A Royal Road to Topology

Convergence of Filters

A Royal Road to Topology
Convergence of Filters

Szymon Dolecki
Mathematical Institute of Burgundy, Dijon, France

World Scientific

NEW JERSEY • LONDON • SINGAPORE • BEIJING • SHANGHAI • HONG KONG • TAIPEI • CHENNAI • TOKYO

Published by

World Scientific Publishing Co. Pte. Ltd.
5 Toh Tuck Link, Singapore 596224
USA office: 27 Warren Street, Suite 401-402, Hackensack, NJ 07601
UK office: 57 Shelton Street, Covent Garden, London WC2H 9HE

Library of Congress Cataloging-in-Publication Data
Names: Dolecki, Szymon, author.
Title: A royal road to topology : convergence of filters / Szymon Dolecki, Mathematical Institute of Burgundy, France.
Description: New Jersey : World Scientific, [2024] | Includes bibliographical references and index.
Identifiers: LCCN 2022013688 | ISBN 9789811232107 (hardcover) | ISBN 9789811232114 (ebook for institutions) | ISBN 9789811232121 (ebook for individuals)
Subjects: LCSH: Topology. | Convergence.
Classification: LCC QA611 .D655 2024 | DDC 514--dc23/eng20220621
LC record available at https://lccn.loc.gov/2022013688

British Library Cataloguing-in-Publication Data
A catalogue record for this book is available from the British Library.

Cover: A. F. Van der Meulen, Ingress of Louis XIV and Marie-Thérèse to Arras on 30 July 1667, Musée des Beaux-Arts, Arras. Photo (C) RMN-Grand Palais / Christian Jean

First published 2024 (hardcover)
Reprinted 2026 (in paperback edition)
ISBN 9789819831296 (pbk)

For any available supplementary material, please visit
https://www.worldscientific.com/worldscibooks/10.1142/12154#t=suppl

Desk Editors: Nandhakumar Krishnan/Tan Rok Ting

Typeset by Stallion Press
Email: enquiries@stallionpress.com

About the Author

Szymon Dolecki was an Emeritus Professor at the Mathematical Institute of Burgundy, Dijon, till 2023. He received his PhD in 1973 from the Institute of Mathematics, Polish Academy of Sciences, where he was employed till 1984. He was a Professor at the University of Limoges (1985–1990) and Burgundy University (1990–2017). He is the author of some hundred publications in convergence theory, topology, optimization, control theory, functional analysis, and history of mathematics. His books are as follows: *Analyse Fondamentale*, Hermann, Paris, 2013 (2nd edition), *Convergence Foundations of Topology*, Frédéric Mynard (coauthor), World Scientific, 2016.

Preface

Topological spaces are a special case of convergence spaces. This textbook introduces topology within a broader context of convergence theory. It comprises the matter usually taught in courses of general topology (Chapters VII to XV).

The book is addressed both to those who wish to learn topology from the start, and to those who, being already knowledgeable, are curious to review topology from a different perspective, which goes well beyond the traditional knowledge. In particular, a topologist will find some arguments ([1]) that become much simpler due to the approach adopted in this book.

As for the beginners, to start up reading this book does not require any knowledge about topologies, not even about metric spaces. It would be enough to have studied convergence of sequences on the real line ([2]), and if you are not familiar with convergence of sequences, which may be the case of those who have studied calculus without sequences, start to read the second chapter by Section II.2. Nevertheless, some level of general mathematical knowledge is presumed, namely familiarity with order, set-theoretic operations, maps, quantifiers, and so on.

We use sequences as a motivation for filters. Having lost their foregoing importance, sequences remain a useful elementary tool to substantiate certain abstract facts involving filters.

---

[1]Let me mention Stone's theorem on paracompactness of metrizable topologies (with a non-technical proof of few lines).

[2]I realize, however, that this seemingly innocuous sentence can have different meaning for a European and an American reader. A traditional European way of teaching analysis is to start by a study of sequences and their convergence on the real line. A typical calculus course in the U. S. A. attains differentiation and integration without mentioning sequences, which is quite an achievement. Indeed, it is hard to conceive for those from the other tradition that it is possible to do this without sequences. The point is that it is feasible, but without defining rigorously the notions of limit and convergence. In other terms, one can be impressively good in calculating derivatives and integrals, without need of knowing what they really are.

Convergence of filters is a natural generalization of convergence of sequences, so that convergence theory is more natural and intuitive to many, perhaps most, students, rather than topology courses based on open and closed sets ([3]).

The framework of convergence is more general than that of topology, and it is exactly the *proper level of generality.* By a "proper level of abstraction" with respect to (an objective of understanding of) a given theory, I mean the level, which is operative on one hand, and intuitive on the other. Of course, this level depends on a student's background, that comprises experience and cognitive pattern, but one can bet that there is some dose of universality in the concept. If an approach is very abstract, its sense is often obscure, and its application arduous; if it is not sufficiently general, its arguments are usually laborious and, most importantly, lacking enough insight.

A knowledgeable reader will recognize some footprints of the *category theory,* outlined in Chapter XXVIII at the end of the book. I refrain from its formal introduction in the body of the book in order not to increase the level of abstraction more than necessary, that is, an objectwise application of concrete categories. This objectwise approach

[3]Filters are either marginal or inexistent in European classics, like [235] of W. SIERPIŃSKI, [169] of K. KURATOWSKI, and [97] of R. ENGELKING, as well as in American standards, like [147] by J. KELLEY, [94] by J. DUGUNDJI, and [191] by J. R. MUNKRES.
The collective oeuvre signed N. BOURBAKI is a vast compendium of mathematics. In the volumes dedicated to General Topology [31] filters are used, but standing back, secondary with respect to open and closed sets. Nevertheless, occasionally they are playing the first roles, like in the presentation of compactness.
The book of L. GILLMAN and M. JERISON [113] constitutes an exception to the low profile of filters, though it is both advanced and specialized to the rings of continuous functions and compactifications of completely regular topologies.
None of these books goes beyond the topological framework. There exists, however, a classical book [40] of E. ČECH, which is developed in a greater generality of pretopologies, in terms of non-idempotent closure.
W. GÄHLER's [107] makes a real difference, but is largely unknown — I believe that it is because of its high level of abstraction and perhaps also because it is written in German, which is not widely known nowadays. There are similarities between the book of GÄHLER and the present one, like use of filters and other relevant families of sets, and structural foundation based on categories and functors. In both, topologies materialize after some 150 pages have been devoted to other themes. We have a common motivation, that is, to base the understanding of topology and analysis on the versatility of the convergence theory, which compensates shortcomings of topology.
A difference lies in an abstract, formal style of the former. It is significant that set-theoretic and category foundations constitute the opening of W. GÄHLER's book, and the appendices of mine.

to categories is an instance of a more general way of looking at things, that is, adapting the tools to the setting in which they are used.

Filters prevail over *nets* in the context of convergence, because, from the convergence viewpoint, each net can be reduced to the filter it determines, and, moreover, because the calculus of nets is intricate in contrast to that, easy, of filters (see Section VII.6).

I have already argued that convergence theory is intuitive. On the other hand, it is very operational in the topological context, because the class of convergences is closed under several natural and essential operations, under which the class of topologies is not!

The class of convergences completes the class of topologies in an analogous way as the field of complex numbers algebraically completes the field of real numbers.

I have been repeating the sentence above, citing myself again and again, but I believe that it grasps the essence. Indeed, various topological quests have solutions only beyond the class of topologies ([4]). Many others can be tackled more straightforwardly and gain on insight when reformulated in convergence terms ([5]).

This sort of fulfillment makes convergence theory far-reaching and easier to comprehend and to implement than topology. Often, it enables a more conscious approach, also to traditional topics of topology. This way we not only observe certain phenomena but also understand why they occur ([6]).

This book attempts to take advantage of these features of convergence theory. It provides essential ingredients of traditional topology courses. But more importantly, it tries to explain the underlying structure and mechanisms.

---

[4]For instance, the coarsest convergence on the set of continuous functions making the natural coupling continuous is always pseudotopological, but not topological, when the underlying topology is not locally compact.

[5]For instance, it becomes clear that the nature of compactness is not topological, but pseudotopological, when we realize that the Tikhonov theorem is a very special case of the commutation of pseudotopological reflectors with products. Another example is that of mixed properties: the image of a Fréchet topology under a quotient map need not be Fréchet, but it is Fréchet under hereditary quotient maps. This becomes clear from the way in which Fréchet topologies and hereditary quotient maps are defined with the aid of the pretopological modifier.

[6]For example, it was long well known that a product of two Fréchet topologies need not to be Fréchet. This fact inspired a search for extra conditions under which a given product is Fréchet. However, a structural explication in convergence-theoretic terms of non-productivity of Fréchet property enabled F. Jordan and F. Mynard to characterize those topologies which product with each Fréchet (respectively, strongly Fréchet) topology is Fréchet [141].

The title of this book alludes to the mentioned advantages of the present approach, which I believe to be more gratifying than many traditional ones: you travel more comfortably and see more.

Sire, there is no royal road to geometry! is attributed by some to Euclid in answering Ptolemy I, and by others to Menaechmus in responding to Alexander the Great, when the monarch asked for a quick introduction to geometry.

Sure enough, this way to topology is not entirely effortless — otherwise it would be boring — but it offers an attractive trade-off: you invest to grasp the world of filters, but soon you will be gladdened, because this world grants you insight.

Perhaps the most significant feature of the class of convergences is *exponentiality* ([7]), that is, there always exists a coarsest convergence on the set of continuous maps (from one convergence to another) that makes the natural coupling jointly continuous. This is not the case of topologies! The fact was discovered seven decades ago by G. CHOQUET ([8]), founder of convergence theory ([9]).

The phenomenon of non-topologicity of spaces of continuous maps occurs also for dual spaces in the Mackey-Arens duality in functional analysis. Also in this case, methods of convergence theory prove to be very fruitful, as witness *Convergence Structures and Applications to Functional Analysis* of R. BEATTIE and H. P. BUTZMANN and *Continuity theory* by D. L. NEL ([10]).

We shall tackle metrizability, which has been one of the oldest topics in topology. Thanks to a novel use of partitions of unity proposed by J. DYDAK ([11]), these — traditionally difficult and technical — investigations become accessible and transparent. Alluding to the title of this book, we can say that they constitute a princely road to metrization.

The subject of this book is similar to that of *Convergence Foundations of Topology* by S. DOLECKI and F. MYNARD ([12]). I tried,

[7]Traditionally called *Cartesian closedness.*

[8]G. CHOQUET [44].

[9]It is interesting that Choquet qualifies convergence theory as an important part of topology.

[10]R. BEATTIE and H. P. BUTZMANN [24], D. L. NEL [204].

[11]J. DYDAK [21] and K. AUSTIN, J. DYDAK [21]. See also S. DOLECKI [64] and S. DOLECKI, F. MYNARD [83].

[12]For practical reasons, I am not always adopting the notation from the previous book [83] by S. DOLECKI and F. MYNARD. If I realize that the notation used in that book is perhaps heavy or not easy to follow, I try another one. In mathematics, the same symbol may have different meanings in different contexts. One should be aware

however, to make it more accessible, by requiring less prerequisites. Its scope is also somewhat altered: while *Convergence Foundations of Topology* tended to derive most of traditional results from more general facts about convergences, *A Royal Road to Topology* develops sooner usual classical topological topics, revisiting them later from a conceptually more advanced perspective of convergence theory.

These objectives implied a method of recurrent introduction of concepts, starting from examples and special cases and ending with full generality and applications.

For accessibility's sake, this book develops on two levels, fundamental, and more advanced. The sections evidenced by asterisk (*) either require more set-theoretic prerequisites, which are supplied in Chapter XXVII, or use more convoluted arguments, or are otherwise complementary. Most chapters include a section called *Supplement*, which — as the name indicates — contains some additional facts, as well as some exercises and their solutions.

For instance, the fundamental level refrains from discussions of cardinalities others than finite, countable, and uncountable without distinction; the cardinality of continuum and other specific uncountable cardinalities may occur, mostly in asterisked sections.

Acknowledgements. I am grateful to Gabriele H. Greco (Trento, Italy), with whom I shared enthusiasm while discovering convergence structures four decades ago. His irreverence to established uses influenced my perception of mathematics and enabled us to transgress the confines of conventional topology.

Frédéric Mynard (New Jersey City University) shares a considerable part in shaping convergence theory. I am grateful to him for several important contributions to this book.

I am beholden to Alois Lechicki (Fürth, Germany) for unfailing and demanding scrutiny of large portions of the manuscript, which incited me to revise certain of its traits.

I am fortunate that Jerzy Wojciechowski (West Virginia University) recently became intrigued by convergence theory. I owe him not only many fine hints and penetrating quests, but also solutions of several open problems.

I appreciate the expertise in set theory of Andrzej Starosolski (Silesian Polytechnic, Gliwice, Poland).

I am indebted to Ahmed Bouziad (University of Rouen) for having detected a mistake in a preliminary version.

---

that an attempt to build a totally precise and consistent system of notation would lead to very complex symbols. Therefore, such a system would be impracticable.

I am thankful to Fadoua Chigr (New Jersey City University) for a diligent perusal of portions of the book.

Finally, I prize suggestions on the presentation made by Zbigniew Lipecki (Mathematical Institute, Polish Academy of Sciences).

Szymon Dolecki
Tavira
October 2023

List of Chapters

Table of Contents

CHAPTER I

Preliminaries

While the ZFC (Zermelo-Fraenkel set theory with the Axiom of Choice) axioms are assumed throughout this book, the Preliminaries are addressed first and foremost to a reader with only basic set-theoretic background. An advanced reader may have a look at it just to get acquainted with notation and terminology. Axiomatic considerations are gathered in Chapter XXVII.

I.1. Relations and maps

In this preliminary section, we shall gather a few basic facts about relations and maps. Mind that we distinguish between the symbol $=$ of *equality* or *equation*, and the symbol $:=$ of *definition*. For instance, in $3 \cdot \frac{1}{3} = 1$, the said symbol denotes an *equality*, in the expression $x + x^2 = 1$, where x is a complex number, an *equation*, which becomes an equality, true or false, when we replace x with a complex number. When we write

$$0! := 1, \text{ and } (n+1)! := (n+1) \cdot n! \text{ for } n > 0,$$

we inductively define the factorial $n!$ for each natural number n.

In mathematics, the notion of *set* is not defined. We use it intuitively, but its sense is made precise by a series of rules (called *axioms*) that are presented in Chapter XXVII. We say that x is an *element* of a set X, or x *belongs* to X, in symbols, $x \in X$. This relation is also undefined, and its meaning can be grasped thanks to the mentioned axioms.

We define, however, a *subset* A of a set X, as follows:

$$A \subset X \iff (x \in A \Longrightarrow x \in X)$$

for each x. The *set-theoretic difference* $X \setminus A$ is defined by

$$x \in X \setminus A \iff (x \in X) \wedge (x \notin A).$$

Let us yet mention that a mathematical definition should be based, in one way or another, on the notions of set and element (see Chapter XXVII for details).

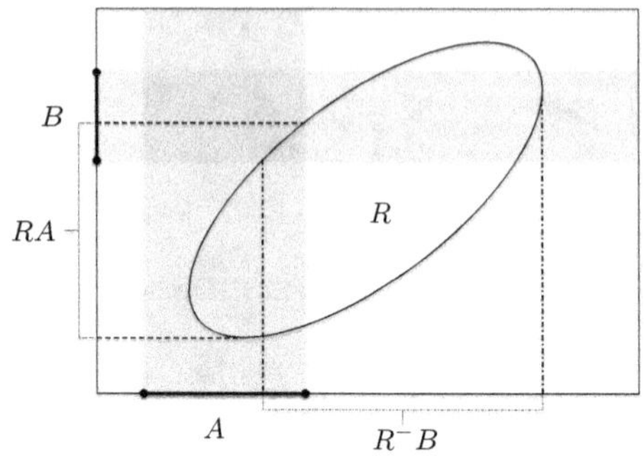

FIGURE I.1.1. The image RA of a set A and the preimage $R^{-}B$ of a set B under a relation R.

If X, Y are sets, then a *relation* R between elements x of X and y of Y is defined as a subset of the product $X \times Y = \{(x, y) : x \in X, y \in Y\}$, that is, $R \subset X \times Y$. In other words, xRy ([1]) if and only if $(x, y) \in R$.

If $R \subset X \times Y, A \subset X$ and $B \subset Y$, then the *image* of A under R and the *preimage* of B under R are respectively defined by

$$RA := \{y \in Y : \underset{x \in A}{\exists} (x, y) \in R\},$$
$$R^{-}B := \{x \in X : \underset{y \in B}{\exists} (x, y) \in R\}.$$

In particular, we abridge

$$Rx := R\{x\} \text{ and } R^{-}y := R^{-}\{y\}. \tag{I.1.1}$$

If $R \subset X \times Y$ and $S \subset Y \times Z$ are two relations, then their *composition* gives rise to a relation $S \circ R \subset X \times Z$, defined by

$$(S \circ R)x := S(Rx).$$

If $\mathcal{A}$ is a family of subsets of a set X, then $\bigcup \mathcal{A}$ is a shorthand for $\bigcup_{A \in \mathcal{A}} A := \{x \in X : \underset{A \in \mathcal{A}}{\exists} x \in A\}$. It is immediate that

$$R(\bigcup \mathcal{A}) = \bigcup_{A \in \mathcal{A}} RA. \tag{I.1.2}$$

On the other hand, an easy and straightforward verification (see Figure I.1.1) yields

[1] x is R-related to y.

PROPOSITION I.1.1. *For $R \subset X \times Y$, $A \subset X$ and $B \subset Y$, the following statements are equivalent:*

$$RA \cap B \neq \varnothing,$$
$$A \cap R^{-}B \neq \varnothing,$$
$$R \cap (A \times B) \neq \varnothing.$$

A relation $Q \subset Y \times X$ is called the *inverse (relation)* of a relation $R \subset X \times Y$ if $x \in Qy$ if and only if $y \in Rx$, and it is denoted by R^{-}.

We recall below a few basic types of relations. For a relation $R \subset X \times X$, its type is defined in terms of a property holding for each x, y, and z.

(reflexive)	$x \in Rx,$
(symmetric)	$y \in Rx \Longrightarrow x \in Ry,$
(antisymmetric)	$y \in Rx$ and $x \in Ry \Longrightarrow y = x,$
(transitive)	$y \in Rx$ and $z \in Ry \Longrightarrow z \in Rx$

The *identity relation* $\triangle_X$ on X is defined by $\triangle_X x := \{x\}$. In other words,

$$\triangle_X = \{(x,x) : x \in X\} \subset X \times X. \tag{I.1.3}$$

Often the *identity relation* on X is called the *diagonal subset* of X.

We know that a *map* $f : X \longrightarrow Y$ associates with every element x of X, a unique element of Y denoted by $f(x)$. The set X is called the *domain* of f, and the set Y is called the *codomain* of f.

The set of all maps from X to Y is denoted by Y^X. We will be using interchangeably

$$f : X \longrightarrow Y \Longleftrightarrow f \in Y^X.$$

This standard definition of a map is not entirely satisfactory for mathematicians, because the term "to associate" has not been defined mathematically. To define a map in terms of sets, let us think of it as of its graph, that is,

$$\mathrm{gr}(f) := \{(x,y) \in X \times Y : y = f(x)\}, \tag{I.1.4}$$

which is a subset of $X \times Y$, hence a relation. A relation of this form is called a *graph*. We will often identify f with $\mathrm{gr}(f)$.

Let us notice that

PROPOSITION I.1.2. *A relation $F \subset X \times Y$ is a graph if and only if $F^{-}Y = X$, and if $F^{-}y_0 \cap F^{-}y_1 \neq \varnothing$ implies $y_0 = y_1$.*

Proposition I.1.2 constitutes, in fact, a definition of a map with the aid of its graph, and is formulated in terms of sets.

A map $f : X \longrightarrow Y$ is called *surjective* (or a *surjection,* or *onto*) if

$$f(X) := \{f(x) : x \in X\} = Y;$$

injective (or an *injection*) if $x_0 \neq x_1$ implies that $f(x_0) \neq f(x_1)$; *bijective* (or a *bijection*) if it is both surjective and injective.

REMARK I.1.3. Note that if $g \circ f$ is surjective, then g is surjective, and if $g \circ f$ is injective, then f is injective.

A bijection $f : X \longrightarrow X$ is also called a *permutation.* Of course, the *identity map* $i_X : X \longrightarrow X$, defined by $i_X(x) := x$ for each $x \in X$, is a permutation.

If a map $f : X \longrightarrow Y$ is bijective, then for each $y \in Y$, there exists a unique element $x \in X$ such that $f(x) = y$. If we denote it by $f^{-1}(y)$, then we define a map $f^{-1} : Y \longrightarrow X$ to the effect that

$$x = f^{-1}(y) \Longleftrightarrow f(x) = y. \tag{I.1.5}$$

We make a distinction between the *inverse relation* f^- of a map $f : X \longrightarrow Y$ and the *inverse map* f^{-1} of f. While f^- exists for every map f (and $f^-(y)$ is a subset of X), the inverse map f^{-1} exists merely for a bijection f (and $f^{-1}(y)$ is an element of X). In this case,

$$\{f^{-1}(y)\} = f^-(y)$$

for each $y \in Y$ ([2]).

If $\varnothing \neq V \subset X$, then $f|_V : V \longrightarrow Y$ is defined by $f|_V(x) := f(x)$ if $x \in V$, and is called the *restriction* of f to V.

DEFINITION I.1.4. If $f : X \longrightarrow Y$ is an injection, then the map $f^{\sim 1} : f(X) \longrightarrow X$ defined by $f^{\sim 1}(f(x)) := x$ for each $x \in X$ is called it the *semi-inverse map* of f.

Of course, the semi-inverse map $f^{\sim 1}$ is a bijection. If f is a bijection, then $f^{\sim 1} = f^{-1}$.

Let us extend the concepts of injectivity and surjectivity to relations.

[2] If $f \in Y^X$ and $g \in X^Y$, and $g \circ f = i_X$ and $f \circ g = i_Y$, then f and g are bijective by Remark I.1.3, and $g = f^{-1}$. On the other hand, if f is surjective, then there exists $h \in X^Y$ such that $f \circ h = i_Y$. Such a map h is called a *quasi-inverse* of f. Indeed, for each $y \in Y$, there exists an element $h(y) \in f^-(y)$. Obviously, $f(h(y)) = y$.

DEFINITION I.1.5. Let V, W be sets. A relation $R \subset V \times W$ is *surjective* if $RV = W$; *injective* if $Rv_0 \cap Rv_1 \neq \varnothing$ implies that $v_0 = v_1$ (3).

Clearly, the graph of f (I.1.4) is surjective (injective) if and only if f is surjective (injective).

If $f : X \longrightarrow Y$, then f^- stands for the inverse relation of (I.1.4), and thus $f^- \subset Y \times X$. It is sometimes called the *fiber relation* of f, and the sets of the form $f^-(y)$, the *fibers* of f.

COROLLARY I.1.6. *A relation $F \subset X \times Y$ is a graph if and only if the inverse relation F^- is surjective and injective.*

PROOF. Indeed, F^- is surjective if and only if $F^-Y = X$, and injective if and only if $F^-y_0 \cap F^-y_1 \neq \varnothing$ implies $y_0 = y_1$, hence, by Proposition I.1.2, if and only if F is a graph. □

I.2. Equivalence relations

A reflexive, symmetric and transitive relation is called an *equivalence (relation)*. Traditionally, an equivalence relation is denoted by $\approx$, or some of its variants, rather than by R, or another capital letter:

$$x \approx y \iff y \in Rx.$$

Let us repeat the axioms of an equivalence relation using this notation:

(reflexive) $\quad x \approx x,$

(symmetric) $\quad x \approx y \Longrightarrow y \approx x,$

(transitive) $\quad (x \approx y) \wedge (y \approx z) \Longrightarrow x \approx z.$

The inverse relation of an equivalence relation $\approx$ is equal to itself, because it is symmetric.

As it is often the case, both notations have their advantages and inconveniences. Therefore, it is sensible to use one or another when it seems to be more suited to a given context.

A family $\mathcal{A}$ of subsets of a set X is called a *partition* of X if $A_0 \cap A_1 = \varnothing$ whenever $A_0 \neq A_1$, and $X = \bigcup \mathcal{A}$. Let E be an equivalence relation on a set X (4). Then

$$\{Ex : x \in X\} \tag{I.2.1}$$

[3]The notion of surjectivity and injectivity was extended from maps to relations by the author in [64] (see also [83]).

[4]In our convention, any capital letter can be used to denote a relation. The choice of E informally indicates that we consider an equivalence relation.

is a family of subsets of X that are either disjoint, or coincide ([5]). Of course, $x \in Ex$ for each $x \in X$.

In (I.2.1), $Ex_0 = Ex_1$ if and only if $x_1 \in Ex_0$. Therefore, (I.2.1) is a partition of X into sets of E-equivalent elements. It is called the *quotient* of X by E, and is denoted by X/E. The map $f_E : X \longrightarrow X/E$ associating Ex to every $x \in X$ is called the *canonical surjection*. Any surjective map $f \in Y^X$ establishes an equivalence relation E_f on X to the effect that $x_1 \in E_f x_0$ whenever $f(x_0) = f(x_1)$. Consequently, $E_f x = f^-(f(x))$ for each $x \in X$. Therefore,

PROPOSITION I.2.1. *There exists a bijection between equivalence relations on a non-empty set X and surjective maps defined on X.*

I.3. Order

A reflexive, antisymmetric and transitive relation is called an *order*. Traditionally an order relation is denoted by $\leq$, or its variants, rather than by R:

$$x \leq y \Longleftrightarrow y \in Rx,$$

and we say y is *greater than or equal to* x (x is *less than or equal to* y).

More generally, a relation $\leq$ on a set X is called a *preorder* if

(reflexive) $\quad x \leq x,$

(transitive) $\quad (x \leq y) \wedge (y \leq z) \Longrightarrow x \leq z.$

If, moreover, $\leq$ fulfills

(antisymmetric) $\quad (x \leq y) \wedge (y \leq x) \Longrightarrow x = y,$

then it is called an *order*.

The inverse relation of $\leq$ is denoted by $\geq$, and is also an order relation, called the *inverse order* of $\leq$. If $A \subset X$ and if there exits an $a_1 \in A$ such that $a \leq a_1$ for each $a \in A$, then we say that a_1 is the *greatest element* of A. An element a_0 of A such that $a_0 \leq a$ for each $a \in A$ is said to be the *least element* of A. None need to exist.

We say that x is *strictly less* than y (equivalently, y is *strictly greater* than x), in symbols,

$$x < y$$

if $x \leq y$ and $x \neq y$. The relation $<$ is that of the *strict order* associated with the order $\leq$.

---

[5]Indeed, suppose that $Ex_0 \cap Ex_1 \neq \varnothing$, and let $x \in Ex_0$. It follows that there exists $w \in Ex_0 \cap Ex_1$, hence $x_0 \in Ew$, because E is (symmetric), and thus $x \in Ew$, because E is (transitive). On the other hand, $w \in Ex_1$. As E is (transitive), $x \in Ex_1$, showing that $Ex_0 = Ex_1$.

EXAMPLE I.3.1. Let $X = 2^{\{0,1\}}$, which is the set of all subsets of $\{0,1\}$, endowed with the inclusion relation $\subset$. In other words, $2^{\{0,1\}} = \{\varnothing, \{0\}, \{1\}, \{0,1\}\}, \varnothing \subset \{0\} \subset \{0,1\}$, and $\varnothing \subset \{1\} \subset \{0,1\}$. We say that $\{0\}$ and $\{1\}$ are *incomparable.*

If $\leq$ is an order X and $\preccurlyeq$ is an order on Y, then a map $f : X \longrightarrow Y$ is called *order-preserving* whenever $x_0 \leq x_1$ implies $f(x_0) \preccurlyeq f(x_1)$ for every $x_0, x_1 \in X$.

EXAMPLE I.3.2 (An order-preserving bijection need not be an order-isomorphism). Let $X = 2^{\{0,1\}}$ be ordered by inclusion and $Y = \{0,1,2,3\}$ by the usual order of $\mathbb{N}$. Let $f \in Y^X$ be given by $f(\varnothing) = 0, f(\{0\}) = 1, f(\{1\}) = 2$, and $f(\{0,1\}) = 3$. Then f is an order-preserving bijection, but $f^{-1}(1)$ and $f^{-1}(2)$ are incomparable.

DEFINITION I.3.3. An order $\leq$ on X is called *linear* or *total* if for any two elements of X either $x_0 \leq x_1$ or $x_1 \leq x_0$. A *linearly ordered set* is also called a *chain.*

The order from Example I.3.1 is not linear. The standard order of the real line is linear ([6]).

EXAMPLE I.3.4 (Diagonal sweeping order). Let us introduce the following relation on $\mathbb{N} \times \mathbb{N}$:

$$(n_0, k_0) \prec (n_1, k_1) \iff (n_0+k_0 < n_1+k_1) \vee (n_0+k_0 = n_1+k_1 \wedge n_0 < n_1).$$

This relation is antisymmetric and transitive, but not reflexive. However, the relation $(n_0, k_0) \preceq (n_1, k_1)$ defined by $(n_0, k_0) \prec (n_1, k_1)$, or $n_0 = n_1$ and $k_0 = k_1$, is antisymmetric, transitive, and reflexive, hence an order ([7]). The product $\mathbb{N} \times \mathbb{N}$ can be viewed as an infinite matrix with the entries of the form (n, k), where n and k are integers.

$(0,0)$	$(1,0)$	$(2,0)$	$(3,0)$
$(0,1)$	$(1,1)$	$(2,1)$	$\cdots$
$(0,2)$	$(1,2)$	$(2,2)$	$\cdots$
$(0,3)$	$\cdots$	$\cdots$	$\cdots$

Let us list the first, second, and few successors in this order:

$$(0,0), (0,1), (1,0), (0,2), (1,1), (2,0), (0,3), (1,2), \ldots$$

This order is linear: all the elements of $\mathbb{N} \times \mathbb{N}$ can be arranged in a sequence ([8]).

---

[6]Incidentally, the term *linear* refers to the (real) line.

[7]In this case, $\prec$ is the *strict order associated* with the order $\preceq$.

[8]For any two elements, (n_0, k_0) and (n_1, k_1), either $(n_0, k_0) \preceq (n_1, k_1)$ or $(n_1, k_1) \preceq (n_0, k_0)$.

EXAMPLE I.3.5 (Lexicographic order). *Lexicographic* order is another type of linear order that can be defined on a product of linear orders. We can arrange the table above by declaring that each line is naturally ordered, and that each element of a line is greater than any element of a preceding line,

$$(0,0),(1,0),(2,0),(3,0)\ldots(0,1),(1,1),(2,1)\ldots(0,2),(1,2)\ldots$$

Formally

$$(x_0,y_0)<(x_1,y_1)\iff\begin{cases}y_0<y_1 & \text{or}\\ y_0=y_1 & \text{and } x_0<x_1\end{cases}.$$

Of course, another lexicographic order be defined so that every column is naturally ordered, and each element of a column is greater than any element of a preceding column.

The name of this order originates from the way of compiling lexicons, where all the words starting with "a" go before those starting with "b", and among those starting with the same letter, they are arranged by the order of the second letter, than by the third, and so on. For example, in the index of this book,

$$\text{open} < \text{order} < \text{ordinal} < \text{paracompact}.$$

Another example is provided by the order in which I arrange sections in this book: for instance, Section III.9 proceeds Section IX.2 followed by Section IX.7 ([9]).

Let us recall some basic notions related to order. Let $(X,\leq)$ be an *ordered set.* For each $A\subset X$, define

$$A^+ := \{x\in X : \underset{a\in A}{\forall}\ a\leq x\},$$
$$A^- := \{x\in X : \underset{a\in A}{\forall}\ a\geq x\}.$$

If A^+ has a least element, then it is denoted by $\bigvee A$ and called the *supremum* (*least upper bound*) of A. Alike, if A^- has a greatest element, then it is denoted by $\bigwedge A$ and called the *infimum* (greatest lower bound) of A. None need exist ([10]).

---

[9]Actually, if A is a finite alphabet, then the alphabetic order is defined on finite sequences $a_1a_2\ldots a_n$, so that

$$b_1b_2\ldots b_nb_{n+1}\ldots b_k > a_1a_2\ldots a_n$$

provided that if $j_0 := \max\{j : a_j = b_j\}$, then $b_{j_0} > a_{j_0}$.

[10]Let $\mathbb{R}$ be the real line with the standard order $\leq$. Then for each $A\subset\mathbb{R}$, which is *bounded above,* which means that $A^+\neq\varnothing$, there exists $\bigvee A$. If we consider, the subset $\mathbb{Q}$ of $\mathbb{R}$, consisting of rational numbers, then $A := \{q\in\mathbb{Q} : q^2 < 2\}$, then

There exist alternative notations for extrema (that is, suprema and infima), especially in the case of the ordered real line $(\mathbb{R}, \le)$, namely

$$\sup A = \bigvee A, \quad \inf A = \bigwedge A.$$

DEFINITION I.3.6. An ordered set $(X, \le)$ is called a *lattice* if $\bigvee A$ and $\bigwedge A$ exist for each non-empty finite subset A of X; a *complete lattice* whenever they exist for each $\varnothing \ne A \subset X$.

DEFINITION I.3.7. A bijective order-preserving map f is said to be an *order isomorphism* if the inverse map f^{-1} is order-preserving.

It follows directly from the definitions that

PROPOSITION I.3.8. *An order isomorphism f preserves the extrema, that is,*

$$f(\bigvee A) = \bigvee f(A), \quad f(\bigwedge A) = \bigwedge f(A).$$

EXAMPLE I.3.9. The *ordered extended real line* $(\overline{\mathbb{R}}, \le)$, where $\overline{\mathbb{R}} := \mathbb{R} \cup \{-\infty, +\infty\}$ is a complete lattice, while the *ordered real line* $(\mathbb{R}, \le)$ is only a relatively complete lattice, that is, each non-empty *bounded* set has a supremum and an infimum ([11]).

The ordered set of Example I.3.1 is a *lattice*, which is necessarily complete, because X is finite. More generally, for each set X, the set $L = 2^X$ of subsets of X is a *complete lattice* with respect to the *inclusion* $\subset$, where

$$A \subset B \Longleftrightarrow (x \in A \Rightarrow x \in B).$$

for $A, B \in 2^X$. Indeed, for any $\mathcal{A} \subset 2^X$, the supremum $\bigvee \mathcal{A}$ and the infimum $\bigwedge \mathcal{A}$ are given by

$$\bigvee \mathcal{A} = \bigcup \mathcal{A} := \bigcup_{A \in \mathcal{A}} A,$$
$$\bigwedge \mathcal{A} = \bigcap \mathcal{A} := \bigcap_{A \in \mathcal{A}} A.$$

If $(L, \le)$ is an ordered set, and M is a subset of L, then $(M, \le)$ constitutes an ordered set with the restriction of the relation $\le$ to M. In other words, if $x, y \in M$, then $x \le y$ with respect to $(M, \le)$ if and only if $x \le y$ with respect to $(L, \le)$.

However, suprema and infima considered in $(M, \le)$ and in $(L, \le)$ are, in general, different!

$A^+ = \{q \in \mathbb{Q} : 2 \le q^2\}$ is non-empty, but does not have a least element (because $\sqrt{2} \notin \mathbb{Q}$), so that the supremum of A does not exist.

[11]More precisely, non-empty bounded above sets admit suprema, and bounded below sets admit infima.

EXAMPLE I.3.10. In Figure I.3.1 the lattice of subsets of $\{0, 1, 2\}$ ordered by the inverse inclusion $\supset$, and its sublattice on the right-hand side. We denote by $\bigvee, \bigwedge$ extrema in the whole lattice, and by $\bigvee_M, \bigwedge_M$ in the sublattice.

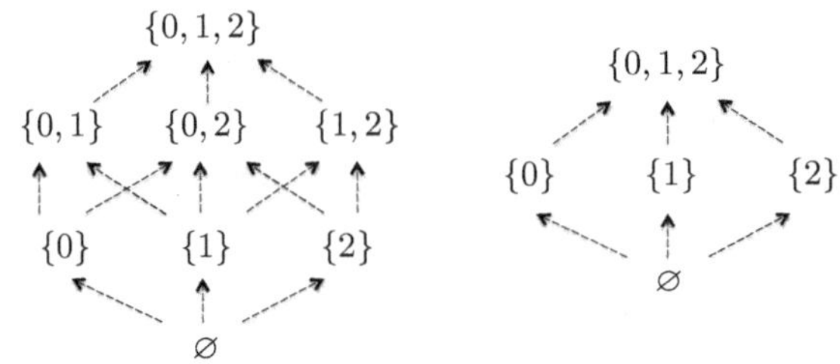

FIGURE I.3.1. An order is defined by arrows $\rightarrow$ and corresponds to $\subset$ on the subsets of $\{0, 1, 2\}$. A sublattice M is placed on the right-hand side.

For instance, $\bigvee\{\{0\}, \{2\}\} = \{0, 2\}$, which is less than $\bigvee_M\{\{0\}, \{2\}\} = \{0, 1, 2\}$.

The existence of $\bigvee A$ in the whole lattice does not imply the existence of $\bigvee_M A$. But, when both exist, then

$$\bigwedge\nolimits_M A \leq \bigwedge A \leq \bigvee A \leq \bigvee\nolimits_M A. \tag{I.3.1}$$

Indeed, if $M \subset L$ and if the suprema exist, then

$$\varnothing \neq \{x \in M : \underset{a \in A}{\forall} a \leq x\} \subset \{x \in L : \underset{a \in A}{\forall} a \leq x\} = A^+,$$

hence the least element $\{x \in M : \underset{a \in A}{\forall} a \leq x\}$ belongs to A^+, and thus is greater than the least element of A^+. Therefore, $\bigvee A \leq \bigvee_M A$, and of course, analogously in the case of infima.

DEFINITION I.3.11. Let $(X, \leq)$ be an ordered set, and let $A \subset X$. Then $a_0 \in A$ is called a *minimal element* of A if $A \ni a \leq a_0$ entails $a = a_0$; a *maximal element* of A if $A \ni a \geq a_0$ entails $a = a_0$.

We denote by

$$\text{Min}A, \quad \text{Max}A$$

the sets of minimal and maximal elements of A, respectively.

If A possesses a least (respectively, greatest) element a_0, then, of course, a_0 is a minimal (respectively, maximal) element of A.

In Example I.3.10, if $A := \{\varnothing, \{0\}, \{1\}, \{2\}\}$, then $\text{Max}A = \{\{0\}, \{1\}, \{2\}\}$ and $\bigvee A = \{0, 1, 2\}$.

I.4. Polarities

If $R \subset X \times Y$ is a relation, then the following subset

$$R^\circ A := \bigcap\nolimits_{x \in A} Rx \tag{I.4.1}$$

of Y is called the *R-polar* of A. We note that

$$A_0 \subset A_1 \Longrightarrow R^\circ A_0 \supset R^\circ A_1, \tag{I.4.2}$$

$$A \subset (R^-)^\circ (R^\circ A). \tag{I.4.3}$$

Indeed, the first formula is obvious. To show the second, let us rewrite $(R^-)^\circ(R^\circ A)$ as

$$(R^-)^\circ(R^\circ A) = \bigcap\nolimits_{y \in R^\circ A} R^- y,$$

and let $x_0 \in A$. By (I.4.1), if $y \in R^\circ A$, then $y \in Rx_0$, equivalently $x_0 \in R^- y$, so that the proof is complete.

It follows that

$$R^\circ((R^-)^\circ(R^\circ A)) = R^\circ A \tag{I.4.4}$$

for every $A \subset X$ and $R \subset X \times Y$. Indeed, $\supset$ holds by (I.4.2) and (I.4.3). When we apply R° to both sides of the equality (I.4.3), then by (I.4.2), we get $\subset$ in (I.4.4). Let us abridge

$$R^{-\circ} := (R^-)^\circ, \text{ and}$$
$$R^{\circ\circ} A := (R^-)^\circ(R^\circ A),$$

and call $R^{\circ\circ}A$, the *R-bipolar* of A. Observe that, for every A_0, A_1, and A,

$$A_0 \subset A_1 \Longrightarrow R^{\circ\circ} A_0 \subset R^{\circ\circ} A_1, \tag{I.4.5}$$

$$A \subset R^{\circ\circ} A, \tag{I.4.6}$$

$$R^{\circ\circ}(R^{\circ\circ} A) = R^{\circ\circ} A. \tag{I.4.7}$$

Indeed, (I.4.5) is a result of a double application of (I.4.2), while (I.4.6) amounts to (I.4.3). Finally, we get (I.4.6) on applying $R^{-\circ}$ to (I.4.4).

PROPOSITION I.4.1. *The pair $(R^\circ, R^{-\circ})$ is an inverse order isomorphism (with respect to $\subset$) between R-bipolars and R^--bipolars.*

PROOF. By (I.4.2), the maps R° and $R^{-\circ}$ are order-inverting. In virtue of Definition I.3.7, it is enough to prove that R° and $R^{-\circ}$ are bijective. Let A_0, A_1 be R-bipolars such that $R^\circ A_0 = R^\circ A_1$. Then

$$A_0 = R^{\circ\circ} A_0 = R^{-\circ}(R^\circ A_0) = R^{-\circ}(R^\circ A_1) = R^{\circ\circ} A_1 = A_1,$$

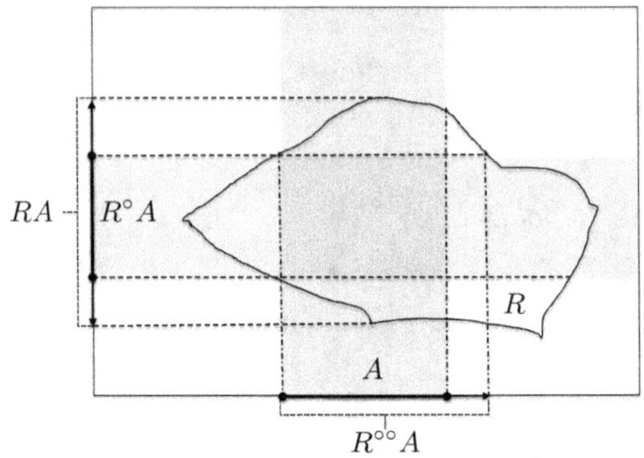

FIGURE I.4.1. A set A, the polar $R^\circ A$, the image RA, and the bipolar $R^{\circ\circ} A$ of A under the relation R.

that is, R° is injective on the set of R-bipolars. If $B = (R^-)^{\circ\circ} B$, then letting $A := R^{-\circ} B$, we get $R^\circ A = R^\circ(R^{-\circ} B) = (R^-)^{\circ\circ} B = B$, that is, R° is surjective. Bijectivity of $R^{-\circ}$ follows *mutatis mutandis*. □

Analogously to Proposition I.1.1,

PROPOSITION I.4.2. *For $R \subset X \times Y$, $A \subset X$ and $B \subset Y$, the following statements are equivalent:*

$$R^\circ A \supset B,$$

$$A \subset R^{-\circ} B,$$

$$A \times B \subset R.$$

I.5. Finite and infinite cardinals

The *cardinality* $\operatorname{card} X$ of a set X is, roughly speaking, the number of elements of X. We shall give a formal definition of cardinality in Chapter XXVII. At this stage, we shall consider only finite and infinite sets, and, among infinite ones, we will distinguish between countable and uncountable ones.

We note that two sets X and Y have the same cardinality

$$\operatorname{card} X = \operatorname{card} Y,$$

that is, are *equipotent*, if there exists a bijection $f : X \to Y$.

EXAMPLE I.5.1. For every $a < b$, the strict interval $]a, b[$ is equipotent with the real line $\mathbb{R}$ ([12]).

[12]Observe first, that any two strict intervals are equipotent, because an appropriate affine map sends one onto another. To conclude, observe that $\arctan :]-\frac{\pi}{2}, \frac{\pi}{2}[\longrightarrow \mathbb{R}$ is a bijection.

On the other hand, the cardinality of X is less than or equal to the cardinality of Y ($\operatorname{card} X \leq \operatorname{card} Y$) whenever there exists an injection ([13]). If $X \neq \varnothing$, then $\operatorname{card} X \leq \operatorname{card} Y$ if and only if there is a surjection $g : Y \longrightarrow X$ ([14]).

We say that Y is *strictly greater* than X, in symbols

$$\operatorname{card} X < \operatorname{card} Y,$$

whenever there is no injection from Y to X. A set X is *finite* whenever each injection $f : X \longrightarrow X$ is a surjection.

We shall write $\operatorname{card} X < \infty$ if X is finite, and $\operatorname{card} X = \infty$ if X is infinite. In particular, $\operatorname{card} \varnothing = 0$, which is the least finite cardinality.

The set of all *natural numbers* is denoted by $\mathbb{N}$. Recall, that $0 \in \mathbb{N}$, and if $n \in \mathbb{N}$ then $n + 1 \in \mathbb{N}$. The map $s : \mathbb{N} \longrightarrow \mathbb{N}$ defined by $s(n) := n + 1$ is an injection that is not a surjection, so that $\mathbb{N}$ is an infinite set. Moreover,

THEOREM I.5.2 (Induction principle). *If $A \subset \mathbb{N}$ has the property that $0 \in A$, and $n \in A$ implies that $n + 1 \in A$, then $A = \mathbb{N}$.*

Theorem I.5.2 follows from a more general Theorem XXVII.4.5. If $m \in \mathbb{N}$, then we denote ([15])

$$\mathbb{N}_m := \{n \in \mathbb{N} : n \geq m\}. \tag{I.5.1}$$

You might find every now and then in the mathematical literature, the notation $n < \omega$ to indicate that n is a natural number ([16]). Hence,

$$n < \omega \Longleftrightarrow n \in \mathbb{N}.$$

By definition, $\operatorname{card} X = \aleph_0$ ([17]) whenever X is equipotent with $\mathbb{N}$. A set X is said to be *countable* if X is the image of the set $\mathbb{N}$. Therefore,

PROPOSITION I.5.3. A set X is countable if and only if it is either finite and non-empty, or $\operatorname{card} X = \aleph_0$.

Let us observe that

[13] This means that X has the same cardinality as a subset of Y.

[14] Indeed, if there is a surjection $g : Y \to X$, then let $f : X \to Y$ be such that $f(x) \in g^{-}(x)$, which assures that f is an injection. Conversely, if $f : X \to Y$ is an injection, then $g(y) := x$ if $y \in f(X)$, and $g(y)$ is arbitrary if $y \notin f(X)$, is an injection.

[15] Giuseppe Peano defined in 1889 the set of natural number in an axiomatic way. The axiom that $\mathbb{N}$ is the least set fulfilling these two conditions, amounts to the principle of induction. The notation that follows is also due to him.

[16] The reason for this notation, to be explained more amply in Chapter XXVII, is that $\omega = \omega_0$ stands for the first infinite ordinal number.

[17] The symbol $\aleph_0$ is pronounced *aleph zero*; $\aleph$ is a letter of the Hebrew alphabet.

PROPOSITION I.5.4. *Each countable union of countable sets is countable.*

PROOF. Indeed, let X_n be a countable set for each $n \in \mathbb{N}$, that is, $X_n = \{x_{n,k} : k \in \mathbb{N}\}$. We do not assume that the sets are distinct, nor the elements of these sets are supposed to be different. This way, we take into account a possibility that the number of sets is finite, and that some, or all among these sets are finite. By Example I.3.4, $\mathbb{N} \times \mathbb{N}$ is countable, and thus $X := \bigcup_{n \in \mathbb{N}} X_n = \{x_{n,k} : n, k \in \mathbb{N}\}$, indexed by the product $\mathbb{N} \times \mathbb{N}$, is countable. By Example I.3.4, there exists a surjective map $\varphi \in (\mathbb{N} \times \mathbb{N})^{\mathbb{N}}$ which arranges the matrix $\mathbb{N} \times \mathbb{N}$. Consequently, X is countable. □

Consequently,

PROPOSITION I.5.5. *The set* $\mathbb{Q}$ *of rational numbers is countable.*

PROOF. Each $q \in \mathbb{Q}$ is of the form $q = \frac{k}{n}$, where k is an integer, and $n \in \mathbb{N}_1$ (non-zero natural number) ([18]). The set $\mathbb{Z}$ of integers is countable, because it can be arranged into a sequence $0, 1, -1, 2, -2, \ldots$ We conclude the proof by using either Proposition I.5.4, or Example I.3.4. □

In course of the proof of Proposition I.5.4, we have demonstrated that

COROLLARY I.5.6. *The product of two countable sets is countable.*

A non-empty set X is called *uncountable* if it is not an image of $\mathbb{N}$, in which case we write $\operatorname{card} X > \aleph_0$.

PROPOSITION I.5.7. *The set* $\mathbb{R}$ *of real numbers is uncountable.*

PROOF. Consider the set $\mathbb{R}$ of real numbers. Suppose that, on the contrary, there exists a surjective map $\varphi \in \mathbb{R}^{\mathbb{N}}$. By assumption, there exists a first natural number n_1 such that $\varphi(0) < \varphi(n_1)$. Again, by assumption, there exists a first natural number n_2 such that $\varphi(0) < \varphi(n_2) < \varphi(n_1)$.

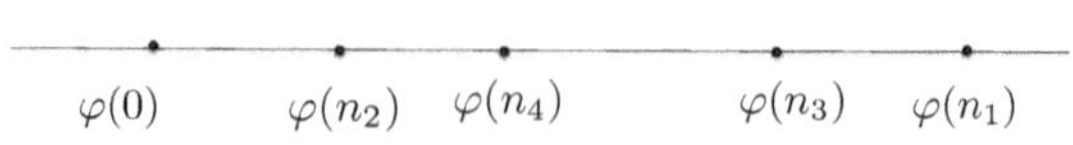

FIGURE I.5.1. Any sequence φ cannot fill the real line.

[18]Each $q \in \mathbb{Q}$ admits many fractional representations; for example, $\frac{1}{3} = \frac{2}{6} = \frac{3}{9} = \frac{11}{33}$.

Accordingly, there is a strictly increasing sequence $(n_k)_k$ such that, for each $k \in \mathbb{N}$,

$$\varphi(n_{2k}) < \varphi(n_{2k+2}) < \ldots < \varphi(n_{2k+3}) < \varphi(n_{2k+1}).$$

As each non-empty bounded subset of $\mathbb{R}$ has an infimum and a supremum, there exists $r \in \mathbb{R}$ such that

$$\sup_{k\in\mathbb{N}} \varphi(n_{2k}) \leq r \leq \inf_{k\in\mathbb{N}} \varphi(n_{2k+1}).$$

By the construction, $r \notin \varphi(\mathbb{N})$, which is a contradiction. □

The cardinality of $\mathbb{R}$ is denoted by $\mathfrak{c}$ and called *continuum.*

We should be aware from the very beginning that there are infinitely many uncountable cardinalities, because

THEOREM I.5.8 (Cantor). *For every set* X,

$$\text{card } X < \text{card } 2^X. \tag{I.5.2}$$

PROOF. The map $h(x) := \{x\}$ is an injection from X to 2^X, hence $\text{card } X \leq \text{card } 2^X$. If $f : X \to 2^X$, then we claim that the set $\{x \in X : x \notin f(x)\}$ does not belong to $f(X)$, that is, f is not surjective.

If on the contrary, there were $x_0 \in X$ such that

$$f(x_0) = \{x \in X : x \notin f(x)\}, \tag{I.5.3}$$

then either $x_0 \in f(x_0)$, that is, $x_0 \notin f(x_0)$ by (I.5.3), or $x_0 \notin f(x_0)$, hence $x_0 \in f(x_0)$ by (I.5.3). We have got a contradiction in any case. Therefore, no $f : X \to 2^X$ is a surjection, and hence (I.5.2). □

Consequently, by an easy induction ([19]), we infer that there are infinitely many infinite cardinals.

By definition, $\lambda^\kappa = \text{card}(Y^X)$, where $\lambda = \text{card } Y$ and $\kappa = \text{card } X$. Therefore,

$$2^\kappa := \text{card } 2^X = \text{card}(\{0,1\}^X)$$

for any set X such that $\text{card } X = \kappa$. Indeed, there exists a bijection between maps $f \in \{0,1\}^X$ and subsets $A \subset X$, realized with the aid of *characteristic functions* χ_A

$$\chi_A(x) := \begin{cases} 1, & x \in A, \\ 0, & x \notin A. \end{cases} \tag{I.5.4}$$

In particular, $\chi_X(x) = 1$ for each $x \in X$, and $\chi_\varnothing(x) = 0$ for each $x \in X$.

[19]Starting from any set X, we get $\text{card } X < \text{card } 2^X < \text{card } 2^{(2^X)}$, and so on.

There exists a widespread convention to denote by $[X]^{<\omega}$ the set of all finite subsets of X. I would rather use $\subset_{\text{fin}}$, that is,

$$A \subset_{\text{fin}} X \iff (A \subset X \wedge \operatorname{card} A < \infty).$$

Along with the symbols $\cup$ and $\bigcup$ (of union), we will also use $\sqcup$ and $\bigsqcup$ of disjoint union, that is,

$$A \cap B = \varnothing \Longrightarrow A \sqcup B := A \cup B.$$

Also $\bigsqcup_{j \in J} A_j := \bigcup_{j \in J} A_j$ provided that $A_j \cap A_k = \varnothing$ if $j \neq k$.

CHAPTER II

From convergence of sequences to the concept of filter

Although filters constitute a principal tool of this book, sequences remain a simple and useful device. The filter approach is more insightful, far-reaching and dexterous than the technique based on sequences. But sequences are designed for enumerating, which is one of the basic procedures in mathematics. We can roughly say that filters are unmatched to formulate and prove theorems, while sequences are unbeaten in constructing counter-examples.

II.1. Convergence of sequences

A *sequence* on a set X is a map from $\mathbb{N}$ to X. As $X^{\mathbb{N}}$ denotes the set of all maps from $\mathbb{N}$ to X, to say that φ is a sequence on X is tantamount to $\varphi \in X^{\mathbb{N}}$. Most commonly, the value of φ at $n \in \mathbb{N}$ is denoted by

$$x_n := \varphi(n),$$

and often a sequence is designated by $(x_n)_{n\in\mathbb{N}}$, or shortly by $(x_n)_n$. By definition,

DEFINITION (A). A sequence φ on the real line $\mathbb{R}$ *converges* to an element x of $\mathbb{R}$ if, for each $\varepsilon > 0$, there exists $n_\varepsilon \in \mathbb{N}$ such that $|\varphi(n) - x| \leq \varepsilon$ for every $n \geq n_\varepsilon$.

Let us reformulate this basic definition in a way that will reveal a novel aspect, which is fundamental for our approach. A strict interval joining a and b will be denoted by ([1])

$$]a, b[:= \{r \in \mathbb{R} : a < r < b\}.$$

As $|\varphi(n) - x| < \varepsilon$ amounts to $\varphi(n) \in]x - \varepsilon, x + \varepsilon[$, we infer that

DEFINITION (B). A sequence φ on the real line *converges* to x if and only if

$$\{n \in \mathbb{N} : \varphi(n) \notin]x - \varepsilon, x + \varepsilon[\}$$

is finite for each $\varepsilon > 0$.

[1]We adopt here the notation $]a, b[$ rather than (a, b), because the latter symbol denotes an ordered pair. This is a traditional French notation.

Let us check that the two definitions are equivalent. Let $\varepsilon > 0$. If $|x_n - x| \leq \varepsilon$ holds for $n \geq n_\varepsilon$, then $\{n : |x_n - x| > \varepsilon\}$ is finite. Conversely, if $\{n : |x_n - x| > \varepsilon\}$ is finite, then letting $n_\varepsilon := \max\{n : |x_n - x| > \varepsilon\} + 1$, we assure that $|x_n - x| \leq \varepsilon$ for each $n \geq n_\varepsilon$.

A subset A of a set N is called *cofinite* if $N \setminus A$ is finite. Accordingly, Definition (B) states that a sequence φ converges to x whenever the set

$$\{n \in \mathbb{N} : \varphi(n) \in]x - \varepsilon, x + \varepsilon[\}$$

is cofinite for each $\varepsilon > 0$.

Notice that Definition (B) does not mention the standard order on the set $\mathbb{N}$ of indices of a sequence, which was used in the standard Definition (A)! It follows, in particular, that any permutation of indices of a sequence does not affect its convergence.

Therefore, the order on the set of indices turns out to be useless, or at least, unnecessary from the convergence point of view. Therefore, instead of using $\mathbb{N}$ for the set of indices, we can take an arbitrary infinite countable set N ([2]). This leads to

Definition (C). A sequence $\varphi \in \mathbb{R}^N$ *converges* to x if and only if, for each $\varepsilon > 0$, there exists a cofinite subset F of N such that

$$\varphi(F) \subset]x - \varepsilon, x + \varepsilon[.$$

We notice that if F is a cofinite subset of N and $F \subset G \subset N$, then G is also a cofinite subset of N. Any finite intersection of cofinite subsets of N is cofinite (because any finite union of finite sets is finite). Finally, the empty subset of N is not cofinite, because N is infinite.

An interval $]x - \varepsilon, x + \varepsilon[$ is a particular (symmetric) neighborhood of x. More generally, a subset V of $\mathbb{R}$ is called a *neighborhood* of x if there exist numbers a and b such that $a < x < b$, and $]a, b[\subset V$. Denote by $\mathcal{N}(x)$ the family of all neighborhoods of x. Notice that

(isotone) $$W \supset V \in \mathcal{N}(x) \Longrightarrow W \in \mathcal{N}(x),$$

(finitely complete) $$V_0, V_1 \in \mathcal{N}(x) \Longrightarrow V_0 \cap V_1 \in \mathcal{N}(x),$$

(proper) $$\varnothing \notin \mathcal{N}(x).$$

Actually, the formula (proper) is a consequence of a stronger property of neighborhoods, that is, $x \in V$ for each $V \in \mathcal{N}(x)$.

Let N be an infinite countable set and $\varphi \in X^N$. Let $\Gamma\varphi$ be the family of subsets of X defined by

$$E \in \Gamma\varphi \Longleftrightarrow \varphi^{-}(E) \text{ is cofinite in } N. \tag{II.1.1}$$

[2]A map from an infinite countable set N to a set X is called a *quence* on X [83].

In other words, $E \in \Gamma\varphi$ if and only if there exists a cofinite subset A of N such that $\varphi(A) \subset E$. Notice that

(isotone) $$F \supset E \in \Gamma\varphi \Longrightarrow F \in \Gamma\varphi,$$

(finitely complete) $$E_0, E_1 \in \Gamma\varphi \Longrightarrow E_0 \cap E_1 \in \Gamma\varphi,$$

(proper) $$\varnothing \notin \Gamma\varphi.$$

PROPOSITION II.1.1. *Let φ be a sequence and $\Gamma\varphi$ be defined by (II.1.1). Then φ converges to x if and only if*

$$\mathcal{N}(x) \subset \Gamma\varphi.$$

PROOF. In fact, suppose that φ converges to x, and let $V \in \mathcal{N}(x)$. By the definition of $\mathcal{N}(x)$, there exists $\varepsilon > 0$ such that $]x-\varepsilon, x+\varepsilon[\subset V$. By definition of $\Gamma\varphi$, there is $E \in \Gamma\varphi$ such that $E \subset V$, hence by (isotone), $V \in \Gamma\varphi$. Conversely, if $]x-\varepsilon, x+\varepsilon[\in \Gamma\varphi$, then $\{n : \varphi(n) \notin]x-\varepsilon, x+\varepsilon[\}$ is finite, hence φ converges to x. □

Subsequences. A sequence $(y_k)_k$ is called a *subsequence* of a sequence $(x_n)_n$ whenever there exists a strictly increasing $(n_k)_k$ such that $y_k = x_{n_k}$ for each $k < \omega$. In other terms, a sequence ψ is a subsequence of a sequence φ if there exists a strictly increasing map $\gamma : \mathbb{N} \longrightarrow \mathbb{N}$ such that $\psi(k) = \varphi(\gamma(k))$ for each $k \in \mathbb{N}$.

As you expect, if ψ is a subsequence of a sequence φ, then

$$\Gamma\varphi \subset \Gamma\psi. \tag{II.1.2}$$

Indeed, $E \in \Gamma\varphi$ if and only if $\varphi^-(E)$ is cofinite. As γ is strictly increasing, ([3]) $\psi^-(E) = \gamma^-(\varphi^-(E))$ is cofinite, hence $E \in \Gamma\psi$.

Hence,

COROLLARY II.1.2. *If a sequence φ converges to x and ψ is a subsequence of φ, then ψ converges to x.*

Of course, we do not need the order on indices to define a subsequence.

PROPOSITION II.1.3. *If $\varphi \in X^N$ and $\psi \in X^K$, where N and K are infinite countable sets, then ψ is a subsequence of φ if and only if there is an injection $\gamma \in N^K$ such that $\psi = \varphi \circ \gamma$.*

It follows that

LEMMA II.1.4. *$\Gamma\varphi \subset \Gamma\psi$ if and only if, for each cofinite subset G of N, the set $\psi^-(\varphi(G))$ is cofinite in K.*

---

[3]Then there is k_0 such that $\{\gamma(k) : k > k_0\} \subset \varphi^-(E)$.

In particular, if ψ is a subsequence of φ, then the condition of Lemma II.1.4 holds.

More generally, if $\psi = \varphi \circ \gamma$ and $\gamma^{-}(G)$ is cofinite for each cofinite subset G of N, then $\Gamma\varphi \subset \Gamma\psi$ ([4]).

From this perspective, $\varphi \subset N \times X$ and $\psi \subset K \times X$, so that $\psi^{-} \subset X \times K$ is the inverse relation of ψ. The composed relation $\psi^{-} \circ \varphi \subset N \times K$, that we call the *correlation* φ with ψ, can be written explicitly

$$\psi^{-} \circ \varphi = \{(n,k) : \varphi(n) = \psi(k)\}. \tag{II.1.3}$$

Now, Lemma II.1.4 can be rephrased: $\Gamma\varphi \subset \Gamma\psi$ if and only if the correlation $\psi^{-} \circ \varphi$ preserves cofinite sets.

The accomplishment of this section is twofold: we have shown that, from the convergence point of view, a sequence $\varphi : \mathbb{N} \longrightarrow X$ can be, not only freed from the order on the set $\mathbb{N}$ of its indices, but ultimately, relieved from the set of indices altogether.

This looks like a miracle, but it is merely a pertinent perspective. Not everything, however, is abandoned during this change of viewpoint, because the structure of cofinite sets is a residue of the natural order on the natural numbers. On the other hand, it is often instrumental to deploy natural numbers in construction or description of sequences.

II.2. The concept of filter

In the preceding section we have observed that the convergence of a sequence to a point of the real line can be characterized by the inclusion between two families of sets, one related to the sequence, the other related to the point. Moreover, these families have some common properties, that is, they are (isotone), (finitely complete), and (proper).

Let X be a non-empty set. A family $\mathcal{F}$ of subsets of X is said to be *isotone* if

$$F \in \mathcal{F} \text{ and } F \subset G \Longrightarrow G \in \mathcal{F}. \tag{isotone}$$

A family $\mathcal{F}$ of subsets of X is called a *filter* (on X) if it is (isotone) and

$$F_0 \in \mathcal{F} \text{ and } F_1 \in \mathcal{F} \Longrightarrow F_0 \cap F_1 \in \mathcal{F}. \tag{finitely complete}$$

A filter $\mathcal{F}$ is said to be *proper* if

$$\varnothing \notin \mathcal{F}. \tag{proper}$$

[4] As we shall see in Section XVI.2, this sufficient condition will become also necessary if we allow γ to be defined on a cofinite subset of K.

Non-empty proper filters are called *non-degenerate*. If the condition (proper) does not hold, then we say that $\mathcal{F}$ is *improper* ([5]), *degenerate* if it is either improper or empty.

The family of neighborhoods of a point, and the family associated with a sequence by (II.1.1) are proper filters.

We denote by $\mathbb{F}$ the class of non-empty proper filters, and by $\overline{\mathbb{F}}$ the class of all filters. If X is a set, then the set of all non-empty proper filters on X is denoted by $\mathbb{F}X$ ([6]), and of all filters on X by $\overline{\mathbb{F}}X$. The only improper filter on a (non-empty) set X is 2^X (the family of all subsets of X), because by (isotone), $\varnothing \in \mathcal{F}$ implies that $G \in \mathcal{F}$ for each $\varnothing \subset G \subset X$. Hence $\overline{\mathbb{F}}X = \mathbb{F}X \cup \{\varnothing, 2^X\}$.

DEFINITION II.2.1 (Filter-base). A subfamily $\mathcal{B}$ of a proper filter $\mathcal{F}$ is called a *base* (or a *filter-base*) of $\mathcal{F}$ if for each $F \in \mathcal{F}$, there exists $B \in \mathcal{B}$ such that $B \subset F$. We say also that $\mathcal{B}$ *generates* $\mathcal{F}$.

In particular, each filter constitutes its own filter-base.

PROPOSITION II.2.2. *A non-empty family $\mathcal{B}$ of subsets of X is a filter-base if and only if $\varnothing \notin \mathcal{B}$ and for each $B_0, B_1 \in \mathcal{B}$, there exists $B \in \mathcal{B}$ such that $B \subset B_0 \cap B_1$.*

PROOF. If $\mathcal{F}$ is a proper filter and $\mathcal{B} \subset \mathcal{F}$, then $B_0 \cap B_1 \in \mathcal{F}$. If $\mathcal{B}$ is a filter-base of $\mathcal{F}$, then there exists $B \in \mathcal{F}$ such that $B \subset B_0 \cap B_1$.

Conversely, if the condition holds, then the family $\mathcal{F}$ such that, for each $F \in \mathcal{F}$ there exists $B \in \mathcal{B}$ with $B \subset F$, is a filter. Obviously, $\mathcal{F}$ is isotone. If now $F_0, F_1 \in \mathcal{F}$, then there exist $B_0, B_1 \in \mathcal{B}$ such that $B_0 \subset F_0$ and $B_1 \subset F_1$, hence there exists $B \in \mathcal{B}$ such that $B \subset B_0 \cap B_1 \subset F_0 \cap F_1$, hence $F_0 \cap F_1 \in \mathcal{F}$. Moreover, $\varnothing \notin \mathcal{F}$ for otherwise there would be $B \in \mathcal{B}$ such that $B \subset \varnothing$. □

If $\mathcal{A} \subset 2^X$, then

$$\mathcal{A}^{\uparrow} := \{F \subset X : \underset{A \in \mathcal{A}}{\exists}\ F \supset A\}. \tag{II.2.1}$$

Sure enough, a family $\mathcal{A}$ is isotone if and only if $\mathcal{A} = \mathcal{A}^{\uparrow}$. It is plain that $\mathcal{B}$ is a filter-base of $\mathcal{F}$ if and only if $\mathcal{B}^{\uparrow} = \mathcal{F}$. If $\mathcal{A} = \{\{x\}\}$, then we abridge

$$x^{\uparrow} := \{\{x\}\}^{\uparrow}.$$

[5] The empty family of subsets of a non-empty set is a proper filter.

[6] We make a distinction between a *class* (of proper filters) and a *set* (of filters on a given set). This is because, by the axioms of set theory (Chapter XXVII), all proper filters do not constitute a set, but the proper filters on any set X do.

DEFINITION II.2.3 (Filter-subbase). A family $\mathcal{D}$ of subsets of X is called *centered* or a *filter-subbase* if

$$\mathcal{D}_0 \subset_{\text{fin}} \mathcal{D} \Longrightarrow \bigcap\nolimits_{D\in\mathcal{D}_0} D \neq \varnothing. \tag{II.2.2}$$

(II.2.2) is called the *finite intersection property.*

If $\mathcal{D}$ is a subbase, then $\{\bigcap \mathcal{D}_0 : \mathcal{D}_0 \subset_{\text{fin}} \mathcal{D}\}$ is a filter-base.

DEFINITION II.2.4 (Principal filter). A filter is called *principal* (or *finitely based*, or else of *finite character*) if it has a finite filter-base.

Consequently, each principal filter on X has a filter-base of the form $\{B\}$, where $B \subset X$, because $\{B\}^{\uparrow} := \{F \subset X : B \subset F\}$ in view of (II.2.1). Of course, $\{B\}^{\uparrow}$ is proper if and only if $B \neq \varnothing$. The class of all proper principal filters is denoted by $\mathbb{F}_0$. Notice that if a proper filter $\mathcal{F}$ is principal, then $\bigcap_{F\in\mathcal{F}} F \in \mathcal{F}$.

Let us extend the concept of *cofinite* set introduced in Section II.1.

DEFINITION II.2.5 (Cofinite set). Let X be a set. A subset F of X is said to be *cofinite* (*in* X) if $X \setminus F$ is finite.

The family of all cofinite subsets of X is denoted by $(X)_0$.

PROPOSITION II.2.6. *The family* $(X)_0$ *of cofinite subsets of a non-empty set* X *is a proper filter if and only if* X *is infinite.*

PROOF. Indeed, if F_0, F_1 are cofinite subsets of X, then $X \setminus (F_0 \cap F_1) = (X \setminus F_0) \cup (X \setminus F_1)$ is finite, hence $F_0 \cap F_1$ is cofinite. Now, X is finite if and only if $\varnothing \in (X)_0$, equivalently $(X)_0$ is improper. □

DEFINITION II.2.7 (Free filter). A filter $\mathcal{F}$ is called *free* if $\bigcap_{F\in\mathcal{F}} F = \varnothing$.

LEMMA II.2.8. *A proper filter* $\mathcal{F}$ *on an infinite set* X *is free if and only if* $\mathcal{F} \supset (X)_0$.

PROOF. By definition, a filter $\mathcal{F}$ on X is free if and only if $\bigcap_{F\in\mathcal{F}} F = \varnothing$, that is, if and only if for each $x \in X$, there is $F \in \mathcal{F}$ such that $x \notin F$, hence $X \setminus \{x\} \in \mathcal{F}$. By Formula (finitely complete), finite intersections of such sets also belong to $\mathcal{F}$, that is, $X \setminus G \in \mathcal{F}$ for each finite subset G of X. In other words, a free filter on X contains all cofinite subsets of X. Conversely, $\bigcap_{H\in(X)_0} H = \varnothing$, because $X \setminus \{x\} \in (X)_0$ for every $x \in X$. Hence, if $\mathcal{F} \supset (X)_0$ then $\bigcap_{F\in\mathcal{F}} F = \varnothing$. □

DEFINITION II.2.9 (Sequential filter). A filter $\mathcal{F}$ on a set X is called *sequential* if there exists a sequence $\varphi \in X^N$ such that $\mathcal{F} = \Gamma\varphi$ (II.1.1).

The class of all *sequential filters* is denoted by $\mathbb{S}$, and the set of all sequential filters on X by $\mathbb{S}X$ or εX. The same definition can be rewritten traditionally, which reads as follows

PROPOSITION II.2.10. *A filter $\mathcal{F}$ on a set X is sequential if and only if there exists a sequence $(x_n)_n$ of elements of X so that*

$$\{\{x_k : k \geq n\} : n \in \mathbb{N}\} \tag{II.2.3}$$

is a filter-base of $\mathcal{F}$.

REMARK II.2.11. Recall that traditionally a sequence $(x_n)_n$ on X is indexed by the set of natural numbers $\mathbb{N}$, and it is identified with a map $\varphi \in X^{\mathbb{N}}$ to the effect that $\varphi(n) := x_n$. Moreover, we have pointed out that the order of $\mathbb{N}$ is irrelevant for convergence questions concerning sequences, so that it is enough to view sequences on X as maps $\varphi \in X^N$, where N is any infinite countable set.

COROLLARY II.2.12. *Each sequential filter contains a countable set.*

We say that a sequence φ is *free* if $\Gamma\varphi$ is a free filter, *principal* if $\Gamma\varphi$ is a principal filter.

By definition, each sequential filter on a set X is of the form $\Gamma\varphi$, where $\varphi \in X^N$ and N is an infinite countable set.

PROPOSITION II.2.13. *A filter $\Gamma\varphi$ is free if and only if $\varphi^-(x)$ is finite for each $x \in \varphi(N)$.*

PROOF. Suppose that there is $x \in \varphi(N \setminus A)$ for each $A \subset_{\text{fin}} N$, that is, $\varphi^-(x) \setminus A \neq \varnothing$ for each $A \subset_{\text{fin}} N$, hence $\varphi^-(x)$ is infinite. Conversely, if for every $x \in \varphi(N)$ there is $A \subset_{\text{fin}} N$ such that $x \notin \varphi(N \setminus A)$, then $\varphi^-(x) \subset A$, thus $\varphi^-(x)$ is finite. $\square$

PROPOSITION II.2.14. *A filter $\Gamma\varphi$ is principal if and only if $\varphi^-(x)$ is infinite for each x in a cofinite subset of $\varphi(N)$.*

PROOF. Let E be a non-empty subset of X such that $\Gamma\varphi = \{E\}^\uparrow$. It means that $E \subset \varphi(F)$ for each cofinite subset F of N. The set $D := \{x \in X : \operatorname{card} \varphi^-(x) < \infty\}$ is disjoint from E, because $E \subset \varphi(N \setminus \varphi^-(x))$ for each $x \in D$. Accordingly, if $x \in E$ then $\varphi^-(x)$ is infinite. On the other hand, D is finite, because otherwise $\varphi^-(D)$ would be an infinite subset of N such that $\varphi^-(D) \cap F \neq \varnothing$ for each cofinite subset F of N. Accordingly, $\psi := \varphi|_{\varphi^-(D)}$ would be a subsequence of φ, which is impossible, because $\psi(\varphi^-(D)) \cap E = \varnothing$.

The filter $\Gamma\varphi$ has a countable base of the form $\{\varphi(N \setminus A) : A \subset_{\text{fin}} N\}$. If $\Gamma\varphi$ is not principal, then the filter-base is infinite, hence there exists a strictly decreasing base $\{B_k : k < \omega\}$. If $x_k \in B_k \setminus B_{k+1}$, then

$(x_k)_k$ is a free subsequence of φ, and $\{x_k : k < \omega\} \cap \bigcap_{k<\omega} B_k \neq \varnothing$. Plainly, $\varphi^-(x_k)$ is finite for each $k < \omega$, for otherwise $x_k \in \bigcap_{m<\omega} B_m$. □

DEFINITION II.2.15. A filter is called *countably carried* if it contains a countable set. We denote by $\mathbb{E}$ the class of countably carried filters.

Consequently, Corollary II.2.12 says that all sequential filters are countably carried.

DEFINITION II.2.16 (Countably based filter). A filter is called *countably based* (or of *countable character*) if it has a countable filter-base.

Notice that if $\mathcal{F}$ is a countably based filter, and $(F_n)_n$ is its filter-base, and

$$G_n := F_0 \cap F_1 \cap \ldots \cap F_n,$$

then $(G_n)_n$ is a decreasing filter-base of $\mathcal{F}$.

The class of countably based filters is denoted by $\mathbb{F}_1$, while the set of all proper countably based filters on X is denoted by $\mathbb{F}_1 X$. Of course, each principal filter is countably based, that is, $\mathbb{F}_0 \subset \mathbb{F}_1$. Every sequential filter is countably based, that is, $\mathbb{S} \subset \mathbb{F}_1$.

DEFINITION II.2.17. The *character* of a filter $\mathcal{F}$, denoted by $\chi(\mathcal{F})$, is the least cardinality of its filter-bases.

If $\chi(\mathcal{F}) < \infty$, then $\chi(\mathcal{F}) = 1$, that is, if and only if $\mathcal{F} \in \mathbb{F}_0$. On the other hand, $\chi(\mathcal{F}) \leq \aleph_0$ if and only if $\mathcal{F} \in \mathbb{F}_1$.

If X is an infinite countable set, then there exists a sequence $(x_n)_{n\in\mathbb{N}}$ such that $X = \{x_n : n < \omega\}$. Then $\{\{x_k : k > n\} : n \in \mathbb{N}\}$ is a filter-base of $(X)_0$.

EXAMPLE II.2.18. Let $\{X_n : n < \omega\}$ be a family of infinite countable sets. Then $X := \bigsqcup_{n<\omega} X_n$ is a countable set. The family $\{\bigcup_{n>k} X_n : k < \omega\}$ is a countable filter-base. The corresponding filter is, of course, countably carried. It will be justified by Proposition XVI.1.2 that it is not sequential.

PROPOSITION II.2.19. *The character of $(X)_0$ is uncountable if and only if X is uncountable.*

PROOF. Assume that the character of $(X)_0$ is uncountable, that is, each base is uncountable. As $\{X \setminus \{x\} : x \in X\}$ is a filter-base of $(X)_0$, so that X is uncountable. If X is uncountable, and by way of contradiction, if $\{F_n : n < \omega\}$ is a family of finite sets, for which $\{X \setminus F_n : n < \omega\}$ is a filter-base of $(X)_0$, then there exists $x_\infty \notin \bigcup_{n<\omega} F_n$, because X is uncountable. Then $X \setminus \{x_\infty\} \in (X)_0$, so that there is $n < \omega$ such that $X \setminus F_n \subset X \setminus \{x_\infty\}$, equivalently $x_\infty \in F_n$, which yields a contradiction. □

Slightly more generally than in Definition II.2.5,

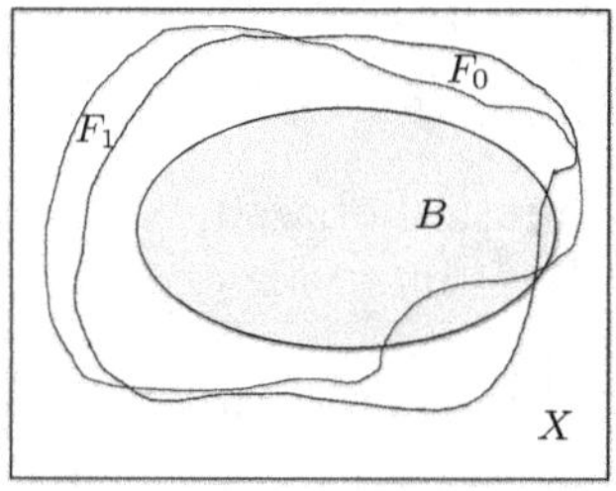

FIGURE II.2.1. Cofinite filter $(B)_0$: $F \in (B)_0$ if and only if $B \setminus F$ is finite.

EXAMPLE II.2.20. If $B \subset X$, then a subset F of X is called *cofinite of* B (or B-*cofinite*) if $B \setminus F$ is finite ([7]). The family $(B)_0$ of all B-cofinite sets is a filter, called the *cofinite filter at* B ([8]). The filter $(B)_0$ is proper if and only if B is infinite, because then $B \setminus \varnothing$ is not finite, and thus $\varnothing \notin (B)_0$. As for $(X)_0$, a filter $(B)_0$ is countably based if and only if B is countable and infinite.

Observe that

PROPOSITION II.2.21. *The cofinite filter of infinite countable set is sequential and free.*

PROOF. Indeed, if B is a countable and infinite subset of X, then $\varphi = i_B \in B^B$ in Definition II.2.9 to the effect that $(B)_0 = \Gamma i_B$. As $X \setminus \{x\} \in (B)_0$ for each $x \in X$, the filter $(B)_0$ is free. □

In particular, $(B)_0$ is a proper sequential filter on X if and only if B is an infinite countable subset of X. Notice that

PROPOSITION II.2.22. *A principal filter* $\{A\}^\uparrow$ *is sequential if and only if* A *is countable.*

PROOF. If $\{A\}^\uparrow$ is a sequential filter, then by Corollary II.2.12, A is countable. Conversely, if A is countable, then if A is finite, say

[7] See also Exercise II.5.7.

[8] Indeed, if $G \supset F$ and $B \setminus F$ is finite, then $B \setminus G$ is finite, thus (isotone) holds. To show (finitely complete), observe that $B \setminus (F_0 \cap F_1) = (B \setminus F_0) \cup (B \setminus F_1)$ is finite, if $F_0, F_1 \in (B)_0$, and thus $F_0 \cap F_1 \in (B)_0$.

$A = \{a_1, \ldots, a_k\}$, then let $\varphi(n) = a_{[n]_k}$, where $[n]_k$ is the rest of the division of n by k, and thus $\{\varphi(n) : n > m\} = A$ for each $m < \omega$.

If A is infinite countable, say $A = \{a_k : k \in \mathbb{N}\}$, then consider a sequence φ defined by

$$a_0, \quad a_0, a_1, \quad a_0, a_1, a_2, \quad a_0, a_1, a_2, a_3, \quad \ldots,$$

and thus $\{\varphi(n) : n > m\} = A$ for each $m < \omega$. □

EXAMPLE II.2.23 (Countably based, non-sequential filter). Let $x \in \mathbb{R}$ and consider the family of intervals of the real line

$$\mathcal{B} := \left\{]x - \tfrac{1}{n}, x + \tfrac{1}{n}[: n \in \mathbb{N}_1 \right\}.$$

This family is evidently countable, but each non-empty strict interval of the real line is uncountable ([9]). Therefore the filter $\mathcal{B}^{\uparrow}$ is not sequential.

EXAMPLE II.2.24 (Cocountable filter). We say that a subset F of X is *cocountable* if $X \setminus F$ is a countable set or empty; the family of all cocountable subsets of X is denoted by $(X)_1$. More generally, if $B \subset X$, then a subset F of X is called *cocountable subset of* B (or B-*co*countable) if $B \setminus F$ is countable or empty. The family $(B)_1$ of all B-cocountable sets is a filter, called the *cocountable filter at* B; it is proper if and only if B is uncountable, because then $B \setminus \varnothing$ is not countable, and thus $\varnothing \notin (B)_1$. Each cocountable filter of an uncountable set is free, because $(B)_0 \subset (B)_1$.

II.3. Order on filters

It is easy to see that the *inclusion* $\subset$ on the set $\overline{\mathbb{F}}X$ is reflexive, antisymmetric, and transitive. Therefore, $(\overline{\mathbb{F}}X, \subset)$ is an ordered set.

It follows from definitions that if $\mathcal{F}_0, \mathcal{F}_1 \in \overline{\mathbb{F}}X$, then

$$\mathcal{F}_0 \vee \mathcal{F}_1 = \{F_0 \cap F_1 : F_0 \in \mathcal{F}_0, F_1 \in \mathcal{F}_1\}, \tag{II.3.1}$$

$$\mathcal{F}_0 \wedge \mathcal{F}_1 = \mathcal{F}_0 \cap \mathcal{F}_1. \tag{II.3.2}$$

If $\mathcal{F}_0, \mathcal{F}_1$ are proper filters, then $\mathcal{F}_0 \wedge \mathcal{F}_1$ is proper, but $\mathcal{F}_0 \vee \mathcal{F}_1$ is proper only if $F_0 \cap F_1 \neq \varnothing$ for each $F_0 \in \mathcal{F}_0$ and $F_1 \in \mathcal{F}_1$.

EXAMPLE II.3.1. If $B_1, \ldots, B_n$ are countable, infinite sets, then $(B_0)_0, \ldots, (B_n)_0$ are free cofinite filters. Then the infimum $(B_0)_0 \wedge \ldots \wedge (B_n)_0 = (B_1 \cup \ldots \cup B_n)_0$. The supremum $(B_0)_0 \vee \ldots \vee (B_n)_0 = (B_1 \cap \ldots \cap B_n)_0$ is proper whenever $B_1 \cap \ldots \cap B_n$ is infinite.

[9] Recall that the real line $\mathbb{R}$ is uncountable. Every non-empty strict interval of $\mathbb{R}$ is uncountable as a bijective image of $\mathbb{R}$.

Recall that an ordered set, in which a supremum and an infimum exist for any two elements, hence for any finite set of elements, is called a *lattice*; a *complete lattice* whenever all non-empty subsets admit suprema and infima (Definition I.3.3).

If $\mathcal{A}$ is a family of subsets of X, then

$$\mathcal{A}^{\cap} := \{\bigcap \mathcal{B} : \mathcal{B} \subset_{\text{fin}} \mathcal{A}\}$$

denotes the family of finite intersections of the elements of $\mathcal{A}$.

PROPOSITION II.3.2. *The set of all filters on a given set, ordered by inclusion, is a complete lattice; if $\boldsymbol{H}$ is a non-empty set of (possibly improper) filters on X, then*

$$\bigwedge \boldsymbol{H} = \bigcap\nolimits_{\mathcal{H} \in \boldsymbol{H}} \mathcal{H}, \quad \bigvee \boldsymbol{H} = (\bigcup\nolimits_{\mathcal{H} \in \boldsymbol{H}} \mathcal{H})^{\cap}.$$

PROOF. The family $\bigcap_{\mathcal{H} \in \boldsymbol{H}} \mathcal{H}$ is included in every filter from $\boldsymbol{H}$, and it is a filter. Indeed, if $F_0, F_1 \in \bigcap_{\mathcal{H} \in \boldsymbol{H}} \mathcal{H}$, then $F_0, F_1 \in \mathcal{H}$ for each $\mathcal{H} \in \boldsymbol{H}$, hence $F_0 \cap F_1 \in \mathcal{H}$ for each $\mathcal{H} \in \boldsymbol{H}$, so that $F_0 \cap F_1 \in \bigcap_{\mathcal{H} \in \boldsymbol{H}} \mathcal{H}$. On the other hand, $\bigcup_{\mathcal{H} \in \boldsymbol{H}} \mathcal{H}$ is the least family of sets including every filter from $\boldsymbol{H}$, but in general it is not a filter. We get the least (possibly improper) filter that includes it, by taking all the finite intersections of the elements of $\bigcup_{\mathcal{H} \in \boldsymbol{H}} \mathcal{H}$. □

COROLLARY II.3.3. *The supremum $\bigvee \boldsymbol{H}$ of a set $\boldsymbol{H}$ of filters is proper if and only if $\bigcap \mathcal{G} \neq \varnothing$ for each finite family $\mathcal{G} \subset \bigcup_{\mathcal{H} \in \boldsymbol{H}} \mathcal{H}$.*

Consequently, the set $\mathbb{F}X$ of proper filters is not even a lattice. Although the intersection of each non-empty set $\boldsymbol{H}$ of proper filters is a proper filter, the supremum exists in $\mathbb{F}X$ if and only if $\bigcup \boldsymbol{H} := \bigcup_{\mathcal{H} \in \boldsymbol{H}} \mathcal{H}$ is closed under finite intersections. Otherwise, the supremum exists in $\overline{\mathbb{F}}X = \mathbb{F}X \cup \{2^X\}$, and is equal to the degenerate filter 2^X.

PROPOSITION II.3.4. *Each infimum of principal filters is a principal filter: $\bigwedge_{A \in \mathcal{A}} \{A\}^{\uparrow} = \{\bigcup_{A \in \mathcal{A}} A\}^{\uparrow}$.*

However,

PROPOSITION II.3.5. *Each filter is a supremum of principal filters.*

PROOF. Let $\mathcal{F}$ be an arbitrary filter. Then $\mathcal{F} = \bigvee_{F \in \mathcal{F}} \{F\}^{\uparrow}$. Of course, $\supset$ holds. On the other hand, if $F_0 \in \mathcal{F}$ then $F_0 \in F_0^{\uparrow}$, hence $F_0 \in \bigvee_{F \in \mathcal{F}} \{F\}^{\uparrow}$. □

PROPOSITION II.3.6. *Each infimum of free filters is free.*

PROOF. Indeed, by Lemma II.2.8, a filter $\mathcal{F}$ on X is free if and only if $\mathcal{F} \supset (X)_0$, hence if $\boldsymbol{F}$ is a set of free filters on X, then $\bigwedge \boldsymbol{F} \supset (X)_0$, hence $\bigwedge \boldsymbol{F}$ is free. □

PROPOSITION II.3.7. *Each countably based filter is an infimum of sequential filters.*

PROOF. Let $\mathcal{F}$ be a countably based filter on a set X, then $\{\mathcal{E} \in \mathbb{S} : \mathcal{E} \supset \mathcal{F}\} \neq \varnothing$. Indeed, let $(F_n)_n$ be a decreasing filter-base of $\mathcal{F}$. If there exists n_0 such that $F_n = F_{n_0}$ for each $n \geq n_0$ then there exists $x \in F_n$ for each $n < \omega$. If $\varphi(n) := x$, then $\Gamma\varphi = x^{\uparrow} \supset \mathcal{F}$. Otherwise, there exists a strictly increasing sequence $(n_k)_k$ and $x_k \in F_{n_k} \setminus F_{n_k+1}$. If now $\varphi(k) := x_k$, then $\Gamma\varphi \supset \mathcal{F}$. Of course, $\mathcal{F} \subset \bigcap\{\mathcal{E} \in \mathbb{S} : \mathcal{E} \supset \mathcal{F}\}$.

Conversely, if $H \notin \mathcal{F}$ and $(F_n)_n$ is a filter-base of $\mathcal{F}$, then $F_n \setminus H \neq \varnothing$ for each $n < \omega$. Accordingly, there exists a sequence $(x_n)_n$ such that $x_n \in F_n \setminus H$. Let $\varphi(n) := x_n$. Then $\Gamma\varphi \supset \mathcal{F}$ and $H \notin \Gamma\varphi \supset \bigcap\{\mathcal{E} \in \mathbb{S} : \mathcal{E} \supset \mathcal{F}\}$. Therefore, $\mathcal{F} \supset \bigcap\{\mathcal{E} \in \mathbb{S} : \mathcal{E} \supset \mathcal{F}\}$. □

If $\mathcal{F}$ is a countably carried filter, then there exists a sequential filter $\mathcal{E} \subset \mathcal{F}$. Indeed, a countable set $E \in \mathcal{F}$, then $\{E\}^{\uparrow} \subset \mathcal{F}$.

The class $\mathbb{E}$ is *isotone*, that is, if $\mathcal{E} \in \mathbb{E}$ and $\mathcal{F} \supset \mathcal{E}$, then $\mathcal{F} \in \mathbb{E}$. The class $\mathbb{S}$ of sequential filters is not isotone, because by Proposition V.4.8, there exist non-sequential filters finer than sequential filters.

DEFINITION II.3.8 (Fréchet filter). A filter is called *Fréchet* if it is an infimum of sequential filters.

Let us warn that the term *Fréchet filters* has been often employed to designate *sequential filters* in our sense ([10])!

Proposition II.3.7 implies that every countably based filter is Fréchet. As we shall see in Example XVI.4.5, there exist Fréchet filters which are not countably based.

If $\mathcal{F}$ is a filter on X and $A \subset X$, then we abridge

$$\mathcal{F} \vee A := \mathcal{F} \vee \{A\}^{\uparrow}, \tag{II.3.3}$$

and call $\mathcal{F} \vee A$ the *restriction* of $\mathcal{F}$ to A. Notice that $\mathcal{F} \vee A$ is proper if and only if $A \cap F \neq \varnothing$ for each $F \in \mathcal{F}$. It is also handy to use another abbreviation, namely

$$\mathcal{F} \setminus A := \mathcal{F} \vee (X \setminus A)^{\uparrow}. \tag{II.3.4}$$

DEFINITION II.3.9 (Ultrafilter). A proper filter $\mathcal{U}$ is called an *ultrafilter* if $\mathcal{H} \supset \mathcal{U}$ implies $\mathcal{H} = \mathcal{U}$ for every proper filter $\mathcal{H}$.

[10]We have adopted the current terminology for several reasons. The main reason is that there is a class of topological spaces traditionally called *Fréchet* spaces. They can be characterized as the spaces, every vicinity filter of which is Fréchet in the sense of Definition II.3.8. Moreover, it is preferable to give a mathematical object a common name evoking its characteristic features, rather than that of a mathematician who contributed to its study. In particular, *sequential filters* are so common and were effectively studied by many a mathematician.

We denote by $\mathbb{U}$ the class of all ultrafilters. For each proper filter $\mathcal{F}$, there exists an ultrafilter $\mathcal{U}$ greater than $\mathcal{F}$ ([11]). The set of all ultrafilters which include a filter $\mathcal{F}$ is denoted by $\beta\mathcal{F}$. Observe that if $\mathcal{F}$ and $\mathcal{G}$ are filters, then $\mathcal{F} \subset \mathcal{G}$ implies $\beta\mathcal{F} \supset \beta\mathcal{G}$. Consequently, the set of all ultrafilters on X is denoted by $\mathbb{U}X$ or by βX (as a shorthand for $\beta\{X\}$).

In other words, ultrafilters on X are maximal elements of $\mathbb{F}X$. If X is a non-empty set, then for any $x \in X$,

$$x^{\uparrow} := \{F \subset X : x \in F\}$$

is an ultrafilter. Indeed, if $\mathcal{H}$ is a filter on X such that $x^{\uparrow} \subset \mathcal{H}$ and $H \in \mathcal{H} \setminus x^{\uparrow}$, then $x \notin H$ and thus $H \cap \{x\} = \varnothing$, contradicting Formula (finitely complete). In view of (II.2.1), $x^{\uparrow} = \{\{x\}\}^{\uparrow}$.

We shall examine ultrafilters in detail in Section V.4. Let us observe right now that

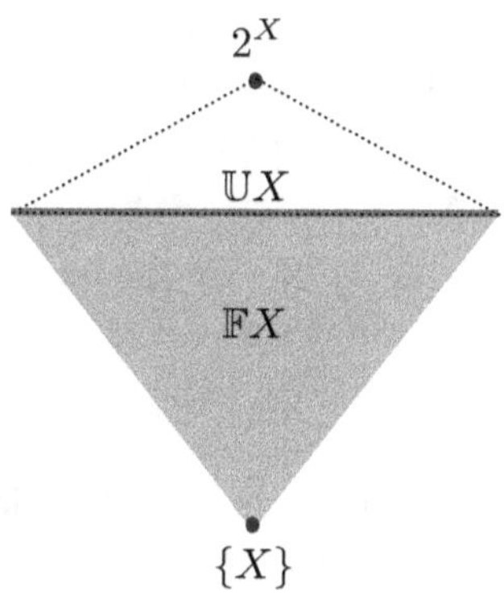

FIGURE II.3.1. The coarsest filter on X is $\{X\}$, the $\mathbb{U}X$ consists of maximal proper filters, and the degenerate filter 2^X is the finest.

PROPOSITION II.3.10. *A filter $\mathcal{F}$ on X is an ultrafilter if and only if either $A \in \mathcal{F}$ or $X \setminus A \in \mathcal{F}$ for every $\varnothing \neq A \subset X$.*

PROOF. If $\mathcal{F}$ is an ultrafilter on X and $\varnothing \neq A \notin \mathcal{F}$, then $F \setminus A \neq \varnothing$ for each $F \in \mathcal{F}$ in view of Formula (isotone). Hence $\{F \setminus A : F \in \mathcal{F}\}$ is a filter-base, so that $\mathcal{F} \setminus A$ is a proper filter. Evidently, $\mathcal{F} \setminus A \supset \mathcal{F}$, and thus $X \setminus A \in \mathcal{F} \setminus A = \mathcal{F}$.

Conversely, if there exists $\varnothing \neq A \notin \mathcal{F}$ so that $A \notin \mathcal{F}$ and $X \setminus A \notin \mathcal{F}$, then $\mathcal{F} \setminus A \supset \mathcal{F}$ and $\mathcal{F} \vee A \supset \mathcal{F}$. These filters are proper and distinct

[11]By Corollary XXVII.4.9 of the Kuratowski-Zorn Theorem XXVII.4.8.

from $\mathcal{F}$ (because $A \in \mathcal{F} \vee A$ and $X \setminus A \in \mathcal{F} \setminus A$), hence $\mathcal{F}$ is not an ultrafilter. □

II.4. Decomposition into free and principal filters

The *kernel* of a family $\mathcal{A}$ of subsets of a set X is defined as the intersection of its elements:

$$\ker \mathcal{A} := \bigcap\nolimits_{A \in \mathcal{A}} A.$$

Therefore, a filter $\mathcal{F}$ is free if and only if $\ker \mathcal{F} = \varnothing$. Hence, a filter $\mathcal{F}$ is principal if and only if $\ker \mathcal{F} \in \mathcal{F}$. The class of free filters is denoted by $\mathbb{F}_*$ and the set of free filters on a set X by $\mathbb{F}_* X$.

For a filter $\mathcal{F}$ on X, the family

$$\mathcal{F}^* := \mathcal{F} \setminus \ker \mathcal{F} \tag{II.4.1}$$

is a free (possibly improper) filter, called the *free part* of $\mathcal{F}$.

Notice that a filter $\mathcal{F}$ is free if and only if $\mathcal{F}^* = \mathcal{F}$. On the other hand, a filter $\mathcal{F}$ is principal if and only if $\mathcal{F}^*$ is improper.

THEOREM II.4.1 (Filter decomposition). *For every filter $\mathcal{F}$ on X, there exists a unique pair $(\mathcal{F}^*, \mathcal{F}^\bullet)$ of (possibly improper) filters such that $\mathcal{F}^*$ is free, $\mathcal{F}^\bullet$ is principal, and*

$$\mathcal{F} = \mathcal{F}^* \wedge \mathcal{F}^\bullet \text{ and } \mathcal{F}^* \vee \mathcal{F}^\bullet = 2^X.$$

PROOF. If $\mathcal{F}$ is a filter on a set X, then let $\mathcal{F}^\bullet := (\ker \mathcal{F})^\uparrow$ and $\mathcal{F}^* := \mathcal{F} \setminus \ker \mathcal{F}$, as in (II.4.1). By definition, $\ker \mathcal{F} \subset F$ for each $F \in \mathcal{F}$, hence $\mathcal{F} \subset (\ker \mathcal{F})^\uparrow = \mathcal{F}^\bullet$ ([12]). As $F \setminus \ker \mathcal{F} \subset F$ for every $F \in \mathcal{F}$, also $\mathcal{F} \subset \mathcal{F} \setminus \ker \mathcal{F} = \mathcal{F}^*$. Therefore, $\mathcal{F} \subset \mathcal{F}^* \wedge \mathcal{F}^\bullet$.

On the other hand, if $F \in \mathcal{F}$, then let $F_* = F \setminus \ker \mathcal{F}$ and $F_\bullet = \ker \mathcal{F}$, so that $F_* \in \mathcal{F}^*, F_\bullet \in \mathcal{F}^\bullet$ and $F_* \cup F_\bullet = F$, proving that $\mathcal{F} \supset \mathcal{F}^* \wedge \mathcal{F}^\bullet$.

It is clear that $\mathcal{F}^* \vee \mathcal{F}^\bullet = 2^X$, because $\ker \mathcal{F} \in \mathcal{F}^\bullet$ and $X \setminus \ker \mathcal{F} \in \mathcal{F}^*$. □

REMARK II.4.2. An ultrafilter $\mathcal{U}$ is either principal or free. As $\mathcal{U} = \mathcal{U}^* \wedge \mathcal{U}^\bullet$, then by the maximality of $\mathcal{U}$, either $\mathcal{U} = \mathcal{U}^*$ and $\mathcal{U}^\bullet$ is improper, or $\mathcal{U} = \mathcal{U}^\bullet$ and $\mathcal{U}^*$ is improper. If $\mathcal{U}$ is a principal ultrafilter on X, then there is $x \in X$ such that $\mathcal{U} = x^\uparrow$. In fact, if $\mathcal{U} = \{A\}^\uparrow$ and $A = A_0 \sqcup A_1$ with $A_0 \neq \varnothing$ and $A_1 \neq \varnothing$, then either $A_0 \in \mathcal{U}$ or $A_1 \in \mathcal{U}$, which contradicts $\mathcal{U} = \{A\}^\uparrow$.

---

[12]If $\ker \mathcal{F} = \varnothing$ then $\mathcal{F}^\bullet = 2^X$; if $\ker \mathcal{F} \in \mathcal{F}$ then $\mathcal{F}^* = 2^X$.

DEFINITION II.4.3. Let $A \subset B \subset X$. The filter

$$(B, A)_0 := \{F \subset X : A \subset F,\ \mathrm{card}(B \setminus F) < \infty\} \tag{II.4.2}$$

is called the *cofinite filter of B centered at A.*

If $A = \varnothing$, then we drop *centered at A* and write $(B)_0$. If $B = A$, then $(A, A)_0 = \{A\}^{\uparrow}$ is a principal filter. The *free part* of $(B, A)_0$ *is* $(B \setminus A)_0$, and the *principal part* is $\{A\}^{\uparrow}$, and thus the decomposition of $(B, A)_0$ of Proposition II.4.1 can be written

$$(B, A)_0 = (B \setminus A)_0 \wedge \{A\}^{\uparrow}. \tag{II.4.3}$$

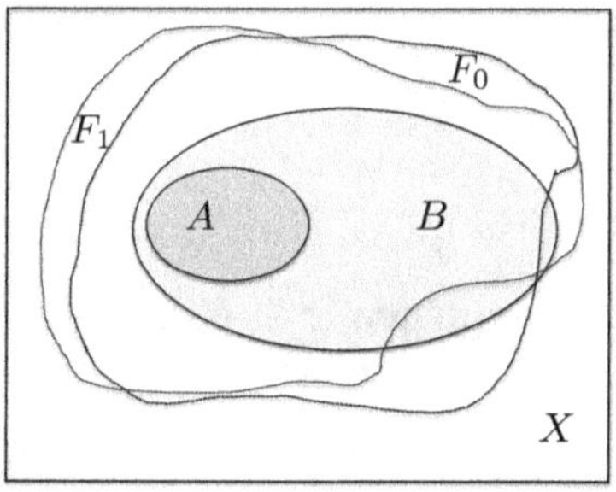

FIGURE II.4.1. The *cofinite filter* $(B, A)_0$ of B centered at A consists of all those F that include A and such that $B \setminus F$ is finite.

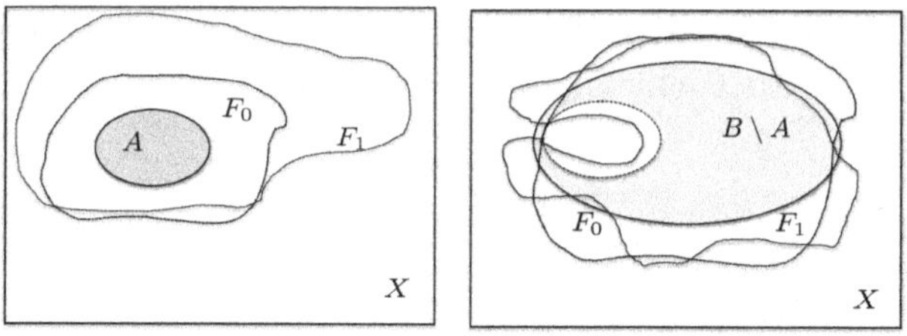

FIGURE II.4.2. Decomposition of the cofinite filter $(B, A)_0$ of B centered at A. The principal part is $\{A\}^{\uparrow} = \{F \subset X : A \subset F\}$, in particular, $A \in \{A\}^{\uparrow}$, and the free part is $(B \setminus A)_0$ the cofinite filter of $B \setminus A$. Generic elements of these filters are represented by F_0 and F_1.

Therefore, $(B, A)_0$ is proper if and only if either $(B \setminus A)_0$ is proper or $\{A\}^\uparrow$ is proper, that is, if either $B \setminus A$ is infinite, or A is non-empty.

If A is a singleton, say $A = \{x_\infty\}$, then we abridge $(B, x_\infty)_0 := (B, \{x_\infty\})_0$ for the *cofinite filter of B centered at x_∞*.

II.5. Supplement

EXAMPLE II.5.1. Consider a sequence of integers $(n_k)_k$. By definition, $(n_k)_k$ tends to infinity,

$$\lim_{k\to+\infty} n_k = +\infty,$$

whenever, for each $n \in \mathbb{N}$, there exists $k(n) \in \mathbb{N}$ such that $n_k \geq n$ for each $k \geq k(n)$. Let us rewrite $(n_k)_k$ as a map $\gamma \in \mathbb{N}^{\mathbb{N}}$, where $\gamma(k) := n_k$. Then, on using the notation introduced in (I.5.1), the condition becomes

$$\forall_{n\in\mathbb{N}} \ \exists_{k(n)\in\mathbb{N}} \ \gamma(\mathbb{N}_{k(n)}) \subset \mathbb{N}_n. \tag{II.5.1}$$

Therefore, $\gamma^-(A)$ is cofinite for each cofinite A.

EXERCISE II.5.2. A sequence $\varphi \in \mathbb{N}^N$ tends to $+\infty$ if and only if $\varphi^-(A)$ is cofinite for each cofinite set A.

SOLUTION. Indeed, φ tends to $+\infty$ if and only if for each cofinite subset A, there is $k(n) \in \mathbb{N}$ such that $\varphi(\mathbb{N}_{k(n)}) \subset A$, that is, $\mathbb{N}_{k(n)} \subset \varphi^-(A)$, equivalently $\varphi^-(A)$ is cofinite.

Accordingly, convergence of a sequence to infinity has been reformulated without employing the order of $\mathbb{N}$. Hence the property can be extended to any map $\varphi \in M^N$ from any infinite countable set N to another infinite countable set M.

EXERCISE II.5.3. Show that the infimum of all sequential filters on $\mathbb{R}$ converging to 0 (in the standard convergence) is generated by $\mathcal{B} := \{(-2^{-n}, 2^{-n}) : n < \omega\}$.

SOLUTION. Denote by $\mathbb{S}(0)$ the set of all sequential filters converging to 0. Of course, $\mathcal{E} \supset \mathcal{B}$ for each $\mathcal{E} \in \mathbb{S}(0)$ by the convergence condition. Consequently, $\bigcap \mathbb{S}(0) \supset \mathcal{B}$. Conversely, if there were $A \in \bigcap \mathbb{S}(0) \setminus \mathcal{B}$ then for every $n \in \mathbb{N}_1$ there would exist

$$r_n \in]-2^{-n}, 2^{-n}[\setminus A. \tag{II.5.2}$$

By construction, the sequence $(r_n)_n$ converges to 0, and thus the filter it generates belongs to $\mathbb{S}(0)$, which implies that there is $m \in \mathbb{N}$ such that $\{r_n : n \geq m\} \subset A$, which contradicts (II.5.2).

EXERCISE II.5.4. Show that $\ker(\bigwedge_{\mathcal{F}\in\boldsymbol{F}} \mathcal{F}) = \bigcup_{\mathcal{F}\in\boldsymbol{F}} \ker \mathcal{F}$.

SOLUTION. Let $x \notin \ker(\bigwedge_{\mathcal{F}\in\boldsymbol{F}}\mathcal{F})$, that is, there is $D \in \bigwedge_{\mathcal{F}\in\boldsymbol{F}}\mathcal{F}$ such that $x \notin D$, but since $D \in \mathcal{F}$ for each $\mathcal{F} \in \boldsymbol{F}$, also $x \notin \ker \mathcal{F}$. Conversely, if $x \notin \bigcup_{\mathcal{F}\in\boldsymbol{F}} \ker \mathcal{F}$, then for each $\mathcal{F} \in \boldsymbol{F}$, there is $F_{\mathcal{F}} \in \mathcal{F}$ such that $x \notin F_{\mathcal{F}}$. Hence $x \notin \ker(\bigwedge_{\mathcal{F}\in\boldsymbol{F}}\mathcal{F})$, because $\bigcup_{\mathcal{F}\in\boldsymbol{F}} F_{\mathcal{F}} \in \bigwedge_{\mathcal{F}\in\boldsymbol{F}}\mathcal{F}$.

EXERCISE II.5.5. Let X be an uncountable set, and let $\mathcal{G}$ be a countably based filter. Show that if $(X)_1 \supset \mathcal{G}$ (where $(X)_1$ is the cocountable filter on X, as in Example II.2.24), then $\ker \mathcal{G} \in (X)_1$.

SOLUTION. Let $\mathcal{G} \approx \{G_n : n < \omega\}$ and $(X)_1 \supset \mathcal{G}$. Then, for each $n < \omega$, there is $W \in (X)_1$ such that $W \subset G_n$, equivalently $G_n \in (X)_1$, which means that G_n is cocountable. As by Proposition I.5.4 each countable union of countable sets is countable, every countable intersection of cocountable sets is cocountable. Hence, $\ker \mathcal{G} = \bigcap_{n<\omega} G_n$ is cocountable, and thus $\ker \mathcal{G} \in (X)_1$.

DEFINITION II.5.6. We call two sets B and D *almost equal* if their *symmetric difference* $B \triangle D := (B \setminus D) \cup (D \setminus B)$ is finite.

EXERCISE II.5.7. Let B and D be infinite subsets of X. Then $(B)_0 = (D)_0$ if and only if $B \triangle D$ is finite.

SOLUTION. Actually, we deduce the result from

$$(B)_0 \leq (D)_0 \Longleftrightarrow \operatorname{card}(D \setminus B) < \infty.$$

Assume that $D_B := D \setminus B$ is finite. Let $F \in (B)_0$, that is, $B_F := B \setminus F$ is finite. Accordingly, $D \subset B \cup D_B$ and $B \subset F \cup B_F$. Hence, $D \subset B \cup D_B \subset F \cup B_F \cup D_B$. As $B_F \cup D_B$ is finite, $F \in (D)_0$. Conversely, if $D \setminus B$ is infinite, then $B \in (B)_0 \setminus (D)_0$.

It turns out that generalized cofinite filters defined in Definition II.4.3 can be characterized in terms of subcofinite filters ([13]).

DEFINITION II.5.8. A filter $\mathcal{F}$ is called *subcofinite* or *almost principal* if there is $F_0 \in \mathcal{F}$ such that $F_0 \setminus F$ is finite for each $F \in \mathcal{F}$.

It follows from the definition that $\mathcal{F} \subset (F_0)_0$. Obviously, each cofinite filter is subcofinite, and each principal filter is subcofinite. It follows from Definition II.4.3 that $(B, A)_0$ is subcofinite for each B and $A \subset B$. An argument like that in Proposition II.2.6 shows that a subcofinite filter containing a countable set, is countably based.

PROPOSITION II.5.9. *Each subcofinite filter is of the form $(B, A)_0$, where B is infinite, and $A \subset B$.*

[13]See *almost principal* filters in [**142**].

PROOF. Indeed, let $\mathcal{F}$ be a subcofinite filter. Then there is $B \in \mathcal{F}$, so that $B \setminus F$ is finite for each $F \in \mathcal{F}$. If $A := \ker \mathcal{F}$, then $\mathcal{F} = (B, A)_0$. □

PROPOSITION II.5.10. *A filter is subcofinite if and only if it has a filter-base of almost equal sets.*

PROOF. Let $\mathcal{F}$ be subcofinite, and let F_0 witness Definition II.5.8. It is clear that $\mathcal{B} := \{F_0 \cap F : F \in \mathcal{F}\}$ is a filter-base of $\mathcal{F}$ such that if $F_1 \in \mathcal{B}$, then $F_1 \setminus F$ is finite for each $F \in \mathcal{F}$ ([14]), and *a fortiori* for each $F \in \mathcal{B}$. On interchanging the roles of F_1 and F, we infer that $F_1 \triangle F = (F_1 \setminus F) \cup (F \setminus F_1)$ is finite.

Conversely, if $\mathcal{D}$ is a filter-base of almost disjoint elements, let $D_0 \in \mathcal{D}$. By assumption, if $F \in \mathcal{D}^{\uparrow}$, then there is $D \in \mathcal{D}$ such that $D \subset F$ and $D_0 \setminus F \subset D_0 \setminus D$ is finite. □

EXAMPLE II.5.11. Let X be an infinite countable set, and let $X = \bigcup_{n<\omega} X_n$ be a partition of X into infinite sets. If we set $F_n := \bigcup_{k>n} X_k$, then the family $\{F_n : n < \omega\}$ is a countable base of a filter that contains a countable set. This filter is not subcofinite.

---

[14]If $F_1 \in \mathcal{B}$ and $F \in \mathcal{F}$, then $F_1 \setminus F = F_0 \setminus (F_0 \cap F_1) \setminus F \subset F_0 \setminus F$, hence is finite.

CHAPTER III

Convergence of filters

III.1. The concept of limit

Let X be a non-empty set, and $\xi \subset \mathbb{F}X \times X$ be a relation between (non-empty, proper) filters $\mathcal{F}$ on X and points of X. We say that x is a *ξ-limit* of $\mathcal{F}$, or that $\mathcal{F}$ *converges to x* in ξ, if $(\mathcal{F}, x) \in \xi$, in symbols

$$x \in \lim_\xi \mathcal{F}. \tag{III.1.1}$$

A relation (III.1.1) is called a *preconvergence*, provided that, for every filters $\mathcal{F}_0$ and $\mathcal{F}_1$ on X,

$$\mathcal{F}_0 \subset \mathcal{F}_1 \Longrightarrow \lim_\xi \mathcal{F}_0 \subset \lim_\xi \mathcal{F}_1. \tag{isotone}$$

A preconvergence ξ is called a *convergence* if for each $x \in X$,

$$x \in \lim_\xi x^\uparrow. \tag{centered}$$

In other terms, a relation ξ is (centered) whenever the principal ultrafilter of x ξ-converges to x, shortly, *each point converges to itself.*

We focus on convergences, but many of our considerations can be easily broaden to preconvergences. If ξ is a convergence on a set, then it is pragmatic to call the pair (X, ξ) a *convergence space* ([1]). Denote by $|\xi|$ the *underlying set* of ξ, that is, $|\xi| = X$. Each convergence determines its underlying set because of (centered).

If a convergence on a set X is implicit, it is sometimes more expedient to write $\lim_X \mathcal{F}$ or $\lim \mathcal{F}$. If $\mathcal{B}$ is a filter-base, then it is often handy to abridge $\lim_\xi \mathcal{B}^\uparrow$ by $\lim_\xi \mathcal{B}$. This convention constitutes an evident extension of limits to filter-bases

$$\lim_\xi \mathcal{B} := \lim_\xi \mathcal{B}^\uparrow. \tag{III.1.2}$$

We shall use the symbol $\mathbf{I}$ to denote the class of all convergences, and $\mathbf{I}_\circ$ that of preconvergences. The set of all convergences on a set X is designed by $\mathbf{I}X$, and that of all preconvergences on X by $\mathbf{I}_\circ X$.

[1] This definition enables an expression *A is a subset of a convergence space (X, ξ)* as an abbreviation of *ξ is a convergence on X, and A is a subset of X.*

The inverse relation ξ^- of ξ associates with every $x \in X$, the set $\xi^-(x)$ of filters that converge to x, that is,

$$\mathcal{F} \in \xi^-(x) \Longleftrightarrow x \in \lim_\xi \mathcal{F}. \tag{III.1.3}$$

The simplest convergences on a given set are the *discrete* and the *chaotic* convergences.

EXAMPLE III.1.1 (Discrete convergence). We denote by $\iota = \iota_X$ the *discrete convergence* (on a set X) defined by $x \in \lim_\iota \mathcal{F}$ implies $\mathcal{F} = x^\uparrow$. In other terms, for each x, the only filter converging to x is the principal ultrafilter of x.

EXAMPLE III.1.2 (Chaotic convergence). The *chaotic convergence* $o = o_X$ on X, also called the *antidiscrete convergence*, is defined by $X = \lim_o \mathcal{F}$ for every $\mathcal{F} \in \mathbb{F}X$, that is, each (proper) filter o-converges to each point.

EXAMPLE III.1.3 (Standard sequential convergence). Consider a convergence $\sigma = \sigma_\mathbb{R}$ on $\mathbb{R}$, defined by

$$x \in \lim_{\sigma_\mathbb{R}} \mathcal{F}$$

if there exists a sequence $\varphi \in \mathbb{R}^N$ such that $x \in \lim \varphi$ and $\varphi[(N_0)] \subset \mathcal{F}$.

EXAMPLE III.1.4 (Standard convergence). The *standard convergence* $\nu = \nu_\mathbb{R}$ of the real line $\mathbb{R}$ is defined by

$$x \in \lim_{\nu_\mathbb{R}} \mathcal{F},$$

provided that $]x - \varepsilon, x + \varepsilon[\in \mathcal{F}$ for each $\varepsilon > 0$, that is, if and only if for each $\varepsilon > 0$, there exists $F \in \mathcal{F}$ such that $|r - x| < \varepsilon$ for each $r \in F$.

DEFINITION III.1.5 (Finitely stable convergences). A convergence ξ is called *finitely stable* ([2]) if for each $\mathcal{F}_0$ and $\mathcal{F}_1$ from $\mathbb{F}|\xi|$,

$$\lim_\xi \mathcal{F}_0 \cap \lim_\xi \mathcal{F}_1 \subset \lim_\xi (\mathcal{F}_0 \cap \mathcal{F}_1). \tag{finitely stable}$$

By the formula (isotone), the converse inclusion always holds. Of course, the property (finitely stable) extends, by easy induction, to arbitrary finite collections of filters.

It is immediate that the convergences from Examples III.1.1 and III.1.2 are finitely stable. To see that the *standard sequential convergence* from Example III.1.3 is finitely stable, observe that finite infima of sequential filters are sequential. The standard convergence from Example III.1.5 is finitely stable, because if $]x - \varepsilon, x + \varepsilon[\in \mathcal{F}_0$ and

[2]Some authors add *finite stability* to the definition of convergence.

$]x-\varepsilon, x+\varepsilon[\in \mathcal{F}_1$ for each $\varepsilon>0$, then $]x-\varepsilon, x+\varepsilon[\in \mathcal{F}_0 \cap \mathcal{F}_1$ for each $\varepsilon>0$.

Non-finitely stable convergences will occur rather seldom in our investigations. Most natural examples of convergences are finitely stable. If ξ is a convergence on a set X, then we define $x \in \lim_{\mathrm{M}\xi} \mathcal{F}$ to the effect that $x \in \lim_{\xi} \mathcal{F}$ and $\mathcal{F}$ is an ultrafilter. It is straightforward that $\mathrm{M}\xi$ is a convergence, called the *ultrafilter modification* of ξ.

PROPOSITION III.1.6. *If ξ is a convergence on a set X, then $\mathrm{M}\xi$ is a non-finitely stable convergence if $\xi \neq \mathrm{M}\xi$.*

PROOF. If there exists a non-maximal filter $\mathcal{F}$ and x such that $x \in \lim_{\xi} \mathcal{F}$, then there exists an ultrafilter $\mathcal{U} \supset \mathcal{F}$ and $\mathcal{U} \neq \mathcal{F}$. Therefore, there exists $U \in \mathcal{U} \backslash \mathcal{F}$, that is, $F \backslash U \neq \varnothing$ for every $F \in \mathcal{F}$, equivalently, $\mathcal{F} \vee (X \setminus U)$ is non-degenerate. Therefore there exists an ultrafilter $\mathcal{W} \supset \mathcal{F} \vee (X \setminus U)$, and obviously $\mathcal{U} \neq \mathcal{W}$. Then $x \in \lim_{\xi} \mathcal{U} \cap \lim_{\xi} \mathcal{W} = \lim_{\mathrm{M}\xi} \mathcal{U} \cap \lim_{\mathrm{M}\xi} \mathcal{W}$, but $\lim_{\mathrm{M}\xi}(\mathcal{U} \wedge \mathcal{W}) = \varnothing$, because $\mathcal{U} \wedge \mathcal{W}$ is not an ultrafilter. □

EXAMPLE III.1.7 (A non-finitely stable convergence). Let θ be a convergence defined on $\mathbb{N} \cup \{\infty\}$ so that $\infty \in \lim_{\theta} \mathcal{F}$ if $\ker \mathcal{F} \subset \{\infty\}$ and either $\mathcal{F}^*$ includes the cofinite filter of the set $\{n \in \mathbb{N} : 2 \nmid n\}$ of odd numbers, or the cofinite filter of $\{n \in \mathbb{N} : 2 \mid n\}$ of even numbers. This is a non-finitely stable convergence, because $\infty \in \lim_{\theta}(2n)_n$, and $\infty \in \lim_{\theta}(2n+1)_n$, but $\lim_{\theta}(\mathbb{N})_0 = \varnothing$.

DEFINITION III.1.8. The *vicinity filter* $\mathcal{V}_{\xi}(x)$ of ξ at x is the infimum of all filters ξ-converging to x, that is,

$$\mathcal{V}_{\xi}(x) := \bigwedge \xi^{-}(x).$$

A vicinity filter $\mathcal{V}_{\xi}(x)$ need not ξ-converge to x.

EXAMPLE III.1.9. Recall the *standard sequential convergence* $\sigma_{\mathbb{R}}$ from Example III.1.3. By virtue of Exercise III.7.1, the vicinity filter $\mathcal{V}_{\sigma_{\mathbb{R}}}(x)$ of $\sigma_{\mathbb{R}}$ at x, is generated by

$$\{]x-\varepsilon, x+\varepsilon[: \varepsilon>0\},$$

hence is not sequential by Example II.2.23. Therefore, $x \notin \lim_{\sigma_{\mathbb{R}}} \mathcal{V}_{\sigma_{\mathbb{R}}}(x)$.

DEFINITION III.1.10 (Pretopologies). A convergence ξ is called a *pretopology* if $\xi^{-}(x)$ has a coarsest filter for each $x \in |\xi|$.

Obviously,

PROPOSITION III.1.11. *A convergence ξ on X is a pretopology if and only $x \in \lim_{\xi} \mathcal{V}_{\xi}(x)$ for each $x \in X$.*

Discrete and chaotic convergences are pretopologies. The *standard convergence* $\nu_{\mathbb{R}}$ from Example III.1.4 is a pretopology, because the vicinity filter $\mathcal{V}_{\nu_{\mathbb{R}}}(x)$ converges to x. The *standard sequential convergence* $\sigma_{\mathbb{R}}$ from Example III.1.3 is not a pretopology, because, by Example III.1.9, its vicinity filters do not converge. Incidentally, observe that

$$\mathcal{V}_{\sigma_{\mathbb{R}}}(x) = \mathcal{V}_{\nu_{\mathbb{R}}}(x). \tag{III.1.4}$$

DEFINITION III.1.12 (Isolated point). An element x of X is called *ξ-isolated* (or, simply, *isolated*, if ξ is implicit) if $x \in \lim_{\xi} \mathcal{F}$ implies that $\mathcal{F} = x^{\uparrow}$.

In other words, a convergence is discrete if and only if all the elements (of the underlying set) are isolated.

DEFINITION III.1.13 (Free and Hausdorff convergences). A convergence ξ is said to be *free* (or T_1) if

$$\lim_{\xi} x^{\uparrow} \subset \{x\}; \tag{III.1.5}$$

it is called *Hausdorff* (or T_2) if $\lim_{\xi} \mathcal{F}$ is at most a singleton for each filter $\mathcal{F}$, that is,

$$\{x_0, x_1\} \subset \lim_{\xi} \mathcal{F} \Longrightarrow x_0 = x_1. \tag{III.1.6}$$

Free convergences are traditionally called T_1. In order to facilitate the reading to those accustomed to the traditional term, I will be recalling the traditional name T_1 — along with the current name *free*.

PROPOSITION III.1.14. *A convergence ξ is free if and only if $x \in \lim_{\xi} \mathcal{F}$ implies that* $\ker \mathcal{F} \subset \{x\}$.

PROOF. If a convergence ξ on X is such that $x_0 \in \lim_{\xi} \mathcal{F}$ and there is $x \neq x_0$ such that $x \in \ker \mathcal{F}$, then $x^{\uparrow} \supset \mathcal{F}$ and thus $x_0 \in \lim_{\xi} x^{\uparrow}$, that is, ξ is not free. Conversely, if ξ is not free, then there exist $x \neq x_0$ such that $x \in \lim_{\xi} x_0^{\uparrow}$. Thus $x_0 \in \ker x_0^{\uparrow} \not\subset \{x\}$. □

By the condition (centered), the inclusion in (III.1.5) is, in fact, the equality. Of course, each Hausdorff convergence is free, because (III.1.6) implies (III.1.5) for $\mathcal{F} = x^{\uparrow}$.

It is clear that the discrete convergence ι_X is Hausdorff for each set X. On the other hand, the chaotic convergence o_X is free (alias T_1) if and only if $\operatorname{card} X = 1$. By the way, $\operatorname{card} X = 1$ implies that $\iota_X = o_X = \xi$ for each convergence ξ on X. Both convergences of the real line from Examples III.1.3 and III.1.4 are Hausdorff.

DEFINITION III.1.15. If ξ is a convergence on X, and $\varnothing \neq V \subset X$, then the *restriction* $\xi|_V$ of ξ to V is defined by

$$\lim_{\xi|_V} \mathcal{F} := V \cap \lim_\xi \mathcal{F}, \tag{III.1.7}$$

where $\mathcal{F}$ is a filter on V.

As in (III.1.7) $\mathcal{F} \in \mathbb{F}V$, it follows that it is a filter-base of a filter on X, that is, $\mathcal{F}^\uparrow \in \mathbb{F}X$. Nonetheless, $\lim_\xi \mathcal{F}$ above makes sense, because of (III.1.2).

It is straightforward that

PROPOSITION III.1.16. *Each restriction of a Hausdorff convergence is Hausdorff, and each restriction of a free convergence is free.*

EXAMPLE III.1.17 (Cofinite convergence). We define the *cofinite convergence* of an infinite set X to be a pretopology θ such that

$$X = \lim_\theta (X)_0.$$

Then $\{x\} = \lim_\theta x^\uparrow$ for each $x \in X$. By (isotone), $x \in \lim_\theta \mathcal{F}$ if and only if $\mathcal{F} \supset (X)_0 \wedge x^\uparrow$ for each $x \in X$. Hence, θ is free (alias T_1), but not Hausdorff.

DEFINITION III.1.18. A convergence ξ is called T_0 if it distinguishes the points, that is, if

$$x_0 \neq x_1 \Longrightarrow \xi^-(x_0) \neq \xi^-(x_1).$$

Of course, each free convergence is T_0, because if ξ is free, then $x_1^\uparrow \in \xi^-(x_1) \setminus \xi^-(x_0)$ for any $x_0 \neq x_1$. The chaotic convergence on a set of cardinality greater than 1 is an example of a convergence that is not T_0.

DEFINITION III.1.19 (Density). Let ξ be a convergence on X. A subset D of X is called ξ-*dense* (or simply, *dense*) if for each $x \in X$ there exists a filter $\mathcal{F}$ such that $D \in \mathcal{F}$ and $x \in \lim_\xi \mathcal{F}$. The *density* $\mathrm{d}(\xi)$ of convergence ξ is the least cardinality of a ξ-dense set. Traditionally a convergence is called *separable* if its density is countable.

Notice that each convergence on a countable set is separable. A subset A of X is ι_X-dense (where ι_X is the discrete convergence of Example III.1.1) if and only if $A = X$. It is o_X-dense (where o_X is the chaotic convergence of Example III.1.2) if and only if $\varnothing \neq A$.

EXAMPLE III.1.20 (A dense subset of the real line). The set $\mathbb{Q}$ of the *rational numbers* is $\nu_\mathbb{R}$-dense, where $\nu_\mathbb{R}$ is the standard convergence on $\mathbb{R}$ of Example III.1.3. Let $r \in \mathbb{R}$. For every $\varepsilon > 0$, there is $n < \omega$, such that $\frac{1}{n} < 2\varepsilon$, so that $]r-\varepsilon, r+\varepsilon[\cap \{\frac{k}{n} : k \in \mathbb{Z}\} \neq \varnothing$. Therefore, $]r-\varepsilon, r+\varepsilon[\cap \mathbb{Q} \neq \varnothing$ for each $\varepsilon > 0$, so that $r \in \lim_{\nu_\mathbb{R}}(\mathcal{V}_{\nu_\mathbb{R}}(r) \vee \mathbb{Q})$.

III.2. A pointwise perspective

Recall that $\xi^-(x)$ is the set of all filters ξ-converging to x. As we have already written (III.1.3),

$$\mathcal{F} \in \xi^-(x) \Longleftrightarrow x \in \lim_\xi \mathcal{F}.$$

It is instructive to look at a convergence at each of its points. Of course, each proposition formulated in terms of $\lim_\xi$ can be reformulated in terms of ξ^-. Although passing from one form to another description provides no new information, it offers a different, sometimes illuminating, perspective. Let us reformulate the axioms of convergence from the pointwise viewpoint.

A relation ξ between filters on X and elements of X is a convergence whenever, for each $x \in X$,

(isotone) $$\mathcal{G} \supset \mathcal{F} \in \xi^-(x) \Longrightarrow \mathcal{G} \in \xi^-(x),$$

(centered) $$x^\uparrow \in \xi^-(x).$$

A convergence ξ is *finitely stable* whenever

(III.2.1) $$\mathcal{F}_0, \mathcal{F}_1 \in \xi^-(x) \Longrightarrow \mathcal{F}_0 \wedge \mathcal{F}_1 \in \xi^-(x).$$

A convergence ξ is Hausdorff if and only if, for each x_0, x_1,

(Hausdorff) $$x_0 \neq x_1 \Longrightarrow \xi^-(x_0) \cap \xi^-(x_1) = \varnothing.$$

Observe that (isotone) and (centered) are *pointwise* properties, that is, they concern each point x of a convergence space separately (from the other points), while (Hausdorff) is not. In contrast, *being free* (alias T_1) is a pointwise property, because ξ is a free convergence on X if and only if $\xi^-(x) \setminus \{x^\uparrow\}$ consists of free filters for each $x \in X$.

Recall that an element x of $|\xi|$ is called *isolated* ([3]) if

(III.2.2) $$\xi^-(x) = \{x^\uparrow\}.$$

EXAMPLE III.2.1. Let ζ be a convergence on the real line such that $\zeta^-(x) := \nu_{\mathbb{R}}^-(x)$ if x is rational, and $\zeta^-(x) := \sigma_{\mathbb{R}}^-(x)$ if x is irrational. We readily see that ζ is a finitely stable convergence. It is *pretopological* at rational points (or rational points are *pretopological*), and not pretopological at irrational points.

DEFINITION III.2.2 (Pavement). A collection of filters $\boldsymbol{D} \subset \xi^-(x)$ is called a *pavement* of ξ at x if for each filter $\mathcal{F} \in \xi^-(x)$, there is $\mathcal{D} \in \boldsymbol{D}$ such that $\mathcal{F} \supset \mathcal{D}$.

It follows from Definition III.1.10 that

---

[3] The term is motivated by the fact that no filter from outside of $\{x\}$ converges to x.

PROPOSITION III.2.3. *A convergence ξ on X is a pretopology if and only if for each $x \in X$, there exists a pavement of ξ at x consisting of one filter.*

COROLLARY III.2.4. *If a finitely stable convergence has a finite pavement at every point, then it is a pretopology.*

EXAMPLE III.2.5 (An uncountable pavement). As the standard sequential convergence $\sigma_{\mathbb{R}}$ from Example III.1.3 is finitely stable, but not a pretopology, it has infinite pavements. Actually, each pavement of $\sigma_{\mathbb{R}}$ is uncountable at every point. Indeed, suppose that $\{\mathcal{E}_n : n < \omega\} \subset \mathbb{S}$ is a $\sigma_{\mathbb{R}}$-pavement at 0. Then for every $n < \omega$, there is a countable set E_n such that $E_n \in \mathcal{E}_n$. Hence, $E := \bigcup_{n<\omega} E_n$ is a countable set. On the other hand, $]-2^{-n}, 2^{-n}[$ is uncountable for each $n < \omega$, hence there exist $x_n \in]-2^{-n}, 2^{-n}[\setminus E$ and thus $(x_n)_n$ is off E and $\sigma_{\mathbb{R}}$-converges to 0, so that $\{\mathcal{E}_n : n < \omega\}$ is not a pavement of $\sigma_{\mathbb{R}}$ at 0.

DEFINITION III.2.6. A convergence ξ on X is said to be of *countable character* at $x \in X$, if there exists a pavement of ξ at x consisting of countably based filters. A convergence is called of *countable character* if it is of countable character at each point.

The discrete and chaotic convergences are of finite character. The standard convergences $\nu_{\mathbb{R}}$ and $\sigma_{\mathbb{R}}$ are of countable character. The cofinite convergence $\pi[x_\infty, (X)_0]$ of Example III.4.3 to come is not of countable character if X is uncountable because of Proposition II.2.19.

III.3. Order on convergences

Let ζ and ξ be two subsets of $\mathbb{F}X \times X$, where X is a non-empty set. We say that ζ is *finer* than ξ, equivalently, ξ is *coarser* than ζ, in symbols

$$\zeta \geq \xi, \tag{III.3.1}$$

if for each $\mathcal{F} \in \mathbb{F}X$,

$$\lim_\zeta \mathcal{F} \subset \lim_\xi \mathcal{F}. \tag{III.3.2}$$

In other words, $\zeta \geq \xi$ if and only if $\zeta \subset \xi$, equivalently $\zeta^- \subset \xi^-$. In particular, (III.3.1) specifies an order on the set $\mathbf{I}_\circ X$ of preconvergences on X, and on the set $\mathbf{I}X$ of convergences on X.

EXAMPLE III.3.1. Observe that $\sigma_{\mathbb{R}} > \nu_{\mathbb{R}}$ (that is, $\sigma_{\mathbb{R}} \geq \nu_{\mathbb{R}}$, but $\sigma_{\mathbb{R}} \neq \nu_{\mathbb{R}}$), where $\sigma_{\mathbb{R}}$ is the *standard sequential convergence* from Example III.1.3, and $\nu_{\mathbb{R}}$ is the *standard convergence* from Example III.1.4.

PROPOSITION III.3.2. *The* discrete convergence ι_X *is the finest, and the* chaotic convergence o_X *is the coarsest convergence on X.*

PROOF. Let ξ be a convergence on X. Indeed, $x^{\uparrow}$ is the only filter that belongs to $\iota^{-}(x)$ for every $x \in X$. By (centered), $\iota^{-}(x) \subset \xi^{-}(x)$ for every $x \in X$ and convergence ξ on X, that is, $\iota \subset \xi$. On the other hand, $\lim_{\xi} \mathcal{F} \subset X = \lim_{o} \mathcal{F}$ for each $\mathcal{F} \in \mathbb{F}X$, that is $\xi \subset o$. □

The order on the set of convergences on X is not linear, unless X is a singleton. Indeed,

EXAMPLE III.3.3. If $x_0, x_1 \in X$, then let $\lim_{\xi_0} x_0^{\uparrow} := \{x_0, x_1\}$ and $\lim_{\xi_0} x^{\uparrow} := \{x\}$ for $x \neq x_0$, and $\lim_{\xi_1} x_1^{\uparrow} := \{x_0, x_1\}$ $\lim_{\xi_1} x^{\uparrow} := \{x\}$ for $x \neq x_1$. Then $x_1 \in \lim_{\xi_0} x_0^{\uparrow} \setminus \lim_{\xi_1} x_0^{\uparrow}$ and $x_0 \in \lim_{\xi_1} x_1^{\uparrow} \setminus \lim_{\xi_0} x_1^{\uparrow}$, so that ξ_0 and ξ_1 are incomparable whenever $x_0 \neq x_1$.

The convergences in the preceding examples are not free. Let us give an example of two free incomparable convergences.

EXAMPLE III.3.4. If X is infinite, $x_0, x_1 \in X$, then let π_0 be a finitely stable convergence on X such that each $x \neq x_0$ is isolated and $x_0 \in \lim_{\pi_0}(X)_0$, and π_1 be a finitely stable convergence such that each $x \neq x_1$ is isolated and $x_1 \in \lim_{\pi_1}(X)_0$. If $x_0 \neq x_1$, then $x_0 \in \lim_{\pi_0}(X)_0 \setminus \lim_{\pi_1}(X)_0$, and $x_1 \in \lim_{\pi_1}(X)_0 \setminus \lim_{\pi_0}(X)_0$. In this case, $\pi_0 \vee \pi_1 = \iota$ and $\pi_0 \wedge \pi_1$ is the cofinite convergence of Example III.1.17.

PROPOSITION III.3.5. *Let ζ and ξ be convergences on a given set so that $\zeta \geq \xi$. If ξ is free (alias T_1), then ζ is free. If ξ is Hausdorff, then ζ is Hausdorff.*

PROOF. If a convergence ξ on X is free, then $\{x\} \supset \lim_{\xi} x^{\uparrow} \supset \lim_{\zeta} x^{\uparrow}$ for each $x \in X$, because $\zeta \geq \xi$, hence ζ is free. If $\{x_0, x_1\} \subset \lim_{\zeta} \mathcal{F}$, then $\{x_0, x_1\} \subset \lim_{\xi} \mathcal{F}$, hence $x_0 = x_1$, because ξ is Hausdorff. Thus ζ is Hausdorff. □

Another example is provided by

EXAMPLE III.3.6 (Sorgenfrey line). Let ς be a convergence on the real line to the effect that $x \in \lim_{\varsigma} \mathcal{F}$ whenever

$$[x, x + \varepsilon[\in \mathcal{F}$$

for each $\varepsilon > 0$. We easily check that ς is a Hausdorff pretopology of countable character, because the coarsest filter ς-convergent to x is the filter generated by the filter-base $\{[x, x + 2^{-n}[: n < \omega\}$. We notice that $\varsigma > \nu_{\mathbb{R}}$, the standard convergence of the real line, as the coarsest filter $\nu_{\mathbb{R}}$-convergent to x is generated by $\{\,]x-\varepsilon, x+\varepsilon[: \varepsilon > 0\}$. This convergence is also called the *right half-open interval convergence*, because it is defined by the intervals closed on the left and open on the right ([4]). We shall sometimes write ς_+ in order to distinguish it from a twin *left half-open interval convergence* ς_-.

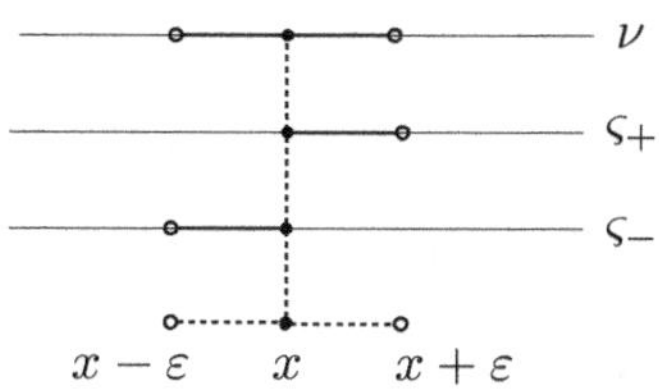

FIGURE III.3.1. Comparison of typical vicinities of x for the *standard convergence* $\nu_{\mathbb{R}}$, and the two variants of the *Sorgenfrey convergence* ς_+ *and* ς_-*, respectively*

$$]x-\varepsilon, x+\varepsilon[,\ [x, x+\varepsilon[,\ ]x-\varepsilon, x].$$

Clearly, the convergence ς_-, defined by $x \in \lim_{\varsigma_-} \mathcal{F}$ by

$$]x-\varepsilon, x] \in \mathcal{F}$$

for each $\varepsilon > 0$, is called the *left half-open interval convergence.* Of course, one of the two Sorgenfrey convergences is transformed to the other by the inversion of the order of the real line ([5]).

Sure enough, neither $\varsigma_+ \geq \varsigma_-$, nor $\varsigma_- \geq \varsigma_+$. In other words, they are *incomparable*. Both are finer than the standard convergence $\nu = \nu_{\mathbb{R}}$. Actually,

$$\varsigma_+ \wedge \varsigma_- = \nu \text{ and } \varsigma_+ \vee \varsigma_- = \iota,$$

where the extrema $\wedge, \vee$ are considered within the finitely stable convergences (See Exercise III.7.3).

PROPOSITION III.3.7. *The supremum and the infimum exist for every non-empty set* Ξ *of convergences on* X*, namely*

$$\lim_{\bigvee \Xi} \mathcal{F} = \bigcap\nolimits_{\xi \in \Xi} \lim_{\xi} \mathcal{F}, \tag{III.3.3}$$

$$\lim_{\bigwedge \Xi} \mathcal{F} = \bigcup\nolimits_{\xi \in \Xi} \lim_{\xi} \mathcal{F}. \tag{III.3.4}$$

[4]L. A. STEEN and J. A. SEEBACH use the term *half-open interval convergence* in [**243**]. This book provides a rich collection of interrelations and counterexamples in topology. It mainly addresses an advanced reader.

[5]We shall discuss in Definition IV.2.8 the notion of *homeomorphic convergences*, of which this situation is a particular case.

PROOF. Clearly, $\bigcap_{\theta\in\Xi}\lim_\theta \mathcal{F} \subset \lim_\xi \mathcal{F}$ for each $\xi \in \Xi$. It is enough to observe that the intersection defines a convergence on X. Similarly, $\lim_\xi \mathcal{F} \subset \bigcup_{\theta\in\Xi}\lim_\theta \mathcal{F}$ for each $\xi \in \Xi$, and the union defines a convergence on X. □

EXAMPLE III.3.8. Let ν_x be a convergence on $\mathbb{R}$ defined by

$$\nu_x(r) = \begin{cases} \{\mathcal{F} : \forall_{\varepsilon>0}\ ]r-\varepsilon, r+\varepsilon[\in \mathcal{F}\}, & r = x, \\ r^\uparrow & r \neq x, \end{cases}$$

for each $x \in \mathbb{R}$. We easily see that $\bigwedge_{x\in\mathbb{R}}\nu_x = \nu$, and $\bigvee_{x\in\mathbb{R}}\nu_x = \iota$.

It is obvious that the supremum of a non-empty set of Hausdorff convergences is Hausdorff. One sees easily that the infimum of (two) Hausdorff convergences need not be Hausdorff.

LEMMA III.3.9. *The supremum and the infimum of each non-empty set of free (alias T_1) convergences is free. The finest free convergence on a set X is the discrete convergence ι_X, and the coarsest free convergence on X is the cofinite convergence of X* ([6]).

PROOF. Let Ξ be a set of free convergences on X, then $\bigvee \Xi$ is free, because if ξ is a free convergence and $\theta \geq \xi$ then $x \in \lim_\theta \mathcal{F} \in \lim_\xi \mathcal{F}$, hence $\ker \mathcal{F} \subset \{x\}$, and thus θ is free. On the other hand, $x \in \lim_{\bigwedge \Xi} \mathcal{F}$, then $\lim_\xi \{x\}^\uparrow \subset \{x\}$ for every $x \in X$ and each $\xi \in \Xi$, thus $\lim_{\bigwedge \Xi}\{x\}^\uparrow \subset \{x\}$ for every $x \in X$. Accordingly, if ξ is a free convergence on X and $\varnothing \neq \lim_\xi \mathcal{F}$, then $\mathcal{F}^* \supset (X)_0$. Therefore, $x \in \lim_{\theta_0} \mathcal{F}$ whenever $\mathcal{F} \supset (X)_0 \wedge x^\uparrow$, then θ_0 is free, and the coarsest one. □

REMARK III.3.10. If ζ and ξ are preconvergences on a set X, then we write

$$\zeta \trianglerighteq \xi$$

whenever (III.3.2) holds for each for each $\mathcal{F} \in \mathbb{F}X$. It is straightforward that $\trianglerighteq$ is an order on the set of preconvergences on X. Of course, if ζ and ξ are convergences on a set X, then $\zeta \trianglerighteq \xi$ if and only if $\zeta \geq \xi$.

If now ξ_0 is a convergence on X_0 and ξ_1 is a convergence on X_1, then

$$\xi_0 \trianglerighteq \xi_1 \tag{III.3.5}$$

means that $\lim_{\xi_0} \mathcal{F} \subset \lim_{\xi_1} \mathcal{F}$ for each $\mathcal{F} \in \mathbb{F}(X_0 \cup X_1)$. Accordingly, (III.3.5) holds if and only if $X_0 \subset X_1$ and $\xi_0 \geq \xi_1|_{X_0}$.

[6]From Example III.1.17.

III.4. Prime convergences

Prime convergences have a particularly simple structure: they are entirely determined by a collection of filters converging to one point.

DEFINITION III.4.1 (Prime convergences). A convergence is called *prime* if it has at most one non-isolated point, which is called the *pole* of that convergence. If x_∞ is a pole of a prime convergence ξ, then we say that ξ is *prime at* x_∞.

A discrete convergence ι_X is prime. A chaotic convergence o_X is prime if and only if X is a singleton.

PROPOSITION III.4.2. *Each prime free (alias T_1) convergence is Hausdorff.*

PROOF. If ξ is prime and discrete, then it is obviously Hausdorff. If ξ is prime non-discrete, then there is a non-isolated point x_∞ in $|\xi|$. If, by way of contradiction, there existed $x \neq x_\infty$ such that $\varnothing \neq \xi^-(x) \cap \xi^-(x_\infty) = \{x^\uparrow\} \cap \xi^-(x_\infty)$, then $x^\uparrow \in \xi^-(x_\infty)$, which means that $x_\infty \in \lim_\xi x^\uparrow$, contrary to (III.1.5). □

We denote by $\pi[x, \boldsymbol{D}]$ the prime convergence ([7]), for which

$$\{\mathcal{D} \wedge x^\uparrow : \mathcal{D} \in \boldsymbol{D}\}$$

is a pavement at the pole x. If, in particular, $\boldsymbol{D} = \{\mathcal{D}\}$, then we abridge $\pi[x, \mathcal{D}]$. Hence, a prime convergence ξ is a pretopology if and only if $\xi = \pi[x, \mathcal{V}_\xi(x)]$ for some $x \in |\xi|$.

EXAMPLE III.4.3. Let X be an infinite set and let $x_\infty \in X$. We define the *prime cofinite convergence* $\pi = \pi[x_\infty, (X)_0]$ on X by declaring isolated each $x \neq x_\infty$, and setting $x_\infty \in \lim_\pi \mathcal{F}$ whenever

$$\mathcal{F} \supset (X, x_\infty)_0,$$

where $(X, x_\infty)_0$ is the cofinite filter centered at x_∞ from Definition II.4.3. In particular, for each free filter $\mathcal{F}$

$$\mathcal{F} \supset (X)_0 \supset (X, x_\infty)_0,$$

hence each free filter on X converges to x_∞.

The cofinite filter $(X)_0$ on an infinite countable set X is free and sequential, while $(X, x_\infty)_0$ is sequential, the free part of which is $(X)_0$ and the principal part is $x_\infty^\uparrow$. If you are accustomed to the language of sequences, it may be more comfortable to see $\pi[x_\infty, (X)_0]$ as a convergence on $X = \{x_\infty\} \cup \{x_n : n < \omega\}$ such that

$$\{x_\infty\} = \lim_\pi (x_n)_n,$$

[7]The underlying set X is implicit.

and $(x_n)_n \wedge x_\infty^\uparrow$ is the coarsest filter-base convergent to x_∞. The prime cofinite convergence is a Hausdorff pretopology.

Each convergence can be decomposed into prime convergences, that is,

PROPOSITION III.4.4. *Each convergence is an infimum of prime convergences.*

PROOF. If θ is a convergence, then for each $x \in |\theta|$, let θ_x be a prime convergence on $|\theta|$ such that $\theta_x^-(x) := \theta^-(x)$, and $\theta_x^-(y) = \{y^\uparrow\}$ for each $y \neq x$. Then $y \in \lim_{\bigwedge_{x\in|\theta|}\theta_x} \mathcal{F}$ if and only if there exists $x \in |\theta|$ such that $y \in \lim_{\theta_x} \mathcal{F}$, equivalently, $\mathcal{F} \in \theta_x^-(y)$, that is, either $y \neq x$, hence $\mathcal{F} = y^\uparrow$, or $y = x$, hence $\mathcal{F} \in \theta_y^-(y) = \theta^-(y)$, which means that $y \in \lim_\theta \mathcal{F}$. □

Similarly to cofinite convergences, we consider ([8])

EXAMPLE III.4.5 (Prime cocountable convergence). Let X be an uncountable set and let $x_\infty \in X$. A *prime cocountable convergence* $\pi[x_\infty, (X)_1]$ is a pretopology, in which the vicinity filter of a non-isolated point x_∞ is the *cocountable filter* of X centered at x_∞. In fact,

$$\mathcal{V}_\pi(x) = (X, x_\infty)_1 := \{F \subset X : x_\infty \in F \wedge \operatorname{card}(X \setminus F) \leq \aleph_0\}.$$

EXAMPLE III.4.6. Let X be an infinite countable set, x_∞ and X_n an infinite subset of X for each $n < \omega$ such that $X = \{x_\infty\} \sqcup \bigsqcup_{n<\omega} X_n$. Let $\{\bigcup_{n>k} X_n : k < \omega\}$ be a filter-base of a filter $\mathcal{G}$. We define a prime convergence γ on X to the effect that $x_\infty \in \lim_\gamma \mathcal{F}$ whenever $\mathcal{F} \supset \mathcal{G} \wedge x_\infty^\uparrow$. Of course, γ is a pretopology, and its vicinity filter is countably based.

Proposition III.4.4 exhibits a decomposition of an arbitrary convergence into prime convergences. On the other hand, each prime convergence can be decomposed into simpler prime convergences, the pavements of which consist of at most two filters each, as in Example III.4.7 below. We shall see in Proposition VI.5.3 that finitely stable convergences can be decomposed into prime pretopologies.

EXAMPLE III.4.7 (Sequential fan). Let $\{x_{n,k} : n, k < \omega\}$ be a set of distinct points, and let $x_\infty \notin \{x_{n,k} : k, < \omega\}$. The set $X = \{x_\infty\} \sqcup \{x_{n,k} : n, k < \omega\}$, endowed with the finest convergence, in which $x_\infty \in \lim_\pi (x_{n,k})_k$ for each $n < \omega$, it is called a *sequential fan*.

---

[8]More generally, if $\operatorname{card} X \geq \kappa \geq \aleph_0$, then a co-$\kappa$ filter consists of subsets F of X such that $\operatorname{card}(X \setminus F) < \kappa$, is proper.

In terms of cofinite filters, let J be an infinite set, and let $\{X_j : j \in J\}$ be a disjoint family of infinite countable subsets of a set X, and let $x_\infty \notin \bigcup_{j\in J} X_j$.

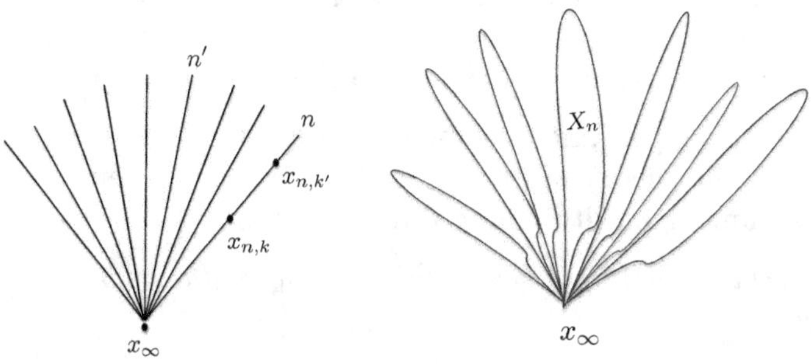

FIGURE III.4.1. Sequential fan (two presentations): (traditional) $x_\infty \longleftarrow_k x_{n,k}$ for each $n < \omega$, and (in terms of filters) for each $n < \omega$, a countable set X_n carries a prime cofinite convergence $(X_n, x_\infty)_0$ of X_n centered at x_∞.

Let $\boldsymbol{X} := \{(X_j, x_\infty)_0 : j \in J\}$ be a pavement of ξ at x_∞. By definition, the convergence $\pi = \pi[x_\infty, \boldsymbol{X}]$ is prime, but not a pretopology, because $\mathcal{V}_\pi(x) \notin \pi^-(x_\infty)$. If J is (infinite) countable, then we recover the description above in terms of sequences. We call $\pi[x_\infty, \boldsymbol{X}]$ a *sequential fan.* We notice that $\pi = \bigwedge_{j\in J} \pi_j$, where $\pi_j := \pi[x_\infty, (X_j)_0]$ is a prime pretopology.

III.5. Finite convergences

A convergence is called *principal* if each convergent filter is principal. It follows from the definition that each filter convergent for a principal convergence is the principal filter of a finite set ([9]).

Hence, each principal, finitely stable convergence is determined by the convergence of principal ultrafilters. Therefore, every principal convergence can be identified with a directed graph, in which the vertices represent both the elements and the corresponding ultrafilters, and the edges, the underlying convergence relation. Namely,

$$x \to y \iff y \in \lim x^\uparrow.$$

A special class of principal convergences is that of *finite convergences*, that is, convergences, the underlying set of which is finite. As

[9]Indeed, there exists a free filter on each infinite set, for instance, the cofinite filter of this set.

the only proper filters on a finite set are principal, each finite convergence is principal.

EXAMPLE III.5.1. Let X be an infinite set and let $\infty \in X$. Define a prime convergence to the effect that $\infty \in \lim_\eta \mathcal{F}$ whenever there exists $A \subset_{\text{fin}} X$ such that $A \cup \{\infty\} \in \mathcal{F}$. Thus η is a finitely stable principal convergence, which is not finite. Notice that η is not a pretopology.

Contrary to appearances, some convergences on finite sets are very important in the theory of convergences, in spite of their simplicity!

As the number of filters on a finite set is finite,

PROPOSITION III.5.2. *Every finitely stable finite convergence is a pretopology.*

On $X = \{0, 1\}$ there are three proper filters: $\{0\}^\uparrow, \{1\}^\uparrow$ and $\{0, 1\}^\uparrow$. Of course, $0 \in \lim \{0\}^\uparrow$ and $1 \in \lim \{1\}^\uparrow$ for every convergence on X. These are the only convergent filters for the discrete convergence ι_X. For the chaotic convergence o_X, all filters converge to all elements. They are both pretopologies, hence finitely stable.

If ξ is a finitely stable convergence on $X = \{0, 1\}$, which is neither chaotic nor discrete, then either $1 \in \lim \{0\}^\uparrow$ or $0 \in \lim \{1\}^\uparrow$.

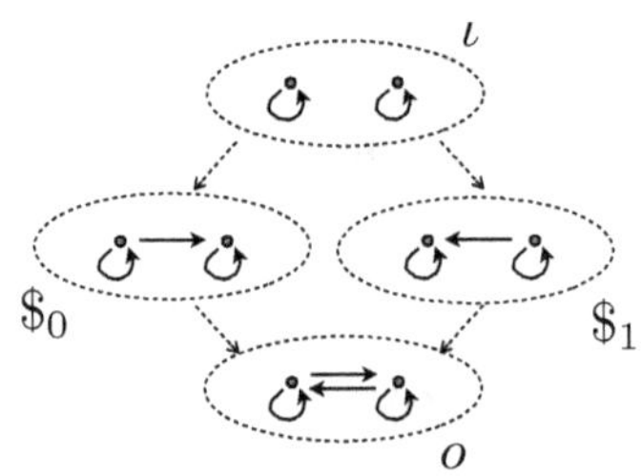

FIGURE III.5.1. The graphs within the ovals represent finitely stable convergences on a two-element set $\{0, 1\}$. We convene that the element on the left is 0, and on the right, is 1.

EXAMPLE III.5.3 (Sierpiński pretopologies). Define a finitely stable convergence $\$_0$ on $X = \{0, 1\}$ by

$$\{0, 1\} = \lim_{\$_0} \{0\}^\uparrow \text{ and } \{1\} = \lim_{\$_0} \{1\}^\uparrow. \tag{III.5.1}$$

Of course, as for every convergence, each point converges to itself. Therefore, we characterize $\$_0$ by saying that 0 converges to 1, but not conversely. For the record, $\$_0$ is this Sierpiński pretopology, for which 1 is isolated, or the *upper Sierpiński* pretopology.

Because $\$_0$ is assumed to be finitely stable, $\lim_{\$_0}\{0,1\}^{\uparrow} = \lim_{\$_0}\{0\}^{\uparrow} \cap \lim_{\$_0}\{1\}^{\uparrow} = \{1\}$.

Similarly, if $\$_1$ is a finitely stable convergence, for which

$$\{0\} = \lim_{\$_1}\{0\}^{\uparrow} \text{ and } \{0,1\} = \lim_{\$_1}\{1\}^{\uparrow}, \tag{III.5.2}$$

then $\lim_{\$_1}\{0,1\}^{\uparrow} = \{0\}$. Remember that $\$_1$ is this Sierpiński pretopology, for which 0 is isolated. We see that $\$_1$ is the convergence on $\{0,1\}$, for which 1 converges to 0, but not conversely, that is the *lower Sierpiński* pretopology. The Sierpiński pretopologies are T_0, but are not free (alias T_1), because $\lim_{\$_0} 0^{\uparrow} \setminus \{0\} \neq \varnothing$, and analogously for $\$_1$. By (III.5.1) and (III.5.2), Sierpiński pretopologies are not comparable. More precisely,

$$\$_0 \vee \$_1 = \iota, \quad \$_0 \wedge \$_1 = o.$$

We notice that the Sierpiński pretopologies are prime: the non-isolated point for $\$_0$ is 1, and for $\$_1$ is 0.

A pretopology on a finite set is determined by its convergent ultrafilters. On an n-point set, there are n ultrafilters (convergent to their defining points). Each of them can converge to any remaining subset, that is, there are 2^{n-1} possibilities for each ultrafilter. Hence there are $2^{(n-1)n}$ distinct pretopologies on an n-point set.

Therefore, on a three-point set there $2^{2\cdot 3} = 64$ pretopologies. As we see in Figure III.5.2, there are sixteen equivalence classes of homeomorphic pretopologies.

III.6. Metrizable convergences

A function $d : X \times X \to \mathbb{R}_+$ is called a *metric* on X if, for each $x, y, z \in X$,

$$d(x,y) = d(y,x), \tag{III.6.1}$$

$$d(x,y) = 0 \Longleftrightarrow x = y, \tag{III.6.2}$$

$$d(x,z) \leq d(x,y) + d(y,z). \tag{III.6.3}$$

If d is a metric on a set X, then

$$x \in \lim_{\mathrm{t}d} \mathcal{F} \Longleftrightarrow \inf_{F\in\mathcal{F}} \sup_{y\in F} d(x,y) = 0 \tag{III.6.4}$$

defines a convergence $\mathrm{t}d$ on X.

In particular, a sequence $(x_n)_n$ of elements of X convergence to $x \in X$ with respect to $\mathrm{t}d$ (that is, $x \in \lim_{\mathrm{t}d}(x_n)_n$) if and only if for each $\varepsilon > 0$, there exists $n_\varepsilon < \omega$ such that $d(x_n, x) < \varepsilon$ for every $n > n_\varepsilon$.

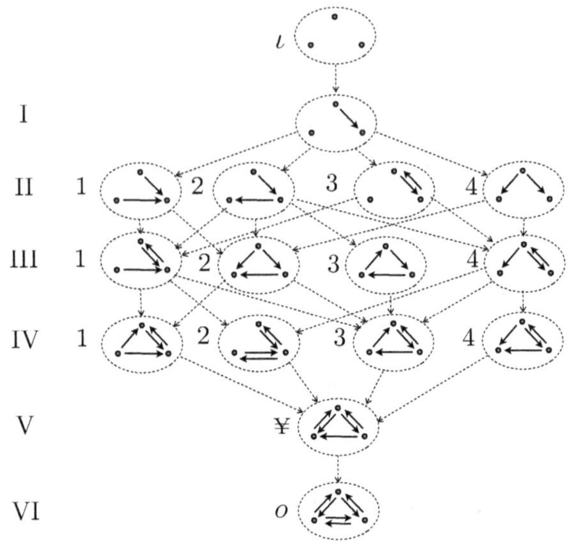

FIGURE III.5.2. There are 16 equivalence classes of pretopologies on a three-point set. A solid arrow inside an oval represents the convergence of a point (that is, the principal ultrafilter) to another point (all points converge also to themselves, because each pretopology is centered). A dotted arrow goes from a finer to the coarser pretopology. The 16 classes are ranged in rows corresponding to the number of solid arrows. We marked the discrete pretopology ι, the chaotic pretopology o, and a Bourdaud pretopology ¥ (discussed in detail in Section VI.7).

DEFINITION III.6.1. If d is a metric on X and $\varepsilon > 0$, then the set

$$B(x,\varepsilon) = B_d(x,\varepsilon) := \{y \in X : d(x,y) < \varepsilon\} \tag{III.6.5}$$

is called a *strict ball* of *radius* ε *centered at* x.

If d is a metric, then $x \in \lim_{\mathrm{t}d} \mathcal{F}$ if and only if $B_d(x,\varepsilon) \in \mathcal{F}$ for each $\varepsilon > 0$. Indeed, as d is non-negative, (III.6.4) means that for each $\varepsilon > 0$ there exists $F \in \mathcal{F}$ such that $F \subset B(x,\varepsilon)$. In other words, the filter

$$\mathcal{V}_{\mathrm{t}d}(x) := \{B_d(x,\varepsilon) : \varepsilon > 0\}^{\uparrow}$$

is the coarsest filter $\mathrm{t}d$-convergent to x.

DEFINITION III.6.2. A convergence ξ on X is called *metrizable* if there exists a metric d such that $\xi = \mathrm{t}d$. Then d is said to be *compatible* with the convergence ξ.

Of course, each metrizable convergence is a pretopology, which is Hausdorff, because of (III.6.2). Let us summarize for future references. Let ξ be a (metrizable) convergence, and d any compatible metric. Then

$$x \in \lim_\xi \mathcal{F} \iff \underset{\varepsilon>0}{\forall}\ B_d(x,\varepsilon) \in \mathcal{F}. \tag{III.6.6}$$

EXAMPLE III.6.3. The convergences from Examples III.1.4 and III.1.1 are metrizable, and those from Examples III.1.3 and III.1.2 are (obviously) not. A metric d on $\mathbb{R}$ that metrizes the standard convergence of Example III.1.4 is

$$d(x_0, x_1) := |x_1 - x_0|. \tag{III.6.7}$$

The discrete convergence on X (of Example III.1.1) can be metrized by a *discrete metric*

$$i_X(x_0, x_1) := \begin{cases} 0, & \text{if } x_0 = x_1, \\ 1, & \text{if } x_0 \neq x_1. \end{cases}$$

Notice that the ball $B_{i_X}(x,\varepsilon) = \{x\}$ if $\varepsilon < 1$ and $B_{i_X}(x,\varepsilon) = X$ if $\varepsilon \geq 1$.

A metrizable convergence can admit several *compatible* metrics, that is, the metrics that metrize it.

EXAMPLE III.6.4. Two metrics compatible with the discrete convergence on $X = \{\frac{1}{n} : n \in \mathbb{N}_1\}$ with $d(r,s) = |r-s|$ and $i_X(r,s) := 1$ if $r \neq s$ and $i_X(r,r) := 0$. Of course, $d \leq i_X$, but there is no $c > 0$ such that $i_X \leq c \cdot d$.

DEFINITION III.6.5 (Euclidean convergence). We call

$$\mathbb{R}^m = \{(x_1, \ldots, x_m) : x_1, \ldots, x_m \in \mathbb{R}\}$$

a *Euclidean space of dimension* m. The functions

$$d(x,y) := \sqrt{\sum_{k=1}^{m} |x_k - y_k|^2},$$
$$s(x,y) := \sum_{k=1}^{m} |x_k - y_k|,$$
$$t(x,y) := \max\{|x_k - y_k| : 1 \leq k \leq m\}$$

are metrics. The first of them is called the *Euclidean metric*.

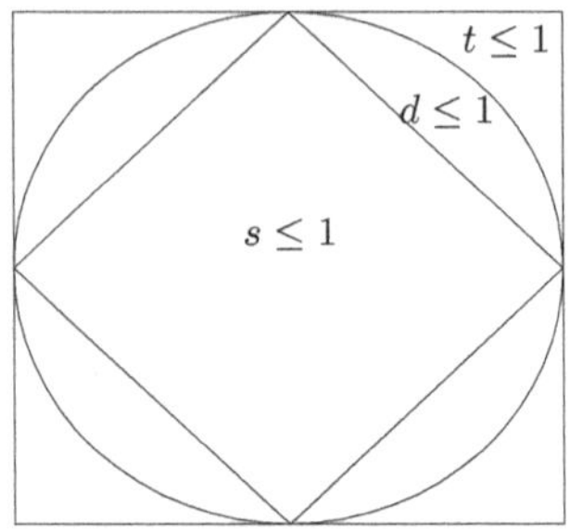

FIGURE III.6.1. Representations of balls of radius 1 with respect to the three metrics in 2-dimensional Euclidean space.

It is straightforward that for each x, y,

$$t\,(x,y) \leq d\,(x,y) \leq s(x,y) \leq m \cdot t\,(x,y)\,, \tag{III.6.8}$$

as one can infer from Figure III.6.1. Consequently, the three metrics above are equivalent, in the sense that

$$\mathrm{t}t = \mathrm{t}d = \mathrm{t}s.$$

Now, it is a simple observation that a filter $\mathcal{F}$ converges to x if and only if $x_k \in \lim_\nu p_k\,[\mathcal{F}]$ for each $1 \leq k \leq m$ (where $p_k(x) := x_k$, the k-th factor of x).

For future references' sake, let us introduce the concept of *norm* in a real vector space X, which is a map $\|\cdot\| : X \longrightarrow \mathbb{R}_+$ such that for each $x, y \in X$, and $\lambda \in \mathbb{R}$,

$$\|x\| = 0 \Longrightarrow x = 0,$$
$$\|\lambda \cdot x\| = |\lambda|\,\|x\|\,,$$
$$\|x + y\| \leq \|x\| + \|y\|\,.$$

Each norm defines a metric by $d(x,y) := \|x - y\|$. In particular, in Euclidean space $\mathbb{R}^m$, where $x = (x_1, \ldots, x_m)$, there are three basic norms:

$$\|x\|_2 := \sqrt{\sum\nolimits_{k=1}^{m} |x_k|^2}, \tag{III.6.9}$$

$$\|x\|_1 := \sum\nolimits_{k=1}^{m} |x_k|\,, \tag{III.6.10}$$

$$\|x\|_\infty := \max\left\{|x_k| : 1 \leq k \leq m\right\}. \tag{III.6.11}$$

They are related to the metrics just defined by

$$d(x,y) = \|x - y\|_2\,, \quad s(x,y) = \|x - y\|_1\,, \quad t(x,y) = \|x - y\|_\infty\,.$$

III.7. Supplement

EXERCISE III.7.1. Show that the filter $\mathcal{V}_{\sigma_{\mathbb{R}}}(x) := \bigwedge_{\mathcal{F}\in\sigma_{\mathbb{R}}^{-}(x)} \mathcal{F}$ has a filter-base consisting of the strict intervals $]x-\varepsilon, x+\varepsilon[$ with $\varepsilon > 0$.

SOLUTION. It follows from Section II.1, that for each sequential filter $\mathcal{F}$ such that $x \in \lim_{\sigma_{\mathbb{R}}} \mathcal{F}$, there exists $\varepsilon > 0$ such that $]x-\varepsilon, x+\varepsilon[\in \mathcal{F}$. If $\mathcal{E} \approx \varphi$ is a sequential filter such that there is $\varepsilon > 0$, for which $]x-\varepsilon, x+\varepsilon[\notin \mathcal{E}$. Consequently, $\{n \in N : \varphi(n) \notin]x-\varepsilon, x+\varepsilon[\}$ is infinite, and thus $x \notin \lim_{\sigma_{\mathbb{R}}} \mathcal{E}$.

EXERCISE III.7.2. Describe $\$_0 \vee \$_1$ and $\$_0 \wedge \$_1$.

SOLUTION. As $(\$_0 \vee \$_1)^-(x) = \$_0^-(x) \cap \$_1^-(x)$, the supremum $\$_0 \vee \$_1$ is the discrete convergence. As $(\$_0 \wedge \$_1)^-(x) = \$_0^-(x) \cup \$_1^-(x)$, the infimum $\$_0 \wedge \$_1$ is the chaotic convergence.

EXERCISE III.7.3. Show that $\varsigma_+ \wedge \varsigma_- = \nu_{\mathbb{R}}$ and $\varsigma_+ \vee \varsigma_- = \iota$, where the extrema $\wedge, \vee$ are considered within the class of finitely stable convergences.

SOLUTION. In fact, by definition, $x \in \lim_{\varsigma_+ \wedge \varsigma_-} \mathcal{F}$ if and only if $x \in \lim_{\varsigma_+} \mathcal{F}$ or $x \in \lim_{\varsigma_-} \mathcal{F}$, that is, either $[x, x+\varepsilon[\in \mathcal{F}$ for each $\varepsilon > 0$, or $]x-\varepsilon, x] \in \mathcal{F}$ for each $\varepsilon > 0$. Or else, the filter-bases $\{[x, x+\varepsilon[: \varepsilon > 0\}$ and $\{]x-\varepsilon, x] : \varepsilon > 0\}$ converge to x in $\varsigma_+ \wedge \varsigma_-$, and since $\varsigma_+ \wedge \varsigma_-$ is supposed to be finitely stable, also $\{]x-\varepsilon, x] \cup [x, x+\varepsilon[: \varepsilon > 0\}$ does. Of course, $]x-\varepsilon, x] \cup [x, x+\varepsilon[=]x-\varepsilon, x+\varepsilon[$, and $\{]x-\varepsilon, x+\varepsilon[: \varepsilon > 0\}$ is the coarsest filter-base $\nu_{\mathbb{R}}$-convergent to x. Now, $x \in \lim_{\varsigma_+ \vee \varsigma_-} \mathcal{F}$ if and only if $x \in \lim_{\varsigma_+} \mathcal{F}$ and $x \in \lim_{\varsigma_-} \mathcal{F}$, that is, $[x, x+\varepsilon[\in \mathcal{F}$ and $]x-\varepsilon, x] \in \mathcal{F}$ for each $\varepsilon > 0$, which amounts to $\{x\} \in \mathcal{F}$, that is, $x \in \lim_{\iota} \mathcal{F}$.

EXERCISE III.7.4. Show that the set of irrational numbers is $\nu_{\mathbb{R}}$-dense in $\mathbb{R}$.

SOLUTION. For each $q \in \mathbb{Q}$ and every $n \in \mathbb{N}_1$, the interval $]q-\frac{1}{n}, q+\frac{1}{n}[$ contains an irrational number r_n. In fact, by Example I.5.1, each open interval is equipotent with the whole real line, hence uncountable by Proposition I.5.7. On the other hand, $\mathbb{Q}$ is countable by Proposition I.5.5. By Definition II.1, $(r_n)_n$ $\nu_{\mathbb{R}}$-converges to q.

EXERCISE III.7.5. Prove (III.6.8).

The following example shows that the property T_0 is nor preserved under restrictions.

EXAMPLE III.7.6. Let υ be a pretopology on $X = \{0,1,2\}$ to the effect that $\lim_\upsilon 0^\uparrow = \lim_\upsilon 1^\uparrow := \{0,1\}$, and $\lim_\upsilon 2^\uparrow := \{0,2\}$. Accordingly,

$$\mathcal{V}_\upsilon(0) = \{0,1,2\}^\uparrow, \mathcal{V}_\upsilon(1) = \{0,1\}^\uparrow, \mathcal{V}_\upsilon(2) = 2^\uparrow,$$

so that υ is T_0. However, the restriction $\upsilon|_{\{0,1\}} = o_{\{0,1\}}$ is not T_0.

REMARK III.7.7. The concept of filter as a family of subsets of a given set, is a special case of a more general notion. Let $(A, \leq)$ be a lattice (see Section II.3). Then $F \subset A$ is called a *lattice-filter* provided that, $F \neq \varnothing$, if $a \in F$ and $x \geq a$, then $x \in F$, and if $a_0, a_1 \in F$ then the infimum $a_0 \wedge a_1 \in F$. Of course, a filter $\mathcal{F}$ on X is a subset of 2^X, for $A_0, A_1 \in \mathcal{F}$ the infimum in $(2^X, \subset)$ is $A_0 \wedge A_1 = A_0 \cap A_1 \in \mathcal{F}$, hence $\mathcal{F}$ is a lattice-filter on $(2^X, \subset)$.

PROPOSITION III.7.8. *If ξ is a finitely stable convergence on X, then $\xi^-(x)$ is a lattice-filter for each $x \in X$.*

Let X be a non-empty set.

PROPOSITION III.7.9. *If $\theta \subset \mathbb{F}X \times X$, then there exists the finest preconvergence $\mathrm{I}_\circ\theta$ that is coarser than θ.*

PROOF. We define $x \in \lim_{\mathrm{I}_\circ\theta} \mathcal{F}$ if there exists $\mathcal{E} \in \mathbb{F}X$ such that $x \in \lim_\theta \mathcal{E}$ and $\mathcal{E} \leq \mathcal{F}$. Then $\mathrm{I}_\circ\theta$ is a preconvergence and $\mathrm{I}_\circ\theta \leq \theta$. If ξ is a preconvergence such that $\xi \leq \theta$, then $\xi = \mathrm{I}_\circ\xi \leq \mathrm{I}_\circ\theta$, so that $\mathrm{I}_\circ\theta$ is the finest preconvergence coarser than θ. □

It is plain that

$$\theta_0 \leq \theta_1 \Longrightarrow \mathrm{I}_\circ\theta_0 \leq \mathrm{I}_\circ\theta_1,$$
$$\mathrm{I}_\circ\theta \leq \theta,$$
$$\mathrm{I}_\circ(\mathrm{I}_\circ\theta) = \mathrm{I}_\circ\theta.$$

A preconvergence ξ on a non-empty set X is a convergence if and only if $\xi \leq \iota_X$. Indeed, $x \in \lim_{\iota_X} \mathcal{F}$ if and only if $\mathcal{F} = x^\uparrow$, hence $\xi \leq \iota_X$ whenever (centered) holds. Hence, for any $\theta \subset \mathbb{F}X \times X$, by Proposition III.7.9, $\mathrm{I}\theta := \mathrm{I}_\circ\theta \wedge \iota_X$ is the finest convergence that is coarser than θ.

LEMMA III.7.10. *If $\theta \subset \mathbb{F}X \times X$, then there exists the finest convergence $\mathrm{I}\theta$ that is coarser than θ.*

We call $\mathrm{I}_\circ$ the *preconvergence modifier*, and I the *convergence modifier*.

EXAMPLE III.7.11. Let $\eta \subset \mathbb{F}\mathbb{R} \times \mathbb{R}$ be defined by $x \in \lim_\eta \mathcal{F}$ whenever there exists a sequence $(x_n)_n$ of elements of $\mathbb{R}$ such that $\mathcal{F} = \{\{x_n : n > k\}\}^\uparrow$ and for each $\varepsilon > 0$ there is $k < \omega$ such that $|x_n - x| < \varepsilon$

for every $n > k$. In other words, this a usual convergence of sequences. Then $\mathrm{I}\eta$ is the standard sequential convergence from Example III.1.3. As $\eta \leq \iota_X$, we infer that $\mathrm{I}\eta = \mathrm{I}_\circ\eta$.

EXAMPLE III.7.12. If $\boldsymbol{F}$ is a set of filters on a set X with the following properties ([10]):

(1) $x^\uparrow \in \boldsymbol{F}$ for each $x \in X$,

(2) if $\mathcal{F} \in \boldsymbol{F}$ and $\mathcal{F} \leq \mathcal{G}$, then $\mathcal{G} \in \boldsymbol{F}$.

The relation $\chi_{\boldsymbol{F}}$ defined by

$$\lim\nolimits_{\chi_{\boldsymbol{F}}} \mathcal{F} := \begin{cases} X & \mathcal{F} \in \boldsymbol{F}, \\ \varnothing & \mathcal{F} \notin \boldsymbol{F}, \end{cases}$$

is a convergence, called a *characteristic convergence* of $\boldsymbol{F}$. If for each pair of filters $\mathcal{F}_0, \mathcal{F}_1 \in \boldsymbol{F}$ implies that $\mathcal{F}_0 \cap \mathcal{F}_1 \in \boldsymbol{F}$, then $\chi_{\boldsymbol{F}}$ is finitely stable; if $\bigwedge \boldsymbol{H} \in \boldsymbol{F}$ for every non-empty subset $\boldsymbol{H}$ of $\boldsymbol{F}$, then $\chi_{\boldsymbol{F}}$ is a pretopology.

[10] The so-called *Cauchy structure* of E. LOWEN-COLEBUNDERS [180].

CHAPTER IV

Continuity

A map is *continuous* if it preserves convergence. Before giving a precise meaning to this vague idea, it is natural to investigate how maps act on filters.

IV.1. Images and preimages

Let X and Y be sets, and let $f \in Y^X$ be a map. The *image* of a subset A of X under f is denoted by $f(A) := \{f(a) : a \in A\}$, and the *preimage* of a subset B of Y under f is denoted by $f^-(B) := \{x \in X : f(x) \in B\}$. Sometimes, in order to avoid too numerous parentheses, we may abridge $fA := f(A)$ and $f^-B := f^-(B)$.

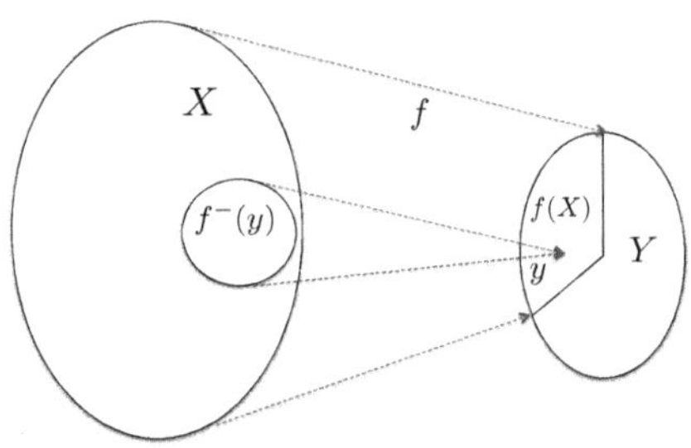

FIGURE IV.1.1. A map $f : X \to Y$, the image $f(X)$ of X, and the preimage $f^-(y)$ of an element y of $f(X)$.

Recall that

$$f(f^-(B)) \subset B, \tag{IV.1.1}$$

$$A \subset f^-(f(A)), \tag{IV.1.2}$$

and the equality holds in (IV.1.1) whenever f is surjective, while the equality holds in (IV.1.2) whenever f is injective.

As $f^-(y_0) \cap f^-(y_1) = \varnothing$ for each $y_0 \neq y_1$,

$$B_0 \cap B_1 = \varnothing \Longrightarrow f^-(B_0) \cap f^-(B_1) = \varnothing,$$

and in particular, for each $B \subset Y$,

(IV.1.3) $$\left(f^{-}\left(B^{c}\right)\right)^{c}=f^{-}(B).$$

If $\mathcal{A}$ and $\mathcal{D}$ are families of subsets of X, then $\mathcal{A}$ is said to be *coarser* than $\mathcal{D}$ (equivalently, $\mathcal{D}$ is *finer* than $\mathcal{A}$), in symbols,

$$\mathcal{A} \leq \mathcal{D}$$

if for each $A \in \mathcal{A}$, there is $D \in \mathcal{D}$ such that $D \subset A$. Notice that if $\mathcal{A} \subset \mathcal{D}$ then $\mathcal{A} \leq \mathcal{D}$.

Recall that a family $\mathcal{A}$ of subsets of a given set X is said to be *isotone* if $A \in \mathcal{A}$ and $A \subset H$ implies that $H \in \mathcal{A}$. Observe that if $\mathcal{D}$ is isotone, and $\mathcal{A} \leq \mathcal{D}$ then $\mathcal{A} \subset \mathcal{D}$.

We say that $\mathcal{B}$ is a *base* of an isotone family $\mathcal{A}$ if

$$\mathcal{B} \subset \mathcal{A} \text{ and } \mathcal{B} \geq \mathcal{A}.$$

If $\mathcal{F}$ is a family of subsets of X, then we write

$$f[\mathcal{F}] := \{f(F) : F \in \mathcal{F}\}.$$

If $\mathcal{G}$ is a family of subsets of Y, then

$$f^{-}[\mathcal{G}] := \{f^{-}(G) : G \in \mathcal{G}\}.$$

LEMMA IV.1.1. *If $f : X \to Y$ and $\mathcal{G} \subset 2^{X}$,*

$$\mathcal{G} \geq f^{-}[f[\mathcal{G}]],$$

and the equality holds if f is injective.

PROOF. By definition, $f^{-}[f[\mathcal{G}]] = \{f^{-}(f(G)) : G \in \mathcal{G}\}$. Now, by (IV.1.2), $G \subset f^{-}(f(G))$, hence $\mathcal{G} \geq f^{-}[f[\mathcal{G}]]$. If f is injective, then $G = f^{-}(f(G))$, so that the equality holds. □

LEMMA IV.1.2. *If $\mathcal{H} \subset 2^{Y}$ and $f : X \longrightarrow Y$, then*

$$f[f^{-}[\mathcal{H}]] \geq \mathcal{H}.$$

If f is surjective, then the equality holds.

PROOF. Indeed, $f^{-}[\mathcal{H}] = \{f^{-}(H) : H \in \mathcal{H}\}$, hence

$$f[f^{-}[\mathcal{H}]] = \{f(f^{-}(H)) : H \in \mathcal{H}\} \geq \mathcal{H},$$

because, by (IV.1.1), $f(f^{-}(H)) \subset H$, and $f(f^{-}(H)) = H$ if f is surjective. Consequently, $f[f^{-}[\mathcal{H}]] \geq \mathcal{H}$, with the equality if f is surjective. □

Lemmas IV.1.1 and IV.1.2 promptly imply that for families $\mathcal{F}$ and $\mathcal{G}$, and a map f,

(IV.1.4) $$f^{-}[\mathcal{H}] \leq \mathcal{G} \Longleftrightarrow \mathcal{H} \leq f[\mathcal{G}].$$

PROPOSITION IV.1.3. *If $\mathcal{F}$ is a proper filter on X and $f \in Y^X$, then $f[\mathcal{F}]$ is a filter-base on Y.*

PROOF. If $B_0, B_1 \in f[\mathcal{F}]$, then there exist $F_0, F_1 \in \mathcal{F}$ such that $B_0 = f(F_0)$ and $B_1 = f(F_1)$. It follows that $B_0 \cap B_1 = f(F_0) \cap f(F_1) \supset f(F_0 \cap F_1) \in f[\mathcal{F}]$ as $\mathcal{F}$ fulfills (finitely complete). On the other hand, if $\varnothing \in f[\mathcal{F}]$ then there would be $F \in \mathcal{F}$ such that $\varnothing = f(F)$, hence $F = \varnothing$ contrary to the assumption that $\mathcal{F}$ is proper. □

Then $f[\mathcal{F}]^\uparrow$ is equal to $f[\mathcal{F}]$ if and only if $f(X) = Y$.

PROPOSITION IV.1.4. *If $\mathcal{G}$ is a proper filter on Y and $f \in Y^X$, then $f^-[\mathcal{G}]$ is a filter-base if and only if $f(X) \cap G \neq \varnothing$ for every $G \in \mathcal{G}$.*

PROOF. As $f^-(G_0 \cap G_1) = f^-(G_0) \cap f^-(G_1)$, in view of Proposition II.2.2, it is enough to note that $f^-(G) \neq \varnothing$ if and only if $f(X) \cap G \neq \varnothing$. □

Note that the filter-base $f^-[\mathcal{G}]$ from Proposition IV.1.4 is a filter if f is injective ([1]).

IV.2. Continuous maps

Let ξ be a convergence on X and τ be a convergence on Y. A map $f : X \to Y$ is said to be *continuous at x* (from ξ to τ) if for every filter $\mathcal{F}$ on X,

$$x \in \lim_\xi \mathcal{F} \Longrightarrow f(x) \in \lim_\tau f[\mathcal{F}]; \tag{IV.2.1}$$

continuous if this holds for each $x \in X$. Mind that $f[\mathcal{F}]$ is in general a filter-base. Nevertheless, in view of (III.1.2), the condition above makes sense.

REMARK IV.2.1. This definition makes also sense for relations. Although we focus on continuity for convergences, we shall occasionally deploy it for preconvergences, as a matter of fact.

We denote by $\mathscr{C}(\xi, \tau)$ the set of all maps that are continuous from ξ to τ (or by $\mathscr{C}(X, Y)$ if convergences on X and Y are implicit). For the sake of brevity, it is more practical to write $f \in \mathscr{C}(\xi, \tau)$, rather than to say: *$f$ is continuous from ξ to τ*. It follows from the definition that $f \in \mathscr{C}(\xi, \tau)$ if and only if

$$f(\lim_\xi \mathcal{F}) \subset \lim_\tau f[\mathcal{F}] \tag{IV.2.2}$$

[1]Conversely, if $x_0, x_1 \in f^-(y_0)$ and $x_0 \neq x_1$, then $X \setminus f^-(y_0) \in f^-[\{Y \setminus \{y_0\}\}^\uparrow]$, but $X \setminus f^-(y_0) \cup \{x_0\} \notin f^-[\{Y \setminus \{y_0\}\}^\uparrow]$.

for each filter $\mathcal{F}$ on X. In terms of inverse relations, a map $f \in \mathscr{C}(\xi, \tau)$ if and only if

$$\mathcal{F} \in \xi^-(x) \Longrightarrow f[\mathcal{F}] \in \tau^-(f(x))$$

for each $x \in X$.

Notice that if ξ and θ are convergences on X, then

$$\xi \geq \theta \Longleftrightarrow i_X \in \mathscr{C}(\xi, \theta),$$

that is, the identity map $i_X : X \to X$ is continuous from ξ to θ. On the other hand, for each convergence ξ on X, and every convergence τ on Y,

$$\mathscr{C}(\iota_X, \tau) = Y^X = \mathscr{C}(\xi, o_Y),$$

that is, each map from a discrete convergence, and every map to a chaotic convergence is continuous. Indeed, since the only ι_X-convergent filters are principal ultrafilters, and for every $f \in Y^X$, the image of a principal ultrafilter is a principal ultrafilter, it is convergent for every convergence: $x \in \lim_\iota x^\uparrow$ implies that $f(x) \in \lim_\tau f[x^\uparrow] = \lim_\tau f(x)^\uparrow$. Every filter converges to each point for a chaotic convergence o, that is, if $x \in \lim_\xi \mathcal{F}$ then $f(x) \in \lim_o f[\mathcal{F}]$.

On the other hand, $\mathscr{C}(o_X, \iota_Y)$ is the set of constant maps. Indeed, $X = \lim_o\{X\}$, that is, $\{X\}$ is a vicinity filter of o at every point. Now, $\overline{y}[\{X\}] = \{y\}$ for each constant map $\overline{y}$, and $\{y\} = \lim_\iota y^\uparrow$.

It is straightforward that

Proposition IV.2.2. *If* $f \in \mathscr{C}(\xi, \tau)$ *and* $g \in \mathscr{C}(\tau, \psi)$, *then* $g \circ f \in \mathscr{C}(\xi, \psi)$.

Example IV.2.3. Let $\nu_\mathbb{R}$ be the standard convergence on the real line as in Example III.1.4. Then $f \in \mathscr{C}(\nu_\mathbb{R}, \nu_\mathbb{R})$ if and only if $\varphi(r) := |f(r) - f(x)|$ is continuous at x for each $x \in \mathbb{R}$. Indeed, $f(x) \in \lim_{\nu_\mathbb{R}} f[\mathcal{F}]$ if and only if for each $\varepsilon > 0$, there exists $F \in \mathcal{F}$ such that $\sup_{r \in F} |f(x) - f(r)| < \varepsilon$ for each $r \in F$.

As $x \in \lim_{\nu_\mathbb{R}} \mathcal{F}$ if and only if $]x-\delta, x+\delta[\, \in \mathcal{F}$ for each $\delta > 0$, we can rephrase the condition above to the effect that for each $\varepsilon > 0$, there exists $\delta > 0$ such that $|r - x| < \delta$ implies that $|f(r) - f(x)| < \varepsilon$.

Example IV.2.4. Consider a metric d compatible with a metrizable convergence ξ on X. Let us observe that $d(\cdot, x) \in \mathscr{C}(\xi, \nu_\mathbb{R})$ for each $x \in X$. Indeed, by (III.6.3) and (III.6.1), $d(y, x) - d(z, x) \leq d(y, z)$, hence $d(z, x) - d(y, x) \leq d(y, z)$, by inverting the role of y and z, so that $|d(y, x) - d(z, x)| \leq d(y, z)$, which implies the continuity of $d(\cdot, x)$ in virtue of Example IV.2.3. If X is a normed vector space, then $\|x\| = d(x, 0)$, where 0 is the zero vector of X. Therefore, the norm is continuous.

LEMMA IV.2.5. *If $f \in \mathscr{C}(\xi,\tau)$, then for each $x \in |\xi|$,*

$$f[\mathcal{V}_\xi(x)] \geq \mathcal{V}_\tau(f(x)). \tag{IV.2.3}$$

PROOF. Indeed, if $W \in \mathcal{V}_\tau(f(x))$, then $f^-(W) \in \mathcal{F}$ for each $\mathcal{F} \in \xi^-(x)$, because $f[\mathcal{F}] \in \tau^-(f(x))$, that is, $f^-(W) \in \mathcal{V}_\xi(x)$, hence $W \supset f(f^-(W)) \in f[\mathcal{V}_\xi(x)]$. □

Consequently, if τ is a pretopology, then $f \in \mathscr{C}(\xi,\tau)$ if and only if (IV.2.3) holds for each $x \in |\xi|$. If ξ and τ are metrizable pretopologies, and d_ξ and d_τ are any respective compatible metrics, then we can rephrase (IV.2.3) as follows:

$$\underset{\varepsilon>0}{\forall}\ \underset{\delta>0}{\exists}\ \underset{w\in|\xi|}{\forall}\ d_\xi(w,x) < \delta \Longrightarrow d_\tau(f(w),f(x)) < \varepsilon. \tag{IV.2.4}$$

On the other hand,

PROPOSITION IV.2.6. *If ξ and τ are metrizable convergences, then $f \in \mathscr{C}(\xi,\tau)$ if and only if*

$$x \in \lim_\xi (x_n)_n \Longrightarrow f(x) \in \lim_\tau (f(x_n))_n \tag{IV.2.5}$$

for each sequence $(x_n)_n$ of elements of $|\xi|$.

PROOF. Let ξ and τ be metrizable, and d_ξ and d_τ are respective compatible metrics, and let ε and δ be as in (IV.2.4). If $x \in \lim_\xi (x_n)_n$, then there exists $n_\delta < \omega$ such that $d_\xi(x_n,x) < \delta$ for each $n > n_\delta$. Hence, by (IV.2.4), $d_\tau(f(x_n),f(x)) < \varepsilon$ for each $n > n_\delta$, so that, $f(x) \in \lim_\tau (f(x_n))_n$.

Conversely, if (IV.2.4) does not hold, then there is $\varepsilon > 0$ such that for every $n < \omega$, there exists x_n such that $d_\xi(x_n,x) < 2^{-n}$ and $d_\tau(f(x_n),f(x)) \geq \varepsilon$ in contradiction with (IV.2.5). □

As one can guess that, in general, sequences are not sufficient to characterize continuity.

EXAMPLE IV.2.7. Consider the convergences $\sigma_\mathbb{R}$ and $\nu_\mathbb{R}$ from Examples III.1.3 and III.1.4. Of course, $i_\mathbb{R} \in \mathscr{C}(\sigma_\mathbb{R},\nu_\mathbb{R})$, but $i_\mathbb{R} \notin \mathscr{C}(\nu_\mathbb{R},\sigma_\mathbb{R})$, because $0 \in \lim_{\nu_\mathbb{R}} \mathcal{V}_{\nu_\mathbb{R}}(0) \setminus \lim_{\sigma_\mathbb{R}} \mathcal{V}_{\nu_\mathbb{R}}(0)$ and $i_\mathbb{R}[\mathcal{V}_{\nu_\mathbb{R}}(0)] = \mathcal{V}_{\nu_\mathbb{R}}(0)$. On the other hand, $\nu_\mathbb{R}$ and $\sigma_\mathbb{R}$ coincide for sequences.

DEFINITION IV.2.8 (Homeomorphism). Convergences ξ and τ are said to be *homeomorphic*, in symbols,

$$\xi \approx \tau$$

if there exists a bijective map $f \in \mathscr{C}(\xi,\tau)$ such that $f^{-1} \in \mathscr{C}(\tau,\xi)$. Then f is called a *homeomorphism*. Denote $\mathscr{H}(\xi,\tau)$ the set of all homeomorphisms between ξ and τ.

Homeomorphic convergences can be seen as identical copies of the same convergence. Therefore, if we identify a convergence, then, in fact, we identify the class of its homeomorphic copies, so that we use to say that a convergence is the required one "up to a homeomorphism".

If $\operatorname{card} X = \operatorname{card} Y$, then $\iota_X \approx \iota_Y$ (the *discrete convergences*) and $o_X \approx o_Y$ (the *chaotic convergences*). In fact, any bijection $h : X \longrightarrow Y$ is a homeomorphism in both cases.

The *Sierpiński pretopologies* of Example III.5.3 are homeomorphic: $\$_0 \approx \$_1$. The map $h : \{0,1\} \longrightarrow \{0,1\}$ given by $h(x) := 1 - x$, is a homeomorphism. Accordingly, by a *Sierpiński pretopology* $\$$ we intend anyone of the homeomorphic variants.

The Sorgenfrey convergences ς_+, ς_- from Example III.3.6 are homeomorphic: $h : \mathbb{R} \longrightarrow \mathbb{R}$ given by $h(x) := -x$, is a homeomorphism.

The standard convergence of the real line (Example III.1.4) and its restriction to an open interval are homeomorphic ([2]).

IV.3. Continuity and order

DEFINITION IV.3.1 (Initial convergence). Let $f : X \to Y$ and let τ be a convergence on Y. The coarsest convergence on X, for which f is continuous is called the *initial convergence* with respect to (f, τ), and is denoted by $f^- \tau$.

PROPOSITION IV.3.2. *A unique initial convergence $f^- \tau$ exists for each f and τ.*

PROOF. Let $\Xi := \{\xi \in \mathbf{I}X : f \in \mathscr{C}(\xi, \tau)\}$. The set Ξ is non-empty, because $\iota_X \in \Xi$. We observe that $f \in \mathscr{C}(\bigwedge \Xi, \tau)$. Indeed, let $x \in \lim_{\bigwedge \Xi} \mathcal{F} = \bigcup_{\xi \in \Xi} \lim_\xi \mathcal{F}$. Hence there is $\xi \in \Xi$ such that $x \in \lim_\xi \mathcal{F}$, hence, by definition, $f(x) \in \lim_\tau f[\mathcal{F}]$. □

The following instrumental formula holds:

$$\lim_{f^- \tau} \mathcal{F} = f^-(\lim_\tau f\,[\mathcal{F}]). \tag{IV.3.1}$$

PROOF. Indeed, $f \in \mathscr{C}(\xi, \tau)$ amounts to $f(\lim_\xi \mathcal{F}) \subset \lim_\tau f[\mathcal{F}]$, hence to $\lim_\xi \mathcal{F} \subset f^-(\lim_\tau f[\mathcal{F}])$. On the other hand, (IV.3.1) defines a convergence. Therefore, $f \in \mathscr{C}(\xi, \tau)$ if and only if $\lim_\xi \mathcal{F} \subset \lim_{f^- \tau} \mathcal{F}$ for each filter $\mathcal{F}$, which means that $f^- \tau$ is the coarsest convergence ξ such that $f \in \mathscr{C}(\xi, \tau)$. □

---

[2]Because $\tan \in \mathscr{H}(\nu_{\mathbb{R}]-\pi/2,\pi/2[}, \nu_{\mathbb{R}})$ is a homeomorphism.

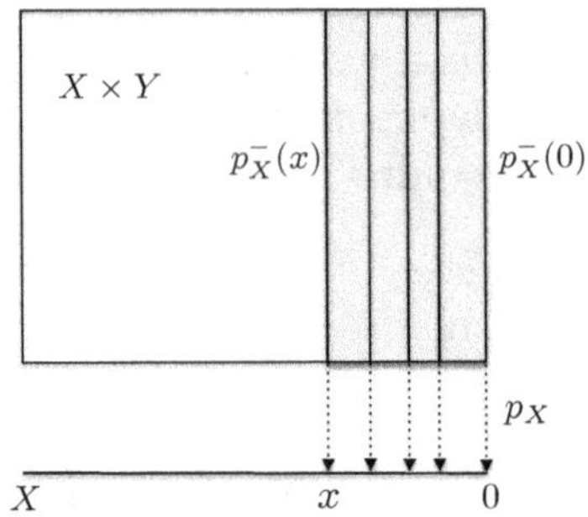

FIGURE IV.3.1. The grey area represents a typical element of the coarsest filter converging to $p_X^-(0)$ for the initial convergence on $X \times Y$ with respect to the projection $p_X : X \times Y \to X$ and a convergence on X.

The *initial convergence* can be seen as a convergence of fibers, rather than that of individual points.

PROPOSITION IV.3.3. *Let $f \in Y^X$ and let τ be a convergence on Y. Then $x \in \lim_{f^-\tau} \mathcal{F}$ implies that*

$$f^-(f(x)) \subset \lim_{f^-\tau} f^-[f[\mathcal{F}]] \subset \lim_{f^-\tau} \mathcal{F}.$$

A plain checking constitutes Exercise IV.10.4. Of course, if f is injective, then $f^-(f(x)) = \{x\}$ is a singleton for each point x, hence $f^-[f[\mathcal{F}]] = \mathcal{F}$.

EXAMPLE IV.3.4. Let η be the pretopology on $X = \{0, 1, 2\}$ to the effect that $\lim_\eta 0^\uparrow = \lim_\eta 1^\uparrow := \{0, 1, 2\}$, and $\lim_\eta 2^\uparrow := \{2\}$. On $Y = \{0, 1\}$ consider the Sierpiński pretopology $\$_0$. Let $f : \{0, 1, 2\} \longrightarrow \{0, 1\}$ be defined by $f(0) = f(1) := 0$ and $f(2) := 1$. Because $\lim_{\$_0} 0^\uparrow = \{0, 1\}$ and $\lim_{\$_0} 1^\uparrow = \{1\}$, we infer that $f^-\$_0 = \upsilon$.

DEFINITION IV.3.5 (Final convergence). Given a convergence ξ on X and a map $f : X \longrightarrow Y$, the finest convergence θ on Y, for which $f \in \mathscr{C}(\xi, \theta)$ is called the *final convergence* with respect to (ξ, f), and is denoted by $f\xi$.

PROPOSITION IV.3.6. *The final convergence always exists.*

PROOF. Let $T := \{\tau \in \mathbf{I}Y : f \in \mathscr{C}(\xi, \tau)\}$. Observe that T is non-empty, because $o_Y \in T$. We shall see that $f\xi = \bigvee T$. Indeed, if $x \in \lim_\xi \mathcal{F}$ then by definition $f(x) \in \lim_\tau f[\mathcal{F}]$ for each $\tau \in T$, hence $f(x) \in \bigcap_{\tau \in T} \lim_\tau \mathcal{F} = \lim_{\bigvee T} \mathcal{F}$, hence $\bigvee T \in T$ is clearly the finest such convergence. $\square$

REMARK IV.3.7. Virtually the same argument shows that there exists a finest preconvergence, for which f is continuous from a convergence ξ. We do not introduce a specific symbol for the latter. The *final preconvergence* is equal to the *final convergence* on the image $f(X)$, that is, the former is equal to $(f\xi)|_{f(X)}$. On the other hand, each $y \notin f(X)$ is $f\xi$-isolated.

It follows from the definition of $f\xi$ that $y \in \lim_{f\xi} \mathcal{G}$ if and only if, either $y \notin f(X)$ and then $(f\xi)^{-}(y) = \{y^{\uparrow}\}$, or there exists a filter $\mathcal{F}$ such that $\lim_{\xi} \mathcal{F} \cap f^{-}(y) \neq \varnothing$ and $\mathcal{G} \geq f[\mathcal{F}]$. Hence ([3])

$$\lim_{f\xi} \mathcal{G} = \bigcup_{f[\mathcal{F}] \leq \mathcal{G}} f(\lim_{\xi} \mathcal{F}) \cup \lim_{\iota} \mathcal{G}.$$

If f is surjective, then

$$\lim_{f\xi} \mathcal{G} = \bigcup_{f[\mathcal{F}] \leq \mathcal{G}} f(\lim_{\xi} \mathcal{F}). \tag{IV.3.2}$$

The following proposition is an immediate consequence of Lemma V.6.4 to come.

PROPOSITION IV.3.8. *If f is surjective, then $y \in \lim_{f\xi} \mathcal{G}$ if and only if there exists a filter $\mathcal{F}_0$ such that $\lim_{\xi} \mathcal{F}_0 \cap f^{-}(y) \neq \varnothing$ and $\mathcal{G} = f[\mathcal{F}_0]$.*

DEFINITION IV.3.9. Let ξ be a convergence on X, and let τ be a convergence on Y. Then a surjective map $f : X \longrightarrow Y$ is called *convergence quotient map* if τ is the finest convergence such that $f \in \mathscr{C}(\xi, \tau)$. It is said to be *finitely stable quotient map* if τ is the finest finitely stable convergence such that $f \in \mathscr{C}(\xi, \tau)$. It is said to be *pretopologically quotient map* if τ is the finest pretopology such that $f \in \mathscr{C}(\xi, \tau)$.

Then $f\xi = \tau$ is called respectively *quotient convergence*, *finitely stable quotient convergence*, and *quotient pretopology*.

A simplest example of a surjective map that is not quotient is the identity map $i_X : X \longrightarrow X$ in the presence of two convergences ξ and τ such that $\xi > \tau$. Then $i \in \mathscr{C}(\xi, \tau)$, but $i\xi = \xi > \tau$.

EXAMPLE IV.3.10. Consider on $[0,1]$ the convergence restricted from the real line, and let us call it μ. Let $h : [0,1] \longrightarrow [0,1[$ be defined by $h(x) := x$ if $x \neq 1$, and $h(1) := 0$. We notice that $h^{-}(r) = \{r\}$ if $0 < r < 1$, and $h^{-}(0) = \{0,1\}$. In other terms, the preimages of points under h are singletons with exception of 0, when it is a doubleton.

[3]By definition, for $f \in Y^X$ and a convergence ξ on X, the final convergence $f\xi$ is defined on Y. If f is not surjective, then $Y \setminus f(X)$ is independent of ξ. Therefore, each $y \in Y \setminus f(X)$ is $f\xi$-isolated.

Then the final convergence $h\mu$ (which is the quotient convergence) differs from μ only at 0. More precisely, $0 \in \lim_{h\mu} \mathcal{F}$ if either $[0, \varepsilon[\in \mathcal{F}$ for every $\varepsilon > 0$, or $\{0\} \cup [1-\varepsilon, 1[\in \mathcal{F}$.

If we denote by τ the pretopology such that $\mathcal{V}_\tau(0) := \mathcal{V}_\mu(0) \wedge (\mathcal{V}_\mu(1) \vee [0,1[)$, and $\mathcal{V}_\tau(r) := \mathcal{V}_\mu(r)$, then τ is the finest pretopology for which $h \in \mathscr{C}(\mu, \tau)$, that is, h is pretopologically quotient with respect to μ and τ. We observe that h is also finitely quotient map (which is a stronger property than that of pretopologically quotient map) with respect to μ and τ, but is not a convergence quotient map, because $h\mu > \tau$.

If $\varnothing \neq V \subset X$, then the map $j_V : V \longrightarrow X$, defined by $j_V(x) := x$ if $x \in V$, is called the *natural injection*.

The notion of *natural injection* is a conceptual artifact, like that of *identity map*. On a subset V of given set X, we define an injection j_V into X so that $j_V(V) = V$. Artificial as it may seem, it is sometimes really useful.

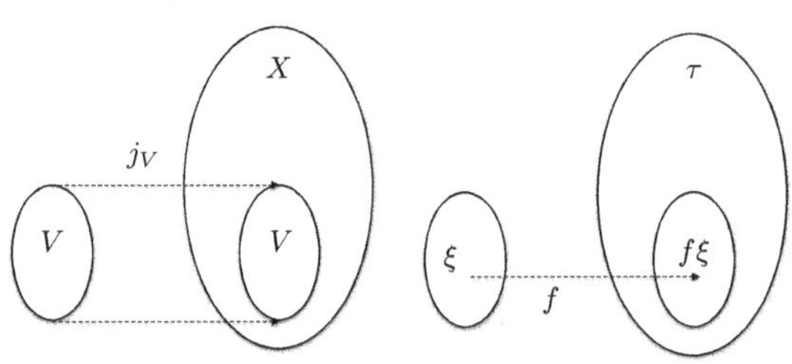

FIGURE IV.3.2. Natural injection $j_V : V \longrightarrow X$, and an embedding $f : \xi \hookrightarrow \tau$.

DEFINITION IV.3.11 (Subconvergence). Let ξ be a convergence on a set X and let $V \subset X$. The initial convergence $j_V^- \xi$ is called a *subconvergence* or a *subspace* of ξ.

The subconvergence $j_V^- \xi$ and the restriction $\xi|_V$ are homeomorphic ([4]), that is, $j_V^- \xi \approx \xi|_V$. Accordingly,

$$\lim_{\xi|_V} \mathcal{F} = V \cap \lim_\xi \mathcal{F}. \tag{IV.3.3}$$

DEFINITION IV.3.12 (Embedding). An injective map is called an *embedding* if its domain and its range are homeomorphic.

[4]By definition, for a filter $\mathcal{F}$ on V,

$$\lim_{j_V^- \xi} \mathcal{F} = j_V^-(\lim_\xi j_V[\mathcal{F}]) = V \cap \lim_\xi \mathcal{F},$$

where the image $j_V[\mathcal{F}]$ in X of a filter $\mathcal{F}$ on V coincides with a filter-base $\mathcal{F}$ in X.

More explicitly, if η in ξ are convergences, then f is an embedding of η in ξ,

$$f : \eta \hookrightarrow \xi$$

if $f \in \mathscr{C}(\eta, \xi)$ and the identity $i \in \mathscr{H}(\eta, f^{-}\xi)$.

Be aware that some authors (5) call a map fulfilling Definition IV.3.12, a *homeomorphic embedding*.

EXAMPLE IV.3.13. Let X_n be an infinite set, $\mathcal{F}_n$ a free filter on X_n, and $x_n \in X_n$ for each $n < \omega$. Let ξ_n be a prime convergence on X_n such that $x_n \in \lim_{\xi_n} \mathcal{F}_n$. Consider disjoint union $X := \bigsqcup_{n<\omega} X_n$, and *sum* $\xi = \bigoplus_{n<\omega} \xi_n$, in which $x \in \lim_\xi \mathcal{F}$ if there is $n < \omega$ such that $x \in X_n \in \mathcal{F}$, and $\mathcal{F} = x^\uparrow$ if $x \neq x_n$, and $\mathcal{F} \geq \mathcal{F}_n$ if $x = x_n$. Therefore, ξ is a pretopology.

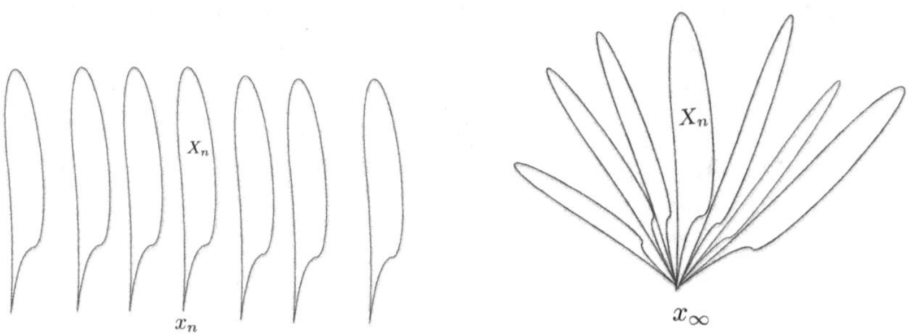

FIGURE IV.3.3. On the left, $\xi = \bigoplus_{n<\omega} \xi_n$, on the right, the fan obtained as the final convergence $f\xi$, where $x_\infty \in \lim_{f\xi} \mathcal{F}$ whenever there exists $n < \omega$ such that $\mathcal{F} \geq \mathcal{F}_n$.

We define $f : X \longrightarrow \{x_\infty\} \cup \bigcup_{n<\omega}(X_n \setminus \{x_n\})$ by

$$f(x) := \begin{cases} x & x \notin \{x_n : n < \omega\} \\ x_\infty & x = x_n \end{cases}.$$

Accordingly, $f\xi$ is the prime convergence, in which $x_\infty \in \lim_{f\xi} \mathcal{F}$ whenever there exists $n < \omega$ such that $\mathcal{F} \geq \mathcal{F}_n$. We notice that $f\xi$ is not a pretopology. A sequential fan from Example III.4.7 is a special case of $f\xi$.

It follows directly from the definitions that

PROPOSITION IV.3.14. *If ζ, ξ are convergences on X, and τ, θ on Y, then*

$$\zeta \geq \xi \text{ and } \tau \geq \theta \Longrightarrow \mathscr{C}(\xi, \tau) \subset \mathscr{C}(\zeta, \theta). \tag{IV.3.4}$$

[5]See R. ENGELKING [97].

On the other hand,

PROPOSITION IV.3.15. *The following statements are equivalent:*

$$f\xi \geq \tau \Longleftrightarrow f \in \mathscr{C}(\xi, \tau) \Longleftrightarrow \xi \geq f^{-}\tau. \tag{IV.3.5}$$

PROOF. Let ξ be a convergence on X, let τ be a convergence on Y, and let $f : X \to Y$ be a map. By definition, $f\xi$ is the finest convergence on Y for which f is continuous from ξ. Hence, $f\xi \geq \tau$ if and only if $f \in \mathscr{C}(\xi, \tau)$. On the other hand, $f^{-}\tau$ is the coarsest convergence on X, for which f is continuous to τ. Therefore, $\xi \geq f^{-}\tau$ if and only if $f \in \mathscr{C}(\xi, \tau)$. □

Let us stress that (IV.3.5) is a very, very useful formula! In fact, usually it is so easier to handle inequalities, rather than to manipulate sets of all continuous maps from one convergence to another. We shall make a frequent usage of it throughout this book.

For more details about sets of continuous maps, see Remark IV.10.10.

By (IV.3.5), convergences ξ and τ are homeomorphic, $\xi \approx \tau$, if and only if there exists a bijective map f such that $\xi = f^{-}\tau$, equivalently $f\xi = \tau$.

Generally,

$$f(f^{-}\tau) \geq \tau, \tag{IV.3.6}$$

$$\xi \geq f^{-}(f\xi). \tag{IV.3.7}$$

Indeed, by definition, $f \in \mathscr{C}(f^{-}\tau, \tau)$, hence (IV.3.6) follows from the left-hand side equivalence of (IV.3.5). Similarly, $f \in \mathscr{C}(\xi, f\xi)$, hence (IV.3.7) follows from the right-hand side equivalence of (IV.3.5).

As it is shown in Propositions IV.10.5 and IV.10.6, $f(f^{-}\tau) = \tau$ whenever f is surjective, and $\xi = f^{-}(f\xi)$ whenever f is injective.

By Proposition IV.3.3, each initial convergence is in fact a convergence of fibers. This is also the case of the initial convergence $f^{-}(f\xi)$ of $f\xi$. However, when f is injective, then the fibers are singletons, so that the convergence of fibers and of points coincide.

EXAMPLE IV.3.16. Let ρ be a convergence on a *circle*

$$S = \{(\cos t, \sin t) : 0 \leq t < 2\pi\} \subset \mathbb{R}^2$$

as a subconvergence of the standard convergence $\nu_{\mathbb{R}}^2$ of $\mathbb{R}^2$, that is, $\rho = j_S^{-}(\nu_{\mathbb{R}}{}^2)$, where $j_S : S \longrightarrow \mathbb{R}^2$ is the standard injection. We notice that ρ is homeomorphic to τ from Example IV.3.10. Indeed, let $g : [0, 1[\longrightarrow S$ be given by

$$g(x) := (\cos 2\pi x, \sin 2\pi x).$$

It is clear that g is a bijection, and $g(]a,b[) =]2\pi a, 2\pi b[$ for every $0 < a < b < 1$. To see that $g \in \mathscr{H}(\rho,\tau)$, it suffices to observe that $g^{-1}[\mathcal{V}_\tau(0)] = \mathcal{V}_\rho(1,0)$, that is, $g([0,\varepsilon[\cup]1-\varepsilon[) =]-2\pi\varepsilon, 2\pi\varepsilon[$ for each $0 < \varepsilon < 1$.

IV.4. Initial convergences

We shall now consider initial convergences with respect to several maps. They are just the suprema of initial convergences with respect to every single map.

Let X be a set, $T = \{\tau_j : j \in J\}$ be a set of convergences, and let for each $j \in J$,

$$f_j : X \longrightarrow |\tau_j|. \tag{IV.4.1}$$

DEFINITION IV.4.1. The coarsest convergence ξ on X, for which $f_j \in \mathscr{C}(\xi,\tau_j)$ for each $j \in J$ is called the *initial convergence* with respect to $\{(f_j,\tau_j) : j \in J\}$.

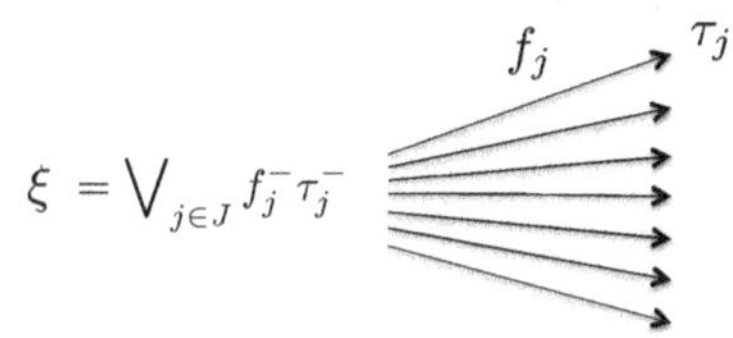

FIGURE IV.4.1. Initial convergence ξ with respect to $\{(f_j,\tau_j) : j \in J\}$.

PROPOSITION IV.4.2. *The initial convergence ξ with respect to every set $\{(f_j,\tau_j) : j \in J\}$ exists.*

PROOF. Indeed, the set of convergences ζ on X, for which all these maps are continuous, that is, $\zeta \geq f_j^- \tau_j$ for each $j \in J$, is not empty, because it contains the discrete convergence ι_X. Therefore,

$$\xi = \bigvee\nolimits_{j\in J} f_j^- \tau_j. \tag{IV.4.2}$$

□

A simplest characterization of initial convergence, that should never be forgotten, is that

LEMMA IV.4.3. *For every x and each filter $\mathcal{F}$,*

$$x \in \lim\nolimits_{\bigvee_{j\in J} f_j^- \tau_j} \mathcal{F}$$

if and only if, for each $j \in J$,

$$f_j(x) \in \lim_{\tau_j} f_j[\mathcal{F}].$$

REMARK IV.4.4. It follows that each supremum of convergences $\bigvee \Xi$ on a set X is the initial convergence with respect to $\{(i, \xi) : \xi \in \Xi\}$, where $i : X \longrightarrow X$ is the identity.

Here is a very useful observation.

PROPOSITION IV.4.5. *If* $\xi = \bigvee_{j \in J} f_j^- \tau_j$ *and* θ *is a convergence, then*

$$h \in \mathscr{C}(\theta, \xi)$$

if and only if, for each $j \in J$,

$$f_j \circ h \in \mathscr{C}(\theta, \tau_j). \tag{IV.4.3}$$

PROOF. By assumption, ξ is given by (IV.4.2). As $h \in \mathscr{C}(\theta, \xi)$ and $f_j \in \mathscr{C}(\xi, \tau_j)$, the composition $f_j \circ h \in \mathscr{C}(\theta, \tau_j)$ for each $j \in J$. Conversely, if for each $j \in J$, (IV.4.3) holds, that is, $\theta \geq (f_j \circ h)^- \tau_j$, equivalently, $\theta \geq h^-(f_j^- \tau_j)$, and else $h\theta \geq f_j^- \tau_j$, then

$$h\theta \geq \bigvee_{j \in J} f_j^- \tau_j = \xi, \tag{IV.4.4}$$

that is, $h \in \mathscr{C}(\theta, \xi)$.

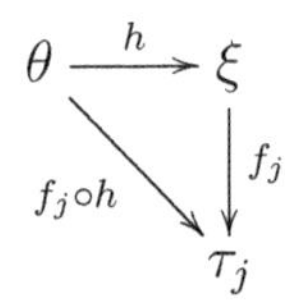

□

Product convergences are instances of initial convergences.

EXAMPLE IV.4.6 (Product of two convergences). If ξ and υ are convergences on X and Y respectively, then the product convergence $\xi \times \upsilon$ on $X \times Y$ is defined by $(x, y) \in \lim_{\xi \times \upsilon} \mathcal{F}$ if and only if there exist filters $\mathcal{G}$ on X and $\mathcal{H}$ on Y such that $x \in \lim_\xi \mathcal{G}, y \in \lim_\upsilon \mathcal{H}$ and $\mathcal{G} \times \mathcal{H} \leq \mathcal{F}$. In other words, a filter $\mathcal{F}$ converges to (x, y) in the *product convergence* if and only if its projections $p_X[\mathcal{F}]$ and $p_Y[\mathcal{F}]$ converge to x and y respectively. Similarly, in finite products of convergences, a filter converges to a point if and only if each projection of that filter converges to the projection of that point.

DEFINITION IV.4.7. Let σ be a convergence. A convergence ξ is called σ-*initial* if there exists a set of maps $\mathscr{F}$ such that $\xi = \bigvee_{f \in \mathscr{F}} f^- \sigma$.

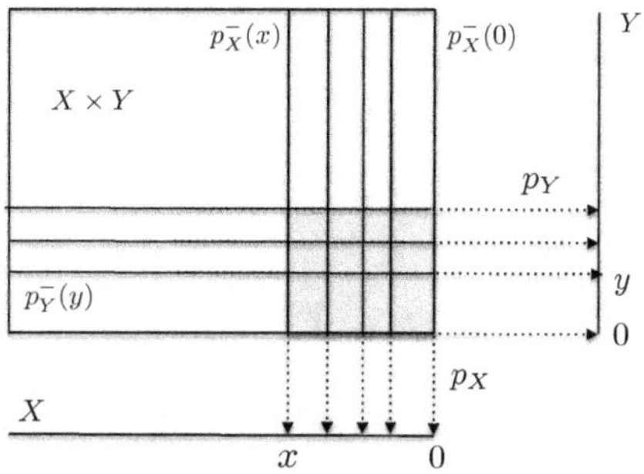

FIGURE IV.4.2. Product of two convergences: an example of the initial convergence on $X \times Y$ with respect of two projections p_X and p_Y. The grey area represents a typical element of the coarsest filter converging to $(0,0)$.

It follows from the definition that if ξ is σ-initial, then $\xi \geq f^-\sigma$ for every $f \in \mathscr{F}$. Consequently, $\mathscr{F} \subset \mathscr{C}(\xi,\sigma)$. Moreover, $\xi \geq \bigvee_{f\in\mathscr{C}(\xi,\sigma)} f^-\sigma$ for any convergences ξ and σ. Therefore,

PROPOSITION IV.4.8. *A convergence ξ is σ-initial if and only if*

$$\xi = \bigvee_{f\in\mathscr{C}(\xi,\sigma)} f^-\sigma.$$

In particular, σ is σ-initial.

EXAMPLE IV.4.9. The convergence $\sigma_\mathbb{R}$ is not $\nu_\mathbb{R}$-initial. We have seen in Example III.1.9 that $0 \notin \lim_{\sigma_\mathbb{R}} \mathcal{V}_{\sigma_\mathbb{R}}(0)$. On the other hand, by (III.1.4), $\mathcal{V}_{\sigma_\mathbb{R}}(x) = \mathcal{V}_{\nu_\mathbb{R}}(x)$ for each $x \in \mathbb{R}$, hence $f[\mathcal{V}_{\sigma_\mathbb{R}}(x)] \geq \mathcal{V}_{\nu_\mathbb{R}}(f(x))$ for every $f \in \mathscr{C}(\sigma_\mathbb{R},\nu_\mathbb{R})$ by Lemma IV.2.5. Therefore, $f(0) \in \lim_{\nu_\mathbb{R}} f[\mathcal{V}_{\sigma_\mathbb{R}}(0)]$, so that $\sigma_\mathbb{R}$ is not $\nu_\mathbb{R}$-initial in virtue of Lemma IV.4.3.

It turns out that there exist non-discrete convergences that are discretely initial, that is, ι_X-initial, where X is a set. More specifically, Example IV.6.8 shows that there exist non-discrete convergences that are $\iota_{\{0,1\}}$-initial.

IV.5. Selections and products

Let $R \subset J \times X$. As already mentioned, the set of all maps from J to X is denoted by X^J ([6]). Consider now maps $\varphi \in X^J$ such that

[6]This notation is related to the fact that there are n^k maps from a set of k elements to a set of n elements. This relation is used in the definition of the cardinality λ^κ when κ and λ are arbitrary cardinal numbers, namely $\lambda^\kappa = \operatorname{card} X^V$, when $\operatorname{card}(V) = \kappa$ and $\operatorname{card}(X) = \lambda$.

$\varphi(j) \in Rj \subset X$ for each $j \in J$. These are those φ whose graph is included in the relation R, that is,

$$\{\varphi \in X^J : \varphi \subset R\}. \tag{IV.5.1}$$

A map $\varphi : J \to X$ is called a *selection* of a relation R if (IV.5.1) holds ([7]). We denote by

$$\prod R = \prod_{j\in J} Rj \tag{IV.5.2}$$

the set of all selections of R, also called the *product* of $\{Rj : j \in J\}$.

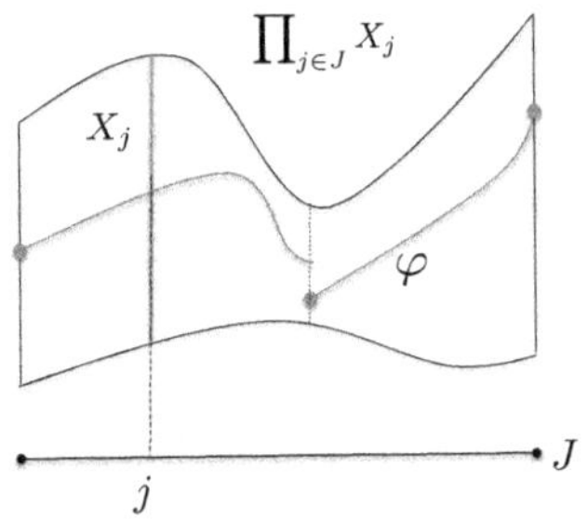

FIGURE IV.5.1. A representation of the product $\prod_{j\in J} X_j$. This is the set of maps $\varphi : J \to \bigcup_{j\in J} X_j$ such that $\varphi(j) \in X_j$ for each $j \in J$, that is, φ are selections.

We have in fact described a construction of the product of a family $\{X_j : j \in J\}$ of sets, indexed by a set J. It is handy to put all these sets together $X := \bigcup_{j\in J} X_j$, so that a relation R from (IV.5.2) fulfills $Rj = X_j$.

Consider first a particular case, where J is finite, say $J = \{1, 2, \ldots, m\}$, and $X_1, X_2, \ldots, X_m$ are sets indexed by J. Then the *product*

$$\prod_{1\le k\le m} X_k := X_1 \times X_2 \times \cdots \times X_m \tag{IV.5.3}$$

consists of m-tuples $(x_1, x_2, \ldots, x_m)$, where $x_1 \in X_1, x_2 \in X_2, \ldots, x_m \in X_m$. The sets $X_1, X_2, \ldots, X_m$ are called *factors* of the product (IV.5.3).

In other words, an element of (IV.5.3) is a finite sequence φ of length m such that $\varphi(k) = x_k \in X_k$ for each $1 \le k \le m$. Such a sequence φ is a map from $\{1, 2, \ldots, m\}$ to $X := X_1 \cup X_2 \cup \cdots \cup X_m$, and thus $\varphi \in \prod_{1\le k\le m} X_k$ if and only if $\varphi \in X^{\{1,2,\ldots,m\}}$ and $\varphi(k) \in X_k$ for each $1 \le k \le m$.

---

[7] The non-emptiness of (IV.5.1), hence of (IV.5.2), under the provision that $Rj \ne \varnothing$ for each $j \in J$, that is, whenever R^- is surjective, is equivalent to the Axiom of Choice (see Chapter XXVII).

The *k-th projection* p_k is defined as the map $p_k : \prod_{1\le j\le m} X_j \to X_k$, such that $p_k(\varphi) := \varphi(k)$. If all the factors are equal, say, $X_k = X$, then we write $X^m = \prod_{1\le k\le m} X$ and call X^m the *m-th power* of X.

If now J is an arbitrary non-empty set, and X_j is a nonempty set for each $j \in J$, then the *product*

$$\prod_{j\in J} X_j \tag{IV.5.4}$$

is defined as the set of maps $\varphi : J \to X := \bigcup_{j\in J} X_j$ such that $\varphi(j) \in X_j$ for each $j \in J$. For each $j \in J$, the set X_j is called a *factor* of (IV.5.4); it is its j-th factor.

Therefore, we recover $\prod R$, introduced at the beginning of this section.

As in the case of finite products, the *j-th projection* p_j is defined as the map

$$p_j : \prod_{k\in J} X_k \to X_j \tag{IV.5.5}$$

such that $p_j(\varphi) := \varphi(j)$. In particular, if all the factors are equal ($X = X_j$ for each $j \in J$), then

$$\prod_{j\in J} X = X^J, \tag{IV.5.6}$$

that is, the set of all maps from J to X is called a *power set* of X or, more precisely, the *J-power* of X.

DEFINITION IV.5.1 (Diagonal map). The *diagonal map*

$$\triangle : X \longrightarrow X^J = \prod_{j\in J} X$$

is defined by $\triangle(x)(j) := x$ for each $j \in J$.

Accordingly, the image $\triangle(X)$ of X under $\triangle$ is the *diagonal subset* of $X^J = \prod_{j\in J} X$, that is,

$$\varphi \in \triangle(X) \Longleftrightarrow \underset{x\in X}{\exists} \underset{j\in J}{\forall} \varphi(j) = x.$$

REMARK IV.5.2. If J consists of two points, then $\triangle : X \longrightarrow X \times X$ is related to the identity relation (I.1.3) by

$$\triangle(X) = \triangle_X = \{(x,x) : x \in X\}.$$

If $f_j : V_j \longrightarrow X_j$ for each $j \in J$, then

DEFINITION IV.5.3 (Product map). The *product map*

$$\prod_{j\in J} f_j : \prod_{j\in J} V_j \longrightarrow \prod_{j\in J} X_j \tag{IV.5.7}$$

is defined factor-wise by

$$(p_j \circ \prod_{k\in J} f_k)(\varphi) := f_j(\varphi(j)).$$

In particular, if $f_j \in X_j^{V_j}$ for $j \in \{0,1\}$, then the product $f_0 \times f_1 : V_0 \times V_1 \longrightarrow X_0 \times X_1$ fulfills

$$(f_0 \times f_1)(v_0, v_1) = (f_0(v_0), f_1(v_1)).$$

IV.6. Product convergences

Product convergences are a special subclass of initial convergences; they are initial with respect to all the projections. We shall see that, conversely, each initial convergence with respect to an injective family of maps is homeomorphic to a subconvergence of a product convergence.

DEFINITION IV.6.1 (Product convergence). Let T be a set of convergences. The *product convergence*

$$\prod T = \prod_{\tau\in T} \tau$$

is the coarsest convergence ξ on $|\prod T|$, for which $p_\tau \in \mathscr{C}(\xi, \tau)$ for each $\tau \in T$.

In other words, the *product convergence* $\prod T$ is the initial convergence with respect to $\{(p_\tau, \tau) : \tau \in T\}$, where $p_\tau : |\prod T| \to |\tau|$, for each $\tau \in T$, that is,

$$\prod T := \bigvee_{\tau\in T} p_\tau^- \tau. \tag{IV.6.1}$$

As we have pointed out in Section IV.5, a set-theoretic product is equal to the set of selections of its factors. In the present case, $\varphi \in \prod_{\tau\in T} |\tau|$ if and only if $p_\tau \circ \varphi \in |\tau|$ for each $\tau \in T$.

REMARK IV.6.2. A set of convergences T can be indexed, for example, $T = \{\tau_j : j \in J\}$, where J is a set of indices. If we denote by Y_j the underlying set of τ_j, then the product convergence $\prod T = \prod_{j\in J} \tau_j$ is a convergence on $Y := \prod_{j\in J} Y_j$. In the same vein, a projection p_j maps Y onto Y_j.

However, instead of picking a set J to index T, we can use T itself as an index set, economizing this way on the symbols used. Then the projection of the product of underlying sets onto the underlying set of a given convergence τ will be denoted by p_τ.

Either way has its advantages and inconveniences. We are going to be flexible in using one or another, according to situations.

In virtue of Lemma IV.4.3,

PROPOSITION IV.6.3. *Let $\mathcal{F}$ be a filter on $\prod_{\tau\in T}|\tau|$. Then*

$$\varphi \in \lim_{\prod T} \mathcal{F}$$

if and only if, for every $\tau \in T$,

$$p_\tau(\varphi) \in \lim_\tau p_\tau[\mathcal{F}].$$

DEFINITION IV.6.4 (Polyhedral filter). A subset of $\prod_{j\in J} X_j$ is said to be a *polyhedron* if it is of the form $D = \prod_{j\in J} D_j$ and $\{j \in J : D_j \neq X_j\}$ is finite. A filter $\mathcal{G}$ on $\prod_{j\in J} X_j$ is called *polyhedral* if it has a filter-base made of polyhedra.

Of course, $D \subset \prod_{j\in J} X_j$ is a polyhedron if and only if there exists a finite set $J_0 \subset J$ and $D_j \subset X_j$ for each $j \in J$ so that $D = \bigcap_{j\in J_0} p_j^-(D_j)$.

If $\mathcal{F}_j \in \mathbb{F}X_j$ for each $j \in J$, then the *product filter* $\prod_{j\in J} \mathcal{F}_j$ is defined by

$$\prod\nolimits_{j\in J} \mathcal{F}_j := \bigvee\nolimits_{j\in J} p_j^-[\mathcal{F}_j],$$

where $p_j : \prod_{k\in J} X_k \longrightarrow X_j$ is the j-th projection. It is clear that each product filter is polyhedral.

PROPOSITION IV.6.5. *A filter is polyhedral if and only if it is a product filter.*

PROOF. If $\mathcal{B}$ is a polyhedral filter-base on $\prod_{j\in J} X_j$, then each $B \in \mathcal{B}$ is of the form $B = \prod_{j\in J} B_j$, where $J_0 := \{j \in J : B_j \neq X_j\}$ is finite, then $p_j(B) = B_j \in p_j[\mathcal{B}]$ for each $j \in J$. Therefore, $B = \bigcap_{j\in J_0} p_j^-(B_j) \in \prod_{j\in J} p_j[\mathcal{B}]$. □

By Proposition IV.4.5,

PROPOSITION IV.6.6. *A map $f \in \mathscr{C}(\xi, \prod T)$ if and only if $p_\tau \circ f \in \mathscr{C}(\xi, \tau)$ for each $\tau \in T$.*

Of course,

PROPOSITION IV.6.7. *Each finite product of discrete convergences is discrete.*

In particular, if $\{0,1\}$ is endowed with the discrete convergence $\iota = \iota_{\{0,1\}}$, then the product $\iota_{\{0,1\}} \times \iota_{\{0,1\}}$ is equal to the discrete convergence $\iota_{\{0,1\}^2}$ (on the product $\{0,1\}^2 = \{0,1\} \times \{0,1\}$). This remains true for any finite product.

EXAMPLE IV.6.8 (Cantor cube). The power set $\{0,1\}^{\mathbb{N}}$ endowed with the product convergence $\iota^{\mathbb{N}} = \prod_{n\in\mathbb{N}} \iota$ is called the *Cantor cube*. Since the only Hausdorff convergence on a finite set is discrete, talking about a Cantor cube $\{0,1\}^{\mathbb{N}}$, implicitly we mean $\iota^{\mathbb{N}}$.

Is the Cantor cube discrete? The answer is no! In fact, let $(\varphi_k)_k$ be the following sequence of elements of $\{0,1\}^{\mathbb{N}}$ ([8]):

$$\varphi_k(n) := \begin{cases} 1 & k = n \\ 0 & k \neq n \end{cases}.$$

By definition, $\varphi \in \lim_{\iota^{\mathbb{N}}}(\varphi_k)_k$ if and only if $\varphi(n) \in \lim_\iota(\varphi_k(n))_k$ for each $n \in \mathbb{N}$. It is clear from the definition above that $\lim_{k\to\infty}\varphi_k(n) = 0$ for each $n \in \mathbb{N}$. Thus, $\varphi(n) := 0$ is a limit of $(\varphi_k(n))_k$ for each $n < \omega$, that is, the free filter generated by $(\varphi_k)_k$ is convergent, so that $\{0,1\}^{\mathbb{N}}$ is not discrete.

EXAMPLE IV.6.9 (Pointwise convergence). Consider the convergence $\nu_{\mathbb{R}}$ on $\mathbb{R}$ from Example III.1.4. Let $\pi = \prod_{j\in J} \nu_j$, where $\nu_j = \nu_{\mathbb{R}}$ for each $j \in J$. Of course, $|\pi| = \prod_{j\in J} \mathbb{R} = \mathbb{R}^J$, that is, the set of all maps from J to $\mathbb{R}$. In other words, $f \in \lim_\pi \mathcal{F}$ if and only if for each $j \in J$,

$$f(j) \in \lim_{\nu_{\mathbb{R}}} \mathcal{F}(j),$$

where

$$\mathcal{F}(j) := \{\{\varphi(j) : \varphi \in F\} : F \in \mathcal{F}\}, \tag{IV.6.2}$$

that is, $\mathcal{F}$ converges *pointwise* to f.

On generalizing Example IV.6.9, consider a convergence σ on a set X. The *J-power convergence* σ^J of σ is defined as

$$\sigma^J := \prod_{j\in J} \sigma,$$

the product convergence, in which $\sigma_j = \sigma$ for each $j \in J$.

DEFINITION IV.6.10 (Pointwise convergence). We say that a filter $\mathcal{F}$ on X^J *σ-pointwise converges* to $f \in X^J$ if $f(j) \in \lim_\sigma \mathcal{F}(j)$ for each $j \in J$, where $\mathcal{F}(j)$ is defined in (IV.6.2).

PROPOSITION IV.6.11. *Every product of Hausdorff convergences is Hausdorff.*

PROOF. If τ_j is a Hausdorff convergence on X_j for each $j \in J$, and $f, g \in \prod_{j\in J} X_j$ and $\mathcal{F}$ is a filter on $\prod_{j\in J} X_j$ such that $\{f, g\} \subset \lim_{\prod_{j\in J}\tau_j} \mathcal{F}$, then $\{f(j), g(j)\} \subset \lim_{\tau_j} p_j[\mathcal{F}]$ for each $j \in J$, hence $f(j) = g(j)$ for each $j \in J$, hence $f = g$. $\square$

[8] As we know, $\varphi \in \{0,1\}^{\mathbb{N}}$ whenever φ is a sequence valued in $\{0,1\}$.

PROPOSITION IV.6.12. *Every product of free (alias T_1) convergences is free.*

PROOF. Let $\{\tau_j : j \in J\}$ be a set of free convergences. By the definition of product, $\varphi \in \lim_{\prod_{j\in J}\tau_j} \mathcal{F}$ if and only if $\varphi(j) \in \lim_{\tau_j} p_j[\mathcal{F}]$ for every $j \in J$. By assumption, $\ker p_j[\mathcal{F}] \subset \{\varphi(j)\}$, hence $\ker \mathcal{F} \subset \ker p_j^-[p_j[\mathcal{F}]] \subset p_j^-(\varphi(j))$ for each $j \in J$, hence

$$\ker \mathcal{F} \subset \bigcap\nolimits_{j\in J} p_j^-(\varphi(j)) = \{\varphi\}.$$

□

IV.7. Diagonal product maps

Let J be a non-empty set, and X_j is a non-empty set for each $j \in J$. If

$$f : V \longrightarrow \prod\nolimits_{j\in J} X_j, \tag{IV.7.1}$$

then, for every $j \in J$,

$$f_j := p_j \circ f \tag{IV.7.2}$$

is the j-th *factor map* $f_j : V \longrightarrow X_j$, where $p_j : \prod_{k\in J} X_k \longrightarrow X_j$ is the j-th *projection* (IV.5.5).

The other way round, if $f_j : V \longrightarrow X_j$ for each $j \in J$, then there exists a unique map $f : V \longrightarrow \prod_{j\in J} X_j$ such that (IV.7.2). Given a family of maps $f_j : V \longrightarrow X_j$ for each $j \in J$,

DEFINITION IV.7.1 (Diagonal product map). The *diagonal product map*

$$\triangle_{k\in J} f_k : V \longrightarrow \prod\nolimits_{j\in J} X_j$$

is the map f such that (IV.7.2) for each $j \in J$.

Thus, explicitly, for each $j \in J$,

$$(\triangle_{k\in J} f_k)(v)(j) = f_j(v). \tag{IV.7.3}$$

We have already considered *diagonal maps* in Definition IV.5.1 and *product maps* in Definition IV.5.3. Therefore, one should be alert not to mix up the concepts. Actually, a *diagonal product map* ([9]) is the composition

---

[9]A *diagonal product map* $\triangle_{j\in J} f_j$ was also called a *diagonal map* in the handbook [97] by R. ENGELKING and in [83]. The *product map* used in [83] was denoted by $\bigotimes_{j\in J} f_j : \prod_{j\in J} X_j \longrightarrow \prod_{j\in J} Y_j$.

$$\triangle_{j\in J} f_j = (\prod_{j\in J} f_j) \circ \triangle,$$

where $\prod_{j\in J} f_j$ is a *product map* (IV.5.7) and $\triangle$ is the *diagonal map*.

In case of two variables, if $\varphi \in X^V$ and $\psi \in Y^V$, then $\varphi \triangle \psi \in (X \times Y)^V$ to the effect that

$$(\varphi \triangle \psi)(v) := (\varphi(v), \psi(v)).$$

Let us observe that

LEMMA IV.7.2. *If either φ or ψ is injective, then the diagonal product map $\varphi \triangle \psi$ is injective.*

PROOF. Indeed, suppose that φ is injective and let $v_0 \neq v_1$. Hence $\varphi(v_0) \neq \varphi(v_1)$, and thus $(\varphi(v_0), \psi(v_0)) \neq (\varphi(v_1), \psi(v_0))$. □

Consequently,

COROLLARY IV.7.3. *If there is $j \in J$ such that f_j is injective, then* $\triangle_{j\in J} f_j$ is injective.

In particular, if $f : X \longrightarrow Y$ is a map, and

$$\mathrm{gr}(f) := \{(x, f(x)) \in X \times Y : x \in X\}$$

is the *graph* of f defined in (I.1.4), then the diagonal product map $i_V \triangle f : X \longrightarrow X \times Y$ is injective and $(i_X \triangle f)(X) = \mathrm{gr}(f)$. Therefore, $i_V \triangle f : X \longrightarrow \mathrm{gr}(f)$ is a bijection, and

$$p_X|_{\mathrm{gr}(f)} = (i_X \triangle f)^{-1},$$

where $p_X : X \times Y \longrightarrow X$ is the projection defined by $p_X((x, y)) := x$ for each $x \in X$ and $y \in Y$.

Of course, the diagonal product map can be injective even if none of its factors is injective.

EXAMPLE IV.7.4. Let $V = [0, 2\pi[$ and consider two maps $\cos : V \longrightarrow \mathbb{R}$ and $\sin : V \longrightarrow \mathbb{R}$. Then $(\cos \triangle \sin)(v) = (\cos v, \sin v) \in \mathbb{R}^2$ for each $v \in V$ defines an injective map, although sin and cos are not injective. The image $(\cos \triangle \sin)(V)$ is the unit circle.

We say that a collection of maps $\{f_j : j \in J\}$ is *injective* or *distinguishes the points* of V if, for each $v_0 \neq v_1$, there exists $j \in J$ such that $f_j(v_0) \neq f_j(v_1)$. Therefore, it is evident that

PROPOSITION IV.7.5. *The diagonal product map $\triangle_{j\in J} f_j$ is injective if and only if $\{f_j : j \in J\}$ is injective.*

In virtue of Proposition IV.6.6, a map f valued in a product is continuous if and only if $p_j \circ f$ for every $j \in J$, where p_j is the j-th projection. Therefore,

PROPOSITION IV.7.6. *The diagonal product map*

$$\triangle_{j\in J} f_j \in \mathscr{C}(\xi, \prod_{j\in J} \tau_j)$$

if and only if $f_j \in \mathscr{C}(\xi, \tau_j)$ *for each* $j \in J$.

We observe that

PROPOSITION IV.7.7. *The initial convergence with respect to several maps is equal to the initial convergence with respect to their product* ([10]), *that is,*

$$\bigvee_{\tau\in T} f_\tau^- \tau = (\triangle_{\tau\in T} f_\tau)^- \prod T.$$

Accordingly,

COROLLARY IV.7.8. *If* ξ *is the initial convergence with respect to an injective collection* $\{(f_\tau, \tau) : \tau \in T\}$, *then the diagonal product map* $\triangle_{\tau\in T} f_\tau$ *is an embedding of* ξ *into* $\prod_{\tau\in T} \tau$.

PROPOSITION IV.7.9. *If* $f \in \mathscr{C}(\xi, \tau)$ *is an embedding, and* $g \in \mathscr{C}(\xi, \theta)$, *then* $f \triangle g \in \mathscr{C}(\xi, \tau \times \theta)$ *is an embedding.*

PROOF. By IV.7.2, $f \triangle g$ is injective. To see that ξ is initial with respect to $f \triangle g$, let $\mathcal{F}$ be a filter such that $(f(x), g(x)) \in \lim_{\tau\times\theta}(f \triangle g)[\mathcal{F}]$. Hence $f(x) \in \lim_\tau p \circ (f \triangle g)[\mathcal{F}]$, where $p \in \mathscr{C}(\tau \times \theta, \tau)$ is a projection. As $p \circ (f \triangle g) = f$, we infer that $f(x) \in \lim_\tau f[\mathcal{F}]$, hence $x \in \lim_\xi \mathcal{F}$, because $f : \xi \hookrightarrow \tau$. □

We recall that if $f \in Y^X$, then $i_X \triangle f : X \longrightarrow \mathrm{gr}(f)$ is a bijection. Hence, Proposition IV.7.9 implies that the graph of a continuous map is homeomorphic to its domain:

COROLLARY IV.7.10. *If* $f \in \mathscr{C}(\xi, \tau)$, *then* ξ *is homeomorphic to the restriction of* $\xi \times \tau$ *to* $\mathrm{gr}(f)$.

IV.8. Final convergences

If we consider a set $\Xi = \{\xi_j : j \in J\}$ of convergences, a set Y and for every $j \in J$ a map $f_j : |\xi_j| \to Y$, then there exists the finest convergence on Y for which f_j is continuous for each $j \in J$.

[10]Indeed, by Lemma IV.4.3, $x \in \lim_{\bigvee_{\tau\in T} f_\tau^- \tau} \mathcal{F}$ if and only if $f_\tau(x) \in \lim_\tau f_\tau[\mathcal{F}]$ for each $\tau \in T$ if and only if $\triangle_{\tau\in T} f_\tau(x) \in \lim_{\prod T}(\triangle_{\tau\in T} f_\tau)[\mathcal{F}]$.

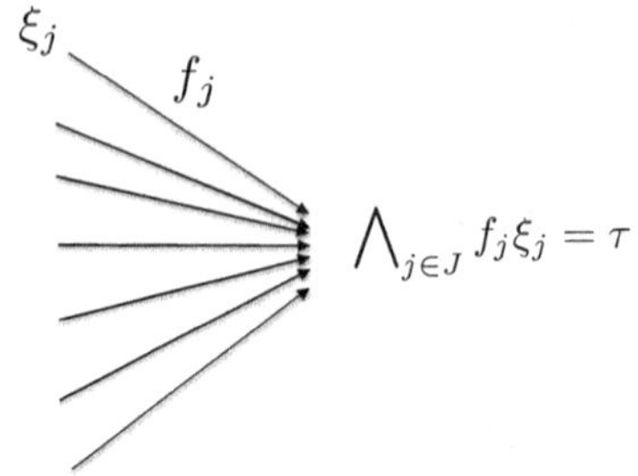

FIGURE IV.8.1. Final convergence

It is called the *final convergence* with respect to $\{(\xi_j, f_j) : j \in J\}$, and is equal to

$$\bigwedge\nolimits_{j\in J} f_j\xi_j.$$

PROPOSITION IV.8.1. *If τ is the final convergence with respect to $\{(\xi_j, f_j) : j \in J\}$, then*

$$h \in \mathscr{C}(\tau, \upsilon)$$

if and only if for each $j \in J$,

$$h \circ f_j \in \mathscr{C}(\xi_j, \upsilon). \tag{IV.8.1}$$

PROOF. If $h \in \mathscr{C}(\tau, \upsilon)$, then for each $j \in J$, (IV.8.1) holds, because each composition of continuous maps is continuous. Conversely, if $h \circ f_j \in \mathscr{C}(\xi_j, \upsilon)$ for each $j \in J$, that is, $(h \circ f_j)\xi_j \geq \upsilon$, equivalently, $f_j\xi_j \geq h^-\upsilon$ for each $j \in J$. Hence

$$\tau = \bigwedge\nolimits_{j\in J} f_j\xi_j \geq h^-\upsilon,$$

that is, $h \in \mathscr{C}(\tau, \upsilon)$.

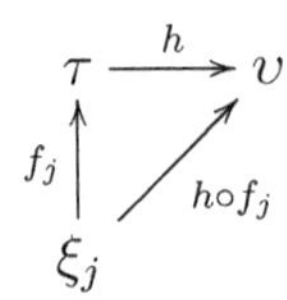

□

Coproducts (simple sums) and quotient convergences are special cases of final convergences.

Let $\{\tau_j : j \in J\}$ be a set of convergences such that $|\tau_j| \cap |\tau_k| = \varnothing$ whenever $j \neq k$.

DEFINITION IV.8.2. The *coproduct* $\bigoplus_{j\in J}\tau_j$ (called also the *sum*) is the final convergence with respect to the family of injections $i_j : |\tau_j| \to \coprod_{k\in J}|\tau_k|$.

In other words,

$$\bigoplus_{j\in J}\tau_j := \bigwedge\nolimits_{j\in J} i_j\tau_j.$$

Therefore, a map from a coproduct is continuous if and only if its composition with each injection is continuous.

PROPOSITION IV.8.3. *Each final convergence with respect to* $\{(\xi, f_j) : j \in J\}$ *is the final convergence with respect to a single map from* $\bigoplus_{j\in J}\xi_j$.

PROOF. Indeed, if we define $f(x) = f_j(x)$ if $x \in |\xi_j|$, then

$$y \in \lim\nolimits_{\bigwedge_{j\in J} f_j\xi_j} \mathcal{G}$$

if there is $j \in J$ such that $y \in \lim_{f_j\xi_j} \mathcal{G}$, that is, if there is a filter $\mathcal{F}$ on $|\xi_j|$ and $x \in |\xi_j|$ such that $x \in \lim_{\xi_j} \mathcal{F}$ and $f(x) = y$. □

Notice that the map f in the proof above is surjective.

A family of maps $\{f_j : j \in J\}$, where $f_j : X_j \longrightarrow Y$, is called *surjective* if $\bigcup_{j\in J} f_j(X_j) = Y$.

We define the *sum map* $\bigsqcup_{j\in J} f_j$ from a disjoint union $\bigsqcup_{j\in J} X_j$ to Y by

$$(\bigsqcup\nolimits_{k\in J} f_k)(x) := f_j(x) \text{ if } x \in X_j.$$

It is obvious that

PROPOSITION IV.8.4. *A family of maps* $\{f_j : j \in J\}$ *is surjective if and only if the sum map* $\bigsqcup_{j\in J} f_j$ *is surjective.*

DEFINITION IV.8.5. A class $\mathbf{W}$ of convergences is called *finally dense* in a class of convergences $\mathbf{G}$ if each convergence $\tau \in \mathbf{G}$, there exists a set $\{\xi_j : j \in J\}$ of $\mathbf{W}$-convergences and a family $\{f_j : j \in J\}$ such that $\tau = \bigwedge_{j\in J} f_j\xi_j$.

IV.9. *Lower semicontinuity

Often the description of a link between convergence spaces takes form of a relation rather than of a map. That is why the concept of continuity of maps has been extended, in various ways, to relations.

Let $R \subset X \times Y$ be a relation. If $\mathcal{A}$ is a family of subsets of X, then

$$R[\mathcal{A}] := \{RA : A \in \mathcal{A}\}.$$

Let ξ be a convergence on X and τ be a convergence on Y.

DEFINITION IV.9.1. A relation R is said to be *lower semicontinuous* at $(x, y) \in X \times Y$ if $x \in \lim_\xi \mathcal{F}$ entails the existence of a filter $\mathcal{G}$ such that $y \in \lim_\tau \mathcal{G}$ so that

$$\mathcal{F} \geq R^-[\mathcal{G}]. \tag{IV.9.1}$$

Notice that if ξ is a pretopology, then it is enough that this inclusion hold for $\mathcal{F} = \mathcal{V}_\xi(x)$, which is the coarsest filter ξ-convergent to x; if τ is a pretopology, then it is enough to take $\mathcal{G} = \mathcal{V}_\tau(y)$. Therefore, if ξ and τ are pretopologies, then R is lower (ξ, τ)-semicontinuous at (x, y) if and only if

$$\mathcal{V}_\xi(x) \geq R^-[\mathcal{V}_\tau(y)].$$

If $Rx = \{f(x)\}$ for $f \in Y^X$, then (IV.9.1) becomes $\mathcal{F} \geq f^-[\mathcal{G}]$.

PROPOSITION IV.9.2. *Let $f \in Y^X$ and $R \subset X \times Y$ be given by $Rx = \{f(x)\}$. Then f is continuous at x if and only if R is lower semicontinuous at $(x, f(x))$.*

PROOF. If f is continuous at x, and $x \in \lim_\xi \mathcal{F}$, then $f(x) \in \lim_\tau f[\mathcal{F}]$. Therefore, $\mathcal{F} \geq f^-[f[\mathcal{F}]]$ in virtue of Lemma IV.1.1. Conversely, for $x \in \lim_\xi \mathcal{F}$, take $\mathcal{G} = f[\mathcal{F}]$ so that $f(x) \in \lim_\tau \mathcal{G}$ and $\mathcal{F} \geq R^-[\mathcal{G}]$. Then by Lemma IV.1.2, $\mathcal{G} \leq f[f^-[\mathcal{G}]] \leq f[\mathcal{F}]$ so that $f(x) \in \lim_\tau f[\mathcal{F}]$, which proves that f is continuous at x. □

Recall that a map $f \in Y^X$ is called a *selection* of a relation $R \subset X \times Y$ whenever $f \subset R$, that is, $f(x) \in Rx$ for each $x \in R^- Y$. Under some additional assumptions, the lower semicontinuity at (x_0, y_0) of a relation R entails the existence of a selection $f \subset R$ that is continuous at x_0 and such that $f(x_0) = y_0$.

Let us consider an abstract version of lower semicontinuity.

DEFINITION IV.9.3. A relation R is said to be *lower semicontinuous from* a filter $\mathcal{F}$ *to* a filter $\mathcal{G}$, in symbols,

$$R \in \mathscr{C}(\mathcal{F}, \mathcal{G}),$$

whenever (IV.9.1) holds, equivalently $R[\mathcal{F}] \geq \mathcal{G}$.

It is clear that this condition implies that there is $F_0 \in \mathcal{F}$ such that $Rx \neq \varnothing$ for every $x \in F_0$.

Abstract lower semicontinuity can be represented as lower semicontinuity between convergences ([11]).

---

[11]If $\mathcal{F}$ is a free filter on X, and $\mathcal{G}$ is a free filter on Y, $x_\infty \in X$, and $y_\infty \in Y$, then a relation $R \subset X \times Y$ is lower semicontinuous from $[x_\infty, \mathcal{F}]$ to $[y_\infty, \mathcal{G}]$ at (x_∞, y_∞) if and only if (IV.9.1) holds.

PROPOSITION IV.9.4 (Selection principle). *Let $\mathcal{F}$ and $\mathcal{G}$ be free countably based filters and $R \in \mathscr{C}(\mathcal{F}, \mathcal{G})$. Then there exist $F \in \mathcal{F}$ and a map defined on F such that $f(x) \in Rx$ for each $x \in F$ and $f \in \mathscr{C}(\mathcal{F}, \mathcal{G})$* ([12]).

PROOF. By (IV.1.4), $\mathcal{G} \leq f[\mathcal{F}]$ amounts to $f^{-}[\mathcal{G}] \leq \mathcal{F}$. Let $\{F_n : n < \omega\}$ be a strictly decreasing base of $\mathcal{F}$, and $\{G_n : n < \omega\}$ be a strictly decreasing base of $\mathcal{G}$. By assumption, there exists n_0 such that $F_{n_0} \subset R^{-}G_0$, so that for each $x \in F_{n_0}$ there exists $f_0(x) \in Rx \cap G_0$. We continue and get a strictly increasing sequence of integers $(n_k)_k$ such that $F_{n_k} \subset R^{-}G_k$. Hence there is a map f_k defined on F_{n_k} so that $f_k(x) \in Rx \cap G_k$ for each $x \in F_{n_k}$. Let

$$f(x) := f_k(x) \text{ if } x \in F_{n_k} \setminus F_{n_{k+1}}.$$

Then $f : F_{n_0} \longrightarrow Y$ is such that $f(x) \in Rx$ for each $x \in F_{n_0}$ and $F_{n_k} \subset R^{-}G_k$ for each $k < \omega$. Consequently, $f^{-}[\mathcal{G}] \leq \mathcal{F}$. □

Let $\mathcal{F}$ be a filter on X, and $\mathcal{G}$ a filter on Y. Let $\psi : X \longrightarrow Z$ and $\varphi : Y \longrightarrow Z$. It is obvious that

$$\varphi[\mathcal{G}] \leq \psi[\mathcal{F}] \tag{IV.9.2}$$

if and only if, for every $G \in \mathcal{G}$, there exists $F \in \mathcal{F}$ such that $\varphi(G) \supset \psi(F)$, that is, $\psi^{-}(\varphi(G)) \supset F$, equivalently $\psi^{-}[\varphi[\mathcal{G}]] \leq \mathcal{F}$, that is,

$$\varphi^{-} \circ \psi \in \mathscr{C}(\mathcal{F}, \mathcal{G}).$$

It other words,

PROPOSITION IV.9.5. *The condition $\varphi[\mathcal{G}] \leq \psi[\mathcal{F}]$ holds if and only if $\varphi^{-} \circ \psi$ is lower semicontinuous from $\mathcal{F}$ to $\mathcal{G}$.*

Actually, by (IV.1.4),

$$\varphi[\mathcal{G}] \leq \psi[\mathcal{F}] \Longleftrightarrow \psi^{-}[\varphi[\mathcal{G}]] \leq \mathcal{F}. \tag{IV.9.3}$$

IV.10. Supplement

Consider a couple of classical instances of continuous and discontinuous maps.

EXAMPLE IV.10.1. Consider the following maps from $\mathbb{R}$ to $\mathbb{R}$:

(1) $f(x) := 0$ if $x < 0$ and $f(x) := 1$ if $x \geq 0$,
(2) $h(x) := \left|\frac{1}{x}\right|$ if $x \neq 0$, and $h(0) := 0$,
(3) $g(0) := 0$, and $g(x) := \sin(\frac{1}{x})$ if $x \neq 0$,
(4) $k(x) := xg(x)$.

[12] A theory of continuous selections of lower semicontinuous relations was initiated by E. MICHAEL [184].

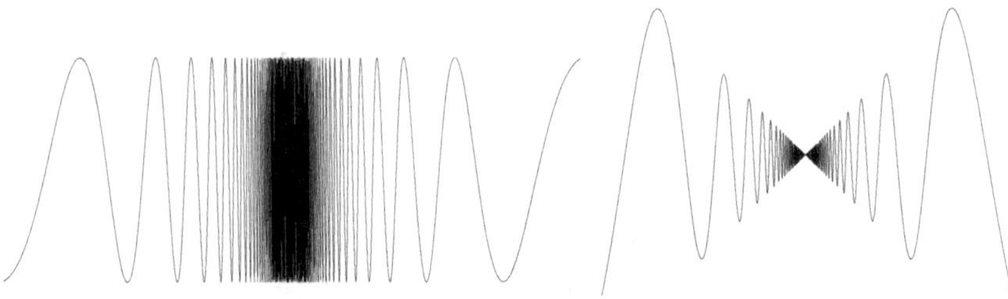

FIGURE IV.10.1. Some examples of discontinuous and continuous real functions: (3) the so called *topologist's sinus* $g(x) = \sin(\frac{1}{x})$ if $x \neq 0$, and its modification (4).

We see that, for the standard convergences on the domain and the range, f in (1) is not continuous, because for the filter-base $\mathcal{E} := \{]-\varepsilon, 0[: \varepsilon > 0\}$, the image $f[\mathcal{E}] = 0^{\uparrow}$, hence

$$\lim_{\nu_{\mathbb{R}}} \mathcal{E} = \{0\}, \lim_{\nu_{\mathbb{R}}} f[\mathcal{E}] = \lim_{\nu_{\mathbb{R}}} 0^{\uparrow} = \{0\} \text{ and } f(0) = 1.$$

As for h in (2), also $\lim_{\nu_{\mathbb{R}}} \mathcal{E} = \{0\}$ and $\lim_{\nu_{\mathbb{R}}} h[\mathcal{E}] = \varnothing$ (we say that $h[\mathcal{E}]$ tends to $+\infty$, but $+\infty \notin \mathbb{R}$), so that h is not continuous. Neither g in (3) is continuous, because $\lim_{\nu_{\mathbb{R}}} g[\mathcal{E}] = \varnothing$. However, k in (4) is continuous: the only point, at which there might be uncertainty, is 0. To see that $\lim_{\nu_{\mathbb{R}}} k[\mathcal{E}] = \{0\}$, observe that $|k(x)| = |k(x) - k(0)| \leq |x| = |x - 0|$.

LEMMA IV.10.2. *Let $\nu_{\mathbb{R}}$ be the standard convergence on the real line. If $f \in \mathscr{C}(\nu_{\mathbb{R}}, \nu_{\mathbb{R}})$ is such that $f(x) \neq 0$ for each $x \in X$, then $1/f$ is continuous.*

PROOF. Let $x \in \mathbb{R}$ be fixed. As f is continuous, there exists $\delta > 0$ and $c > 0$ such that $|f(r)| > c$ for each $r \in]x - \delta, x + \delta[$. Then $\left|\frac{1}{f(r)} - \frac{1}{f(x)}\right| = \frac{|f(x)-f(r)|}{|f(r)f(x)|} < \frac{1}{c^2}|f(x) - f(r)|$. Accordingly, for each $\varepsilon > 0$, there exists $\eta > 0$ such that $r - x < \eta$ implies $|f(x) - f(r)| < c^2\varepsilon$, hence $\left|\frac{1}{f(r)} - \frac{1}{f(x)}\right| < \varepsilon$. □

EXAMPLE IV.10.3. Let τ be the convergence on a one-point set, say $\{0\}$. It follows that $\mathbb{F}\{0\}$ consists of a single proper filter $0^{\uparrow} = \{\{0\}\}$, and $\lim_{\tau} 0^{\uparrow} = \{\{0\}\} = \{0\}$. It is the only convergence on $\{0\}$, which is simultaneously the discrete and chaotic topology. Moreover, it is both free and principal. If X is an infinite set and $f : X \longrightarrow \{0\}$ is a map, then this map is unique, that is, $f(x) = 0$ for each $x \in X$. Accordingly, $f^{-}\tau$ is a pretopology, and, by definition, $x \in \lim_{f^{-}\tau} \mathcal{F}$ whenever $0 = f(x) \in \lim_{\tau} f[\mathcal{F}] = \lim_{\tau} 0^{\uparrow} = \{\{0\}\}$, hence $f^{-}\tau = o_X$ is the chaotic pretopology.

EXERCISE IV.10.4. Prove Proposition IV.3.3.

SOLUTION. Indeed, $x \in \lim_{f^{-}\tau} \mathcal{F} = f^{-}(\lim_{\tau} f[\mathcal{F}])$ amounts to $f(x) \in \lim_{\tau} f[\mathcal{F}]$, and thus $f^{-}(f(x)) \subset f^{-}(\lim_{\tau} f[\mathcal{F}]) = \lim_{f^{-}\tau} \mathcal{F}$. On the other hand, $f(f^{-}(f(A))) = f(A)$ for each (not necessarily surjective) map f and each A, and thus

$$\lim_{f^{-}\tau} f^{-}[f[\mathcal{F}]] = f^{-}(\lim_{\tau} f[f^{-}[f[\mathcal{F}]]]) = f^{-}(\lim_{\tau} f[\mathcal{F}]) = \lim_{f^{-}\tau} \mathcal{F}.$$

Note that

$$\xi \geq f^{-}(f\xi). \tag{IV.10.1}$$

Indeed, by definition, $f \in C(\xi, f\xi)$, which is equivalent to (IV.10.1) by the second equivalence in (IV.3.5).

PROPOSITION IV.10.5. *If f is injective, then*

$$f^{-}(f\xi) = \xi.$$

PROOF. If $x \in \lim_{f^{-}(f\xi)} \mathcal{F}$ then $f(x) \in \lim_{f\xi} f[\mathcal{F}]$, that is, by Proposition IV.3.8, there is a filter $\mathcal{F}_0$ such that $x \in \lim_{\xi} \mathcal{F}_0$ and $f[\mathcal{F}] = f[\mathcal{F}_0]$ and, since f is injective, $\mathcal{F} = \mathcal{F}_0$. □

On the other hand,

$$f(f^{-}\tau) \geq \tau, \tag{IV.10.2}$$

because by definition, $f \in C(f^{-}\tau, \tau)$, which is equivalent to (IV.10.2) by the first equivalence in (IV.3.5).

PROPOSITION IV.10.6. *If f is surjective, then*

$$f(f^{-}\tau) = \tau. \tag{IV.10.3}$$

PROOF. Indeed, by Proposition IV.3.8, $y \in \lim_{ff^-\tau} \mathcal{G}$ if and only if there exists $\mathcal{F}$ such that $\mathcal{G} = f[\mathcal{F}]$ and $y \in f(\lim_{f^-\tau} \mathcal{F})$. As f is surjective,

$$f(\lim_{f^-\tau} \mathcal{F}) = ff^-(\lim_\tau f[\mathcal{F}]) = \lim_\tau f[\mathcal{F}] = \lim_\tau \mathcal{G}.$$

□

LEMMA IV.10.7. *For any collection $\boldsymbol{F}$ of filters and a map h,*

$$h[\bigwedge \boldsymbol{F}] = \bigwedge\{h[\mathcal{F}] : \mathcal{F} \in \boldsymbol{F}\}.$$

PROOF. Note that $W \in h[\bigwedge \boldsymbol{F}]$ whenever, for each $\mathcal{F} \in \boldsymbol{F}$, there exists $F_\mathcal{F} \in \mathcal{F}$ such that $\bigcup_{\mathcal{F}\in\boldsymbol{F}} h(F_\mathcal{F}) = h(\bigcup_{\mathcal{F}\in\boldsymbol{F}} F_\mathcal{F}) \subset W$, equivalently $W \in \bigwedge\{h[\mathcal{F}] : \mathcal{F} \in \boldsymbol{F}\}$. □

By Lemma IV.10.7, if $\mathcal{F}$ is a filter on X and $f \in Y^X$, then

$$f[\mathcal{F}] = f[\mathcal{F}^*] \wedge f[\mathcal{F}^\bullet].$$

It follows that $f[\mathcal{F}^\bullet]$ is principal. However, the image of a free filter need not be free, for example, if X is infinite, and $f : X \longrightarrow \{0\}$, then $(X)_0$ is free, and $f[(X)_0] = \{0\}$ is principal. Nevertheless,

$$f(\ker \mathcal{F}) \subset \ker f[\mathcal{F}]. \tag{IV.10.4}$$

EXERCISE IV.10.8. Show that

$$f(\bigwedge\nolimits_{\xi\in\Xi} \xi) = \bigwedge\nolimits_{\xi\in\Xi} f\xi, \quad f^-(\bigvee\nolimits_{\tau\in T} \tau) = \bigvee\nolimits_{\tau\in T} f^-\tau,$$
$$g(\bigwedge\nolimits_{j\in J} f_j\xi_j) = \bigwedge\nolimits_{\xi\in\Xi}(g \circ f_j)\xi_j, \quad f^-(\bigvee\nolimits_{j\in J} g_j^-\tau_j) = \bigvee\nolimits_{j\in J}(g_j \circ f)^-\tau_j.$$

SOLUTION. Indeed, $f(\bigwedge_{\xi\in\Xi} \xi) \le f\theta$ for each $\theta \in \Xi$, hence $\le$ holds in the first line. To show $\ge$, let $\bigwedge_{\xi\in\Xi} f\xi \ge \tau$, hence $f\theta \ge \tau$ for each $\theta \in \Xi$, equivalently, $\theta \ge f^-\tau$ for each $\theta \in \Xi$, that is, $\bigwedge_{\theta\in\Xi} \theta \ge f^-\tau$, equivalently, $f(\bigwedge_{\xi\in\Xi} \xi) \ge \tau$. Other equalities can be proved analogously.

EXERCISE IV.10.9. Show that ρ from Example IV.3.16 and μ from Example IV.3.10 are not homeomorphic.

REMARK IV.10.10. Recall of Remark III.3.10 that if ζ and ξ are preconvergences on X, then $\zeta \trianglerighteq \xi$ if and only if $\lim_\zeta \mathcal{F} \subset \lim_\xi \mathcal{F}$ for each filter $\mathcal{F}$ on X. Similarly, if τ is a preconvergence on Y, then we write $f \in \mathscr{C}(\xi, \tau)$ whenever $f(\lim_\xi \mathcal{F}) \subset \lim_\tau f[\mathcal{F}]$ for each filter $\mathcal{F}$ on X.

Notice that (IV.3.5) extends to preconvergences:

$$f\xi \trianglerighteq \tau \Longleftrightarrow f \in \mathscr{C}(\xi,\tau) \Longleftrightarrow \xi \trianglerighteq f^{-}\tau.$$

Therefore, it is straightforward that (IV.3.4) still holds if the considered objects are preconvergences. Observe that

$$(\zeta \trianglerighteq \xi) \wedge (\tau \trianglerighteq \theta) \Longrightarrow \mathscr{C}(\xi,\tau) \subset \mathscr{C}(\zeta,\theta).$$

Moreover,

(IV.10.5)

$$\mathscr{C}\left(\bigwedge\nolimits^{\trianglelefteq} \Xi,\tau\right) = \bigcap\nolimits_{\xi\in\Xi} \mathscr{C}(\xi,\tau), \qquad \mathscr{C}\left(\xi,\bigvee\nolimits^{\trianglelefteq} T\right) = \bigcap\nolimits_{\tau\in T} \mathscr{C}(\xi,\tau).$$

Indeed, $\bigwedge^{\trianglelefteq} \Xi \trianglelefteq \xi$ implies $\mathscr{C}(\bigwedge^{\trianglelefteq} \Xi,\tau) \subset \mathscr{C}(\xi,\tau)$ for each $\xi \in \Xi$. Conversely, if $f\xi \trianglerighteq \tau$ for each $\xi \in \Xi$, then $\bigwedge^{\trianglelefteq} \Xi \trianglelefteq \xi$. As for the second equality, $\mathscr{C}(\xi, \bigvee^{\trianglelefteq} T) \subset \mathscr{C}(\xi,\tau)$ for each $\tau \in T$. Conversely, if $f\xi \trianglerighteq \tau$ for each $\tau \in T$, then $f\xi \trianglerighteq \bigvee^{\trianglelefteq} T$.

Of course, in particular, (IV.10.5) holds for convergences. On the other hand,

$$\bigcup\nolimits_{\xi\in\Xi} \mathscr{C}(\xi,\tau) \subset \mathscr{C}(\bigvee\nolimits^{\trianglelefteq} \Xi,\tau), \qquad \bigcup\nolimits_{\tau\in T} \mathscr{C}(\xi,\tau) \subset \mathscr{C}(\xi,\bigwedge\nolimits^{\trianglelefteq} T).$$

EXAMPLE IV.10.11. Let $\pi = \pi[x_\infty, (X)_1]$ be the *prime cocountable convergence* on an uncountable set X, and let $(X)_1$ stand for the cocountable filter on X (Example II.2.24). All the elements $x \neq x_\infty$ are π-isolated, and $x_\infty \in \lim_\pi \mathcal{F}$ if and only if $\ker \mathcal{F} \subset \{\infty\}$ and $\mathcal{F}^* \geq (X)_1$.

We will see that if $f \in \mathscr{C}(\pi, \nu_{\mathbb{R}})$, then there is $W \in (X)_1$ such that f is constant on W (that is, $f(W)$ is a singleton). Without loss of generality, assume that $f(0) = 0$. Then let $G_n := f^{-}(]2^{-n}, 2^{-n}[)$ be the preimage under f of the strict interval $]2^{-n}, 2^{-n}[$ for each $n \in \mathbb{N}$. By Exercise II.5.5, $f^{-}(0) = \bigcap_{n<\omega} G_n \in (X)_1$, which means that f is constant on an element of $(X)_1$.

REMARK IV.10.12. Notice that the projection $p_X : X \times Y \longrightarrow X$ restricted to the graph

$$\mathrm{gr}(f) := \{(x, f(x)) : x \in X\}$$

is a homeomorphism, that is, $p_X|_{\mathrm{gr}(f)} \in \mathscr{H}(\,(\xi \times \tau)|_{\mathrm{gr}(f)}\,, \xi)$. Indeed, $p_X \circ (i_X \triangle f) = i_X$. There is still another useful representation of the graph. If Y is the underlying set of τ, that is, $Y = |\tau|$, then

$$\mathrm{gr}(f) = (f \times i_Y)^{-} \triangle(Y),$$

where $\triangle(Y)$ is the *diagonal* of $Y \times Y$. In our case, $(f \times i_Y)(x, y) = (f(x), y)$. Indeed, $(f \times i_Y)(x, y) \in \triangle(Y) = \{(y, y) : y \in Y\}$ if and only if $f(x) = y$.

EXERCISE IV.10.13. Let $\{\xi_j : j \in J\}$ be a set of convergences on X. Then

$$(\prod_{j\in J} \xi_j)\Big|_{\triangle(X)} = \bigvee_{j\in J} \xi_j.$$

SOLUTION. Indeed, $\varphi \in \lim_{(\prod_{j\in J}\xi_j)|_{\triangle(X)}} \mathscr{F}$ whenever $\varphi \in \triangle(X) \cap \lim_{\prod_{j\in J}\xi_j} \mathscr{F}$ and $\triangle(X) \in \mathscr{F}$. Therefore, there is $x \in X$ such that $\varphi = \triangle(x)$ and a filter $\mathcal{F}$ on X such that $\mathscr{F} = \triangle[\mathcal{F}]$ so that $p_j(\varphi) = x$ and $p_j[\mathscr{F}] = \mathcal{F}$ and $x \in \lim_{\xi_j} \mathcal{F}$ for each $j \in J$,

$$x \in \bigcap_{j\in J} \lim_{\xi_j} \mathcal{F} = \lim_{\bigvee_{j\in J}\xi_j} \mathcal{F}.$$

DEFINITION IV.10.14. A convergence ξ is said to be *homogeneous* if for every $x_0, x_1 \in |\xi|$, there exists $h \in \mathscr{H}(\xi, \xi)$ such that $h(x_0) = x_1$.

For each set X, the convergences ι_X, o_X are homogeneous. The *standard sequential convergence* $\sigma_{\mathbb{R}}$ (Example III.1.3), and the standard convergence $\nu_{\mathbb{R}}$ (Example III.1.4) are homogeneous. Also the convergences ρ from Example IV.3.16 and thus $h\mu$ from Example IV.3.10 are homogeneous. The *Sierpiński pretopology* $\$_0$ is not homogeneous, hence $\$_1$ is not homogeneous.

EXAMPLE IV.10.15. Let $\nu_{\mathbb{R}}$ be the standard convergence on $\mathbb{R}$ and let $f : \mathbb{R} \longrightarrow \mathbb{R}$ be defined by

$$f(x) := \begin{cases} x & x \notin \mathbb{Z}, \\ 0 & x \in \mathbb{Z}. \end{cases}$$

The standard convergence $\nu_{\mathbb{R}}$ is a pretopology: for each $x \in \mathbb{R}$ there exists the coarsest filter $\mathcal{N}_{\nu_{\mathbb{R}}}(x)$ that converges to x. Notice that $f\nu_{\mathbb{R}}$ is not a pretopology. Indeed,

$$0 \in \lim_{f\nu_{\mathbb{R}}} f[\mathcal{N}_{\nu_{\mathbb{R}}}(n)]$$

for each integer $n \in \mathbb{Z}$. However, $0 \notin \lim_{f\nu_{\mathbb{R}}} \bigcap_{n\in\mathbb{Z}} f[\mathcal{N}_{\nu_{\mathbb{R}}}(n)]$, because the filter $\bigcap_{n\in\mathbb{Z}} \mathcal{N}_{\nu_{\mathbb{R}}}(n)$ is not $\nu_{\mathbb{R}}$-convergent.

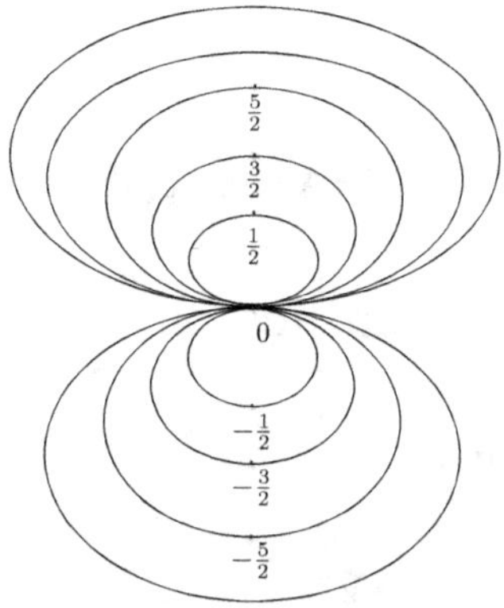

FIGURE IV.10.2. The image of $\mathbb{R}$ under the quotient map of Example IV.10.15.

CHAPTER V

Families of sets

Let us revisit here more systematically Section IV.1 on families of sets. Recall that

$$A \in 2^X \iff A \subset X. \tag{V.0.1}$$

Accordingly, if $\mathcal{A}$ is a *family* ([1]) of subsets of X, then $\mathcal{A} \subset 2^X$. If $\mathcal{A}, \mathcal{D} \subset 2^X$, then

$$\mathcal{A} \subset \mathcal{D}$$

means that $A \in \mathcal{A}$ implies $A \in \mathcal{D}$.

REMARK V.0.1. On the other hand, if A and B are subsets of X, then $A \subset B$ means that $x \in A$ implies $x \in B$. Hence, the meaning of the relation $\subset$ depends on the context, which disambiguates the use. In particular, $(2^{(2^X)}, \subset)$ is an order relation ([2]). Accordingly, $\cap, \bigcap$ stand for infima, and $\cup, \bigcup$ for suprema with respect to this order.

DEFINITION V.0.2. We say that a family $\mathcal{A}$ is *coarser* than a family $\mathcal{D}$ ($\mathcal{D}$ is *finer* than $\mathcal{A}$)

$$\mathcal{A} \leq \mathcal{D},$$

whenever for each $A \in \mathcal{A}$ there exists $D \in \mathcal{D}$ such that $D \subset A$.

Of course, if $\mathcal{A} \subset \mathcal{D}$ then $\mathcal{A} \leq \mathcal{D}$, but not vice versa ([3]). The relation $\leq$ is *transitive* and *reflexive*, but not *antisymmetric*, that is, $\leq$ is a *preorder*. Families $\mathcal{A}$ and $\mathcal{D}$ are said to be *equivalent*

$$\mathcal{A} \approx \mathcal{D}$$

whenever $\mathcal{A} \leq \mathcal{D}$ and $\mathcal{D} \leq \mathcal{A}$.

---

[1]Families of subsets of a given set are sets. The term *family* is used for practical purpose of distinguishing between *sets* and *sets of sets*. For the same reason, we shall use the term a *collection* for a set of families of subsets.

[2]Let us write it explicitly in the case of families of subsets of X (that is, a relation on $2^{(2^X)}$): the relation $\subset$ is *reflexive* ($\mathcal{A} \subset \mathcal{A}$), *antisymmetric* (if $\mathcal{A} \subset \mathcal{B}$ and $\mathcal{B} \subset \mathcal{A}$, then $\mathcal{A} = \mathcal{B}$), and *transitive* (if $\mathcal{A}_0 \subset \mathcal{A}_1$ and $\mathcal{A}_1 \subset \mathcal{A}_2$, then $\mathcal{A}_0 \subset \mathcal{A}_2$).

[3]For example, $\mathcal{B}$ is a base of a filter $\mathcal{F}$ whenever $\mathcal{B} \subset \mathcal{F} \leq \mathcal{B}$, but $\mathcal{B}$ is not necessarily equal to $\mathcal{F}$.

Consider now two operations $(\cdot)_c : 2^X \to 2^X$, given by

$$\mathcal{H}_c := \{H^c : H \in \mathcal{H}\}, \tag{V.0.2}$$

where $H^c := X \setminus H$, and $(\cdot)^c : 2^X \to 2^X$, given by

$$\mathcal{H}^c := \{D : D \notin \mathcal{H}\}. \tag{V.0.3}$$

We observe that $(\cdot)_c$ and $(\cdot)^c$ are *involutive*, that is,

$$(\mathcal{H}_c)_c = \mathcal{H}, \qquad (\mathcal{H}^c)^c = \mathcal{H},$$

in particular, they are bijections ([4]). Observe that, for a collection $\boldsymbol{H}$ of families,

$$(\bigcup\nolimits_{\mathcal{H}\in\boldsymbol{H}} \mathcal{H})_c = \bigcup\nolimits_{\mathcal{H}\in\boldsymbol{H}} \mathcal{H}_c, \quad (\bigcap\nolimits_{\mathcal{H}\in\boldsymbol{H}} \mathcal{H})_c = \bigcap\nolimits_{\mathcal{H}\in\boldsymbol{H}} \mathcal{H}_c,$$
$$(\bigcup\nolimits_{\mathcal{H}\in\boldsymbol{H}} \mathcal{H})^c = \bigcap\nolimits_{\mathcal{H}\in\boldsymbol{H}} \mathcal{H}^c.$$

The composition of $(\cdot)_c$ and $(\cdot)^c$, in any order, results in the so called *conjugation*. The *conjugate family* $\mathcal{A}^\star$ of $\mathcal{A} \subset 2^X$ is defined by

$$\mathcal{A}^\star := \{H : H^c \notin \mathcal{A}\}. \tag{V.0.4}$$

In other words, $H \in \mathcal{A}^\star$ if and only if $H^c \notin \mathcal{A}$. As a composition of two involutions,

$$\mathcal{A}^{\star\star} = \mathcal{A}$$

the operation $(\cdot)^\star$ is an involution ([5]).

LEMMA V.0.3. *For each family* $\mathcal{A}$,

$$\mathcal{A}^\star = (\mathcal{A}_c)^c = (\mathcal{A}^c)_c.$$

PROOF. Let $H \in (\mathcal{A}_c)^c$, that is, $H \notin \mathcal{A}_c$, equivalently $H^c \notin \mathcal{A}$. The latter means that $H^c \in \mathcal{A}^c$, thus $H \in (\mathcal{A}^c)_c$. □

Accordingly,

$$(\bigcup\nolimits_{\mathcal{H}\in\boldsymbol{H}} \mathcal{H})^\star = \bigcap\nolimits_{\mathcal{H}\in\boldsymbol{H}} \mathcal{H}^\star, \qquad (\bigcap\nolimits_{\mathcal{H}\in\boldsymbol{H}} \mathcal{H})^\star = \bigcup\nolimits_{\mathcal{H}\in\boldsymbol{H}} \mathcal{H}^\star, \tag{V.0.5}$$

the second equality following from the first by involution.

---

[4]A map $h : W \longrightarrow W$ is an *involution* if $h(h(w)) = w$ for every $w \in W$. In other words, h is an involution if $h \circ h = i_W$ the identity on W. Hence, by Remark I.1.3, each involution is a bijection.

[5]A direct checking is also immediate: $B \in \mathcal{A}^{\star\star}$ if and only if $B^c \notin \mathcal{A}^\star$, equivalently $B \in \mathcal{A}$.

V.1. Isotone and antitone families

As we have said, a family $\mathcal{A} \subset 2^X$ is said to be *isotone* if

$$D \supset A \in \mathcal{A} \Longrightarrow D \in \mathcal{A}.$$

For each $\mathcal{C} \subset 2^X$, there exists ([6]) the least isotone family that includes $\mathcal{C}$, namely

$$\mathcal{C}^{\uparrow} := \{H \subset X : \underset{C \in \mathcal{C}}{\exists} C \subset H\}.$$

We denote by $\mathbb{I}$ the class of isotone families, and by $\mathbb{I}X$ the set of isotone families of subsets X.

The relations $\subset$ and $\leq$ coincide on the set of isotone families. In fact, $\mathcal{A} \subset \mathcal{D}$ implies $\mathcal{A} \leq \mathcal{D}$, and

PROPOSITION V.1.1. *If $\mathcal{D}$ is isotone and $\mathcal{A} \leq \mathcal{D}$, then $\mathcal{A} \subset \mathcal{D}$.*

PROOF. Indeed, if $A \in \mathcal{A} \leq \mathcal{D}$, then there exists $D \in \mathcal{D}$ such that $D \subset A$, hence $A \in \mathcal{D}$, because $\mathcal{D}$ is isotone. □

If $\boldsymbol{A} \subset \mathbb{I}X \subset 2^{(2^X)}$ is a collection of isotone families, then $\bigcup \boldsymbol{A} = \bigcup_{\mathcal{A} \in \boldsymbol{A}} \mathcal{A}$ and $\bigcap \boldsymbol{A} = \bigcap_{\mathcal{A} \in \boldsymbol{A}} \mathcal{A}$ are isotone. Therefore, the union and the intersection are respectively the supremum and the infimum in the ordered set $(\mathbb{I}X, \subset)$.

EXAMPLE V.1.2. Let A, B be subsets of X. Then $\{A\} \cap \{B\} = \varnothing$ if and only if $A \neq B$, while $\{A\}^{\uparrow} \cap \{B\}^{\uparrow} = \{F \subset X : (A \subset F) \wedge (B \subset F)\} = (A \cup B)^{\uparrow}$. Now, $\{A\} \cup \{B\} = \{A, B\}$, while $\{A\}^{\uparrow} \cup \{B\}^{\uparrow} = \{F \subset X : (A \subset F) \vee (B \subset F)\} = \{A, B\}^{\uparrow}$.

DEFINITION V.1.3. If $\mathcal{A}$ is an isotone family, then $\mathcal{B} \subset \mathcal{A}$ is called a *base* of $\mathcal{A}$ if $\mathcal{B} \geq \mathcal{A}$.

A family $\mathcal{P} \subset 2^X$ is called *antitone* if

$$R \subset P \in \mathcal{P} \Longrightarrow R \in \mathcal{P}.$$

For each $\mathcal{C} \subset 2^X$, there exists the least antitone family that includes $\mathcal{C}$, namely

$$\mathcal{C}^{\downarrow} := \{H \subset X : \underset{C \in \mathcal{C}}{\exists} H \subset C\}.$$

We denote by $\mathbb{A}$ the class of antitone families, and by $\mathbb{A}X$ the set of antitone families of subsets X.

[6]The existence follows from two facts, first, the intersection of any collection of isotone families is isotone; second, there is the largest isotone family on each set X, namely 2^X.

Notice that the operations $(\cdot)_c$ and $(\cdot)^c$ defined by (II.2.2) and (V.0.3) are *antitone*, that is, if $\mathcal{H}$ is isotone, then $\mathcal{H}_c$ and $\mathcal{H}^c$ are antitone, and *vice versa*. On the other hand, $(\cdot)^\star$ is *isotone* ([7]), as the composition of two antitone operations.

DEFINITION V.1.4 (Refinement). Let $\mathcal{P}, \mathcal{R} \subset 2^X$. We say that $\mathcal{R}$ *refines* $\mathcal{P}$ (or is a *refinement* of $\mathcal{P}$), in symbols,

$$\mathcal{R} \lhd \mathcal{P},$$

if for every $R \in \mathcal{R}$ there is $P \in \mathcal{P}$ such that $R \subset P$.

If $\mathcal{R} \subset \mathcal{P}$ then $\mathcal{R} \lhd \mathcal{P}$. On the other hand,

PROPOSITION V.1.5. *If $\mathcal{P}$ is antitone and $\mathcal{R} \lhd \mathcal{P}$, then $\mathcal{R} \subset \mathcal{P}$.*

PROOF. Let $R \in \mathcal{R}$ and $\mathcal{R} \lhd \mathcal{P}$, then there is $P \in \mathcal{P}$ such that $R \subset P$. As $\mathcal{P}$ is antitone, $R \in \mathcal{P}$. □

If $\boldsymbol{A} \subset \mathbb{A}X \subset 2^{(2^X)}$, then $\bigcup \boldsymbol{A} = \bigcup_{\mathcal{A} \in \boldsymbol{A}} \mathcal{A}$ and $\bigcap \boldsymbol{A} = \bigcap_{\mathcal{A} \in \boldsymbol{A}} \mathcal{A}$ are antitone. Therefore, the union and the intersection are respectively the supremum and the infimum in the ordered set $(\mathbb{A}X, \subset)$.

PROPOSITION V.1.6. *For every $\mathcal{P}$ and $\mathcal{R}$,*

$$\mathcal{R} \lhd \mathcal{P} \Longleftrightarrow \mathcal{R}_c \leq \mathcal{P}_c. \tag{V.1.1}$$

PROOF. In fact, $A \in \mathcal{R}_c$ if and only if $A^c \in \mathcal{R}$. If now $\mathcal{R} \lhd \mathcal{P}$, then there is $P \in \mathcal{P}$ such that $A^c \subset P$, equivalently, $P^c \subset A$. Since $P^c \in \mathcal{P}_c$, we infer that $\mathcal{R}_c \leq \mathcal{P}_c$. The converse argument is analogous. □

V.2. Ideals, filters, distributive families

Filters are isotone families, stable under finite intersections, that is,

$$F_0 \in \mathcal{F} \text{ and } F_1 \in \mathcal{F} \Longrightarrow F_0 \cap F_1 \in \mathcal{F}. \tag{V.2.1}$$

Evidently, (V.2.1) extends by induction to all finite intersections of elements of $\mathcal{F}$.

A dual notion with respect to filter is that of ideal.

DEFINITION V.2.1. An antitone family $\mathcal{I} \subset 2^X$ is called an *ideal* whenever

$$I_0 \in \mathcal{I} \text{ and } I_1 \in \mathcal{I} \Longrightarrow I_0 \cup I_1 \in \mathcal{I}. \tag{V.2.2}$$

An ideal $\mathcal{I}$ is said to be *proper* if $X \notin \mathcal{I}$. We say that $\mathcal{B} \subset \mathcal{P}$ is an *ideal-base* of an ideal $\mathcal{P}$ (or $\mathcal{P}$ is *generated* by $\mathcal{B}$) if for each $P \in \mathcal{P}$, there exists $B \in \mathcal{B}$ so that $P \subset B$.

---

[7]That is, if $\mathcal{H}$ is isotone, then $\mathcal{H}^\star$ is isotone, and if $\mathcal{H}$ is antitone, then $\mathcal{H}^\star$ is antitone.

PROPOSITION V.2.2. *A family $\mathcal{I}$ is an ideal if and only if $\mathcal{I}_c$ is a filter. The ideal $\mathcal{I}$ is proper if and only if the filter $\mathcal{I}_c$ is proper.*

PROOF. Indeed, we know already that $(\cdot)_c$ is antitone. Now, by setting $F_0 = (I_0)^c, F_1 = (I_1)^c$, and $\mathcal{F} = \mathcal{I}_c$ in (V.2.1), and noting that $F_0 \cap F_1 = (I_0 \cup I_1)^c$, we get equivalence with (V.2.2). On the other hand, $\varnothing \in \mathcal{F}$ amounts to $X = \varnothing^c \in \mathcal{F}_c = \mathcal{I}$. □

DEFINITION V.2.3. An isotone family $\mathcal{D} \subset 2^X$ is called *distributive* whenever

$$D_0 \cup D_1 \in \mathcal{D} \Longrightarrow D_0 \in \mathcal{D} \text{ or } D_1 \in \mathcal{D}. \tag{V.2.3}$$

PROPOSITION V.2.4. *A family $\mathcal{D}$ is distributive if and only if $\mathcal{D}^\star$ is a filter.*

PROOF. In fact, $\mathcal{D}$ is isotone if and only if $\mathcal{D}^\star$ is isotone: suppose that $\mathcal{D}$ is isotone, and let $B \supset A \in \mathcal{D}^\star$. Then $A^c \notin \mathcal{D}$ and $B^c \subset A^c$, hence $B^c \notin \mathcal{D}$, because $\mathcal{D}$ is isotone, hence $B \in \mathcal{D}^\star$. The converse fact follows from involution.

Suppose that $\mathcal{D}$ is a distributive family, and let $F_0 \in \mathcal{D}^\star$ and $F_1 \in \mathcal{D}^\star$. Equivalently, $(F_0)^c \notin \mathcal{D}$ and $(F_1)^c \notin \mathcal{D}$, hence, by (V.2.3), $(F_0 \cap F_1)^c = (F_0)^c \cup (F_1)^c \notin \mathcal{D}$, that is, $F_0 \cap F_1 \in \mathcal{D}^\star$.

Conversely, suppose that $\mathcal{F} = \mathcal{D}^\star$ is a filter, and $D_0 \cup D_1 \in \mathcal{D}$. Then $(D_0 \cup D_1)^c = (D_0)^c \cap (D_1)^c \notin \mathcal{F}$, thus by (V.2.1) either $(D_0)^c \notin \mathcal{F}$ or $(D_1)^c \notin \mathcal{F}$, equivalently, either $D_0 \in \mathcal{D}$ or $D_1 \in \mathcal{D}$. □

As * is an involution, the family $\mathcal{F}^\star$ is distributive provided that $\mathcal{F}$ is a filter. Notice that if $\mathcal{F}$ is a proper filter, then

$$\mathcal{F} \subset \mathcal{F}^\star.$$

Indeed, if $F \in \mathcal{F}$, then $F^c \notin \mathcal{F}^\star$. As $\mathcal{F}$ is proper, $\mathcal{F}^\star \neq \varnothing$ and thus $F \cup F^c = X \in \mathcal{F}^\star$, hence by (V.2.3), $F \in \mathcal{F}^\star$.

We illustrate the notions just discussed in case of finite underlying sets.

EXAMPLE V.2.5. Let $X = \{0,1,2\}$, and let $\mathcal{F} = \{\{2\},\{0,2\},\{1,2\}, \{0,1,2\}\}$. Then $\mathcal{F}$ is an ultrafilter. We notice that $\mathcal{F}^\star = \mathcal{F}$. The family of the complements of the elements of $\mathcal{F}$, that is, $\mathcal{F}_c = \{\varnothing, \{1\}, \{0\}, \{0,1\}\}$ is an ideal.

EXAMPLE V.2.6. Now, $\mathcal{F} = \{\{0,2\},\{0,1,2\}\}$ is a non-maximal filter. By definition, the conjugate family $\mathcal{F}^\star$ is the complementary set of the complements of elements of $\mathcal{F}$.

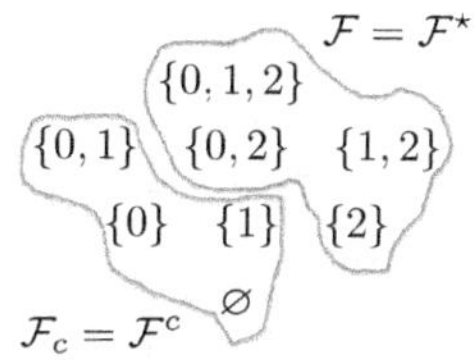

FIGURE V.2.1. Here $\mathcal{F}$ is a maximal filter, and $\mathcal{F}^\star = \mathcal{F}$.

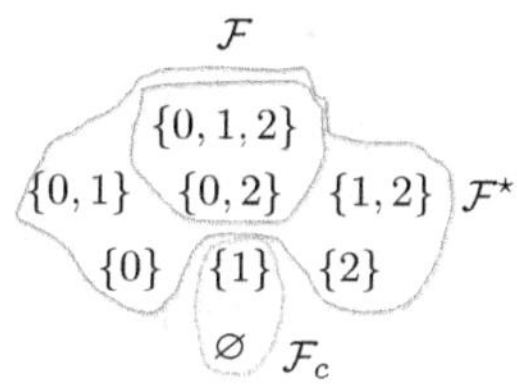

FIGURE V.2.2. Here $\mathcal{F}$ is a non-maximal filter, $\mathcal{F}_c$ is the complementary ideal, and $\mathcal{F}^\star$ is the conjugate family of $\mathcal{F}$, which is strictly larger than $\mathcal{F}$.

Let us recapitulate the action of involutions $(\cdot)_c, (\cdot)^c$, and $(\cdot)^\star$ between the classes of (possibly improper or empty) families, $\overline{\mathbb{F}}$, $\overline{\mathbb{I}}$, and $\overline{\mathbb{D}}$.

If a filter on a non-empty set X is improper, that is, equal to 2^X, then the conjugate family $(2^X)^\star = \varnothing$. On the other hand, the family 2^X is distributive, and $(2^X)^\star = \varnothing$.

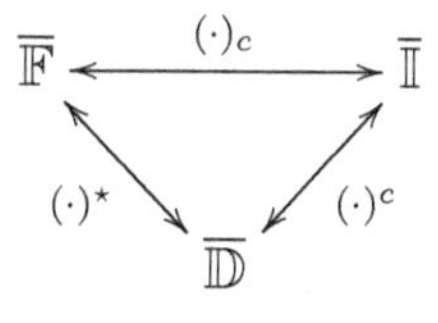

REMARK V.2.7. Of course, the operation $(\cdot)^c$ transforms $\overline{\mathbb{F}}$ onto another class of families of sets ([8]). It consists of antitone families $\mathcal{H}$ such that

$$H_0 \cap H_1 \in \mathcal{H} \Longrightarrow H_0 \in \mathcal{H} \text{ or } H_1 \in \mathcal{H}.$$

Of course, in Example V.2.6, $\mathcal{F}^c$ consists of all the sets that are not in the filter $\mathcal{F}$. Properties of filters can be translated to equivalent properties concerning this class.

[8]Victor Porton informed me by E-mail about such classes, which he calls *mixers*.

V.3. The notion of grill

If X is a set, then the *grill* is a relation on 2^X defined by

$$A\#B \Longleftrightarrow A \cap B \neq \varnothing. \tag{V.3.1}$$

It can be thought as a shorthand, but its extension to families of subsets of X provides a very useful notation. If $\mathcal{A}, \mathcal{B} \subset 2^X$, then we say that $\mathcal{A}$ and $\mathcal{B}$ *mesh*, in symbols,

$$\mathcal{A}\#\mathcal{B} \tag{V.3.2}$$

if $A\#B$ for each $A \in \mathcal{A}$ and each $B \in \mathcal{B}$. We abridge $\{A\}\#\mathcal{B}$ and $\{A\}^{\uparrow}\#\mathcal{B}$ to $A\#\mathcal{B}$.

If a family $\mathcal{A}$ is centered (that is, a filter-subbase; see Definition II.2.3), then $\mathcal{A}\#\mathcal{A}$, but not conversely (see Exercise V.6.2).

If $\mathcal{A}$ and $\mathcal{B}$ do not mesh, in symbols

$$\mathcal{A}\perp\mathcal{B}, \mathcal{A}\neg\#\mathcal{B} \text{ or } \neg(\mathcal{A}\#\mathcal{B}),$$

then they are called *dissociated*. A collection $\boldsymbol{A}$ of families is called *dissociated* if $\mathcal{A}_0$ and $\mathcal{A}_1$ are dissociated for each pair $\mathcal{A}_0, \mathcal{A}_1 \in \boldsymbol{A}$. A family $\mathcal{A}$ is said to be *dissociated* from a collection $\boldsymbol{B}$ whenever $\mathcal{A}\perp\mathcal{B}$ for each $\mathcal{B} \in \boldsymbol{B}$.

Finally, if $\mathcal{A} \subset 2^X$, then the *grill* $\mathcal{A}^{\#}$ is defined by

$$\mathcal{A}^{\#} := \{H \subset X : H\#\mathcal{A}\}.$$

REMARK V.3.1. Notice that $H \in \mathcal{A}^{\#}$ if and only if for each $A \in \mathcal{A}$ there exists $x_{A,H} \in H \cap A$. Therefore, $\{x_{A,H} : A \in \mathcal{A}\} \subset H$ and $\{x_{A,H} : A \in \mathcal{A}\} \in \mathcal{A}^{\#}$ for each $A \in \mathcal{A}$ and $H \in \mathcal{H}$, and thus $\{\{x_{A,H} : A \in \mathcal{A}\} : H \in \mathcal{A}^{\#}\}$ is a base of $\mathcal{A}^{\#}$.

It is plain that

$$\mathcal{A} \leq \mathcal{B} \Longrightarrow \mathcal{A}^{\#} \supset \mathcal{B}^{\#}. \tag{V.3.3}$$

On the other hand, it is clear that for each $\mathcal{A}$, the grill $\mathcal{A}^{\#}$ is *isotone*, that is, if $H \in \mathcal{A}^{\#}$ and $H \subset J$, then $J \in \mathcal{A}^{\#}$. On the other hand,

$$\mathcal{A} \subset \mathcal{A}^{\#\#}.$$

Indeed, if $A \in \mathcal{A}$ and $H \in \mathcal{A}^{\#}$, then, by definition, $A\#H$, hence $A \in \mathcal{A}^{\#\#}$.

PROPOSITION V.3.2. *$\mathcal{A}^{\#} \subset \mathcal{A}^{\star}$ for every $\mathcal{A}$. Moreover, $\mathcal{A}$ is isotone if and only if*

$$\mathcal{A}^{\#} = \mathcal{A}^{\star}.$$

PROOF. If $H \notin \mathcal{A}^\star$, equivalently $H^c \in \mathcal{A}$, then $H \notin \mathcal{A}^\#$, because $H \cap H^c = \varnothing$.

Suppose that $\mathcal{A}$ is isotone and that $H \notin \mathcal{A}^\#$. By definition, there exists $A \in \mathcal{A}$ such that $A \cap H = \varnothing$, equivalently $A \subset H^c$, so that $H^c \in \mathcal{A}$, that is, $H \notin \mathcal{A}^\star$.

If $\mathcal{A}$ is not isotone, that is, there exists $A \in \mathcal{A}$ and $B \notin \mathcal{A}$ such that $A \subset B$, then $A \cap B^c \subset A \cap A^c = \varnothing$, proving that $B^c \notin \mathcal{A}^\#$, though $B^c \in \mathcal{A}^\star$. □

COROLLARY V.3.3. *For each isotone family $\mathcal{A}$,*

$$\mathcal{A}^{\#\#} = \mathcal{A}.$$

Therefore, if $\mathcal{A}$ is isotone, then

$$H \notin \mathcal{A}^\# \Longleftrightarrow H^c \in \mathcal{A}. \tag{V.3.4}$$

REMARK V.3.4. Notice that

$$(2^X)^\# = \varnothing, \text{ and } \varnothing^\# = 2^X$$

([9]). Indeed, $(2^X)^\# = \varnothing$, because $\varnothing \in 2^X$ ([10]) and $\neg(\varnothing \# A)$ for each $A \in 2^X$. On the other hand, $\varnothing^\# = 2^X$, because $\varnothing^\# = ((2^X)^\#)^\# = 2^X$ by Proposition V.3.2, since 2^X is isotone.

We have already observed that if $\mathcal{F}$ is a proper filter, then

$$\mathcal{F} \subset \mathcal{F}^\#. \tag{V.3.5}$$

By Proposition V.2.4,

COROLLARY V.3.5. *A family is distributive if and only if it is the grill of a filter.*

Let us recollect now Proposition I.1.1 rewriting it with the aid of #.

$$RA\#B \Longleftrightarrow A\#R^- B \Longleftrightarrow (A \times B)\#R. \tag{V.3.6}$$

This very simple fact appeared in Preliminaries, because it is basic for our relation calculus. Its extension to families of sets is very easy, and very important fo that calculus.

PROPOSITION V.3.6. *Let $R \subset X \times Y$, and let $\mathcal{A} \subset 2^X$ and $\mathcal{B} \subset 2^Y$. Then*

$$R[\mathcal{A}]\#\mathcal{B} \Longleftrightarrow \mathcal{A}\#R^-[\mathcal{B}] \Longleftrightarrow (\mathcal{A} \times \mathcal{B})\#R. \tag{V.3.7}$$

[9]Here, $\varnothing \in 2^{(2^X)}$ denotes the empty family of subsets of X.

[10]Here, $\varnothing \in 2^X$ denotes the empty subset of X.

It is enough to observe that $R[\mathcal{A}]\#\mathcal{B}$ if and only if (V.3.6) holds for each $A \in \mathcal{A}$ and $B \in \mathcal{B}$. In particular, if R corresponds to a map $f \in Y^X$, then

COROLLARY V.3.7. *If $f : X \to Y, \mathcal{A} \subset 2^X$, and $\mathcal{B} \subset 2^Y$, then*

$$f[\mathcal{A}]\#\mathcal{B} \Longleftrightarrow \mathcal{A}\#f^-[\mathcal{B}]. \tag{V.3.8}$$

Let $\boldsymbol{A}$ be a set of isotone families. Then

$$(\bigcup \boldsymbol{A})^\# = \bigcap\nolimits_{\mathcal{A}\in\boldsymbol{A}} \mathcal{A}^\#, \quad (\bigcap \boldsymbol{A})^\# = \bigcup\nolimits_{\mathcal{A}\in\boldsymbol{A}} \mathcal{A}^\#. \tag{V.3.9}$$

This fact follows from (V.0.5) and Proposition V.3.2 and Corollary V.3.3. Nevertheless, it is advisable to check the formulae directly (Exercise V.6.3).

In the case of finite collections of filters, the second relation in (V.3.9) can be sharpened.

PROPOSITION V.3.8. *Let $\mathcal{H}$ be a filter, and $\boldsymbol{G}$ be a finite set of isotone families. Then*

$$\mathcal{H}\#\bigcap\nolimits_{\mathcal{G}\in\boldsymbol{G}} \mathcal{G} \Longleftrightarrow \underset{\mathcal{G}\in\boldsymbol{G}}{\exists} \mathcal{H}\#\mathcal{G}.$$

PROOF. The implication $\Leftarrow$ holds, because $\bigcap \boldsymbol{G}$ is coarser than any $\mathcal{G} \in \boldsymbol{G}$. To prove $\Rightarrow$, suppose that $\mathcal{H}$ is a filter such that $\mathcal{H}\perp\mathcal{G}$ for each $\mathcal{G} \in \boldsymbol{G}$. In other terms, for each $\mathcal{G} \in \boldsymbol{G}$, there exists $G_\mathcal{G} \in \mathcal{G}$ so that $(G_\mathcal{G})^c \in \mathcal{H}$. Hence $\bigcup_{\mathcal{G}\in\boldsymbol{G}} G_\mathcal{G} \in \bigcap_{\mathcal{G}\in\boldsymbol{G}} \mathcal{G}$, and $\bigcap_{\mathcal{G}\in\boldsymbol{G}} (G_\mathcal{G})^c \in \mathcal{H}$, that is, $\bigcup_{\mathcal{G}\in\boldsymbol{G}} G_\mathcal{G} \notin \mathcal{H}^\#$. □

The assumption of finiteness in Proposition V.3.8 cannot be dropped.

EXAMPLE V.3.9. Let $\mathcal{H}$ be a free filter on an (infinite) set A. Then $\{A\}^\uparrow = \bigcap_{a\in A} a^\uparrow$ and $\mathcal{H}\#\{A\}^\uparrow$, but $a^\uparrow \notin \mathcal{H}^\#$ for each $a \in A$.

V.4. Ultrafilters

Recall that maximal filters are called *ultrafilters* (Definition II.3.9). If $\mathcal{F}$ is a filter on X, then we denote by $\beta\mathcal{F}$ the set of all ultrafilters on X that are finer than $\mathcal{F}$. It follows from Corollary XXVII.4.9 that $\beta\mathcal{F} \neq \varnothing$ for every proper filter $\mathcal{F}$.

If $\mathcal{F} = \{F\}^\uparrow$, then we abridge $\beta F := \beta\{F\}^\uparrow$. If $\mathcal{U}$ is an ultrafilter, then by the convention above, $\beta\mathcal{U} = \{\mathcal{U}\}$.

By Remark II.4.2, an ultrafilter $\mathcal{U}$ on a set X is either free or principal, and in the latter case, $\mathcal{U} = x^\uparrow$ for some $x \in X$. It is clear that there are as many principal ultrafilters on a set as there are points of this set. It is not difficult to see that there are infinitely many free ultrafilters on each infinite set ([11]); actually Proposition XII.6.6 will establish the cardinality of βX in terms of the cardinality of X.

PROPOSITION V.4.1. *A filter $\mathcal{F}$ is an ultrafilter if and only if $\mathcal{F} = \mathcal{F}^{\#}$.*

PROOF. If a filter $\mathcal{F}$ is not an ultrafilter, then there exist a filter $\mathcal{G} \supset \mathcal{F}$ and $G_0 \in \mathcal{G} \setminus \mathcal{F}$. Therefore $G_0 \in \mathcal{G} \subset \mathcal{G}^{\#} \subset \mathcal{F}^{\#}$. Conversely, if there is $H \in \mathcal{F}^{\#} \setminus \mathcal{F}$, then $\mathcal{F} \vee H$ is a proper filter finer than $\mathcal{F}$ and that contains H. Since $H \notin \mathcal{F}$, the filter $\mathcal{F}$ is not maximal. $\square$

By Proposition V.4.1, each ultrafilter is distributive (V.2.3), which improves Proposition II.3.10.

COROLLARY V.4.2. *If $\mathcal{U}$ is an ultrafilter and $D_0 \cup D_1 \in \mathcal{U}$, then either $D_0 \in \mathcal{U}$ or $D_1 \in \mathcal{U}$.*

PROPOSITION V.4.3. *If $\mathcal{U}_0$ and $\mathcal{U}_1$ are distinct ultrafilters on a set X, then there exist $X_0 \in \mathcal{U}_0$ and $X_1 \in \mathcal{U}_1$ such that $X_0 \sqcup X_1 = X$.*

PROOF. Indeed, by maximality, $\mathcal{U}_0 \setminus \mathcal{U}_1 \neq \varnothing$ and $\mathcal{U}_1 \setminus \mathcal{U}_0 \neq \varnothing$, hence in particular there exists $X_0 \in \mathcal{U}_0 \setminus \mathcal{U}_1$. As $X = X_0 \cup (X \setminus X_0) \in \mathcal{U}_1$, by Corollary V.4.2, $X_1 := X \setminus X_0 \in \mathcal{U}_1$. $\square$

Observe that for every proper filter $\mathcal{F}$,

$$\mathcal{F} = \bigcap\nolimits_{\mathcal{U} \in \beta\mathcal{F}} \mathcal{U}, \text{ and } \mathcal{F}^{\#} = \bigcup\nolimits_{\mathcal{U} \in \beta\mathcal{F}} \mathcal{U}. \tag{V.4.1}$$

Indeed, the inclusion $\subset$ is obvious. Conversely, if $H \notin \mathcal{F}$, then $H^c \in \mathcal{F}^{\#}$, hence $\mathcal{F} \vee H^c$ is a proper filter. If $\mathcal{U}$ is an ultrafilter finer than $\mathcal{F} \vee H^c$, then of course $\mathcal{U} \in \beta\mathcal{F}$ and $H^c \in \mathcal{U}$. It follows that $H \notin \mathcal{U}$, hence $H \notin \bigcap_{\mathcal{U} \in \beta\mathcal{F}} \mathcal{U}$. The second equality follow from (V.3.9) and the fact that $\mathcal{U}^{\#} = \mathcal{U}$ for each ultrafilter $\mathcal{U}$.

PROPOSITION V.4.4. *An ultrafilter is sequential if and only if it is the principal filter of a singleton.*

PROOF. We have seen that an ultrafilter is either principal or free, and we know from Proposition XVI.5.3, that each principal filter of a countable set is sequential. Suppose that $\mathcal{E}$ is a free sequential filter, that is, by Proposition XVI.5.3, $\mathcal{E} = (E)_0$, where E is an infinite countable set. Let $E_0 \cup E_1 = E$ be a partition of E into two infinite sets. Of course, $E_0 \cup E_1 \in \mathcal{E}$ but $E_0 \notin \mathcal{E}$ and $E_1 \notin \mathcal{E}$, because E_0 and E_1 are not cofinite in E. By Proposition V.4.1, $\mathcal{E}$ is not an ultrafilter. $\square$

PROPOSITION V.4.5. *If $f : X \to Y$ and $\mathcal{U}$ is an ultrafilter, then $f[\mathcal{U}]$ is an ultrafilter-base.*

[11]Because such sets can be partitioned into infinitely many disjoint infinite sets.

PROOF. Let $\mathcal{U}$ be an ultrafilter, and $H \in f[\mathcal{U}]^{\#}$, equivalently, $f^{-}(H)\#\mathcal{U}$ and thus by Proposition V.4.1, $f^{-}(H) \in \mathcal{U}$, hence $f(f^{-}(H)) \in f[\mathcal{U}]$ and $f(f^{-}(H)) \subset H$, hence $H \in f[\mathcal{U}]^{\uparrow}$ is an ultrafilter by Proposition V.4.1. □

The following useful property is sometimes called the *compactness property of filters* ([12]).

PROPOSITION V.4.6 (Compactness of filters). *Let $U_{\mathcal{U}} \in \mathcal{U}$ for every $\mathcal{U} \in \beta\mathcal{F}$. Then there exists a finite subcollection $\boldsymbol{G}$ of $\beta\mathcal{F}$ such that $\bigcup_{\mathcal{U}\in\boldsymbol{G}} U_{\mathcal{U}} \in \mathcal{F}$.*

PROOF. If by way of contradiction, $\bigcup_{\mathcal{U}\in\boldsymbol{G}} U_{\mathcal{U}} \notin \mathcal{F}$ for every finite subcollection $\boldsymbol{G}$ of $\beta\mathcal{F}$, hence $\mathcal{B} = \{\bigcap_{\mathcal{U}\in\boldsymbol{G}} U_{\mathcal{U}}^{c} : \boldsymbol{G} \subset_{\text{fin}} \beta\mathcal{F}\}$ is a filter-base that meshes with $\mathcal{F}$. If $\mathcal{W}$ is an ultrafilter finer than $\mathcal{B} \vee \mathcal{F}$, then $U_{\mathcal{W}} \in \mathcal{W}$ and $U_{\mathcal{W}}^{c} \in \mathcal{B} \subset \mathcal{W}$, yielding a contradiction. □

Let us illustrate applicability of Proposition V.4.6 in the case of countably carried filters.

PROPOSITION V.4.7. *If every $\mathcal{U} \in \beta\mathcal{F}$ is countably carried, then $\mathcal{F}$ is countably carried.*

PROOF. By definition, for each $\mathcal{U} \in \beta\mathcal{F}$, there is a countable set $U_{\mathcal{U}} \in \mathcal{U}$, hence, by Proposition V.4.6, there is $\boldsymbol{F} \subset_{\text{fin}} \beta\mathcal{F}$ such that $\bigcup_{\mathcal{U}\in\boldsymbol{F}} U_{\mathcal{U}} \in \mathcal{F}$. □

PROPOSITION V.4.8. *The character of each free ultrafilter is uncountable.*

PROOF. By way of contradiction let $\mathcal{B} := \{B_n : n < \omega\}$ be a strictly decreasing filter-base of an ultrafilter $\mathcal{U}$. If we pick $x_n \in B_n \setminus B_{n+1}$, then $\{x_n : n < \omega\}$ consists of distinct elements, and intersects B_n for each $n < \omega$, hence $\{x_n : n < \omega\} \in \mathcal{U}^{\#} = \mathcal{U}$, because $\mathcal{U}$ is an ultrafilter. As

$$\{x_n : n < \omega\} = \{x_{2n} : n < \omega\} \cup \{x_{2n+1} : n < \omega\},$$

either $\{x_{2n} : n < \omega\} \in \mathcal{U}$ or $\{x_{2n+1} : n < \omega\} \in \mathcal{U}$. In the first case, $x_{2m+1} \in B_m \setminus \{x_{2n} : n < \omega\}$ for each $m < \omega$, so that $\mathcal{B}$ is not a filter-base of $\mathcal{U}$. In the second case, $x_{2m} \in B_m \setminus \{x_{2n+1} : n < \omega\}$ for each $m < \omega$, and thus $\mathcal{B}$ is not a filter-base of $\mathcal{U}$. □

By the way, Proposition V.4.4 is a consequence of the proposition above.

---

[12]This terminology will be later justified in terms of the Stone topology of the space of ultrafilters.

V.5. Various classes of filters

We have already considered several classes of proper filters in Section II.2. Let us recall that $\mathbb{F}$ stands for the class of all filters, and $\mathbb{F}X$ denotes the set of all filters on a set X ([13]). More generally, we adopt the convention that if $\mathbb{A}$ is a subclass of $\mathbb{F}$, then $\mathbb{A}X$ stands for the set of all filters on a set X that are in $\mathbb{A}$. Of course, $\mathbb{A} \subset \mathbb{B}$ means that $\mathbb{A}X \subset \mathbb{B}X$ for each set X.

The class of *principal filters* $\mathbb{F}_0$ coincides with the filters of character equal to 1 (Definition II.2.4), that of *countably based* filters $\mathbb{F}_1$ with the filters of countable character. The class of *free filters* is denoted by $\mathbb{F}_*$, that of *ultrafilters* by $\mathbb{U}$, and of free ultrafilters $\mathbb{U}_*$.

Of course,

$$\mathbb{F}_0 \subset \mathbb{F}_1 \subset \mathbb{F}.$$

Notice that the subclasses $\mathbb{F}_0$ and $\mathbb{F}_1$ of $\mathbb{F}$ were singled out in terms of the least cardinality of filter-bases. Because of Proposition V.4.8, $\mathbb{U} \cap \mathbb{F}_1 \subset \mathbb{F}_0$, in other words, countably based ultrafilters are principal.

Let us add to the triad above the class of countably complete filters.

DEFINITION V.5.1. A filter $\mathcal{F}$ is called *countably complete* or *countably deep* if $\bigcap \mathcal{D} \in \mathcal{F}$ for each countable family $\mathcal{D} \subset \mathcal{F}$.

The class of countably complete filters is denoted by $\mathbb{F}_{\wedge 1}$.

PROPOSITION V.5.2. *Each countably complete filter on a countable set is principal.*

PROOF. Let $\mathcal{F} \in \mathbb{F}_{\wedge 1}X$, where X is countable, say $X = \{x_n : n < \omega\}$. I claim that $\ker \mathcal{F} \neq \varnothing$. Otherwise, for each $n < \omega$, there is $F_n \in \mathcal{F}$ such that $x_n \notin F_n$, so that $\{x_n : n < \omega\} \cap \bigcap_{k<\omega} F_k = \varnothing$, a contradiction.

Let us show that $\ker \mathcal{F} \in \mathcal{F}$. Otherwise, $\mathcal{F}^* = \mathcal{F} \setminus \ker \mathcal{F}$ is a proper free filter, hence there is a strictly descending sequence $(F_n)_n$ of elements of $\mathcal{F}$ such that $F_n \setminus \ker \mathcal{F} \in \mathcal{F}^*$.

Now, $\bigcap_{n<\omega} F_n \in \mathcal{F}$, because $\mathcal{F} \in \mathbb{F}_{\wedge 1}$, hence $\bigcap_{n<\omega}(F_n \setminus \ker \mathcal{F}) = \bigcap_{n<\omega} F_n \setminus \ker \mathcal{F} \in \mathcal{F}^*$ by definition, so that $\mathcal{F}^* \in \mathbb{F}_{\wedge 1}$. Therefore, by the first part of the proof, $\ker \mathcal{F}^* \neq \varnothing$, which is a contradiction, because by definition $\mathcal{F}^*$ is free. We have proved that $\ker \mathcal{F} \in \mathcal{F}$, that is, $\mathcal{F} \in \mathbb{F}_0$. □

The cocountable filter from Example II.2.24 is an instance of a countably complete filter.

---

[13]From the point of view of set theory, the class of all filters does not constitute a set, but all the filters on a given set is a set (see Chapter XXVII).

PROPOSITION V.5.3. *The cocountable filter on an uncountable set is countably complete.*

PROOF. Let X be an uncountable set, and let $F_n \in (X)_1$ for each $n < \omega$. Accordingly, $X \setminus F_n$ is countable for each $n < \omega$, hence $X \setminus \bigcap_{n<\omega} F_n = \bigcup_{n<\omega}(X \setminus F_n)$ is countable as a countable union of countable sets, and thus $\bigcap_{n<\omega} F_n \in (X)_1$. □

Of course, $\mathbb{F}_0 \subset \mathbb{F}_{\wedge 1} \subset \mathbb{F}$. In fact,

PROPOSITION V.5.4. *Each countably complete, countably based filter is principal:*

$$\mathbb{F}_1 \cap \mathbb{F}_{\wedge 1} \subset \mathbb{F}_0.$$

PROOF. Indeed, if $(F_n)_n$ is a filter-base of a countably complete filter $\mathcal{F}$, then $\bigcap_{n<\omega} F_n \in \mathcal{F}$. On the other hand, for each $F \in \mathcal{F}$, there exists $n < \omega$ such that $F_n \subset F$, so that $\bigcap_{n<\omega} F_n = \ker \mathcal{F}$, hence $\mathcal{F}$ is principal. □

REMARK V.5.5. It turns out that the existence of free countably complete ultrafilters cannot be assured without additional set-theoretic assumptions. The problem gave rise to deep investigations ([14]). A similar concept is discussed in XIX.5.

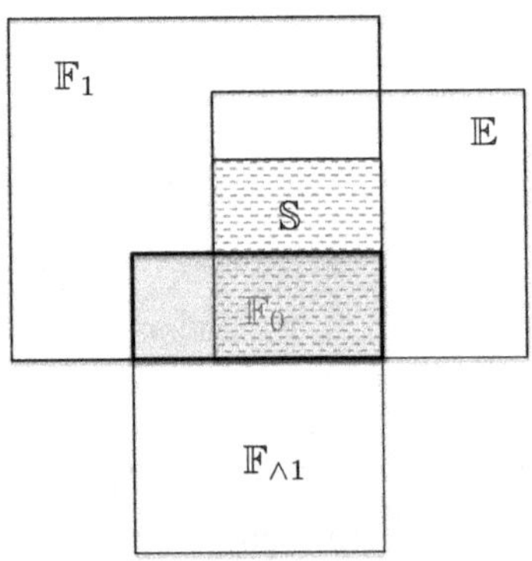

FIGURE V.5.1. Filter classes: principal filters $\mathbb{F}_0 = \mathbb{F}_1 \cap \mathbb{F}_{\wedge 1}$ (the junction of countably based filters $\mathbb{F}_1$ and countably complete filters $\mathbb{F}_{\wedge 1}$); sequential filters $\mathbb{S} \subset \mathbb{F}_1 \cap \mathbb{E}$ are countably based $\mathbb{F}_1$ and countably carried $\mathbb{E}$, but a countably carried, countably based filter need not be sequential.

[14]See T. Jech [**136**, p. 56].

Another classification of filters is founded on the cardinality of its elements.

DEFINITION V.5.6. A filter is called *uniform* if all its elements have the same cardinality.

In other words, a filter $\mathcal{F}$ on a set X is uniform if and only if $\operatorname{card}(F) = \operatorname{card}(X)$ for every $F \in \mathcal{F}$. For example, the cofinite filter $(X)_0$ is uniform for each infinite set X, and the cocountable filter $(X)_1$ is uniform for each uncountable set X ([15]).

PROPOSITION V.5.7 (*). *For each infinite set X, each uniform filter $\mathcal{F}$ on X, there exists a uniform ultrafilter $\mathcal{U} \in \beta\mathcal{F}$.*

PROOF. Let $\mathcal{F}$ be a uniform filter on X. Suppose that none of its ultrafilters is uniform, that is, for each $\mathcal{U} \in \beta\mathcal{F}$ there exists $U_{\mathcal{U}} \in \mathcal{U}$ such that $\operatorname{card} U_{\mathcal{U}} < \operatorname{card} X$. By Proposition V.4.6, there is a finite subset $\boldsymbol{G}$ of $\beta\mathcal{F}$ such that $\bigcup_{\mathcal{U}\in\boldsymbol{G}} U_{\mathcal{U}} \in \mathcal{F}$. By Corollary XXVII.6.5, $\operatorname{card} \bigcup_{\mathcal{U}\in\boldsymbol{G}} U_{\mathcal{U}} < \operatorname{card} X$, a contradiction. □

Of course, the principal filter $\{X\}$ on X is uniform. Hence, by Proposition V.5.7, on each infinite set there exist uniform ultrafilters.

REMARK V.5.8. A uniform ultrafilter on an uncountable set cannot be finer than any sequential filter. It follows that in the standard convergence of the real line $\nu_{\mathbb{R}}$, there exist ultrafilters that converge to 0 that are not finer than any sequential filter. On the other hand, each ultrafilter on a countable set is finer than a sequential filter ([16]).

DEFINITION V.5.9 (Countably tight filter). A filter $\mathcal{F}$ is called *countably tight* if for every $H \in \mathcal{F}^{\#}$ there exists a countable subset A of H such that $A \in \mathcal{F}^{\#}$. The class of countably tight filters is denoted by $\mathbb{F}_{\#1}$.

REMARK V.5.10. As we have seen in Proposition V.2.2, $\mathcal{F} \subset 2^X$ is a filter if and only if $\mathcal{F}_c = \{X \setminus F : F \in \mathcal{F}\}$ is an ideal. Let us adopt a convention:

$$\boldsymbol{F}_{\neg} := \{\mathcal{F}_c : \mathcal{F} \in \boldsymbol{F}\}.$$

Accordingly, $\boldsymbol{F}$ is a set of filters, then $\boldsymbol{F}_{\neg}$ is the set of the complementary ideals, and conversely. We extend this notation to classes of filters and ideals, that is, if $\mathbb{H}$ is a class of filters, then $\mathbb{H}_{\neg}$ is the class of complementary ideals.

[15]This follows from Corollary XXVII.6.5.

[16]An ultrafilter on a countable set is either principal, hence sequential, or free; in the latter case it is finer than the cofinite filter of that countable set, which is sequential.

Consequently, $\mathbb{F}_\neg$ is the class of all ideals, $(\mathbb{F}_1)_\neg$ is the class of countably based ideals (called *σ-ideals*), $(\mathbb{F}_0)_\neg$ is the class of principal ideals, and so on.

As we know from Section II.3, the set of proper filters on a given set is closed under infima with respect to the order given by inclusion (on isotone families). On the other hand, a supremum of a set of filters $\boldsymbol{F}$ exists whenever $\bigcup_{\mathcal{F}\in\boldsymbol{F}}\mathcal{F}$ is a filter-subbase (see Definition II.2.3).

PROPOSITION V.5.11. *Every infimum of principal filters is principal.*

PROOF. Let $\boldsymbol{F} \subset \mathbb{F}_0 X$. Hence, for each $\mathcal{F} \in \boldsymbol{F}$, there is $F_\mathcal{F} \neq \varnothing$ such that $\mathcal{F} = F_\mathcal{F}^\uparrow$. Then $A \in \bigwedge \boldsymbol{F}$ if and only if $A \supset \bigcup_{\mathcal{F}\in\boldsymbol{F}} F_\mathcal{F}$, that is, $\bigwedge \boldsymbol{F} = (\bigcup_{\mathcal{F}\in\boldsymbol{F}} F_\mathcal{F})^\uparrow$. □

On the other hand, by Proposition II.3.5, each filter is a supremum of principal filters, that is, for every filter $\mathcal{F}$,

$$\mathcal{F} = \bigvee_{F\in\mathcal{F}}\{F\}^\uparrow.$$

It follows, in particular, that the supremum of a set of countably based filters need not be countably based (because principal filters are countably based). However,

PROPOSITION V.5.12. *Each proper countable supremum of countably based filters is countably based.*

PROOF. Indeed, if $\{F_{n,k} : k < \omega\}$ is a filter-base of a filter $\mathcal{F}_n$ for $n < \omega$, then the family of finite intersections of

$$\{F_{n,k} : n, k < \omega\}$$

is a countable filter-base (of $\bigvee_{n<\omega}\mathcal{F}_n$). □

An infimum of countably many countably based filters need not be countably based.

By Definition II.3.8, a filter is called a Fréchet filter if it is an infimum of sequential filters.

PROPOSITION V.5.13. *A filter is Fréchet if and only if it is an infimum of countably based filters.*

PROOF. As each sequential filter is countably based, every Fréchet filter fulfills the condition. Conversely, a filter is an infimum of countably based filters, then it is Fréchet, because each countably based filter is an infimum of sequential filters by Proposition II.3.7. □

Therefore, if we set

$$\mathbb{F}_1(\mathcal{F}) := \{\mathcal{G} \in \mathbb{F}_1 : \mathcal{G} \geq \mathcal{F}\},$$

then a filter $\mathcal{F}$ is Fréchet if and only if

$$\mathcal{F} = \bigcap\nolimits_{\mathcal{G}\in\mathbb{F}_1(\mathcal{F})} \mathcal{G}. \tag{V.5.1}$$

V.6. Supplement

EXERCISE V.6.1. Let $\mathcal{A}_j$ be an isotone family of subsets of X for each $j \in J$. Show that if $f : X \longrightarrow Y$, then

$$f[\bigcap\nolimits_{j\in J} \mathcal{A}_j] = \bigcap\nolimits_{j\in J} f[\mathcal{A}_j].$$

SOLUTION. Indeed, $B \in f[\bigcap_{j\in J} \mathcal{A}_j]$ whenever for each $j \in J$, there is $A_j \in \mathcal{A}_J$ such that $B \supset f(\bigcup_{j\in J} A_j) = \bigcup_{j\in J} f(A_j) \in \bigcap_{j\in J} f[\mathcal{A}_j]$, and conversely, $B \in \bigcap_{j\in J} f[\mathcal{A}_j]$ provided that for each $j \in J$, there is A_j such that $B \supset \bigcup_{j\in J} f(A_j)$.

EXERCISE V.6.2. Show that if a family $\mathcal{A}$ is centered, then $\mathcal{A}\#\mathcal{A}$, but not conversely.

SOLUTION. If $\mathcal{A}$ is centered, then, in particular, $A_0\#A_1$ for each $A_0, A_1 \in \mathcal{A}$. To see that the converse does not hold, let $X := \{0,1,2\}$, and let $\mathcal{A} := \{\{0,1\},\{1,2\},\{2,0\},\{0,1,2\}\}$. Indeed, $\{0,1\} \cap \{1,2\} \cap \{2,0\} = \varnothing$.

EXERCISE V.6.3. Check (V.3.9)

$$(\bigcup \boldsymbol{A})^{\#} = \bigcap\nolimits_{\mathcal{A}\in\boldsymbol{A}} \mathcal{A}^{\#}, \quad (\bigcap \boldsymbol{A})^{\#} = \bigcup\nolimits_{\mathcal{A}\in\boldsymbol{A}} \mathcal{A}^{\#}.$$

SOLUTION. Indeed, $H\#\bigcup_{\mathcal{A}\in\boldsymbol{A}} \mathcal{A}$ if and only if $H\#A$ for each $\mathcal{A} \in \boldsymbol{A}$ and each $A \in \mathcal{A}$, that is, $H \in \mathcal{A}^{\#}$ for each $\mathcal{A} \in \boldsymbol{A}$. As well, $H \in (\bigcap \boldsymbol{A})^{\#}$, equivalently $H^c \notin \bigcap \boldsymbol{A}$, that is, there exists $\mathcal{A} \in \boldsymbol{A}$ such that $H^c \notin \mathcal{A}$, equivalently $H \in \mathcal{A}^{\#} \subset \bigcup_{\mathcal{A}\in\boldsymbol{A}} \mathcal{A}^{\#}$.

LEMMA V.6.4. *Let $f : X \longrightarrow Y$, let $\mathcal{F}$ be a filter on X and $\mathcal{G}$ a filter on Y such that $f[\mathcal{F}] \leq \mathcal{G}$. Then*

$$f[\mathcal{F} \vee f^-[\mathcal{G}]] \approx \mathcal{G}.$$

PROOF. On one hand, $f[\mathcal{F} \vee f^-[\mathcal{G}]] \geq f[f^-[\mathcal{G}]] \geq \mathcal{G}$. On the other, if $\mathcal{U} \in \beta\mathcal{G}$, then by assumption, $\mathcal{F}\#f^-[\mathcal{U}]$. If $\mathcal{W}_{\mathcal{U}} \in \beta(\mathcal{F} \vee f^-[\mathcal{U}])$, then $f[\mathcal{W}_{\mathcal{U}}]\#\mathcal{U}$, hence $f[\mathcal{W}_{\mathcal{U}}] = \mathcal{U}$, because $f[\mathcal{W}_{\mathcal{U}}]$ is an ultrafilter. Therefore,

$$f[\mathcal{F} \vee f^-[\mathcal{G}]] \leq f[\bigcap\nolimits_{\mathcal{U}\in\beta\mathcal{G}} \mathcal{W}_{\mathcal{U}}] = \bigcap\nolimits_{\mathcal{U}\in\beta\mathcal{G}} f[\mathcal{W}_{\mathcal{U}}] = \mathcal{G},$$

the first equality resulting from Exercise V.6.1 and the second from (V.4.1). □

We are now in a position to provide a proof of Proposition IV.3.8:

PROOF. By (IV.3.2), $y \in \lim_{f\xi} \mathcal{G}$ is tantamount to the existence of a filter $\mathcal{F}$ such that $f^-(y)\# \lim_\xi \mathcal{F}$ and $f[\mathcal{F}] \leq \mathcal{G}$. By setting $\mathcal{F}_0 := \mathcal{F} \vee f^-[\mathcal{G}]$, we ensure that $f^-(y)\# \lim_\xi \mathcal{F}_0$ and $f[\mathcal{F}_0] = \mathcal{G}$ by Lemma V.6.4. □

PROPOSITION V.6.5 (*). *If $\mathcal{U}$ is an ultrafilter on an infinite set X, then* $\operatorname{card}(\mathcal{U}) = 2^{\operatorname{card} X}$.

PROOF. If $\mathcal{U} \in \beta X$ and X is infinite, then for each $A \in 2^X \setminus \{\varnothing, X\}$, either $A \in \mathcal{U}$ or $A^c := X \setminus A \in \mathcal{U}$. Thus $\mathcal{U} \sqcup \mathcal{U}_c = 2^X$, and by (XXVII.6.1), $\operatorname{card}(\mathcal{U}) = \operatorname{card}(\mathcal{U}_c) = 2^{\operatorname{card} X}$ ([17]). □

On the other hand, it follows from Corollary XXVII.4.9 that there are as many ultrafilters on X as there are collections of subsets of X.

We know by Lemma II.2.8 that a filter $\mathcal{F}$ on an infinite set X is free if and only if $\mathcal{F} \geq (X)_0$. Let us say that a filter $\mathcal{F}$ on an uncountable set X is *hyperfree* whenever $\mathcal{F} \geq (X)_1$, where $(X)_1$ is the cocountable filter of X from Example II.2.24. For each filter on an uncountable set X, there is a hyperfree part $\mathcal{F}^\sharp := \mathcal{F}^* \vee (X)_1$. However, the remaining free part $\bigwedge(\beta\mathcal{F}^* \setminus \beta\mathcal{F}^\sharp)$ generally meshes $\mathcal{F}^\sharp$. Therefore, the decomposition of a filter into a hyperfree part and the other part, like in Theorem II.4.1 is not possible.

A free filter has no hyperfree ultrafilters if and only if it contains a countable set.

EXERCISE V.6.6. Check that $\mathcal{A}_j$ is an isotone family for each $j \in J$, then

$$\{\bigcup_{j\in J} A_j : A_j \in \mathcal{A}_j, j \in J\}$$

is a base of $\bigcap_{j\in J} \mathcal{A}_j$.

Saturation. A set $\boldsymbol{F}$ of filters is called *saturated* if for each $\mathcal{U} \in \beta(\bigwedge \boldsymbol{F})$, there is $\mathcal{F} \in \boldsymbol{F}$ such that $\mathcal{F} \leq \mathcal{U}$. We say that a set $\boldsymbol{G}$ of filters is *saturation stable* if $\bigwedge \boldsymbol{F} \in \boldsymbol{G}$ for each saturated subset $\boldsymbol{F}$ of $\boldsymbol{G}$.

PROPOSITION V.6.7. *The class $\mathbb{S}$ of sequential filters is saturation stable.*

[17]The proof communicated by A. STAROSOLSKI.

PROOF. Let $\mathcal{F}$ be a free filter such that for each $\mathcal{U} \in \beta\mathcal{F}$, there exists a free sequential filter $\mathcal{E}_{\mathcal{U}} \leq \mathcal{U}$. Then there is a finite set $\boldsymbol{U} \subset \beta\mathcal{F}$ such that

$$\bigwedge\nolimits_{\mathcal{U}\in \boldsymbol{U}} \mathcal{E}_{\mathcal{U}} \leq \mathcal{F}.$$

Indeed, let $\mathcal{E}_{\mathcal{U}}$ be a cofinite filter of $E_{\mathcal{U}}$. Hence $E_{\mathcal{U}} \leq \mathcal{U}$ for each $\mathcal{U} \in \beta\mathcal{F}$, thus by Proposition V.4.6, there is a finite subset $\boldsymbol{U}$ of $\beta\mathcal{F}$ such that $E = \bigcup_{\mathcal{U}\in\boldsymbol{U}} E_{\mathcal{U}} \in \mathcal{F}$, that is, $(E)_0 \leq \mathcal{F}$. But $(E)_0 = (\bigcup_{\mathcal{U}\in\boldsymbol{U}} E_{\mathcal{U}})_0 = \bigwedge_{\mathcal{U}\in\boldsymbol{U}}(E_{\mathcal{U}})_0 = \bigwedge_{\mathcal{U}\in\boldsymbol{U}} \mathcal{E}_{\mathcal{U}}$. □

***Criteria of classification.** We have singled out various classes of filters that we have considered in diverse circumstances of convergence theory. They can be classified following several reasoned criteria. Classification of filters will be instrumental in classification of convergences.

Let $\mathcal{F}$ be a filter.

(1) The *character* of $\mathcal{F}$ is the least cardinality of a filter-base of $\mathcal{F}$, that is,

$$\chi(\mathcal{F}) := \min\{\operatorname{card}\mathcal{B} : \mathcal{B} \subset \mathcal{F}, \mathcal{B} \geq \mathcal{F}\}.$$

The class $\mathbb{F}_\gamma$ consists of all filters such that $\chi(\mathcal{F}) < \aleph_\gamma$. In particular,
$\mathbb{F}_0$ is the class of filters $\mathcal{F}$ with $\chi(\mathcal{F}) < \aleph_0$, that is, of filters of *finite character*, hence of character 1, called also *principal* filters.
$\mathbb{F}_1$ is the class of filters $\mathcal{F}$ with $\chi(\mathcal{F}) < \aleph_1$, that is, of filters of *countable character*, also called *countably based*.

(2) The *depth* of $\mathcal{F}$ is the supremum of cardinalities of $\mathcal{G} \subset \mathcal{F}$, the intersection of which is in $\mathcal{F}$, that is,

$$\varphi(\mathcal{F}) := \sup\{\kappa : \mathcal{G} \subset \mathcal{F}, \operatorname{card}\mathcal{G} < \kappa \Rightarrow \bigcap \mathcal{G} \in \mathcal{F}\}.$$

We note that $\varphi(\mathcal{F}) = \aleph_0$ for every filter $\mathcal{F}$, because if $\operatorname{card}\mathcal{G} < \aleph_0$, then $\mathcal{G}$ is finite. The class $\mathbb{F}_{\wedge 1}$ consists of those filters $\mathcal{F}$, for which $\bigcap \mathcal{G} \in \mathcal{F}$ if $\mathcal{G}$ is a countable subfamily of $\mathcal{F}$. Such filters are called *σ-complete*.
If the defining supremum does not exist, in other words, if $\bigcap \mathcal{G} \in \mathcal{F}$ for each $\mathcal{G} \subset \mathcal{F}$, then $\ker \mathcal{F} = \bigcap \mathcal{F} \in \mathcal{F}$, then $\mathcal{F} \in \mathbb{F}_0$, that is, the filter is principal.

(3) The robustness of $\mathcal{F}$ is the least cardinality elements of $\mathcal{F}$, that is,

$$\|\mathcal{F}\| := \min\{\operatorname{card} F : F \in \mathcal{F}\}.$$

Of course, $1 \leq \|\mathcal{F}\|$ for every $\mathcal{F} \in \mathbb{F}$, and $\aleph_0 \leq \|\mathcal{F}\|$ for each free filter $\mathcal{F}$. If $\mathcal{F} \in \mathbb{F}X$ and $\|\mathcal{F}\| = \operatorname{card} X$, then $\mathcal{F}$ is called *uniform*.

(4) The *tightness* is the least cardinal such that

$$\tau(\mathcal{F}) := \min\{\kappa : \operatorname{card} H < \kappa \Rightarrow H \notin \mathcal{F}^{\#}\}.$$

In particular, a filter $\mathcal{F}$ is called tight if $\tau(\mathcal{F}) = \aleph_0$.

There are some easy interrelations between the listed numbers. For example,

PROPOSITION V.6.8. *If the character or the weight does not exceed the depth, then the filter is principal, that is,*

$$\chi(\mathcal{F}) \leq \varphi(\mathcal{F}) \Longrightarrow \mathcal{F} \in \mathbb{F}_0,$$
$$\|\mathcal{F}\| \leq \varphi(\mathcal{F}) \Longrightarrow \mathcal{F} \in \mathbb{F}_0.$$

On the other hand,

PROPOSITION V.6.9. *The character is preserved under finite suprema, and the weight is subadditive, more precisely, for each filters $\mathcal{F}_0$ and $\mathcal{F}_1$,*

$$\chi(\mathcal{F}_0 \vee \mathcal{F}_1) = \chi(\mathcal{F}_0) \vee \chi(\mathcal{F}_1),$$
$$\|\mathcal{F}_0 \wedge \mathcal{F}_1\| \leq \|\mathcal{F}_0\| + \|\mathcal{F}_1\|.$$

On the other hand,

PROPOSITION V.6.10. *The depth is preserved under infima, more precisely,*

$$\varphi(\bigwedge \boldsymbol{F}) \geq \min\{\varphi(\mathcal{F}) : \mathcal{F} \in \boldsymbol{F}\}.$$

PROOF. If $\lambda \leq \min\{\varphi(\mathcal{F}) : \mathcal{F} \in \boldsymbol{F}\}$, if $\mathcal{G} \subset \bigwedge \boldsymbol{F}$ and $\operatorname{card} \mathcal{G} \leq \lambda$, then $\bigcap \mathcal{G} \in \mathcal{F}$ for each $\mathcal{F} \in \boldsymbol{F}$, hence $\bigcap \mathcal{G} \subset \bigwedge \boldsymbol{F}$. Thus $\lambda \leq \varphi(\bigwedge \boldsymbol{F})$. □

PROPOSITION V.6.11. *The character is preserved under composition, that is,*

$$\chi(\mathcal{H}) \vee \chi(\mathcal{D}) \geq \chi(\mathcal{H}[\mathcal{D}]).$$

PROOF. If $\mathcal{H}_0$ is a filter-base of $\mathcal{H}$ of cardinality λ and $\mathcal{D}_0$ is a filter-base of $\mathcal{D}$ of cardinality κ, then $\{HD : H \in \mathcal{H}_0, D \in \mathcal{D}_0\}$ is a filter-base of $\mathcal{H}[\mathcal{D}]$, and its cardinality is equal to $\lambda \cdot \kappa = \max(\lambda, \kappa)$ by (XXVII.6.1). The inequality follows. □

CHAPTER VI

Pretopologies

A convergence is a *pretopology* if for each point (of the underlying set), there is a coarsest filter converging to that point. Pretopologies embrace topologies, which has been so far the most known and studied class of convergences. The class of pretopologies, however, is structurally more versatile than that of topologies.

Pretopologies have already been considered in the first half of twentieth century, under various names, by RIESZ, SIERPIŃSKI, ČECH, HAUSDORFF, but not systematically explored. CHOQUET investigated them in relation with pseudotopologies, a central class of convergences.

VI.1. Basic facts about pretopologies

Let ξ be a convergence. Recall that the infimum of all filters ξ-converging to x

$$\mathcal{V}_{\xi}(x) := \bigwedge \xi^{-}(x), \tag{VI.1.1}$$

is called the *vicinity filter of* x for ξ ([1]). We call each element V of $\mathcal{V}_{\xi}(x)$, a *vicinity* of ξ at x.

In general, a vicinity filter $\mathcal{V}_{\xi}(x)$ does not converge to x for ξ. For instance, the vicinity filter $\mathcal{V}_{\sigma_{\mathbb{R}}}(r)$ for the standard sequential convergence $\sigma_{\mathbb{R}}$ on $\mathbb{R}$ from Example III.1.3 does not converge to r.

In Definition III.1.10, a convergence ξ is called a *pretopology* if for every $x \in |\xi|$, there is a coarsest filter among the filters ξ-convergent to x. This filter is, of course, the vicinity filter $\mathcal{V}_{\xi}(x)$. Therefore,

PROPOSITION VI.1.1. *A convergence ξ is a pretopology if and only if for each $x \in |\xi|$,*

$$x \in \lim_{\xi} \mathcal{V}_{\xi}(x). \tag{VI.1.2}$$

Of course,

$$\zeta \geq \xi \Longrightarrow \mathcal{V}_{\zeta}(x) \geq \mathcal{V}_{\xi}(x). \tag{VI.1.3}$$

[1]Of course, $\bigwedge \xi^{-}(x) = \bigcap \{\mathcal{F} : x \in \lim_{\xi} \mathcal{F}\}$.

If moreover, ζ and ξ are pretopologies, then the converse holds as well in (VI.1.3), by virtue of Propositions VI.1.1. Indeed, if $\mathcal{V}_\zeta(x) \geq \mathcal{V}_\xi(x)$ and $x \in \lim_\zeta \mathcal{F}$, then $\mathcal{F} \geq \mathcal{V}_\zeta(x) \geq \mathcal{V}_\xi(x)$, hence $x \in \lim_\xi \mathcal{F}$, because ξ is a pretopology, that is, $\zeta \geq \xi$.

We have already come across numerous pretopologies (discrete and chaotic convergences, the standard convergence of the real line, the cofinite and prime cocountable convergences). We have also encountered non-pretopological convergences, like the standard sequential convergence of the real line.

PROPOSITION VI.1.2. *A convergence ξ is a pretopology if and only if*

$$\bigcap\nolimits_{\mathcal{G}\in\boldsymbol{G}} \lim_\xi \mathcal{G} \subset \lim_\xi\left(\bigwedge\nolimits_{\mathcal{G}\in\boldsymbol{G}} \mathcal{G}\right) \tag{VI.1.4}$$

for every set $\boldsymbol{G}$ of filters.

PROOF. If ξ is a pretopology, and $x \in \lim_\xi \mathcal{G}$ for each $\mathcal{G} \in \boldsymbol{G}$, equivalently, $\mathcal{V}_\xi(x) \leq \mathcal{G}$ for each $\mathcal{G} \in \boldsymbol{G}$, then $\mathcal{V}_\xi(x) \leq \bigwedge \boldsymbol{G}$, hence $x \in \lim_\xi(\bigwedge \boldsymbol{G})$. Conversely, assume (VI.1.4) and recall (VI.1.1). □

It turns out that pretopologies are determined by their adherences. The *adherence* with respect to a convergence ξ of a subset A of $|\xi|$ is defined by

$$\operatorname{adh}_\xi A := \bigcup\nolimits_{\mathcal{F}\#A} \lim_\xi \mathcal{F}. \tag{VI.1.5}$$

REMARK VI.1.3. Notice that

$$x \in \operatorname{adh}_\xi A$$

if and only if there exists a filter $\mathcal{H}$ such that $x \in \lim_\xi \mathcal{H}$ and $A \in \mathcal{H}$. Indeed, by definition, $x \in \lim_\xi A$ if and only if there exists a filter $\mathcal{F}$ such that $A \in \mathcal{F}^\#$ and $x \in \lim_\xi \mathcal{F}$. Hence $\mathcal{H} := \mathcal{F} \vee A$ is a proper filter, $A \in \mathcal{H}$ and $x \in \lim_\xi \mathcal{H}$.

PROPOSITION VI.1.4. *For each convergence ξ, every $A \subset |\xi|$, and each $x \in |\xi|$,*

$$x \in \operatorname{adh}_\xi A \Longleftrightarrow A \in \mathcal{V}_\xi(x)^\#.$$

PROOF. By definition, $x \in \operatorname{adh}_\xi A$ if and only if there exists a filter $\mathcal{F}$ such that $x \in \lim_\xi \mathcal{F}$ and $A \in \mathcal{F}^\# \subset \mathcal{V}_\xi(x)^\#$. Conversely, if $x \notin \operatorname{adh}_\xi A$, then $A \notin \mathcal{F}^\#$ for every filter $\mathcal{F}$ such that $x \in \lim_\xi \mathcal{F}$. By (V.3.9), $A \notin \mathcal{V}_\xi(x)^\#$. □

PROPOSITION VI.1.5. *A convergence ξ is a pretopology if and only if for each filter $\mathcal{F}$,*

$$\lim_\xi \mathcal{F} = \bigcap_{H \in \mathcal{F}^\#} \operatorname{adh}_\xi H. \tag{VI.1.6}$$

PROOF. If $x \in \lim_\xi \mathcal{F}$, then, by (VI.1.5), $x \in \operatorname{adh}_\xi A$ for each $A \in \mathcal{F}^\#$, that is, the inclusion $\subset$ holds for every convergence ξ.

If ξ is a pretopology, and $x \in \operatorname{adh}_\xi A$ for each $A \in \mathcal{F}^\#$, equivalently $A \in \mathcal{V}_\xi(x)^\#$ for each $A \in \mathcal{F}^\#$, that is, $\mathcal{F}^\# \subset \mathcal{V}_\xi(x)^\#$, equivalently, $\mathcal{V}_\xi(x) \leq \mathcal{F}$, hence $x \in \lim_\xi \mathcal{F}$. □

EXAMPLE VI.1.6 (Missing-a-family pretopology). Let X be an infinite set, $x_\infty \in X$, and let $\mathcal{A} \subset 2^X$. Let η be a prime convergence such that $x_\infty \in \lim_\eta \mathcal{F}$ whenever $\ker \mathcal{F} \subset \{x_\infty\}$ and $\mathcal{F} \perp \mathcal{A}$. Then η is a pretopology.

Indeed, if $\mathcal{F}$ is a filter and $x_\infty \notin \lim_\eta \mathcal{F}$, then, by the definition of η, there is $A \in \mathcal{A}$ such that $A \in \mathcal{F}^\#$. Therefore, $x_\infty \notin \operatorname{adh}_\eta A$, because if $\mathcal{G}$ a filter such that $A \in \mathcal{G}^\#$, then $x_\infty \notin \lim_\eta \mathcal{G}$ by the definition of η. Thus by (VI.1.6), η is a pretopology.

Here is an example of non-pretopological convergence.

EXAMPLE VI.1.7 (Missing-an-ultrafilter convergence). Let X be an infinite set, $\mathcal{U}$ a free ultrafilter on X, and let $x_\infty \in X$. Define a prime convergence ξ on X, for which $x_\infty \in \lim_\xi \mathcal{F}$ whenever $\mathcal{U} \notin \beta\mathcal{F}$ and $\ker \mathcal{F} \subset \{x_\infty\}$. Observe that $x_\infty \in \operatorname{adh}_\xi H$ for each $H \in \mathcal{U}$, because $x_\infty \in \lim_\xi \mathcal{W}$ for each ultrafilter $\mathcal{W}$ on H such that $\mathcal{W} \neq \mathcal{U}$. Therefore, by Proposition VI.1.5, ξ is not a pretopology.

Propositions VI.1.4 and VI.1.5 indicate that pretopologies are entirely determined by their adherences. Therefore, we shall study in detail the operation of adherence, defined in (VI.1.5) for arbitrary convergences, and will infer its particular role in the case of pretopologies.

EXAMPLE VI.1.8 (Bisequence pretopology). Consider $X := \{x_\infty\} \sqcup \{x_n : n < \omega\} \sqcup \bigsqcup_{n<\omega} \{x_{n,k} : k < \omega\}$ endowed with the finest convergence θ, in which $x_\infty \in \lim_\theta (x_n)_n$ and $x_n \in \lim_\theta (x_{n,k})_k$ for each $n < \omega$.

More precisely, let $X_\infty := \{x_n : n < \omega\}$ and let $X_n := \{x_{n,k} : k < \omega\}$ for each $n < \omega$. Then the vicinity filters $\mathcal{V}_\theta(x_\infty) = (X_\infty)_0 \wedge x_\infty^\uparrow$ and $\mathcal{V}_\theta(x_n) = (X_n)_0 \wedge x_n^\uparrow$ are the respective cofinite filters (centered at x_∞ and x_n) for each $n < \omega$, while all other points are isolated. This is the *bisequence pretopology*.

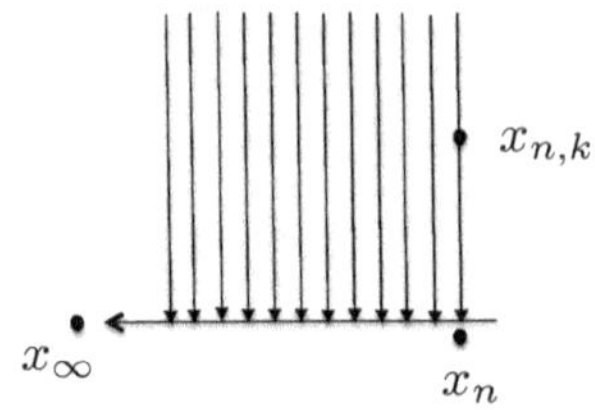

FIGURE VI.1.1. A *bisequence* $x_\infty \longleftarrow_n x_n \longleftarrow_k x_{n,k}$ defines the *bisequence pretopology.*

Then

$$x_\infty \in \mathrm{adh}_\theta(\mathrm{adh}_\theta\{x_{n,k} : n, k < \omega\}) \setminus \mathrm{adh}_\theta\{x_{n,k} : n, k < \omega\}.$$

Here is another example:

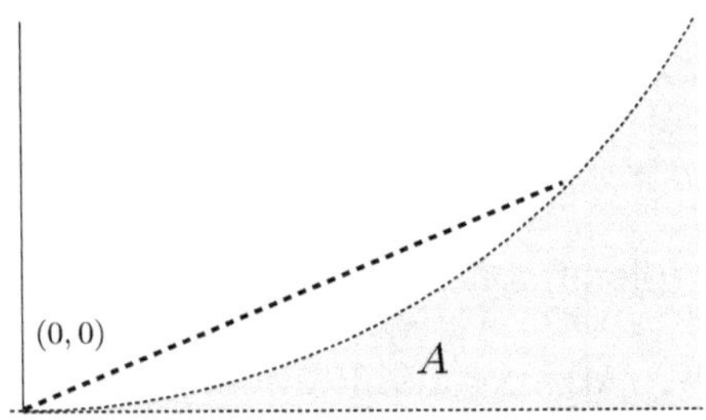

FIGURE VI.1.2. The point $(0,0)$ is not in $\mathrm{adh}_\rho A$, because the distance between $(0,0)$ and A along each straight line is strictly positive.

EXAMPLE VI.1.9 (Radial pretopology). The *radial pretopology* ρ is defined on the plane $\mathbb{R}^2$ as follows: $x \in \mathrm{adh}_\rho A$ if there exists a vector h and a strictly decreasing real sequence $(t_n)_n$ converging to 0 such that $x + t_n h \in A$. Let $A := \{(r,s) \in \mathbb{R}^2 : 0 < s \leq r^2, 0 < r\}$. Then $(0,0) \in \mathrm{adh}_\rho(\mathrm{adh}_\rho A) \setminus \mathrm{adh}_\rho A$.

VI.2. Adherence

We have just defined in (VI.1.5) the adherence of a convergence. Observe that if ζ and ξ are convergences (on the same set), then

$$\zeta \geq \xi \Longrightarrow \mathrm{adh}_\zeta A \subset \mathrm{adh}_\xi A$$

for each A. If moreover, ζ and ξ are pretopologies, then the converse holds as well.

PROPOSITION VI.2.1. *If ζ and ξ are pretopologies on X, and* $\operatorname{adh}_\zeta A \subset \operatorname{adh}_\xi A$ *for each $A \subset X$, then $\zeta \geq \xi$.*

PROOF. If $\operatorname{adh}_\zeta A \subset \operatorname{adh}_\xi A$ for each A, and $x \in \lim_\zeta \mathcal{F}$, then $x \in \operatorname{adh}_\zeta A \subset \operatorname{adh}_\xi A$ for each $A \# \mathcal{F}$, hence, by Proposition VI.1.5, $x \in \lim_\xi \mathcal{F}$, that is, $\zeta \geq \xi$. □

EXAMPLE VI.2.2. The adherence for the discrete pretopology ι on X fulfills

$$\operatorname{adh}_\iota A = A,$$

since the only filter that converges to x is the principal filter $x^\uparrow$ of x.

EXAMPLE VI.2.3. The adherence for the chaotic pretopology o on X fulfills

$$\operatorname{adh}_o A = \begin{cases} X & \text{if } A \neq \varnothing, \\ \varnothing & \text{if } A = \varnothing, \end{cases}$$

because every proper filter converges to each point.

Given a convergence ξ on a set X, the adherence is a map $\operatorname{adh}_\xi : 2^X \longrightarrow 2^X$.

PROPOSITION VI.2.4. *Let ξ be a convergence, and let A, A_0, A_1 and B be subsets of $|\xi|$. Then*

(a) $$\operatorname{adh}_\xi \varnothing = \varnothing,$$

(b) $$A \subset B \Longrightarrow \operatorname{adh}_\xi A \subset \operatorname{adh}_\xi B,$$

(c) $$A \subset \operatorname{adh}_\xi A,$$

(d) $$\operatorname{adh}_\xi (A_0 \cup A_1) = \operatorname{adh}_\xi A_0 \cup \operatorname{adh}_\xi A_1.$$

PROOF. (a) Because $\varnothing \notin \mathcal{F}^\#$ for every filter $\mathcal{F}$. (b) Because if $A \in \mathcal{F}^\#$ and $A \subset B$, then $B \in \mathcal{F}^\#$. (c) Because $A \in x^\uparrow = (x^\uparrow)^\#$ for each $x \in A$. (d) Because if $A_0 \cup A_1 \in \mathcal{F}^\#$ then either $A_0 \in \mathcal{F}^\#$ or $A_1 \in \mathcal{F}^\#$; the inverse inclusion follows from (b). □

Of course, (b) and (d) extend, by a simple induction, to

$$\operatorname{adh}_\xi (\bigcup \mathcal{A}) = \bigcup_{A \in \mathcal{A}} \operatorname{adh}_\xi A \tag{VI.2.1}$$

for each finite family $\mathcal{A}$. The equality (VI.2.1) also holds in a more general context.

DEFINITION VI.2.5. A family $\mathcal{A}$ of subsets of a convergence space is called *locally finite* if for each convergent filter $\mathcal{F}$ there exists $F \in \mathcal{F}$ such that $\{A \in \mathcal{A} : A \# F\}$ is finite; $\mathcal{A}$ is called *discrete* if for each convergent filter $\mathcal{F}$ there exists $F \in \mathcal{F}$ such that $\operatorname{card}\{A \in \mathcal{A} : A \# F\} \leq 1$. A subset C is called *discrete* if the family $\{\{x\} : x \in C\}$ is discrete.

It is straightforward that if ξ is a pretopology on X, then a family $\mathcal{A}$ is locally finite if and only if for every $x \in X$, there exists a vicinity $V \in \mathcal{V}_\xi(x)$ that intersects at most finitely many $A \in \mathcal{A}$. A family $\mathcal{A}$ is discrete if and only if for every $x \in X$, there exists a vicinity $V \in \mathcal{V}_\xi(x)$ that intersects at most a single $A \in \mathcal{A}$.

The elements of a discrete family $\mathcal{A}$ are pairwise disjoint, because if $x \in A \in \mathcal{A}$, then $x \in \lim x^\uparrow$ and $\{x\} \cap A \neq \varnothing$. Of course, each discrete family is locally finite. Any finite family of overlapping sets ([2]) is locally finite, but not discrete.

PROPOSITION VI.2.6. *If $\mathcal{A}$ is a locally finite family of subsets of $|\xi|$, then (VI.2.1) holds.*

PROOF. By Proposition VI.2.4 (b), the inclusion $\supset$ holds for an arbitrary family $\mathcal{A}$. If $x \in \operatorname{adh}_\xi(\bigcup_{A\in\mathcal{A}} A)$, then there exists a filter $\mathcal{F}$ on $\bigcup_{A\in\mathcal{A}} A$ such that $x \in \lim_\xi \mathcal{F}$. As $\mathcal{A}$ is locally finite, there exist $F \in \mathcal{F}$ and a finite subfamily $\mathcal{A}_0$ of $\mathcal{A}$ such that $F \cap \bigcup_{A\in\mathcal{A}\setminus\mathcal{A}_0} A = \varnothing$. Therefore, $x \in \operatorname{adh}_\xi(\bigcup_{A\in\mathcal{A}_0} A) = \bigcup_{A\in\mathcal{A}_0} \operatorname{adh}_\xi A$ by Proposition VI.2.4 (d). □

COROLLARY VI.2.7. *If ξ is a free (alias T_1) convergence and C is a ξ-discrete subset, then* $\operatorname{adh}_\xi C = C$.

PROOF. Indeed, as C is ξ-discrete,

$$\operatorname{adh}_\xi C = \operatorname{adh}_\xi\Big(\bigcup\nolimits_{x\in C}\{x\}\Big) = \bigcup\nolimits_{x\in C} \operatorname{adh}_\xi\{x\} = \bigcup\nolimits_{x\in C}\{x\},$$

because ξ is free. □

REMARK VI.2.8. If C is a ξ-discrete subset, then $\xi|_C = \iota_C$ (the restriction of ξ to C is a discrete convergence). The converse does not hold. For example, for each prime non-discrete pretopology ξ on X with a pole x_∞, then $\operatorname{adh}_\xi(X \setminus \{x_\infty\}) = X$, hence $X \setminus \{x_\infty\}$ is not a ξ-discrete subset, but the restriction of ξ to $X \setminus \{x_\infty\}$ is a discrete convergence. Therefore, in order to avoid possible misunderstanding, one says that C is a ξ-closed discrete subspace, meaning that $\xi|_C = \iota_C$ and $\operatorname{adh}_\xi C = C$.

On the other hand, the inclusion

$$\operatorname{adh}_\xi(A_0 \cap A_1) \subset \operatorname{adh}_\xi A_0 \cap \operatorname{adh}_\xi A_1$$

cannot be reversed. A simple example is provided by $A_0 := [-1, 0[$ and $A_1 :=]0, 1]$ with the standard topology of the real line. Another instance is that of the sets of rational and irrational numbers, in which

[2]If $\{0,1,2\}$ is endowed with the discrete convergence, then the family $\{\{0,1\},\{1,2\}\}$ is not discrete.

case the intersection is empty, hence its adherence is empty, while the intersection of the adherences is the whole real line.

EXAMPLE VI.2.9. The two standard convergences of the real line $\sigma_{\mathbb{R}}$ and $\nu_{\mathbb{R}}$ from Examples III.1.3 and III.1.4 have the same adherence

$$\operatorname{adh}_{\sigma_{\mathbb{R}}} A = \operatorname{adh}_{\nu_{\mathbb{R}}} A$$

for each $A \subset \mathbb{R}$. The inclusion $\operatorname{adh}_{\sigma_{\mathbb{R}}} A \subset \operatorname{adh}_{\nu_{\mathbb{R}}} A$ holds, because $\sigma_{\mathbb{R}} \geq \nu_{\mathbb{R}}$. Conversely, if $x \in \operatorname{adh}_{\nu_{\mathbb{R}}} A$, that is, $A \in \mathcal{V}_{\xi}(x)^{\#}$, which means that for every $n \in \mathbb{N}_1$, there exists $x_n \in \left(x - \frac{1}{n}, x + \frac{1}{n}\right) \cap A$. The sequential filter $\mathcal{E}$ generated by $(x_n)_n$ converges to x in $\nu_{\mathbb{R}}$, hence in $\sigma_{\mathbb{R}}$. Therefore $x \in \operatorname{adh}_{\nu_{\mathbb{R}}} A$, because $A \in \mathcal{E}$.

VI.3. Inherence

The *inherence* $\operatorname{inh}_{\xi} A$ of a subset A of $|\xi|$ is defined by

$$x \in \operatorname{inh}_{\xi} A \Longleftrightarrow A \in \mathcal{V}_{\xi}(x). \tag{VI.3.1}$$

In other words, x belongs to the inherence of A if and only if A is a vicinity of x. It follows that, if ζ and ξ are convergences on a set X, then

$$\zeta \geq \xi \Longrightarrow \operatorname{inh}_{\zeta} A \supset \operatorname{inh}_{\xi} A.$$

In particular, $\operatorname{inh}_{\iota} A = A$, and $\operatorname{inh}_{o} X = X$ and $\operatorname{inh}_{o} A = \varnothing$ if $A \neq X$. Observe that

PROPOSITION VI.3.1. *For each $A \subset |\xi|$,*

$$\operatorname{inh}_{\xi} A = (\operatorname{adh}_{\xi} A^{c})^{c}. \tag{VI.3.2}$$

PROOF. Indeed, $x \in (\operatorname{adh}_{\xi} A^{c})^{c}$, that is, $x \notin \operatorname{adh}_{\xi} A^{c}$. By Proposition VI.1.4, this is equivalent to $A^{c} \notin \mathcal{V}_{\xi}(x)^{\#}$, which is $A \in \mathcal{V}_{\xi}(x)$ in virtue of (V.3.4). Finally, this amounts to $x \in \operatorname{inh}_{\xi} A$ by (VI.3.1). □

The following properties are respectively equivalent to those listed in Proposition VI.2.4 with the aid of (VI.3.2):

$$\begin{gathered}
\operatorname{inh}_{\xi} X = X, \\
A \subset B \Longrightarrow \operatorname{inh}_{\xi} A \subset \operatorname{inh}_{\xi} B, \\
\operatorname{inh}_{\xi} A \subset A, \\
\operatorname{inh}_{\xi}(A_0 \cap A_1) = \operatorname{inh}_{\xi} A_0 \cap \operatorname{inh}_{\xi} A_1,
\end{gathered}$$

for every A, A_0, A_1 and B.

VI.4. Continuity

Because continuity generalizes order, it is not surprising that continuous maps between pretopologies can be characterized in terms of images and preimages of vicinity filters and of adherences.

Therefore, if we stay within the class of pretopologies, we can argue entirely in terms of adherences.

PROPOSITION VI.4.1. *Let ξ and τ be convergences on X and Y, respectively, and let $f : X \to Y$. The following statements are equivalent* ([3]):

$$f[\mathcal{V}_\xi(x)] \geq \mathcal{V}_\tau(f(x)), \tag{1}$$

$$\mathcal{V}_\xi(x) \geq f^-[\mathcal{V}_\tau(f(x))], \tag{2}$$

$$f(\mathrm{adh}_\xi A) \subset \mathrm{adh}_\tau f(A), \tag{3}$$

$$\mathrm{adh}_\xi f^-(B) \subset f^-(\mathrm{adh}_\tau B). \tag{4}$$

Moreover,

$$f \in \mathscr{C}(\xi, \tau) \tag{5}$$

implies any of them, and if τ is a pretopology, each of them implies (5).

PROOF. By applying f^- to (1), we get (2), because by Lemma IV.1.1, $\mathcal{G} \geq f^-[f[\mathcal{G}]]$ for any filter $\mathcal{G}$.

(2) $\Rightarrow$ (3). If $y \in f(\mathrm{adh}_\xi A)$, then there exists $x \in f^-(y) \cap \mathrm{adh}_\xi A$, and thus $A \in \mathcal{V}_\xi(x)^\#$ by Proposition VI.1.4. By (2), $A \# f^-[\mathcal{V}_\tau(f(x))]$ because of (V.3.3), equivalently, $f(A) \# \mathcal{V}_\tau(f(x))$, that is, $f(x) \in \mathrm{adh}_\tau f(A)$ by Proposition VI.1.4.

By setting $A = f^-(B)$ in (3), we get $f(\mathrm{adh}_\xi f^-(B)) \subset \mathrm{adh}_\tau f(f^-(B)) \subset \mathrm{adh}_\tau B$. By applying f^- to the last inclusion, we get

$$\mathrm{adh}_\xi f^-(B) \subset f^-(f(\mathrm{adh}_\xi f^-(B))) \subset f^-(\mathrm{adh}_\tau B),$$

that is, (4).

(4) $\Rightarrow$ (1). Suppose that (1) does not hold, that is, there exists $H \in \mathcal{V}_\tau(f(x)) \setminus f[\mathcal{V}_\xi(x)]$. Equivalently, $B = H^c \notin \mathcal{V}_\tau(f(x))^\#$ and $B = H^c \in f[\mathcal{V}_\xi(x)]^\#$. In other words, $f(x) \notin \mathrm{adh}_\tau B$ and $x \in \mathrm{adh}_\xi f^-(B)$, that is, $x \in \mathrm{adh}_\xi f^-(B) \setminus f^-(\mathrm{adh}_\tau B)$, which contradicts (4).

Now, suppose (5), that is, if $\mathcal{F} \in \xi^-(x)$ then $f[\mathcal{F}] \in \tau^-(f(x))$. Hence

$$f[\mathcal{V}_\xi(x)] = f[\bigwedge\nolimits_{\mathcal{F}\in\xi^-(x)} \mathcal{F}] = \bigwedge\nolimits_{\mathcal{F}\in\xi^-(x)} f[\mathcal{F}] \geq \mathcal{V}_\tau(f(x)).$$

[3] For each $A \subset X, B \subset Y, x \in X$.

Finally, if $x \in \lim_\xi \mathcal{F}$ then $\mathcal{F} \geq \mathcal{V}_\xi(x)$, hence by (1), $f[\mathcal{F}] \geq \mathcal{V}_\tau(f(x))$, which means that $f(x) \in \lim_\tau f[\mathcal{F}]$, because τ is a pretopology. □

Let us now write explicit formulas for adherences corresponding to the final and initial convergences. Let ξ and τ be arbitrary convergences, and let $f : |\xi| \longrightarrow |\tau|$.

PROPOSITION VI.4.2. *The adherence of the initial convergence with respect to* τ *by a map* f, *is given by*

$$\operatorname{adh}_{f^-\tau} H = f^-(\operatorname{adh}_\tau f(H)). \tag{VI.4.1}$$

PROOF. By definition, $x \in \operatorname{adh}_{f^-\tau} H$ whenever there is a filter $\mathcal{F}$ such that $H \in \mathcal{F}^\#$ and $x \in \lim_{f^-\tau} \mathcal{F}$, equivalently $f(x) \in \lim_\tau f[\mathcal{F}]$ and $f(H) \in f[\mathcal{F}]^\#$, thus $f(x) \in \operatorname{adh}_\tau f(H)$, that is, $x \in f^-(\operatorname{adh}_\tau f(H))$.

On the other hand, if $f(x) \in \operatorname{adh}_\tau f(H)$ then there is a filter $\mathcal{G}$ such that $f(x) \in \lim_\tau \mathcal{G}$ and $f(H) \in \mathcal{G}^\#$, which amounts to $H \# f^-[\mathcal{G}]$, by (V.3.8). As $\mathcal{G} \leq f[f^-[\mathcal{G}]]$,

$$x \in f^-(\lim_\tau \mathcal{G}) \subset f^-\left(\lim_\tau f[f^-[\mathcal{G}]]\right) = \lim_{f^-\tau} f^-[\mathcal{G}],$$

thus $x \in \operatorname{adh}_{f^-\tau} H$. □

PROPOSITION VI.4.3. *The adherence of the final convergence of* ξ *by* f *is given by*

$$\operatorname{adh}_{f\xi} G = G \cup f(\operatorname{adh}_\xi f^-(G)). \tag{VI.4.2}$$

PROOF. By definition, $y \in f(\operatorname{adh}_\xi f^-(G))$ if and only if there exists a filter $\mathcal{F}$ such that $f^-(G) \in \mathcal{F}^\#$ and $y \in f(\lim_\xi \mathcal{F}) \subset \lim_{f\xi} f[\mathcal{F}]$, because $f \in \mathscr{C}(\xi, f\xi)$. Hence $G \in f[\mathcal{F}]^\#$ and thus $y \in \operatorname{adh}_{f\xi} G$.

Conversely, $y \in \operatorname{adh}_{f\xi} G$ means that there is a filter $\mathcal{D}$ such that $y \in \lim_{f\xi} \mathcal{D}$ and $G \in \mathcal{D}^\#$. If $y \notin f(X)$ then $\mathcal{D} = y^\uparrow$ and $y \in G$. Otherwise, we can assume that $f(X) \in \mathcal{D}$ so that there is a filter $\mathcal{F}$ on X such that $\mathcal{D} = f[\mathcal{F}]^\uparrow$ and $y \in \lim_{f\xi} f[\mathcal{F}] = f(\lim_\xi f^-[f[\mathcal{F}]]) \subset f(\lim_\xi \mathcal{F})$. Of course, $f^-(G) \in \mathcal{F}^\#$ by (V.3.8), hence $y \in f(\operatorname{adh}_\xi f^-(G))$. □

Of course, we could have used the two propositions above to establish the equivalences between continuity and two last conditions in Proposition VI.4.1.

In the special case when f is surjective (VI.4.2) becomes

$$\operatorname{adh}_{f\xi} G = f(\operatorname{adh}_\xi f^-(G)). \tag{VI.4.3}$$

One can expect that each proposition formulated in terms of adherences can be translated, via (VI.3.2) to an equivalent proposition in terms of inherences. This expectation is almost, but not entirely, fulfilled. For example, Proposition VI.4.1 can be completed by

PROPOSITION VI.4.4. *Let ξ and τ be convergences on X and Y, respectively, and let $f : X \to Y$. If $f \in \mathscr{C}(\xi, \tau)$, then for every $D \subset Y$,*

$$f^{-}(\operatorname{inh}_{\tau} D) \subset \operatorname{inh}_{\xi} f^{-}(D). \tag{VI.4.4}$$

If, moreover τ is a pretopology, then the converse also holds.

PROOF. Indeed, by (4), $\operatorname{adh}_{\xi} f^{-}(B) \subset f^{-}(\operatorname{adh}_{\tau} B)$. By applying the complementation, we get by (VI.3.2) and (IV.1.3),

$$\operatorname{inh}_{\xi} f^{-}(B^{c}) \supset f^{-}((\operatorname{adh}_{\tau} B)^{c}) = f^{-}(\operatorname{inh}_{\tau} B^{c}),$$

and thus with $D = B^{c}$, we get (VI.4.4). By applying the complementation to (VI.4.4), we get by (IV.1.3),

$$f^{-}(\operatorname{adh}_{\tau} D^{c}) \supset \operatorname{adh}_{\xi} f^{-}(D^{c}),$$

and we recover (4) for $B = D^{c}$.

The last statement follows directly from Proposition VI.4.1. □

REMARK VI.4.5. However, a counterpart of (3), that is, $f(\operatorname{inh}_{\xi} C) \supset \operatorname{inh}_{\tau} f(C)$ is not equivalent to (VI.4.4). The point is that, in general, $(f(A))^{c} \neq f(A^{c})$, for example if $f : \{0,1\} \longrightarrow \{0\}$ is constant and $A = \{0\}$.

VI.5. Structural aspects of pretopologies

PROPOSITION VI.5.1. *Each supremum of pretopologies is a pretopology.*

PROOF. Let Ξ be a non-empty set of pretopologies on X, and let $\boldsymbol{D}$ be a non-empty collection of filters on X. Then, by (III.3.3),

$$\begin{aligned}\lim_{\bigvee \Xi}(\bigwedge_{\mathcal{D}\in\boldsymbol{D}} \mathcal{D}) &= \bigcap_{\xi\in\Xi} \lim_{\xi}(\bigwedge_{\mathcal{D}\in\boldsymbol{D}} \mathcal{D}) = \bigcap_{\xi\in\Xi}\bigcap_{\mathcal{D}\in\boldsymbol{D}} \lim_{\xi} \mathcal{D} \\ &= \bigcap_{\mathcal{D}\in\boldsymbol{D}}\bigcap_{\xi\in\Xi} \lim_{\xi} \mathcal{D} = \bigcap_{\mathcal{D}\in\boldsymbol{D}} \lim_{\bigvee \Xi} \mathcal{D},\end{aligned}$$

hence, $\bigvee \Xi$ is a pretopology. □

The situation is completely different in case of infima.

PROPOSITION VI.5.2. *If $\{\pi_n : n < \omega\}$ is a sequence of pretopologies such that $\pi_n \geq \pi_{n+1}$ for each $n < \omega$, and there is x such that $\mathcal{V}_{\pi_n}(x) > \mathcal{V}_{\pi_{n+1}}(x)$ for each $n < \omega$, then $\pi = \bigwedge_{n<\omega} \pi_n$ is not a pretopology.*

PROOF. We observe that $\mathcal{V}_{\pi}(x) = \bigcap_{n<\omega} \mathcal{V}_{\pi_n}(x) < \mathcal{V}_{\pi_m}(x)$ for each $m < \omega$. If $x \in \lim_{\pi} \mathcal{V}_{\pi}(x)$, then, by definition of infimum, there exists $m < \omega$ such that $x \in \lim_{\pi_m} \mathcal{V}_{\pi}(x)$, which yields a contradiction. □

Proposition III.4.4, saying that each convergence is an infimum of prime convergences, can be refined to the effect that

Proposition VI.5.3. *Each finitely stable convergence is an infimum of prime pretopologies.*

Proof. Let ξ be a finitely stable convergence on a set X. For each $x \in X$ and each filter $\mathcal{F}$ such that $x \in \lim_\xi \mathcal{F}$ let $\xi_{(x,\mathcal{F})}$ be a pretopology on X, defined by

$$\mathcal{V}_{\xi_{(x,\mathcal{F})}}(x) := \mathcal{F} \wedge x^\uparrow,$$

and $\mathcal{V}_{\xi_{(x,\mathcal{F})}}(y) := y^\uparrow$ if $y \neq x$. Then $\xi_{(x,\mathcal{F})}$ is a prime pretopology, and

$$\xi = \bigwedge\{\xi_{(x,\mathcal{F})} : (\mathcal{F}, x) \in \xi\}.$$

□

Corollary VI.5.4. *Each finitely stable convergence is a final convergence with respect to pretopologies.*

Proof. Let ξ be a finitely stable convergence as in the proof of Proposition VI.5.3. Let $f_{(x,\mathcal{F})} : \left|\xi_{(x,\mathcal{F})}\right| \longrightarrow X$ is the identity, and thus $f_{(x,\mathcal{F})}(x) = x$ and $f_{(x,\mathcal{F})}[\mathcal{F}] = \mathcal{F}$. Accordingly, the final convergence with respect to

$$\{f_{(x,\mathcal{F})} : (\mathcal{F}, x) \in \xi\}$$

is homeomorphic to $\bigwedge\{\xi_{(x,\mathcal{F})} : (\mathcal{F}, x) \in \xi\}$.

Consider a pretopology obtained as a simple sum $\pi = \bigoplus_{(\mathcal{F},x)\in\xi} \xi_{(x,\mathcal{F})}$, and $f : |\pi| \longrightarrow |\xi|$, defined by $f(p) := f_{(x,\mathcal{F})}(p)$ if $p \in \left|\xi_{(x,\mathcal{F})}\right|$. By Proposition IV.8.3, $f\pi = \xi$. □

We can rephrase Corollary VI.5.4 in terms of final density from Definition IV.8.5 as follows:

Corollary VI.5.5. *The class of pretopologies is finally dense in the class of finitely stable convergences.*

Pretopologies with the order inherited from convergences form a complete lattice. More precisely, denote by $\mathbf{S}_0$ the class of pretopologies, and by $\mathbf{S}_0 X$, the set of all pretopologies on a set X.

Proposition VI.5.6. *The set of pretopologies on a given set is a complete lattice.*

Proof. Proposition VI.5.1 ensures that each non-empty set of pretopologies has a supremum. Let Π be a non-empty set of pretopologies on X. Consider $\Xi := \{\xi \in \mathbf{S}_0 X : \forall_{\pi\in\Pi}\ \xi \leq \pi\}$. By Proposition VI.5.1, $\bigvee \Xi$ exists and $\bigvee \Xi \in \Xi$. Accordingly, $\bigwedge_{\mathbf{S}_0} \Pi = \bigvee \Xi$. □

PROPOSITION VI.5.7. *If τ is a pretopology, then $f^{-}\tau$ is also a pretopology.*

PROOF. With a view to apply Proposition VI.1.5, let

$$x \in \bigcap_{A \in \mathcal{F}^{\#}} \operatorname{adh}_{f^{-}\tau} A,$$

and let $H \# f[\mathcal{F}]$. Therefore, $f^{-}(H) \# \mathcal{F}$, so that $x \in \operatorname{adh}_{f^{-}\tau} f^{-}(H)$, that is, $f(x) \in \operatorname{adh}_{\tau} f(f^{-}(H)) \subset \operatorname{adh}_{\tau} H$. As τ is a pretopology, $f(x) \in \lim_{\tau} f[\mathcal{F}]$ by (VI.1.6), hence by (IV.3.1) $x \in \lim_{f^{-}\tau} \mathcal{F}$. By Proposition VI.1.5, $f^{-}\tau$ is a pretopology. □

It follows from Propositions VI.5.1 and VI.5.7 that

COROLLARY VI.5.8. *Each product of pretopologies is a pretopology, and each subconvergence of a pretopology is a pretopology.*

DEFINITION VI.5.9. A property of convergences is called *hereditary* (respectively, *productive*) if it is preserved under restrictions (respectively, under products).

For example, by Proposition III.1.16, the Hausdorff property is hereditary. Accordingly, Corollary VI.5.8 states that pretopologicity is productive and hereditary.

Let τ_j be a pretopology on X_j. By Definition IV.6.1, the product convergence $\tau := \prod_{j \in J} \tau_j$ is the coarsest convergence such that the projection $p_j \in \mathscr{C}(\tau, \tau_j)$ for each $j \in J$.

Let us describe vicinity filters of τ. If $\varphi \in \prod_{j \in J} X_j$, then it follows from the definition that

$$\mathcal{V}_{\tau}(\varphi) = \bigvee_{j \in J} p_j^{-}[\mathcal{V}_{\tau_j}(\varphi(j))].$$

Hence $V \in \mathcal{V}_{\tau}(\varphi)$ if and only if there exist $m < \omega$, and $j_0, j_1, \ldots, j_m \in J$ such that

$$\bigcap_{0 \leq k \leq m} p_{j_k}^{-}(V_{j_k}) \subset V.$$

In other words, $V \in \mathcal{V}_{\tau}(\varphi)$ if and only if there exists a finite subset J_0 of J and $V_j \in \mathcal{V}_{\tau_j}(\varphi(j))$ such that $V_j = X_j$ for each $j \notin J_0$ and

$$\prod_{j \in J} V_j \subset V.$$

Accordingly, a vicinity in the product is determined by vicinities of finitely many factors.

Let us illustrate the situation in a special case, where $X_j = X$ for each $j \in J$.

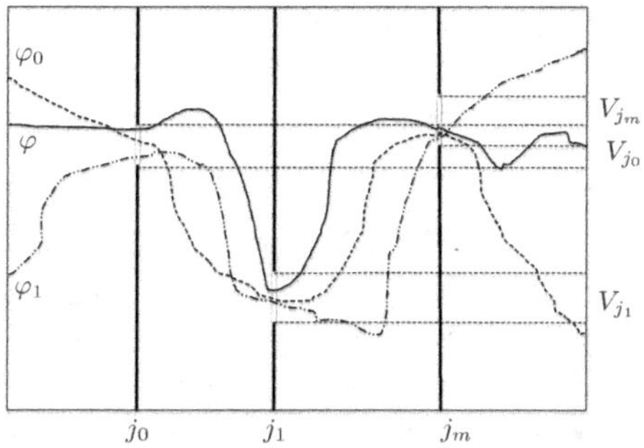

FIGURE VI.5.1. A typical vicinity V of φ consists of those φ_0, for which $\varphi_0(j) \in V_j \in \mathcal{V}_{\tau_j}(\varphi(j))$ for a finite number of j, say, $j_0, j_1, \ldots, j_m$.

We already know that each pretopology can be described also in terms of adherences, and in terms of inherences.

EXAMPLE VI.5.10. It is straightforward that

$$\mathrm{adh}_{\prod_{j\in J}\tau_j} \prod\nolimits_{j\in J} A_j = \prod\nolimits_{j\in J} \mathrm{adh}_{\tau_j} A_j,$$

Indeed, let $\tau = \prod_{j\in J}\tau_j$. Then $\varphi \in \mathrm{adh}_\tau \prod_{j\in J} A_j$ if and only if $\prod_{j\in J} A_j \# \mathcal{V}_\tau(\varphi)$. In particular, $A_j \# \mathcal{V}_j(\varphi(j))$ for each $j \in J$, hence $\varphi(j) \in \mathrm{adh}_{\tau_j} A_j$.

A formula for $\mathrm{adh}_{\prod_{j\in J}\tau_j} A$ for an arbitrary A is rather complex (see Propositions VI.8.5 and VI.8.6).

VI.6. Pretopologizer

PROPOSITION VI.6.1. *For each convergence θ, there exists the finest pretopology that is coarser than θ.*

PROOF. Let θ be a convergence on a set X. The set

$$A_\theta := \{\pi \in \mathbf{S}_0 X : \pi \leq \theta\} \tag{VI.6.1}$$

is not empty, because the chaotic convergence o_X is the coarsest pretopology on X, hence $o_X \in A_\theta$. By Proposition VI.5.1, $\bigvee A_\theta \in A_\theta$, hence

$$\mathrm{S}_0\theta := \bigvee A_\theta$$

is the finest pretopology coarser than θ ([4]). □

Accordingly, we have defined an operation S_0 on convergences, that we will call the *pretopologizer*.

[4]Hence, $\mathrm{S}_0\theta := \bigvee A_\theta = \bigvee\{\pi \in \mathbf{S}_0 X : \pi \leq \theta\}$.

PROPOSITION VI.6.2. *The pretopologizer is*

(increasing) $\zeta \leq \xi \Longrightarrow \mathrm{S}_0\zeta \leq \mathrm{S}_0\xi,$

(descending) $\mathrm{S}_0\xi \leq \xi,$

(idempotent) $\mathrm{S}_0(\mathrm{S}_0\xi) = \mathrm{S}_0\xi.$

PROOF. With the notation (VI.6.1), if $\zeta \leq \xi$ then $A_\zeta \subset A_\xi$, hence $\mathrm{S}_0\zeta = \bigvee A_\zeta \leq \bigvee A_\xi = \mathrm{S}_0\xi$. Now $\mathrm{S}_0\xi \leq \xi$, because $\mathrm{S}_0\xi \in A_\xi$. As $\mathrm{S}_0\xi$ is a pretopology, the supremum of pretopologies coarser than $\mathrm{S}_0\xi$ is equal $\mathrm{S}_0\xi$. □

It follows that

$$\bigvee_{\mathbf{S}_0}\Xi = \bigvee \Xi, \text{ and } \bigwedge_{\mathbf{S}_0}\Xi = \mathrm{S}_0(\bigwedge \Xi)$$

for each set $\Xi \subset \mathbf{S}_0 X$.

THEOREM VI.6.3. *For each convergence τ and every f,*

$$\mathrm{S}_0\left(f^-\tau\right) \geq f^-\left(\mathrm{S}_0\tau\right). \tag{VI.6.2}$$

PROOF. By applying f^- to $\tau \geq \mathrm{S}_0\tau$, we get $f^-\tau \geq f^-\left(\mathrm{S}_0\tau\right)$. By Proposition VI.5.7, $f^-\left(\mathrm{S}_0\tau\right)$ is a pretopology, so that $\mathrm{S}_0\left(f^-\tau\right) \geq \mathrm{S}_0\left(f^-\left(\mathrm{S}_0\tau\right)\right) = f^-(\mathrm{S}_0\tau)$, that is, (VI.6.2). □

PROPOSITION VI.6.4. *For each convergence ξ,*

$$f(\mathrm{S}_0\xi) \geq \mathrm{S}_0(f\xi). \tag{VI.6.3}$$

PROOF. By applying f to (VI.6.2), we get $f(\mathrm{S}_0(f^-\tau)) \geq f(f^-(\mathrm{S}_0\tau)) \geq \mathrm{S}_0\tau$, because of (IV.3.6). We set $\tau = f\xi$ to get

$$f(\mathrm{S}_0\xi) \geq f(\mathrm{S}_0(f^-f\xi)) \geq \mathrm{S}_0(f\xi),$$

because by (IV.3.7), $\xi \geq f^-f\xi$. □

As $f\xi \geq \tau$ implies $\mathrm{S}_0(f\xi) \geq \mathrm{S}_0\tau$, by (VI.6.3),

$$\mathscr{C}(\xi,\tau) \subset \mathscr{C}(\mathrm{S}_0\xi, \mathrm{S}_0\tau). \tag{VI.6.4}$$

Let us establish an explicit formula for the pretopologizer. As pretopologies have been characterized in several ways, there are several possible approaches to describe the pretopologizer.

PROPOSITION VI.6.5. *If ξ is a convergence, then $x \in \lim_{\mathrm{S}_0\xi}\mathcal{F}$ if and only if $\mathcal{F} \geq \mathcal{V}_\xi(x)$.*

PROOF. Suppose that ξ is a convergence on X. Define $x \in \lim_\eta \mathcal{F}$ whenever $\mathcal{F} \geq \mathcal{V}_\xi(x)$ for each $x \in X$. Obviously, the relation η is a convergence such that $\xi \geq \eta$. In fact, η is a pretopology, because $\mathcal{V}_\eta(x) = \bigcap_{\mathcal{F}\in\eta^-(x)} \mathcal{F} \geq \mathcal{V}_\xi(x)$, so that $x \in \lim_\eta \mathcal{V}_\eta(x)$. If θ is a pretopology such that $\theta \leq \xi$, then $\mathcal{V}_\theta(x) \leq \mathcal{V}_\xi(x) \leq \mathcal{V}_\eta(x)$, hence $\theta \leq \eta$ in virtue of Proposition VI.2.1. Accordingly, η is the finest among the pretopologies that are coarser than ξ, so that $\eta = \mathrm{S}_0\xi$ is the pretopologization of ξ. □

EXAMPLE VI.6.6 (Standard convergence). Recall Examples III.1.3 and III.1.4. We have observed in (III.1.4) that $\mathcal{V}_{\nu_{\mathbb{R}}}(x) = \mathcal{V}_{\sigma_{\mathbb{R}}}(x)$, and thus $\nu_{\mathbb{R}} = \mathrm{S}_0\sigma_{\mathbb{R}}$.

COROLLARY VI.6.7. *If ξ is a convergence, then for each filter $\mathcal{F}$,*

$$\lim\nolimits_{\mathrm{S}_0\xi} \mathcal{F} = \bigcap\nolimits_{A\in\mathcal{F}^{\#}} \mathrm{adh}_\xi\, A. \tag{VI.6.5}$$

PROOF. Indeed, $\mathcal{F} \geq \mathcal{V}_\xi(x)$ if and only if $\mathcal{F}^{\#} \subset \mathcal{V}_\xi(x)^{\#}$, that is, if each $A \in \mathcal{F}^{\#}$ implies that $x \in \mathrm{adh}_\xi\, A$, equivalently, $x \in \lim_{\mathrm{S}_0\xi} \mathcal{F}$. □

We conclude that vicinity filters, adherence and inherence do not depend on a convergence, but on its pretopologization, that is, for each x and each A,

$$\begin{aligned}\mathcal{V}_\xi(x) &= \mathcal{V}_{\mathrm{S}_0\xi}(x),\\ \mathrm{adh}_\xi\, A &= \mathrm{adh}_{\mathrm{S}_0\xi}\, A,\\ \mathrm{inh}_\xi\, A &= \mathrm{inh}_{\mathrm{S}_0\xi}\, A.\end{aligned}$$

EXAMPLE VI.6.8 (Missing-an-ultrafilter). Recall Example VI.1.7: X is an infinite set, $\mathcal{U}$ a free ultrafilter on X, and $x_\infty \in X$. A prime convergence ξ on X is defined by $x_\infty \in \lim_\xi \mathcal{F}$ whenever $\mathcal{U} \notin \beta\mathcal{F}$ and $\ker \mathcal{F} \subset \{x_\infty\}$.

We notice that $\mathcal{V}_\xi(x) = (X)_0 \wedge x_\infty^{\uparrow}$, where $(X)_0$ is the cofinite filter of X. Indeed, $(X)_0$ is the infimum of all free ultrafilters on X. On the other hand, the only free ultrafilter $\mathcal{U}$ that is not in $\xi^-(x_\infty)$ fulfills $\mathcal{U} \geq \bigwedge \xi^-(x_\infty)$. Otherwise there would be $U \in \mathcal{U}$ such that $X \setminus U \in \mathcal{W}$ for each free ultrafilter $\mathcal{W} \neq \mathcal{U}$, which is not possible, as U is infinite. Accordingly, the pretopologization $\mathrm{S}_0\xi$ of ξ is the prime cofinite pretopology $\pi[x_\infty, (X)_0]$.

PROPOSITION VI.6.9. *If a convergence ξ is free (alias T_1), then $\mathrm{S}_0\xi$ is free.*

PROOF. Because $\mathrm{adh}_{\mathrm{S}_0\xi}\{x\} = \mathrm{adh}_\xi\{x\} \subset \{x\}$. □

The inequality (VI.6.2) can be improved. In fact,

PROPOSITION VI.6.10. *For each f and τ,*

$$\mathrm{S}_0\left(f^-\tau\right) = f^-\left(\mathrm{S}_0\tau\right). \tag{VI.6.6}$$

PROOF. Because of (VI.6.2), it is enough to show that $f^{-}(\mathrm{S}_0\tau) \geq \mathrm{S}_0(f^{-}\tau)$. Let $x \in \lim_{f^{-}(\mathrm{S}_0\tau)} \mathcal{F}$, that is, $f(x) \in \lim_{\mathrm{S}_0\tau} f[\mathcal{F}]$. Equivalently, $f(x) \in \mathrm{adh}_\tau H$ for each $H \in f[\mathcal{F}]^{\#}$. If now $A \in \mathcal{F}^{\#}$, then $f(A) \in f[\mathcal{F}]^{\#}$, hence $f(x) \in \mathrm{adh}_\tau f(A)$, that is, $x \in \mathrm{adh}_{f^{-}\tau} A$ for each $A \in \mathcal{F}^{\#}$. Consequently, $x \in \lim_{\mathrm{S}_0(f^{-}\tau)} \mathcal{F}$. □

As we can check in Exercise VI.8.3, the inequality (VI.6.2) is equivalent to two other characterizations of the functorial character of S_0. The equality (VI.6.6) embraces an additional property, that is, heredity of the pretopologizer. In particular, Proposition VI.6.10 implies that

$$\mathrm{S}_0(\xi|_V) = (\mathrm{S}_0\xi)|_V, \tag{VI.6.7}$$

where $\xi|_V$ is the restriction of a convergence ξ to a non-empty subset V of $|\xi|$, and $(\mathrm{S}_0\xi)|_V$ is the restriction of $\mathrm{S}_0\xi$ to V.

The pretopologizer commutes neither with suprema, nor with products.

EXAMPLE VI.6.11 ($\mathrm{S}_0\xi \vee \tau \not\geq \mathrm{S}_0(\xi \vee \tau)$). Let ξ be a countable sequential fan of Example III.4.7, that is, $\{X_n : n < \omega\}$ is a disjoint family of infinite countable sets not containing ∞, and $X := \{\infty\} \cup \bigcup_{n<\omega} X_n$. By definition, ξ is a prime convergence on X such that $\infty \in \lim_\xi \mathcal{F}$ whenever there is $n < \omega$ such that $\{\infty\} \cup N \in \mathcal{F}$ for each cofinite subset N of X_n. Notice that $V \in \mathcal{V}_\xi(\infty)$ if and only if $\infty \in V$ and $X_n \setminus V$ is finite for each $n < \omega$.

Let τ be a prime convergence on the same underlying set such that $\infty \in \lim_\tau \mathcal{F}$ provided that $V_n := \{\infty\} \cup \bigcup_{k>n} X_k \in \mathcal{F}$ for each $n < \omega$. Thus $\{V_n : n < \omega\}$ is a filter-base of the vicinity filter of τ at ∞. Therefore, τ is a pretopology, hence $\mathrm{S}_0\tau = \tau$.

Accordingly, $\mathrm{S}_0\xi \vee \tau$ is a prime pretopology, for which W is a vicinity at ∞ if $\infty \in W$, and $X_n \setminus V$ is finite for all but finitely many n.

On the other hand, $\xi \vee \tau$ is discrete, because if $\mathcal{F}$ is a free filter and $\infty \in \lim_{\xi\vee\tau} \mathcal{F}$, then there is $n < \omega$ such that $\{\infty\} \cup \bigcup_{k\leq n} X_n \in \mathcal{F}$, but also $\{\infty\} \cup \bigcup_{k>n} X_n \in \mathcal{F}$, which yields a *contradiction*. We conclude that $\mathrm{S}_0(\xi \vee \tau) > \mathrm{S}_0\xi \vee \tau$.

The same convergences can be used to prove that the pretopologizer is not productive.

EXAMPLE VI.6.12 ($\mathrm{S}_0\xi \times \tau \not\geq \mathrm{S}_0(\xi \times \tau)$). Let ξ and τ be the convergences from Example VI.6.11. By Exercise IV.10.13 and by (VI.6.7),

$$(\mathrm{S}_0\xi \times \tau)|_{\Delta X} = \mathrm{S}_0\xi \vee \tau \not\geq \mathrm{S}_0(\xi \vee \tau) = \mathrm{S}_0((\xi \times \tau)|_{\Delta X}) = \mathrm{S}_0(\xi \times \tau)|_{\Delta X}.$$

As the inequality of two convergences does not hold on a subset ΔX, it does not hold on the whole $X \times X$, hence $\mathrm{S}_0\xi \times \tau \not\geq \mathrm{S}_0(\xi \times \tau)$.

As we have observed in VI.5.4, $f\xi$ need not be a pretopology when ξ is a pretopology. By Defintion IV.3.9, a surjective map $f \in \mathscr{C}(\xi, \tau)$ is called a *pretopologically quotient* whenever

$$\tau \geq \mathrm{S}_0(f\xi).$$

In other words, a pretopology τ is a pretopological quotient if is the finest pretopology, for which f is continuous from ξ.

EXAMPLE VI.6.13. An evident instance of a continuous map that is not pretopologically quotient is the identity i_X on a set X, on which a non-pretopological convergence ξ is defined. Then $i \in \mathscr{C}(\xi, \mathrm{S}_0\xi)$, but is not pretopologically quotient, because $\xi > \mathrm{S}_0\xi$.

VI.7. Initial density of the Bourdaud pretopology

We will consider here a special, but very important, pretopology on a three point set, which is called the *Bourdaud pretopology,* and is denoted by ¥.

EXAMPLE VI.7.1 (Bourdaud pretopology). Let ¥ be a pretopology defined on $\{0,1,2\}$ by convergence of ultrafilters as follows:

$$\lim\nolimits_{¥} 0^{\uparrow} = \{0,1\},\ \lim\nolimits_{¥} 1^{\uparrow} = \{0,1,2\} = \lim\nolimits_{¥} 2^{\uparrow}.$$

([5]) As for vicinities,

$$\mathcal{V}_{¥}(0) = \mathcal{V}_{¥}(1) = \{\{0,1,2\}\},\ \mathcal{V}_{¥}(2) = \{\{1,2\},\{0,1,2\}\}.$$

Accordingly, $\mathrm{adh}_{¥}\{0\} = \{0,1\}, \mathrm{adh}_{¥}\{1\} = \{0,1,2\} = \mathrm{adh}_{¥}\{2\}$, hence

$$\mathrm{adh}_{¥}(\mathrm{adh}_{¥}\{0\}) \setminus \mathrm{adh}_{¥}\{0\} \neq \varnothing.$$

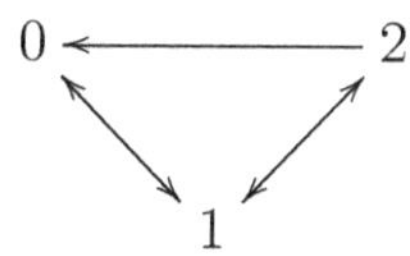

LEMMA VI.7.2. *If ξ is a pretopology on X, then $f \in \mathscr{C}(\xi, ¥)$ if and only if* $\mathrm{adh}_\xi f^-(0) \subset f^-(\{0,1\})$.

PROOF. Let us use one of the conditions of Proposition VI.4.1, say (4). The only significant case is when $B = \{0\}$, that is, $\mathrm{adh}_\xi f^-(0) \subset f^-(\mathrm{adh}_{¥}\{0\}) = f^-(\{0,1\})$. □

THEOREM VI.7.3. *Each pretopology is a supremum of initial convergences with respect to the Bourdaud pretopology.*

[5]Therefore, $\lim_{¥}\{\{0,1\}\}^{\uparrow} = \{0,1\} = \lim_{¥}\{\{0,2\}\}^{\uparrow}, \lim_{¥}\{\{1,2\}\}^{\uparrow} = \{0,1,2\}$, and $\lim_{¥}\{\{0,1,2\}\}^{\uparrow} = \{0,1\}$.

PROOF. For every $A \subset X$, define a map $f_A : X \longrightarrow \{0,1,2\}$ by

$$f_A(x) := \begin{cases} 0 & x \in A, \\ 1 & x \in \operatorname{adh}_\xi A \setminus A, \\ 2 & x \notin \operatorname{adh}_\xi A. \end{cases} \tag{VI.7.1}$$

Notice that f_A is continuous. Indeed, as $f_A^-(0) = A$ and $f_A^-(\{0,1\}) = \operatorname{adh}_\xi A$, the continuity follows from Lemma VI.7.2. Therefore, $\xi \geq (f_A)^-¥$, and thus $\xi \geq \bigvee_{A \subset X}(f_A)^-¥$.

To see that the equality holds, suppose that $x \notin \lim_\xi \mathcal{F}$, that is, by Proposition VI.1.5, there is $A_0 \subset X$ such that $A_0 \in \mathcal{F}^\#$ and $x \notin \operatorname{adh}_\xi A_0$. We shall show that $x \notin \operatorname{adh}_{f_{A_0}^- ¥} A_0$, equivalently by (VI.4.1), $f_{A_0}(x) \notin \operatorname{adh}_¥ f_{A_0}(A_0)$. Indeed, by definition of f_{A_0} above, $f_{A_0}(A_0) = \{0\}$, $f_{A_0}(x) = 2$, and $\operatorname{adh}_¥\{0\} = \{0,1\}$, that is, $f_{A_0}(x) \notin \operatorname{adh}_¥ f_{A_0}(A_0)$.

As $A_0 \in \mathcal{F}^\#$, also $f_{A_0}(A_0)\# f_{A_0}[\mathcal{F}]$, hence $f_{A_0}(x) \notin \lim_¥ f_{A_0}[\mathcal{F}]$ by (VI.1.6). Therefore,

$$x \notin \lim_{f_{A_0}^- ¥} \mathcal{F} \supset \lim_{\bigvee_{A\subset X} f_A^- ¥} \mathcal{F}.$$

□

Let us extend Definition IV.4.7:

DEFINITION VI.7.4 (Initial density). A convergence σ is called *initially dense* in a class $\mathbf{X}$ of convergences if each $\xi \in \mathbf{X}$ is σ-*initial*, that is, there exists a set $\mathscr{F} \subset \mathscr{C}(\xi, \sigma)$ such that $\xi = \bigvee_{f \in \mathscr{F}} f^- \sigma$.

A class $\mathbf{X}$ is said to be *simple* if there exists in $\mathbf{X}$ an initially dense convergence.

If σ is a convergence initially dense in $\mathbf{X}$, then $\sigma \in \mathbf{X}$, because $\sigma = i^-\sigma$, where i stands for the identity map.

COROLLARY VI.7.5. *The Bourdaud pretopology is initially dense in the class of pretopologies.*

COROLLARY VI.7.6. *Each pretopology is the initial convergence with respect to a product of Bourdaud pretopologies.*

PROOF. If ξ is a pretopology on X, then for every $A \subset X$, let f_A be given by (VI.7.1). Let Y_A be a copy of $\{0,1,2\}$ and $¥_A$ be the Bourdaud pretopology on Y_A. Define $F : X \longrightarrow \prod_{A\subset X} Y_A$ by

$$(p_A \circ F)(x) := f_A(x),$$

where p_A is the projection of $\prod_{A\subset X} Y_A$ onto Y_A for each $A \subset X$. Then, by Theorem VI.7.3,

$$F^-(\prod_{A\subset X} ¥_A) = \bigvee_{A\subset X} {f_A}^- ¥_A = \xi.$$

□

In view of Definition VI.7.4,

COROLLARY VI.7.7. *The class of pretopologies is simple.*

VI.8. Supplement

PROPOSITION VI.8.1. *If ρ is a metrizable pretopology, then $x \in \mathrm{adh}_\rho A$ if and only if there exists a sequential filter $\mathcal{E}$ such that $x \in \lim_\rho \mathcal{E}$ and $A \in \mathcal{E}$.*

PROOF. If the condition holds, then $x \in \mathrm{adh}_\rho A$, by definition of adherence. Conversely, if $x \in \mathrm{adh}_\rho A$ then there is a filter $\mathcal{F}$ such that $A \in \mathcal{F}^\#$ and $x \in \lim_\rho \mathcal{F}$. As each metrizable topology is a pretopology, this implies that $\mathcal{V}_\rho(x) \subset \mathcal{F}$ hence $\mathcal{F}^\# \subset \mathcal{V}_\rho(x)^\#$. Let d be one of the metrics compatible with ρ, and let $B_d(x, \varepsilon)$ be the ball defined by d, as in (III.6.5). As

$$\left\{B_d\left(x, \tfrac{1}{n}\right) : n \in \mathbb{N}_1\right\}$$

is a filter-base of $\mathcal{V}_\rho(x)$, for each $n \in \mathbb{N}_1$, there exists $x_n \in B_d\left(x, \frac{1}{n}\right) \cap A$. Consequently, $x \in \lim_\rho (x_n)_n$ and $A \in (x_n)_n$. □

PROPOSITION VI.8.2. *Each metrizable pretopology has idempotent adherence.*

PROOF. Suppose that π is metrizable (by a metric d) and let $x \in \mathrm{adh}_\pi(\mathrm{adh}_\pi A)$. By Proposition VI.8.1, there exists a sequence $(x_n)_n$ on $\mathrm{adh}_\pi A$ such that $x \in \lim_\pi (x_n)_n$. Hence, for each $n \in \mathbb{N}_1$, there exists a sequence $(x_{n,k})_k$ on A such that $x_n \in \lim_\pi (x_{n,k})_k$. If $p \in \mathbb{N}_1$, then, in particular, there exists n_p such that $d(x_{n_p}, x) < \frac{1}{p}$. If $r_p := \frac{1}{p} - d(x_n, x)$, then there exists k_p such that $d(x_{n_p,k_p}, x_{n_p}) < r_p$. Therefore

$$d(x_{n_p,k_p}, x) < d(x_{n_p,k_p}, x_{n_p}) + d(x_{n_p}, x) < \tfrac{1}{p}.$$

Accordingly, $x \in \lim_\pi \left(x_{n_p,k_p}\right)_p$ and $x_{n_p,k_p} \in A$ for each $p \in \mathbb{N}_1$, hence $x \in \mathrm{adh}_\pi A$. □

EXERCISE VI.8.3. Show that formulae (VI.6.2) , (VI.6.3), and (VI.6.4) are equivalent (for every variable that occurs):

$$\mathrm{S}_0(f^-\tau) \geq f^-(\mathrm{S}_0\tau), \tag{1}$$

$$f(\mathrm{S}_0\xi) \geq \mathrm{S}_0(f\xi), \tag{2}$$

$$\mathscr{C}(\xi, \tau) \subset \mathscr{C}(\mathrm{S}_0\xi, \mathrm{S}_0\tau). \tag{3}$$

SOLUTION. Assume (1) and let $\tau = f\xi$, so that $f^-(\mathrm{S}_0(f\xi)) \leq \mathrm{S}_0(f^- f\xi) \leq \mathrm{S}_0\xi$ by (IV.3.6). Applying f to the obtained inequality, we get $\mathrm{S}_0(f\xi) \leq ff^-(\mathrm{S}_0(f\xi)) \leq f(\mathrm{S}_0\xi)$, that is, (2).

If $f \in \mathscr{C}(\xi, \tau)$, that is, $f\xi \geq \tau$, then, by (2), $f(\mathrm{S}_0\xi) \geq \mathrm{S}_0(f\xi) \geq \mathrm{S}_0\tau$, hence $f \in \mathscr{C}(\mathrm{S}_0\xi, \mathrm{S}_0\tau)$, that is, (3).

By (3), $f \in \mathscr{C}(f^-\tau, \tau) \subset \mathscr{C}(\mathrm{S}_0(f^-\tau), \mathrm{S}_0\tau)$, hence from the second equivalence in (IV.3.5) we get (1).

PROPOSITION VI.8.4. *If Ξ is a set of pretopologies on a set X, then*

$$\mathcal{V}_{\bigvee \Xi}(x) = \bigvee\nolimits_{\xi \in \Xi} \mathcal{V}_\xi(x). \tag{VI.8.1}$$

PROOF. As $\bigvee \Xi \geq \xi$ for each $\xi \in \Xi$,

$$\mathcal{V}_{\bigvee \Xi}(x) \geq \bigvee\nolimits_{\xi \in \Xi} \mathcal{V}_\xi(x).$$

As $x \in \lim_\xi \mathcal{V}_\xi(x) \subset \lim_\xi \bigvee_{\xi \in \Xi} \mathcal{V}_\xi(x)$ for each $\xi \in \Xi$, hence $x \in \lim_{\bigvee \Xi} \bigvee_{\xi \in \Xi} \mathcal{V}_\xi(x)$, and since $\bigvee \Xi$ is a pretopology, $\mathcal{V}_{\bigvee \Xi}(x) \leq \bigvee_{\xi \in \Xi} \mathcal{V}_\xi(x)$. □

Therefore,

PROPOSITION VI.8.5. *If Ξ is a set of pretopologies on a set X, then*

$$\mathrm{inh}_{\bigvee \Xi} A = \bigcup_{\Sigma \subset_{\mathrm{fin}} \Xi} \ \bigcup_{\bigcap_{\sigma \in \Sigma} A_\sigma \subset A} \ \bigcap_{\sigma \in \Sigma} \mathrm{inh}_\sigma A_\sigma.$$

PROOF. Indeed, $x \in \mathrm{inh}_{\bigvee \Xi} A$ if and only if $A \in \mathcal{V}_{\bigvee \Xi}(x)$, and by (VI.8.1), there exists a finite subset Σ of Ξ such that for each $\sigma \in \Sigma$, there exists $A_\sigma \in \mathcal{V}_\sigma(x)$ such that $\bigcap_{\sigma \in \Sigma} A_\sigma \subset A$. As $A_\sigma \in \mathcal{V}_\sigma(x)$ amounts to $x \in \mathrm{inh}_\sigma A_\sigma$, the equality holds. □

PROPOSITION VI.8.6. *If Ξ is a set of pretopologies on a set X, then*

$$\mathrm{adh}_{\bigvee \Xi} A = \bigcap_{\Sigma \subset_{\mathrm{fin}} \Xi} \ \bigcap_{A \subset \bigcup_{\sigma \in \Sigma} A_\sigma} \ \bigcup_{\sigma \in \Sigma} \mathrm{adh}_\sigma A_\sigma.$$

Any assignment of filters $\{\mathcal{W}(x) : x \in X\}$ such that $x \in W$ for every $W \in \mathcal{W}(x)$, determines the vicinities of a unique pretopology. Indeed,

LEMMA VI.8.7. *If $\mathcal{W}(x)$ is a filter on X such that $x \in \ker \mathcal{W}(x)$ for each $x \in X$, then there exists a unique pretopology λ on X such that $\mathcal{V}_\lambda(x) = \mathcal{W}(x)$ for every $x \in X$.*

PROOF. Let $x \in \lim_\lambda \mathcal{F}$ provided that $\mathcal{F} \geq \mathcal{W}(x)$ for each $x \in X$. Then λ is obviously a pretopology on X and $x \in \lim_\lambda \mathcal{W}(x)$. Accordingly, $\mathcal{W}(x) \geq \mathcal{V}_\lambda(x) = \bigwedge_{\mathcal{F} \geq \mathcal{W}(x)} \mathcal{F} \geq \mathcal{W}(x)$. □

PROPOSITION VI.8.8. *If* ah $: 2^X \longrightarrow 2^X$ *is such that for all* $A, A_0, A_1 \subset X$,

$$\text{ah } \varnothing = \varnothing, \tag{VI.8.2}$$

$$A \subset \text{ah } A, \tag{VI.8.3}$$

$$\text{ah}(A_0 \cup A_1) = \text{ah } A_0 \cup \text{ah } A_1, \tag{VI.8.4}$$

then there exists a unique pretopology λ *on* X *such that for every* $A \subset X$,

$$\text{adh}_\lambda A = \text{ah}\, A.$$

PROOF. For each $x \in X$ define an isotone family $\mathcal{W}(x)$ of subsets of X by setting $A \in \mathcal{W}(x)^\#$ if and only if $x \in \text{ah } A$. To see that $\mathcal{W}(x)$ is a filter, equivalently $\mathcal{W}(x)^\#$ is a filter-grill, let $A_0 \cup A_1 \in \mathcal{W}(x)^\#$. By definition, $x \in \text{ah}(A_0 \cup A_1)$, hence by (VI.8.4), either $x \in \text{ah } A_0$ or $x \in \text{ah } A_1$, so that $\mathcal{W}(x)^\#$ is a filter-grill. By (VI.8.3), $x \in \text{ah}\,\{x\}$, hence $\{x\} \in \mathcal{W}(x)^\#$ and thus $x \in \ker \mathcal{W}(x)$. By Lemma VI.8.7, there exists a unique pretopology λ such that $\mathcal{V}_\lambda(x) = \mathcal{W}(x)$, and thus $\text{adh}_\lambda A = \text{ah}\, A$ in view of Proposition VI.1.4. $\square$

CHAPTER VII

Topological structures

A topology is a pretopology, whose adherence is idempotent. Topologies are by far the most known and studied convergences. Here they come out as special pretopologies. As we anticipated, the class of topologies is harder to handle than that of all convergences or that of pretopologies. This is why we tackle topologies from a broader perspective. Nevertheless, in spite of structural deficiencies of the class of topologies, it remains a core of convergence theory.

VII.1. Closed sets, closure

Let ξ be a convergence on a set X. A subset A of X is called *ξ-closed* or simply, *closed* if $\lim_\xi \mathcal{F} \subset A$ for every filter $\mathcal{F}$ such that $A \in \mathcal{F}$. Denote by $\mathcal{C}_\xi$ the family of all ξ-closed sets.

PROPOSITION VII.1.1. *A set A is ξ-closed if and only if*

$$\operatorname{adh}_\xi A \subset A.$$

PROOF. Suppose that A is ξ-closed and $x \in \operatorname{adh}_\xi A$. By Remark VI.1.3, there exists a filter $\mathcal{H}$ such that $A \in \mathcal{H}$, and $x \in \lim_\xi \mathcal{H} \subset A$, by the assumption. Conversely, if $\operatorname{adh}_\xi A \subset A$ and $A \in \mathcal{F}$, then $\lim_\xi \mathcal{F} \subset \operatorname{adh}_\xi \mathcal{F} \subset A$ by the definition of adherence. □

REMARK VII.1.2. Mind that ξ-closed sets are defined for an arbitrary convergence ξ, and, in general, $\operatorname{adh}_\xi A$ is not ξ-closed.

Observe that

$$\zeta \geq \xi \Longrightarrow \mathcal{C}_\zeta \supset \mathcal{C}_\xi. \tag{VII.1.1}$$

In particular, for the discrete convergence ι_X, and chaotic convergence o_X on a set X,

$$\mathcal{C}_{\iota_X} = 2^X \text{ and } \mathcal{C}_{o_X} = \{\varnothing, X\}.$$

PROPOSITION VII.1.3. *If ξ is a convergence on X, then*

$$\varnothing, X \in \mathcal{C}_\xi, \tag{VII.1.2}$$

$$\mathcal{H} \subset \mathcal{C}_\xi \Longrightarrow \bigcap \mathcal{H} \in \mathcal{C}_\xi, \tag{VII.1.3}$$

$$\mathcal{H} \subset_{\text{fin}} \mathcal{C}_\xi \Longrightarrow \bigcup \mathcal{H} \in \mathcal{C}_\xi. \tag{VII.1.4}$$

PROOF. Clearly, $\lim_\xi \mathcal{F} \subset X$ for each filter $\mathcal{F} \in \mathbb{F}X$. No (proper) filter $\mathcal{F}$ fulfills $\varnothing \in \mathcal{F}$, hence $\lim_\xi \mathcal{F} \subset \varnothing$ is emptily fulfilled, proving (VII.1.2). To see (VII.1.3), let $x \in \operatorname{adh}_\xi(\bigcap \mathcal{H})$, where $\mathcal{H} \subset \mathcal{C}_\xi$. As $\bigcap \mathcal{H} \subset H$ for each $H \in \mathcal{H}$, Proposition VI.2.4 (b) implies that $x \in \operatorname{adh}_\xi H$ for each $H \in \mathcal{H}$, and thus $x \in H$, because each $H \in \mathcal{H}$ is ξ-closed. Consequently, $x \in \bigcap \mathcal{H}$. If $\mathcal{H}$ is finite, and $x \in \operatorname{adh}_\xi (\bigcup \mathcal{H})$, then it follows by simple induction from Proposition VI.2.4 (d) that there exists $H \in \mathcal{H}$ for which $x \in \operatorname{adh}_\xi H$, hence $x \in H \subset \bigcup \mathcal{H}$, because H is ξ-closed, that is, (VII.1.4) is fulfilled. □

EXAMPLE VII.1.4. For the standard pretopology ν of the real line, the intervals $[r,s] := \{x \in \mathbb{R} : r \leq x \leq s\}$ and $[r,+\infty[:= \{x \in \mathbb{R} : r \leq x\}$ and $]-\infty, s] := \{x \in \mathbb{R} : x \leq s\}$ are ν-closed, as well as their finite unions and locally finite unions (see Definition VI.2.5).

PROPOSITION VII.1.5. *Let ξ be a convergence on a set X. For every subset A of X, there exists the least ξ-closed set that includes A.*

PROOF. Indeed, the family $\mathcal{C}_\xi(A) := \{H \in \mathcal{C}_\xi : A \subset H\}$ is non-empty, because $X \in \mathcal{C}_\xi(A)$. By (VII.1.3),

$$\operatorname{cl}_\xi A := \bigcap \{H \in \mathcal{C}_\xi : A \subset H\}$$

is ξ-closed, and, of course, $A \subset \operatorname{cl}_\xi A$. Hence, $\operatorname{cl}_\xi A \in \mathcal{C}_\xi(A)$, and thus, $\operatorname{cl}_\xi A$ is the least ξ-closed set including A. □

The set $\operatorname{cl}_\xi A$ is called the *closure* of A. Observe that A is ξ-closed if and only if

$$\operatorname{cl}_\xi A \subset A.$$

PROPOSITION VII.1.6. *The closure operation fulfills (for each A, B, A_0, A_1)*

(a) $$\operatorname{cl}_\xi \varnothing = \varnothing,$$

(b) $$A \subset B \Longrightarrow \operatorname{cl}_\xi A \subset \operatorname{cl}_\xi B,$$

(c) $$A \subset \operatorname{cl}_\xi A,$$

(d) $$\operatorname{cl}_\xi(A_0 \cup A_1) = \operatorname{cl}_\xi A_0 \cup \operatorname{cl}_\xi A_1,$$

(e) $$\operatorname{cl}_\xi(\operatorname{cl}_\xi A) = \operatorname{cl}_\xi A.$$

PROOF. (a-c) follow immediately from definitions. By (VII.1.4), $\mathrm{cl}_\xi A_0 \cup \mathrm{cl}_\xi A_1$ is ξ-closed, and obviously includes $A_0 \cup A_1$, thus includes the least ξ-closed set $\mathrm{cl}_\xi(A_0 \cup A_1)$ including $A_0 \cup A_1$. The converse inclusion follows immediately from (b), hence (d) holds. As $\mathrm{cl}_\xi A$ is ξ-closed, it is equal to the least ξ-closed set that includes it, that is $\mathrm{cl}_\xi(\mathrm{cl}_\xi A) = \mathrm{cl}_\xi A$, in particular, (e) holds. □

COROLLARY VII.1.7. *The family $\mathcal{C}_\xi$ of all ξ-closed sets constitutes a complete lattice with respect to the inclusion, so that if $\varnothing \neq \mathcal{H} \subset \mathcal{C}_\xi$, then*

$$\bigwedge \mathcal{H} = \bigcap\nolimits_{H\in\mathcal{H}} H, \quad \bigvee \mathcal{H} = \mathrm{cl}_\xi \bigcup\nolimits_{H\in\mathcal{H}} H. \tag{VII.1.5}$$

PROOF. By (VII.1.3) the infimum is obviously equal to the intersection. As a union of closed sets is not necessarily closed, the supremum is the least closed set including it, that is, its closure. □

For each $A \subset |\xi|$,

$$\mathrm{adh}_\xi A \subset \mathrm{cl}_\xi A.$$

Indeed, as $A \subset \mathrm{cl}_\xi A$, by (b), $\mathrm{adh}_\xi A \subset \mathrm{adh}_\xi(\mathrm{cl}_\xi A) \subset \mathrm{cl}_\xi A$, because $\mathrm{cl}_\xi A$ is ξ-closed.

COROLLARY VII.1.8. *The adherence* adh_ξ *is idempotent if and only if* $\mathrm{adh}_\xi = \mathrm{cl}_\xi$.

PROOF. By Proposition VII.1.6, cl_ξ is idempotent. Conversely, if adh_ξ is idempotent, then in virtue of Proposition VII.1.1, $\mathrm{adh}_\xi(\mathrm{adh}_\xi A) \subset \mathrm{adh}_\xi A$ means that $\mathrm{adh}_\xi A$ is ξ-closed, hence $\mathrm{cl}_\xi A \subset \mathrm{cl}_\xi(\mathrm{adh}_\xi A) \subset \mathrm{adh}_\xi A$. □

VII.2. Open sets, interior, neighborhoods

Let ξ be a convergence on a set X. A subset O of X is called *ξ-open* if $X \setminus O$ is ξ-closed. Let us denote by $\mathcal{O}_\xi$ the set of all ξ-open sets.

It follows from the definition that

PROPOSITION VII.2.1. *A subset O of X is ξ-open if and only if for each filter $\mathcal{F}$ on X,*

$$O \cap \lim\nolimits_\xi \mathcal{F} \neq \varnothing \Longrightarrow O \in \mathcal{F}. \tag{VII.2.1}$$

PROOF. Indeed, if O is ξ-open and $O \notin \mathcal{F}$ amounts to $X \setminus O \in \mathcal{F}^{\#}$, hence $\lim_\xi \mathcal{F} \subset X \setminus O$, because $X \setminus O$ is ξ-closed. □

It follows immediately from Proposition VII.1.3 that

$$X, \varnothing \in \mathcal{O}_\xi, \tag{VII.2.2}$$

$$\mathcal{A} \subset \mathcal{O}_\xi \Longrightarrow \bigcup \mathcal{A} \in \mathcal{O}_\xi, \tag{VII.2.3}$$

$$\mathcal{A} \subset_{\text{fin}} \mathcal{O}_\xi \Longrightarrow \bigcap \mathcal{A} \in \mathcal{O}_\xi. \tag{VII.2.4}$$

If $\zeta \geq \xi$ then $\mathcal{O}_\zeta \supset \mathcal{O}_\xi$. When ξ is a convergence on X, it is convenient to denote, for each subset A of X,

$$\mathcal{O}_\xi(A) := \{O \in \mathcal{O}_\xi : A \subset O\}, \tag{VII.2.5}$$

and, in particular, $\mathcal{O}_\xi(x) := \mathcal{O}_\xi(\{x\})$ for each $x \in X$. Hence,

$$\mathcal{O}_\xi(A) = \bigcap\nolimits_{x \in A} \mathcal{O}_\xi(x).$$

For instance, $\mathcal{O}_{\iota X} = \mathcal{C}_{\iota X} = 2^X$, and $\mathcal{O}_{o X} = \mathcal{C}_{o X} = \{\varnothing, X\}$.

EXAMPLE VII.2.2. A set is $\nu_{\mathbb{R}}$-open if and only if it is a union of strict intervals $]r, s[$, where $r \leq s$. Indeed, $x \in \lim_{\nu_{\mathbb{R}}} \mathcal{F}$ if and only if $]r, s[\in \mathcal{F}$ for each r and s such that $r < x < s$.

EXAMPLE VII.2.3. For the (upper) Sorgenfrey line ς_+ (that is, the right half open interval convergence) from Example III.3.6, a set A is open if and only if for each $r \in A$, there exists $s > r$ so that $[r, s[\subset A$. Indeed, $r \in \lim_{\varsigma_+} \mathcal{F}$ if and only if $[r, s[\in \mathcal{F}$ for each s such that $r < s$. It is straightforward that for $r < s$, the intervals $]-\infty, r[$, $[r, s[$, and $[s, +\infty[$ are both ς_+-open and ς_+-closed; the intervals $]r, s[$ and $]s, +\infty[$ are ς_+-open, but not ς_+-closed.

Proposition VII.1.1 says that a set A is ξ-closed if and only if $\operatorname{adh}_\xi A \subset A$. As $\operatorname{inh}_\xi A = (\operatorname{adh}_\xi A^c)^c$, we infer that

PROPOSITION VII.2.4. *A set A is ξ-open if and only if*

$$A \subset \operatorname{inh}_\xi A.$$

Recall (VI.3.1) $x \in \operatorname{inh}_\xi A$ if and only if $A \in \mathcal{V}_\xi(x)$. Therefore,

COROLLARY VII.2.5. *A set A is open if and only if it is a vicinity of each of its elements.*

REMARK VII.2.6. Observe that Corollary VII.2.5 is a particular form of Proposition VII.2.1 to the effect that the condition on convergent filters is replaced by that on vicinity filters. The latter are infima of the former, but generally need not converge.

For each subset A of $|\xi|$, there exists the largest ξ-open set included in A. This set is called the *interior* of A, and is denoted by $\operatorname{int}_\xi A$. It is clear that, similarly to (VI.3.2),

$$\operatorname{int}_\xi A = (\operatorname{cl}_\xi A^c)^c. \tag{VII.2.6}$$

In general, $\operatorname{int}_\xi A \subset \operatorname{inh}_\xi A \subset A$ for each A. Moreover, inh_ξ is idempotent if and only if $\operatorname{int}_\xi A = \operatorname{inh}_\xi A$ for each A. It immediately follows from Proposition VII.1.6 via (VII.2.6) that

PROPOSITION VII.2.7. *The operation of interior fulfills (for each A, B, A_0, A_1)*

(a) $$\operatorname{int}_\xi X = X,$$

(b) $$A \subset B \Longrightarrow \operatorname{int}_\xi A \subset \operatorname{int}_\xi B,$$

(c) $$\operatorname{int}_\xi A \subset A,$$

(d) $$\operatorname{int}_\xi A_0 \cap \operatorname{int}_\xi A_1 = \operatorname{int}_\xi(A_0 \cap A_1),$$

(e) $$\operatorname{int}_\xi A = \operatorname{int}_\xi(\operatorname{int}_\xi A).$$

By Corollary VII.1.7 and (VII.2.6),

COROLLARY VII.2.8. *The family $\mathcal{O}_\xi$ of all ξ-open sets is a complete lattice with respect to inclusion, so that if $\varnothing \neq \mathcal{A} \subset \mathcal{O}_\xi$, then*

$$\bigwedge \mathcal{A} = \operatorname{int}_\xi\Big(\bigcap\nolimits_{A\in\mathcal{A}} A\Big), \quad \bigvee \mathcal{A} = \bigcup\nolimits_{A\in\mathcal{A}} A.$$

DEFINITION VII.2.9 (Neighborhood). Let ξ be a convergence on a set X. A subset V of X is called a *ξ-neighborhood* (or simply, *neighborhood*) of an element x if there exists a ξ-open set O such that $x \in O \subset V$.

The set of all ξ-neighborhoods of x, is denoted by $\mathcal{N}_\xi(x)$. By Corollary VII.2.5, $\mathcal{N}_\xi(x) \subset \mathcal{V}_\xi(x)$ for every convergence ξ, and each $x \in |\xi|$. Notice that

$$V \in \mathcal{N}_\xi(x) \Longleftrightarrow x \in \operatorname{int}_\xi V. \tag{VII.2.7}$$

Therefore, each property of interior, listed in Proposition VII.2.7, corresponds to an equivalent property in terms of neighborhood. In fact, (VII.2.7) translates the first four of them to the effect that $\mathcal{N}_\xi(x)$ is a filter, and x belongs to its kernel. The condition (e) of Proposition VII.2.7 is tantamount to

$$V \in \mathcal{N}_\xi(x) \Longrightarrow \underset{W\in\mathcal{N}_\xi(x)}{\exists} \ \underset{w\in W}{\forall} \ V \in \mathcal{N}_\xi(w). \tag{VII.2.8}$$

Indeed, $V \in \mathcal{N}_\xi(x)$ amounts to $x \in \operatorname{int}_\xi V \subset \operatorname{int}_\xi(\operatorname{int}_\xi V)$, that is, $W := \operatorname{int}_\xi V \in \mathcal{N}_\xi(x)$, and thus $V \in \mathcal{N}_\xi(w)$ for each $w \in W$.

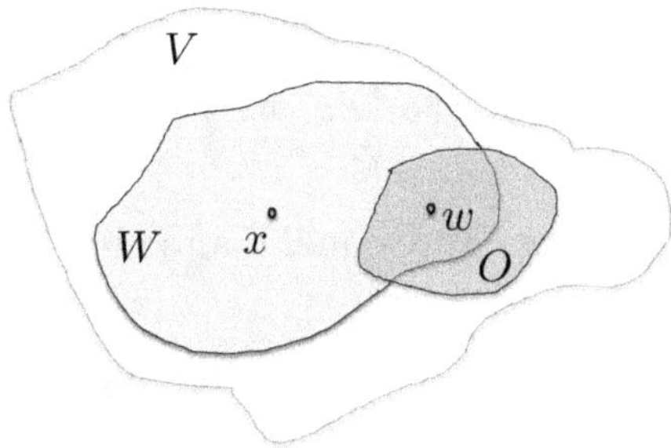

FIGURE VII.2.1. For each neighborhood V of x, there is a neighborhood W of x such that V is a neighborhood of every $w \in W$.

In other terms,

$$\mathcal{N}_\xi(x) \subset \bigcup_{W \in \mathcal{N}_\xi(x)} \bigcap_{w \in W} \mathcal{N}_\xi(w). \tag{VII.2.9}$$

VII.3. Topologies

DEFINITION VII.3.1. A convergence is called a topology if it is a pretopology whose adherence is idempotent.

More explicitly, a convergence θ is a topology if it is a pretopology such that, for each A,

$$\operatorname{adh}_\theta(\operatorname{adh}_\theta A) \subset \operatorname{adh}_\theta A, \tag{VII.3.1}$$

equivalently, if $\operatorname{inh}_\theta B \subset \operatorname{inh}_\theta(\operatorname{inh}_\theta B)$ for each B. It follows from Proposition VI.1.5 that

PROPOSITION VII.3.2. *If a convergence θ is a topology then, for every filter $\mathcal{F}$,*

$$\lim_\theta \mathcal{F} = \bigcap_{A \in \mathcal{F}^\#} \operatorname{cl}_\theta A. \tag{VII.3.2}$$

PROOF. If θ is a topology, hence a pretopology, then by Proposition VI.1.5, $\lim_\theta \mathcal{F} = \bigcap_{A \in \mathcal{F}^\#} \operatorname{adh}_\theta A$, and since $\operatorname{adh}_\theta$ is idempotent, $\operatorname{adh}_\theta A = \operatorname{cl}_\theta A$ for every set A. □

COROLLARY VII.3.3. *If θ is a topology, then*

$$x \in \lim_\theta \mathcal{F}$$

if and only if $x \in C$ for each θ-closed set $C \in \mathcal{F}^\#$.

PROOF. By (VII.3.2), $\lim_\theta \mathcal{F} = \bigcap_{A \in \mathcal{F}^\#} \operatorname{cl}_\theta A \subset \bigcap_{C \in \mathcal{F}^\# \cap \mathcal{C}_\theta} C$, but in fact the equality holds, because $\operatorname{cl}_\theta A \in \mathcal{C}_\theta$ for each A. □

Recall (VII.1.1) that if $\zeta \geq \xi$, then $\mathcal{C}_\zeta \supset \mathcal{C}_\xi$. By Proposition VII.3.2,

COROLLARY VII.3.4. *If* ζ *and* ξ *are topologies and* $\mathcal{C}_\zeta \supset \mathcal{C}_\xi$, *then* $\zeta \geq \xi$.

Therefore, two topologies are equal if and only if they have the same closed sets.

COROLLARY VII.3.5. *If* ζ *and* ξ *are topologies and* $\mathcal{O}_\zeta \supset \mathcal{O}_\xi$, *then* $\zeta \geq \xi$.

Therefore, two topologies are equal if and only if they have the same open sets. Recall (VII.2.5) that $\mathcal{O}_\theta(x)$ denotes the family of θ-open sets that contain x. Rephrasing Corollary VII.3.3,

COROLLARY VII.3.6. *If* θ *is a topology, then*

$$x \in \lim_\theta \mathcal{F}$$

if and only if $O \in \mathcal{F}$ *for each* $O \in \mathcal{O}_\theta(x)$.

In other terms,

$$x \in \lim_\theta \mathcal{F} \Longleftrightarrow \mathcal{N}_\theta(x) \leq \mathcal{F}.$$

As we have mentioned in the previous section, if $\zeta \geq \xi$, then $\mathcal{O}_\zeta \supset \mathcal{O}_\xi$.

LEMMA VII.3.7. *The initial convergence with respect to a topology is a topology.*

PROOF. Let τ be a topology. We want to show that $f^-\tau$ is also a topology. By Proposition VI.5.7, $f^-\tau$ is a pretopology. It is then enough to show that $\mathrm{adh}_{f^-\tau}$ is idempotent. Indeed, by (VI.4.1)

$$\begin{aligned}&\mathrm{adh}_{f^-\tau}(\mathrm{adh}_{f^-\tau} A)\\ &= f^-(\mathrm{adh}_\tau f f^-(\mathrm{adh}_\tau f(A)) \subset f^-(\mathrm{adh}_\tau \mathrm{adh}_\tau f(A))\\ &\subset f^-(\mathrm{adh}_\tau f(A)) = \mathrm{adh}_{f^-\tau} A,\end{aligned}$$

because $f(f^-(D)) \subset D$ and adh_τ is idempotent by assumption. □

COROLLARY VII.3.8. *Every subconvergence of a topology is a topology.*

Continuity between topological spaces can be characterized in terms of closed and open sets.

PROPOSITION VII.3.9. *If* ξ *and* τ *are topologies, then* $f \in \mathscr{C}(\xi, \tau)$ *is equivalent to each of the statements:*

(1) $f^-(O) \in \mathcal{O}_\xi$ *for every* $O \in \mathcal{O}_\tau$,
(2) $f^-(C) \in \mathcal{C}_\xi$ *for every* $C \in \mathcal{C}_\tau$.

PROOF. Let $f \in \mathscr{C}(\xi, \tau)$ and $O \in \mathcal{O}_\tau$ and let $x \in f^-(O)$, equivalently $f(x) \in O$. If $x \in \lim_\xi \mathcal{F}$, then by continuity $f(x) \in \lim_\tau f[\mathcal{F}]$, hence by Proposition VII.2.1, $O \in f[\mathcal{F}]$, that is, there is $F \in \mathcal{F}$ such that $f(F) \subset O$, hence $F \subset f^-(O)$ and thus $f^-(O) \in \mathcal{F}$, which proves that $f^-(O) \in \mathcal{O}_\xi$, that is, (1).

Let $C \in \mathcal{C}_\tau$, then $C^c \in \mathcal{O}_\tau$, hence by (1), $f^-(C)^c = f^-(C^c) \in \mathcal{O}_\xi$, thus $f^-(C) \in \mathcal{C}_\xi$.

If $f \notin \mathscr{C}(\xi, \tau)$, then there is $x \in \lim_\xi \mathcal{F}$ so that $f(x) \notin \lim_\tau f[\mathcal{F}]$. Then, by Corollary VII.3.3, there is $C \in \mathcal{C}_\tau(f(x))$ such that $C \notin f[\mathcal{F}]^\#$, equivalently $f^-(C) \notin \mathcal{F}^\#$, and $f^-(C) \in \mathcal{C}_\xi(x)$. Consequently, $x \notin \lim_\xi \mathcal{F}$ in virtue of Corollary VII.3.3, a contradiction. □

Analogously to Proposition VI.4.1,

PROPOSITION VII.3.10. *Let ξ and τ be convergences on X and Y, respectively, and let $f : X \to Y$. The following statements are equivalent (for each $A \subset X, B \subset Y, x \in X$):*

$$f[\mathcal{N}_\xi(x)] \geq \mathcal{N}_\tau(f(x)), \tag{1}$$

$$\mathcal{N}_\xi(x) \geq f^-[\mathcal{N}_\tau(f(x))], \tag{2}$$

$$f(\mathrm{cl}_\xi A) \subset \mathrm{cl}_\tau f(A), \tag{3}$$

$$\mathrm{cl}_\xi f^-(B) \subset f^-(\mathrm{cl}_\tau B), \tag{4}$$

$$f^-(\mathrm{int}_\tau B) \subset \mathrm{int}_\xi f^-(B). \tag{5}$$

Moreover,

$$f \in \mathscr{C}(\xi, \tau) \tag{6}$$

implies any of them, and if τ is a topology, each of them implies (6).

The only difference with Propositions VI.4.1 and VI.4.4 is idempotency of the operations. Although the arguments to prove these formulae are essentially the same, it is advisable to check them in order to get used to the formalism.

REMARK VII.3.11. Remember that $f(\mathrm{int}_\xi C) \supset \mathrm{int}_\tau f(C)$ is not equivalent to the conditions above.

By Corollary VII.3.8, if τ is a topology on a set Y, and $X \subset Y$, then the restriction $\tau|_X$ of τ to X is a topology. Let us describe it in terms of closed sets, and open sets, respectively.

PROPOSITION VII.3.12. *A subset P of X is $\tau|_X$-open if and only if there is a τ-open subset O of Y such that $P = O \cap X$.*

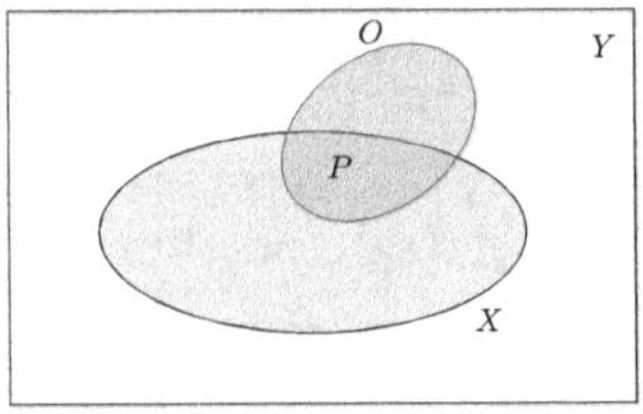

FIGURE VII.3.1. A subset P of X is open if and only if there is an open subset O of Y such that $P = O \cap X$.

PROOF. Let O be a τ-open subset of Y. To see that $X \cap O$ is $\tau|_X$-open, let $x \in X \cap O \in \mathcal{N}_\tau(x) \vee X = \mathcal{N}_{\tau|_X}(x)$ by definition of a subspace, proving that $X \cap O$ is $\tau|_X$-open.

Let a subset P of X be $\tau|_X$-open. By definition, $P \in \mathcal{N}_{\tau|_X}(x) = \mathcal{N}_\tau(x) \vee X$ for each $x \in P$, hence there is $O_x \in \mathcal{O}_\tau(x)$ such that $O_x \cap X \subset P$, because $\mathcal{O}_\tau(x)$ is a filter-base of $\mathcal{N}_\tau(x)$. Therefore, $O := \bigcup_{x\in P} O_x$ is τ-open, $P \subset O$ and $O \cap X = \bigcup_{x\in P}(O_x \cap X) \subset P$. □

By taking the complements of P and O in the proof above, we get

COROLLARY VII.3.13. *A subset F of X is $\tau|_X$-closed if and only if there is a τ-closed subset C of Y such that $F = C \cap X$.*

PROPOSITION VII.3.14. *Each prime pretopology is a topology.*

PROOF. Let π be a prime pretopology on a set X, and let x_∞ be a distinguished point (if π is discrete, then x_∞ is any point). Either $\operatorname{adh}_\pi A = A$ and thus $\operatorname{adh}_\pi(\operatorname{adh}_\pi A) = A$, or $x_\infty \in \operatorname{adh}_\pi A \setminus A$, and then $\operatorname{adh}_\pi A = A \cup \{x_\infty\}$, hence $\operatorname{adh}_\pi(\operatorname{adh}_\pi A) \subset \operatorname{adh}_\pi A \cup \{x_\infty\} = A \cup \{x_\infty\} = \operatorname{adh}_\pi A$. □

Because in the proof of Proposition VI.5.3, we used prime pretopologies, we have already proved that

PROPOSITION VII.3.15. *Each finitely stable convergence is an infimum of prime topologies.*

COROLLARY VII.3.16. *Each finitely stable convergence is a convergence quotient of a topology.*

PROOF. It is straightforward that any sum of topologies is a topology. Let $\xi = \bigwedge_{j\in J} \tau_j$, where τ_j is a topology on $X = |\xi|$ for each $j \in J$, and let $\varphi_j : \{j\} \times X \longrightarrow X$ be given by $\varphi_j(j, x) := x$. Hence $\varphi_j^- \tau_j$ is homeomorphic to τ_j, hence a topology. Then the sum $\bigoplus_{j\in J} \varphi_j^- \tau_j$ (Definition IV.8.2) is a topology on $\bigsqcup_{j\in J}(\{j\} \times X)$. If now $f : \bigsqcup_{j\in J}(\{j\} \times$

$X) \longrightarrow X$ is defined by $f(j,x) := x$, then $f(\bigoplus_{j\in J}\varphi_j\tau_j) = \xi$, that is, the convergence quotient of a topology is equal to ξ. $\square$

VII.4. Structure of the class of topologies

Let us denote by $\mathbf{T}$ the class of topologies. Consequently, $\mathbf{T} \subset \mathbf{S}_0 \subset \mathbf{I}$, where $\mathbf{S}_0$ stands for the class of pretopologies, and $\mathbf{I}$ for that of all convergences. According to our convention, $\mathbf{T}X$ stands for the set of all topologies on a given set X.

THEOREM VII.4.1. *For each convergence ξ, there exists the finest topology coarser than ξ.*

PROOF. Define

$$\lim_\eta \mathcal{F} := \bigcap\nolimits_{A\in\mathcal{F}^\#} \mathrm{cl}_\xi A,$$

and observe that η is a convergence coarser than ξ ([1]). We claim that $\mathcal{C}_\eta = \mathcal{C}_\xi$. Indeed, $\mathcal{C}_\eta \subset \mathcal{C}_\xi$, because $\xi \geq \eta$. Conversely, if $C \in \mathcal{C}_\xi$ and $C \in \mathcal{F}^\#$, then $\lim_\eta \mathcal{F} \subset \mathrm{cl}_\xi C = C$, by the definition of η, and thus $C \in \mathcal{C}_\eta$. Therefore,

$$\lim_\eta \mathcal{F} = \bigcap\nolimits_{A\in\mathcal{F}^\#} \mathrm{cl}_\eta A,$$

that is, η is a topology, coarser than ξ, and such that $\mathcal{C}_\eta = \mathcal{C}_\xi$. If now τ is any topology coarser than ξ, then by (VII.1.1), $\mathcal{C}_\tau \subset \mathcal{C}_\xi = \mathcal{C}_\eta$, hence $\tau \leq \eta$ in virtue of Corollary VII.3.4, showing that η is the finest among the topologies that are coarser than ξ. $\square$

We denote by $\mathrm{T}\xi$ the finest topology among all topologies that are coarser than ξ, and call it the *topologization* of ξ. It follows that $\mathrm{T}\xi = \eta$ from the proof above, hence

$$\lim_{\mathrm{T}\xi} \mathcal{F} = \bigcap\nolimits_{A\in\mathcal{F}^\#} \mathrm{cl}_\xi A = \bigcap\nolimits_{F\in\mathcal{C}_\xi\cap\mathcal{F}^\#} F; \tag{VII.4.1}$$

the second equality is straightforward (Exercise VII.7.3).

It follows from the proof and the definition above that $\mathcal{C}_\xi = \mathcal{C}_{\mathrm{T}\xi}$ for every convergence ξ, hence, for every A,

$$\mathrm{cl}_\xi A = \mathrm{cl}_{\mathrm{T}\xi} A.$$

[1]Indeed, $\mathcal{F}_0 \leq \mathcal{F}_1$ implies that $\lim_\eta \mathcal{F}_0 \subset \lim_\eta \mathcal{F}_1$, and $x \in \lim_\eta x^\uparrow$. Moreover, if $x \in \lim_\xi \mathcal{F}$ then $x \in \mathrm{cl}_\xi A$ for every $A \in \mathcal{F}^\#$.

It is clear that the resulting operation T, called the *topologizer*, fulfills

$$\zeta \geq \xi \Longrightarrow \mathrm{T}\zeta \geq \mathrm{T}\xi, \tag{VII.4.2}$$

$$\mathrm{T}\xi \leq \xi, \tag{VII.4.3}$$

$$\mathrm{T}(\mathrm{T}\xi) \geq \mathrm{T}\xi. \tag{VII.4.4}$$

COROLLARY VII.4.2. *Each supremum of topologies is a topology.*

PROOF. Let Ξ be a non-empty set of topologies on a set X. By (VII.4.2)

$$\mathrm{T}(\bigvee \Xi) \geq \bigvee_{\xi \in \Xi} \mathrm{T}\xi = \bigvee_{\xi \in \Xi} \xi = \bigvee \Xi,$$

because $\mathrm{T}\xi = \xi$ for each $\xi \in \Xi$. □

COROLLARY VII.4.3. *The class of topologies is a complete lattice.*

PROOF. It remains to show that each non-empty set Ξ of topologies on a given set X has an infimum. This holds, because $A_\Xi := \{\tau \in \mathbf{T}X : \forall_{\xi \in \Xi}\, \tau \leq \xi\}$ is non-empty, because $o \in A_\Xi$, and by Corollary VII.4.2, $\bigvee A_\Xi \in A_\Xi$. Hence, by definition, $\bigwedge_{\mathbf{T}} \Xi = \bigvee A_\Xi$, where $\bigwedge_{\mathbf{T}}$ denotes the infimum with respect to the lattice $\mathbf{T}X$ □

As a result, if $\Xi \neq \varnothing$, then

$$\bigwedge\nolimits_{\mathbf{T}} \Xi = \mathrm{T}(\bigwedge \Xi).$$

Because a product is the supremum of initial convergences with respect to the projections, by Proposition VII.3.7,

COROLLARY VII.4.4. *Each product of topologies is a topology.*

EXAMPLE VII.4.5 (Bourdaud pretopology). We have considered the *Bourdaud pretopology* ¥ in Example VI.7.1. Clearly, ¥ is not a topology, because its adherence is not idempotent. It turns out that the topologization of the Bourdaud pretopology is the chaotic topology: $\mathrm{T}¥ = o$, because $\mathrm{cl}_¥ A \supset \mathrm{adh}_¥(\mathrm{adh}_¥ A) = \{0, 1, 2\} = \mathrm{adh}_o A$ for every $A \neq \varnothing$.

By Lemma VII.3.7, $f^-\theta$ is a topology for any map f and each topology θ. Therefore, for each convergence τ,

$$\mathrm{T}(f^-\tau) \geq f^-(\mathrm{T}\tau). \tag{VII.4.5}$$

Indeed, by (VII.4.3) $\tau \geq \mathrm{T}\tau$, hence $f^-\tau \geq f^-(\mathrm{T}\tau)$. By (VII.4.2), $\mathrm{T}(f^-\tau) \geq \mathrm{T}f^-(\mathrm{T}\tau) = f^-(\mathrm{T}\tau)$, because $f^-(\mathrm{T}\tau)$ is a topology in virtue of Lemma VII.3.7.

Like Exercise VI.8.3, this amounts to

$$f(\mathrm{T}\xi) \geq \mathrm{T}(f\xi) \tag{VII.4.6}$$

that for any map f and every convergence ξ, as well as to

$$\mathscr{C}(\xi,\tau) \subset \mathscr{C}(\mathrm{T}\xi,\mathrm{T}\tau)$$

for any ξ and τ.

However, unlike for pretopologizers, (VII.4.5) cannot be replaced by equality. In particular, there are θ and V such that $\mathrm{T}(\theta|_V) \neq (\mathrm{T}\theta)|_V$.

EXAMPLE VII.4.6 (Arens topology). The bisequence pretopology from Example VI.1.8 is not a topology. Indeed, let $A := \bigcup_{n<\omega} X_n \setminus X_\infty$. Then $x_\infty \notin \operatorname{adh}_\theta A = \bigcup_{n<\omega} X_n$, while $x_\infty \in \operatorname{adh}^2_\theta A := \operatorname{adh}_\theta(\operatorname{adh}_\theta A) = X$.

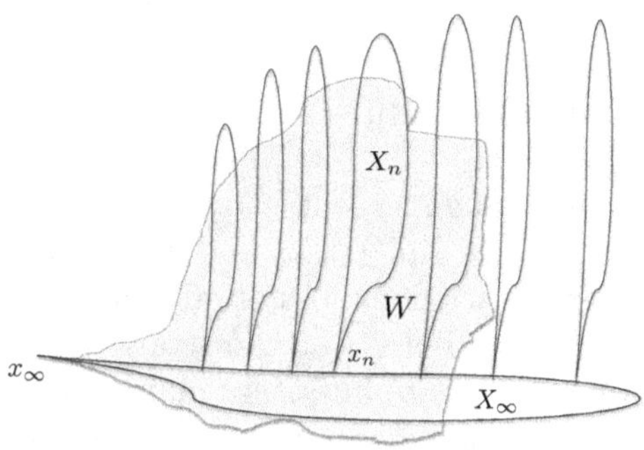

FIGURE VII.4.1. The Arens topology.

Actually, $\operatorname{cl}_\theta D = \operatorname{adh}^2_\theta D$ for each set D. A subset W is a θ-neighborhood of x_∞ whenever $x_\infty \in W$ and there is $n < \omega$ such that $X_k \setminus W$ is finite for each $n < k < \omega$.

Let $V := A \cup \{x_\infty\}$. Then $\theta|_V$ is the discrete topology, hence $\mathrm{T}(\theta|_V)$ is discrete. On the other hand, $(\mathrm{T}\theta)|_V$ is not discrete. Indeed, $U \subset V$ is a $(\mathrm{T}\theta)|_V$-neighborhood of x_∞, provided that $x_\infty \in U$ and there is $n < \omega$ such that $X_k \setminus U$ is finite for each $n < k < \omega$. This topology is called the *Arens topology*. Therefore, $(\mathrm{T}\theta)|_V < \mathrm{T}(\theta|_V)$.

The topologizer commutes neither with suprema, nor with products. Indeed,

EXAMPLE VII.4.7. Examples VI.6.11 and VI.6.12 exhibit a convergence ξ and a pretopology τ such that $\mathrm{S}_0\xi \vee \tau \ngeq \mathrm{S}_0(\xi \vee \tau)$, and consequently $\mathrm{S}_0\xi \times \tau \ngeq \mathrm{S}_0(\xi \times \tau)$. As ξ, τ, and $\xi \vee \tau$ are prime, $\tau, \mathrm{S}_0\xi$

and $\mathrm{S}_0(\xi \vee \tau)$ are prime pretopologies, hence topologies by Proposition VII.3.14. Therefore, $\mathrm{T}\xi \vee \tau \ngeq \mathrm{T}(\xi \vee \tau)$ and τ is a topology.

Although the convergence $\xi \times \tau$ on $X \times X$ is not prime, its restriction to the diagonal set $\triangle(X)$ is prime. Moreover, $\triangle(X)$ is $(\xi \times \tau)$-closed. Therefore, $(x_\infty, x_\infty) \in \lim_{\mathrm{T}\xi \times \tau} \triangle[\mathcal{F}] \setminus \lim_{\mathrm{T}(\xi \times \tau)} \triangle[\mathcal{F}]$, proving that there exists a prime convergence ξ and a prime topology such that $\mathrm{T}\xi \times \tau \ngeq \mathrm{T}(\xi \times \tau)$.

PROPOSITION VII.4.8. *If a convergence ξ is free (alias T_1), then $\mathrm{T}\xi$ is free.*

PROOF. From $\mathrm{adh}_\xi\{x\} \subset \{x\}$, we get that $\mathrm{cl}_\xi\{x\} = \mathrm{adh}_\xi\{x\} = \{x\}$, hence $\lim_{\mathrm{T}\xi} x^\uparrow = \mathrm{cl}_\xi\{x\} \subset \{x\}$. □

EXAMPLE VII.4.9 (Radial topology). Let $X := \mathbb{R}^2$. We define the *radial topology* λ by declaring that a set A is λ-closed whenever for every straight line $L \subset \mathbb{R}^2$ the intersection $A \cap L$ is closed in L endowed with the standard topology $\nu_{\mathbb{R}}$ of $\mathbb{R}$. It is easy to check that the family $\mathcal{C}_\lambda$ (of λ-closed sets) fulfills (VII.1.2)(VII.1.2) and (VII.1.4), hence defines a topology.

We claim that each λ-convergent sequence lies on a finite union of straight lines. Let $x_\infty \in \lim_\lambda (x_n)_n$ and $(x_n)_n$ does not fulfill the condition. We can assume without loss of generality ([2]) that the terms of the sequence are distinct. Let L_0 be the straight line through x_0 and x_1. By the assumption, there are infinitely many terms of the sequence off L_0, so let n_2 be the least index such that $x_{n_2} \notin L_0$.

We proceed by induction. If each straight line has at most two points common with $\{x_0, x_1, \ldots, x_{n_k}\}$, then by inductive assumption, there are infinitely many terms of the sequence that are off the affine hull $\mathrm{aff}\{x_0, x_1, \ldots, x_{n_k}\}$ of $\{x_0, x_1, \ldots, x_{n_k}\}$ ([3]). So let n_{k+1} be the first index such that $x_{n_{k+1}} \notin \mathrm{aff}\{x_0, x_1, \ldots, x_{n_k}\}$.

The subsequence $(x_{n_k})_k$ constructed above has the property that each straight line meets at most its two terms, hence the set $S_m := \{x_{n_k} : k > m\}$ is closed for each $m < \omega$. Accordingly,

$$\lim\nolimits_\lambda (x_{n_k})_k \subset \bigcap_{m<\omega} \mathrm{cl}_\lambda S_m = \bigcap_{m<\omega} S_m = \varnothing,$$

which completes the proof.

[2]If a term occurs a finite number of times, we can remove all its occurrences but one without modifying its limit. If a term occurs infinitely many times, then this term is necessarily equal to the limit. Then either the number of the remaining terms is finite, and the theorem is proved, or it is enough to prove for the sequence of the remaining terms.

[3]By definition, the *affine hull* of a set A in a vector space is the union of all straight lines through at least two elements of the set.

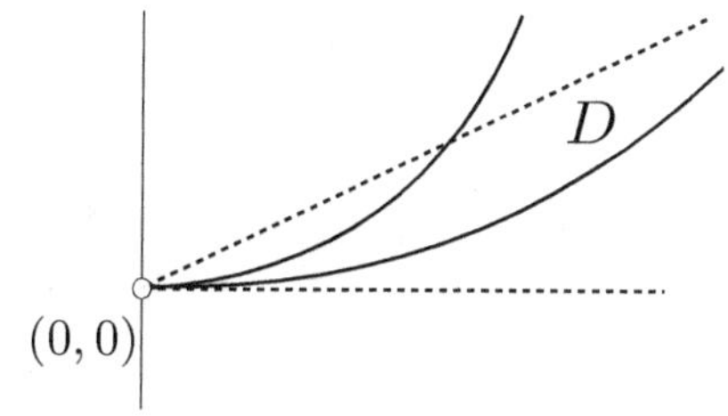

FIGURE VII.4.2. The set $D := \{(r,s) : r^2 \le s \le 2r^2, r > 0\}$ is λ-closed, but $(0,0) \in \operatorname{cl}_{\nu_{\mathbb{R}}^2} D$.

Recall the *radial pretopology* ρ from Example VI.1.9, for which $x \in \operatorname{adh}_\rho A$ whenever there is h and a sequence $(t_n)_n$ converging to 0 such that $x + t_n h \in A$, that is, there is a sequence $(x_n)_n$ in A of the form $x_n = x + t_n h$ converging to x. In other terms, if h is non-zero, then $L := \{x + th : t \in \mathbb{R}\}$ is a straight line such that $x \in \operatorname{adh}_\rho (L \cap A)$.

Accordingly, A is ρ-closed whenever $\operatorname{adh}_\rho A \subset A$, that is, if $L \cap A$ is closed in L. Therefore, the *radial topology* λ is the topologization of the *radial pretopology* ρ, that is, $\lambda = \mathrm{T}\rho$. Of course, any $\nu_{\mathbb{R}}^2$-closed set is ρ-closed, so that $\mathrm{T}\rho \ge \nu_{\mathbb{R}}^2$. In fact, the inequality is strict, as shows Figure VII.4.2.

VII.5. Initial density of the Sierpiński topology

The *Sierpiński topology* \$ was defined in Example III.5.3 on a set of cardinality 2, so that one singleton is \$-closed, and the other is not. Therefore there are two homeomorphic Sierpiński topologies $\$_0$ and $\$_1$ on $\{0,1\}$. We set $\mathcal{C}_{\$_0} := \{\varnothing, \{1\}, \{0,1\}\}$, and $\mathcal{C}_{\$_1} := \{\varnothing, \{0\}, \{0,1\}\}$. The term *Sierpiński topology* \$ refers to any of two variants, while we use specifically the upper variant $\$_0$ and the lower variant $\$_1$ to fix the setting.

Let us take $\$_0$ here, so that $\lim_{\$_0} 0^\uparrow = \{0,1\}$, and $\lim_{\$_0} 1^\uparrow = \{1\}$.

Recall (I.5.4) that there exists a bijection between maps $f \in \{0,1\}^X$ and subsets $A \subset X$. This bijection can be realized with the aid of *characteristic functions* χ_A, or *indicator functions* ψ_A, and, of course, in many other ways. Let

$$\chi_A(x) := \begin{cases} 1, & x \in A, \\ 0, & x \notin A. \end{cases} \qquad \psi_A(x) := \begin{cases} 0, & x \in A, \\ 1, & x \notin A. \end{cases} \tag{VII.5.1}$$

In particular, $\chi_X(x) = 1$ for each $x \in X$, and $\chi_\varnothing(x) = 0$ for each $x \in X$. Of course, $\psi_A = \chi_{A^c}$ and vice versa. Moreover, for each $f : X \longrightarrow \{0,1\}$,

$$f = \chi_{f^-(1)} = \psi_{f^-(0)}.$$

We notice that

LEMMA VII.5.1. *If ξ is a convergence on X, then $f \in \mathscr{C}(\xi, \$_0)$ if and only if $f^-(1)$ is ξ-closed, and $f \in \mathscr{C}(\xi, \$_1)$ if and only if $f^-(1)$ is ξ-open.*

PROOF. Indeed, the only proper $\$_0$-closed set is $\{1\}$. □

In other words, $f \in \mathscr{C}(\xi, \$_0)$ if and only if there is a ξ-closed set H such that $f = \chi_H$. Therefore, we have evidenced a bijection between continuous functions valued in the Sierpiński topology and closed sets.

THEOREM VII.5.2. *Each topology is a supremum of preimages of the Sierpiński topology.*

PROOF. Let ξ be a topology on a set X. As for every $H \in \mathcal{C}_\xi$, the map $\chi_H \in \mathscr{C}(\xi, \$_0)$, equivalently, $\xi \geq \chi_H^-(\$_0)$,

$$\xi \geq \bigvee_{H \in \mathcal{C}_\xi} \chi_H^-(\$_0).$$

Conversely, if $x \notin \lim_\xi \mathcal{F}$, then by (VII.4.1) there exists a ξ-closed set $H \in \mathcal{F}^\#$ such that $x \notin H$, that is, $\{1\} \in \chi_H[\mathcal{F}]$ and $\chi_H(x) = 0$. As $0 \notin \lim_{\$_0} 1^\uparrow$, we infer that $x \notin \lim_{\chi_H^- \$_0} \mathcal{F}$ ([4]). □

In other words, each topology is the initial convergence with respect to a product of Sierpiński topologies. By Definition VI.7.4,

COROLLARY VII.5.3. *The Sierpiński topology is initially dense in the class of topologies, hence the class of topologies is simple.*

By Definition III.1.18, a convergence ξ is T_0 if $\xi^-(x_0) \neq \xi^-(x_1)$ for all $x_0 \neq x_1$. It follows that

PROPOSITION VII.5.4. *A topology ξ is T_0 if and only if for all $x_0 \neq x_1$, there exists an open set that contains exactly one of them.*

It follows from Example III.7.6 that the property T_0 is not preserved under restriction ([5]). However, it is straightforward that

PROPOSITION VII.5.5. *Each restriction of a T_0-topology is T_0.*

---

[4]It follows from a general formula (IV.3.1) for initial convergences that $x \in \lim_{\chi_H^- \$_0} \mathcal{F}$ if and only if $\chi_H(x) \in \lim_{\$_0} \chi_H[\mathcal{F}]$.

[5]It is straightforward that the topologization of the T_0-convergence from Example III.7.6 is not T_0.

Actually, by Proposition IV.7.7, each topology is initial with respect to the product $\prod_{f\in\mathscr{C}(\xi,\$)}\$$, called *Sierpiński cube*. Moreover, if a topology ξ is T_0, then the corresponding diagonal map $\triangle_{f\in\mathscr{C}(\xi,\$)}f$ is injective, because by Proposition VII.5.4, for all $x_0\neq x_1$, there exists $f\in\mathscr{C}(\xi,\$)$ such that $f(x_0)\neq f(x_1)$.

COROLLARY VII.5.6. *Every T_0-topology ξ is embeddable in $\prod_{f\in\mathscr{C}(\xi,\$)}\$$.*

Theorem IX.5.1 to come provides an estimate of the dimension of the Sierpiński cube from Corollary VII.5.6.

Mind that, by Proposition X.1.2 to come, the lower and the upper topologies ν_- and ν_+ of the real line are also initially dense in the class of topologies.

VII.6. *Remarks on convergence of nets

I provide here a short outline of nets, a device that was conceived to generalize sequences ([6]) in view for applications in topology and analysis. As we shall see, nets can be reduced to filters in most cases. On the other hand, calculus of nets is convoluted, which makes them rather difficult to use.

Recall that a transitive reflexive relation $\leq$ on a non-empty set V is called a *preorder*, and $(V,\leq)$ is said to be a *preordered set*. Accordingly an order is an antisymmetric preorder.

A *directed set* is a preordered set $(V,\leq)$ such that, for any $v_0,v_1\in V$, there exists $v\in V$ so that $v_0\leq v$ and $v_1\leq v$. More extensively, $(V,\leq)$ is a directed set if V is non-empty, and for every v,v_0,v_1,v_2 in V,

$$v\leq v, \tag{VII.6.1}$$

$$(v_0\leq v_1)\wedge(v_1\leq v_2)\Longrightarrow v_0\leq v_2, \tag{VII.6.2}$$

$$\underset{v\in V}{\exists}\,(v_0\leq v)\wedge(v_1\leq v). \tag{VII.6.3}$$

Denote by

$$V^{\uparrow}(v):=\{w\in V:v\leq w\}$$

the set of successors of v in V (with respect to $\leq$). It is plain that if $v_0,v_1\in V$, then, by (VII.6.3), there is $v\in V$ such that

$$V^{\uparrow}(v)\subset V^{\uparrow}(v_0)\cap V^{\uparrow}(v_1).$$

Therefore,

[6]Introduced by E. H. MOORE and L. H. SMITH in [187] in 1922; employed by J. KELLEY in [147].

PROPOSITION VII.6.1. *If $(V, \leq)$ is a directed set, then the family*

$$\mathcal{V} := \{V^{\uparrow}(v) : v \in V\} \tag{VII.6.4}$$

is a filter-base.

In other words, a directed set $(V, \leq)$ generates on V a filter $\mathcal{V}^{\uparrow}$.

DEFINITION VII.6.2. A *net* on a set X is a map $\varphi \in X^V$, where $(V, \leq)$ is a directed set.

Accordingly, if $\varphi : V \longrightarrow X$ is a net, then $\varphi[\mathcal{V}]$ is a filter-base on X.

Sequences constitute a special class of nets, as well as transfinite sequences, introduced in Section XXVII.4.

If τ is a topology on X, then we say that a net $\varphi \in X^V$ converges to x in τ if for every $O \in \mathcal{O}_\tau(x)$ there is $v \in V$ such that $\varphi(w) \in O$ for every $w \geq v$.

PROPOSITION VII.6.3. *Let τ be a topology on X. A net $\varphi \in X^V$ converges to x in τ if and only if*

$$x \in \lim\nolimits_\tau \varphi[\mathcal{V}]. \tag{VII.6.5}$$

If τ is a convergence, then we say that net φ converges to x in τ whenever (VII.6.5) holds.

Proposition VII.6.3 reduces convergence of a net $\varphi \in X^V$ to convergence of $\varphi[\mathcal{V}]$, where $\mathcal{V}$ is the filter-base determined on V by the preorder defined on V.

Conversely, ([7])

PROPOSITION VII.6.4. *If $\mathcal{F}$ is a filter on X, then there exists a net $\varphi \in X^V$ such that $\mathcal{F} = \varphi[\mathcal{V}]^{\uparrow}$, where $\mathcal{V}$ is a filter-base (VII.6.4).*

PROOF. Indeed, for any filter-base $\mathcal{B}$ of $\mathcal{F}$ let $V := \{(x, B) : x \in X, B \in \mathcal{B}\}$ and

$$(x_0, B_0) \leq (x_1, B_1) \Longleftrightarrow B_0 \supset B_1,$$

and let $\varphi(x, B) := x$. Then the family $\mathcal{V}$ consisting of $V^{\uparrow}(x, B) = \{(w, W) : B \supset W\}$ is a filter-base on V. On the other hand, $\varphi(V^{\uparrow}(x, B)) = B$, so that $\varphi[\mathcal{V}] = \mathcal{F}$. □

REMARK VII.6.5. Any directed set $(V, \leq)$ determines, in a canonical way, via (VII.6.4), a filter $\mathcal{V}^{\uparrow}$ on V. This is this filter that specifies convergence properties of the net. Therefore, we can forget about the preorder $\leq$ and argue only in terms of $\varphi[\mathcal{V}]$, as we have done in the case of sequences in Section II.1. This leads to a concept of external filters.

[7]See J. Kelley [147, p. 83].

DEFINITION VII.6.6. An *external filter* of a set X is a pair $(\varphi, \mathcal{V})$ such that $\mathcal{V}$ is a filter on a set V and $\varphi \in X^V$.

If ξ is a convergence on X, then an external filter $(\varphi, \mathcal{V})$ ξ-*converges* to x whenever $x \in \lim_\xi \varphi[\mathcal{V}]$. There exist some situations where a use of external filters is handy.

DEFINITION VII.6.7. A net $(\psi \mid W, \leq)$ on X is said to be a *subnet* of $(\varphi \mid V, \leq)$ if there exists a map $f : W \longrightarrow V$ such that $\psi = \varphi \circ f$, and for each $v \in V$ there exists $w \in W$ such that $f(W^{\uparrow}(w)) \subset V^{\uparrow}(v)$.

It is straightforward that

PROPOSITION VII.6.8. *If a net ψ is a subnet of a net φ, then*

$$\varphi[\mathcal{V}] \leq \psi[\mathcal{W}]. \tag{VII.6.6}$$

You may recognize a familiar relation (IV.9.2), discussed in Section IV.9 because of its pertinence to *subquences*, that is, generalized subsequences.

Subsequences constitute a special case of subnets. Already in that particular case, Proposition VII.6.8 cannot be reversed. This is why the concept of subquence emerged. A potential generalization of subquences in the context of nets becomes a bit tricky.

The things become quite complex, when we pass to directed products ([8]) and *Theorem on iterated limits.* It is hard to find *maximal nets* in the literature, and listed in MathSciNet, the titles containing *ultra-nets* are mentioned only a couple of times. Although J. KELLEY founds his *General Topology* ([9]) on nets, neither *maximal nets* nor *ultra-nets* can be found in the Index, but we find *ultrafilters* instead, even if filters are treated marginally. Is this an implicit confession of difficulties?

VII.7. Supplement

Traditionally, topologies are defined in terms of open sets, or in terms of the closure operation ([10]). Let X be a non-empty set, and let

---

[8] J. Kelley [147, pp. 68–69]

[9] See J. KELLEY [147]

[10] W. SIERPIŃSKI [235], K. KURATOWSKI [169], R. ENGELKING [97], J. KELLEY [147], and J. DUGUNDJI [94].

$\mathcal{O}$ be a family of subsets of X fulfilling

$$\varnothing, X \in \mathcal{O},$$

$$\mathcal{A} \subset \mathcal{O} \Longrightarrow \bigcup \mathcal{A} \in \mathcal{O},$$

$$\mathcal{A} \subset_{\text{fin}} \mathcal{O} \Longrightarrow \bigcap \mathcal{A} \in \mathcal{O}.$$

A *topological space* is a pair $(X, \mathcal{O})$, and the $\mathcal{O}$ is said to be the family of *open sets* or the *topology* on X. Any topological space $(X, \mathcal{O})$ defines a convergence $\tau_{\mathcal{O}}$ on X to the effect that

$$x \in \lim_{\tau_{\mathcal{O}}} \mathcal{F} \Longleftrightarrow \mathcal{O}(x) \subset \mathcal{F}, \tag{VII.7.1}$$

where $\mathcal{O}(x) := \{O \in \mathcal{O} : x \in O\}$. It is obviously a convergence. Moreover,

PROPOSITION VII.7.1. *If $(X, \mathcal{O})$ is a topological space, then the open sets associated with the convergence $\tau_{\mathcal{O}}$ coincide with $\mathcal{O}$, that is, $\mathcal{O}_{\tau_{\mathcal{O}}} = \mathcal{O}$.*

PROOF. If $O \in \mathcal{O}$, then by (VII.7.1), $O \in \mathcal{F}$ for each filter $\mathcal{F}$ such that $O \cap \lim_{\tau_{\mathcal{O}}} \mathcal{F} \neq \varnothing$. Hence, $\mathcal{O} \subset \mathcal{O}_{\tau_{\mathcal{O}}}$. If $P \in \mathcal{O}_{\tau_{\mathcal{O}}}$, then by definition of $\tau_{\mathcal{O}}$, for every $x \in P$, there is $O_x \in \mathcal{O}(x)$ such that $O_x \subset P$, hence $P = \bigcup_{x \in P} O_x \subset P$, and thus $P \in \mathcal{O}$. Therefore, $\mathcal{O}_{\tau_{\mathcal{O}}} \subset \mathcal{O}$. $\square$

EXAMPLE VII.7.2. If we consider the *upper* variant $\$_0$ of $\$$, then by Lemma VII.5.1, $f \in \mathscr{C}(\xi, \$_0)$ if and only if $f^-(1)$ is ξ-closed, which is the characteristic function $f = \chi_A$ of a ξ-closed set A. Therefore, the set $\mathcal{C}_\xi$ of all ξ-closed sets can be identified with $\mathscr{C}(\xi, \$_0)$ by

$$A \in \mathcal{C}_\xi \Longleftrightarrow \chi_A \in \mathscr{C}(\xi, \$_0).$$

As for the *lower* variant $\$_1$ of $\$$, by the same lemma, $f \in \mathscr{C}(\xi, \$_1)$ if and only if $f^-(1)$ is ξ-open, which is the characteristic function $f = \chi_A$ of a ξ-open set A. Therefore, the set $\mathcal{O}_\xi$ of ξ-open sets can be identified with $\mathscr{C}(\xi, \$_1)$ by

$$A \in \mathcal{O}_\xi \Longleftrightarrow \chi_A \in \mathscr{C}(\xi, \$_1).$$

Accordingly, the fact that if $f \in \mathscr{C}(\xi, \sigma)$ and $O \in \mathcal{O}_\sigma$, then $f^-(O) \in \mathcal{O}_\xi$, acquires an interpretation in terms of composition of continuous maps: if $\chi_O \in \mathscr{C}(\sigma, \$_1)$ and $f \in \mathscr{C}(\xi, \sigma)$, then $\chi_{f^-(O)} = \chi_O \circ f \in \mathscr{C}(\xi, \$_1)$, because $1 = \chi_{f^-(O)}(x)$ if and only if $x \in f^-(O)$, equivalently $f(x) \in O$, that is, $1 = \chi_O(f(x)) = (\chi_O \circ f)(x)$.

EXERCISE VII.7.3. Show that $\lim_{\mathrm{T}\xi} \mathcal{F} = \bigcap_{F \in \mathcal{C}_\xi \cap \mathcal{F}^\#} F$.

SOLUTION. Clearly, $\bigcap_{F\in\mathcal{C}_\xi\cap\mathcal{F}^\#} F = \bigcap_{F\in\mathcal{C}_\xi\cap\mathcal{F}^\#} \mathrm{cl}_\xi F$, because if $F \in \mathcal{C}_\xi$ then $F = \mathrm{cl}_\xi F$. Therefore, $\bigcap_{A\in\mathcal{F}^\#} \mathrm{cl}_\xi A \subset \bigcap_{F\in\mathcal{C}_\xi\cap\mathcal{F}^\#} \mathrm{cl}_\xi F$. Conversely, if $x \notin \bigcap_{A\in\mathcal{F}^\#} \mathrm{cl}_\xi A$, then there is $A \in \mathcal{F}^\#$ such that $x \notin \mathrm{cl}_\xi A$, hence $\mathrm{cl}_\xi A \in \mathcal{C}_\xi\cap\mathcal{F}^\#$ and $x \notin \mathrm{cl}_\xi A$.

We know that the ordered set $(\mathbf{I}X, \leq)$ of convergences on X is a complete lattice. It turns out that the set $\mathbf{T}X$ of topologies on X, with the order induced from $(\mathbf{I}X, \leq)$, is a complete lattice too.

Indeed, let $\bigvee, \bigwedge$ be the extrema ([11]) with respect to $(\mathbf{I}X, \leq)$, then $\bigvee = \bigvee_{\mathbf{T}}$, because a (convergence) supremum of topologies is a topology. If $\Xi \subset \mathbf{T}X \subset \mathbf{I}X$, then $\bigwedge \Xi$ need not be a topology. Nevertheless, $\mathrm{T}(\bigwedge \Xi)$ is the finest topology coarser than $\bigwedge \Xi$, so that

$$\bigwedge\nolimits_{\mathbf{T}} \Xi = \mathrm{T}(\bigwedge \Xi).$$

If we look at the families of ξ-closed sets $\mathcal{C}_\xi$ and ξ-open sets $\mathcal{O}_\xi$, then we observe that $\mathcal{C}_{\bigwedge_{\mathbf{T}}\Xi} \subset \bigcap_{\xi\in\Xi} \mathcal{C}_\xi$, because of (VII.1.1. However, since $\bigcap_{\xi\in\Xi} \mathcal{C}_\xi$ fulfills (VII.1.2), (VII.1.3), and (VII.1.4),

$$\mathcal{C}_{\bigwedge_{\mathbf{T}}\Xi} = \bigcap\nolimits_{\xi\in\Xi} \mathcal{C}_\xi.$$

Indeed, if $\varnothing, X \in \mathcal{C}_\xi$ for each $\xi \in \Xi$, hence $\varnothing, X \in \bigcap_{\xi\in\Xi} \mathcal{C}_\xi$; if $\mathcal{D} \subset \bigcap_{\xi\in\Xi} \mathcal{C}_\xi \subset \mathcal{C}_\eta$ for each $\eta \in \Xi$, hence $\bigcap \mathcal{D} \in \mathcal{C}_\eta$ (and $\bigcap \mathcal{D} \in \mathcal{C}_\eta$ if $\mathcal{D}$ is finite) for each $\eta \in \Xi$, thus $\bigcap \mathcal{D} \in \bigcap_{\xi\in\Xi} \mathcal{C}_\xi$, and $\bigcup \mathcal{D} \in \bigcap_{\xi\in\Xi} \mathcal{C}_\xi$ if $\mathcal{D}$ is finite.

On the other hand, $\bigcup_{\xi\in\Xi} \mathcal{C}_\xi \subset \mathcal{C}_{\bigvee\Xi}$, but $\bigcup_{\xi\in\Xi} \mathcal{C}_\xi$ is not closed under intersections. Therefore, $\mathcal{C}_{\bigvee\Xi}$ is the least family including $\bigcup_{\xi\in\Xi} \mathcal{C}_\xi$ and fulfilling (VII.1.3) and (VII.1.4).

DEFINITION VII.7.4 (Pointwise property). A convergence property $\mathfrak{P}$ is said to be *pointwise* if there exists a class of filters $\mathbb{P}$ such that $\xi \in \mathfrak{P}$ provided that $\xi^-(x) \subset \mathbb{P}$ for each $x \in |\xi|$.

In other words, pointwise properties depend only on filters convergent to individual points, one by one. For example, pretopologicity or character are pointwise properties, but topologicity or compactness are not.

PROPOSITION VII.7.5. *Being free (alias T_1) is a pointwise property.*

PROOF. Indeed, a convergence ξ is free, whenever that $\xi^-(x) \cap \mathbb{F}_0$ is a singleton for each $x \in |\xi|$. Since ξ is (centered), $x^\uparrow \in \xi^-(x)$, and since $\mathcal{F} \in \xi^-(x)$ implies that $\ker \mathcal{F} \subset \{x\}$. Therefore, $x^\uparrow$ is the only principal filter in $\xi^-(x)$. □

[11] That is, suprema and infima.

We have considered certain properties at individual points. For example, a map can be continuous at certain points, but not at some others. A convergence can be of countable character at some points, of finite character at some others (isolated points), and of uncountable character at some others. For example, a convergence ξ on X is *pretopological* at x_0 if $x_0 \in \lim_\xi \mathcal{V}_\xi(x_0)$. But certain non-pointwise properties can be considered at individual points. For example, a pretopology ξ on X is said to be *topological* at $x_0 \in X$ if $\int_{\mathcal{N}_\xi(x_0)} \mathcal{N}_\xi(\cdot) \geq \mathcal{N}_\xi(x_0)$ (see (VII.2.9)).

EXAMPLE VII.7.6. Let $X = \{\infty\} \sqcup \bigsqcup_{n<\omega} X_n$, where X_n is an infinite set for each $n < \omega$. Let $\infty \in \lim_\xi \mathcal{F}$ whenever $\ker \mathcal{F} \subset \{\infty\}$ and there is $k < \omega$ such that $\{\infty\} \sqcup \bigsqcup_{n<k} X_n \in \mathcal{F}$. Then every $x \in X \setminus \{\infty\}$ is topological, while ∞ is not.

CHAPTER VIII

Adherences, covers, and compactness

We have employed so far adherences of subsets of convergence spaces. We are going now to extend this concept to arbitrary families of subsets of convergence spaces, and in particular, to filters.

VIII.1. Adherences

Recall from Section V.3 that families $\mathcal{A}$ and $\mathcal{B}$ *mesh*, in symbols, $\mathcal{A}\#\mathcal{B}$ whenever $A\#B$, that is, $A \cap B \neq \varnothing$ for each $A \in \mathcal{A}$ and $B \in \mathcal{B}$. We write $\mathcal{A}\perp\mathcal{B}$ if $\mathcal{A}$ and $\mathcal{B}$ do not mesh.

The *adherence* $\operatorname{adh}_\xi \mathcal{A}$ of a family $\mathcal{A}$ of subsets of $|\xi|$ is defined by

$$\operatorname{adh}_\xi \mathcal{A} := \bigcup\nolimits_{\mathbb{F}\ni\mathcal{F}\#\mathcal{A}} \lim\nolimits_\xi \mathcal{F}, \tag{VIII.1.1}$$

that is, $x \in \operatorname{adh}_\xi \mathcal{A}$ whenever there exists a filter $\mathcal{F}$ such that $\mathcal{F}\#\mathcal{A}$ and $x \in \lim_\xi \mathcal{F}$. A family $\mathcal{A}$ is called *ξ-adherent* (or simply, *adherent*) if $\operatorname{adh}_\xi \mathcal{A} \neq \varnothing$.

In particular, if $\{A\} \subset \mathcal{A} \subset \{A\}^\uparrow$, then

$$\operatorname{adh}_\xi A = \operatorname{adh}_\xi \mathcal{A},$$

that is, (VIII.1.1) is an extension of the operation of *principal adherence* defined in (VI.1.5).

Generalizing Proposition VI.2.4, for each convergence ξ and for any families $\mathcal{A}, \mathcal{B}$,

$$\operatorname{adh}_\xi 2^X = \varnothing, \tag{VIII.1.2}$$

$$\mathcal{A} \leq \mathcal{B} \Longrightarrow \operatorname{adh}_\xi \mathcal{A} \supset \operatorname{adh}_\xi \mathcal{B}, \tag{VIII.1.3}$$

$$\operatorname{adh}_\xi(\mathcal{A} \cap \mathcal{B}) = \operatorname{adh}_\xi \mathcal{A} \cup \operatorname{adh}_\xi \mathcal{B}. \tag{VIII.1.4}$$

Indeed, (VIII.1.2) holds, as $\mathcal{F}\perp 2^X$ for every filter on X (because $\varnothing \in 2^X$). (VIII.1.3) is straightforward. To see (VIII.1.4), let $x \in \operatorname{adh}_\xi(\mathcal{A} \cap \mathcal{B})$. Accordingly, there exists a filter $\mathcal{H}$ such that $x \in \lim_\xi \mathcal{H}$ and $\mathcal{H}\#(\mathcal{A} \cap \mathcal{B})$. By Proposition V.3.8, either $\mathcal{H}\#\mathcal{A}$ or $\mathcal{H}\#\mathcal{B}$, hence $x \in \operatorname{adh}_\xi \mathcal{A} \cup \operatorname{adh}_\xi \mathcal{B}$.

In particular, $\operatorname{adh}_\xi \mathcal{A}^\uparrow = \operatorname{adh}_\xi \mathcal{A}$, because $\mathcal{A} \subset \mathcal{A}^\uparrow \leq \mathcal{A}$ for any family $\mathcal{A}$ of subsets of $|\xi|$. Moreover, for every $x \in |\xi|$,

$$x \in \operatorname{adh}_\xi x^\uparrow. \tag{VIII.1.5}$$

In fact, this means that there exists $\mathcal{F} \# x^\uparrow$ such that $x \in \lim_\xi \mathcal{F}$. But, $x^\uparrow \# x^\uparrow$ and $x \in \lim_\xi x^\uparrow$, because each convergence ξ is (centered).

REMARK VIII.1.1. As $\{A\}^\uparrow \cap \{B\}^\uparrow = \{A \cup B\}^\uparrow$, (VIII.1.4) generalizes (d) of Proposition VI.2.4. In general, the inequality $\operatorname{adh}_\xi(\mathcal{A} \vee \mathcal{B}) \subset \operatorname{adh}_\xi \mathcal{A} \cap \operatorname{adh}_\xi \mathcal{B}$ cannot be reversed. A particular case $\operatorname{adh}_\xi(F \cap G) \subset \operatorname{adh}_\xi F \cap \operatorname{adh}_\xi G$ was discussed in Section VI.2.

LEMMA VIII.1.2. *If $\mathcal{F}$ is a filter, then*

$$\operatorname{adh}_\xi \mathcal{F} = \bigcup\nolimits_{\mathbb{F} \ni \mathcal{G} \supset \mathcal{F}} \lim_\xi \mathcal{G} = \bigcup\nolimits_{\mathcal{U} \in \beta\mathcal{F}} \lim_\xi \mathcal{U}. \tag{VIII.1.6}$$

PROOF. Indeed, if $\mathcal{F}$ is a filter and $x \in \operatorname{adh}_\xi \mathcal{F}$, then by (VIII.1.1), there is a filter $\mathcal{H}$ such that $\mathcal{H} \# \mathcal{F}$ and $x \in \lim_\xi \mathcal{H}$. Then $\mathcal{G} := \mathcal{F} \vee \mathcal{H} \supset \mathcal{F}$ and $x \in \lim_\xi \mathcal{H} \subset \lim_\xi \mathcal{G} \subset \lim_\xi \mathcal{U}$ for any ultrafilter $\mathcal{U}$ finer than $\mathcal{G}$, hence finer than $\mathcal{F}$.

If now $x \notin \operatorname{adh}_\xi \mathcal{F}$, then by (VIII.1.1), for each $x \notin \lim_\xi \mathcal{H}$ for each $\mathcal{H} \# \mathcal{F}$, in particular, $x \notin \lim_\xi \mathcal{G}$ for each filter $\mathcal{G}$ such that $\mathcal{G} \supset \mathcal{F}$, so that $x \notin \lim_\xi \mathcal{U}$ for each $\mathcal{U} \in \beta\mathcal{F}$. □

Formula (VIII.1.6) does not hold for arbitrary families $\mathcal{F}$, for if $F_0, F_1 \in \mathcal{F}$ are non-empty disjoint, then $\beta\mathcal{F} = \varnothing$. On the other hand, $\operatorname{adh}_\xi\{F_0, F_1\} = \operatorname{adh}_\xi F_0 \cup \operatorname{adh}_\xi F_1$ is non-empty.

Obviously, $\mathcal{F} \subset \mathcal{F}^\#$ for each filter $\mathcal{F}$. Therefore

$$\lim_\xi \mathcal{F} \subset \operatorname{adh}_\xi \mathcal{F} \tag{VIII.1.7}$$

for each filter $\mathcal{F}$, and by Lemma VIII.1.2,

$$\operatorname{adh}_\xi \mathcal{U} = \lim_\xi \mathcal{U}$$

for every ultrafilter $\mathcal{U}$.

PROPOSITION VIII.1.3. *If ξ is a Hausdorff convergence and $\lim_\xi \mathcal{F} \neq \varnothing$, then*

$$\operatorname{adh}_\xi \mathcal{F} = \lim_\xi \mathcal{F}.$$

PROOF. If $\lim_\xi \mathcal{F} \neq \varnothing$, then there is $x \in \lim_\xi \mathcal{F} \subset \lim_\xi \mathcal{U} = \{x\}$ for each $\mathcal{U} \in \beta\mathcal{F}$, because ξ is Hausdorff. Therefore, by Lemma VIII.1.2, $\operatorname{adh}_\xi \mathcal{F} \subset \{x\}$. □

The Hausdorff property is essential in Proposition VIII.1.3.

EXAMPLE VIII.1.4. Consider the Sierpiński topology $\$_0$ from Example III.5.3. As we have seen, the filter $\{\{0,1\}\}^{\uparrow}$ is convergent, and $\{1\} = \lim_{\$_0}\{\{0,1\}\}^{\uparrow}$, while $0 \in \lim_{\$_0}\{0\} \subset \mathrm{adh}_{\$_0}\{\{0,1\}\}^{\uparrow}$.

Clearly, for each family $\mathcal{A}$ on $|\xi| = |\theta|$,

$$\xi \geq \theta \Longrightarrow \mathrm{adh}_{\xi}\mathcal{A} \subset \mathrm{adh}_{\theta}\mathcal{A}. \tag{VIII.1.8}$$

PROPOSITION VIII.1.5. *If ξ is a pretopology, then for each family $\mathcal{A}$,*

$$\mathrm{adh}_{\xi}\,\mathcal{A} = \bigcap\nolimits_{A\in\mathcal{A}} \mathrm{adh}_{\xi}\,A. \tag{VIII.1.9}$$

PROOF. Indeed, if $x \in \mathrm{adh}_{\xi}\,A$ for each $A \in \mathcal{A}$, then for each A there exists a filter $\mathcal{F}_A \# A$ such that $x \in \lim_{\xi}\mathcal{F}_A$. Accordingly $\mathcal{F}_A \geq \mathcal{V}_{\xi}(x)$, that is, $A \in \mathcal{F}_A^{\#} \leq \mathcal{V}_{\xi}(x)^{\#}$ for each $A \in \mathcal{A}$, and thus $\mathcal{A}\#\mathcal{V}_{\xi}(x)$, which means that $x \in \mathrm{adh}_{\xi}\,\mathcal{A}$. □

Actually, Proposition XVIII.10.2 to come is a kind of converse of Proposition VIII.1.5.

COROLLARY VIII.1.6. *If ξ is a topology, then* $\mathrm{adh}_{\xi}\,\mathcal{A} = \bigcap_{A\in\mathcal{A}} \mathrm{cl}_{\xi}\,A$.

REMARK VIII.1.7. If $\mathcal{A}$ is a family of subsets of X, then $\mathcal{A}^{\cap}$ denotes the family of all finite intersections of elements of $\mathcal{A}$. As $\mathcal{A} \subset \mathcal{A}^{\cap}$, hence $\mathcal{A} \leq \mathcal{A}^{\cap}$, by (VIII.1.3), $\mathrm{adh}_{\xi}\,\mathcal{A}^{\cap} \subset \mathrm{adh}_{\xi}\,\mathcal{A}$. This inclusion cannot be reversed, even if $\mathcal{A}$ is centered.

EXAMPLE VIII.1.8. Let S be the circle from Example IV.3.16. Define $S(t) := (\cos t, \sin t) \in \mathbb{R}^2$, and let $A_0 := S([0,\pi[), A_1 := S(]\pi, 2\pi])$, and $\mathcal{A} := \{A_0, A_1\}$. As ν^2 is a topology, $\mathrm{adh}_{\nu^2}\,\mathcal{A} = \mathrm{adh}_{\nu^2}\,A_0 \cap \mathrm{adh}_{\nu^2}\,A_1 = \{S(0), S(\pi)\}$. On the other hand, $\mathcal{A}^{\cap}$ contains $A_0 \cap A_1 = \{S(0)\}$, and thus $\mathrm{adh}_{\nu^2}\,\mathcal{A}^{\cap} = \{S(0)\}$.

PROPOSITION VIII.1.9. *If ξ is a finitely stable convergence and $\mathcal{A}$ is finite, then*

$$\mathrm{adh}_{\xi}\,\mathcal{A} = \bigcap\nolimits_{A\in\mathcal{A}} \mathrm{adh}_{\xi}\,A.$$

PROOF. If $x \in \bigcap_{A\in\mathcal{A}} \mathrm{adh}_{\xi}\,A$, then for each $A \in \mathcal{A}$, there is a filter $\mathcal{F}_A$ such that $x \in \lim_{\xi}\mathcal{F}_A$ and $A \in \mathcal{F}_A^{\#}$. Hence $x \in \lim_{\xi}\bigcap_{A\in\mathcal{A}}\mathcal{F}_A$ and $\mathcal{A}\#(\bigcap_{A\in\mathcal{A}}\mathcal{F}_A)$. □

Let us provide formulas for the adherences with respect to initial and final convergences. They generalize the formulas (VI.4.1) and (VI.4.2), established for principal adherences. Let $\mathcal{H}$ and $\mathcal{G}$ be isotone families. Then

$$\operatorname{adh}_{f^- \tau} \mathcal{H} = f^-(\operatorname{adh}_\tau f[\mathcal{H}]), \tag{VIII.1.10}$$

$$\operatorname{adh}_{f\xi} \mathcal{G} = \ker \mathcal{G} \cup f(\operatorname{adh}_\xi f^-[\mathcal{G}]). \tag{VIII.1.11}$$

PROOF. By definition, $x \in \operatorname{adh}_{f^-\tau} \mathcal{H}$ if and only if there exists $\mathcal{F}$ such that $\mathcal{F}\#\mathcal{H}$ such that $x \in \lim_{f^-\tau} \mathcal{F} = f^-(\lim_\tau f[\mathcal{F}])$, that is, $f(x) \in \lim_\tau f[\mathcal{F}]$. It follows that $f[\mathcal{F}]\#f[\mathcal{H}]$, and thus $f(x) \in \operatorname{adh}_\tau f[\mathcal{H}]$, that is, $x \in f^-(\operatorname{adh}_\tau f[\mathcal{H}])$. Conversely, this holds if and only if there exists $\mathcal{F}$ such that $\mathcal{F}\#f[\mathcal{H}]$ and $f(x) \in \lim_\tau \mathcal{F}$. By Lemma IV.1.2, $\mathcal{F} \leq f[f^-[\mathcal{F}]]$, hence $f(x) \in \lim_\tau f[f^-[\mathcal{F}]]$, that is, $x \in \lim_{f^-\tau} f^-[\mathcal{F}]$. On the other hand, $\mathcal{F}\#f[\mathcal{H}]$ amounts to $f^-[\mathcal{F}]\#\mathcal{H}$, and thus $x \in \operatorname{adh}_{f^-\tau} \mathcal{H}$.

By definition, $y \in f(\operatorname{adh}_\xi f^-[\mathcal{G}])$ whenever there exists $\mathcal{F}\#f^-[\mathcal{G}]$ such that $y \in f(\lim_\xi \mathcal{F}) \subset f(\lim_\xi f[f^-[\mathcal{F}]]) = \lim_{f\xi} f^-[\mathcal{F}]$. As $\mathcal{F}\#f^-[\mathcal{G}]$ amounts to $f^-[\mathcal{F}]\#\mathcal{G}$, we conclude that $y \in \operatorname{adh}_{f\xi} \mathcal{G}$. Conversely if $y \in \operatorname{adh}_{f\xi} \mathcal{G}$, then there is a filter $\mathcal{F}$ such that $\mathcal{F}\#\mathcal{G}$ and $y \in \lim_{f\xi} \mathcal{F}$. If $y \notin f(X)$ and $y \in \lim_{f\xi} \mathcal{F}$, then $\mathcal{F} = y^\uparrow$, hence $\{y\}\#\mathcal{G}$ is equivalent to $y \in \ker \mathcal{G}$. Otherwise, there exists a filter $\mathcal{D}$ such that $f^-(y) \cap \lim_\xi \mathcal{D} \neq \varnothing$ and $\mathcal{F} \geq f[\mathcal{D}]$. It follows that $f[\mathcal{D}]\#\mathcal{G}$, equivalently $\mathcal{D}\#f^-[\mathcal{G}]$, and thus $f^-(y) \cap \operatorname{adh}_\xi f^-[\mathcal{G}] \neq \varnothing$, that is, $y \in f(\operatorname{adh}_\xi f^-[\mathcal{G}])$. □

Accordingly, if f is surjective, then

$$\operatorname{adh}_{f\xi} \mathcal{G} = f(\operatorname{adh}_\xi f^-[\mathcal{G}]). \tag{VIII.1.12}$$

COROLLARY VIII.1.10. *If $f \in \mathscr{C}(\xi, \tau)$ and $\mathcal{H}$ is a family of sets, then*

$$x \in \operatorname{adh}_\xi \mathcal{H} \Longrightarrow f(x) \in \operatorname{adh}_\tau f[\mathcal{H}].$$

Of course, Corollary VIII.1.10 can be readily proved directly.

PROOF. If $x \in \operatorname{adh}_\xi \mathcal{H}$, then there is $\mathcal{F}\#\mathcal{H}$ such that $x \in \lim_\xi \mathcal{F}$, hence $f[\mathcal{F}]\#f[\mathcal{H}]$ and $f(x) \in \lim_\tau f[\mathcal{F}]$, hence $f(x) \in \operatorname{adh}_\tau f[\mathcal{H}]$. □

VIII.2. Covers and inherences

The *inherence* of a family $\mathcal{D}$ with respect to a convergence ξ is defined by

$$\operatorname{inh}_\xi \mathcal{D} := (\operatorname{adh}_\xi \mathcal{D}_c)^c. \tag{VIII.2.1}$$

If $\mathcal{D} = \{D\}$, then $\operatorname{inh}_\xi D = \operatorname{inh}_\xi\{D\}$ coincides with the *principal inherence* from Section VI.3. If ξ is a pretopology, then in virtue of Proposition VIII.1.5,

$$\operatorname{inh}_\xi \mathcal{D} = \bigcup\nolimits_{D \in \mathcal{D}} \operatorname{inh}_\xi D. \tag{VIII.2.2}$$

It follows from Proposition VIII.1.9 that if ξ is a finitely stable convergence and $\mathcal{D}$ is finite, then (VIII.2.2) holds.

Recall Definition V.1.4 that a family $\mathcal{R}$ *refines* a family $\mathcal{P}$, in symbols $\mathcal{R} \lhd \mathcal{P}$, if for every $R \in \mathcal{R}$ there is $P \in \mathcal{P}$ such that $R \subset P$.

It follows from (VIII.1.2), (VIII.1.3), and (VIII.1.4) that, for antitone families $\mathcal{P}, \mathcal{R}$,

$$\operatorname{inh}_\xi 2^X = X, \tag{VIII.2.3}$$

$$\mathcal{R} \lhd \mathcal{P} \Longrightarrow \operatorname{inh}_\xi \mathcal{R} \supset \operatorname{inh}_\xi \mathcal{P}, \tag{VIII.2.4}$$

$$\operatorname{inh}_\xi(\mathcal{P} \cap \mathcal{R}) = \operatorname{inh}_\xi \mathcal{P} \cap \operatorname{inh}_\xi \mathcal{R}. \tag{VIII.2.5}$$

Hence, in particular, if $R \subset P$, then $\operatorname{inh}_\xi R \subset \operatorname{inh}_\xi P$, and $\operatorname{inh}_\xi(P \cap R) = \operatorname{inh}_\xi P \cap \operatorname{inh}_\xi R$.

Definition VIII.2.1 (Cover). Let ξ be a convergence on X. A family $\mathcal{P}$ of sets is called a ξ*-cover* of a set $A \subset X$, in symbols,

$$\mathcal{P} \succ_\xi A,$$

if $\mathcal{P} \cap \mathcal{F} \neq \varnothing$ for every filter $\mathcal{F}$ such that $A \cap \lim_\xi \mathcal{F} \neq \varnothing$.

Remark VIII.2.2. In a special case, where a convergence is a discrete topology ι, then ι-covers coincide with set-theoretic covers, that is,

$$\mathcal{P} \succ_\iota A \Longleftrightarrow A \subset \bigcup\nolimits_{P \in \mathcal{P}} P.$$

Indeed, the condition $x \in A \cap \lim_\iota \mathcal{F}$ entails $\mathcal{F} = x^\uparrow$, hence $\varnothing \neq \mathcal{P} \cap \mathcal{F} = \mathcal{P} \cap x^\uparrow$, that is, there is $P \in \mathcal{P}$ such that $x \in P$.

Lemma VIII.2.3. *Let $\mathcal{P}$ be a family of sets. The following statements are equivalent:*

$$\mathcal{P} \succ_\xi A,$$
$$\operatorname{adh}_\xi \mathcal{P}_c \cap A = \varnothing,$$
$$A \subset \operatorname{inh}_\xi \mathcal{P}.$$

Proof. A family $\mathcal{P}$ is not a ξ-cover of a set A if and only if there exists a filter $\mathcal{F}$ such that $A \cap \lim_\xi \mathcal{F} \neq \varnothing$ and $P \notin \mathcal{F}$ for each $P \in \mathcal{P}$, equivalently $\mathcal{P}_c \# \mathcal{F}$ by (V.3.4), that is, $A \cap \operatorname{adh}_\xi \mathcal{P}_c \neq \varnothing$. Finally, $\operatorname{adh}_\xi \mathcal{P}_c \cap A = \varnothing$ amounts to $\operatorname{adh}_\xi \mathcal{P}_c \subset A^c$, that is, $A \subset \operatorname{inh}_\xi \mathcal{P}$. □

Let $\mathcal{P}$ and $\mathcal{R}$ be two families of subsets of a convergence space. We observe that

Proposition VIII.2.4. *If $\mathcal{R}$ is a cover of A and refines $\mathcal{P}$, then $\mathcal{P}$ is a cover of A.*

PROOF. Indeed, if $\mathcal{R} \succ_\xi A$ and $A \cap \lim_\xi \mathcal{F} \neq \varnothing$ then there is $R \in \mathcal{R}$ such that $R \in \mathcal{F}$. As $\mathcal{R} \triangleleft \mathcal{P}$, there is $P \in \mathcal{P}$ such that $R \subset P$, hence $P \in \mathcal{F}$. $\square$

Proposition VIII.2.4 can be written down: if $\mathcal{R} \succ_\xi A$ and $\mathcal{R} \triangleleft \mathcal{P}$, then $\mathcal{P} \succ_\xi A$. In terms of inherences, it is rephrased:

$$(A \subset \operatorname{inh}_\xi \mathcal{R}) \wedge (\mathcal{R} \triangleleft \mathcal{P}) \Longrightarrow A \subset \operatorname{inh}_\xi \mathcal{P}.$$

Perhaps it becomes most evident written in terms of adherences:

$$(\operatorname{adh}_\xi \mathcal{R}_c \cap A = \varnothing) \wedge (\mathcal{R}_c \leq \mathcal{P}_c) \Longrightarrow \operatorname{adh}_\xi \mathcal{P}_c \cap A = \varnothing.$$

Let us mention that

LEMMA VIII.2.5. *If ξ is a convergence on X and $\mathcal{P}$ is a ξ-cover, then $\mathcal{P}$ is a ξ-cover of any $W \subset X$.*

In virtue of Proposition VIII.1.2,

PROPOSITION VIII.2.6. *If ξ is a pretopology, then $\mathcal{P}$ is a ξ-cover of A if and only if*

$$A \subset \bigcup_{P \in \mathcal{P}} \operatorname{inh}_\xi P. \tag{VIII.2.6}$$

If ξ is a topology, then each family $\mathcal{R}$ of ξ-open sets, such that $A \subset \bigcup \mathcal{R}$, is a ξ-cover of A ([1]). Conversely,

COROLLARY VIII.2.7. *If ξ is a topology, then for each ξ-cover $\mathcal{P}$ of A, there is a refinement $\mathcal{R} \subset \mathcal{O}_\xi$ of $\mathcal{P}$ such that $A \subset \bigcup \mathcal{R}$.*

PROOF. Let $\mathcal{P}$ be a cover of A. As ξ is, in particular, a pretopology, (VIII.2.6) holds. By topologicity, $\operatorname{inh}_\xi P = \operatorname{int}_\xi P$ is ξ-open for each P, hence $\mathcal{R} := \{\operatorname{inh}_\xi P : P \in \mathcal{P}\} \subset \mathcal{O}_\xi$ and $\mathcal{R} \triangleleft \mathcal{P}$. $\square$

Let us relate the notion of convergence cover to the traditional concept of *open cover* in a topological space.

DEFINITION VIII.2.8 (Open cover). A family $\mathcal{P}$ of subsets of X is called a *cover* of a set A if $A \subset \bigcup_{P \in \mathcal{P}} P$. If ξ is a topology on X, then a cover is said to be *ξ-open* or simply *open* if it consists of ξ-open sets.

Accordingly, each (set-theoretic) cover of a subset A of X is a ι_X-open cover of A. On the other hand, if $\mathcal{P}$ is a ξ-open cover of A whenever $A \subset \bigcup_{P \in \mathcal{P}} P = \bigcup_{P \in \mathcal{P}} \operatorname{int}_\xi P$, hence when $\mathcal{P}$ is a ξ-cover of A.

More generally,

[1] In fact, if $x \in A$ then there is $R \in \mathcal{R}$ such that $x \in R \in \mathcal{V}_\xi(x)$.

DEFINITION VIII.2.9. A family $\mathcal{P}$ is a ξ-*cover* of a family $\mathcal{A}$, in symbols,

$$\mathcal{P} \succ_{\xi} \mathcal{A}$$

if there is $A \in \mathcal{A}$ such that $\mathcal{P}$ is a ξ-*cover* of A.

THEOREM VIII.2.10. *If $\mathcal{A}$ is an isotone family, then the following statements are equivalent:*

$$\mathcal{P} \succ_{\xi} \mathcal{A}, \tag{VIII.2.7}$$

$$\operatorname{adh}_{\xi} \mathcal{P}_c \notin \mathcal{A}^{\#}, \tag{VIII.2.8}$$

$$\operatorname{inh}_{\xi} \mathcal{P} \in \mathcal{A}. \tag{VIII.2.9}$$

PROOF. $\mathcal{P} \succ_{\xi} \mathcal{A}$ whenever there exists $A \in \mathcal{A}$ such that $\mathcal{P} \succ_{\xi} A$. By Lemma VIII.2.3, this is equivalent to $\operatorname{adh}_{\xi} \mathcal{P}_c \cap A = \varnothing$, that is, $\operatorname{adh}_{\xi} \mathcal{P}_c \notin \mathcal{A}^{\#}$. This amounts to $(\operatorname{adh}_{\xi} \mathcal{P}_c)^c \in \mathcal{A}$, and by (VIII.2.1), to $\operatorname{inh}_{\xi} \mathcal{P} \in \mathcal{A}$. □

REMARK. Actually, the symbol $\succ_{\xi}$ becomes redundant, because $\mathcal{P} \succ_{\xi} \mathcal{A}$ if and only if $\operatorname{inh}_{\xi} \mathcal{P} \in \mathcal{A}$ for each isotone family $\mathcal{A}$.

DEFINITION VIII.2.11 (Pseudocover). A family $\mathcal{P}$ of subsets of a convergence space X is called a *pseudocover* of a subset A if every ultrafilter convergent to an element of A contains an element of $\mathcal{P}$.

By Proposition V.4.6, a family $\mathcal{P}$ is a pseudocover of A if and only if for every filter $\mathcal{F}$ that converges to an element of A there is a finite subfamily $\mathcal{P}_0$ of $\mathcal{P}$ such that $\bigcup \mathcal{P}_0 \in \mathcal{F}$. Of course, every cover is a pseudocover, and each ideal pseudocover is a cover.

EXAMPLE VIII.2.12. If ξ is a topology on X, and $P \subset X$ such that $\operatorname{cl}_{\xi} P \cap \operatorname{cl}_{\xi}(X \setminus P) \neq \varnothing$, then $\{P, X \setminus P\}$ is a ξ-pseudocover, but not a ξ-cover, of X, because the neighborhood filter of $x \in \operatorname{cl}_{\xi} P \cap \operatorname{cl}_{\xi}(X \setminus P)$ contains neither P nor $X \setminus P$.

An open pseudocover in a topological space is a cover. Indeed, let $\mathcal{P}$ be an open pseudocover of A in a topological space, that is, for each $x \in A$ there is a finite subfamily $\mathcal{P}_0$ of $\mathcal{P}$ such that $x \in \operatorname{int}(\bigcup \mathcal{P}_0)$, but since the elements of $\mathcal{P}_0$ are open, there is $P \in \mathcal{P}_0$ such that $x \in P$. It is straightforward that

PROPOSITION VIII.2.13. *$\mathcal{P}$ is a ξ-pseudocover of A if and only $\lim_{\xi} \mathcal{U} \cap A = \varnothing$ for every $\mathcal{U} \in \beta \mathcal{P}_c$.*

VIII.3. Compact convergences

A convergence is called compact if every filter is adherent. Explicitly, a convergence ξ on X is *compact* if

$$\operatorname{adh}_{\xi}\mathcal{H}\neq\varnothing \tag{VIII.3.1}$$

for each $\mathcal{H}\in\mathbb{F}X$. It follows that the chaotic topology o_X is compact for each non-empty set X, and the discrete topology ι_X is compact if and only if X is finite.

By Lemma VIII.1.2, a convergence ξ is compact if and only if

$$\lim\nolimits_{\xi}\mathcal{U}\neq\varnothing$$

for each ultrafilter $\mathcal{U}$ on X. It follows that every convergence on a finite set is compact.

A subset A of X is called *ξ-compact* if the convergence $\xi|_A$ is compact.

PROPOSITION VIII.3.1. *A subset A of X is ξ-compact if and only if, for each $\mathcal{H}\in\mathbb{F}X$,*

$$A\in\mathcal{H}^{\#}\Longrightarrow\operatorname{adh}_{\xi}\mathcal{H}\cap A\neq\varnothing.$$

PROOF. If A is ξ-compact and $\mathcal{G}$ is a filter on A, then $\varnothing\neq\operatorname{adh}_{\xi|_A}\mathcal{G}=A\cap\operatorname{adh}_{\xi}\mathcal{G}$. Conversely, if the condition holds and $A\in\mathcal{H}^{\#}$, then $\mathcal{H}\vee A$ contains A, hence $A\in(\mathcal{H}\vee A)^{\#}$, and $\operatorname{adh}_{\xi|_A}(\mathcal{H}\vee A)=A\cap\operatorname{adh}_{\xi}(\mathcal{H}\vee A)\neq\varnothing$. □

Consequently, A is ξ-compact if and only if $\lim_{\xi}\mathcal{U}\cap A\neq\varnothing$ for each $\mathcal{U}\in\beta A$.

PROPOSITION VIII.3.2. *An interval of the real line is compact if and only if it is closed and bounded.*

PROOF. Let $a<b$ and let $\mathcal{U}\in\beta[a,b]$. Then either $[a,\frac{a+b}{2}]\in\mathcal{U}$ or $[\frac{a+b}{2},b]\in\mathcal{U}$. Let $[a_1,b_1]$ be that that belongs to $\mathcal{U}$. By induction, construct a sequence of intervals

$$[a,b]\supset[a_1,b_1]\supset\ldots\supset[a_k,b_k]$$

that belong to $\mathcal{U}$. Then $(a_k)_k$ is an increasing sequence, $(b_k)_k$ is a decreasing sequence, $|b_k-a_k|=2^k\,|b_0-a_0|$ tends to 0 as k goes to ∞. Let $x\in[a,b]$ be such that

$$\sup\nolimits_{k<\omega}a_k\leq x\leq\inf\nolimits_{k<\omega}b_k.$$

Accordingly, $[a_k,b_k]\in\mathcal{N}_{\nu}(x)^{\#}$ for each $k<\omega$, so that $\mathcal{U}\geq\mathcal{N}_{\nu}(x)$.

If an interval is not closed, say $[a,b[$, then any ultrafilter containing $]b-\varepsilon,b[$ with $0<\varepsilon<b-a$, converges to $b\notin[a,b[$, which proves that $[a,b[$ is not compact.

If an interval is not bounded, say $[a, +\infty[$, then any ultrafilter containing $[n, +\infty[$ with $a < n$, is not convergent, and thus $[a, +\infty[$ is not compact. $\square$

PROPOSITION VIII.3.3. *If $f \in \mathscr{C}(\xi, \tau)$ and A is ξ-compact, then $f(A)$ is τ-compact.*

PROOF. Let $\mathcal{U} \in \beta f(A)$, equivalently $f^{-}[\mathcal{U}] \# A$. Hence there exists $\mathcal{W} \in \beta(f^{-}[\mathcal{U}] \vee A)$, and $A \# \lim_{\xi} \mathcal{W}$ by the ξ-compactness of A. From $f \in \mathscr{C}(\xi, \tau)$ it follows that $f(A) \# \lim_{\tau} f[\mathcal{W}]$. On the other hand, $f[\mathcal{W}] \geq f[f^{-}[\mathcal{U}]] \geq \mathcal{U}$ by Lemma IV.1.2, and since $\mathcal{W}$ is an ultrafilter, $f[\mathcal{W}] \approx \mathcal{U}$ by Proposition V.4.5. $\square$

THEOREM VIII.3.4 (Tikhonov). *Let Ξ be a non-empty set of convergences. Then $\prod \Xi$ is compact if and only if ξ is compact for each $\xi \in \Xi$.*

PROOF. Let ξ_i be a convergence on X_i for each $i \in I$ and let $\xi := \prod_{i \in I} \xi_i$. If ξ is compact, then ξ_i is compact for each $i \in I$, because $p_i \in \mathscr{C}(\xi, \xi_i)$, where $p_i : \prod_{k \in I} \xi_k \longrightarrow X_i$ is the canonical projection.

Conversely, let ξ_i be compact for each $i \in I$, and let $\mathcal{U}$ be an ultrafilter on $\prod_{i \in I} X_i$. Then $p_i[\mathcal{U}]$ is an ultrafilter on X_i for each $i \in I$, and since ξ_i is compact, there exists $x_i \in \lim_{\xi_i} p_i[\mathcal{U}]$ for each $i \in I$. If $\gamma \in \prod_{i \in I} X_i$ be given by $\gamma(i) := x_i$, then $\gamma \in \lim_{\xi} \mathcal{U}$, hence $\xi = \prod_{i \in I} \xi_i$ is compact. $\square$

It follows from Lemma VIII.2.3 that

PROPOSITION VIII.3.5. *A convergence ξ on X is compact if and only if $X = \operatorname{inh}_{\xi} \mathcal{P}$ implies that $X \in \mathcal{P}$ for every ideal $\mathcal{P}$ of subsets of X.*

PROOF. Indeed, $\mathcal{P}$ is an ideal if and only if the family of complements $\mathcal{P}_c$ (of the complements of elements of $\mathcal{P}$) is a filter. By definition, ξ is compact if and only if $\operatorname{adh}_{\xi} \mathcal{P}_c = \varnothing$ implies that $\varnothing \in \mathcal{P}_c$ for every ideal $\mathcal{P}$. By (VIII.2.1), this is means that if $X = \operatorname{inh}_{\xi} \mathcal{P}$, then $X \in \mathcal{P}$. $\square$

COROLLARY VIII.3.6. *Let ξ be a convergence on X. A subset A of X is ξ-compact if and only if, for each ideal $\mathcal{P}$ on X,*

$$A \subset \operatorname{inh}_{\xi} \mathcal{P} \Longrightarrow A \in \mathcal{P}.$$

Hence, by (VIII.2.2), if ξ is a pretopology, then a set A is ξ-compact if and only if for each ideal $\mathcal{P}$,

$$A \subset \bigcup_{P \in \mathcal{P}} \operatorname{inh}_{\xi} P \Longrightarrow A \in \mathcal{P}. \tag{VIII.3.2}$$

Finally, if ξ is a topology, then $\operatorname{inh}_{\xi} = \operatorname{int}_{\xi}$.

PROPOSITION VIII.3.7. *If ξ is a topology, then a set A is ξ-compact if and only if for each family $\mathcal{Q} \subset \mathcal{O}_\xi$ such that $A \subset \bigcup_{Q\in Q} Q$, there exists $\mathcal{Q}_0 \subset_{\text{fin}} \mathcal{Q}$ so that $A \subset \bigcup_{Q\in\mathcal{Q}_0} Q$.*

PROOF. Suppose that A is ξ-compact and $\mathcal{Q} \subset \mathcal{O}_\xi$ such that $A \subset \bigcup_{Q\in Q} Q = \bigcup_{Q\in Q} \operatorname{int}_\xi Q$. If $\mathcal{P}$ is the ideal generated by $\mathcal{Q}^\cup$, then by (VIII.3.2), there exists $\mathcal{Q}_0 \subset_{\text{fin}} \mathcal{Q}$ such that $A \subset \bigcup_{Q\in\mathcal{Q}_0} Q$. Conversely, if the condition holds, and $\mathcal{P}$ is an ideal such that $A \subset \bigcup_{P\in\mathcal{P}} \operatorname{int}_\xi P$, then $\mathcal{Q} := \{\operatorname{int}_\xi P : P \in \mathcal{P}\}$ is a family of ξ-open sets such that $A \subset \bigcup_{Q\in Q} Q$. Therefore, there exists a finite subfamily $\mathcal{P}_0$ of $\mathcal{P}$ such that $A \subset \bigcup_{P\in\mathcal{P}_0} \operatorname{int}_\xi P \subset \bigcup_{P\in\mathcal{P}_0} P \in \mathcal{P}$, and thus $A \in \mathcal{P}$. □

In view of Definition VIII.2.8,

COROLLARY VIII.3.8. *A subset A of a topological space is compact if and only if every open cover of A has a finite subcover of A.*

REMARK VIII.3.9. It turns out that the characterization of Corollary VIII.3.8 of compactness in topological spaces can be generalized in various ways. One is, of course, the original definition, but there exist also other possibilities, which are not equivalent to the definition for arbitrary convergences. They will be discussed in Section XX.1.

DEFINITION VIII.3.10. A convergence ξ is said to be *countably compact* if (VIII.3.1)

$$\operatorname{adh}_\xi \mathcal{H} \neq \varnothing$$

holds for every countably based filter $\mathcal{H}$.

Of course, each compact convergence is countably compact, but not conversely. Although countable compactness is preserved under continuous maps, the product of two countably compact topologies need not be countably compact ([2]). Analogously to the foregoing considerations,

PROPOSITION VIII.3.11. *A subset A of a topological space is countably compact if and only if every countable open cover of A has a finite subcover of A.*

PROOF. Let ξ be a countably compact convergence on X. If $\{H_n : n < \omega\}$ is a decreasing filter-base of $\mathcal{H}$, then $\operatorname{adh}_\xi \mathcal{H} \neq \varnothing$. Equivalently, if $X = \operatorname{inh}_\xi \mathcal{H}_c$ then there exists $n < \omega$ such that $H_n^c = X$. If now ξ is a topology and $\mathcal{P} = \{P_n : n < \omega\}$ is an increasing family of ξ-open sets such that $X = \bigcup_{n<\omega} P_n$, then $\bigcup_{n<\omega} P_n = \operatorname{inh}_\xi \mathcal{P}^{\cup\downarrow}$ and $\mathcal{P}^{\cup\downarrow}$ is an ideal. By the assumption of countable compactness, there is $n < \omega$ such that $P_n^c = \varnothing$, that is, $X = P_n = \bigcup_{k<n} P_k$. Conversely, if ξ is a

[2]See Examples XX.4.3 and XII.3.2.

topology and $\mathcal{H} = \{H_n : n < \omega\}$ is a decreasing filter-base, then by Corollary VIII.1.6, $\operatorname{adh}_\xi \mathcal{H} = \bigcap_{n<\omega} \operatorname{cl}_\xi H_n = \operatorname{adh}_\xi \mathcal{H}^{\uparrow\cap}$. If $\operatorname{adh}_\xi \mathcal{H} = \varnothing$ then $\{\operatorname{int}_\xi H_n^c : n < \omega\}$ is an increasing countable ξ-open cover of X, hence by assumption there is $n < \omega$ such that $\operatorname{int}_\xi H_n^c = X$, that is, $\operatorname{cl}_\xi H_n = \varnothing$, a contradiction. □

DEFINITION VIII.3.12. A convergence ξ is said to be *Lindelöf* if (VIII.3.1)

$$\operatorname{adh}_\xi \mathcal{H} \neq \varnothing$$

holds for every countably complete filter $\mathcal{H}$.

The Lindelöf property is preserved under continuous maps. The product of two Lindelöf topologies need not be Lindelöf ([3]). Moreover,

PROPOSITION VIII.3.13. *A subset A of a topological space is Lindelöf if and only if every open cover of A has a countable subcover of A.*

PROOF. Let ξ be a convergence on X. Notice that if $\operatorname{adh}_\xi \mathcal{H} \neq \varnothing$ for each countably complete filter $\mathcal{H}$, then $\operatorname{adh}_\xi \mathcal{G} \neq \varnothing$ for each filter $\mathcal{G}$ with countable intersection property ([4]). Indeed, if $\mathcal{G}$ has this property, then $\mathcal{H} := \{\bigcap \mathcal{F} : \mathcal{F} \subset \mathcal{G}, \operatorname{card} \mathcal{F} \leq \aleph_0\}$ is a countably complete filter finer than $\mathcal{G}$. By assumption, $\varnothing \neq \operatorname{adh}_\xi \mathcal{G} \subset \operatorname{adh}_\xi \mathcal{H}$. Therefore, ξ is Lindelöf if and only if $\operatorname{adh}_\xi \mathcal{G} \neq \varnothing$ for each filter $\mathcal{G}$ with countable intersection property. Equivalently, if $\operatorname{inh}_\xi \mathcal{G}_c = X$ implies that there exists a countable subfamily $\mathcal{F}$ of $\mathcal{G}$ such that $\bigcup \mathcal{F}_c = X$.

To see the converse let ξ be a topology and $\mathcal{P}$ be a family of ξ-open sets that covers X. Then by Corollary VIII.1.6, $X = \bigcup_{P \in \mathcal{P}} P = \operatorname{inh}_\xi \mathcal{P}^{\cup\downarrow}$ and $\mathcal{P}^{\cup\downarrow}$ is an ideal, hence there is a countable subfamily $\mathcal{P}_0$ of $\mathcal{P}$, thus also of $\mathcal{P}^{\cup\downarrow}$, such that $\bigcup \mathcal{P}_0 = X$. An obvious contraposition completes the proof. □

The discrete topology on an infinite countable set is Lindelöf but is not compact. Indeed, if X is infinite, then $\{\{x\} : x \in X\}$ is an ι_X-open cover without any finite subcover.

There exist countably compact topologies that are not compact, but commonly known examples, although not very difficult, require some preparation and will be presented later (see Example XII.4.5).

PROPOSITION VIII.3.14. *Every closed subset of a compact (countably compact, Lindelöf) convergence space is compact (respectively, countably compact, Lindelöf).*

[3] See Example XI.3.12.

[4] That is, if $\{G_n : n < \omega\} \subset \mathcal{G}$ then $\bigcap_{n<\omega} G_n \neq \varnothing$.

PROOF. Let ξ be a compact convergence on X, and let $C \in \mathcal{C}_\xi$. If $\mathcal{F}$ is a filter such that $C \in \mathcal{F}$, then $\varnothing \neq \mathrm{adh}_\xi \mathcal{F} \subset C$, because ξ is compact, and C is ξ-closed. By choosing $\mathcal{F}$ to be countably based and countably deep, we prove the two variants. □

Mind that no separation conditions are required in Proposition IX.3.1!

A compact set need not be closed.

EXAMPLE VIII.3.15. The set $\{0\}$ is $\$_0$-compact, but is not $\$_0$-closed, because $\lim_{\$_0} 0^{\uparrow} = \{0, 1\}$.

However,

PROPOSITION VIII.3.16. *Every compact subset of a Hausdorff convergence space is closed.*

PROOF. Let ξ be a Hausdorff convergence on X. Let C be ξ-compact, $\mathcal{F}$ be a filter on C, and let $\{x\} = \lim_\xi \mathcal{F}$. It follows that $C \cap \mathrm{adh}_\xi \mathcal{F} \neq \varnothing$ by the ξ-compactness of C. As ξ is Hausdorff, $\mathrm{adh}_\xi \mathcal{F} = \lim_\xi \mathcal{F}$ by Proposition VIII.1.3, because $\lim_\xi \mathcal{F} \neq \varnothing$. We conclude that $x \in C$, so that C is ξ-closed. □

REMARK VIII.3.17. Let τ be a convergence on Y and ξ be a convergence on a subset X of Y, so that $\xi = \tau|_X$. In other words, ξ is a subconvergence of τ. Let $A \subset X \subset Y$. Then A is τ-compact if and only if A is ξ-compact. Indeed, if $\mathcal{U} \in \beta A$, then, by (IV.3.3), $\lim_\xi \mathcal{U} = X \cap \lim_\tau \mathcal{U}$. Therefore,

$$A \cap \lim_\xi \mathcal{U} = A \cap X \cap \lim_\tau \mathcal{U} = A \cap \lim_\xi \mathcal{U} = A \cap \lim_\tau \mathcal{U}.$$

A property $P(\xi)$ of subsets of a convergence ξ is *absolute* whenever if $A \in P(\xi)$ and $\xi|_A = \tau|_A$, then $A \in P(\tau)$. Therefore,

COROLLARY VIII.3.18. *Compactness of sets is an absolute property.*

We shall discuss compactness, countable compactness and Lindelöf property in topological context (Section IX.3, Chapter XII) before tackling them in full generality in Chapter XX.

VIII.4. Supplement

LEMMA VIII.4.1. *If ξ is a free (alias T_1) convergence, then for any sequence $(x_n)_n$,*

$$\mathrm{adh}_\xi\{x_n : n < \omega\} = \{x_n : n < \omega\} \cup \mathrm{adh}_\xi (x_n)_n.$$

PROOF. If ξ is any convergence, then $x \in \mathrm{adh}_\xi\{x_n : n < \omega\}$ if and only if there exists $\mathcal{U} \in \beta\{x_n : n < \omega\}$ such that $x \in \lim_\xi \mathcal{U}$. If $\mathcal{U}$ is principal, there is $n < \omega$ such that $\mathcal{U} = x_n^\uparrow$, and thus $x \in \lim_\xi x_n^\uparrow$. Therefore, if ξ is free, then $x = x_n$. If $\mathcal{U}$ is free, then $\mathcal{U} \geq \mathcal{E}$, where $\mathcal{E}$ stands for the cofinite filter of $\{x_n : n < \omega\}$, that is, $x \in \mathrm{adh}_\xi \mathcal{E} = \mathrm{adh}_\xi (x_n)_n$ in virtue of Lemma VIII.1.2. □

PROPOSITION VIII.4.2. *Let ξ be a convergence. A set A is ξ-compact if and only if for each family $\mathcal{Q}$ of sets, $A \in \mathrm{inh}_\xi \mathcal{Q}$ implies the existence of $\mathcal{Q}_0 \subset_{\mathrm{fin}} \mathcal{Q}$ such that $A \subset \bigcup \mathcal{Q}_0$.*

PROOF. Let A be ξ-compact and $A \in \mathrm{inh}_\xi \mathcal{Q}$. Then $\mathcal{Q}$ refines the ideal $\mathcal{Q}^{\cup\downarrow}$ and thus by Proposition VIII.2.4, $A \in \mathrm{inh}_\xi \mathcal{Q}^{\cup\downarrow}$. As A is ξ-compact, $A \in \mathcal{Q}^{\cup\downarrow}$, that is, there exitsts $\mathcal{Q}_0 \subset_{\mathrm{fin}} \mathcal{Q}$ such that $A \subset \bigcup \mathcal{Q}_0$. Conversely, if the condition holds and $\mathcal{P}$ is an ideal such that $A \in \mathrm{inh}_\xi \mathcal{P}$, then there is $\mathcal{P}_0 \subset_{\mathrm{fin}} \mathcal{P}$ such that $A \subset \bigcup \mathcal{P}_0 \in \mathcal{P}$. □

EXAMPLE VIII.4.3. If $X = Y = \mathbb{R}$ endowed with the standard pretopology $\nu = \nu_\mathbb{R}$. The function f, defined by $f(x) := \frac{1}{x}$ if $x \neq 0$ and $f(0) := 0$, is continuous at each $x \neq 0$, and is not continuous at 0, because $\lim_\nu(\frac{1}{n})_n = \{0\}$, while $\lim_\nu(f(\frac{1}{n}))_n = \varnothing$ since $f(\frac{1}{n}) = n$ for each $n \in \mathbb{N}$. The corresponding relation R, defined by $Rx = \{\frac{1}{x}\}$ if $x \neq 0$ and $R0 = \{0\}$, is graph-closed. This follows from the continuity of f at each $x \neq 0$. On the other hand, the vicinity filter $\mathcal{V}_\nu(0)$ is the coarsest filter ν-convergent to 0, and its image $f[\mathcal{V}_\nu(0)]$ has a filter-base of the form $\{]-\infty, -n] \cup \{0\} \cup [n, +\infty[: n \in \mathbb{N}\}$. Therefore, if $\mathcal{U} \in \beta f[\mathcal{V}_\nu(0)]$, then by Proposition II.3.10, either $\{0\} \in \mathcal{U}$, hence $\mathcal{U} = 0^\uparrow$ and thus $\lim_\nu 0^\uparrow = \{0\}$, or $]-\infty, -n] \in \mathcal{U}$ for every $n < \omega$, or else $[n, +\infty[\in \mathcal{U}$ for every $n < \omega$, so that in these two cases $\lim_\nu \mathcal{U} = \varnothing$.

REMARK VIII.4.4. If oper : $2^{(2^X)} \longrightarrow 2^{(2^X)}$, then for $A \subset X$, we abridge

$$\mathrm{oper} A := \mathrm{oper}\{A\}.$$

Then, the superscript *natural* ♮ modifies the meaning of oper to the effect that

$$\mathrm{oper}^\natural \mathcal{A} := \{\mathrm{oper}\, A : A \in \mathcal{A}\}. \tag{VIII.4.1}$$

For example, $\mathrm{adh}_\xi^\natural \mathcal{A} = \{\mathrm{adh}_\xi A : A \in \mathcal{A}\}$ ([5]) denotes the family of ξ-adherences of the elements of $\mathcal{A}$.

[5]If oper is isotone, that is, $A_0 \subset A_1$ entails oper $A_0 \subset$ oper A_1, and $\mathcal{F}$ is a filter, then $\mathrm{oper}^\natural \mathcal{F}$ is (possibly improper) filter-base.

CHAPTER IX

Topological concepts

IX.1. Topological separation

A topology is said to be T_0 if for any two different points, there is an open set that contains exactly one of them ([1]).

It follows from Definition III.1.13 that a topology is free (alias T_1) if and only if each singleton is closed.

By Definition III.1.13, a convergence ξ is *Hausdorff* (alias T_2) if $\{x_0, x_1\} \subset \lim_\xi \mathcal{F}$ implies $x_0 = x_1$. Consequently, a topology ξ is *Hausdorff* if and only if ([2])

$$\mathcal{N}_\xi(x_0) \# \mathcal{N}_\xi(x_1) \Longrightarrow x_0 = x_1.$$

Since each neighborhood filter has a filter-base consisting of open sets,

PROPOSITION IX.1.1. *A topology ξ is Hausdorff if and only if $x_0 \neq x_1$ implies the existence of ξ-open sets O_0 and O_1 such that $x_0 \in O_0, x_1 \in O_1$ and $O_0 \cap O_1 = \varnothing$.*

By Proposition III.1.16, Hausdorff and T_1 properties are hereditary, and by Corollary VII.3.8, topologicity is hereditary, so that each restriction of a Hausdorff topology is a Hausdorff topology, and each restriction of a free topology is a free topology.

It is sometimes helpful to remember that

PROPOSITION IX.1.2. *A convergence ξ on X is Hausdorff if and only if the diagonal $\triangle(X) = \{(x,x) : x \in X\}$ is $\xi \times \xi$-closed.*

PROOF. By definition, $\triangle(X)$ is $\xi \times \xi$-closed if and only if whenever a filter $\mathcal{G}$ is such that $\triangle(X) \in \mathcal{G}$ and $(x_0, x_1) \in \lim_{\xi\times\xi} \mathcal{G}$, then $(x_0, x_1) \in \triangle(X)$, that is, $x_0 = x_1$. As $\triangle$ is injective, there is a filter $\mathcal{F}$ on X such that $\mathcal{G} \approx \triangle[\mathcal{F}]$. Now, $(x_0, x_1) \in \lim_{\xi\times\xi} \mathcal{G}$ if and only if $x_0 \in \lim_\xi p_0[\mathcal{G}]$ and $x_1 \in \lim_\xi p_1[\mathcal{G}]$, where $p_0, p_1 : X \times X \longrightarrow X$ are the canonical

[1] We have come accross with the property T_0 for arbitrary convergences in Definition III.1.18, and in Proposition VII.5.4 in the case of topologies. Proposition VII.5.5 says that each restriction of a T_0-topology is T_0 (although this is not true for arbitrary convergences).

[2] If $\mathcal{F} := \mathcal{N}_\xi(x_0) \vee \mathcal{N}_\xi(x_1)$ is proper, then $\{x_0, x_1\} \subset \lim_\xi \mathcal{F}$.

projections. On the other hand, $p_0[\mathcal{G}] = p_1[\mathcal{G}] = \mathcal{F}$, so that $x_0, x_1 \in \lim_\xi \mathcal{F}$. Therefore, ξ is Hausdorff if and only if $x_0 = x_1$. $\square$

The identity map $i \in \mathscr{C}(\xi, \xi)$ is always continuous, but its graph $\mathrm{gr}(i)$ is $\xi \times \xi$-closed if and only if ξ is Hausdorff.

LEMMA IX.1.3 (Closed coincidence set). *Let ξ be a convergence, and $f, g \in \mathscr{C}(\xi, \theta)$, where θ is a Hausdorff convergence. Then $\{f = g\}$ is ξ-closed.*

PROOF. Let ξ be a convergence on X and θ a convergence on Y. The diagonal product map $f \bigtriangleup g : X \longrightarrow Y \times Y$ is continuous $f \bigtriangleup g \in \mathscr{C}(\xi, \theta \times \theta)$ by Proposition IV.7.6. Therefore, by Proposition IX.1.2, $\triangle(Y)$ is $\theta \times \theta$-closed, and by Proposition VII.3.10, $\{f = g\} = (f \bigtriangleup g)^-(\triangle(Y))$ is ξ-closed. $\square$

COROLLARY IX.1.4. *If ξ is a convergence, τ is a Hausdorff convergence, and $f_0, f_1 \in \mathscr{C}(\xi, \tau)$ are equal on a ξ-dense set, then $f_0 = f_1$.*

PROOF. Let D be a ξ-dense set such that $D \subset \{f_0 = f_1\}$. Then $|\xi| = \mathrm{cl}_\xi D \subset \mathrm{cl}_\xi\{f_0 = f_1\} = \{f_0 = f_1\}$ by Lemma IX.1.3. $\square$

DEFINITION IX.1.5 (Regular topology). A topology ξ is called *regular* if for each ξ-closed set F and every $x \notin F$, there exist ξ-open sets O_0 and O_1 such that $x \in O_0, F \subset O_1$ and $O_0 \cap O_1 = \varnothing$. A free (alias T_1) regular topology is called T_3.

Therefore, ξ is regular if and only if for each point x and a ξ-closed set F,

$$\mathcal{N}_\xi(x) \# \mathcal{N}_\xi(F) \Longrightarrow x \in F.$$

REMARK IX.1.6. Each regular free topology is Hausdorff. If ξ is free and $x_0 \neq x_1$, then $\{x_1\}$ is ξ-closed and $x_0 \notin \{x_1\}$, and if ξ is regular, there are disjoint ξ-open sets O_0 and O_1 such that $x_0 \in O_0$ and $\{x_1\} \subset O_1$.

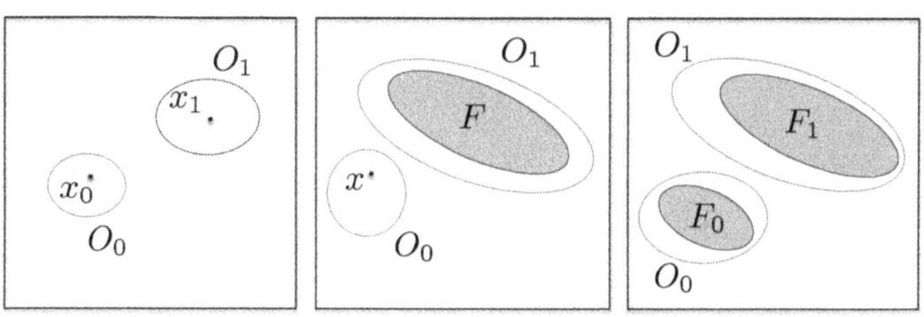

FIGURE IX.1.1. Three kinds of separation in topological spaces: Hausdorff property, regularity, and normality.

Observe that if F is ξ-closed and $x \notin F$, then $F^c \in \mathcal{O}_\xi(x) \subset \mathcal{N}_\xi(x)$. If O_0 and O_1 are ξ-open sets as above then $x \in O_0 \subset O_1^c \subset F^c$, hence $O_0 \in \mathcal{N}_\xi(x_0)$ and $\mathrm{cl}_\xi O_0 \subset F^c$. Accordingly,

COROLLARY IX.1.7. *A topology ξ is regular if and only if for each x and every $V \in \mathcal{N}_\xi(x)$, there exists $W \in \mathcal{N}_\xi(x)$ such that $\mathrm{cl}_\xi W \subset V$.*

In other words, a topology ξ is regular if and only if for each x,

$$\{\mathrm{cl}_\xi W : W \in \mathcal{N}_\xi(x)\} \geq \mathcal{N}_\xi(x).$$

Accordingly, a topology is regular if and only if every neighborhood filter has a filter-base consisting of closed sets. In particular, metrizable topologies are regular.

DEFINITION IX.1.8 (Boundary). The *boundary* $\partial_\xi A$ of A is defined as

$$\partial_\xi A = \mathrm{cl}_\xi A \cap \mathrm{cl}_\xi(X \setminus A),$$

where X is the underlying set of ξ.

Of course, $\partial_\xi A = \partial_\xi(X \setminus A)$ for every $A \subset X$. On the other hand,

$$\partial_\xi A = \mathrm{cl}_\xi A \setminus \mathrm{int}_\xi A, \tag{IX.1.1}$$

Notice that $\partial_\iota A = \varnothing$ for each $A \subset X$, where ι is the discrete topology on X, and $\partial_o A = X$ if $\varnothing \neq A \neq X$, where o is the chaotic topology on X.

EXAMPLE IX.1.9. We notice that $\partial_{\nu_\mathbb{R}}[0,1] = \partial_{\nu_\mathbb{R}}[0,1[= \partial_{\nu_\mathbb{R}}]0,1[= \{0,1\}$. Observe that $\partial_{\nu_\mathbb{R}^2} A = \{(x,y) : \|(x,y)\|_2 = 1\}$ if $\{(x,y) : \|(x,y)\|_2 < 1\} \subset A \subset \{(x,y) : \|(x,y)\|_2 \leq 1\}$.

DEFINITION IX.1.10. A subset of a convergence space is called *clopen* if it is both closed and open.

The notion of clopen set is, of course, topological. We notice that if ξ is a topology, then A is ξ-clopen whenever

$$\mathrm{cl}_\xi A \subset A \subset \mathrm{int}_\xi A. \tag{IX.1.2}$$

In particular, each subset of a discrete topology is clopen. On the other hand, the only clopen sets for the chaotic topology o on X are $\varnothing$ and X.

By (IX.1.2) and (IX.1.1),

PROPOSITION IX.1.11. *A set is clopen if and only if its boundary is empty.*

EXAMPLE IX.1.12 (A Hausdorff, non-regular topology). Let ρ be a convergence defined on the real line by $\rho^-(x) := \nu_{\mathbb{R}}^-(x)$ if $x \neq 0$, and $0 \in \lim_\rho \mathcal{F}$ provided that $0 \in \lim_{\nu_{\mathbb{R}}} \mathcal{F}$ and $\{2^{-n} : n < \omega\} \notin \mathcal{F}^\#$.

Of course, $\nu_{\mathbb{R}} \leq \rho$, hence, $\nu_{\mathbb{R}} \leq \mathrm{T}\rho$, in particular, $\mathrm{T}\rho$ is Hausdorff. To see that ρ is a topology, notice that $\{2^{-n} : n < \omega\}$ is ρ-closed, and $0 \notin \{2^{-n} : n < \omega\}$. The neighborhood filter $\mathcal{N}_\rho(0)$ has a filter-base of the form

$$\{V \setminus \{2^{-n} : n < \omega\} : V \in \mathcal{O}_{\nu_{\mathbb{R}}}(0)\},$$

hence $0 \in \lim_\rho \mathcal{N}_\rho(0)$ and thus $\mathrm{T}\rho = \rho$. Since $\mathrm{cl}_\rho(]-r, r[\setminus\{2^{-n} : n < \omega\}) = [-r, r]$ is not included in $V \setminus \{2^{-n} : n < \omega\}$ for any $V \in \mathcal{N}_{\nu_{\mathbb{R}}}(0)$, it follows that ρ is not regular.

Regularity is hereditary.

PROPOSITION IX.1.13. *Each subspace of a regular topology is regular.*

PROOF. Let τ be a regular topology on Y and let $\varnothing \neq X \subset Y$. If $V \in \mathcal{O}_{\tau|_X}(x)$, then by Proposition VII.3.12, there exists $O \in \mathcal{O}_\tau(x)$ such that $V = O \cap X$. By regularity of τ, there exists $Q \in \mathcal{O}_\tau(x)$ such that $\mathrm{cl}_\tau Q \subset O$. Hence $Q \cap X \in \mathcal{O}_{\tau|_X}(x)$, and

$$V = O \cap X \supset \mathrm{cl}_\tau Q \cap X \subset Q \cap X \in \mathcal{O}_{\tau|_X}(x).$$

As $\mathrm{cl}_\tau Q \cap X$ is $\tau|_X$-closed, the proof is complete. □

PROPOSITION IX.1.14. *Each product of regular topologies is a regular topology.*

PROOF. Let ξ_j be a regular topology on X_j for each $j \in J$. Then $V \in \mathcal{N}_{\prod_{j\in J}\xi_j}(\varphi)$ if there is a finite subset J_0 and $V_j \in \mathcal{N}_{\xi_j}(\varphi(j))$ for each $j \in J$ such that $V_j = X_j$ if $j \notin J_0$ and

$$\prod\nolimits_{j\in J} V_j \subset V.$$

By Corollary IX.1.7, for each $j \in J_0$, there is $W_j \in \mathcal{N}_{\xi_j}(\varphi(j))$ such that $\mathrm{cl}_{\xi_j} W_j \subset V_j$. On setting $W_j := X_j$ if $j \notin J_0$, we get $\prod_{j\in J} W_j \in \mathcal{N}_{\prod_{j\in J}\xi_j}(\varphi)$, the $\prod_{j\in J}\xi_j$-closure of which is included in V. □

DEFINITION IX.1.15. We say that a topology is *normal* if for any pair F_0, F_1 of disjoint ξ-closed sets, there exist open sets O_0, O_1 such that $F_0 \subset O_0, F_1 \subset O_1$ and $O_0 \cap O_1 = \varnothing$.

Consequently, a topology is normal if and only if for any pair of ξ-closed sets F_0, F_1,

$$\mathcal{N}_\xi(F_0)\#\mathcal{N}_\xi(F_1) \Longrightarrow F_0\#F_1.$$

REMARK IX.1.16 (*). If ξ is not normal, then there exists a pairwise disjoint family $\mathcal{A} \subset \mathcal{C}_\xi$ such that $\operatorname{card} \mathcal{A} > 1$ and $\bigvee_{A \in \mathcal{A}} \mathcal{N}_\xi(A)$ is a proper filter. The least cardinal λ such that there exists a family $\mathcal{A}$ of cardinality λ such that $\bigvee_{A \in \mathcal{A}} \mathcal{N}_\xi(A)$ is proper is called the *non-normality number of* ξ ([3]).

By Exercise IX.5.9,

PROPOSITION IX.1.17. *A topology ξ is normal if and only if for each $F \in \mathcal{C}_\xi$ and every $V \in \mathcal{O}_\xi(F)$, there exists $W \in \mathcal{O}_\xi(F)$ such that $\operatorname{cl}_\xi W \subset V$.*

In other words, a topology ξ is normal if and only if for each $F \in \mathcal{C}_\xi$,

$$\{\operatorname{cl}_\xi W : W \in \mathcal{N}_\xi(F)\} \geq \mathcal{N}_\xi(F).$$

We notice that each normal free (alias T_1) topology is regular, because each singleton is closed, so that the condition of Proposition IX.1.17 implies that of Corollary IX.1.7.

As we shall see in Exercise XI.5.1, each metrizable topology is normal. In particular, the standard topology of the real line $\nu_\mathbb{R}$ is normal, as well as the standard topologies $\nu_\mathbb{R}^n$ of n-dimensional Euclidean space.

EXAMPLE IX.1.18 (Prime cofinite topology is normal). Consider the prime cofinite topology $\pi = \pi[x_\infty, (X)_0]$ on an infinite set X from Example III.4.3. As each x distinct from x_∞ is π-isolated, a set A is π-open provided that $x_\infty \notin A$ or $X \setminus A$ is finite. Accordingly, F is π-closed if $x_\infty \in F$ or F is finite. Therefore, π is regular, because each $V \in \mathcal{N}_\pi(x_\infty)$ is π-closed. To see that π is normal, let F_0 and F_1 be disjoint π-closed sets. If $x_\infty \in F_0$, then F_1 is finite, hence $\operatorname{int}_\pi F_1 = F_1$, so that F_1 and $X \setminus F_1$ are π-open disjoint. Otherwise, both F_0 and F_1 are finite and π-open.

Here is a classical example of a Hausdorff regular topology that is not normal.

EXAMPLE IX.1.19 (*Niemytzki plane*). Consider the closed upper half-plane $X := \{(x, y) \in \mathbb{R}^2 : y \geq 0\}$ endowed with the following pretopology ζ: if $y > 0$ then $\mathcal{V}_\zeta(x, y) := \mathcal{N}_{\nu_\mathbb{R}^2}(x, y)$, where $\nu_\mathbb{R}^2$ stands for the standard topology of the real plane. A filter-base of $\mathcal{V}_\zeta(x, 0)$ is composed of $V_n := B((x, 2^{-n}), 2^{-n}) \cup \{(x, 0)\}$ with $n < \omega$.

It is immediate that each V_n is ζ-open, $\operatorname{cl}_\zeta V_n \subset V_{n-1}$ for $n \in \mathbb{N}_1$, and thus ζ is a Hausdorff, regular topology. However, it is not normal.

[3]See S. DOLECKI, [57] and S. DOLECKI, T. NOGURA, R. PEIRONE, and G. M. REED [86].

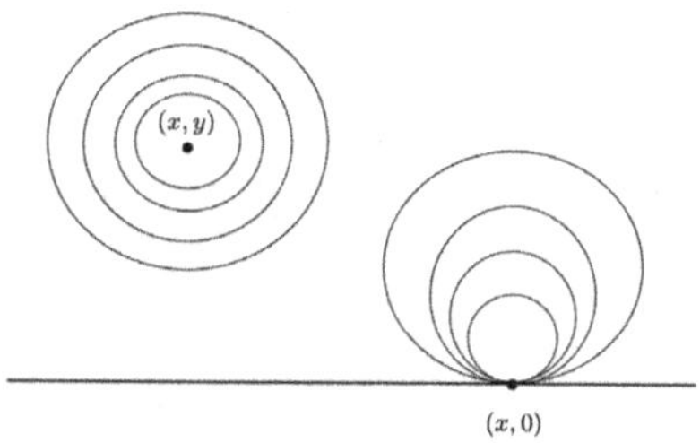

FIGURE IX.1.2. Niemytzki plane: typical neighborhoods for $y = 0$ and $y > 0$.

Start by observing that each subset F of $L := \{(x, 0) : x \in \mathbb{R}\}$ is ζ-closed. If ζ were normal, for each $F \subset L$, there would be $U_F \in \mathcal{O}_\zeta(F)$ and $W_F \in \mathcal{O}_\zeta(L \setminus F)$ such that $U_F \cap W_F = \varnothing$.

We shall get a contradiction with the aid of Theorem I.5.8 of Cantor, by showing that the assignment $F \mapsto U_F \cap (\mathbb{Q} \times \mathbb{Q}_+)$ is injective.

Indeed, on one hand, there are $2^{\mathfrak{c}}$ subsets F of the real line L, and on the other, there are $2^{\aleph_0} = \mathfrak{c}$ of the rational upper half-plane. This would mean that $2^{\mathfrak{c}} \leq \mathfrak{c}$, which is false by Theorem I.5.8. Hence, such an assignment does not exist.

Let us suppose that $F_1 \setminus F_0 \neq \varnothing$, say $x \in F_1 \setminus F_0$. As $F_1 \subset U_{F_1}$ and $L \setminus F_0 \subset W_{F_0}$, we infer that the ζ-open set $U_{F_1} \cap W_{F_0}$ is non-empty, because $x \in U_{F_1} \cap W_{F_0}$, hence

$$\varnothing \neq U_{F_1} \cap W_{F_0} \cap (\mathbb{Q} \times \mathbb{Q}_+) = U_{F_1} \cap (\mathbb{Q} \times \mathbb{Q}_+) \setminus (U_{F_0} \cap (\mathbb{Q} \times \mathbb{Q}_+)),$$

showing the injectivity of $F \mapsto U_F \cap (\mathbb{Q} \times \mathbb{Q}_+)$.

In Example X.7.14 we will prove once more that the Niemytzki plane is not normal with the aid of Baire Category Theorem X.4.15 ([4]).

EXAMPLE IX.1.20. Consider the Sierpiński topology $\$_0$ (Example III.5.3), that is, the topology on $\{0, 1\}$, in which $\mathcal{C}_{\$_0} = \{\varnothing, \{1\}, \{0, 1\}\}$. Obviously, this topology is normal, because in each pair of disjoint closed sets, one is necessarily empty, hence simultaneously closed and open, that is, clopen. However, $\$_0$ is not regular. In fact, $\{0\}$ is a least open neighborhood of 0, so that if $W \in \mathcal{O}_{\$_0}(0)$, then $\mathrm{cl}_{\$_0} W = \mathrm{cl}_{\$_0}\{0\} = \{0, 1\}$ is not included in $\{0\}$.

[4]Actually it was proved in [57] that the non-normality (see Remark IX.1.16) of the Niemytzki plane is continuum, and more generally in [86] that the non-normality number of every separable regular topology with a closed discrete subspace of cardinality continuum is at least continuum.

Here is an instrumental condition of normality in terms of open covers:

LEMMA IX.1.21. *A free (alias T_1) topology on X is normal if for each closed set F and each $O \in \mathcal{O}_X(F)$, there exists a countable open cover $\{O_n : n < \omega\}$ of F such that $\bigcup_{n<\omega} \mathrm{cl}_X O_n \subset O$.*

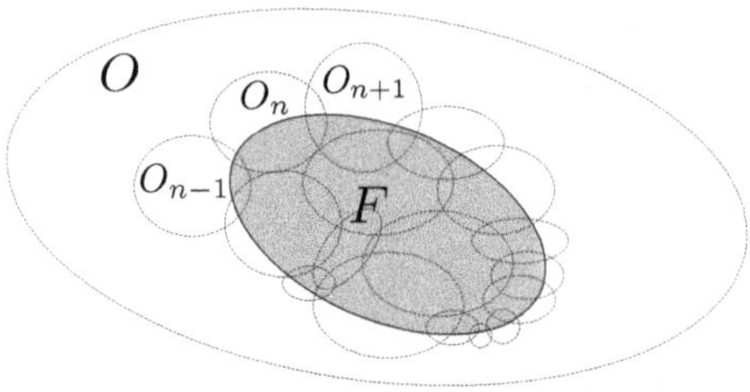

FIGURE IX.1.3. A criterion for normality: if for each closed set F and $O \in \mathcal{O}(F)$, there exists a countable cover $\{O_n : n < \omega\}$ of F such that $\bigcup_{n<\omega} \mathrm{cl}\, O_n \subset O$, then the topology is normal.

PROOF. Suppose that a free topology on X fulfills the condition. Let F_0 and F_1 be disjoint closed sets. Then $X \setminus F_1 \in \mathcal{O}_X(F_0)$, and there exists $\{O_n : n < \omega\} \subset \mathcal{O}_X$ such that

$$F_0 \subset \bigcup_{n<\omega} O_n \text{ and } \bigcup_{n<\omega} \mathrm{cl}_X O_n \cap F_1 = \varnothing.$$

On the other hand, $X \setminus F_0 \in \mathcal{O}_X(F_1)$, so that there exists a family $\{V_n : n < \omega\}$ of open sets such that $F_1 \subset \bigcup_{n<\omega} V_n$ and $\bigcup_{n<\omega} \mathrm{cl}_X V_n \cap F_0 = \varnothing$. Let

$$G_n := O_n \setminus \bigcup_{k\leq n} \mathrm{cl}_X V_k, \text{ and } H_n := V_n \setminus \bigcup_{k\leq n} \mathrm{cl}_X O_k.$$

Then $\bigcup_{n<\omega} G_n \in \mathcal{O}_X(F_0)$ and $\bigcup_{n<\omega} H_n \in \mathcal{O}_X(F_1)$ are disjoint. Indeed, let $m \geq n$, and let $x \in G_n \cap H_m$. Then on one hand, $x \in O_n$, and on the other, $x \notin \mathrm{cl}_X O_n$, which is a contradiction. We proceed similarly, if $m \leq n$. □

COROLLARY IX.1.22. *Each regular free (alias T_1) topology on a countable set is normal.*

PROOF. Let F be a closed set and $O \in \mathcal{O}(F)$. By regularity, for each $x \in F$ there exists $O_x \in \mathcal{O}(x)$ such that $\mathrm{cl}\, O_x \subset O$. As F is countable, we can apply Lemma IX.1.21 to conclude. □

DEFINITION IX.1.23. If ξ is a topology, then two sets A, B are called *ξ-separated* if

$$\mathrm{cl}_\xi A \cap B = A \cap \mathrm{cl}_\xi B = \varnothing.$$

Two disjoint closed sets are separated. Two disjoint open sets are separated.

EXAMPLE IX.1.24. Let $\nu_{\mathbb{R}}$ be the standard topology of $\mathbb{R}$. Then, for instance, $]0,1[, [1,2]$ are not $\nu_{\mathbb{R}}$-separated. On the other hand, $]0,1[,]1,2]$ are $\nu_{\mathbb{R}}$-separated.

LEMMA IX.1.25. *If ξ is a topology on X, and $\varnothing \neq W \subset X$. Two $\xi|_W$-closed disjoint subsets of W are ξ-separated.*

PROOF. Let $F_0, F_1 \in \mathcal{C}_{\xi|_W}$ be disjoint. By Corollary VII.3.13, there exist $C_0, C_1 \in \mathcal{C}_\xi$ such that $F_0 = C_0 \cap W$ and $F_1 = C_1 \cap W$. Then

$$\begin{aligned}\mathrm{cl}_\xi F_0 \cap F_1 &\subset C_0 \cap F_1 = C_0 \cap (C_1 \cap W)\\ &= (C_0 \cap W) \cap (C_1 \cap W) = F_0 \cap F_1 = \varnothing,\end{aligned}$$

and similarly $F_0 \cap \mathrm{cl}_\xi F_1 = \varnothing$. □

COROLLARY IX.1.26. *Two sets A, B are separated if and only if they are closed disjoint in the subspace $A \cup B$.*

PROOF. One implication has been already proved in Lemma IX.1.25. Let ξ be a topology on X, and let A, B be ξ-separated. Let $W := A \cup B$. Then $\mathrm{cl}_{\xi|_W} A = \mathrm{cl}_\xi A \cap (A \cup B) = A$, hence A is $\xi|_W$-closed. Analogously, B is $\xi|_W$-closed. □

Normality is not hereditary (see Example IX.1.28). By Proposition IV.6.11, every product of Hausdorff topologies is Hausdorff. By Proposition IX.1.14, each product of regular topologies is regular. A product of two normal topologies need not be normal, as Example X.7.22 will show.

In view of Definition VI.5.9, a topology is called *hereditarily normal* if every restriction is normal.

Of course, each hereditarily normal topology is normal, but there exist normal topologies that are not hereditarily normal (Example IX.1.28).

PROPOSITION IX.1.27. *The following statements are equivalent:*

1. *a topology ξ is hereditarily normal,*
2. *every open subspace of ξ is normal,*
3. *for any two ξ-separated sets A_0, A_1 there exist $O_0 \in \mathcal{O}_\xi(A_0)$ and $O_1 \in \mathcal{O}_\xi(A_1)$ such that $O_0 \cap O_1 = \varnothing$.*

PROOF. (1) $\Rightarrow$ (2) is obvious. (2) $\Rightarrow$ (3). Let A, B such that $\mathrm{cl}_\xi A \cap B = \varnothing = A \cap \mathrm{cl}_\xi B$, and let

$$W := X \setminus (\mathrm{cl}_\xi A \cap \mathrm{cl}_\xi B).$$

Then $A \subset W$ and $B \subset W$, and $\mathrm{cl}_{\xi|_W} A \cap \mathrm{cl}_{\xi|_W} B = \varnothing$. Hence by (2), there exist disjoint $\xi|_W$-open sets O_0 and O_1 such that $A \subset O_0$ and $B \subset O_1$. As W is ξ-open, O_0 and O_1 are ξ-open.

(3) $\Rightarrow$ (1). Let $W \subset X$ and assume that $F_0, F_1 \in \mathcal{C}_{\xi|_W}$ are disjoint, hence ξ-separated by Lemma IX.1.25. By assumption, there exist $O_0 \in \mathcal{O}_\xi(F_0)$ and $O_1 \in \mathcal{O}_\xi(F_1)$ such that $O_0 \cap O_1 = \varnothing$. Therefore, $O_0 \cap W \in \mathcal{O}_{\xi|_W}(F_0)$ and $O_1 \cap W \in \mathcal{O}_{\xi|_W}(F_1)$, so that $\xi|_W$ is normal. $\square$

EXAMPLE IX.1.28 (A normal, non-hereditarily normal topology). Let X be a countable set with the prime cofinite topology at x_∞. Let Y be an uncountable set with the prime cofinite topology at y_∞. By Proposition IX.3.4 to come, their product is normal. However, $Z := X \times Y \setminus \{(x_\infty, y_\infty)\}$ is not normal.

Indeed, $F_0 := (X \setminus \{x_\infty\}) \times \{y_\infty\}$ and $F_1 := \{x_\infty\} \times (Y \setminus \{y_\infty\})$ are Z-closed and disjoint. Let O_0, O_1 be Z-open sets such that $F_0 \subset O_0$ and $F_1 \subset O_1$. For each $x \in X \setminus \{x_\infty\}$ there exists $U_x \in \mathcal{O}_Y(y_\infty)$ such that $U_x \subset O_1$. Hence,

$$O_0 \supset \bigcup \{U_x \times \{y_\infty\} : x \in F_0\} \in \mathcal{O}_Z(F_0).$$

On the other hand, $\bigcap_{x \in X \setminus \{x_\infty\}} U_x$ is a cocountable subset of Y, because U_x is a cofinite subset of Y for each $x \in X \setminus \{x_\infty\}$. If $y \in \bigcap_{x \in X \setminus \{x_\infty\}} U_x$, then there is a cofinite subset V of X such that $(x_\infty, y) \in \{x_\infty\} \times V \subset O_0 \cap O_1$, proving that Z is not normal.

IX.2. Bases

DEFINITION IX.2.1. A subfamily $\mathcal{B}$ of open sets is called a *base for the open sets* if each open set is a union of elements of $\mathcal{B}$.

Dually,

DEFINITION IX.2.2. A subfamily of closed sets is called a *base for the closed sets* if each closed set is an intersection of its elements.

REMARK IX.2.3. The notion of *base for the closed sets* is deployed less often than that of *base for the open sets*. Hence a *base* is usually understood as that for the open sets, unless specified otherwise.

Let us rephrase Definition IX.2.1 in more detail: a family $\mathcal{B}$ of θ-open sets is a *base* for the open sets of θ if for each $x \in |\theta|$ and every

$O \in \mathcal{O}_\theta(x)$, there exists $B \in \mathcal{B}$ such that

$$x \in B \subset O.$$

DEFINITION IX.2.4. A subfamily $\mathcal{S}$ of $\mathcal{O}_\theta$ is called a *subbase* if the family $\mathcal{S}^\cap$ of finite intersections is a base of θ.

LEMMA IX.2.5. *If ξ is a convergence on X, $\mathcal{S} \subset \mathcal{O}_\xi$ with $\bigcup \mathcal{S} = X$, then ξ is a topology with subbase $\mathcal{S}$ if and only if*

$$\mathcal{S} \cap \{x\}^\uparrow \subset \mathcal{F} \Longrightarrow x \in \lim_\xi \mathcal{F} \tag{IX.2.1}$$

for every $x \in X$ and $\mathcal{F} \in \mathbb{F}X$.

PROOF. If ξ is a topology with subbase $\mathcal{S}$ then $\mathcal{S}^\cap$ is a base for the topology ξ and $\mathcal{S} \cap \{x\}^\uparrow \subset \mathcal{F}$ implies $\mathcal{S}^\cap \cap \{x\}^\uparrow \subset \mathcal{F}$, hence $\mathcal{O}_\xi(x) \subset \mathcal{F}$ and the conclusion follows.

Conversely, if (IX.2.1) then $x \in \lim_\xi \mathcal{F}$ whenever $\mathcal{O}_\xi(x) \subset \mathcal{F}$ and ξ is topological. Moreover, $\mathcal{S}$ is a subbase for the topology, for $\mathcal{S}^\cap \cap \{x\}^\uparrow$ is a filter-base of $\mathcal{N}_\xi(x)$ composed of open sets. □

Of course, the family of all open sets is itself a base. Here is a criterion to recognize that a family of subsets constitutes a base for the open sets of a topology.

PROPOSITION IX.2.6. *A family $\mathcal{B}$ of sets is a base for the open sets of a topology on X if and only if, for each $x \in X$, there is $B \in \mathcal{B}$ such that $x \in B$, and if $B_0, B_1 \in \mathcal{B}$ and $x \in B_0 \cap B_1$, then there exists $B \in \mathcal{B}$ such that $x \in B \subset B_0 \cap B_1$.*

PROOF. In fact, $\mathcal{O}$ is defined as the family of all unions of elements of $\mathcal{B}$. Then for each $x \in X$, there is $B \in \mathcal{B}$ such that $x \in B$, because $X \in \mathcal{O}$. On the other hand, if $O_0, O_1 \in \mathcal{O}$ and $x \in O_0 \cap O_1$, then by definition there exist $B_0, B_1 \in \mathcal{B}$ such that $x \in B_0 \subset O_0$ and $x \in B_1 \subset O_1$, hence $x \in B_0 \cap B_1 \in \mathcal{O}$, and thus there is $B \in \mathcal{B}$ such that $x \in B \subset B_0 \cap B_1$. □

Dually,

PROPOSITION IX.2.7. *A family $\mathcal{C}$ of subsets of X is a base for the closed sets of a topology on X if and only if for each $x \in X$, there is $C \in \mathcal{C}$ such that $x \notin C$, and if $C_0, C_1 \in \mathcal{C}$ and $x \notin C_0 \cup C_1$, then there exists $C \in \mathcal{C}$ such that $x \notin C \supset C_0 \cup C_1$.*

EXAMPLE IX.2.8. The family $\{\,]a,b[\,: a,b \in \mathbb{R}\}$ constitutes a base for the standard topology $\nu_\mathbb{R}$ on $\mathbb{R}$ ([5]). Also $\{\,]a,b[\,: a,b \in \mathbb{Q}\}$ is a base of $\nu_\mathbb{R}$. It follows that $\nu_\mathbb{R}$ has a countable base. Of course, each base of $\nu_\mathbb{R}$ is infinite, so that the least cardinality of its bases is $\aleph_0$.

---

[5]Of course, $]a,b[\, \neq \varnothing$ if and only if $a < b$.

REMARK IX.2.9. If $\mathcal{B}$ is a base for the θ-open sets, then for each $x \in |\theta|$, the family

$$\mathcal{B}(x) := \{B \in \mathcal{B} : x \in B\}$$

is a filter-base of the neighborhood filter $\mathcal{N}_\theta(x)$. Conversely, if $\mathcal{W}_\theta(x)$ is a filter-base of $\mathcal{N}_\theta(x)$ for each $x \in |\theta|$, then $\mathcal{W} := \bigcup_{x\in|\theta|} \mathcal{W}_\theta(x)$ is a base for the θ-open sets.

EXAMPLE IX.2.10. In the discrete topology ι_X on a set X, each subset of X is open. Therefore, $\{\{x\} : x \in X\}$ is a base for the open sets of ι_X.

It follows from the definitions that

PROPOSITION IX.2.11. *Let ξ_j be a topology on X_j for each $j \in J$, and let $\mathcal{B}_j$ be a base for the ξ_j-open sets. Then the family of all $\prod_{j\in J} V_j$ such that there is $J_0 \subset_{\text{fin}} J$ so that $V_j \in \mathcal{B}_j$ for $j \in J_0$ and $V_j = X_j$ if $j \in J \setminus J_0$ is a base for the open sets of $\prod_{j\in J} \xi_j$.*

DEFINITION IX.2.12 (Weight). The least cardinality of a topological base of τ is called the *weight* of τ, and is denoted by $\mathrm{w}(\tau)$.

Traditionally a topology of *countable weight* is called *second countable.*

The weight of the discrete topology ι_X on X is $\operatorname{card} X$, because each base of ι_X must include $\{\{x\} : x \in X\}$. The weight of the chaotic topology o_X is one, because $\{X\}$ is the only base of o_X.

EXAMPLE IX.2.13. The prime cofinite pretopology on an infinite set X from Example III.4.3 is a topology because of Proposition VII.3.14. Its weight is equal to $\operatorname{card} X$. Indeed, the least base for $\pi[x_\infty, (X)_0]$ consists of all singletons but $\{x_\infty\}$ and of $X \setminus \{x\}$, where $x \neq x_\infty$.

It follows from Proposition IX.2.11 and Corollary XXVII.6.5 to come that

PROPOSITION IX.2.14. *If κ is an infinite cardinal,* $\operatorname{card} J \leq \kappa$ *and $\mathrm{w}(\xi_j) \leq \kappa$ for each $j \in J$, then $\mathrm{w}(\prod_{j\in J} \xi_j) \leq \kappa$.*

Let us notice that if $\mathcal{B}$ is a base for a topology ξ on X, and $B\cap D \neq \varnothing$ for each $B \in \mathcal{B}$, then a subset D of X is ξ-dense by Definition III.1.19. In particular, if $x_B \in B$ for each $B \in \mathcal{B}$ then $\{x_B : B \in \mathcal{B}\}$ is ξ-dense. Therefore,

COROLLARY IX.2.15. *The density is less than or equal to the weight.*

Hence, if ξ is a topology, then

$$\mathrm{d}(\xi) \leq \mathrm{w}(\xi).$$

For metrizable topologies, the converse also holds.

PROPOSITION IX.2.16. *Each metrizable topology of countable density has countable weight.*

PROOF. Indeed, if W is a ξ-dense set for a topology ξ on a set X metrized by a metric d, then

$$\mathcal{B} := \{B_d(w, \tfrac{1}{2^n}) : w \in W, n < \omega\}$$

is a base for the ξ-open sets. Indeed, let O be a ξ-open set and $x \in O$. Then there is $n < \omega$ such that $B_d(x, \frac{1}{2^n}) \subset O$. If $w \in W \cap B_d(x, \frac{1}{2^{n+1}})$, then $x \in B_d(w, \frac{1}{2^{n+1}}) \subset B_d(x, \frac{1}{2^n}) \subset O$. □

DEFINITION IX.2.17. A topology is called *zero-dimensional* if it has a base consisting of clopen sets.

The discrete topologies are zero-dimensional. If a topology is zero-dimensional, then each neighborhood filter has a filter-base of clopen sets. Therefore, zero-dimensional topologies are regular.

REMARK IX.2.18. The reasoning of the proof of Proposition IX.2.16 does not extend to non-metrizable topologies.

EXAMPLE IX.2.19 (The Sorgenfrey line has uncountable weight). The *Sorgenfrey line topology* ς is defined on $\mathbb{R}$ by a base for the open sets $\mathcal{B} := \{[a, b[: a < b\}$. Accordingly, $[b, +\infty[$ and $]-\infty, a[$ are open as (countable) unions of intervals from $\mathcal{B}$. Therefore, $[a, b[$ is closed for each $a < b$, so that a base for the open sets consists of clopen sets, and thus ς is *zero-dimensional* (Definition IX.2.17). As every interval $[a, b[$ with $a < b$ contains a rational number, ς is of *countable density* (that is, separable). The topology ς is of *countable character*, because $\{[a, a + 2^{-n}[: n < \omega\}$ is a filter-base of $\mathcal{N}_\varsigma(a)$ for each $a \in \mathbb{R}$.

From $\varsigma > \nu_{\mathbb{R}}$, it follows that ς is Hausdorff. However, the weight of ς is equal to $\mathfrak{c}$ ([6]). Suppose that there exists a countable base $\mathcal{B} = \{B_n : n < \omega\}$. We can assume without loss of generality that $b_n := \inf B_n > -\infty$ for each $n < \omega$. Indeed, if there is $k < \omega$ such that $b_k = -\infty$, then $\mathcal{B} \setminus \{B_k\}$ is still a base for the open sets of ς, because if $x \in B_k$, then there is $n < \omega$ and $B_{x,n} \in \mathcal{B}$ such that $x \in B_{x,n} \subset [x, \frac{1}{n}[$, and thus $B_{x,n} \neq B_k$. As $A := \{b_n : n < \omega\}$ is countable, there exists $x_0 \in \mathbb{R} \setminus A$. If $r > x_0$, then there is no $n < \omega$ such that $x_0 \in B_n \subset [x_0, r[$. In fact,

[6]Recall that $\mathfrak{c} := \operatorname{card} \mathbb{R}$, and is called *continuum*.

if $b_n < x_0$, then $B_n \setminus [x_0, r[\neq \varnothing$, and if $x_0 < b_n$, then $[x_0, b_n[\cap B_n = \varnothing$. Hence, by Proposition IX.2.16, ς is not metrizable.

LEMMA IX.2.20. *Let ξ be a topology. For any family $\mathcal{S} \subset \mathcal{O}_\xi$, there is $\mathcal{S}_0 \subset \mathcal{S}$ such that $\bigcup \mathcal{S} = \bigcup \mathcal{S}_0$ and* $\operatorname{card} \mathcal{S}_0 \leq \mathrm{w}(\xi)$.

PROOF. Consider a ξ-base $\mathcal{W}$ such that $\operatorname{card} \mathcal{W} = \mathrm{w}(\xi)$. Let $\mathcal{S} \subset \mathcal{O}_\xi$ and let $\mathcal{W}_\mathcal{S} \subset \mathcal{W}$ consists of those $W \in \mathcal{W}$, for which there exists $S \in \mathcal{S}$ such that $W \subset S$. For every $W \in \mathcal{W}_\mathcal{S}$, let $S_W \in \mathcal{S}$ be such that $W \subset S_W$. Then $\mathcal{S}_0 := \{S_W : W \in \mathcal{W}_\mathcal{S}\}$ is of cardinality $\mathrm{w}(\xi)$ and $\bigcup \mathcal{S} = \bigcup \mathcal{S}_0$. Indeed, if $x \in \bigcup \mathcal{S}$, then there exist $S \in \mathcal{S}$ and $W \in \mathcal{W}$ such that $x \in W \subset S$, and thus $W \in \mathcal{W}_\mathcal{S}$. Hence, by construction, there is $S_W \in \mathcal{S}_0$ such that $W \subset S_W$. As $x \in W$, we have $x \in S_W$, and thus $x \in \bigcup \mathcal{S}_0$. □

THEOREM IX.2.21. *Let ξ be a topology of infinite weight* $\mathrm{w}(\xi)$. *Then, for each base $\mathcal{B}$ for the open sets of ξ, there exists $\mathcal{S} \subset \mathcal{B}$ such that $\mathcal{S}$ is a base of cardinality* $\mathrm{w}(\xi)$.

PROOF. Let $\mathcal{B}$ and $\mathcal{W}$ be two bases for the open sets of a topology ξ so that $\operatorname{card} \mathcal{W} = \mathrm{w}(\xi)$. For each $W \in \mathcal{W}$, let $\mathcal{B}_W := \{B \in \mathcal{B} : B \subset W\}$. As $\mathcal{B}$ is a ξ-base, and W is ξ-open, $W = \bigcup \mathcal{B}_W$. By Lemma IX.2.20, there is $\mathcal{S}_W \subset \mathcal{B}_W$ of cardinality $\mathrm{w}(\xi)$ and such that $W = \bigcup \mathcal{S}_W$.

Then the family $\mathcal{S} := \bigcup_{W \in \mathcal{W}} \mathcal{S}_W$ is a base of cardinality $\mathrm{w}(\xi)$. Indeed, if $O \in \mathcal{O}_\xi(x)$, then there is $W \in \mathcal{W}$ such that $x \in W \subset O$, because $\mathcal{W}$ be a ξ-base. Then, by construction, there is $S \in \mathcal{S}_W \subset \mathcal{B}_W$ such that $x \in S \subset W \subset O$, which proves that $\mathcal{S}$ is a ξ-base. As $\operatorname{card} \mathcal{W} = \mathrm{w}(\xi)$ and $\operatorname{card} \mathcal{S}_W = \mathrm{w}(\xi)$, we conclude that $\operatorname{card} \mathcal{S} \leq \mathrm{w}(\xi) \cdot \mathrm{w}(\xi) = \mathrm{w}(\xi)$ by Corollary XXVII.6.5. □

LEMMA IX.2.22. *If $\mathcal{P}$ is a ξ-open cover and $\mathcal{B}$ is a ξ-base, then there is a covering refinement $\mathcal{P}_\mathcal{B}$ of $\mathcal{P}$ such that $\mathcal{P}_\mathcal{B} \subset \mathcal{B}$.*

PROOF. For each x in the underlying set X and each $P \in \mathcal{P}$ such that $x \in P$, there is $B_{x,P} \in \mathcal{B}$ such that $x \in B_{x,P} \subset P$. Then $\mathcal{P}_\mathcal{B} := \{B_{x,P} : x \in X, P \in \mathcal{P}\}$ is a required refinement. □

By Lemma IX.2.20,

COROLLARY IX.2.23. *If $\mathcal{P}$ is an open cover of ξ, then there exists $\mathcal{P}_0 \subset \mathcal{P}$ with* $\operatorname{card} \mathcal{P}_0 \leq \mathrm{w}(\xi)$ *and $\mathcal{P}_0$ is a cover of ξ.*

IX.3. Compact, countably compact, and Lindelöf topologies

Recall that by Corollary VIII.3.8, Proposition VIII.3.11, and Proposition VIII.3.13,

PROPOSITION IX.3.1. *A topology is compact if and only if each open cover has a finite subcover, countably compact if and only if each open countable cover has a finite subcover; Lindelöf if and only if every open cover has a countable subcover.*

It follows from Corollary IX.2.23 that

COROLLARY IX.3.2. *Each topology of countable weight is Lindelöf.*

It follows from Examples IX.2.19 and IX.3.13 (to come) that the converse need not hold.

PROPOSITION IX.3.3. *Every Hausdorff compact topology is regular.*

PROOF. Let ξ be a Hausdorff compact topology on X, let C be a ξ-closed subset of X, and $x_0 \notin C$. Then by Proposition IX.3.3, for each $x \in C$, there exist $P_x \in \mathcal{O}_\xi(x_0)$ and $O_x \in \mathcal{O}_\xi(x)$ such that $P_x \cap O_x = \varnothing$. By the compactness of C, there is a finite subset F of C such that $C \subset \bigcup_{x\in F} O_x$. On the other hand, $\bigcap_{x\in F} P_x \in \mathcal{O}_\xi(x_0)$ and $\bigcup_{x\in F} O_x \cap \bigcap_{x\in F} P_x = \varnothing$, that is, ξ is regular. □

PROPOSITION IX.3.4. *Every Hausdorff compact topology is normal.*

PROOF. Let ξ be a Hausdorff compact topology on X, and let C_0 and C_1 be disjoint ξ-closed sets. Then by Proposition IX.3.3, for each $x \in C_1$, there exist $P_x \in \mathcal{O}_\xi(C_0)$ and $O_x \in \mathcal{O}_\xi(x)$ such that $P_x \cap O_x = \varnothing$. By the compactness of C_1, there is a finite subset F of C_1 such that $C_1 \subset \bigcup_{x\in F} O_x$. On the other hand, $\bigcap_{x\in F} P_x \in \mathcal{O}_\xi(C_0)$ and $\bigcup_{x\in F} O_x \cap \bigcap_{x\in F} P_x = \varnothing$, that is, ξ is normal. □

We shall see in Corollary XI.3.11 to come that every Hausdorff regular Lindelöf topology is normal.

EXAMPLE IX.3.5 (Linear order topology). If $(Z, \leq)$ is a linearly ordered set, then the sets of the form $\{z \in Z : z < s\}$ and $\{z \in Z : z > r\}$ form a subbase for the open sets of a *linear order topology*. We shall denote

$$]r, s[:= \{z \in Z : r < z < s\}, \ [r, s[:= \{z \in Z : r \leq z < s\},$$

and so on. For instance, the real line $\mathbb{R}$, the rational line $\mathbb{Q}$, and the interval $[0, 1]$ with the standard order are linearly ordered.

THEOREM IX.3.6. *A linearly ordered topology is compact if and only if it is a complete lattice.*

PROOF. If $(Z, \leq)$ is not complete, then take a subset A of Z without supremum, that is, the set $A^+ := \{z \in Z : \forall_{a \in A} a \leq z\}$ does not have a least element. Then the sets $\{z \in Z : z < a\}$ with $a \in A$ and $\{z \in Z : z > b\}$ with $b \in A^+$ is an open cover of Z. Indeed, if $x \in A$, then x is not the greatest element of A, and thus there is $a \in A$ such that $x < a$. Therefore each element of A or less than an element of A belongs to $\{z \in Z : z < a\}$ for some $a \in A$. If $x \in A^+$, then by assumption there is $b \in A^+$ such that $b < x$, so that $x \in \{z \in Z : z > b\}$. If there were a finite subcover of this cover, then either A would have a greatest element, or there would be a least element of A^+, contrary to the assumption.

Suppose that $(Z, \leq)$ is a complete lattice, and let $\mathcal{P}$ be an open cover of Z. Let S be the set of those $s \in Z$ for which $[0, s[$ has a finite subcover of $\mathcal{P}$. Let $s_+ := \sup S$. If $s_+ = 1$, the proof is completed.

Suppose that $s_+ < 1 = \sup Z$. Then there exist $P \in \mathcal{P}$ and $r < s_+ < t$ such that $s_+ \in]r, t[\subset P$.

If $s_+ \notin S$, then either $S \cap]r, t[\neq \varnothing$, and thus $]r, t[\subset S$, because $[0, t[$ has a finite subcover of $\mathcal{P}$, in particular, $s_+ \in S$, a contradiction, or $S \cap]r, t[= \varnothing$, so that $r \in S$, hence $r \geq s$ for each $s \in S$, that is to say, $r = \sup S$, a contradiction with $r < s_+$. Thus $s_+ \in S$ so that $]r, t[\subset S$. It follows that $]s_+, t[= \varnothing$, that is, $t \in S$ and $s_+ < t$, a contradiction. □

EXAMPLE IX.3.7 (*). Consider the order topology on $X := \omega_1 \sqcup \{\omega_1\}$ (the set of ordinals less than or equal to the first uncountable ordinal ω_1). A base of open sets for this topology consists of $]x, y]$, where $x < y \leq \omega_1$. This topology is compact by Theorem IX.3.6.

Let us provide a direct proof. Let $\mathcal{Q}$ be an open cover of X. Let A be the set of x, for which there exists a finite subfamily $\mathcal{Q}_x$ of $\mathcal{Q}$ such that $[0, x] \subset \bigcup \mathcal{Q}_x$. It suffices to show that $X \setminus A = \varnothing$. If by way of contradiction x_0 is the least element of $X \setminus A$, then let $Q_0 \in \mathcal{Q}$ be such that $x_0 \in Q_0$, so that there is $x < x_0$ such that $]x, x_0] \subset Q_0$. By definition, $x \in A$, so that there is a finite family $\mathcal{Q}_x$ such that $[0, x] \subset \bigcup \mathcal{Q}_x$. Consequently, $[0, x_0] \subset \bigcup_{Q \in \mathcal{Q}_x} Q \cup Q_0$, that is, $x_0 \in A$, which yields a contradiction.

The notion of *isolated point* can be easily relativized:

DEFINITION IX.3.8. Let ξ be a pretopology on X. An element x is called a *ξ-accumulation point* of a subset A of X if

$$x \in \operatorname{adh}_\xi(A \setminus \{x\}).$$

Otherwise, x is called *ξ-isolated at A*.

If every point is ξ-isolated at A, then for each $x \in X$ there exists $V \in \mathcal{V}_\xi(x)$ such that $V \cap A$ is at most a singleton, in other words, A is a ξ-discrete set by Definition VI.2.5, that is a ξ-closed discrete subspace.

PROPOSITION IX.3.9. *A free (alias T_1) topology is countably compact if and only if each infinite (countable) set has an accumulation point.*

PROOF. The claim is equivalent to the fact that a free topology τ on X is not countably compact if and only if there is a countable τ-closed discrete subspace. By definition, τ is not countably compact if and only if there is a strictly ascending sequence $\{O_n : n < \omega\} \subset \mathcal{O}_\tau$ such that $\bigcup_{n<\omega} O_n = X$. Equivalently, for each $n < \omega$, there is $x_n \in O_{n+1} \setminus O_n$ so that $\bigcap_{n<\omega} \mathrm{cl}_\tau\{x_k : k > n\} = \varnothing$, because $\mathrm{cl}_\tau\{x_k : k > n\} \cap O_n = \varnothing$ for each $n < \omega$. In other words, for each $x \in X$, there is $O \in \mathcal{O}_\tau(x)$ and $n < \omega$ such that $O \cap \{x_k : k > n\} = \varnothing$. On the other hand, for each $m \leq n$ such that $x_n \neq x$, the set $X \setminus \{x_n\} \in \mathcal{O}_\tau(x)$, because τ is free. Hence a τ-neighborhood $O \setminus \{x_0, \ldots, x_n\}$ of x has at most one common element with $\{x_n : n < \omega\}$. □

PROPOSITION IX.3.10. *A free topology is countably compact if and only if each locally finite family of non-empty sets is finite.*

PROOF. If τ is a free countably compact topology on X and let $\mathcal{A}$ be a family of non-empty subsets of X. If the claim does not hold, then there exists $\{A_n : n < \omega\} \subset \mathcal{A}$ such that $A_{n+1} \setminus \bigcup_{k=0}^{n} A_k \neq \varnothing$ for each $n < \omega$. Then $F_n := \mathrm{cl}_\tau(\bigcup_{k>n} A_k)$ is a decreasing sequence of non-empty τ-closed sets, thus there exists $x_\infty \in \bigcap_{n<\omega} F_n$ in virtue of countable compactness of τ. By Proposition VI.2.6, $F_n = \bigcup_{k>n} \mathrm{cl}_\tau A_k$, so that there is an infinite sequence $(k_n)_n$ such that $x_\infty \in \mathrm{cl}_\tau A_{k_n}$ for each $n < \omega$, contrary to the local finiteness of $\mathcal{A}$.

Conversely, if there exists an infinite countable set $A = \{a_n : n < \omega\}$ of distinct points without τ-accumulation points, then the family $\{\{a_n\} : n < \omega\}$ is locally finite, but not finite. □

In virtue of Lemma IX.1.21,

PROPOSITION IX.3.11. *Every regular free (alias T_1) topology of countable weight is normal.*

PROOF. Let $\mathcal{B}$ be a countable base for the open sets. Let F be a closed set and let $O \in \mathcal{O}_X(F)$. By regularity, for each $x \in F$, there exists $B_x \in \mathcal{B}$ such that $x \in B_x \subset \operatorname{cl}_X B_x \subset O$. Thus X is normal by Lemma IX.1.21. □

EXAMPLE IX.3.12. Each subspace of the *standard topology* of the real line $\nu_{\mathbb{R}}$ is Lindelöf, because it has countable weight. The *standard topology*, however, is not countably compact. Indeed, $\{]-n,n[: n<\omega\}$ is an open cover of $\mathbb{R}$ without any finite subcover. The *standard topology* is *σ-compact*, that is, a countable union of compact sets.

Coming back to the Sorgenfrey line topology from Example III.7.3,

EXAMPLE IX.3.13 (Sorgenfrey line). Each compact subset of the *Sorgenfrey line topology* $\varsigma = \varsigma_+$ is countable. Indeed, if K is ς-compact, then for each $r \in K$, there exists $x_r < r$ such that $]x_r, r[\cap K = \varnothing$. Otherwise, there would be an increasing sequence $(r_n)_n$ of elements of K such that $\lim_{n\to\infty} |r - r_n| = 0$. Therefore

$$K \subset]-\infty, r_0[\cup \bigcup\nolimits_{n<\omega} [r_n, r_{n+1}[\cup [r, \infty[,$$

that is, a countable open cover of K stops being a cover of K if we remove from it any interval of the form $[r_n, r_{n+1}[$. Accordingly, there is no finite subcover. We conclude that if K is ς-compact, then $\bigcup_{r\in K}]x_r, r[\cap K = \varnothing$, where $x_r < r$ for each $r \in K$, and, moreover, $]x_r, r[\cap]x_s, s[= \varnothing$ if $s \neq r$. As for each $r \in K$, there is $q_r \in]x_r, r[\cap \mathbb{Q}$, we witness an injection of K into $\mathbb{Q}$, and thus K is countable.

It follows that ς is not σ-compact. Nevertheless, ς is Lindelöf. Indeed, if $\mathcal{P}$ is a ς-open cover, then there exists a countable subfamily $\mathcal{P}_0$ of $\mathcal{P}$ such that $A := \bigcup_{P\in\mathcal{P}} \operatorname{int}_{\nu_{\mathbb{R}}} P = \bigcup_{P\in\mathcal{P}_0} \operatorname{int}_{\nu_{\mathbb{R}}} P$, because A is $\nu_{\mathbb{R}}$-Lindelöf, as a subspace of $\nu_{\mathbb{R}}$. If $x \notin A$, then there exists $r_x > x$ such that $]x, r_x[\subset A$. As these intervals are disjoint, they are at most countably many. Hence $\mathbb{R} \setminus A$ is countable and thus Lindelöf.

EXAMPLE IX.3.14 (A compact, non-separable topology of countable character). Consider the unit square $X \times Y$, where $X = Y = [0,1]$, with the topology of lexicographic order.

More precisely, $(x_0,y_0) < (x_1,y_1)$ if either $x_0 < x_1$, or $x_0 = x_1$, and $y_0 < y_1$ for every $(x_0,y_0), (x_1,y_1) \in I^2$. We note that this ordered set is a complete lattice. Indeed, if $\varnothing \neq R \subset I^2$, then R is a relation, and $R^-Y \neq \varnothing$. Let $x_+ := \sup\{x \in X : x \in R^-Y\}$ be the supremum of the

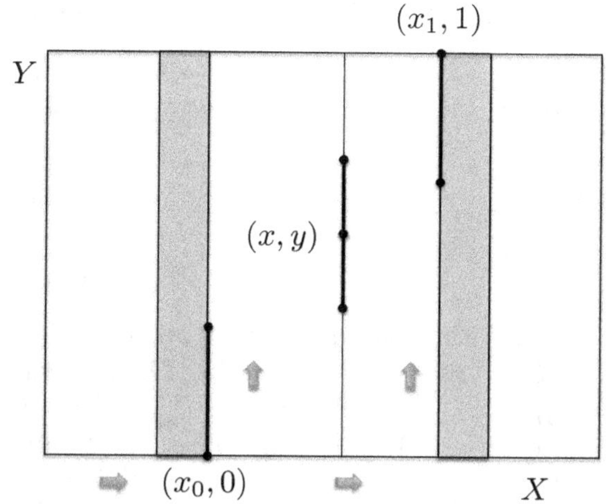

FIGURE IX.3.1. The unit square with lexicographic order (indicated by horizontal arrows first, vertical arrows afterwards). Generic neighborhoods of points of three types are marked.

subset $R^{-}Y$ of $X = [0,1]$. There are two cases: either $Rx_+ \neq \varnothing$, so we let $y_+ := \sup Rx_+$, and then $(x_+, y_+) = \sup R$, or $Rx_+ = \varnothing$ and then $(x_+, 0) = \sup R$ ([7]).

We realize that a neighborhood base of $\alpha = (x_0, y_0)$ is of the form

$$([r, x_0[\times[0,1]) \cup (\{x_0\} \times [0, s]) \text{ if } y_0 = 0,$$
$$\{x_0\}\times]y_0 - \varepsilon, y_0 + \varepsilon[\text{ if } 0 < x_0, y_0 < 1,$$
$$(\{x_0\} \times [r, 1]) \cup ([x_0, s[\times[0,1]) \text{ if } y_0 = 1.$$

It is plain that the topology is Hausdorff and of countable character. By Theorem IX.3.6, this topology is compact, hence normal because of Proposition IX.3.4.

The lexicographic order topology is not separable, because if D is dense, then $D \cap \{x\}\times]\frac{1}{4}, \frac{3}{4}[\neq \varnothing$ for each $x \in X$, and thus $\mathrm{card} D \geq \mathfrak{c}$.

IX.4. Open, closed, topologically quotient, and perfect maps

Let ξ be a topology on X, and τ be a topology on Y.

DEFINITION IX.4.1. A surjective map $f \in \mathscr{C}(\xi, \tau)$ is called *topologically quotient* if τ is the finest topology for which f is continuous.

[7] In fact, for every $(\alpha, \beta) < (x_+, 0)$, there is $\alpha < \gamma < x_+$ such that $R\gamma \neq \varnothing$.

In other words, f is topologically quotient whenever $\tau = \mathrm{T}(f\xi)$.

PROPOSITION IX.4.2. *Let ξ and τ be topologies. If $f \in C(\xi, \tau)$ is topologically quotient, then a set P is τ-open if and only if $f^-(P)$ is ξ-open.*

PROOF. Let ξ be a topology on X, and τ a topology on Y. The family

$$f^{\rightarrow}[\mathcal{O}_\xi] := \{P \subset Y : f^-(P) \in \mathcal{O}_\xi\} \tag{IX.4.1}$$

fulfills the axioms for the open sets of a topology, namely $f^-(Y) = X \in \mathcal{O}_\xi$ and $f^-(\varnothing) = \varnothing \in \mathcal{O}_\xi$; if $f^-(P) \in \mathcal{O}_\xi$ for each $P \in \mathcal{P}$, then $f^-(\bigcup_{P\in\mathcal{P}} P) = \bigcup_{P\in\mathcal{P}} f^-(P) \in \mathcal{O}_\xi$, and $f^-(\bigcap_{P\in\mathcal{P}} P) = \bigcap_{P\in\mathcal{P}} f^-(P) \in \mathcal{O}_\xi$ if $\mathcal{P}$ is a finite family. If θ is a topology on Y and $f \in C(\xi, \theta)$, then $f^-(P) \in \mathcal{O}_\xi$ for each $P \in \mathcal{O}_\theta$, so that $f^{\rightarrow}[\mathcal{O}_\xi] \supset \mathcal{O}_\theta$. Therefore, $f^{\rightarrow}[\mathcal{O}_\xi] = \mathcal{O}_{\mathrm{T}(f\xi)} = \mathcal{O}_{f\xi}$. □

DEFINITION IX.4.3. A map is called *open* if the image of each open set is open. A map is called *closed* if the image of every closed set is closed.

PROPOSITION IX.4.4. *Each surjective continuous open map is topologically quotient* ([8]).

PROOF. Let ξ be a topology on X, and τ a topology on Y. If $f \in \mathscr{C}(\xi, \tau)$ is surjective and open, and $P \subset Y$ is such that $f^-(P) \in \mathcal{O}_\xi$, then by the assumption, $P = f(f^-(P)) \in \mathcal{O}_\tau$, because f is surjective. □

PROPOSITION IX.4.5. *Each surjective continuous closed map is topologically quotient.*

PROOF. If $P \subset Y$ is such that $f^-(P) \in \mathcal{O}_\xi$, then $X \setminus f^-(P) = f^-(Y \setminus P)$ is ξ-closed. Hence, by the assumption, $f(f^-(Y \setminus P)) \in \mathcal{C}_\tau$. As f is surjective, $f(f^-(Y \setminus P)) = Y \setminus P$, and thus $P \in \mathcal{O}_\tau$. □

The following two examples show that a topologically quotient map need be neither open nor closed.

EXAMPLE IX.4.6 (An open map that is not closed). Let $f : \mathbb{R} \to S^1$ be given by

$$f(x) := (\cos 2\pi x, \sin 2\pi x).$$

To see that f is open, note that f maps strict intervals onto strict intervals. As each open subset O of $\mathbb{R}$ is a union of strict intervals, $f(O)$ is the union of the images of these intervals under f by (I.1.2),

[8] We will see in Section XVIII.9 that each open map is (convergence) quotient.

hence is open. However, the set $N := \{n+\frac{1}{n} : n \in \mathbb{N}\}$ is closed, but its image under f is not closed. Indeed, as n tends to $+\infty$, the sequence $f(n+\frac{1}{n}) = (\cos\frac{2\pi}{n}, \sin\frac{2\pi}{n})$ converges to $(1,0)$, which does not belong to $f(N)$.

Consider the restriction of f from Example IX.4.6 to $[0,1]$:

EXAMPLE IX.4.7 (A closed map that is not open). Let $f : [0,1] \to S^1$ be given by $f(x) := (\cos 2\pi x, \sin 2\pi x)$. This is not an open map. Indeed, $[0,\frac{1}{2}[$ is open in $[0,1]$ (with respect to $\nu|_{[0,1]}$), but $f([0,\frac{1}{2}[)$ is not open in S^1, because, for example, $((\cos\frac{-2\pi}{n}, \sin\frac{-2\pi}{n}))_n$ converges to $f(0)$.

This map is, however, closed. Indeed, if $C \subset [0,1]$ is closed and $y \notin f(C)$, equivalently $f^-(y) \cap C = \varnothing$, either $f^-(y) = \{0,1\}$ and thus there is $\varepsilon > 0$ such that $([0,\varepsilon[\cup]1-\varepsilon]) \cap C = \varnothing$, hence the segment of the circle corresponding to the angles from $]-2\pi\varepsilon, 2\pi\varepsilon[$ is disjoint from $f(C)$, or there is $0 < x < 1$ such that $f^-(y) = \{x\}$, thus there is $\varepsilon > 0$ such that $]x-\varepsilon, x+\varepsilon[\cap C = \varnothing$, hence the segment of the circle corresponding to the angles from $]y-2\pi\varepsilon, y+2\pi\varepsilon[$ is disjoint from $f(C)$.

Consider a relation $R \subset X \times Y$, a topology ξ on X, and a topology τ on Y. A relation is called *upper semicontinuous* at x_0 whenever for each $O \in \mathcal{O}_\tau(Rx_0)$, there exists $V \in \mathcal{N}_\xi(x_0)$ such that $RV \subset O$.

It is called *upper semicontinuous* if it is upper semicontinuous at every element of X ([9]). Actually,

PROPOSITION IX.4.8. *A relation R is upper semicontinuous if and only if R^-C is closed for each closed set C.*

PROOF. Let R be upper semicontinuous and C be τ-closed. If $y \notin R^-C$, equivalently $Ry \cap C = \varnothing$, then by the upper semicontinuity assumption, there is $V \in \mathcal{O}_\xi(y)$ such that $RV \cap C = \varnothing$, equivalently, $V \cap R^-C = \varnothing$, that is, R^-C is ξ-closed.

Conversely, let $O \in \mathcal{O}_\tau(Ry)$. Therefore, $C := (X \setminus O)$ is τ-closed and $C \cap Ry = \varnothing$, equivalently, $y \notin R^-C$. By assumption, R^-C is ξ-closed, hence $V := X \setminus R^-C$ is a ξ-open neighborhood of y and $V \cap R^-C = \varnothing$, equivalently $RV \cap C = \varnothing$, that is, $RV \subset O$. □

It follows from Proposition IX.4.8 that

COROLLARY IX.4.9. *A map f is closed if and only if the relation f^- is upper semicontinuous.*

[9]It follows that R^-Y is ξ-closed. Indeed, if $x \notin R^-Y$ then $\varnothing \in \mathcal{O}_\tau(Rx)$ so that there exists $V \in \mathcal{O}_\xi(x)$ such that $RV = \varnothing$, that is $V \cap R^-Y = \varnothing$.

A surjective map f is called *perfect* if f is closed, and $f^-(y)$ is ξ-compact for each $y \in |\tau|$ ([10]).

It is plain that the map f from Example IX.4.7 is perfect.

PROPOSITION IX.4.10. *If $f : |\xi| \longrightarrow |\tau|$ is perfect, and K is τ-compact, then $f^-(K)$ is ξ-compact.*

PROOF. Let $\mathcal{P}$ be a family of ξ-open sets such that $\bigcup_{P\in\mathcal{P}} P \supset f^-(K)$. As $f^-(y)$ is ξ-compact for every $y \in Y$, there exists $\mathcal{P}_y \subset_{\text{fin}} \mathcal{P}$ such that $Q_y := \bigcup_{P\in\mathcal{P}_y} P \supset f^-(y)$, and thus $y \in (f(Q_y^c))^c$ and $(f(Q_y^c))^c$ is τ-open for each $y \in Y$. Hence $\bigcup_{y\in K}(f(Q_y^c))^c \supset K$, and by compactness, there exists a finite subset A of K such that $\bigcup_{y\in A}(f(Q_y^c))^c \supset K$. Therefore,

$$\varnothing = \bigcap\nolimits_{y\in A} f(Q_y^c) \cap K \supset f(\bigcap\nolimits_{y\in A} Q_y^c) \cap K,$$

and thus $\bigcap_{y\in A} Q_y^c \cap f^-(K) = \varnothing$, equivalently $\bigcup_{y\in A} Q_y \supset f^-(K)$. It follows that $\bigcup_{y\in A} \mathcal{P}_y$ is a finite ξ-subcover of $\mathcal{P}$ of $f^-(K)$. □

PROPOSITION IX.4.11. *If $f : |\xi| \longrightarrow |\tau|$ is perfect, and $\mathcal{A}$ is a ξ-locally finite family, then $f[\mathcal{A}]$ is τ-locally finite.*

PROOF. As $\mathcal{A}$ is ξ-locally finite, for each $y \in Y$ and each $x \in f^-(y)$ there exists $O_x \in \mathcal{O}_\xi(x)$ such that $\{A \in \mathcal{A} : A\#O\}$ is finite. As $f^-(y)$ is ξ-compact, there exists a finite subset F_y of $f^-(y)$ such that $Q_y := \bigcup_{x\in F_y} O_x \supset f^-(y)$ and $\{A \in \mathcal{A} : A\#Q_y\}$ is finite. As f is closed, by Proposition IX.4.8, for every $y \in Y$, there exists $O_y \in \mathcal{O}_\tau(y)$ such that $f^-(O_y) \subset Q_y$. As $f^-(O_y)\#A$ if and only if $O_y\#f(A)$, the family $f[\mathcal{A}]$ is τ-locally finite. □

REMARK IX.4.12. Observe that we have not assumed continuity in the two propositions above.

PROPOSITION IX.4.13. *Let $f \in \mathscr{C}(\xi,\tau)$ be a perfect map. If ξ is Hausdorff, then τ is Hausdorff. If ξ is Hausdorff and regular, then τ is (Hausdorff) regular.*

PROOF. If $y_0 \neq y_1$, then $f^-(y_0)$ and $f^-(y_1)$ are ξ-compact and disjoint. Hence there exist $O_0 \in \mathcal{O}_\xi(f^-(y_0))$ and $O_1 \in \mathcal{O}_\xi(f^-(y_1))$ such that $O_0 \cap O_1 = \varnothing$. As f is closed, the sets $Y\setminus f(X\setminus O_0)$ and $Y\setminus f(X\setminus O_1)$ are τ-open neighborhoods of y_0 and y_1 respectively. Let us show that they are disjoint:

$$\begin{aligned}[Y \setminus f(X \setminus O_0)] \cap [Y \setminus f(X \setminus O_1)] &= Y \setminus [f(X \setminus O_0) \cup f(X \setminus O_1)] \\ &= Y \setminus f[(X \setminus O_0) \cup (X \setminus O_1)] = Y \setminus f(X \setminus (O_0 \cap O_1)) = \varnothing.\end{aligned}$$

[10] Traditional definition comprises continuity.

If F is τ-closed, and $y \notin F$, then $f^{-}(y) \cap f^{-}(F) = \varnothing$. As $f^{-}(y)$ is ξ-compact and $f^{-}(F)$ is ξ-closed, there exist $O_0 \in \mathcal{O}_\xi(y)$ and $O_1 \in \mathcal{O}_\xi(f^{-}(F))$ such that $O_0 \cap O_1 = \varnothing$. As before, $(f(O_0^c))^c$ and $(f(O_1^c))^c$ are τ-open and disjoint neighborhoods of y and F respectively. $\square$

IX.5. Supplement

For any cardinal κ, let $\$^\kappa \approx \prod_{j\in\kappa} \$$ be the *Sierpiński cube* of dimension κ ([11]).

Showing that a topology ξ is \$-initial in Theorem VII.5.2, we could have taken only the characteristic functions of the members of a base for the ξ-closed sets. As there is a base of cardinality $\mathrm{w}(\xi)$, we conclude that

THEOREM IX.5.1. *Each T_0-topology of weight κ is embeddable in a Sierpiński cube of dimension κ.*

DEFINITION IX.5.2. A convergence is said to be *universal* for a class of convergences if each convergence from that class is embeddable into that convergence.

Therefore,

COROLLARY IX.5.3. *The Sierpiński κ-cube is universal for T_0-topologies of weight κ.*

PROPOSITION IX.5.4. *Let ξ be a convergence on X and $f \in Y^X$. Then D is $f\xi$-closed if and only if $f^{-}(D)$ is ξ-closed.*

PROOF. By definition, D is $f\xi$-closed if and only if $D \supset \operatorname{adh}_{f\xi} D = D \cup f(\operatorname{adh}_\xi f^{-}(D))$ if and only if $f(\operatorname{adh}_\xi f^{-}(D)) \subset D$. By (IV.1.2) $\operatorname{adh}_\xi f^{-}(D) \subset f^{-}(D)$, that is, $f^{-}(D)$ is ξ-closed. Conversely, by applying f and (IV.1.1), $f(f^{-}(D)) \subset D$, we conclude. $\square$

COROLLARY IX.5.5. *If ξ and τ are convergences, τ is Hausdorff, and $f \in \mathscr{C}(\xi, \tau)$, then $\operatorname{gr}(f)$ is $(\xi \times \tau)$-closed.*

PROOF. By Remark IV.10.12, $\operatorname{gr}(f) = (f \times i_Y)^{-}\triangle(Y)$, hence by Proposition IX.1.2, $\triangle(Y)$ is $(\tau \times \tau)$-closed, and since $f \times i_Y \in \mathscr{C}(\xi \times \tau, \tau \times \tau)$, the preimage $(f \times i_Y)^{-}\triangle(Y)$ is $(\xi \times \tau)$-closed. $\square$

EXERCISE IX.5.6. Show that if τ is a regular topology, R is upper semicontinuous at x_0 and Rx_0 is τ-closed, then R is graph-closed at x_0.

---

[11]The *Sierpiński cube* is called the *Alexandrov cube* by R. ENGELKING [97].

SOLUTION. If $x_0 \in \lim_\xi \mathcal{F}$ and $y \notin Rx_0$, then there is a τ-closed neighborhood V of y such that $V \cap Rx_0 \neq \varnothing$. By upper semicontinuity, there is $F \in \mathcal{F}$ such that $RF \cap V = \varnothing$. If $\mathcal{U} \in \beta R[\mathcal{F}]$ is such that $y \in \lim_\tau \mathcal{U}$, then $V \in \mathcal{U}$ and thus $\lim_\tau \mathcal{U} \subset V$, but since $R[\mathcal{F}] \leq \mathcal{U}$, also $Y \setminus V \in \mathcal{U}$, which yields a contradiction.

EXERCISE IX.5.7. Using closed and open sets, show that every compact subset of a Hausdorff topological space is closed.

SOLUTION. Let ξ be a Hausdorff topology on X. If C is ξ-compact and $x_0 \notin C$, then for every $x \in C$ there exist $O_x \in \mathcal{O}_\xi(x)$ and $P_x \in \mathcal{O}_\xi(x_0)$ such that $O_x \cap P_x = \varnothing$. As C is ξ-compact, there is a finite subset F of C such that $C \subset \bigcup_{x\in F} O_x$. On the other hand, $\bigcap_{x\in F} P_x \in \mathcal{O}_\xi(x_0)$ and

$$C \cap \bigcap\nolimits_{x\in F} P_x \subset \bigcup\nolimits_{x\in F} O_x \cap \bigcap\nolimits_{x\in F} P_x = \varnothing,$$

proving that C is ξ-closed.

EXERCISE IX.5.8. Each closed subspace of a normal topology is normal.

SOLUTION. Let A be a closed subspace of a normal topology on X. If F_0, F_1 are disjoint closed subsets of A, hence of X, then there exist disjoint X-open sets O_0 and O_1 such that $F_0 \subset O_0$ and $F_1 \subset O_1$. By Proposition VII.3.12, $O_0 \cap A$ and $O_1 \cap A$ are A-open.

EXERCISE IX.5.9. A topological space ξ is normal if and only if for each ξ-closed set A and $U \in \mathcal{O}_\xi(A)$ there exists $W \in \mathcal{O}_\xi(A)$ such that

$$\operatorname{cl}_\xi W \subset U. \tag{IX.5.1}$$

SOLUTION. As A is ξ-closed and $U \in \mathcal{O}_\xi(A)$, the sets A and $X \setminus U$ are ξ-closed and disjoint. By normality, there exist $W \in \mathcal{O}_\xi(A)$ and $V \in \mathcal{O}_\xi(X \setminus U)$ such that $W \cap V = \varnothing$. Therefore, $\operatorname{cl}_\xi W \cap V = \varnothing$, hence $X \setminus \operatorname{cl}_\xi W \supset V \supset X \setminus U$, so that $\operatorname{cl}_\xi W \subset U$.

Conversely, if A, B are disjoint ξ-closed sets, then $A \subset X \setminus B$. Hence there is $W \in \mathcal{O}_\xi(A)$ such that $\operatorname{cl}_\xi W \subset X \setminus B$, which means that $X \setminus \operatorname{cl}_\xi W \in \mathcal{O}_\xi(B)$ and $W \cap (X \setminus \operatorname{cl}_\xi W) = \varnothing$.

EXERCISE IX.5.10. A topology is compact if and only if each family of closed sets with finite intersection property has non-empty intersection.

SOLUTION. Let τ be a topology on X. By definition, τ is compact if for each $\mathcal{P} \subset \mathcal{O}_\tau$ such that $\bigcup_{P\in\mathcal{P}} P = X$, there exists a finite subfamily $\mathcal{P}_0$ of $\mathcal{P}$ such that $\bigcup_{P\in\mathcal{P}_0} P = X$. If we set $\mathcal{D} := \mathcal{P}_c$, then we get an equivalent condition: if for each $\mathcal{D} \subset \mathcal{C}_\tau$ such that $\bigcap_{D\in\mathcal{D}_0} D \neq \varnothing$ for every finite subfamily $\mathcal{D}_0$ of $\mathcal{D}$, then $\bigcap_{D\in\mathcal{D}} D \neq \varnothing$.

EXERCISE IX.5.11 (Cantor condition). A topology is countably compact if and only if

$$\bigcap_{n<\omega} C_n \neq \varnothing$$

for each decreasing sequence $\{C_n : n < \omega\}$ of non-empty closed sets.

SOLUTION. Let τ be a topology on X. By definition, τ is countably compact if for every sequence $\{O_n : n < \omega\} \subset \mathcal{O}_\tau$ such that $\bigcup_{n<\omega} O_n = X$, there exists $n_0 < \omega$ such that $\bigcup_{n \leq n_0} O_n = X$. By setting $Q_n := \bigcup_{k \leq n} O_n$, we obtain an equivalent condition: for each increasing sequence $\{Q_n : n < \omega\}$ of τ-open sets such that $\bigcup_{n<\omega} Q_n = X$, there exists $n_0 < \omega$ such that $Q_0 = X$. It is enough to set $C_n := X \setminus Q_n$ to conclude.

DEFINITION IX.5.12. We say that a convergence ξ is *pretopologically Hausdorff* if $\mathrm{S}_0\xi$ is Hausdorff; *topologically Hausdorff* if $\mathrm{T}\xi$ is Hausdorff.

Of course, a Hausdorff pretopology is pretopologically Hausdorff, and a Hausdorff topology is topologically Hausdorff.

It follows immediately from Definition III.4.1 (see also Proposition III.4.2),

PROPOSITION IX.5.13. *Every prime free (alias T_1) convergence is topologically Hausdorff.*

As $\lim_\xi \mathcal{F} \subset \lim_{\mathrm{S}_0} \mathcal{F} \subset \lim_{\mathrm{T}\xi} \mathcal{F}$ for each $\mathcal{F}$, a topologically Hausdorff convergence is pretopologically Hausdorff, which in turn is Hausdorff. The converse implications do not hold in general.

EXAMPLE IX.5.14 (Hausdorff, not S_0-Hausdorff). Let $\{X_n : n < \omega\}$ be disjoint infinite countable sets, and let $0, \infty \notin \bigcup_{n<\omega} X_n$. We define a convergence ζ on $X := \{0, \infty\} \sqcup \bigsqcup_{n<\omega} X_n$ by declaring isolated all the elements of $\bigcup_{n<\omega} X_n$, and letting for each $n < \omega$,

$$\{0\} = \lim_\zeta (X_n)_0, \text{ and } \{\infty\} = \lim_\zeta \mathcal{X}$$

where

$$\mathcal{X} := \{\bigcup_{k>m} X_k : m < \omega\}^\uparrow. \tag{IX.5.2}$$

It follows from the definition that ζ is Hausdorff. On the other hand, if $V \in \mathcal{V}_\zeta(0)$ then $X_n \setminus V$ is finite for each $n < \omega$, while $W \in \mathcal{V}_\zeta(\infty)$ implies that there is $m < \omega$ such that $X_n \subset W$ for each $n < m$. Consequently, $\mathcal{V}_\zeta(0) \# \mathcal{V}_\zeta(\infty)$, hence $\{0, \infty\} \subset \lim_\zeta(\mathcal{V}_\zeta(0) \vee \mathcal{V}_\zeta(\infty))$, so that $\mathrm{S}_0\zeta$ is not Hausdorff.

EXAMPLE IX.5.15 (S_0-Hausdorff, not T-Hausdorff). Let $Y := X \cup \{x_n : n < \omega\}$, where X is defined in Example IX.5.14, and $X \cap \{x_n : n < \omega\} = \varnothing$. We define a convergence η, for which all the elements of $\bigcup_{n<\omega} X_n$ are isolated, and for each $n < \omega$,

$$\{0\} = \lim_\eta (x_n)_n, \ \{x_n\} = \lim_\eta (X_n)_0, \ \text{and } \{\infty\} = \lim_\eta \mathcal{X},$$

where $\mathcal{X}$ is defined in (IX.5.2). To see that $S_0\eta$ is Hausdorff, let us notice that $X_n \notin \mathcal{X}$ for every $n < \omega$. However, $T\eta$ is not Hausdorff. Indeed, if $O \in \mathcal{O}_\eta(0)$ then there is $m < \omega$ such that $X_n \setminus O$ is finite for each $n > m$. Hence $O \in \mathcal{X}^\#$, which means that $\mathcal{N}_\eta(0)\#\mathcal{N}_\eta(\infty)$, so that η is not T-Hausdorff.

A family $\mathcal{A}$ of subsets of a set X is called *point-finite* if the set $\{A \in \mathcal{A} : x \in A\}$ is finite for every $x \in X$.

THEOREM IX.5.16 (*). *For every point-finite open cover $\mathcal{P}$ of a normal topological space, there exists an open cover whose closures constitute a refinement of $\mathcal{P}$.*

PROOF. Let ξ be a normal topology on X, and let $\{P_j : j \in J\}$ be a point-finite ξ-open cover. Consider the collection $\boldsymbol{Q}$ of ξ-open covers of the form $\mathcal{Q} = \{Q(j) : j \in J\}$ such that either $Q(j) = P_j$ or $\mathrm{cl}_\xi Q(j) \subset P_j$ for each $j \in J$.

We order $\boldsymbol{Q}$ to the effect that $\mathcal{Q}_0 \succeq \mathcal{Q}_1$ if $Q_0(j) = Q_1(j)$ whenever $Q_0(j) \neq P_j$. If $\boldsymbol{Q}_0 \subset \boldsymbol{Q}$ is a chain, then let $\mathcal{Q}_\infty$ defined by $Q_\infty(j) := \bigcap_{\mathcal{Q}\in\boldsymbol{Q}_0} Q(j)$. Then $\mathcal{Q}_\infty \in \boldsymbol{Q}$. Indeed, it is clear that either $Q_\infty(j) = P_j$ or there exists $\mathcal{Q} \in \boldsymbol{Q}_0$ such that $Q_\infty(j) = Q(j)$ and $\mathrm{cl}_\xi Q(j) \subset P_j$. It is enough to show that $X = \bigcup_{j\in J} Q_\infty(j)$. If $x \in X$, then since $\{P_j : j \in J\}$ is point-finite, there exists a finite subset J_0 of J such that $x \in \bigcap_{j\in J_0} P_j$ and $x \notin P_j$ for each $j \in J \setminus J_0$. If there exists $j_0 \in J_0$ such that $Q_\infty(j_0) = P_{j_0}$, then $x \in Q_\infty(j_0) \subset \bigcup_{j\in J} Q_\infty(j)$. If $Q_\infty(j) \neq P_j$ for each $j \in J_0$, we can assume that for each $j \in J_0$, there exists $\mathcal{Q}_j \in \boldsymbol{Q}_0$ such that $Q_j(j) \neq P_j$, and since $\boldsymbol{Q}_0$ is a chain, there exists $k \in J_0$ such that $\mathcal{Q}_k \preceq \mathcal{Q}_j$ for each $j \in J_0$. As $\mathcal{Q}_k$ is a cover, there is $j \in J_0$ such that $x \in Q_k(j) = Q_\infty(j)$.

By Theorem XXVII.4.8 of Kuratowski-Zorn, there exists a minimal element $\mathcal{Q}_0$ of $\boldsymbol{Q}$. It remains to prove that $\mathrm{cl}_\xi Q_0(j) \subset P_j$ for each $j \in J$. If by way of contradiction, there exists $j_0 \in J$ such that $\mathrm{cl}_\xi Q_0(j_0) \setminus P_{j_0} \neq \varnothing$, then $X \setminus \bigcup_{j\neq j_0} Q_0(j)$ is a ξ-closed set included in $Q_0(j_0)$. By normality, there exists a ξ-open set O such that $X \setminus \bigcup_{j\neq j_0} Q_0(j) \subset O \subset \mathrm{cl}_\xi O \subset Q_0(j_0)$. It follows from the construction that $Q_0(j_0) = P_{j_0}$. By setting $Q_1(j_0) := O$ and $Q_1(j) := Q_0(j)$ if $j \neq j_0$, we obtain a family

$\mathcal{Q}_1 \in \boldsymbol{Q}$ such that $\mathcal{Q}_0 \succeq \mathcal{Q}_1$, hence $\mathcal{Q}_0 = \mathcal{Q}_1$ by the minimality of $\mathcal{Q}_0$, a contradiction. □

LEMMA IX.5.17. *If every open cover of a topological space has a refining locally finite closed cover, then every open cover has a refining locally finite open cover.*

PROOF. Let ξ be a topology on X, and let $\mathcal{P}$ be a ξ-open cover. Let $\{A_j : j \in J\}$ be any ξ-locally finite cover of X which is a refinement of $\mathcal{P}$, and for every $x \in X$, let $V_x \in \mathcal{N}_\xi(x)$ such that $\{j \in J : A_j \# V_x\}$ is finite. Let $\mathcal{Q}$ be a ξ-closed cover of X which is a ξ-locally finite refinement of $\{V_x : x \in X\}$. For every $j \in J$, define

$$W_j := X \setminus \bigcup\{Q \in \mathcal{Q} : Q \cap A_j = \varnothing\}.$$

Then W_j is a ξ-open set and $A_j \subset W_j$. For each $Q \in \mathcal{Q}$ and every $j \in J$,

$$W_j \# Q \Longleftrightarrow A_j \# Q.$$

For each $j \in J$, let $P(j) \in \mathcal{P}$ be such that $A_j \subset P(j)$, and let $U_j := W_j \cap P(j)$. Then $\{U_j : j \in J\}$ is a ξ-open cover which is a refinement of $\mathcal{P}$. For every $x \in X$, there exists $N_x \in \mathcal{N}_\xi(x)$ such that $\{Q \in \mathcal{Q} : N_x \# Q\}$ is finite, and for each $Q \in \mathcal{Q}$, the set $\{j \in J : Q \# A_j\}$ is finite, $\{U_j : j \in J\}$ is ξ-locally finite. □

Density related terms. We have observed that a pretopology is determined by its adherence, hence also by its inherence. Therefore, the concepts defined with the aid of adherence or inherence are pretopological, for example, density. There exists a plethora of terms related to density.

Observe that a subset A of X is *ξ-dense* for a convergence ξ on X (Definition III.1.19) if and only if $\operatorname{adh}_\xi A = X$.

A subset of a convergence space is called *codense* if its complement is dense. Hence, a set is ξ-codense if and only if $\operatorname{inh}_\xi A = \varnothing$ ([12]). By Example III.1.20 and Exercise III.7.4, the sets $\mathbb{Q}$ and $\mathbb{R} \setminus \mathbb{Q}$ are both dense and codense.

DEFINITION IX.5.18. A set A is called *nowhere dense* or *rare* if

$$\operatorname{inh}_\xi(\operatorname{adh}_\xi A) = \varnothing.$$

A countable union of rare sets is called a *meager* set or of *first category* set. A non-meager set is called of *second category*. A complement of a meager set is called *residual*.

[12]In fact, A is ξ-codense if and only if $X = \operatorname{adh}_\xi(X \setminus A)$, equivalently, $\varnothing = \operatorname{inh}_\xi A$.

EXAMPLE IX.5.19 (A nowhere dense set). A subset $A = \{2^{-n} : n < \omega\}$ of the real line is nowhere dense. In fact, $\operatorname{adh}_{\nu_\mathbb{R}} A = \operatorname{cl}_{\nu_\mathbb{R}} A = A \cup \{0\}$, hence is countable. On the other hand, each countable subset of the real line has empty interior. In fact, if $\operatorname{int}_{\nu_\mathbb{R}} B \neq \varnothing$, then there exist $a < b$ such that $]a, b[\subset B$, hence B is not countable.

CHAPTER X

Functional study of topologies

The word *function* is used as a synonym of *map*. Yet, often a *function* is understood in a narrower sense, as a real-valued map, and we will follow here this convention.

The term *functional study of topologies* will be used here to designate the interplay between topologies and the corresponding real-valued continuous functions.

In Section III.6 we have briefly described metrizable topologies. In fact, a topology ξ on X is metrizable if there exists a metric $d : X \times X \longrightarrow \mathbb{R}_+$ such that the family $\{d(\cdot, x) : x \in X\}$ of functions determines ξ, more precisely, if $x \in \lim_\xi \mathcal{F}$ whenever $\inf_{F\in\mathcal{F}} \sup_{y\in F} d(x,y) = 0$.

Of course, any given convergence ξ specifies $\mathscr{C}(\xi, \nu_\mathbb{R})$, and vice versa the family $\mathscr{C}(\xi, \nu_\mathbb{R})$ determines the corresponding $\nu_\mathbb{R}$-initial topology on the underlying set $|\xi|$ of ξ.

The initial topology is coarser than the original convergence, and if it is equal, it is necessarily a topology, called *functionally initial.* We shall show in Section X.9 that functionally initial topologies coincide with *functionally regular topologies*, traditionally called *completely regular.*

X.1. The real line

In addition to the standard topology $\nu = \nu_\mathbb{R}$ we define two other convergences on $\mathbb{R}$. The *lower convergence* ν_- on the real line by $r \in \lim_{\nu_-} \mathcal{F}$ whenever

$$r \leq \sup_{F\in\mathcal{F}} \inf F. \tag{X.1.1}$$

Consequently, if $r \in \lim_{\nu_-} \mathcal{F}$ and $s \leq r$, then $s \in \lim_{\nu_-} \mathcal{F}$. In other terms, each convergent filter converges to a lower half-line. Actually,

$$\lim_{\nu_-} \mathcal{F} =]-\infty, \sup_{F\in\mathcal{F}} \inf F].$$

The *upper convergence* ν_+ is defined analogously: $r \in \lim_{\nu_+} \mathcal{F}$ whenever $r \geq \inf_{F\in\mathcal{F}} \sup F$. This is a topology, and the family of its open sets consists of $\varnothing, \mathbb{R}$ and $]-\infty, t[$, where $t \in \mathbb{R}$.

Of course, ν_+ and ν_- are homeomorphic ([1]).

PROPOSITION X.1.1. *The lower and the upper convergences are topologies.*

PROOF. Indeed, let $\mathcal{O}_{\nu_-}$ consists of $\varnothing, \mathbb{R}$ and $]s, +\infty[$ for each $s \in \mathbb{R}$. It is straightforward that O_{ν_-} fulfills the condition for open sets of a convergence. On the other hand, (X.1.1) means that $\mathcal{F}$ lower converges to r if and only if for each $s < r$, there exists $F \in \mathcal{F}$ such that $F \subset]s, +\infty[$. Therefore ν_- is a topology, so is its homeomorphic image ν_+. □

PROPOSITION X.1.2. *The lower and the upper topologies are initially dense in the class of topologies.*

PROOF. In virtue of Theorem VII.5.2, the Sierpiński topology is initially dense in the class of topologies. Hence, it is enough to prove that there exists $f \in \mathscr{C}(\$_0, \nu_+)$ such that $\$_0 = f^-\nu_+$. Let $f(0) = 0$ and $f(1) = 1$.

As $\mathcal{C}_{\$_0} = \{\varnothing, \{1\}, \{0, 1\}\}$, and a proper ν_+-closed set is of the form $[r, +\infty[$, we need to consider three cases: (1) $f(1) < r$, then $f^-([r, +\infty[) = \varnothing$; (2) $f(0) < r \leq f(1)$, then $f^-([r, +\infty[) = \{1\}$, and (3) $r \leq f(0)$, then $f^-([r, +\infty[) = \{0, 1\}$. □

One can prove Proposition X.1.2 directly, without using the fact that the Sierpiński topology is initially dense in the class of topologies (see Exercise X.10.3).

The supremum $\nu_- \vee \nu_+$ is a topology in virtue of Proposition VII.4.2. Actually, it is equal to the standard topology:

$$\nu_- \vee \nu_+ = \nu.$$

Indeed, by the definition of supremum, all the strict intervals $]s, t[$ are open in $\nu_- \vee \nu_+$, and we have seen that they constitute a base for the open sets of ν.

LEMMA X.1.3. *Each bounded above increasing sequence converges to its supremum, and each bounded below decreasing sequence converges to its infimum.*

PROOF. If a sequence $(r_n)_n$ on $\mathbb{R}$ is bounded above, its supremum exists. Let $r := \sup\{r_n : n < \omega\}$. By definition, for each $s < r$, there exists $n_0 < \omega$ such that $s < r_{n_0} \leq r$, and since the sequence is increasing, $r_n \in]s, r]$ for each $n \geq n_0$. Similarly for the second claim. □

[1]For example, by $r \longmapsto -r$.

A series of positive terms is said to be *convergent* if

$$\lim_\nu (\sum_{k=0}^{n} a_k)_n \neq \varnothing,$$

and we denote the limit by $\sum_{n=0}^{\infty} a_n$. Otherwise, we set $\sum_{n=0}^{\infty} a_n = \infty$. Note that if $\sum_{n=0}^{\infty} a_n < \infty$, then its value is independent of the order of the terms (because the terms are positive), and

$$\lim_{n\to\infty} \sum_{k=n}^{\infty} a_k = 0, \tag{X.1.2}$$

hence $\lim_{n\to\infty} a_n = 0$.

Among criteria that ensure convergence of series, we will make use of a simplest one, concerning *geometric series*. Namely,

LEMMA X.1.4. *A geometric series is convergent* $\sum_{n=0}^{\infty} q^n < \infty$ *if and only if* $0 \leq q < 1$, *and in this case* ([2])

$$\sum_{n=0}^{\infty} q^n = \frac{1}{1-q}. \tag{X.1.3}$$

X.2. Real-valued functions

Let ξ be a convergence on X. A function f is called *lower semicontinuous* at x if for every filter $\mathcal{F} \in \xi^-(x)$,

$$f(x) \leq \sup_{F\in\mathcal{F}} \inf_{w\in F} f(w); \tag{X.2.1}$$

f is called *upper semicontinuous* at x if for every filter $\mathcal{F} \in \xi^-(x)$,

$$f(x) \geq \inf_{F\in\mathcal{F}} \sup_{w\in F} f(w). \tag{X.2.2}$$

Recall that ν_- and ν_+ stand for the lower and the upper topologies respectively. It is straightforward that

PROPOSITION X.2.1. *A function f is lower ξ-semicontinuous if and only if $f \in \mathscr{C}(\xi, \nu_-)$; upper ξ-semicontinuous if and only if $f \in \mathscr{C}(\xi, \nu_+)$.*

Notice that

$$\mathscr{C}(\xi, \nu_-) \cap \mathscr{C}(\xi, \nu_+) = \mathscr{C}(\xi, \nu).$$

COROLLARY X.2.2. *A function f is lower ξ-semicontinuous if and only if $\{f \leq r\}$ is ξ-closed for each $r \in \mathbb{R}$. A function f is upper ξ-semicontinuous if and only if $\{f \geq r\}$ is ξ-closed for each $r \in \mathbb{R}$.*

---

[2]We consider here the series with positive terms. For arbitrary terms, the condition for the convergence of the geometric series $\sum_{n=0}^{\infty} q^n$ is $|q| < 1$. Indeed, $(1-q)\sum_{k=0}^{n} q^k = \sum_{k=0}^{n} q^k - \sum_{k=0}^{n} q^{k+1} = 1 - q^{n+1}$, hence if $|q| < 1$, then

$$\sum_{n=0}^{\infty} q^n = \lim_{n\to\infty} \sum_{k=0}^{n} q^k = \lim_{n\to\infty} \tfrac{1-q^{n+1}}{1-q} = \frac{1}{1-q}.$$

Note that $0^0 = 1$ in virtue of Proposition XXVII.2.8.

PROOF. Let f be lower ξ-semicontinuous and let $x \in \lim_\xi \mathcal{F}$. Now $x \notin \{f \leq r\}$ whenever $r < f(x)$, which by (X.2.1) is equivalent to the existence of $F \in \mathcal{F}$ such that $r < f(w)$ for each $w \in F$. In other words, $X \setminus \{f \leq r\} = \{f > r\} \in \mathcal{F}$, that is, $\{f \leq r\}$ is ξ-closed. □

COROLLARY X.2.3. *A function f is continuous if and only if it is both lower and upper semicontinuous.*

The extrema of sets of functions are defined pointwise, that is,

$$\sup \mathscr{F}(x) := \sup\{f(x) : f \in \mathscr{F}\},$$
$$\inf \mathscr{F}(x) := \inf\{f(x) : f \in \mathscr{F}\}.$$

In general, extrema can admit infinite values.

LEMMA X.2.4. *If $\mathscr{F}$ is a set of lower semicontinuous functions, then* $\sup \mathscr{F}$ *is lower semicontinuous. If $\mathscr{F}$ is a set of upper semicontinuous functions, then* $\inf \mathscr{F}$ *is upper semicontinuous.*

PROOF. Let us abridge

$$F(x) := \sup \mathscr{F}(x) = \sup\{f(x) : f \in \mathscr{F}\}, \tag{X.2.3}$$

where $x \in X$. Let $x \in X$ and $r < F(x)$. Then there exists $f \in \mathscr{F}$ such that $f(x) > r$. As f is lower semicontinuous, there exists $U \in \mathcal{N}_X(x)$ such that $r < f(u) \leq F(u)$ for each $u \in U$, hence $\{F > r\}$ is open. □

DEFINITION X.2.5. A set $\mathscr{F} \subset \mathbb{R}^X$ is said to be *equicontinuous* if for each $x \in X$ and every $\varepsilon > 0$, there exists $U \in \mathcal{N}_X(x)$ such that $|f(u) - f(x)| < \varepsilon$ for each $u \in U$ and every $f \in \mathscr{F}$.

Of course, each element of an equicontinuous family is a continuous function.

PROPOSITION X.2.6. *If $\mathscr{F}$ is equicontinuous, then* $\sup \mathscr{F}$ *and* $\inf \mathscr{F}$ *are continuous, provided that they do not have infinite values.*

PROOF. By Lemma X.2.4, $F = \sup \mathscr{F}$ is lower semicontinuous. Let $x \in X$ and $\varepsilon > 0$. In particular, there exists $U \in \mathcal{N}_X(x)$ such that $f(u) < f(x) + \varepsilon \leq F(x) + \varepsilon$ for each $u \in U$. Hence $F(u) = \sup \mathscr{F}(u) \leq F(x) + \varepsilon$, that is, F is upper semicontinuous, hence continuous. □

Actually, if we adjoin the supremum and the infimum to an equicontinuous family, then the resulting family is also equicontinuous. In the sequel we will use an equivalent fact only for suprema.

LEMMA X.2.7. *If* $\mathscr{F}$ *is equicontinuous and* $\sup\{f(x) : f \in \mathscr{F}\} < +\infty$ *for each* $x \in X$*, then the family*

$$\{\sup_{f\in\mathscr{T}} f : \varnothing \neq \mathscr{T} \subset \mathscr{F}\}$$

is equicontinuous.

PROOF. Let us abridge $f_{\mathscr{T}} := \sup_{f\in\mathscr{T}} f$ for each (not necessarily finite) subfamily $\mathscr{T}$ of $\mathscr{F}$. By assumption, for each $x \in X$ and every $\varepsilon > 0$, there exists $U \in \mathcal{N}_X(x)$ such that $|f(u) - f(x)| < \frac{\varepsilon}{2}$ for each $u \in U$ and $f \in \mathscr{T}$. If $\mathscr{T} \subset \mathscr{F}$, then there is $f \in \mathscr{T}$ such that $f_{\mathscr{T}}(x) - \frac{\varepsilon}{2} < f(x)$, thus $f_{\mathscr{T}}(x) - \varepsilon < f(x) - \frac{\varepsilon}{2} < f(u) \le f_{\mathscr{T}}(u)$ for each $u \in U$. On the other hand, if $f \in \mathscr{T}$ then $f_{\mathscr{T}}(x) + \varepsilon \ge f(x) + \varepsilon > f(u)$ for each $u \in U$, hence $f_{\mathscr{T}}(x) + \varepsilon \ge f_{\mathscr{T}}(u)$ for each $u \in U$. □

X.3. More on metrics and metrizable topologies

Recall that a *metric* on a non-empty set X is a function $d : X \times X \longrightarrow \mathbb{R}_+$ such that (III.6.1), (III.6.2), and (III.6.3) for each x, y, and z,

$$d(x, y) = d(y, x),$$
$$d(x, y) = 0 \Longleftrightarrow x = y,$$
$$d(x, z) \le d(x, y) + d(y, z).$$

Recollect as well that a topology ξ is called *metrizable* if there exists a metric such that $\xi = \mathrm{t}d$, that is, (III.6.4) $x \in \lim_\xi \mathcal{F}$ provided that

$$\inf_{F\in\mathcal{F}} \sup_{y\in F} d(x, y) = 0.$$

Accordingly, a metric determines a topology altogether. Although a metric is a single function, but being a function of two variables, it actually operates as a family of functions: if d is a metric on X, then $\{d(\cdot, x) : x \in X\}$ is the family of functions that specifies the topology. More precisely, if $\xi = \mathrm{t}d$, then, for every $x \in X$, the ξ-vicinity filter of x is given by

$$\mathcal{V}_\xi(x) = \{\{w \in X : d(w, x) < \varepsilon\} : \varepsilon > 0\}^\uparrow. \tag{X.3.1}$$

Metrics are called *equivalent* if they define the same topology. Note that if d is a metric, then $\min(d, 1)$ is an equivalent metric. Therefore, each metrizable topology admits a bounded metric.

LEMMA X.3.1. *If* $d : X \times X \longrightarrow \mathbb{R}_+$ *is a metric, then for each* $y \in X$,

$$|d(x_0, y) - d(x_1, y)| \le d(x_0, x_1). \tag{X.3.2}$$

PROOF. By (III.6.3), $d(x_0,y) \leq d(x_0,x_1)+d(x_1,y)$, hence $d(x_0,y)-d(x_1,y) \leq d(x_0,x_1)$. By interchanging x_0 with x_1, we conclude. □

Consequently, each metric is continuous. More precisely, for each $y \in X$, the function $x \mapsto d(x,y)$ is continuous. It follows that every strict ball $B(x,\varepsilon) := \{y \in X : d(x,y) < \varepsilon\}$ is open. Therefore, each vicinity filter of a metrizable convergence consists of open sets. It results that each metrizable convergence is a topology.

If d is a ξ-compatible metric, then

$$\xi = \bigvee_{x\in|\xi|} d(\cdot,x)^{-}\nu,$$

that is,

COROLLARY X.3.2. *If d is a compatible metric of a metrizable topology ξ, then ξ is initial with respect to $\{d(\cdot,x) : x \in |\xi|\}$ and the standard topology ν of the real line.*

COROLLARY X.3.3. *Each metrizable convergence is a Hausdorff topology of countable character.*

In particular, if ξ is metrizable, then $\mathrm{adh}_\xi A = \mathrm{cl}_\xi A$, because ξ is a topology.

LEMMA X.3.4. *If d is a metric, then*

$$|d(x_0,y_0) - d(x_1,y_1)| \leq d(x_0,x_1) + d(y_0,y_1). \tag{X.3.3}$$

PROOF. Applying (III.6.3) twice, we get

$$d(x_0,y_0) \leq d(x_0,x_1) + d(x_1,y_1) + d(y_1,y_0),$$

hence

$$d(x_0,y_0) - d(x_1,y_1) \leq d(x_0,x_1) + d(y_1,y_0),$$

and, by inverting the role of x_0 and x_1 above, where it is useful, we get (X.3.3). □

COROLLARY X.3.5. *If d is a ξ-compatible metric, then $d \in \mathscr{C}(\xi \times \xi,\nu)$.*

For a metric d on a set X, and for a non-empty subset A of X, we define a *distance function* from A,

$$d(x,A) := \inf\{d(x,a) : a \in A\}. \tag{X.3.4}$$

The triangular inequality (III.6.3) generalizes to

$$|d(x_0,A) - d(x_1,A)| \leq d(x_0,x_1). \tag{X.3.5}$$

PROOF. By definition (X.3.4), and by (III.6.3), $d(x_0, A) \leq d(x_0, a) \leq d(x_0, x_1) + d(x_1, a)$ for each $a \in A$. Therefore, $d(x_0, A) - d(x_1, a) \leq d(x_0, x_1)$ for each $a \in A$. Hence,

$$d(x_0, A) - d(x_1, A) = d(x_0, A) - \inf_{a \in A} d(x_1, a) \leq d(x_0, x_1).$$

On inverting the role of x_0 and x_1, we conclude that (X.3.5) holds. □

COROLLARY X.3.6. *The distance function $d(\cdot, A)$ is continuous, and if $\xi = \mathrm{t}d$, then*

$$\{x : d(x, A) = 0\} = \mathrm{cl}_\xi A.$$

PROOF. As $d(\cdot, A)$ is continuous, $\{x : d(x, A) = 0\}$ is ξ-closed, and includes A. Therefore, $\mathrm{cl}_\xi A \subset \{x : d(x, A) = 0\}$. On the other hand, if $x \notin \mathrm{cl}_\xi A$, then there exists $\varepsilon > 0$ such that $B(x, \varepsilon) \cap A = \varnothing$, equivalently $d(x, A) > \varepsilon$, so that $x \notin \mathrm{cl}_\xi A$. □

The *distance* between two subsets A_0 and A_1 of a metric space (X, d) is defined by

$$\mathrm{dist}(A_1, A_0) := \inf_{x \in A_1} d(x, A_0) = \inf\{(x_0, x_1) : x_0 \in A_0, x_1 \in A_1\}.$$

PROPOSITION X.3.7. *If K is a compact subset, and F is a closed subset of a metric space (X, d) and $K \cap F = \varnothing$, then $\mathrm{dist}_d(K, F) > 0$.*

PROOF. By Corollary X.3.6, the function $d(\cdot, F)$ is continuous, hence there exists $x_0 \in K$ such that $\mathrm{dist}_d(K, F) = d(x_0, F) > 0$, by Corollary X.3.6, because $x_0 \notin F$. □

PROPOSITION X.3.8. *Each countable product of metrizable topologies is metrizable.*

PROOF. Let τ_n be a topology on a set X_n metrized by a metric d_n. As $\max(d_n, 1)$ is topologically equivalent to d_n, we assume without loss of generality that $|d_n| \leq 1$ for each $n < \omega$. Then

$$d(f, g) := \sum\nolimits_{n=0}^{\infty} \frac{1}{2^n} d_n(f(n), g(n)) \tag{X.3.6}$$

for every $f, g \in \prod_{n<\omega} X_n$, is a metric on $\prod_{n<\omega} X_n$, as one easily checks.

We need to prove that convergence in the product topology $\tau := \prod_{n<\omega} \tau_n$ coincides with that with respect to d. Let $(f_k)_k$ be a sequence of elements of $\prod_{n<\omega} X_n$ such that $\lim_{k\to\infty} d(f_k, f)$. It follows that $\lim_{k\to\infty} d_n(f_k(n), f(n))$ for each $n < \omega$, hence $f(n) \in \lim_{\tau_n}(f_k(n))_k$ for each $n < \omega$, and thus $f \in \lim_\tau (f_k)_k$ by definition.

Conversely, if $f \in \lim_\tau (f_k)_k$ then $\lim_{k\to\infty} d_n(f_k(n), f(n))$ for each $n < \omega$, because d_n is a metric compatible with τ_n for each $n < \omega$. If

$\varepsilon > 0$, then there exists n_ε such that $\sum_{n=n_\varepsilon}^{\infty} \frac{1}{2^n} < \frac{\varepsilon}{2}$. Therefore, for every $f, g \in \prod_{n<\omega} X_n$,

$$\sum_{n=n_\varepsilon}^{\infty} \frac{1}{2^n} d_n(f(n), g(n)) < \frac{\varepsilon}{2},$$

because $d_n(x, y) \leq 1$ for each $n < \omega$. On the other hand, for each $n < n_\varepsilon$, there exists $k_n(\varepsilon) < \omega$ such that for $k > \max\{k_n(\varepsilon) : n < n_\varepsilon\}$ and $n < n_\varepsilon$,

$$d_n(f_k(n), f(n)) < \frac{\varepsilon}{4}.$$

Therefore, for $k > \max\{k_n(\varepsilon) : n < n_\varepsilon\}$,

$$d(f_k, f) = \frac{\varepsilon}{4} \sum_{n<n_\varepsilon}^{\infty} \frac{1}{2^n} + \frac{\varepsilon}{2} < \varepsilon.$$

□

EXAMPLE X.3.9 (Baire space). The *Baire space* B is defined on $\mathbb{N}^{\mathbb{N}}$ as an infinite countable product of the discrete topology $\iota_{\mathbb{N}}$. Recall that a sequence $(\varphi_k)_k$ converges to φ in $\prod_{n\in\mathbb{N}} \iota_X$ whenever, for each $n \in \mathbb{N}$, there exists $k_n < \omega$ such that $\varphi_k(n) = \varphi(n)$ for each $k > k_n$.

Each discrete topology is metrizable. The standard compatible metric is the *discrete metric,* that is, $i_X : X \times X \longrightarrow \{0, 1\}$, given by $i_X(x_0, x_1) = 0$ if $x_0 = x_1$, and 1 otherwise. Accordingly, the Baire space is metrizable, and (X.3.6) is a compatible metric when $d_n = i_X$ for each $n < \omega$. We shall define later other compatible metrics of particular interest for the countable products of discrete topologies.

Let d be a metric on a set Z and X be a set. A filter $\mathcal{F}$ on Z^X *d-uniformly converges* to $f_\infty \in \mathbb{R}^X$, in symbols, $f_\infty \in \lim_{\mathfrak{u}(d)} \mathcal{F}$, if

$$\sup_{F\in\mathcal{F}} \inf_{f\in F} \sup_{x\in X} d(f(x), f_\infty(x)) = 0. \tag{X.3.7}$$

PROPOSITION X.3.10. *Uniform convergence is metrizable.*

PROOF. For $f, g \in Z^X$ and a metric d on Z, let

$$D(f, g) := \sup\nolimits_{x\in X} d(f(x), g(x)).$$

It is straightforward that D is a metric. Then (X.3.7) amounts to

$$\sup\nolimits_{F\in\mathcal{F}} \inf\nolimits_{f\in F} D(f, f_\infty) = 0,$$

and since D is non-negative, to $\lim_\nu D(\mathcal{F}, f_\infty) = \{0\}$, where

$$D(\mathcal{F}, h) := \{\{D(f, h) : f \in F\} : F \in \mathcal{F}\}.$$

Hence the *uniform convergence* $\mathfrak{u}$ is a metrizable topology. □

The *norm* $\|h\|$ of $h \in \mathbb{R}^X$ is defined by

$$\|h\| := \sup_{x\in X} |h(x)| .$$

It is straightforward that $\|h\| \geq 0$ for each $h \in \mathbb{R}^X$, and $\|h\| = 0$ if and only if $h = 0$; if $\lambda \in \mathbb{R}$ then $\|\lambda \cdot h\| = |\lambda| \cdot \|h\|$. Finally, for any $h_0, h_1 \in \mathbb{R}^X$,

$$\|h_0 + h_1\| \leq \|h_0\| + \|h_1\| . \tag{X.3.8}$$

It is clear that $\|h\| < \infty$ if and only if there is $r \in \mathbb{R}$ such that $|h(x)| < r$ for all $x \in X$, that is, whenever h is a *bounded function*. Then $D(h_0, h_1) := \|h_0 - h_1\|$ is a metric on $\mathbb{R}^X$.

PROPOSITION X.3.11. *For each topology ξ on X ([3]), $\mathscr{C}(\xi,\nu)$ is closed for the uniform topology $\mathfrak{u}$.*

PROOF. Let $f_\infty \in \lim_{\mathfrak{u}} \mathcal{F}$ and $\mathscr{C}(\xi,\nu) \in \mathcal{F}$. Let $x \in X$ and $\varepsilon > 0$. By definition, there is $F \in \mathcal{F}$ such that $\|f - f_\infty\| < \frac{\varepsilon}{3}$ for each $f \in F$. Pick any $f \in F$, and a neighborhood V of x so that $|f(v) - f(x)| < \frac{\varepsilon}{3}$ for each $v \in V$. Consequently, for each $v \in V$,

$$|f_\infty(v) - f_\infty(x)| \leq |f_\infty(v) - f(v)| + |f(v) - f(x)| + |f(x) - f_\infty(x)| < \varepsilon,$$

proving that $f_\infty \in \mathscr{C}(\xi,\nu)$. □

A metric d on X is called *totally bounded* if, for each $\varepsilon > 0$, there exists a finite subset of X such that $X \subset B_d(F,\varepsilon)$.

PROPOSITION X.3.12. *Each compatible metric of a countably compact metrizable topology is totally bounded.*

PROOF. By way of contradiction suppose that there exists $\varepsilon > 0$ such that $X \setminus B_d(F,\varepsilon) \neq \varnothing$ for each $F \subset_{\text{fin}} X$. Hence, by induction, there exists a sequence $(x_n)_n$ of elements of X such that $d(x_n, x_k) \geq \varepsilon$ for each $n \neq k$. Then $\{x_n : n < \omega\}$ is a closed discrete subspace hence is not compact by Proposition IX.3.9. □

COROLLARY X.3.13. *Every countably compact metrizable topology is separable.*

PROOF. Let d be a compact metric on X. Then, by (1), for each $n < \omega$, there is a finite set F_n such that $X \subset B_d(F_n, 2^{-n})$. Therefore, the countable set $\bigcup_{n<\omega} F_n$ is d-dense. □

COROLLARY X.3.14. *Every countably compact metrizable topology is compact.*

[3] *This is true for each convergence ξ, but since $\mathscr{C}(\xi,\nu) = \mathscr{C}(\mathrm{T}\xi,\nu)$, it is enough to consider only topologies ξ.*

PROOF. By Corollary X.3.13 and Proposition IX.2.16, each countably compact metrizable topology is of countable weight, hence Lindelöf by Corollary IX.3.2. Finally each countably compact Lindelöf convergence is compact. □

X.4. Completely metrizable topologies

A filter $\mathcal{F}$ on a metric space (X, d) is said to be *d-fundamental* or a *Cauchy* filter if for each $\varepsilon > 0$, there exists $F \in \mathcal{F}$ such that the *diameter* $\operatorname{diam} F < \varepsilon$, that is,

$$\inf_{F \in \mathcal{F}} \operatorname{diam} F = 0.$$

LEMMA X.4.1. *If $\mathcal{F}$ is a d-fundamental filter such that $x \in \operatorname{adh}_{\mathrm{t}d} \mathcal{F}$, then $x \in \lim_{\mathrm{t}d} \mathcal{F}$.*

PROOF. Let $x_\infty \in \operatorname{adh}_{\mathrm{t}d} \mathcal{F}$. If $\varepsilon > 0$, then for each $F \in \mathcal{F}$, there is $x_F \in F$ such that $d(x_F, x_\infty) < \frac{\varepsilon}{2}$. As $\mathcal{F}$ is d-fundamental, there exists $F \in \mathcal{F}$ such that $d(x, x_F) < \frac{\varepsilon}{2}$ for each $x \in F$. Therefore, if $x \in F$, then

$$d(x, x_\infty) \leq d(x, x_F) + d(x_F, x_\infty) < \varepsilon,$$

that is, $B_d(x_\infty, \varepsilon) \in \mathcal{F}$, which proves that $x_\infty \in \lim_{\mathrm{t}d} \mathcal{F}$. □

In particular, a sequence $(x_n)_n$ is called *Cauchy* if the corresponding sequential filter is fundamental. Lemma X.4.1 specializes to sequences, as below. As adherence and sequential adherence coincide in metrizable spaces, this specialization is a simple corollary. Nevertheless, we provide a proof for the reader's convenience.

COROLLARY X.4.2. *If a Cauchy sequence has a convergent subsequence, then it is convergent.*

PROOF. Let $(x_n)_n$ be a Cauchy sequence, and let $(x_{n_k})_k$ converge to x. If $\varepsilon > 0$, then there exists $n_\varepsilon < \omega$ such that $d(x_n, x_m) < \frac{\varepsilon}{2}$ for each $n, m > n_\varepsilon$, and there exists $k_\varepsilon < \omega$ such that $d(x_{n_k}, x) < \frac{\varepsilon}{2}$ for $k > k_\varepsilon$. Consequently, for $n > n_\varepsilon$ and $m = n_k > n_\varepsilon$ with $k > k_\varepsilon$,

$$d(x_n, x) \leq d(x_n, x_{n_k}) + d(x_{n_k}, x) < \varepsilon.$$

□

A metric is said to be *complete* if each fundamental filter is adherent ([4]).

PROPOSITION X.4.3. *The following statements are equivalent for a metric space:*

[4]Recall that a filter $\mathcal{F}$ is ξ-adherent if $\operatorname{adh}_\xi \mathcal{F} \neq \varnothing$.

(1) *a metric is complete,*
(2) *each fundamental filter converges,*
(3) *each Cauchy sequence converges,*
(4) *each Cauchy sequence is adherent.*

PROOF. (1) $\Rightarrow$ (2). If d is a complete metric on X, then, by definition, $\operatorname{adh}_{td} \mathcal{F} \neq \varnothing$ for each d-fundamental. By Lemma X.4.1, $\lim_{td} \mathcal{F} \neq \varnothing$. (2) $\Rightarrow$ (3). In particular, this holds for every sequential filter. (3) $\Rightarrow$ (4). Every convergent filter is adherent. (4) $\Rightarrow$ (1). Let $\mathcal{F}$ be d-fundamental. Then for each $n < \omega$, there is $F_n \in \mathcal{F}$ such that $\operatorname{diam}(F_n) < 2^{-n}$. We can assume that the sequence $(F_n)_n$ is decreasing, for otherwise, we replace F_{n+1} by $F_{n+1} \cap F_n$. If $x_n \in F_n$, then $(x_n)_n$ is a Cauchy sequence such that $(x_n)_n \geq (F_n)_n$, hence by assumption $\varnothing \neq \operatorname{adh}(x_n)_n \subset \operatorname{adh}(F_n)_n$. As $(F_n)_n$ is a filter-base of a fundamental filter $\varnothing \neq \lim(F_n)_n$ by Lemma X.4.1. As $(F_n)_n \leq \mathcal{F}$, also $\varnothing \neq \lim \mathcal{F} \subset \operatorname{adh} \mathcal{F}$. □

It is straightforward that

PROPOSITION X.4.4. *Each closed subset of a complete metric space is complete.*

Here is a classical characterization of completeness.

THEOREM X.4.5 (Cantor). *A metric space is complete if and only if every decreasing sequence of non-empty closed sets with the diameter tending to 0 has non-empty intersection.*

PROOF. Assume completeness, and let $(F_n)_n$ be a decreasing sequence of non-empty closed sets with $\lim_{n\to\infty} \operatorname{diam} F_n = 0$. If $x_n \in F_n$, then for $\varepsilon > 0$ there is $n < \omega$ such that $\operatorname{diam} F_n < \varepsilon$. As $x_k \in F_n$ for each $k \geq n$, the sequence $(x_n)_n$ is Cauchy, hence convergent. Consequently $\lim(x_k)_k \subset F_n$, because F_n is closed for each $n < \omega$. Therefore $\varnothing \neq \bigcap_{n<\omega} F_n$.

Conversely, if a metric is not complete, then there exists a Cauchy non-convergent sequence $(x_n)_n$. Then $F_m := \{x_n : n > m\}$ is closed for each $m < \omega$. Indeed, if $x \in \operatorname{cl} F_{m_0}$, then there is a strictly increasing subsequence $(n_k)_k$ such that $x \in \lim(x_{n_k})_k$, which is contradictory by Lemma X.4.1. On the other hand, $\bigcap_{m<\omega} F_m = \varnothing$. □

REMARK X.4.6. Actually, the non-empty intersection in the first part of the proof is a singleton, because its diameter must be 0.

A topology is called *completely metrizable* if there exists a compatible complete metric.

LEMMA X.4.7. *The standard topology $\nu_{\mathbb{R}}$ of the real line is completely metrizable.*

PROOF. We shall see that the Euclidean metric $d(r,s) = |r-s|$ is complete. If $(r_n)_n$ is a d-Cauchy sequence, then $\{r_n : n < \omega\}$ is d-bounded. By Bolzano-Weierstraß Theorem XII.1.1, there exists a convergent subsequence, hence d is complete by Proposition X.4.3. □

By definition, a completely metrizable topology admits a complete metric. However, it can admit as well non-complete metrics.

EXAMPLE X.4.8 (Complete and a non-complete metrics). Like in Example III.6.4, consider $X := \{2^{-n} : n < \omega\}$ with the subspace topology of $\mathbb{R}$. The resulting topology ν_X is discrete. The restriction d_X of the Euclidean metric d of $\mathbb{R}$ is not complete. Indeed, $\{\{2^{-n} : n > m\} : m < \omega\}$ is a filter-base of the cofinite filter of X. As $\operatorname{diam}\{2^{-n} : n > m\} = 2^{-m-1}$, the filter is d_X-fundamental, but it is not convergent. Let $i_X(2^{-n}, 2^{-k}) = 1$ if $n \neq k$, and $i_X(2^{-n}, 2^{-n}) = 0$ for each $n, k < \omega$. This is also a compatible metric of the discrete topology on X. Therefore, the only i_X-fundamental filters are principal ultrafilters, which are convergent. Therefore, the topology is completely metrizable. Therefore, the identity map $i \in \mathscr{C}(\nu_X, \nu_X)$ maps a complete metric space onto a non-complete metric space.

It follows that the continuous image of a complete metric space need not be complete; for the same reason, Cauchy sequences are not preserved under continuous maps.

On the other hand,

PROPOSITION X.4.9. *Each compatible metric of a metrizable compact topology is complete.*

PROOF. If d is such a metric, and $(x_n)_n$ is a Cauchy sequence, then by compactness, there exists a convergent subsequence $(x_{n_k})_k$. By Corollary X.4.2, $(x_n)_n$ is convergent. □

PROPOSITION X.4.10. *The product metric (X.3.6)*

$$d(f,g) := \sum_{n=0}^{\infty} \frac{1}{2^n} d_n(f(n), g(n))$$

on $\prod_{n\in\mathbb{N}} X_n$ is complete if and only if d_n is complete for each $n < \omega$.

PROOF. Let d be complete, and let $(x_k)_k$ be a Cauchy sequence in X_n. For an arbitrary $\psi \in \prod_{m\in\mathbb{N}} X_m$, define a sequence $(\varphi_k)_k$ by

$$\varphi_k(m) := \begin{cases} x_k & m = n, \\ \psi(m) & m \neq n. \end{cases}$$

As $d(\varphi_k, \varphi_l) = 2^{-n} d_n(x_k, x_l)$, the sequence $(\varphi_k)_k$ is Cauchy, hence convergent to an element φ of $\prod_{m\in\mathbb{N}} X_m$. Therefore, $(x_k)_k$ converges to $\varphi(n)$.

Conversely, let d_n be complete for each $n < \omega$, and let $(\varphi_k)_k$ be a Cauchy sequence in the product, hence by (X.3.6), $(\varphi_k(n))_k$ is d_n-Cauchy, hence convergent to some $x_n \in X_n$ for every $n < \omega$. Consequently, $(\varphi_k)_k$ converges to φ, given by $\varphi(n) := x_n$. □

COROLLARY X.4.11. *A countable product of metrizable topologies is completely metrizable if and only if every factor is completely metrizable.*

In particular, the Baire space of Example X.3.9 is completely metrizable.

By Proposition X.4.4, each closed subset of a completely metrizable space is completely metrizable. Each open subset of a completely metrizable space is completely metrizable. Indeed, by Lemma X.10.4 to come, each open subset of a topology ξ is homeomorphic to a closed subset of $\xi \times \nu_{\mathbb{R}}$, which is completely metrizable by Corollary X.4.11 and Lemma X.4.7.

DEFINITION X.4.12. A set is called a $G_{\aleph_0}$*-set* (traditionally, G_δ-set) if it is a countable intersection of open sets. A set is called a $F_{\aleph_0}$*-set* (traditionally, F_σ-set) if it is a countable union of closed sets.

PROPOSITION X.4.13. *Each $G_{\aleph_0}$-subset of a completely metrizable topology is completely metrizable.*

PROOF. By Lemma X.10.5 to come, each $G_{\aleph_0}$-subset of a topology ξ is homeomorphic to a closed subset of $\xi \times \prod_{n<\omega} \nu_{\mathbb{R}}$, which is completely metrizable by Corollary X.4.11 and Lemma X.4.7. □

PROPOSITION X.4.14. *The topology of uniform convergence is completely metrizable.*

PROOF. In fact, $d(f, g) := \|f - g\| = \sup_{x\in X} |f(x) - g(x)|$ is a complete metric. Indeed, if $(f_n)_n$ is a Cauchy sequence, that is, for each $\varepsilon > 0$ there is $n_\varepsilon < \omega$ such that $\|f_n - f_k\| < \varepsilon$ for each $n, k > n_\varepsilon$. In particular,

$$|f_n(x) - f_k(x)| < \varepsilon \tag{X.4.1}$$

for each $n, k > n_\varepsilon$ and for each $x \in X$. Therefore, the sequence $(f_n(x))_n$ is Cauchy. By Lemma X.4.7, there exists a subsequence $(f_{n_p}(x))_p$ convergent to an element $f(x)$ of $\mathbb{R}$. By replacing in (X.4.1) k by n_p, we conclude that $|f_n(x) - f(x)| \leq \varepsilon$ for every x and each $n > n_\varepsilon$.

Consequently,

$$\|f_n - f\| = \sup_{x\in X} |f_n(x) - f(x)| \leq \varepsilon$$

and each $n > n_\varepsilon$. Therefore, $(f_n)_n$ uniformly converges to f. □

The following theorem is one of the most instrumental results in topology and analysis. The term *category theorem* refers to the sets of *first category* that are countable unions of nowhere dense sets (Definition IX.5.18 to come).

In these terms, the following theorem says that completely metrizable spaces are not of first category.

THEOREM X.4.15 (Baire category theorem). *The intersection of a sequence of open dense subsets of a completely metrizable space is dense.*

PROOF. Fix a complete metric compatible on X. For each $n < \omega$, let Q_n be an open dense subset of X. Let $x \in X$ and let $r > 0$. By density, there exists $x_1 \in Q_0 \cap B(x, r)$, and since Q_0 is open, there is $0 < r_1 < \frac{r}{2}$ such that $\operatorname{cl} B(x_1, r_1) \subset Q_0 \cap B(x, r)$. Suppose that we have found sequences $r_0 = r, r_1, \ldots, r_n$ and $x_0 := x, x_1, \ldots, x_n$ such that $0 < r_{k+1} < \frac{r_k}{2}$ and

$$\operatorname{cl} B(x_{k+1}, r_{k+1}) \subset Q_k \cap B(x_k, r_k) \tag{X.4.2}$$

for $k < n < \omega$. Then there exist x_{n+1} and r_{n+1} such that (X.4.2) for $k = n$. As $(r_n)_n$ tends to 0, by Theorem X.4.5 of Cantor,

$$\varnothing \neq \bigcap_{n<\omega} \operatorname{cl} B(x_n, r_n) \subset \bigcap_{n<\omega} Q_n,$$

and thus $B(x, r) \cap \bigcap_{n<\omega} Q_n \neq \varnothing$. □

COROLLARY X.4.16. *If τ is a completely metrizable topology on X, and $(F_n)_n$ is a sequence of τ-closed sets such that $X = \bigcup_{n<\omega} F_n$, then there is $n < \omega$ such that $\operatorname{int}_\tau F_n \neq \varnothing$.*

PROOF. Let $Q_n := X \setminus F_n$. Then $X = \bigcup_{n<\omega} F_n$ amounts to $\varnothing = \bigcap_{n<\omega} Q_n$, hence by Theorem X.4.15, there is $n < \omega$ such that Q_n is not dense, that is, $\operatorname{cl}_\tau Q \neq X$, equivalently $\operatorname{int}_\tau F_n \neq \varnothing$. □

EXAMPLE X.4.17 ($\mathbb{Q}$ *is not completely metrizable*). The standard topology ν restricted to the set $\mathbb{Q}$ of rational numbers is not completely metrizable. Indeed, if $\mathbb{Q} = \{q_n : n < \omega\}$ then $\mathbb{Q} = \bigcup_{n<\omega}\{q_n\}$, where each singleton is closed and $\operatorname{int}_{\nu_\mathbb{Q}}\{q_n\} = \varnothing$ for each $n < \omega$, hence in virtue of Corollary X.4.16, $\nu_\mathbb{Q}$ is not completely metrizable.

X.5. Completion

We shall see that each metric space can be isometrically embedded into a complete metric space, that is, if (X,d) is a metric space, then there exists a complete metric space (Y,h) and an injection $j : X \longrightarrow Y$ so that

$$h(j(x_0), j(x_1)) = d(x_0, x_1).$$

Actually, one can isometrically embed (X,d) into the space $\mathscr{C}_b(X,\mathbb{R})$ of continuous bounded functions equipped with the metric of uniform convergence.

Let (X,d_X) and (Y,d_Y) be metric spaces. A map $f \in Y^X$ is called an *isometry* if for any $x_0, x_1 \in X$,

$$d_Y(f(x_0), f(x_1)) = d_X(x_0, x_1).$$

Let (X,d) be a metric space.

Definition X.5.1 (Completion). A complete metric space $(\widehat{X}, \widehat{d})$ such that (X,d) is isometric with a dense subspace of $(\widehat{X}, \widehat{d})$ is called a *completion* of (X,d).

Theorem X.5.2 (Hausdorff). *Each metric space has a completion.*

Proof. Let (X,d) be a metric space. Fix any $a \in X$, and define

$$f_x(y) := d(x,y) - d(y,a).$$

By (X.3.2), $|f_x(y)| \leq d(x,a)$, so that $f_x \in \mathbb{R}^X$ is bounded for each $x \in X$. On the other hand,

$$|f_x(y_0) - f_x(y_1)| = |d(x,y_0) - d(y_0,a) - d(x,y_1) + d(y_1,a)| \leq 2d(y_0,y_1),$$

that is, $f_x \in \mathscr{C}$ for every $x \in X$. This way, $x \mapsto f_x$ maps X into the space $\widehat{X}$ of continuous, bounded functions on X. The function $D : \widehat{X} \times \widehat{X} \longrightarrow \mathbb{R}_+$ defined by

$$D(f,g) := \|f - g\| = \sup\{|f(x) - g(x)| : x \in X\} \tag{X.5.1}$$

is easily seen to be a metric. We observe now that

$$\begin{aligned} f_x(y) - f_w(y) &= d(x,y) - d(y,a) - d(w,y) + d(y,a) \\ &= d(x,y) - d(w,y) \leq d(x,w). \end{aligned}$$

If $y = w$, then $f_x(w) - f_w(w) = d(x,w)$, hence

$$d(x,w) \leq D(f_x, f_w) = \sup\{|f_x(y) - f_w(y)| : y \in X\} \leq d(x,w),$$

which proves that $x \mapsto f_x$ is an isometry. Therefore, (X,d) is isometrically embedded in the space of continuous bounded functions endowed with a complete metric compatible with the uniform convergence. Accordingly, the closure of the image of this embedding is a completion of (X, d). □

Subsequently, we shall construct a completion using a method of F. HAUSDORFF ([5]).

PROOF. Let $\mathcal{S}(X, d)$ be the set of d-Cauchy sequences on X. Two sequences $\varphi, \psi \in \mathcal{S}(X, d)$ are equivalent, in symbols $\varphi \approx \psi$, whenever

$$\lim_{n\to\infty} d(\varphi(n), \psi(n)) = 0.$$

This limit exists. Indeed, by Lemma X.3.4,

$$|d(\varphi(n), \psi(n)) - d(\varphi(m), \psi(m))| \leq d(\varphi(n), \varphi(m)) + d(\psi(n), \psi(m))$$

fro each $n, m < \omega$, and thus the real-valued sequence $\{d(\varphi(n), \psi(n)) : n < \omega\}$ is Cauchy. As the metric $|\cdot|$ on $\mathbb{R}$ is complete, the sequence converges.

The function

$$\widehat{d}(\varphi, \psi) := \lim_{n\to\infty} d(\varphi(n), \psi(n))$$

is symmetric, fulfills the triangle inequality, and $\widehat{d}(\varphi, \psi) = 0$ if and only if $\varphi \approx \psi$. Hence $\widehat{d}$ a metric on the set $\widehat{X}$ of $\approx$-equivalence classes of $\mathcal{S}(X, d)$ ([6]).

We shall see that the metric space $(\widehat{X}, \widehat{d})$ is complete. To this end, we will show that each $\widehat{d}$-Cauchy sequence on $\widehat{X}$ is $\widehat{d}$-convergent. Let us take a $\widehat{d}$-Cauchy sequence $(\widehat{\varphi}_k)_k$ of elements of $\widehat{X}$, where $\widehat{\varphi}_k$ is the equivalence class of $\varphi_k \in \mathcal{S}(X, d)$ for each $k < \omega$.

Because $(\widehat{\varphi}_k)_k$ is $\widehat{d}$-Cauchy, there is a strictly increasing sequence $(k(m))_m$ such that $\widehat{d}(\varphi_k, \varphi_l) = \lim_{n\to\infty} d(\varphi_k(n), \varphi_l(n)) < 2^{-m}$ for all $k, l \geq k(m)$. Therefore, for all $m < \omega$ and $k, l \geq k(m)$, there exists $n_m(k, l) \geq k(m)$ such that for each $n \geq n_m(k, l)$,

$$d(\varphi_k(n), \varphi_l(n)) < 2^{-m}. \tag{X.5.2}$$

We shall construct $\zeta \in \mathcal{S}(X, d)$ that is a limit of the sequence $(\varphi_n)_n$ ([7]). We start by picking a strictly increasing sequence $(a_m)_{m<\omega}$ of natural numbers.

[5]The completion theorem was proved by F. HAUSDORFF [127], but the proof below is due to K. KURATOWSKI [167], and is reported in R. ENGELKING [97, p. 276].

[6]Formally, $\widehat{D}(\widehat{\varphi}, \widehat{\psi}) := \widehat{d}(\varphi, \psi)$, where $\widehat{\varphi}$ stands for the class of $\approx$-equivalence containing φ.

[7]Of course, there are multiple representations of each $\approx$-equivalence class, so there are many equivalent limits.

Let $0 < a_0 < \ldots a_{m-1} < a_m$ and

$$a_m \geq \max\{n_m(k,l) : k(m) \leq k, l \leq k(m+1)\}.$$

This way, (X.5.2) holds for $k(m) \leq k, l \leq k(m+1)$ and $n > a_m$.

We define ζ for n fulfilling $a_m \leq n < a_{m+1}$ for each $m < \omega$, as follows: let $\zeta(n) := \varphi_0(n)$ for each $n < a_0$, and $\zeta(n) := \varphi_{k(m)}(n)$ for $a_m \leq n < a_{m+1}$.

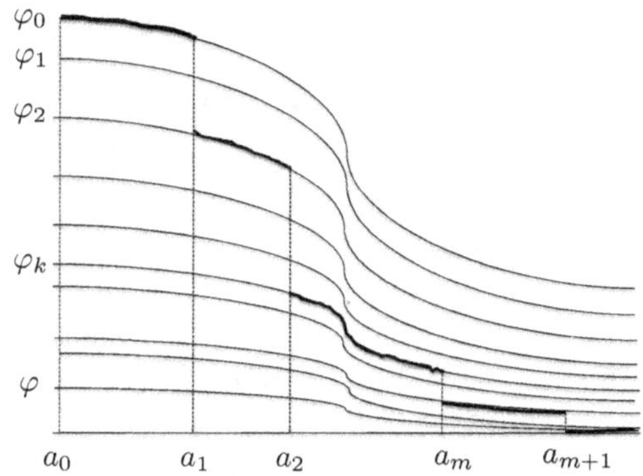

FIGURE X.5.1. Construction of a limit sequence ζ for the sequence of sequences $(\varphi_n)_n$. The graph of ζ is glued from the bold pieces.

To see that ζ is Cauchy, we notice that if $a_m \leq n < a_{m+1}$, then $d(\zeta(n), \zeta(a_{m+1})) = d(\varphi_{k(m)}(n), \varphi_{k(m+1)}(a_{m+1})) < 2^{-m}$, hence for any $j > n$,

$$d(\zeta(n), \zeta(j)) \leq \sum\nolimits_{p=0}^{\infty} 2^{-m-p} \leq 2^{-m+1},$$

that is, $\zeta \in \mathcal{S}(X, d)$. We shall prove that $\lim_{k\to\infty} \widehat{d}(\varphi_k, \zeta) = 0$. If $a_m \leq n < a_{m+1}$, and $k \geq k(m)$, then $d(\varphi_k(n), \zeta(n)) < 2^{-m}$, and thus for $k \geq k(m)$,

$$\widehat{d}(\varphi_k, \zeta) = \lim\nolimits_{n\to\infty} d(\varphi_k(n), \zeta(n)) < 2^{-m}.$$

Now, for each $x \in X$, we denote by $j(x)$ the constant sequence, that is, $j(x)(n) := x$ for each $n < \omega$, which is of course convergent, hence Cauchy. The corresponding equivalence class is that of almost constant sequences ([8]). Sure enough, $\widehat{d}(j(x_0), j(x_1)) = d(x_0, x_1)$, and thus $j : (X, d) \longrightarrow (\widehat{X}, \widehat{d})$ is an isometric embedding. Moreover, $j(X)$

[8]A sequence $\varphi \in X^N$ is called almost constant if $x \in X$ such that $\{n \in N : \varphi(n) \neq x\}$ is finite

is $\widehat{d}$-dense, because for every $\varphi \in \mathcal{S}(X,d)$ and each $\varepsilon > 0$, there is n_ε such that $d(\varphi(n), \varphi(n_\varepsilon)) < \varepsilon$, and so that $\widehat{d}(\varphi, j(\varphi(n_\varepsilon))) \leq \varepsilon$. $\square$

Let d be a metric on Y compatible with the topology τ, and let $f \in Y^V$, where V is a ξ-dense subset of X. The *oscillation* $\omega_V(f,x)$ of f at x is defined by

$$\omega_V(f,x) := \inf\{\operatorname{diam} f(O) : O \in \mathcal{N}_\xi(x)\}.$$

Mind that, by our convention, $f(O) = \{f(w) : w \in V \cap O\}$. Then the set

$$\Omega_V(f) := \{x \in X : \omega_V(f,x) = 0\} = \{\omega_V(f,\cdot) = 0\}$$

is a $G_{\aleph_0}$-subset of X, because $\Omega_V(f) = \bigcap_{n<\omega}\{\omega_V(f,\cdot) < 2^{-n}\}$.

PROPOSITION X.5.3. *If (Y,d) is complete, V is a ξ-dense subset of X, and $f \in \mathscr{C}(\xi|_V, \mathrm{t}d)$, then there exists $F \in \mathscr{C}(\xi|_{\Omega_V(f)}, \mathrm{t}d)$ such that $F|_V = f$.*

PROOF. If $x \in \Omega_V(f)$, then $f[\mathcal{N}_\xi(x)]$ is a Cauchy filter on (Y,d), hence there exists a unique $F(x) \in Y$ such that for each $\varepsilon > 0$, there exists $O \in \mathcal{O}_\xi(x)$ such that $\mathrm{cl}_{\mathrm{t}d} f(O) \subset B(F(x),\varepsilon)$. If $x \in V$, then $F(x) = f(x)$, so that F is an extension of f. If $w \in O \cap \Omega_V(f)$, then $F(w) \in \mathrm{cl}_{\mathrm{t}d} f(O)$, and thus $d(w,x) < \varepsilon$, so that $F \in \mathscr{C}(\xi|_{\Omega_V(f)}, \mathrm{t}d)$. $\square$

DEFINITION X.5.4 (Uniform continuity). Let d_X and d_Y be respective metrics on X and Y. A map $f \in Y^X$ is called *uniformly continuous*, in symbols $f \in \mathscr{U}(d_X, d_Y)$, if for every $\varepsilon > 0$, there exists $\delta > 0$ such that, for each $x_0, x_1 \in X$,

$$d_X(x_0,x_1) < \delta \Longrightarrow d_Y(f(x_0), f(x_1)) < \delta.$$

It is straightforward that a uniformly continuous map is continuous, that is, $\mathscr{U}(d_X, d_Y) \subset \mathscr{C}(\mathrm{t}d_X, \mathrm{t}d_Y)$.

Of course, each isometry is uniformly continuous. More generally, each *Lipschitz* map f, that is, such that there is a constant $l > 0$ so that, for each $x_0, x_1 \in X$,

$$d_Y(f(x_0), f(x_1)) \leq l \cdot d_X(x_0,x_1)$$

is uniformly continuous.

EXAMPLE X.5.5. The function $f_0(x) := x^2$ is a continuous function that is not uniformly continuous. The function $f_1(x) := \sqrt{|x|}$ is uniformly continuous, but not Lipschitz.

PROPOSITION X.5.6. *If V is a d_X-dense subset of X and $f \in \mathscr{U}(d_X|_V, d_Y)$, then $\Omega_V(f) = X$.*

PROOF. Let $x \in X$ and let $\varepsilon > 0$. Let $\delta > 0$ correspond to ε in the definition of uniform continuity of f. By the density of V, there is $O \in \mathcal{O}_X(x)$ such that $d_X(x_0, x_1) < \delta$ for each $x_0, x_1 \in O \cap V$. Hence, $d_Y(f(x_0), f(x_1)) < \varepsilon$ for each $x_0, x_1 \in O \cap V$, and thus $\omega_V(f, x) < \varepsilon$. □

We can quantify uniform continuity. Let d_X and d_Y be respective metrics on X and Y, and $f : Y^X$. The *rate of uniform continuity* $m_f : \mathbb{R}_+ \longrightarrow \mathbb{R}_+$ of f is defined by

$$m_f(\varepsilon) := \sup\{d_X(x_0, x_1) : d_Y(f(x_0), f(x_1)) < \varepsilon, x_0, x_1 \in X\}.$$

Accordingly, a mapping f is uniformly continuous if and only if m_f is increasing and strictly positive. We say that a function F has *almost the same rate* as f if $d_X(w_0, w_1) < m_f(\varepsilon)$ implies $d_Y(F(w_0), F(w_1)) \leq \varepsilon$.

THEOREM X.5.7. *Each uniformly continuous map from a dense subset of a metric space to a complete metric space can be extended to a uniformly continuous map on the whole space at almost the same rate.*

PROOF. By Propositions X.5.3 and X.5.6, f can be extended to a continuous map F. Let $\varepsilon > 0$ and let $x_0, x_1 \in X$ such that $d_X(x_0, x_1) < m_f(\varepsilon)$. By the density of V, for each $\delta > 0$, there exist v_0 and v_1 on V converging to x_0 and x_1 respectively, such that $d_X(v_0, x_0) < \delta$ and $d_X(v_1, x_1) < \delta$. Accordingly,

$$d_Y(F(x_0), F(x_1)) \leq d_Y(F(x_0), f(v_0)) + d_Y(f(v_0), f(v_1)) + d_Y(f(v_0), F(x_1)).$$

When δ tends to 0, the inequality above yields $d_Y(F(x_0), F(x_1)) \leq \varepsilon$, because F is a continuous extension of f and d_Y is continuous. □

COROLLARY X.5.8. *Each isometry from a dense subset of a metric space to a complete metric space can be extended to a unique isometry on the whole space.*

PROOF. If f is an isometry on a dense subset V of X, then $V \times V$ is dense in $X \times X$. Hence a continuous function $d_Y(f(v_0), f(v_1)) = d_X(v_0, v_1)$ has a unique extension to $X \times X$ in virtue of Corollary IX.1.4. □

X.6. Functionally closed and open sets

It is instrumental to use here an alternative notation for the preimages of various intervals. For example, if $f \in \mathbb{R}^X$, $a, b, r \in \mathbb{R}$, then

$$\{a < f \leq b\} := \{x \in X : a < f(x) \leq b\} = f^-(]a, b]).$$

Other instances

$$\{f > r\}, \{f \leq r\}, \{f = r\},$$

designate respectively $f^-(]r, +\infty[), f^-(]-\infty, r]), f^-(r)$.

If $f, g \in \mathbb{R}^X$, then

$$\{f < g\} := \{x \in X : f(x) < g(x)\}.$$

Finally, if $r \in \mathbb{R}$ then $\widehat{r} \in \mathbb{R}^X$ is the constant function defined by $\widehat{r}(x) := r$ for each $x \in X$. Hence $\{f < \widehat{r}\} = \{f < r\}$.

If a convergence ξ on a set X is fixed, as will be the case in this section, then we often abridge

$$\mathscr{C} := \mathscr{C}(\xi, \nu) = \mathscr{C}(\mathrm{T}\xi, \nu).$$

Notice that if $f, f_0, f_1 \in \mathscr{C}$, then

$$|f|, f_0 + f_1, f_0 \cdot f_1, \max(f_0, f_1) = f_0 \vee f_1$$

also belong to $\mathscr{C}$. In particular, $-f, f + r, f_0 \wedge f_1 \in \mathscr{C}$.

DEFINITION X.6.1. Let ξ be a convergence on X. A subset $A \subset X$ is called *functionally open* if there exists $f \in \mathscr{C}(\xi, \nu)$ such that

$$A = \{f > 0\};$$

it is called *functionally closed* if there exists $f \in \mathscr{C}(\xi, \nu)$ such that

$$A = \{f = 0\}.$$

We shall use $\mathcal{C}^I_\xi$ to denote the family of functionally closed sets, and $\mathcal{O}^I_\xi$ to denote the family of functionally open sets (with respect to $\mathscr{C}(\xi, \nu)$). Of course, for each convergence ξ,

$$\mathcal{C}^I_\xi \subset \mathcal{C}_\xi, \qquad \mathcal{O}^I_\xi \subset \mathcal{O}_\xi.$$

The converse does not hold in general.

EXAMPLE X.6.2. Each continuous (real-valued) function on a Sierpiński topology is constant, which means that the only non-empty functionally closed set is the whole $\{0, 1\}$. In fact, if there existed a continuous non-constant function, then there would be $f \in \mathscr{C}(\$_0, \nu)$ such that $f(0) = 0$ and $f(1) = 1$. Hence $\{0\} = \{f = 0\} \in \mathcal{C}_{\$_0}$, which is a contradiction.

A set is functionally closed if and only if its complementary set is functionally open. Indeed, if a set is functionally closed, hence of the form $\{f = 0\}$, where $f \in \mathscr{C}$, then $X \setminus \{f = 0\} = \{|f| > 0\}$, and $|f| \in \mathscr{C}$. On the other hand, a functionally open set is of the form $\{f > 0\}$ with $f \in \mathscr{C}$, and $X \setminus \{f > 0\} = \{f \leq 0\} = \{f \vee \widehat{0} = 0\}$, and $f \vee 0 \in \mathscr{C}$.

Each functionally closed set is a countable intersection of functionally open sets. In fact, $\{f = 0\} = \bigcap_{n<\omega}\{|f| < 2^{-n}\}$. Accordingly, every functionally open set is a countable union of functionally closed sets.

Each finite union of functionally closed sets is functionally closed. Indeed, if $f_0, f_1 \in \mathscr{C}$ then $|f_0| \wedge |f_1| \in \mathscr{C}$, and

$$\{f_0 = 0\} \cup \{f_1 = 0\} = \{|f_0| \wedge |f_1| = 0\}.$$

Arbitrary intersections of functionally closed sets need not be functionally closed. For instance, in a prime cocountable topology on an uncountable set (see Example X.7.23), the distinguished singleton is an intersection of functionally closed sets, but is not functionally closed.

Countable intersections of functionally closed sets are functionally closed.

LEMMA X.6.3. *If* $\{f_n : n < \omega\} \subset \mathscr{C}$ *and* $\sum_{n<\omega} \sup |f_n| < \infty$, *then* $\sum_{n<\omega} f_n \in \mathscr{C}$.

PROOF. Let $F := \sum_{n<\omega} f_n$ and let

$$F_n := \sum\nolimits_{0 \le k \le n} f_k$$

be a partial sum of the first $n+1$ terms of the series. Of course, $F_n \in \mathscr{C}$ for each $n < \omega$. To prove the continuity of F at x, take $\varepsilon > 0$. Then, by (X.1.2), there is n_ε such that $\sum_{n>n_\varepsilon} \sup |f_n| < \frac{\varepsilon}{3}$. On the other hand, there is a neighborhood W of x such that $|F_{n_\varepsilon}(w) - F_{n_\varepsilon}(x)| < \frac{\varepsilon}{3}$ for each $w \in W$. Consequently, for each $w \in W$,

$$|F(w) - F(x)| \\ \le |F(w) - F_{n_\varepsilon}(w)| + |F_{n_\varepsilon}(w) - F_{n_\varepsilon}(x)| + |F_{n_\varepsilon}(x) - F(x)| < \varepsilon.$$

□

PROPOSITION X.6.4. *Each countable intersection of functionally closed sets is functionally closed.*

PROOF. Let $\{f_n : n < \omega\} \subset \mathscr{C}$. We need to show that there exists $f \in \mathscr{C}$ such that

$$\{f = 0\} = \bigcap\nolimits_{n<\omega} \{f_n = 0\}. \tag{X.6.1}$$

Let $f := \sum_{n<\omega}(|f_n| \wedge \frac{1}{2^n})$, and observe that $f \in \mathscr{C}$ in virtue of Lemma X.6.3. As each term of the series is non-negative, $f(x) = 0$ if and only if $|f_n(x)| \wedge \frac{1}{2^n} = 0$ for each $n < \omega$, hence (X.6.1) holds. □

All the facts established above have their counterparts for functionally open sets, in particular, the family of functionally open sets is stable under finite intersections, and countable unions.

X.7. Functional separation

In Section IX.1 we studied various types of separation properties for topologies. Here we will consider diverse functional separation properties, comparing them with those already discussed. We remind that if ξ is a fixed convergence, we abridge $\mathscr{C}(\xi,\nu)$ to $\mathscr{C}$.

DEFINITION X.7.1. Two sets A_0 and A_1 are called *functionally separated* if there exists $f \in \mathscr{C}$ such that $A_0 \subset \{f = 0\}$ and $A_1 \subset \{f = 1\}$ ([9]); *functionally exactly separated* if there exists $f \in \mathscr{C}$ such that $A_0 = \{f = 0\}$ and $A_1 = \{f = 1\}$.

Notice that A_0 and A_1 are functionally separated if and only if there exists $f \in \mathscr{C}$ such that $\sup_{A_0} f < \inf_{A_1} f$ (Exercise X.10.2). On the other hand,

LEMMA X.7.2. *Each pair of disjoint functionally closed sets is functionally exactly separated.*

PROOF. If $A_0 = \{f_0 = 0\}$ and $A_1 = \{f_1 = 0\}$ are disjoint, then the function

$$f(x) := \frac{|f_0(x)|}{|f_0(x)| + |f_1(x)|}$$

is well-defined for each x and belongs to $\mathscr{C}$. Indeed, if $x \in A_0$ then $f_0(x) = 0$ and $f_1(x) \neq 0$, and vice versa if $x \in A_1$. Otherwise, $|f_0(x)| + |f_1(x)| > 0$ for each x. □

A convergence ξ is *functionally Hausdorff* if for any pair of distinct elements x_0, x_1, there exists $f \in \mathscr{C}$ such that $f(x_0) \neq f(x_1)$. Sure enough, this is equivalent to the existence of $h \in \mathscr{C}$ such that $h(x_0) = 0$ and $h(x_1) = 1$.

Each functionally Hausdorff convergence is topologically Hausdorff, because if h is as above, then

$$\{h < \tfrac{1}{2}\} \in \mathcal{O}_\xi(x_0), \{h > \tfrac{1}{2}\} \in \mathcal{O}_\xi(x_1), \text{ and } \{h < \tfrac{1}{2}\} \cap \{h > \tfrac{1}{2}\} = \varnothing.$$

The converse does not hold, because there exist Hausdorff topologies, in which each continuous function is constant, as we shall see later.

Of course, functionally exactly separated sets are functionally separated and closed, while functionally separated sets are disjoint.

DEFINITION X.7.3 (Functionally regular). A topology is called *functionally regular* if for every closed set F and each $x \notin F$, there exist $f \in \mathscr{C}$ such that $\sup_F f < f(x)$.

[9] GILLMAN and JERISON [113] and R. ENGELKING [97] use the term *completely separated.*

In other words, a topology ξ is functionally regular, if for each ξ-closed set, and $x \notin F$ there exists $f \in \mathscr{C}(\xi, \nu)$ such that $F \subset \{f = 0\}$ and $f(x) = 1$.

Functionally regular topologies are also called *completely regular;* Hausdorff functionally regular are called *Tikhonov* or *$T_{3\frac{1}{2}}$-topologies* ([10]).

REMARK X.7.4. This abundance of names dedicated to a property suggests that this property has been considered as important. This is in fact the case, because several consequential constructions, like the Čech-Stone compactification, are possible for the class of Hausdorff, functionally regular topologies.

We have adopted ([11]) the term *functionally regular topology* because of the conviction that it describes the property more suggestively than the other terms. Nevertheless, I shall be often recalling traditional terms in order to facilitate the orientation to those accustomed to the traditional terminology.

PROPOSITION X.7.5. *A topology is functionally regular (alias completely regular) if and only if it has a base for the open sets consisting of functionally open sets.*

PROOF. This is only a reformulation of the definition: if $O \in \mathcal{O}_\xi(x)$, then there is $f \in \mathscr{C}$ such that $x \in \{f > 0\} \subset O$. □

Zero-dimensional topologies are functionally regular, because the characteristic functions of clopen sets are continuous. It is clear that each functionally regular topology is regular. We will give an example of topology, for which the converse is not true.

It was observed in Example III.4.5 that each continuous function on a prime cocountable topology is constant on a neighborhood of the distinguished point.

LEMMA X.7.6. *If $\mathcal{V}_\xi(x)$ is a cofinite filter of an uncountable set V_0 centered at x and $f \in \mathscr{C}(\xi, \nu)$, then $\{f = f(x)\}$ is a cocountable subset of V_0.*

---

[10]The terminology slightly varies from a paper to another. For instance, R. Engelking in [97] calls a topology completely regular if it is functionally regular and Hausdorff.

[11]S. DOLECKI and F. MYNARD [83]. Moreover, some definitions comprise additionally the Hausdorff property. Incidentally, the property was introduced earlier by P. URYSOHN [252] and studied by A. TIKHONOV [251]. There exist several Latin transliterations of Тихонов from Russian: Tikhonov (most recent), Tychonoff, Tichonoff, Tichonov.

PROOF. By assumption, $f^-B(f(x), 2^{-n})$ is cofinite in V_0 for each $n < \omega$. Therefore, $\{f = f(x)\} = \bigcap_{n<\omega} f^-B(f(x), 2^{-n})$ is cocountable in V_0, as a countable intersection of cofinite sets. □

EXAMPLE X.7.7 (*A Hausdorff, regular, non-functionally regular topology*). In a preliminary step we define on $Z := \{(x, y) : 0 \le y \le 2\}$ a topology so that each (x, y) with $y > 0$ is isolated, and V is a neighborhood of $(x, 0)$ if $I_x \cup J_x \setminus V$ is finite, where

$$I_x := \{(x, y) \in Z : y \ge 0\}, \quad J_x := \{(x - y, y) \in Z : y \ge 0\}.$$

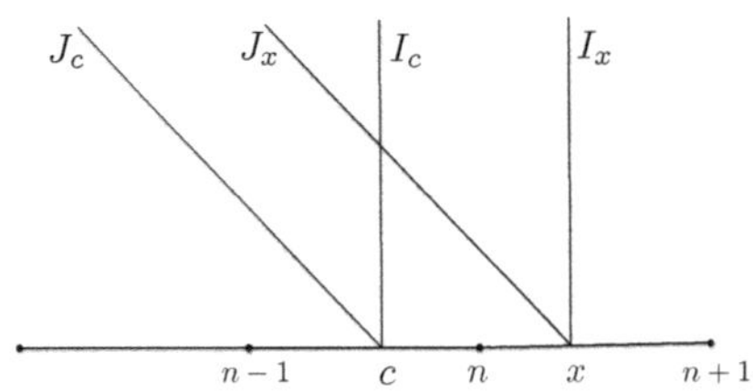

FIGURE X.7.1. The set of $x \in [n, n+1]$ such that J_x intersect of $I_c \cap f^-(0)$ is cocountable.

It follows that the topology defined is Hausdorff and zero-dimensional, hence functionally regular (alias completely regular). We will show that if $f \in \mathscr{C}$ and $f(x, 0) = 0$ for each $x \le 0$, then $([n-1, n] \times \{0\}) \cap f^-(0)$ is infinite for each $n < \omega$.

We proceed by induction on n. This is obvious for $n = 0$. Suppose that inductive hypothesis holds for $n \ge 0$, and consider $x \in [n, n+1]$. By assumption, there is an infinite subset C of $[n-1, n]$ such that $f(c, 0) = 0$ for each $c \in C$. Consequently, by Lemma X.7.6, $f^-(0)$ is cocountable in I_c. Mind that $I_c \cap J_x = \{(c, x-c)\}$ is at most a singleton. Consequently, the set of $x \in [n, n+1]$ such that $I_c \cap J_x \cap f^-(0) \ne \varnothing$ is cocountable, hence

$$\Lambda_C := \bigcap_{c \in C} \{x \in [n, n+1] : I_c \cap J_x \cap f^-(0) \ne \varnothing\}$$

is cocountable. Therefore, $V \cap f^-(0) \ne \varnothing$ for each neighborhood V of $(x, 0)$, because V is cofinite in J_x. Therefore, $f(x, 0) = 0$, and thus $f^-(0)$ is cocountable in $[n, n+1] \times \{0\}$.

Let now $Z_\infty := Z \cup \{\infty\}$ and

$$U_n := \{(x, y) \in Z : x \ge n\} \cup \{\infty\}$$

for $n \in \mathbb{Z}$, be a neighborhood base at ∞. Then $\mathrm{cl}_{Z_\infty} U_n \subset U_{n-2}$, and thus the topology of Z_∞ is Hausdorff and regular. On the other hand,

$F := \{(x,0) : x \leq 0\}$ is closed, $\infty \notin F$, but $f(\infty) = 0$ for each continuous map f such that $F \subset f^{-}(0)$. We conclude that the topology defined on Z_∞ is not functionally regular.

The situation is entirely different for normality:

THEOREM X.7.8 (Urysohn). *Each pair of disjoint closed sets in a normal topological space is functionally separated.*

PROOF. Let ξ be a normal topology on X, and let A_0 and A_1 be disjoint closed subsets of X. We will construct inductively a family of open sets $\{U_r : r \in [0,1] \cap \mathbb{Q}\}$ such that $A_0 \subset U_0, U_1 \subset X \setminus A_1$, and

$$r < t \Longrightarrow \mathrm{cl}_\xi U_r \subset U_t. \tag{X.7.1}$$

Let $[0,1] \cap \mathbb{Q} = \{r_n : n < \omega\}$ so that $r_0 = 0$ and $r_1 = 1$. Let $U_1 := X \setminus A_1$. By Exercise IX.5.9, there exists $U_0 \in \mathcal{O}(A_0)$ such that

$$\mathrm{cl}_\xi U_0 \subset U_1.$$

If $U_{r_0}, \ldots, U_{r_n}$ have been already defined, then there exist $l, m \leq n$ such that r_l is the greatest and r_m is the least so that $r_l < r_{n+1} < r_m$. It follows that $\mathrm{cl}_\xi U_{r_l} \subset U_{r_m}$. By inductive assumption, there exists an open set $U_{r_{n+1}}$ such that

$$\mathrm{cl}_\xi U_{r_l} \subset U_{r_{n+1}} \subset \mathrm{cl}_\xi U_{r_{n+1}} \subset U_{r_m}.$$

Let

$$f(x) := \inf\{r : x \in U_r\}$$

if $x \notin A_1$ and $f(x) := 1$ if $x \in A_1$. Of course, $f(x) = 0$ for each $x \in A_0$. We claim that $f \in \mathscr{C}(\xi, \nu)$. If $f(x) < s$, then there exists $r < s$ such that $x \in U_r \subset \{f < s\} = f^{-}([0, s[)$. If $s < f(x)$ then there is $t > s$ such that $x \notin U_t$, and by (X.7.1), there is $s < r < t$ such that $x \notin \mathrm{cl}_\xi U_r$. Therefore $x \in X \setminus \mathrm{cl}_\xi U_r \subset \{s < f\} = f^{-}(]s, 1])$. □

LEMMA X.7.9. *Let C be a closed subset for a normal topology on X and let f be a continuous function on C such that $\sup_C |f| \leq c$. Then there exists $h \in \mathscr{C}$ such that $\sup_X |h| \leq \frac{1}{3}c$ and $\sup_C |f - h| \leq \frac{2}{3}c$.*

PROOF. Indeed, the sets

$$C_0 := \{x \in C : f(x) \leq -\frac{1}{3}c\} \text{ and } C_1 := \{x \in C : \frac{1}{3}c \leq f(x)\}$$

are closed and disjoint, hence, by normality, there is $g \in \mathscr{C}$ such that $C_0 \subset \{g = 0\}$ and $C_1 \subset \{g = 1\}$. Let $h(x) := \frac{2}{3}c(g(x) - \frac{1}{2})$ for each $x \in X$. Then $|h(x)| \leq \frac{1}{3}c$ for each $x \in X$ and $|f(x) - h(x)| \leq \frac{2}{3}c$ for each $x \in C$. □

THEOREM X.7.10 (Tietze extension theorem). *Every continuous function on a closed subset of a normal topological space can be continuously extended to the whole space so that the bounds are preserved.*

PROOF. Let C be a closed subset of a normal topological space X and let f be a continuous function on C.

Assume first that $f(C) \subset [-1,1]$. Then, by Lemma X.7.9, there exists $h_0 \in \mathscr{C}$ such that $|h_0(x)| \leq \frac{1}{3}$ for each $x \in X$ and $|f(x) - h_0(x)| \leq \frac{2}{3}$ for each $x \in C$. We construct by induction a sequence $(h_n)_n \subset \mathscr{C}$ such that

$$\sup_X |h_n| \leq \tfrac{1}{3}(\tfrac{2}{3})^n, \text{ and } \sup_C \left| f - \sum_0^n h_k \right| \leq (\tfrac{2}{3})^{n+1}$$

for each $n < \omega$. This has been already done for $n = 0$. If we have constructed $h_0, \ldots, h_n$, then let $f_n(x) := f(x) - \sum_0^n h_k(x)$ for each $x \in C$. By Lemma X.7.9, there exists $h_{n+1} \in \mathscr{C}$ such that $\sup_X |h_{n+1}| \leq \frac{1}{3}(\frac{2}{3})^{n+1}$ and $\sup_C |f_n - h_{n+1}| \leq \frac{2}{3}(\frac{2}{3})^{n+1}$. By Proposition X.3.11, the function $h(x) := \sum_{n=0}^{\infty} h_n(x)$ for each $x \in X$, is continuous, $h(x) = f(x)$ for each $x \in C$, and $\sup_X |\sum_{n=0}^{\infty} h_n| \leq \frac{1}{3}\sum_{n=0}^{\infty}(\frac{2}{3})^n = 1$.

Moreover, if $f(C) \subset]-1,1[$, then there exists a continuous extension F such that also $F(X) \subset]-1,1[$. Indeed, we have proved the existence of an extension h of f that fulfills $h(X) \subset [-1,1]$. The set $\{h = 1\}$ is closed and disjoint from X, so that by Theorem X.7.8 of Urysohn, there exists a continuous map $g : X \longrightarrow [0,1]$ such that $C \subset \{g = 1\}$ and $\{h = 1\} \subset \{g = 0\}$. Therefore, $F := g \cdot h$ is a continuous extension of f such that $F(X) \subset]-1,1[$.

Let $f : C \longrightarrow \mathbb{R}$ be continuous. Take any homeomorphism $\varphi : \mathbb{R} \longrightarrow]-1,1[$, and extend $\varphi \circ f$ to a continuous function $F : X \longrightarrow]-1,1[$. Then $\varphi^{-1} \circ F$ is continuous on X, and if $x \in C$, then $(\varphi^{-1} \circ F)(x) = \varphi^{-1}(\varphi(f(x))) = f(x)$, so that $\varphi^{-1} \circ F$ is an extension of f. $\square$

PROPOSITION X.7.11. *Each closed $G_{\aleph_0}$-set of a normal topological space is functionally closed, and each open $F_{\aleph_0}$-set is functionally open.*

PROOF. If F is a closed $G_{\aleph_0}$-set in a normal topological space, then let $\{G_n : n < \omega\}$ be a sequence of open sets such that $F = \bigcap_{n<\omega} G_n$. Then there is a sequence $\{g_n \in [0, 2^{-n}] : n < \omega\} \subset \mathscr{C}$ such that $F \subset \{g_n = 0\}$ and $\{g_n = 2^{-n}\} \subset G_n$. Then $\gamma_n = \inf_{0 \leq k \leq n} g_k \in \mathscr{C}$ and $(\gamma_n)_n$ converges uniformly to a function γ such that $F = \bigcap_{n<\omega}\{\gamma_n = 0\} = \{\gamma = 0\}$, and, by Proposition X.3.11, $\gamma \in \mathscr{C}$. $\square$

Of course, each normal free (alias T_1) topology is functionally regular (alias completely regular). In the context of Hausdorff functionally regular topologies, it is possible to functionally separate two closed sets (like in Theorem X.7.8 of Urysohn), provided that one of them is compact.

PROPOSITION X.7.12. *Two disjoint closed sets in a Hausdorff functionally regular (alias completely regular) topological space are functionally separated, provided that one of them is compact.*

PROOF. Let ξ be a Hausdorff functionally regular topology on X, let K be a ξ-compact subset, and C be a ξ-closed subset of X such that $K \cap C = \varnothing$. As ξ is functionally regular, for each $x \in K$, there exists $f_x \in \mathscr{C}(\xi, \nu_{\mathbb{R}})$ such that $f_x(x) > 1$ and $C \subset \{f_x = 0\}$. Accordingly, $K \subset \bigcup_{x \in K}\{f_x > 1\}$. By compactness, there exists a finite subset F of K such that $K \subset \bigcup_{x\in F}\{f_x > 1\}$. Consequently, $f := \max_{x \in F} f_x \in \mathscr{C}(\xi, \nu_{\mathbb{R}})$ and $K \subset \{f > 1\}$. On the other hand, $C \subset \{f = 0\}$. □

Observe that Lemma X.7.9 also holds for a compact subset F of a Hausdorff functionally regular (alias completely regular) topological space. As each continuous function on a compact set is bounded, the proof of Theorem X.7.10 works if C is a compact subset for a Hausdorff functionally regular topology. Hence,

COROLLARY X.7.13. *Every continuous function on a compact subset of a Hausdorff functionally regular (alias completely regular) topological space can be continuously extended to the whole space so that the bounds are preserved.*

EXAMPLE X.7.14 (*A functionally regular, non-normal topology*). We have seen that the topology ζ of the Niemytzki plane from Example IX.1.19 is not normal. Here we show that it is functionally regular (alias completely regular), and we provide another proof of non-normality, based on the Baire Category Theorem X.4.15.

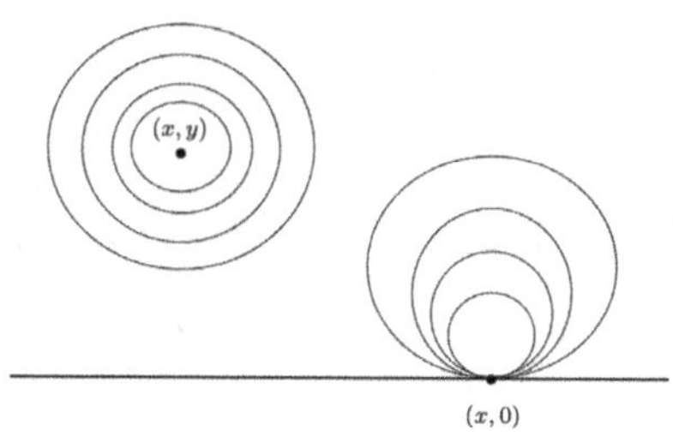

FIGURE X.7.2. Niemytzki plane: typical neighborhoods for $y = 0$ and $y > 0$.

Indeed, if F is a ζ-closed set, and $(x, y) \notin F$, then either $y > 0$ and there is $r > 0$ such that $B((x, y), r) \cap F = \varnothing$, or $y = 0$ and there is $r > 0$ such that $B((x, r), r) \cap F = \varnothing$. In the first case, let $f(v, w) :=$

$\frac{1}{r}(|v-x|^2+|w-y|^2)^{-2}\wedge 1$ be continuous, because ζ coincides with ν^2 within $B((x,y),r)$. Also $f(x,y)=0$, and $\inf_F f = 1$. In the latter case, let $f(v,w) := \frac{s}{r}$ if (v,w) lies on the circle of radius $0<s<r$ centered at (x,s) ([12]), $f(x,0):=0$ and $f(v,w):=1$ if $(v,w)\notin B((x,r),r)$.

Recall Example IX.1.19 that L is discrete in ζ. In particular, the set $F := \{(x,0) : x \notin \mathbb{Q}\}$ is ζ-closed. On the other hand, F is a subspace of the real line that has the Baire property. If ζ were normal, then there would be $U \in \mathcal{O}_\zeta(F)$ and $W \in \mathcal{O}_\zeta(L\setminus F)$ such that $U \cap W = \varnothing$. Hence, for each $x \notin \mathbb{Q}$, there is $r(x)>0$ such that $B((x,r(x)),r(x)) \subset U$. If

$$A_n := \{x \in \mathbb{R}\setminus\mathbb{Q} : r(x) > 2^{-n}\},$$

then $\mathbb{R}\setminus\mathbb{Q} \subset \bigcup_{n<\omega} \mathrm{cl}_{\mathbb{R}\setminus\mathbb{Q}} A_n$, hence, by the Baire Category Theorem X.4.15, there exist $n<\omega$ and $a<b$ such that $]a,b[$ is ν-dense in $\mathrm{cl}_{\mathbb{R}\setminus\mathbb{Q}} A_n$. If $x \in]a,b[\cap\mathbb{Q}$, then by assumption, there is $s>0$ such that $B((x,s),s) \subset W$, which implies that $W \cap U \neq \varnothing$ a contradiction.

DEFINITION X.7.15 (Perfectly normal). A topology is called *perfectly normal* if each pair of disjoint closed sets can be functionally exactly separated.

By Definition X.7.1,

PROPOSITION X.7.16. *A topology is perfectly normal if and only if for each pair F_0, F_1 of disjoint closed sets, there exists $f \in \mathscr{C}$ such that $F_0 = \{f=0\}$ and $F_1=\{f=1\}$.*

LEMMA X.7.17. *A topology is perfectly normal if and only if it is normal and each open set is $F_{\aleph_0}$ (equivalently, each closed set is $G_{\aleph_0}$).*

PROOF. The necessity is an immediate consequence of the definition. Conversely, if a topology is normal and the condition holds, then, by Proposition X.7.11, all the closed sets are functionally closed. By Lemma X.7.2 and Proposition X.7.16, the proof is complete. □

LEMMA X.7.18. *Each Hausdorff regular topology with a σ-locally finite base is perfectly normal.*

PROOF. Consider a Hausdorff regular topology on X. Let $(\mathcal{W}_n)_n$ be a sequence of locally finite families of open subsets of X such that $\bigcup_{n<\omega}\mathcal{W}_n$ is a base for the topology. Let O be an open set. Then, for each $x \in O$, there exist $n(x)<\omega$ and $W(x) \in \mathcal{W}_{n(x)}$ such that $x \in W(x) \subset \mathrm{cl}_X W(x) \subset O$, because X is regular. Let X_n consist of those

[12]That is, if $(|v-x|^2+|w-s|^2)^{-2}=s$.

$x \in O$, for which $n(x) = n$. Then $\{W(x) : x \in X_n\} \subset \mathcal{W}_n$ is locally finite, hence

$$O = \bigcup_{n<\omega} \bigcup_{x \in X_n} \mathrm{cl}_X W(x) = \bigcup_{n<\omega} \mathrm{cl}_X(\bigcup_{x \in X_n} W(x)),$$

that is, O is $F_{\aleph_0}$. By Lemma X.7.17, the topology is perfectly normal. $\square$

By Exercise XI.5.1, metrizable topologies are perfectly normal. On the other hand, each normal topology of countable weight is perfectly normal. In fact, if $\mathcal{B}$ is a countable base, then, for every open set O,

$$O = \bigcup\{\mathrm{cl}_X B : \mathcal{B} \ni B \subset O\},$$

so that O is $F_{\aleph_0}$.

THEOREM X.7.19 (Vedenisov). *A free (alias T_1) topology is perfectly normal if and only if each closed set is functionally closed.*

PROOF. If two disjoint closed sets are functionally closed, then they are functionally exactly separated by virtue of Lemma X.7.2, that is, the topology is perfectly normal. If a perfectly normal topology is free, then for each proper closed subset F there exists $x \notin F$. As x is closed, there exists $f \in \mathscr{C}$ such that $F = \{f = 0\}$ and $f(x) = 1$, hence F is functionally closed. $\square$

EXAMPLE X.7.20 (*Sorgenfrey line*). The *Sorgenfrey line topology* ς is perfectly normal. Let us start by making a more general observation: if O is a ς-open set, then $O \setminus \mathrm{int}_\nu O$ is countable ([13]).

Now, if $a \in O \setminus \mathrm{int}_\nu O$, then there exists $x(a) > a$ such that $[x, a(x)[\subset O$, hence $]a, x(a)[\subset O$, that is, a is a ν-boundary point of O. If now $a, b \in O \setminus \mathrm{int}_\nu O$, and $x(a)$ and $x(b)$ are such that $[a, x(a)[\subset O$ and $[b, x(b)[\subset O$, then $[a, x(a)[\cap [b, x(b)[= \varnothing$. Since $[a, x(a)[\cap \mathbb{Q} \neq \varnothing$, the difference $O \setminus \mathrm{int}_\nu O$ is countable.

Indeed, if $O \in \mathcal{O}_\varsigma$ then $\mathrm{int}_\nu O$ is a countable union of ν-open intervals of type $]a, b[$, each of which is a $F_{\aleph_0}(\varsigma)$-set, because $]a, b[= \bigcup_{n<\omega} [a_n, b[$. Therefore, $\mathrm{int}_\nu O$ is a $F_{\aleph_0}(\varsigma)$-set, too. It follows that O is a $F_{\aleph_0}(\varsigma)$-set, hence by Corollary X.7.11, O is functionally open, thus by Theorem X.7.19 of Vedenisov, ς is perfectly normal.

We shall show that the square of the Sorgenfrey topology is not normal. This implies that

COROLLARY X.7.21. *Normality is not preserved under finite products.*

[13]Where ν stands for the standard topology of real line

EXAMPLE X.7.22 (*Sorgenfrey plane*). The Sorgenfrey line ζ is normal. We will see now that $\zeta \times \zeta$ is not normal. Consider the variant ζ_+, in which $[x, x+\varepsilon[$ are basic open sets. Then $\mathcal{V}_{\zeta^2}(x,y) \approx \{S_\varepsilon(x,y) : \varepsilon > 0\}$, where $S_\varepsilon(x,y) := [x, x+\varepsilon[\times [y, y+\varepsilon[$.

To show that ζ^2 is not normal, consider a subset $L := \{(x,y) \in \mathbb{R}^2 : y = -x\}$, which is a closed discrete subspace in the Sorgenfrey plane. Indeed, L is ν^2-closed, hence *a fortiori* ζ^2-closed. On the other hand, if $l \in L$, then there is x such that $l = (x, x)$, and thus $S_\varepsilon(x,x) \cap L = \{(x,x)\}$ for each $\varepsilon > 0$. Therefore, $L_{\mathbb{Q}} := \{(q,q) \in L : q \in \mathbb{Q}\}$ is closed, and $L \setminus L_{\mathbb{Q}}$ is closed. Let O_0, O_1 be ζ^2-open sets including respectively $L_{\mathbb{Q}}$ and $L \setminus L_{\mathbb{Q}}$. For each $(x,x) \in L \setminus L_{\mathbb{Q}}$ there is $r_x > 0$ such that $S_{r_x}(x,x) \subset O_1$.

Let $Q_n := \{x \in \mathbb{R} \setminus \mathbb{Q} : r_x \geq 2^{-n}\}$ for each $n < \omega$. Then $\mathbb{R} = \bigcup_{n<\omega} \operatorname{cl}_\nu Q_n$, hence by Theorem X.4.15 of Baire, there exists $n < \omega$ such that $\operatorname{int}_\nu \operatorname{cl}_\nu Q_n \neq \varnothing$. If $q \in \operatorname{int}_\nu \operatorname{cl}_\nu Q_n$, then there exists $t > 0$ so that $S_t(q,q) \subset O_0$. On the other hand, there is $x \in \operatorname{int}_\nu \operatorname{cl}_\nu Q_n \setminus \mathbb{Q}$ such that $|x - q| < \min(t, 2^{-n})$, hence $\varnothing \neq S_t(x,x) \cap S_{2^{-n}}(q,q) \subset O_1 \cap O_0$. A contradiction.

EXAMPLE X.7.23 (*A Hausdorff, compact, not perfectly normal topology*). A prime cofinite topology $\pi[x_\infty, (X)_0]$ on an uncountable set is normal by Example IX.1.18. If $f \in \mathscr{C}(\pi, \nu)$ and $f(x_\infty) = 0$, then for each $n < \omega$, the set $\{|f| < 2^{-n}\}$ is cofinite, hence $\{f = 0\} = \bigcap_{n<\omega}\{|f| < 2^{-n}\}$ is cocountable. Consequently, $x_\infty \in \{f = 0\}$, but $\{x_\infty\} \neq \{f = 0\}$, because the underlying set is uncountable. Therefore, the closed set $\{x_\infty\}$ is not functionally closed. If $\mathcal{P}$ is an open cover, and $P_\infty \in \mathcal{P}$ is such that $x_\infty \in P_\infty$, then $X \setminus P_\infty$ is finite. Hence there is a finite subfamily $\mathcal{P}_0$ of $\mathcal{P}$ such that $X \setminus P_\infty \subset \bigcup \mathcal{P}_0$, and thus $\mathcal{P}_0 \cup \{P_\infty\}$ is a finite cover of X.

An akin example with an interesting additional property, is provided by a cocountable topology.

EXAMPLE X.7.24 (*A normal, not perfectly normal topology*). A prime cocountable topology $\pi[x_\infty, (X)_1]$ on an uncountable set X was considered in Exercise IV.10.11. Let us recall that F is closed if either $x_\infty \in F$ or F is countable. Therefore, if F_0, F_1 are disjoint closed sets, then either they are both countable, and then $F_0 \in \mathcal{O}(F_0)$ and $F_1 \in \mathcal{O}(F_1)$, or F_0 is countable and $x_\infty \in F_1$. In this case, $F_0 \in \mathcal{O}(F_0)$ and $X \setminus F_0 \in \mathcal{O}(F_1)$. Consequently, this topology is normal.

As we have concluded in Exercise IV.10.11, for each continuous function f there exists a neighborhood W of x_∞ such that $f(W) = \{f(x_\infty)\}$, that is, f is constant on W. Accordingly, this topology is not perfectly normal.

We have encountered several separation axioms. I am gathering them here for the reader's convenience. As their meaning vary from one author to another, one should be alert. I have been employing a personalized system, because of the conviction that a properly chosen term can be helpful in memorizing and comprehension (if the name is evocative). On the other hand, when using this personalized terminology, additionally I am recalling most usual terms to help keeping track.

As some definitions of regularities and normalities do not require the Hausdorff property, certain authors add Hausdorffness in order to keep the sequence of T_n ascending. Let us note, however, that most often the addition of the T_1 axiom has the same effect. Indeed, a regular T_1 topology is T_2.

	This book terminology	Traditional
T_0	T_0	T_0
T_1	free	T_1
T_2	Hausdorff	Hausdorff
T_3	regular + free	regular + T_1
$T_{3\frac{1}{2}}$ or Tikhonov	functionally regular + free	completely regular + T_1
T_4	normal + free	normal + T_1
T_5	completely normal + free	completely normal + T_1

TABLE X.7.1. Separation axioms.

X.8. *A non-trivial regular topology with each continuous function constant

Not only there exist Hausdorff, regular, non-functionally regular topologies, but even non-trivial Hausdorff, regular topologies on which each continuous function is constant [**130**].

LEMMA X.8.1. *Let A be an uncountable set, and B be a countable set. Let*

$$H := (A \times B) \cup A \cup B$$

be endowed with the topology, for which the points in $A \times B$ are isolated, while $V \in \mathcal{N}(a)$ whenever $a \in V$ and $(\{a\} \times B) \setminus V$ is finite, and $W \in \mathcal{N}(b)$ whenever $b \in W$ and $(A \times \{b\}) \setminus W$ is finite. Then, for each continuous function f on H, there is a cocountable subset C of A such that f is constant on $(C \times B) \cup C \cup B$.

PROOF. If $f \in \mathscr{C}$, then for each $b \in B$, the set

$$\Lambda_b := \{x \in A : f(x,b) = f(b)\}$$
$$= \bigcap_{n<\omega} \{x \in A : |f(x,b) - f(b)| < 2^{-n}\}$$

is cocountable. As B is countable, $\Lambda := \bigcap_{b \in B} \Lambda_b$ is also cocountable. I claim that f is constant on Λ. Otherwise, there are $a_0, a_1 \in \Lambda$ such that $f(a_0) < t < f(a_1)$.

Therefore, there exist $V_0 \in \mathcal{N}(a_0)$ and $V_1 \in \mathcal{N}(a_1)$ such that $V_0 \subset \{f < t\}$ and $V_1 \subset \{f > t\}$. Thus $(\{a_0\} \times B) \setminus \{f < t\}$ is finite, and $(\{a_1\} \times B) \setminus \{f > t\}$ is finite, hence there is $b \in B$ such that $(a_0, b) \in \{f < t\}$ and $(a_1, b) \in \{f > t\}$, a contradiction.

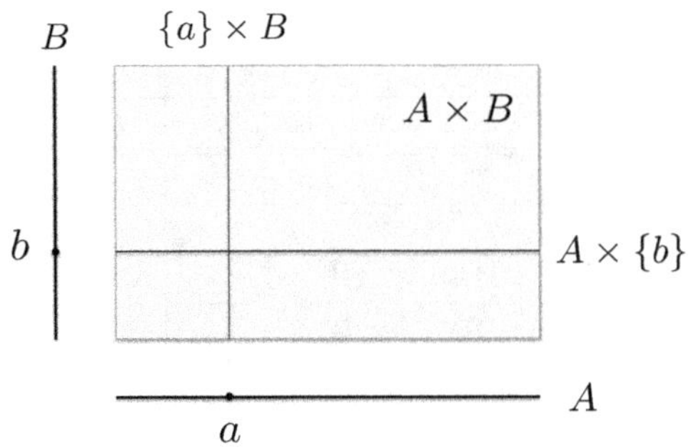

FIGURE X.8.1. As Λ_b are cocountable in A, and B is countable, $\Lambda = \bigcap_{b \in B} \Lambda_b$ is cocountable in A. The trace of a function $\{f = t\}$ illustrates the argument by contradiction to the fact that $f(a_0) = f(a_1)$.

□

The following Example X.8.2 adopts and refines the idea of Example X.7.6.

EXAMPLE X.8.2 (A Hausdorff, regular, non-functionally regular topology). ([14]) Let $X = \mathbb{R}^2 \cup \{-\infty, +\infty\}$, let $\triangle := \{(r,r) : r \in \mathbb{R}\}$ be the diagonal set, and, for each $r \in \mathbb{R}$, let

$$\varGamma_r := \{(r,t) : r - 3 < t \leq r\} \cup \{(s,r) : r \leq s < r + 3\}.$$

We define a topology θ on X to the effect that all the elements of $X \setminus \triangle$ are isolated, V is a neighborhood of (r,r) whenever $\varGamma_r \setminus V$ is finite. Accordingly, the topology is Hausdorff and zero-dimensional, hence functionally regular (alias completely regular).

[14] Appeared in [45] by K. CIESIELSKI and J. WOJCIECHOWSKI, who refined a method of A. MYSIOR [202].

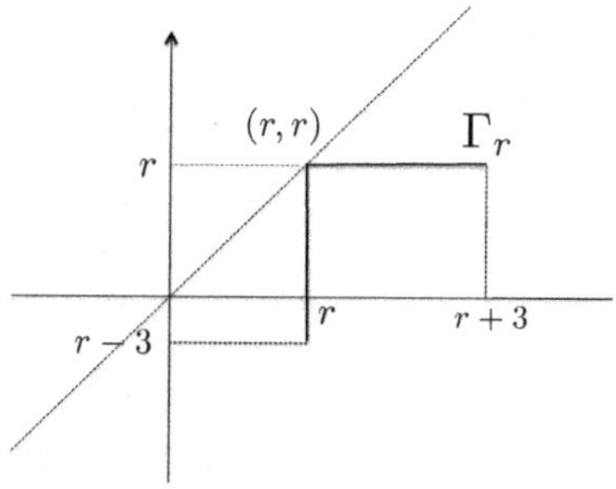

FIGURE X.8.2. A neighborhood filter of (r, r) is the cofinite filter of Γ_r centered at (r, r).

Given $r \in \mathbb{R}$, consider a rectangle $A \times B$, where $A = [r, r+2]$ and B is an infinite countable subset of $[r-1, r]$. Then the restriction of θ on $(A \times B) \cup A \cup B$ is homeomorphic to the topology of Lemma X.8.1, and thus, on identifying A with $A' = \Delta A$, and B with $B' = \Delta^- B$, we infer that each continuous function is constant on $B' \cup C'$, where C' is a cocountable subset of A'.

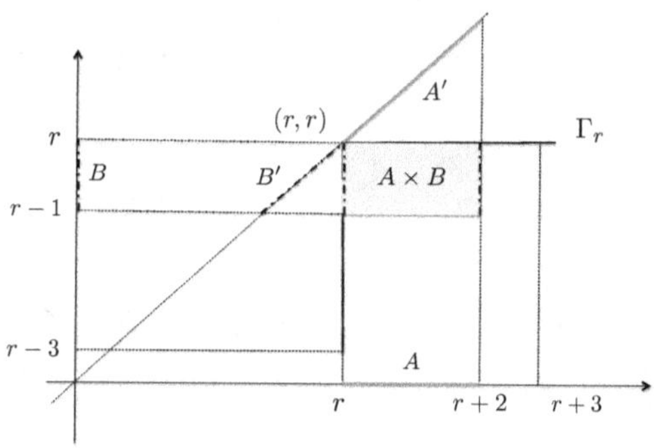

FIGURE X.8.3. Each continuous function is constant on a cocountable subset of $\Delta[r, +\infty[$.

Consequently, if f is a continuous function on X, then f is constant on a cocountable subset of $\Delta[r, +\infty[$ for every $r \in \mathbb{R}$. This follows by induction on $k < \omega$ applied to $\Delta[r-1+k, r+k]$.

Let Q be an extension of X, in which a filter-bases of $\mathcal{N}_Q(+\infty)$ is composed of $V_n^{\uparrow} := \{(x, y) : x \geq n\}$ and of $\mathcal{N}_Q(-\infty)$ is composed of $V_n^{\downarrow} := \{(x, y) : x \leq -n\}$, respectively, for each $n < \omega$. As $\mathrm{cl}_Q V_n^{\uparrow} \subset V_{n-2}^{\uparrow}$ and $\mathrm{cl}_Q V_n^{\downarrow} \subset V_{n+2}^{\downarrow}$, the topology of Q is Hausdorff and regular. If $f \in \mathscr{C}$ then for each $\varepsilon > 0$, there exists $n < \omega$ such that $|f(-\infty) - f(x, y)| < \varepsilon$

for each $(x, y) \in V_n^{\downarrow}$, and $|f(+\infty) - f(x, y)| < \varepsilon$ for each $(x, y) \in V_n^{\uparrow}$. By taking $r < n - 1$, we infer that $|f(+\infty) - f(-\infty)| < 2\varepsilon$.

PROPOSITION X.8.3. *For every Hausdorff, regular topological space Z, there exists a Hausdorff, regular topological space $Q(Z)$ such that Z is closed in $Q(Z)$, and each continuous function on $Q(Z)$ is constant on Z.*

PROOF. Let Q be a Hausdorff regular topological space and two points a and b such that

$$f \in \mathscr{C}(Q, \nu) \Longrightarrow f(a) = f(b). \tag{X.8.1}$$

Let Z be another Hausdorff regular topological space. We topologize the product $Z \times Q$ as follows: $O \in \mathcal{O}_{Z\times Q}(z, q)$, then there is $U \in \mathcal{O}_Q(q)$ so that $\{z\} \times U \subset O$, and if $O \in \mathcal{O}_{Z\times Q}(z, a)$, then there is $W \in \mathcal{O}_Z(z)$ that $W \times \{a\} \subset O$.

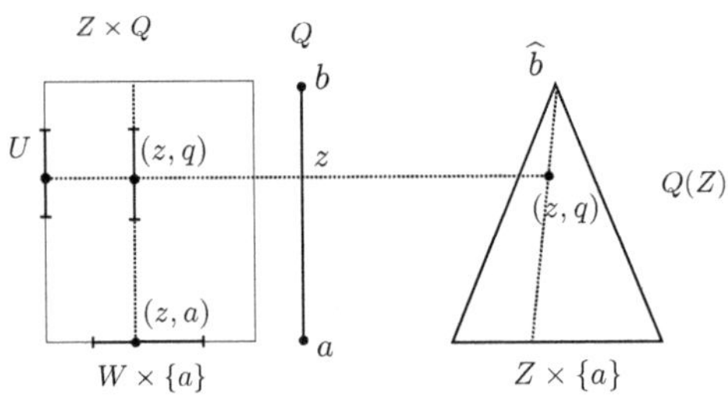

FIGURE X.8.4. The quotient $Q(Z)$ of $Z \times Q$ by identifying $(z, b) \approx \widehat{b}$.

Now, the quotient topology $Q(Z)$ is obtained by identifying $z \times \{b\} \approx \widehat{b}$: let $h : Z \times Q \longrightarrow Q(Z)$ be defined by

$$h(z, q) := \begin{cases} (z, q) & z \neq b, \\ \widehat{b} & z = b. \end{cases}$$

Observe that if $F \in \mathscr{C}$ then $\{F(\widehat{b})\} = F(Z \times \{a\})$. Indeed, for $z \in Z$ the function F_z restricted to $\{z\} \times Q$ is continuous, hence by (X.8.1), $f(z, a) = f_z(a) = f(\widehat{b})$ for each $z \in Z$.

Hence, f is constant on $Z \times \{a\}$. Moreover, $Z \times \{a\}$ is closed in $Q(Z)$. To see that the quotient topology is regular, we only need to check neighborhoods of $\widehat{b}$. If $B \in \mathcal{N}(\widehat{b})$, then for each $z \in Z$, there is

$B_z \in \mathcal{N}_Q(b)$ such that $h(\{z\} \times B_z) \subset B$, hence by the regularity of Q, there exists $V_z \in \mathcal{N}_Q(b)$ such that $\mathrm{cl}_Q V_z \subset B_z$, and thus

$$\mathrm{cl}_{Q(Z)} \bigcup_{z \in Z} h(\{z\} \times V_z) \subset \bigcup_{z \in Z} h(\{z\} \times \mathrm{cl}_Q V_z) \subset \bigcup_{z \in Z} h(\{z\} \times B_z) \subset B.$$

□

LEMMA X.8.4 (Herrlich). *If $\{X_n : n < \omega\}$ is a sequence of Hausdorff, regular topological spaces such that X_n is a closed subspace of X_{n+1} so that each continuous function on X_{n+1} is constant on X_n, then $X_\infty := \bigcup_{n<\omega} X_n$ with the final topology is Hausdorff regular and such that every continuous function on X_∞ is constant*

PROOF. Let X_0 be a singleton with the discrete topology. Then $X_n \subset X_{n+1}$ is a sequence of Hausdorff regular topologies such that X_n is closed in X_{n+1}, and each $f \in \mathscr{C}(X_{n+1}, \nu)$ is constant on X_n for each $n < \omega$. Then $X_\infty := \bigcup_{n<\omega} X_n$ with the final topology $\tau := \bigwedge j_n \tau_n$, where $j_n : X_n \longrightarrow X$ is the canonical injection, τ_n is the topology of X_n.

Observe that τ is Hausdorff and regular. If a and b are distinct elements of X_∞, there exists $n < \omega$ such that $a, b \in X_n$. Then for each $k \geq n$ there exist disjoint sets $A_k \in \mathcal{O}_{\tau_k}(a)$ and $B_k \in \mathcal{O}_{\tau_k}(b)$. On the other hand, by Proposition VII.3.12, for each $k < \omega$, there exist disjoint $D_{k+1}, G_{k+1} \in \mathcal{O}_{\tau_k}$ such that

$$A_k = D_{k+1} \cap X_k, \quad B_k = D_{k+1} \cap X_k.$$

Therefore, $\bigcup_{k \geq n} D_k$ and $\bigcup_{k \geq n} G_k$ are disjoint τ-open sets. if $x \notin F$, where F is τ-closed, then for each n such that $x \in X_n$ there exist disjoint sets $A_n \in \mathcal{O}_{\tau_n}(x)$ and $B_n \in \mathcal{O}_{\tau_n}(F \cap X_n)$ such that $A_n \subset A_{n+1}$ and $B_n \subset B_{n+1}$. Therefore, $\bigcup_{n<\omega} A_n \in \mathcal{O}_\tau(x)$ and $\bigcup_{n<\omega} B_n \in \mathcal{O}_\tau(F)$ are disjoint.

Finally, $f \in \mathscr{C}(\tau, \nu)$ if and only if $f \circ j_n \in \mathscr{C}(\tau_n, \nu)$ for each $n < \omega$. Accordingly, f is constant on X_n for each $n < \omega$, hence constant on the union. □

THEOREM X.8.5. *There exists a Hausdorff regular topology on an infinite set such that each continuous function is constant.*

PROOF. Let Q be the topology from Example X.8.2. Then, by Proposition X.8.3, for any Hausdorff regular topology X_0, the sequence defined by $X_{n+1} := Q(X_n)$ for each $n < \omega$, fulfills the conditions of Lemma X.8.4, so that X_∞ defined in the lemma is a required space. □

X.9. Functionally initial topologies

By Definition IV.4.7, a convergence ξ is ν-*initial* if there exists $\mathscr{F} \subset \mathscr{C}(\xi,\nu)$ such that $\xi = \bigvee_{f\in\mathscr{F}} f^{-}\nu$. If ξ is ν-initial, then the condition holds also for $\mathscr{F} = \mathscr{C}(\xi,\nu)$, because $f \in \mathscr{C}(\xi,\nu)$ if and only if $\xi \geq f^{-}\nu$.

Therefore, according to our terminology from Definition IV.4.1, a convergence is ν-initial whenever it is initial with respect to $\mathscr{C}(\xi,\nu)$. For this reason, we also call it *functionally initial*.

By Definition VI.7.4, ν is initially dense in the class of ν-initial topologies. Let

$$\mathrm{R}^{\nu}\xi := \bigvee_{f\in\mathscr{C}(\xi,\nu)} f^{-}\nu,$$

and observe that $\mathrm{R}^{\nu}\xi$ is the coarsest convergence on $|\xi|$, for which each $f \in \mathscr{C}(\xi,\nu)$ is continuous. Therefore, ξ is functionally initial whenever

$$\xi = \mathrm{R}^{\nu}\xi.$$

In particular, each functionally initial convergence is a topology by Lemma VII.3.7, and thus $\mathrm{T}\xi \geq \mathrm{R}^{\nu}\xi$ for every convergence ξ.

It is straightforward that

$$\zeta \geq \xi \Longrightarrow \mathrm{R}^{\nu}\zeta \geq \mathrm{R}^{\nu}\xi,$$
$$\xi \geq \mathrm{R}^{\nu}\xi,$$
$$\mathrm{R}^{\nu}(\mathrm{R}^{\nu}\xi) = \mathrm{R}^{\nu}\xi.$$

It follows immediately from the definition (and was already observed in Lemma IV.4.3) that

PROPOSITION X.9.1. *A topology ξ is functionally initial if and only if*

$$x \in \lim\nolimits_{\xi} \mathcal{F}$$

provided that, for each $f \in \mathscr{C}(\xi,\nu)$,

$$f(x) \in \lim\nolimits_{\nu} f[\mathcal{F}].$$

PROPOSITION X.9.2. *For each convergence τ and each map valued in $|\tau|$,*

$$\mathrm{R}^{\nu}(f^{-}\tau) \geq f^{-}(\mathrm{R}^{\nu}\tau). \tag{X.9.1}$$

PROOF. Indeed, by Exercise IV.10.8,

$$f^{-}(\mathrm{R}^{\nu}\tau) = f^{-}\Big(\bigvee_{h\in\mathscr{C}(\tau,\nu)} h^{-}\nu\Big) = \bigvee_{h\in\mathscr{C}(\tau,\nu)} (h\circ f)^{-}\nu$$
$$\leq \bigvee_{g\in\mathscr{C}(f^{-}\tau,\nu)} g^{-}\nu = \mathrm{R}^{\nu}(f^{-}\tau),$$

because $f \in \mathscr{C}(f^{-}\tau,\tau)$ and thus $h\circ f \in \mathscr{C}(f^{-}\tau,\nu)$. □

Similarly to Proposition VI.6.4, (X.9.1) is equivalent to

$$f(\mathrm{R}^{\nu}\xi) \geq \mathrm{R}^{\nu}(f\xi)$$

for every convergence ξ and each map f from $|\xi|$. Analogously to (VI.6.4), each of the inequalities above is equivalent to

$$\mathscr{C}(\xi,\tau) \subset \mathscr{C}(\mathrm{R}^{\nu}\xi, \mathrm{R}^{\nu}\tau)$$

for all convergences ξ and τ.

COROLLARY X.9.3. *Each product of functionally regular topologies is functionally regular.*

LEMMA X.9.4. *Each functionally initial topology has a base for the open sets that is composed of functionally open sets.*

PROOF. As $\{]a,b[: a < b, a, b \in \mathbb{R}\}$ is a base for the open sets for the standard topology ν on $\mathbb{R}$,

$$\{\{a < f < b\} : a < b, f \in \mathscr{C}(\xi,\nu)\} \tag{X.9.2}$$

is a subbase for the open sets of $\mathrm{R}^{\nu}\xi$ (Definition IX.2.1), that is, the family of finite intersections of (X.9.2) constitutes a base for the open sets. As we shall see, however, (X.9.2) itself is already a base. Indeed,

$$\{a < f < b\} = \{a < f\} \cap \{f < b\}$$
$$= \{0 < -f + a\} \cap \{0 < f - b\} = \{0 < (-f + a) \wedge (f - b)\},$$

and $(-f+a) \wedge (f-b) \in \mathscr{C}(\xi,\nu)$. Consequently, each element of (X.9.2) is of the form $\{f > 0\}$ for some $f \in \mathscr{C}(\xi,\nu)$. On the other hand, each finite intersection of functionally open sets is functionally open, because $\{f_0 > 0\} \cap \{f_1 > 0\} = \{f_0 \wedge f_1 > 0\}$. Hence,

$$\{\{f > 0\} : f \in \mathscr{C}(\xi,\nu)\}$$

is a base for the open sets of $\mathrm{R}^{\nu}\xi$, and ξ is functionally initial if and only if $\xi = \mathrm{R}^{\nu}\xi$. □

By Definition X.7.3, a topology ξ is *functionally regular* (alias completely regular) if for each $C \in \mathcal{C}_{\xi}$ and every $x \notin C$, there exists $f \in \mathscr{C}(\xi,\nu)$ such that $f(x) > \sup_C f$. Proposition X.7.5 says that a topology is functionally regular if and only if it admits a base for the open sets consisting of functionally open sets.

THEOREM X.9.5. *A topology is functionally regular (alias completely regular) if and only if it is functionally initial.*

PROOF. By Proposition X.7.5 and Lemma X.9.4, each functionally initial topology is functionally regular. To prove the converse we shall use Proposition X.9.1. If $x \notin \lim_\xi \mathcal{F}$, that is, by Corollary VII.3.3, there exists $C \in \mathcal{C}_\xi$ such that $x \notin C$ and $C \in \mathcal{F}^\#$. By functional regularity, there exists $f \in \mathscr{C}(\xi, \nu)$ such that $f(x) = 1$ and $C \subset \{f = 0\}$. Hence $0 \in \bigcap_{F \in \mathcal{F}} f(F)$ so that $1 \notin \lim_\nu f[\mathcal{F}]$, hence ξ is functionally initial. □

DEFINITION X.9.6. A *pseudometric* on X is a function $d : X \times X \longrightarrow \mathbb{R}_+$ such that for every $x, y, z \in X$,

$$d(x,x) = 0, \tag{X.9.3}$$

$$d(y,x) = d(x,y), \tag{X.9.4}$$

$$d(x,z) \leq d(x,y) + d(y,z). \tag{X.9.5}$$

A pseudometric is a metric if for every $x, y \in X$,

$$d(x,y) = 0 \Longrightarrow x = y.$$

A convergence ξ on X is called *pseudometrizable* if there exists a family $\mathcal{D}$ of pseudometrics on X such that $x \in \lim_\xi \mathcal{F}$ whenever for each $d \in \mathcal{D}$,

$$\inf_{F \in \mathcal{F}} \sup_{y \in F} d(x,y) = 0. \tag{X.9.6}$$

As $\{\{y : d(x,y) < \varepsilon\} : \varepsilon > 0\}$ is a filter-base of $\mathcal{V}_\xi(x)$ consisting of ξ-open sets, every pseudometrizable convergence is a topology. Unlike metrizable topologies, a pseudometrizable topology need not be Hausdorff.

PROPOSITION X.9.7. *A topology is functionally initial if and only if it is pseudometrizable.*

PROOF. Let ξ be pseudometrizable, and let $\mathcal{D}$ be a family of pseudometrics on X witnessing the property. For each $d \in \mathcal{D}$ and $x \in X$, let $d_x(y) := d(x,y)$. Then, $d_x[\mathcal{F}] = \{d_x(F) : F \in \mathcal{F}\}$. Then by defintion $x \in \lim_\xi \mathcal{F}$ provided that (X.9.6), that is, $d_x(x) = 0 \in \lim_\nu d_x[\mathcal{F}]$ for each $d \in \mathcal{D}$. Therefore, ξ is functionally initial by Proposition X.9.1.

Let ξ be a topology on X. For any $F \subset_{\text{fin}} \mathscr{C}(\xi,\nu)$, let

$$d_F(x,y) := \max\{|f(x) - f(y)| : f \in F\}.$$

It is clear that d_F fulfills (X.9.3) and (X.9.4). Let us check (X.9.5). Indeed,

$$\begin{aligned} d_F(x,z) = \max\nolimits_{f\in F} |f(x) - f(z)| &= \max\nolimits_{f\in F} |f(x) - f(y) + f(y) - f(z)| \\ &= \max\nolimits_{f\in F}(|f(x) - f(y)| + |f(y) - f(z)|) \\ \leq \max\nolimits_{f\in F} |f(x) - f(y)| + \max\nolimits_{f\in F} |f(y) - f(z)| &= d_F(x,y) + d_F(y,z). \end{aligned}$$

Accordingly, d_F is a jointly ξ-continuous pseudometric. The family $\{d_F(\cdot, y) : F \subset_{\min} \mathscr{C}(\xi, \nu), y \in X\}$ determines $\mathrm{R}^{\nu}\xi$. Indeed, for each ξ-open set O and each $x \in O$, there is $F \subset_{\text{fin}} \mathscr{C}(\xi, \nu)$ such that $x \in \{d_F(\cdot, x) < \varepsilon\} \subset O$. □

COROLLARY X.9.8. *A topology is functionally regular if and only if it is pseudometrizable.*

We have observed in our separation considerations for a convergence ξ that we can either use $\mathscr{C}(\xi, \nu)$ or $\mathscr{C}(\xi, I)$. Therefore,

COROLLARY X.9.9. *A topology is functionally regular if and only if it is I-initial.*

Hence by Proposition IV.7.7, ξ is I-initial if and only if it is initial with respect to $(\prod \mathscr{C}, \prod_{f\in\mathscr{C}} I)$, where $\mathscr{C} := \mathscr{C}(\xi, I)$. The *diagonal product map* of all maps, continuous from ξ to I with the standard topology, is valued in the product of copies of I, called the *Tikhonov cube*

$$\prod\nolimits_{f\in\mathscr{C}} I \approx I^{\mathscr{C}}$$

endowed with the product topology.

Of course, this topology depends merely on the cardinality of $\mathscr{C}(\xi, I)$, that is, on the number of maps continuous from ξ to I. We denote by I^{κ}, the Tikhonov cube of *dimension* κ.

In order to see that each Hausdorff functionally regular (alias completely regular) topology can be embedded in a Tikhonov cube, we need a criterion for a map between topologies to be an embedding. Corollary IV.7.8 yields a condition for general convergences, which we shall specify for topologies.

LEMMA X.9.10. *Let ξ and τ be topologies, then an injective map $f \in \mathscr{C}(\xi, \tau)$ is an embedding if and only if $f(x) \notin \mathrm{cl}_{\tau} f(C)$ for each ξ-closed set C and every $x \notin C$.*

PROOF. Let $f \in \mathscr{C}(\xi, \tau)$ be an injective map. Then $f^{\sim 1} \in \mathscr{C}(\tau|_{f(X)}, \xi)$ [15] if and only if $f(C)$ is $\tau\,|_{f(X)}$-closed, that is,

$$\mathrm{cl}_{\tau} f(C) \cap f(X) \subset f(C),$$

for every ξ-closed set C. In other terms, if $y \in f(X) \setminus f(C)$, equivalently, $\varnothing = C \cap f^{-}(y)$, then $y \notin \mathrm{cl}_{\tau} f(C)$. As f is injective, for each $y \in f(X)$, there is $x \in X$ such that $f^{-}(y) = \{x\}$, that is, $x = f^{\sim 1}(y)$, so that the proof is complete. □

[15] For $f^{\sim 1}$ see Definition I.1.4.

Let ξ and τ_j be topologies on V and X_j respectively, and let $f_j \in {X_j}^V$ for each $j \in J$. We say that the family $\{f_j : j \in J\}$ separates points from ξ-closed sets whenever for each ξ-closed set C and every $x \notin C$, there exists $j \in J$ such that $f_j(x) \notin \mathrm{cl}_{\tau_j} f_j(C)$.

COROLLARY X.9.11. *Let ξ and τ_j be free (alias T_1) topologies and let $f_j \in \mathscr{C}(\xi, \tau_j)$ for each $j \in J$. If $\{f_j : j \in J\}$ separates points from ξ-closed sets, then $\triangle_{j\in J} f_j$ is an embedding.*

PROOF. As ξ is free, each singleton is ξ-closed, so that $\{f_j : j \in J\}$ is injective. Let us show that $F := \triangle_{j\in J} f_j$ separates points from ξ-closed sets. By way of contradiction, suppose that there is a ξ-closed set C and $x \notin C$ such that $F(x) \in \mathrm{cl}_{\prod_{j\in J}\tau_j} F(C)$. This means that there exists a filter $\mathcal{F}$ on C such that $F(x) \in \lim_{\prod_{j\in J}\tau_j} F[\mathcal{F}]$, equivalently $f_j(x) \in \lim_{\tau_j} f_j[\mathcal{F}]$ for each $j \in J$, that is, $f_j(x) \in \mathrm{cl}_{\tau_j} f_j(C)$ for every $j \in J$, contrary to the assumption. Consequently, $F = \triangle_{j\in J} f_j \in \mathscr{C}(\xi, \prod_{j\in J}\tau_j)$ is an injection separating points from ξ-closed sets, hence an embedding in virtue of Lemma X.9.10. □

If ξ is a free functionally regular topology, then it is Hausdorff, and by the very definition, the set $\mathscr{C}(\xi, I)$ of continuous functions valued in $I = [0, 1]$ endowed with ν_I, separates points from ξ-closed sets.

Therefore, by Corollary X.9.11, the diagonal product map

$$\triangle\mathscr{C} = \triangle_{f\in\mathscr{C}(\xi,I)} f$$

is an embedding of ξ into $\prod_{f\in\mathscr{C}} I$.

COROLLARY X.9.12. *Every free (alias T_1) functionally regular (alias completely regular) topology can be embedded in a Tikhonov cube.*

THEOREM X.9.13. *The Tikhonov cube I^κ is a universal space for free functionally regular topologies of infinite weight κ.*

PROOF. As each base for the open sets of a functionally regular topology ξ has a refining base consisting of functionally open sets, ξ is initial with respect to continuous functions which determine the elements of that base. Hence, by Theorem IX.2.21, there is a base consisting of functionally open sets of cardinality equal to the weight.

By Lemma III.3.9 and Corollary X.9.3, each Tikhonov cube is free functionally regular. It follows from Proposition IX.2.14 that $\mathrm{w}(I^\kappa) \leq \kappa$ for every infinite κ. As the discrete topology on a set of cardinality κ is a functionally regular topology of weight κ, and thus embeddable in I^κ, we infer that $\kappa \leq \mathrm{w}(I^\kappa)$. □

THEOREM X.9.14 (*Urysohn metrization theorem*). *A free (alias T_1) regular topology of countable weight is metrizable.*

PROOF. By Proposition IX.3.11, a topology τ on X fulfilling the conditions is normal. Let $\mathcal{B}$ be a countable base, and let

$$\mathcal{W} := \{(B, W) \in \mathcal{B} \times \mathcal{B} : \mathrm{cl}_\tau B \subset W\}.$$

By Theorem X.7.8 of Urysohn, for each $(B, W) \in \mathcal{V}$ there exists $f_{B,W} \in \mathscr{C}(\tau, I)$ such that $B \subset \{f_{B,W} = 0\}$ and $X \setminus W \subset \{f_{B,W} = 1\}$. As τ is free, the family

$$\mathscr{F} := \{f_{B,W} : (B, W) \in \mathcal{B} \times \mathcal{B}\}$$

distinguishes the points from the closed sets, and thus τ is initial with respect to $\mathscr{F}$ and the standard topology of $[0, 1]$, that is, is functionally initial. As τ is free, and $\mathscr{F}$ distinguishes the points, the diagonal map $\prod \mathscr{F}$ is an embedding into the Tikhonov cube $I^{\mathscr{F}}$, which is metrizable, because $\mathscr{F}$ is countable, in virtue of Proposition X.3.8. □

X.10. Supplement

EXERCISE X.10.1. Show that each lower semicontinuous function on a countably compact topological space attains its infima.

SOLUTION. Let f be lower semicontinuous function on a countably compact topological space X. Let $(r_n)_n$ be a strictly decreasing sequence of reals such that $\inf\{r_n : n < \omega\} = \inf_X f$. By lower semicontinuity, each level set $\{f \le r\}$ is closed, hence, by countable compactness, there is $x \in \bigcap_{n<\omega}\{f \le r_n\}$, and thus $f(x) = \inf_X f$.

EXERCISE X.10.2. A_0 and A_1 are functionally separated if and only if there exists $f \in \mathscr{C}$ such that $\sup_{A_0} f < \inf_{A_1} f$.

SOLUTION. If $A_0 \subset \{f = 0\}$ and $A_1 \subset \{f = 1\}$, then $0 = \sup_{A_0} f < \inf_{A_1} f = 1$. Conversely, if $a = \sup_{A_0} f < \inf_{A_1} f = b$, then let $h := \frac{1}{b-a}((f \vee a) \wedge b) - a$. We check that $h(x) = 0$ for $x \in A_0$, and $h(x) = 1$ for $x \in A_1$.i

Let ν_- be the lower topology on the real line (Section X.1).

EXERCISE X.10.3. Show that a convergence τ is a topology if and only if $\tau = \mathrm{R}^{\nu_-}\tau$.

SOLUTION. For each $F \in \mathcal{C}_\tau$, the function $\psi_F(x) := 0$ if $x \in F$ and $\psi_F(x) := 1$ if $x \notin F$ (VII.5.1) is lower τ-semicontinuous, hence $\psi_F \in \mathscr{C}(\tau, \nu_-)$. Therefore,

$$\tau \ge \mathrm{R}^{\nu_-}\tau = \bigvee_{f \in \mathscr{C}(\tau,\nu_-)} f^- \nu_- \ge \bigvee_{F \in \mathcal{C}_\tau} \psi_F^-(\nu_-).$$

On the other hand, $\$_1$ on $\{0, 1\}$ is embedded in ν_- on $\mathbb{R}$ via the natural injection $j \in \mathbb{R}^{\{0,1\}}$. Hence, $\psi_F^-(\nu_-) = \psi_F^-(\$)$, and thus $\mathrm{R}^{\nu_-}\tau \ge \mathrm{R}^{\$}\tau = \tau$.

Lemma X.10.4. *Each open set of a metrizable topology is homeomorphic to a closed subspace of* $\xi \times \nu_{\mathbb{R}}$.

Proof. Let g be a metric on X compatible with ξ, let O be a ξ-open set. Then

$$f(x) := \frac{1}{g(x, X \setminus O)}$$

is well defined, and $i_X \triangle f \in \mathscr{C}(\xi, \xi \times \nu_{\mathbb{R}})$ is a bijection, hence an embedding. Because of (I.1.4), the image $i_X \triangle f$ is the graph $\mathrm{gr}(f)$ of f, which is closed, because of Corollary IX.5.5. □

More generally,

Lemma X.10.5. *Each* $G_{\aleph_0}$*-subspace of a metrizable topology* ξ *is homeomorphic to a closed subspace of* $\xi \times \prod_{n<\omega} \nu_{\mathbb{R}}$.

Proof. Let $\eta = \xi \times \prod_{n<\omega} \nu_{\mathbb{R}}$ and let D be a $G_{\aleph_0}(\eta)$-subset of $|\xi| \times \prod_{n<\omega} \mathbb{R}$, that is, $D = \bigcap_{n<\omega} O_n$, where $\{O_n = n < \omega\}$ is a (decreasing) sequence of η-open sets such that. Let g be a metric a metric on X compatible with ξ. Then

$$f_n(x) := \frac{1}{g(x, X \setminus O_n)}$$

is well defined on X for each $n < \omega$, because if $x \in O_n$, then the distance $g(x, X \setminus O_n) > 0$, as $X \setminus O_n$ is ξ-closed. Hence, $f_n \in \mathscr{C}(\xi, \nu_{\mathbb{R}})$ for each $n < \omega$. Therefore $f := i_X \triangle (\triangle_{n<\omega} f_n)$ is an embedding. Indeed, $p_n \circ f = f_n$ is continuous, and if p_X stands for the projection on X, then $p_X \circ f = i_X \in \mathscr{C}(\xi, \xi)$, so that f is a bijection and $f \in \mathscr{C}(\xi, \eta)$. Moreover, $f^{-}|_{f(X)} \in \mathscr{C}(\eta|_{f(X)}, \xi)$. Of course, the graph of f is η-closed because of Corollary IX.5.5. □

Lemma X.10.6. *In a metrizable space, if* $(x_n)_n$ *converges to* x_∞, *and for each* $n < \omega$, *there exists a sequence* $(x_{n,k})_k$ *converging to* x_n, *then there is a map* $f : \mathbb{N} \longrightarrow \mathbb{N}$ *such that for each* $B \in \mathcal{N}(x_\infty)$ *there is* $n_B < \omega$ *such that* $\{(x_{n,k}) : n > n_B, k > f(n)\} \subset B$.

Proof. Let d be a compatible metric, and let $V_p := B(x_\infty, 2^{-p})$ for each $p < \omega$. Then for each $p < \omega$ there exists n_p such that $x_n \in V_p$ for each $n \geq n_p$, and since V_p is open, there exists $f_p : \mathbb{N} \longrightarrow \mathbb{N}$ such that $x_{n,k} \in V_p$ for each $k \geq f_p(n)$ and $n \geq n_p$, and $f_p(n) = 0$ for $n < n_p$. Let $f(n) := \sup_{p<\omega} f_p(n)$. As $\{n : x_n \notin V_p\}$ is finite for each $p < \omega$, the set $F_n := \{p : f_p(n) > 0\}$ is finite for each $n < \omega$, and thus $f(n) < \infty$ for each $n < \omega$. Then f is a needed map. Indeed, if $p < \omega$, then for $n \geq n_p$,

$$k \geq f(n) \geq f_p(n),$$

thus $x_{n,k} \in V_p$. □

COROLLARY X.10.7. *In a metrizable space, if* $(x_n)_n$ *converges to* x, *and for each* $n < \omega$ *there exists a sequence* $(x_{n,k})_k$ *converging to* x_n, *then there is a sequence* $(k_n)_n$ *such that* $(x_{n,k_n})_n$ *converges to* x.

EXERCISE X.10.8. Let f be a real-valued infinitely differentiable function on the real line such that for each $x \in \mathbb{R}$, there exists $n < \omega$ so that the n-th derivative $f^{(n)}(x) = 0$. Show that

(1) there exists an open interval, on which f is equal to a polynomial function,
(2) f is a polynomial function on the whole real line ([16]).

SOLUTION. (1) For each $n < \omega$, the set $F_n := \{f^{(n)} = 0\}$ is closed, because all the derivatives are continuous. As $\mathbb{R} = \bigcup_{n<\omega} F_n$, by Corollary X.4.16 of Baire, there exists $n < \omega$ such that $\operatorname{int} F_n \neq \varnothing$, so that there exist $a < b$ such that $]a, b[\subset F_n$, that is, $f^{(n)}(x) = 0$ for each $a < x < b$. By integrating $f^{(n)}$ n times, we get a polynomial function, the coefficients of which we can choose to make equal to f.

(2) As each open subset of the real line is completely metrizable, adapting the argument above, we conclude that for each open interval I, there exist $n < \omega$ and $a < b$ such that $]a, b[\subset I$ and $f^{(n)}(x) = 0$ for each $a < x < b$.

Let $\mathcal{J}$ be the set of open intervals such that for each $J \in \mathcal{J}$, the function $f|_J$ is polynomial.

The set $\bigcup_{J\in\mathcal{J}} J$ is open and dense. Let us call L the complementary set. Of course, L is closed, hence completely metrizable. Moreover, L has no isolated points, for if $x \in L$ were an isolated point, then there would be $a < b$ such that f would be polynomial on $]a, x]$ and on $[x, b[$. This means that there exists $m < \omega$ such that $f^{(m)}$ is zero in $]a, b[$ by continuity. Hence $]a, b[$ would be an element of $\mathcal{J}$, which is a contradiction.

If $L \neq \varnothing$, then by Corollary X.4.16 of Baire, there exists $k < \omega$ and $a < b$ such that such $]a, b[\cap L \neq \varnothing$ and $f^{(k)}(x) = 0$ for each $x \in]a, b[\cap L$. By definition, $\bigcup_{J\in\mathcal{J}} J$ covers $]a, b[\setminus L$. If $J \in \mathcal{J}$ is such that $J \#]a, b[$, then $]a, b[\setminus J \neq \varnothing$, for otherwise $J \# L$, which is a contradiction. Therefore at least one end-point x_J of J belongs to $]a, b[\cap L$.

By definition of $\mathcal{J}$, there exists $m < \omega$ such that $f^{(m)}(x) = 0$ for each $x \in J$. If $m \leq k$, then $f^{(k)}(x) = 0$ for each $x \in J$. If $m > k$, then integrate $f^{(m)}$ over J for $m - k$ times with the boundary condition $0 = f^{(k)}(x_J) = f^{(k+1)}(x_J) = \ldots$, so that $f^{(k)}(x) = 0$ for each $x \in J$. Therefore, $f^{(k)}(x) = 0$ for each $x \in J$ and every $J \in \mathcal{J}$ such that $J \#]a, b[$. Consequently, $f^{(k)}(x) = 0$ for each $x \in]a, b[$, so that $]a, b[$ is

[16]H. D. BRUNK and R. P. BOAS [34]

an element of $\mathcal{J}$, in contradiction with $]a,b[\cap L \neq \varnothing$. It follows that $L = \varnothing$, that is, f is a polynomial on the whole real line.

EXERCISE X.10.9 (Dini theorem). Let τ be a countably compact topology on X. If for an increasing sequence $\{f_n : n < \omega\} \subset \mathscr{C}(\tau, \nu)$, the function $f(x) := \lim_\nu (f_n(x))_n \in \mathscr{C}(\tau, \nu)$, then $(f_n)_n$ converges uniformly to f.

SOLUTION. Let $\varepsilon > 0$, and let $F_n := \{x \in X : f_n(x) \geq f(x) - \varepsilon\}$. Then $\{F_n : n < \omega\} \subset \mathcal{C}_\tau$ is a descending sequence. By assumption, for each $x \in X$, there is $n < \omega$ such that $x \notin F_n$, that is, $\bigcap_{n<\omega} F_n = \varnothing$. Therefore, by the Cantor condition from Exercise IX.5.11, there exists $n_0 < \omega$ such that $C_{n_0} = \varnothing$, hence $C_n = \varnothing$ for each $n \geq n_0$. This means that $\sup\{|f(x) - f_n(x)| : x \in X\} < \varepsilon$ for each $n \geq n_0$.

LEMMA X.10.10. *The square root on* $[0,1]$ *is the uniform limit of a sequence of continuous functions.*

PROOF. Define a sequence $\{p_n : n < \omega\} \subset \mathbb{R}^{[0,1]}$ so that $p_0(x) := 0$ for each $x \in [0,1]$, and if $n < \omega$, then

$$p_{n+1}(x) := p_n(x) + 2^{-1}(x - p_n(x)^2) \tag{X.10.1}$$

for each $x \in [0,1]$. We shall show by induction that

$$0 \leq p_n(x) \leq \sqrt{x} \tag{X.10.2}$$

for each $x \in [0,1]$. This is obvious for $n = 0$. So, assume that (X.10.2) holds for $n \geq 0$, and observe that

$$\begin{aligned}\sqrt{x} - p_{n+1}(x) &= \sqrt{x} - p_n(x) - \tfrac{1}{2}(\sqrt{x} - p_n(x))(\sqrt{x} + p_n(x)) \\ &= (\sqrt{x} - p_n(x))(1 - \tfrac{1}{2}(\sqrt{x} + p_n(x))) \geq 0\end{aligned}$$

for each $x \in [0,1]$. It follows immediately from the inductive definition that $(p_n)_n$ is increasing and bounded by the square root. Hence $p(x) := \lim_\nu (p_n(x))_n$ exists for each $x \in [0,1]$. By applying the limit to (X.10.1), we obtain $p(x) = p(x) + 2^{-1}(x - p(x)^2)$, so that $p(x) = \sqrt{x}$ for each $x \in [0,1]$. By the Dini theorem (Exercise X.10.9), $(p_n)_n$ converges uniformly to the square root. □

A subset R of $\mathbb{R}^X$ is called a *ring* if $f+g, f-g$ and $f \cdot g$ belong to R provided that $f, g \in R$.

COROLLARY X.10.11. *If a ring* R *of continuous functions on a compact topological space* X *contains all the constant functions and is closed under the uniform convergence, then* $|f|, \max(f,g)$, *and* $\min(f,g)$ *belong to* R *provided that* $f, g \in R$. *In particular,* R *is a lattice for the pointwise order.*

PROOF. There exists $r > 0$ such that $\max|f(x)| \leq r$ for each $x \in X$. Thus $g := \frac{1}{r}f$ is bounded by 1. As $|t| = \sqrt{t^2}$, the function $|g|$ is the pointwise limit of $\{p_n \circ g^2 : n < \omega\} \subset R$, and by Lemma X.10.10, $|g|$ is the pointwise limit of elements of R, hence belongs to R by assumption, and thus $|f| \in R$. It is enough to note that

$$\max(f, g) = \tfrac{1}{2}(f + g + |f - g|),$$
$$\min(f, g) = \tfrac{1}{2}(f + g - |f - g|).$$

□

THEOREM X.10.12 (Stone-Weierstraß). *If τ is a Hausdorff compact topology, and $R \subset \mathscr{C}(\tau, \nu)$ is a ring containing all the constant functions, that separates the points, and is closed for the topology of uniform convergence, then $R = \mathscr{C}(\tau, \nu)$.*

PROOF. It is enough to prove that for every $f \in \mathscr{C}(\tau, \nu)$ and each $\varepsilon > 0$, there exists $f_\varepsilon \in R$ such that $|f(x) - f_\varepsilon(x)| < \varepsilon$ for each $x \in X := |\tau|$. For each pair of distinct points $a, b \in X$, there exists $f_{a,b} \in R$ such that $f_{a,b}(a) = f(a)$ and $f_{a,b}(b) = b$. Indeed, by assumption, there exists $g \in R$ such that $g(a) \neq g(b)$. Then the function

$$f_{a,b}(x) = f(a) + \frac{g(x) - g(a)}{g(b) - g(a)}(f(b) - f(a))$$

is a required function. If $a = b$, then let $f_{a,b}(x) = f(a) = f(b)$ for each $x \in X$, that is, a constant function. For each $a, b \in X$, let

$$U_{a,b} := \{x \in X : f_{a,b}(x) < f(x) + \varepsilon\},$$
$$W_{a,b} := \{x \in X : f_{a,b}(x) > f(x) - \varepsilon\}.$$

For each $a \in X$, the family $\{U_{a,b} : b \in X\}$ is a τ-open cover, hence by compactness, there exists a finite subset X_a of X such that $\bigcup_{b \in X_a} U_{a,b} = X$. The function $f_a := \min\{f_{a,b} : b \in X_a\}$ belongs to R by Corollary X.10.11, and $f_a(x) < f(x) + \varepsilon$ for each $x \in X$, and

$$x \in W_a := \bigcap\nolimits_{b \in X_a} W_{a,b} \Longrightarrow f(x) - \varepsilon < f_a(x).$$

It follows that $\{W_a : a \in X\}$ is a τ-open cover, hence by compactness, there exists a finite subset A of X so that $\bigcup_{a \in A} W_a = X$. The function $f_\varepsilon := \max\{f_a : a \in A\}$ is a required function by Corollary X.10.11. □

The family of all polynomials is a ring that contains all the constant functions, and the identity function separates the points. By taking all the uniform limits of sequences of polynomials, we get a ring that contains all the constant functions, separates the points, and is closed under uniform convergence. Thus, by Theorem X.10.12 of Stone-Weierstraß,

COROLLARY X.10.13. *Each continuous function on a bounded closed interval is a uniform limit of polynomials.*

A linear order $\leq$ is called *dense* if for each $a < b$ there exists c such that $a < c < b$. It is called *unbounded* if for each c there exist a and b such that $a < c < b$.

THEOREM X.10.14 (Cantor). *Any two countable dense linear unbounded orders are isomorphic.*

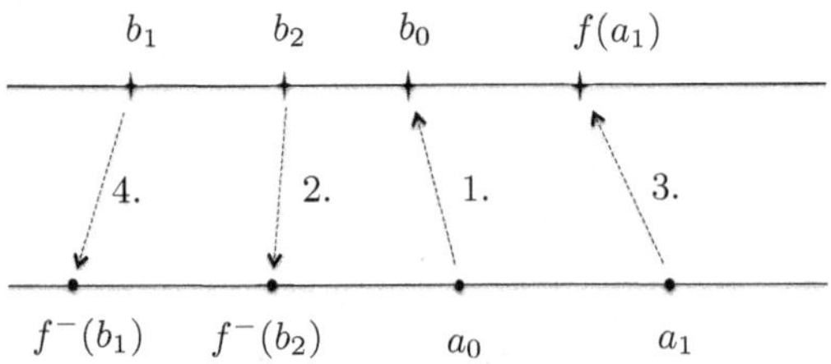

FIGURE X.10.1. Every countable unbounded dense linearly ordered set is isomorphic with the set of rationals. Assign b_0 to a_0, then to b_1 the first a_n less than a_0. Then assign to a_1 the first b_k greater than b_0 and b_1, then to b_2 the first a_{n+p} that is between $f^-(b_1)$ and a_0, and so on, maintaining the order.

PROOF. Let $A := \{a_n : n < \omega\}$ and $B := \{b_n : n < \omega\}$ be dense unbounded linearly ordered sets. We define an isomorphism $f : A \longrightarrow B$ by induction, by picking alternatively $f(a_n)$ and $f^-(b_n)$. Let $f(a_0) := b_0$. If $b_1 < b_0$, then let $f^-(b_1) := a_n$ where $n := \min\{k : a_k < a_0\}$, and if $b_1 > b_0$, then let $f^-(b_1) := a_n$ where $n := \min\{k : a_k > a_0\}$. This is possible, because the order is unbounded. Let I be a strict interval joining b_0 with b_1. If $a_1 \neq f^-(b_1)$, then let $f(a_1) := b_n$, where $n := \min\{k : b_k \in I\}$, which is possible, because the order is dense; if $a_1 = f^-(b_1)$, then $f(a_2)$ is defined as above. In general, if we have already defined f for a finite number of elements, and n is the least index, for which $f(a_n)$ is yet not defined, then we pick for it b_m preserving the order and such that m is the least index yet not assigned. $\square$

COROLLARY X.10.15. *Each countable linear dense unbounded order is isomorphic with $\mathbb{Q}$.*

A linear order is called *conditionally complete* if each non-empty bounded above set has a supremum. This implies that each non-empty bounded below set has an infimum. It can be shown that each dense unbounded linear order, has a unique (up to isomorphism) conditional completion.

It follows that $\mathbb{R}$ constitutes a unique (up to isomorphism) order completion of $\mathbb{Q}$. On the other, the standard order of $\mathbb{R}$ determines its standard topology, which is homeomorphic to $]0,1[$. Consequently, $\mathbb{Q}$ is homeomorphic to $]0,1[\cap\mathbb{Q}$, and $\mathbb{R}\setminus\mathbb{Q}$ to $]0,1[\setminus\mathbb{Q}$.

CHAPTER XI

Functional partitions and metrization

In this chapter we shall consider *functional covers* of topologies, like *partitions of unity*, which are convenient in our functional study, in particular, in metrizability quest.

XI.1. Infinite sums of positive real numbers

If $\varnothing \neq A \subset \mathbb{R}_+$ is possibly uncountable, then we define

$$\sum A := \sup\{\sum R : R \subset_{\text{fin}} A\}.$$

In other words, $\sum A$ is the supremum of all finite sums of elements of A. Let us observe that $\sum A$ is either a non-negative real number, or $+\infty$. Of course, if $\sum A < \infty$, then for each $\varepsilon > 0$, there exists a finite subset R of A such that

$$\sum A - \varepsilon < \sum R.$$

Consequently, the remaining $A \setminus R$ is negligible, that is, $\sum(A \setminus R) < \varepsilon$. Moreover,

LEMMA XI.1.1. *If $\sum A < \infty$, then $\{a \in A : a > 0\}$ is countable.*

PROOF. Suppose that $\{a \in A : a > 0\}$ is uncountable. Then there is $n \in \mathbb{N}_1$ such that $\{a \in A : \frac{1}{n} \leq a\}$ is uncountable. Otherwise, $\{a \in A : a > 0\} = \bigcup_{n<\omega}\{a \in A : \frac{1}{n} \leq a\}$, and a countable union of countable sets is countable. In particular, $\sum A \geq \frac{k}{n}$ for each $k < \omega$, hence $\sum A = \infty$. □

As the sums of series of positive terms do not depend on the order of the terms, we infer that

COROLLARY XI.1.2. *If $A = \{a_n : n < \omega\} \subset \mathbb{R}_+$, then $\sum A = \sum_{n=0}^{\infty} a_n$.*

XI.2. Partitions of unity

We shall consider families $\mathscr{F} \subset {\mathbb{R}_+}^X$ of positive real-valued functions. As $\mathscr{F}$ can be uncountable, the meaning of the sum $\sum \mathscr{F}$ is that derived pointwise from the preceding section. If $\mathscr{F}(x) := \{f(x) : f \in \mathscr{F}\}$ for each $x \in X$, then $\sum \mathscr{F}(x) := \sum\{f(x) : f \in \mathscr{F}\}$.

Therefore, if $\mathscr{F}_{\text{fin}\sum}$ stands for the family of finite sums of elements of $\mathscr{F}$, then

$$\sum \mathscr{F}(x) = \sup\{h(x) : h \in \mathscr{F}_{\text{fin}\sum}\}. \tag{sum}$$

We fix a topology on X, and denote by

$$\mathscr{C}_+ := \{f \in \mathbb{R}_+{}^X : f \in \mathscr{C}(X, \nu)\}$$

the set of all continuous positive functions. We assume tacitly that from now on $\mathscr{F} \subset \mathscr{C}_+$, occasionally recalling this fact. Of course, $\sum \mathscr{F}$ need not be a real function, because it can take values $+\infty$. In other terms, in general, $\sum \mathscr{F} \in \overline{\mathbb{R}}_+{}^X$ is an *extended real function*, where $\overline{\mathbb{R}}_+ := \mathbb{R} \cup \{+\infty\}$ is the *extended upper real line*. If $\sum \mathscr{F}(x) < \infty$ for each $x \in X$, then we say that $\sum \mathscr{F}$ is *limited* ([1]).

DEFINITION XI.2.1 (Partition). A family $\mathscr{F}$ is called a *partition of a function* F if $F(x) = \sum \mathscr{F}(x)$ for each $x \in X$; it is called a *partition of unity* on X if $\sum \mathscr{F}(x) = 1$ for each $x \in X$.

DEFINITION XI.2.2 (Functional cover). A family $\mathscr{F}$ of functions is called a *functional cover* of X if for each $x \in X$, there is $f \in \mathscr{F}$ such that $f(x) > 0$.

Accordingly, $\mathscr{F}$ is a functional cover of X whenever, for each $x \in X$,

$$\sup \mathscr{F}(x) := \sup\{f(x) : f \in \mathscr{F}\} > 0.$$

DEFINITION XI.2.3. The *support* of $\mathscr{F}$ is defined by

$$\partial \mathscr{F} := \{\{f > 0\} : f \in \mathscr{F}\}. \tag{support}$$

Sure enough, the support of a family of continuous functions consists of functionally open sets. Clearly,

PROPOSITION XI.2.4. *A family $\mathscr{F} \subset \mathscr{C}_+$ is a functional cover if and only if $\partial \mathscr{F}$ is a cover.*

If a family $\mathscr{F} \subset \mathscr{C}_+$ is finite, then $\sup \mathscr{F}$ and $\sum \mathscr{F}$ are (limited) continuous functions, but otherwise they need be neither limited, nor continuous. In any case, $\sup \mathscr{F}(x) \leq \sum \mathscr{F}(x)$ for each $x \in X$.

Recall (Definition V.1.4) that a family $\mathcal{A}$ of subsets of a given set is a *refinement* of a family $\mathcal{D}$ (or $\mathcal{A}$ *refines* $\mathcal{D}$) if for each $A \in \mathcal{A}$,

---

[1]The term *finite function*, used by some authors ([95]), has a disadvantage of evoking finite sets. In order to avoid misunderstandings, we reserve the term partition of F for limited functions, unlike [95].

there exists $D \in \mathcal{D}$ such that $A \subset D$. It is instrumental to extend this concept to families of functions.

DEFINITION XI.2.5. We say that a family of functions $\mathscr{H}$ *refines* a family of functions $\mathscr{F}$ whenever $\partial\mathscr{H}$ refines $\partial\mathscr{F}$.

Let us make explicit this definition and its hybrid variants: if $\mathscr{H}$, $\mathscr{F} \subset \mathscr{C}_+$ and $\mathcal{A} \subset \mathcal{O}$, then ([2])

(1) $\mathscr{H}$ *refines* $\mathscr{F}$ if for each $h \in \mathscr{H}$, there exists $f \in \mathscr{F}$ such that

$$\{h > 0\} \subset \{f > 0\},$$

(2) $\mathscr{H}$ *refines* $\mathcal{A}$ if for each $h \in \mathscr{H}$, there exists $A \in \mathcal{A}$ such that

$$\{h > 0\} \subset A,$$

(3) $\mathcal{A}$ *refines* $\mathscr{F}$ if for each $A \in \mathcal{A}$, there exists $f \in \mathscr{F}$ such that

$$A \subset \{f > 0\}.$$

We shall revisit some separation properties in terms of functional covers. By Theorem X.7.19 of Vedenisov, a free topology on X is perfectly normal if and only if each open cover $\mathcal{W}$ of X, there exists a functional cover $\mathscr{W}$ such that $\mathcal{W} = \partial\mathscr{W}$. As for normality,

PROPOSITION XI.2.6. *A Hausdorff topology is normal if and only if each finite open cover has a refining partition of unity.*

PROOF. Suppose that the property holds, and let F_0, F_1 be closed subsets of X such that $F_0 \cap F_1 = \varnothing$. Then $(X \setminus F_0) \cup (X \setminus F_1) = X$, hence there is a partition of unity $\{f_0, f_1\}$ such that $\{f_0 > 0\} \subset X \setminus F_0$ and $\{f_1 > 0\} \subset X \setminus F_1$, equivalently $F_0 \subset \{f_0 = 0\}$ and $F_1 \subset \{f_1 = 0\} = \{f_0 = 1\}$, because $f_0(x) + f_1(x) = 1$ for each $x \in X$, proving normality.

Conversely, suppose that X is normal, and let $\mathcal{P}$ be a finite open cover of X. We proceed by induction on the cardinality n of $\mathcal{P}$. If $n = 1$, then $\mathcal{P} = \{X\}$ and the constant function 1 is a partition of unity refining X. If $n = 2$, then the equivalent conditions of the first part of the proof yield the result. If the condition has been established for n, and $\mathcal{P}$ has $n+1$ elements, say, $\mathcal{P} = \{P_0, P_1, \ldots, P_n, P_{n+1}\}$, then $\{\bigcup_{0 \le k \le n} P_k, P_{n+1}\}$ is an open cover, hence there exists a partition of unity $\{f_0, f_1\}$ refining this cover. As $\{P_0, P_1, \ldots, P_n\}$ is an open cover of a closed, hence normal, subset $X \setminus P_{n+1}$ of X, by inductive assumption,

[2]Various authors use for (2) the term: *$\mathscr{H}$ is subordinated to $\mathcal{A}$*, or $\mathscr{H}$ is *$\mathcal{A}$-small.*

there is a refining partition of unity $\{g_0, g_1, \ldots, g_n\}$. It follows that $\{f_0 \cdot g_0, f_0 \cdot g_1, \ldots, f_0 \cdot g_n, f_1\}$ is a partition of unity refining $\mathcal{P}$. □

Under what conditions the sum $\sum \mathscr{F}$ is continuous? A necessary condition is that it be limited. We will give a simple characterization of the continuity of $\sum \mathscr{F}$ in terms of the family of finite sums of the elements of $\mathscr{F}$.

PROPOSITION XI.2.7. *Let $\mathscr{F} \subset \mathscr{C}_+$ be a partition of F. Then F is continuous if and only if for each $\varepsilon > 0$ and $x \in X$, there exist $U \in \mathcal{N}_X(x)$ and a finite subset $\mathscr{T}$ of $\mathscr{F}$ such that*

$$\sup_{u \in U} \sum\nolimits_{f \in \mathscr{F} \setminus \mathscr{T}} f(u) < \varepsilon. \tag{XI.2.1}$$

PROOF. Let $\varepsilon > 0$ and $x \in X$. Suppose that (XI.2.1) holds. By definition of (sum), there exists a finite subset $\mathscr{T}$ of $\mathscr{F}$ such that $F(x) - \varepsilon < \sum \mathscr{T}(x)$, hence by the continuity of $\sum_{f \in \mathscr{T}} f$, there exists $W \in \mathcal{N}_X(x)$ such that $F(x) - \varepsilon < \sum \mathscr{T}(u) \leq F(u)$ for each $u \in W$. Let $U \in \mathcal{N}_X(x)$ and $\mathscr{T}$ be such that (XI.2.1) holds. Then $F(u) - \sum \mathscr{T}(u) < \frac{\varepsilon}{2}$ for each $u \in U$. Hence, by the continuity of $\sum \mathscr{T}$, for each $u \in U$,

$$F(x) + \varepsilon \geq \sum \mathscr{T}(x) + \varepsilon > \sum \mathscr{T}(u) + \varepsilon/2 > F(u).$$

Conversely, if F is continuous, $\varepsilon > 0$, and $x \in X$, then there exists $U \in \mathcal{N}_X(x)$ and a finite subset $\mathscr{T}$ of $\mathscr{F}$ such that $F(x) - \sum \mathscr{T}(x) < \frac{\varepsilon}{2}$, and, by the continuity of F and each $f \in \mathscr{T}$, also $F(u) - \sum \mathscr{T}(u) < \varepsilon$ for each $u \in U$. Hence, $\sum_{f \in \mathscr{F} \setminus \mathscr{T}} f(u) = F(u) - \sum_{f \in \mathscr{T}} f(u) < \varepsilon$ for each $u \in U$. □

THEOREM XI.2.8. *Let $\mathscr{F} \subset \mathscr{C}_+$ be a partition of a function F. Then F is continuous if and only if $\mathscr{F}_{\mathrm{fin}\sum}$ is equicontinuous.*

PROOF. If F is continuous, $x \in X$ and $\varepsilon > 0$, then, by (XI.2.1), there exists a finite subset $\mathscr{T}$ of $\mathscr{F}$ and $U \in \mathcal{N}_X(x)$ such that $\sum(\mathscr{F} \setminus \mathscr{T})(u) < \varepsilon$ for each $u \in U$. If $\mathscr{S}$ is another finite subset of $\mathscr{F}$, then $\sum_{f \in \mathscr{F} \setminus \mathscr{T}} f(u) < \frac{\varepsilon}{3}$ for each $u \in U$. On the other hand, $\{\sum(\mathscr{S} \cap \mathscr{T}) : \mathscr{S} \subset_{\mathrm{fin}} \mathscr{F}\}$ is semicontinuous. Therefore, there exists $W \in \mathcal{N}_X(x)$ such that $\left|\sum_{f \in \mathscr{S} \cap \mathscr{T}} (f(x) - f(u))\right| < \frac{\varepsilon}{3}$ for each $u \in W$ and each finite $\mathscr{S} \subset \mathscr{F}$. As a result, for each $u \in W \cap U$ and each finite $\mathscr{S} \subset \mathscr{F}$,

$$\left|\sum\nolimits_{f \in \mathscr{S}} f(x) - \sum\nolimits_{f \in \mathscr{S}} f(u)\right| \leq$$

$$\left|\sum\nolimits_{f \in \mathscr{S} \cap \mathscr{T}} (f(x) - f(u))\right| + \left|\sum\nolimits_{f \in \mathscr{S} \setminus \mathscr{T}} (f(u) - f(x))\right| < \varepsilon.$$

Conversely, if $\mathscr{F}_{\text{fin}\sum}$ is equicontinuous, then by Lemma X.2.6, (sum) is continuous. □

Of course, $\mathscr{F} \subset \mathscr{F}_{\text{fin}\sum}$, and thus

COROLLARY XI.2.9. *If* $F = \sum \mathscr{F}$ *is continuous, then* $\mathscr{F}$ *is equicontinuous.*

A family $\mathscr{F}$ is called *strongly equicontinuous* if for every $x \in X$ and $\varepsilon > 0$, there exist $U \in \mathcal{N}_X(x)$ and a finite $\mathscr{T} \subset \mathscr{F}$ such that

$$\sup_{u \in U} f(u) < \varepsilon \tag{XI.2.2}$$

for each $f \in \mathscr{F} \setminus \mathscr{T}$.

Each strongly equicontinuous family is equicontinuous. Sure enough, (XI.2.2) implies that $|f(u) - f(x)| < \varepsilon$ for each $u \in U$ and $f \in \mathscr{F} \setminus \mathscr{T}$, so that $\mathscr{F} \setminus \mathscr{T}$ is equicontinuous. As $\mathscr{T}$ is a finite family of continuous functions, it is equicontinuous, and thus $\mathscr{F} = (\mathscr{F} \setminus \mathscr{T}) \cup \mathscr{T}$ is equicontinuous.

On recalling that a family is a partition of a function valued in positive reals, then by definition its sum is limited,

PROPOSITION XI.2.10. *Each equicontinuous partition is strongly equicontinuous.*

PROOF. By definition, for each $x \in X$ and $\varepsilon > 0$, there is a finite subfamily $\mathscr{T}$ of $\mathscr{F}$ such that $\sum \mathscr{F}(x) < \sum \mathscr{T}(x) + \frac{\varepsilon}{2}$. Hence, $\sum(\mathscr{F} \setminus \mathscr{T})(x) < \frac{\varepsilon}{2}$. On the other hand, by equicontinuity, there exists $U \in \mathcal{N}_X(x)$ such that $\sup_{u \in U} f(u) < f(x) + \frac{\varepsilon}{2}$ for each $f \in \mathscr{F}$. Therefore, if $f \in \mathscr{F} \setminus \mathscr{T}$, then

$$\sup_{u \in U} f(u) < f(x) + \tfrac{\varepsilon}{2} \leq \sum(\mathscr{F} \setminus \mathscr{T})(x) + \tfrac{\varepsilon}{2} < \varepsilon.$$

□

Let us recall Definition VI.2.5 that a family $\mathcal{A}$ of subsets of a topological space on X is *locally finite* if for each $x \in X$, there exists $O \in \mathcal{O}(x)$ such that $\{A \in \mathcal{A} : A \# O\}$ is finite; *discrete* if for each $x \in X$, there exists $O \in \mathcal{O}(x)$ such that $\operatorname{card}\{A \in \mathcal{A} : A \# O\} \leq 1$.

These notions are adapted to a family of continuous functions $\mathscr{F}$ with the aid of $\partial \mathscr{F}$, the family of supports of $\mathscr{F}$, namely $\mathscr{F}$ is *locally finite* (respectively, *discrete*) whenever for each $x \in X$, there exists $O \in \mathcal{O}(x)$ such that

$$\{f \in \mathscr{F} : \sup_O f > 0\}$$

is finite (respectively, is at most a singleton).

A family of subsets of a topological space is called *σ-locally finite* (respectively, *σ-discrete*) if it is a countable union of locally finite (respectively, discrete) families.

PROPOSITION XI.2.11. *A functional cover $\mathscr{F}$ with limited $\sum \mathscr{F}$ is equicontinuous if and only if $\{\max(0, f-\varepsilon) : f \in \mathscr{F}\}$ is locally finite for each $\varepsilon > 0$.*

PROOF. As $\sum \mathscr{F}$ is limited, $\mathscr{F}$ is equicontinuous if and only if it is strongly equicontinuous by Proposition XI.2.10, that is, for each $x \in X$ and every $\varepsilon > 0$, there exist $U \in \mathcal{N}_X(x)$ and a finite subset $\mathscr{T}$ of $\mathscr{F}$ such that $\sum(\mathscr{F} \setminus \mathscr{T})(u) < \varepsilon$ for each $u \in U$. Equivalently, $f(u) - \varepsilon < 0$ for each $f \in \mathscr{F} \setminus \mathscr{T}$ and $u \in U$, or else $\{\max(0, f-\varepsilon) : f \in \mathscr{F}\}$ is locally finite. $\square$

PROPOSITION XI.2.12. *A functional cover $\mathscr{F}$ with limited (sum) is equicontinuous if and only if $\{\max(0, f-h) : f \in \mathscr{F}\}$ is locally finite for each strictly positive $h \in \mathscr{C}_+$.*

PROOF. If $\mathscr{F}$ is equicontinuous and $\sum \mathscr{F}$ is limited, then $\mathscr{F}$ is strongly equicontinuous, and thus, for each $x \in X$, there exist $U \in \mathcal{N}_X(x)$ and a finite subset $\mathscr{T}$ of $\mathscr{F}$ such that $\sum(\mathscr{F} \setminus \mathscr{T})(u) < \frac{1}{2}h(x)$ for each $u \in U$. On the other hand, there exists $W \in \mathcal{N}_X(x)$ such that $W \subset U$, and such that $h(w) > \frac{1}{2}h(x)$ for each $w \in W$. Consequently, $f(w) - h(w) < 0$ for each $w \in W$ and $f \in \mathscr{F} \setminus \mathscr{T}$. Proposition XI.2.11 implies the converse. $\square$

LEMMA XI.2.13. *Each equicontinuous functional cover, that is a partition of a limited function, has a locally finite, regular, refining functional cover.*

PROOF. Let $\mathscr{F}$ be as required, and let $h(x) := \sup\{f(x) : f \in \mathscr{F}\} \leq \sum \mathscr{F}(x)$ for each $x \in X$. By Proposition X.2.6, h is continuous and strictly positive. Let $g_f(x) := \max(0, f(x) - \frac{1}{2}h(x))$. For each $x \in X$ there is $f \in \mathscr{F}$ such that $f(x) > \frac{1}{2}h(x)$, hence $g_f(x) > 0$. Therefore, $\{g_f : f \in \mathscr{F}\}$ is a functional cover, and $\mathrm{cl}_X\{g_f > 0\} = \{f \geq \frac{1}{2}h\} \subset \{f > 0\}$, that is regular and refining $\mathscr{F}$. By Proposition XI.2.12, $\{g_f : f \in \mathscr{F}\}$ is locally finite (hence $g := \sum_{f \in \mathscr{F}} g_f$ is limited and continuous). $\square$

As g is continuous and strictly positive, $\{\frac{g_f}{g} : f \in \mathscr{F}\}$ is a partition of unity. Therefore,

COROLLARY XI.2.14. *Each equicontinuous functional cover, that is a partition of a limited function, has a refining partition of unity.*

PROPOSITION XI.2.15. *Each partition of unity has a locally finite regular refining partition of unity.*

PROOF. If $\mathscr{F}$ is a partition of unity, then, by Theorem XI.2.8, $\mathscr{F}$ is equicontinuous. Then, by Proposition XI.2.11, $\{\max(0, f - \frac{1}{2}) : f \in \mathscr{F}\}$ is locally finite, and $\{f - \frac{1}{2} > 0\} \subset \mathrm{cl}_X\{f - \frac{1}{2} > 0\} \subset \{f > 0\}$. Hence $\{2 \cdot \max(0, f - \frac{1}{2}) : f \in \mathscr{F}\}$ is a regular refinement of $\mathscr{F}$, and a partition of unity. □

XI.3. Paracompactness

DEFINITION XI.3.1 (Paracompact topology). A topology is called *paracompact* if every open cover has a locally finite refining cover.

Obviously, every compact topology is paracompact. By Proposition IX.3.10,

PROPOSITION XI.3.2. *Each free paracompact countably compact topology is compact.*

LEMMA XI.3.3. *If F_0, F_1 are closed subsets of a paracompact topological space X such that $\mathcal{N}(x) \perp \mathcal{N}(F_1)$ for each $x \in F_0$, then $\mathcal{N}(F_0) \perp \mathcal{N}(F_1)$.*

PROOF. By hypothesis, for each $x \in F_0$, there exist $U_x \in \mathcal{N}(x)$ and $V_x \in \mathcal{N}(F_1)$ with $U_x \cap V_x = \varnothing$. Let $\mathcal{W}$ be a locally finite refining cover of $\{U_x : x \in F_0\} \cup \{X \setminus F_0\}$. Consider a family $\mathcal{W}_0$ consisting of those $W \in \mathcal{W}$, for which $W \# F_0$. In particular, for each $W \in \mathcal{W}_0$, there exists $x \in F_0$ such that $W \subset U_x$, hence $\mathrm{cl}\, W \cap F_1 = \varnothing$. Therefore,

$$F_0 \subset \bigcup\nolimits_{W \in \mathcal{W}_0} W \subset \mathrm{cl}(\bigcup\nolimits_{W \in \mathcal{W}_0} W) = \bigcup\nolimits_{W \in \mathcal{W}_0} \mathrm{cl}\, W,$$

by virtue of Proposition VI.2.6. Hence $\bigcup_{W \in \mathcal{W}_0} W \in \mathcal{N}(F_0)$, and $X \setminus \bigcup_{W \in \mathcal{W}_0} \mathrm{cl}\, W \in \mathcal{N}(F_1)$. □

COROLLARY XI.3.4. *Each Hausdorff paracompact topology is regular.*

PROOF. Let F_0 be a closed set and $x_1 \notin F_0$. Apply Lemma XI.3.3 to F_0 and $F_1 := \{x_1\}$. □

COROLLARY XI.3.5. *Each Hausdorff paracompact topology is normal.*

PROOF. Let F_0, F_1 be disjoint closed sets. By Corollary XI.3.4, $\mathcal{N}(x) \perp \mathcal{N}(F_1)$ for each $x \in F_0$. Apply Lemma XI.3.3 to conclude that $\mathcal{N}(F_0)$ and $\mathcal{N}(F_1)$ do not mesh. □

PROPOSITION XI.3.6. *For a Hausdorff topological space X, the following conditions are equivalent:*

(1) *X is paracompact,*
(2) *for each open cover of X, there exists a refining locally finite partition of unity,*
(3) *for each open cover of X, there exists a refining partition of unity.*

PROOF. (1) $\Rightarrow$ (2). Let X be Hausdorff and paracompact, and let $\mathcal{W}$ be an open cover of X. Since, by Corollary XI.3.4, X is regular, for every $x \in W$ there exists $V_x \in \mathcal{O}(x)$ such that $\operatorname{cl} V_x \subset W$. Therefore, $\mathcal{V} := \{V_x : x \in W, W \in \mathcal{W}\}$ is an open cover, such that $\{\operatorname{cl} V : V \in \mathscr{V}\}$ refines $\mathcal{W}$. As X is paracompact, there exists an open locally finite cover $\mathcal{D}$ refining $\mathcal{V}$. Consequently, for each $D \in \mathcal{D}$, there exists $V \in \mathcal{V}$ and $W \in \mathcal{W}$ so that $\operatorname{cl} D \subset \operatorname{cl} V \subset W$. Now, by Theorem X.7.8 of Urysohn, there exists $f_D \in \mathscr{C}$ such that $D \subset \{f = 0\}$ and $\{f < 1\} \subset W$. As a result, $\{f_D : D \in \mathcal{D}\}$ is a locally finite functional refinement of $\mathcal{W}$, hence

$$\{\frac{f_D}{\sum_{B \in \mathcal{D}} f_B} : D \in \mathcal{D}\}$$

is a locally finite partition of unity, refining $\mathcal{W}$. It is obvious that (2) $\Rightarrow$ (3).

Suppose that (3) holds and let $\mathcal{W}$ be an open cover of X. Then there exists a partition of unity $\mathscr{F}$ refining $\mathcal{W}$. In particular, $\mathscr{F}$ is equicontinuous, hence by Proposition XI.2.15, has a locally finite partition of unity refining $\mathscr{F}$, hence refining $\mathcal{W}$. □

LEMMA XI.3.7. *For each equicontinuous functional cover bounded by $c > 0$ of a topological space, there exists a refining equicontinuous partition of a function bounded by c.*

PROOF. * *This proof necessitates some concepts off our prerequisites. We shall describe them briefly here, leaving precise definitions and observations to Chapter XXVII:*

An ordered set $(S, \leq)$ is said to be *well-ordered* if each non-empty subset has a least element. Each well-ordered set is linearly ordered ([3]). We shall use below the fact that *each set can be well-ordered* ([4]).

Let $\mathscr{F}$ be an equicontinuous functional cover of X, bounded by $c > 0$. Let us well order $\mathscr{F} = \{f_s : s \in S\}$. Let

$$F_s(x) := \max(0, f_s(x) - \sup_{t<s} f_t(x)).$$

[3]Because a subset $\{s_0, s_1\}$, either s_0 or s_1 is the least element, hence either $s_0 \leq s_1$ or $s_1 \leq s_0$.

[4]This is the wording of the *Zermelo Theorem*, which is equivalent to the *Axiom of Choice*, discussed in Chapter XXVII.

By Lemma X.2.7, $\{\sup_{t<s} f_t : s \in S\}$ is equicontinuous, hence $\{F_s : s \in S\}$ is equicontinuous. We notice that $t < s$ implies that

$$F_s(x) \leq f_s(x) - f_t(x) \tag{XI.3.1}$$

for each $x \in X$. If $s(x)$ is a least $s \in S$ such that $f_s(x) > 0$, then $F_{s(x)}(x) = f_{s(x)}(x)$, and thus $\sum_{s\in S} F_s(x) > 0$. If $T = \{t_0, t_1, \ldots, t_n\}$ is a strictly increasing finite sequence, then, by (XI.3.1),

$$\begin{aligned}\sum_{t\in T} F_t(x) &\leq f_{t_0}(x) + (f_{t_1}(x) - f_{t_0}(x)) + \cdots + (f_{t_n}(x) - f_{t_{n-1}}(x)) \\ &= f_{t_n}(x) \leq c.\end{aligned}$$

Hence, by definition (sum), $F(x) := \sum_{s\in S} F_s(x) \leq c$, so that $\{F_s : s \in S\}$ is refining $\mathscr{F}$, and is an equicontinuous partition of a bounded function. □

THEOREM XI.3.8 (Stone). *Each metrizable topology is paracompact.*

PROOF. ([5]) If d is a compatible metric on X bounded by 1, and if $\mathcal{P}$ is an open cover of X, then, by (X.3.4), $\mathscr{P} = \{d(\cdot, X\backslash P) : P \in \mathcal{P}\}$ is an equicontinuous functional cover bounded by 1. Sure enough, $\mathscr{P}$ refines $\mathcal{P}$, because $\partial\mathscr{P} = \mathcal{P}$ (in fact, $\{d(\cdot, X \setminus P) > 0\} = P$ for each $P \in \mathcal{P}$). By Lemma XI.3.7, there exists a refining equicontinuous functional cover that is a partition of a function bounded by 1. Hence, by Lemma XI.2.13, there exists a refining locally finite (functional) cover. □

COROLLARY XI.3.9. *Each metrizable topology has a σ-locally finite base.*

PROOF. For a compatible metric on X, the family $\mathcal{B}_n := \{B(x, 2^{-n}) : x \in X\}$ (balls with respect to that metric) is an open cover, hence, by Theorem XI.3.8 of Stone, there exists a locally finite refinement $\mathcal{D}_n$. As $\bigcup_{n<\omega} \mathcal{B}_n$ is a base of the topology, $\bigcup_{n<\omega} \mathcal{D}_n$ is a σ-locally finite base refining $\bigcup_{n<\omega} \mathcal{B}_n$. □

PROPOSITION XI.3.10. *Each regular Lindelöf topology is paracompact.*

PROOF. Let $\mathcal{P}$ be an open cover of a regular Lindelöf topological space X. By regularity, for each $x \in X$, there exist $V_x \in \mathcal{O}(x)$ and $P_x \in \mathcal{P}$ such that $\operatorname{cl} V_x \subset P_x$. In other words, $\{V_x : x \in X\}$ is an open

[5]This famous theorem was proved by A. H. STONE in [244], but the proof was quite long and difficult. A short proof was published by M.E. RUDIN [226] but was very technical, based on involved calculations. The proof presented above is due to J. DYDAK [95].

refinement of $\mathcal{P}$. It follows from the Lindelöf property that there exists a sequence $(x_n)_n$ of elements of X such that $\{V_{x_n} : n < \omega\}$ is a cover. If we set

$$Q_n := P_{x_n} \setminus \mathrm{cl}(V_{x_0} \cup V_{x_1} \cup \cdots \cup V_{x_{n-1}}),$$

then $\{Q_n : n < \omega\}$ consists of open sets. For each $x \in X$, there is a least $n < \omega$ such that $x \in P_{x_n}$, and thus $x \in Q_n$, but $x \notin Q_{n+k}$ for each $k < \omega$. Consequently, $\{Q_n : n < \omega\}$ is a locally finite refinement of $\mathcal{P}$. □

In virtue of Corollary XI.3.5,

COROLLARY XI.3.11. *Each Hausdorff regular Lindelöf topology is normal.*

EXAMPLE XI.3.12. *The Sorgenfrey line topology* is Lindelöf, hence paracompact, because it is regular by Example X.7.20. If $\mathcal{P}$ is a ς-open cover of $\mathbb{R}$, then there is a countable set $\{x_n : n < \omega\}$ such that $\mathbb{R} = \{x_n : n < \omega\} \cup \bigcup_{P \in \mathcal{P}} \mathrm{int}_\nu P$. As ν has countable weight, by Lemma IX.2.23, there exists a countable subfamily $\mathcal{P}_0$ of $\mathcal{P}$ such that $\bigcup_{P \in \mathcal{P}_0} \mathrm{int}_\nu P = \bigcup_{P \in \mathcal{P}} \mathrm{int}_\nu P$. For each $n < \omega$ such that $x_n \notin \bigcup_{P \in \mathcal{P}_0} P$, let $P_n \in \mathcal{P}$ be such that $x_n \in P_n$. Then $\mathcal{P}_0 \cup \{P_n : n < \omega\}$ is a countable ς-open cover of $\mathbb{R}$.

REMARK XI.3.13. In Example X.7.22 we noticed that the Sorgenfrey plane ζ^2 is not normal, although the Sorgenfrey line ζ is normal. By Proposition IX.1.13, ζ^2 is regular. Also ζ^2 is not Lindelöf, because of Corollary XI.3.11. Finally, by Example XI.3.12, the product of two Lindelöf topologies need not to be Lindelöf.

XI.4. Metrization theorems

The class of metrizable topologies is an earliest abstract class of topologies studied in mathematics. Metric spaces appeared at the beginning of the twentieth century ([6]), as an abstraction of Euclidean spaces, which marked the ninetieth century, arising as a generalization of the real line.

In contrast to concrete objects like the standard topology of the real line, metric spaces are defined abstractly by a set of axioms.

One of the central themes of research in topology at its early axiomatic period, was a *metrizability quest*:

PROBLEM. How can one recognize that a given topology is metrizable?

[6] The PhD thesis of MAURICE FRÉCHET [103] was a foundational work introducing metric spaces.

In other terms, what are the topological properties that characterize metrizable topologies? For example, we know that each metrizable topology is of countable character, perfectly normal, and paracompact. Is the converse true? The answer is no, as we can infer from Examples IX.2.19, X.7.20, and XI.3.12 of the *Sorgenfrey line*.

This problem has a long history of solutions, which are sometimes formulated in rather complex terms ([7]). For the class of free topologies of countable weight, regularity characterizes metrizability by Urysohn Metrization Theorem X.9.14.

I am going to present first a most recent result, the *Dydak theorem*, because it has a clear proof, and other classical theorems can be lucidly deduced from it.

As we shall see, the *Dydak theorem* is not entirely a topological characterization, because it uses partition of unity, which is a functional concept.

THEOREM XI.4.1 (Dydak). *A Hausdorff topology is metrizable if and only if it has a partition $\mathscr{F}$ of unity, the support $\partial\mathscr{F}$ of which is a base for the open sets.*

PROOF. If a topology on X is metrizable, then fix any compatible metric on X, and let $\mathcal{B}_n := \{B(x, 2^{-n}) : x \in X\}$ for each $n \in \mathbb{N}_1$. Then $\mathcal{B} := \bigcup_{n \in \mathbb{N}_1} \mathcal{B}_n$ is a base for the open sets. By Proposition XI.3.6, for each $n \in \mathbb{N}_1$, there exists a partition of unity $\{f_{n,x} : x \in X\}$ such that $\{f_{n,x} > 0\} \subset B(x, 2^{-n})$ for each $x \in X$. Accordingly,

$$\{2^{-n} f_{n,x} : x \in X, n \in \mathbb{N}_1\}$$

is a partition of unity, the support of which is a base (refining $\mathcal{B}$).

Conversely, let $\mathscr{F}$ be a partition of unity of X such that $\{\{f > 0\} : f \in \mathscr{F}\}$ is a base of the topology τ on X. Then

$$d(x, y) = \sum_{f \in \mathscr{F}} |f(x) - f(y)| \qquad \text{(XI.4.1)}$$

is a metric (Exercise XI.5.3) that is compatible with τ. Indeed, for each $x \in X$, the map $d(x, \cdot) \in \mathscr{C}(\tau, \nu)$ by Theorem XI.2.8, because $\mathscr{F}$ is equicontinuous as a partition of a continuous function. Therefore, $\tau \geq \mathrm{t}d$. To see that $\mathrm{t}d \geq \tau$, let $O \in \mathcal{O}_\tau(x)$. As $\partial\mathscr{F}$ is a base for the open sets of τ, there exists $f \in \mathscr{F}$ such that $x \in \{f > 0\} \subset O$. Notice that if $y \notin O$, then $f(y) = 0$, hence $d(x, y) \geq |f(x) - f(y)| = f(x) > 0$. Therefore, the $\mathrm{t}d$-open set $B_d(x, f(x)) \subset O$, and thus $\mathrm{t}d \geq \tau$. □

[7] I report after R. ENGELKING [97, p. 334] a list of papers containing principal metrization theorems: P. ALEXANDROFF and P. URYSOHN [3], R. L. Moore [188], A. H. STONE [244], R. H. BING [28], K. MORITA [189], A. V. ARHANGEL'SKII [8], [18], P. ALEXANDROFF [2], A. H. STONE [246].

THEOREM XI.4.2 (Nagata-Smirnov). *A topology is metrizable if and only if it is Hausdorff regular and has a σ-locally finite base for the open sets.*

PROOF. The necessity follows from Corollary XI.3.9. Conversely, suppose that a Hausdorff regular topology on X has a base for the open sets $\mathcal{B} = \bigcup_{n<\omega} \mathcal{B}_n$, where $\mathcal{B}_n$ is a locally finite cover of X ([8]). By Lemma X.7.18, the topology is perfectly normal, so that for each $B \in \mathcal{B}$ there exists $f_B \in \mathscr{C}_+$ such that $B = \{f_B > 0\}$. For each $n > 0$, the (locally finite) family $\{f_B : B \in \mathcal{B}_n\}$ is a partition of a continuous function. By Corollary XI.2.14, for each $n < \omega$, there exists a partition of unity $\mathscr{F}_n$ refining $\mathcal{B}_n$, hence $\bigcup_{n<\omega}\{2^{-n-1}f : f \in \mathscr{F}_n\}$ is a partition of unity refining $\mathcal{B}$, so that the topology is metrizable by Theorem XI.4.1 of Dydak. □

THEOREM XI.4.3 (Bing). *A topology is metrizable if and only if it is Hausdorff regular and has a σ-discrete base for the open sets.*

Since each σ-discrete family is σ-locally finite, the sufficiency follows from Theorem XI.4.2 of Nagata and Smirnov. The necessity follows from the following ([9])

THEOREM XI.4.4 (Stone theorem). *Every open cover of a metrizable space has a refining σ-discrete, locally finite open cover.*

PROOF. (*) Let $\mathcal{P}$ be an open cover of (X, d), and let $B(A, r) := \{x \in X : \operatorname{dist}(x, A) < r\}$ be a ball of radius $r > 0$ around a set A. Let $\mathcal{P} = \{P_\alpha : \alpha < \delta\}$ be well-ordered. We define for $n < \omega$, a family $\mathcal{V}_n = \{V_{\alpha,n} : \alpha < \delta\}$ of open sets. Let

$$V_{\alpha,n} := B(A_{\alpha,n}, 2^{-n}),$$

where $A_{\alpha,n}$ consists of x such that α is the least ordinal for which $x \in P_\alpha$, and $B(x, 3 \cdot 2^{-n}) \subset P_\alpha$, but $x \notin V_{\gamma,m}$ for each $\gamma < \delta$ and $0 \le m < n$.

It follows that $V_{\alpha,n}$ is an open set included in P_α for each $n < \omega$ and every $\alpha < \delta$. The family $\mathcal{V} := \bigcup_{n<\omega} \mathcal{V}_n$ is an open refinement of $\mathcal{P}$. We shall see that $\mathcal{V}$ is a cover of X. Indeed, if $x \in X$, then take the least $\alpha < \delta$ such that $x \in P_\alpha$ and $n < \omega$ such that $B(x, 3 \cdot 2^{-n}) \subset P_\alpha$. Then either $x \in V_{\alpha,n}$ or there exists $m < n$ and $\gamma < \delta$ such that $x \in V_{\gamma,m}$.

---

[8]If a locally finite $\mathcal{B}_n$ were not a cover, we adjoin to it the whole space X.

[9]The proof of the Stone theorem (A. H. STONE [245]) follows that of R. ENGELKING [97, p. 280]. A proof following Dydak's approach can be found in [83, p. 493].

We shall prove that $d(x_\alpha, x_\gamma) > 2^{-n}$ for $x_\alpha \in V_{\alpha,n}$ and $x_\gamma \in V_{\gamma,n}$ each $n < \omega$ and for $\alpha \neq \gamma$, showing that $\mathcal{V}_n$ is a discrete family for each $n < \omega$.

In fact, if $\gamma > \alpha$ then there exist $x \in A_{\alpha,n}$ and $w \in A_{\gamma,n}$ such that $x_\alpha \in B(x, 2^{-n}) \subset V_{\alpha,n}$ and $x_\gamma \in B(w, 2^{-n}) \subset V_{\gamma,n}$. As $B(x, 3 \cdot 2^{-n}) \subset P_\alpha$, and $w \notin P_\alpha$, we infer that $d(x, w) \geq 3 \cdot 2^{-n}$, and thus

$$d(x_\alpha, x_\gamma) \geq d(x, w) - d(x_\alpha, x) - d(x_\gamma, w) > 2^{-n}.$$

To prove that $\mathcal{V}$ is locally finite, observe that for $\alpha < \delta$ and $k, j < \omega$, if $B(w, 2^{-k}) \subset V_{\alpha,j}$, then $B(w, 2^{-j-k}) \cap V_{\gamma,n} = \varnothing$ for $n \geq j + k$ and $\gamma \in A$. Indeed, from the definition of $x \in A_{\gamma,n}$, it follows that $x \notin V_{\alpha,j}$ whenever $n \geq j + k$, and as $B(w, 2^{-k}) \subset V_{\alpha,j}$, we infer that $d(x, w) > 2^{-k}$ for each such x. Hence $j + k \geq k + 1$ and $n \geq k + 1$ entail $B(w, 2^{-j-k}) \cap B(x, 2^{-n}) = \varnothing$. Consequently, $B(w, 2^{-j-k})$ meets at most $j + k - 1$ members of $\mathcal{V}$. □

XI.5. Supplement

EXERCISE XI.5.1. If F_0, F_1 be disjoint closed subsets of a metrizable space X, then the function

$$f(x) := \frac{d(x, F_0)}{d(x, F_0) + d(x, F_1)}$$

is continuous, $f(F_0) = \{0\}$ and $f(F_1) = \{1\}$.

SOLUTION. The function $d(x, F) = 0$ if and only if $x \in \operatorname{cl} F$. As F_0, F_1 are disjoint, either $x \notin F_0$ or $x \notin F_1$, hence $d(x, F_0) + d(x, F_1) > 0$ for each $x \in X$. By Corollary X.3.6, the distance function f is continuous, hence the quotient is continuous. It is clear that $f(x) = 0$ whenever $x \in F_0$, and $f(x) = 1$ whenever $x \in F_1$.

EXERCISE XI.5.2. Show that a metrizable topology of countable density has countable weight.

SOLUTION. Indeed, if d is a compatible metric, and A is a countable dense set, then

$$\{B(a, 2^{-n}) : a \in A, n < \omega\}$$

is a countable base for the open sets. Indeed, if $O \in \mathcal{O}(x)$, then there is n such that $x \in B(x, 2^{-n}) \subset O$. As A is dense, there is $a \in A$ such that $d(a, x) < 2^{-n+1}$. Hence $x \in B(a, 2^{-n+1}) \subset O$.

We shall see that each compatible metric is jointly continuous.

EXERCISE XI.5.3. Check that (XI.4.1) is a metric.

SOLUTION. Symmetry (III.6.1) is evident. If $d(x_0, x_1) = 0$ then $f(x_0) = f(x_1) = 0$ for each $f \in \mathscr{F}$, hence $x_0 = x_1$, because otherwise, by assumption, there would be $f \in \mathscr{F}$ such that $f(x_0) > 0$ and $f(x_1) = 0$. Thus (III.6.2) holds. (III.6.3) follows from

$$\begin{aligned} d(x,z) &= \sum\nolimits_{f\in\mathscr{F}} |f(x) - f(z)| \le \sum\nolimits_{f\in\mathscr{F}} (|f(x) - f(y)| + |f(y) - f(z)|) \\ &= \sum\nolimits_{f\in\mathscr{F}} |f(x) - f(y)| + \sum\nolimits_{f\in\mathscr{F}} |f(y) - f(z)| = d(x,y) + d(y,z). \end{aligned}$$

PROPOSITION XI.5.4. *A Hausdorff compact topology τ on X is metrizable if and only if the diagonal $\Delta(X)$ is $G_{\aleph_0}$ in $\tau \times \tau$.*

PROOF. ([10]) If τ is a metrizable topology on X, then so is $\tau \times \tau$, and hence is perfectly normal. Therefore every closed subset is $G_{\aleph_0}$, and in particular $\Delta(X) = \{(x,x) : x \in X\}$.

Assume that $G_n \in \mathcal{O}_{\tau\times\tau}$ for each $n < \omega$ and that $\Delta(X) = \bigcap_{n<\omega} G_n$. Being Hausdorff and compact, $\tau\times\tau$ is normal by Proposition IX.3.4 and the diagonal $\Delta(X)$ is $\tau \times \tau$-closed. Therefore, there exists a sequence $(P_n)_n \subset \mathcal{O}_{\tau\times\tau}$ such that $\Delta(X) \subset P_{n+1} \subset \mathrm{cl}_{\tau\times\tau} P_{n+1} \subset P_n \cap G_n$ for each $n < \omega$, so that $\bigcap_{n<\omega} P_n = \Delta(X)$.

Note that for every $x \in X$ and for each $O \in \mathcal{O}_\tau(x)$, there is $n < \omega$ such that $P_n x \subset O$. If, by way of contradiction, $x_n \in P_n x \setminus O$, that is, $(x, x_n) \in P_n$ for each $n < \omega$, and thus $x \in \lim_\tau (x_n)_n$.

Of course, for each $n < \omega$ and every $x \in X$, there is $B_{n,x} \in \mathcal{O}_\tau(x)$ such that $B_{n,x} \times B_{n,x} \subset P_n$. By compactness, for each $n < \omega$, there is a finite subset F_n of X such that $\mathcal{B}_n := \{B_{n,x} : x \in F_n\}$ is a τ-cover of X. Consequently, $\mathcal{B} := \bigcup_{n<\omega} \mathcal{B}_n$ is countable. To see that $\mathcal{B}$ is a base, let $O \in \mathcal{O}_\tau(x)$. Then there is $n < \omega$ and $w \in F_n$ so that $x \in B_{n,w} \subset P_n x \subset O$.

Apply Theorem XI.4.2 of Nagata and Smirnov to infer that τ is metrizable. □

LEMMA XI.5.5. *If ξ is a metrizable topology and $f \in \mathscr{C}(\xi, \tau)$ is perfect, then every τ-open cover has a τ-locally finite refining τ-open cover.*

PROOF. If $\mathcal{P}$ is a τ-open cover, then $f^-[\mathcal{P}]$ is a ξ-open cover. By Theorem XI.3.8 of Stone, there exists a ξ-locally finite ξ-open covering refinement $\{R_j : j \in J\}$ of $f^-[\mathcal{P}]$ which is a ξ-cover. In virtue of Theorem IX.5.16, there exists a ξ-open cover $\{Q_j : j \in J\}$ such that $\mathrm{cl}_\xi Q_j \subset R_j$ for each $j \in J$. Then by Proposition IX.4.11,

[10] Due to V. E. SCHNEIDER [233]. The argument used is from the blog of STEVEN CLONTZ (9.02.2014), University of Southern Alabama.

$\{f(\mathrm{cl}_\xi Q_j) : j \in J\}$ is a τ-locally finite, τ-closed cover which is a refinement of $\mathcal{P}$. □

THEOREM XI.5.6. *Metrizability is preserved under continuous perfect maps.*

PROOF. ([11]) Let ξ a metrizable topology on X, τ be a topology on Y, and $f \in \mathscr{C}(\xi, \tau)$ be perfect. We will use Lemma XI.5.5 to construct a σ-locally finite base for the open sets of τ. We shall consider balls with respect to a fixed metric d compatible with ξ. For each $y \in Y$ and every $n < \omega$, let

$$U_n(y) := B(f^-(y), 2^{-n}), \text{ and } W_n(y) := Y \setminus f(X \setminus U_n(y)).$$

Accordingly, $f^-(y) \subset f^-(W_n(y)) \subset U_n(y)$ for each $y \in Y$. Notice that for each y and t in Y,

$$f^-(t) \subset U_n(y) \Longrightarrow t \in W_n(y). \tag{XI.5.1}$$

For each $n < \omega$, the family $\mathcal{W}_n := \{W_n(y) : y \in Y\}$ is obviously a τ-open cover. On the other hand, for each $y \in Y$, the family $\{W_n(y) : n < \omega\}$ is a filter-base of $\mathcal{N}_\tau(y)$. Indeed, if $O \in \mathcal{O}_\tau(y)$, then $f^-(O) \supset f^-(y)$, and as $f^-(y)$ is ξ-compact, by Proposition X.3.7, there is $n < \omega$ such that $U_n(y) \subset f^-(O)$, and thus $W_n(y) \subset f(U_n(y)) \subset O$.

We shall prove that for every $y \in Y$ and each $n < \omega$, there exists $k < \omega$ such that for each $z \in Y$,

$$y \in W_k(z) \Longrightarrow W_k(z) \subset W_n(y). \tag{XI.5.2}$$

Indeed, by Proposition X.3.7, there is $k \geq n+1$ such that $U_k(y) \subset f^-(W_{n+1}(y))$.

If $y \in W_k(z)$, then $f^-(y) \subset f^-(W_k(z)) \subset U_k(z) = B(f^-(z), 2^{-k})$. In particular, $\varnothing \neq f^-(z) \cap U_k(y) \subset f^-(z) \cap f^-(W_{n+1}(y))$, hence $f^-(z) \subset \bigcup_{w \in W_{n+1}(y)} f^-(w) = f^-(W_{n+1}(y)) \subset U_{n+1}(y)$ ([12]).

Let $t \in W_k(z)$, then $f^-(t) \subset f^-(W_k(z)) \subset U_k(z) = B(f^-(z), 2^{-k})$. Hence for every $x_0 \in f^-(t)$, there exists $x_1 \in f^-(z)$ such that $d(x_0, x_1) < 2^{-k} < 2^{-n-1}$, and there exists $x_2 \in f^-(y)$ such that $d(x_1, x_2) < 2^{-n-1}$, and thus $f^-(t) \subset U_n(y)$, hence by (XI.5.1), $t \in W_n(y)$.

By Lemma XI.5.5, for each $n < \omega$, the τ-open cover $\mathcal{W}_n$ has a locally finite refinement $\mathcal{B}_n$ which is a τ-cover, so that $\bigcup_{n<\omega} \mathcal{B}_n$ is a σ-locally finite family, which is a base for the open sets of τ. Indeed, if $O \in \mathcal{O}_\tau$ and $y \in O$, then there exists $n < \omega$ such that $W_n(y) \subset O$. Thus there exists $k < \omega$ such that (XI.5.2) holds. Let $B \in \mathcal{B}_k$ be such

[11]K. MORITA and S. HANAI [190], A. H. STONE [245] (see R. ENGELKING [97]).

[12]Because if $f^-(z) \# f^-(w)$, then $z = w$.

that $y \in B$. By assumption, there exists $z \in Y$ such that $B \subset W_k(z)$, so that

$$y \in B \subset W_k(z) \subset W_n(y) \subset O.$$

As, by Proposition IX.4.13, τ is Hausdorff and regular, by Theorem XI.4.2 of Nagata and Smirnov, τ is metrizable. $\square$

CHAPTER XII

Compact topologies

Compactness is one of the most important concepts of topology and analysis. In this chapter, we will study compactness in topological context in terms of adherent filters, reviewing several prominent examples of compact topologies. We need to know more about convergences before we will be in a position to revisit compactness in full generality in Chapter XX.

XII.1. A short story of compactness

The significance of compactness prompts me to delineate here ideas that lead to the concept in its present form.

In Section II.1, we considered convergence of sequences on the real line in order to motivate the introduction of filters. Passing from sequences to filters in investigating the notion of compactness is also quite illuminating. It is not surprising that the idea of compactness was first studied in terms of sequences. A modern approach that emerged at the end of the nineteenth century took time before becoming commonplace.

FIGURE XII.1.1. Bernard Bolzano (1781–1848), Karl Weierstraß (1815–1897), Émile Borel (1871–1956), and Henri Lebesgue (1875–1941).

All started on the real line at the very beginning of nineteenth century. B. BOLZANO proved in 1817 ([1]), and K. WEIERSTRASS independently later, that

[1] B. BOLZANO **[30]**.

THEOREM XII.1.1 (Bolzano-Weierstraß). *Each bounded sequence of elements of the real line has a convergent subsequence.*

Although we have already proved a variant of the Bolzano–Weierstraß theorem (Proposition VIII.3.2), we present here a more traditional proof.

PROOF. If $(x_n)_n$ is a bounded sequence, then there exist $a_0 < b_0$ such that $x_n \in [a_0, b_0]$ for each $n < \omega$. Choose any interval $[a_0, \frac{a_0+b_0}{2}]$ or $[\frac{a_0+b_0}{2}, b_0]$ provided that it contains infinitely many elements of $(x_n)_n$, and call it $[a_1, b_1]$. By induction, construct a sequence of intervals

$$[a_0, b_0] \supset [a_1, b_1] \supset \ldots \supset [a_k, b_k]$$

containing infinitely many elements of $(x_n)_n$, and such that $|b_j - a_j| = 2\,|b_{j+1} - a_{j+1}|$ for every $0 \leq j < k$. Sure enough, $[a_{k+1}, b_{k+1}]$ is that half of $[a_k, b_k]$ that contains infinitely many elements of $(x_n)_n$. Let $x_{n_k} \in [a_k, b_k]$ for every $k < \omega$. Then $(a_k)_k$ is an increasing sequence, $(b_k)_k$ is a decreasing sequence, $|b_k - a_k| = 2^k\,|b_0 - a_0|$ tends to 0 as k goes to ∞, and

$$a_k \leq x_{n_k} \leq b_k.$$

Therefore, by Lemma X.1.3, there exists (a unique) x_∞ such that $\lim_{k\to\infty} x_{n_k} = x_\infty$. □

It was in 1895 that E. BOREL proved ([2]) that

THEOREM XII.1.2 (Borel). *Each countable open cover of a bounded, closed interval has a finite subcover.*

PROOF. Let I a bounded, closed interval, and let $\{O_n : n < \omega\}$ be a sequence of open sets such that $\bigcup_{n<\omega} O_n \supset I$. If, by way of contradiction, for each $n < \omega$, there exists $x_n \in I \setminus \bigcup_{m\leq n} O_m$, then by Theorem XII.1.1, there exists a convergent subsequence $(x_{n_k})_k$ of $(x_n)_n$. Let $x_\infty \in \lim_\nu (x_{n_k})_k$. As I is closed, $x_\infty \in I$, but on the other hand, $x_\infty \notin \bigcup_{n<\omega} O_n$, which is a contradiction. □

In fact, Theorem XII.1.2 asserts that bounded, closed intervals are countably compact. In 1903, H. LEBESGUE observed that

THEOREM XII.1.3 (Lebesgue). *Each open cover of a bounded, closed interval has a finite subcover.*

PROOF. Let $\mathcal{P}$ be a family of open sets such that $\bigcup_{P\in\mathcal{P}} P \supset I$. Let $\mathcal{B} = \{B_n : n < \omega\}$ be a base for the open sets of I ([3]). Let N be the

[2] In quite a complicated way, by the way.

[3] For example, $\{I \cap]q - 2^{-k}, q + 2^{-k}[: q \in \mathbb{Q} \cap I, k < \omega\}$.

set of $n < \omega$ for which there exists $P \in \mathcal{P}$ such that $B_n \subset P$. For each $n \in N$, let $P_n \in \mathcal{P}$ be such that $B_n \subset P_n$.

By assumption, for each $x \in I$, there exists $P \in \mathcal{P}$ so that $x \in P$, and since $\mathcal{B}$ is a base, there exists $n < \omega$ such that $x \in B_n \subset P$. Hence $n \in N$ and thus $B_n \subset P_n$. Therefore, $I \subset \bigcup_{n \in N} P_n$. To conclude, apply Theorem XII.1.2 of Borel. □

REMARK XII.1.4. We also could use the fact that the weight of the real line is countable, hence by Lemma IX.2.20, there is a countable subfamily $\mathcal{P}_0$ of $\mathcal{P}$ such that $\bigcup_{P \in \mathcal{P}_0} P \supset I$.

In other words, from each open cover of I, one can extract a countable open cover, that is, I is a *Lindelöf* topological space.

The fact that a subset of the real line is compact if and only if it is closed and bounded is usually called either the *Borel-Lebesgue theorem* or the *Heine-Borel theorem*.

However, already in 1884, G. CANTOR proved that ([4])

THEOREM XII.1.5 (Cantor). *Let $\mathcal{D}$ be a family of subsets of Euclidean space with the property that if $X_1, \ldots, X_n$ are any sets such that $\bigcup_{1 \le k \le n} X_k \in \mathcal{D}$, then there exists $1 \le k \le n$ such that $X_k \in \mathcal{D}$. If A is a bounded set such that $A \in \mathcal{D}$, then there exists a point x_∞ such that each $V \in \mathcal{N}_\nu(x_\infty)$ belongs to $\mathcal{D}$.*

Recall from Definition V.2.3, that the family $\mathcal{D}$ described in Theorem XII.1.5 is called *distributive*. This nomenclature is due to G. PEANO, who reformulated *Cantor theorem* also in terms of ideals in 1887 ([5]).

THEOREM XII.1.6 (Peano). *If A is a bounded subset of Euclidean space E, and $\mathcal{I}$ is an ideal of subsets of E such that $\mathcal{I} \cap \mathcal{N}_\nu(x) \neq \varnothing$ for each $x \in E$, then $A \in \mathcal{I}$.*

Notice that the condition $\mathcal{I} \cap \mathcal{N}_\nu(x) \neq \varnothing$ for each $x \in E$, amounts to

$$\bigcup_{I \in \mathcal{I}} \operatorname{int}_\nu I \supset E,$$

that is, $\mathcal{I}$ is an ν-open cover of the Euclidean space E. Therefore, Theorem XII.1.6 says that

COROLLARY XII.1.7. *If A is a bounded subset of Euclidean space, then $A \in \mathcal{I}$ for each ideal openly based cover $\mathcal{I}$ of E.*

[4]See G. CANTOR [36]. I have here slightly reworded the Cantor theorem.

[5]See G. CANTOR [36], G. PEANO [215], and [71].

Of course, if $\mathcal{P}$ is an open cover of A, and A is closed, then $\mathcal{P} \cup \{E \setminus A\}$ is an open cover of E. The ideal $(\mathcal{P} \cup \{E \setminus A\})^{\cup\downarrow}$ generated is a cover of E. Hence, by Corollary XII.1.7, there is a finite subfamily $\mathcal{P}_0$ of $\mathcal{P}$ such that $A \subset \bigcup_{P \in \mathcal{P}_0} P \cup E \setminus A$.

COROLLARY XII.1.8. *If A is a bounded, closed subset of Euclidean space, then each open cover of A has a finite subcover.*

FIGURE XII.1.2. Georg Cantor (1845–1918), Giuseppe Peano (1858–1932), Leopold Vietoris (1891–2002), and Henri Cartan (1904–2008).

As we see, the results of CANTOR and PEANO not only precede and surpass those of BOREL and LEBESGUE, but also are formulated in an accomplished, surprisingly contemporary language. To conclude this overview, let us recollect Lemma VIII.2.3, to get a counterpart of Corollary XII.1.7.

PROPOSITION XII.1.9. *If A is a bounded subset of Euclidean space, then for each filter $\mathcal{H}$,*

$$A \in \mathcal{H}^{\#} \Longrightarrow \operatorname{adh}_\nu \mathcal{H} \neq \varnothing.$$

PROOF. By Lemma VIII.2.3, $\mathcal{H}$ is a filter such that $A \cap \operatorname{adh}_\nu \mathcal{H} = \varnothing$ if and only if $\mathcal{H}_c$ is an ideal ν-cover of A. By noting that $A \in \mathcal{H}_c$ amounts to $A^c \in \mathcal{H}$, that is, we get that if $A \in \mathcal{H}^{\#}$ then $\operatorname{adh}_\nu \mathcal{H} \neq \varnothing$. □

COROLLARY XII.1.10. *A subset A of Euclidean space is closed bounded if and only if, for each filter $\mathcal{H}$,*

$$A \in \mathcal{H}^{\#} \Longrightarrow A \cap \operatorname{adh}_\nu \mathcal{H} \neq \varnothing. \qquad \text{(XII.1.1)}$$

PROOF. We easily see that the conditions above are necessary. Indeed, if subset A of Euclidean space is not bounded, then there exists a sequence $(x_n)_n$ of elements of A such that $\|x_n\| > n$, and thus non-adherent. On the other hand, if A is not closed, then there exists an ultrafilter $\mathcal{U}$ on A with $\lim_\nu \mathcal{U} \cap A = \varnothing$, as ν is Hausdorff. □

Although filters were first formally defined by H. CARTAN in 1937 ([6]), L. VIETORIS defined a filter-base and used it in the context of compactness ([7]) already in 1921.

As a purpose of this overview is to merely sketch developments of the concept of compactness, it is far from being exhaustive. Let us resume.

PROPOSITION XII.1.11. *Let A be a subset of a Euclidean space E with its standard topology. A is compact if and only if one of the following equivalent statements holds.*

(1) *If a filter $\mathcal{F}$ meshes A, then* $\operatorname{adh} \mathcal{F} \cap A \neq \varnothing$,
(2) *Each sequence of elements of A has a subsequence convergent to an element of A,*
(3) *Each countable open cover of A has a finite subcover,*
(4) *Each open cover of A has a finite subcover.*

It is instructive and recommended to prove these equivalences (Exercise XII.10.3). In topological spaces, only (1) and (4) are equivalent, and in general convergences spaces, they are all different! In the sequel, we shall detail their interrelations.

THEOREM XII.1.12 (Weierstraß). *Each continuous function on a non-empty closed bounded subset of Euclidean space attains its extrema.*

Surprisingly, extension of the Weierstraß theorem, from closed intervals to closed bounded subsets of Euclidean space, was not obvious for the mathematicians of nineteenth century. For instance, an attempt of G. DARBOUX from 1872 ([8]) comprised imprecise formulation and proof, and unnecessary assumptions. It was G. PEANO who achieved this extension in a masterly way, as a Corollary of Theorem XII.1.6. The following proof is perhaps most elementary.

PROOF OF THEOREM XII.1.12. Let f be a continuous function on a non-empty closed bounded set A. Let $(r_n)_n$ be a strictly decreasing sequence of reals such that $\inf_{n<\omega} r_n = \inf_A f$. Thus, for each $n < \omega$, there exists $x_n \in A$ such that $f(x_n) < r_n$. By Proposition XII.1.11, there exists a sequence $(n_k)_k$ and x_∞ such that $x_\infty \in \lim_\nu (x_{n_k})_k$. If $r > \inf_A f$, then $r > r_n$ for all but possibly finitely many n, so that $\inf_A f = \lim_\nu f(x_{n_k})_k = f(x_\infty) > -\infty$. □

Recall Definition X.5.4 that if d_X and d_Y are metrics on X and Y respectively, then a map $f : X \longrightarrow Y$ is called *uniformly continuous*

[6] H. CARTAN [**38, 37**]; see also G. CHOQUET [**43**].
[7] In German, *der Kranz* in [**259**].
[8] G. DARBOUX [**51**]

if for every $\varepsilon > 0$ there is $\delta > 0$ such that $d_X(x_0, x_1) < \delta$ implies $d_Y(f(x_0), f(x_1)) < \varepsilon$ for each $x_0, x_1 \in X$.

In 1872 EDUARD HEINE published ([9]) the following result.

THEOREM XII.1.13 (Heine). *A continuous map between metric spaces is uniformly continuous provided that the domain is compact.*

PROOF. Let (X, d_X) and (Y, d_Y) be metric spaces, and $f : X \longrightarrow Y$. Suppose that X is compact, f is continuous, but not uniformly continuous. Then there exists $\varepsilon > 0$ such that, for each $n < \omega$, there exist $x_n, w_n \in X$ such that $d_X(x_n, w_n) < 2^{-n}$ and $d_Y(f(x_n), f(w_n)) \geq \varepsilon$. As in metrizable spaces compactness and sequential compactness coincide, there exist $x_\infty \in X$ and an infinite $A \subset \mathbb{N}$ such that $x_\infty \in \lim_{dd_X}(x_n)_{n\in A}$, and there exist $w_\infty \in X$ and an infinite $B \subset A$ such that $w_\infty \in \lim_{dd_X}(w_n)_{n\in B}$. As for each $n \in B$,

$$d_X(x_\infty, w_\infty) \leq d_X(x_\infty, w_n) + d_X(x_n, w_n) + d_X(w_n, w_\infty),$$

we infer that $x_\infty = w_\infty$. Hence $f(x_\infty) = f(w_\infty)$ by the continuity of f, and on the hand, by the continuity of f and d_Y, also $d_Y(f(x_\infty), f(w_\infty)) \geq \varepsilon$, which yields a contradiction. □

XII.2. Compact sets in topological spaces

Recall from Section VIII.3 that if ξ is a convergence on X, and $A \subset X$, then A is *ξ-compact* whenever the restriction $\xi|_A$ is compact. By Proposition VIII.3.1, A is ξ-compact whenever, for each filter $\mathcal{H}$,

$$A \in \mathcal{H}^\# \Longrightarrow A \cap \operatorname{adh}_\xi \mathcal{H} \neq \varnothing. \tag{XII.2.1}$$

DEFINITION XII.2.1 (Compactoid). Let ξ be a topology on X. A subset A of X is said to be *ξ-compactoid* or *relatively ξ-compact* if $\operatorname{adh}_\xi \mathcal{H} \neq \varnothing$ for each filter $\mathcal{H}$ such that $A \in \mathcal{H}^\#$.

Of course, each compactoid closed set is compact. However, the closure of a compactoid set need not be even compactoid.

EXAMPLE XII.2.2. Consider $B := \{(x, y) \in \mathbb{R}^2 : x^2 + y^2 \leq 1\}$ and $B_< := \{(x, y) \in \mathbb{R}^2 : x^2 + y^2 < 1\}$. Let η be a topology on B defined so that $\mathcal{N}_\eta(b) := \mathcal{N}_{\nu^2}(b) \vee B$ if $b \in B_<$, and $\mathcal{N}_\eta(b) := \mathcal{N}_{\nu^2}(b) \vee (B_< \cup \{b\})$ if $b \in B \setminus B_<$. Then $B_<$ is η-compactoid, $\operatorname{cl}_\eta B_< = B$, but B is not η-compactoid, because $\{B_< \cup \{b\} : b \in B \setminus B_<\}$ is an η-cover of B without any finite subcover.

PROPOSITION XII.2.3. *If ξ is a compact topology, and τ is a Hausdorff topology such that $\xi \geq \tau$, then $\xi = \tau$.*

[9] E. HEINE [129].

PROOF. Let $x \in \lim_\tau \mathcal{F}$. As τ is Hausdorff, $\lim_\tau \mathcal{U} = \{x\}$ for every $\mathcal{U} \in \beta\mathcal{F}$. As ξ is compact and $\xi \geq \tau$,

$$\varnothing \neq \mathrm{adh}_\xi \mathcal{U} = \lim_\xi \mathcal{U} \subset \lim_\tau \mathcal{U} = \{x\}$$

for each $\mathcal{U} \in \beta\mathcal{F}$. By (V.4.1), $\mathcal{F} = \bigwedge \beta\mathcal{F}$ so that $x \in \bigcap_{\mathcal{U}\in\beta\mathcal{F}} \lim_\xi \mathcal{U} \subset \lim_\xi \mathcal{F}$ because of (VI.1.4). □

COROLLARY XII.2.4. *If ξ is a compact topology, $f \in \mathscr{C}(\xi, \tau)$ is a bijection, and τ is a Hausdorff topology, then $f \in \mathscr{H}(\xi, \tau)$.*

PROOF. Let F be ξ-closed. Then, by Proposition VIII.3.3, $f(F)$ is τ-compact, hence τ-closed by Proposition VIII.3.16. Hence, $f^{-1} \in \mathscr{C}(\tau, \xi)$ by Proposition VII.3.9. □

By Proposition IX.3.4, each Hausdorff compact topology is normal. Therefore,

COROLLARY XII.2.5. *Each Hausdorff compact topology is functionally regular (alias completely regular).*

Hence, by Corollary X.9.12,

COROLLARY XII.2.6. *Each Hausdorff compact topology is homeomorphic to a closed subset of a Tikhonov cube.*

PROOF. By Corollary XII.2.5, a Hausdorff compact topology is functionally regular, hence, in virtue of Proposition X.9.12, it is embedded in a Tikhonov cube, which is compact and Hausdorff by Proposition IV.6.11 and Theorem VIII.3.4. By Proposition VIII.3.3 and Corollary XII.2.4, it is homeomorphic to its image under the embedding. □

XII.3. Weaker variants of compactness

By Definition VIII.3.10, a topology is countably compact if and only if every countably based filter has non-empty adherence. By Proposition IX.3.1, a topology is *countably compact* if every open countable cover has a finite subcover.

PROPOSITION XII.3.1. *Every continuous real-valued function on a countably compact set attains its extrema.*

Actually, Exercise X.10.1 goes a step further, extending the result above to semicontinuous functions.

Here is an example of a countably compact, non-compact topology.

EXAMPLE XII.3.2 (*). On the well ordered set ω_1 (of countable ordinal numbers), consider the order topology. The topology is not compact, because ω_1 is dense in the Hausdorff compact space $\omega_1 \sqcup \{\omega_1\}$ of Example IX.3.7. To show that it is countably compact, let $\{\alpha_n : n < \omega\}$ be such that $\alpha_n < \alpha_{n+1}$ for each $n < \omega$, that is, an infinite countable set. Then $\alpha := \sup\{\alpha_n : n < \omega\} < \omega_1$ is an accumulation point, and thus ω_1 is countably compact in virtue of Proposition IX.3.9.

We call a topology ξ *functionally compact* if each $f \in \mathscr{C}(\xi, \nu)$ is bounded. Traditionally functionally compact, functionally regular topologies are called *pseudocompact*.

PROPOSITION XII.3.3. *Every countably compact topology is functionally compact (alias pseudocompact).*

PROOF. Let ξ be a countably compact topology on X, and let $f \in \mathscr{C}(\xi, \nu)$. Then $X \subset \bigcup_{n<\omega}\{|f| < n\}$, which is a union of open sets, hence by countable compactness, there is $n < \omega$ such that $|f(x)| < n$ for each $x \in X$. □

LEMMA XII.3.4. *Each Hausdorff functionally compact normal topology is countably compact.*

PROOF. Suppose that a topology τ on X is a not countably compact. By Proposition IX.3.9, there exists a τ-closed discrete subspace $A := \{a_n : n < \omega\}$ of X. The function $f(a_n) := n$ is τ_A-continuous, because τ_A is discrete. As τ is normal, by Theorem X.7.10 of Tietze, there exists $F \in \mathscr{C}(\tau, \nu)$ such that $F_A = f$. Accordingly, F_A is unbounded, hence τ is not functionally compact. □

Like in Proposition VIII.3.3,

PROPOSITION XII.3.5. *If $f \in \mathscr{C}(\xi, \tau)$ and A is countably ξ-compact, then $f(A)$ is countably τ-compact.*

PROOF. Let $\mathcal{H}$ be a countably based filter such that $\mathcal{H} \# f(A)$, equivalently $f^{-}[\mathcal{H}] \# A$, hence $A \# \operatorname{adh}_\xi f^{-}[\mathcal{H}]$ by the countable ξ-compactness of A. As $f \in \mathscr{C}(\xi, \tau)$, by Corollary VIII.1.10, $f(A) \# \operatorname{adh}_\tau \mathcal{H}$. □

REMARK XII.3.6. Unlike for compact topologies, the product of two countably compact topologies need not be countably compact! See Example XX.4.3.

DEFINITION XII.3.7. A topology is called *locally compactoid* (*locally compact*) if each point has a compactoid (compact) neighborhood; *hereditarily locally compact* if every neighborhood filter has a filter-base consisting of compact sets.

Each discrete topology is hereditarily locally compact, and every compact topology is locally compact.

EXAMPLE XII.3.8. The spaces of rational and irrational numbers are not locally compactoid. There exists $\nu_{\mathbb{Q}}$-base of the form $]a,b[\cap\mathbb{Q}$, where $a<b$. None of these sets is $\nu_{\mathbb{Q}}$-compactoid, because if $t\in\,]a,b[\setminus\mathbb{Q}$, then there exists a sequence $(q_n)_n$ on $]a,b[\cap\mathbb{Q}$ such that $\{t\}=\lim_\nu(q_n)_n$, so that $\mathrm{adh}_{\nu_{\mathbb{Q}}}(q_n)_n=\varnothing$. Therefore, no neighborhood filter contains a compactoid set. The case of $\nu_{\mathbb{R}\setminus\mathbb{Q}}$ is analogous.

PROPOSITION XII.3.9. *Every regular locally compactoid topology is hereditarily locally compact.*

PROOF. Indeed, if $V\in\mathcal{N}_\xi(x)$ is ξ-compactoid, then by regularity, there exists a ξ-closed $C\in\mathcal{N}_\xi(x)$ such that $C\subset V$. Then C is ξ-compact, as a closed compactoid set. □

Here is an example of a Hausdorff regular topology that is not locally compact.

EXAMPLE XII.3.10. The *prime cocountable topology* $\pi=\pi[x_\infty,(X)_1]$ on an uncountable set X (from Examples III.4.5 and X.7.24) is regular but not locally compactoid. Indeed, each π-neighborhood V of x_∞ is cocountable (or the whole X), hence uncountable. If $\mathcal{U}\in\beta V$ is free and contains a countable set, then $\lim_\pi\mathcal{U}=\varnothing$.

The discrete topology ι_X is locally compact, and the identity map $i\in\mathscr{C}(\iota_X,\xi)$ for every topology ξ on X, in particular if ξ is not locally compact. Therefore,

COROLLARY XII.3.11. *Local compactness is not preserved under continuous maps.*

We will see later when a map preserves local compactness (Corollary XXI.7.6).

EXAMPLE XII.3.12. Consider a prime pretopology $\gamma=\pi[x_\infty,\mathcal{G}]$ (hence a topology by Proposition VII.3.14) on an infinite countable set X from Example III.4.6, such that $\mathcal{G}$ is countably based, and there is a filter-base $\{G_n:n<\omega\}$ of $\mathcal{G}$ such that $G_n\setminus G_{n+1}$ is infinite for each $n<\omega$. Notice that γ is a Hausdorff regular topology of countable character that is not locally compact. Indeed, if $\mathcal{U}\in\beta(G_n\setminus G_{n+1})$, then $\lim_\gamma\mathcal{U}=\varnothing$.

It is not difficult to see that a Hausdorff locally compact topology is functionally regular (alias completely regular) (Exercise XII.10.8).

A topology is called *σ-compact* if it is a countable union of compact sets.

A topology is called *hemicompact* if there exists an increasing sequence $\{K_n : n < \omega\}$ of compact sets such that for each compact set K, there is $n < \omega$ for which $K \subset K_n$.

Obviously, each hemicompact topology is σ-compact.

PROPOSITION XII.3.13. *A Hausdorff topology of countable weight is hemicompact if and only if it is locally compact.*

PROOF. Let X be a Hausdorff topological space of countable weight. If X is locally compact, then by Theorem IX.2.21, there exists a sequence $\{B_n : n < \omega\}$ of compact subsets of X such that $\{\operatorname{int} B_n : n < \omega\}$ is a base of X. If K is a compact subset of X, then there exists $n < \omega$ such that $K \subset \bigcup_{m \leq n} \operatorname{int} B_m \subset \bigcup_{m \leq n} B_m$. Then $K_n := \bigcup_{m \leq n} B_m$ defines the sequence of compact sets with the required property.

Conversely, let X be hemicompact with an increasing sequence $\{K_n : n < \omega\}$ witnessing hemicompactness. Let x be a point without compact neighborhood, and $\{V_n : n < \omega\}$ be a neighborhood base of x. Then for each $n < \omega$ there exists $x_n \in V_n \setminus K_n$. On one hand, $\{x\} = \lim (x_n)_n$, thus $K := \{x_m : m < \omega\} \cup \{x\}$ is compact, and $x_n \in K \setminus K_n$ for each $n < \omega$, which is a contradiction. □

EXAMPLE XII.3.14. The space $\mathbb{Q}$ of rational numbers (with the standard topology) is σ-compact and not hemicompact. Indeed, $\mathbb{Q}$ is a Hausdorff countable topological space, hence σ-compact. Since $\mathbb{Q}$ is of countable weight (as a subspace of $\mathbb{R}$), and (nowhere) locally compact, it is not hemicompact by Proposition XII.3.13.

XII.4. Sequentially compact topologies

DEFINITION XII.4.1. A topology ξ is said to be *sequentially compact* if every sequence has a ξ-convergent subsequence. If ξ is a topology on X, then a subset A of X is called *sequentially compact* if $\xi|_A$ is a sequentially compact topology.

PROPOSITION XII.4.2. *Each sequentially compact topology is countably compact.*

PROOF. Let ξ be a sequentially compact topology on X, and let $\{H_n : n < \omega\}$ be a descending sequence of non-empty subsets of X. Then, for each $n < \omega$, there is $x_n \in H_n$. By hypothesis, there exists a subsequence $(x_{n_k})_k$ such that $\varnothing \neq \lim_\xi (x_{n_k})_k \subset \bigcap_{n<\omega} \operatorname{cl}_\xi H_n$. □

Therefore, both compactness and sequential compactness imply countable compact. We shall see that that neither compactness entails sequential compactness, nor conversely.

EXAMPLE XII.4.3 (A compact, non-sequentially compact topology). Consider the product $\eta := \prod_{r\in[0,1]} \iota(r)$, where $\iota(r)$ stands for the discrete topology of $\{0,1\}$ for each $r \in [0,1]$. The underlying set of the Hausdorff topology η is

$$X := \{0,1\}^{[0,1]},$$

as in (IV.5.6). Of course, η is homeomorphic to the pointwise convergence.

By Theorem VIII.3.4 of Tikhonov, η is a compact topology. To see that η is not sequentially compact, consider, for each $r \in [0,1]$, a binary representation of the form

$$r = \sum_{n=1}^{\infty} \frac{r_n}{2^n},$$

where $r_n \in \{0,1\}$, and if r has two representations ([10]), then there is n_0 such that $r_n = 0$ for each $n > n_0$. Define a sequence $(f_n)_{n\in\mathbb{N}_1}$ of elements of X by

$$f_n(r) := r_n.$$

We claim that there is no convergent subsequence. Indeed, if $f \in X$ is a limit of $(f_{n_k})_k$, then for each $r \in [0,1]$, there is $k(r)$ such that $f(r) = f_{n_k}(r)$ for each $k > k(r)$. To see the contradiction, consider r such that $r_{n_k} = 0$ if k is even, and $r_{n_k} = 1$ if k is odd.

On the other hand, Example XII.3.2 is a Hausdorff sequentially compact, non-compact topology. Let us first observe that

PROPOSITION XII.4.4. *Each countable product of sequentially compact topologies is sequentially compact.*

PROOF. Let ξ_m be a sequentially compact topology on X_m for each $m < \omega$, and let $\xi := \prod_{m<\omega} \xi_m$. If $X := \prod_{m<\omega} X_m$, and $f_n \in X^N$, then $(f_n)_n$ is a sequence of elements of X, that is, $f_n(m) \in X_m$ for each $m < \omega$. As ξ_0 is sequentially compact, there exists an infinite subset N_0 of N such that $(f_n(0))_{n\in N_0}$ ξ_0-converges to an element x_0 of X_0. By induction, if $N_0 \supset N_1 \supset \ldots \supset N_m$ are infinite subsets of N, and x_k is a ξ_k-limit of $(f_n(k))_{n\in N_k}$ for $0 \le k \le m$, then by the sequential compactness of ξ_{m+1}, there exists an infinite subset N_{m+1} of N_m and $x_{m+1} \in X_{m+1}$ such that $(f_n(m+1))_{n\in N_{m+1}}$ ξ_{m+1}-converges to x_{m+1}.

[10]For example, for $r = \frac{1}{2}$ we take $r_1 = 1$, and $r_n = 0$ for $n > 1$; the other representation is $r_1 = 0$, and $r_n = 1$ for $n > 1$.

By Lemma XVI.2.1 to come, there exists a subset N_∞ of N such that $(f_n(m))_{n\in N_\infty}$ ξ_m-converges to x_m for each $m < \omega$. Therefore, if we set $f(m) := x_m$, then $f \in \lim_\xi (f_n)_{n\in N_\infty}$. □

EXAMPLE XII.4.5 (A sequentially compact, non-compact topology). Let Y be a subset of $X = \{0,1\}^{[0,1]}$ from Example XII.4.3 consisting of those f for which

$$s(f) := \{r \in [0,1] : f(r) \neq 0\}$$

is countable. This is an example of *Σ-product*. Then $Y \neq X$ and Y is η-dense. Indeed, if $f_0 \in X$ and $V \in \mathcal{N}_\eta(f_0)$, then there is a finite subset F of $[0,1]$ such that $\{f \in X : \underset{r\in F}{\forall} f(r) = f_0(r)\} \subset V$. Thus, if $f_F(r) := f_0(r)$ if $r \in F$, and $f_F(r) := 0$ otherwise, then $f_F \in V \cap Y$. Therefore, the restricted topology $\eta|_Y$ is not compact, because Y is not an η-closed subset of a Hausdorff, compact topological space. On the other hand, if $(f_n)_n$ is a sequence on Y, then $S := \bigcup_{n<\omega} s(f_n)$ is a countable set. Therefore, $Z := \{0,1\}^S$ is an η-closed subset of Y, and by Proposition XII.4.4, there exists a η_Y-convergent subsequence of $(f_n)_n$.

Example XII.4.5 is also an instance of a countably compact, non-compact topology.

XII.5. The Cantor set

For each sequence $\varphi \in \{0,1,2\}^{\mathbb{N}_1}$, there exists $r \in [0,1]$ such that

$$r = \sum_{n=1}^{\infty} \frac{\varphi(n)}{3^n} \tag{XII.5.1}$$

and conversely, each $r \in [0,1]$ admits $\varphi \in \{0,1,2\}^{\mathbb{N}_1}$ — its *ternary representation* — such that (XII.5.1) holds. Some admit two ternary representations, for instance, $\frac{1}{3} = \sum_{n=2}^{\infty} \frac{2}{3^n}$.

By definition, the *Cantor set* C consists of those $r \in [0,1]$, for which there exists a representation (XII.5.1) with $\varphi_r(n) \in \{0,2\}$ for each $n \in \mathbb{N}_1$.

PROPOSITION XII.5.1. *Each element of the Cantor set has a unique ternary representation.*

PROOF. Let $\varphi, \psi \in \{0,2\}^{\mathbb{N}_1}$ be two representations of the same number from C. If n_0 is the first integer such that $\psi(n) \neq \varphi(n)$, say $\varphi(n_0) = 2$ and $\psi(n_0) = 0$, then by a formula for the geometric series (X.1.3),

$$\frac{2}{3^{n_0}} = \sum_{n=n_0+1}^{\infty} \frac{\psi(n)-\varphi(n)}{3^n} \leq \sum_{n=n_0+1}^{\infty} \frac{2}{3^n} = \frac{2}{3^{n_0+1}} \frac{1}{1-\frac{1}{3}} = \frac{1}{3^{n_0}},$$

a contradiction. It follows that $\psi(n) = \varphi(n)$ for each $n \in \mathbb{N}_1$. $\square$

Accordingly, for each $r \in C$, there is a unique $\varphi \in \{0, 2\}^{\mathbb{N}_1}$ such that (XII.5.1) holds.

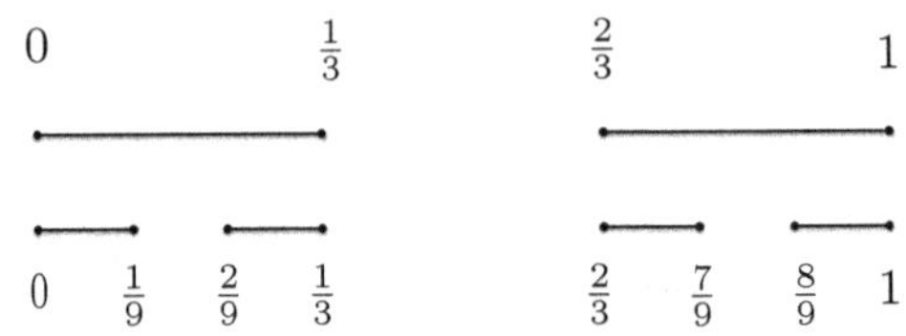

FIGURE XII.5.1. Two levels of construction of the Cantor set: C_1 and C_2.

The Cantor set can be represented as the intersection of a sequence of the following sets: $C_0 := [0, 1]$, and recursively, with the notation of (XII.5.1),

$$C_{n+1} := \{r \in C_n : \varphi(n+1) \neq 1\}.$$

Clearly, C_n is ν-compact for every $n < \omega$. Hence the intersection $C = \bigcap_{n<\omega} C_n$ is ν-compact and non-empty. We notice that C_n consists of 2^n disjoint closed intervals of length 3^{-n}. The sum of the lengths of the intervals composing C_n is equal to $(\frac{2}{3})^n$.

The topology of the Cantor set is the restriction $\nu|_C$ of the standard topology ν. Therefore, the Cantor set is completely metrizable as a non-empty closed subset of a completely metrizable topology; we shall use on C the restriction of the standard metric.

The difference $[0, 1] \setminus C$ is a union of disjoint ν-open intervals. The sum of their lengths is equal to 1.

Let us denote by $C_{n,k}$, the k-th interval composing C_n, counting from the left. Then C is homeomorphic to

$$Q_{n,k} := C \cap C_{n,k}$$

for each $1 \leq k \leq 2^n$ and $n < \omega$. Moreover, each $Q_{n,k}$ is $\nu|_C$-clopen ($\nu|_C$-closed as an intersection of ν-closed sets, $\nu|_C$-open as the complement of a finite union of $\nu|_C$-closed sets).

For every sequence $(k_n)_n$ such that $1 \leq k_n \leq 2^n$, the intersection $\bigcap_{n<\omega} Q_{n,k_n}$ is a singleton $\{x\}$, and $\{Q_{n,k_n} : n < \omega\}$ is a filter-base of the neighborhood filter of x with respect to $\nu|_C$. On the other hand, for each $n \in \mathbb{N}_1$, the family $\{Q_{n,k} : 1 \leq k \leq 2^n\}$ is a $\nu|_C$-open cover. Therefore,

$$\{Q_{n,k} : 1 \leq k \leq 2^n, n < \omega\} \tag{XII.5.2}$$

is a base for the open sets of $\nu|_C$, called the *canonical base* of C.

LEMMA XII.5.2. *A sequence* $(\varphi_m)_m$ *converges to* φ *in the Cantor set* C *if and only if for each* $n \in \mathbb{N}_1$ *there is* m_n *such that* $\varphi_m(n) = \varphi(n)$ *for each* $m \geq m_n$.

PROOF. If the condition holds, then for each $\varepsilon > 0$, there is $n \in \mathbb{N}_1$ such that $\frac{1}{3^n} < \varepsilon$, hence there exists m_n such that $\varphi_m(k) = \varphi(k)$ for each $k \leq n$ and every $m \geq m_n$, hence $\sum_{k=n+1}^{\infty} \frac{2}{3^k} = \frac{1}{3^n} < \varepsilon$ for each $m \geq m_n$. Conversely, if the condition does not hold, and n_0 is the first natural number such that $\varphi_m(n_0) \neq \varphi(n_0)$ for infinitely many m, then $\frac{|\varphi_m(n_0)-\varphi(n_0)|}{3^{n_0}} = \frac{2}{3^{n_0}} \leq \sum_{n=n_0+1}^{\infty} \frac{2}{3^n} = \frac{1}{3^{n_0}}$ gives a contradiction. □

In Example IV.6.8 we investigated the product convergence $\iota^{\mathbb{N}}$ on $\{0,1\}^{\mathbb{N}} \approx \prod_{n \in \mathbb{N}_1} \{0,1\}$, called a *Cantor cube*. By Theorem VIII.3.4 of Tikhonov, the Cantor cube is compact. Actually,

PROPOSITION XII.5.3. *The Cantor set* C *and the Cantor cube are homeomorphic.*

PROOF. The map $F : \{0,1\}^{\mathbb{N}} \longrightarrow C$, given by

$$F(f) := \sum_{n=0}^{\infty} \frac{2f(n)}{3^{n+1}},$$

is surjective by (XII.5.1), and injective in virtue of Proposition XII.5.1. To see that F is continuous, recall that a sequence convergences in a product space if and only it converges factor-wise. As in our case the factors are discrete, $(f_m)_m$ converges to f in the Cantor cube if and only if for each $n < \omega$, there exists m_n such that $f_m(n) = f(n)$ for each $m \geq m_n$. Hence, in virtue of Lemma XII.5.2, $(F(f_m))_m$ converges to $F(f)$ in the Cantor set. By Corollary XII.2.4, F is a homeomorphism. □

It follows in particular that $\{0,1\}^{\mathbb{N}}$ and C are equipotent. Hence, by Exercise XVI.5.1, they are equipotent to $[0,1]$, which is equipotent to the real line by Example I.5.1. Recall that the cardinality of the real line is called the *continuum*, and is denoted by $\mathfrak{c}$. As $\{0,1\}^{\mathbb{N}}$ is equipotent with the set of all subsets of $\mathbb{N}$, as shown in (I.5.4), and $\operatorname{card} \mathbb{N} = \aleph_0$, we have proved that

$$\mathfrak{c} = 2^{\aleph_0}. \tag{XII.5.3}$$

Recall that a topology is said to be zero-dimensional if it has a base of clopen sets (Definition IX.2.17).

COROLLARY XII.5.4. *The Cantor set and the Cantor cube are zero-dimensional.*

On the other hand, it turns out that ([11])

PROPOSITION XII.5.5. *The interval* $[0,1]$ *is a continuous image of the Cantor set* C.

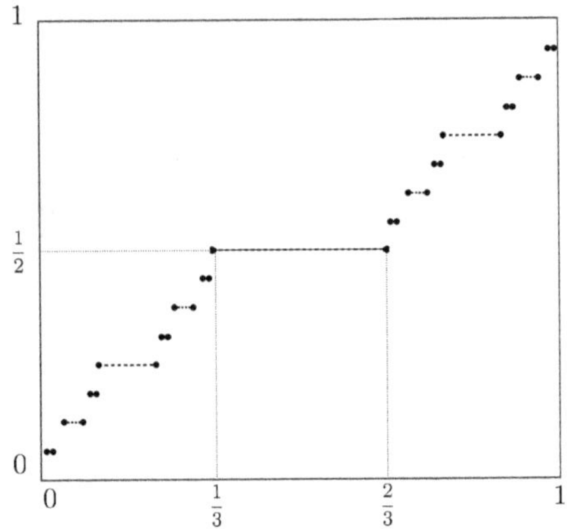

FIGURE XII.5.2. The unit interval is a continuous image of the Cantor set.

PROOF. Indeed, each element r of the Cantor set C has a unique representation (XII.5.1), where $\varphi(n) \in \{0,2\}$ for each $n \in \mathbb{N}_1$. Then

$$f(r) := \frac{1}{2}\sum_{n=1}^{\infty} \frac{\varphi(n)}{2^n}$$

represents an element of $[0,1]$, because $0 \leq f(r) \leq \sum_{n=1}^{\infty} \frac{1}{2^n} \leq 1$. The map f is surjective, because each $t \in [0,1]$ has a binary representation $t = \sum_{n=1}^{\infty} \frac{\varphi(n)}{2^n}$, where $\varphi(n) \in \{0,1\}$ for each $n \in \mathbb{N}_1$, hence $2\sum_{n=1}^{\infty} \frac{\varphi(n)}{3^n} \in f^{-}(t)$. We extend f to the whole interval $[0,1]$ by setting the constant values on the excluded intervals, for instance, $\frac{1}{2}$ on $]\frac{1}{3}, \frac{2}{3}[$, which yields a continuous map. □

REMARK XII.5.6. The map f is not injective, because otherwise it would be a homeomorphism. We can explicitly mark out the points t of $[0,1]$, where $f^{-}(t)$ is not a singleton (in these cases it has 2 elements), as in Figure XII.5.2. If $t \in [0,1]$ has a binary representation $\varphi(n) = 0$ for almost all n, then let $\#t$ be the greatest natural number such that $\varphi(n) = 0$ for each $n > \#t$. Consequently,

$$t = \sum_{n=1}^{\#t} \frac{\varphi(n)}{2^n} = \sum_{n=1}^{\#t-1} \frac{\varphi(n)}{2^n} + \sum_{n=\#t+1}^{\infty} \frac{1}{2^n}.$$

[11]The figure after [168, p. 165].

Hence $f^{-}(t) = \{r_0, r_1\}$, where $r_0 = \sum_{n=1}^{\#t} \frac{2\varphi(n)}{3^n}$ and $r_1 = \sum_{n=1}^{\#t-1} \frac{2\varphi(n)}{3^n} + \sum_{n=\#t+1}^{\infty} \frac{2}{3^n}$.

It follows from Corollary XII.2.6 that every compact metrizable topology is a continuous image of a closed subset of a Cantor set. This result can be strengthened.

PROPOSITION XII.5.7. *Each compact metrizable topology is a continuous image of the Cantor set.*

Indeed, this is a consequence of the following

LEMMA XII.5.8. *Every non-empty closed subset of the Cantor set is a continuous image of the Cantor set.*

PROOF. Consider the Euclidean metric on C. Let $F \subset C$ be non-empty and closed. Let $\{G_m : m < \omega\}$ be a subsequence of the canonical base (XII.5.2) of C such that $C \setminus F = \bigcup_{m<\omega} G_m$. In particular, each G_m is clopen and $\lim_{m\to\infty} \operatorname{diam} G_m = 0$. As F and G_m are compact for each $m < \omega$, there exist two points $p_m \in G_m$ and $q_m \in F$ such that $|p_m - q_m| = \operatorname{dist}(G_m, F)$. We define a map $f : C \longrightarrow F$ by

$$f(x) := \begin{cases} x & x \in F, \\ q_m & x \in G_m. \end{cases}$$

It is clearly a surjective map. We shall show that f is continuous. The restriction of f to F is the identity, and its restriction to G_m is constant for each $m < \omega$. We need only to check that if $(x_k)_k$ is a sequence on $C \setminus F$ and $x = \lim_{k\to\infty} x_k$, then $x = f(x) = \lim_{k\to\infty} f(x_k)$.

For each $k < \omega$, there is m_k such that $x_k \in G_{m_k}$. On the other hand, for each m, the number of k such that $x_k \in G_m$ is finite, because $x \notin G_m$ and G_m is closed. By definition, $f(x_k) = q_{m_k}$ and $f(x) = x$.

$$\begin{aligned} |f(x) - f(x_k)| = |x - q_{m_k}| &\le |x - x_k| + |x_k - p_{n_k}| + |p_{n_k} - q_{n_k}| \\ &\le |x - x_k| + \operatorname{diam} G_{m_k} + \operatorname{dist}(G_{m_k}, F) \\ &\le |x - x_k| + \operatorname{diam} G_{m_k} + |x - x_k| \,. \end{aligned}$$

As $\lim_{m\to\infty} \operatorname{diam} G_m = 0$, we conclude that f is continuous. □

XII.6. *Cardinality aspects

As the asterisk * indicates, this section is more difficult, and can be omitted without affecting the comprehension of the sequel.

Here we use the convention $A \in [X]^{<\omega}$ instead of $A \subset_{\text{fin}} X$, whenever A is finite and $A \subset X$.

LEMMA XII.6.1. *If κ is infinite, then the weight of the Cantor cube ι^κ is κ.*

PROOF. Indeed, $\{0,1\}^\kappa$ has a base of cardinality κ, for example,

$$\mathcal{B} := \{\{f \in \{0,1\}^\kappa : f|_A = h\} : h \in \{0,1\}^A, A \in [\kappa]^{<\omega}\}.$$

As $\operatorname{card}([\kappa]^{<\omega}) = \kappa$ and $\operatorname{card}(\{0,1\}^A) = 2^{\operatorname{card} A}$ is finite, $\operatorname{card} \mathcal{B} = \kappa$. □

LEMMA XII.6.2. *If κ is an infinite cardinal, and $X_t = X$ is a set endowed with the discrete topology and* $\operatorname{card} X \leq \kappa$, *then the density of $\prod_{t \in 2^\kappa} X_t$ is equal to κ.*

PROOF. Let $T := \{0,1\}^\kappa$, and let $\mathcal{B}$ be a base of $\{0,1\}^\kappa$ of cardinality κ. Then, there exists a base of $\prod_{t\in T} X_t$ consisting of

$$\{\varphi \in \prod\nolimits_{t \in T} X_t : \varphi(t_k) \in X, 1 \leq k \leq n\},$$

where $n < \omega$, $\{t_1, \ldots, t_n\} \subset T$ for $1 \leq k \leq n$, and t_k are distinct. As the product topology is Hausdorff, there exist disjoint $B_k \in \mathcal{B}$ such that $t_k \in B_k$. Let Φ consists of those $\varphi \in \prod_{t\in T} X$ be such that $\varphi^-(\varphi(x_k)) := B_k$, and constant out of $B_1 \cup \ldots \cup B_n$.

Then for any $\psi \in \prod_{t \in T} X$, any $B_1, \ldots, B_n \in \mathcal{B}$, there exists $\varphi \in \Phi$ that coincides with ψ on $B_1, \ldots, B_n$. Therefore Φ is dense in $\prod_{t\in T} X$.

As each $\{B_1, \ldots, B_n\} \in [\mathcal{B}]^{<\omega}$ there are κ of them, and for each such n-tuple, there are κ piecewise-constant selection $\varphi \in \prod_{t\in T} X$, so that $\operatorname{card} \Phi = \kappa$. Therefore, the density of $\prod_{t\in 2^\kappa} X$ is κ. □

THEOREM XII.6.3 (Hewitt-Marczewski-Pondiczery). *If κ is an infinite cardinal, and τ_t is a topology with $\mathrm{d}(\tau_t) \leq \kappa$ for each $t \in 2^\kappa$, then $\mathrm{d}(\prod_{t\in 2^\kappa} \tau_t) \leq \kappa$.*

PROOF. ([12]) Let τ_t be a topology on X_t, and A_t be a τ_t-dense subset of X_t such that $\operatorname{card} A_t \leq \kappa$ for each $t \in T = 2^\kappa$. Then $\prod_{t\in T} A_t$ is dense in $\tau := \prod_{t\in T} \tau_t$, as $\operatorname{cl}_\tau \prod_{t\in T} A_t = \prod_{t\in T} \operatorname{cl}_{\tau_t} A_t$. By Lemma XII.6.2, the density of $\prod_{t\in T} \iota_t$, where ι_t is the discrete topology on A_t, is less than or equal to κ, thus there is a subset D of cardinality less than or equal to κ of $\prod_{t\in T} A_t$ such that such that $\operatorname{cl}_{\prod_{t\in T} \iota_t} D = \prod_{t\in T} A_t$, hence $\operatorname{cl}_\tau D = \prod_{t\in T} X_t$. □

If $\kappa = \aleph_0$, then $2^{\aleph_0} = \mathfrak{c}$, then Theorem XII.6.3 yields

COROLLARY XII.6.4. *The density of the Cantor cube $\iota^\mathfrak{c}$ is $\aleph_0$, the weight is $\mathfrak{c}$ and the cardinality $2^\mathfrak{c}$.*

[12]Based on [97, Theorem 2.3.15] by R. ENGELKING.

Let X be an infinite set. If $\mathcal{A} \subset 2^X$, then we convene that

$$A^0 := A, \text{ and } A^1 := X \setminus A.$$

We say that $\mathcal{A} \subset 2^X$ is *independent* if for each finite subfamily $\mathcal{A}_0$ of $\mathcal{A}$, and each function $f : \mathcal{A} \longrightarrow \{0,1\}$

$$\bigcap\nolimits_{A \in \mathcal{A}_0} A^{f(A)} \neq \varnothing. \tag{XII.6.1}$$

THEOREM XII.6.5 (Hausdorff). *For each infinite set X, there exists an independent family of cardinality $2^{\operatorname{card} X}$.*

PROOF. By Lemma XII.6.2, there exists a dense subset X of cardinality κ of $\prod_{t \in T}\{0,1\}$ with the product topology, where $T = 2^\kappa$. Then $\mathcal{A} := \{p_t^-(0) \cap X : t \in T\}$ is an independent family of subsets of X, and its cardinality is 2^κ.

Indeed, $\{p_t^-(0), p_t^-(1)\}$ is an open partition of $\prod_{t \in T}\{0,1\}$ for each $t \in T$. On the other hand, for each finite sequence $t_1, \ldots, t_n$ of elements of T and $\alpha_0, \ldots, \alpha_n \in \{0,1\}$, the set $\bigcap_{k=0}^n p_{t_k}^-(\alpha_k)$ is a non-empty open set, hence $X \cap \bigcap_{k=1}^n p_{t_k}^-(\alpha_k) \neq \varnothing$ because of the density of X. If $A_0, \ldots, A_n$ are elements of $\mathcal{A}$, then there exist $t_1, \ldots, t_n \in T$ such that $A_k = p_{t_k}^-(0)$, so that for any $\alpha_0, \ldots, \alpha_n \in \{0,1\}$,

$$\bigcap\nolimits_{k=1}^n A^{\alpha_k} = X \cap \bigcap\nolimits_{k=1}^n p_{t_k}^-(\alpha_k) \neq \varnothing,$$

proving that $\mathcal{A}$ is independent. It is obvious that $\operatorname{card}\mathcal{A} = 2^\kappa$. □

PROPOSITION XII.6.6. *Let X be an infinite set of cardinality κ. Then the cardinality of βX is 2^{2^κ}.*

PROOF. By Theorem XII.6.5, on each set of infinite cardinality κ, there exists an independent family $\mathcal{A}$ of cardinality 2^κ. For each $f : \mathcal{A} \longrightarrow \{0,1\}$, the family $\mathcal{A}_f := \{A^{f(A)} : A \in \mathcal{A}\}$ is a filter-subbase, because of (XII.6.1). Moreover, if $f_0, f_1 : \mathcal{A} \longrightarrow \{0,1\}$ are distinct, then $\mathcal{A}_{f_0}$ and $\mathcal{A}_{f_1}$ do not mesh, because then there is $A \in \mathcal{A}$ such that $f_0(A) + f_1(A) = 1$, hence $A^{f_0(A)} \cap A^{f_1(A)} = \varnothing$. Therefore $\beta(\mathcal{A}_{f_0}^\cap) \cap \beta(\mathcal{A}_{f_1}^\cap) = \varnothing$ if $f_0 \neq f_1$, and thus $\operatorname{card} \beta X \geq 2^{2^\kappa}$. On the other hand, $\operatorname{card} \beta X \leq 2^{2^\kappa}$, because $\beta X \subset 2^{(2^X)}$. □

COROLLARY XII.6.7. *The cardinality of $\beta\mathbb{N}$ is $2^{\mathfrak{c}}$.*

XII.7. The Stone topology

We are going to define a canonical Hausdorff compact topology on the set βX of all ultrafilters on X, the so-called *Stone topology*.

Considering a convergence on a set of ultrafilters is a bit intricate, not only conceptually, but also notation-wise, as each ultrafilter $\mathcal{U}$ plays two roles: it is at the same time a family of subsets of X and a point of βX,

$$X \in \mathcal{U} \in \beta X.$$

Moreover, we shall consider ultrafilters on βX, that is, elements of $\beta(\beta X)$.

Therefore, it is sensible to give two different names to the same ultrafilter, according to its current role, in order to avoid misinterpretations.

Let us adopt the following convention: if $p \in \beta X$ is considered as a point, then we denote by $\mathcal{U}(p)$ the same ultrafilter considered as a family of subsets of X, that is, if $\mathcal{U}(p) \in 2^{(2^X)}$. If we want to indicate or stress this latter role, then we shall use $\mathbb{U}X$ to denote the set of all ultrafilters considered as families. For example,

$$p \in \beta X \Longleftrightarrow \mathcal{U}(p) \in \mathbb{U}X.$$

Following the proposed convention, if $\mathcal{F} \in \mathbb{F}X$, then

$$\beta\mathcal{F} := \{p \in \beta X : \mathcal{U}(p) \geq \mathcal{F}\}. \tag{XII.7.1}$$

If in particular $\mathcal{F} = \{F\}^{\uparrow}$ is a principal filter, where $F \subset X$, then we abridge

$$\beta F := \beta\{F\}^{\uparrow} = \{p \in \beta X : F \in \mathcal{U}(p)\}.$$

We also denote by

$$\beta^* X := \beta(X)_0$$

the set of all free ultrafilters on X.

Stone transform. Accordingly, (XII.7.1) defines a map

$$\beta : \mathbb{F}X \longrightarrow 2^{\beta X},$$

called the *Stone transform*.

Let us observe that if $\mathcal{F}_0, \mathcal{F}_1 \in \mathbb{F}X$, then

$$\mathcal{F}_0 \leq \mathcal{F}_1 \Longrightarrow \beta\mathcal{F}_0 \supset \beta\mathcal{F}_1, \tag{XII.7.2}$$

and

$$\beta(\mathcal{F}_0 \vee \mathcal{F}_1) = \beta\mathcal{F}_0 \cap \beta\mathcal{F}_1, \quad \beta(\mathcal{F}_0 \wedge \mathcal{F}_1) = \beta\mathcal{F}_0 \cup \beta\mathcal{F}_1.$$

In particular, if A_0, A_1 are subsets of X, then

$$\beta(A_0 \cap A_1) = \beta A_0 \cap \beta A_1, \quad \beta(A_0 \cup A_1) = \beta A_0 \cup \beta A_1.$$

We notice also that

$$\beta^* A_0 = \beta^* A_1$$

if and only if A_0 and A_1 are *almost equal*, that is, such that

$$A_0 \triangle A_1 = (A_0 \setminus A_1) \cup (A_1 \setminus A_0)$$

is finite (Definition II.5.6).

There are as many ultrafilters as elements on a finite set. For an infinite set X, however, by Proposition XII.6.6,

$$\operatorname{card}(\beta X) = \operatorname{card}(2^{(2^X)}).$$

If X is infinite countable, like the set $\mathbb{N}$ of natural numbers, then in virtue of (XII.5.3), $\operatorname{card}(2^{\mathbb{N}}) = \operatorname{card}\mathbb{R} = \mathfrak{c}$, thus $\operatorname{card}(\beta\mathbb{N}) = 2^{\mathfrak{c}}$.

Stone topology. If X is infinite, then the *Stone topology* $\beta = \beta\iota_X$ on βX is defined by declaring as a base for the open sets the family

$$\mathcal{B} := \{\beta A : A \subset X\}. \tag{XII.7.3}$$

Indeed, $\mathcal{B}$ fulfills the requirements of Proposition IX.2.6 for a base of the open sets of a topology, because the whole space $\beta X \in \mathcal{B}$ and if $A_0, A_1 \subset X$, and if $p \in \beta A_0 \cap \beta A_1$, equivalently $A_0 \in \mathcal{U}(p)$ and $A_1 \in \mathcal{U}(p)$, hence $A_0 \cap A_1 \in \mathcal{U}(p)$, and thus $p \in \beta(A_0 \cap A_1) = \beta A_0 \cap \beta A_1$. Therefore,

$$\{\beta U : U \in \mathcal{U}(p)\} \tag{XII.7.4}$$

is a filter-base of the neighborhood filter $\mathcal{N}_\beta(p)$ of $p \in \beta X$.

PROPOSITION XII.7.1. *The Stone topology β is Hausdorff and zero-dimensional.*

PROOF. If $p_0, p_1 \in \beta X$ and $p_0 \neq p_1$, then there is $A \in \mathcal{U}(p_0) \setminus \mathcal{U}(p_1)$. Hence, by Proposition II.3.10, $X \setminus A \in \mathcal{U}(p_1)$. Thus $p_0 \in \beta A$ and $p_1 \in \beta(X \setminus A) = \beta X \setminus \beta A$. Therefore, β is Hausdorff, and the family (XII.7.3) consists of β-clopen sets. □

If $x \in X$, then $x^\uparrow = \{F \subset X : x \in F\}$ is a principal ultrafilter.

PROPOSITION XII.7.2. *The subset $\{x^\uparrow : x \in X\}$ of βX is open and dense. The restriction of the topology β to $\{x^\uparrow : x \in X\}$ is discrete.*

PROOF. By definition, $V \in \mathcal{N}_\beta(x^\uparrow)$ whenever there is $A \subset X$ such that $x \in A$ and $\beta A \subset V$. The least such A is $\{x\}$. Consequently, $\{x^\uparrow : x \in X\}$ is β-open.

Let $p \in \beta X$. For every $W \in \mathcal{N}_\beta(p)$, there is $U \in \mathcal{U}(p)$ such that $\beta U \subset W$. It follows from $\varnothing \neq U \subset X$ that $\{x^\uparrow : x \in X\} \cap \beta U \neq \varnothing$. □

It follows that ι_X is embedded in β. On recalling Definition III.1.19, $\mathrm{d}(\beta X) = \operatorname{card} X$, as we have already established in Proposition XII.6.6, and in particular,

COROLLARY XII.7.3. *If X is countable, then βX is separable.*

As we have observed, if X is infinite, then βX is a huge set in comparison to X, and X is β-dense. On the other hand, we will see with Proposition XII.8.4 that

COROLLARY XII.7.4. *The β-density of $\beta^* X$ is not less than $2^{\operatorname{card} X}$.*

PROPOSITION XII.7.5. *The Stone topology is compact.*

PROOF. Let $\mathcal{P}$ be an open cover of βX. As the family (XII.7.3) is a base for β, there is a family $\mathcal{A}$ of subsets of X such that $\{\beta A : A \in \mathcal{A}\}$ is a covering refinement of $\mathcal{P}$. Accordingly, for each $p \in \beta X$, there exists $A_p \in \mathcal{A}$ such that $p \in \beta A_p$, equivalently $A_p \in \mathcal{U}(p)$. By Proposition V.4.6, there is a finite subfamily $\mathcal{A}_0$ of $\mathcal{A}$ such that $\bigcup_{A_p \in \mathcal{A}_0} A_p \in X^{\uparrow}$, that is, $\bigcup_{A_p \in \mathcal{A}_0} \beta A_p = \beta X$. □

PROPOSITION XII.7.6. *The character of the Stone topology is uncountable at every point of $\beta^* X = \beta X \setminus X$.*

PROOF. Let us show that if $p \in \beta^* X$ and $\{W_n : n < \omega\}$ is a filter-base of $\mathcal{N}_\beta(p)$, then $\operatorname{int}_\beta \bigcap_{n<\omega} W_n \neq \varnothing$, which is a contradiction. Indeed, by (XII.7.4), there is a decreasing sequence $\{U_n : n < \omega\} \subset \mathcal{U}(p)$ such that $p \in \beta U_n \subset W_n$. By Lemma XVI.2.1 to come, there exists an infinite set U_∞ that is almost included in U_n for every $n < \omega$. Therefore, $\beta U_\infty \subset \bigcap_{n<\omega} \beta U_n \subset \bigcap_{n<\omega} W_n$. As βU_∞ is β-open, the proof is complete. □

PROPOSITION XII.7.7. *The set $\beta \mathcal{F}$ is β-closed for every filter $\mathcal{F}$.*

PROOF. Let $\mathcal{F}$ be a filter on X. If $p \notin \beta \mathcal{F}$, then there exists $U \in \mathcal{U}(p)$ such that $X \setminus U \in \mathcal{F}$, and thus $\beta X \setminus \beta U = \beta(X \setminus U) \subset \beta \mathcal{F}$, that is, $p \in \beta U$ and $\beta U \cap \beta \mathcal{F} = \varnothing$. □

PROPOSITION XII.7.8. *If $D \subset \beta X$, then*

$$\operatorname{cl}_\beta D = \beta\Big(\bigcap\nolimits_{p \in D} \mathcal{U}(p)\Big). \tag{XII.7.5}$$

PROOF. As $\bigcap_{p \in D} \mathcal{U}(p)$ is a filter, by Proposition XII.7.7, $\beta(\bigcap_{p \in D} \mathcal{U}(p))$ is β-closed and includes D. Therefore, $\operatorname{cl}_\beta D \subset \beta(\bigcap_{p \in D} \mathcal{U}(p))$.

Conversely, let $q \notin \operatorname{cl}_\beta D$, that is, there exists $U \in \mathcal{U}(q)$ such that $\beta U \cap D = \varnothing$, so that $X \setminus U \in \mathcal{U}(p)$ for each $p \in D$, hence $X \setminus W \in \bigcap_{p \in D} \mathcal{U}(p)$, and thus $q \notin \beta(\bigcap_{p \in D} \mathcal{U}(p))$. □

COROLLARY XII.7.9. *A subset D of βX is β-closed if and only if there exists a filter $\mathcal{F}$ such that $D = \beta \mathcal{F}$.*

Consequently, $\beta \mathcal{F}$ is β-compact for each filter $\mathcal{F}$ on X. In particular, if $\mathcal{D}$ is a family of subsets of X such that $\beta \mathcal{F} \subset \bigcup_{D \in \mathcal{D}} \beta D$, then there exists a finite subfamily $\mathcal{D}_0$ of $\mathcal{D}$ such that $\beta\mathcal{F} \subset \bigcup_{D\in\mathcal{D}_0} \beta D$. In other words, if for each $\mathcal{U} \in \beta\mathcal{F}$ there is $D \in \mathcal{D}$ such that $D \in \mathcal{U}$, then there exists a finite subfamily $\mathcal{D}_0$ of $\mathcal{D}$ such that $\bigcup \mathcal{D}_0 = \bigcup_{D\in\mathcal{D}_0} D \in \mathcal{F}$. Therefore,

COROLLARY XII.7.10. *If $U_{\mathcal{U}} \in \mathcal{U}$ for each $\mathcal{U} \in \beta\mathcal{F}$, then there exists a finite subset $\boldsymbol{A}$ of $\beta\mathcal{F}$ such that $\bigcup_{\mathcal{U}\in\boldsymbol{A}} U_{\mathcal{U}} \in \mathcal{F}$.*

This way, we have recovered Proposition V.4.6.

THEOREM XII.7.11. *The Stone transform is an inverse-order isomorphism between $\mathbb{F}X$ and the family $\mathcal{C}_\beta$ of β-closed sets.*

PROOF. If $\mathcal{F}_0 \leq \mathcal{F}_1$ then $\beta\mathcal{F}_0 \supset \beta\mathcal{F}_1$. By Corollary XII.7.9, the Stone transform is surjective, and it is obviously injective. Finally, if $D_0, D_1 \in \mathcal{C}_\beta$ are such that $D_0 \subset D_1$, then $\bigwedge D_0 = \bigcap_{p\in D_0} \mathcal{U}(p) \geq \bigcap_{p\in D_1} \mathcal{U}(p) = \bigwedge D_1$. □

COROLLARY XII.7.12. *If $\boldsymbol{F} \subset \mathbb{F}X$, then*

$$\beta(\bigvee \boldsymbol{F}) = \bigcap_{\mathcal{F}\in\boldsymbol{F}} \beta\mathcal{F},$$

$$\beta(\bigwedge \boldsymbol{F}) = \mathrm{cl}_\beta(\bigcup_{\mathcal{F}\in\boldsymbol{F}} \beta\mathcal{F}).$$

PROOF. By Proposition I.3.8 on preservation of suprema and infima by order isomorphisms, the Stone transform inverts the extrema. In the complete lattice of closed sets for a topology, the infimum is the closure of the union. □

REMARK XII.7.13 (*). We shall see in Section XXI.2 that Theorem XII.7.11 is an immediate consequence of Theorem XXI.2.5. Indeed, if $\bigwedge D := \bigcap_{\mathcal{U}\in D} \mathcal{U}$ for each $D \subset 2^{\beta X}$, then the pair of decreasing maps $\beta, \bigwedge$ is a Galois connection. Indeed, by (V.4.1), $\bigwedge(\beta\mathcal{F}) = \mathcal{F}$ for each $\mathcal{F} \in \mathbb{F}X$, and $\beta(\bigwedge D) \supset D$ for every $D \subset 2^{\beta X}$. As $\bigwedge \beta$ and $\beta \bigwedge$ are ascending,

$$\mathbb{F}X \underset{\bigwedge}{\overset{\beta}{\rightleftarrows}} 2^{\beta X}$$

is a Galois connection.

Convergence. We are going to characterize convergence of filters in the Stone topology with the aid of the contour operation. Let Y be a set, and for each $y \in Y$, let $\mathcal{S}(y)$ be a family of subsets of X.

DEFINITION XII.7.14. The *contour* of $\{\mathcal{S}(y) : y \in Y\}$ along a filter $\mathcal{F}$ on Y is defined by

$$\int_{\mathcal{F}} \mathcal{S}(\cdot) = \bigcup_{F\in\mathcal{F}} \bigcap_{y\in F} \mathcal{S}(y). \qquad \text{(XII.7.6)}$$

In particular, if $\mathcal{F} \in \mathbb{F}(\beta X)$ and $\mathcal{U}(p) \in \mathbb{U}X$ for each $p \in \beta X$, then $\int_{\mathcal{F}} \mathcal{U}(\cdot)$ is a filter on X. In other words, $A \in \int_{\mathcal{F}} \mathcal{U}(\cdot)$ if and only if there exists $F \in \mathcal{F}$ such that $A \in \mathcal{U}(y)$ for each $y \in F$, that is,

$$A \in \textstyle\int_{\mathcal{F}} \mathcal{U}(\cdot) \Longleftrightarrow \beta A \in \mathcal{F}. \tag{XII.7.7}$$

PROPOSITION XII.7.15. *A filter $\mathcal{F}$ on βX converges to $p \in \beta X$, that is, $p \in \lim_\beta \mathcal{F}$ if and only if*

$$\mathcal{U}(p) = \textstyle\int_{\mathcal{F}} \mathcal{U}(\cdot). \tag{XII.7.8}$$

PROOF. By definition of the Stone topology, $p \in \lim_\beta \mathcal{F}$ if and only if $\beta A \in \mathcal{F}$ for each $A \in \mathcal{U}(p)$, by (XII.7.7), equivalently $\mathcal{U}(p) \subset \int_{\mathcal{F}} \mathcal{U}(\cdot)$, hence (XII.7.8) by maximality of ultrafilters. □

COROLLARY XII.7.16. *If $\mathcal{W} \in \beta(\beta X)$ and $p \in \beta X$, then*

$$p \in \lim\nolimits_\beta \mathcal{W} \Longleftrightarrow \mathcal{U}(p) = \textstyle\int_{\mathcal{W}} \mathcal{U}(\cdot).$$

As for each $p \in \beta X$, the neighborhood filter $\mathcal{N}_\beta(p)$ converges to p, by (XII.7.8),

$$\mathcal{U}(p) = \textstyle\int_{\mathcal{N}_\beta(p)} \mathcal{U}(\cdot). \tag{XII.7.9}$$

The Stone topology β on βX has a property that every map from X to a compact Hausdorff topological space τ has a unique continuous extension to a map from βX to τ.

THEOREM XII.7.17. *For every Hausdorff compact topology τ and each $f \in |\tau|^X$, there exists a unique map $F \in \mathscr{C}(\beta, \tau)$ such that $F|_X = f$. It is given by the formula*

$$F(p) \in \lim\nolimits_\tau f[\mathcal{U}(p)] \tag{XII.7.10}$$

for every ultrafilter $p \in \beta X$.

PROOF. Since τ is Hausdorff and compact, the map F in (XII.7.10) is well defined. It is unique by Lemma IX.1.3, because X is β-dense and τ is Hausdorff. We need to show that F is continuous, that is,

$$u_0 \in \lim\nolimits_\beta \mathcal{G} \Longrightarrow F(u_0) \in \lim\nolimits_\tau F[\mathcal{G}] \tag{XII.7.11}$$

for every $u_0 \in \beta X$ and $\mathcal{G} \in \mathbb{F}(\beta X)$. By Proposition XII.7.15, $u_0 \in \lim_\beta \mathcal{G}$ if and only if $\mathcal{U}(u_0) = \int_{\mathcal{G}} \mathcal{U}(\cdot)$. It follows that $f[\mathcal{U}(u_0)] = f[\int_{\mathcal{G}} \mathcal{U}(\cdot)]$, and thus

$$F(u_0) \in \lim\nolimits_\tau f[\mathcal{U}(u_0)] = \lim\nolimits_\tau f[\textstyle\int_{\mathcal{G}} \mathcal{U}(\cdot)].$$

The proof will be complete if we show that

$$\lim\nolimits_\tau F[\mathcal{G}] = \lim\nolimits_\tau f[\textstyle\int_{\mathcal{G}} \mathcal{U}(\cdot)]. \tag{XII.7.12}$$

If $y \in \lim_\tau F[\mathcal{G}]$ and $O \in \mathcal{O}_\tau(y)$, then there is $G \in \mathcal{G}$ such that $F(w) \in O$ for each $w \in G$. By (XII.7.10), there is $U_w \in \mathcal{U}(w)$ so that $f(U_w) \subset O$, equivalently $f(\bigcup_{w\in G} U_w) \subset O$. As $\bigcup_{w\in G} U_w \in \int_{\mathcal{G}} \mathcal{U}(\cdot)$, it follows that $O \in f[\int_{\mathcal{G}} \mathcal{U}(\cdot)]$, and thus $y \in \lim_\tau f[\int_{\mathcal{G}} \mathcal{U}(\cdot)]$. As τ is Hausdorff and compact, both limits are singletons, so that the equality follows. $\square$

EXAMPLE XII.7.18. If $f \in X^X$ is a permutation, then its extension $F = f^*$ fulfills $f^*(p) \in \lim_\beta f[\mathcal{U}(p)]$, hence by Proposition XII.7.15, $\mathcal{U}(f^*(p)) = \int_{f[\mathcal{U}(p)]} \mathcal{U}(\cdot)$. By definition of contour, $A \in \int_{f[\mathcal{U}(p)]} \mathcal{U}(\cdot)$ if and only if there exists $U \in \mathcal{U}(p)$ such that $A \in \mathcal{U}(x) = x^\uparrow$ for each $x \in f(U)$, that is, $f(U) \subset A$. In other words, $\int_{f[\mathcal{U}(p)]} \mathcal{U}(\cdot) = f[\mathcal{U}(p)]$, and thus

$$\mathcal{U}(f^*(p)) = f[\mathcal{U}(p)].$$

We shall now complete Theorem XII.7.17 by a sort of converse.

THEOREM XII.7.19. *If ξ is a Hausdorff compact topology on Y such that X is ξ-densely embedded, and for each Hausdorff compact topology τ and every $f \in |\tau|^X$, there exists a unique $F \in \mathscr{C}(\xi, \tau)$ such that $F|_X = f$, then ξ is homeomorphic to β.*

PROOF. Let X be a ξ-dense subset of Y, where ξ has the properties listed above. As ξ is densely embedded in βX, there exists a unique map $F \in \mathscr{C}(\xi, \beta)$ so that $F|_X = i_X$, the identity on X. On the other hand, by Theorem XII.7.17, there exists a unique map $G \in \mathscr{C}(\beta, \xi)$ so that $G|_X = i_X$. Therefore, $G \circ F \in \mathscr{C}(\xi, \xi)$ coincides with the identity on the dense subset X of Y, hence by Corollary IX.1.4 on uniqueness of extensions, $G \circ F = i_Y$. Therefore, by Remark I.1.3, G is surjective and F is injective. On the other hand, $F \circ G \in \mathscr{C}(\beta, \beta)$ coincides with the identity on the dense subset X of βX, thus $F \circ G = i_{\beta X}$, and thus G is injective and F is surjective. hence by Corollary XII.2.4 on bijections from Hausdorff compact topologies, $F \in \mathscr{H}(\xi, \beta)$ is a homeomorphism. Therefore, $G = F^{-1}$. $\square$

COROLLARY XII.7.20. *Each function $f \in [0,1]^X$ has a unique continuous extension to a function $F \in \mathscr{C}(\beta, \nu|_{[0,1]})$.*

THEOREM XII.7.21. *Let X be infinite countable. Each infinite β-closed subset of βX includes a subset homeomorphic to βX.*

PROOF. Let C be an infinite β-closed subset of βX and let $p_0, p_1 \in C$ be distinct. Then, as β is Hausdorff, there exist $O_0 \in \mathcal{O}_\beta(p_0)$ and $O_1 \in \mathcal{O}_\beta(p_1)$ such that $O_0 \cap O_1 = \varnothing$. We construct by induction, $P = \{p_n : n < \omega\} \subset C$ and $\{O_n : n < \omega\}$ such that $O_n \in \mathcal{O}_\beta(p_n)$ and $O_k \cap O_n = \varnothing$ for every $k \neq n$. If $k > 0$ and such sequences are constructed for $0 \leq n < k$, then, as β is normal by Proposition IX.3.4, there exists $O_k \in \mathcal{O}_\beta(p_k)$ such that $(O_0 \cup \cdots \cup O_{k-1}) \cap O_k = \varnothing$.

Let $f \in \mathscr{C}(\beta|_P, [0,1]\}$. Define $f_0 : X \longrightarrow [0,1]$ by

$$f_0(x) := \begin{cases} f(p_n) & x \in O_n, \\ 0 & x \in X \setminus \bigcup_{n<\omega} O_n. \end{cases}$$

As X is a β-discrete subspace, f_0 is continuous, hence has an extension $F_0 \in \mathscr{C}(\beta, [0,1])$. Since X is β-dense and $[0,1]$ is Hausdorff, by Lemma IX.1.3, the function F_0 coincides with f on $\mathrm{cl}_\beta(X \cap O_n)$ for each $n < \omega$. Hence, F_0 coincides with f on P.

Therefore, for each $f \in \mathscr{C}(\beta|_P, [0,1])$, there exists an extension $F \in \mathscr{C}(\beta|_{\mathrm{cl}_\beta P}, [0,1])$, and thus by Corollary XII.7.20, $\mathrm{cl}_\beta P = \beta P$. And since P is homeomorphic to the discrete topology on the countable set X, also $\mathrm{cl}_\beta P$ is homeomorphic to $\mathrm{cl}_\beta X = \beta X$. □

XII.8. *Almost disjoint families

Two subsets A and B of an infinite set X are called *almost disjoint* if $A \cap B$ is finite. Notice that two sets are almost disjoint if and only if their cofinite filters do not mesh.

DEFINITION XII.8.1. A family $\mathcal{A}$ of infinite pairwise almost disjoint subsets of X is called *almost disjoint* (shortly, *AD*).

An almost disjoint family $\mathcal{A}$ is said to be *maximal* (shortly, *MAD*) if $\mathcal{A} \subset \mathcal{D}$ implies $\mathcal{A} = \mathcal{D}$ for any almost disjoint family $\mathcal{D}$.

PROPOSITION XII.8.2. *A family $\mathcal{A}$ of subsets of X is* MAD *if and only if for each infinite $B \subset X$, there exists $A \in \mathcal{A}$ such that $A \cap B$ is infinite.*

PROOF. Indeed, if B is infinite and $A \cap B$ is finite for each $A \in \mathcal{A}$, then $\mathcal{A} \cup \{B\}$ is almost disjoint. □

It is clear that each family of disjoint sets is almost disjoint. A *MAD* family $\mathcal{A}$ can be finite ([13]), however,

PROPOSITION XII.8.3. *Every infinite* MAD *family on an infinite set is uncountable.*

PROOF. Suppose that $\mathcal{A} = \{A_n : n < \omega\}$ is *MAD*, where $A_n \neq A_k$ if $n \neq k$. Let $x_0 \in A_0$. Then by induction we select a sequence $(x_n)_n$ such that $x_n \in A_n \setminus (A_0 \cup \ldots \cup A_{n-1})$, and we infer that $\{x_n : n < \omega\} \cup \mathcal{A}$ is *AD*, contradicting maximality. □

[13]For example, a family $\{A_0, A_1\}$ on $\mathbb{N}$, where A_0 is the set of even, and A_1 of odd numbers.

Moreover,

PROPOSITION XII.8.4. *On each infinite countable set X, there exists a* MAD *family of cardinality* $\mathfrak{c}$.

PROOF. It is enough to construct such a family on the set of rational numbers $\mathbb{Q}$. By Example III.1.20, for each $r \in \mathbb{R} \setminus \mathbb{Q}$ there exists $\varphi_r \in \mathbb{Q}^{\mathbb{N}}$ such that $r \in \lim_\nu \varphi_r$. Then $\mathcal{R} := \{\varphi_r(\mathbb{N}) : r \in \mathbb{R} \setminus \mathbb{Q}\}$ is almost disjoint, because if $r_0 < s < r_1$, then the sets $\{n \in \mathbb{N} : s < \varphi_{r_0}(n)\}$ and $\{n : \varphi_{r_1}(n) > s\}$ are finite. The cardinality of $\mathcal{R}$ is equal to that of $\mathbb{R} \setminus \mathbb{Q}$, that is, $\mathfrak{c} = 2^{\aleph_0}$. By Theorem XXVII.4.8 of Zorn, there exists a *MAD* family $\mathcal{A}$ such that $\mathcal{R} \subset \mathcal{A}$, and thus $\operatorname{card}\mathcal{A} \geq \mathfrak{c}$. As $\mathcal{A} \subset 2^{\mathbb{N}}$, its cardinality is not greater than $\operatorname{card}(2^X) = 2^{\aleph_0} = \mathfrak{c}$. □

PROPOSITION XII.8.5. *If X is an infinite countable set, then there exists a family of cardinality $2^{\mathfrak{c}}$ of mutually disjoint β-closed subsets of $\beta^* X$.*

PROOF. Let $\{X_n : n < \omega\}$ be a partition of X into infinite sets. By Proposition XII.8.4, for each $n < \omega$, there is a *MAD* family $\mathcal{A}_n$ of cardinality $\mathfrak{c}$ on X_n. There exist $2^{\mathfrak{c}}$ selections $\varphi : \mathbb{N} \longrightarrow \beta^* X_n$ so that $\varphi(n) \in A_{\varphi(n)} \in \mathcal{A}_n$ and $A_{\varphi(n)} \neq A_{\psi(n)}$ if $\varphi \neq \psi$ for every $n < \omega$. Therefore, if $\varphi \neq \psi$, then $\bigcup_{n<\omega} A_{\varphi(n)} \in \int_{(n)} \mathcal{U}(\varphi(\cdot))$ and $\bigcup_{n<\omega}(A_{\psi(n)} \setminus A_{\varphi(n)}) \in \int_{(n)} \mathcal{U}(\psi(\cdot))$. Hence $\beta^* \int_{(n)} \mathcal{U}(\varphi(\cdot)) \cap \beta^* \int_{(n)} \mathcal{U}(\psi(\cdot)) = \varnothing$. Consequently, there are $2^{\mathfrak{c}}$ mutually disjoint β-closed subsets of $\beta^* X$. □

Cardinality aspects of almost disjoint families are studied in detail in Section XXVII.9.

Almost disjoint topology. With every almost disjoint family $\mathcal{A}$ on an infinite set X, we associate a canonical topology, called the *almost disjoint topology* of $\mathcal{A}$ and denoted by $\zeta_{\mathcal{A}}$.

Let $\mathcal{A}$ is parametrized by a set Y, as follows

$$\mathcal{A} = \{A_y : y \in Y\}.$$

We define on $X \sqcup Y$, the topology $\zeta_{\mathcal{A}}$ to the effect that each $x \in X$ is $\zeta_{\mathcal{A}}$-isolated, and for each $y \in Y$,

$$\mathcal{N}_{\zeta_{\mathcal{A}}}(y) := y^{\uparrow} \wedge (A_y)_0,$$

where, as usual, $(A_y)_0$ stands for the cofinite filter of A_y.

We observe that the *almost disjoint topology* $\zeta_{\mathcal{A}}$ is Hausdorff, because if $y_0, y_1 \in Y$ are distinct, then either one of them is in X, hence is isolated, or $y_0, y_1 \in Y \setminus X$ so that $A_{y_0} \cap A_{y_1}$ is finite, hence $\mathcal{N}_{\zeta_{\mathcal{A}}}(y_0)$ does not mesh with $\mathcal{N}_{\zeta_{\mathcal{A}}}(y_1)$.

It is clear that X is $\zeta_{\mathcal{A}}$-open and $\zeta_{\mathcal{A}}$-dense, and that its complement Y is $\zeta_{\mathcal{A}}$-closed and $\zeta_{\mathcal{A}}$-discrete. Moreover, $\zeta_{\mathcal{A}}$ is hereditarily locally compact (Definition XII.3.7), because the cofinite subsets of $A_y \cup \{y\}$ centered at y constitute a filter-base of $\mathcal{N}_{\zeta_{\mathcal{A}}}(y)$ for each $y \in Y$, hence each free filter on any cofinite subset of A_y converges to y in $\zeta_{\mathcal{A}}$.

Let us focus on $\zeta_{\mathcal{A}}$, for which $\mathcal{A}$ is an almost disjoint family on an infinite countable set. In this case, each element of $\mathcal{A}$ is countable and infinite, hence $\mathcal{N}_{\zeta_{\mathcal{A}}}(y)$ is a sequential filter. In particular, as a Hausdorff locally compact topology, $\zeta_{\mathcal{A}}$ is functionally regular (alias completely regular) (Exercise XII.10.8). If $\mathcal{A}$ is finite, then obviously $\zeta_{\mathcal{A}}$ is compact. If $\mathcal{A}$ is infinite, then $\zeta_{\mathcal{A}}$ is not even countably compact, because the set Y is infinite, discrete and closed. We conclude that

PROPOSITION XII.8.6. *If $\mathcal{A}$ is an almost disjoint family on an infinite countable set, then $\zeta_{\mathcal{A}}$ is a Hausdorff, separable, functionally regular (alias completely regular), locally compact topology.*

On the other hand,

PROPOSITION XII.8.7. *If $\mathcal{A}$ is an infinite* MAD *family on an infinite countable set, then $\zeta_{\mathcal{A}}$ is functionally compact, but not countably compact.*

PROOF. Let $\mathcal{A}$ be *MAD*, and $f \in \mathscr{C}(\zeta_{\mathcal{A}}, \nu)$. Suppose f is not bounded. Then, for every $k \in \mathbb{N}$, there $y_k \in Y$ such that $|f(y_k)| > k$. As X is $\zeta_{\mathcal{A}}$-dense, there exists $x_k \in X$ such that $|f(x_k)| > k$. Since $\{x_k : k < \omega\}$ is infinite and $\mathcal{A}$ is maximal, there exists $y \in Y$ such that $\{x_k : k < \omega\} \cap A_y$ is infinite. This means that there is a subsequence of $(x_k)_k$ that is finer than $(A_y)_0$ and thus convergent to y, which yields a contradiction. □

By Lemma XII.3.4,

COROLLARY XII.8.8. *If $\mathcal{A}$ is an infinite* MAD *family on an infinite countable set, then $\zeta_{\mathcal{A}}$ is a functionally compact, non-normal topology.*

If $\mathcal{A}$ is an infinite almost disjoint family of infinite subsets of X, then the family $\mathcal{A}_c$ is a filter-subbase. Indeed, if $A_0, \ldots, A_n \in \mathcal{A}$,

$$A_0^c \cap \ldots \cap A_n^c = (A_0 \cup \ldots \cup A_n)^c \neq \varnothing.$$

DEFINITION XII.8.9 (Residual filter). Let $\mathcal{A}$ be an *AD* family of infinite subsets of X, then the filter generated by $\{X \setminus A : A \in \mathcal{A}\}$ is called the *residual filter* of $\mathcal{A}$, and is denoted by $\mathrm{c}\mathcal{A}$.

PROPOSITION XII.8.10. *The residual filter of a AD family is proper if and only if the family is infinite.*

PROOF. Indeed, the residual filter $\mathrm{c}\mathcal{A}$ of an almost disjoint family $\mathcal{A}$ is improper whenever $\beta^*X = \bigcup_{A\in\mathcal{A}} \beta^*A$, and since $\{\beta^*A : A \in \mathcal{A}\}$ is a disjoint cover by β-open sets of a compact topological space, $\mathcal{A}$ is finite. $\square$

It follows from the definition that

PROPOSITION XII.8.11. *If $\mathcal{A}$ is an infinite AD family on X, then*

$$\beta^*(\mathrm{c}\mathcal{A}) = \beta^*X \setminus \bigsqcup_{A\in\mathcal{A}} \beta^*A. \tag{XII.8.1}$$

PROOF. By definition, $\mathcal{U} \in \beta^*(\mathrm{c}\mathcal{A})$ if and only if $X \setminus A \in \mathcal{U}$ for each $A \in \mathcal{A}$. $\square$

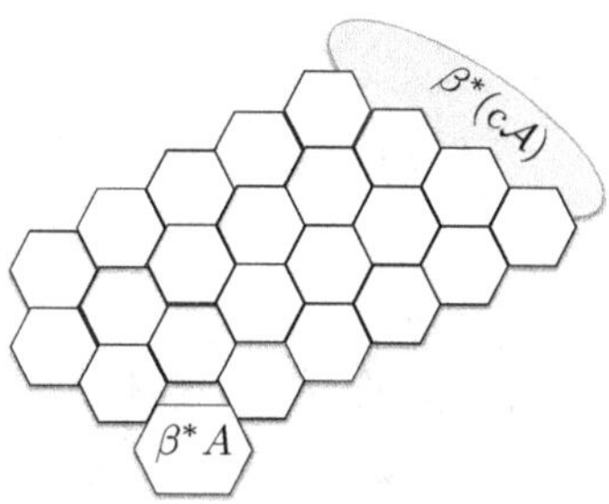

FIGURE XII.8.1. If $\mathcal{A}$ is an infinite almost disjoint family, then β^*A is β-clopen for each $A \in \mathcal{A}$. The Stone image $\beta^*(\mathrm{c}\mathcal{A})$ of the residual filter has empty β-interior if and only if $\mathcal{A}$ is maximal.

As for each $y \in Y$, the set $A_y \cup \{y\}$ is $\zeta_{\mathcal{A}}$-compact, $\mathrm{c}\mathcal{A}$ is the $\zeta_{\mathcal{A}}$-cocompact filter.

PROPOSITION XII.8.12. *An AD family $\mathcal{A}$ is maximal if and only if* $\mathrm{int}_\beta\, \beta^*(\mathrm{c}\mathcal{A}) = \varnothing$.

PROOF. Indeed, $\mathrm{int}_\beta\, \beta^*(\mathrm{c}\mathcal{A}) \neq \varnothing$ if and only if there exists $B \subset X$ such that $\beta^*B \subset \beta^*(\mathrm{c}\mathcal{A})$, equivalently whenever $\mathcal{A} \cup \{B\}$ is AD. $\square$

Therefore, an AD family is maximal if and only if $\bigcup_{A\in\mathcal{A}} \beta^*A$ is dense in β^*X.

If X is infinite countable, then $(A)_0$ is a sequential filter for each infinite $A \subset X$. Therefore, $\varepsilon(\mathrm{c}\mathcal{A}) = \varnothing$ if and only if $\mathcal{A}$ is maximal.

Other useful filters associated with almost disjoint families are contours $\int_{\mathcal{F}} \mathcal{N}_{\zeta_{\mathcal{A}}}(\cdot)$, where $\mathcal{F}$ is a free filter on Y. Applying the Stone transform to $\int_{\mathcal{F}} \mathcal{N}_{\zeta_{\mathcal{A}}}(\cdot) = \bigcup_{F\in\mathcal{F}} \bigcap_{y\in F} \mathcal{N}_{\zeta_{\mathcal{A}}}(\cdot)$, we get

$$\beta(\textstyle\int_{\mathcal{F}} \mathcal{N}_{\zeta_{\mathcal{A}}}(\cdot)) = \bigcap_{F\in\mathcal{F}} \mathrm{cl}_\beta(\bigcup_{y\in F} \beta\mathcal{N}_{\zeta_{\mathcal{A}}}(\cdot)).$$

LEMMA XII.8.13. *Let $\mathcal{A}$ be an AD family on X. If $\mathcal{F}$ is a free filter on Y, then $\int_{\mathcal{F}} \mathcal{N}_{\zeta_{\mathcal{A}}}(\cdot)$ is dissociated from $\mathcal{N}_{\zeta_{\mathcal{A}}}(\cdot)$ for each $y \in Y$.*

PROOF. As $\mathcal{F}$ is a free filter on Y, for each $y_0 \in Y$, there is $F_0 \in \mathcal{F}$ such that $y_0 \notin F_0$. As every $\beta^*\mathcal{N}_{\zeta_{\mathcal{A}}}(y)$ is β-clopen, $\beta^*\mathcal{N}_{\zeta_{\mathcal{A}}}(y_0) \cap \mathrm{cl}_\beta(\bigcup_{y\in F_0} \beta^*\mathcal{A}(y)) = \varnothing$, hence a fortiori $\beta^*\mathcal{N}_{\zeta_{\mathcal{A}}}(y_0) \cap \beta \int_{\mathcal{F}} \mathcal{N}_{\zeta_{\mathcal{A}}}(\cdot) = \varnothing$. □

COROLLARY XII.8.14. *Let $\mathcal{A}$ be an AD family on X. Then*

$$\beta(\textstyle\int_{(Y)_0} \mathcal{N}_{\zeta_{\mathcal{A}}}(\cdot)) = \mathrm{cl}_\beta(\bigcup_{y\in F} \beta^*\mathcal{N}_{\zeta_{\mathcal{A}}}(y)) \setminus \bigcup_{y\in F} \beta^*\mathcal{N}_{\zeta_{\mathcal{A}}}(y)$$

the boundary of $\bigcup_{y\in F} \beta^*\mathcal{N}_{\zeta_{\mathcal{A}}}(y)$.

By (XII.8.1),

COROLLARY XII.8.15. *Let $\mathcal{A}$ be an AD family on X. If $\mathcal{F}$ is a free filter on Y, then $\int_{\mathcal{F}} \mathcal{N}_{\zeta_{\mathcal{A}}}(\cdot) \geq \mathrm{c}\mathcal{A}$.*

Hence, in particular,

$$\textstyle\int_{(Y)_0} \mathcal{N}_{\zeta_{\mathcal{A}}}(\cdot) \geq \mathrm{c}\mathcal{A}. \tag{XII.8.2}$$

COROLLARY XII.8.16. *The contour $\int_{(Y)_0} \mathcal{N}_{\zeta_{\mathcal{A}}}(\cdot)$ is equal to the residual filter $\mathrm{c}\mathcal{A}$ if and only if $\mathcal{A}$ is maximal.*

PROOF. By Corollary XII.8.14, $\beta \int_{(Y)_0} \mathcal{N}_{\zeta_{\mathcal{A}}}(\cdot)$ is the β-boundary of the β-open set $\bigcup_{y\in F} \beta^*\mathcal{N}_{\zeta_{\mathcal{A}}}(y)$. By Proposition XII.8.12, $\beta^*(\mathrm{c}\mathcal{A})$ is the β-boundary of $\bigcup_{y\in F} \beta^*\mathcal{N}_{\zeta_{\mathcal{A}}}(y)$ if and only if $\mathcal{A}$ is maximal. □

Compact almost disjoint topology. On $Y_\infty := X \sqcup Y \sqcup \{\infty\}$ we define a pretopology $\zeta_{\mathcal{A}}^\infty$ which coincides on $X \sqcup Y$ with $\zeta_{\mathcal{A}}$, and $\infty \in \lim_{\zeta_{\mathcal{A}}^\infty} \mathcal{U}$ for each $\mathcal{U} \in \beta^*(X \sqcup Y)$ such that $\lim_{\zeta_{\mathcal{A}}} \mathcal{U} = \varnothing$.

PROPOSITION XII.8.17. *Let $\mathcal{A}$ be an almost disjoint family. The extension $\zeta_{\mathcal{A}}^\infty$ of $\zeta_{\mathcal{A}}$ to Y_∞ is a Hausdorff compact topology.*

PROOF. The compactness of $\zeta_{\mathcal{A}}^\infty$ follows from the fact that if $\mathcal{U} \in \beta^*Y_\infty$ then either $Y \in \mathcal{U}$ and hence $\mathcal{U} \geq (Y)_0$, or $X \in \mathcal{U}$, in which case, either there is $y \in Y$ such that $A_y \in \mathcal{U}$, so that $y \in \lim_{\zeta_{\mathcal{A}}^\infty} \mathcal{U}$, or $\infty \in \lim_{\zeta_{\mathcal{A}}^\infty} \mathcal{U}$ by construction. As $\mathrm{c}\mathcal{A} = \bigcap\{\mathcal{U} \in \beta^*X : \mathcal{U} \perp \mathcal{A}\}$, it follows from pretopologicity that $\infty \in \lim_{\zeta_{\mathcal{A}}^\infty} \mathrm{c}\mathcal{A}$. From (XII.8.2) it follows that $\infty \in \lim_{\zeta_{\mathcal{A}}^\infty} \int_{(Y)_0} \mathcal{N}_{\zeta_{\mathcal{A}}}(\cdot)$, and since, by construction, $\mathcal{N}_{\zeta_{\mathcal{A}}}(y) = \mathcal{N}_{\zeta_{\mathcal{A}}^\infty}(y)$, the pretopology $\zeta_{\mathcal{A}}^\infty$ is a topology. □

In particular, the $\zeta_{\mathcal{A}}^\infty$-neighborhood filter at ∞ is equal to

$$\mathcal{N}_{\zeta_{\mathcal{A}}^\infty}(\infty) = \mathrm{c}\mathcal{A} \wedge (Y)_0 \wedge \infty^\uparrow. \tag{XII.8.3}$$

REMARK XII.8.18. Let us note that the residual filter $\mathrm{c}\mathcal{A}$ is the restriction to X of the $\zeta_{\mathcal{A}}$-cocompactoid filter. Indeed, $K \in \mathcal{K}(\zeta_{\mathcal{A}})$ if and only if it is a finite union of the elements of $\{A_y \cup \{y\} : y \in Y\}$ by convening that $A_x = \{x\}$ if $x \in X$. Hence $\mathcal{K}^\circ(\zeta_{\mathcal{A}}) \vee X = \mathcal{A}^{\cup\downarrow}$.

REMARK XII.8.19. A compact maximal almost disjoint topology is a quotient topology of the Stone topology, the corresponding topologically quotient map is in fact a perfect map.

Let X be an infinite countable set, and let $\beta = \beta\iota_X$ be the Stone topology on βX. On the other hand, let $\mathcal{A} = \{A_y : y \in Y\}$ be a maximal almost disjoint family on X.

Let $h : \beta X \longrightarrow Y_\infty$ be defined so that $h(x) := x$ for each $x \in X$, and $h(p) := y$ if $A_y \in \mathcal{U}(p)$, and $h(p) := \infty$ if $\mathcal{U}(p) \geq \mathrm{c}\mathcal{A}$. In other terms, $h(p) := y$ if $p \in \beta^* A_y$, and $h(p) := \infty$ if $p \in \beta^*(\mathrm{c}\mathcal{A})$. If $p \in \lim_\beta \mathcal{W}$ where $\mathcal{W} \in \beta^*(\beta X)$, then by Corollary XII.7.16, $\mathcal{U}(p) = \int_{\mathcal{W}} \mathcal{U}(\cdot)$.

If $p \in \beta^* A_y$, then $A_y \in \mathcal{U}(p)$, hence by (XII.7.7), $\beta A_y \in \mathcal{W}$. Therefore, $h(\beta A_y) \subset A_y \cup \{y\}$. In particular, if $q \in \beta^* A_y$, then $\mathcal{U}(q) \geq (A_y)_0$, and thus $y \in \lim_{\zeta_{\mathcal{A}}^\infty} h[\mathcal{W}]$. Similarly, if $p \in \beta^*(\mathrm{c}\mathcal{A})$, then $\mathcal{U}(p) \geq \mathrm{c}\mathcal{A}$ and $\infty \in \lim_{\zeta_{\mathcal{A}}^\infty} h[\mathcal{W}]$. Accordingly, $h \in \mathscr{C}(\beta, \zeta_{\mathcal{A}}^\infty)$, and as βX is compact, h is perfect.

XII.9. *Hyperconvergences and semicontinuities

In the case of the *upper* variant $\$_0$ of $\$$, by Lemma VII.5.1, $f \in \mathscr{C}(\xi, \$_0)$ if and only if $f^-(1)$ is ξ-closed, that is, $f = \chi_A$ is the characteristic function of a ξ-closed set A. Therefore, the *hyperset* $\mathcal{C}_\xi$ of all ξ-closed sets can be identified with $\mathscr{C}(\xi, \$_0)$ by

$$A \in \mathcal{C}_\xi \Longleftrightarrow \chi_A \in \mathscr{C}(\xi, \$_0).$$

As for the *lower* variant $\$_1$ of $\$$, by the same lemma, $f \in \mathscr{C}(\xi, \$_1)$ if and only if $f^-(1)$ is ξ-open, which is the characteristic function $f = \chi_A$ of a ξ-open set A. The *hyperset* $\mathcal{O}_\xi$ of ξ-open sets can be identified with $\mathscr{C}(\xi, \$_1)$ by

$$A \in \mathcal{O}_\xi \Longleftrightarrow \chi_A \in \mathscr{C}(\xi, \$_1).$$

Accordingly, the fact that if $f \in \mathscr{C}(\xi, \sigma)$ and $O \in \mathcal{O}_\sigma$, then $f^-(O) \in \mathcal{O}_\xi$, has an interpretation in terms of composition of continuous maps: if $\chi_O \in \mathscr{C}(\sigma, \$_1)$ and $f \in \mathscr{C}(\xi, \sigma)$, then $\chi_{f^-(O)} = \chi_O \circ f \in \mathscr{C}(\xi, \$_1)$ ([14]).

Therefore, convergences on sets of closed or open subsets of a convergence space constitute a special case of convergences on spaces of continuous maps.

[14]Because $1 = \chi_{f^-(O)}(x)$ if and only if $x \in f^-(O)$, equivalently $f(x) \in O$, that is, $1 = \chi_O(f(x)) = (\chi_O \circ f)(x)$.

If ξ is a topology on a set X and $\mathscr{F} \in \mathbb{F}2^X$ is a filter on the set of subsets of X, then

$$\text{(XII.9.1)} \qquad \mathrm{Ls}_\xi \mathscr{F} := \bigcap\nolimits_{F \in \mathscr{F}} \mathrm{cl}_\xi (\bigcup\nolimits_{A \in F} A),$$

$$\text{(XII.9.2)} \qquad \mathrm{Li}_\xi \mathscr{F} := \bigcap\nolimits_{H \in \mathscr{F}^\#} \mathrm{cl}_\xi (\bigcup\nolimits_{A \in H} A)$$

are the *upper* and *lower hyperlimits* of $\mathscr{F}$ with respect to ξ.

Note that both hyperlimits are ξ-closed subsets of X. As $\mathscr{F}$ is a filter, $\mathscr{F} \subset \mathscr{F}^\#$, and thus

$$\mathrm{Li}_\xi \mathscr{F} \subset \mathrm{Ls}_\xi \mathscr{F}.$$

If $\xi = \iota_X$ is the discrete topology on X, then they become

$$\text{(XII.9.3)} \qquad \mathrm{Ls}\mathscr{F} = \bigcap\nolimits_{F \in \mathscr{F}} \bigcup\nolimits_{A \in F} A, \qquad \mathrm{Li}\mathscr{F} = \bigcup\nolimits_{F \in \mathscr{F}} \bigcap\nolimits_{A \in F} A,$$

which are the *set-theoretic upper* and *lower* hyperlimits, respectively ([15]).

For a topology ξ on X, let $[\xi, \$_0]$ and $[\xi, \$_0]^\#$ be relations be defined by

$$A_0 \in \lim_{[\xi,\$_0]} \mathscr{F} \iff \mathrm{Ls}_\xi \mathscr{F} \subset A_0,$$
$$A_0 \in \lim_{[\xi,\$_0]^\#} \mathscr{F} \iff A_0 \subset \mathrm{Li}_\xi \mathscr{F}$$

for $A_0 \in 2^X$ and a filter $\mathscr{F}$ on 2^X. Observe that if $\mathscr{F} \subset \mathscr{G}$ then

$$\lim_{[\xi,\$_0]} \mathscr{F} \subset \lim_{[\xi,\$_0]} \mathscr{G}, \quad \lim_{[\xi,\$_0]^\#} \mathscr{F} \subset \lim_{[\xi,\$_0]^\#} \mathscr{G},$$

that is, $[\xi, \$_0]$ and $[\xi, \$_0]^\#$ are preconvergences on 2^X.

Another observation is that if $A_1 \supset A_0$ and $A_0 \in \lim_{[\xi,\$_0]} \mathscr{F}$, then $A_1 \in \lim_{[\xi,\$_0]} \mathscr{F}$. Therefore, if $A_0 \in \lim_{[\xi,\$_0]} \{A\}^\uparrow$, then $\mathrm{cl}_\xi A \subset A_0$. This means that $\mathrm{cl}_\xi A$ is the least subset of X that belongs to $\lim_{[\xi,\$_0]} \{A\}^\uparrow$, and that X belongs to every $[\xi, \$_0]$-limit. Therefore, typically $[\xi, \$_0]$ is not free (alias T_1).

Consequently, the restriction of a preconvergence $[\xi, \$_0]$ on 2^X to $\mathcal{C}_\xi$ is a convergence. Using here the term "convergence", we implicitly indicate that we talk about the restriction to the set of closed sets.

Note that if ξ is a compact topology, then $\varnothing \notin \lim_{[\xi,\$_0]} \mathscr{F}$ for every filter $\mathscr{F}$ such that $\{\varnothing\} \notin \mathscr{F}$; in other words, if ξ is a compact topology, then $\varnothing$ is $[\xi, \$_0]$-isolated.

---

[15]Traditionally, upper and lower hyperconvergences have been called respectively the *upper* and *lower Kuratowski* convergences; $[\xi, \$_0]^\#$ has been also called the *lower Vietoris* topology (see Section XX.10 for terminological and historical comments).

A useful concept in this study is the *reduced filter* $\mathrm{r}\mathscr{F}$ of a filter $\mathscr{F}$ on $\mathcal{C}_\xi$ ([16]), that is,

$$\mathrm{r}\mathscr{F} := \{\mathrm{r}(\mathrm{F}) : \mathrm{F} \in \mathscr{F}\}^{\uparrow},$$

where

$$\mathrm{r}(\mathrm{F}) := \bigcup\nolimits_{A \in \mathrm{F}} A.$$

As $\mathrm{F}_0 \cap \mathrm{F}_1 \in \mathscr{F}$ for any $\mathrm{F}_0, \mathrm{F}_1 \in \mathscr{F}$, and $\mathrm{r}(\mathrm{F}_0 \cap \mathrm{F}_1) = \mathrm{r}\mathrm{F}_0 \cap \mathrm{r}\mathrm{F}_1$, we conclude that $\mathrm{r}(\mathscr{F})$ is indeed a filter on $|\xi|$. However, the reduced filter of a proper filter need not to be proper. For example, since $\varnothing \in \mathcal{C}_\xi$, the proper filter $\{\varnothing\}^{\uparrow \mathcal{C}_\xi}$ on $\mathcal{C}_\xi$, but $\varnothing = \bigcup_{A \in \varnothing} A = \varnothing \in \mathrm{r}(\{\varnothing\}^{\uparrow \mathcal{C}_\xi})$.

Filters on hypersets are called *hyperfilters*. We notice that if $\mathscr{F}$ and $\mathscr{G}$ are hyperfilters, then

$$\mathscr{F} \leq \mathscr{G} \Longrightarrow \mathrm{r}\mathscr{F} \leq \mathrm{r}\mathscr{G}.$$

More generally, if $\boldsymbol{F}$ is a collection of hyperfilters, then

$$\mathrm{r}(\bigwedge \boldsymbol{F}) = \bigwedge\nolimits_{\mathscr{F} \in \boldsymbol{F}} \mathrm{r}\mathscr{F},$$

because $H \in \bigwedge_{\mathscr{F} \in \boldsymbol{F}} \mathrm{r}\mathscr{F}$ whenever $H \in \mathrm{r}\mathscr{F}$ for each $\mathscr{F} \in \boldsymbol{F}$. If $\mathrm{F}_{\mathscr{F}} \in \mathscr{F}$ implies $\mathrm{r}(\mathrm{F}_{\mathscr{F}}) \subset H$, then $\mathrm{r}(\bigcup_{\mathscr{F} \in \boldsymbol{F}} \mathrm{F}_{\mathscr{F}}) = \bigcup_{\mathscr{F} \in \boldsymbol{F}} \mathrm{r}(\mathrm{F}_{\mathscr{F}}) \subset H$. On the other hand, $\mathrm{F} \in \bigwedge \boldsymbol{F}$ whenever there is a selection $\mathrm{F}_{\mathscr{F}} \in \mathscr{F}$ such that $\bigcup_{\mathscr{F} \in \boldsymbol{F}} \mathrm{F}_{\mathscr{F}} \subset \mathrm{F}$.

On recalling that if ξ is a topology, then $\mathrm{adh}_\xi \mathcal{F} = \bigcap_{F \in \mathcal{F}} \mathrm{cl}_\xi F$, we infer from (XII.9.1) that

LEMMA XII.9.1. *If ξ is a topology, $\mathscr{F}$ is a hyperfilter on $\mathcal{C}_\xi$ and $A_0 \in \mathcal{C}_\xi$, then*

$$A_0 \in \lim\nolimits_{[\xi,\$_0]} \mathscr{F} \Longleftrightarrow \mathrm{adh}_\xi \mathrm{r}\mathscr{F} \subset A_0. \tag{XII.9.4}$$

If $A_1 \supset A_0$ and $A_1 \in \lim_{[\xi,\$_0]^\#} \mathscr{F}$, then $A_0 \in \lim_{[\xi,\$_0]^\#} \mathscr{F}$. Moreover, $\mathrm{cl}_\xi A$ is the largest subset of X that belongs to $\lim_{[\xi,\$_0]^\#} \{A\}^\uparrow$. In particular, $A \in \lim_{[\xi,\$_0]^\#} \{A\}^\uparrow$ for every $A \subset X$, and thus $[\xi, \$_0]^\#$ is a convergence on 2^X.

PROPOSITION XII.9.2. *If ξ is a topology, then $[\xi, \$_0]^\#$ is a topology* ([17]).

PROOF. Let ([18])

$$O^\# := \{A \in \mathcal{C}_\xi : A \# O\}. \tag{XII.9.5}$$

We will show that $\{O^\# : O \in \mathcal{O}_\xi\}$ is a subbase for the open sets of $[\xi, \$_0]^\#$ (Definition IX.2.4) and that $\mathrm{T}([\xi, \$_0]^\#) = [\xi, \$_0]^\#$.

[16] The notion of *reduced* and *erected* filters was introduced by A. LECHICKI in [**172**].
[17] And if ξ is a pretopology, then $[\xi, \$_0]^\#$ is a pretopology.
[18] Recall that $A \# O$ means that $A \cap O \neq \varnothing$.

By definition, $A_0 \in \lim_{[\xi,\$_0]^\#} \mathscr{F}$ if and only if, for each $x \in A_0$ every $\mathrm{H} \in \mathscr{F}^\#$ and every $O \in \mathcal{O}_\xi(x)$, there exists $A \in \mathrm{H}$ such that $A \cap O \neq \varnothing$. Equivalently, for each $x \in A_0$ and every $O \in \mathcal{O}_\xi(x)$, there exists $\mathrm{F} \in \mathscr{F}$ such that $A \cap O \neq \varnothing$ for each $A \in \mathrm{F}$, that is to say, $A_0 \in O^\# \in \mathscr{F}$. This proves that $A_0 \in \lim_{[\xi,\$_0]^\#} \mathscr{F}$ if and only if $O^\# \in \mathscr{F}$ for every $O \# A_0$ (In particular, $O^\#$ is $[\xi, \$_0]^\#$-open for every ξ-open O). Therefore, $\{O^\# : A_0 \# O, O \in \mathcal{O}_\xi\}$ is a filter-subbase (Definition II.2.3) of the neighborhood filter $\mathcal{N}_{[\xi,\$_0]^\#}(A_0)$. □

Of course, hyperconvergences just discussed have their isomorphic counterparts on hyperspaces of open sets, that is, $[\xi, \$_1]$ is a convergence on $\mathcal{O}_\xi \approx \mathscr{C}(\xi, \$_1)$ and $[\xi, \$_1]^\#$ is a convergence on the whole of 2^X ([19]). This time,

$$A_0 \in \lim_{[\xi,\$_1]} \mathscr{F} \iff A_0 \subset \bigcup_{\mathrm{F}\in\mathscr{F}} \mathrm{int}_\xi(\bigcap_{A\in\mathrm{F}} A),$$

$$A_0 \in \lim_{[\xi,\$_1]^\#} \mathscr{F} \iff \bigcup_{\mathrm{H}\in\mathscr{F}^\#} \mathrm{int}_\xi(\bigcap_{A\in\mathrm{H}} A) \subset A_0.$$

Lower hyperconvergences is intimately related to lower semicontinuity, which we discussed in Section IV.9.

Consider a relation $R \subset X \times Y$, a convergence ξ on X and a topology τ on Y. In virtue of our convention, $Rx := \{y \in Y : (x, y) \in R\}$ and $RA := \bigcup_{x\in A} Rx$ for any $A \subset X$.

A map $R^\bullet : X \longrightarrow 2^Y$ given by $R^\bullet(x) := Rx$ is called a *multifunction* corresponding to R. Therefore, for a subset F of X,

$$R^\bullet(F) = \{R^\bullet(x) \in 2^Y : x \in F\},$$

and for a family $\mathcal{F}$ of subsets of X,

$$R^\bullet[\mathcal{F}] = \{R^\bullet(F) : F \in \mathcal{F}\}.$$

A relation R is said to be *lower semicontinuous* at x_0 if it is lower semicontinuous at (x_0, y_0) for every $y_0 \in Rx_0$ (Definition IV.9.1).

PROPOSITION XII.9.3. *The relation R is (ξ, τ)-lower semicontinuous at x_0 if and only if $R^\bullet(x_0) \in \lim_{[\tau,\$_0]^\#} R^\bullet[\mathcal{F}]$ for every filter $\mathcal{F}$ such that $x_0 \in \lim_\xi \mathcal{F}$.*

PROOF. By definition, R is (ξ, τ)-lower semicontinuous at x_0 if and only if for every τ-open set O such that $O \cap Rx_0 \neq \varnothing$ there is $F \in \mathcal{F}$ such that $O \cap Rx \neq \varnothing$ for each $x \in F$. Equivalently, if $R^\bullet(x_0) \in O^\#$ implies that $R^\bullet(F) \subset O^\#$. □

[19]The former has been traditionally called the *Scott convergence*; the latter has not been much used.

COROLLARY XII.9.4. *Let ξ and τ be topologies on X and Y respectively. Then $R \subset X \times Y$ is (ξ, τ)-lower semicontinuous if and only if $R^{-}O \in \mathcal{O}_{\xi}$ for each $O \in \mathcal{O}_{\tau}$.*

If ξ is a topology on X and τ is a topology on Y, then a relation $R \subset X \times Y$ is said to be *graph-closed* at $x_0 \in X$ whenever $\operatorname{adh}_{\tau} R[\mathcal{N}_{\xi}(x_0)] \subset Rx_0$.

PROPOSITION XII.9.5. *A relation R is (ξ, τ)-graph-closed at x_0 if and only if $Rx_0 \in \lim_{[\tau,\$_0]} R^{\bullet}[\mathcal{F}]$ for every filter $\mathcal{F}$ such that $x_0 \in \lim_{\xi} \mathcal{F}$.*

PROOF. If $x_0 \in \lim_{\xi} \mathcal{F}$ implies that $R^{\bullet}(x_0) \in \lim_{[\tau,\$_0]} R^{\bullet}[\mathcal{F}]$ if and only if

$$\bigcap\nolimits_{F \in \mathcal{F}} \operatorname{cl}_{\tau} RF \subset Rx_0.$$

$\square$

If $f \in \mathscr{C}(\xi, \tau)$ and τ is Hausdorff, then the relation R, defined by $Rx := \{f(x)\}$ for each x, is graph-closed. In fact, $x \in \lim_{\xi} \mathcal{F}$ implies that $\{f(x)\} = \lim_{\tau} f[\mathcal{F}] = \operatorname{adh}_{\tau} f[\mathcal{F}]$ by Proposition VIII.1.3. The converse implication need not hold (Example VIII.4.3).

XII.10. Supplement

LEMMA XII.10.1. *Let ξ be a compact topology on X and let $\mathcal{H}$ be a filter on X. If $O \in \mathcal{O}_{\xi}(\operatorname{adh}_{\xi} \mathcal{H})$, then $O \in \mathcal{H}$.*

PROOF. Suppose that $O \notin \mathcal{H}$, that is, $X \setminus O \in \mathcal{H}^{\#}$. Then $X \setminus O$ is ξ-closed, hence ξ-compact, because ξ is compact. Consequently, $\operatorname{adh}_{\xi} \mathcal{H} \cap (X \setminus O) \neq \varnothing$, contrary to the assumption that $\operatorname{adh}_{\xi} \mathcal{H} \subset O$. $\square$

This fact will be generalized in Lemma XX.6.8.

COROLLARY XII.10.2. *If ξ is a compact topology on X, and $(H_n)_n$ is a decreasing sequence of non-empty closed sets, and $O \in \mathcal{O}_{\xi}(\bigcap_{n<\omega} H_n)$, then there is $n_O < \omega$ such that $H_n \subset O$ for each $n > n_O$.*

EXERCISE XII.10.3. Prove the equivalences of Proposition XII.1.11.

(1) If $\mathcal{U} \in \beta A$, then $\lim \mathcal{U} \cap A \neq \varnothing$,
(2) If a filter $\mathcal{F}$ meshes A, then $\operatorname{adh} \mathcal{F} \cap A \neq \varnothing$,
(3) Each sequence of elements of A has a subsequence convergent to an element of A,
(4) Each countable open cover of A has a finite subcover,
(5) Each open cover of A has a finite subcover.

SOLUTION. Let ξ standard topology of a Euclidean space E. If $A \in \mathcal{F}^{\#}$, and $\mathcal{U} \in \beta(\mathcal{F} \vee A)$, then, by (1), $\varnothing \neq \lim_{\xi} \mathcal{U} \cap A \subset \operatorname{adh}_{\xi} \mathcal{F} \cap A$. Let $(x_n)_n$ be a sequence of elements of A. Then, by (2), there exists $x \in \operatorname{adh}_{\xi} \mathcal{F} \cap A$, where $\mathcal{F}$ is the filter generated by $(x_n)_n$. Therefore

$$x \in A \cap \bigcap\nolimits_{F \in \mathcal{F}} \operatorname{adh}_{\xi} F = A \cap \bigcap\nolimits_{k<\omega} \operatorname{adh}_{\xi}\{x_n : n > k\},$$

hence, by Lemma XXI.1.8, there exists a subsequence $(x_{n_k})_k$ such that $x \in \lim_{\xi}(x_{n_k})_k$, which yields (3).

Suppose that $(Q_n)_n$ is an increasing open cover of A such that $A \setminus Q_n \neq \varnothing$ for each $n < \omega$. Hence, there exists a sequence $(x_n)_n$ such that $x_n \in A \setminus Q_n$. By (3), there exists a subsequence $(x_{n_k})_k$ such that $\{x\} = \lim_{\xi}(x_{n_k})_k$. Therefore, $x \in A \setminus \bigcup_{n<\omega} Q_n$, hence $(Q_n)_n$ is not a cover of A, contrary to the assumption.

If $\mathcal{Q}$ is an arbitrary open cover of A, and $E_{\mathbb{Q}}$ is the set of all elements of E having all rational factors, then for each $x \in A$ and each $Q \in \mathcal{Q}$ such that $x \in Q$ there exists $x_Q \in Q \cap E_{\mathbb{Q}}$. As $E_{\mathbb{Q}}$ is countable, there a countable subfamily of $\mathcal{Q}$ that is a cover of A. By (4), there is a finite subfamily of $\mathcal{Q}$ that covers A.

Let $\mathcal{U} \in \beta A$. If $\varnothing = A \cap \lim_{\xi} \mathcal{U} = A \cap \bigcap_{U \in \mathcal{U}} \operatorname{cl}_{\xi} U$, equivalently, $A \subset \bigcup_{U \in \mathcal{U}} \operatorname{int}_{\xi} U^c$, then, by (5), there a finite family $\mathcal{D} \subset \mathcal{U}$ such that $A \cap \bigcap_{U \in \mathcal{D}} \operatorname{cl}_{\xi} U = \varnothing$. Hence, *a fortiori*, $A \cap \bigcap_{U \in \mathcal{D}} U = \varnothing$, and since $\bigcap_{U \in \mathcal{D}} U \in \mathcal{U}$, we conclude that $A \notin \mathcal{U}^{\#} = \mathcal{U}$, contrary to the assumption.

EXERCISE XII.10.4. Show that there exists a Hausdorff compact topology that is not separable.

SOLUTION. For an infinite countable set X, the set $\beta^* X$ of free ultrafilters endowed with the Stone topology is compact but not separable. As X is β-open, $\beta|_{\beta^* X}$ is compact. If, by way of contradiction, $\{p_n \in \beta^* X : n < \omega\}$ is β-dense. If $p \notin \boldsymbol{U}$ then for each $n < \omega$, there is an infinite $A_n \subset X$ such that $p \in \beta^* A_n$ and $p_n \in \beta^*(X \setminus A_n)$. By Lemma XVI.2.1, there is an infinite $A_\infty \subset X$ that is almost included in A_n for each $n < \omega$, that is, $\beta^* A_\infty \subset \bigcap_{n<\omega} \beta^* A_n$. Therefore, $\varnothing \neq \operatorname{int}_{\beta} \bigcap_{n<\omega} \beta^* A_n$, hence $\beta^* X \neq \operatorname{cl}_{\beta}(\bigcup_{n<\omega} \beta^*(X \setminus A_n)) \supset \operatorname{cl}_{\beta}\{p_n : n < \omega\}$, in contradiction of density.

EXAMPLE XII.10.5. The *lexicographic order topology* of the unit square from Example IX.3.14 is a Hausdorff compact topology of countable character. In view of Exercise XII.10.4, it is not metrizable, because it is not separable.

A topology is called *hypercompact* if every filter converges. Each chaotic topology is hypercompact. An example of non-chaotic hypercompact topology is the Sierpiński topology $.

EXERCISE XII.10.6. If a free topology is hypercompact, then its underlying set is a singleton.

SOLUTION. Let ξ be a free topology on X, that is, $\lim_\xi x^\uparrow \subset \{x\}$ for each $x \in X$. If there exist $x_0, x_1 \in X$ such that $x_0 \neq x_1$ and ξ is hypercompact, then

$$\varnothing \neq \lim_\xi \{\{x_0, x_1\}\}^\uparrow \subset \lim_\xi x_0^\uparrow \cap \lim_\xi x_1^\uparrow \subset \{x_0\} \cap \{x_1\} = \varnothing,$$

a contradiction.

If ξ is a topology on X, then $x \in X$ is called a *point of local compactness* if there exists a ξ-compact ξ-neighborhood of x.

EXERCISE XII.10.7. Show that the points of local compactness of a topology constitute an open set.

SOLUTION. Let ξ be a topology on X. If x is a point of local compactness of ξ, then there is a ξ-compact set K such that $x \in \operatorname{int}_\xi K$. Each element of the ξ-open set $\operatorname{int}_\xi K$ is of local compactness of ξ.

EXERCISE XII.10.8. Show that a Hausdorff locally compact topology is functionally regular.

SOLUTION. Let ξ a Hausdorff locally compact topology on a set X, $F \in \mathcal{C}_\xi$ and $x \notin F$. Then there exists a ξ-compact set K such that $x \in \operatorname{int}_\xi K \setminus F$. Let $O \in \mathcal{O}_\xi(x)$ such that $O \subset K \setminus F$. As $\xi|_K$ is a Hausdorff compact topology, it is, in particular, functionally regular, hence there exists $f \in \mathscr{C}(\xi_K, I)$ such that $f(x) = 1$ and $f(K \setminus O) = 0$. Let $h \in [0,1]^X$ coincide with f on K and $h(w) = 0$ for $w \in X \setminus K$. Then $h \in \mathscr{C}(\xi, I)$, because $h^-(0)$ is ξ-closed and $\{f > 0\} \subset K$, and $F \subset h^-(0)$.

EXAMPLE XII.10.9. Let $j_X : X \longrightarrow \beta X$ be the *canonical injection* given by $j_X(x) := x^\uparrow$. In other words, j_X associates with each $x \in X$, the principal ultrafilter $x^\uparrow$. By Theorem XII.7.17, there is a unique $J_X \in \mathscr{C}(\beta X, \beta X)$ such that $J_X(x) = x^\uparrow$, namely $J_X(p) \in \lim_\beta j_X[\mathcal{U}(p)]$. By Corollary XII.7.16, this means that $\mathcal{U}(J_X(p)) = \int_{j_X[\mathcal{U}(p)]} \mathcal{U}(\cdot)$. Therefore, $A \in \mathcal{U}(J_X(p))$ if and only if there is $U \in \mathcal{U}(p)$ such that $A \in x^\uparrow$ for each $x \in U$, that is, $U \subset A$. As a result, $\mathcal{U}(J_X(p)) = \mathcal{U}(p)$, that is, $J_X(p) = p$ for each $p \in \beta X$.

Actually, by interpreting conclusions of a general theorem, we have recovered a known fact: as for the identification bijection $I_\beta : \beta X \longrightarrow$

$\mathbb{U}X$ given $I(p) := \mathcal{U}(p)$, the inverse map I_β^{-1} coincides with J_X on a dense subset of βX ([20]), by Lemma IX.1.3, $J_X = I_\beta^{-1}$.

Actually, every Hausdorff compact set is a continuous image of a Cantor cube. The dimension of the cube depends on the weight of the considered compact set.

THEOREM XII.10.10. *Each Hausdorff compact topology of infinite weight κ is a continuous image of a closed subspace of the κ-dimensional Cantor cube.*

PROOF. We introduce here a specific notation with a view to apply Lemma XIV.1.6. For each Hausdorff compact topology τ, there is a cardinal κ such that τ can be embedded into a Sierpiński cube $\$^\kappa$ in virtue of Corollary IX.5.3. This amounts to $\tau = (\$^\kappa)|_Y$, where Y is a subset of $\{0,1\}^\kappa$. Let $i : \{0,1\}^\kappa \longrightarrow \{0,1\}^\kappa$ be the identity map. As $\iota_{\{0,1\}} \geq \$$, we conclude that $i \in \mathscr{C}(\iota_{\{0,1\}}{}^\kappa, \$^\kappa)$. Let $A := i^-(Y)$ and $X := \mathrm{cl}_{\iota_{\{0,1\}}{}^\kappa} A$, and finally set $\xi := (\iota_{\{0,1\}}{}^\kappa)\big|_X$.

Denote $f := i_A$, the restriction to A of the identity map i. Then $f \in \mathscr{C}(\xi_A, \tau)$ and to conclude, we need to extend it to $\mathscr{C}(\xi, \tau)$. In order to apply Lemma XIV.1.6, consider disjoint τ-closed sets B_0 and B_1. By construction, there exist $\$^\kappa$-closed sets K_0 and K_1 such that $B_0 = K_0 \cap Y$ and $B_1 = K_1 \cap Y$. As a result, $(K_0 \cap K_1) \cap Y = \varnothing$.

We proceed carefully, having in mind that $\$^\kappa$ is not Hausdorff. If $\mathcal{D}$ is a standard base for the $\$^\kappa$-open sets, then D is $\iota_{\{0,1\}}{}^\kappa$-clopen for each $D \in \mathcal{D}$. Then, for each $y \in Y$ there is $y \in D_y \in \mathcal{D}$ such that $D_y \cap K_0 \cap K_1 = \varnothing$, and since Y is $\$^\kappa$-compact, there is a finite subset Y_0 of Y such that the $\iota_{\{0,1\}}{}^\kappa$-clopen $O := \bigcup_{y \in Y_0} D_y$ includes $Y = A$, and is disjoint from $K_0 \cap K_1$. Therefore,

$$X = \mathrm{cl}_{\iota_{\{0,1\}}{}^\kappa} A \subset O \setminus (K_0 \cap K_1). \tag{XII.10.1}$$

Accordingly, $f^-(B_0) = K_0 \cap f^-(Y) = K_0 \cap A$, and since K_0 is *a fortiori* $\iota_{\{0,1\}}{}^\kappa$-closed, $\mathrm{cl}_\xi f^-(B_0) \subset K_0 \cap X$, and of course, $\mathrm{cl}_\xi f^-(B_1) \subset K_1 \cap X$. As a result, by (XII.10.1),

$$\mathrm{cl}_\xi f^-(B_0) \cap \mathrm{cl}_\xi f^-(B_1) \subset K_0 \cap K_1 \cap X = \varnothing.$$

By Lemma XIV.1.6, there exists $F \in \mathscr{C}(\xi, \tau)$ such that $F_A = f$, so that $F\xi = \tau$ and ξ is a closed subspace topology of a Cantor cube. □

[20]We abuse here slightly the notation, by identifying x with $x^\uparrow$.

CHAPTER XIII

Connected and disconnected topologies

A convergence is connected if the empty set and the whole underlying set are the only clopen sets. As we see, connectedness is defined in purely topological terms, and thus is a topological property. Consequently, we will discuss connectedness uniquely in topological setting. Connectedness for non-topological convergences is outlined in [83].

A study of a class implies to some extent a study of its complementary class. The title of this chapter evokes explicitly connected and disconnected topologies, because we shall be equally interested by both-classes.

XIII.1. Connected topologies

DEFINITION XIII.1.1. A topology ξ is called *connected* if the empty set and the whole space are the only sets that are both open and closed:

$$\mathcal{C}_\xi \cap \mathcal{O}_\xi = \{\varnothing, |\xi|\}.$$

A topology is called *disconnected* if it is not connected.

Accordingly, each chaotic topology is connected. On the other hand, a discrete topology ι_X is connected if and only if X is a singleton, say $X = \{x\}$. If ζ is connected and $\zeta \geq \xi$, then $\mathcal{C}_\xi \cap \mathcal{O}_\xi \subset \mathcal{C}_\zeta \cap \mathcal{O}_\zeta = \{\varnothing, X\}$, hence ξ is connected.

A topology ξ on X is disconnected if and only if there exists $A \subset X$ such that $\varnothing \neq A \neq X$ and $\{A, X \setminus A\} \subset \mathcal{O}_\xi$, in other words, if there exists a proper ξ-clopen set.

LEMMA XIII.1.2. *A topology is connected if and only if every continuous map to* $\{0,1\}$ *endowed with the discrete topology is constant.*

PROOF. Let ξ be a topology on X and $f \in \mathscr{C}(\xi, \iota_{\{0,1\}})$. If $f(X) = \{0,1\}$, then $f^-(0), f^-(1)$ are non-empty ξ-closed and complementary, so that ξ is not connected. If ξ is not connected, then there exists a non-trivial ξ-clopen set A, then $\chi_A \in \mathscr{C}(\xi, \iota_{\{0,1\}})$ is not constant. □

More generally,

PROPOSITION XIII.1.3. *If* ζ *is a connected topology and* $f \in \mathscr{C}(\zeta, \xi)$ *is surjective, then* ξ *is connected.*

PROOF. If, ξ is disconnected, then by Lemma XIII.1.2, there exists a surjection $h \in \mathscr{C}(\xi, \iota_{\{0,1\}})$. Then $h \circ f \in \mathscr{C}(\zeta, \iota_{\{0,1\}})$ is surjective, hence ζ is disconnected. □

EXAMPLE XIII.1.4. The Sierpiński topology is connected. Indeeed, $\mathcal{O}_{\$_0} = \{\varnothing, \{0\}, \{0,1\}\}$, so that there is no proper $\$_0$-clopen set.

DEFINITION XIII.1.5. A subset of a topological space is called *connected* if the restricted topology is connected.

PROPOSITION XIII.1.6. *A subset of a real line is connected if and only if it is an interval.*

Recall that an interval I is a subset of $\mathbb{R}$ such that if $r_0, r_1 \in I$ with $r_0 \leq r_1$, then $[r_0, r_1] \subset I$.

PROOF. Let I be an interval. Suppose that by way of contradiction I is not connected. Then there exist $a, b \in I$ such that $a < b$ and disjoint closed sets A, B such that $a \in A, b \in B$ and $A \cup B = I$. Let

$$c := \sup([a, b] \cap A).$$

As A is closed, $c \in A$, hence $]c, b] \subset B$, hence $c \in B$ since B is closed in I, which is a contradiction.

Suppose that $A \subset \mathbb{R}$ is not an interval. Then there exists $r \notin A$ and the sets $]-\infty, r[\cap A$ and $]r, +\infty[\cap A$ are both non-empty. By Proposition VII.3.12, they are open in the topology restricted to A. As they are complementary in A, they are also closed, and thus A is disconnected. □

COROLLARY XIII.1.7 (Bolzano). *The image under a continuous function of a connected topological space to the real line is an interval.*

PROPOSITION XIII.1.8. *If ξ is a functionally Hausdorff connected topology on a set X with* $\operatorname{card} X > 1$. *Then* $\operatorname{card} X \geq \mathfrak{c}$.

PROOF. Indeed, if ξ is a functionally Hausdorff topology on X and $x_0 \neq x_1$, then there is $f \in \mathscr{C}(\xi, \nu)$ such that $f(x_0) = 0$ and $f(x_1) = 1$, then $[0, 1] \subset f(X)$, and thus $\mathfrak{c} \leq \operatorname{card} f(X) \leq \operatorname{card} X$. □

COROLLARY XIII.1.9. *Every connected Hausdorff functionally regular (alias completely regular) free (alias T_1) topology which is not a singleton is of cardinality greater than $\mathfrak{c}$.*

If ξ is a topology, then two sets A, B are called *ξ-separated* if

$$\operatorname{cl}_\xi A \cap B = A \cap \operatorname{cl}_\xi B = \varnothing$$

(see Definition IX.1.23). Separated sets are necessarily disjoint, but not conversely (see Example IX.1.24). If, however, two disjoint sets are either both open or both closed, then they are separated.

LEMMA XIII.1.10. *Each connected subset of the union of two separated sets is included in one of them.*

PROOF. By definition, a set in a topological space is connected if it is connected as a subspace. By Corollary IX.1.26, two separated sets A and B are closed and disjoint in $A \cup B$, so that if a connected set $D \subset A \cup B$, either $D \subset A$ or $D \subset B$. □

COROLLARY XIII.1.11. *If two connected subsets are not separated, then their union is connected.*

More generally,

PROPOSITION XIII.1.12. *If $\{A_t : t \in T\}$ is a family of ξ-connected sets, and there is $t_0 \in T$ such that A_t, A_{t_0} are not separated for each $t \in T$, then $\bigcup_{t\in T} A_t$ is ξ-connected.*

PROOF. Let C_0, C_1 be ξ-separated sets such that $\bigcup_{t\in T} A_t = C_0 \cup C_1$. As A_{t_0} is ξ-connected, by Lemma XIII.1.10, A_{t_0} is included in one of them, say $A_{t_0} \subset C_0$. By assumption, A_t, A_{t_0} are not separated for every $t \neq t_0$, hence $A_t \cup A_{t_0}$ is connected, and thus $A_t \cup A_{t_0} \subset C_0$ for each $t \in T$. Hence $C_1 = \varnothing$. By Corollary XIII.1.11, $\bigcup_{t\in T} A_t$ is ξ-connected. □

In particular, if $\bigcap_{t\in T} A_t \neq \varnothing$ and A_t is connected for each $t \in T$, then $\bigcup_{t\in T} A_t$ is connected.

COROLLARY XIII.1.13. *If for each pair $a_0, a_1 \in A$, there exists a ξ-connected set C such that $\{a_0, a_1\} \subset C \subset A$, then A is ξ-connected.*

PROOF. Let $a_0 \in A$. By assumption, for each $a \in A$, there exists a ξ-connected set C_a such that $\{a_0, a\} \subset C_a$. Then A is ξ-connected by Proposition XIII.1.12. □

PROPOSITION XIII.1.14. *If A is ξ-connected and $A \subset B \subset \mathrm{cl}_\xi A$, then B is ξ-connected.*

PROOF. Let $f \in \mathscr{C}(\xi_B, \iota_{\{0,1\}})$. As A is ξ-connected, f is constant on A, hence by Lemma IX.1.3, f is constant on $\mathrm{cl}_\xi A$. □

It follows that if a set A includes a dense connected subset, then A is connected.

COROLLARY XIII.1.15. *If $(F_n)_n$ is a decreasing sequence of non-empty, compact, connected sets in a Hausdorff topological space, then $\bigcap_{n<\omega} F_n$ is connected.*

PROOF. We can assume without loss of generality that a topology on a set X is compact ([1]). Then $F := \bigcap_{n<\omega} F_n$ is non-empty and compact. By way of contradiction, let K_0 and K_1 be disjoint non-empty compact sets such that $F = K_0 \cup K_1$. By Proposition IX.3.4, X is normal, hence there exist disjoint open sets O_0 and O_1 such that $K_0 \subset O_0$ and $K_1 \subset O_1$. For every $n < \omega$,

$$F_{0,n} := F_n \setminus O_1, \qquad F_{1,n} := F_n \setminus O_0.$$

Then $\bigcap_{n<\omega} F_{0,n} = K_0$ and $\bigcap_{n<\omega} F_{1,n} = K_1$. By Corollary XII.10.2, there exists $N < \omega$ such that $F_{0,N} \subset O_0$ and $F_{1,N} \subset O_1$. Consequently, $F_N = F_{N,0} \sqcup F_{N,1}$ is a disjoint union of closed subsets of X, hence is disconnected, contrary to the assumption. □

REMARK XIII.1.16. The intersection of a sequence of non-empty closed connected sets need not to be connected (see Exercise XIII.7.2).

THEOREM XIII.1.17. *The product of topologies is connected if and only if each factor is connected.*

PROOF. If X, Y are connected topological spaces, then for each pair $(x_0, y_0), (x_1, y_1) \in X \times Y$, the sets $\{x_0\} \times Y$ and $X \times \{y_1\}$ are homeomorphic to X and Y respectively, hence connected. $(x_0, y_1) \in (\{x_0\} \times Y) \cap (X \times \{y_1\})$, hence $\{x_0\} \times Y) \cup (X \times \{y_1\}$ is connected by Corollary XIII.1.11. By Corollary XIII.1.13, $X \times Y$ is connected. It follows by induction that each product of finitely many connected spaces is connected.

Let now X_j be a topological space for each $j \in J$. If $\prod_{j \in J} X_j$ is connected, then X_j is connected in virtue of Proposition XIII.1.3. Conversely, let X_j be connected for each $j \in J$. Pick any $\varphi \in \prod_{j\in J} X_j$, and for each finite F subset of J, define $A_j := X_j$ if $j \in F$, and $A_j := \{\varphi(j)\}$ if $j \notin F$. Then $A_F := \prod_{j \in J} A_j$ is connected, because it is homeomorphic to $\prod_{j\in F} X_j$. As $\varphi \in \bigcap_{F \subset_{\text{fin}} J} A_F$, the union $A := \bigcup_{F\subset_{\text{fin}} J} A_F$ is connected.

On the other hand, A is dense in $\prod_{j\in J} X_j$, hence $\prod_{j \in J} X_j$ is connected. Indeed, if $x \in \prod_{j\in J} X_j$ and $V \in \mathcal{N}_{\prod_{j\in J} X_j}(x)$, then there exist $F \subset_{\text{fin}} J$ and $V_j \in \mathcal{N}_{X_j}(x(j))$ for each $j \in J$ such that $V_j := X_j$ if $j \notin F$ and $V_F := \prod_{j\in J} V_j \subset V$. Therefore, $V \cap A \supset V_F \cap A_F \neq \varnothing$, because if $j \in F$, then $V_j \cap A_j = V_j \cap X_j$, and if $j \notin F$, then $V_j \cap A_j = \{x(j)\}$. □

As a result, the Tikhonov cube of any dimension is connected, and the Cantor cube of any dimension is disconnected.

[1] Set $X := F_0$.

XIII.2. Components and quasi-components

DEFINITION XIII.2.1. The *component* $C(x)$ of a point x of a topological space X is the union of all connected sets that contain x.

It is plain that each component is connected and closed by Proposition XIII.1.14. Two components are either disjoint or coincide. Therefore, all the components constitute a partition of the space. Obviously, a topology on X is connected whenever X is a component of each $x \in X$.

DEFINITION XIII.2.2. The *quasi-component* $Q(x)$ of a point x of a topological space X is the intersection of all clopen sets that contain x.

It follows that each quasi-component is closed.

PROPOSITION XIII.2.3. *Two quasi-components are either disjoint or coincide.*

PROOF. Let X be a topological space. If $x_1 \notin Q(x_0)$, then by definition, there is a clopen set A such that $x_0 \in A$ and $x_1 \in X \setminus A$. As $X \setminus A$ is clopen, $x_1 \in Q(x_1) \subset X \setminus A$. □

As a result, all the quasi-components form a partition of X.

If D is a connected set and $x \in D$, then $D \subset A$ for each clopen set A such that $x \in A$. Otherwise, $D \cap A$ and $D \setminus A$ would be separated. Therefore, $D \subset Q(x)$, hence

$$C(x) \subset Q(x), \tag{XIII.2.1}$$

for each $x \in X$.

Obviously, if $Q(x)$ is connected, then $Q(x) = C(x)$.

EXAMPLE XIII.2.4 (A quasi-component that is not a component). Let $X_n := [-1,1] \times \{2^{-n}\}$ for each $n < \omega$, $X_\infty := [-1,1] \times \{0\}$, and let $X := \bigcup_{n<\omega} X_n \cup X_\infty \setminus \{(0,0)\}$ with the topology restricted from the standard plane topology. It is straightforward that the components of X are X_n for each $n < \omega$, $[-1,0[\times\{0\}$ and $]0,1] \times \{0\}$.

We will see that the quasi-component of $(1,0)$ is $X_\infty \setminus \{(0,0)\}$. Indeed, if $Q \in \mathcal{O}(1,0)$, then there is n_Q such that $Q \# X_n$ for each $n > n_Q$. If, moreover, Q is closed, then $Q \cap X_n$ is a clopen subset of a connected set X_n, and if $n > n_Q$, then it is non-empty. Therefore, $X_n \subset Q$ for $n > n_Q$. And since $\bigcap_{n_Q<\omega} \operatorname{cl} \bigcup_{n>n_Q} X_n = X_\infty \setminus \{(0,0)\}$, the set $X_\infty \setminus \{(0,0)\}$ is an intersection of clopen sets containing $(1,0)$.

THEOREM XIII.2.5. *In each Hausdorff compact topological space the components and the quasi-components coincide.*

PROOF. Let ξ be a Hausdorff compact topology on X. It is enough to show that each quasi-component is ξ-connected. Suppose that there exist two ξ-closed disjoint sets F_0, F_1 such that $Q(x) = F_0 \cup F_1$, and $x \in F_0$. By Proposition IX.3.4, ξ is normal, hence there exist disjoint ξ-open sets O_0 and O_1 such that $F_0 \subset O_0$ and $F_1 \subset O_1$. The family $\mathcal{G}(x)$ of ξ-clopen sets that contain x is closed under finite intersections, hence is a filter-base, and $Q(x) \in \mathcal{G}(x)$. Since $Q(x)$ is ξ-compact, by Lemma XII.10.1, there exists $G \in \mathcal{G}(x)$ such that $Q(x) \subset G \subset O_0 \cup O_1$. As $\mathrm{cl}_\xi O_0 \cap O_1 = \varnothing$,

$$\mathrm{cl}_\xi(O_0 \cap G) \subset \mathrm{cl}_\xi O_0 \cap G = \mathrm{cl}_\xi O_0 \cap G \cap (O_0 \cup O_1) = O_0 \cap G,$$

which shows that $O_0 \cap G$ is ξ-clopen and $x \in O_0 \cap G$, and thus $Q(x) \subset O_0 \cap G$. Therefore, $F_1 \subset Q(x) \subset O_0$, which yields a contradiction. □

LEMMA XIII.2.6. *Let ξ be a Hausdorff compact connected topology on X. If $\varnothing \neq F \neq X$ is ξ-closed, and C is a component of F, then $C \cap \partial_\xi F \neq \varnothing$.*

PROOF. Suppose that C is a component of F such that $C \cap \partial_\xi F = \varnothing$. Hence $F \setminus \partial_\xi F \in \mathcal{O}_\xi(C)$. By Theorem XIII.2.5, C is a quasi-component, that is, $C = \bigcap_{D \in \mathcal{D}} D$, where $\mathcal{D}$ denotes the family of $\xi|_F$-clopen sets that include C. By Lemma XII.10.1, there is $D \in \mathcal{D}$ such that $D \subset F \setminus \partial_\xi F$. Therefore, there exists $O \in \mathcal{O}_\xi$ such that $D = O \cap F$, hence $O \cap \partial_\xi F = \varnothing$, and $O \cap \mathrm{int}_\xi F$ is ξ-open. On the other hand, D is $\xi|_F$-closed, hence ξ-closed. Since ξ is a connected topology on X, and D is ξ-clopen, $D = X$, which contradicts $F \neq X$. □

It is obvious that a Hausdorff connected topological space cannot be represented as a finite union of closed disjoint sets, because these subsets are in fact clopen. Is it possible to represent it as a countable union of closed disjoint subsets? The answer is no if the space is compact ([2]), but there exist (non-compact) Hausdorff connected spaces that are unions of countably many closed disjoint sets like that of Example XIII.7.5.

THEOREM XIII.2.7 (Sierpiński). *A Hausdorff compact connected topology is not equal to an infinite countable union of disjoint non-empty closed sets.*

PROOF. Let ξ be a Hausdorff compact and connected topology on X, and $X = \bigsqcup_{n<\omega} X_n$, where $\{X_n : n < \omega\}$ are ξ-closed (and mutually disjoint). We need to show that at most one of these sets is non-empty.

[2]These facts were proved for metric spaces by W. SIERPIŃSKI in **[234]**; see also **[169**, p. 175].

By way of contradiction, let X_0 and X_1 be non-empty. By normality, there exist disjoint $O_0, O_1 \in \mathcal{O}_\xi$ such that $X_0 \subset O_0$ and $X_1 \subset O_1$. Let $x \in X_1$ and let C_0 be the component of x in $\operatorname{cl}_\xi O_1$. Of course, $C_0 \cap O_0 = \varnothing$. By Lemma XIII.2.6, $C_0 \cap \partial_\xi(\operatorname{cl}_\xi O_1) \neq \varnothing$. As $X_1 \subset \operatorname{int}_\xi \operatorname{cl}_\xi O_1$, there exists $k \neq 1$ such that $C_0 \cap X_k \neq \varnothing$. Therefore, C_0 is a compact and connected set such that $C_0 \cap X_0 = \varnothing$, and $C_0 \cap X_n \neq \varnothing$ for at least two $n < \omega$.

By induction, we construct a decreasing sequence $\{C_n : n < \omega\}$ of non-empty closed connected sets such that $X_n \cap C_n = \varnothing$ for each $n < \omega$. Therefore,

$$\bigcup\nolimits_{n<\omega} X_n \cap \bigcap\nolimits_{n<\omega} C_n = \varnothing,$$

and, on the other hand, $\bigcap_{n<\omega} C_n \neq \varnothing$, because $(C_n)_n$ is a decreasing sequence of non-empty ξ-compact sets. A contradiction. □

XIII.3. Locally and arcwise connected topologies

DEFINITION XIII.3.1 (Locally connected). A topology is said to be *locally connected at* x if each neighborhood of x includes a connected neighborhood of x. It is called *locally connected* if this holds for every element x.

Each discrete topology is locally connected. Each chaotic topology is both connected and locally connected.

EXAMPLE XIII.3.2 (A broom). Let $x_n > x_{n+1}$ converge to 0 in $\nu_{\mathbb{R}}$. Denote $L := \operatorname{cl} \bigcup_{n<\omega} \{((1-y)x_n, y) : 0 \leq y \leq 1\}$. Then L is a connected, locally disconnected subspace of $\mathbb{R}^2$. Indeed, it is not locally connected at each point of the form $(0, y)$, where $0 \leq y < 1$.

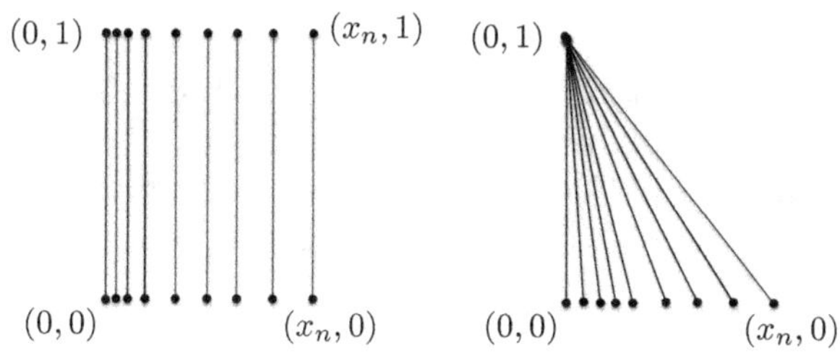

FIGURE XIII.3.1. The bundle of segments (with the horizontal component topology discrete) is locally connected, and the broom is a continuous image of the bundle, and is not locally connected.

Consider now the union $X := \bigcup_{n<\omega}\{(x_n, y) : 0 \leq y \leq 1\} \cup \{(0, y) : 0 \leq y \leq 1\}$ endowed with the product topology $\iota \times \nu$ (horizontally discrete, vertically standard). The map $f : X \longrightarrow L$ defined by $f(x, y) := ((1-y)x, y)$ for each $(x, y) \in L$ is continuous. Observe that X is locally connected, disconnected, while L is connected and locally disconnected.

It follows from Example XIII.3.2 that the continuous image of a locally connected topology need not be locally connected. It follows directly from the definition that

LEMMA XIII.3.3. *Each component of a locally connected topology is open.*

Therefore, in a locally connected topology, the components are clopen, and thus coincide with the quasi-components, that is, for each $x \in X$,

$$C(x) = Q(x).$$

On the other hand, each open subspace of a locally connected topology is locally connected. The converse also holds.

PROPOSITION XIII.3.4. *A topology is locally connected if and only if all components of open subspaces are open.*

PROOF. If the property holds, and $O \in \mathcal{O}_\xi(x)$, then an O-component V of x is a ξ-open subset of O, so that ξ is locally connected. □

DEFINITION XIII.3.5. A topology τ on X is called *pathwise connected* if for any $x_0, x_1 \in X$ there exists a continuous map $f : I \longrightarrow X$ such that $f(0) = x_0$ and $f(1) = x_1$; it is said to be *arcwise connected* if for any $x_0, x_1 \in X$ there exists an embedding $f : I \longrightarrow X$ such that $f(0) = x_0$ and $f(1) = x_1$.

Each arcwise connected topology is pathwise connected. Every pathwise connected topology is connected. Indeed, the unit interval I is connected and a path $f(I)$ is connected as a continuous image of I. As each pair of points is linked by a path, the topology is connected by Corollary XIII.1.13.

Notice that if τ is an arcwise connected topology on a set X such that $\operatorname{card} X > 1$, then $\operatorname{card} X \geq \mathfrak{c}$. For instance, the Sierpiński topology $\$$ is pathwise connected, but not arcwise connected ([3]).

[3]Indeed, if $f : [0, 1] \longrightarrow \{0, 1\}$ is defined by $f(x) := 0$ if $0 \leq x < 1$ and $f(1) := 1$, then $f \in \mathscr{C}(I, \$_0)$, but of course I is not embedded in $\$_0$.

EXAMPLE XIII.3.6. In Example IV.10.1, we considered the so called *topologist's sinus*, that is a map $g \in \mathbb{R}^{\mathbb{R}}$ defined by $g(x) = \sin \frac{1}{x}$ if $x \neq 0$, and $g(0) = 0$.

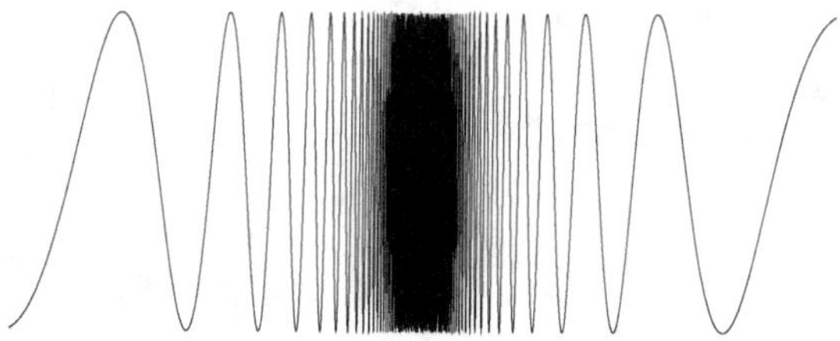

FIGURE XIII.3.2. Topologist's sinus

Consider now the graph $\operatorname{gr}(g) := \{(x, g(x)) : x \in \mathbb{R}\}$ with the topology restricted from $\mathbb{R}^2$. This is a connected topology that is not pathwise connected. Indeed, the sets $G_+ := \{(x, g(x)) : x > 0\}$ and $G_- := \{(x, g(x)) : x < 0\}$ are connected, because g is continuous at each $x \neq 0$. Hence $\operatorname{cl}_{\nu^2} G_+$ and $\operatorname{cl}_{\nu^2} G_-$ are connected by Proposition XIII.1.14, hence $G_+ \subset G_+ \cup \{(0,0)\} \subset \operatorname{cl}_{\nu^2} G_+$ and $G_- \subset G_- \cup \{(0,0)\} \subset \operatorname{cl}_{\nu^2} G_-$, hence the graph $\operatorname{gr}(g)$ is connected. On the other hand, if $f \in \mathscr{C}(I, \operatorname{gr}(g))$ is such that $f(0) = (0,0)$ and $f(1) \in G_+$, then there would be $t > 0$ such that $f([0,t]) \subset \operatorname{gr}(g) \cap \{(x,y) : |y| \leq \frac{1}{2}\}$, which is impossible.

Before the question of which topologies are continuous images of the closed interval was fully answered by Theorem XIII.3.11 of Hahn-Mazurkiewicz, PEANO constructed in 1890 his celebrated curve that fills a square ([4]). Earlier on, in 1884 CANTOR ([5]) showed that a square is equipotent with an interval, and he was amazed by this fact writing this famous sentence *"Ich sehe es, aber ich glaube es nicht"* ([6]). This fact is proved in Chapter XXVII, but it has a direct elementary proof — that is that of CANTOR (Exercise XIII.7.8).

THEOREM XIII.3.7 (Peano). *There exists a continuous map of the unit interval onto the unit square.*

PEANO proved Theorem XIII.3.7 providing an explicit formula for that map. The figures below, represent successive interpolations of

[4]See G. PEANO [**217**]

[5]See G. CANTOR [**36**]

[6]In German, *I see it, but I don't believe it!*

Peano's curve: the marked dots lie on Peano's curve. A detailed description and discussion of Peano's proof is available ([7]).

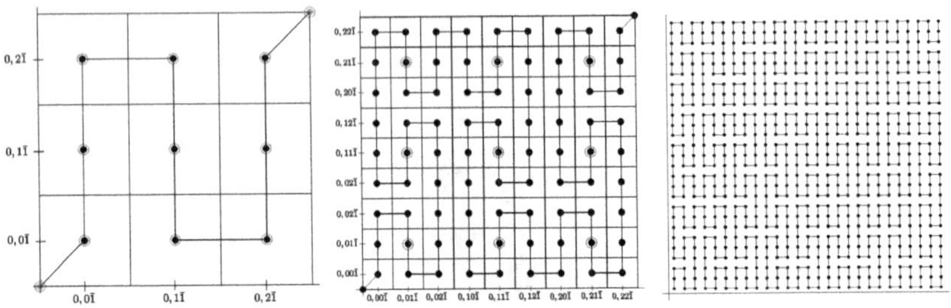

FIGURE XIII.3.3. Three polygonal lines inscribed in Peano's curve. They correspond respectively to two, four and eight digits of the ternary representation of the curve.

Consequently, D. HILBERT provided in 1891 a shorter but non-constructive proof of existence ([8]).

PROOF. As usual, I denotes $[0,1]$ with the standard topology. For each $n \in \mathbb{N}_1$, let $f_n \in \mathscr{C}(I, I^2)$ be a curve joining the centers of 9^n squares, as on Figure XIII.3.3. As $\|f_n(t) - f_{n+k}(t)\| \leq 3^{-n}$ for each $n, k \in \mathbb{N}$ and $t \in [0,1]$, the sequence is uniformly convergent, and the limit is a continuous map in view of Proposition X.3.11. As the union of the centers of consecutive squares is dense, the image $f(I)$ is dense in I^2, and since I is compact, $f(I)$ is compact and closed, hence equal to I^2. □

A sequence of sets $G_0, G_1, \ldots, G_n$ is called a *chain* joining points v and w if $v \in G_0, w \in G_n$ and $G_j \cap G_k \neq \varnothing$ for $|j-k| \leq 1$.

REMARK XIII.3.8. If there is a chain joining two points, then there a subchain $G_0, G_1, \ldots, G_n$ joining these points and such that $G_j \cap G_k \neq \varnothing$ if and only if $|j-k| \leq 1$. Indeed, if $J := \{0 \leq j \leq n : G_j \cap G_k \neq \varnothing\}$ contains other j than $k-1, k$ and $k+1$, then G_k joins $G_{\min J}$ with $G_{\max J}$. Therefore we can remove G_j for each $\min J < j < \max J$ from the chain without effecting the connection. This, of course, can be done properly by induction on the length of chains.

LEMMA XIII.3.9. *If $\mathcal{G}$ is an open cover of a connected topological space X, then each pair of elements of X can be joined by a chain of elements of $\mathcal{G}$.*

[7] G. PEANO [71, Section 5.1]; see also [64, p. 318].

[8] Using binary representation to describe consecutive approximations (D. HILBERT [131]).

PROOF. Given $v \in X$, the set $J(v)$ of elements w of a connected space X that can be joined with v by a chain of elements of $\mathcal{G}$ is obviously open. It is also closed, for if $w \in \operatorname{cl}_X J(v)$, then there exists $G_w \in \mathcal{G}$ such that $w \in G_w$. Let $x \in G_w \cap J(v)$. then there is a chain $G_0, G_1, \ldots, G_n$ such that $v \in G_0$ and $x \in G_n$. Hence, $G_0, G_1, \ldots, G_n, G_w$ is a required chain. It follows that $J(v) = X$ for each $v \in X$. □

LEMMA XIII.3.10. *Each completely metrizable, connected and locally connected space is arcwise connected.*

PROOF. Let $v, w \in X$, and let $\varepsilon > 0$. As X is locally connected, there exists a cover $\mathcal{G}_1$ of X consisting of open connected sets of diameter less than ε. By Lemma XIII.3.9, there exists an irreducible chain $G_0, G_1, \ldots, G_n$ of elements of $\mathcal{G}_1$ such that $v \in G_0$ and $w \in G_n$. Of course, $G_0 \cup \cdots \cup G_n$ is connected. Let $\mathcal{G}_2$ be a cover of $G_0 \cup \cdots \cup G_n$ by connected open sets of diameter less than $\frac{\varepsilon}{2}$ and such that for each $G \in \mathcal{G}_2$, there is $0 \leq k \leq n$ such that $\operatorname{cl}_X G \subset G_k$. By Lemma XIII.3.9, there exists an irreducible chain of elements of $\mathcal{G}_1$ that joins v with w. By induction, we construct a sequence $(\mathcal{L}_n)_n$ of irreducible chains consisting of open connected sets of diameter less than $\frac{\varepsilon}{n}$ that connect v with w, and such that for each n and every $L_{n+1} \in \mathcal{L}_{n+1}$, there is $L_n \in \mathcal{L}_n$ such that $\operatorname{cl}_X L_{n+1} \subset L_n$.

Let $\mathcal{L}_n = \{L_{n,k} : 0 \leq k \leq m_n\}$ and let $(\mathcal{P}_n)_n$ be a sequence of open covers of $[0,1]$ such that $\mathcal{P}_n = \{P_{n,k} : 0 \leq k \leq m_n\}$ is a chain joining 0 with 1, and moreover $\operatorname{cl} P_{n+1,k} \subset P_{n,n(k)}$ if and only if $\operatorname{cl}_X L_{n+1,k} \subset L_{n,n(k)}$. Then for $t \in [0,1]$, there is a sequence $(P_{n,k_n})_n$ such that $\{t\} = \bigcap_{n<\omega} \operatorname{cl} P_{n,k_n} = \bigcap_{n<\omega} P_{n,k_n}$. On the other hand, by completeness, there is a unique element $h(t)$ of $\bigcap_{n<\omega} \operatorname{cl}_X L_{n,k_n}$. Accordingly, $\{h(t) : 0 \leq t \leq 1\}$ is connected. As $\{h(t) : 0 \leq t \leq 1\}$ is a continuous image of a compact set under a bijection, it is homeomorphic to $[0,1]$. □

THEOREM XIII.3.11 (Hahn-Mazurkiewicz). *Each metrizable, compact, connected, and locally connected topology is a continuous image of* $[0,1]$.

PROOF. Let τ be a topology on X with the listed properties. By Theorem XII.10.10 and Lemma XII.5.8, there is a continuous map f from the Cantor set C onto X. We shall continuously extend f to $F : [0,1] \longrightarrow X$. We know that $[0,1] \setminus C = \bigcup_{n<\omega}]a_n, b_n[$ of disjoint open intervals such that $\lim_{n\to\infty} |b_n - a_n| = 0$. By Theorem XII.1.13 of Heine, f is uniformly continuous, so that $\lim_{n\to\infty} \operatorname{diam} |f(b_n) - f(a_n)| = 0$. By Lemma XIII.3.10, τ is arcwise connected, hence for each $n < \omega$, let F map $[a_n, b_n]$ either bijectively onto the arc joining $f(a_n)$ with $f(b_n)$, or onto $f(a_n) = f(b_n)$ if the equality holds. Finally, for $x \in C$, we set $F(x) := f(x)$. Then F is the sought map. □

XIII.4. Disconnected topologies

It is plain to see that each free (alias T_1) topology of cardinality > 1 that has an isolated point is disconnected, because the singleton consisting of that point is a proper clopen set.

EXAMPLE XIII.4.1. The spaces $\mathbb{Q}$ of *rational numbers*, and $\mathbb{R} \setminus \mathbb{Q}$ of *irrational numbers* are disconnected without isolated points. To see that $\mathbb{Q}$ is disconnected, take two irrational numbers $r < s$, and notice that $\{q \in \mathbb{Q} : s > q > r\} = \{q \in \mathbb{Q} : s \geq q \geq r\}$ are $\nu_{\mathbb{Q}}$-clopen. Therefore, the topology $\nu_{\mathbb{Q}}$ has a base of clopen sets. An analogous argument works for $\mathbb{R} \setminus \mathbb{Q}$.

By Definition IX.2.17 a topology is *zero-dimensional* if it has a base consisting of clopen sets, hence $\mathbb{Q}$ and $\mathbb{R} \setminus \mathbb{Q}$ are zero-dimensional.

We have already noticed that Cantor cubes are disconnected. More generally, any product of discrete topologies of cardinality > 1 is disconnected by virtue of Theorem XIII.1.17. In particular, the *Baire space* $\prod_{n<\omega} \mathbb{N} \approx \mathbb{N}^{\mathbb{N}}$ is disconnected. We shall see in Theorem XIII.5.16 that the Baire space is homeomorphic to the space of irrational numbers.

It was already observed that every zero-dimensional topology is regular. Actually,

PROPOSITION XIII.4.2. *Each zero-dimensional free (alias T_1) topology is functionally regular (alias completely regular).*

PROOF. Indeed, if F is a closed set and $x \notin F$, then there exists a clopen set C such that $F \cap C = \varnothing$ and $x \in C$. Then χ_C is continuous, $\chi_C(x) = 1$, and $F \subset \{\chi_C = 0\}$. □

Of course, the discrete topologies are zero-dimensional. Notice that all prime topologies are zero-dimensional.

EXAMPLE XIII.4.3 (A zero-dimensional topology). Let $\pi = \pi[x_\infty, \mathcal{G}]$ be a *prime topology* on an infinite set X. Accordingly, $\ker \mathcal{G} = \{x_\infty\}$ and $\mathcal{G} \setminus \{x_\infty\}$ is a free filter on X. Of course, $\mathcal{G} = \mathcal{N}_\pi(x_\infty)$. Accordingly $O \in \mathcal{O}_\pi$ whenever either $x_\infty \notin O$ or $O \in \mathcal{G}$. Therefore, $\mathrm{cl}_\pi O = O$ for each $O \in \mathcal{O}_\pi$, so that every point has a neighborhood filter-base consisting of π-clopen sets.

We have noticed in Example IX.2.19 that the Sorgenfrey line is zero-dimensional.

PROPOSITION XIII.4.4. *If a free topology is zero-dimensional, then all its quasi-components are singletons.*

PROOF. If $x \neq x_0$, then there exists $O \in \mathcal{O}(x_0)$ such that $x \notin O$, because $\{x\}$ is closed. By assumption, there exists a clopen set C such that $x_0 \in C \subset O$. Therefore, $x \notin C \supset Q(x)$, where $Q(x)$ denotes the quasi-component of x. □

By (XIII.2.1), all the components of each zero-dimensional free topology are singletons.

PROPOSITION XIII.4.5. *The product of topologies is zero-dimensional if and only if each factor is zero-dimensional.*

PROOF. Let $\xi := \prod_{j\in J} \xi_j$ be a product of topologies on $\prod_{j\in J} X_j$. For each $\varphi \in \prod_{j\in J} X_j$ and every a neighborhood V of φ, there exists a finite subset J_0 of J and $V_j \in \mathcal{N}_{\xi_j}(\varphi(j))$ such that $\prod_{j\in J} V_j \subset V$ and $V_j = X_j$ for each $j \in J \setminus J_0$. Consequently, $\prod_{j\in J} V_j$ is ξ-clopen if and only if V_j is ξ_j-clopen for each $j \in J$. □

THEOREM XIII.4.6. *Each Hausdorff zero-dimensional topology ξ can be embedded in the Cantor cube of dimension* $\mathrm{w}(\xi)$.

PROOF. Let ξ be a zero-dimensional topology on X. Consider a base $\mathcal{B}$ of ξ of cardinality $\mathrm{w}(\xi)$ consisting of ξ-clopen sets. For every $B \in \mathcal{B}$, let $\chi_B : X \longrightarrow \{0,1\}$. Sure enough, $\chi_B \in \mathscr{C}(\xi, \{0,1\})$ and thus $\prod_{B\in\mathcal{B}} \chi_B : X \longrightarrow \prod_{B\in\mathcal{B}}\{0,1\}$ is injective and separating points from closed sets, and thus an embedding by Corollary VII.3.9. □

COROLLARY XIII.4.7. *The Cantor cube of infinite dimension κ is universal for all zero-dimensional topologies of weight $\leq \kappa$.*

XIII.5. Ultrametric and ultrametrizable spaces

A function $d : X \times X \longrightarrow \mathbb{R}_+$ is called an *ultrametric* (or *non-Archimedean metric*) on X if, for each $x, y, z \in X$,

$$d(x,y) = d(y,x), \tag{XIII.5.1}$$

$$d(x,y) = 0 \Longleftrightarrow x = y, \tag{XIII.5.2}$$

$$d(x,z) \leq \max(d(x,y), d(y,z)). \tag{XIII.5.3}$$

Since, $\max(a,b) \leq a + b$ for any positive reals a, b, each ultrametric is a metric, as defined in Section X.3.6.

Ultrametrics are quite special metrics, because

PROPOSITION XIII.5.1. *If d is an ultrametric on X and $x, y, z \in X$, then at least two elements of $\{d(x,y), d(y,z), d(x,z)\}$ are equal.*

PROOF. Indeed, if for instance $d(x,y) < d(y,z) < d(x,z)$, then (XIII.5.3) would not hold. □

EXAMPLE XIII.5.2. A *discrete metric* on X as in Example III.6.3, that is,

$$i_X(x_0, x_1) := \begin{cases} 0, & \text{if } x_0 = x_1, \\ 1, & \text{if } x_0 \neq x_1, \end{cases}$$

is obviously an ultrametric.

PROPOSITION XIII.5.3. *Let d be an ultrametric on X, and let*

$$B_r(x) := \{y \in X : d(x,y) < r\}.$$

Then for $x, y \in X$ and $r, s > 0$,

(1) *If $y \in B_r(x)$, then $B_r(y) = B_r(x)$,*
(2) *If $r \leq s$ and $B_r(x) \cap B_s(y) \neq \varnothing$, then $B_r(x) \subset B_s(y)$,*
(3) $\mathrm{cl}_{td}\, B_r(x) \subset B_r(x)$.

PROOF. (1) If $y \in B_r(x)$ and $z \in B_r(x)$, then

$$d(z,y) \leq \max(d(z,x), d(x,y)) < r,$$

hence $z \in B_r(y)$. To show the converse, mind that $y \in B_r(x)$ amounts to $x \in B_r(y)$.

(2) If $z \in B_r(x) \cap B_s(y)$, then by (1), $x \in B_r(z) \subset B_s(z)$ and since $z \in B_s(y)$, again by (1), $B_s(z) = B_s(y)$.

(3) If $y \notin B_r(x)$ and $0 < s < r$, then $B_s(y) \cap B_r(x) = \varnothing$, which implies that $y \notin \mathrm{cl}_{td}\, B_r(x)$. Indeed, if $B_s(y) \cap B_r(x) \neq \varnothing$, hence by (2), $B_s(y) \subset B_r(x)$, a contradiction. □

COROLLARY XIII.5.4. *Each ultrametrizable topology is zero-dimensional.*

DEFINITION XIII.5.5. A metrizable topology has *covering dimension zero* whenever for each open cover there exists a refining cover of pairwise disjoint open sets ([9]).

It is clear that the refining cover above consists of clopen sets.

PROPOSITION XIII.5.6. *If a topology is ultrametrizable, then it has covering dimension zero.*

PROOF. Consider an ultrametric on a set X and let $\mathcal{P}$ be an open cover. For each $x \in X$, let $r(x)$ be the largest number such that there exists $P \in \mathcal{P}$ such that $B_{r(x)}(x) \subset P$. It follows that $\{B_{r(x)}(x) : x \in X\}$ is a cover of X refining $\mathcal{P}$ and consisting of clopen sets. If $B_{r(x)}(x) \cap B_{r(y)}(y) \neq \varnothing$, say $r(x) \leq r(y)$, then by Proposition

[9]See Remark XIII.7.10.

XIII.5.3 (2), $B_{r(x)}(x) \subset B_{r(y)}(y)$. As $x \in B_{r(y)}(y)$, by by Proposition XIII.5.3 (1), $B_{r(y)}(x) = B_{r(y)}(y)$, hence $r(x) = r(y)$, which shows that $B_{r(x)}(x) = B_{r(y)}(y)$. □

PROPOSITION XIII.5.7. *Each separable zero-dimensional topology has covering dimension zero.*

PROOF. Let θ be a zero-dimensional separable topology, and let $\mathcal{B}$ be a base of θ consisting of θ-clopen sets. If $\mathcal{P}$ is an θ-open cover, then by Lemma IX.2.22, there exists a refinement $\mathcal{R}$ of $\mathcal{P}$ which is a θ-open cover and $\mathcal{R} \subset \mathcal{B}$. By Corollary IX.2.23, there exists a θ-open cover $\mathcal{R}_0 \subset \mathcal{R}$ of cardinality $\aleph_0$, say $\mathcal{R}_0 = \{R_n : n < \omega\}$. Then $S_0 := R_0$ and $S_n := R_n \setminus \bigcup_{0 \leq k < n} R_k$ constitutes a θ-open cover of $\mathcal{P}$ consisting of disjoint θ-clopen sets. □

Recall that in a metric space (X, d), the *diameter* of a set A is defined by $\operatorname{diam} A := \sup\{d(a_0, a_1) : a_0, a_1 \in A\}$, and for a family $\mathcal{A}$ of subsets of X, by

$$\operatorname{diam} \mathcal{A} := \sup\{\operatorname{diam} A : A \in \mathcal{A}\}.$$

THEOREM XIII.5.8. *A metrizable topology is ultrametrizable if and only if it has covering dimension zero.*

PROOF. If a topology is ultrametrizable, then by Proposition XIII.5.6 the condition holds.

Conversely, let (X, d) be a metric space and the condition holds. There exists a sequence $(\mathcal{P}_n)_n$ of disjoint clopen covers such that $\mathcal{P}_{n+1}$ is a refinement of $\mathcal{P}_n$, and $\operatorname{diam} \mathcal{P}_n < 2^{-n}$. Indeed, by assumption, for an open cover by unit balls, there exists a refining cover $\mathcal{P}_0$ of clopen sets of diameter less than 1. We proceed by induction. If $\mathcal{P}_0, \mathcal{P}_1, \ldots, \mathcal{P}_n$ fulfill the inductive assumptions, then let $\mathcal{P}_{n+1}$ be a refining cover of $\{P \cap B_d(x, 2^{-(n+1)}) : P \in \mathcal{P}_n, x \in X\}$. Then $\bigcup_{n<\omega} \mathcal{P}_n$ is a base for the topology compatible with d. We need to construct an ultrametric equivalent to d.

For $x, y \in X$, let $\#(x, y)$ be the first $n < \omega$ such that x and y are not in the same $P \in \mathcal{P}_n$. Let $s : \mathbb{N} \longrightarrow \mathbb{R}_+$ be strictly decreasing and $\lim_{n\to\infty} s(n) = 0$. Now, define

$$m(x, y) := \begin{cases} s(\#(x, y)), & x \neq y, \\ 0, & x = y. \end{cases} \tag{XIII.5.4}$$

It is clear that m is valued in positive reals, is symmetric, and $m(x, y) = 0$ implies that $x = y$. We need to show that

$$m(x, z) \leq \max(m(x, y), m(y, z))$$

for each $x, y, z \in X$. If $n = \min\{\#(x,z), \#(x,y), \#(y,z)\}$, then either x, y and z belong to three different elements of $\mathcal{P}_n$, that is, $n = \#(x,z) = \#(x,y) = \#(y,z)$, or two are in the same element P of $\mathcal{P}_n$ and one is in another, say, $x, y \in P$ and $z \notin P$. In the latter case, $n = \#(x,z) = \#(y,z) < \#(x,y)$, so that $m(x,y) \leq m(x,z) = m(y,z)$. Therefore, each of these values is less than or equal to the supremum of the two others.

To see that m is a compatible metric, let $M_r(x) := \{y \in X : m(x,y) < r\}$ and notice that if $s(n) \leq r < s(n-1)$, then $M_r(x) = M_{s(n)}(x)$. As every $\mathcal{P}_n$ is a partition, let $P_{n,x}$ denote $P \in \mathcal{P}_n$ such that $x \in P$. Then $y \in M_{s(n)}(x)$ if and only if $s(\#(x,y)) < s(n)$ if and only if $\#(x,y) > n$ if and only if $y \in P_{n,x}$. Hence $M_{s(n)}(x) = P_{n,x}$. □

REMARK XIII.5.9. Theorem XIII.5.8 is due to J. DE GROOT [**52**]. The proof above follows that of J. VAUGHAN in [**258**]. For a list of necessary and sufficient conditions for ultrametrizability, see [**22**, Theorem 1.13].

EXAMPLE XIII.5.10. The *Cantor set* C was studied in detail in Section XII.5. For two elements r, s of C, we defined by $\#(r,s)$, the first $n \in \mathbb{N}_1$ such that the n-th coefficients of the ternary representations (XII.5.1) are different, that is, $r^{[n]} \neq s^{[n]}$. Then for every $r \in C$, there exists $P \in \mathcal{P}_n$, for which $r \in P$. Then $P = \{t \in C : t^{[n]} = r^{[n]}\}$. It follows that if $P, R \in \mathcal{P}_n$, then $\operatorname{diam} P = 3^{-n} = \operatorname{diam} R$ and $\sup\{d(x,y) : x \in P, y \in R\} \geq 3^{-n}$.

Therefore, if we set $m(r,s) = 3^{\#(r,s)}$, like in (XIII.5.4), then

$$d(r,s) \leq m(r,s) \leq 3d(r,s).$$

By Corollary XII.5.3, the Cantor set is homeomorphic to the Cantor cube of dimension $\aleph_0$. Of course, any homeomorphism sends covers $\mathcal{P}_n$ for $n < \omega$ onto the corresponding covers of the Cantor cube. For example, the homeomorphism $F : \{0,1\}^{\mathbb{N}} \longrightarrow C$ defines an ultrametric $m_F(f,g) := m(F^{-1}(f), F^{-1}(g))$.

REMARK XIII.5.11. If X is any set endowed with the discrete topology ι_X, then the product topology on $X^{\mathbb{N}}$ is zero-dimensional, as a product of zero-dimensional topologies. For any $f, g \in X^{\mathbb{N}}$, let $\#(f,g)$ be the least $n < \omega$ such that $f(n) \neq g(n)$, and set $\#(f,g) = 0$ if $f = g$. Then (XIII.5.4) defines un ultrametric on $X^{\mathbb{N}}$.

THEOREM XIII.5.12. *A topology is ultrametrizable if and only if it is homeomorphic to a subset of a countable product of discrete topologies.*

PROOF. [10] Let m be an ultrametric on a set X and let ξ be a compatible topology. Let $r_n > r_{n+1}$ for each $n < \omega$ and $\lim_{n\to\infty} r_n = 0$. The set $\mathcal{M}_n$ of balls of radius r_n forms a partition of X made of ξ-clopen sets.

Let $f_n : X \longrightarrow \mathcal{M}_n$ assign to every $x \in X$ the ball counting x. It is clear that $f_n \in \mathscr{C}(\xi, \iota_n)$, where ι_n stands for the discrete topology on $\mathcal{M}_n$. The diagonal product map $\Delta_{n<\omega} f_n$ is an embedding by Corollary IV.7.8, that is, a homeomorphism of ξ with a subspace of $\prod_{n<\omega} \iota_n$.

Conversely, if τ_n is the discrete topology on X_n for each $n < \omega$, then

$$m(x, y) := \inf\{2^{-n} : \underset{k<n}{\forall}\ x(k) = y(k)\}$$

is an ultrametric on $\prod_{n<\omega} X_n$ that is compatible with $\prod_{n<\omega} \tau_n$. □

COROLLARY XIII.5.13. *Each countable product of ultrametrizable topologies is ultrametrizable.*

In particular,

COROLLARY XIII.5.14. *The Baire space is ultrametrizable.*

EXAMPLE XIII.5.15 (Baire space). The set $\mathbb{N}^{\mathbb{N}}$ with the topology of pointwise convergence on the power of discrete topologies on $\mathbb{N}$ is called the *Baire space* $\mathbb{B}$. We shall see that the Baire space is homeomorphic to the space $\mathbb{R} \setminus \mathbb{Q}$ of *irrational numbers.* Actually, it is handy to prove that it is homeomorphic to $]0, 1[\setminus\mathbb{Q}$. For any finite sequence $(a_n)_n \subset \mathbb{N}_1$, we define by induction the *continued fraction* by

$$r_1 = \frac{1}{a_1},\ r_2 = \frac{1}{a_1 + \frac{1}{a_2}},\ r_3 = \frac{1}{a_1 + \frac{1}{a_2 + \frac{1}{a_3}}}, \ldots \tag{XIII.5.5}$$

and we notice that for any $(a_n)_{n\in\mathbb{N}_1} \subset \mathbb{N}_1$,

$$0 < r_1 < r_3 < r_5 < \cdots < r_4 < r_2 < 1 \tag{XIII.5.6}$$

and $r_n \in \mathbb{Q}$ for each $n < \omega$. Conversely, by the Euclid algorithm, each rational number from $]0, 1]$ has a finite representation (XIII.5.5). A rough estimate yields (Exercise XIII.7.7)

$$|r_n - r_{n+1}| \leq \frac{1}{n}, \tag{XIII.5.7}$$

hence $\lim_{n\to\infty} |r_n - r_{n+1}| = 0$, and thus $(r_n)_n$ is a Cauchy sequence. Then, by completeness of the real line, $r := \lim_{n\to\infty} r_n$ exists. We shall

[10] See A. C. M. VAN ROOIJ [256].

see that it is always an irrational number. Denote ([11]) $\|a_0\| := a_0$, and

$$\|a_0, a_1, \ldots, a_n\| := a_0 + \|a_1, \ldots, a_n\|^{-1},$$

where $a_0 \in \mathbb{Z}$ and $a_k \in \mathbb{N}_1$ for $1 \leq k < \omega$. Therefore, every rational number is represented this way. In our case $a_0 = 0$, because we consider merely the rationals between 0 and 1.

THEOREM XIII.5.16. *The Baire space is homeomorphic to the space of irrational numbers.*

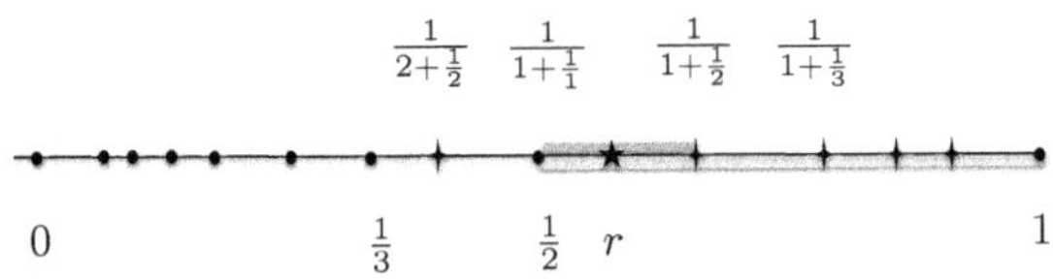

FIGURE XIII.5.1. Baire space: for example, an irrational point r belongs to the interval between $[0, 1, 1]$ and $[0, 1, 2]$.

PROOF. It is more practical to consider $\mathbb{N}_1^{\mathbb{N}_1}$ rather than (a homeomorphic) $\mathbb{N}^{\mathbb{N}}$. We define $F : \mathbb{N}_1^{\mathbb{N}_1} \longrightarrow]0, 1[$ by

$$F(\varphi) := \lim_{n\to\infty} \|0, \varphi(1), \varphi(2), \ldots, \varphi(n)\| .$$

Then $F(\varphi)$ is an irrational number. Each basic neighborhood V_N of $\varphi \in \mathbb{N}_1^{\mathbb{N}_1}$ is determined by a finite subset $\{1, \ldots, N\}$ of $\mathbb{N}_1$ to the effect that

$$V_N(\varphi) := \{\psi \in \mathbb{N}_1^{\mathbb{N}_1} : \underset{1\leq n\leq N}{\forall} \psi(n) = \varphi(n)\}.$$

The image $F(V_N(\varphi))$ is the interval with the endpoints $\|0, \ldots, \varphi(N-1)\|$ and $\|0, \ldots, \varphi(N)\|$, their mutual position depends on the parity of N (XIII.5.6). Of course, $F(V_N(\varphi)) \cap \mathbb{Q} = \varnothing$ for every $N \in \mathbb{N}_1$. Consequently, $\{F(V_N(\varphi)) : 1 \leq N < \infty\}$ is a sequence of basic neighborhoods of $F(\varphi)$. □

REMARK XIII.5.17. For an odd N, the collection of intervals

$$\mathcal{P}_N := \{] \|0, \ldots, \varphi(N), k\| , \|0, \ldots, \varphi(N)\| [\setminus \mathbb{Q} : k \in \mathbb{N}_1\}$$

is a cover of $]0, 1[\setminus \mathbb{Q}$ by clopen intervals, and

$$\mathcal{P}_{N+1} := \{] \|0, \ldots, \varphi(N+1)\| , \|0, \ldots, \varphi(N+1), k\| [\setminus \mathbb{Q} : k \in \mathbb{N}_1\}$$

is its subsequent refinement.

---

[11] For the sake of brevity, we shall use the traditional notation, modified to avoid a confusion with the symbols of intervals.

XIII.6. Cantor-Bendixon theorem

Recall Definition IX.3.8 that an element x of a subset A of a topological space is called *isolated* if there exists an open set O such that $O \cap A = \{x\}$. As we have pointed out in Section XIII.4, if a subset of a topological space has isolated points, then it is disconnected.

EXAMPLE XIII.6.1. Consider the following subset

$$X := \{x \in \mathbb{R} : x \leq 0\} \cup \{2^{-n} : n < \omega\} \cup \{2^{-n} + 2^{-k} : n < \omega, 2n < k < \omega\}$$

of the real line. The isolated points of X are those in the component on the right. When we remove them, then we get $X_1 := \{x \in \mathbb{R} : x \leq 0\} \cup \{2^{-n} : n < \omega\}$. All the point in the component on the right are isolated points of X_1, and when we remove them, there remains $\{x \in \mathbb{R} : x \leq 0\}$, which has no isolated points.

LEMMA XIII.6.2. *Every subset of the real line has at most countably many isolated points.*

PROOF. Let $A \subset \mathbb{R}$ and let $\{I_n : n < \omega\}$ be the family of intervals of the form $]a, b[$ with $a, b \in \mathbb{Q}$. If x be an isolated point of A, then there exists a least $n(x) < \omega$ such that $I_{n(x)} \cap A = \{x\}$. If $y \neq x$ is another isolated point of A, then $I_{n(y)} \neq I_{n(x)}$, and the proof is complete. □

On the other hand, a set without isolated points need not be connected.

A subset of a convergence space is also called *dense in itself* if it is *without isolated points*. A closed set without isolated points is said to be *perfect*.

For example, $\mathbb{Q}$ and $\mathbb{R} \setminus \mathbb{Q}$ are without isolated points and disconnected. Of course, they are not perfect subsets of $\mathbb{R}$. The *Cantor set* is an example of a perfect, disconnected (even zero-dimensional) subset of the real line.

LEMMA XIII.6.3. *Each union of sets without isolated points and the closure of a set without isolated points are without isolated points.*

PROOF. Let $\mathcal{A}$ consists of sets without isolated points. If $x \in \bigcup_{A \in \mathcal{A}} A$, then there exists $A_0 \in \mathcal{A}$ such that, by assumption,

$$x \in \mathrm{cl}_\xi(A_0 \setminus \{x\}) \subset \mathrm{cl}_\xi(\bigcup_{A \in \mathcal{A}} A \setminus \{x\}).$$

We prove that if A is without isolated points, then $\mathrm{cl}_\xi A$ is without isolated points too. We need to show that if $y \in \mathrm{cl}_\xi A$, then

$$y \in \mathrm{cl}_\xi(\mathrm{cl}_\xi A \setminus \{y\}).$$

If $y \in A$, then, by assumption, $y \in \mathrm{cl}_\xi(A \setminus \{y\}) \subset \mathrm{cl}_\xi(\mathrm{cl}_\xi A \setminus \{y\})$. If $y \in \mathrm{cl}_\xi A \setminus A$, then for each $O \in \mathcal{O}_\xi(y)$, there exists $x_O \neq y$ such that $x_O \in O \cap A \subset O \cap \mathrm{cl}_\xi A$, hence $O \cap (\mathrm{cl}_\xi A \setminus \{y\}) \neq \varnothing$. □

We remember that the cardinality of the Cantor set is *continuum* ([12]).

LEMMA XIII.6.4 (*). *The cardinality of each non-empty perfect subset of the real line is equal to continuum.*

PROOF. Let $A \subset \mathbb{R}$ be perfect. We shall construct an injection of $\{0,1\}^{\mathbb{N}}$ to A. Let Λ be the *binary tree*, the set of finite sequences valued in $\{0,1\}$ ordered by inclusion, that is, the empty sequence $\varnothing \in \Lambda$ is the least element, and $(a_1, \ldots, a_n) \leq (b_1, \ldots, b_k)$ whenever $n \leq k$ and $a_j = b_j$ for every $1 \leq j \leq n$. We define $F : \Lambda \longrightarrow A$ by induction on the length ([13]) of elements of Λ. Let $F(\varnothing) \in A$ be arbitrary. Suppose that F is defined up to the length $n < \omega$. Let $\varepsilon_0 = 1$, and

$$\varepsilon_n := \tfrac{1}{4} \min\{|F(t_0) - F(t_1)| : t_0 \neq t_1, l(t_0) = l(t_1) = n\}. \tag{XIII.6.1}$$

If $l(t) = n$, let $F(t,0) := F(t)$, and $F(t,1)$ be any element of $\{x \in A : |x - F(t)| < \varepsilon_n\} \cap A$ different from $F(t)$, which is possible, because A is without isolated points. Then, for each $\{t_n : n < \omega\}$ such that $l(t_n) = n$, the sequence $(F(t_n))_n$ is a Cauchy sequence because of (XIII.6.1), and thus converges to an element of A, because A is closed. Moreover, if $\varphi, \psi \in \{0,1\}^{\mathbb{N}}$ and $\varphi \neq \psi$, then

$$F(\varphi) := \lim_{n\to\infty} F(\varphi|_n) \neq \lim_{n\to\infty} F(\psi|_n) =: F(\psi)$$

because of (XIII.6.1). Accordingly, we have defined an injective map $F : \{0,1\}^{\mathbb{N}} \longrightarrow A$, and thus $\mathrm{card}\, A \geq 2^{\aleph_0} = \mathfrak{c}$. □

DEFINITION XIII.6.5. A set is called *scattered* if every non-empty subset contains an isolated point.

Accordingly, a set is scattered if and only if it does not include any non-empty set without isolated points.

Of course, each discrete set is scattered. A free convergent sequence with its limit (on a real line) is a scattered non-discrete set.

The following result is a consequence of Lemma XIII.6.3.

COROLLARY XIII.6.6. *Each topological space is a disjoint union of a perfect set and a scattered set.*

---

[12]The following lemma and other facts of this section about subsets of real line are taken from [**136**, pp. 33–35].

[13]By definition, the length $l(t) = n$ if t is of the form $t = (a_1, \ldots, a_n)$.

PROOF. If $\mathcal{A}$ be the family of all without ξ-isolated points subsets of X. Then $X_0 := \mathrm{cl}_\xi(\bigcup \mathcal{A})$ is without ξ-isolated points and ξ-closed. If a non-empty set A is without isolated points, then $A \in \mathcal{A}$, hence $A \subset X_0$. Then $X \setminus X_0$ is scattered. □

For example, there are no accumulation points for the discrete topology ι_X, and each point is an accumulation point for the chaotic topology o_X, provided that $\operatorname{card} X > 1$. Each point is an accumulation point for the standard sequential convergence on $\mathbb{R}$, and for the standard topology on $\mathbb{R}$. For the Sierpiński topologies from Example III.5.3, 1 is an accumulation point for $\$_0$, and 0 for $\$_1$.

We denote by $\mathrm{d}_\xi A$ the set of ξ-accumulation points of A, and call it the *derived set* of A. Accordingly, $x \in A \setminus \mathrm{d}_\xi A$ if and only if x is an *ξ-isolated point of A.*

Observe that $\mathrm{d}_\xi A$ is ξ-closed and that $\mathrm{adh}_\xi A = A \cup \mathrm{d}_\xi A$ for each $A \subset |\xi|$. On the other hand,

$$A \subset B \Longrightarrow \mathrm{d}_\xi A \subset \mathrm{d}_\xi B, \text{ and } \mathrm{d}_\xi (A \cup B) \subset \mathrm{d}_\xi A \cup \mathrm{d}_\xi B.$$

A set A is without ξ-isolated points if and only if $A \subset \mathrm{d}_\xi A$.

REMARK XIII.6.7 (*). One can iterate the operation of derivation d_ξ. By definition, $\mathrm{d}_\xi^0 A := A$ for any subset A of $|\xi|$. If $\alpha > 0$ is an ordinal, then

$$\mathrm{d}_\xi^\alpha A := \mathrm{d}_\xi(\bigcap\nolimits_{\gamma<\alpha} \mathrm{d}_\xi^\gamma A).$$

Therefore, $\mathrm{d}_\xi^{\alpha+1} A = \mathrm{d}_\xi(\mathrm{d}_\xi^\alpha A)$. In particular, $\mathrm{d}_\xi^1 A = \mathrm{d}_\xi A$. In other words, at every iteration we remove isolated points. By Theorem XXVII.4.6 of Zermelo, there exists a least α such that $\mathrm{d}_\xi^\alpha A = \mathrm{d}_\xi^{\alpha+1} A$, called the *Cantor-Bendixon rank* of A.

LEMMA XIII.6.8 (*). *Each strictly descending transfinite sequence of closed subsets of the real line is countable.*

PROOF. Let $\{A_\alpha : \alpha < 2^{\mathfrak{c}}\}$ be a strictly descending of $\nu_\mathbb{R}$-closed sets. Then for each α, there are $q, r \in \mathbb{Q}$ such that $]q, r[\cap A_\alpha \neq \varnothing$ and $]q, r[\cap A_{\alpha+1} = \varnothing$. This describes an injection into $\mathbb{Q} \times \mathbb{Q}$,which is countable. □

THEOREM XIII.6.9 (Cantor-Bendixon). *Each closed subset of the real line is a disjoint union of a perfect set and of a countable set.*

PROOF. Let A be a closed subset of the real line. Let $A_0 := A$, and $A_{\alpha+1} := \mathrm{d}_\nu A_\alpha$ if α is a non-limit ordinal, and

$$A_\alpha := \mathrm{d}_\nu(\bigcap\nolimits_{\beta<\alpha} A_\beta)$$

otherwise. As the (transfinite) sequence is descending, there exists a least ordinal $\alpha(A)$ such that $A_\alpha = A_{\alpha(A)}$ for each $\alpha \geq \alpha(A)$. If $A_{\alpha(A)} \neq \varnothing$, then it is perfect. By Lemma XIII.6.8, $\alpha(A)$ is countable, and by Lemma XIII.6.2, $A_\alpha \setminus A_{\alpha+1}$ is countable for each $\alpha < \alpha(A)$, so that $A \setminus A_{\alpha(A)}$ is countable. □

EXAMPLE XIII.6.10. Each compact subset of rational numbers is scattered. Indeed, if $K \subset \mathbb{Q}$ is $\nu_{\mathbb{Q}}$-compact, then it is $\nu_{\mathbb{R}}$-compact, hence $\nu_{\mathbb{R}}$-closed. By Corollary XIII.6.6, $K = K_0 \sqcup K_1$, where K_0 is $\nu_{\mathbb{R}}$-perfect and K_1 is $\nu_{\mathbb{R}}$-scattered. By Lemma XIII.6.4, if K_0 is non-empty, then the cardinality of K_0 is $\mathfrak{c}$, and since K is countable, we infer $K_0 = \varnothing$.

REMARK XIII.6.11. There exist compact subsets of the irrational numbers of cardinality $\mathfrak{c}$. Indeed, for any $\varepsilon > 0$, there exists a $\nu_{\mathbb{R}}$-open subset O of length less than ε such that $\mathbb{Q} \subset O$. Indeed, if $\mathbb{Q} = \{q_n : n < \omega\}$, then let I_n be an open interval of length $\varepsilon 2^{-(n+1)}$ such that $q_n \in I_n$. Then $O = \bigcup_{n<\omega} I_n$ has is of length less than ε. Therefore, if $\varepsilon < 1$, then $[0,1] \setminus O$ is a compact subset of cardinality continuum of irrational numbers.

XIII.7. Supplement

EXERCISE XIII.7.1. For each $f \in \mathscr{C}(I, I)$, there exists $x_\infty \in I$ such that $f(x_\infty) = x_\infty$, that is, x_∞ is a fixed point of f.

SOLUTION. If $h(x) := f(x) - x$, then $h \in \mathscr{C}(I, \mathbb{R})$ is continuous and such that $h(0) \geq 0 \geq h(1)$. Hence $0 \in [h(1), h(0)] \subset h(I)$, by Corollary XIII.1.7. Therefore, there exists $x_\infty \in I$ such that $0 = h(x_\infty) = f(x_\infty) - x_\infty$.

This elementary fact was generalized to continuous maps from I^n into I^n by L. BROUWER ([14]); there is no easy proof of the existence of fixed points for $f \in \mathscr{C}(I^n, I^n)$ for $n > 1$. The proof cited, uses *dimension theory.*

EXERCISE XIII.7.2. Find a decreasing sequence of non-empty closed connected sets such that the intersection is not connected.

SOLUTION. For example, in $\mathbb{R}^2$, let $L := \{(x,y) : y = 0\}, H := \{(x,y) : xy = 1, x > 0\}$, and $F_n := \{(x,y) : 0 \leq y \leq \frac{1}{x}, x \geq n + 1\} \cup L \cup H$. Then F_n is connected and closed for each $n < \omega$, but $\bigcap_{n<\omega} F_n = L \cup H$ is disconnected.

[14]See for instance R. Engelking [97, Theorem 7.3.18].

EXERCISE XIII.7.3. Prove that

(1) $\mathbb{R}^n$ is arcwise connected for each $n < \omega$,
(2) If $n > 1$, then $\mathbb{R}^n \setminus \{a\}$ is arcwise connected for each $a \in \mathbb{R}^n$,
(3) If $n > 1$ and $A \subset \mathbb{R}^n$ is countable, then $\mathbb{R}^n \setminus A$ is arcwise connected.

SOLUTION. (1) If $x_0, x_1 \in \mathbb{R}^n$ and $x_0 \neq x_1$, then

$$[x_0, x_1] := \{(1-\alpha)x_0 + \alpha x_1 : \alpha \in [0,1]\}$$

is homomorphic with $[0,1]$. (2) If $x_0, x_1 \in \mathbb{R}^n \setminus \{a\}$, then either $a \notin [x_0, x_1]$ and thus $[x_0, x_1] \subset \mathbb{R}^n \setminus \{a\}$, or $a \in]x_0, x_1[$, hence $[0,1]$ is homeomorphic to $[x_0, x] \cup [x, x_1] \subset \mathbb{R}^n \setminus \{a\}$ for any $x \notin [x_0, x_1]$. (3) If $x_0, x_1 \in \mathbb{R}^n \setminus A$, then there exists a straight line L in $\mathbb{R}^n$ such that $x_0, x_1 \notin L$, because $n > 1$. On the other hand,

$$\{x \in L : A \cap ([x_0, x] \cup [x, x_1]) \neq \varnothing\}$$

is countable, because A is countable. Hence, there exists $x \in L$ such that $A \cap ([x_0, x] \cup [x, x_1]) = \varnothing$.

EXAMPLE XIII.7.4 (A Hausdorff regular connected topology on a countable set). Consider the set $X := \mathbb{Q} \times \mathbb{Q}_+$ endowed with a topology ρ, defined as follows: for each $x \in X$, consider the equilateral triangle, for which x is a vertex and the two remaining vertices called $x_0 = (r_0, 0), x_1 = (r_1, 0)$ lie on $\mathbb{Q} \times \{0\}$. Then $V \in \mathcal{N}_\rho(x)$ if there exists $\varepsilon > 0$ such that

$$\{x\} \cup]r_0 - \varepsilon, r_0 + \varepsilon[\times \{0\} \cup]r_1 - \varepsilon, r_1 - \varepsilon[\times \{0\} \subset V.$$

In case when $x \in \mathbb{Q} \times \{0\}$, the triangle degenerates to one point $x_0 = x = x_1$, so that the formula above reduces to $]r - \varepsilon, r + \varepsilon[\times \{0\} \subset V$. We easily see that if $x \neq y$, then x_0, y_0, x_1, y_1 are distinct, if both x, y are in the strict upper plane; x_0, y, x_1 are distinct if x is in the strict upper plane, and $y \in \mathbb{Q} \times \{0\}$, thus ρ is Hausdorff.

It is not difficult to see that the closure of a basic neighborhood of z is represented by a gray figure on Figure XIII.7.1. In particular, ρ is not regular.

On the other hand, for two arbitrary points x, y lying in the strict upper plane, $O_x \in \mathcal{O}_\rho(x)$ and $O_y \in \mathcal{O}_\rho(y)$, the closures $\operatorname{cl}_\rho O_x \# \operatorname{cl}_\rho O_y$. Therefore, the space cannot be a union of two non-empty, disjoint ρ-clopen sets A and B, for if $x \in A$ and $y \in B$, then there exist $O_x \in \mathcal{O}_\rho(x)$ and $O_y \in \mathcal{O}_\rho(y)$ such that $\varnothing \neq \operatorname{cl}_\rho O_x \cap \operatorname{cl}_\rho O_y \subset A \cap B$.

There exist many examples of Hausdorff connected topologies that are countable unions of disjoint closed sets, perhaps the oldest being given by W. SIERPIŃSKI in [**234**]. The following is one of the simplest.

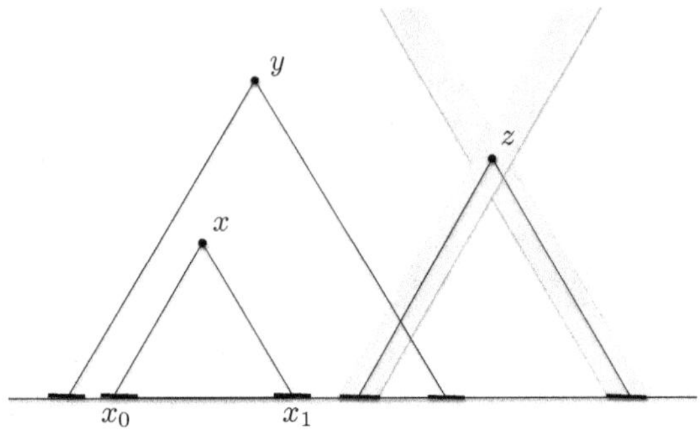

FIGURE XIII.7.1. A Hausdorff countable connected topology. The gray figure represents the closure of a neighborhood of z.

EXAMPLE XIII.7.5. Let L_n, M_n for $0 < n < \omega$ be subsets of the real plane, defined by

$$L_n := \{(x, \tfrac{1}{n} - (n-1)x) : x \geq 0\}, \quad M_n := \{(x, -\tfrac{1}{n} - (n-1)x) : x \leq 0\}.$$

Let $L := \bigcup_{0<n<\omega} L_n$, $M := \bigcup_{0<n<\omega} M_n$, and $X := L \cup M$. Then X is connected. Indeed,

$$L_n \# \operatorname{cl} M, \quad \operatorname{cl} L \# M_k$$

for each $0 < n, k < \omega$. Thus if $f \in \{0,1\}^X$ is continuous, then it is constant on each L_n and on each M_n. Let us show that $f(L_n) = f(M_k)$ for each $0 < n, k < \omega$.

As $L_n \# \operatorname{cl} M$, then by continuity, the set $A_n := \{k < \omega : f(L_n) = f(M_k)\}$ is cofinite, and for each $k \in A_n$, the set $B_k := \{p < \omega : f(L_p) = f(M_k)\}$ is cofinite. Therefore, there is such $m < \omega$ and $j \in \{0,1\}$ such that $f(L_n) = \{j\} = f(M_k)$ for each $n, k > m$. If now $l < m$, then, by the argument just deployed, $f(L_l) = f(M_k)$ for infinitely many $k < \omega$ and thus $f(L_l) = \{j\}$. Similarly, $f(M_l) = f(L_k)$ for infinitely many $k < \omega$ and thus $f(M_l) = \{j\}$.

PROPOSITION XIII.7.6. *Each Hausdorff sequentially founded convergence that is a topology is scattered.*

PROOF. Let X be a topological, sequentially founded convergence. Let A be a non-empty subset of X without isolated points. Pick $a \in A$. By assumption, $a \in \operatorname{cl}(A \setminus \{a\})$ and, since X is sequentially founded, there is a free sequence $(a_n)_n$ of elements of A converging to a. For the same reason, for each n there is a sequence $(a_{n,k})_k$ of elements of A converging to a_n. By topologicity, the contour converges to a, but there is no coarser sequential filter that does. $\square$

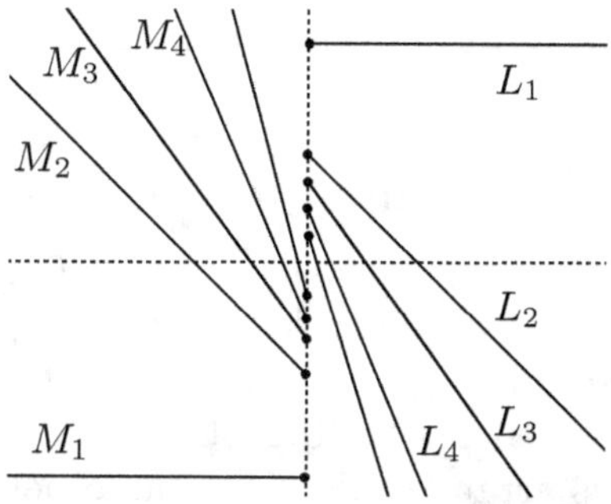

FIGURE XIII.7.2. A connected subspace of the real plane that is a union of countably many disjoint closed sets.

EXERCISE XIII.7.7. Let $(b_n)_n \subset \mathbb{N}_1$. Set $[b_0] := b_0$, and for $n > 0$,

$$[b_0, b_1, \ldots, b_n] := b_0 + \frac{1}{[b_1, \ldots, b_n]}.$$

Therefore $[b_0, b_1, \ldots, b_n, b_{n+1}] = [b_0, b_1, \ldots, b_n + \frac{1}{b_{n+1}}]$. Show that for $1 \leq x < y$,

$$|[b_0, b_1, \ldots, b_n, b_{n+1}] - [b_0, b_1, \ldots, b_n]| \leq \frac{1}{b_{n+1}} \frac{1}{1+n}.$$

SOLUTION. As $[b_0, b_1, \ldots, b_n, b_{n+1}] = [b_0, b_1, \ldots, b_n + \frac{1}{b_{n+1}}]$, we show that for $1 \leq x < y$,

$$|[b_0, b_1, \ldots, b_n, x] - [b_0, b_1, \ldots, b_n, y]| \leq \frac{|y-x|}{xy+n}.$$

For $n = 0$, we get $[b_0, x] - [b_0, y] = \frac{y-x}{xy}$. By inductive hypothesis, for $n+1$,

$$\left|[b_0, b_1, \ldots, b_n, b_{n+1} + \frac{1}{x}] - [b_0, b_1, \ldots, b_n, b_{n+1} + \frac{1}{y}]\right|$$

$$\leq \frac{\left|\frac{1}{y} - \frac{1}{x}\right|}{(b_{n+1} + \frac{1}{x})(b_{n+1} + \frac{1}{y}) + n} \leq \frac{\left|\frac{1}{y} - \frac{1}{x}\right|}{1 + \frac{1}{x} + \frac{1}{y} + \frac{1}{xy} + n}$$

$$\leq \frac{|x-y|}{xy+y+x+1+n} \leq \frac{|y-x|}{xy+n}.$$

EXERCISE XIII.7.8. Show that (1) $\operatorname{card}([0,1]^2) = \operatorname{card}([0,1])$, (2) $\operatorname{card}([0,1]^{\mathbb{N}}) = \operatorname{card}([0,1])$.

SOLUTION. (1) Each $r \in]0,1]$ has an infinite decimal representation of the form

$$0.a_1a_2\ldots a_n\ldots = \frac{a_1}{10} + \frac{a_2}{100} + \cdots + \frac{a_n}{10^n} + \cdots,$$

where $a_k \in \{0,1,\ldots,0\}$ ([15]). For each $(s,t) \in]0,1]^2$, let $s = 0.b_1b_2\ldots b_n\ldots$ and $t = 0.c_1c_2\ldots c_n\ldots$. Then let $0.a_1a_2\ldots a_n\ldots \mapsto (0.a_1a_3a_5\ldots, 0.a_2a_4a_6\ldots)$ is a bijection.

(2) If $\varphi \in]0,1]^{\mathbb{N}_1}$, then $\varphi(n) = 0.\varphi_{n1}\varphi_{n2}\varphi_{n3}\ldots$is an infinite decimal representation. Let us arrange them in a table below.

$$\begin{array}{llllll} \varphi_{11} & \varphi_{12} & \varphi_{13} & \cdots & \varphi_{1k} & \cdots \\ \varphi_{21} & \varphi_{22} & \varphi_{23} & \cdots & \varphi_{2k} & \cdots \\ \cdots \\ \varphi_{n1} & \varphi_{n2} & \varphi_{n3} & \cdots & \varphi_{nk} & \cdots \\ \cdots \end{array}$$

For a given $r = 0.a_1a_2\ldots a_k\ldots$, let $f(r) \in]0,1]^{\mathbb{N}}$, like in Example I.3.4, be given by

$$\begin{array}{llll} a_1 & a_3 & a_6 & a_{10} \\ a_2 & a_5 & a_9 & \cdots \\ a_4 & a_8 & \cdots & \cdots \\ a_7 & \cdots & \cdots & \cdots \end{array}.$$

Then f is surjective, proving (2).

EXERCISE XIII.7.9. Let (X,d) be a metric space, and let $\varepsilon > 0$. A finite sequence $x_0, x_1, \ldots, x_n$ is called an *ε-chain* if $d(x_k, x_{k+1}) < \varepsilon$ for each $0 \le k < n$. A metric space is called *well-chained* if for any two elements v, w of X and every $\varepsilon > 0$ there exist $n < \omega$ and an *ε-chain* $x_0, x_1, \ldots, x_n$ such that $v = x_0$ and $w = x_n$.

(1) Each metric connected space is well-chained.
(2) Every metric compact well-chained space is connected.

SOLUTION. (1) For $v \in X$, let $J_\varepsilon(v)$ be the set of all those $w \in X$, for which there is $n < \omega$ and an *ε-chain* $x_0, x_1, \ldots, x_n$ such that $v = x_0$ and $w = x_n$. We observe that $J_\varepsilon(v)$ is open. Indeed, if $w \in J_\varepsilon(v)$, then for every $x \in B(w,\varepsilon)$ we set $x_{n+1} := x$, obtaining an ε-chain $x_0, x_1, \ldots, x_n, x_{n+1}$ joining v and x. But, on the other hand, $J_\varepsilon(v)$ is also closed. Indeed, if $(w_k)_k$ is a sequence of elements of $J_\varepsilon(v)$ and $w \in \lim_{k\to\infty} w_k$, then there is $k_\varepsilon < \omega$ such that $d(w, w_{k_\varepsilon}) < \varepsilon$, so that v and w are connected by an ε-chain. As the space is connected, and

[15] Some reals have two such representation, and in this case one is finite and the other infinite, like $0.1 = 0.0999\ldots$.

$J_\varepsilon(v)$ is clopen and non-empty, $J_\varepsilon(v) = X$ for every $\varepsilon > 0$ and each $v \in X$.

(2) Suppose that (X, d) is compact and well-chained, but disconnected. Then there exist two disjoint closed subsets X_0 and X_1. By Corollary X.3.5 and Corollary XII.3.1, there exist $x_0 \in X_0$ and $x_1 \in X_1$ such that

$$\delta := \operatorname{dist}(X_0, X_1) = d(x_0, x_1) > 0.$$

Consequently, there is no δ-chain joining x_0 with x_1. A contradiction.

Metrizable topologies of covering dimension zero were introduced in Definition XIII.5.5. More generally, a (functionally regular) topology ξ on X is *strongly zero-dimensional* or has *covering dimension zero* if for each finite functionally ξ-open cover $\mathcal{Q}$, there exists a ξ-clopen refinement $\mathcal{P}$ of $\mathcal{Q}$ that is a partition of X.

REMARK XIII.7.10. By Theorem XI.3.8, each metrizable topology is paracompact. Hence, by the Dowker theorem ([16]), the definition above and Definition XIII.5.5 are equivalent for metrizable topologies.

[16]See [97, Theorem 7.2.4].

CHAPTER XIV

Extensions and compactifications

An *extension* of a topology ξ is a topology τ, in which ξ is densely embedded. If f is an embedding, that is, ξ and $f\xi$ are homeomorphic, then it is often practical to see ξ as a dense subspace of its extension τ. A Hausdorff compact extension of a topology ξ is called a *compactification* of ξ ([1]).

After having established necessary and sufficient conditions for a Hausdorff topology to admit compactifications, we will proceed to construct some of them. Our investigation resembles that of a paleontologist, attempting to describe a prehistoric creature on the base of its partial residues.

XIV.1. Extensions of spaces and maps

By Corollaries X.9.12 and XII.2.5, we know that

COROLLARY XIV.1.1. *A topology has a compactification if and only if it is Hausdorff and functionally regular (alias completely regular).*

Accordingly, if ξ is a topology on X densely embedded in a Hausdorff compact topology τ on Y, then

(1) for each point $y \in Y \setminus X$, there exists a coarsest filter $\mathcal{H}(y)$ on X such that $\{y\} = \lim_\tau \mathcal{H}(y)$. Actually,

$$\mathcal{H}(y) = \mathcal{N}_\tau(y) \vee X.$$

(2) As τ is Hausdorff and ξ is its dense subspace, $\operatorname{adh}_\xi \mathcal{H}(y) = \varnothing$, and if $y_0 \neq y_1$ then $\mathcal{H}(y_0) \perp \mathcal{H}(y_1)$.

(3) As τ is compact, for every $\mathcal{U} \in \beta X$ such that $\lim_\xi \mathcal{U} = \varnothing$, there exists a unique $y \in Y$ such that $\mathcal{U} \in \beta_X \mathcal{H}(y)$.

(4) As τ is a topology, by (VII.2.9), for each $y \in Y \setminus X$,

$$\mathcal{H}(y) \subset \textstyle\int_{\mathcal{H}(y)} \mathcal{H}(\cdot).$$

(5) If $Y \setminus X$ is infinite, then by compactness of τ, for each $\mathcal{W} \in \beta_X(Y \setminus X)$ there is a unique $y \in Y$ such that $y \in \lim_\tau \mathcal{W}$.

[1]The restriction to Hausdorff compactifications is pragmatic, as the traditional results presented here have been carried out in that framework.

All these conditions are in general not sufficient for a reconstruction of a Hausdorff compactification, because functional regularity must be involved at some stage.

(1) For each $f \in \mathscr{C}(\xi, I)$ there exists $F \in \mathscr{C}(\tau, I)$ such that $F|_X = f$.
(2) For each $C \in \mathcal{C}_\tau$ and $y \notin C$, there is $h \in \mathscr{C}(\tau, I)$ such that $h(x) > \sup_C h$.

Actually, an extension of a Hausdorff, functionally regular (alias completely regular) topology to some of its compactifications will be achieved through an extension of continuous maps. Moreover, we have already made our first steps in this direction, embedding Hausdorff, functionally regular topologies into Tikhonov cubes (Corollary X.9.12).

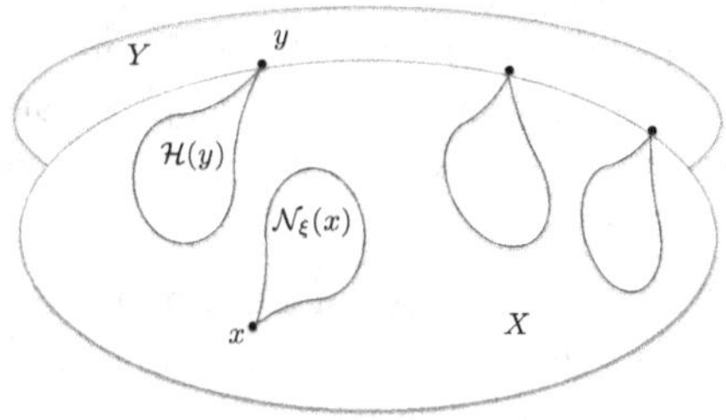

FIGURE XIV.1.1. An extension τ of a Hausdorff topology ξ on X to Y. If $y \in Y \setminus X$, then there is a filter $\mathcal{H}(y)$ on X that τ-converges to y, because X is τ-dense. We will see that $Y \setminus X \in \mathcal{N}_\tau(x)^\#$, that is, X need not be τ-open.

EXAMPLE XIV.1.2 (Alexandrov compactification). A simplest example of compactification that one can think of is the following: if ξ is a Hausdorff non-compact topology on X, then let ξ_∞ be defined on $X \sqcup \{\infty\}$ to the effect that $\infty \in \lim_{\xi_\infty} \mathcal{U}$ for each $\mathcal{U} \in \beta X$ such that $\lim_\xi \mathcal{U} = \varnothing$. As ξ_∞ is supposed to be a topology, $\mathcal{N}_{\xi_\infty}(\infty) \vee X = \bigcap\{\mathcal{U} : \lim_\xi \mathcal{U} = \varnothing\}$. Since ξ_∞ is supposed to be Hausdorff, we need to assure that $\mathcal{N}_\xi(x) \perp \mathcal{N}_{\xi_\infty}(\infty)$ for each $x \in X$. In other words, for each $x \in X$, there is $V \in \mathcal{N}_\xi(x)$ such that $X \setminus V \in \mathcal{N}_{\xi_\infty}(\infty) \vee X$, which means that if $\mathcal{U} \in \beta V$ then $\lim_\xi \mathcal{U} \neq \varnothing$. In other terms, V is ξ-compactoid.

REMARK XIV.1.3. By Example XIV.1.2, local compactoidness is a necessary condition for a Hausdorff topology ξ to have a Hausdorff compact one-point extension ξ_∞. As ξ_∞ is a Hausdorff compact topology, it is regular in virtue of Proposition IX.3.3, hence ξ is regular, because regularity is hereditary. By Proposition XII.3.9, ξ is (hereditarily) locally compact.

PROPOSITION XIV.1.4. *A Hausdorff topology has a one-point compactification if and only if it is locally compact.*

PROOF. By Example XIV.1.2 and Remark XIV.1.3, if a Hausdorff topology has a one-point compactification, then it is locally compact. By Exercise XII.10.8, every Hausdorff locally compact topology is functionally regular, hence has a compactification by Corollary XIV.1.1. Then the construction from Example XIV.1.2 yields a one-point compactification. We see that ξ_∞ is a topology. Indeed, the family of complements in $X \sqcup \{\infty\}$ of ξ-compact sets is a filter-base of the ξ_∞-neighborhood filter of ∞. $\square$

EXAMPLE XIV.1.5. Let ι_X be a discrete topology on an infinite set X. As $\lim_{\iota_X} \mathcal{U} = \varnothing$ for each free ultrafilter $\mathcal{U}$ on X, we conclude that $\mathcal{N}_{(\iota_X)_\infty}(\infty) \vee X = (X)_0$, that is, the *cofinite filter* on X. Therefore, the prime cofinite topology $\pi[\infty, (X)_0]$ is a one-point compactification of the discrete topology ι_X.

LEMMA XIV.1.6 (Extension of continuous maps). *Let ξ be a topology on X, and τ a Hausdorff compact topology on Y. If A is a ξ-dense subset of X, and $f \in \mathscr{C}(\xi|_A, \tau)$, then there exists $F \in \mathscr{C}(\xi, \tau)$ such that $F|_A = f$ if and only if for each pair of τ-closed sets B_0, B_1,*

$$B_0 \cap B_1 = \varnothing \Longrightarrow \mathrm{cl}_\xi f^-(B_0) \cap \mathrm{cl}_\xi f^-(B_1) = \varnothing. \tag{XIV.1.1}$$

PROOF. If such F exists, and $B_0, B_1 \in \mathcal{C}_\tau$ with $B_0 \cap B_1 = \varnothing$, hence $F^-(B_0) \cap F^-(B_1) = \varnothing$. On the other hand, $\mathrm{cl}_\xi f^-(B_0) \subset F^-(B_0)$ and $\mathrm{cl}_\xi f^-(B_1) \subset F^-(B_1)$, because $F^-(B_0) \in \mathcal{C}_\xi$ and $F^-(B_1) \in \mathcal{C}_\xi$. Consequently, (XIV.1.1).

Conversely, by assumption of density, $\mathcal{N}_\xi(x) \vee A$ is proper for each $x \in X$, and since τ is compact, for each $x \in X$, there exists $F(x) \in \mathrm{adh}_\tau f[\mathcal{N}_\xi(x) \vee A]$. Actually, $F(x)$ is the only such element, for if $y_0 \neq y_1$ were elements of $\mathrm{adh}_\tau f[\mathcal{N}_\xi(x) \vee A]$, then there would be $W_0 \in \mathcal{N}_\tau(y_0)$ and $W_1 \in \mathcal{N}_\tau(y_1)$ such that $\mathrm{cl}_\tau W_0 \cap \mathrm{cl}_\tau W_1 = \varnothing$, because τ is Hausdorff and regular. Therefore, (XIV.1.1) holds for $B_0 = \mathrm{cl}_\tau W_0$ and $B_1 = \mathrm{cl}_\tau W_1$. Therefore, $B_j \# f[\mathcal{N}_\xi(x) \vee A]$, equivalently $(A \cap f^- B_j) \# \mathcal{N}_\xi(x)$ for $j \in \{0, 1\}$, which means that $x \in \mathrm{cl}_\xi f^-(B_0) \cap \mathrm{cl}_\xi f^-(B_1)$, a contradiction.

As τ is compact, $\lim_\tau \mathcal{U} \neq \varnothing$ for each $\mathcal{U} \in \beta f[\mathcal{N}_\xi(x) \vee A]$, hence $\varnothing \neq \lim_\tau \mathcal{U} \subset \mathrm{adh}_\tau f[\mathcal{N}_\xi(x) \vee A] = \{F(x)\}$. If $W \in \mathcal{N}_\tau(F(x))$, then $W \in \mathcal{U}$ for each $\mathcal{U} \in \beta f[\mathcal{N}_\xi(x) \vee A]$, and thus $W \in f[\mathcal{N}_\xi(x) \vee A]$. Therefore, $F(x) \in \lim_\tau f[\mathcal{N}_\xi(x) \vee A]$. In particular, if $a \in A$, then $F(a) \in \lim_\tau f[\mathcal{N}_\xi(a) \vee A] = \{f(a)\}$, so that $F|_A = f$. $\square$

XIV.2. Compactifications

DEFINITION XIV.2.1. A Hausdorff compact topology τ is called a *compactification* of a topology ξ if ξ is *densely embedded* in τ ([2]).

We denote by $\xi \hookrightarrow \tau$ the fact that ξ is densely embedded in τ, and when the embedding j is explicit, we write

$$j : \xi \hookrightarrow \tau.$$

The set-theoretic difference between the underlying set of a compactification and that of the embedded topology $|\tau| \setminus |j\xi|$ is called the *remainder* of the compactification.

As ξ and $j\xi$ are homeomorphic, it is sensible to imagine that a topology ξ is a subspace of its compactification τ.

EXAMPLE XIV.2.2. In Section XII.7, we studied the Stone topology $\beta\iota_X$ on the set βX of all ultrafilters on an infinite set X. As $\beta\iota_X$ is Hausdorff and compact, and X is a $\beta\iota_X$-dense subset of βX, we infer that $\beta\iota_X$ is a compactification of ι_X.

If we consider an infinite set X, then the following topologies are compactifications of the *discrete topology* ι_X on X:

(1) the *prime cofinite topology* $\pi[\infty, (X)_0]$ on $X \sqcup \{\infty\}$ (Example XIV.1.5),
(2) the *compact MAD topology* $\zeta_{\mathcal{A}}^{\infty}$, where $\mathcal{A} \subset 2^X$ (Section XII.8),
(3) the *Stone topology* $\beta\iota_X$ topology on βX (Section XII.7).

Let ξ be a Hausdorff functionally regular topology on X, and let θ be a compactification of ξ on Y. By definition, there exists a $j \in \mathscr{C}(\xi, \theta)$ such that $j \in \mathscr{H}(\xi, \theta|_{j(X)})$ and $\mathrm{cl}_\theta\, j(X) = Y$. We denote by

$$\mathfrak{C}(\xi)$$

the class of all compactifications of ξ.

Two compactifications $j_0 : \xi \hookrightarrow \theta_0$ and $j_1 : \xi \hookrightarrow \theta_1$ are said to be *equivalent*

$$\theta_0 \approx \theta_1$$

if there exists a homeomorphism $h \in \mathscr{H}(\theta_0, \theta_1)$ such that $h \circ j_0 = j_1$.

---

[2]Often the embedding takes part of a formal definition of a compactification, that is to say, (j, τ) is said a compactification of ξ if the conditions above hold.

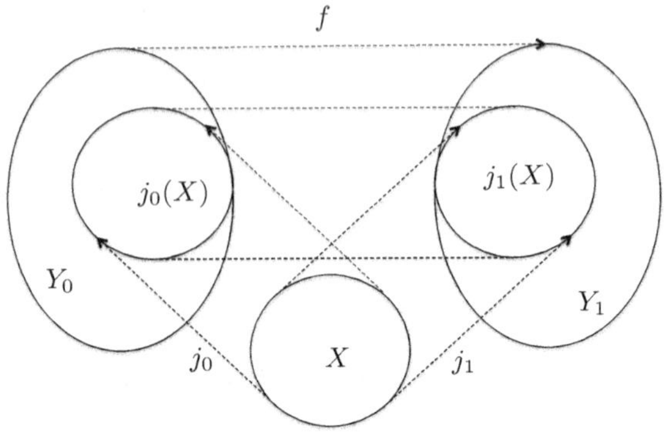

FIGURE XIV.2.1. A topology on X and its two compactifications on Y_0 and Y_1 realized with the aid of respective injections j_0 and j_1. A continuous map $f : Y_0 \longrightarrow Y_1$ (from a larger compactification to a smaller one) is an extension of a bijection $j_1 j_0^- : j_0(X) \longrightarrow j_1(X)$.

We say that $j_0 : \xi \twoheadrightarrow \theta_0$ is *larger* than $j_1 : \xi \twoheadrightarrow \theta_1$, in symbols

$$\theta_0 \succeq \theta_1 \tag{XIV.2.1}$$

if there exists $f \in \mathscr{C}(\theta_0, \theta_1)$ such that $f \circ j_0 = j_1$.

On recalling that ξ is a topology defined on X, the condition above means that f maps every point of X, considered as a common subspace of θ_0 and θ_1, onto itself.

The relation (XIV.2.1) is clearly transitive and reflexive. We shall see that it is also antisymmetric on the $\approx$ equivalence classes.

PROPOSITION XIV.2.3. *If $\theta_0 \succeq \theta_1$ and $\theta_1 \succeq \theta_0$, then $\theta_0 \approx \theta_1$.*

PROOF. Let ξ be a topology on X, and $j_0 : \xi \twoheadrightarrow \theta_0$ and $j_1 : \xi \twoheadrightarrow \theta_1$ be compactifications such that there exist $f_0 \in \mathscr{C}(\theta_0, \theta_1)$ and $f_1 \in \mathscr{C}(\theta_1, \theta_0)$ so that $f_0 \circ j_0 = j_1$ and $f_1 \circ j_1 = j_0$. Then $f_1 \circ f_0 \in \mathscr{C}(\theta_0, \theta_0)$ and $f_1 \circ f_0 \circ j_0 = f_1 \circ j_1 = j_0$, that is, $f_1 \circ f_0$ coincides on a dense subset with a restriction of the identity, hence is bijective by Lemma IX.1.3. Similarly, $f_0 \circ f_1 \in \mathscr{C}(\theta_1, \theta_1)$ is a bijection. Hence, f_0 and f_1 are bijections, by Remark I.1.3. By Corollary XII.2.4, each continuous bijection from a compact topology to a Hausdorff topology is a homeomorphism, so that $f_0 \in \mathscr{H}(\theta_0, \theta_1)$ and $f_1 = f_0^{-1}$. □

PROPOSITION XIV.2.4. *Let $\theta_0, \theta_1 \in \mathfrak{C}(\xi)$. If $\theta_0 \succeq \theta_1$, then for each pair of A, B of ξ-closed sets,*

$$\mathrm{cl}_{\theta_1} j_1(A) \cap \mathrm{cl}_{\theta_1} j_1(B) = \varnothing \Longrightarrow \mathrm{cl}_{\theta_0} j_0(A) \cap \mathrm{cl}_{\theta_0} j_0(B) = \varnothing.$$

PROOF. Let $f \in \mathscr{C}(\theta_0, \theta_1)$ be such that $f \circ j_0 = j_1$. Let A, B be ξ-closed sets, and assume that $\mathrm{cl}_{\theta_1} j_1(A) \cap \mathrm{cl}_{\theta_1} j_1(B) = \varnothing$, then $\mathrm{cl}_{\theta_0} f^-(j_1 A) \cap \mathrm{cl}_{\theta_0} f^-(j_1 B) = \varnothing$, because $f \in \mathscr{C}(\theta_0, \theta_1)$. By assumption, $f \circ j_0 = j_1$, hence $j_0(A) \subset f^- f(j_0(A)) = f^-(j_1(A))$, hence $\mathrm{cl}_{\theta_0} j_0(A) \cap \mathrm{cl}_{\theta_0} j_1(B) = \varnothing$. □

LEMMA XIV.2.5. *Every non-empty subfamily of compactifications has a greatest element.*

PROOF. Let ξ be a topology on X and let $\Theta \subset \mathfrak{C}(\xi)$. As $j_\theta : \xi \hookrightarrow \theta$, by Corollary IV.7.8, the diagonal product map

$$k := \triangle_{\theta \in \Theta} j_\theta : \xi \hookrightarrow \prod \Theta,$$

is an embedding, so that $\alpha\xi := \mathrm{cl}_{\prod \Theta} k(X)$ is Hausdorff and compact, and $k : \xi \hookrightarrow \alpha\xi$ is a compactification of ξ. As the projection $p_\theta \in \mathscr{C}(\prod \Theta, \theta)$ and $p_\theta \circ k = j_\theta$, we conclude that $\alpha\xi = \bigvee_{\mathfrak{C}} \Theta$. □

COROLLARY XIV.2.6. *Each Hausdorff, functionally regular (alias completely regular) topology has a largest compactification, unique up to a homeomorphism.*

PROOF. The existence follows from Lemma XIV.2.5 and the uniqueness from Proposition XIV.2.3. □

By Proposition XIV.1.4,

PROPOSITION XIV.2.7. *A Hausdorff, functionally regular (alias completely regular) topology has a least compactification if and only if it is locally compact.*

REMARK XIV.2.8 (*). As each compactification of a topology ξ on X has an equivalent compactification which is a subspace of a Tikhonov cube of dimension depending on X, we can consider $\mathfrak{C}(\xi)$ as a set. If θ_0 is larger than θ_1, then a map f defining $\theta_0 \succeq \theta_1$ is perfect (see Proposition XX.11.13), because θ_1 is compact, and the fibers are compact subspaces of θ_0. In particular, f is a topological quotient map, which is injective if and only if it is a homeomorphism.

REMARK XIV.2.9 (*Cardinalities). A Hausdorff, functionally regular topology on an infinite set X is embedded in a Tikhonov cube of dimension equal to the cardinality of $\mathscr{C}$ — a set of continuous functions that separate the topology. Therefore this dimension is not greater than $\mathfrak{c}^{\mathrm{card}\,X} = (2^{\aleph_0})^{\mathrm{card}\,X} = 2^{\mathrm{card}\,X}$. Hence, the cardinality of a compactification is not greater than the cardinality of a Tikhonov cube of dimension $2^{\mathrm{card}\,X}$, that is, by Corollary XXVII.6.5, $\mathfrak{c}^{2^{\mathrm{card}\,X}} = 2^{\aleph_0 2^{\mathrm{card}\,X}} = 2^{2^{\mathrm{card}\,X}}$. In the case of $X = \mathbb{N}$, this is $2^{\mathfrak{c}}$ and it is the cardinality of $\beta\mathbb{N}$.

XIV.3. Filters based in lattices

In the case of discrete topology ι_X on an infinite set X, the compactification $\beta\iota_X$ of Example XIV.2.2 was constructed on the set βX of all ultrafilters on X.

An analogous construction can be carried out for any Hausdorff, functionally regular topology. This time, arbitrary ultrafilters need be replaced by maximal filters among those based in functionally closed sets. Properties of such maximal filters are a bit more intricate than those of ultrafilters.

Consider a *lattice* $\mathcal{Z}$ *of subsets* of X. Explicitly, $\mathcal{Z} \subset 2^X$ is such that

$$Z_0, Z_1 \in \mathcal{Z} \Longrightarrow (Z_0 \cap Z_1 \in \mathcal{Z}) \wedge (Z_0 \cup Z_1 \in \mathcal{Z}). \qquad \text{(XIV.3.1)}$$

This implies, by an easy induction, that if $\mathcal{Z}_0 \subset_{\text{fin}} \mathcal{Z}$, then $\bigcap \mathcal{Z}_0 \in \mathcal{Z}$ and $\bigcup \mathcal{Z}_0 \in \mathcal{Z}$.

DEFINITION XIV.3.1. A proper filter $\mathcal{F}$ on X is called a $\mathcal{Z}$*-filter* if it has a filter-base consisting of elements of $\mathcal{Z}$ ([3]).

For example, if ξ is a convergence on X, then ξ, the family of all ξ-closed sets, is a lattice. The family of functionally closed sets $\mathcal{C}_\xi^I$ is also a lattice.

If $\boldsymbol{F}$ is a collection of $\mathcal{Z}$-filters linearly ordered by inclusion, then $\bigcup \boldsymbol{F}$ is a $\mathcal{Z}$-filter. Indeed, let $Z_0, Z_1 \in \bigcup \boldsymbol{F}$, then there exists $\mathcal{F}_0, \mathcal{F}_1 \in \boldsymbol{F}$ such that $Z_0 \in \mathcal{F}_0$ and $Z_1 \in \mathcal{F}_1$. As $\boldsymbol{F}$ is a linearly ordered, $\mathcal{F} := \mathcal{F}_0 \cup \mathcal{F}_1 \in \mathcal{F}_1$ and thus $Z_0 \cap Z_1 \in \mathcal{F} \subset \bigcup \boldsymbol{F}$. Also if $F \in \bigcup \boldsymbol{F}$ and $\mathcal{Z} \ni G \supset F$, then there is $\mathcal{F} \in \bigcup \boldsymbol{F}$ such that $F \in \mathcal{F}$, hence $G \in \mathcal{F} \subset \bigcup \boldsymbol{F}$.

Hence, by the Kuratowski-Zorn Theorem XXVII.4.8,

PROPOSITION XIV.3.2. *For every $\mathcal{Z}$-filter $\mathcal{F}$, there exists a maximal $\mathcal{Z}$-filter $\mathcal{U} \supset \mathcal{F}$.*

A maximal $\mathcal{Z}$-filter is called a $\mathcal{Z}$*-ultrafilter.* If $\mathcal{Z}$ is a lattice of subsets of X, then in general a $\mathcal{Z}$-ultrafilter is not an ultrafilter-base on X.

PROPOSITION XIV.3.3. *A $\mathcal{Z}$-filter $\mathcal{F}$ is a $\mathcal{Z}$-ultrafilter if and only if $\mathcal{Z} \cap \mathcal{F}^\# \subset \mathcal{F}$.*

---

[3] A filter on a lattice $\mathcal{Z}$ is a subfamily $\mathcal{F}$ of $\mathcal{Z}$ fulfilling (XIV.3.1), for example, they are called *z-filters* in [**113**] of L. GILLMAN and M. JERISON, in reference of zero sets, that is, the sets of the form $\{f = 0\}$, where f is a continuous functions. That approach was adopted in [**83**]. Here I consider instead filters based in a lattice $\mathcal{Z}$. These two techniques are equivalent.

PROOF. If $\mathcal{F}$ is a $\mathcal{Z}$-ultrafilter and $Z \in \mathcal{Z} \cap \mathcal{F}^{\#}$, then $Z\#F$ for every $F \in \mathcal{Z} \cap \mathcal{F}$. As $F \cap Z \in \mathcal{Z}$, the family $\{F \cap Z : F \in \mathcal{F}\}$ is a filter-base consisting of elements of $\mathcal{Z}$. Then, there exists a $\mathcal{Z}$-ultrafilter $\mathcal{U} \geq \{F \cap Z : F \in \mathcal{F}\} \geq \mathcal{F}$, and thus $\mathcal{U} = \mathcal{F}$.

Conversely, if a $\mathcal{Z}$-filter $\mathcal{F}$ is not maximal, then there exists a $Z \in \mathcal{Z} \setminus \mathcal{F}$ such that $\mathcal{F} \cup \{Z\}$ is a filter-subbase. Consequently, $Z \in \mathcal{F}^{\#}$. □

COROLLARY XIV.3.4. *If $\mathcal{U}$ and $\mathcal{V}$ are distinct $\mathcal{Z}$-ultrafilters, then there exist $U \in \mathcal{U}$ and $V \in \mathcal{V}$ such that $U \cap V = \varnothing$.*

PROOF. Indeed, if $\mathcal{U}\#\mathcal{V}$, then $\{U \cap V : U \in \mathcal{U}, V \in \mathcal{V}\}$ includes $\mathcal{U}$ and $\mathcal{V}$, hence $\mathcal{U}$ and $\mathcal{V}$ are not distinct $\mathcal{Z}$-ultrafilters. □

REMARK XIV.3.5. If $\mathcal{Z}$ is a lattice of subsets of X, then $\mathcal{Z}(x) := \{Z \in \mathcal{Z} : x \in Z\}$ is a $\mathcal{Z}$-filter. It is a $\mathcal{Z}$-ultrafilter if and only if for each $Z \in \mathcal{Z}$ and each $x \notin Z$, there exists $W \in \mathcal{Z}$ such that $x \in W$ and $W \cap Z = \varnothing$. Indeed, if $Z \in \mathcal{Z} \setminus \mathcal{Z}(x)$, then $Z \notin \mathcal{Z}(x)^{\#}$ if and only if the condition holds, and we apply Proposition XIV.3.3.

On the other hand, $\mathcal{Z}(x_0) \neq \mathcal{Z}(x_1)$ if and only if there exists $Z \in \mathcal{Z}$ that contains one point and the other. If besides the condition above holds, then by Corollary XIV.3.4, there exist $Z_0 \in \mathcal{Z}(x_0)$ and $Z_1 \in \mathcal{Z}(x_1)$ and $Z_0 \cap Z_1 = \varnothing$. This property can be deduced directly, because if there is $Z_0 \in \mathcal{Z}$ such that, say, $x_0 \in Z_0$ and $x_1 \notin Z_0$, then by the first property, there is $Z_1 \in \mathcal{Z}$ such that $x_1 \in Z_1$ and $Z_0 \cap Z_1 = \varnothing$.

Notice that if ξ is a Hausdorff functionally regular topology, then $\mathcal{C}^I_\xi$ has both properties.

DEFINITION XIV.3.6. Let us call a lattice $\mathcal{Z}$ of subsets of X *separated* if for each $x \in X$, the family $\mathcal{Z}(x) := \{Z \in \mathcal{Z} : x \in Z\}$ is a $\mathcal{Z}$-ultrafilter such that $\mathcal{Z}(x_0) \neq \mathcal{Z}(x_1)$ if $x_0 \neq x_1$.

A $\mathcal{Z}$-filter $\mathcal{F}$ is called *prime* if for each pair $Z_0, Z_1 \in \mathcal{Z}$,

$$Z_0 \cup Z_1 \in \mathcal{F} \Longrightarrow (Z_0 \in \mathcal{F}) \vee (Z_1 \in \mathcal{F}).$$

PROPOSITION XIV.3.7. *Each $\mathcal{Z}$-ultrafilter is prime.*

PROOF. Let $\mathcal{U}$ be a $\mathcal{Z}$-ultrafilter, and let $Z_0 \notin \mathcal{U}$ and $Z_1 \notin \mathcal{U}$. By Proposition XIV.3.3, there exist $U_0, U_1 \in \mathcal{U}$ such that $Z_0 \cap U_0 = \varnothing$ and $Z_1 \cap U_1 = \varnothing$. Hence $(Z_0 \cup Z_1) \cap (U_0 \cap U_1) = \varnothing$, and as $U_0 \cap U_1 \in \mathcal{U}$, we infer that $Z_0 \cup Z_1 \notin \mathcal{U}^{\#}$, and since $\mathcal{U}$ is a $\mathcal{Z}$-ultrafilter, $Z_0 \cup Z_1 \notin \mathcal{U}$ by Proposition XIV.3.3. □

PROPOSITION XIV.3.8. *If $\mathcal{U}$ is an ultrafilter, then $\mathcal{U} \cap \mathcal{Z}$ is prime.*

PROOF. Let $H_0, H_1 \in \mathcal{Z}$ and $H_0 \cup H_1 \in \mathcal{U}$, then either $H_0 \in \mathcal{U}$ or $H_1 \in \mathcal{U}$, because $\mathcal{U}$ is an ultrafilter. □

EXAMPLE XIV.3.9. Consider $\mathcal{C}_\nu$, the lattice of ν-closed subsets of the real line. If $\mathcal{U}_+$ is an ultrafilter such that $0 \in \lim_\nu \mathcal{U}_+$ and $]0, +\infty[\in \mathcal{U}_+$, then $\mathcal{U}_+ \cap \mathcal{C}_\nu$ is a prime $\mathcal{C}_\nu$-filter. If $Z \in \mathcal{U}_+ \cap \mathcal{C}_\nu$, then $0 \in Z$. We take an analogous ultrafilter $\mathcal{U}_-$ with $]-\infty, 0[\in \mathcal{U}_-$ and $0 \in \lim_\nu \mathcal{U}_-$, and notice that if $Z \in \mathcal{U}_- \cap \mathcal{C}_\nu$, then $0 \in Z$. Accordingly, $(\mathcal{U}_+ \cap \mathcal{C}_\nu) \# (\mathcal{U}_- \cap \mathcal{C}_\nu)$, so that they are not $\mathcal{C}_\nu$-ultrafilters. They included in $\{Z \in \mathcal{C}_\nu : 0 \in Z\}$, which is a $\mathcal{C}_\nu$-ultrafilter, because if $0 \notin W \in \mathcal{C}_\nu$, then there exist $a < 0 < b$ such that $[a, b] \cap W = \varnothing$, hence $W \notin \{Z \in \mathcal{C}_\nu : 0 \in Z\}^\#$.

DEFINITION XIV.3.10. A lattice $\mathcal{Z}$ of subsets of X is called *normal* whenever for any two disjoint $U_0, U_1 \in \mathcal{Z}$, there exist $Z_0, Z_1 \in \mathcal{Z}$ such that

$$U_0 \cap Z_1 = \varnothing, \quad U_1 \cap Z_0 = \varnothing, \text{ and } Z_0 \cup Z_1 = X. \tag{XIV.3.2}$$

EXAMPLE XIV.3.11. The family $\mathcal{C}_\xi$ (of all ξ-closed sets) is a normal lattice if and only if ξ is normal.

PROPOSITION XIV.3.12. *Each prime $\mathcal{Z}$-filter in a normal lattice $\mathcal{Z}$ is included in a unique $\mathcal{Z}$-ultrafilter.*

PROOF. Let $\mathcal{Z}$ be a normal lattice on X. Let $\mathcal{F}$ be a prime $\mathcal{Z}$-filter and let $\mathcal{U}_0, \mathcal{U}_1$ be $\mathcal{Z}$-ultrafilters such that $\mathcal{U}_0 \cap \mathcal{U}_1 \supset \mathcal{F}$. If $\mathcal{U}_0 \neq \mathcal{U}_1$, then there exist $U_0 \in \mathcal{U}_0$ and $U_1 \in \mathcal{U}_1$ such that $U_0 \cap U_1 = \varnothing$. By the normality of $\mathcal{Z}$, there exist $Z_0, Z_1 \in \mathcal{Z}$ such that (XIV.3.2) holds. Since $Z_0 \cup Z_1 = X \in \mathcal{F}$, either $Z_1 \in \mathcal{F} \subset \mathcal{U}_0$ or $Z_0 \in \mathcal{F} \subset \mathcal{U}_1$. Hence, either $\varnothing = Z_1 \cap U_0 \in \mathcal{U}_0$ or $\varnothing = Z_0 \cap U_1 \in \mathcal{U}_1$, which yields a contradiction. □

EXAMPLE XIV.3.13. A topology ξ is Hausdorff, functionally regular (alias completely regular) if and only if the set $\mathcal{C}_\xi^I$ of functionally ξ-closed sets is a base. Let ξ be a Hausdorff functionally regular topology. Then by Lemma X.7.2 on functional separation, the set $\mathcal{C}_\xi^I$ of all functionally closed sets is a normal lattice.

LEMMA XIV.3.14. *If $\mathcal{U}$ is a prime* $\mathcal{C}_\xi^I$-filter *and $f \in \mathscr{C}(\xi, \zeta)$, then $f[\mathcal{U}]$ is a prime* $\mathcal{C}_\zeta^I$-filter .

PROOF. If $H_0, H_1 \in \mathcal{C}_\zeta^I$ and $f \in \mathscr{C}(\xi, \zeta)$, then $H_0 \cup H_1 \in f[\mathcal{U}]$ implies that $f^-(H_0) \cup f^-(H_1) = f^-(H_0 \cup H_1) \in \mathcal{U}$ and $f^-(H_0), f^-(H_1) \in \mathcal{C}_\xi^I$, hence either $f^-(H_0) \in \mathcal{U}$ or $f^-(H_1) \in \mathcal{U}$, thus either $H_0 \supset ff^-(H_0) \in f[\mathcal{U}]$ or $H_1 \supset ff^-(H_1) \in f[\mathcal{U}]$. □

Of course, if ξ is Hausdorff functionally regular (alias completely regular), then $\mathcal{N}_\xi(x)$ is $\mathcal{C}_\xi^I$-based and $\mathcal{O}_\xi^I$-based.

REMARK XIV.3.15. As ξ is a topology, $\operatorname{adh}_\xi \mathcal{F} = \bigcap_{F \in \mathcal{F}} \operatorname{cl}_\xi F$. Accordingly, if $\mathcal{F}$ is a $\mathcal{C}_\xi^I$-filter, then $\operatorname{adh}_\xi \mathcal{F} = \bigcap_{F \in \mathcal{F}} F = \ker \mathcal{F}$, hence if $x \in \lim_\xi \mathcal{F}$, then $\{x\} = \bigcap_{F \in \mathcal{F}} F$.

Recollect from Example X.7.23 that the *prime cofinite topology* $\pi[x, (X)_0]$ on an uncountable set X is Hausdorff and compact, hence normal, but not perfectly normal. Indeed, each continuous function is constant on a cocountable set containing the distinguished point x_∞, so that the closed set $\{x_\infty\}$ is not functionally closed. In Example X.7.24 we showed that also the *prime cocountable topology* $\pi[x_\infty, (X)_1]$ on an uncountable set X is normal, but not perfectly normal for the same reasons.

PROPOSITION XIV.3.16. *Let ξ be a Hausdorff, functionally regular (alias completely regular) topology, and let $\mathcal{F}$ be a prime $\mathcal{C}_\xi^I$-filter. Then the following statements are equivalent:*

(1) $x \in \operatorname{adh}_\xi \mathcal{F}$,
(2) $x \in \lim_\xi \mathcal{F}$,
(3) $\{x\} = \bigcap_{F \in \mathcal{F}} F$.

PROOF. Let X be the underlying set of ξ. We show that (1) implies (2). Let V be a functionally ξ-closed ξ-neighborhood of x. Then there is a functionally ξ-open ξ-neighborhood O of x included in V. Hence $X \setminus O$ is ξ-closed, and $V \cup (X \setminus O) = X \in \mathcal{F}$, and since $\mathcal{F}$ is prime, either $V \in \mathcal{F}$ or $X \setminus O \in \mathcal{F}$. As $x \notin X \setminus O$, we infer that $V \in \mathcal{F}$, and thus (2) holds. (2) implies (3) by Remark XIV.3.15, and (3) implies (1) in virtue of Proposition VIII.1.3. □

LEMMA XIV.3.17. *If ξ is a Hausdorff functionally regular topology, then*

$$\mathcal{U}(x) := \{W \in \mathcal{C}_\xi^I : x \in W\}$$

is a filter-base of a unique $\mathcal{C}_\xi^I$-ultrafilter ξ-convergent to x.

PROOF. The family $\mathcal{U}(x)$ is obviously a $\mathcal{C}_\xi^I$-filter-base. If $x \notin W \in \mathcal{C}_\xi^I$, then there is $f \in \mathscr{C}(\xi, I)$ such that $f(x) = 0$ and $W \subset \{f = 1\}$. Thus $\{f \leq \frac{1}{2}\} \in \mathcal{U}(x)$ and $\{f \leq \frac{1}{2}\} \cap W = \varnothing$, that is, $W \notin \mathcal{U}(x)^\#$. It follows that $\mathcal{U}(x)$ is a filter-base of a $\mathcal{C}_\xi^I$-ultrafilter. □

XIV.4. Čech-Stone compactification

The largest compactification of a convergence ξ on X is called the *Čech-Stone compactification* and is denoted by $\beta\xi$ ([4]).

REMARK XIV.4.1. In the literature on compactifications, one writes βX to design $\beta\xi$ when it is clear which topology ξ is considered on X; for instance, $\beta\mathbb{N}$ is a usual notation for the Čech-Stone compactification of the discrete topology on $\mathbb{N}$, which we denote by $\beta\iota_{\mathbb{N}}$.

[4] E. ČECH [39] and M. H. STONE [247].

By Corollaries XIV.1.1 and XIV.2.6, a topology has a Čech-Stone compactification if and only if it is Hausdorff and functionally regular (alias completely regular).

PROPOSITION XIV.4.2. *Let ξ be a Hausdorff, functionally regular topology on X, σ a Hausdorff compact topology on Z, and $f \in \mathscr{C}(\xi, \sigma)$. Then there exists a map $F \in \mathscr{C}(\beta\xi, \sigma)$ such that $F|_X = f$.*

PROOF. Let $f \in \mathscr{C}(\xi, \sigma)$ and let $b : \xi \looparrowright \beta\xi$ be a maximal compactification, where $b \in |\beta\xi|^X$. By Proposition IV.7.9, $f \times b : \xi \looparrowright \sigma \times \beta\xi$ is an embedding, and since σ is Hausdorff and compact on Z, a compactification of ξ. As $\beta\xi$ is maximal, there exists $g \in \mathscr{C}(\beta\xi, \sigma \times \beta\xi)$ such that $g \circ b = f \times b$. Therefore, if $F = p_Z \circ g \in \mathscr{C}(\beta\xi, \sigma)$, then $F \circ b = p_Z \circ g \circ b = f$. □

PROPOSITION XIV.4.3. *A compactification τ of ξ is maximal if and only if for each Hausdorff compact topology ζ, every $f \in \mathscr{C}(\xi, \zeta)$ can be extended to $F \in \mathscr{C}(\tau, \zeta)$.*

PROOF. If τ is a maximal compactification, then f can be extended to F by Proposition XIV.4.2. Conversely, let $g : \xi \looparrowright \tau$ and suppose that the extension property above holds for τ. Then there exists an extension $B \in \mathscr{C}(\tau, \beta\xi)$ of $b : \xi \looparrowright \beta\xi$. Hence $B \circ g = b$, and thus by definition $\tau \succeq \beta\xi$, so that $\tau \approx \beta\xi$ in virtue of Proposition XIV.2.3. □

Given a Hausdorff, functionally regular (alias completely regular) topology, we shall construct its maximal compactification following the guidelines from Section XIV.1.

So let ξ be a Hausdorff functionally regular topology. It turns out that the non-ξ-adherent filters $\mathcal{H}(y)$, corresponding to new points y of the maximal compactification $\beta\xi$, are maximal filters based in functionally ξ-closed sets.

Given a topology ξ, a filter is called *functionally regular* if it is a $\mathcal{C}^I_\xi$*-filter*, that is, by Definition XIV.3.1, a filter having a base of $\mathscr{C}(\xi, I)$-closed sets.

REMARK XIV.4.4. If $W \in \mathcal{C}^I_\xi$ and $y \in \mathrm{cl}_\tau W$, then there is a $\mathcal{C}^I_\xi$-ultrafilter $\mathcal{U}$ on W such that $y \in \lim_\tau \mathcal{U}$. Indeed, by assumption, $\mathcal{N}_\tau(y) \vee W$ is proper and $\mathcal{C}^I_\xi$-based, hence there is a $\mathcal{C}^I_\xi$-ultrafilter finer than $\mathcal{N}_\tau(y) \vee W$.

By Propositions XIV.4.2 and XIV.4.3, the Čech-Stone compactification of a Hausdorff functionally regular topology ξ is a topology τ fulfilling any condition of Theorem XIV.4.5 below.

THEOREM XIV.4.5. *Let τ be a Hausdorff, functionally regular (alias completely regular) topology, and let ξ be a dense subspace of τ. Then the following statements are equivalent:*

(1) *For each Hausdorff compact topology ζ and $f \in \mathscr{C}(\xi, \zeta)$, there exists an extension $F \in \mathscr{C}(\tau, \zeta)$ of f.*
(2) *For each $f \in C(\xi, I)$, there exists $F \in \mathscr{C}(\tau, I)$ such that $F|_{|\xi|} = f$.*
(3) *If Z_0, Z_1 are functionally ξ-closed sets and $Z_0 \cap Z_1 = \varnothing$, then*

$$\mathrm{cl}_\tau Z_0 \cap \mathrm{cl}_\tau Z_1 = \varnothing.$$

(4) *If Z_0, Z_1 are functionally ξ-closed sets, then $\mathrm{cl}_\tau(Z_0 \cap Z_1) = \mathrm{cl}_\tau Z_0 \cap \mathrm{cl}_\tau Z_1$.*
(5) *For each $y \in |\tau|$, there exists a unique C^I_ξ-ultrafilter $\mathcal{U}$ such that $y \in \lim_\tau \mathcal{U}$.*

PROOF. ([5]) (1) $\Rightarrow$ (2), because I is a Hausdorff compact topological space.

(2) $\Rightarrow$ (3). As Z_0, Z_1 are functionally ξ-closed and disjoint, by Lemma X.7.2, there exists $f \in \mathscr{C}(\xi, I)$ such that $Z_0 \subset \{f = 0\}$ and $Z_1 \subset \{f = 1\}$. By (2), there exists an extension $F \in \mathscr{C}(\tau, I)$ of f. Clearly, $Z_0 \subset \{F = 0\}$ and $Z_1 \subset \{F = 1\}$, and thus $\mathrm{cl}_\tau Z_0 \cap \mathrm{cl}_\tau Z_1 = \varnothing$.

(3) $\Rightarrow$ (4). If $y \in \mathrm{cl}_\tau Z_0 \cap \mathrm{cl}_\tau Z_1$, then $V \cap Z_0 \neq \varnothing$ and $V \cap Z_1 \neq \varnothing$ for each functionally τ-closed $V \in \mathcal{N}_\tau(y)$, because τ is functionally regular. Then $V \cap Z_0$ and $V \cap Z_1$ are functionally ξ-closed, and of course, $W \cap V \cap Z_0 \neq \varnothing$ and $W \cap V \cap Z_1 \neq \varnothing$ for each $W \in \mathcal{N}_\tau(y)$, so that $y \in \mathrm{cl}_\tau(V \cap Z_0) \cap \mathrm{cl}_\tau(V \cap Z_1)$, hence by (3), $V \cap Z_0 \cap Z_1 \neq \varnothing$ for each functionally τ-closed $V \in \mathcal{N}_\tau(y)$, so that $y \in \mathrm{cl}_\tau(Z_0 \cap Z_1)$.

(4) $\Rightarrow$ (5). Let $y \in |\tau|$. By τ-density of $|\xi|$ and by functional regularity of τ, for each functionally τ-closed $V \in \mathcal{N}_\tau(y)$, the intersection $V \cap |\xi|$ is functionally ξ-closed and non-empty. Hence $\mathcal{N}_\tau(y) \vee |\xi|$ is a functionally ξ-regular filter τ-convergent to y, hence there exists a C^I_ξ-ultrafilter $\mathcal{U} \geq \mathcal{N}_\tau(y) \vee |\xi|$, and thus $y \in \lim_\tau \mathcal{U}$. If $\mathcal{W}$ is another C^I_ξ-ultrafilter finer than $\mathcal{N}_\tau(y) \vee |\xi|$, then $\mathcal{U} \# \mathcal{W}$, hence $\mathcal{U} = \mathcal{W}$ by virtue of Proposition XIV.3.3.

(5) $\Rightarrow$ (1). Let ζ be a Hausdorff compact topology, and let $f \in \mathscr{C}(\xi, \zeta)$. If $y \in |\tau|$, then there exists a unique C^I_ξ-ultrafilter $\mathcal{U}_y$ such that $y \in \lim_\tau \mathcal{U}_y$.

As ζ is compact, $\mathrm{adh}_\zeta f[\mathcal{U}_y] \neq \varnothing$, and since, by Lemma XIV.3.14, $f[\mathcal{U}_y]$ is ζ-prime, by XIV.3.16, $\lim_\zeta f[\mathcal{U}_y]$ is a singleton that we shall denote by $F(y)$. In particular, if $y \in |\xi|$, then $\mathcal{U}_y = y^\uparrow$ and $F(y) = f(y)$, which means that the function F just defined is an extension of f.

[5]Based on [113, Theorem 6.4] of L. GILLMAN and M. JERISON.

We notice that if $W \in \mathcal{C}^I_\zeta$, then

$$y \in \mathrm{cl}_\tau f^-(W) \Longrightarrow F(y) \in W. \tag{XIV.4.1}$$

Indeed, by Remark XIV.4.4, this implies that there is a $\mathcal{C}^I_\xi$-ultrafilter $\mathcal{U}$ on $f^-(W)$ such that $y \in \lim_\tau \mathcal{U}$. As the only $\mathcal{C}^I_\xi$-ultrafilter convergent to y is $\mathcal{U}_y$, we infer that $f^-(W) \in \mathcal{U}_y$, hence $W \in f[\mathcal{U}_y]$ and thus $F(y) \in W$.

We need to prove that $F \in \mathscr{C}(\tau, \zeta)$. Let $y \in |\tau|$ and W_0 be a functionally ζ-closed neighborhood of $F(y)$. Let W_1 be a functionally ζ-closed set such that $|\zeta| \setminus W_1 \in \mathcal{N}_\zeta(F(y))$ and $|\zeta| \setminus W_1 \subset W_0$. In particular, $F(y) \notin W_1$.

Then $W_0 \cup W_1 = |\zeta|$, thus $f^-(W_0) \cup f^-(W_1) = |\xi|$, so that $\mathrm{cl}_\tau f^-(W_0) \cup \mathrm{cl}_\tau f^-(W_1) = |\tau|$. As $F(y) \notin W_1$, we conclude that $y \notin \mathrm{cl}_\tau f^-(W_1)$, hence $V := |\tau| \setminus \mathrm{cl}_\tau f^-(W_1) \in \mathcal{N}_\tau(y)$. We observe that if $v \in V$, then $v \in \mathrm{cl}_\tau f^-(W_0)$, so that by (XIV.4.1) with y replaced by v, we conclude that $F(v) \in W_0$. □

By Proposition XIV.4.3 and Theorem XIV.4.5,

Corollary XIV.4.6. *An extension τ of a Hausdorff, functionally regular (alias completely regular) topology ξ is a maximal compactification if and only if it fulfills any of the equivalent conditions of Theorem XIV.4.5.*

Corollary XIV.4.7. *For any compactification τ of ξ fulfilling any condition of Theorem XIV.4.5, there is $h \in \mathscr{H}(\beta\xi, \tau)$ such that $h(x) = x$ for each $x \in |\xi|$.*

Before describing explicitly $\beta\xi$ for a given ξ, let us make some observations. An immediate consequence of Theorem XIV.4.5 is that

Proposition XIV.4.8. *If A is ξ-clopen, then $\mathrm{cl}_{\beta\xi} A$ is $\beta\xi$-clopen.*

Corollary XIV.4.9. *The Čech-Stone compactification $\beta\xi$ is connected if and only if ξ is connected.*

Proof. By Proposition XIII.1.3, each continuous image of a connected set is connected, and by Proposition XIII.1.14 the closure of a connected set is connected. Therefore if ξ is connected then $\beta\xi$ is connected. Proposition XIV.4.8 implies that if ξ is disconnected then $\beta\xi$ is disconnected. □

Remark XIV.4.10. It can be proved ([6]) that the Čech-Stone compactification $\beta\xi$ is strongly zero-dimensional if and only if ξ is strongly zero-dimensional.

[6]See R. Engelking [97, Theorem 6.2.12].

DEFINITION XIV.4.11. Let $\xi \trianglerighteq \tau$ be topologies (see Remark III.3.10), that is, $\xi \geq \tau|_{|\xi|}$. Then ξ is said to be *functionally extendable* to τ whenever every ξ-continuous bounded function can be extended to a τ-continuous bounded function.

This is the property (2) of Theorem XIV.4.5 without the density assumption.

REMARK XIV.4.12. Traditionally, *functionally extendable* topologies have been called *C^*-embedded* ([7]).

Theorem X.7.10 of Tietze asserts that-every closed subset of a normal topological space is functionally extendable to the whole space. Each compact subspace of a functionally regular topology is functionally extendable, by Corollary X.7.13.

PROPOSITION XIV.4.13. *Let ξ and τ are Hausdorff, functionally regular (alias completely regular) topologies and $\xi \trianglerighteq \tau$. Then ξ is functionally extendable in τ if and only if*

$$\tau|_{\mathrm{cl}_{\beta\tau}|\xi|} = \beta\xi.$$

PROOF. If ξ is functionally extendable to τ, and τ is functionally extendable to $\beta\tau$, then ξ is functionally extendable to $\beta\tau$. If ξ is functionally extendable to $\beta\tau$, then obviously, ξ is functionally extendable to τ. As each continuous bounded function on ξ is extendable to $\mathrm{cl}_{\beta\tau}|\xi|$, we conclude that $\tau|_{\mathrm{cl}_{\beta\tau}|\xi|} = \beta\xi$. □

PROPOSITION XIV.4.14. *The following statements are equivalent:*

(1) $\xi \trianglerighteq \tau \trianglerighteq \beta\xi$,
(2) $\beta\xi = \beta\tau$.

PROOF. (1) implies (2). As ξ is densely embedded in $\beta\xi$, also τ is $\beta\xi$-dense. Therefore, for each $f \in \mathscr{C}(\tau, I)$, the restriction $f|_{|\xi|} \in \mathscr{C}(\xi, I)$ and there exists $F \in \mathscr{C}(\beta\xi, I)$ such that $f|_{|\xi|} = F|_{|\xi|}$, because ξ is functionally extendable to $\beta\xi$. Hence $F|_{|\tau|} = f$, so that τ is functionally extendable to $\beta\xi$, thus $\beta\tau$ is homeomorphic to $\beta\xi$ by Theorem XIV.4.5 (2).

(2) implies (1), because $\xi \trianglerighteq \tau \trianglerighteq \beta\tau = \beta\xi$. □

[7] L. GILLMAN and M. JERISON [113]. I do not keep the traditional notation, because in convergence theory, * is used to indicate objects relative to free filters, and the term *functionally embeddable* refers to the exponential hull of Hausdorff functionally regular topologies.

XIV.5. A construction of maximal compactifications

In this section, the Čech-Stone compactification $\beta\xi$ of a Hausdorff, functionally regular (alias completely regular) topology ξ will be constructed as a Hausdorff compact topology on the set of all $\mathcal{C}_\xi^I$-ultrafilters.

We shall proceed, however, in slightly more general terms of $\mathcal{Z}$-ultrafilters with respect to a normal lattice $\mathcal{Z}$, because this more abstract argument appears to be simpler.

By Definiton XIV.3.6, a lattice $\mathcal{Z}$ of subsets of X is *separated* if for each $x \in X$, the family $\{Z \in \mathcal{Z} : x \in Z\}$ is a distinct $\mathcal{Z}$-ultrafilter.

THEOREM XIV.5.1. *If $\mathcal{Z}$ is a separated normal lattice of subsets of X, then on the set of all $\mathcal{Z}$-ultrafilters there exists a Hausdorff compact topology so that X is densely embedded and the limit of each $\mathcal{Z}$-ultrafilter on X is its defining point.*

PROOF. Let $\{\mathcal{H}(y) : y \in Y\}$ be the set of all $\mathcal{Z}$-ultrafilters on X. In particular, for each $x \in X \subset Y$,

$$\mathcal{H}(x) := \{Z \in \mathcal{Z} : x \in Z\},$$

is a $\mathcal{Z}$-ultrafilter, and since $\mathcal{Z}$ is separated, $x_0 \neq x_1$ implies that $\mathcal{H}(x_0) \neq \mathcal{H}(x_1)$. We define $\mathrm{ah} : \mathcal{Z} \longrightarrow 2^Y$ by setting $\mathrm{ah}Z := \{y \in Y : Z \in \mathcal{H}(y)\}$ for every $Z \in \mathcal{Z}$. In other words, by Proposition XIV.3.3,

$$y \in \mathrm{ah}Z \Longleftrightarrow Z \in \mathcal{H}(y)^{\#} \Longleftrightarrow Z \in \mathcal{H}(y). \tag{XIV.5.1}$$

Then, for every $Z, Z_0, Z_1 \in \mathcal{Z}$,

$$Z \subset \mathrm{ah}Z, \tag{XIV.5.2}$$

$$\mathrm{ah}(Z_0 \cup Z_1) = \mathrm{ah}Z_0 \cup \mathrm{ah}Z_1, \tag{XIV.5.3}$$

$$\mathrm{ah}(Z_0 \cap Z_1) = \mathrm{ah}Z_0 \cap \mathrm{ah}Z_1. \tag{XIV.5.4}$$

If needed, proofs are in ([8]). It follows that, for any $Z_0, Z_1 \in \mathcal{Z}$,

$$Z_0 \subset Z_1 \Longrightarrow \mathrm{ad}Z_0 \subset \mathrm{ad}Z_1,$$

$$Z_0 \cap Z_1 = \varnothing \Longrightarrow \mathrm{ad}Z_0 \cap \mathrm{ad}Z_1 = \varnothing. \tag{XIV.5.5}$$

[8](XIV.5.2). If $y \in Z$, then there exists $x \in X$ such that $y = x$, so that $Z \in \mathcal{H}(x)$ and thus $y = x \in \mathrm{ah}Z$. (XIV.5.3). If $y \in \mathrm{ah}(Z_0 \cup Z_1)$, equivalently $Z_0 \cup Z_1 \in \mathcal{H}(y)$, then either $Z_0 \in \mathcal{H}(y)$ or $Z_1 \in \mathcal{H}(y)$, because $\mathcal{H}(y)$ is prime, that is, either $y \in \mathrm{ah}Z_0$ or $y \in \mathrm{ah}Z_1$. (XIV.5.4). If $y \in \mathrm{ah}Z_0 \cap \mathrm{ah}Z_1$, then $y \in \mathrm{ah}Z_0$ and $y \in \mathrm{ah}Z_1$, thus $Z_0 \in \mathcal{H}(y)$ and $Z_1 \in \mathcal{H}(y)$, and thus $Z_0 \cap Z_1 \in \mathcal{H}(y)$ (because $\mathcal{H}(y)$ is a $\mathcal{Z}$-filter), equivalently, $y \in \mathrm{ah}(Z_0 \cap Z_1)$.

By (XIV.5.3),

$$\text{(XIV.5.6)} \qquad \mathrm{ah}^{\natural}\mathcal{Z} := \{\mathrm{ah}Z : Z \in \mathcal{Z}\}$$

is an ideal-base of subsets of Y. Accordingly, by Proposition IX.2.7, the family (XIV.5.6) is a base for the closed sets of a topology η on Y. In particular, if $Z \in \mathcal{Z}$ then by (XIV.5.2), $Z \subset \mathrm{ah}Z \in \mathrm{ah}^{\natural}\mathcal{Z}$, hence $\mathrm{cl}_\eta Z \subset \mathrm{ah}Z$.

On the other hand, if $Z, W \in \mathcal{Z}$ are such that $Z \subset \mathrm{ah}W$, then $Z \subset \mathrm{ah}W \cap X = W$, so that $\mathrm{ah}Z \subset \mathrm{ah}W$, hence $\mathrm{ah}Z \subset \mathrm{cl}_\eta Z$. Therefore,

$$\text{(XIV.5.7)} \qquad \mathrm{cl}_\eta Z = \mathrm{ah}Z$$

for each $Z \in \mathcal{Z}$.

Since $\mathrm{cl}_\eta Z \cap X = Z$ for each $Z \in \mathcal{Z}$, the family $\mathcal{Z}$ is a base for the closed sets of the restriction $\eta|_X$.

By Corollary XIV.3.4, if $y_0 \neq y_1$, then there exist $Z_0, Z_1 \in \mathcal{Z}$ such that $Z_0 \in \mathcal{H}(y_0)$ and $Z_1 \in \mathcal{H}(y_1)$, and $Z_0 \cap Z_1 = \varnothing$. As $\mathcal{Z}$ is normal, there exist $W_0, W_1 \in \mathcal{Z}$ such that $W_0 \cup W_1 = X$ as well as $Z_0 \cap W_1 = \varnothing$ and $Z_1 \cap W_0 = \varnothing$. Hence, $\mathrm{cl}_\eta W_0 \cup \mathrm{cl}_\eta W_1 = Y$, that is, $(\mathrm{cl}_\eta W_0)^c \cap (\mathrm{cl}_\eta W_1)^c = \varnothing$, as well as $y_0 \in \mathrm{cl}_\eta Z_0 \subset (\mathrm{cl}_\eta W_1)^c$ and $y_1 \in \mathrm{cl}_\eta Z_1 \subset (\mathrm{cl}_\eta W_0)^c$. Therefore η is a Hausdorff topology.

As $y \in \bigcap_{Z \in \mathcal{H}(y)} \mathrm{cl}_\eta Z$, the set X is η-dense. Actually,

$$\text{(XIV.5.8)} \qquad \{y\} = \lim\nolimits_\eta \mathcal{H}(y)$$

for each $y \in Y$. Indeed, if $V \in \mathcal{O}_\eta(y)$, then there exists $Z \in \mathcal{Z}$ such that $y \in (\mathrm{cl}_\eta Z)^c \subset V$. Hence $Z \notin \mathcal{H}(y)$, and thus $Z \notin \mathcal{H}(y)^{\#}$. Therefore, there is $H \in \mathcal{H}(y)$ such that $Z \cap H = \varnothing$. By (XIV.5.5) and (XIV.5.7), $\mathrm{cl}_\eta Z \cap \mathrm{cl}_\eta H = \varnothing$, so that $H \subset \mathrm{cl}_\eta H \subset (\mathrm{cl}_\eta Z)^c \subset V$, and thus $V \in \mathcal{H}(y)$.

To see that η is compact, let $\mathcal{F}$ be a filter on Y. Consider

$$\widehat{\mathcal{F}} := \{Z \in \mathcal{Z} : \underset{F \in \mathcal{F}}{\exists} F \subset \mathrm{cl}_\eta Z\},$$

and observe that $\widehat{\mathcal{F}}$ is a $\mathcal{Z}$-filter-base on X because of (XIV.5.4) and (XIV.5.7). Hence by Proposition XIV.3.2, there exists $y_0 \in Y$ such that $\widehat{\mathcal{F}} \subset \mathcal{H}(y_0)$. Therefore,

$$y_0 \in \mathrm{adh}_\eta \widehat{\mathcal{F}} = \bigcap_{F \in \mathcal{F}} \bigcap_{\mathrm{cl}_\eta Z \supset F} \mathrm{cl}_\eta Z = \bigcap_{F \in \mathcal{F}} \mathrm{cl}_\eta F = \mathrm{adh}_\eta \mathcal{F}.$$

□

In particular, when $\mathcal{Z} = 2^X$, then each filter on X is a $\mathcal{Z}$-filter, so that each ultrafilter is a $\mathcal{Z}$-ultrafilter. Accordingly, the topology constructed coincides with $\beta\iota_X$.

More generally,

COROLLARY XIV.5.2. *If ξ is a Hausdorff, functionally regular (alias completely regular) topology, then the topology constructed in Theorem XIV.5.1 for the lattice $\mathcal{Z}$ of functionally closed sets is the Čech-Stone compactification of ξ.*

PROOF. If ξ is a Hausdorff functionally regular topology, then by Example XIV.3.13, the family $\mathcal{C}_\xi^I$ of functionally closed sets is a separated normal lattice, and is a base for the ξ-closed sets. The topology η constructed in Theorem XIV.5.1 fulfills $\eta|_X = \xi$. Hence η is a compactication of ξ such that (XIV.5.8) holds for each $\mathcal{C}_\xi^I$-ultrafiler $\mathcal{H}(y)$. By (5) of Theorem XIV.4.5, η is the Čech-Stone compactification of ξ. $\square$

XIV.6. Supplement

EXAMPLE XIV.6.1 ($\beta\iota_{\mathbb{N}}$ on $\beta\mathbb{N}$). The Čech-Stone compactification $\beta\iota_{\mathbb{N}}$ of the natural numbers is disconnected, actually strongly zero-dimensional, because $\iota_{\mathbb{N}}$ has these properties.

If A is an infinite subset of $\mathbb{N}$, then $\iota_A \subset \iota_{\mathbb{N}}$, and obviously, ι_A is functionally extendable to $\iota_{\mathbb{N}}$. By Proposition XIV.4.13, $\beta\iota_A \subset \beta\iota_{\mathbb{N}}$ and $\beta\iota_A = \beta\iota_{\mathbb{N}}|_{\mathrm{cl}_{\beta\iota_{\mathbb{N}}} A}$. On the other hand, $\beta\iota_A$ and $\beta\iota_{\mathbb{N}}$ are homeomorphic. If $\mathbb{N} \setminus A$ is finite, then $\beta^*\iota_A = \beta^*\iota_{\mathbb{N}}$.

Let us split $\mathbb{N} = \mathbb{N}_0 \sqcup \mathbb{N}_1$, where $\mathbb{N}_0$ stands for the set of even integers, and $\mathbb{N}_1$ for the set of odd integers, then $\beta\iota_{\mathbb{N}_0} \oplus \beta\iota_{\mathbb{N}_1} = \beta\iota_{\mathbb{N}}$, that is the Stone topology on the set of all ultrafilters on $\mathbb{N}$ is a simple sum of its two homeomorphic copies. Indeed, $\mathbb{N}_0$ and $\mathbb{N}_1$ are disjoint and ι-closed, hence by Theorem XIV.4.5, $\mathrm{cl}_{\beta\iota_{\mathbb{N}}} \beta\mathbb{N}_0 \cap \mathrm{cl}_{\beta\iota_{\mathbb{N}}} \beta\mathbb{N}_1 = \varnothing$, and each $\mathcal{U} \in \beta\mathbb{N}$ contains either $\mathbb{N}_0$ or $\mathbb{N}_1$.

If $\mathcal{A}$ is an infinite *MAD* family on $\mathbb{N}$ (see Section XII.8), then $\bigoplus_{A\in\mathcal{A}} \beta\iota_A \subset \beta\iota_{\mathbb{N}}$, but the inclusion is strict. As $\mathcal{A}$ is uncountable, $\beta\iota_{\mathbb{N}}$ includes uncountably many disjoint copies of itself; actually, by Proposition I.5.4, it includes a simple sum of $\mathfrak{c}$ copies of itself.

EXAMPLE XIV.6.2. The Čech-Stone compactification $\beta\nu_{\mathbb{R}}$ of the reals is connected. If $f : \mathbb{N} \longrightarrow \mathbb{Q}$ is a bijection, then $f(\mathbb{N})$ is dense in $\beta\nu_{\mathbb{R}}$ and $f \in \mathscr{C}(\iota_{\mathbb{N}}, \beta\nu_{\mathbb{R}})$. By Theorem XIV.4.5, there exists an extension $F \in \mathscr{C}(\beta\iota_{\mathbb{N}}, \beta\nu_{\mathbb{R}})$ of f. As $\mathbb{R}$ is $\beta\nu_{\mathbb{R}}$-dense, F is a surjection, $\beta\nu_{\mathbb{R}}$ is a continuous image of $\beta\iota_{\mathbb{N}}$. Therefore,

$$\mathrm{card}\,|\beta\nu_{\mathbb{R}}| = \mathrm{card}\,|\beta\iota_{\mathbb{N}}| = 2^{\mathfrak{c}}.$$

As $\mathbb{N}$ is a $\nu_{\mathbb{R}}$-closed subset of $\mathbb{R}$, it follows from Proposition XIV.4.13, that

$$\beta\nu_{\mathbb{R}}|_{\mathrm{cl}_{\beta\nu_{\mathbb{R}}}\mathbb{N}} = \beta\iota_{\mathbb{N}}.$$

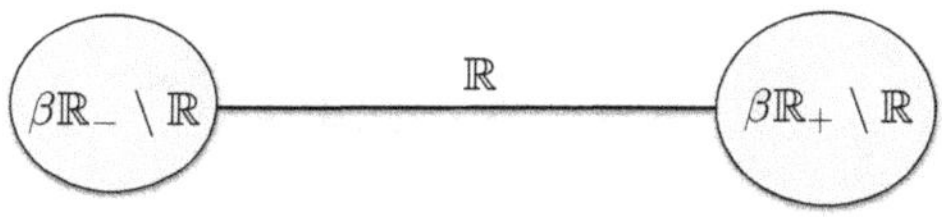

FIGURE XIV.6.1. The Čech-Stone compactification of the real line.

As $\nu_{\mathbb{R}}$ is locally compact, and $\mathbb{R}$ is $\beta_{\mathbb{R}}$-dense, $\mathbb{R}$ is $\beta\nu_{\mathbb{R}}$-open in virtue of Exercise XII.10.7.

A more precise picture of $\beta\nu_{\mathbb{R}}$ appears when we perceive that each bounded ultrafilter on $\mathbb{R}$ is $\nu_{\mathbb{R}}$-convergent ([9]). If an ultrafilter $\mathcal{U}$ on $\mathbb{R}$ is unbounded, then either $\{r \in \mathbb{R} : r \geq n\} \in \mathcal{U}$ for each $n < \omega$, or $\{r \in \mathbb{R} : r \leq -n\} \in \mathcal{U}$ for each $n < \omega$. Hence, we can consider $\beta\mathbb{R}_+ = \bigcap_{n<\omega} \beta\{r \in \mathbb{R} : r \geq n\}$ and $\beta\mathbb{R}_- = \bigcap_{n<\omega} \beta\{r \in \mathbb{R} : r \leq -n\}$ and represent $\beta\nu_{\mathbb{R}}$ as in Figure XIV.6.1.

Observe that, although $\beta\nu_{\mathbb{R}}$ is connected, the remainder is $\beta\nu_{\mathbb{R}}$-disconnected ([10]).

EXAMPLE XIV.6.3. The Čech-Stone compactification $\beta\nu_{\mathbb{Q}}$ of the rational numbers is disconnected, actually strongly zero-dimensional.

The remainder of $\beta\nu_{\mathbb{Q}}$ is $\beta\nu_{\mathbb{Q}}$-dense. Let $q \in \mathbb{Q}$. As $\mathcal{N}_{\beta\nu_{\mathbb{Q}}}(q)$ has a filter-base consisting of compact sets, $V \setminus \mathbb{Q} \neq \varnothing$ for every $V \in \mathcal{N}_{\beta\nu_{\mathbb{Q}}}(q)$, because $\nu_{\mathbb{Q}}$ is not locally compact. Therefore, $\mathbb{Q}$ is not $\beta\nu_{\mathbb{Q}}$-open.

Also $\beta\nu_{\mathbb{Q}}$ is a continuous image of $\beta\iota_{\mathbb{N}}$. If $f : \mathbb{N} \longrightarrow \mathbb{Q}$ is a bijection, then $f(\mathbb{N})$ is dense in $\beta\nu_{\mathbb{Q}}$ and $f \in \mathscr{C}(\iota_{\mathbb{N}}, \beta\nu_{\mathbb{Q}})$. By Theorem XIV.4.5, there exists an extension $F \in \mathscr{C}(\beta\iota_{\mathbb{N}}, \beta\nu_{\mathbb{Q}})$ of f. As $\mathbb{Q}$ is $\beta\nu_{\mathbb{Q}}$-dense, F is a surjection, $\beta\nu_{\mathbb{Q}}$ is a continuous image of $\beta\iota_{\mathbb{N}}$. Therefore,

$$\operatorname{card}|\beta\nu_{\mathbb{Q}}| = \operatorname{card}|\beta\iota_{\mathbb{N}}| = 2^{\mathfrak{c}}.$$

On the other hand, by Proposition XIV.4.13,

$$\beta\nu_{\mathbb{Q}}|_{\mathrm{cl}_{\beta\nu_{\mathbb{Q}}}\mathbb{N}} = \beta\iota_{\mathbb{N}}.$$

[9] A filter on $\mathbb{R}$ is said to be bounded if it contains a bounded set.

[10] Indeed, $\arctan \in \mathscr{C}(\nu_{\mathbb{R}}, [-\frac{\pi}{2}, \frac{\pi}{2}])$ extends to $a \in \mathscr{C}(\beta\nu_{\mathbb{R}}, [-\frac{\pi}{2}, \frac{\pi}{2}])$, and $a(\beta\mathbb{R} \setminus \mathbb{R}) = \{-\frac{\pi}{2}, \frac{\pi}{2}\}$.

CHAPTER XV

Uniform structures

Usually topology courses comprise uniform structures ([1]). This chapter delineates basic facts on the subject.

Uniform continuity and uniform convergence have been defined with the aid of metrics. In order to enlarge the scope of uniform structures from metrizable to functionally regular topologies a recourse was made to collections of pseudometrics. To enclose all topologies, one needed to employ quasi-pseudometrics.

On the other hand, there emerged a concept of using a filter on the square $X \times X$ to gauge a convergence on X. In this approach, uniformities play the role of collections of pseudometrics, and quasi-uniformities that of collections of quasi-pseudometrics.

We shall go a step further by considering preuniformities, which enable us to tackle pretopologies. And finally, we shall comment on extension of these ideas to more general convergences.

The first abstract uniform structures considered in the literature are uniformities. They are suitable to characterize functionally regular topologies ([2]). More recently, quasi-uniformities were introduced to fit all topologies ([3]).

XV.1. Preuniformities

Recall that $\triangle_X := \{(x,x) : x \in X\}$ stands for the identity relation on X. In this chapter we shall use the abbreviation $\triangle := \triangle_X$.

DEFINITION XV.1.1. A *preuniformity* on X is a filter $\mathcal{R}$ on $X \times X$ such that $\triangle \subset \ker \mathcal{R}$ ([4]).

A preuniformity $\mathcal{R}_1$ is said to be *finer* than a preuniformity $\mathcal{R}_0$ whenever $\mathcal{R}_1 \geq \mathcal{R}_0$. Each set of preuniformities on X has an infimum

[1]For example, R. ENGELKING [97], J. KELLEY [147]. See also L. NACHBIN [203] and W. GÄHLER [107].

[2]The concept of uniformity and the related main results are due to A. WEIL [260].

[3]P. FLETCHER and W. F. LINDGREN [101], H.-P. KÜNZI [164].

[4]The term *semi-quasi-uniformity* was used in [175] and in the literature cited therein.

and a supremum. The filter $\Delta^{\uparrow} = \{F \subset X \times X : \Delta \subset F\}$ is called the *discrete preuniformity* on X, and the filter $\{X \times X\}$ is said to be the *chaotic preuniformity*. They are respectively the finest and the coarsest preuniformities on X.

As $\mathcal{R} \in \mathbb{F}(X \times X)$, it follows that each $R \in \mathcal{R}$ is a binary relation. Using our convention (I.1.1), $Rx = \{y \in X : (x, y) \in R\}$ is the image of $x \in X$ under R. For a preuniformity $\mathcal{R}$, the family

$$\mathcal{R}(x) := \{Rx : R \in \mathcal{R}\}$$

is a proper filter on X ([5]). If $\mathcal{R}$ is a preuniformity on X, then we define

$$x \in \lim_{\lambda\mathcal{R}} \mathcal{F} \Longleftrightarrow \mathcal{F} \geq \mathcal{R}(x) \tag{XV.1.1}$$

for each $x \in X$. We observe that $\lambda\mathcal{R}$ is a pretopology. Indeed, $x \in \lim_{\lambda\mathcal{R}} x^{\uparrow}$ because $x \in Rx$ for each $R \in \mathcal{R}$ and every $x \in X$, and $\mathcal{R}(x)$ is the coarsest filter convergent to x. Specifically,

$$\mathcal{V}_{\lambda\mathcal{R}}(x) = \mathcal{R}(x)$$

for each $x \in X$. It is obvious that $\lambda\Delta^{\uparrow}$ is the discrete topology, and $\lambda\{X \times X\}$ is the chaotic topology on X.

Let us observe that for any $A \subset X$,

$$\operatorname{adh}_{\lambda\mathcal{R}} A = \bigcap_{R\in\mathcal{R}} R^{-}A. \tag{XV.1.2}$$

Indeed, by definition, $x \in \operatorname{adh}_{\lambda\mathcal{R}} A$ if and only if $A \cap Rx \neq \varnothing$, equivalently $x \in R^{-}A$, for each $R \in \mathcal{R}$.

PROPOSITION XV.1.2. *The pretopology $\lambda\mathcal{R}$ is free (alias T_1) if and only if* $\ker \mathcal{R} = \bigcap_{R\in\mathcal{R}} R = \Delta$.

PROOF. Indeed, $\lambda\mathcal{R}$ is free if and only if $x_1 \in \operatorname{adh}_{\lambda\mathcal{R}}\{x_0\}$ implies that $x_0 = x_1$, that is, by (XV.1.2), $\bigcap_{R\in\mathcal{R}} R^{-} \subset \Delta$, equivalently $\bigcap_{R\in\mathcal{R}} R \subset \Delta$. □

If $\mathcal{R}$ is a preuniformity, then $\mathcal{R}^{-} := \{R^{-} : R \in \mathcal{R}\}$ is also a preuniformity, called the *conjugate preuniformity* of $\mathcal{R}$. Two preuniformities, one conjugate of the other, are in general incomparable.

EXAMPLE XV.1.3. Let $X = \mathbb{R}$ be the real line. If $B_r := \{(x, y) : |y - x| < r\}$, then $\{B_r : r < \infty\}$ is a filter-base of a preuniformity $\mathcal{V}$ on X. It is plain that $\lambda\mathcal{V} = \nu$ is the standard topology. Clearly, $\mathcal{V}^{-} = \mathcal{V}$

[5]Indeed, $\{x\} = \Delta x \subset Rx$ for each $x \in X$, and if $D_0, D_1 \in \mathcal{R}(x)$, then there exist $R_0, R_1 \in \mathcal{R}$ such that $D_0 = R_0x$ and $D_1 = R_1x$, so that

$$D_0 \cap D_1 = R_0x \cap R_1x = (R_0 \cap R_1)x \in \mathcal{R}(x),$$

because $R_0 \cap R_1 \in \mathcal{R}$; finally, if $D \supset R(x)$, then $R \subset R \cup (\{x\} \times D) \in \mathcal{R}$, hence $D = [R \cup (\{x\} \times D)]x \in \mathcal{R}(x)$.

is symmetric, so that $\lambda\mathcal{V}^- = \lambda\mathcal{V}$. If $B_r^+ := \{(x, y) : \max(y - x, 0) < r\}$, then $\{B_r^+ : r < \infty\}$ is a filter-base of a preuniformity $\mathcal{V}_+$ on X; analogously, $B_r^- := \{(x, y) : \max(x - y, 0) < r\}$, then $\{B_r^- : r < \infty\}$ is a filter-base of a preuniformity $\mathcal{V}_-$ on X. They are conjugate one of the other, and incomparable. Observe that $\lambda\mathcal{V}_+ = \varsigma_+$ is the *upper Sorgenfrey line* of Example III.3.6, and $\lambda\mathcal{V}_- = \varsigma_-$ is the *lower Sorgenfrey line.*

A preuniformity $\mathcal{R}$ on X is called *Hausdorff* if for any pair x_0, x_1 of distinct elements of X, there is $R \in \mathcal{R}$ such that $Rx_0 \cap Rx_1 = \varnothing$, that is, if $\lambda\mathcal{R}$ is Hausdorff.

PROPOSITION XV.1.4. *The pretopology $\lambda\mathcal{R}$ is Hausdorff if and only if* [6]

$$\bigcap\nolimits_{R\in\mathcal{R}} R^- \circ R = \triangle.$$

PROOF. For every preuniformity $\mathcal{R}$ and each $R \in \mathcal{R}$, we have $R \supset \triangle$, and thus $R^- \supset \triangle^- = \triangle$. Therefore, $R^- \circ R \supset \triangle \circ \triangle = \triangle$. On the other hand, if $(x_0, x_1) \notin \triangle$, that is, $x_0 \neq x_1$. Now $\mathcal{R}$ is Hausdorff if and only if there exists $R \in \mathcal{R}$ such that $Rx_0 \cap Rx_1 = \varnothing$. This means that $x_1 \notin R^- Rx_0$, that is, $(x_0, x_1) \notin \bigcap_{R\in\mathcal{R}} R^- \circ R$. □

EXAMPLE XV.1.5. For each $t \in \mathbb{R}$, let $R_t := \{(x, y) \in \mathbb{R} \times \mathbb{R} : x \geq t, y \leq x\}$. Then $\{R_t \cup \triangle : t \in \mathbb{R}\}$ is a filter-base of a preuniformity $\mathcal{R}$. We observe that

$$R_t x = \begin{cases} \{x\} & x < t, \\ ]-\infty, x] & x \geq t. \end{cases}$$

Therefore, $\lambda\mathcal{R}$ is the discrete topology on $\mathbb{R}$, hence Hausdorff. On the other hand,

$$R_t^- y = \begin{cases} \{y\} \cup [t, +\infty[& y < t, \\ [y, +\infty[& y \geq t. \end{cases}$$

Therefore, $\lambda\mathcal{R}^-$ is not Hausdorff.

If π is a pretopology and $\mathcal{P}$ is a preuniformity such that $\pi = \lambda\mathcal{P}$, then we say that $\mathcal{P}$ is *compatible* with π. It is known [7] that

PROPOSITION XV.1.6. *For each pretopology, there exists a compatible preuniformity.*

---

[6] The condition given by P. FLETCHER and W. F. LINDGREN [101, 1.9 (e)] is different from the present one due to a non-standard definition of the composition of two relations adopted by these authors.

[7] A. LECHICKI AND J. ZIEMIŃSKA [175]

PROOF. If π is a pretopology on X, then the filter

$$\mathcal{P} := \bigwedge_{x\in X} \mathcal{V}_\iota(x) \times \mathcal{V}_\pi(x)$$

is a required preuniformity. Indeed, $P \in \mathcal{P}$ if for each $x \in X$, there is $V_x \in \mathcal{V}_\pi(x)$ such that $P \supset \{x\} \times V_x$. Therefore, $\triangle \subset \ker \mathcal{P}$, and $\mathcal{P}(x) = \mathcal{V}_\pi(x)$ for every $x \in X$. More explicitly, $P \in \mathcal{P}$ if and only if for each $w \in X$ there exists $V_w \in \mathcal{V}_\pi(w)$ such that $P \supset \bigcup_{w\in X}\{w\} \times V_w$. Hence, $Px \supset \bigcup_{w\in X}(\{w\} \times V_w)x = V_x$. □

If $\mathbf{U}X$ stands for the set of all preuniformities on X, the map $\lambda : \mathbf{U}X \longrightarrow \mathbf{S}_0X$ associates to each $\mathcal{R} \in \mathbf{U}X$ a pretopology $\lambda\mathcal{R}$ on X. Of course, $\mathcal{R}_0 \leq \mathcal{R}_1$ implies that $\lambda\mathcal{R}_0 \leq \lambda\mathcal{R}_1$. Actually, λ preserves suprema: if $\{\mathcal{R}_j : j \in J\}$ is a set of preuniformities on X, then $\lambda(\bigvee_{j\in J} \mathcal{R}_j) = \bigvee_{j\in J} \lambda(\mathcal{R}_j)$.

Therefore, for every $\pi \in \mathbf{S}_0X$ and $\varnothing \neq \{\mathcal{R}_j : j \in J\} \subset \lambda^-\pi$, since $\mathcal{V}_\pi(x) = \mathcal{R}_jx$ for each $j \in J$, also $\mathcal{V}_\pi(x) = (\bigvee_{j\in J} \mathcal{R}_j)(x)$. This means that

PROPOSITION XV.1.7. *For each pretopology, there exists the finest compatible preuniformity.*

DEFINITION XV.1.8 (Semi-uniformity). A preuniformity $\mathcal{R}$ is called a *semi-uniformity* if $R \in \mathcal{R}$ implies $R^- \in \mathcal{R}$.

PROPOSITION XV.1.9. *A preuniformity $\mathcal{R}$ is a semi-uniformity if and only if $y \in \lim_{\lambda\mathcal{R}} x^\uparrow$ implies that $x \in \lim_{\lambda\mathcal{R}} y^\uparrow$ for every x and y.*

PROOF. By definition, $y \in \lim_{\lambda\mathcal{R}} x^\uparrow$ if and only if $y \in Rx$ for every $R \in \mathcal{R}$. If $\mathcal{R}$ is a semi-uniformity, then $R \in \mathcal{R}$ implies $R^- \in \mathcal{R}$, hence $x \in Ry$ for every $R \in \mathcal{R}$, that is, $x \in \lim_{\lambda\mathcal{R}} y^\uparrow$. The converse is obtained by rephrasing the conditions. □

Actually, the condition of Proposition XV.1.9 is equivalent to

$$y \in \mathrm{adh}_{\lambda\mathcal{R}}\{x\} \Longrightarrow x \in \mathrm{adh}_{\lambda\mathcal{R}}\{y\}$$

for every x and y.

If $\mathcal{R}$ is a preuniformity, then ([8])

$$\mathcal{R}^\flat := \mathcal{R} \vee \mathcal{R}^-$$

is a semi-uniformity.

---

[8]I depart from the usual notation $\mathcal{R}^\star$, because the asterisk has been already adopted to design the conjugate family (V.0.4).

It is interesting that $\lambda\mathcal{R}$ is T_0 if and only if $\lambda\mathcal{R}^\flat$ is Hausdorff ([9]). It is straightforward that the supremum and the infimum of any set of semi-uniformities are semi-uniformities. Actually, $\mathcal{R}^\flat$ is the coarsest semi-uniformity that is finer than a preuniformity $\mathcal{R}$.

DEFINITION XV.1.10 (Quasi-uniformity). A preuniformity $\mathcal{R}$ is called a *quasi-uniformity* if for every $R \in \mathcal{R}$, there exists $S \in \mathcal{R}$ such that $S^2 := S \circ S \subset R$ ([10]).

If $\mathcal{R}$ is a quasi-uniformity on X, then the relation $\ker \mathcal{R}$ in $X \times X$ is a partial order. Indeed, the relation is reflexive, because $\triangle \subset \ker \mathcal{R}$. On the other hand, if $(x_0, x_1) \in \ker \mathcal{R}$ and $(x_1, x_2) \in \ker \mathcal{R}$, and $R \in \mathcal{R}$, then there exists $V \in \mathcal{R}$ such that $V \circ V \subset R$. Hence $(x_0, x_1) \in V$ and $(x_1, x_2) \in V$, so that $(x_0, x_2) \in V \circ V \subset R$, proving that $(x_0, x_2) \in \ker \mathcal{R}$.

DEFINITION XV.1.11 (Locally symmetric). A quasi-uniform space $(X, \mathcal{R})$ is said to be *locally symmetric* if for each $R \in \mathcal{R}$ and $x \in X$, there exists a symmetric $P \in \mathcal{R}$ such that $P^2(x) \subset Rx$.

PROPOSITION XV.1.12. *If $\mathcal{R}$ is a quasi-uniformity, then the associated convergence $\lambda\mathcal{R}$ is a topology.*

PROOF. As $\lambda\mathcal{R}$ is a pretopology, it is enough to show that if $V \in \mathcal{V}_{\lambda\mathcal{R}}(x)$, then there exists $W \in \mathcal{V}_{\lambda\mathcal{R}}(x)$ such that $V \in \mathcal{V}_{\lambda\mathcal{R}}(W)$. By definition, there is $R \in \mathcal{R}$ such that $Rx \subset V$, and there is $S \in \mathcal{R}$ such that $S^2 \subset R$, hence $S(Sx) \subset Rx$. Accordingly, $W := Sx \in \mathcal{V}_{\lambda\mathcal{R}}(x)$ and $V(w) := Sw$ for each $w \in W$, we get $V \supset Rx \supset S(Sx) = \bigcup_{w \in W} V(w)$. □

EXAMPLE XV.1.13. A *quasi-pseudometric* on X is a function $d : X \times X \longrightarrow \mathbb{R}_+$ such that $d(x, x) = 0$, and $d(x, z) \leq d(x, y) + d(y, z)$ whenever $x, y, z \in X$. If d is a quasi-pseudometric on X, then the function $d^\star : X \times X \longrightarrow \mathbb{R}_+$, defined by $d^\star(x, y) := d(y, x)$ for each $x, y \in X$, is also a quasi-pseudometric, called the *conjugate quasi-pseudometric* of d. A quasi-pseudometric d is called a *pseudometric* if $d = d^\star$. A pseudometric d is called a *metric* if $d(x, y) = 0$ implies $x = y$ for every $x, y \in X$.

[9]Suppose that $\lambda\mathcal{R}$ is T_0 and $x_1 \neq x_0$. Then either there is $R_0 \in \mathcal{R}$ such that $x_1 \notin R_0 x_0$ or there is $R_1 \in \mathcal{R}$ such that $x_0 \notin R_1 x_1$, that is, $x_1 \notin R_0 \cap R_1^- x_0$. On the other hand, $\mathcal{R}^\flat$ is generated by $\{R_0 \cap R_1^- : R_0, R_1 \in \mathcal{R}\}$.

[10]The original term *semi-uniformity* of L. NACHBIN [**203**, p. 57] has been more recently replaced in the literature by that of *quasi-uniformity* used by M. G. MURDESHWAR and S. A. NAIMPALLY [**192**]. The term *semi-uniformity* was employed by E. Čech with a different meaning.

If d is a quasi-pseudometric on X, let $U_\varepsilon := \{(x,y) \in X \times X : d(x,y) < \varepsilon\}$. Then $\{U_\varepsilon : \varepsilon > 0\}$ is a filter-base on $X \times X$ such that

$$\triangle \subset U_0 := \{(x,y) \in X \times X : d(x,y) = 0\} = \bigcap_{\varepsilon>0} U_\varepsilon.$$

On the other hand, $U_{\frac{\varepsilon}{2}} \circ U_{\frac{\varepsilon}{2}} \subset U_\varepsilon$ for each $\varepsilon > 0$, so that the filter $\mathcal{U}$ generated by $\{U_\varepsilon : \varepsilon > 0\}$ is a quasi-uniformity, and thus $\lambda\mathcal{U}$, defined by (XV.1.1), is a topology of countable character.

It turns out ([11]) that

PROPOSITION XV.1.14 (Pervin quasi-uniformity). *For each topology τ, there exists a quasi-uniformity $\mathcal{P}_\tau$ such that $\tau = \lambda\mathcal{P}_\tau$.*

PROOF. Let τ be a topology on X. For every $O \in \mathcal{O}_\tau$, let

$$S_O := (O \times O) \cup ((X \setminus O) \times X).$$

Then $\{S_O : O \in \mathcal{O}_\tau\}$ is a filter-subbase on $X \times X$. As $\triangle \subset S_O$ for every $O \in \mathcal{O}_\tau$, the corresponding filter $\mathcal{P}_\tau$ fulfills $\triangle \subset \ker \mathcal{P}_\tau$.

On the other hand, $S_O \circ S_O \subset S_O$. If $(x,z) \in S_O \circ S_O$, then there exists $y \in X$ such that $(x,y) \in S_O$ and $(y,z) \in S_O$. Indeed, if $x \in O$, then $(x,y) \in O \times O$, hence $y \in O$, thus $(y,z) \in O \times O$; if $x \in X \setminus O$, then $y \in X \setminus O$, hence $z \in X \setminus O$.

Consequently, if $P \in \mathcal{P}_\tau$, then there exist $n \in \mathbb{N}_1$ and $O_1, O_2, \ldots, O_n \in \mathcal{O}_\tau$ such that $P \supset S := S_{O_1} \cap \cdots \cap S_{O_n}$ and $S \circ S \subset S \subset P$. Therefore, $\mathcal{P}_\tau$ is a quasi-uniformity.

If $O \in \mathcal{O}_\tau$ and $x \in O$, then $S_O x = O$, so that $\mathcal{O}_\tau \subset \mathcal{O}_{\lambda\mathcal{P}_\tau}$. Conversely, if $G \in \mathcal{O}_{\lambda\mathcal{P}_\tau}$ and $x \in G$, then there is $P \in \mathcal{P}_\tau$ such that $Px \subset G$. Hence, there exist $n \in \mathbb{N}_1$ and $O_1, O_2, \ldots, O_n \in \mathcal{O}_\tau$ such that $Px \supset (S_{O_1} \cap \cdots \cap S_{O_n})x = S_{O_1}x \cap \cdots \cap S_{O_n}x \ni x$. On the other hand, for each $1 \leq k \leq n$, either $S_{O_k}x = O_k$ or $S_{O_k}x = X$, so that $(S_{O_1} \cap \cdots \cap S_{O_n})x$ is τ-open. This proves that $G \in \mathcal{O}_\tau$. □

The quasi-uniformity of a topology τ, constructed in the proof of Proposition XV.1.14 is not the only quasi-uniformity compatible with τ.

EXAMPLE XV.1.15. Let ν_I be the standard topology of the interval $I = [0,1]$. The standard metric $d : I \times I$ defined by $d(x,y) := |y - x|$ gives rise to a quasi-uniformity $\mathcal{N}$ ([12]), for which $\{\{(x,y) \in I \times I : |y - x| < 2^{-n}\} : n < \omega\}$ is a filter-base. By Proposition XV.1.12, $\lambda\mathcal{N} = \lambda\mathcal{P}_{\nu_I} = \nu_I$, but the filters $\mathcal{N}$ and $\mathcal{P}_{\nu_I}$ are not comparable.

[11] A proof in [50] by Á. CSÁSZÁR is based on the theory of syntopogenic structures. We follow here the straightforward proof of [220] by W. J. PERVIN.

[12] We shall see that it is a *uniformity*, which we will define later.

Proposition XV.1.16. *For each preuniformity $\mathcal{R}$ on a set X, there is a quasi-uniformity, denoted by* $\mathrm{T}\mathcal{R}$, *that is the finest among all quasi-uniformities on X coarser than $\mathcal{R}$.*

Proof. If $\boldsymbol{T}$ is a non-empty set of quasi-uniformities on X, and $A \in \bigvee \boldsymbol{T}$, then there exist $\boldsymbol{F} \subset_{\text{fin}} \boldsymbol{T}$ and $T_{\mathcal{T}} \in \mathcal{T}$ for each $\mathcal{T} \in \boldsymbol{F}$ such that $A \supset \bigcap_{\mathcal{T} \in \boldsymbol{F}} T_{\mathcal{T}}$. By assumption, for each $\mathcal{T} \in \boldsymbol{F}$, there exists $B_{\mathcal{T}} \in \mathcal{T}$ such that $B_{\mathcal{T}} \circ B_{\mathcal{T}} \subset T_{\mathcal{T}}$. If $B := \bigcap_{\mathcal{T} \in \boldsymbol{F}} B_{\mathcal{T}}$, then $\triangle \subset B$, hence $B \in \bigvee \boldsymbol{T}$, and $B \circ B \subset A$. Obviously, the chaotic preuniformity $\mathcal{U}_o = \{X \times X\}$ is a quasi-uniformity. Accordingly, the set of quasi-uniformities $\boldsymbol{T}_{\mathcal{R}}$ that are coarser than a preuniformity $\mathcal{R}$ is non-empty, and $\mathrm{T}\mathcal{R} = \bigvee \boldsymbol{T}_{\mathcal{R}}$ is the sought quasi-uniformity. □

We used the same symbol T in the context of convergences to denote the *topologizer* (the reflector on the class of topologies). Although T can appear in the same formula in two different roles, there would be no confusion, provided that these roles are clearly distinguished, as in the following (XV.1.3).

The supremum of a set of quasi-uniformities coincides with its supremum in the lattice of preuniformilties, while $\bigwedge_{\mathrm{T}} \boldsymbol{R} = \mathrm{T}(\bigwedge \boldsymbol{R})$ for any set $\boldsymbol{R}$ of preuniformities on the same underlying set. Therefore, by Proposition XV.1.14,

Corollary XV.1.17. *For each topology, there exists a finest compatible quasi-uniformity.*

As $\lambda(\mathrm{T}\mathcal{R})$ is a topology coarser than the pretopology $\lambda\mathcal{R}$, we infer that

$$\mathrm{T}(\lambda\mathcal{R}) \geq \lambda(\mathrm{T}\mathcal{R}). \tag{XV.1.3}$$

As we shall see in Example XV.1.18, this inequality can be strict.

It is obvious that a subset O of X is $\lambda\mathcal{R}$-open if for each $x \in O$, there exists $R \in \mathcal{R}$ such that $Rx \subset O$.

Example XV.1.18. Consider on

$$X = \{x_\infty\} \sqcup \{x_n : n < \omega\} \sqcup \{x_{n,k} : k > n, n < \omega\}.$$

the *bisequence pretopology* θ from Example VI.1.8. For each $m < \omega$, let $P_m \subset X \times X$ defined by

$$P_m x_\infty := \{x_\infty\} \cup \{x_n : n \geq m\},$$
$$P_m x_n := \{x_n\} \cup \{x_{n,k} : k \geq m\},\ P_m x_{n,k} := \{x_{n,k}\}.$$

Then $\{P_m : m < \omega\}$ is a filter-base of a preuniformity $\mathcal{P}$. The preuniformity $\mathrm{T}\mathcal{P}$ contains the sets of the form $P_m^2 = P_m \circ P_m$ with $m < \omega$.

In particular,

$$P_m^2 x_\infty = \{x_\infty\} \cup \{x_n : n \geq m\} \cup \{x_{n,k} : n \geq m, k \geq m\}.$$

It follows that $P_m^{2+l} = P_m^2$ for each $m, l < \omega$. Accordingly, $\{P_m^2 x_\infty : m < \omega\}$ is a filter-base of $\mathcal{V}_{\lambda(\mathrm{T}\mathcal{P})}(x_\infty)$, while $V \in \mathcal{N}_{\lambda\mathcal{P}}(x_\infty) = \mathcal{N}_\theta(x_\infty)$ whenever there exists $m < \omega$, and, for each $n \geq m$ there is $k_n < \omega$ such that

$$\{x_\infty\} \cup \{x_n : n \geq m\} \cup \{x_{n,k} : k \geq k_n, n \geq m\} \subset V.$$

Therefore, $\mathrm{T}(\lambda\mathcal{P}) > \lambda(\mathrm{T}\mathcal{P})$.

REMARK XV.1.19. If $\mathcal{R}$ is a preuniformity such that $\lambda\mathcal{R} = \xi$, then, for each $M \subset X \times X$,

$$\mathrm{adh}_{\xi\times\xi} M = \bigcap\nolimits_{R\in\mathcal{R}} R^- \circ M \circ R.$$

In fact, $(x, y) \in \mathrm{adh}_{\xi\times\xi} M$ if and only if $(Rx \times Ry)\# M$ for each $R \in \mathcal{R}$, that is, $Ry\# M(Rx)$ or else $y \in R^-(M(Rx))$, equivalently $(x, y) \in R^- \circ M \circ R$ for each $R \in \mathcal{R}$.

XV.2. Uniformities

DEFINITION XV.2.1 (Uniformity). A filter $\mathcal{R}$ on $X \times X$ such that $\ker \mathcal{R} \supset \Delta$ is called a *uniformity* if for every $R \in \mathcal{R}$, there exists $S \in \mathcal{R}$ such that $S \circ S \subset R$, and if $\mathcal{R}^- = \mathcal{R}$.

Accordingly, a preuniformity is a uniformity if and only if it is both a semi-uniformity and a quasi-uniformity.

Notice that if $\mathcal{R}$ is a uniformity on X, then for each $A \subset X$ and $R \in \mathcal{R}$,

$$A \times A \subset R \Longrightarrow \mathrm{cl}_{\lambda\mathcal{R}} A \times \mathrm{cl}_{\lambda\mathcal{R}} A \subset R^3. \tag{XV.2.1}$$

Indeed, if $x_0, x_1 \in \mathrm{cl}_{\lambda\mathcal{R}} A$, then there exist $a_0, a_1 \in A$ such that $a_0 \in Rx_0$ and $a_1 \in Rx_1$. By assumption, $a_1 \in Ra_0$. Therefore, $x_1 \in Ra_1 \subset R^2 a_0 \subset R^3 x_0$.

EXAMPLE XV.2.2. Let $\mathcal{D}$ be a family of pseudometrics on X such that for any $x_0 \neq x_1$, there exists $d \in \mathcal{D}$ such that $d(x_0, x_1) > 0$. Then the family

$$\{(x, y) \in X \times X : d(x, y) < 2^{-n}\}, \tag{XV.2.2}$$

where $n < \omega$ and $d \in \mathcal{D}$ is a base for a uniformity $\mathcal{U}_\mathcal{D}$ induced by (XV.2.2). Let $\mathcal{U}_\nu$ stand for the standard uniformity on $\mathbb{R}$. If $\mathcal{R}$ is a uniformity, then a pseudometric d is said to be $\mathcal{R}$-uniform if for every $\varepsilon > 0$, there is $R \in \mathcal{R}$ such that $R \subset \{d < \varepsilon\}$.

If $\mathcal{P}$ is a quasi-uniformity, then $\mathcal{P}^\flat = \mathcal{P} \vee \mathcal{P}^-$ is a uniformity. It is straightforward that the set of uniformities on a given set is a complete lattice, and

PROPOSITION XV.2.3. *If $\mathcal{P}$ is a quasi-uniformity, then $\mathcal{P}^\flat$ is the coarsest uniformity finer than $\mathcal{P}$.*

LEMMA XV.2.4. *Let $\mathcal{V}$ be a uniformity on X, and for each $V \in \mathcal{V}$, let $\{V_n : n < \omega\} \subset \mathcal{V}$ be such that $V_n^- = V_n$ and $V_0 = X \times X, V_1 = V$, and $V_{n+1}^3 \subset V_n$ for $n < \omega$. Then there exists a pseudometric $d : X \times X \longrightarrow \mathbb{R}_+$ such that for each $n \in \mathbb{N}_1$,*

$$V_n \subset \{d \leq 2^{-n}\} \subset V_{n-1}. \tag{XV.2.3}$$

PROOF. Let $f(x, y) := 0$ if $(x, y) \in \bigcap_{n<\omega} V_n$, and

$$(x, y) \in V_n \setminus V_{n+1} \Longrightarrow f(x, y) := 2^{-n}. \tag{XV.2.4}$$

Of course, $f(y, x) = f(x, y)$ for each $x, y \in X$. Let

$$d(x, y) := \inf\{\sum_{m=1}^{k} f(x_{m-1}, x_m) : x = x_0, y = x_k, k \in \mathbb{N}_1\}. \tag{XV.2.5}$$

It is clear that $d(x, x) = 0$ and $d(y, x) = d(x, y)$ for each $x, y \in X$. Moreover, by definition (XV.2.5), for every $x, y, z \in X$,

$$d(x, z) \leq d(x, y) + d(y, z),$$

so that d is a pseudometric. As $d(x, y) \leq f(x, y)$ for each $x, y \in X$, we conclude that $V_n \subset \{d \leq 2^{-n}\}$.

It remains to prove that $f(x, y) \leq 2d(x, y)$ for each $x, y \in X$ to ensure that $\{d \leq 2^{-n}\} \subset V_{n-1}$ for $n \in \mathbb{N}_1$. In other words, we need to show that

$$f(x, y) \leq 2 \sum_{m=1}^{k} f(x_{m-1}, x_m) \tag{XV.2.6}$$

for every $k \in \mathbb{N}_1$ and each sequence $x = x_0, \ldots, x_k = y$. Let us proceed by induction on k. This is obvious for $k = 1$. If $l > 1$ and (XV.2.6) holds for every $k < l$, let $x_0, \ldots, x_l$ with $x_0 = x, x_l = y$, and

$$r := \sum_{m=1}^{l} f(x_{m-1}, x_m).$$

Since $f(x, y) \leq 1$, it is enough to consider $r < \frac{1}{2}$. If $0 < r$, then either $f(x_0, x_1) \leq \frac{1}{2}r$ or $f(x_{l-1}, x_l) \leq \frac{1}{2}r$. Let $f(x_0, x_1) \leq \frac{1}{2}r$. Let j be the least index such that $\sum_{m=1}^{j} f(x_{m-1}, x_m) > \frac{1}{2}r$. Then $\sum_{m=j+1}^{l} f(x_{m-1}, x_m) \leq \frac{1}{2}r$. By the inductive assumption, $f(x_0, x_{j-1}) \leq r$ and

$f(x_j, x_l) \le r$, and by the definition of r, also $f(x_{j-1}, x_j) \le r$. If p is the least integer such that $2^{-p} \le r$, then by (XV.2.4), $(x_0, x_{j-1}) \in V_l, (x_j, x_l) \in V_p$ and $(x_{j-1}, x_j) \in V_p$. By assumption, $(x, y) = (x_0, x_1) \in V_{p-1}$ and thus, by (XV.2.4), $f(x, y) \le 2^{1-p} \le 2r$.

If $r = 0$, then $(x_{m-1}, x_m) \in V_j$ for $m = 1, \ldots, l$ and $j < \omega$, hence $(x, y) \in V_j^l$ for $j < \omega$, and thus $f(x, y) = 0$. □

COROLLARY XV.2.5. *For each uniformity, there exists a family of (uniformly continuous) pseudometrics that determines the uniformity.*

PROOF. Let $\mathcal{U}$ be a uniformity on X. By definition and induction, for each $U \in \mathcal{U}$, there exists $W \in \mathcal{U}$ such that $W^3 \subset U$ ([13]). Therefore, for each $U \in \mathcal{U}$, there exists $\{V_n : n < \omega\} \subset \mathcal{U}$ fulfilling the assumptions of Lemma XV.2.4. Consequently, there is a pseudometric $d = d_U$ such that (XV.2.3). Hence

$$V_{n+1}x \subset \{w \in X : d(w, x) < 2^{-n}\} \subset V_n x$$

for every $x \in X$. Therefore, by taking all $\{d_U(\cdot, x) < 2^{-n}\}$ with $x \in X$ and $n < \omega$, and for each $U \in \mathcal{U}$, we get a base for the open sets of $\lambda\mathcal{U}$. □

We say that a topology is *uniformizable* if it has a compatible uniformity. By Corollary XV.2.5, uniformizable topologies are pseudometrizable.

On the other hand, it follows from Proposition X.9.7 that a topology is pseudometrizable if and only if it is functionally regular. Therefore,

THEOREM XV.2.6 (Weil). *A Hausdorff topology is uniformizable if and only if it is functionally regular (alias completely regular).*

It is clear that each uniformity $\mathcal{R}$ has a filter-base consisting of $\lambda\mathcal{R} \times \lambda\mathcal{R}$-open sets, and a filter-base consisting of $\lambda\mathcal{R} \times \lambda\mathcal{R}$-closed sets. This is because, for each $R \in \mathcal{R}$, there is a pseudometric d_R such that $\{d_R < 1\} \subset \{d_R \le 1\} \subset R$.

PROPOSITION XV.2.7. *A uniformity is induced by a metric if and only if it is countably based.*

PROOF. If a uniformity is induced by a metric d, then $\{\{d < 2^{-n}\} : n < \omega\}$ is a filter-base. Conversely, if $\mathcal{R}$ is a uniformity and $\{R_n : n < \omega\}$ is a filter-base, then for each $n < \omega$ there is $V_n \in \mathcal{R}$ such that $V_n^3 \subset R_n$, hence $\bigcap_{n<\omega} V_n = \Delta$. Thus, the pseudometric from the proof of Lemma XV.2.4 is a metric. □

[13]Indeed, there is $V_0 \in \mathcal{U}$ such that $V_0^2 \subset U$, and there is $V_1 \in \mathcal{U}$ such that $V_1^2 \subset V_0$. Hence, $V_1^2 \circ V_0 \subset U$, thus $W := V_0 \cap V_1$ is a good choice.

XV.3. Operations

DEFINITION XV.3.1 (Uniform continuity). Let $\mathcal{P}$ and $\mathcal{R}$ be preuniformities on X and Y respectively. We say that a map $f \in Y^X$ is *uniformly continuous*, in symbols $f \in \mathscr{U}(\mathcal{P}, \mathcal{R})$, whenever $(f \times f)[\mathcal{P}] \geq \mathcal{R}$.

Accordingly,

PROPOSITION XV.3.2. *If $\mathcal{P}$ and $\mathcal{R}$ are preuniformities, then $f \in \mathscr{U}(\mathcal{P}, \mathcal{R})$ if and only if, for each $R \in \mathcal{R}$, there is $P \in \mathcal{P}$ such that $(x_0, x_1) \in P$ implies $(f(x_0), f(x_1)) \in R$* for all $(x_0, x_1) \in X \times X$.

DEFINITION XV.3.3 (Uniform isomorphism). If $\mathcal{P}$ and $\mathcal{R}$ are preuniformities, then a bijective map f is called a *uniform isomorphism* if $f \in \mathscr{U}(\mathcal{P}, \mathcal{R})$ and $f^{-1} \in \mathscr{U}(\mathcal{R}, \mathcal{P})$.

As expected,

PROPOSITION XV.3.4. *If $\mathcal{P}$ and $\mathcal{R}$ are preuniformities, then $\mathscr{U}(\mathcal{P}, \mathcal{R}) \subset \mathscr{C}(\lambda\mathcal{P}, \lambda\mathcal{R})$.*

PROOF. Let $f \in \mathscr{U}(\mathcal{P}, \mathcal{R})$ and $x \in X$. Let $V \in \mathcal{V}_{\lambda\mathcal{R}}(f(x))$. By definition, there exists $R \in \mathcal{R}$ such that $V \supset R(f(x))$, hence there is $P \in \mathcal{P}$ such that $f(w) \in R(f(x))$ for each $w \in Px \in \mathcal{V}_{\lambda\mathcal{P}}(x)$. $\square$

See Example X.5.5 of continuous non-uniformly continuous functions.

PROPOSITION XV.3.5. *Let $\mathcal{P}$ and $\mathcal{R}$ be preuniformities. If $f \in \mathscr{C}(\lambda\mathcal{P}, \lambda\mathcal{R})$ and $\lambda\mathcal{P}$ is compact, then $f \in \mathscr{U}(\mathcal{P}, \mathcal{R})$.*

PROOF. Let $\mathcal{P}$ be a preuniformity on X and $\mathcal{R}$ a preuniformity on Y, so that $\lambda\mathcal{P}$ is compact, and let $f \in \mathscr{C}(\lambda\mathcal{P}, \lambda\mathcal{R})$. If $R \in \mathcal{R}$, then by assumption, for every $x \in X$, there exists $P_x \in \mathcal{P}$ such that $P_x x \subset f^-(Rf(x))$, that is, $x \in \operatorname{inh}_{\lambda\mathcal{P}} f^-(Rf(x))$. In other words, $\{P_x x : x \in X\}$ is a $\lambda\mathcal{P}$-cover of X. By compactness, there is a finite subset F of X such that $\{P_x x : x \in F\}$ is a $\lambda\mathcal{P}$-cover of X. On the other hand, $P := \bigcap_{x \in F} P_x \in \mathcal{P}$, so that $Px \subset f^-(Rf(x))$ for every $x \in X$. Hence, if $w \in Px$, then $f(w) \in Rf(x)$, and by Proposition XV.3.2, $f \in \mathscr{U}(\mathcal{P}, \mathcal{R})$. $\square$

Let $\mathcal{R}_j$ be a preuniformity on X_j for each $j \in J$.

DEFINITION XV.3.6. The coarsest preuniformity $\mathcal{R}$ on $\prod_{j \in J} X_j$, for which $p_j \in \mathscr{U}(\mathcal{R}, \mathcal{R}_j)$ for each $j \in J$ is called the *product preuniformity* of $\{\mathcal{R}_j : j \in J\}$.

It is straightforward that

PROPOSITION XV.3.7. *The product preuniformity with respect to* $\{\mathcal{R}_j : j \in J\}$ *is equal to the product filter* $\prod_{j\in J}\mathcal{R}_j$ *on* $\prod_{j\in J}(X_j \times X_j)$.

PROOF. Indeed, by definition, $\mathcal{R} \geq p_j^-[\mathcal{R}_j]$ for every $j \in J$, and thus $\mathcal{R} := \bigvee_{j\in J} p_j^-[\mathcal{R}_j]$ is the coarsest filter with this property. □

In other terms, $R \in \prod_{j\in J}\mathcal{R}_j$ whenever there is a finite set J_0 of J and $R_j \in \mathcal{R}_j$ such that $R_j = X_j$ for each $j \in J \setminus J_0$ and $\prod_{j\in J} R_j \subset R$.

Consequently,

$$\lambda(\prod_{j\in J}\mathcal{R}_j) = \prod_{j\in J}\lambda\mathcal{R}_j.$$

Indeed, $\varphi \in \lim_{\prod_{j\in J}\lambda\mathcal{R}_j}\mathcal{F}$ whenever $\varphi(j) \in \lim_{\lambda\mathcal{R}_j} p_j[\mathcal{F}]$, that is, $\mathcal{V}_{\lambda\mathcal{R}_J}(\varphi(j)) \leq p_j[\mathcal{F}]$ for every $j \in J$.

Moreover, each product of semi-uniformities, quasi-uniformities, and uniformities is, respectively, a semi-uniformity, a quasi-uniformity, and a uniformity.

It is straightforward that

PROPOSITION XV.3.8. *Let* $\mathcal{P}$ *be a uniformity on* X*, and* $\mathcal{R}_j$ *a uniformity on* Y_j *for each* $j \in J$*. If* $f : X \longrightarrow \prod_{j\in J} Y_j$*, then* $f \in \mathscr{U}(\mathcal{P}, \prod_{j\in J}\mathcal{R}_j)$ *if and only if* $p_j \circ f \in \mathscr{U}(\mathcal{P}, \mathcal{R}_j)$ *for each* $j \in J$.

On the other hand ([14]),

THEOREM XV.3.9. *Each Hausdorff uniform space is uniformly isomorphic with a subspace of a product of metrizable uniform spaces.*

PROOF. Let $\mathcal{V}$ be a uniformity on X. By Lemma XV.2.4, for each $V \in \mathcal{V}$, there exists $\{V_n : n < \omega\} \subset \mathcal{V}$ be such that $V_n^- = V_n$ and $V_0 = X \times X, V_1 = V$, and $V_{n+1}^3 \subset V_n$ for $n < \omega$, and there exists a pseudometric d_V such that (XV.2.3)

$$V_n \subset \{d_V \leq 2^{-n}\} \subset V_{n-1}$$

for each $n \in \mathbb{N}_1$. Let $E_V \subset X \times X$ be defined by $(x_0, x_1) \in E_V$ if and only if $d_V(x_0, x_1) = 0$. It turns out that

$$(x_0, x_1), (y_0, y_1) \in E_V \Longrightarrow d_V(x_0, y_0) = d_V(x_1, y_1).$$

Indeed, $d_V(x_1,y_1) \leq d_V(x_1,x_0) + d_V(x_0,y_0) + d_V(y_0,y_1) = d_V(x_0,y_0)$, and similarly $d_V(x_0,y_0) \leq d_V(x_1,y_1)$. In particular, if $y \in E_V x$ and $z \in E_V y$, then $d_V(x, z) \leq d_V(x, y) + d_V(y, z) = 0$, hence $z \in E_V x$. Therefore, E_V is an equivalence relation.

[14]See R. ENGELKING [97, Theorem 8.2.3].

For any $x \in X$, let $E_V x$ be the E_V-equivalence class of x, and X_V stands for the set of E_V-equivalence classes. Let $f_V : X \longrightarrow X_V$ be defined by $f_V(x) := E_V x$. Finally

$$D_V(f_V(x_0), f_V(x_1)) := d_V(x_0, x_1)$$

defines a metric on X_V. If now $\mathcal{V}_V$ is the uniformity on X_V defined by D_V, then obviously $f_V \in \mathscr{U}(\mathcal{V}, \mathcal{V}_V)$.

By Proposition XV.3.8, the diagonal product map $f := \Delta_{V \in \mathcal{V}} f_V$ fulfills $f \in \mathscr{U}(\mathcal{V}, \prod_{V \in \mathcal{V}} \mathcal{V}_V)$. As $\ker \mathcal{V} = \Delta$, for each distinct $x, y \in X$, there exists $V \in \mathcal{V}$ such that $d_V(x,y) > 0$, hence $f(x) \neq f(y)$. Therefore, the semi-inverse $f^{\sim 1} : f(X) \longrightarrow X$ of f is well defined. Now $f^{\sim 1} \in \mathscr{U}((\prod_{V \in \mathcal{V}} \mathcal{V}_V)\big|_{f(X)}, \mathcal{V})$, because for each $V \in \mathcal{V}$ there is $\{V_n : n < \omega\}$ so that $V_1 = V$ and $V_{n+1}^3 \subset V_n$ for each $n \in \mathbb{N}_1$, and if $D_V(f_V(x), f_V(y)) = d_V(x,y) < 2^{-2}$, then in virtue of (XV.2.3) $(x,y) \in V_1 = V$. $\square$

If X is a set, and $\mathcal{R}$ is a preuniformity on a set Y, then we define the *uniform convergence* on Y^X to the effect that $f_\infty \in \lim_{\mathfrak{u}\mathcal{R}} \mathcal{F}$ whenever for each $R \in \mathcal{R}$, there is $F \in \mathcal{F}$ such that $(f_\infty(x), f(x)) \in R$ for each $x \in X$ and $f \in F$.

We observe that each $R \in \mathcal{R}$ defines a relation $\widehat{R}$ on Y^X so that $(f,g) \in \widehat{R}$ provided that $(f(x), g(x)) \in R$ for each $x \in X$. It is plain that

$$\widehat{R_0 \cap R_1} = \widehat{R_0} \cap \widehat{R_1}$$

for any $R_0, R_1 \in \mathcal{R}$. Consequently, $\widehat{\mathcal{R}} := \{\widehat{R} : R \in \mathcal{R}\}$ is a preuniformity whenever $\mathcal{R}$ is a preuniformity. Moreover, $\widehat{R^-} = (\widehat{R})^-$ and $\widehat{S \circ R} = \widehat{S} \circ \widehat{R}$ for every $R, S \in \mathcal{R}$, so that $\widehat{\mathcal{R}}$ is a quasi-uniformity if $\mathcal{R}$ is a quasi-uniformity, and $\widehat{\mathcal{R}}$ is a uniformity if $\mathcal{R}$ is a uniformity. In particular, $\lambda\widehat{\mathcal{R}}$ coincides with the uniform convergence. See Example X.5.5.

XV.4. Completeness

Let $\mathcal{R}$ be a preuniformity on X.

DEFINITION XV.4.1 (Cauchy filter). A filter $\mathcal{F}$ on X is said to be an *$\mathcal{R}$-Cauchy filter* if for every $R \in \mathcal{R}$ there exists $x \in X$ such that $Rx \in \mathcal{F}$.

It is obvious that each $\lambda\mathcal{R}$-convergent filter is $\mathcal{R}$-Cauchy. If $\mathcal{P}$ and $\mathcal{R}$ are preuniformities such that $\mathcal{P} \leq \mathcal{R}$, then each $\mathcal{R}$-Cauchy filter is a $\mathcal{P}$-Cauchy filter. In particular, every $\mathcal{R}^\flat$-Cauchy filter is an $\mathcal{R}$-Cauchy filter and an $\mathcal{R}^-$-Cauchy filter.

PROPOSITION XV.4.2. *Let $\mathcal{R}$ be a preuniformity on X. A filter $\mathcal{F}$ on X is $\mathcal{R}^\flat$-Cauchy if and only if $\mathcal{F} \times \mathcal{F} \geq \mathcal{R}$.*

PROOF. Suppose that $\mathcal{F} \times \mathcal{F} \geq \mathcal{R}$. Let $R \in \mathcal{R}$ and let $F \in \mathcal{F}$ be such that $F \times F \subset R$. If $x \in F$, then $F = (F \times F)x \subset Rx$ and analogously $F = (F \times F)^- x \subset R^- x$, hence $F \subset (R \cap R^-)x$, that is, $(R \cap R^-)x \in \mathcal{F}$.

Conversely, if $\mathcal{F}$ is $\mathcal{R}^\flat$-Cauchy filter, and let $R \in \mathcal{R}$. Pick $P \in \mathcal{R}$ such that $P \circ P \subset R$. By assumption, there exists $F \in \mathcal{F}$ and $x \in X$ such that $F \subset (P \cap P^-)x$. If $(a, b) \in F \times F$, then $a \in P^- x$ and $b \in Px$, so that $(a, x) \in P$ and $(x, b) \in P$, hence $(a, b) \in P \circ P \subset R$ so that $F \times F \subset R$, that is, $\mathcal{F} \times \mathcal{F} \geq \mathcal{R}$. $\square$

EXAMPLE XV.4.3. Let $X := \{0\} \cup \{\frac{1}{n} : n \in \mathbb{N}_1\}$ and let $R \in \mathcal{R}$ whenever there is $m \in \mathbb{N}_1$ such that $R_m := \{(0, \frac{1}{n}) : n > m\} \cup \triangle \subset R$. Then $\mathcal{R}$ is a quasi-uniformity, because $R_m \circ R_m = R_m$. We see that $\lambda\mathcal{R}$ is the prime topology, where the $\lambda\mathcal{R}$-neighborhood filter of 0 is the cofinite filter centered at 0. On the other hand, $\mathcal{R}^\flat = \{\triangle\}^\uparrow$, so that each $\mathcal{R}^\flat$-Cauchy filter is a principal ultrafilter. Consequently, the $\lambda\mathcal{R}$-neighborhood filter is not $\mathcal{R}^\flat$-Cauchy.

PROPOSITION XV.4.4. *If $\mathcal{P}$ and $\mathcal{R}$ are preuniformities and $f \in \mathscr{U}(\mathcal{P}, \mathcal{R})$ and $\mathcal{F}$ be a $\mathcal{P}$-Cauchy filter, then $f[\mathcal{F}]$ is a $\mathcal{R}$-Cauchy filter-base.*

PROOF. If $R \in \mathcal{R}$, then there is $P \in \mathcal{P}$ such that $(x, y) \in P$ implies that $(f(x), f(y)) \in R$. As $\mathcal{F}$ is a $\mathcal{P}$-Cauchy filter, there exists $x \in X$ and $F \in \mathcal{F}$ such that $F \subset Px$. Hence, $(x, y) \in P$ for each $y \in F$, and thus $(f(x), f(y)) \in R$ and $f(y) \in f(F)$. Consequently, $f(P) \subset Rf(x)$ proving that $f[\mathcal{F}]$ is a $\mathcal{R}$-Cauchy filter-base. $\square$

Similarly,

PROPOSITION XV.4.5. *For each $j \in J$, let $\mathcal{R}_j$ be a preuniformity on Y_j and $f_j \in Y_j^X$. Let $\mathcal{P} := \bigvee_{j \in J} f_j^- [\mathcal{R}_j]$ be the coarsest quasi-uniformity, for which $f_j \in \mathscr{U}(\mathcal{P}, \mathcal{R}_j)$ for each $j \in J$. Then a filter $\mathcal{F}$ is a $\mathcal{P}$-Cauchy if and only if $f_j[\mathcal{F}]$ is a $\mathcal{R}_j$-Cauchy filter-base for each $j \in J$.*

Therefore, if $(X, \mathcal{P})$ is a subspace of $(Y, \mathcal{R})$, then the filter induced by a $\mathcal{R}^\flat$-Cauchy filter is a $\mathcal{P}^\flat$-Cauchy filter. By Propositions XV.4.4 and XV.4.5,

PROPOSITION XV.4.6. *Let $\mathcal{R}_j$ be a preuniformity for each $j \in J$. If $\mathcal{R} := \prod_{j \in J} \mathcal{R}_j$ is the product preuniformity, then a filter $\mathcal{F}$ is $\mathcal{R}$-Cauchy if and only if $p_j[\mathcal{F}]$ is a $\mathcal{R}_j$-Cauchy filter for each $j \in J$.*

DEFINITION XV.4.7. A preuniformity $\mathcal{R}$ is called *complete* if every $\mathcal{R}$-Cauchy filter is $\lambda\mathcal{R}$-adherent; it is called *convergence-complete* if every $\mathcal{R}$-Cauchy filter is $\lambda\mathcal{R}$-convergent.

EXAMPLE XV.4.8. Let $X := [0,1]$ and for every $\varepsilon > 0$, let $V_\varepsilon \subset X \times X$ be given by

$$V_\varepsilon := \triangle \cup (\{0\} \times [0, \varepsilon[\cup \{1\} \times]1 - \varepsilon, 1]).$$

Notice that $V_\varepsilon \circ V_\varepsilon \subset V_\varepsilon$. Consider the quasi-uniformity $\mathcal{V}$ generated by $\{V_\varepsilon : \varepsilon > 0\}$. Then $\lambda\mathcal{V}^-$ is the discrete quasi-uniformity. To see that $\mathcal{V}$ is complete, let $\mathcal{F}$ be a $\mathcal{V}$-Cauchy ultrafilter. If $\mathcal{F}$ converges neither to 0 nor to 1, then there exists $\varepsilon > 0$ such that $[\varepsilon, 1-\varepsilon] \in \mathcal{F}$. If $x \in X$ is such that $V_\varepsilon x \in \mathcal{F}$, then $V_\varepsilon x = \{x\}$ and thus $\mathcal{F}$ converges to x. Therefore, $\mathcal{V}$ is a complete qusi-uniformity. On the other hand, the filter-base $\{]0, \varepsilon] \cup [1-\varepsilon, 1[: \varepsilon > 0\}$ is a $\mathcal{V}$-Cauchy non-convergent filter-base.

PROPOSITION XV.4.9. *Each locally symmetric complete quasi-uniformity is convergence-complete.*

PROOF. Let $\mathcal{F}$ be a $\mathcal{R}$-Cauchy filter and $p \in \operatorname{adh}_{\lambda\mathcal{R}} \mathcal{F}$. Let $R \in \mathcal{R}$ and let $P \in \mathcal{R}$ be symmetric and such that $P^3(p) \subset Rp$. On the other hand, there exists $x \in X$ such that $Px \in \mathcal{F}$, hence $Px \# Pp$, because $Pp \in \mathcal{F}^\#$. As P is symmetric, $x \in P^2(p)$, hence $Px \subset P^3(p) \subset Rp$ and thus $Rp \in \mathcal{F}$, which proves that $p \in \lim_{\lambda\mathcal{R}} \mathcal{F}$. □

It is straightforward that a closed subspace of a complete quasi-uniform space is complete. On the other hand, a complete subspace of a Hausdorff quasi-uniform space need not to be closed.

EXAMPLE XV.4.10. In Example XV.4.3, the subset $\{\frac{1}{n} : n \in \mathbb{N}_1\}$ is not closed, and the restriction of the quasi-uniformity $\mathcal{R}$ is discrete, hence complete.

However,

PROPOSITION XV.4.11. *If $\mathcal{R}$ is a complete uniformity on X and $M \subset X$, then the restriction $\mathcal{R}_M$ of $\mathcal{R}$ to $M \times M$ is complete if and only if M is $\lambda\mathcal{R}$-closed.*

PROOF. Assume that $\mathcal{R}_M$ is complete, and let $x_0 \in cl_{\lambda\mathcal{R}} M$. Then $\mathcal{B} := \{Rx_0 \cap M : \operatorname{cl}_{\lambda\mathcal{R}\times\lambda\mathcal{R}} R \subset R \in \mathcal{R}\}$ is an $\mathcal{R}_M$-Cauchy filter-base consisting of $\lambda\mathcal{R}_M$-closed sets, hence $\varnothing \neq \bigcap \mathcal{B} \subset \{x_0\}$ and $x_0 \in M$ by the completeness of $\mathcal{R}_M$. □

PROPOSITION XV.4.12. *If $\mathcal{R}_j$ is a quasi-uniformity on X_j for each $j \in J$, then $\mathcal{R} := \prod_{j\in J} \mathcal{R}_j$ is complete if and only if $\mathcal{R}_j$ is complete for each $j \in J$.*

PROOF. Assume that $\mathcal{R}_j$ is a complete quasi-uniformity on X_j for each $j \in J$. If $\mathcal{F}$ is a $\mathcal{R}$-Cauchy filter, then by Proposition XV.4.6, $p_j[\mathcal{F}]$ is a $\mathcal{R}_j$-Cauchy filter for each $j \in J$, hence $\lambda\mathcal{R}_j$-adherent for each $j \in J$, and thus $\mathcal{F}$ is $\lambda\mathcal{R}$-adherent. Conversely, if $\mathcal{R}$ is complete, then $\mathcal{R}_j$ is complete for each $j \in J$, because X_j is $\lambda\mathcal{R}$-closed for each $j \in J$. □

THEOREM XV.4.13. *Let $(X, \mathcal{P})$ and $(Y, \mathcal{R})$ be uniform spaces, and let D be a $\lambda\mathcal{P}$-dense subset of X. If $\mathcal{R}$ is complete and Hausdorff, and $f \in \mathscr{U}(\mathcal{P}|_D, \mathcal{R})$, then there exists $F \in \mathscr{U}(\mathcal{P}, \mathcal{R})$ such that $F|_D = f$.*

PROOF. As $f \in \mathscr{U}(\mathcal{P}|_D, \mathcal{R})$, we infer from Proposition XV.4.4 that $\{f(Px \cap D) : P \in \mathcal{P}\}$ is an $\mathcal{R}$-Cauchy filter-base, so that, by completeness and the Hausdorff property,

$$\{F(x)\} = \bigcap\nolimits_{P \in \mathcal{P}} \mathrm{cl}_{\lambda\mathcal{R}} f(Px \cap D)$$

defines a mapping. It is immediate that $F(x) = f(x)$ for each $x \in D$.

To see that F is uniformly continuous, let $R \in \mathcal{R}$ and let $R_0 \in \mathcal{R}$ such that $R_0^6 \subset R$. By uniform continuity, there exists $P \in \mathcal{P}$ that is $(\lambda\mathcal{P}\times\lambda\mathcal{P})$-open and such that $(f(w_0), f(w_1)) \in R_0$ for every $w_0, w_1 \in D$ such that $(w_0, w_1) \in P^2$. By (XV.2.1), for every $A \subset X$,

$$\text{(XV.4.1)} \quad A \times A \subset P^2 \Longrightarrow \mathrm{adh}_{\lambda\mathcal{R}} f(A \cap D) \times \mathrm{adh}_{\lambda\mathcal{R}} f(A \cap D) \subset R_0^3.$$

If now $x_0, x_1 \in X$ are such that $(x_0, x_1) \in P$, then $Px_0 \# Px_1$, so that there exists $w \in Px_0 \cap Px_1 \cap D$. It follows that

$$f(w) \in \mathrm{adh}_{\lambda\mathcal{R}} f(Px_0 \cap D) \cap \mathrm{adh}_{\lambda\mathcal{R}} f(Px_1 \cap D),$$

and by (XV.4.1), $\mathrm{adh}_{\lambda\mathcal{R}} f(Px_0 \cap D) \cup \mathrm{adh}_{\lambda\mathcal{R}} f(Px_1 \cap D) \subset R_0^6 \subset R$, so that $(F(x_0), F(x_1)) \in R$. □

COROLLARY XV.4.14. *Let $\mathcal{P}$ and $\mathcal{R}$ be complete uniformities. Let A be a $\lambda\mathcal{P}$-dense and B be a $\lambda\mathcal{R}$-dense, and $\mathcal{P}_A$ and $\mathcal{R}_B$ be the respective restrictions. If $f \in \mathscr{U}(\mathcal{P}_A, \mathcal{R}_B)$ and $f^{-1} \in \mathscr{U}(\mathcal{R}_B, \mathcal{P}_A)$, then there exists (a unique) $F \in \mathscr{U}(\mathcal{P}, \mathcal{R})$ such that $F^{-1} \in \mathscr{U}(\mathcal{R}, \mathcal{P})$ and $F|_A = f$.*

A detailed justification of the following remarks can be found in [**97**] by R. ENGELKING. By Hausdorff Theorem X.5.2, each metric space is homeomorphic to a subspace of a complete metric space. By Proposition X.4.3, a metric is complete if and only if each filter containing arbitrarily small sets is adherent. It follows that a uniformity induced by a metric is complete if and only if the metric is complete.

By Theorem XV.3.9, each Hausdorff uniform space is uniformly isomorphic with a subspace of a product of metrizable uniform spaces.

On the other hand, by Proposition XV.4.12, a product uniformity is complete if and only if each factor uniformity is complete. Therefore,

COROLLARY XV.4.15. *For every uniform space, there exists a complete uniform space (unique up to uniform isomorphism) so that the former is a dense subset of the latter.*

It suffices to note that uniqueness up to uniform isomorphism follows from Corollary XV.4.14.

XV.5. Supplement

Let me just hint how the concept of uniform structure extends to general convergences.

We defined a preuniformity on a set X as a filter $\mathcal{R}$ on $X \times X$ such that $\bigcap_{R\in\mathcal{R}} R \supset \Delta$. Then $\mathcal{R}$ determines a pretopology $\lambda\mathcal{R}$ on X to the effect that $x \in \lim_{\lambda\mathcal{R}} \mathcal{F}$ whenever $\mathcal{R}(x) := \{Rx : R \in \mathcal{R}\} \leq \mathcal{F}$.

More generally, a *uniform convergence structure* on a set X is a collection $\boldsymbol{R}$ of filters on $X \times X$ containing the principal filter $\Delta^{\uparrow}$ of the diagonal ([15]), and such that $R^{-}X = X$ for each $R \in \mathcal{R}$ and every $\mathcal{R} \in \boldsymbol{R}$ ([16]).

A convergence $\lambda\boldsymbol{R}$ on X is called *compatible* with a uniform convergence structure $\boldsymbol{R}$ if $x \in \lim_{\lambda\boldsymbol{R}} \mathcal{F}$ whenever there is $\mathcal{R} \in \boldsymbol{R}$ such that $\mathcal{R}(x) \leq \mathcal{F}$.

Accordingly, $\{\mathcal{R}(x) : R \in \boldsymbol{R}\}$ is a pavement of $\lambda\boldsymbol{R}$ at x. A convergence $\lambda\boldsymbol{R}$ is finitely stable whenever $\boldsymbol{R}$ is stable under finite infima.

PROPOSITION XV.5.1. *Every convergence admits a compatible uniform convergence structure* ([17]).

PROOF. Let ξ be a convergence on X. If $\{\mathcal{F}(x) : x \in X\}$ is a selection of ξ, then the filter

$$\mathscr{F} := \bigwedge\nolimits_{x\in X} (x^{\uparrow} \times \mathcal{F}(x))$$

is well defined on $X \times X$. Taking all possible selections, valued in a system of pavements, we get a uniform convergence structure, that we call Ξ. As $\mathscr{F}(x) = \mathcal{F}(x)$ for each $\mathscr{F}$ and $x \in X$, we infer that $\lambda\Xi = \xi$. □

[15]The approach of A. LECHICKI and J. ZIEMIŃSKA [175] is even more general.

[16]The latter condition is designed to assure that $\mathcal{R}(x)$ is a proper filter for every $\mathcal{R} \in \boldsymbol{R}$ and each $x \in X$.

[17]This is essentially a result A. LECHICKI and J. ZIEMIŃSKA [175].

CHAPTER XVI

Sequentially founded convergences

Sequentially founded convergences refer to one of the most basic concepts, that of convergence of sequences. Let us revisit here convergence of sequences from a convergence theoretic viewpoint.

XVI.1. Sequential filters

Recall that a *sequence* φ on a set X is a map from an infinite countable set N to X. A filter on a set X is called *sequential* whenever it is of the form $\Gamma\varphi$ for some $\varphi \in X^N$ (II.1.1), where $E \in \Gamma\varphi$ if and only if $\varphi^-(E)$ is cofinite in N. A sequence φ on X is called *stationary* if there exists $x \in X$ such that $\varphi(n) = x$ for almost all $n \in N$, equivalently if $\Gamma\varphi = x^\uparrow$.

Remember that $(N)_0$ stands for the cofinite filter of N.

PROPOSITION XVI.1.1. *A filter $\mathcal{E}$ on X is sequential if and only if there exists $\varphi \in X^N$ such that $\varphi[(N)_0]$ is a filter-base of $\mathcal{E}$.*

PROOF. By definition, $E \in \mathcal{E} = \Gamma\varphi$ if and only if $\varphi^-(E) \in (N)_0$, hence $E \supset \varphi(\varphi^-(E)) \in \varphi[(N)_0]$. □

Explicitly,

$$\Gamma\varphi = \varphi[(N)_0]^\uparrow. \tag{XVI.1.1}$$

The class of sequential filters is denoted by $\mathbb{S}$. For a non-empty family $\mathcal{A}$ of subsets of a given set X, let us denote

$$\mathbb{S}\mathcal{A} := \{\mathcal{E} \in \mathbb{S} : \mathcal{E} \geq \mathcal{A}\}, \tag{XVI.1.2}$$

the set of all sequential filters that are finer than $\mathcal{A}$.

Sequential filters can be characterized without recourse to sequences. We have already observed that each sequential filter contains a countable set (infinite or finite), and is countably based.

PROPOSITION XVI.1.2. *A filter is sequential if and only if it contains a countable set and has a countable filter-base of almost equal sets.*

PROOF. A filter is sequential if and only if it has a filter-base of the form $\varphi[(N)_0]$. This filter-base is countable, consisting of countable sets that are almost equal. Indeed, $\varphi(F_1)\backslash\varphi(F_0) \subset \varphi(N)\backslash\varphi(F_0) \subset \varphi(N\backslash F_0)$ is finite ([1]). Conversely, if $\{F_k : k < \omega\}$ is a decreasing filter-base of almost equal countable sets of a filter $\mathcal{F}$, then let $F_\infty := \ker \mathcal{F}$. If $F_\infty = \varnothing$, then let φ be a bijection from $\mathbb{N}$ onto F_0. If $F_0 = F_\infty$, then let $\varphi \in F_\infty^{\mathbb{N}}$ be such that $\{\varphi^-(x) : x \in F_\infty\}$ is a partition of $\mathbb{N}$ into infinite sets; in this case, $\varphi^-(F_k) = \mathbb{N}$, hence is cofinite in $\mathbb{N}$ for each $k < \omega$. If $F_0 \neq F_\infty \neq \varnothing$, then $F_0 \setminus F_\infty$ is infinite; let $N = N_0 \sqcup N_\infty$ be a partition of N into countable infinite sets, and let $\varphi \in X^N$ be such that the restriction φ_0 to N_0 is a bijection, and $\{\varphi^-(x) : x \in F_\infty\}$ be a partition of N_∞ into infinite sets. □

COROLLARY XVI.1.3. *If $\mathcal{E}$ is a sequential filter and $H \in \mathcal{E}^\#$, then $\mathcal{E} \vee H$ is sequential.*

LEMMA XVI.1.4. *If $\mathcal{F}$ is a free filter such that for every $\mathcal{U} \in \beta\mathcal{F}$, there exists a sequential filter $\mathcal{E}_\mathcal{U} \leq \mathcal{U}$, then there is a finite subset $\boldsymbol{F}$ of $\beta\mathcal{F}$ such that $\bigwedge_{\mathcal{U}\in\boldsymbol{F}} \mathcal{E}_\mathcal{U} \leq \mathcal{F}$.*

PROOF. As each $\mathcal{U} \in \beta\mathcal{F}$ is free, there is a countable set $U_\mathcal{U} \in \mathcal{U}$ such that $\mathcal{E}_\mathcal{U} \leq (U_\mathcal{U})_0 \leq \mathcal{U}$. Thus, by Proposition V.4.6, there is $\boldsymbol{F} \subset_{\text{fin}} \beta\mathcal{F}$ such that $U := \bigcup_{\mathcal{U}\in\boldsymbol{F}} U_\mathcal{U} \in \mathcal{F}$. Hence $\bigwedge_{\mathcal{U}\in\boldsymbol{F}} \mathcal{E}_\mathcal{U} \leq (U)_0 \leq \mathcal{F}$. □

PROPOSITION XVI.1.5. *The image of each sequential filter is sequential.*

PROOF. If $f : X \longrightarrow Y$ and $\mathcal{E} \in \mathbb{S}X$, then by Definition II.2.9, there exists $\varphi \in X^N$ such that $\mathcal{E} = \varphi[(N)_0]$. Therefore, $f[\mathcal{E}] = f[\varphi[(N)_0]] = (f \circ \varphi)[(N)_0]$ is a sequential filter-base. □

In general, a product of two sequential filters is not sequential.

EXAMPLE XVI.1.6. Let X and Y be infinite countable sets. Then $(X)_0$ and $(Y)_0$ are free sequential filters, but their product $\mathcal{D} := (X)_0 \times (Y)_0$ is not sequential. We will show that this product is not a cofinite filter, hence it is not sequential in view of Proposition XVI.5.3. Indeed, if $D \in \mathcal{D}$, then there are $A \subset X$ and $B \subset Y$ such that $D \supset A \times B \in \mathcal{D}$. If $a_0 \in A$, then $(A \setminus \{a_0\}) \times B$ belongs to $\mathcal{D}$, but is not cofinite.

REMARK XVI.1.7. Notice that if $p_Y : X \times Y \longrightarrow Y$ is the projection on Y (see Example XVI.1.6), then $p_Y^-(Y)_0$ is not cofinite, hence is not sequential, but is a preimage of a sequential filter $(Y)_0$.

[1]If $y \notin \varphi(N \setminus F_0)$ then $\varphi^-(y) \subset F_0$, hence $y \in \varphi(F_0)$, thus $y \notin \varphi(N) \setminus \varphi(F_0)$.

XVI.2. Subsequences and subquences

In traditional notation, a sequence $(y_k)_k$ is a *subsequence* of a sequence $(x_n)_n$ whenever there exists a strictly increasing sequence $(n_k)_k$ such that, for each $k < \omega$,

$$y_k = x_{n_k}. \tag{XVI.2.1}$$

As a result, a subsequence $(y_k)_{k \in N}$ of $(x_n)_{n \in N}$, defined via (XVI.2.1) with $n_k = \gamma(k)$, could be conveniently represented by

$$(x_n)_{n \in \gamma(N)},$$

or simply by $(x_n)_{n \in N_0}$, where N_0 stands for an infinite subset of N. This convention facilitates a treatment of several, possibly infinitely many, consecutive subsequences,

$$(x_n)_{n \in N_1}, \ldots, (x_n)_{n \in N_m}, \ldots, \tag{XVI.2.2}$$

where $N \supset N_1 \supset \cdots \supset N_m, \ldots$. It has, however, some limitations. For example, it does not enable to design a common subsequence $(x_n)_{n \in N_\infty}$ of $(x_n)_{n \in N_m}$ for $m < \omega$. What one can get, however, is an almost common subsequence.

LEMMA XVI.2.1 (Pseudointersection). *If $\{A_n : n < \omega\}$ is a decreasing sequence of infinite sets, then there exists an infinite set A_∞ such that $A_\infty \setminus A_n$ is finite for each $n < \omega$.*

PROOF. If there is $n_0 < \omega$ such that $A_{n+1} = A_n$ for every $n \geq n_0$, then $A_\infty = A_n$. Otherwise, there exists a strictly increasing sequence $(n_k)_k$ such that for each $k < \omega$, there exists $x_k \in A_{n_k} \setminus A_{n_k+1}$. Therefore, $A_\infty := \{x_k : k < \omega\}$ is an infinite countable set such that $A_\infty \setminus A_n$ is finite for each $n < \omega$. □

More generally, if $\mathcal{A}$ is a family of subsets of a set X, then a subset A_∞ is called a *pseudointersection* of $\mathcal{A}$ if $A_\infty \setminus A$ is finite for each $A \in \mathcal{A}$ ([2]).

COROLLARY XVI.2.2. *If (XVI.2.2) is such that $N_m \supset N_{m+1}$ for each $m < \omega$, then there exists N_∞ such that $(x_n)_{n \in N_\infty \cap N_m}$ is a subsequence of $(x_n)_{n \in N_m}$ for each $m < \omega$.*

We observed in (II.1.2) that if ψ is a subsequence of φ, then

$$\Gamma\psi \geq \Gamma\varphi. \tag{XVI.2.3}$$

Let me recall a justification. By definition, $\psi \in X^K$ is a subsequence of $\varphi \in X^N$ if and only if there exists an injective map $\gamma : K \longrightarrow N$ such that $\psi = \varphi \circ \gamma$. If $G \in \Gamma\varphi$, then by (XVI.1.1), $\varphi^-(G)$ is cofinite

[2] A pseudointersection is not unique.

in N, and since γ is injective, $\gamma^-(\varphi^-(G))$ is cofinite in K, so that $\psi^-(G) = (\varphi \circ \gamma)^-(G)$ is cofinite in K, that is, $G \in \Gamma\psi$ by (XVI.1.1).

In order to a characterize (XVI.2.3), we are going to extend the concept of *subsequence* to that of *subquence*.

A sequence $\psi \in X^K$ is called a *subquence* of a sequence $\varphi \in X^N$ if there exists a map $\gamma : K_0 \longrightarrow N$, where $K_0 \subset_{\text{cofin}} K$, such that $\gamma^-(G) \subset_{\text{cofin}} K$ for each $G \subset_{\text{cofin}} N$.

On setting $\mathcal{F} := [(K)_0]^\uparrow$ and $\mathcal{G} := [(N)_0]^\uparrow$, so that $\Gamma\psi = \psi[\mathcal{F}]$ and $\Gamma\varphi = \varphi[\mathcal{G}]$, and thus (XVI.2.3) becomes

$$\text{(XVI.2.4)} \qquad \psi[\mathcal{F}] \geq \varphi[\mathcal{G}].$$

It follows from (IV.9.3) that (XVI.2.3) is equivalent to

$$\varphi^- \circ \psi \in \mathscr{C}(\mathcal{F}, \mathcal{G}),$$

that is, to the lower semicontinuity of $\varphi^- \circ \psi$ from $\mathcal{F}$ to $\mathcal{G}$ (Definition IV.9.3).

Recall (II.1.3) the correlation between φ and ψ is given by

$$\psi^- \circ \varphi = \{(n, k) \in N \times K : \varphi(n) = \psi(k)\}.$$

Let X, Y be sets. A relation $f \subset X \times Y$ is called a *cofinitely partial map* if there exists $X_0 \subset_{\text{cofin}} X$ such that $f \in Y^{X_0}$.

As $\mathcal{F}$ and $\mathcal{G}$ are cofinite filters on countable sets, hence free countably based filters, we can apply Proposition IV.9.4 to the effect that

COROLLARY XVI.2.3. *If $\mathcal{F}$ and $\mathcal{G}$ are cofinite filters such that (XVI.2.4) holds, then there exists a cofinitely partial map γ such that*

$$\text{(XVI.2.5)} \qquad \varphi^- \circ \psi \supset \gamma \in \mathscr{C}(\mathcal{F}, \mathcal{G}).$$

Consequently, $\psi \supset \varphi \circ \gamma$, and since ψ is a map, $\psi =_{\text{cofin}} \varphi \circ \gamma$. Therefore,

PROPOSITION XVI.2.4. *Let φ and ψ be sequences on X. Then $\Gamma\psi \geq \Gamma\varphi$ if and only if ψ is a subquence of φ.*

PROOF. Let $\mathcal{F} = (N)_0$ and $\mathcal{G} = (K)_0$, If $\psi \in X^K$ is a subquence of $\varphi \in X^N$, then there is $\gamma \in \mathscr{C}(\mathcal{F}, \mathcal{G})$ such that $\gamma \subset \varphi^- \circ \psi$. Hence $\varphi^- \circ \psi \in \mathscr{C}(\mathcal{F}, \mathcal{G})$, that is, (XVI.2.3), that is, $\Gamma\psi \geq \Gamma\varphi$.

Conversely, $\Gamma\psi \geq \Gamma\varphi$ is equivalent to (XVI.2.3), so that ψ is a subquence of φ by Corollary XVI.2.3. □

COROLLARY XVI.2.5. *Two sequences φ and ψ are equivalent, that is,*

$$\Gamma\psi = \Gamma\varphi$$

if and only if there exists a relation $\Delta \subset \varphi^{-} \circ \psi$ *such that* Δn *and* $\Delta^{-} k$ *are finite for every* $n, k < \omega$, *and* ΔN *and* $\Delta^{-} K$ *are cofinite* ([3]).

PROOF. If $\Delta \subset \varphi^{-} \circ \psi$ and $\Delta \in \mathscr{C}(\mathcal{F}, \mathcal{G})$ and $\Delta^{-} \in \mathscr{C}(\mathcal{G}, \mathcal{F})$, then $\varphi[\mathcal{F}] \leq \psi[\mathcal{G}] \leq \varphi[\mathcal{F}]$. Conversely, there exist $\gamma_0 \in \mathscr{C}(\mathcal{F}, \mathcal{G})$ and $\gamma_1 \in \mathscr{C}(\mathcal{G}, \mathcal{F})$, then $\Delta := \gamma_0 \cup \gamma_1^{-}$ fulfills the conditions. $\square$

XVI.3. Sequentially founded convergences

A *convergence of sequences* θ on a non-empty set X is defined as a relation between sequences of elements of X and elements of X, that is, a subset θ of $X^N \times X$ such that

$$\Gamma\psi \geq \Gamma\varphi \Longrightarrow \lim_{\theta} \varphi \subset \lim_{\theta} \psi \tag{XVI.3.1}$$

and $x \in \lim_{\theta} x^{\uparrow}$, where $x^{\uparrow}$ is the filter generated by the stationary sequence of x.

On assigning to every sequence φ on X, the corresponding sequential filter $\Gamma\varphi$ given by (II.1.1), we define a subset $\widehat{\theta}$ of $\mathbb{S}X \times X \subset \mathbb{F}X \times X$ to the effect that $(\Gamma\varphi, x) \in \widehat{\theta}$ whenever $(\varphi, x) \in \theta$.

The relation $\widehat{\theta}$ is well defined, because two sequences correspond to the same filter if and only if they are equivalent. In other words, (XVI.3.1) implies that

$$\Gamma\psi = \Gamma\varphi \Longrightarrow \lim_{\theta} \varphi = \lim_{\theta} \psi.$$

In order to extend $\widehat{\theta}$ to a convergence of filters in our sense, it is enough to apply the modifier I (Lemma III.7.10). In other terms, we start with a sequence convergence θ, we map it into a relation $\widehat{\theta}$ in $\mathbb{F}X \times X$, and then extend it to a convergence $\mathrm{I}\widehat{\theta}$.

$$x \in \lim_{\mathrm{I}\widehat{\theta}} \mathcal{F} \Longleftrightarrow \underset{\varphi \in X^N}{\exists} (x \in \lim_{\theta} \varphi) \wedge (\mathcal{F} \geq \Gamma\varphi).$$

DEFINITION XVI.3.1. A convergence ξ is called *sequentially founded* if $x \in \lim_{\xi} \mathcal{F}$ implies the existence of a sequential filter $\mathcal{E}$ such that $x \in \lim_{\xi} \mathcal{E}$ and $\mathcal{E} \leq \mathcal{F}$.

In other words,

PROPOSITION XVI.3.2. *A convergence is sequentially founded if and only if it has at every point a pavement consisting of sequential filters.*

We denote by **Seq** the class of sequentially founded convergences, and by **Seq**X the set of sequentially founded convergences on X.

[3] S. DOLECKI and G. H. GRECO [69].

For example, the discrete convergence ι_X on any set X ([4]) is sequentially founded, because $x^{\uparrow}$ is a sequential filter for each $x \in X$. The chaotic convergence o_X on X is sequentially founded if and only if X is countable ([5]).

Finite convergences are sequentially founded, because each filter on a finite set is principal and sequential. A simplest infinite Hausdorff non-discrete sequentially founded convergence is that of Example XVI.3.3 below.

EXAMPLE XVI.3.3. Let X be an infinite countable set, and let $x_\infty \in X$. We define a prime pretopology σ on X by setting $x_\infty \in \lim_\sigma \mathcal{F}$ whenever $\mathcal{F} \geq (X)_0 \wedge x_\infty^{\uparrow}$. This is a sequentially founded convergence, because $(X)_0$ is a sequential filter.

Here is a slightly more complex prime sequentially founded convergence.

EXAMPLE XVI.3.4. Let $(X_n)_n$ be a sequence of disjoint infinite countable sets, let $X := \bigcup_{n<\omega} X_n$, and let $x_\infty \in X$. We define a prime convergence η on X by declaring $\{(X_n)_0 : n < \omega\} \cup \{x_\infty^{\uparrow}\}$ a pavement of η at x_∞, and, of course, $\{x^{\uparrow}\}$ is a pavement of η at each $x \neq x_\infty$. This is a sequentially founded convergence, which is not a pretopology. Indeed, the filter

$$\mathcal{V}_\eta(x_\infty) = \bigwedge_{n<\omega} (X_n)_0 \wedge x_\infty^{\uparrow},$$

is not sequential. In fact, if $A \in \bigwedge_{n<\omega}(X_n)_0$, then $X_n \setminus A$ is finite for each $n < \omega$. Let $x_n \in A \cap X_n$ for each $n < \omega$. Then $A \setminus \{x_n : n < \omega\} \in \bigwedge_{n<\omega}(X_n)_0$, but is not cofinite in A.

The *standard sequential convergence* $\sigma_{\mathbb{R}}$ form Example III.1.3 is of course sequentially founded.

PROPOSITION XVI.3.5. *Each non-empty infimum of sequentially founded convergences is sequentially founded.*

PROOF. Let Ξ be a non-empty set of sequentially founded convergences on a given set X. Then $x \in \lim_{\bigwedge \Xi} \mathcal{F}$ means that there exists $\xi \in \Xi$ such that $x \in \lim_\xi \mathcal{F}$. By definition, there exists a sequential filter $\mathcal{E} \leq \mathcal{F}$ such that $x \in \lim_\xi \mathcal{E} \subset \lim_{\bigwedge \Xi} \mathcal{E}$, hence, $\bigwedge \Xi$ is sequentially founded. □

PROPOSITION XVI.3.6. *For each convergence ξ, there exists a coarsest sequentially founded convergence that is finer than ξ.*

[4] Of course, non-empty, by the definition of convergence.

[5] Indeed, if X is countable, then the coarsest filter $\{X\}$ converges to each point and is sequential.

PROOF. Let ξ be a convergence on a set X. Consider the set of all sequentially founded convergences that are finer than ξ,

$$B_\xi := \{\theta \in \mathbf{Seq}X : \theta \geq \xi\}. \tag{XVI.3.2}$$

The set B_ξ is non-empty, because the discrete convergence $\iota_X \in B_\xi$. By Proposition XVI.3.5, $\bigwedge B_\xi \in B_\xi$, and thus is its coarsest element. $\square$

The *sequential modification* $\mathrm{Seq}\,\xi$ of ξ stands for the coarsest sequentially founded convergence that is finer than ξ. The resulting operation Seq is called the *sequential modifier.*

It is easy to describe the sequential modification of a given convergence.

$$x \in \lim_{\mathrm{Seq}\xi} \mathcal{F} \iff \underset{\mathcal{E}\in\mathbb{S}}{\exists} (x \in \lim_\xi \mathcal{E}) \wedge (\mathcal{F} \geq \mathcal{E}), \tag{XVI.3.3}$$

that is, it is the restriction of a convergence to sequential filters and, of course, to all filters that are finer than these sequential filters.

EXAMPLE XVI.3.7. If $\nu_\mathbb{R}$ is the *standard convergence* of the real line from Example III.1.4, then $\mathrm{Seq}\nu_\mathbb{R} = \sigma_\mathbb{R}$, the *standard sequential convergence* $\sigma_\mathbb{R}$ from Example III.1.3.

PROPOSITION XVI.3.8. *The sequential modifier* Seq *is*

(increasing) $$\zeta \geq \xi \Longrightarrow \mathrm{Seq}\,\zeta \geq \mathrm{Seq}\,\xi,$$

(ascending) $$\mathrm{Seq}\,\xi \geq \xi,$$

(idempotent) $$\mathrm{Seq}\,(\mathrm{Seq}\,\xi) = \mathrm{Seq}\,\xi.$$

PROOF. Let B_ξ be defined as in (XVI.3.2). Clearly, if $\zeta \geq \xi$ then $B_\zeta \subset B_\xi$, and thus $\mathrm{Seq}\,\zeta = \bigwedge B_\zeta \geq \bigwedge B_\xi = \mathrm{Seq}\,\xi$. As $\mathrm{Seq}\,\xi = \bigwedge B_\xi \in B_\xi$, we infer that $\mathrm{Seq}\,\xi \geq \xi$. Finally, as $\mathrm{Seq}\,\xi$ is sequentially founded, $\bigwedge B_{\mathrm{Seq}\,\xi} = \mathrm{Seq}\,\xi$. $\square$

PROPOSITION XVI.3.9. *If $f : X \longrightarrow Y$ and ξ is a sequentially founded convergence on X, then the final convergence $f\xi$ is sequentially founded.*

PROOF. A convergence at isolated points is sequentially founded. Thus, it is enough to consider surjective maps. Let $y \in \lim_{f\xi} \mathcal{G}$. By (IV.3.2), there exists a filter $\mathcal{F}$ (on X) such that $f[\mathcal{F}] \leq \mathcal{G}$ and $y \in f(\lim_\xi \mathcal{F})$. Hence there is $x \in f^-(y)$ such that $x \in \lim_\xi \mathcal{F}$, and since ξ is sequentially founded, there is $\mathcal{E} \in \mathbb{S}$ such that $x \in \lim_\xi \mathcal{E}$ and $\mathcal{E} \leq \mathcal{F}$. Therefore, $f[\mathcal{E}] \leq f[\mathcal{F}] \leq \mathcal{G}$ and $y = f(x) \in f(\lim_\xi \mathcal{E})$, so that $y \in \lim_{f\xi} f[\mathcal{E}]$. By virtue of Proposition XVI.1.5, $f[\mathcal{E}] \in \mathbb{S}$, and thus the convergence $f\xi$ is sequentially founded. $\square$

PROPOSITION XVI.3.10. *For every* $f : X \longrightarrow Y$ *and each convergence* ξ *on* X,

$$f(\operatorname{Seq}\xi) \geq \operatorname{Seq}(f\xi). \tag{XVI.3.4}$$

PROOF. It follows from $\operatorname{Seq}\xi \geq \xi$ that $f(\operatorname{Seq}\xi) \geq f\xi$. By Proposition XVI.3.9, $f(\operatorname{Seq}\xi)$ is sequentially founded, hence is finer than $\operatorname{Seq}(f\xi)$, the coarsest sequentially founded convergence finer than $f\xi$. □

PROPOSITION XVI.3.11. *For every* $f : X \longrightarrow Y$ *and each convergence* τ *on* Y,

$$\operatorname{Seq}(f^-\tau) \geq f^-(\operatorname{Seq}\tau). \tag{XVI.3.5}$$

PROOF. Applying f^- to (XVI.3.4) and setting $\xi = f^-\tau$, we get

$$\operatorname{Seq}(f^-\tau) \geq f^-f(\operatorname{Seq}(f^-\tau)) \geq f^-(\operatorname{Seq}(ff^-\tau)) \geq f^-(\operatorname{Seq}\tau),$$

because of (IV.3.7) and (IV.3.6). □

PROPOSITION XVI.3.12. *For any convergences* ξ *and* τ,

$$\mathscr{C}(\xi,\tau) \subset \mathscr{C}(\operatorname{Seq}\xi, \operatorname{Seq}\tau).$$

PROOF. If $f \in \mathscr{C}(\xi,\tau)$, that is, $f\xi \geq \tau$, then $\operatorname{Seq}(f\xi) \geq \operatorname{Seq}\tau$, hence by (XVI.3.4), $f(\operatorname{Seq}\xi) \geq \operatorname{Seq}\tau$, that is, $f \in \mathscr{C}(\operatorname{Seq}\xi, \operatorname{Seq}\tau)$. □

An initial convergence with respect to a sequentially founded convergence need not to be sequentially founded.

EXAMPLE XVI.3.13. Let X and Y be infinite countable sets. Consider the projection $p_Y : X \times Y \longrightarrow Y$ ([6]). Let $y_\infty \in Y$ and define a convergence σ_Y as a prime convergence such that $y_\infty \in \lim_{\sigma_Y}(Y)_0$. Then σ_Y is a sequentially founded convergence ([7]), but $p_Y^-\sigma_Y$ is not sequentially founded. In fact, $\{X\} \times (Y)_0$ is a coarsest filter-base converging in $p_Y^-\sigma_Y$ to (x, y_∞) (for any x), but is not a cofinite filter of any of its elements, because the sets of the form $X \times \{y\}$ are infinite. Hence $\{X\} \times (Y)_0$ is the vicinity filter of any (x, y_∞), but is not sequentially founded.

A product of sequentially founded convergences need not to be sequentially founded.

EXAMPLE XVI.3.14. Let X and Y be infinite countable sets, let $x_\infty \in X, y_\infty \in Y$, and let σ_X and σ_Y be prime convergences defined by $x_\infty \in \lim_{\sigma_X}(X)$ and $y_\infty \in \lim_{\sigma_Y}(Y)_0$, respectively. Then σ_X and σ_Y

[6]That is, $p_Y(x,y) := y$ for each $(x,y) \in X \times Y$.

[7]Which is also a Hausdorff pretopology.

are sequentially founded, but $\sigma_X \times \sigma_Y$ is not, because $(X)_0 \times (Y)_0$ is the coarsest filter converging to (x_∞, y_∞) in $\sigma_X \times \sigma_Y$, and we have seen in Example XVI.1.6 that $(X)_0 \times (Y)_0$ is not sequentially founded.

DEFINITION XVI.3.15 (Urysohn convergence). A sequentially founded convergence ξ on X is said to be *Urysohn* if for every $x \in X$ and each sequential filter $\mathcal{E}$ such that $x \notin \lim_\xi \mathcal{E}$ there exists $\mathcal{F} \in \mathbb{S}\mathcal{E}$ such that $x \notin \lim_\xi \mathcal{G}$ for each $\mathcal{G} \in \mathbb{S}\mathcal{F}$.

In other words, a convergence $\xi = \mathrm{Seq}\xi$ on X is Urysohn if and only if, for each $\mathcal{E} \in \mathbb{S}X$,

$$\lim_{\mathrm{U}\xi} \mathcal{E} := \bigcap_{\mathcal{F} \in \mathbb{S}\mathcal{E}} \bigcup_{\mathcal{G} \in \mathbb{S}\mathcal{F}} \lim_\xi \mathcal{G} \subset \lim_\xi \mathcal{E}. \tag{XVI.3.6}$$

Let us translate the definition in terms of sequences:

PROPOSITION XVI.3.16. *A sequentially founded convergence is Urysohn if and only if whenever $(x_n)_n$ does not converge to x, then there exists a subsequence $(x_{n_k})_k$ such that no subsequence of $(x_{n_k})_k$ converges to x.*

EXAMPLE XVI.3.17. The *standard sequential convergence* $\sigma_{\mathbb{R}}$ (Examples III.1.3 and III.1.9) is Urysohn. Indeed, $x \notin \lim_{\sigma_{\mathbb{R}}} (x_n)_n$ if and only if there exists $\varepsilon > 0$ and a subsequence $(x_{n_k})_k$ such that $x_{n_k} \notin]x-\varepsilon, x+\varepsilon[$. Obviously, no subsequence of $(x_{n_k})_k$ converges to x in $\sigma_{\mathbb{R}}$.

A non-Urysohn convergence is defined in the following

EXAMPLE XVI.3.18. Let $\mathcal{A}$ be a *MAD* family on an infinite countable set X, and let $X_\infty = X \sqcup \{\infty\}$. Define a prime convergence η such that $\infty \in \lim_\eta \mathcal{F}$ whenever there exists $\mathcal{A}_0 \subset_{\text{fin}} \mathcal{A}$ so that $\mathcal{F} \geq (\bigcup_{A \in \mathcal{A}_0} A)_0 \cup \infty^\uparrow$. Then η is a sequentially founded finitely stable convergence, which is not Urysohn. In fact, $\infty \notin \lim_\eta (X)_0$, but for every subsequence $(B)_0$ of $(X)_0$, there is $A \in \mathcal{A}$ such that $A \cap B$ is infinite, hence $(A \cap B)_0 \geq (B)_0 \vee (A)_0$, and thus $\infty \in \lim_\eta (A \cap B)_0$.

We shall supply a characterization of Urysohn convergences in Proposition XXI.5.1.

XVI.4. *Extrema of sets of sequential filters

Arbitrary infima of sequential filters need not be sequential, because every countably based filter is an infimum of sequential filters. Actually, by Definition II.3.8, a filter is called Fréchet if it is an infimum of sequential filters.

PROPOSITION XVI.4.1. *A filter $\mathcal{F}$ is Fréchet if and only if for each $H \in \mathcal{F}^{\#}$ there exists a sequential filter $\mathcal{E} \in \mathbb{S}(\mathcal{F} \vee H)$.*

PROOF. Let $\mathcal{F} = \bigcap_{\mathcal{E}\in\mathbb{S}\mathcal{F}} \mathcal{E}$ and let $H \in \mathcal{F}^{\#}$. Then by (V.3.9), $H \in \bigcup_{\mathcal{E}\in\mathbb{S}\mathcal{F}} \mathcal{E}^{\#}$, hence there exists $\mathcal{E} \in \mathbb{S}\mathcal{F}$ such that $H \in \mathcal{E}^{\#}$, that is, $\mathcal{E} \vee H \geq \mathcal{F} \vee H$, and, by Corollary XVI.1.3, $\mathcal{E} \vee H$ is sequential.

Conversely, if a filter $\mathcal{F}$ on X is not Fréchet, then there exists $A \in \bigcap_{\mathcal{E}\in\mathbb{S}\mathcal{F}} \mathcal{E}$ and $A \notin \mathcal{F}$, equivalently $A^c \in \mathcal{F}^{\#}$. Thus $\mathcal{F} \vee A^c$ is a proper filter. If $\mathcal{E} \in \mathbb{S}(\mathcal{F} \vee A^c) \subset \mathbb{S}\mathcal{F}$ then $A^c \in \mathcal{E}$ in contradiction with $A \in \bigcap_{\mathcal{E}\in\mathbb{S}\mathcal{F}} \mathcal{E}$. Therefore, $\mathbb{S}(\mathcal{F} \vee A) = \varnothing$. □

A sequence $(\mathcal{F}_n)_n$ of filters on a given set is called *monotone* if either it is *increasing*, that is, $\mathcal{F}_n \leq \mathcal{F}_{n+1}$ for each $n < \omega$, or *decreasing*, that is, $\mathcal{F}_n \geq \mathcal{F}_{n+1}$ for each $n < \omega$.

Let us remind that if a countable supremum of countably based filters is proper, then it is countably based.

However, a countable supremum of sequential filters is in general not sequential.

PROPOSITION XVI.4.2. *If $(\mathcal{E}_n)_n$ is an increasing sequence of free sequential filters such that $\bigvee_{n<\omega} \mathcal{E}_n$ is sequential, then there is $n_0 < \omega$ such that $\mathcal{E}_{n_0} = \mathcal{E}_n$ for every $n \geq n_0$.*

PROOF. Suppose that $(\mathcal{E}_n)_n$ is strictly increasing. Then there is a sequence $(E_n)_n$, of decreasing sets, such that $\mathcal{E}_n$ is the cofinite filter of E_n and $E_n \setminus E_{n+1}$ is infinite for every n. We shall see that $(E_n)_n$ is a filter-base of $\mathcal{F} = \bigvee_{n<\omega} \mathcal{E}_n$, and thus by Proposition XVI.5.3, $\mathcal{F}$ constitutes not sequential. Suppose that $(E_n)_n$ is not a filter-base of $\mathcal{F}$: there is $F \in \mathcal{F}$ and a sequence $(x_n)_n$ such that $x_n \in E_n \setminus F$ for each $n < \omega$. Then the set $\{x_n : n < \omega\}$ is almost included in E_n for every n, hence $(x_n)_n \geq \mathcal{E}_n$ for every n, and thus $(x_n)_n \geq \mathcal{F}$. But on the other hand, $\{x_n : n < \omega\} \cap F = \varnothing$, which yields a contradiction. □

However,

LEMMA XVI.4.3. *For each increasing sequence of free sequential filters $(\mathcal{E}_n)_n$ there exists a sequential filter $\mathcal{E} \geq \bigvee_{n<\omega} \mathcal{E}_n$.*

PROOF. By assumption, there exists a decreasing sequence $(E_n)_n$ of infinite countable sets such that $\mathcal{E}_n = (E_n)_0$. By Lemma XVI.2.1, there exists an infinite set E such that $E \setminus E_n$ is finite for each $n < \omega$, and thus $\mathcal{E} := (E)_0 \geq (E_n)_0 = \mathcal{E}_n$ for every $n < \omega$. □

Similarly,

PROPOSITION XVI.4.4. *If $(\mathcal{E}_n)_n$ is a decreasing sequence of free sequential filters such that $\bigwedge_{n<\omega}\mathcal{E}_n$ is sequential, then there is $n_0 < \omega$ such that $\mathcal{E}_n = \mathcal{E}_{n_0}$ for every $n \geq n_0$.*

PROOF. The filter $\bigwedge_{n<\omega}\mathcal{E}_n$ is free, as an infimum of free filters on a given set. If $\bigwedge_{n<\omega}\mathcal{E}_n$ is sequential, then it is the cofinite filter of an infinite countable set E. Because $(\mathcal{E}_n)_n$ is decreasing, there exists a increasing sequence $(E_n)_n$ of subsets of E such that $\mathcal{E}_n$ is the cofinite filter of E_n for every n. If $\bigwedge_{n<\omega}\mathcal{E}_n$ is not equal to any $\mathcal{E}_n$, then $E \setminus E_n$ is infinite for every n. Hence there exists a sequence (x_n) of distinct terms such that $x_n \notin E_n$ for every n. The infinite set $H = \{x_k : k < \omega\}$ has the property that $H \cap E_n$ is finite for every n, so that $E \setminus H \in \mathcal{E}_n$ for every n and thus $E \setminus H \in \bigwedge_{n<\omega}\mathcal{E}_n$ contrary to the fact that $\bigwedge_{n<\omega}\mathcal{E}_n$ is the cofinite filter of E. □

On the other hand, the infimum of a decreasing sequence of sequential filters need not be even countably based.

EXAMPLE XVI.4.5 (Non-countably based Fréchet filter). Let $(E_n)_n$ be a sequence of disjoint infinite countable sets, and $\mathcal{E}_n$ be the cofinite filter of E_n. If $\mathcal{F}_n := \bigwedge_{0\leq k\leq n}\mathcal{E}_k$, then $\mathcal{F}_n \geq \mathcal{F}_{n+1}$ for every $n < \omega$. Let $\mathcal{F} := \bigwedge_{n<\omega}\mathcal{F}_n = \bigwedge_{n<\omega}\mathcal{E}_n$. Arrange $E_n = \{x_{n,k} : k < \omega\}$ for every $n < \omega$, so that

$$\{\{x_{n,k} : k > m\} : m < \omega\}$$

is a filter-base of $\mathcal{E}_n = (E_n)_0$. Therefore, if

$$W_f := \bigcup\nolimits_{n<\omega}\{x_{n,k} : k > f(n)\},$$

then $\mathcal{W} := \{W_f : f \in \mathbb{N}^{\mathbb{N}}\}$ constitutes a filter-base of $\mathcal{F} = \bigwedge_{n<\omega}\mathcal{E}_n$ by Exercise V.6.6. We shall see that it is not possible to extract from $\mathcal{W}$ a countable filter-base of $\mathcal{F}$.

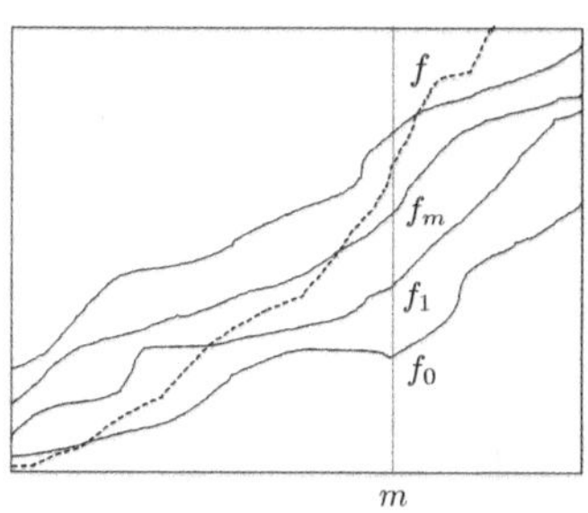

FIGURE XVI.4.1. Any filter-base of $\mathcal{F} = \bigwedge_{n<\omega}\mathcal{F}_n$ is uncountable.

By way of contradiction, suppose that there is a sequence $f_m \in \mathbb{N}^{\mathbb{N}}$ such that $(W_{f_m})_m$ is a filter-base of $\mathcal{F}$. For every $n \in \mathbb{N}$, let

$$f(n) := \max_{m \leq n} f_m(n) + 1. \tag{XVI.4.1}$$

Then $W_f \in \mathcal{F}$, hence there exists $m \in \mathbb{N}$ such that $W_{f_m} \subset W_f$, that is $f(n) \leq f_m(n)$ for each $n \in \mathbb{N}$. But by (XVI.4.1), $f(m) > f_m(m)$, which yields a contradiction.

However, for every sequence $(\mathcal{E}_n)_n$ of sequential filters, there is a sequential filter $\mathcal{E}$ such that $\mathcal{E} \leq \bigwedge_{n<\omega} \mathcal{E}_n$. Indeed, if they are all principal, that is, if $\mathcal{E}_n = \{E_n\}^{\uparrow}$, where E_n is a countable set for each $n < \omega$, then $\mathcal{E} := \{\bigcup_{n<\omega} E_n\}^{\uparrow}$; if they are all free, then $\mathcal{E}_n = (E_n)_0$, where E_n is an infinite countable set for each $n < \omega$, then $\mathcal{E} := (\bigcup_{n<\omega} E_n)_0$. More precisely,

LEMMA XVI.4.6. *If $(\mathcal{E}_n)_n$ is a sequence of sequential filters, and $\mathcal{F}$ is a countably based filter such that $\mathcal{E}_n \geq \mathcal{F}$ for every $n < \omega$, then there is a sequential filter $\mathcal{E}$ for which $\bigwedge_{n<\omega} \mathcal{E}_n \geq \mathcal{E} \geq \mathcal{F}$.*

PROOF. Because $\bigwedge_{n<\omega} \mathcal{E}_n = \bigwedge_{n<\omega}(\mathcal{E}_n^{\bullet} \wedge \mathcal{E}_n^{*}) = \bigwedge_{n<\omega} \mathcal{E}_n^{\bullet} \wedge \bigwedge_{n<\omega} \mathcal{E}_n^{*}$, we consider separately the free and the principal parts of $\mathcal{E}_n$. It follows from the assumptions that $\mathcal{E}_n^{\bullet} \geq \mathcal{F}^{\bullet}$, and thus the principal sequential filter $\bigwedge_{n<\omega} \mathcal{E}_n^{\bullet} = (\bigwedge_{n<\omega} \mathcal{E}_n)^{\bullet}$ is finer than $\mathcal{F}^{\bullet} \geq \mathcal{F}$.

Let $\mathcal{E}_n := (E_n)_0$, where E_n is an infinite countable set E_n and $E_n \subset E_{n+1}$ for each $n < \omega$. Let $\{F_k : k < \omega\}$ be a decreasing filter-base of $\mathcal{F}$. We observe that $F_k \cap E_n \in \mathcal{E}_n$, equivalently $E_n \setminus F_k$ is finite, for each $n, k < \omega$, because $\mathcal{E}_n \geq \mathcal{F}$ for each $n < \omega$. Hence $E := \bigcup_{n<\omega} F_n \cap E_n \in \bigwedge_{n<\omega} \mathcal{E}_n$, so that $\mathcal{E} := (E)_0 \leq \bigwedge_{n<\omega} \mathcal{E}_n$. On the other hand, $E_n \setminus F_k$ is finite for each $n, k < \omega$, so that for each $k < \omega$,

$$E \setminus F_k = \bigcup_{n<k} (F_n \cap E_n) \setminus F_k \subset \bigcup_{n<k} (E_n \setminus F_k)$$

is finite, and thus $\mathcal{E} \geq \mathcal{F}$. □

We conclude that

COROLLARY XVI.4.7. *If $(\mathcal{F}_n)_n$ is an increasing sequence of free sequential filters, and $(\mathcal{E}_k)_k$ is a decreasing sequence of free sequential filters, such that $\mathcal{F}_n \leq \mathcal{E}_k$ for every n and k, then there is a sequential filter $\mathcal{E}$ such that*

$$\bigvee_{n<\omega} \mathcal{F}_n \leq \mathcal{E} \leq \bigwedge_{k<\omega} \mathcal{E}_k.$$

PROOF. As the filter $\bigvee_{n<\omega} \mathcal{F}_n$ is countably based, we can use Lemma XVI.4.6 to conclude. □

XVI.5. Supplement

EXERCISE XVI.5.1. Show that the number of convergent sequences with rational terms on the real line is continuum.

SOLUTION. By Example III.1.20, for each $r \in \mathbb{R} \setminus \mathbb{Q}$ there exists an infinite set $\{q_n(r) : n < \omega\}$ such that $(q_n(r))_n$ converges to r. Then $\mathcal{R} := \{\{q_n(r) : n < \omega\} : r \in \mathbb{R} \setminus \mathbb{Q}\}$ is almost disjoint. Indeed, if $r_0 < s < r_1$, then almost all elements of $\{q_n(r_0) : n < \omega\}$ and finitely many elements of $\{q_n(r_1) : n < \omega\}$ are less than s. The cardinality of $\mathcal{R}$ is equal to that of $\mathbb{R} \setminus \mathbb{Q}$, hence is $2^{\aleph_0}$.

EXERCISE XVI.5.2. Finite suprema and finite infima of sequential filters are sequential.

SOLUTION. Let $\mathcal{E}, \mathcal{F}$ be sequential filters. Let us use Proposition XVI.1.2 to prove that $\mathcal{E} \vee \mathcal{F}$ and $\mathcal{E} \wedge \mathcal{F}$ are sequential. It is clear that $\mathcal{E} \vee \mathcal{F}$ and $\mathcal{E} \wedge \mathcal{F}$ contain countable sets. If $\{E_n : n < \omega\}$ and $\{F_n : n < \omega\}$ are respective filter-bases of $\mathcal{E}$ and $\mathcal{F}$, then $\{E_n \cap F_n : n < \omega\}$ and $\{E_n \cup F_n : n < \omega\}$ are respective filter-bases of $\mathcal{E} \vee \mathcal{F}$ and $\mathcal{E} \wedge \mathcal{F}$. Let us see that they consist of almost equal sets. Let $A := E_n \setminus E_k$ and $B := E_n \setminus E_k$ are finite. Then

$$E_n \cap F_n \subset (E_k \cup A) \cap (F_k \cup B) \subset (E_k \cap F_k) \cup A \cup B,$$
$$E_n \cup F_n \subset (E_k \cup F_k) \cup A \cup B,$$

are finite. The proof is completed by obvious induction.

In Section II.4 the *kernel* of a family of sets was defined as the intersection of its elements.

PROPOSITION XVI.5.3. *A filter $\mathcal{E}$ is sequential if and only if it is a proper cofinite filter of a countable set centered at* $\ker \mathcal{E}$, *that is, if there is $E \in \mathcal{E}$ such that*

$$\mathcal{E} = (E, \ker \mathcal{E})_0$$

PROOF. Indeed, let $\varphi \in X^N$ such that $\varphi[(N)_0]$ is a filter-base of a sequential filter $\Gamma\varphi$. Then $\ker \Gamma\varphi = \{x \in \varphi(N) : \operatorname{card} \varphi^-(x) = \infty\}$ and $\Gamma\varphi = (\varphi(N), \ker \Gamma\varphi)_0$. Thus $x \in \ker \Gamma\varphi$ if and only if $\varphi^-(x) \in (N)_0^\#$, equivalently $\operatorname{card} \varphi^-(x) = \infty$. Indeed, $x \in \ker \Gamma\varphi = \bigcap_{F \subset_{\text{fin}} N} \varphi(N \setminus F)$, that is, $x \in \varphi(N \setminus F)$ for each $F \subset_{\text{fin}} N$, which means that $\varphi^-(x)$ is infinite. $\square$

EXAMPLE XVI.5.4. Consider

(1) The sequence $1, \frac{1}{2}, \frac{1}{3}, \ldots, \frac{1}{n}, \ldots$ generates the cofinite filter of $\{\frac{1}{n} : n \in \mathbb{N}_0\}$ (8); therefore it is a free filter.

[8] Where $\mathbb{N}_0 := \mathbb{N} \setminus \{0\}$.

(2) A constant sequence with $x_n = 1$ for each n, generates a cofinite filter of $\{1\}$ at $\{1\}$; it is a principal filter.
(3) The sequence $1, \frac{1}{2}, \frac{1}{2}, \frac{1}{3}, \frac{1}{3}, \frac{1}{3}, \ldots, \underbrace{\frac{1}{n}, \ldots, \frac{1}{n}}_{n \text{ times}}$ generates the same filter as the sequence $1, \frac{1}{2}, \frac{1}{3}, \ldots, \frac{1}{n}, \ldots$.
(4) Finally, the sequence $1, 1, \frac{1}{2}, 1, \frac{1}{2}, \frac{1}{3}, \ldots, \underbrace{1, \ldots, \frac{1}{n}}_{n \text{ terms}}$ is a cofinite filter of $\{\frac{1}{n} : n \in \mathbb{N}_0\}$ centered at $\{\frac{1}{n} : n \in \mathbb{N}_0\}$; it generates the principal filter $\{F \subset X : \{\frac{1}{n} : n \in \mathbb{N}_0\} \subset F\}$.

CHAPTER XVII

Structural aspects

([1]) The class **Seq** of *sequentially founded convergences* is closed under infima, and contains the finest convergence, which is the discrete convergence ι. These two properties of **Seq** are equivalent to the existence of an increasing, idempotent, ascending map Seq, associating with each convergence ξ the coarsest convergence $\mathrm{Seq}\,\xi$ of **Seq** that is finer than ξ.

On the other hand, the classes of *pretopologies* $\mathbf{S}_0$ and *topologies* $\mathbf{T}$ are closed under suprema, and contain the coarsest convergence, which is the chaotic convergence o. This amounts to the existence of increasing, idempotent, descending maps S_0 and T, associating with each convergence ξ the finest pretopology $\mathrm{S}_0\xi$ and the finest topology $\mathrm{T}\xi$ that are coarser than ξ.

These are instances of a general principle that associates certain complete sublattices of convergences with some specific increasing maps on sets of convergences.

DEFINITION XVII.0.5. A map F, assigning to a convergence ξ a convergence $F\xi$, is called a *modifier* if for any convergences ξ and ζ,

(concrete) $$|F\xi| = |\xi|,$$
(increasing) $$\zeta \leq \xi \Longrightarrow F\zeta \leq F\xi.$$

We call a convergence θ an F*-convergence* if $F\theta = \theta$. Accordingly,

$$\mathrm{fix}F := \{\theta : \theta = F\theta\}$$

is the class of F-convergences.

XVII.1. Projective and coprojective classes

If a subclass $\mathbf{J}$ of convergences is closed under suprema and contains all the chaotic convergences, then we call it *projective*.

[1]Although Sections XXI.2 and XXI.3 also concern structural aspects, they are postponed in order to allow an accumulation of motivating examples.

More precisely, $\mathbf{J}$ is projective whenever for each (non-empty) set X and every non-empty $\Xi \subset \mathbf{J}X$,

$$\bigvee \Xi \in \mathbf{J}X,$$

and $o_X \in \mathbf{J}X$.

For example, the classes of pretopologies and topologies are projective.

PROPOSITION XVII.1.1. *If* $\mathbf{J}$ *is a projective class, then for every convergence* ξ *there exists the finest convergence* $J\xi$ *from* $\mathbf{J}$ *coarser than* ξ*. Moreover, the modifier* J *is a* projector, *that is,*

(increasing) $\quad \zeta \leq \xi \Longrightarrow J\zeta \leq J\xi,$

(descending) $\quad J\xi \leq \xi,$

(idempotent) $\quad J(J\xi) = J\xi.$

Although it would be enough to adapt, *mutatis mutandis,* the proofs of combined Proposition VI.6.1 and Proposition VI.6.2, we shall provide a proof for the reader's convenience.

PROOF. Let $\mathbf{J}$ be a projective class. For any a convergence ξ on a set X, define

$$A_\xi := \{\theta \in \mathbf{J}X : \theta \leq \xi\}, \tag{XVII.1.1}$$

and notice that $A_\xi \neq \varnothing$, because $o_X \in \mathbf{J}X$. Since $\mathbf{J}X$ is closed under suprema, $\bigvee A_\xi \in A_\xi$, hence $J\xi := \bigvee A_\xi$ is the finest element of $\mathbf{J}X$ that is coarser than ξ.

Of course, if $\zeta \leq \xi$ then $A_\zeta \subset A_\xi$, hence $J\zeta = \bigvee A_\zeta \leq \bigvee A_\xi = J\xi$, that is, J is (increasing). Now $J\xi \leq \xi$, because $J\xi \in A_\xi$. As $J\xi \in \mathbf{J}X$, the supremum of convergences from $\mathbf{J}X$ that are coarser than $\mathbf{J}X$ is equal $J\xi$, hence J is (idempotent). □

Analogously, if a subclass $\mathbf{E}$ of convergences is closed under infima and contains all the discrete convergences, then we call it *coprojective.*

More precisely, a class $\mathbf{E}$ is coprojective whenever for each (non-empty) set X and every non-empty $\Xi \subset \mathbf{E}X$,

$$\bigwedge \Xi \in \mathbf{E}X,$$

and $\iota_X \in \mathbf{E}X$.

An example of a coprojective class is that of sequentially founded convergences.

PROPOSITION XVII.1.2. *If* $\mathbf{E}$ *is a coprojective class, then for every convergence* ξ *there exists the coarsest convergence* $E\xi$ *from* $\mathbf{E}$ *finer than* ξ*. Moreover, the so defined modifier* E *is a* coprojector, *that is,*

(increasing) $\zeta \le \xi \Longrightarrow E\zeta \le E\xi,$

(ascending) $E\xi \ge \xi,$

(idempotent) $E(E\xi) = E\xi.$

PROOF. Adapt the preceding proof *mutatis mutandis.* □

PROPOSITION XVII.1.3. *Projective and coprojective classes constitute complete lattices.*

PROOF. Let $\mathbf{J}$ be a reflective class of convergences and X a non-empty set. We need to show that each non-empty subset Ξ of $\mathbf{J}X$ has a supremum and an infimum (with respect to $\mathbf{J}X$). By definition of reflective class, $\bigvee \Xi \in \mathbf{J}X$, hence $\bigvee_{\mathbf{J}} \Xi = \bigvee \Xi$ ([2]).

On the other hand, the infimum $\bigwedge \Xi$ with respect to the set $\mathbf{I}X$ of all convergences always exists, but need not belong to $\mathbf{J}X$. However, $J(\bigwedge \Xi)$, where J is the projector on $\mathbf{J}X$, is the finest convergence in $\mathbf{J}X$ coarser than $\bigwedge \Xi$. Therefore,

$$\bigwedge\nolimits_{\mathbf{J}} \Xi = J(\bigwedge \Xi)$$

is the infimum of Ξ with respect to $\mathbf{J}X$. The variant for coprojective classes is obtained by inverting the order. Of course, if $\varnothing \neq \Xi \subset \mathbf{E}X$, where $\mathbf{E}$ is a coprojective class of convergences, X a non-empty set, and E is the corresponding coprojector, then

$$\bigvee\nolimits_{\mathbf{E}} \Xi = E(\bigvee \Xi).$$

□

Let us observe that the identity modifier I (such that $I\xi := \xi$ for each ξ) is simultaneously a projector and a coprojector.

If G is a descending modifier, then for every non-empty set X, and each $\xi \in \mathbf{I}X$, we define $G^0\xi := \xi$, and for any ordinal number $\alpha > 0$,

$$G^\alpha \xi := G(\bigwedge\nolimits_{\beta<\alpha} G^\beta \xi)$$

is a descending modifier. In particular, $G^1\xi = G(G^0\xi) = G\xi$.

PROPOSITION XVII.1.4. *If* G *is a descending modifier, then for every non-empty set* X*, there exists an ordinal* $\gamma > 0$ *such that* G^γ *is a projector in* $\mathbf{I}X$*.*

---

[2]In other words, suprema with respect to $\mathbf{J}X$ exist, because they coincide with the corresponding suprema with respect to $\mathbf{I}X$, the set of all convergences on X.

PROOF. (*) It follows from the definition that if $\alpha_0 < \alpha_1$ then $G^{\alpha_0}\xi \geq G^{\alpha_1}\xi$ for each $\xi \in \mathbf{I}X$. Hence, by Zermelo Theorem XXVII.4.6, there exists a least ordinal γ, for which $G^{\gamma+1}\xi = G^{\gamma}\xi$ for each $\xi \in \mathbf{I}X$, and thus G^{γ} is an idempotent descending modifier, that is, a projector on $\mathbf{I}X$. □

Similarly, if H is an ascending modifier, then for every non-empty set X, and each $\xi \in \mathbf{I}X$, we define $H^0\xi := \xi$, and $H^{\alpha}\xi = H(\bigvee_{\beta<\alpha} H^{\beta}\xi)$ is an ascending modifier for any ordinal number $\alpha > 0$. An adaptation *mutatis mutandis* of the preceding proof yields

COROLLARY XVII.1.5. *If H is an ascending modifier, then for every non-empty set X, there exists an ordinal $\gamma > 0$ such that G^{γ} is a coprojector in $\mathbf{I}X$.*

PROPOSITION XVII.1.6. *If F is a modifier, then the class*

$$\{F \geq \mathrm{I}\} := \{\xi : F\xi \geq \xi\} \tag{XVII.1.2}$$

is projective, and the class

$$\{\mathrm{I} \geq F\} := \{\xi : \xi \geq F\xi\} \tag{XVII.1.3}$$

is coprojective.

PROOF. As the chaotic convergence o is the coarsest convergence on every given set, $Fo \geq o$. Let Ξ be a set of convergences ξ (on the same underlying set). If (XVII.1.2) for each $\xi \in \Xi$, then $F(\bigvee \Xi) \geq F\xi \geq \xi$ for each $\xi \in \Xi$, hence $F(\bigvee \Xi) \geq \bigvee \Xi$.

As the discrete convergence ι is the finest convergence on every given set, $\iota \geq F\iota$. If (XVII.1.3) for each $\xi \in \Xi$, then $F(\bigwedge \Xi) \leq F\xi \leq \xi$ for each $\xi \in \Xi$, hence $F(\bigwedge \Xi) \leq \bigwedge \Xi$. □

Notice that if F is an ascending modifier, then $\{F \geq \mathrm{I}\}$ is the class of all convergences; similarly, if F is a descending modifier, then $\{\mathrm{I} \geq F\}$ is the class of all convergences.

REMARK XVII.1.7 (*). If F is any concrete map, then $\{F = \mathrm{I}\} \subset \{F \geq \mathrm{I}\}$. If F is a descending modifier, then

$$\{F = \mathrm{I}\} = \{F \geq \mathrm{I}\}.$$

Indeed, let $\xi \in \{F \geq \mathrm{I}\}$, that is, $F\xi \geq \xi$. As F is descending, $\xi \geq F\xi$, hence $F\xi = \xi$. It is clear that for any modifier F, the modifier $F \wedge \mathrm{I}$ is descending, and moreover,

$$\{F \wedge \mathrm{I} = \mathrm{I}\} = \{F \geq \mathrm{I}\}. \tag{XVII.1.4}$$

We can construct the projector on $\{F \geq \mathrm{I}\}$, by iterating the descending modifier $G := F \wedge \mathrm{I}$. By Proposition XVII.1.4, for each non-empty set

X, there exists an ordinal $\gamma > 0$ such that G^γ is a projector in $\mathbf{I}X$. Moreover, $\{G = \mathrm{I}\} = \{F \geq \mathrm{I}\}$ by (XVII.1.4). On the other hand, if $G^\beta\xi = \xi$ for each $0 < \beta < \alpha$, then $G^\alpha\xi = G(\bigwedge_{\beta<\alpha} G^\beta\xi) \geq G\xi \geq \xi$, hence $G^\alpha\xi = \xi$, because G^α is a descending modifier. Consequently, $G^\gamma = (F \wedge \mathrm{I})^\gamma$ is the projector on $\{F = \mathrm{I}\}$ in $\mathbf{I}X$.

Similarly, if F is an ascending modifier, then

$$\{F = \mathrm{I}\} = \{F \leq \mathrm{I}\},$$

and we can construct the corresponding coprojector, by iterating $F \vee \mathrm{I}$.

XVII.2. Functors

We enter here the realm of *category theory*, and hurry to notify those in the know that we define here functors only on objects! In fact, on the level of generality of our considerations, functors in the sense of category theory can be recovered from their values on objects, as explained in Chapter XXVIII.

DEFINITION XVII.2.1. A modifier F is called a *functor* if it preserves continuity, that is,

(functorial) $$\mathscr{C}(\xi,\tau) \subset \mathscr{C}(F\xi, F\tau)$$

for every ξ and τ.

Therefore, a functor is a map $F : \mathbf{I} \longrightarrow \mathbf{I}$ such that $F(\mathbf{I}X) \subset \mathbf{I}X$ for every (non-empty) set X, that is, if $|F\xi| = |\xi|$ for every convergence ξ. In other words, F is (concrete).

Note that (functorial) implies (increasing). Indeed, if $\xi \geq \tau$, then ξ and τ have the same underlying set, say X, and the identity $i_X \in \mathscr{C}(\xi,\tau)$, hence by (functorial), $i_X \in \mathscr{C}(F\xi, F\tau)$, equivalently $F\xi \geq F\tau$.

Sure enough, if F and G are functors, then

$$\mathscr{C}(\xi,\tau) \subset \mathscr{C}(F\xi, F\tau) \subset \mathscr{C}(G(F\xi), G(F\tau)),$$

hence

PROPOSITION XVII.2.2. *Every finite composition of functors is a functor.*

Recall the following characterizations of continuity (IV.3.5):

$$f\xi \geq \tau \iff f \in \mathscr{C}(\xi,\tau) \iff \xi \geq f^-\tau.$$

PROPOSITION XVII.2.3. *A modifier F is a functor if and only if for every f and τ,*

(XVII.2.1) $$F(f^-\tau) \geq f^-(F\tau),$$

if and only if for every f and ξ,

$$f(F\xi) \geq F(f\xi). \tag{XVII.2.2}$$

PROOF. Suppose that (functorial) holds for every ξ and τ. By the definition of initial convergence, $f^-\tau \geq \tau$, hence, by (IV.3.5) and (functorial),

$$f \in \mathscr{C}(f^-\tau, \tau) \subset \mathscr{C}(F(f^-\tau), F\tau).$$

Therefore, by the right-hand equivalence of (IV.3.5), we get (XVII.2.1).

If we set $\tau = f\xi$ in (XVII.2.1), then, by (IV.3.7), we get

$$F\xi \geq F(f^-f\xi) \geq f^-(Ff\xi).$$

By applying f to the inequality above, in view of (IV.3.7), we infer

$$f(F\xi) \geq ff^-(Ff\xi) \geq F(f\xi),$$

which is (XVII.2.2).

Let $f \in \mathscr{C}(\xi, \tau)$, that is, $f\xi \geq \tau$, hence $F(f\xi) \geq F\tau$ in virtue of (increasing), and of (XVII.2.2),

$$f(F\xi) \geq F(f\xi) \geq F\tau,$$

that is, $f \in \mathscr{C}(F\xi, F\tau)$. □

PROPOSITION XVII.2.4. *If $\mathfrak{F}$ is a collection of functors, then*

$$\bigvee \mathfrak{F} := \bigvee_{F\in\mathfrak{F}} F, \qquad \bigwedge \mathfrak{F} := \bigwedge_{F\in\mathfrak{F}} F$$

are functors.

PROOF. It is straightforward that $\bigvee \mathfrak{F}$ and $\bigwedge \mathfrak{F}$ are (concrete) and (increasing) provided that each $F \in \mathfrak{F}$ is (concrete) and (increasing). If $f \in \mathscr{C}(\xi, \tau)$ then $f \in \mathscr{C}(F\xi, F\tau)$, equivalently $f(F\xi) \geq F\tau$ for each functor $F \in \mathfrak{F}$. Then $f(\bigvee_{F\in\mathfrak{F}} F\xi) \geq G\tau$ for each $G \in \mathfrak{F}$, hence $f(\bigvee_{F\in\mathfrak{F}} F\xi) \geq \bigvee_{F\in\mathfrak{F}} F\tau$, that is, $f \in \mathscr{C}(\bigvee_{F\in\mathfrak{F}} F\xi, \bigvee_{F\in\mathfrak{F}} F\tau)$. Similarly, $f \in \mathscr{C}(F\xi, F\tau)$ amounts to $F\xi \geq f^-(F\tau)$, hence, by assumption $G\xi \geq f^-(\bigwedge_{F\in\mathfrak{F}} F\tau)$ for each $G \in \mathfrak{F}$, and thus $\bigwedge_{F\in\mathfrak{F}} F\xi \geq f^-(\bigwedge_{F\in\mathfrak{F}} F\tau)$, that is, $f \in \mathscr{C}(\bigwedge_{F\in\mathfrak{F}} F\xi, \bigwedge_{F\in\mathfrak{F}} F\tau)$. □

PROPOSITION XVII.2.5. *For any functor F,*

$$F(\xi \times \tau) \geq F\xi \times F\tau$$

for each pair ξ and τ of convergences.

PROOF. Indeed, let ξ be a convergence on X and τ a convergence on Y. Then $\xi \times \tau = p_X^-\xi \vee p_Y^-\tau$, and thus by (XVII.2.1), $F(\xi \times \tau) \geq F(p_X^-\xi) \geq p_X^-F\xi$, and $F(\xi \times \tau) \geq F(p_Y^-\tau) \geq p_Y^-F\tau$. Hence, $F(\xi \times \tau) \geq p_X^-F\xi \vee p_Y^-F\tau = F\xi \times F\tau$. □

The converse inequality

$$F\xi \times F\tau \geq F(\xi \times \tau),$$

for each pair of convergences ξ and τ, does not hold for all functors F. If F is such that the inequality above holds, then $F\xi \times F\tau = F(\xi \times \tau)$, and we say that F *commutes with finite* products.

Commutation with finite products of a functor, or its failure, have important consequences for the class of convergences the functor determines.

XVII.3. Reflective and coreflective classes

A class $\mathbf{J}$ of convergences is called *reflective* if it is projective and closed under initial convergences, that is,

$$\tau \in \mathbf{J} \Longrightarrow f^{-}\tau \in \mathbf{J} \tag{XVII.3.1}$$

for each f and τ. Therefore,

PROPOSITION XVII.3.1. *Each reflective class is closed under products.*

PROOF. Let $\mathbf{J}$ be a reflective class. If Ξ is a set convergences on a given set, then by definition $\prod \Xi = \bigvee_{\xi \in \Xi} p_{|\xi|}^{-}\xi$,where $p_{|\xi|}$ is the projection onto $|\xi|$. If $\xi \in \mathbf{J}$ for each $\xi \in \Xi$, then $p_{|\xi|}^{-}\xi \in \mathbf{J}$, because $\mathbf{J}$ is reflective. Hence their supremum is in $\mathbf{J}$, because each reflective class is in particular projective. $\square$

PROPOSITION XVII.3.2. *A projective class* $\mathbf{J}$ *is reflective if and only if the associated projector* J *is a functor.*

PROOF. Let a class $\mathbf{J}$ be reflective, and let $f : X \longrightarrow Y$. If τ is a convergence on Y then $J\tau \in \mathbf{J}$, hence by (XVII.3.1), $f^{-}(J\tau) \in \mathbf{J}$. On the other hand, $\tau \geq J\tau$, because J is a projector, and thus $f^{-}\tau \geq f^{-}(J\tau)$. Therefore, (XVII.2.1) holds

$$J(f^{-}\tau) \geq f^{-}(J\tau),$$

because $J(f^{-}\tau)$ is the finest convergence in $\mathbf{J}$ that is coarser than $f^{-}\tau$.

Conversely, if a projector J is a functor, and if $\tau \in \mathbf{J}Y$, that is, $J\tau \geq \tau$, then $J(f^{-}\tau) \geq f^{-}(J\tau) \geq f^{-}\tau$, hence $f^{-}\tau \in \mathbf{J}X$, proving (XVII.3.1). $\square$

DEFINITION XVII.3.3. A functor is called a *reflector* if it is the projector on a reflective class.

We have already thoroughly studied two reflectors, the *pretopologizer* S_0, and the *topologizer* T.

EXAMPLE XVII.3.4. A convergence θ on a set X is called *characteristic* if $\lim_\theta \mathcal{F} \neq \varnothing$ implies that $\lim_\theta \mathcal{F} = X$. Of course, the chaotic convergence o_X is characteristic. On the other hand, if Θ is a non-empty set of characteristic convergences on X, then $\bigvee \Theta$ is characteristic. Indeed, if $\varnothing \neq \lim_{\bigvee \Theta} \mathcal{F} = \bigcap_{\theta\in\Theta} \lim_\theta \mathcal{F}$, then $\lim_\theta \mathcal{F} \neq \varnothing$, hence $\lim_\theta \mathcal{F} = X$, for every $\theta \in \Theta$, thus $X = \lim_{\bigvee \Theta} \mathcal{F}$. Accordingly, the class of characteristic convergences is projective. Let χ be the corresponding projector. Sure enough, if ξ is a convergence on X, then

$$\lim_{\chi\xi} \mathcal{F} = \begin{cases} X & \lim_\xi \mathcal{F} \neq \varnothing \\ \varnothing & \lim_\xi \mathcal{F} = \varnothing \end{cases}.$$

If

$$\varnothing = \lim_{f^-(\chi\tau)} \mathcal{F} = f^-(\lim_{\chi\tau} f[\mathcal{F}]),$$

then $\lim_\tau f[\mathcal{F}] = \varnothing$, hence $\varnothing = \lim_{f^-\tau} \mathcal{F} = \lim_{\chi_{f^-\tau}} \mathcal{F}$. Therefore

$$\chi_{f^-\tau} \geq f^-\chi_\tau,$$

proving that χ is a functor, hence a reflector. Accordingly, the class of characteristic convergences is reflective.

A class $\mathbf{E}$ is called *coreflective* if it is coprojective and closed under final convergences, that is,

$$\xi \in \mathbf{E} \Longrightarrow f\xi \in \mathbf{E} \tag{XVII.3.2}$$

for each f and ξ.

PROPOSITION XVII.3.5. *A coprojective class* $\mathbf{E}$ *is coreflective if and only if the associated coprojector* E *is a functor.*

PROOF. Let a class $\mathbf{E}$ be reflective, and let $f : X \longrightarrow Y$. If ξ is a convergence on X then $E\xi \in \mathbf{E}$, hence by (XVII.3.2), $f(E\xi) \in \mathbf{E}$. On the other hand, $E\xi \geq \xi$, because E is a coprojector, and thus $f(E\xi) \geq f\xi$. Therefore, (XVII.2.2) holds

$$f(E\xi) \geq E(f\xi),$$

because $E(f\xi)$ is the coarsest convergence in $\mathbf{J}$ that is finer than $f\xi$.

Conversely, if a coprojector E is a functor, and $\xi \in \mathbf{E}X$, that is, $E\xi \leq \xi$, then $E(f\xi) \leq f(E\xi) \leq f\xi$, hence $f\xi \in \mathbf{E}Y$, proving (XVII.3.2). □

DEFINITION XVII.3.6. A functor is called a *coreflector* if it is the coprojector on a coreflective class.

By Lemma III.3.9,

PROPOSITION XVII.3.7. *The class of free (alias T_1) convergences is coprojective.*

The map $\xi \longmapsto \xi^*$ associating with each convergence ξ, the coarsest free convergence finer than ξ, is a coprojector. It can be easily described by

$$x \in \lim_{\xi^*} \mathcal{F} \Longleftrightarrow (x \in \lim_{\xi} \mathcal{F} \wedge \ker \mathcal{F} \subset \{x\}).$$

However, the class of free convergences is not coreflective.

EXAMPLE XVII.3.8. Let X be an infinite countable set, and let $x_\infty \in X$. Consider the prime topology $\pi = \pi(x_\infty, (X)_0)$ and a map $f : X \longrightarrow \{0, 1\}$ defined by $f(x) := 0$ if $x \neq x_\infty$ and $f(x_\infty) := 1$. Of course, π is free, while the final convergence $f\pi$ is not. Indeed, $\lim_{f\pi} 0^\uparrow = \bigcup_{f[\mathcal{F}]\leq 0^\uparrow} f(\lim_\pi \mathcal{F})$. As $f[(X, \{x_\infty\})_0] = 0^\uparrow$ and $x_\infty \in \lim_\pi (X, \{x_\infty\})_0$, we infer that $1 \in \lim_{f\pi} 0^\uparrow$. In other terms, $f\pi = \$_0$ is the upper Sierpiński topology, which is not free.

REMARK XVII.3.9. By Proposition XVII.2.3, the inequalities (XVII.2.1) and (XVII.2.2) are equivalent for every modifier F, and are tantamount to (functorial). Nevertheless, if F is a reflector, then in general $\xi \in \mathrm{fix}F$ does not imply that $f\xi \in \mathrm{fix}F$, for example, in the case of topologies, $\xi \in \mathbf{T}$ does not imply that $f\xi \in \mathbf{T}$. Similarly, if F is a coreflector, then in general $\tau \in \mathrm{fix}F$ does not imply that $f^-\tau \in \mathrm{fix}F$, for example, in the case of sequentially founded convergences, $\tau \in \mathbf{Seq}$ does not imply that $f^-\tau \in \mathbf{Seq}$.

Let us observe that if F and G are functors, then the compositions $G \circ F$ is a functor. Furthermore, if both F and G are reflectors (respectively, coreflectors), then $G \circ F$ is a reflector (respectively, coreflector).

PROPOSITION XVII.3.10. *If $\mathfrak{F}$ consists of reflectors, then $\bigvee \mathfrak{F}$ is a reflector.*

PROOF. By Proposition XVII.2.4, $\bigvee \mathfrak{F}$ is a functor. If $F\xi \leq \xi$ for each $F \in \mathfrak{F}$, then $\bigvee_{F\in\mathfrak{F}} F\xi \leq \xi$. If each $F \in \mathfrak{F}$ is idempotent, that is, $F\xi \leq F(F\xi)$, then

$$\bigvee\nolimits_{F\in\mathfrak{F}} F(\bigvee\nolimits_{H\in\mathfrak{F}} H\xi) \geq \bigvee\nolimits_{F\in\mathfrak{F}} F(F\xi) \geq \bigvee\nolimits_{F\in\mathfrak{F}} F\xi$$

which proves that $\bigvee \mathfrak{F}$ is idempotent, hence a reflector. □

Analogously,

PROPOSITION XVII.3.11. *If $\mathfrak{F}$ consists of coreflectors, then $\bigwedge \mathfrak{F}$ is a coreflector.*

REMARK XVII.3.12. Proposition XVII.1.4 states that successive iterations of a descending modifier in a set of convergences results in a projector. If, moreover, this modifier is a functor, then by Propositions XVII.2.2 and XVII.2.4, the resulting projector is a reflector. Similarly, successive iterations of an ascending functor in a set of convergences results in a coreflector.

EXAMPLE XVII.3.13 (discretization modifier). Let $\mathrm{Dis}\,\xi := \iota_{|\xi|}$, that is, Dis associates with every convergence ξ the discrete convergence on the same underlying set. The modifier Dis is obviously (concrete) and (increasing). As $\mathscr{C}(\mathrm{Dis}\,\xi, \mathrm{Dis}\,\tau) = \mathscr{C}(\iota_{|\xi|}, \iota_{|\tau|}) \supset \mathscr{C}(\xi, \tau)$, it is a functor. In fact, $\mathrm{fix}(\mathrm{Dis})X = \{\iota_X\}$ for every non-empty set X. Therefore, the *discretization modifier* Dis is a coreflector. We notice that $F \leq \mathrm{Dis}$ for each functor F.

Another special case is provided by

EXAMPLE XVII.3.14. The *identity functor* I is defined by $\mathrm{I}\xi := \xi$ for each convergence ξ. As $\mathrm{fix}(\mathrm{I}) = \mathbf{I}$ is the class of all convergences, I is simultaneously reflective and coreflective. An idempotent functor F is reflective whenever $F \leq \mathrm{I}$, and coreflective if $\mathrm{I} \leq F$.

We already know that regular topologies are preserved by products (Proposition IX.1.14) and by subspaces (Proposition IX.1.13). We shall see later (Theorem XIX.3.11) that the class of regular topologies is reflective.

The class of normal topologies is not reflective, because subspaces of normal topologies are not necessarily normal (Example IX.1.28).

Nor is the class of normal topologies coreflective. Indeed, each Hausdorff topology of countable character is an infimum of metrizable, hence normal, topologies. The Hausdorff non-regular topology of Example IX.1.12 is of countable character, but not normal.

XVII.4. Initial and final functors

If ξ and τ are arbitrary convergences, then $\mathscr{C}(\xi, \tau) \neq \varnothing$, because each constant map is continuous ([3]).

Consequently, for any convergences ξ and σ, the convergence

$$\mathrm{R}^{\sigma}\xi := \bigvee\nolimits_{f \in \mathscr{C}(\xi,\sigma)} f^{-}\sigma \tag{XVII.4.1}$$

is well-defined. On recalling that $f \in \mathscr{C}(\xi, \sigma)$ whenever $\xi \geq f^{-}\sigma$, we infer that $\xi \geq \mathrm{R}^{\sigma}\xi$ for each convergence ξ. Therefore,

[3] More explicitly, if $y \in |\tau|$ and $\widehat{y}$ stands for the constant map $\widehat{y}(x) := y$ for each $x \in |\xi|$, then $(\widehat{y})^{-}\tau$ is the chaotic convergence on $|\xi|$, hence $\xi \geq (\widehat{y})^{-}\tau$ for each convergence ξ.

LEMMA XVII.4.1. *Let ξ and σ be convergences. Then $\mathrm{R}^{\sigma}\xi$ is the coarsest convergence θ on $|\xi|$, for which $\mathscr{C}(\theta,\sigma)=\mathscr{C}(\xi,\sigma)$.*

In particular,

$$\mathscr{C}(\mathrm{R}^{\sigma}\xi,\sigma)=\mathscr{C}(\xi,\sigma). \tag{XVII.4.2}$$

By Definition IV.4.7, a convergence ξ is called σ-initial if $\xi=\mathrm{R}^{\sigma}\xi$. A special case of functionally initial topologies was considered in Section X.9.

PROPOSITION XVII.4.2. *For each convergence σ, the modifier R^{σ} is a reflector.*

PROOF. If $\zeta\geq\xi$ then $\mathscr{C}(\xi,\sigma)\subset\mathscr{C}(\zeta,\sigma)$, hence by definition $\mathrm{R}^{\sigma}\zeta\geq\mathrm{R}^{\sigma}\xi$. We have already observed that $\xi\geq\mathrm{R}^{\sigma}\xi$ for each convergence ξ. It follows from (XVII.4.2) that $\mathrm{R}^{\sigma}(\mathrm{R}^{\sigma}\xi)=\mathrm{R}^{\sigma}\xi$. To see that R^{σ} is a functor, take a convergence τ and a map valued in $|\tau|$. Then, by Exercise IV.10.8,

$$\begin{aligned}f^{-}(\mathrm{R}^{\sigma}\tau)=f^{-}(\bigvee\nolimits_{h\in\mathscr{C}(\tau,\sigma)}h^{-}\sigma)&=\bigvee\nolimits_{h\in\mathscr{C}(\tau,\sigma)}(h\circ f)^{-}\sigma\\&\leq\bigvee\nolimits_{g\in\mathscr{C}(f^{-}\tau,\sigma)}g^{-}\sigma=\mathrm{R}^{\sigma}(f^{-}\tau),\end{aligned}$$

because $f\in\mathscr{C}(f^{-}\tau,\tau)$ and thus $h\circ f\in\mathscr{C}(f^{-}\tau,\sigma)$. □

Moreover, $\xi=\mathrm{R}^{\sigma}\xi$ if and only if ξ is an initial convergence with respect to $\mathscr{C}(\xi,\sigma)$, that is, ξ is σ-initial. We call $\mathrm{R}^{\sigma}\xi$ the *σ-initial reflection* of ξ.

In particular, obviously $\sigma=\mathrm{R}^{\sigma}\sigma$, because the identity $i_{|\sigma|}\in\mathscr{C}(\sigma,\sigma)$. In fact, $\mathrm{fix}(\mathrm{R}^{\sigma})$ is the *reflective hull* of σ, that is,

PROPOSITION XVII.4.3. *The class $\mathrm{fix}(\mathrm{R}^{\sigma})$ is the least reflective class of convergences containing σ.*

PROOF. Indeed, let $\mathbf{A}$ be a reflective class such that $\sigma\in\mathbf{A}$. Accordingly, $f^{-}\sigma\in\mathbf{A}$ for each map f valued in $|\sigma|$. In particular, each chaotic topology $o_X\in\mathbf{A}X$ as the initial convergence of σ with respect to constant maps. If now $\Xi\subset\mathbf{A}X$ is such that, for each $\xi\in\Xi$, there exists a map $f_{\xi}:X\longrightarrow|\sigma|$ such that $\xi=f_{\xi}^{-}\sigma$, then

$$\bigvee\Xi=\bigvee\nolimits_{\xi\in\Xi}f_{\xi}^{-}\sigma=\bigvee\nolimits_{\xi\in\Xi}\mathrm{R}^{\sigma}\xi\leq\mathrm{R}^{\sigma}(\bigvee\Xi),$$

so that $\bigvee\Xi\in\mathbf{A}X$. □

By Corollary VII.5.3, the Sierpiński topology \$ is initially dense in the class of topologies, and by Corollary VI.7.5, the Bourdaud pretopology ¥ is initially dense in the class of pretopologies.

More generally, if $\mathbf{D}$ is a class of convergences and ξ is a convergence, then by Proposition XVII.3.10,

$$\mathrm{R}^{\mathbf{D}}\xi := \bigvee\nolimits_{\sigma\in\mathbf{D}} \mathrm{R}^{\sigma}\xi$$

defines a reflector $\mathrm{R}^{\mathbf{D}}$ ([4]).

In virtue of Definition VI.7.4 a convergence ξ is σ-initial whenever $\xi = \mathrm{R}^{\sigma}\xi$.

More generally, by Definition VI.7.4, a class $\mathbf{D}$ is *initially dense* in a class $\mathbf{X}$ whenever $\mathbf{X} \subset \mathrm{fix}(\mathrm{R}^{\mathbf{D}})$. A class $\mathbf{X}$ of convergences is called *simple* if there exists a set $\boldsymbol{D}$ of convergences that is initially dense in $\mathbf{X}$.

In view of Proposition IV.7.7, if a convergence is initial with respect to a set $\boldsymbol{D}$ of convergences, then it is $\prod \boldsymbol{D}$-initial. Therefore,

PROPOSITION XVII.4.4. *A class of convergences is simple if and only if it contains an initially dense convergence.*

This concept enables us to infer certain facts about $\mathrm{fix}(\mathrm{R}^{\mathbf{D}})$ by checking them only for $\mathbf{D}$.

It follows from the definition of initial density that

COROLLARY XVII.4.5. *If $\mathbf{D}$ is a class of convergences, then $\mathbf{D}$ is initially dense in $\mathrm{fix}(\mathrm{R}^{\mathbf{D}})$, equivalently $\mathrm{fix}(\mathrm{R}^{\mathbf{D}})$ is the reflective hull of $\mathbf{D}$.*

PROPOSITION XVII.4.6. *If $\mathbf{D} \subset \mathbf{S}$ and $\mathbf{D}$ is initially dense in $\mathbf{S}$, then $\mathrm{R}^{\mathbf{S}} = \mathrm{R}^{\mathbf{D}}$.*

PROOF. It is clear that $\mathrm{R}^{\mathbf{S}}\xi \geq \mathrm{R}^{\mathbf{D}}\xi$ for each convergence ξ. On the other hand, $\mathrm{R}^{\mathbf{S}}\xi = \bigvee_{\sigma\in\mathbf{S}} \bigvee_{f\in\mathscr{C}(\xi,\sigma)} f^{-}\sigma$, and $\sigma = \bigvee_{\delta\in\mathbf{D}} \bigvee_{g\in\mathscr{C}(\sigma,\delta)} g^{-}\delta$ for each $\sigma \in \mathbf{S}$. Therefore, by Exercise IV.10.8,

$$\begin{aligned}\mathrm{R}^{\mathbf{S}}\xi &= \bigvee\nolimits_{\sigma\in\mathbf{S}} \bigvee\nolimits_{f\in\mathscr{C}(\xi,\sigma)} \bigvee\nolimits_{\delta\in\mathbf{D}} \bigvee\nolimits_{g\in\mathscr{C}(\sigma,\delta)} (g\circ f)^{-}\delta \\ &\leq \bigvee\nolimits_{\sigma\in\mathbf{S}} \bigvee\nolimits_{\delta\in\mathbf{D}} \bigvee\nolimits_{h\in\mathscr{C}(\xi,\delta)} h^{-}\delta = \bigvee\nolimits_{\sigma\in\mathbf{S}} \mathrm{R}^{\mathbf{D}}\xi = \mathrm{R}^{\mathbf{D}}\xi.\end{aligned}$$

□

If J is a modifier, then we abridge

$$\mathrm{R}^{J} := \mathrm{R}^{\mathrm{fix}J}.$$

PROPOSITION XVII.4.7. *If J is a reflector, then $\mathrm{R}^{J} = J$.*

[4]Although the supremum above is formulated in terms of a class of convergences, it is, in fact, a supremum over a set of convergences. Indeed, given a convergence ξ on a set X, if $f \in \mathscr{C}(\xi,\sigma)$ then $f^{-}\sigma$ is a convergence on X, for which $f^{-}\sigma = i_{X}^{-}(f^{-}\sigma)$.

PROOF. Indeed, by definition,

$$\mathrm{R}^J\xi = \bigvee\nolimits_{\sigma=J\sigma} \bigvee\nolimits_{f\in\mathscr{C}(\xi,\sigma)} f^-\sigma. \tag{XVII.4.3}$$

Since J is reflective, $f^-\sigma \in \mathrm{fix}J$ for any map f provided that $\sigma \in \mathrm{fix}J$. As any supremum of elements of fix$J$ is in fix$J$, we infer that $J(\mathrm{R}^J\xi) = \mathrm{R}^J\xi$. Applying J to $\mathrm{R}^J\xi \leq \xi$ and using the preceding equality, we get $\mathrm{R}^J\xi \leq J\xi$. On the other hand, setting $\sigma = J\xi$, and taking the identity $i \in \mathscr{C}(\xi, J\xi)$, (XVII.4.3) yields $\mathrm{R}^J\xi \geq J\xi$. □

PROPOSITION XVII.4.8. *Let J be a reflector, and let $\mathbf{D}$ be initially dense in* fix(J). *The following formulas are equivalent:*

(1) $\theta \geq J\xi$,
(2) $\mathscr{C}(\xi,\sigma) \subset \mathscr{C}(\theta,\sigma)$ *for each* $\sigma = J\sigma$,
(3) $\mathscr{C}(\xi,\sigma) \subset \mathscr{C}(\theta,\sigma)$ *for each* $\sigma \in \mathrm{D}$.

PROOF. (1) implies (2). Let $\sigma = J\sigma$. As J is a reflector,

$$\mathscr{C}(\xi,\sigma) \subset \mathscr{C}(J\xi, J\sigma) = \mathscr{C}(J\xi,\sigma) \subset \mathscr{C}(\theta,\sigma)$$

by (IV.3.4), because of (1). It is obvious that (2) implies (3). If $\sigma \in \mathbf{D}$ and $f \in \mathscr{C}(\xi,\sigma)$, then by virtue of (3), $\theta \geq f^-\sigma$. Hence

$$\theta \geq \bigvee\nolimits_{\sigma\in \boldsymbol{D}} \bigvee\nolimits_{f\in\mathscr{C}(\xi,\sigma)} f^-\sigma = \mathrm{R}^{\mathbf{D}}\xi = J\xi.$$

□

Analogously, for any two convergences ξ and τ,

$$\mathrm{C}^\xi\tau := \bigwedge\nolimits_{f\in\mathscr{C}(\xi,\tau)} f\xi$$

represents the finest convergence θ on $|\tau|$ such that $f\xi \geq \theta$ for each $f \in \mathscr{C}(\xi,\tau)$. Consequently, for each non-empty class $\mathbf{W}$ of convergences, the convergence

$$\mathrm{C}^{\mathbf{W}}\tau := \bigwedge\nolimits_{\xi\in\mathbf{W}} \mathrm{C}^\xi\tau$$

is the finest among θ for which $\mathscr{C}(\xi,\tau) = \mathscr{C}(\xi,\theta)$.

Similarly, $\mathrm{C}^{\mathbf{W}}$ is a coreflector for any class $\mathbf{W}$ of convergences. Moreover, fix$(\mathrm{C}^{\mathbf{W}})$ is the least coreflective class of convergences including $\mathbf{W}$, that is, fix$(\mathrm{C}^{\mathbf{W}})$ is the *coreflective hull* of $\mathbf{W}$.

A class $\mathbf{W}$ of convergences is called *finally dense* in a class $\mathbf{G}$ whenever, for each $\tau \in \mathbf{G}$, there is a set Ξ of $\mathbf{W}$ and a collection of maps $f_\xi \in \mathscr{C}(\xi,\tau)$ such that

$$\tau = \bigwedge\nolimits_{\xi\in\Xi} f_\xi\xi. \tag{XVII.4.4}$$

Accordingly,

PROPOSITION XVII.4.9. *Let* $\mathbf{W}$ *and* $\mathbf{G}$ *be classes of convergences. Then* $\mathbf{W}$ *is finally dense in* $\mathbf{G}$ *if and only if* $\mathbf{G} \subset \mathrm{fix}(\mathrm{C}^{\mathbf{W}})$.

For example, the class of pretopologies $\mathbf{S}_0$ is *finally dense* in the class $\mathbf{F}_0$ of *finitely stable* convergences, that is, each $\mathbf{F}_0$-convergence is final with respect to a set of pretopologies.

Corollary VI.5.4 ascertains that each finitely stable convergence is a final convergence with respect to pretopologies. In other terms,

COROLLARY XVII.4.10. *The class of pretopologies is finally dense in the class of finitely stable convergences.*

Roughly speaking, each free sequence is finally dense in the class of sequentially founded convergences. Let us give this statement a precise meaning.

EXAMPLE XVII.4.11 (Prime cofinite topology). Like in Example III.4.3, for any infinite countable set N, the cofinite filter $(N)_0$ is a sequential free filter. The *prime cofinite topology* σ on $\overline{N} := N \sqcup \{\infty\}$ is defined by setting $\mathcal{N}_\sigma(\infty) := (N)_0 \wedge \infty^\uparrow$. A map $\varphi \in X^{\overline{N}}$ is called an *extended sequence* on X. If ξ is a convergence on X, then $\varphi \in \mathscr{C}(\sigma, \xi)$ if and only if $\varphi(\infty) \in \lim_\xi \varphi[(N)_0]$. If, moreover, ξ is sequentially founded, then for each $x \in \lim_\xi \mathcal{F}$, there exists an extended sequence φ such that $x = \varphi(\infty)$ and $\mathcal{F} \geq \varphi[(N)_0]$.

Therefore,

PROPOSITION XVII.4.12. *The prime cofinite topology on an infinite countable set is finally dense in the class of sequentially founded convergences.*

If E is a modifier, then we abridge

$$\mathrm{C}^E := \mathrm{C}^{\mathrm{fix}E}.$$

PROPOSITION XVII.4.13. *If* E *is a coreflector, then* $\mathrm{C}^E = E$.

PROOF. By definition, $\mathrm{C}^E\tau = \bigwedge_{\zeta=E\zeta} \bigwedge_{f\in\mathscr{C}(\zeta,\tau)} f\zeta$. If $\zeta \in \mathrm{fix}E$, then by assumption, $f\zeta \in \mathrm{fix}E$, and as $\mathrm{fix}E$ is stable under all infima, we infer that $E(\mathrm{C}^E\tau) = \mathrm{C}^E\tau$. Therefore, $\mathrm{C}^E\tau \geq E\tau$. By setting $\zeta = E\tau$ and taking $i \in \mathscr{C}(E\tau, \tau)$, we conclude that $\mathrm{C}^E\tau \leq E\tau$. □

PROPOSITION XVII.4.14. *If* $\mathbf{W} \subset \mathbf{G}$ *and* $\mathbf{W}$ *is finally dense in* $\mathbf{G}$, *then* $\mathrm{C}^{\mathbf{W}} = \mathrm{C}^{\mathbf{G}}$.

PROOF. It is clear that $\mathrm{C}^{\mathbf{W}} \geq \mathrm{C}^{\mathbf{G}}$. Now $\mathrm{C}^{\mathbf{G}}\tau = \bigwedge_{\zeta \in \mathbf{G}} \bigwedge_{f \in \mathscr{C}(\zeta,\tau)} f\zeta$, and $\zeta = \bigwedge_{\xi \in \mathbf{W}} \bigwedge_{g \in \mathscr{C}(\xi,\zeta)} g\xi$ for every $\zeta \in \mathbf{G}$. Therefore, by Exercise IV.10.8,

$$\mathrm{C}^{\mathbf{G}}\tau = \bigwedge_{\zeta \in \mathbf{G}} \bigwedge_{f \in \mathscr{C}(\zeta,\tau)} \bigwedge_{\xi \in \mathbf{W}} \bigwedge_{g \in \mathscr{C}(\xi,\zeta)} f(g\xi) \geq \bigwedge_{\xi \in \mathbf{W}} \bigwedge_{h \in \mathscr{C}(\xi,\tau)} h\xi = \mathrm{C}^{\mathbf{W}}\tau.$$

□

PROPOSITION XVII.4.15. *Let E be a coreflector, and let $\mathbf{W}$ be finally dense in* fix(E). *The following formulas are equivalent:*

(1) $E\sigma \geq \tau$,
(2) $\mathscr{C}(\xi,\sigma) \subset \mathscr{C}(\xi,\tau)$ *for each* $\xi = E\xi$,
(3) $\mathscr{C}(\xi,\sigma) \subset \mathscr{C}(\xi,\tau)$ *for each* $\xi \in \mathbf{W}$.

PROOF. If $E\sigma \geq \tau$ and $f\xi \geq \sigma$, then $f(E\xi) \geq E(f\xi) \geq E\sigma \geq \tau$, and if $\xi = E\xi$, then $f \in \mathscr{C}(\xi,\tau)$. As $\mathbf{W}$ is finally dense in fix(E), (2) implies (3). If $f \in \mathscr{C}(\xi,\sigma)$ and $\xi \in \mathbf{W}$, then by (3), $f\xi \geq \tau$ for each $\xi \in \mathbf{W}$. Hence, by Proposition XVII.4.13,

$$E\sigma = \mathrm{C}^{\mathbf{W}}\sigma = \bigwedge_{\xi \in \mathbf{W}} \bigwedge_{f \in \mathscr{C}(\xi,\sigma)} f\xi \geq \tau.$$

□

XVII.5. Supplement

REMARK XVII.5.1 (Weak topologies). If ξ and θ are convergences on X and Z respectively, and $\mathscr{F} \subset \mathscr{C}(\xi,\theta)$, then there exists the coarsest convergence $\sigma_{\mathscr{F}}(\xi,\theta)$ on X such that $f(x) \in \lim_\theta f[\mathcal{F}]$ for every $f \in \mathscr{F}$. Of course, $\xi \geq \sigma_{\mathscr{F}}(\xi,\theta)$, and it may happen that the inequality is strict (for example, see Theorem X.8.5). In other words, $\sigma_{\mathscr{F}}(\xi,\theta)$ is the $\mathscr{F}$-initial reflection of ξ.

$\mathscr{F}$-initial convergences are studied in functional analysis, where θ is a topology on the set of scalars, ξ is a vector topology, and $\mathscr{F}$ is a subset of $\mathscr{C}(\xi,\theta)$ consisting of linear maps. Then $\sigma_{\mathscr{F}}(\xi,\theta)$ is the $\mathscr{F}$*-weak topology* on X.

We proved in Section III.7 that a map I_0 associating with any $\theta \subset \mathbb{F}X \times X$, the preconvergence $\mathrm{I}_0\theta$, is concrete, increasing, descending, and idempotent. Therefore, I_0 is a generalized projector. Similarly, the map I associating with each $\theta \subset \mathbb{F}X \times X$ the convergence $\mathrm{I}\theta$, is a generalized projector. We shall see that they generalize reflectors. Indeed, if $\theta \subset \mathbb{F}X \times X$ and $\tau \subset \mathbb{F}Y \times Y$, then

(XVII.5.1) $$\mathscr{C}(\theta,\tau) \subset \mathscr{C}(\mathrm{I}_\circ\theta, \mathrm{I}_\circ\tau) \subset \mathscr{C}(\mathrm{I}\theta, \mathrm{I}\tau).$$

Indeed, if $f \in \mathscr{C}(\theta, \tau)$ and $x \in \lim_{\mathrm{I}_\circ \theta} \mathcal{F}$, then there exists a filter $\mathcal{E} \leq \mathcal{F}$ such that $x \in \lim_\theta \mathcal{E}$. It follows that $f(x) \in \lim_\tau f[\mathcal{E}]$. Moreover, $f[\mathcal{E}] \leq f[\mathcal{F}]$, so that $f(x) \in \lim_{\mathrm{I}_\circ \tau} f[\mathcal{F}]$. To prove the second inclusion, if $x \in \lim_{\mathrm{I}\theta} \mathcal{F}$, then there is $\mathcal{E} \in \mathbb{F}X$ such that $x \in \lim_{\mathrm{I}_\circ \theta} \mathcal{E}$ and $\mathcal{F} \geq \mathcal{E}$ or $\mathcal{F} = x^\uparrow$. Hence $f(x) \in \lim_{\mathrm{I}_\circ \tau} f[\mathcal{E}] \subset \lim_{\mathrm{I}\theta} f[\mathcal{F}]$ and $f(x) \in \lim_{\mathrm{I}\theta} f(x)^\uparrow$.

CHAPTER XVIII

Fundamental classes

In this chapter, we classify convergences in several manners according to how filters determine limits in one way or another.

Stability indicates the cardinality of a set of filters convergent to a point, for which the intersection also converges to that point. We have already encountered *finitely stable convergences*, and *pretopologies*, which are stable for arbitrary sets of filters.

The *paving number* is the least cardinality of pavements; it is equal to 1 in the case of pretopologies.

A *foundation* of a convergence is a class of filters such that every point has a pavement consisting of filters from that class. In Chapter XVI we studied *sequentially founded convergences*. The classes of principal, countably based, countably complete, and countably carried filters are other examples of foundations.

A convergence is *adherence-determined* by a class of filters if the limit of every filter is the intersection of the adherences of the filters from the class meshing with that filter. For example, pretopologies are determined by the adherences of principal filters. We shall study other important classes, like *pseudotopologies* (determined by the class of all filters), *paratopologies* (countably based filters), and *hypotopologies* (countably complete filters).

XVIII.1. Stability and pavements

By definition, a convergence is non-pretopological if it has at least one point without a coarsest convergent filter, like in Figure XVIII.1.1.

For a convergence ξ on a set X and $x \in X$, consider the following implication

$$\boldsymbol{A} \subset \xi^{-}(x) \Longrightarrow (\bigwedge \boldsymbol{A}) \in \xi^{-}(x). \tag{XVIII.1.1}$$

We have already considered convergences that are *finitely stable at* x, that is, when (XVIII.1.1) holds for each finite collection $\boldsymbol{A}$ of filters.

A convergence is called *stable at* x if (XVIII.1.1) holds for each $\boldsymbol{A}$. Therefore, a convergence is stable at x if and only if it is pretopological at x (Recall Proposition VI.1.2).

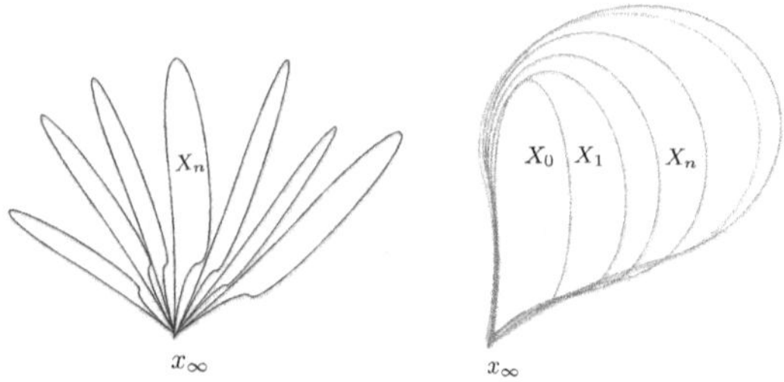

FIGURE XVIII.1.1. The fan represents a pavement of a convergence that is not finitely stable. The second figure illustrates a finitely stable, non-pretopological convergence. A sequence $(X_n)_n$ of infinite countable sets, X_n (almost) included in X_{n+1} for each $n < \omega$, and such that $X_{n+1} \setminus X_n$ is infinite for each $n < \omega$. Let $x_\infty \in \lim \mathcal{F}$ if there exists $n < \omega$ such that $\mathcal{F}$ is finer than the cofinite filter of X_n centered at x_∞.

We shall also consider convergences that are *countably stable at* x, that is, such that (XVIII.1.1) holds for each countable collection $\boldsymbol{A}$.

EXAMPLE XVIII.1.1 (Countably stable convergence which is not stable). The *standard sequential convergence* σ of the real line is defined by $r \in \lim_\sigma \mathcal{F}$ if there exists a sequential filter $\mathcal{E} \leq \mathcal{F}$ so that $\mathcal{E}$ converges to r in the *standard convergence.* The convergence σ is not a pretopology, because by Proposition II.3.7, $\mathcal{N}_\sigma(r) = \mathcal{N}_\nu(r)$, which is not sequential.

However, σ is countably stable. In fact, if $\{\mathcal{F}_n : n < \omega\} \subset \sigma^-(r)$, then there exist a deceasing sequence of sequential filters $\{\mathcal{E}_n : n < \omega\} \subset \sigma^-(r)$ such that $\bigwedge_{k\leq n} \mathcal{F}_k \geq \mathcal{E}_n \geq \mathcal{N}_\nu(r)$. By Lemma XVI.4.6, there exists a sequential filter $\mathcal{E}$ such that $\bigwedge_{n<\omega} \mathcal{E}_n \geq \mathcal{E} \geq \mathcal{N}_\nu(r)$.

More generally, a convergence ξ is said to be κ*-stable at* x if (XVIII.1.1) holds for each collection $\boldsymbol{A}$ such that card $\boldsymbol{A} \leq \kappa$. A convergence is called κ*-stable if it is* κ-stable at each point.

We denote by $\mathbf{P}_0$ the class of *finitely stable* convergences, and by $\mathbf{P}_1$ the class of *countably stable* convergences. Of course, $\mathbf{S}_0 \subset \mathbf{P}_1 \subset \mathbf{P}_0$, where $\mathbf{S}_0$ stands for the class of pretopologies.

PROPOSITION XVIII.1.2. *Let* κ *be a cardinal number. If* Ξ *is a set of* κ*-stable convergences (on a common set), then* $\bigvee \Xi$ *is* κ*-stable.*

PROOF. If $\boldsymbol{F}$ is a collection of filters such that card$\boldsymbol{F} \leq \kappa$ and $x \in \lim_\xi \mathcal{F}$ for each $\mathcal{F} \in \boldsymbol{F}$ and every $\xi \in \Xi$, then $x \in \lim_\xi(\bigwedge \boldsymbol{F})$ for each ξ, hence $x \in \lim_{\bigvee \Xi}(\bigwedge \boldsymbol{F})$. $\square$

As the chaotic convergence on any set is stable, hence κ-stable for each cardinal κ, we infer that

COROLLARY XVIII.1.3. *For each convergence θ, there exists a finest κ-stable convergence that is coarser than θ.*

PROPOSITION XVIII.1.4. *Let us denote by $F_\kappa\theta$ the finest κ-stable convergence that is coarser than θ. Then (for every ζ, ξ, τ and f)*

(increasing) $$\zeta \leq \xi \Longrightarrow F_\kappa\zeta \leq F_\kappa\xi,$$

(descending) $$F_\kappa\xi \leq \xi,$$

(idempotent) $$F_\kappa(F_\kappa\xi) = F_\kappa\xi,$$

(XVIII.1.2) $$F_\kappa(f^-\tau) = f^-(F_\kappa\tau).$$

PROOF. Indeed, let $x \in \lim_{F_\kappa(f^-\tau)} \mathcal{G}$, then there is a collection $\boldsymbol{F}$ of filters such that card $\boldsymbol{F} \leq \kappa, \mathcal{G} \geq \bigwedge \boldsymbol{F}$, and $x \in \lim_{f^-\tau} \mathcal{F}$, equivalently, $f(x) \in \lim_\tau f[\mathcal{F}]$, for each $\mathcal{F} \in \boldsymbol{F}$. It follows that $f[\mathcal{G}] \geq f[\mathcal{F}]$ for each $\mathcal{F} \in \boldsymbol{F}$, hence $f(x) \in \lim_{F_\kappa\tau} f[\mathcal{G}]$, that is, $x \in \lim_{f^-(F_\kappa\tau)} \mathcal{G}$. Conversely, if $x \in \lim_{f^-(F_\kappa\tau)} \mathcal{G}$, then there exists a collection $\boldsymbol{H}$ of filters, such that card $\boldsymbol{H} \leq \kappa, f[\mathcal{G}] \geq \mathcal{H}$, and $f(x) \in \lim_\tau \mathcal{H} \subset \lim_\tau(\mathcal{H} \vee f(X))$, and thus $x \in f^-(\lim_\tau f[f^-[\mathcal{H}]]) = \lim_{f^-\tau} f^-[\mathcal{H}]$ for each $\mathcal{H} \in \boldsymbol{H}$. As $\mathcal{G} \geq f^-[f[\mathcal{G}]] \geq f^-[\mathcal{H}]$ for each $\mathcal{H} \in \boldsymbol{H}$, we infer that $x \in \lim_{F_\kappa(f^-\tau)} \mathcal{G}$. $\square$

Of course, if $\lambda \geq \kappa$, then $F_\lambda \leq F_\kappa$. It follows from Proposition XVIII.1.4 that

COROLLARY XVIII.1.5. *The class of κ-stable convergences is reflective.*

Actually, (XVIII.1.2) is more than needed for the projector F_κ to be a reflector: $F_\kappa(f^-\tau) \geq f^-(F_\kappa\tau)$. The equality (XVIII.1.2) indicates that the reflector F_κ is hereditary (Definition XVIII.5.10 to come).

By Definition III.2.2, a collection of filters $\boldsymbol{D} \subset \xi^-(x)$ is called a *pavement* of ξ at x if for each filter $\mathcal{F} \in \xi^-(x)$, there is $\mathcal{D} \in \boldsymbol{D}$ such that $\mathcal{F} \geq \mathcal{D}$.

DEFINITION XVIII.1.6. A collection of filters $\boldsymbol{D} \subset \xi^-(x)$ is called a *pseudopavement* of ξ at x if for each ultafilter $\mathcal{U}$ such that $x \in \lim_\xi \mathcal{U}$, there exists $\mathcal{D} \in \boldsymbol{D}$ such that $\mathcal{U} \geq \mathcal{D}$.

PROPOSITION XVIII.1.7. *For $\boldsymbol{D} \subset \xi^-(x)$, the following statements are equivalent:*

(1) *$\boldsymbol{D}$ is a pseudopavement,*
(2) *for each filter $\mathcal{F}$ such that $x \in \operatorname{adh}_\xi \mathcal{F}$, there exists $\mathcal{D} \in \boldsymbol{D}$ such that $\mathcal{F} \# \mathcal{D}$,*

(3) *for each filter $\mathcal{F}$ such that $x \in \lim_\xi \mathcal{F}$, there exists $\mathcal{D} \in \boldsymbol{D}$ such that $\mathcal{F}\#\mathcal{D}$.*

PROOF. (1) $\Rightarrow$ (2). If $x \in \mathrm{adh}_\xi \mathcal{F}$, then there exists $\mathcal{U} \in \beta\mathcal{F}$ such that $x \in \lim_\xi \mathcal{U}$, thus by (1), there exists $\mathcal{D} \in \boldsymbol{D}$ such that $\mathcal{U} \geq \mathcal{D}$, hence $\mathcal{F}\#\mathcal{D}$. (2) $\Rightarrow$ (3). Obvious. (3) $\Rightarrow$ (1). Because if $\mathcal{U}$ is an ultrafilter such that $\mathcal{U}\#\mathcal{D}$, then $\mathcal{U} \geq \mathcal{D}$. □

Of course, each pavement is a pseudopavement. Let κ be a cardinal number.

REMARK XVIII.1.8. If $\boldsymbol{D}$ is a pavement of ξ at x and for every $\mathcal{D} \in \boldsymbol{D}$ there is $P_\mathcal{D} \in \mathcal{D}$, then $\{P_\mathcal{D} : \mathcal{D} \in \boldsymbol{D}\}$ is a ξ-cover of $\{x\}$ (Definition VIII.2.1). Indeed, if $x \in \lim_\xi \mathcal{F}$, then there exists $\mathcal{D} \in \boldsymbol{D}$ such that $\mathcal{F} \geq \mathcal{D}$, hence $P_\mathcal{D} \in \mathcal{F}$. Similarly, if $\boldsymbol{D}$ is a pseudopavement of ξ at x and for every $\mathcal{D} \in \boldsymbol{D}$ there is $P_\mathcal{D} \in \mathcal{D}$, then $\{P_\mathcal{D} : \mathcal{D} \in \boldsymbol{D}\}$ is a ξ-pseudocover of $\{x\}$ (Definition VIII.2.11).

DEFINITION XVIII.1.9. We say that a convergence ξ is *κ-paved at x* if there is a pavement at x of cardinality κ; a convergence ξ is *κ-pseudopaved at x* if there is a pseudopavement at x of cardinality κ.

Each κ-paved convergence is κ-pseudopaved. In particular, a convergence is *finitely paved at x* if there exists a finite pavement at x, *countably paved at x* if there exists a countable pavement at x.

The *paving number* $\mathfrak{p}(\xi, x)$ is the least cardinality of a pavement. The *paving number* $\mathfrak{p}(\xi)$ of a convergence ξ is defined as

$$\mathfrak{p}(\xi) := \sup\{\mathfrak{p}(\xi, x) : x \in |\xi|\}.$$

The *pseudopaving number* $\mathfrak{p}_\#(\xi, x)$ is the least cardinality of a pseudopavement. The *pseudopaving number* $\mathfrak{p}_\#(\xi)$ of a convergence ξ is defined as

$$\mathfrak{p}_\#(\xi) := \sup\{\mathfrak{p}_\#(\xi, x) : x \in |\xi|\}.$$

Since each pavement is a pseudopavement, $\mathfrak{p}_\#(\xi, x) \leq \mathfrak{p}(\xi, x)$. An instance of the strict inequality will be given in Example XXV.7.4.

It is clear that

PROPOSITION XVIII.1.10. *If a convergence is λ-stable at x and κ-paved at x, and $\lambda \geq \kappa$, then it is pretopological at x.*

PROOF. Indeed, if $\boldsymbol{D}$ is a pavement of ξ at x and $\mathrm{card}\, \boldsymbol{D} = \kappa$ and ξ is κ-stable at x, then by λ-stability, $\{\bigwedge \boldsymbol{D}\}$ is a pavement of ξ at x, that is, the vicinity filter $\mathcal{V}_\xi(x) = \bigwedge \boldsymbol{D}$. □

We recover Corollary III.2.4 to the effect that each finitely stable finitely paved convergence is pretopological. Similarly,

COROLLARY XVIII.1.11. *If a countably stable convergence is countably paved, then it is a pretopology.*

EXAMPLE XVIII.1.12 (Countably stable convergence that is not countably paved). The *standard sequential convergence* σ from Example XVIII.1.1 is not countably paved at any point, because by Proposition XVIII.1.10, it is countably stable and non-pretopological at each point.

EXAMPLE XVIII.1.13 (Countably stable convergence that is not countably paved). Let X be an uncountable set, and let $x_\infty \in X$. We define a convergence ζ by declaring isolated each $x \neq x_\infty$, and $x_\infty \in \lim_\zeta \mathcal{F}$ whenever $\ker \mathcal{F} \subset \{x_\infty\}$, and there is a countable set $F \in \mathcal{F}$.

The convergence ζ is countably stable, for if $\{\mathcal{F}_n : n < \omega\} \subset \zeta^-(x_\infty)$ then $\ker(\bigwedge_{n<\omega} \mathcal{F}_n) \subset \zeta^-(x_\infty)$, and for any $F_n \in \mathcal{F}_n$, the set $\bigcup_{n<\omega} F_n \in (\bigwedge_{n<\omega} \mathcal{F}_n)$.

On the other hand, $\mathcal{V}_\zeta(x_\infty)$ does not contain countable sets, and thus ζ is not a pretopology.

EXAMPLE XVIII.1.14 (Countably paved, finitely stable convergence that is not countably stable). Let $\{X_n : n < \omega\}$ be a set of infinite countable sets, and x_∞ such that $x_\infty \in X_n \cap X_k$ for every $n, k \in \mathbb{N}$. Define a convergence η on $X := \bigcup_{n<\omega} X_n$ by declaring isolated each $x \neq x_\infty$, and $x_\infty \in \lim_\eta \mathcal{F}$ whenever $\ker \mathcal{F} \subset \{x_\infty\}$ and there is m such that $\bigcup_{n\leq m} X_n \in \mathcal{F}^*$. Then $\{x_\infty^\uparrow \wedge (\bigcup_{n\leq m} X_n)_0 : n < \omega\}$ is a countable pavement of η at x_∞. On the other hand, it is clear that finite infima of filters convergent to a given point, converge to that point, but

$$x_\infty \notin \lim_\eta \bigwedge_{n<\omega} (X_n)_0,$$

so that τ is not countably stable. We notice that $\{(X_n \setminus \{x_\infty\}, x_\infty)_0 : n < \omega\} \cup \{x_\infty^\uparrow\}$ is a minimal pavement.

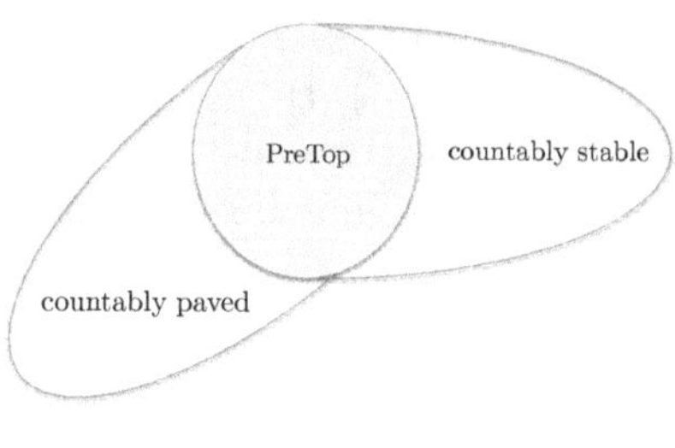

FIGURE XVIII.1.2. A convergence is a pretopology if and only if it is countably paved and countably stable.

XVIII.2. Foundations

We shall classify convergences in terms of types of filters forming pavements.

DEFINITION XVIII.2.1. Let $\mathbb{H}$ be a class of filters. A convergence ξ on X is said to be $\mathbb{H}$-*founded* provided that $x \in \lim_\xi \mathcal{F}$ implies the existence of $\mathcal{H} \in \mathbb{H}X$ such that $\mathcal{H} \leq \mathcal{F}$ and $x \in \lim_\xi \mathcal{H}$ ([1]).

In such case we say that $\mathbb{H}$ is a *foundation* of ξ.

We denote by $\mathbf{B}_{\mathbb{H}}$ the class of $\mathbb{H}$-founded convergences. Clearly,

$$\mathbb{H}_0 \subset \mathbb{H}_1 \Longrightarrow \mathbf{B}_{\mathbb{H}_0} \subset \mathbf{B}_{\mathbb{H}_1}.$$

In particular, every convergence is $\mathbb{F}$-founded, while a convergence is principal if and only if it is $\mathbb{F}_0$-founded. On the other hand, if a convergence is free (alias T_1), then it is $(\mathbb{F}_* \cup \mathbb{U}_\bullet)$-founded, where $\mathbb{U}_\bullet$ stands for the class of principal ultrafilters.

Accordingly, a convergence ξ is $\mathbb{S}$-*founded* (or *sequentially founded*) if $x \in \lim_\xi \mathcal{F}$ implies that there exists a sequential filter $\mathcal{E}$ (that is, $\mathcal{E} \in \mathbb{S}$) such that $\mathcal{E} \leq \mathcal{F}$ and $x \in \lim_\xi \mathcal{E}$.

EXAMPLE XVIII.2.2. A convergence is called *countably carried if it is* $\mathbb{E}$-*founded* convergences, where $\mathbb{E}$ stands for the class of *countably carried* filters, that is, the filters that contain a countable set (Definition II.2.15).

Of course, each $\mathbb{S}$-founded convergence is $\mathbb{E}$-founded, because every filter finer than a sequential filter is countably carried. In fact, a filter is countably carried if and only if it is finer than a sequential filter. However, this does not mean that each $\mathbb{E}$-founded convergence is $\mathbb{S}$-founded (see Example II.2.18).

In Example III.7.12 we defined, on a set X, a characteristic convergence $\chi_{\boldsymbol{F}}$ of an isotone set $\boldsymbol{F}$ of filters containing principal ultrafilters to the effect that $X = \lim_{\chi_{\boldsymbol{F}}} \mathcal{F}$ if $\mathcal{F} \in \boldsymbol{F}$ and $\varnothing = \lim_{\chi_{\boldsymbol{F}}} \mathcal{F}$ if $\mathcal{F} \notin \boldsymbol{F}$.

The class $\mathbb{E}$ of countably carried filters is isotone and, moreover, contains all principal ultrafilters ([2]). Hence $\chi_{\mathbb{E}X}$ is a characteristic convergence on X. Therefore, a convergence ξ on X is countably carried if and only if $\xi \geq \chi_{\mathbb{E}X}$. It follows that for any convergence θ on X,

$$\mathrm{B}_{\mathbb{E}}\theta = \theta \vee \chi_{\mathbb{E}X}.$$

PROPOSITION XVIII.2.3. *If a class* $\mathbb{H}$ *of filters contains all principal ultrafilters, then the class of* $\mathbb{H}$*-founded convergences is coprojective.*

[1] In other words, if $\mathbb{H} \cap \xi^-(x)$ is a pavement of ξ at x.

[2] Actually, $\mathbb{E}$ contains all principal filters of countable sets.

PROOF. As all principal ultrafilters belong to $\mathbb{H}$, the discrete convergence ι_X on X is $\mathbb{H}$-founded. On the other hand, if Ξ is a non-empty set of $\mathbb{H}$-founded convergences on a set X, then

$$x \in \lim_{\bigwedge \Xi} \mathcal{F} = \bigcup_{\xi \in \Xi} \lim_\xi \mathcal{F}$$

implies that there exists $\xi \in \Xi$ such that $x \in \lim_\xi \mathcal{F}$, hence there exists $\mathcal{H} \in \mathbb{H}$ such that $\mathcal{H} \leq \mathcal{F}$ and $x \in \lim_\xi \mathcal{H} \subset \lim_{\bigwedge \Xi} \mathcal{H}$, proving that $\bigwedge \Xi$ is $\mathbb{H}$-founded. □

As a result, for every convergence, there exists the coarsest among finer $\mathbb{H}$-founded convergences.

DEFINITION XVIII.2.4. A class $\mathbb{H}$ of filters is called *final* if ([3])

$$\mathcal{H} \in \mathbb{H} \Longrightarrow f[\mathcal{H}] \in \mathbb{H}. \tag{XVIII.2.1}$$

PROPOSITION XVIII.2.5. *If $\mathbb{H}$ is a final class containing all principal ultrafilters, then the class of $\mathbb{H}$-founded convergences is coreflective.*

PROOF. By Proposition XVIII.2.3, the class of $\mathbb{H}$-founded convergences is coprojective. Let $f : X \longrightarrow Y$, and let $\xi \in \mathbf{B}_{\mathbb{H}}X$. Let $y \in \lim_{f\xi} \mathcal{F}$, that is, either $\mathcal{F} = y^{\uparrow}$, or there exist a filter $\mathcal{D}$ and an element $x \in \lim_\xi \mathcal{D} \cap f^-(y) \neq \varnothing$ and $f[\mathcal{D}] \leq \mathcal{F}$. By assumption, $y^{\uparrow} \in \mathbb{H}Y$. On the other hand, since $\xi \in \mathbf{B}_{\mathbb{H}}X$, there exists $\mathcal{H} \in \mathbb{H}X$ such that $x \in \lim_\xi \mathcal{H}$ and $\mathcal{H} \leq \mathcal{D}$. Consequently, $y = f(x) \in \lim_{f\xi} f[\mathcal{H}]$, and $f[\mathcal{H}] \leq f[\mathcal{D}] \leq \mathcal{F}$. As by assumption $f[\mathcal{H}] \in \mathbb{H}Y$, the class of $\mathbb{H}$-founded convergences is coreflective. □

We denote by $\mathrm{B}_{\mathbb{H}}$ the *coreflector* on the class of $\mathbb{H}$-founded convergences. It is straightforward that

$$\lim_{\mathrm{B}_{\mathbb{H}}\xi} \mathcal{F} = \bigcup_{\mathbb{H} \ni \mathcal{H} \leq \mathcal{F}} \lim_\xi \mathcal{H}.$$

Among basic classes of convergences founded by a class of filters are already studied in Section XVI.3 *sequentially founded*, of *finite character* also called *finitely generated*, of *countable character*, *countably stable*, and *countably carried.*

Later on, we shall also consider convergences founded in a class that depends on convergences, for example, *locally compact* convergences.

DEFINITION XVIII.2.6. Let $\mathbb{H}$ and $\mathbb{L}$ be classes of filters. Then $\mathbb{H}$ is said to be $\mathbb{L}$-*composable* if, for each $\mathcal{H} \in \mathbb{H}X$ and every $\mathcal{L} \in \mathbb{L}(X \times Y)$,

$$\mathcal{L}[\mathcal{H}] \in \mathbb{H}Y.$$

In particular, we say that $\mathbb{H}$ is *composable* or *self-composable* if it is $\mathbb{H}$-composable.

[3]Of course, this means that if $f : X \longrightarrow Y$ and $\mathcal{H} \in \mathbb{H}X$, then $f[\mathcal{H}] \in \mathbb{H}Y$.

Foundation	Convergence class		Alternative name
$\mathbb{F}_0$	of finite character	$\mathbf{I}_0$	finitely generated
$\mathbb{F}_1$	of countable character	$\mathbf{I}_1$	first countable
$\mathbb{F}_{\wedge 1}$	countably complete	$\mathbf{I}_{\wedge 1}$	P-space
$\mathbb{S}$	sequentially founded	**Seq**	
$\mathbb{E}$	countably carried	**E**	

TABLE XVIII.2.1. Some basic foundations.

If $\mathbb{F}_0 \subset \mathbb{H} \subset \mathbb{F}$ and $\mathbb{H}$ is self-composable, then it is obviously $\mathbb{F}_0$-composable. On the other hand, $\mathbb{H}$ is $\mathbb{F}_0$-composable if and only if $R[\mathcal{H}] \in \mathbb{H}$ for every relation R and each $\mathcal{H} \in \mathbb{H}$. Hence,

COROLLARY XVIII.2.7. *Each $\mathbb{F}_0$-composable class of filters is initial and final.*

It is straightforward that the classes $\mathbb{F}_0, \mathbb{F}_1, \mathbb{F}_{\wedge 1}$, and $\mathbb{E}$ are composable.

PROPOSITION XVIII.2.8. *Each composable class of filters including $\mathbb{F}_0$ is closed under finite suprema and finite products, in extenso,*

(1) $\mathcal{F}, \mathcal{G} \in \mathbb{H}X \Longrightarrow \mathcal{F} \vee \mathcal{G} \in \mathbb{H}X$,
(2) $(\mathcal{F} \in \mathbb{H}X) \wedge (\mathcal{G} \in \mathbb{H}Y) \Longrightarrow \mathcal{F} \times \mathcal{G} \in \mathbb{H}(X \times Y)$.

PROOF. (1) Let $d : X \longrightarrow X \times X$ be the *diagonal map* given by $d(x) := (x, x)$. If $\mathcal{F} \in \mathbb{H}X$, then $d[\mathcal{F}] \in \mathbb{H}(X \times X)$, and thus if $\mathcal{G} \in \mathbb{H}X$, then $\mathcal{F} \vee \mathcal{G} = d[\mathcal{F}][\mathcal{G}] \in \mathbb{H}X$. (2) If $\mathcal{F} \in \mathbb{H}X$ and $\mathcal{G} \in \mathbb{H}Y$, then $\mathcal{F} \times \mathcal{G} = p_X^-[\mathcal{F}] \vee p_Y^-[\mathcal{G}] \in \mathbb{H}(X \times Y)$. □

COROLLARY XVIII.2.9. *If $\mathbb{H}$ includes $\mathbb{F}_0$ and is composable, then $\mathrm{B}_{\mathbb{H}}$ commutes with finite products.*

PROOF. Indeed, we need show that for each pair ξ and τ of convergences,

$$\mathrm{B}_{\mathbb{H}}\xi \times \mathrm{B}_{\mathbb{H}}\tau \geq \mathrm{B}_{\mathbb{H}}(\xi \times \tau).$$

If $(x, y) \in \lim_{\mathrm{B}_{\mathbb{H}}\xi \times \mathrm{B}_{\mathbb{H}}\tau} \mathcal{F}$ then there exist $\mathcal{H}_0, \mathcal{H}_1 \in \mathbb{H}$ such that $x \in \lim_\xi \mathcal{H}_0$ and $y \in \lim_\tau \mathcal{H}_1$ and $\mathcal{H}_0 \leq p_X[\mathcal{F}]$ and $\mathcal{H}_1 \leq p_Y[\mathcal{F}]$. Hence $\mathcal{F} \geq \mathcal{H}_0 \times \mathcal{H}_1 \in \mathbb{H}$. As by Proposition XVII.2.5 the inverse inequality holds for every functor,

$$\mathrm{B}_{\mathbb{H}}\xi \times \mathrm{B}_{\mathbb{H}}\tau = \mathrm{B}_{\mathbb{H}}(\xi \times \tau).$$

□

Let us review in more detail the second class, that of convergences of countable character. It includes the class of sequentially founded convergences, and that of metrizable topologies.

As we have already said, a convergence ξ is of *countable character* if $x \in \lim_\xi \mathcal{F}$ implies the existence of $\mathcal{D} \in \mathbb{F}_1$ such that $\mathcal{D} \leq \mathcal{F}$ and $x \in \lim_\xi \mathcal{D}$.

The class of convergences of countable character is denoted by $\mathbf{I}_1$. We have concluded that $\mathbf{I}_1$ is a coreflective class, and the associated coreflector is denoted by I_1. In particular, the quotient of an $\mathbf{I}_1$-convergence is an $\mathbf{I}_1$-convergence.

By Propositiopn V.5.12,

LEMMA XVIII.2.10. *The class $\mathbb{F}_1$ of countably based filters is closed under countable suprema.*

COROLLARY XVIII.2.11. *Each countable supremum of convergences of countable character is of countable character.*

PROOF. Let Ξ be a countable set of $\mathbf{I}_1$-convergences, and let $x \in \lim_{\bigvee \Xi} \mathcal{F}$, that is, $x \in \lim_\xi \mathcal{F}$ for each $\xi \in \Xi$. Accordingly, for each $\xi \in \Xi$ there exists $\mathcal{I}_\xi \in \mathbb{F}_1$ such that $\mathcal{I}_\xi \leq \mathcal{F}$ and $x \in \lim_\xi \mathcal{I}_\xi$. By Lemma XVIII.2.10, $\mathcal{I} := \bigvee_{\xi \in \Xi} \mathcal{I}_\xi \in \mathbb{F}_1$, and $x \in \lim_\xi \mathcal{I}$ for each $\xi \in \Xi$, hence $x \in \lim_{\bigvee \Xi} \mathcal{I}$. As $\mathcal{I} \leq \mathcal{F}$, we conclude that $\bigvee \Xi \in \mathbf{I}_1$. $\square$

PROPOSITION XVIII.2.12. *The initial convergence of a convergence of countable character is of countable character.*

PROOF. If τ is a convergence of countable character, then $f^- \tau$ is. Indeed, if $x \in \lim_{f^- \tau} \mathcal{F}$, that is, $f(x) \in \lim_\tau f[\mathcal{F}]$. By assumption, there is $\mathcal{G} \in \mathbb{F}_1$ such that $f(x) \in \lim_\tau \mathcal{G}$ and $\mathcal{G} \leq f[\mathcal{F}]$. Therefore, $f^-[\mathcal{G}] \leq f^-[f[\mathcal{F}]] \leq \mathcal{F}$, and $f^-[\mathcal{G}] \in \mathbb{F}_1$. $\square$

COROLLARY XVIII.2.13. *Each countable product of convergences of countable character is of countable character.*

Let us extend our considerations to foundations depending on convergence. We shall consider now a map $\mathbb{H}(\cdot)$ associating with each convergence ξ on a set X, a collection $\mathbb{H}(\xi)$ of filters on X that includes $\{x^\uparrow : x \in X\}$. Then we define a modifier $\mathrm{B}_\mathbb{H}$ to the effect that

$$\lim_{\mathrm{B}_\mathbb{H} \xi} \mathcal{F} = \bigcup_{\mathbb{H}\xi \ni \mathcal{G} \leq \mathcal{F}} \lim_\xi \mathcal{G}. \tag{XVIII.2.2}$$

DEFINITION XVIII.2.14 (Generalized foundation). We shall call a map $\mathbb{H}(\cdot)$ a *generalized foundation* if for all convergences ζ and ξ and for each map f,

(antitone) $$\zeta \leq \xi \Longrightarrow \mathbb{H}\zeta \supset \mathbb{H}\xi,$$

(idempotent) $$\mathbb{H}(\mathrm{B}_{\mathbb{H}}\xi) = \mathbb{H}\xi,$$

(final) $$\mathcal{H} \in \mathbb{H}\xi \Longrightarrow f[\mathcal{H}] \in \mathbb{H}(f\xi).$$

Mind that here the term *antitone* refers to a map $\mathbb{H}(\cdot)$, and not to its particular value $\mathbb{H}(\xi)$ at ξ.

PROPOSITION XVIII.2.15. *For each convergence ξ, let $\mathbb{H}(\xi)$ be a collection of filters on $|\xi|$ containing all principal ultrafilters.*

(1) *If $\mathbb{H}$ is (antitone), then $\mathrm{B}_{\mathbb{H}}$ is an ascending modifier.*
(2) *If $\mathbb{H}$ is (antitone) and (idempotent), then $\mathrm{B}_{\mathbb{H}}$ is a coprojector.*
(3) *If $\mathbb{H}$ is (antitone) and (final), then $\mathrm{B}_{\mathbb{H}}$ is a functor.*

PROOF. (1) If $x \in \lim_{\mathrm{B}_{\mathbb{H}}\xi} \mathcal{F}$, that is, there exists a filter $\mathcal{G} \leq \mathcal{F}$ such that $\mathcal{G} \in \mathbb{H}\xi \subset \mathbb{H}\zeta$ and $x \in \lim_{\xi} \mathcal{G} \subset \lim_{\zeta} \mathcal{G}$, hence $x \in \lim_{\mathrm{B}_{\mathbb{H}}\zeta} \mathcal{F}$, proving that $\mathrm{B}_{\mathbb{H}}$ is a modifier. As $\lim_{\xi} \mathcal{G} \subset \lim_{\xi} \mathcal{F}$ for $\mathcal{G} \leq \mathcal{F}$, we infer that $\lim_{\mathrm{B}_{\mathbb{H}}\xi} \mathcal{F} \subset \lim_{\xi} \mathcal{F}$ for each $\mathcal{F}$ and ξ, proving that $\mathrm{B}_{\mathbb{H}}$ is ascending.

(2) It suffices to prove that $\mathrm{B}_{\mathbb{H}}$ is idempotent, equivalently, $\lim_{\mathrm{B}_{\mathbb{H}}\xi} \mathcal{F} \subset \lim_{\mathrm{B}_{\mathbb{H}}(\mathrm{B}_{\mathbb{H}}\xi)} \mathcal{F}$. If $x \in \lim_{\mathrm{B}_{\mathbb{H}}\xi} \mathcal{F}$, then there exists $\mathcal{H} \in \mathbb{H}\xi$ such that $\mathcal{H} \leq \mathcal{F}$ and $x \in \lim_{\xi} \mathcal{H}$. As $\mathbb{H}(\mathrm{B}_{\mathbb{H}}\xi) = \mathbb{H}\xi$, we infer that $x \in \lim_{\mathrm{B}_{\mathbb{H}}(\mathrm{B}_{\mathbb{H}}\xi)} \mathcal{F}$.

(3) We only need to show that $f(\mathrm{B}_{\mathbb{H}}\xi) \geq \mathrm{B}_{\mathbb{H}}(f\xi)$ for every map f and each convergence ξ on its domain. If $y \in \lim_{f(\mathrm{B}_{\mathbb{H}}\xi)} \mathcal{G}$, then there exists $x \in f^{-}(y)$ and a filter $\mathcal{F}$ such that $x \in \lim_{\mathrm{B}_{\mathbb{H}}\xi} \mathcal{F}$ and $f[\mathcal{F}] \leq \mathcal{G}$. Therefore, there exists a filter $\mathcal{H} \in \mathbb{H}\xi$ such that $\mathcal{H} \leq \mathcal{F}$ and $x \in \lim_{\xi} \mathcal{H}$. Then $f[\mathcal{H}] \leq \mathcal{G}$ and $f[\mathcal{H}] \in \mathbb{H}(f\xi)$, because $\mathbb{H}$ is (final). On the other hand, $y = f(x) \in \lim_{f\xi} f[\mathcal{H}]$, so that $y \in \lim_{\mathrm{B}_{\mathbb{H}}(f\xi)} \mathcal{G}$. □

COROLLARY XVIII.2.16. *If $\mathbb{H}$ is a generalized foundation, then $\mathrm{B}_{\mathbb{H}}$ is a coreflector.*

XVIII.3. Pseudotopologies

Each convergence determines an adherence of filters, that is, if ξ is a convergence on X and $\mathcal{H}$ is a filter on X, then by Lemma VIII.1.2,

$$\mathrm{adh}_{\xi} \mathcal{H} = \bigcup_{\mathcal{U} \in \beta\mathcal{H}} \lim_{\xi} \mathcal{U} \tag{XVIII.3.1}$$

is the ξ-adherence of $\mathcal{H}$. In particular, for every convergence ξ,

$$\mathrm{adh}_{\xi} \mathcal{U} = \lim_{\xi} \mathcal{U} \tag{XVIII.3.2}$$

for every ultrafilter $\mathcal{U}$ on X.

It turns out that among all the convergences θ such that $\mathrm{adh}_\theta = \mathrm{adh}_\xi$, there exists the coarsest one. Let X be a non-empty set. Let $\mathrm{ah} : \mathbb{F}X \longrightarrow 2^X$ be such that, for each x and every filter $\mathcal{H}$,

$$x \in \mathrm{ah}\, x^\uparrow, \tag{XVIII.3.3}$$

$$\mathrm{ah}\, \mathcal{H} = \bigcup\nolimits_{\mathcal{U} \in \beta\mathcal{H}} \mathrm{ah}\, \mathcal{U}. \tag{XVIII.3.4}$$

Then $\mathrm{ah}\, 2^X = \varnothing$, and $\mathrm{ah}\, \mathcal{H} \subset \mathrm{ah}\, \mathcal{G}$ provided that $\mathcal{G} \leq \mathcal{H}$.

LEMMA XVIII.3.1. *If (XVIII.3.3) and (XVIII.3.4) hold, then*

$$\lim\nolimits_\eta \mathcal{F} = \bigcap\nolimits_{\mathcal{U} \in \beta\mathcal{F}} \mathrm{ah}\, \mathcal{U}$$

is a convergence such that $\mathrm{adh}_\eta \mathcal{H} = \mathrm{ah}\, \mathcal{H}$ for every filter $\mathcal{H}$.

PROOF. If $\mathcal{F}_0 \leq \mathcal{F}_1$, then $\beta\mathcal{F}_0 \supset \beta\mathcal{F}_1$, so that $\lim_\eta \mathcal{F}_0 \subset \lim_\eta \mathcal{F}_1$. Moreover, $\lim_\eta \mathcal{U} = \mathrm{ah}\, \mathcal{U}$ for every $\mathcal{U} \in \beta X$, hence $\lim_\eta x^\uparrow = \mathrm{ah}\, x^\uparrow \ni x$. We conclude that $\lim_\eta$ represents a convergence. Let now $\mathcal{H} \in \mathbb{F}X$. Then, by (XVIII.3.4),

$$\mathrm{adh}_\eta \mathcal{H} = \bigcup\nolimits_{\mathcal{U} \in \beta\mathcal{H}} \lim\nolimits_\eta \mathcal{U} = \bigcup\nolimits_{\mathcal{U} \in \beta\mathcal{H}} \mathrm{ah}\, \mathcal{U} = \mathrm{ah}\, \mathcal{H}.$$

□

DEFINITION XVIII.3.2. A convergence ξ is called a *pseudotopology* if

$$\lim\nolimits_\xi \mathcal{F} = \bigcap\nolimits_{\mathbb{F} \ni \mathcal{H} \# \mathcal{F}} \mathrm{adh}_\xi \mathcal{H}. \tag{XVIII.3.5}$$

By Lemma XVIII.3.1, for every convergence ξ, there exists a pseudotopology η such that $\mathrm{adh}_\eta = \mathrm{adh}_\xi$. It easily follows from (XVIII.3.5) that

$$\lim\nolimits_\xi \mathcal{F} = \bigcap\nolimits_{\mathcal{U} \in \beta\mathcal{F}} \lim\nolimits_\xi \mathcal{U}. \tag{XVIII.3.6}$$

Indeed, as $\mathcal{U} \in \beta\mathcal{F}$ implies $\mathcal{U} \# \mathcal{F}$, we infer that $\bigcap_{\mathbb{F} \ni \mathcal{H} \# \mathcal{F}} \mathrm{adh}_\xi \mathcal{H} \subset \bigcap_{\mathcal{U} \in \beta\mathcal{F}} \lim_\xi \mathcal{U}$. On the other hand, if $x \notin \mathrm{adh}_\xi \mathcal{H}$ for $\mathcal{H} \# \mathcal{F}$, then in particular $x \notin \mathrm{adh}_\xi \mathcal{U}$ for any $\mathcal{U} \in \beta(\mathcal{H} \vee \mathcal{F})$ by (XVIII.3.1).

PROPOSITION XVIII.3.3. *For each convergence ξ and every filter $\mathcal{F}$, let*

$$\lim\nolimits_{\mathrm{S}\xi} \mathcal{F} := \bigcap\nolimits_{\mathcal{U} \in \beta\mathcal{F}} \lim\nolimits_\xi \mathcal{U}.$$

Then S *is a reflector.*

PROOF. By Lemma XVIII.3.1, if we set $\mathrm{ah}\, \mathcal{U} = \mathrm{adh}_\xi \mathcal{U}$ for every $\mathcal{U} \in \beta\mathcal{F}$, then $\mathrm{S}\xi = \eta$ in virtue of (XVIII.3.2). We infer that $\xi \geq \mathrm{S}\xi$, because $\lim_\xi \mathcal{F} \subset \lim_\xi \mathcal{U}$ for each $\mathcal{U} \in \beta\mathcal{F}$. It follows from (XVIII.3.6) that ξ is a pseudotopology if and only if $\mathrm{S}\xi = \xi$. It is evident that $\zeta \geq \xi$ implies $\mathrm{S}\zeta \geq \mathrm{S}\xi$. To see that S is idempotent, notice that, by definition, $\lim_{\mathrm{S}\xi} \mathcal{U} = \lim_{\xi} \mathcal{U}$ for every ultrafilter.

Finally, if τ is a pseudotopology on Y, and $f : X \longrightarrow Y$, then $f^{-}\tau$ is a pseudotopology. Indeed, let $x \in \lim_{f^{-}\tau} \mathcal{U}$, that is, $f(x) \in \lim_{\tau} f[\mathcal{U}]$, for every $\mathcal{U} \in \beta\mathcal{F}$. If now $\mathcal{W} \in \beta f[\mathcal{F}]$, then $\mathcal{F}\# f^{-}[\mathcal{W}]$ and thus there is $\mathcal{U} \in \beta(\mathcal{F} \vee f^{-}[\mathcal{W}])$. It follows that $f[\mathcal{U}] \geq f[f^{-}[\mathcal{W}]] \geq \mathcal{W}$, hence $f[\mathcal{U}] = \mathcal{W}$ because $f[\mathcal{U}]$ is an ultrafilter. As we assume that τ is a pseudotopology, this means that $f(x) \in \lim_{\tau} f[\mathcal{F}]$, equivalently $x \in \lim_{f^{-}\tau} \mathcal{F}$. $\square$

Of course, for each convergence ξ and each filter $\mathcal{F}$,

(XVIII.3.7) $$\lim_{\mathrm{S}\xi} \mathcal{F} = \bigcap_{\mathbb{F} \ni \mathcal{H} \# \mathcal{F}} \operatorname{adh}_{\xi} \mathcal{H},$$

and, for each filter $\mathcal{H}$,

(XVIII.3.8) $$\operatorname{adh}_{\mathrm{S}\xi} \mathcal{H} = \operatorname{adh}_{\xi} \mathcal{H},$$

and thus $\lim_{\mathrm{S}\xi} \mathcal{U} = \lim_{\xi} \mathcal{U}$ for each ultrafilter $\mathcal{U}$. In other words, the convergence of a filter for a pseudotopology is determined by the convergence of (finer) ultrafilters.

It follows from Section XVIII.5 that the *class of pseudotopologies* **S** is reflective, and its reflector S, called the *pseudotopologizer*, fulfills additionally, for each f and each τ,

(XVIII.3.9) $$\mathrm{S}(f^{-}\tau) = f^{-}(\mathrm{S}\tau).$$

PROOF. Suppose that $x \in \lim_{f^{-}(\mathrm{S}\tau)} \mathcal{F}$, that is, $f(x) \in \lim_{\mathrm{S}\tau} f[\mathcal{F}] \subset \operatorname{adh}_{\tau} \mathcal{W}$ for every $\mathcal{W} \in \beta f[\mathcal{F}]$. We need to show that $x \in \lim_{\mathrm{S}(f^{-}\tau)} \mathcal{F}$. If $\mathcal{U} \in \beta\mathcal{F}$, then $f[\mathcal{U}] \in \beta f[\mathcal{F}]$. Therefore, $f(x) \in \operatorname{adh}_{\tau} f[\mathcal{U}]$, equivalently $x \in \operatorname{adh}_{f^{-}\tau} \mathcal{U}$ by (VIII.1.10), as needed. $\square$

LEMMA XVIII.3.4. *If ξ is a convergence, τ is a pseudotopology, and*

$$f(\lim_{\xi} \mathcal{U}) \subset \lim_{\tau} f[\mathcal{U}]$$

for every ultrafilter $\mathcal{U}$, then $f \in \mathscr{C}(\xi, \tau)$.

PROOF. Obviously, continuity implies the condition. Let $x \in |\xi|$ and $x \in \lim_{\xi} \mathcal{F}$, and let $\mathcal{W} \in \beta(f[\mathcal{F}])$ be such that $f(x) \in \lim_{\tau} \mathcal{W}$. Then $f^{-}[\mathcal{W}]\#\mathcal{F}$, and thus there exists $\mathcal{U} \in \beta(f^{-}[\mathcal{W}] \vee \mathcal{F})$, so that $x \in \lim_{\xi} \mathcal{U}$ and $\mathcal{U} \geq f^{-}[\mathcal{W}]$. Therefore, $f(x) \in \lim_{\tau} f[\mathcal{U}]$ and $f[\mathcal{U}]\#\mathcal{W}$, that is, $f[\mathcal{U}] = \mathcal{W}$, because $f[\mathcal{U}]$ is an ultrafilter by Proposition V.4.5. As τ is a pseudotopology, and the argument holds for each $\mathcal{W} \in \beta(f[\mathcal{F}])$, we infer that $f(x) \in \lim_{\tau} f[\mathcal{F}]$. $\square$

EXAMPLE XVIII.3.5 (Missing-filters pseudotopology). Let X be an infinite set, and for every $x \in X$, let $\boldsymbol{H}_x$ be a set of free filters on X. Let $x \in \lim_\xi \mathcal{F}$ whenever $\ker \mathcal{F} \subset \{x\}$ and $\mathcal{F} \perp \mathcal{H}$ for each $\mathcal{H} \in \boldsymbol{H}_x$. By definition, $x \notin \operatorname{adh}_\xi \mathcal{H}$ for each $\mathcal{H} \in \boldsymbol{H}_x$. The convergence ξ is a free pseudotopology. Indeed, if $x \notin \lim_\xi \mathcal{F}$, then either $\ker \mathcal{F} \setminus \{x\} \neq \varnothing$ or there is $\mathcal{H} \in \boldsymbol{H}_x$ such that $\mathcal{F} \# \mathcal{H}$ since $x \notin \operatorname{adh}_\xi \mathcal{H}$, so that (XVIII.3.5) holds.

Therefore, the convergence ξ from Example VI.1.7 is a pseudotopology.

An example of a non-pretopological pseudotopology is provided by the familiar standard sequential convergence of the real line.

EXAMPLE XVIII.3.6 (A non-pretopological pseudotopology: standard sequential convergence). The *standard sequential convergence* σ of the real line is defined by $r \in \lim_\sigma \mathcal{F}$ if there exists a sequential filter $\mathcal{E} \leq \mathcal{F}$ so that $\mathcal{E}$ converges to r in the *standard topology*. In Example XVIII.1.1, we have seen that σ is not a pretopology.

The convergence σ is a pseudotopology. Indeed, it is enough to consider a free filter $\mathcal{F}$ such that $r \in \lim_\sigma \mathcal{U}$ for each $\mathcal{U} \in \beta\mathcal{F}$. Accordingly, for each $\mathcal{U} \in \beta\mathcal{F}$ there is an infinite countable set $E_\mathcal{U} \in \mathcal{U}$, hence $\mathcal{E}_\mathcal{U} \subset \mathcal{U}$ and $r \in \lim_\sigma \mathcal{E}_\mathcal{U}$, where $\mathcal{E}_\mathcal{U}$ is the cofinite filter of $E_\mathcal{U}$. By Proposition V.4.6, there is a finite subset $\boldsymbol{F}$ of $\beta\mathcal{F}$ such that $\bigcup_{\mathcal{U} \in \boldsymbol{F}} E_\mathcal{U} \in \mathcal{F}$, and thus $\mathcal{E} := \bigwedge_{\mathcal{U} \in \boldsymbol{F}} \mathcal{E}_\mathcal{U}$ is the cofinite filter of $\bigcup_{\mathcal{U} \in \boldsymbol{F}} E_\mathcal{U}$, hence $\mathcal{E} \subset \mathcal{F}$ ([4]) and $r \in \lim_\sigma \mathcal{E} \subset \lim_\sigma \mathcal{F}$.

Here is an example of non-pseudotopological convergence.

EXAMPLE XVIII.3.7 (A Hausdorff, finitely stable, non-pseudotopological convergence). Let X be an infinite set and $x_\infty \in X$. Consider a prime convergence ζ such that $x_\infty \in \lim_\zeta \mathcal{F}$ whenever there is a finite subset $\boldsymbol{F}$ of $\beta^* X \cup \{x_\infty^\uparrow\}$ such that $\mathcal{F} = \bigwedge \boldsymbol{F}$. The convergence ζ is finitely stable, but not pseudotopological, because the cofinite filter $(X)_0$ is the infimum of all free ultrafilters on X, that is,

$$(X)_0 = \bigcap\nolimits_{\mathcal{U} \in \beta^* X} \mathcal{U}$$

does not ζ-converge to x_∞, contrary to (XVIII.3.6). We conclude that the pseudotopologization $\mathrm{S}\zeta$ of ζ is the *prime cofinite convergence* on X at x_∞ from Example III.4.3. As $\mathrm{S}\zeta$ is a topology, we infer that $\mathrm{S}\zeta = \mathrm{T}\zeta$.

Of course, as for every reflective class, the class of pseudotopologies is closed under suprema. Moreover,

[4]Indeed, every free filter $\mathcal{F}$ is finer than the cofinite filter of any $F_0 \in \mathcal{F}$.

PROPOSITION XVIII.3.8. *If Ξ is a set of convergences, then*

$$\mathrm{S}(\bigvee \Xi) = \bigvee_{\xi \in \Xi} \mathrm{S}\xi.$$

PROOF. By definition, for every filter $\mathcal{F}$,

$$\begin{aligned}\lim_{\mathrm{S}(\bigvee \Xi)} \mathcal{F} &= \bigcap_{\mathcal{U} \in \beta \mathcal{F}} \lim_{\bigvee \Xi} \mathcal{U} = \bigcap_{\mathcal{U} \in \beta \mathcal{F}} \bigcap_{\xi \in \Xi} \lim_{\xi} \mathcal{U} \\ &= \bigcap_{\xi \in \Xi} \bigcap_{\mathcal{U} \in \beta \mathcal{F}} \lim_{\xi} \mathcal{U} = \bigcap_{\xi \in \Xi} \lim_{\mathrm{S}\xi} \mathcal{U} = \lim_{\bigvee_{\xi \in \Xi} \mathrm{S}\xi} \mathcal{F}.\end{aligned}$$

□

A straightforward corollary of this fact is of great importance.

THEOREM XVIII.3.9. *The pseudotopologizer commutes with each product, that is, for each set of convergences Ξ,*

$$\mathrm{S}(\prod \Xi) = \prod_{\xi \in \Xi} \mathrm{S}\xi.$$

PROOF. In virtue of (IV.6.1), $\prod \Xi = \bigvee_{\xi \in \Xi} p_\xi^- \xi$, where $p_\xi : \prod_{\zeta \in \Xi} |\zeta| \to |\xi|$ is the ξ-projection. Therefore, by Proposition XVIII.3.8 and (XVIII.3.9),

$$\mathrm{S}(\prod \Xi) = \mathrm{S}(\bigvee_{\xi \in \Xi} p_\xi^- \xi) = \bigvee_{\xi \in \Xi} \mathrm{S}(p_\xi^- \xi) = \bigvee_{\xi \in \Xi} p_\xi^- (\mathrm{S}\xi) = \prod_{\xi \in \Xi} \mathrm{S}\xi.$$

□

The pseudotopological infimum of a set Ξ of pseudotopologies fulfills

$$\bigwedge_{\mathrm{S}} \Xi = \mathrm{S}(\bigwedge \Xi).$$

PROPOSITION XVIII.3.10. *If Ξ is a set of pseudotopologies on a set X, then $x \in \lim_{\bigwedge_{\mathrm{S}} \Xi} \mathcal{F}$ whenever*

$$\beta_X \mathcal{F} \subset \beta X \cap \bigcup_{\xi \in \Xi} \xi^-(x). \tag{XVIII.3.10}$$

PROOF. Indeed,

$$\lim_{\mathrm{S}(\bigwedge \Xi)} \mathcal{F} = \bigcap_{\mathcal{U} \in \beta \mathcal{F}} \bigcup_{\xi \in \Xi} \lim_{\xi} \mathcal{U},$$

that is, $x \in \lim_{\bigwedge_{\mathrm{S}} \Xi} \mathcal{F}$ whenever for each $\mathcal{U} \in \beta_X \mathcal{F}$ there is $\xi \in \Xi$ such that $x \in \lim_{\xi} \mathcal{U}$, equivalently $\mathcal{U} \in \xi^-(x)$, that is, (XVIII.3.10). □

XVIII.4. *Initial density, exponential hulls

In virtue of Corollary VI.7.5, the *Bourdaud pretopology* is initially dense in the class of pretopologies. By Corollary VII.5.3, the *Sierpiński topology* is initially dense in the class of topologies. We will see that no pseudotopology is *initially dense* in the class of pseudotopologies.

A class of convergences is said to be *simple* if there exists a convergence that is initially dense in this class. Consequently, the classes of topologies and of pretopologies are simple. We will prove that the classes of all convergences and of pseudotopologies are not simple.

LEMMA XVIII.4.1. *Let X, Y be sets such that X is infinite and $\operatorname{card} X > \operatorname{card} Y$. Then for each uniform ultrafilter $\mathcal{U}$ on X and each $f : X \longrightarrow Y$, there exists a uniform ultrafilter $\mathcal{W} \neq \mathcal{U}$ such that $f[\mathcal{U}] = f[\mathcal{W}]$.*

PROOF. Consider the family $\mathcal{P} := \{f^-(y) : y \in f(X)\}$ of subsets of X, a subfamily $\mathcal{P}_0$ of $\mathcal{P}$ consisting of finite sets, and $\mathcal{P}_1 := \mathcal{P} \setminus \mathcal{P}_0$. As $\operatorname{card} \mathcal{P} = \operatorname{card} Y$, we deduce that $\operatorname{card}(\bigcup_{P\in\mathcal{P}_0} P) < \operatorname{card} X$. It follows that $\mathcal{P}_1$ is not empty. Let us partition each $P \in \mathcal{P}_1$ into two disjoint subsets A_P and B_P of the same cardinality ([5]), set $P_0 := \bigcup_{P\in\mathcal{P}_0} P$, and

$$A := P_0 \cup \bigcup\nolimits_{P\in\mathcal{P}_1} A_P, \text{ and } B := P_0 \cup \bigcup\nolimits_{P\in\mathcal{P}_1} B_P.$$

Then $X = A \cup B \in \mathcal{U}$, hence either $A \in \mathcal{U}$ or $B \in \mathcal{U}$, say $A \in \mathcal{U}$. Consider a map $h : A \longrightarrow B$, such that $h|_{P_0}$ is the identity and be a bijection between A_P and B_P for each $P \in \mathcal{P}_1$. Then $\mathcal{W} := h[\mathcal{U}]$ is an ultrafilter (as an image of an ultrafilter) different from $\mathcal{U}$. Indeed, $A \in \mathcal{U}, B \in \mathcal{W}$, and $A \cap B = P_0 \notin \mathcal{U}$, because $\mathcal{U}$ is uniform, and $\operatorname{card} P_0 < \operatorname{card} X = \min\{\operatorname{card} U : U \in \mathcal{U}\}$. Hence $A \setminus P_0 \in \mathcal{U}$ and is disjoint from B. Finally, $f[\mathcal{U}] \approx f[\mathcal{W}] = (f \circ h)[\mathcal{U}]$, and $\mathcal{W} = h[\mathcal{U}]$ is uniform, because h restricted to $A \setminus P_0 \in \mathcal{U}$ is a bijection. □

THEOREM XVIII.4.2 (Lowen-Colebunders and Lowen). *The class of convergences is not simple* ([6]).

PROOF. We need to show that for every convergence τ on Y, there exists a convergence ξ such that $\xi > \bigvee_{f\in\mathscr{C}(\xi,\tau)} f^-\tau$. Let X an infinite set such that $\operatorname{card} X > \operatorname{card} Y$ and let $\mathcal{U}$ be a uniform ultrafilter on X ([7]). Let $x_\infty \in X$ and let ξ be the prime convergence such that $x_\infty \in \lim_\xi \mathcal{F}$ whenever $\ker \mathcal{F} \subset \{x_\infty\}$ and $\mathcal{U} \notin \beta\mathcal{F}$. By Lemma XVIII.4.1, for each

[5]That is, $P = A_P \cup B_P$ and $A_P \cap B_P = \varnothing$.

[6]E. LOWEN-COLEBUNDERS and R. LOWEN [177].

[7]This is possible, because by Proposition V.5.7 there exist uniform filters on each infinite set.

map $f \in Y^X$ there exists an ultrafilter $\mathcal{W}_f \neq \mathcal{U}$ such that $f[\mathcal{U}] = f[\mathcal{W}_f]$. It follows from the definition of ξ that $\lim_\xi \mathcal{U} = \varnothing$ and $x_\infty \in \lim_\xi \mathcal{W}_f$ for each $f \in Y^X$. If $f \in \mathscr{C}(\xi, \tau)$, then $f(x_\infty) \in \lim_\tau f[\mathcal{W}_f] = \lim_\tau f[\mathcal{U}]$, and thus

$$x_\infty \in \lim_{\bigvee_{f\in\mathscr{C}(\xi,\tau)} f^-\tau} \mathcal{U} \neq \lim_\xi \mathcal{U} = \varnothing.$$

□

As ξ used in the proof of Theorem XVIII.4.2 is a pseudotopology, we infer that

COROLLARY XVIII.4.3. *The class of pseudotopologies is not simple.*

XVIII.5. Adherence-determined convergences

Let $\mathbb{H}$ be a class of filters including all principal filters, that is, $\mathbb{F}_0 \subset \mathbb{H} \subset \mathbb{F}$. It follows from this assumption that $\mathbb{H}X \neq \varnothing$ for each non-empty set X.

DEFINITION XVIII.5.1. A convergence ξ is said to be $\mathbb{H}$-*adherence-determined* if for each filter $\mathcal{F}$ ([8]),

$$\lim_\xi \mathcal{F} = \bigcap_{\mathbb{H}\ni\mathcal{H}\#\mathcal{F}} \operatorname{adh}_\xi \mathcal{H}. \tag{XVIII.5.1}$$

We denote by $\mathbf{A}_\mathbb{H}$ the class of $\mathbb{H}$-adherence-determined convergences. Let us observe that if $\mathbb{H}_0, \mathbb{H}_1$ are initial classes such that $\mathbb{F}_0 \subset \mathbb{H}_0 \subset \mathbb{H}_1 \subset \mathbb{F}$, then $\mathbf{A}_{\mathbb{H}_0} \subset \mathbf{A}_{\mathbb{H}_1}$. In particular, $\mathbf{A}_{\mathbb{F}_0}$ is the least and $\mathbf{A}_\mathbb{F}$ is the largest class of adherence-determined convergences ([9]).

Accordingly, $\mathbf{A}_{\mathbb{F}_0}$ is the class of pretopologies, and $\mathbf{A}_\mathbb{F}$ the class of pseudotopologies.

Recall that a class $\mathbb{H}$ is called *initial* if

$$\mathcal{H} \in \mathbb{H}Y \Longrightarrow f^-[\mathcal{H}] \in \mathbb{H}X \tag{XVIII.5.2}$$

for every $f : X \longrightarrow Y$ and each $\mathcal{H}$.

PROPOSITION XVIII.5.2. *If $\mathbb{H}$ is an initial class of filters, then $\mathbf{A}_\mathbb{H}$ is reflective.*

PROOF. Indeed, $o_X \in \mathbf{A}_\mathbb{H}X$, because o_X is a pretopology, that is $o_X \in \mathbf{A}_{\mathbb{F}_0} \subset \mathbf{A}_\mathbb{H}$.

Let $\Xi \subset \mathbf{A}_\mathbb{H}X$, and let $x \in \operatorname{adh}_{\bigvee\Xi} \mathcal{H}$ for every $\mathcal{H} \in \mathbb{H}X$ such that $\mathcal{H}\#\mathcal{F}$. Thus $x \in \operatorname{adh}_\xi \mathcal{H}$ for every $\mathcal{H} \in \mathbb{H}X$ and for every $\xi \in \Xi$,

[8] In [**83**, Definition XIV.3.1], adherence-determination was defined in a more general sense, which admitted that a class $\mathbb{H}$ depend on the convergence, and not necessarily includes $\mathbb{F}_0$. Topologies are adherence-determined in that broader sense.

[9] F. MYNARD and J. WOJCIECHOWSKI established a characterization of simplicity in the case of adherence-determined classes [**201**].

because $\bigvee \Xi \geq \xi$. Hence, by (XVIII.5.1), $x \in \lim_{\xi} \mathcal{F}$ for every $\xi \in \Xi$, thus $x \in \lim_{\bigvee \Xi} \mathcal{F}$.

Let $\tau \in \mathbf{A}_{\mathbb{H}} Y$ and let $f : X \longrightarrow Y$. We will see that $f^{-}\tau \in \mathbf{A}_{\mathbb{H}} X$. So let $x \in \operatorname{adh}_{f^{-}\tau} \mathcal{H}$, equivalently $f(x) \in \operatorname{adh}_{\tau} f[\mathcal{H}]$ by (VIII.1.10), for each $\mathcal{H} \in \mathbb{H}X$ such that $\mathcal{H}\#\mathcal{F}$. We need to show that $x \in \lim_{f^{-}\tau} \mathcal{F}$, that is, $f(x) \in \lim_{\tau} f[\mathcal{F}]$.

Let $\mathcal{G} \in \mathbb{H}Y$ be such that $\mathcal{G}\#f[\mathcal{F}]$. Therefore $f^{-}[\mathcal{G}]\#\mathcal{F}$, and since $\mathbb{H}$ is initial, $f^{-}[\mathcal{G}] \in \mathbb{H}X$, so that $f(x) \in \operatorname{adh}_{\tau} f[f^{-}[\mathcal{G}]]$. As $ff^{-}[\mathcal{G}]\#f[\mathcal{F}]$, and by Lemma VIII.1.4, $f[f^{-}[\mathcal{G}]] \geq \mathcal{G}$, we infer that $f(x) \in \operatorname{adh}_{\tau} \mathcal{G}$. By (XVIII.5.1), $f(x) \in \lim_{\tau} f[\mathcal{F}]$, equivalently, $x \in f^{-}(\lim_{\tau} f[\mathcal{F}]) = \lim_{f^{-}\tau} \mathcal{F}$. □

It follows that, for each initial class $\mathbb{H}$, there exists a reflector $\mathrm{A}_{\mathbb{H}}$ on $\mathbb{H}$-adherence-determined convergences. As a functor, it preserves continuity, that is, $\mathscr{C}(\xi, \tau) \subset \mathscr{C}(\mathrm{A}_{\mathbb{H}}\xi, \mathrm{A}_{\mathbb{H}}\tau)$ for each convergences ξ and τ, equivalently, $\mathrm{A}_{\mathbb{H}}(f^{-}\tau) \geq f^{-}(\mathrm{A}_{\mathbb{H}}\tau)$ for each f and τ.

It is useful to establish an explicit formula for this reflector.

PROPOSITION XVIII.5.3. *If $\mathbb{H}$ is initial, then*

$$\lim_{\mathrm{A}_{\mathbb{H}}\xi} \mathcal{F} = \bigcap_{\mathbb{H} \ni \mathcal{H}\#\mathcal{F}} \operatorname{adh}_{\xi} \mathcal{H} \tag{XVIII.5.3}$$

for every filter $\mathcal{F}$.

PROOF. Notice that $\lim_{\xi} \mathcal{F} \subset \lim_{\mathrm{A}_{\mathbb{H}}\xi} \mathcal{F}$, because $\lim_{\xi} \mathcal{F} \subset \operatorname{adh}_{\xi} \mathcal{H}$ whenever $\mathcal{F}\#\mathcal{H}$. Hence $\xi \geq \mathrm{A}_{\mathbb{H}}\xi$. Since $\zeta \geq \xi$ implies $\operatorname{adh}_{\zeta} \mathcal{A} \subset \operatorname{adh}_{\xi} \mathcal{A}$ for every family $\mathcal{A}$, it follows that

$$\zeta \geq \xi \Longrightarrow \mathrm{A}_{\mathbb{H}}\zeta \geq \mathrm{A}_{\mathbb{H}}\xi.$$

To see that $\mathrm{A}_{\mathbb{H}}$ is idempotent, it is enough to show that

$$\operatorname{adh}_{\mathrm{A}_{\mathbb{H}}\xi} \mathcal{H} = \operatorname{adh}_{\xi} \mathcal{H} \tag{XVIII.5.4}$$

for every $\mathcal{H} \in \mathbb{H}$. If $\mathcal{H}_0 \in \mathbb{H}$ and $x \in \operatorname{adh}_{\mathrm{A}_{\mathbb{H}}\xi} \mathcal{H}_0$, then there is a filter $\mathcal{F}\#\mathcal{H}_0$ such that $x \in \lim_{\mathrm{A}_{\mathbb{H}}\xi} \mathcal{F}$. Therefore, $x \in \operatorname{adh}_{\xi} \mathcal{H}$ for every filter $\mathcal{H} \in \mathbb{H}$ such that $\mathcal{F}\#\mathcal{H}$, in particular, $x \in \operatorname{adh}_{\xi} \mathcal{H}_0$. □

As a matter of fact, we have already exhibited a special case of (XVIII.5.3) for $\mathbb{H} = \mathbb{F}_0$, resulting in $\mathrm{A}_{\mathbb{F}_0} = \mathrm{S}_0$, that is, the pretopologizer, and for $\mathbb{H} = \mathbb{F}$, resulting in $\mathrm{A}_{\mathbb{F}} = \mathrm{S}$, that is, the pseudotopologizer.

EXAMPLE XVIII.5.4 (Missing-a-filter prime convergence). Let $\mathbb{H}$ be an initial class, X be an infinite set, and $\mathcal{H}_{\infty} \in \mathbb{H}X$. Let η be a prime free convergence to the effect that $x_{\infty} \in \lim_{\eta} \mathcal{F}$ whenever $\ker \mathcal{F} \subset \{x_{\infty}\}$ and $\mathcal{F} \perp \mathcal{H}_{\infty}$. Notice that $\operatorname{adh}_{\eta} \mathcal{H}_{\infty} = \varnothing$, because if $\mathcal{G}\#\mathcal{H}_{\infty}$, then $\lim_{\eta} \mathcal{G} = \varnothing$. The convergence η is $\mathbb{H}$-adherence-determined. Indeed, if

$x_\infty \notin \lim_\eta \mathcal{F}$ and $\ker \mathcal{F} \subset \{x_\infty\}$, then $\mathcal{F}\#\mathcal{H}_\infty$ and $x_\infty \notin \operatorname{adh}_\eta \mathcal{H}_\infty$, which proves that η is $\mathbb{H}$-adherence-determined.

EXAMPLE XVIII.5.5 (Missing-filters convergence). Let X be an infinite set, and for each $x \in X$, let $\boldsymbol{H}_x \subset \mathbb{H}X$. Then we define a free convergence ζ to the effect that $x \in \lim_\zeta \mathcal{F}$ whenever $\ker \mathcal{F} \subset \{x\}$ such that $\mathcal{F} \perp \mathcal{H}$ for each $\mathcal{H} \in \boldsymbol{H}_x$. Then ζ is an $\mathbb{H}$-adherence-determined convergence. Indeed, if $x \notin \lim_\zeta \mathcal{F}$ then, by definition of ζ, either $\ker \mathcal{F} \setminus \{x\} \neq \varnothing$, or there exists is $\mathcal{H}_0 \in \boldsymbol{H}_x$ such that $\mathcal{F}\#\mathcal{H}_0$. If $\mathcal{G}\#\mathcal{H}_0$ then $x \notin \lim_\zeta \mathcal{G}$ by definition of ζ. Therefore, $x \notin \operatorname{adh}_\zeta \mathcal{H}_0$, that is $x \notin \lim_{\mathrm{A}_\mathbb{H}\zeta} \mathcal{F}$, which proves that $\zeta = \mathrm{A}_\mathbb{H}\zeta$.

COROLLARY XVIII.5.6. *Let $\mathbb{H}$ be a class of filters. Then $\zeta \geq \mathrm{A}_\mathbb{H}\xi$ if and only if*

$$\operatorname{adh}_\zeta \mathcal{H} \subset \operatorname{adh}_\xi \mathcal{H}$$

for each $\mathcal{H} \in \mathbb{H}$.

PROOF. One implication follows from (VIII.1.8). Assume the condition, let $x \in \lim_\zeta \mathcal{F}$ and let $\mathcal{H}\#\mathcal{F}$. Then $x \in \operatorname{adh}_\zeta \mathcal{H} \subset \operatorname{adh}_\xi \mathcal{H}$. Therefore, $x \in \bigcap_{\mathbb{H}\ni\mathcal{H}\#\mathcal{F}} \operatorname{adh}_\xi \mathcal{H} = \lim_{\mathrm{A}_\mathbb{H}\xi} \mathcal{F}$. □

The formula just established enables us to infer that

THEOREM XVIII.5.7. *If $\mathbb{H}$ is composable, then*

$$\mathrm{A}_\mathbb{H}(f^-\tau) = f^-(\mathrm{A}_\mathbb{H}\tau),$$

for each f and τ.

PROOF. Suppose that $x \in \lim_{f^-(\mathrm{A}_\mathbb{H}\tau)} \mathcal{F}$, that is, $f(x) \in \lim_{\mathrm{A}_\mathbb{H}\tau} f[\mathcal{F}] \subset \operatorname{adh}_\tau \mathcal{D}$ for every filter $\mathcal{D} \in \mathbb{H}$ such that $\mathcal{D}\#f[\mathcal{F}]$. We need to show that $x \in \lim_{\mathrm{A}_\mathbb{H}(f^-\tau)} \mathcal{F}$. So let $\mathcal{H} \in \mathbb{H}X$ be such that $\mathcal{H}\#\mathcal{F}$. It follows that $f[\mathcal{H}]\#f[\mathcal{F}]$ and $f[\mathcal{H}] \in \mathbb{H}$. Therefore, $f(x) \in \operatorname{adh}_\tau f[\mathcal{H}]$, equivalently $x \in \operatorname{adh}_{f^-\tau} \mathcal{H}$, as needed. □

COROLLARY XVIII.5.8. *If $\mathbb{H}$ is composable, then $A_\mathbb{H}(\xi|_B) = (A_\mathbb{H}\xi)|_B$ for each $B \subset |\xi|$.*

Mind that the class of topologies is not adherence-determined. In particular, there exists a convergence θ on X and $B \subset X$ so that

$$\mathrm{T}(\theta|_B) \neq (\mathrm{T}\theta)|_B\,.$$

Indeed,

EXAMPLE XVIII.5.9. Consider the pretopology θ from Example VI.1.8. If $B := X \setminus X_\infty$, then $\theta|_B = \iota_B$, hence $\mathrm{T}(\theta|_B) = \iota_B$, while $x_\infty \in \operatorname{cl}_\theta(B \setminus \{x_\infty\})$, showing that $(\mathrm{T}\theta)|_B \neq \iota_B$.

Nevertheless, if B is τ-closed, then $\mathrm{T}(\tau|_B) = (\mathrm{T}\tau)|_B$ (Exercise XVIII.10.3).

DEFINITION XVIII.5.10. We say that a reflector J is *hereditary* if for each τ and every $B \subset |\tau|$,

$$(J\tau)|_B = J(\tau|_B).$$

LEMMA XVIII.5.11. *If $J \geq \mathrm{T}$ is hereditary, then $J \geq \mathrm{S}_0$.*

PROOF. Let J be a hereditary reflector greater than T. Suppose that there exists a convergence τ such that $J\tau \ngeq \mathrm{S}_0\tau$, that is, there exists $x \in \lim_{J\tau} \mathcal{F} \setminus \lim_{\mathrm{S}_0\tau} \mathcal{F}$. Accordingly, there exists $B \subset |\tau|$ such that $B \in \mathcal{F}^\#$ and $x \notin \mathrm{adh}_\tau B$. If $X := B \cup \{x\}$, then $x \in \lim_{J\tau}(\mathcal{F} \vee B)$, hence

$$x \in \lim\nolimits_{(J\tau)|_X}(\mathcal{F} \vee B),$$

but $x \notin \mathrm{adh}_{\tau|_X} B$. As $\mathrm{adh}_{\tau|_X} B \subset B \in (\mathcal{F} \vee B)^\#$, so that B is $\tau|_X$-closed, and thus $x \notin \lim_{\mathrm{T}(\tau|_X)}(\mathcal{F} \vee B) \supset \lim_{J(\tau|_X)}(\mathcal{F} \vee B)$, which means that J is not hereditary. □

Because S_0 is hereditary by (XVIII.5.8),

THEOREM XVIII.5.12 (Mynard). *The pretopologies constitute the reflective hereditary hull of topologies.*

Recall (Proposition XVIII.1.4) that for each cardinal κ, the *κ-stable modifier* F_κ is a hereditary reflector. In particular, $F_\kappa \geq \mathrm{S}_0$.

DEFINITION XVIII.5.13. A convergence is called a

(1) *paratopology* if it is $\mathbb{F}_1$-adherence-determined. We denote the class of paratopologies by $\mathbf{S}_1$ and the reflector, called the *paratopologizer*, by S_1.
(2) *hypotopology* if it is $\mathbb{F}_{\wedge 1}$-adherence-determined. We denote the class of hypotopologies by $\mathbf{S}_{\wedge 1}$ and the reflector, called the *hypotopologizer*, by $\mathrm{S}_{\wedge 1}$.

We shall examine them in detail in the forthcoming sections. At the present time, let us gather the mentioned adherence-determined classes in a table.

Class name	reflector	filter class	filter class name
pseudotopology	S	$\mathbb{F}$	all
hypotopology	$\mathrm{S}_{\wedge 1}$	$\mathbb{F}_{\wedge 1}$	countably complete
paratopology	S_1	$\mathbb{F}_1$	countably based
pretopology	S_0	$\mathbb{F}_0$	principal

TABLE XVIII.5.1. Basic adherence-determined classes of convergences.

The classes $\mathbb{F}_0, \mathbb{F}_1, \mathbb{F}_{\wedge 1}, \mathbb{F}$, are composable, in particular initial, the case of $\mathbb{F}$, consisting of all filters, being obvious. More generally, the class of filters of character less than a given cardinal is initial ([10]). Surely, $\mathbb{F}_0 \subset \mathbb{F}_1 \subset \mathbb{F}$ and $\mathbb{F}_0 \subset \mathbb{F}_{\wedge 1} \subset \mathbb{F}$, and thus, in virtue of (XVIII.5.1),

$$\mathrm{S}_0 \leq \mathrm{S}_1 \leq \mathrm{S}, \text{ and } \mathrm{S}_0 \leq \mathrm{S}_{\wedge 1} \leq \mathrm{S}.$$

Therefore, the class of pretopologies is included in those of paratopologies and hypotopologies, and the latter are included in that of pseudotopologies.

XVIII.6. Paratopologies

A convergence ξ is a *paratopology* if $x \in \lim_\xi \mathcal{F}$ provided that $x \in \operatorname{adh}_\xi \mathcal{H}$ for each countably based filter $\mathcal{H}$ such that $\mathcal{H} \# \mathcal{F}$. As we have seen in Section XVIII.5, the class $\mathbf{S}_1$ of paratopologies is adherence-determined by the class $\mathbb{F}_1$ of countably based filters, hence reflective and hereditary. It follows also that the *paratopologizer* S_1 fulfills

$$\lim_{\mathrm{S}_1 \xi} \mathcal{F} = \bigcap_{\mathbb{F}_1 \ni \mathcal{H} \# \mathcal{F}} \operatorname{adh}_\xi \mathcal{H}. \tag{XVIII.6.1}$$

EXAMPLE XVIII.6.1 (A non-pretopological paratopology). Let $\{Y_n : n < \omega\}$ be an increasing family of sets such that $Y_{n+1} \setminus Y_n$ is infinite countable for each $n < \omega$, and let $Y_0 := \{\infty\}$. We define a convergence υ on $Y := \bigcup_{n<\omega} Y_n$ to the effect that

$$y \in \lim_\upsilon \mathcal{F}$$

whenever $\mathcal{F} = y^\uparrow$, or if $y = \infty$ and there is $n < \omega$ so that $\mathcal{F} \geq (Y_n)_0 \wedge \infty^\uparrow$.

[10]Indeed, the *character* $\chi(\mathcal{F})$ of a filter $\mathcal{F}$, that is, the least cardinality of filter-bases of $\mathcal{F}$ (see Definition XX.2.16) is stable under images and preimages. In fact, if $\mathcal{B}$ is a filter-base of a filter $\mathcal{F}$, then $f[\mathcal{B}]$ is a filter-base of $f[\mathcal{F}]^\uparrow$, and $f^-[\mathcal{B}]$ is a base of $f^-[\mathcal{F}]$. Moreover, $\operatorname{card}(f[\mathcal{B}]) \leq \operatorname{card}(\mathcal{B}) = \operatorname{card}(f^-[\mathcal{B}])$.

This is a finitely stable convergence, which is prime, and thus Hausdorff. It is obviously not a pretopology, because $\mathcal{V}_\upsilon(\infty) = \bigwedge_{n<\omega}(Y_n)_0$ does not converge. We notice that if $\mathcal{H}$ is a free filter, then $\infty \in \operatorname{adh}_\upsilon \mathcal{H}$ whenever there is $n < \omega$ such that $Y_n \cap H$ is infinite for every $H \in \mathcal{H}$, equivalently $\mathcal{H} \vee Y_n \geq (Y_n)_0$.

To see that υ is a paratopology, let $\mathcal{F}$ be a filter such that $x_\infty \in \operatorname{adh}_\upsilon \mathcal{H}$ for each countably based filter such that $\mathcal{H}\#\mathcal{F}$. We need to show that there is $n < \omega$ such that $Y_n \in \mathcal{F}$. If, on the contrary, $Y \setminus Y_n \in \mathcal{F}^\#$ for each $n < \omega$, then $\mathcal{H} \approx (Y \setminus Y_n)_n$ is a countably based filter such that $\mathcal{H}\#\mathcal{F}$ and $\operatorname{adh}_\upsilon \mathcal{H} = \varnothing$, a contradiction.

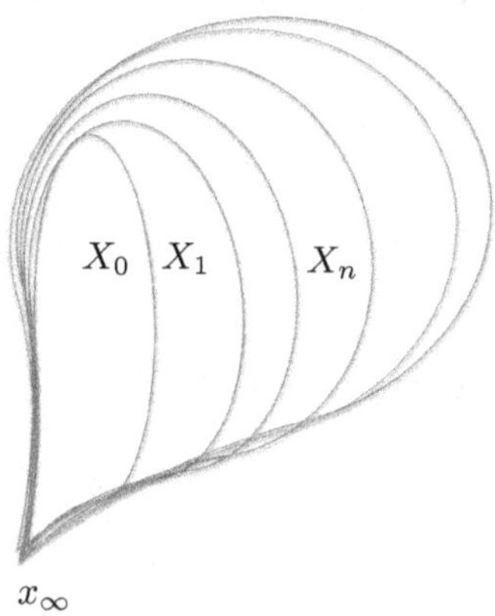

FIGURE XVIII.6.1. A finitely stable, non-pretopological convergence. An increasing sequence $(X_n)_n$ of sets such that $X_{n+1} \setminus X_n$ is infinite for each $n < \omega$ and $x_\infty \in \bigcap_{n<\omega} X_n$. Let τ_n be a prime pretopology at x_∞ so that τ_n is the restriction of τ_{n+1} to X_n. Then the final convergence on $X = \bigcup_{n<\omega} X_n$ is a paratopology.

EXAMPLE XVIII.6.2. Recall a *bisequence pretopology* from Example VI.1.8. Here is an alternative presentation: $\{X_n : n < \omega\}$ is a disjoint family of infinite countable sets, and $x_n \in X_n$ for each $n < \omega$.

A pretopology θ is defined on $X := \bigcup_{n<\omega} X_n \cup \{x_\infty\}$, so that $\mathcal{V}_\theta(x_n) = (X_n, x_n)_0$ is the cofinite filter of X_n centered at x_n, and $\mathcal{V}_\theta(x_\infty) = (X_\infty, x_\infty)_0$ is the cofinite filter of X_∞ centered at x_∞, where $X_\infty := \{x_n : n < \omega\} \cup \{x_\infty\}$, and $\mathcal{V}_\theta(x) = x^\uparrow$ otherwise. Consider now

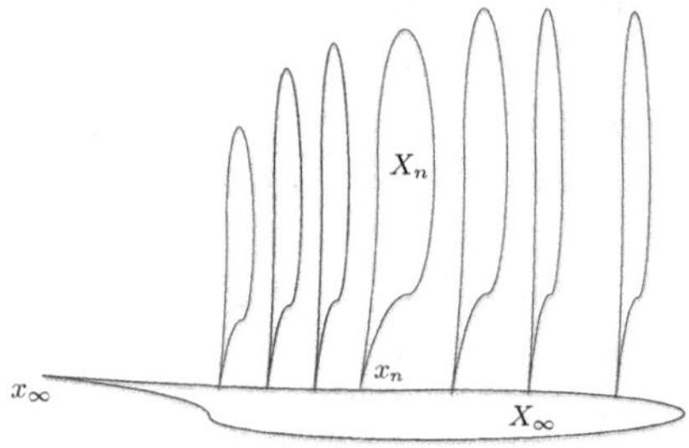

FIGURE XVIII.6.2. A bisequence pretopology: $x_n \in \lim_\theta(X_n, x_n)_0$ for each $n < \omega$, and $x_\infty \in \lim_\theta(X_\infty, x_\infty)_0$, where $X_\infty := \{x_\infty\} \cup \{x_n : n < \omega\}$.

a map $f : X \longrightarrow \bigcup_{n<\omega}(X_n \setminus \{x_n\}) \cup \{x_\infty\}$ defined by

$$f(x) := \begin{cases} x & x \notin X_\infty, \\ x_\infty & x \in X_\infty, \end{cases}$$

and consider the quotient $f\theta$, that is, the final convergence of θ under f.

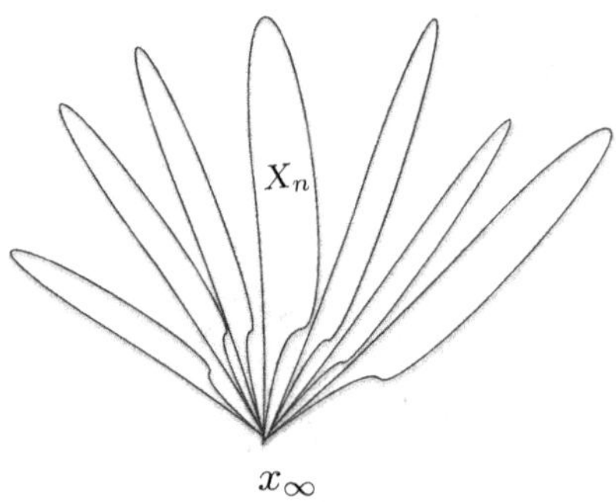

FIGURE XVIII.6.3. A fan: $x_\infty \in \lim(X_n, x_\infty)_0$ for each $n < \omega$.

We shall see that the finitely stable modification of $\mathrm{P}_0(f\theta)$ of $f\theta$ is a non-pretopological paratopology. Indeed, all the points but x_∞ are isolated, and $x_\infty \in \lim_{\mathrm{P}_0(f\theta)} \mathcal{F}$ whenever there is $n < \omega$ such that

$$\mathcal{F} \geq (\bigcup_{k \leq n} f(X_k), x_\infty)_0,$$

where $(B, x_\infty)_0$ denotes the cofinite filter of B centered at $\{x_\infty\}$ (see Definition II.4.3). It follows that $\mathrm{P}_0(f\theta)$ is of countable character. Clearly, $\mathrm{P}_0(f\theta)$ is not finitely paved at x_∞, so that it is not a pretopology.

At this point, we realize that $\mathrm{P}_0(f\theta)$ is actually the paratopology υ described in Example XVIII.6.1 (an evident homeomorphism with $Y_n := \bigcup_{k\leq n} f(X_k)$ and $\infty := x_\infty$).

Since paratopologies form a reflective class of convergences, suprema, products, and initial convergences of paratopologies are paratopological. Moreover, as paratopologies are adherence-determined,

$$\mathrm{S}_1(f^-\tau) = f^-(\mathrm{S}_1\tau)$$

for every convergence τ and each map f. *A priori* nothing assures paratopologicity of infima. Nevertheless, some results on infima of paratopologies are a consequence of specific traits of this class.

Recall that if $\{\pi_n : n < \omega\}$ is a sequence of convergences on X such that $\pi_n \geq \pi_{n+1}$ for each $n < \omega$, then by (III.3.4),

$$x \in \lim_{\bigwedge_{n<\omega}\pi_n} \mathcal{F} \Longleftrightarrow \underset{n<\omega}{\exists}\ x \in \lim_{\pi_n} \mathcal{F}$$

for every $x \in X$ and each filter $\mathcal{F}$.

Of course, the set of paratopologies on a given non-empty set is a complete lattice, as any other set of reflective convergences. As $\mathbf{S}_1 \subset \mathbf{S} \subset \mathbf{I}$ for a set Ξ of paratopologies on a given set, $\bigwedge_{\mathbf{S}_1} \Xi \leq \bigwedge_{\mathbf{S}} \Xi \leq \bigwedge \Xi$.

We shall see in Corollary XXII.6.7 that if Ξ is a countable set of paratopologies, then the pseudotopological infimum of Ξ is a paratopology, that is, $\bigwedge_{\mathbf{S}} \Xi \in \mathbf{S}_1$, and thus $\bigwedge_{\mathbf{S}_1} \Xi = \bigwedge \Xi$.

Proposition XVIII.6.3. *Each countably stable paratopology is a pretopology.*

Proof. Let us first prove that

$$\mathrm{S}_1\mathrm{P}_1 = \mathrm{S}_0. \tag{XVIII.6.2}$$

Indeed, $\mathrm{S}_1 \geq \mathrm{S}_0$ and $\mathrm{P}_1 \geq \mathrm{P} = \mathrm{S}_0$, so that $\mathrm{S}_1\mathrm{P}_1 \geq \mathrm{S}_0$. Conversely, let $x \in \lim_{\mathrm{S}_0\xi} \mathcal{F}$, and let $\mathcal{H} \in \mathbb{F}_1$ such that $\mathcal{H}\#\mathcal{F}$. Let $\{H_n : n < \omega\}$ be a filter-base of $\mathcal{H}$, hence $H_n\#\mathcal{F}$ for each $n < \omega$, hence $x \in \operatorname{adh}_{\mathrm{S}_0\xi} H_n = \operatorname{adh}_\xi H_n$ for each $n < \omega$. Therefore, for each $n < \omega$, there exists $\mathcal{U}_n \in \beta H_n$ such that $x \in \lim_\xi \mathcal{U}_n$, and thus $x \in \lim_{\mathrm{P}_1\xi}(\bigwedge_{n<\omega} \mathcal{U}_n)$. On the other hand, $(\bigwedge_{n<\omega} \mathcal{U}_n)\#\mathcal{H}$, so that $x \in \operatorname{adh}_{\mathrm{P}_1\xi} \mathcal{H}$. Consequently, $x \in \lim_{\mathrm{S}_1\mathrm{P}_1\xi} \mathcal{F}$.

If now ξ is a countably stable paratopology, then $\mathrm{S}_0\xi = \mathrm{S}_1(\mathrm{P}_1\xi) = \mathrm{S}_1\xi = \xi$, hence ξ is a pretopology. □

In Example XVIII.1.1 we showed that the standard sequential convergence of the real line σ is countably stable, but not stable, that is, is not a pretopology. As σ is a pseudotopology by Example XVIII.3.6, σ is not a paratopology by Proposition XVIII.6.3.

Here is another example of a countably stable pseudotopology which is not a paratopology.

EXAMPLE XVIII.6.4. Consider a prime convergence ξ on an uncountable set X such that $x_\infty \in \lim_\xi \mathcal{F}$ whenever $\ker\mathcal{F} \subset \{x_\infty\}$ and $\mathcal{F}$ contains a countable set. The convergence ξ is countably stable, because if $x_\infty \in \bigcap_{n<\omega} \lim_\xi \mathcal{F}_n$ then for each $n < \omega$, there is a countable $F_n \in \mathcal{F}_n$, hence $\bigcup_{n<\omega} F_n$ is countable and belongs to $\bigwedge_{n<\omega} \mathcal{F}_n$. Moreover $\ker(\bigwedge_{n<\omega} \mathcal{F}_n) \subset \{x_\infty\}$ by Exercise II.5.4.

To see that ξ is a pseudotopology, suppose that $x_\infty \in \lim_\xi \mathcal{U}$ for each $\mathcal{U} \in \beta\mathcal{F}$, that is there is a countable $U_\mathcal{U} \in \mathcal{U}$ and $\ker\mathcal{U} \subset \{x_\infty\}$. By Proposition V.4.6, there is a finite subset $\boldsymbol{A}$ of $\beta\mathcal{F}$ such that the countable set $\bigcup_{\mathcal{U}\in\boldsymbol{A}} U_\mathcal{U} \in \mathcal{F}$. On the other hand, $\ker\mathcal{F} = \bigcup_{\mathcal{U}\in\beta\mathcal{F}} \ker\mathcal{U} \subset \{x_\infty\}$, hence $x_\infty \in \lim_\xi \mathcal{F}$.

On the other hand, ξ is not a pretopology, for if we split X into an uncountable family of infinite countable sets $\{X_j : j \in J\}$, then $x_\infty \in \lim_\xi (X_j)_0$ for each $j \in J$, but $\bigwedge_{j\in J}(X_j)_0$ does not contain a countable set, hence $x_\infty \notin \lim_\xi \bigwedge_{j\in J}(X_j)_0$. By Proposition XVIII.6.3, a countably stable paratopology is a pretopology, and thus ξ is not a paratopology.

What is the paratopologization of ξ? Consider a free countably based filter $\mathcal{H}$ such that $x_\infty \in \operatorname{adh}_\xi \mathcal{H}$. Accordingly, there exists a filter $\mathcal{D} \geq \mathcal{H}$ such that $x_\infty \in \lim_\xi \mathcal{D}$, so that there exists a countable set $D_0 \in \mathcal{D}$, and, of course, $\ker\mathcal{D} \subset \{x_\infty\}$. Then $x_\infty \in \lim_\xi (D_0)_0$ and $(D_0)_0 \# \mathcal{H}$ ([11]).

Of course, $(C)_0 \geq (X)_0$ for each infinite subset C of X. In particular, if C is infinite countable, then $x_\infty \in \lim_\xi (C)_0$. Hence, if H is an infinite subset of X, then $H \in (X)_0^\#$ and $x_\infty \in \operatorname{adh}_\xi (H)_0$ and thus $x_\infty \in \operatorname{adh}_\xi (C)_0 = \lim_\xi (C)_0$ for each countable, infinite subset C of H. Therefore, $\mathcal{V}_\xi(x_\infty) = (X, x_\infty)_0$, and thus $\mathrm{S}_1\xi = \mathrm{S}_0\xi = \pi[x_\infty, (X)_0] = \mathrm{T}\xi$, because the convergence is prime.

PROPOSITION XVIII.6.5. *Each countably pseudopaved pseudotopology is a paratopology.*

PROOF. Let ξ be a pseudotopology, and let $\{\mathcal{G}_n : n < \omega\}$ be a pseudopavement of ξ at x. If $x \notin \lim_\xi \mathcal{F}$, then we need to find a countably

[11]Hence, there is a sequential filter $\mathcal{E} \geq \mathcal{H} \vee (C)_0$ with $x_\infty \in \lim_\xi \mathcal{E}$. If $x_\infty \in \operatorname{adh}_\xi \mathcal{H}$ for each countably based filter $\mathcal{H}$ with $\mathcal{H} \# \mathcal{F}$, then $x_\infty \in \lim_{\mathrm{S}_1\xi} \mathcal{F}$.

based filter $\mathcal{H}$ such that $\mathcal{H}\#\mathcal{F}$ and such that $x \notin \operatorname{adh}_\xi \mathcal{H}$. As ξ is a pseudotopology, there exists $\mathcal{U} \in \beta\mathcal{F}$ such that $x \notin \lim_\xi \mathcal{U}$. Hence, $\mathcal{U} \notin \beta\mathcal{G}_n$ for each $n < \omega$, that is, there exists $U_n \in \mathcal{U}$ such that $(U_n)^c \in \mathcal{G}_n$ for each $n < \omega$. Accordingly, $\{U_n : n < \omega\}\#\mathcal{F}$ and $x \notin \operatorname{adh}_\xi\{U_n : n < \omega\}$. Indeed, if $\mathcal{W} \in \beta\{U_n : n < \omega\}$ and $x \in \lim_\xi \mathcal{W}$, then there exists $n_0 < \omega$ such that $\mathcal{W} \geq \mathcal{G}_{n_0}$, which is impossible, because $(U_{n_0})^c \in \mathcal{G}_{n_0}$. □

The convergence ξ of Example XVIII.6.4 can be used to construct a counterexample of non-commutation of S_1 with suprema and products.

Example XVIII.6.6. There exist a prime pseudotopology ξ and a prime topology τ such that

$$\mathrm{T}\xi \vee \tau = \mathrm{S}_0\xi \vee \tau = \mathrm{S}_1\xi \vee \tau < \mathrm{S}_1(\xi \vee \tau) = \mathrm{S}_0(\xi \vee \tau) = \mathrm{T}(\xi \vee \tau).$$

Take ξ from Example XVIII.6.4 on an uncountable set X, and let τ be the prime cocountable topology centered at x_∞, that is, $\mathcal{V}_\tau(x_\infty) = (X, x_\infty)_1$, the cocountable filter centered at x_∞. Of course, τ is a topology, because it is prime. We see that $\xi \vee \tau = \iota$, because $x_\infty \in \lim_{\xi\vee\tau} \mathcal{F}$ whenever $\ker \mathcal{F} \subset \{x_\infty\}$ and $\mathcal{F}$ contains a countable set, and the free part of $\mathcal{F}$ contains all cocountable sets. It follows that $\mathcal{F} = \{x_\infty\}$.

It follows that none of the three reflectors $\mathrm{S}_0, \mathrm{S}_1$, and T commutes with suprema. We have already given Examples VI.6.11, VI.6.12 and VII.4.7 that witness non-commutation of S_0 and T with suprema and products.

Proposition XVIII.6.7. *The paratopologizer does not commute with finite products.*

Proof. Let ξ and τ be convergences on a set X from Example XVIII.6.6. We have seen that $\mathrm{S}_1\xi \vee \tau \ngeq \mathrm{S}_1(\xi \vee \tau)$. By Corollary XVIII.5.8, the paratopologizer is hereditary. Therefore, recalling Exercise IV.10.13, we get

$$(\mathrm{S}_1\xi \times \tau)|_{\triangle X} = \mathrm{S}_1\xi \vee \tau \ngeq \mathrm{S}_1(\xi \vee \tau) = \mathrm{S}_1((\xi \times \tau)|_{\triangle X}) = \mathrm{S}_1(\xi \times \tau)|_{\triangle X},$$

so that $\mathrm{S}_1\xi \times \tau \ngeq \mathrm{S}_1(\xi \times \tau)$. □

We shall consider now a paratopology of special interest.

Example XVIII.6.8 (Wojciechowski paratopology). Let $\beth$ be a pseudotopology on

$$W := \mathbb{N} \sqcup \{\infty_0, \infty_1, \infty_2\}$$

such that every filter $\daleth$-converges to $W \setminus \{\infty_2\}$, and $\infty_2 \in \lim_\daleth \mathcal{F}$ whenever $\mathcal{F}$ contains $F \subset_{\text{fin}} W \setminus \{\infty_0\}$.

Obviously, all the points but ∞_2 are pretopological. To see that $\daleth$ is a paratopology, assume that $\mathcal{U} \in \beta W$ such that $\infty_2 \in \operatorname{adh}_\daleth \mathcal{H}$ for each $\mathcal{H} \in \mathbb{F}_1$ such that $\mathcal{H} \leq \mathcal{U}$. Accordingly, there is a finite set N such that $\infty_0 \notin N$ and $N \in \mathcal{H}^\#$, so that $\mathcal{H} \vee N \leq \mathcal{U} \vee N$ are proper filters. Consequently, $\mathcal{U}$ is a principal ultrafilter different from $\infty_0^\uparrow$, and thus $\infty_2 \in \lim_\daleth \mathcal{U}$.

THEOREM XVIII.6.9 (Wojciechowski). *The Wojciechowski paratopology is initially dense in the class of paratopologies.*

PROOF. Let ξ be a paratopology on a set X. For each sequence $\mathcal{B} = \{B_n : n < \omega\}$ of subsets of X such that $B_n \supset B_{n+1}$ for each $n < \omega$, let $f_\mathcal{B} : X \longrightarrow W$, where W is the set from Example XVIII.6.8, be defined by

$$f_\mathcal{B}(w) := \begin{cases} n & w \in B_n \setminus B_{n+1}, \\ \infty_0 & w \in \bigcap_{n<\omega} B_n, \\ \infty_1 & w \in \operatorname{adh}_\xi \mathcal{B} \setminus B_0, \\ \infty_2 & \text{otherwise.} \end{cases}$$

Let us prove that $f_\mathcal{B} \in \mathscr{C}(\xi, \daleth)$ for each sequence $\mathcal{B}$. Let $w \in \lim_\xi \mathcal{F}$. If $\mathcal{B} \perp \mathcal{F}$, that is, there exist $n < \omega$ and $F \in \mathcal{F}$ such that $F \cap B_m = \varnothing$ for each $m > n$, hence $\{0, \ldots, n, \infty_1, \infty_2\} \in f_\mathcal{B}[\mathcal{F}]$, and thus $f_\mathcal{B}(w) \in \lim_\daleth f_\mathcal{B}[\mathcal{F}]$; if $\mathcal{B} \# \mathcal{F}$, then $w \in \operatorname{adh}_\xi \mathcal{B}$, hence $f_\mathcal{B}(w) \neq \infty_2$ and thus $f_\mathcal{B}(w) \in \lim_\daleth f_\mathcal{B}[\mathcal{F}]$.

By Proposition XVII.4.2, $\xi \geq \mathrm{R}^\daleth \xi = \bigvee_{f \in \mathscr{C}(\xi,\daleth)} f^- \daleth$. To see that the equality holds, let $w \notin \lim_\xi \mathcal{F}$. By paratopologicity, there exists $\mathcal{H} \in \mathbb{F}_1 X$ such that $\mathcal{H} \# \mathcal{F}$ and $w \notin \operatorname{adh}_\xi \mathcal{H}$. Therefore, there is $H_0 \in \mathcal{H}$ such that $w \notin H_0$. Let $\mathcal{B}$ be a non-increasing sequence of sets such that $B_0 = H_0$ and $\mathcal{B}$ is a filter-base of $\mathcal{H}$. Then $f_\mathcal{B}(w) = \infty_2$ and either for each $F \in \mathcal{F}$, there are infinitely many n in $f_\mathcal{B}(F)$ or $\infty_0 \in f_\mathcal{B}(F)$. Therefore, $f_\mathcal{B}(w) \notin \lim_\daleth f_\mathcal{B}[\mathcal{F}]$. □

COROLLARY XVIII.6.10. *The class of paratopologies is simple.*

XVIII.7. Hypotopologies

A convergence ξ is a *hypotopology* if $x \in \lim_\xi \mathcal{F}$ provided that $x \in \operatorname{adh}_\xi \mathcal{H}$ for each countably complete filter $\mathcal{H}$ such that $\mathcal{H} \# \mathcal{F}$. As we have seen in Section XVIII.5, the class $\mathbf{S}_{\wedge 1}$ of hypotopologies is adherence-determined by the class $\mathbb{F}_{\wedge 1}$ of countably complete filters, hence is reflective and hereditary. It follows also that the *hypotopologizer* $\mathrm{S}_{\wedge 1}$ fulfills

$$\lim_{\mathrm{S}_{\wedge 1}\xi} \mathcal{F} = \bigcap_{\mathbb{F}_{\wedge 1} \ni \mathcal{H} \# \mathcal{F}} \mathrm{adh}_{\xi} \mathcal{H}. \tag{XVIII.7.1}$$

Notice that

PROPOSITION XVIII.7.1. *If ξ is a convergence on a countable set, then $\mathrm{S}_{\wedge 1}\xi = \mathrm{S}_0\xi$.*

PROOF. By Proposition V.5.2, countably complete and principal filters coincide on countable sets. Hence (XVIII.7.1) coincides with the pretopologizer (VI.6.5). □

COROLLARY XVIII.7.2. *The hypotopologizer does not commute with finite products and suprema.*

PROOF. It was shown in Examples VI.6.11 and VI.6.12 that the pretopologizer does not commute with finite products and suprema. As the underlying set in these examples is countable, the pretopologizer and the hypotopologizer coincide by Proposition XVIII.7.1. □

LEMMA XVIII.7.3. *The cocountable filter on an uncountable set is the coarsest free countably complete filter on that set.*

PROOF. Let X be an uncountable set. Then $(X)_1$ is countably complete. Indeed, if $\{F_n : n < \omega\} \subset (X)_1$ then $X \setminus F_n$ is countable for each $n < \omega$, hence $\bigcup_{n<\omega}(X \setminus F_n)$ is countable, and thus $\bigcap_{n<\omega} F_n = X \setminus (\bigcup_{n<\omega}(X \setminus F_n))$ is cocountable. If $\mathcal{F}$ is free, then $\mathcal{F} \geq (X)_0$, and if D is a countable subset of X, then $X \setminus \{d\} \in \mathcal{F}$ for each $d \in D$, hence if moreover $\mathcal{F}$ is countably complete, then $X \setminus D = \bigcap_{d\in D}(X \setminus \{d\}) \in \mathcal{F}$. □

PROPOSITION XVIII.7.4. *Each hypotopology is countably stable.*

PROOF. We shall prove that

$$\mathrm{P}_1 \geq \mathrm{S}_{\wedge 1}, \tag{XVIII.7.2}$$

which already implies the statement. Let $x \in \lim_{\mathrm{P}_1\xi} \mathcal{F}$. By definition, there exists a sequence $\{\mathcal{F}_n : n < \omega\} \subset \xi^-(x)$ such that $\mathcal{F} \geq \bigwedge_{n<\omega} \mathcal{F}_n$. If $\mathcal{H}$ is a countably complete filter such that $\mathcal{H}\#\mathcal{F}$, then there there is $n < \omega$ such that $\mathcal{H}\#\mathcal{F}_n$, thus $x \in \mathrm{adh}_\xi \mathcal{H}$, hence $x \in \lim_{\mathrm{S}_{\wedge 1}\xi} \mathcal{F}$.

Indeed, if, by way of contradiction, for each $n < \omega$, there is $F_n \in \mathcal{F}_n$ such that $F_n \notin \mathcal{H}^\#$, that is, $F_n^c \in \mathcal{H}$, then $\bigcap_{n<\omega} F_n^c \in \mathcal{H}$, equivalently $\bigcup_{n<\omega} F_n \notin \mathcal{H}^\#$ and $\bigcup_{n<\omega} F_n \in \bigwedge_{n<\omega} \mathcal{F}_n$, which yields a contradiction. □

COROLLARY XVIII.7.5. *If a hypotopology is a paratopology, then it is a pretopology.*

PROOF. If ξ is a hypotopology, that is, $\xi \leq \mathrm{S}_{\wedge 1}\xi$, then by (XVIII.7.2), $\xi \leq \mathrm{P}_1\xi$. By applying S_1 to this inequality, we get $\mathrm{S}_1\xi \leq \mathrm{S}_1\mathrm{P}_1\xi$. As ξ is a paratopology, $\xi \leq \mathrm{S}_1\xi$, hence by (XVIII.6.2),

$$\xi \leq \mathrm{S}_1\xi \leq \mathrm{S}_1\mathrm{S}_{\wedge 1}\xi \leq \mathrm{S}_1\mathrm{P}_1\xi = \mathrm{S}_0\xi.$$

□

Actually, more is known:

$$\mathrm{S}_1\mathrm{S}_{\wedge 1} = \mathrm{S}_{\wedge 1}\mathrm{S}_1 = \mathrm{S}_0. \tag{XVIII.7.3}$$

We have already proved that $\mathrm{S}_1\mathrm{S}_{\wedge 1} \leq \mathrm{S}_0$. As $\mathrm{S}_0 \leq \mathrm{S}_1$ and $\mathrm{S}_0 \leq \mathrm{S}_{\wedge 1}$, we infer that $\mathrm{S}_1\mathrm{S}_{\wedge 1} = \mathrm{S}_0$ and $\mathrm{S}_0 \leq \mathrm{S}_{\wedge 1}\mathrm{S}_1$. Therefore, it remains to prove that $\mathrm{S}_0 \geq \mathrm{S}_{\wedge 1}\mathrm{S}_1$, which will be done in Proposition XXII.6.5.

EXAMPLE XVIII.7.6 (A non-pretopological hypotopology). The countably stable pseudotopology from Example XVIII.6.4 is a hypotopology. Notice that $\mathrm{adh}_\xi(X)_1 = \varnothing$, because if a filter $\mathcal{D}$ meshes $(X)_1$, then each $D \in \mathcal{D}$ is uncountable, hence $\lim_\xi \mathcal{D} = \varnothing$. Therefore, $\mathrm{adh}_\xi(X)_1 = \varnothing$, and thus $\mathrm{adh}_\xi \mathcal{H} = \varnothing$ for each free countably complete filter $\mathcal{H}$, by virtue of Lemma XVIII.7.3. As ξ is prime, so is $\mathrm{S}_{\wedge 1}\xi$. Let $x_\infty \in \lim_{\mathrm{S}_{\wedge 1}\xi} \mathcal{F}$, then $x_\infty \in \mathrm{adh}_\xi \mathcal{H}$ for each countably complete filter $\mathcal{H}\#\mathcal{F}$. If $x_\infty \notin \lim_\xi \mathcal{F}$, then either $\mathcal{F}\#(X)_1$, or there exists $x \neq x_\infty$ such that $x \in \ker \mathcal{F}$ and thus $\mathcal{F} = x^\uparrow$, so that $\{x\} = \lim_\xi \mathcal{F} \subset \lim_{\mathrm{S}_{\wedge 1}\xi} \mathcal{F}$, which yields a contradiction.

A special case of Example XVIII.5.4 yields an example of hypotopology.

EXAMPLE XVIII.7.7. Let $\mathcal{H}_\infty$ be a countably complete filter. A prime pseudotopology ξ is defined so that $x_\infty \in \lim_\xi \mathcal{F}$ whenever $\mathcal{H}_\infty \perp \mathcal{F}$ and $\ker \mathcal{F} \subset \{x_\infty\}$. Then ξ is a hypotopology. Indeed, first we note that $\mathrm{adh}_\xi \mathcal{H}_\infty = \varnothing$. If now $\ker \mathcal{F} \subset \{x_\infty\}$ and $x_\infty \notin \lim_\xi \mathcal{F}$, then $\mathcal{H}_\infty\#\mathcal{F}$ and $x \notin \mathrm{adh}_\xi \mathcal{H}_\infty$. As by assumption, $\mathcal{H}_\infty \in \mathbb{F}_{\wedge 1}$, the convergence ξ is a hypotopology.

The class of hypotopologies is not simple ([12]). This fact will be proved with the aid of families of *closed unbounded* subsets of cardinals in Section XXVII.7. By Theorem XXVII.7.2, the family of closed unbounded subsets of a regular uncountable cardinal is, in particular, countably complete. Let $\mathcal{C}_\lambda$ be the family of closed unbounded subsets of a regular uncountable cardinal λ.

[12]First proved by J. WOJCIECHOWSKI. This is also a consequence of a more general result of F. MYNARD and J. WOJCIECHOWSKI [201].

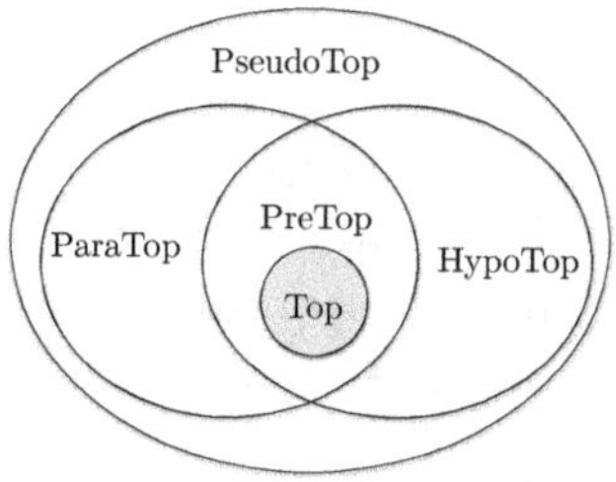

FIGURE XVIII.7.1. Adherence-determined classes of convergences: pretopologies, paratopologies, hypotopologies, and pseudotopologies.

LEMMA XVIII.7.8 (*). *Let Y be a set and λ be an uncountable regular cardinal such that $\lambda > \operatorname{card} Y$. Then for every map $f : \lambda \longrightarrow Y$, there exists a uniform λ-complete filter $\mathcal{F}$ such that $\mathcal{F} \perp \mathcal{C}_\lambda$ and $f[\mathcal{F}] = f[\mathcal{C}_\lambda{}^\uparrow]$.*

PROOF. Consider the family $\mathcal{P} := \{f^-(y) : y \in f(\lambda)\}$ of subsets of λ, a subfamily $\mathcal{P}_0 := \{f^-(y) : \operatorname{card} f^-(y) < \lambda\}$, and $\mathcal{P}_1 := \mathcal{P} \setminus \mathcal{P}_0$. Let

$$X_0 := \bigcup \mathcal{P}_0, \quad X_1 := \bigcup \mathcal{P}_1.$$

As $\operatorname{card} \mathcal{P} = \operatorname{card} Y$, we deduce that $\operatorname{card} X_0 < \lambda$, hence $\operatorname{card} X_1 = \lambda$. To show that there exists $C \in \mathcal{C}_\lambda$ such that

$$\operatorname{card}(f^-(y) \cap C) = \operatorname{card}(f^-(y) \setminus C) = \lambda,$$

for every $y \in f(\lambda)$ such that $f^-(y) \in \mathcal{P}_1$, let us write

$$\mathcal{P}_1 = \{f^-(y) : y < \kappa\},$$

where κ is a cardinal identified with a subset of Y. We shall now construct by transfinite induction two families $\{A_\alpha : \alpha < \lambda\}$ and $\{B_\alpha : \alpha < \lambda\}$ of subsets of λ such that, any two distinct members of $\{A_\alpha : \alpha < \lambda\} \cup \{B_\alpha : \alpha < \lambda\}$ are disjoint, and for each $y < \kappa$ and every $\alpha < \lambda$, there exist distinct elements $a_\alpha(y)$ and $b_\alpha(y)$ of λ so that

$$\{a_\alpha(y)\} = A_\alpha \cap f^-(y), \qquad \{b_\alpha(y)\} = B_\alpha \cap f^-(y), \tag{XVIII.7.4}$$

and

$$\operatorname{cl}(\bigcup\nolimits_{\alpha<\lambda} A_\alpha) \cap \bigcup\nolimits_{\alpha<\lambda} B_\alpha = \varnothing.$$

If $\beta < \lambda$ and for each $\alpha < \beta$, we have constructed A_α and B_α such that (XVIII.7.4) and $\operatorname{cl}(\bigcup_{\alpha<\beta} A_\alpha) \cap \bigcup_{\alpha<\beta} B_\alpha = \varnothing$, then for each $y < \kappa$, let $a_\beta(y) \in f^-(y)$ fulfill

$$a_\beta(y) > \sup(\operatorname{cl} \bigcup\nolimits_{\alpha<\beta} A_\alpha \cup \bigcup\nolimits_{\alpha<\beta} B_\alpha).$$

Let $A_\beta := \{a_\beta(y) : y < \kappa\}$. Now, for each $y < \kappa$, let $b_\beta(y) \in f^-(y)$ fulfill $b_\beta(y) > \sup A_\beta$. Let $B_\beta := \{b_\beta(y) : y < \kappa\}$.

Then $C := \mathrm{cl}(\bigcup_{\alpha<\lambda} A_\alpha)$ is a closed unbounded subset of λ, disjoint from $\bigcup_{\alpha<\lambda} B_\alpha$. Moreover, for each $y < \kappa$, both $C \cap f^-(y)$ and $f^-(y) \setminus C$ have cardinality λ, so that there exists a bijection $h : X_0 \cup C \longrightarrow X_0 \cup (X_1 \setminus C)$ such that $h(x) := x$ if $x \in X_0$, and $h(f^-(y) \cap C) = f^-(y) \setminus C$ for each $y < \kappa$. Let $\mathcal{F} := h[\mathcal{C}_\lambda^\uparrow]$. As, by Theorem XXVII.7.2, $\mathcal{C}_\lambda^\uparrow$ is λ-complete and uniform, so is $\mathcal{F}$. By construction, $\mathcal{F} \perp \mathcal{C}_\lambda$ and $f[\mathcal{F}] = f[\mathcal{C}_\lambda^\uparrow]$. ☐

THEOREM XVIII.7.9 (Wojciechowski). *The class of hypotopologies is not simple.*

PROOF. Let τ be a convergence on a set Y and let $\lambda > \operatorname{card} Y$. Let ξ be the missing $\mathcal{C}_\lambda$ prime hypotopology on λ from Example XVIII.5.4. By Lemma XVIII.7.8, for every $f \in Y^\lambda$, there is a λ-complete filter $\mathcal{F}_f$ on λ, dissociated from $\mathcal{C}_\lambda$ and such that $f[\mathcal{F}_f] = f[\mathcal{C}_\lambda^\uparrow]$. Then $\lim_\xi \mathcal{C}_\lambda = \varnothing$ and $\infty \in \lim_\xi \mathcal{F}_f$. If $f \in \mathscr{C}(\xi, \tau)$, then $f(\infty) \in \lim_\tau f[\mathcal{F}_f] = \lim_\tau f[\mathcal{C}_\lambda]$. Therefore,

$$\infty \in \lim\nolimits_{\bigvee_{f \in \mathscr{C}(\xi,\tau)} f^- \tau} \mathcal{C}_\lambda \neq \lim\nolimits_\xi \mathcal{C}_\lambda = \varnothing.$$

☐

XVIII.8. Commutation of functors with products

Theorem XVIII.3.9 shows that the pseudotopologizer S commutes with arbitrary products, thus *a fortiori* $\mathrm{S}\xi \times \mathrm{S}\tau = \mathrm{S}(\xi \times \tau)$. On the other hand, Example XVIII.6.6 demonstrates that none of other basic reflectors (topologizer T, pretopologizer S_0, and paratopologizer S_1) does.

EXAMPLE XVIII.8.1. The coreflector $\mathrm{I}_1 = B_{\mathbb{F}_1}$ on the class of convergences of *countable character*. Then $\mathrm{I}_1\xi \times \mathrm{I}_1\tau \geq \mathrm{I}_1(\xi \times \tau)$. In fact, if $(x, y) \in \lim_{\mathrm{I}_1\xi \times \mathrm{I}_1\tau} \mathcal{F}$, that is, $x \in \lim_{\mathrm{I}_1\xi} p_X[\mathcal{F}]$ and $y \in \lim_{\mathrm{I}_1\tau} p_Y[\mathcal{F}]$. Hence there exist $\mathcal{E}_X, \mathcal{E}_Y \in \mathbb{F}_1$ such that $x \in \lim_\xi \mathcal{E}_X$ and $y \in \lim_\tau \mathcal{E}_Y$ and $\mathcal{E}_X \leq p_X[\mathcal{F}]$ and $\mathcal{E}_Y \leq p_Y[\mathcal{F}]$. Consequently, $\mathcal{E}_X \times \mathcal{E}_Y \in \mathbb{F}_1$, $(x, y) \in \lim_{\xi\times\tau}(\mathcal{E}_X \times \mathcal{E}_Y)$, and $\mathcal{E}_X \times \mathcal{E}_Y \leq \mathcal{F}$, proving that $(x, y) \in \lim_{\mathrm{I}_1(\xi\times\tau)} \mathcal{F}$ ([13]).

[13]This extends to countable products, because if $\mathcal{E}_n \in \mathbb{F}_1$ for $n < \omega$, then $\prod_{n<\omega} \mathcal{E}_n \in \mathbb{F}_1$. Indeed, if $\mathcal{B}_n = \{B_{n,k} : k < \omega\}$ is a filter-base of $\mathcal{E}_n$ for each $n < \omega$, then $\mathcal{B} = \{B_{n,k} : n, k < \omega\}$ is a subbase of $\bigvee_{n<\omega} \mathcal{E}_n$, hence $\bigvee_{n<\omega} \mathcal{E}_n \in \mathbb{F}_1$, thus $\prod_{n<\omega} \mathcal{E}_n \in \mathbb{F}_1$.

PROPOSITION XVIII.8.2. *An adherence-determined (with respect to a composable class) reflector commutes with (finite) suprema if and only if it does with (finite) products.*

PROOF. Let $J = \mathrm{A}_{\mathbb{J}}$, where $\mathbb{J}$ is a composble class of filters. By Theorem XVIII.5.7, $J(f^{-}\tau) = f^{-}(J\tau)$. Suppose that J commutes with finite suprema, and let ξ be a convergence on X and τ a convergence on Y. Then

$$\begin{aligned} J(\xi \times \tau) &= J(p_X^{-}\xi \vee p_Y^{-}\tau) \leq J(p_X^{-}\xi) \vee J(p_Y^{-}\tau) \\ &= p_X^{-}(J\xi) \vee p_Y^{-}(J\tau) = J\xi \times J\tau. \end{aligned}$$

Conversely, let $J(\xi \times \tau) \leq J\xi \times J\tau$ for each ξ and τ, and let θ_0 and θ_1 be convergences on a set V. Consider two copies of V, that is, V_0 and V_1, and let identify V with the diagonal

$$\triangle = \{(v_0, v_1) \in V_0 \times V_1 : v_0 = v_1\}.$$

Let $\varphi_0 : \triangle \longrightarrow V_0, \varphi_1 : \triangle \longrightarrow V_1$ defined by $\varphi_0(v_0, v_1) := v_0$, and $\varphi_1(v_0, v_1) := v_1$. Then $\varphi_0\theta_0 \times \varphi_1\theta_1 = \varphi_0^{-}\varphi_0\theta_0 \vee \varphi_1^{-}\varphi_1\theta_1 = \theta_0 \vee \theta_1$. Hence,

$$\begin{aligned} J(\theta_0 \vee \theta_1) &= J(\varphi_0^{-}(\varphi_0\theta_0) \vee \varphi_1^{-}(\varphi_1\theta_1)) = J(\varphi_0\theta_0 \times \varphi_1\theta_1) \\ &\leq J(\varphi_0\theta_0) \times J(\varphi_1\theta_1) \\ &= \varphi_0^{-}J(\varphi_0\theta_0) \vee \varphi_1^{-}J(\varphi_1\theta_1) = J\varphi_0^{-}(\varphi_0\theta_0) \vee J\varphi_1^{-}(\varphi_1\theta_1) = J\theta_0 \vee J\theta_1. \end{aligned}$$

□

By Example XVIII.6.6 and Corollary XVIII.7.2, $\mathrm{S}_0, \mathrm{S}_1$, and $\mathrm{S}_{\wedge 1}$ do not commute with finite products.

EXAMPLE XVIII.8.3 ($\mathrm{T}\xi \times \tau \ngeq \mathrm{T}(\xi \times \tau)$). Prime convergences ξ and τ from Example VI.6.11 are such that $\mathrm{S}_0\xi \vee \tau \ngeq \mathrm{S}_0(\xi \vee \tau)$ and $\mathrm{S}_0\xi \times \tau \ngeq \mathrm{S}_0(\xi \times \tau)$. Accordingly, $\mathrm{S}_0\xi$ is a prime pretopology, hence a topology, so that $\mathrm{T}\xi = \mathrm{S}_0\xi$. As $\xi \vee \tau = \iota$ is the discrete convergence, hence the discrete topology, $\iota = \mathrm{T}(\xi \vee \tau) \nleq \mathrm{T}\xi \vee \tau$. In Example VI.6.12, it was shown that $(\xi \times \tau)|_{\triangle} = \xi \vee \tau$

XVIII.9. Classification of quotient maps

As we have said in Definition IX.4.1, a surjective map $f \in \mathscr{C}(\xi, \tau)$ is called a *topologically quotient map* if ξ and τ are topologies, and if $\tau = \mathrm{T}(f\xi)$.

The definition above comprises the continuity $f\xi \geq \tau$ on one hand, and the condition that $\tau \geq \mathrm{T}(f\xi)$ on the other. It is rational to consider these two facets separately, because some arguments hinge on one, and some others on another facet.

This is why, in the study of various concepts of quotient map for convergences, we shall not assume continuity ([14]).

Therefore, from now on a surjective map $f : |\xi| \longrightarrow |\tau|$ is called *topologically quotient* if $\tau \geq \mathrm{T}(f\xi)$. More generally,

DEFINITION XVIII.9.1. Let J be a reflector. A surjective map $f : |\xi| \longrightarrow |\tau|$ is called a *J-quotient map*, in symbols $f \in Q_J(\xi, \tau)$, whenever

$$\tau \geq J(f\xi). \tag{XVIII.9.1}$$

Notice that (XVIII.9.1) can be interpreted as a sort of inverse continuity.

REMARK XVIII.9.2. It is immediate that if $J \leq L$ are reflectors, then each L-quotient map is J-quotient. On the other hand, if a convergence θ fulfills $J\theta < \theta \leq L\theta$, then the identity $i \notin Q_L(\theta, J\theta)$ is not an L-quotient, but $i \in Q_J(\theta, J\theta)$ is a J-quotient.

If $J = \mathrm{I}$, the identity reflector, then (XVIII.9.1) becomes

$$\tau \geq f\xi, \tag{XVIII.9.2}$$

and is called a (*convergence*) *quotient map*. In other words, a surjective map $f \in Q(\xi, \tau)$ if and only if τ is finer than the final convergence $f\xi$. It follows from Proposition IV.3.8 that

PROPOSITION XVIII.9.3. *A surjective map f is quotient, that is $\tau \geq f\xi$, if and only if for each filter $\mathcal{G}$ and each point y such that $y \in \lim_\tau \mathcal{G}$,*

$$\underset{x \in f^-(y)}{\exists} \; \underset{\mathcal{F}}{\exists} \, (x \in \lim\nolimits_\xi \mathcal{F}) \wedge (\mathcal{G} = f[\mathcal{F}]). \tag{XVIII.9.3}$$

Such maps are traditionally called (in the context of topologies) *almost open* ([15]). The term suggests that they are akin to open maps. In topology, a continuous map between topological spaces is called *open* if it maps each open set onto an open set (Definition IX.4.3).

---

[14]As in [56], [83], and other papers on this theme.

[15]See A. V. ARHANGEL'SKII [7].

DEFINITION XVIII.9.4. Let ξ and τ be convergences. A map $f : |\xi| \longrightarrow |\tau|$ is called *open* if for each filter $\mathcal{G}$ and a point y such that $y \in \lim_\tau \mathcal{G}$,

$$\forall_{x\in f^-(y)} \exists_{\mathcal{F}} (x \in \lim_\xi \mathcal{F}) \wedge (\mathcal{G} \approx f[\mathcal{F}]). \tag{XVIII.9.4}$$

Obviously, each surjective open map is almost open. Let us relate Definition XVIII.9.4 to Definition IX.4.3 (formulated in the framework of topologies).

LEMMA XVIII.9.5. *If a map is open (Definition XVIII.9.4), then the image of each open set is open.*

PROOF. Let O be a ξ-open set, let $y \in f(O)$, and let $y \in \lim_\tau \mathcal{G}$. Consequently, there exists $x \in O \cap f^-(y)$ and by (XVIII.9.4), there exists a filter $\mathcal{F}$ such that $x \in \lim_\xi \mathcal{F}$ and $\mathcal{G} = f[\mathcal{F}]$. Thus $O \in \mathcal{F}$, because $O \in \mathcal{O}_\xi(x)$, so that $f(O) \in f[\mathcal{F}] = \mathcal{G}$, proving that $f(O)$ is τ-open. □

PROPOSITION XVIII.9.6. *If the images of open sets are open under a map between topological spaces, then the map is open (Definition XVIII.9.4).*

PROOF. Let ξ and τ be topologies. If $O \in \mathcal{O}_\xi(x)$, then by the condition, $f(O) \in \mathcal{O}_\tau(f(x))$, hence $f[\mathcal{N}_\xi(x)] \leq \mathcal{N}_\tau(y)$ for each x and y such that $f(x) = y$. If $y \in \lim_\tau \mathcal{G}$, that is, $f[\mathcal{N}_\xi(x)] \leq \mathcal{G}$ for each $x \in f^-(y)$, then by Lemma V.6.4, the filter $\mathcal{F} := \mathcal{N}_\xi(x) \vee f^-[\mathcal{G}]$ is such that $\mathcal{G} \approx f[\mathcal{F}]$ and, of course, $x \in \lim_\xi \mathcal{F}$. □

REMARK XVIII.9.7. The traditional definition of open map includes continuity. Therefore, $f \in \mathscr{C}(\xi, \tau)$ is open if $f[\mathcal{N}_\xi(x)] \geq \mathcal{N}_\tau(f(x))$ (by continuity), and $\mathcal{N}_\tau(f(x)) \geq f[\mathcal{N}_\xi(x)]$ (because the images of open sets are open) for each x. As a result, $f[\mathcal{N}_\xi(x)] = \mathcal{N}_\tau(f(x))$ for each x.

LEMMA XVIII.9.8. *Let ξ be a convergence on X and $f : X \longrightarrow Y$ be surjective. Then a set O is $f\xi$-open if and only if $f^-(O)$ is ξ-open.*

PROOF. If O is $f\xi$-open, then $f^-(O)$ is ξ-open, because $f \in \mathscr{C}(\xi, f\xi)$. Conversely, let $f^-(O)$ be ξ-open, and let $y \in O \cap \lim_{f\xi} \mathcal{G}$. By Proposition IV.3.8, there exists $x \in f^-(y)$ and a filter $\mathcal{F}$ such that $x \in \lim_\xi \mathcal{F}$ and $f[\mathcal{F}] \approx \mathcal{G}$. It follows from assumptions that $f^-(O) \in \mathcal{F}$, hence $f(f^-(O)) \in f[\mathcal{F}] \approx \mathcal{G}$, hence $O \in \mathcal{G}$, which proves that O is $f\xi$-open. □

Therefore,

PROPOSITION XVIII.9.9. *Let f be a surjection. Then $\tau \geq \mathrm{T}(f\xi)$ if and only if O is τ-open whenever $f^-(O)$ is ξ-open.*

REMARK XVIII.9.10. As traditionally, *quotient* signifies *topologically quotient* and continuous (and moreover the convergences considered are topologies), a classical counterpart of Proposition XVIII.9.9 says that a map is quotient whenever O is τ-open if and only if $f^{-}(O)$ is ξ-open.

The cases of S_0-quotient, S_1-quotient, and S-quotient maps, all correspond to adherence-determined reflectors, and will be described jointly.

If a reflector is adherence-determined, that is, of the form $A_{\mathbb{H}}$, where $\mathbb{H}$ is an initial class of filters, then we say that a map is $\mathbb{H}$-*quotient* whenever it is $A_{\mathbb{H}}$-quotient. By Corollary XVIII.5.6,

PROPOSITION XVIII.9.11. *Let $\mathbb{H}$ be an initial class of filters. A surjective map $f : |\xi| \to |\tau|$ is $\mathbb{H}$-quotient if and only if for each $\mathcal{H} \in \mathbb{H}$,*

$$\text{(XVIII.9.5)} \qquad \operatorname{adh}_{\tau} \mathcal{H} \subset \operatorname{adh}_{f\xi} \mathcal{H}.$$

We notice that (XVIII.9.5) holds if and only if, for each $\mathcal{H} \in \mathbb{H}$ and each $y \in |\tau|$,

$$\text{(XVIII.9.6)} \qquad y \in \operatorname{adh}_{\tau} \mathcal{H} \Longrightarrow f^{-}(y) \cap \operatorname{adh}_{\xi} f^{-}[\mathcal{H}] \neq \varnothing.$$

Indeed, as f is surjective, by (VIII.1.12), $\operatorname{adh}_{\tau} \mathcal{H} \subset \operatorname{adh}_{f\xi} \mathcal{H} = f(\operatorname{adh}_{\xi} f^{-}[\mathcal{H}])$, so that (XVIII.9.5) amounts to (XVIII.9.6).

The following table reports the traditional names of the quotient maps ([16]) discussed:

J	reflector	J-quotient	traditional name
I	*identity*	$\tau \geq f\xi$	*almost open*
S	*pseudotopologizer*	$\tau \geq \mathrm{S}(f\xi)$	*biquotient*
$\mathrm{S}_{\wedge 1}$	*hypotopologizer*	$\tau \geq \mathrm{S}_{\wedge 1}(f\xi)$	*weakly biquotient*
S_1	*paratopologizer*	$\tau \geq \mathrm{S}_1(f\xi)$	*countably biquotient*
S_0	*pretopologizer*	$\tau \geq \mathrm{S}_0(f\xi)$	*hereditarily quotient*
T	*topologizer*	$\tau \geq \mathrm{T}(f\xi)$	*quotient*

TABLE XVIII.9.1. A surjective map $f : |\xi| \longrightarrow |\tau|$ is *J-quotient whenever $\tau \geq J(f\xi)$*. The corresponding traditional name is used when ξ and τ are topologies and, moreover, $f \in \mathscr{C}(\xi, \tau)$.

All these variants of quotient are distinct because of Remark XVIII.9.2.

[16] *Weakly biquotiont maps* were introduced by C. LIU [**176**].

Proposition VIII.2.10 enables us to reformulate the characterizations above in terms of covers or inherences, which is important, because traditionally the discussed variants of quotient were considered in terms of covers.

PROPOSITION XVIII.9.12. *Let* $\mathbb{H}$ *be an initial class of filters. A surjective map* f *is* $\mathbb{H}$*-quotient if and only if for each* $\mathcal{P}$ *such that* $\mathcal{P}_c \in \mathbb{H}$ *and for each* y*,*

$$\text{inh}_\xi \mathcal{P} \supset f^-(y) \Longrightarrow y \in \text{inh}_\tau f[\mathcal{P}]. \tag{XVIII.9.7}$$

PROOF. Indeed, set $\mathcal{P} = f^-[\mathcal{H}_c] = f^-[\mathcal{H}]_c$ in (XVIII.9.6). Then $\mathcal{H}_c = f[f^-[\mathcal{H}_c]] = f[\mathcal{P}]$. □

We always keep in mind that in general the final convergence $f\xi$ of a topology ξ is not a topology, and thus the condition $\tau = J(f\xi)$ for a reflector $J \geq \mathrm{T}$ is not trivial. This particular topic is one of forthright examples how non-topological methods apply in purely topological context.

Recollect that if θ is a pretopology, then $\mathcal{P}$ is a θ-cover of A, that is, $A \subset \text{inh}_\theta \mathcal{P}$ is equivalent to $A \subset \bigcup_{P \in \mathcal{P}} \text{inh}_\theta P$, and thus if θ is a topology, also to $A \subset \bigcup_{P \in \mathcal{P}} \text{int}_\theta P$.

Hereditarily quotient. A surjective map f between topological spaces is $\mathbb{F}_0$-*quotient* (*hereditarily quotient*) if and only if $y \in \text{int} f(O)$ for each open $O \supset f^-(y)$. In fact, as $\mathbb{F}_0$ is the class of principal filters, the family of complements $(\mathbb{F}_0)_c$ is that of principal ideals. Therefore, the fact that $O^\downarrow$ is an open cover of $f^-(y)$ amounts to $f^-(y) \subset \text{int}\, O$. It is enough to take O open, that is, $\text{int}\, O = O$.

Countably biquotient. A surjective map f between topological spaces is $\mathbb{F}_1$-*quotient* (*countably biquotient*) if and only if for any sequence $(O_n)_n$ of open sets such that $f^-(y) \subset \bigcup_{n<\omega} O_n$, there exists $n < \omega$ such that $y \in \text{int} f(O_n)$. Indeed, as $\mathbb{F}_1$ is the class of countably based filters, the family of complements $(\mathbb{F}_1)_c$ is that of countably based ideals. Therefore, $f^-(y) \in \text{inh}\{O_n : n < \omega\}$ amounts to $f^-(y) \subset \bigcup_{n<\omega} \text{int}\, O_n$. It is enough to consider the ideals that are based in open sets, so that we can take $O_n = \text{int}\, O_n$ without loss of generality.

Biquotient. A surjective map f between topological spaces is $\mathbb{F}$-*quotient* (*biquotient*) if and only if for any family $\mathcal{P}$ of open sets such that $f^-(y) \subset \bigcup_{P \in \mathcal{P}} P$, there exists $P \in \mathcal{P}$ such that $y \in \text{int} f(P)$. As $\mathbb{F}$ is the class of all filters, the family of complements $(\mathbb{F})_c$ is that of all ideals. Similarly to the preceding cases, it is enough to consider openly based ideals.

What is important about quotient maps corresponding to adherence-determined reflectors is their hereditariness.

DEFINITION XVIII.9.13. A class of maps $\mathscr{G}$ is called *hereditary* if for each $g \in \mathscr{G}(\xi, \tau)$, and for each subset B of the range of g, the restriction

$$g|_{g^- B} : g^- B \longrightarrow B$$

belongs to $\mathscr{G}(\xi|_{g^- B}, \tau|_B)$.

The class of continuous maps is obviously hereditary.

PROPOSITION XVIII.9.14. *The class of (convergence) quotient maps is hereditary.*

PROOF. Let $f \in Y^X$ be surjective, and let ξ be a convergence on Y. Then

$$(f\xi)|_B = f|_{fB}(\xi|_{fB}). \tag{XVIII.9.8}$$

In fact, by definition of restriction, $y \in \lim_{(f\xi)|_B} \mathcal{G}$ if $y \in B \in \mathcal{G}$ and $y \in \lim_{f\xi} \mathcal{G}$. As f is surjective, by (IV.3.2), there exists $\mathcal{F}$ and x such that $x \in \lim_\xi \mathcal{F}$, $f(x) = y$, and $f[\mathcal{F}] = \mathcal{G}$. Accordingly, $f^- B \in f^-[f[\mathcal{F}]] \leq \mathcal{F}$ and $x \in f^- B$, that is, $x \in \lim_{\xi|_{f^- B}} \mathcal{F}$. As a result, $y \in \lim_{f|_{f^- B}(\xi|_{f^- B})} \mathcal{G}$. Conversely, as $f|_{f^- B} : f^- B \longrightarrow B$ is surjective, the last formula means that there exist $x \in f^- B$ and a filter $\mathcal{F}$ on $f^- B$ such that $f|_{f^- B}(x) = y$ and $f|_{f^- B}[\mathcal{F}] = \mathcal{G}$. In particular, $x \in \lim_\xi \mathcal{F}$, where $\mathcal{F}$ is a filter-base on X. Accordingly, $y \in \lim_{f\xi} \mathcal{G}$. As $f(f^- B) = B$ by the surjectivity of f, $B \in \mathcal{G}$ and $f(x) \in B$. Therefore, $y \in \lim_{(f\xi)|_B} \mathcal{G}$, and if $\tau = f\xi$, then $\tau|_B \geq (f\xi)|_B = f|_{f^- B}(\xi|_{f^- B})$. □

We have observed that adherence-determined reflectors J fulfill

$$J(f^- \tau) = f^-(J\tau)$$

for each map f and every convergence τ. The equality above implies in particular that

$$J(\theta|_V) = (J\theta)|_V \tag{XVIII.9.9}$$

for any convergence θ and each non-empty subset V of its underlying set.

THEOREM XVIII.9.15. *If J is a reflector commuting with restrictions, then the class of J-quotient maps is hereditary.*

PROOF. If $\tau \geq J(f\xi)$, then by (XVIII.9.9), $\tau|_B \geq (J(f\xi))|_B = J(f\xi)|_B$, hence by (XVIII.9.8), $\tau|_B \geq J(f|_{f^- B}\, \xi|_{f^- B})$. □

COROLLARY XVIII.9.16. *For each initial class $\mathbb{H}$ of filters, the class of $\mathbb{H}$-quotient maps is hereditary.*

In general, T-quotient maps are not hereditary. Indeed, we have observed in Remark XVIII.9.2 that if $\mathrm{S}_0\xi = \xi > \mathrm{T}\xi$, then the identity from $\tau = \mathrm{T}\xi$ to ξ is a T-quotient, but not an S_0-quotient. On the other hand,

THEOREM XVIII.9.17. *A* T*-quotient map is hereditary if and only if it is an* S_0*-quotient.*

PROOF. By Theorem XVIII.9.15, each S_0-quotient is hereditary. Suppose that f is a T-quotient, which is not an S_0-quotient, that is $\tau \geq \mathrm{T}(f\xi)$ but there is there exists a filter $\mathcal{G}$ and a point y such that $y \in \lim_\tau \mathcal{G} \setminus \lim_{\mathrm{S}_0(f\xi)} \mathcal{G}$. Therefore, there exists $A \in \mathcal{G}^{\#}$ such that $y \in \mathrm{cl}_{f\xi} A \setminus \mathrm{adh}_{f\xi} A$. In particular, $f^-(y) \cap \mathrm{adh}_\xi f^-(A) = \varnothing$ by (VI.4.3).
Let $B := A \cup \{y\}$. Indeed, A is not $\tau|_B$-closed by construction. On the other hand, $f^-(B) = f^-(A) \cup f^-(y)$, so that $f^-(A)$ is closed in $\xi|_{f^-B}$. Thus $f|_{f^-B}$ is not T-quotient by Proposition XVIII.9.9. □

XVIII.10. Supplement

EXAMPLE XVIII.10.1 (Standard sequential convergence). The collection of all sequential filters σ-converging to x, is a pavement of σ at x. Notice that σ is finitely stable, but is not a pretopology. Indeed, it is not difficult to check (see Exercise II.5.3) that

$$\mathcal{V}_\sigma(x) = \{]x - \tfrac{1}{n}, x + \tfrac{1}{n}[: n \in \mathbb{N}_1\}^{\uparrow}. \tag{XVIII.10.1}$$

As we have observed in Example II.2.23, $\mathcal{V}_\sigma(x)$ is not (finer than) a sequential filter, thus $x \notin \lim_\sigma \mathcal{V}_\sigma(x)$. Consequently, σ does not admit finite pavings, hence is not a pretopology. The convergence σ is Hausdorff. Indeed, if $x_0 < x_1$ and $n > \frac{1}{|x_1 - x_0|}$, then $x_0 + \frac{1}{n} < x_1 - \frac{1}{n}$, hence

$$]x_0 - \tfrac{1}{n}, x_0 + \tfrac{1}{n}[\cap]x_1 - \tfrac{1}{n}, x_1 + \tfrac{1}{n}[= \varnothing,$$

and thus $\mathcal{V}_\sigma(x_0)$ does not mesh $\mathcal{V}_\sigma(x_1)$. Therefore, $\{x_0, x_1\} \subset \lim_\sigma \mathcal{F}$ implies that $\mathcal{F} \supset \mathcal{V}_\sigma(x_0) \vee \mathcal{V}_\sigma(x_1) = 2^X$, which is a contradiction.

PROPOSITION XVIII.10.2. *If ξ is a pseudotopology such that*

$$\mathrm{adh}_\xi \mathcal{F} \supset \bigcap_{F \in \mathcal{F}} \mathrm{adh}_\xi F$$

for each filter $\mathcal{F}$, then ξ is a pretopology.

PROOF. If ξ is not a pretopology, then there exists x such that $x \notin \lim_\xi \mathcal{V}_\xi(x)$. Let $\mathcal{U} \in \beta\mathcal{V}_\xi(x)$ be such that $x \notin \lim_\xi \mathcal{U} = \mathrm{adh}_\xi \mathcal{U}$.

Then $U \in \mathcal{V}_\xi(x)^\#$ for each $U \in \mathcal{U}$, that is, $x \in \operatorname{adh}_\xi U$ for each $U \in \mathcal{U}$, so that $\operatorname{adh}_\xi \mathcal{U} \neq \bigcap_{U \in \mathcal{U}} \operatorname{adh}_\xi U$. $\square$

EXERCISE XVIII.10.3. Show that if B is τ-closed, then $\mathrm{T}(\tau|_B) = (\mathrm{T}\tau)|_B$.

SOLUTION. It suffices to show that each $\tau|_B$-closed set is $(\mathrm{T}\tau)|_B$-closed. If C is a $\tau|_B$-closed subset of B, that is, if $C \in \mathcal{F}$, then $\lim_{\tau|_B} \mathcal{F} \subset C$. But $\lim_{\tau|_B} \mathcal{F} = B \cap \lim_\tau \mathcal{F} \subset C = B \cap C$, which is τ-closed.

PROPOSITION XVIII.10.4. *Let J be a reflector. Then the composition of two J-quotient maps is J-quotient.*

PROOF. Let $f \in Q_J(\xi, \tau)$ and $g \in Q_J(\tau, \theta)$. Then $\tau \geq J(f\xi)$ implies $g\tau \geq gJ(f\xi) \geq J(g(f\xi))$ by (XVII.2.2). Hence, $\theta \geq J(J(g \circ f(\xi))) = J(g \circ f(\xi))$, that is, $g \circ f \in Q_J(\xi, \theta)$. $\square$

CHAPTER XIX

Diagonality and regularity

As we shall see, diagonality and regularity are characterized by the same contour operation, but in two antithetic ways.

Let us start by establishing some equalities between set-theoretic limits. They readily entail analogous equalities regarding certain contours, which will be instrumental in the sequel ([1]).

Let Y, Z be non-empty sets, $\mathcal{G}$ an isotone family of subsets of Y, and $S : Y \longrightarrow 2^Z$. Then the *lower* and the *upper set-theoretic limits* are defined, as in (XII.9.3), by

$$\mathrm{Li}_{\mathcal{G}} S(\cdot) := \bigcup\nolimits_{G \in \mathcal{G}} \bigcap\nolimits_{y \in G} S(y), \ \mathrm{Ls}_{\mathcal{G}} S(\cdot) := \bigcap\nolimits_{G \in \mathcal{G}} \bigcup\nolimits_{y \in G} S(y).$$

It turns out that

LEMMA XIX.0.5. *If $\mathcal{G}$ is an isotone family, then*

$$\mathrm{Li}_{\mathcal{G}} S = \mathrm{Ls}_{\mathcal{G}^\#} S, \quad \mathrm{Li}_{\mathcal{G}^\#} S = \mathrm{Ls}_{\mathcal{G}} S. \tag{XIX.0.2}$$

PROOF. By definition, $z \in \mathrm{Li}_{\mathcal{G}} S$ whenever there exists $G \in \mathcal{G}$ such that $x \in S(y)$ for each $y \in G$. Therefore, for every $H \in \mathcal{G}^\#$ and every $G \in \mathcal{G}$, there is $y \in H \cap G$ such that $x \in S(y)$, equivalently $x \in \bigcap_{H \in \mathcal{G}^\#} \bigcup_{y \in H} S(y)$. Replacing $\mathcal{G}$ by $\mathcal{G}^\#$ in the first equality of (XIX.0.2), we get the second equality, because $\mathcal{G}^{\#\#} = \mathcal{G}$ for each isotone family $\mathcal{G}$. □

XIX.1. Contour operation

We have defined the *contour operation* in (XII.7.6) for filters. Although it extends naturally to arbitrary families, it is enough to restrict our considerations of *contour* to isotone families of sets. Greater generality, useless for our purposes, would slightly complicate the calculus without adding anything meaningful!

If $\mathcal{S}(y)$ is an isotone family of subsets of X for each $y \in G$, then we write

$$\mathcal{S}(G) := \bigcap\nolimits_{y \in G} \mathcal{S}(y). \tag{XIX.1.1}$$

[1]These equalities constitute a special case of some laws valid for complete completely distributive lattices (G. H. GRECO [118]).

This is a handy notation, which we have already informally used in previous chapters.

EXAMPLE XIX.1.1. In particular, if ξ is a convergence on X and $Y = X$, then $\mathcal{V}_\xi(G)$ is the set of all vicinities of G, and $\mathcal{N}_\xi(G)$ stands for the set of all neighborhoods of G.

Let us extend Definition XII.7.14.

DEFINITION XIX.1.2. If $\mathcal{G} \in \mathbb{I}Y$ and $\mathcal{S} : Y \longrightarrow \mathbb{I}X$, then

$$\mathcal{S}\{\mathcal{G}\} = \textstyle\int_{\mathcal{G}} \mathcal{S}(\cdot) := \bigcup_{G\in\mathcal{G}} \bigcap_{y\in G} \mathcal{S}(y)$$

is called a *contour* of $\mathcal{S}$ ([2]).

REMARK XIX.1.3. The double notation is introduced in order to help those who might use also [83], where a symbol similar to the first of our symbols was adopted. Depending on context, either one or the other symbol might be easier to implement.

For instance, if ξ is a convergence on X and $\mathcal{V}_\xi : X \longrightarrow \mathbb{F}X$ is the vicinity map, which associates the vicinity filter $\mathcal{V}_\xi(x)$ on X with each element x of X, then $\mathcal{V}_\xi\{\mathcal{V}_\xi(x)\} = \int_{\mathcal{V}_\xi(x)} \mathcal{V}_\xi \in \mathbb{F}X$.

It is immediate that $\int_{\mathcal{G}} \mathcal{S} \in \mathbb{I}X$. If $\mathcal{G}$ is a principal filter on Y, that is, there is $G \subset Y$ such that $\mathcal{G} = G^\uparrow = \{D \subset Y : G \subset D\}$, then $\int_{G^\uparrow} \mathcal{S} = \mathcal{S}(G)$. If $\mathcal{G}$ is an isotone family of subsets of $Y = \mathbb{I}X$, and $\mathcal{I}(y) := y^\uparrow$ for each $y \in Y$, then

$$\textstyle\int_{\mathcal{G}} \mathcal{I}(\cdot) = \mathcal{G}. \tag{XIX.1.2}$$

EXAMPLE XIX.1.4 (Sequential contour). If $(x_{n,k})_k$ is a sequence on X for each $n < \omega$ (that is, a *bisequence* on X) then A belongs to

$$\int_{(n)} (x_{n,k})_k \tag{XIX.1.3}$$

whenever there exists $n_0 < \omega$ such that for each $n > n_0$, there is $k_n < \omega$, for which $x_{n,k} \in A$ for each $k > k_n$. If $\mathcal{E}_n$ is a sequential filter $\mathcal{E}_n \approx (x_{n,k})_k$, then $\int_{(n)} (x_{n,k})_k = \int_{(\mathbb{N})_0} \mathcal{E}_n$,where $(\mathbb{N})_0$ is the cofinite filter of $\mathbb{N}$. Sequential contours play an important role in the study of P-points (see Section XIX.5).

[2]Introduced by H. J. KOWALSKY in [161] without any name. C. COOK and H. FISCHER in [48] call it the *compression operator*. Independently, Z. FROLÍK defines the contour for ultrafilters in [106]. Let us observe that $\int_{\mathcal{G}} \mathcal{S} = \mathrm{Li}_{\mathcal{G}}\, \mathcal{S}$, the *set-theoretic lower limit* defined in (XII.9.3). The latter symbol was used in [91], while in [83] we used $\mathcal{S}(\mathcal{G})$.

PROPOSITION XIX.1.5. *There is no sequential filter finer than a sequential contour.*

PROOF. Let $X := \{(n,k) : n,k < \omega\}$, that is, $X = \mathbb{N} \times \mathbb{N}$, and let $\mathcal{C} := \int_k (n,k)_k$ stand for the sequential contour (XIX.1.3). Suppose that a sequence $y_p := (n_p, k_p)_p$ is finer than $\mathcal{C}$. Then $(n_p)_p$ tends to ∞, because otherwise there is N such that $\{p < \omega : n_p \leq N\}$ is infinite, and thus $\{y_p : n_p \leq N\} \cap \{(n,k) : n > N\} = \varnothing$, so that there are infinitely many terms of the sequence lying off $\{(n,k) : n > N\} \in \mathcal{C}$.

Let $N_p := \max\{n_0, \ldots, n_p\}$ and $K_p := \max\{k_0, \ldots, k_p\}$, and

$$f(n) := \begin{cases} 0 & n < N_0, \\ K_p & N_p \leq n < N_{p+1}. \end{cases}$$

Then the strict epigraph $\{(n,k) : f(n) < k\}$ of f belongs to $\mathcal{C}$, and is disjoint from $\{y_p : p < \omega\}$. □

Notice that if $\mathcal{G}(y) = \mathcal{G}$ is constant, then $\int_{\mathcal{F}} \mathcal{G}(\cdot) = \mathcal{G}$ for every filter $\mathcal{G}$ on Y. Indeed, $A \in \int_{\mathcal{F}} \mathcal{G}(\cdot)$ whenever there exists $F \in \mathcal{F}$ such that $A \in \mathcal{G}$ for each $y \in F$, and since $F \neq \varnothing$ for every $F \in \mathcal{F}$, we infer that $\int_{\mathcal{F}} \mathcal{G}(\cdot) = \mathcal{G}$.

We shall review some basic properties of the operation of contour. Let us assume that all considered families are isotone. It is straightforward that if $\mathcal{F}$ and $\mathcal{G}(y)$ are isotone for each y, then $\int_{\mathcal{F}} \mathcal{G}$ is isotone.

LEMMA XIX.1.6. *If $\mathcal{F}_0 \subset \mathcal{F}_1$ and $\mathcal{G}_0(y) \subset \mathcal{G}_1(y)$ for each y, then*

$$\int_{\mathcal{F}_0} \mathcal{G}_0(\cdot) \subset \int_{\mathcal{F}_1} \mathcal{G}_1(\cdot).$$

PROOF. If $A \in \int_{\mathcal{F}_0} \mathcal{G}_0(\cdot)$, that is, there is $F \in \mathcal{F}_0 \subset \mathcal{F}_1$ such that $A \in \mathcal{G}_0(y) \subset \mathcal{G}_1(y)$ for each $y \in F$, hence $A \in \int_{\mathcal{F}_1} \mathcal{G}_1(\cdot)$. □

PROPOSITION XIX.1.7. *If $\mathcal{F}$ and $\mathcal{G}(y)$ are filters for each y, then $\int_{\mathcal{F}} \mathcal{G}(\cdot)$ is a filter.*

PROOF. Let $A_0, A_1 \in \int_{\mathcal{F}} \mathcal{G}(\cdot)$, that is, there exist $F_0, F_1 \in \mathcal{F}$ such that $A_0 \in \mathcal{G}(y)$ for each $y \in F_0$ and $A_1 \in \mathcal{G}(y)$ for each $y \in F_1$. Hence $A_0 \cap A_1 \in \mathcal{G}(y)$ for each $y \in F_0 \cap F_1 \in \mathcal{F}$. □

PROPOSITION XIX.1.8. *If $\mathcal{G}(y)$ is a free filter on X for every $y \in Y$ and $\mathcal{F}$ is a filter, then $\int_{\mathcal{F}} \mathcal{G}(\cdot)$ is a free filter.*

PROOF. By assumption, $\mathcal{G}(y) \geq (X)_0$ for each $y \in Y$, hence by Lemma XIX.1.6, $\int_{\mathcal{F}} \mathcal{G}(\cdot) \geq \int_{\mathcal{F}} (X)_0 = (X)_0$, that is, $\int_{\mathcal{F}} \mathcal{G}(\cdot)$ is a free filter on X. □

LEMMA XIX.1.9. *Let $\mathcal{G}^{\#}(y) := (\mathcal{G}(y))^{\#}$. Then*

$$(\text{XIX.1.4}) \qquad \left(\int_{\mathcal{F}} \mathcal{G}(\cdot)\right)^{\#} = \int_{\mathcal{F}^{\#}} \mathcal{G}(\cdot)^{\#}.$$

PROOF. Indeed, by (V.3.9),

$$(\textstyle\int_{\mathcal{F}} \mathcal{G}(\cdot))^{\#} = (\bigcup_{F\in\mathcal{F}} \bigcap_{y\in F} \mathcal{G}(y))^{\#}$$

$$\bigcap_{F\in\mathcal{F}} \bigcup_{y\in F} \mathcal{G}^{\#}(y) = \bigcup_{H\in\mathcal{F}^{\#}} \bigcap_{y\in H} \mathcal{G}^{\#}(y) = \textstyle\int_{\mathcal{F}^{\#}} \mathcal{G}(\cdot)^{\#}.$$

□

COROLLARY XIX.1.10. *If $\mathcal{F}$ and $\mathcal{G}(y)$ are filter-grills for each y, then $\int_{\mathcal{F}} \mathcal{G}$ is a filter-grill.*

PROOF. It follows from the assumption that $\mathcal{F}^{\#} \in \mathbb{F}$ and $\mathcal{G}^{\#}(y) \in \mathbb{F}$ for each y. Hence, by (XIX.1.4) and by Proposition XIX.1.7, $(\int_{\mathcal{F}} \mathcal{G}(\cdot))^{\#} \in \mathbb{F}$, thus $\int_{\mathcal{F}} \mathcal{G}(\cdot)$ is a filter-grill. □

By Proposition V.4.1,

COROLLARY XIX.1.11. *If $\mathcal{F}$ and $\mathcal{G}(y)$ are ultrafilters for each y, then $\int_{\mathcal{F}} \mathcal{G}$ is an ultrafilter.*

Observe that

$$\textstyle\int_{\bigcap_{j\in J}\mathcal{F}_j} \mathcal{G}(\cdot) = \bigcap_{j\in J} \int_{\mathcal{F}_j} \mathcal{G}(\cdot). \tag{XIX.1.5}$$

Indeed, by Lemma XIX.1.6, $\int_{\bigcap_{j\in J}\mathcal{F}_j} \mathcal{G}(\cdot) \subset \bigcap_{j\in J} \int_{\mathcal{F}_j} \mathcal{G}(\cdot)$, because $\bigcap_{k\in J} \mathcal{F}_k \subset \mathcal{F}_j$ for each $j \in J$. If $A \in \bigcap_{j\in J} \int_{\mathcal{F}_j} \mathcal{G}(\cdot)$, then $A \in \int_{\mathcal{F}_j} \mathcal{G}(\cdot)$ for each $j \in J$. Hence there is $F_j \in \mathcal{F}_j$ with $A \in \bigcap_{y\in F_j} \mathcal{G}(y)$ for each $j \in J$. Therefore, $A \in \bigcap_{j\in J} \bigcap_{y\in F_j} \mathcal{G}(y) = \mathcal{G}(\bigcup_{j\in J} F_j)$ and $\bigcup_{j\in J} F_j \in \bigcap_{j\in J} \mathcal{F}_j$, thus $A \in \int_{\bigcap_{j\in J}\mathcal{F}_j} \mathcal{G}(\cdot)$.

DEFINITION XIX.1.12 (Precontour). Consider sets X and Y. A map $\mathcal{S} : Y \longrightarrow \mathbb{I}X$ from Y to the set of isotone families of subsets of X is called a *precontour*.

In a special case, where $\mathcal{S}(x)$ is a vicinity filter at x and $X = Y$, a precontour corresponds to a pretopology.

As in the pretopological case, for $y \in Y$ and $A \subset X$, let

$$y \in \mathrm{ah}_{\mathcal{S}} A \Longleftrightarrow A \in \mathcal{S}(y)^{\#},$$
$$y \in \mathrm{ih}_{\mathcal{S}} A \Longleftrightarrow A \in \mathcal{S}(y),$$

where $\mathrm{ah}_{\mathcal{S}}$ is the *preadherence* and $\mathrm{ih}_{\mathcal{S}}$ is the *preinherence* of $\mathcal{S}$.

We use now the notation introduced in Remark VIII.4.4 to the effect that

$$\mathrm{ah}_{\mathcal{S}}^{\natural} \mathcal{A} = \{\mathrm{ah}_{\mathcal{S}} A : A \in \mathcal{A}\}.$$

LEMMA XIX.1.13. *Let $\mathcal{A} \in \mathbb{I}X$ and $\mathcal{D} \in \mathbb{I}Y$. Then*

$$(\textstyle\int_{\mathcal{D}} \mathcal{S}(\cdot))\#\mathcal{A} \Longleftrightarrow \mathcal{D}\# \operatorname{ah}^{\natural}_{\mathcal{S}} \mathcal{A}.$$

PROOF. Let $\mathcal{D}\# \operatorname{ah}^{\natural}_{\mathcal{S}} \mathcal{A}$, that is, for each $A \in \mathcal{A}$ and $D \in \mathcal{D}$, there is $y \in D \cap \operatorname{ah}_{\mathcal{S}} A$, that is, $A \in \mathcal{S}^{\#}(y)$, equivalently $A \in \bigcup_{y\in D} \mathcal{S}^{\#}(y) = \mathcal{S}(D)^{\#}$, that is,

$$\mathcal{A} \subset \bigcap\nolimits_{D\in\mathcal{D}} \mathcal{S}(D)^{\#} = (\bigcup\nolimits_{D\in\mathcal{D}} \mathcal{S}(D))^{\#} = (\textstyle\int_{\mathcal{D}} \mathcal{S})^{\#}.$$

□

In virtue of Lemma XIX.1.13,

COROLLARY XIX.1.14 (Adherence-vicinity duality). *If $\mathcal{A}$ and $\mathcal{D}$ are isotone families of subsets of X and ξ is a convergence on X, then*

$$(\textstyle\int_{\mathcal{D}} \mathcal{V}_{\xi}(\cdot))\#\mathcal{A} \Longleftrightarrow \mathcal{D}\# \operatorname{adh}^{\natural}_{\xi} \mathcal{A}.$$

In this specific case, $Y = X$, and moreover $x \in V$ for every $V \in \mathcal{V}_{\xi}(x)$. Consequently, for any isotone family,

$$\textstyle\int_{\mathcal{D}} \mathcal{V}_{\xi}(\cdot) \leq \mathcal{D}. \tag{XIX.1.6}$$

Indeed, if $A \in \int_{\mathcal{D}} \mathcal{V}_{\xi}(\cdot)$, then by definition there exists $D \in \mathcal{D}$ such that $A \in \mathcal{V}_{\xi}(x)$ for each $x \in D$. As $A \in \mathcal{V}_{\xi}(x)$ amounts to $x \in \operatorname{inh}_{\xi} A$, it follows that $D \subset \operatorname{inh}_{\xi} A \subset A$, which proves (XIX.1.6).

XIX.2. Diagonality

DEFINITION XIX.2.1. Let ξ be a convergence on a set X. A map $\mathcal{S} : X \longrightarrow \mathbb{F}X$ is called a *ξ-selection* or a *convergence selection* of ξ, if for each $x \in X$,

$$x \in \lim_{\xi} \mathcal{S}(x).$$

In particular, $\mathcal{S}(x) := x^{\uparrow}$ is a convergence selection for every convergence on X.

EXAMPLE XIX.2.2. If ξ is a pretopology, then $\mathcal{S}(x) := \mathcal{V}_{\xi}(x)$ is a ξ-selection. Of course, this ξ-selection determines ξ.

DEFINITION XIX.2.3. A convergence ξ is called *diagonal* if, for each filter $\mathcal{F}$ and every ξ-selection $\mathcal{S}$,

$$\lim_{\xi} \mathcal{F} \subset \lim_{\xi} \textstyle\int_{\mathcal{F}} \mathcal{S}. \tag{XIX.2.1}$$

We have already observed that a pretopology ξ on X is a topology if and only if

$$\textstyle\int_{\mathcal{V}_{\xi}(x)} \mathcal{V}_{\xi} \geq \mathcal{V}_{\xi}(x)$$

for each $x \in X$. In particular, in case of pretopologies, it is enough to check diagonality conditions for vicinity filters.

PROPOSITION XIX.2.4. *Every topology is diagonal.*

PROOF. Let ξ be a topology on X, and let $\mathcal{S}$ be a ξ-selection, that is, $\mathcal{S}(x) \geq \mathcal{N}_\xi(x)$ for each $x \in X$. To see that ξ is diagonal, let $x_0 \in \lim_\xi \mathcal{F}$, that is, $\mathcal{F} \geq \mathcal{N}_\xi(x_0)$. Consequently, $\int_{\mathcal{F}} \mathcal{S} \geq \int_{\mathcal{N}_\xi(x_0)} \mathcal{N}_\xi \geq \mathcal{N}_\xi(x_0)$ in virtue of (VII.2.9). □

Here is a basic example of a Hausdorff non-diagonal pretopology.

EXAMPLE XIX.2.5 (A non-diagonal pretopology). Let $(X_n)_n$ be a sequence of disjoint infinite countable sets, let $X_\infty := \{x_n : n < \omega\}$ be such that $x_n \in X_n$ for each $n < \omega$, and let $x_\infty \notin X_n$ for each $n < \omega$. Let

$$X := \bigcup\nolimits_{n<\omega} X_n \cup \{x_\infty\}.$$

Consider on X a pretopology θ given by setting $\mathcal{V}_\theta(x_n)$ to be the cofinite filter of X_n centered at x_n, and $\mathcal{V}_\theta(x_\infty)$ to be the cofinite filter of $\{x_\infty, x_0, \ldots\}$ centered at x_∞, while all the other points be isolated.

This pretopology is not diagonal, because

$$\int_{(x_n)_n} (X_n)_0 \ngeq \mathcal{V}_\theta(x_\infty).$$

Indeed, if $A \in \int_{(x_n)_n} (X_n)_0$ then A is cofinite in X_n for almost all n, while there is $V \in \mathcal{V}_\theta(x_\infty)$, which is not cofinite in any X_n.

DEFINITION XIX.2.6. A convergence ξ is called *pseudodiagonal* or *weakly diagonal* if $\mathrm{adh}_\xi \mathcal{F}$ is ξ-closed, that is,

$$\mathrm{adh}_\xi(\mathrm{adh}_\xi \mathcal{F}) \subset \mathrm{adh}_\xi \mathcal{F} \tag{XIX.2.2}$$

for each filter $\mathcal{F}$.

PROPOSITION XIX.2.7. *Each diagonal convergence is pseudodiagonal.*

PROOF. Let $x_0 \in \mathrm{adh}_\xi(\mathrm{adh}_\xi \mathcal{F})$. Accordingly, there exists $\mathcal{U} \in \beta(\mathrm{adh}_\xi \mathcal{F})$ such that $x_0 \in \lim_\xi \mathcal{U}$. For each $x \in \mathrm{adh}_\xi \mathcal{F}$, there exists $\mathcal{S}(x) \in \beta\mathcal{F}$ such that $x \in \lim_\xi \mathcal{S}(x)$. If we set $\mathcal{S}(x) := x^\uparrow$ for every $x \notin \mathrm{adh}_\xi \mathcal{F}$, then $\mathcal{S}$ is an ultrafilter ξ-selection. As ξ is diagonal, $x_0 \in \lim_\xi \int_{\mathcal{U}} \mathcal{S}(\cdot)$. By Corollary XIX.1.11, $\int_{\mathcal{U}} \mathcal{S}(\cdot)$ is an ultrafilter, and

$$\int_{\mathcal{U}} \mathcal{S}(\cdot) \geq \bigcap\nolimits_{x\in \mathrm{adh}_\xi \mathcal{F}} \mathcal{S}(x) \geq \mathcal{F}.$$

Consequently, $x_0 \in \lim_\xi \int_{\mathcal{U}} \mathcal{S}(\cdot) \subset \mathrm{adh}_\xi \mathcal{F}$. □

REMARK XIX.2.8. In the proof above, adh_ξ appears in two forms, that is, first it assigns to a filter $\mathcal{F}$ a set $\mathrm{adh}_\xi \mathcal{F}$, then it assigns to a set $\mathrm{adh}_\xi \mathcal{F}$, a set $\mathrm{adh}_\xi(\mathrm{adh}_\xi \mathcal{F})$. In the second employment, adh_ξ is the principal adherence, or else, the adherence of a principal filter $(\mathrm{adh}_\xi \mathcal{F})^\uparrow$.

DEFINITION XIX.2.9. A convergence ξ on X is called *subdiagonal* if for each $A \subset X$,

$$\operatorname{adh}_\xi(\operatorname{adh}_\xi A) \subset \operatorname{adh}_\xi A. \tag{XIX.2.3}$$

Condition (XIX.2.3) is a special case of (XIX.2.2), where $\mathcal{F} = \{A\}^\uparrow$ is a principal filter. Therefore, each pseudodiagonal convergence is subdiagonal.

Of course, ξ is a subdiagonal convergence whenever $\operatorname{cl}_\xi A \subset \operatorname{adh}_\xi A$ for every A, that is,

COROLLARY XIX.2.10. *A convergence ξ is subdiagonal if and only if*

$$\mathrm{T}\xi \geq \mathrm{S}_0\xi. \tag{XIX.2.4}$$

Therefore,

COROLLARY XIX.2.11. *A pretopology is subdiagonal if and only if it is a topology.*

Actually, although no implication below can be reversed,

$$\text{topological} \Longrightarrow \text{diagonal} \Longrightarrow \text{pseudodiagonal} \Longrightarrow \text{subdiagonal},$$

all of them coincide for pretopologies, because, by Proposition XIX.2.4, each topology is diagonal.

On the other hand, there exist diagonal convergences that are not topologies.

EXAMPLE XIX.2.12 (A diagonal, non-topological paratopology). Let $\{X_n : n < \omega\}$ be sets such that $X_n \subset X_{n+1}$ for each $n < \omega$, and let $x_\infty \in \bigcap_{n<\omega} X_n$. Let τ_n be a prime topology at x_∞ on a set X_n so that τ_n is the restriction of τ_{n+1} to X_n for each $n < \omega$. The final convergence τ on $X := \bigcup_{n<\omega} X_n$ with respect to the natural embeddings is prime. On the other hand, τ is a paratopology (see Example XVIII.6.1). If we assume that τ does not coincide with any τ_n, then τ is not a pretopology. Let us check that τ is diagonal. If $x_\infty \in \lim_\tau \mathcal{F}$, then there is $n < \omega$ such that $\mathcal{F} \geq \mathcal{N}_{\tau_n}(x_\infty)$, and if $\mathcal{S} : X \longrightarrow \mathbb{F}X$ is a τ-selection, then there exists $n \leq k < \omega$ such that $\mathcal{S}(x_\infty) \geq \mathcal{N}_{\tau_k}(x_\infty)$. Therefore, $\int_{\mathcal{F}} \mathcal{S} \geq \int_{\mathcal{N}_{\tau_n}(x_\infty)} \mathcal{N}_{\tau_k} = \mathcal{N}_{\tau_k}(x_\infty)$, hence $x_\infty \in \lim_\tau \int_{\mathcal{F}} \mathcal{S}$.

Actually, in Example XIX.2.12 we used the following fact.

LEMMA XIX.2.13. *Each prime, finitely stable convergence is diagonal.*

PROOF. Let π be a prime convergence at x_∞ on X. Let $x_\infty \in \lim_\pi \mathcal{F}$ and $\mathcal{S}$ be a π-selection, in particular $x_\infty \in \lim_\pi \mathcal{S}(x_\infty)$ and $\mathcal{S}(x) = x^\uparrow$

for $x \neq x_\infty$. If $A \in \int_{\mathcal{F}} \mathcal{S}$, then there is $F \in \mathcal{F}$ such that $A \in \mathcal{S}(x)$ for each $x \in F$, that is, $F \setminus \{x_\infty\} \subset A$ and $A \in \mathcal{S}(x_\infty)$ if $\ker \mathcal{F} \neq \varnothing$, and in this case $\ker \mathcal{F} = \{x_\infty\}$. So that if $\mathcal{F}$ is free, then $\int_{\mathcal{F}} \mathcal{S} = \mathcal{F}$ or $\int_{\mathcal{F}} \mathcal{S} = \mathcal{F} \wedge \mathcal{S}(x_\infty)$. □

The following characterization of pseudodiagonal convergences will be instrumental ([3]).

PROPOSITION XIX.2.14. *A convergence ξ is pseudodiagonal if and only if for every x, each filter $\mathcal{F}$ such that $x \in \lim_\xi \mathcal{F}$, and each ξ-selection $\mathcal{S}$,*

$$x \in \bigcap\nolimits_{F \in \mathcal{F}} \operatorname{adh}_\xi \mathcal{S}(F).$$

PROOF. If ξ is pseudodiagonal, $x \in \lim_\xi \mathcal{F}$ and $\mathcal{S}$ is a ξ-selection, then $F \subset \operatorname{adh}_\xi \mathcal{S}(F)$, because $w \in \lim_\xi \mathcal{S}(w)$ for each $F \in \mathcal{F}$ and every $w \in F$. Therefore, $\operatorname{adh}_\xi \mathcal{S}(F) \in \mathcal{F}$, and thus, by assumption,

$$x \in \lim_\xi \mathcal{F} \subset \operatorname{adh}_\xi(\operatorname{adh}_\xi \mathcal{S}(F)) \subset \operatorname{adh}_\xi \mathcal{S}(F).$$

Conversely, if $x \in \operatorname{adh}_\xi(\operatorname{adh}_\xi \mathcal{H})$, then there exists a filter $\mathcal{F}$ such that $x \in \lim_\xi \mathcal{F}$ and $\operatorname{adh}_\xi \mathcal{H} \in \mathcal{F}$. For each $w \in \operatorname{adh}_\xi \mathcal{H}$, let $\mathcal{S}(w) \geq \mathcal{H}$ and such that $w \in \lim_\xi \mathcal{S}(w)$, and $\mathcal{S}(w) := w^\uparrow$ for each $w \notin \operatorname{adh}_\xi \mathcal{H}$, constitutes a ξ-selection. By assumption, $x \in \operatorname{adh}_\xi \mathcal{S}(\operatorname{adh}_\xi \mathcal{H}) \subset \operatorname{adh}_\xi \mathcal{H}$. □

EXAMPLE XIX.2.15 (A pseudodiagonal, non-diagonal pseudotopology). Let σ be the *standard sequential convergence* from Example XVIII.1.1. Let $x_\infty \in \lim_\sigma \mathcal{F}$ and let $\mathcal{S}$ be a σ-selection. Hence $x_\infty \in \lim_\sigma (x_n)_n$ and $\mathcal{F} \geq (x_n)_n \vee x_\infty^\uparrow$ and $\mathcal{S}(x_n) \geq (x_{n,k})_k$ with $x_n \in \lim_\sigma (x_{n,k})_k$ and $\mathcal{S}(x_\infty) \geq (x_{\infty,k})_k$ with $x_\infty \in \lim_\sigma (x_{\infty,k})_k$. By Lemma X.10.6, there is a map $f : \mathbb{N} \longrightarrow \mathbb{N}$ such that $x_\infty \in \lim_\nu (x_{n,k_n})_n$ if $k_n \geq f(n)$. Consequently, $x_\infty \in \lim_\sigma (x_{n,k_n})_n$, hence $x_\infty \in \operatorname{adh}_\sigma \mathcal{S}(\{x_n : n > p\})$ for each $p < \omega$. As $x_\infty \in \lim_\sigma (x_n)_n$, we conclude that $x_\infty \in \operatorname{adh}_\sigma \mathcal{S}(F)$ for each $F \in \mathcal{F}$, so that σ is pseudodiagonal by Proposition XIX.2.14.

To see that σ is not diagonal, let $\mathcal{F} \approx (x_n)_n$ be a strictly decreasing sequence convergent to 0, and $\mathcal{S}(x_n)$ be a strictly decreasing sequence convergent to x_n for each $n < \omega$. Then $0 \notin \lim_\sigma \int_{\mathcal{F}} \mathcal{S}$, because there is no sequential filter convergent to 0 coarser than $\int_{\mathcal{F}} \mathcal{S}$. By way of contradiction suppose that there exists such a filter. Then it is necessarily free, and there is a strictly decreasing sequence $(y_p)_p$ that determines this filter. For each $q < \omega$, there exists $A \in \int_{\mathcal{F}} \mathcal{S}$ such that $A \subset \{y_p : p > q\}$. In particular, there is $n < \omega$ such that almost all terms of $(x_{n,k})_k$ lie in

[3] Both the definition of pseudodiagonal convergences and this characterization are due to EVA LOWEN-COLEBUNDERS [179] who used the term *weakly diagonal convergences*.

$\{y_p : p > q\}$. As $(x_{n,k})_k$ converges to x_n, a subsequence of $(y_p)_p$ would converge to x_n, which yields a contradiction.

Finally,

EXAMPLE XIX.2.16 (A Hausdorff subdiagonal, non-pseudodiagonal pseudotopology). ([4]) In Example VI.1.8 we considered a bisequence pretopology defined on

$$X := \{x_\infty\} \sqcup X_\infty \sqcup \bigsqcup\nolimits_{n<\omega} X_n.$$

where $X_\infty := \{x_n : n < \omega\}$, and $X_n := \{x_{n,k} : k < \omega\}$. We shall use X to define a non-pretopological pseudotopology η to the effect that $x_{n,k}$ is isolated for each $n, k < \omega$, and if $\mathcal{U} \in \beta^* X$, then $x_n \in \lim_\eta \mathcal{U}$ whenever $X_n \in \mathcal{U}$, and $x_\infty \in \lim_\eta \mathcal{U}$ if either $X_\infty \in \mathcal{U}$ or there is an (infinite) set $A \in \mathcal{U}$ such that $A \cap X_n$ is finite for each $n < \omega$. The pseudotopology η is defined by extending this convergence to all filters via (XVIII.3.6).

It is clear that η is Hausdorff. To see that it is subdiagonal, let $A \subset X$ and

$$x \in \operatorname{adh}_\eta(\operatorname{adh}_\eta A).$$

The only non-trivial case is when $A \subset \bigsqcup_{n<\omega} X_n$ ([5]). Then either x is isolated, hence $x \in A$, or there is $n < \omega$ such that $x = x_n$ and then $A \cap X_n$ is infinite, hence $x_n \in \operatorname{adh}_\eta A$, or $x = x_\infty$. So suppose that $x_\infty \notin \operatorname{adh}_\eta A$, thus $A \cap X_n \neq \varnothing$ for finitely many n, hence $X_\infty \cap \operatorname{adh}_\eta A$ is also finite, so that $x \notin \operatorname{adh}_\eta(\operatorname{adh}_\eta A)$.

We define a filter $\mathcal{F}$ on $\bigsqcup_{n<\omega} X_n$ so that $F \in \mathcal{F}$ whenever $X_n \setminus F$ is finite for every $n < \omega$. Then $x_\infty \notin \operatorname{adh}_\eta \mathcal{F}$, because if $A \cap X_n$ is finite for each $n < \omega$, then $A \notin \mathcal{F}^\#$, and also $X_\infty \notin \mathcal{F}^\#$. However, $X_\infty \subset \operatorname{adh}_\eta \mathcal{F}$, hence

$$x_\infty \in \operatorname{adh}_\eta(\operatorname{adh}_\eta \mathcal{F}) \setminus \operatorname{adh}_\eta \mathcal{F},$$

contrary to (XIX.2.2). Therefore, η is not pseudodiagonal, and thus, by Corollary XIX.2.11, it is not a pretopology.

PROPOSITION XIX.2.17. *Diagonal convergences constitute a projective class.*

PROOF. Obviously, each chaotic convergence is diagonal. Suppose that Ξ is a set of diagonal convergences on a given set. We shall see that $\bigvee \Xi$ is diagonal. If $x \in \lim_{\bigvee \Xi} \mathcal{F}$and $\mathcal{S}$ is a $\bigvee \Xi$-selection, then by definition of supremum, $x \in \lim_\xi \mathcal{F}$ for each $\xi \in \Xi$, and $\mathcal{S}$ is a

[4]This is a variant of an example of J. WOJCIECHOWSKI [**262**].

[5]As the adherence of a finite union is equal to the union of these adherences.

ξ-selection for each $\xi \in \Xi$. Therefore, $x \in \lim_\xi \int_{\mathcal{F}} \mathcal{S}$ for each $\xi \in \Xi$, hence $x \in \lim_{\bigvee \Xi} \int_{\mathcal{F}} \mathcal{S}$. □

The corresponding projector D is called the *diagonalizer*. As each topology is diagonal, $\mathrm{D} \geq \mathrm{T}$, and since each diagonal convergence ξ is subdiagonal, that is, $\mathrm{S}_0\xi = \mathrm{T}\xi$,

$$\mathrm{S}_0\mathrm{D} = \mathrm{T}.$$

We shall see that D is not a functor, hence is not a reflector.

EXAMPLE XIX.2.18. Let θ be a *convergent bisequence* from Example VI.1.8, and let $f : X \longrightarrow Y := \{x_{n,k} : n, k < \omega\} \cup \{\infty\}$ such that $f^-(\infty) = \{x_n : n < \omega\} \cup \{x_\infty\}$. Then $\tau := \mathrm{P}_0 f\theta$, where P_0 is the finitely stable modifier, is diagonal by Lemma XIX.2.13. By construction, $f \in \mathscr{C}(\theta, \tau)$. Of course, $x_\infty \in \lim_{\mathrm{T}\theta} \mathcal{N}_\theta(x_\infty)$; on the other hand, $\mathcal{N}_\theta(x_\infty) = \int_{\mathcal{V}_\theta(x_\infty)} \mathcal{V}_\theta$ and thus $\mathrm{D}\theta = \mathrm{T}\theta$. On the other hand, $\infty = f(x_\infty) \notin \lim_\tau f[\mathcal{N}_\theta(x_\infty)]$, because $f(\{x_{n,k} : n > 0\}) \cap f(\{x_{n,k} : n \leq n_0\}) = \varnothing$ for each $n_0 < \omega$. Accordingly, $f \notin \mathscr{C}(\mathrm{D}\theta, \tau)$.

XIX.3. Regularity

Formula (VIII.4.1) in the case of the operation of adherence becomes

$$\mathrm{adh}^\natural_\xi \mathcal{F} := \{\mathrm{adh}_\xi F : F \in \mathcal{F}\}.$$

Of course, when $\mathcal{F} \in \mathbb{F}$, the family $\mathrm{adh}^\natural_\xi \mathcal{F}$ is, in general, a filter-base, but we know that the limit of a filter-base is that of the filter it generates. Of course, $\mathrm{adh}^\natural_\xi \mathcal{F} \leq \mathcal{F}$.

DEFINITION XIX.3.1. A convergence ξ is called *regular* if for every filter $\mathcal{F}$,

$$\lim_\xi \mathcal{F} \subset \lim_\xi \mathrm{adh}^\natural_\xi \mathcal{F}. \tag{XIX.3.1}$$

A Hausdorff regular convergence is called a T_3-*convergence*.

PROPOSITION XIX.3.2. *A pretopology ξ is regular if and only if for each x and every $V \in \mathcal{V}_\xi(x)$, there exists $W \in \mathcal{V}_\xi(x)$ such that* $\mathrm{adh}_\xi W \subset V$.

PROOF. For a pretopology ξ, the condition (XIX.3.1) becomes

$$\mathcal{V}_\xi(x) \leq \mathrm{adh}^\natural_\xi \mathcal{V}_\xi(x)$$

for every x, which, written *in extenso,* yields the wording of the proposition. □

DEFINITION XIX.3.3. A convergence ξ is called *topologically regular*, or T-*regular* if

$$\lim_\xi \mathcal{F} \subset \lim_\xi \mathrm{cl}_\xi^\natural \mathcal{F}$$

for each filter $\mathcal{F}$.

In particular, a topologically regular convergence ξ has at every point a pavement consisting of ξ-*closedly based* filters ([6]).

Of course, a regular topology is topologically regular, because for topologies, the closure and the adherence coincide ([7]). Example IX.1.12 is an elementary instance of a Hausdorff, non-regular topology. A Hausdorff, regular pretopology that is not topologically regular is described in Example XXII.3.2.

COROLLARY XIX.3.4. *Every regular subdiagonal convergence is topologically regular.*

PROOF. A convergence ξ is subdiagonal whenever $\mathrm{adh}_\xi A = \mathrm{cl}_\xi A$ for each A. It follows that $\mathrm{adh}_\xi^\natural \mathcal{F} = \mathrm{cl}_\xi^\natural \mathcal{F}$ for each filter $\mathcal{F}$. Hence, if $x \in \lim_\xi \mathcal{F}$, then $x \in \lim_\xi \mathrm{cl}_\xi^\natural \mathcal{F}$. □

Recall that each prime free convergence is topologically Hausdorff.

PROPOSITION XIX.3.5. *Each prime finitely stable convergence is topologically regular.*

PROOF. Let ξ be a prime convergence on X, let x_∞ be a distinguished point. As all the points but possibly x_∞ are isolated, $\mathrm{adh}_\xi A = \mathrm{cl}_\xi A \subset A \cup \{x_\infty\}$ for each $A \subset X$. Therefore if $x_\infty \in \lim_\xi \mathcal{F}$, then $\mathrm{cl}_\xi^\natural \mathcal{F} \geq \mathcal{F} \wedge \{x_\infty\}$, hence $x_\infty \in \lim_\xi \mathrm{cl}_\xi^\natural \mathcal{F}$. □

EXAMPLE XIX.3.6. Let $A_\infty := \{x_n : n < \omega\}$ and $A_n := \{x_{n,k} : k < \omega\}$. On $X = \{x_\infty\} \sqcup A_\infty \sqcup \bigsqcup_{n<\omega} A_n$ define a pretopology η to the effect that $x_{n,k}$ is isolated for each $n, k < \omega$, $\mathcal{V}_\eta(x_\infty) \approx \int_{(n)} (x_{n,k})_k \wedge x_\infty^\uparrow$ and $\mathcal{V}_\eta(x_n) \approx (x_{n,k})_k \wedge x_n^\uparrow$ for each $n < \omega$.

EXAMPLE XIX.3.7. We observe that η is a topology. However, η is not regular, because for each $V \in \int_{(n)} (x_{n,k})_k$, there is n_0 and a map $\varphi \in \mathbb{N}^\mathbb{N}$ such that $\{x_{n,k} : k > \varphi(n), n > n_0\} \subset V$, hence $\{x_n : n > n_0\} \subset \mathrm{adh}_\eta V$, but $\lim_\eta (x_n)_n = \varnothing$.

---

[6]A filter $\mathcal{F}$ is ξ-*closedly based* whenever $\mathcal{F} \cap \mathcal{C}_\xi$ is a filter-base of $\mathcal{F}$.

[7]Topological regularity is an example of regularity with respect to an operation different from the self-adherence. If θ and ξ are convergences on the same underlying set, then ξ is said θ-*regular* whenever $\lim_\xi \mathcal{F} \subset \lim_\xi \mathrm{adh}_\theta^\natural \mathcal{F}$. In particular, ξ is T-*regular* if ξ is Tξ-regular.

LEMMA XIX.3.8. *Each supremum of regular convergences is regular.*

PROOF. Let Ξ be a set of regular convergence on X and let $x \in \lim_{\bigvee \Xi} \mathcal{F}$. Therefore, for each $\xi \in \Xi$,

$$x \in \lim_\xi \mathcal{F} \subset \lim_\xi \operatorname{adh}_\xi^\natural \mathcal{F} \subset \lim_\xi \operatorname{adh}_{\bigvee \Xi}^\natural \mathcal{F},$$

because each ξ is regular, and $\operatorname{adh}_\xi^\natural \mathcal{F} \leq \operatorname{adh}_{\bigvee \Xi}^\natural \mathcal{F}$, as $\operatorname{adh}_{\bigvee \Xi} A \subset \operatorname{adh}_\xi A$ for each A. It follows that $x \in \lim_{\bigvee \Xi} \operatorname{adh}_{\bigvee \Xi}^\natural \mathcal{F}$. $\square$

As the chaotic convergences are regular, we infer that

PROPOSITION XIX.3.9. *The class of regular convergences is projective.*

The corresponding projector Reg is called the *regularizer*.

LEMMA XIX.3.10. *The initial convergence with respect to a regular convergence is regular.*

PROOF. Indeed, suppose that τ is regular, and let $x \in \lim_{f^- \tau} \mathcal{F}$. Then

$$f(x) \in \lim_\tau f[\mathcal{F}] \subset \lim_\tau \operatorname{adh}_\tau^\natural f[\mathcal{F}],$$

because τ is regular. By Lemma IV.1.2,

$$\operatorname{adh}_\tau^\natural f[\mathcal{F}] \leq f[f^-[\operatorname{adh}_\tau^\natural f[\mathcal{F}]]] \approx f[\{f^-(\operatorname{adh}_\tau f(F)) : F \in \mathcal{F}\}].$$

By (VI.4.1), $\operatorname{adh}_{f^- \tau} F = f^-(\operatorname{adh}_\tau f(F))$, so that $f(x) \in \lim_\tau f[\operatorname{adh}_{f^-\tau}^\natural \mathcal{F}]$, that is, $x \in \lim_{f^-\tau} \operatorname{adh}_{f^-\tau}^\natural \mathcal{F}$. $\square$

Consequently, by (XVII.3.1)

THEOREM XIX.3.11. *The class* **Reg** *of regular convergences is reflective.*

COROLLARY XIX.3.12. *Each subconvergence of a regular convergence is regular, and each product of regular convergences is regular.*

Regarding the contour operation, recall from (XIX.1.6) that $\mathcal{V}_\xi\{\mathcal{H}\} = \int_{\mathcal{H}} \mathcal{V}_\xi(\cdot) \leq \mathcal{H}$ for any isotone family of sets $\mathcal{H}$. Therefore,

$$\operatorname{adh}_\xi \mathcal{H} \subset \operatorname{adh}_\xi \textstyle\int_{\mathcal{H}} \mathcal{V}_\xi(\cdot) = \operatorname{adh}_\xi \mathcal{V}_\xi\{\mathcal{H}\}$$

for each filter $\mathcal{H}$ and every convergence ξ.

PROPOSITION XIX.3.13. *If a convergence ξ is regular, then for every filter $\mathcal{H}$,*

$$\operatorname{adh}_\xi \mathcal{V}_\xi\{\mathcal{H}\} = \operatorname{adh}_\xi \textstyle\int_{\mathcal{H}} \mathcal{V}_\xi(\cdot) \subset \operatorname{adh}_\xi \mathcal{H}. \tag{XIX.3.2}$$

If moreover ξ is a pseudotopology and (XIX.3.2) holds for each filter $\mathcal{H}$, then ξ is regular.

PROOF. If $x \in \mathrm{adh}_\xi \int_{\mathcal{H}} \mathcal{V}_\xi(\cdot)$, then there exists a filter $\mathcal{F}$ such that $x \in \lim_\xi \mathcal{F}$ and $\mathcal{F}\#\int_{\mathcal{H}} \mathcal{V}_\xi(\cdot)$, equivalently $\mathrm{adh}_\xi^\natural \mathcal{F}\#\mathcal{H}$ in virtue of Corollary XIX.1.14. By regularity, $x \in \lim_\xi(\mathrm{adh}_\xi^\natural \mathcal{F})$, hence $x \in \mathrm{adh}_\xi \mathcal{H}$. Suppose that ξ is a pseudotopology, and let $x \notin \lim_\xi(\mathrm{adh}_\xi^\natural \mathcal{F})$, hence, by (XVIII.3.5), there exists a filter $\mathcal{H}$ such that $\mathcal{H}\# \mathrm{adh}_\xi^\natural \mathcal{F}$ and $x \notin \mathrm{adh}_\xi \mathcal{H}$. Then, again by Corollary XIX.1.14, $\int_{\mathcal{H}} \mathcal{V}_\xi(\cdot)\#\mathcal{F}$, and by (XIX.3.2) $x \notin \mathrm{adh}_\xi \int_{\mathcal{H}} \mathcal{V}_\xi(\cdot)$, hence $x \notin \lim_\xi \mathcal{F}$ by (VIII.1.1). □

DEFINITION XIX.3.14 (Regular refinement). Let ξ be a convergence on X, and let $\mathcal{P}, \mathcal{S}$ be families of subsets of X. Then $\mathcal{S}$ is called a *ξ-regular refinement* of $\mathcal{P}$ if $\mathrm{adh}_\xi^\natural \mathcal{S}$ is a refinement of $\mathcal{P}$.

Now $\mathrm{adh}_\xi^\natural \mathcal{S} \triangleleft \mathcal{P}$ whenever for each $S \in \mathcal{S}$ there is $P \in \mathcal{P}$ such that $\mathrm{adh}_\xi S \subset P$.

LEMMA XIX.3.15. *If ξ is a regular convergence and $A \subset \mathrm{inh}_\xi \mathcal{P}$, then there exists a ξ-regular refinement $\mathcal{S}$ of $\mathcal{P}$ such that $A \subset \mathrm{inh}_\xi \mathcal{S}$.*

PROOF. Let $\mathcal{F}$ be a filter on A such that $\varnothing \neq A \cap \lim_\xi \mathcal{F} \subset A \cap \lim_\xi \mathrm{adh}_\xi^\natural \mathcal{F}$. Consequently, there exists $P \in \mathcal{P}$ and $F_{\mathcal{F}} \in \mathcal{F}$ such that $\mathrm{adh}_\xi F_{\mathcal{F}} \subset P$. Then $\mathcal{S} := \{F_{\mathcal{F}} : A \cap \lim_\xi \mathcal{F} \neq \varnothing\}$ is a regular refinement $\mathcal{P}$ with $A \subset \mathrm{inh}_\xi \mathcal{S}$. □

In other words, on recalling Lemma VIII.2.3 saying that $\mathcal{P}$ is a ξ-cover of A whenever $A \subset \mathrm{inh}_\xi \mathcal{P}$, we infer that each cover for a regular convergence has a regular covering refinement. Dually,

COROLLARY XIX.3.16. *If ξ is a regular convergence on X, and $A \cap \mathrm{adh}_\xi \mathcal{H} = \varnothing$, then there exists a filter $\mathcal{G}$ on X such that $\mathrm{inh}_\xi^\natural \mathcal{G} \leq \mathcal{H}$ and $A \cap \mathrm{adh}_\xi \mathcal{G} = \varnothing$.*

PROOF. Indeed, by (V.1.1), $\mathrm{adh}_\xi^\natural \mathcal{S} \triangleleft \mathcal{P}$ whenever $(\mathrm{adh}_\xi^\natural \mathcal{S})_c \leq \mathcal{P}_c$. On the other hand,

$$\begin{aligned}(\mathrm{adh}_\xi^\natural \mathcal{S})_c &= \{\mathrm{adh}_\xi S : S \in \mathcal{S}\}_c = \{(\mathrm{adh}_\xi S)^c : S \in \mathcal{S}\}\\ &= \{\mathrm{inh}_\xi S^c : S \in \mathcal{S}\} = \mathrm{inh}_\xi^\natural \mathcal{S}_c,\end{aligned}$$

and thus $\mathrm{adh}_\xi^\natural \mathcal{S} \triangleleft \mathcal{P} \Longleftrightarrow \mathrm{inh}_\xi^\natural \mathcal{S}_c \leq \mathcal{P}_c$. □

Let us relate Proposition XIX.3.13 to Corollary XIX.3.16.

LEMMA XIX.3.17. *Let ξ be a convergence, and $\mathcal{G}$ and $\mathcal{H}$ be families of subsets of $|\xi|$. Then*

$$\mathcal{G} \leq \int_{\mathcal{H}} \mathcal{V}_\xi(\cdot) \Longleftrightarrow \mathrm{inh}_\xi^\natural \mathcal{G} \leq \mathcal{H}.$$

PROOF. The first inequality means that for each $G \in \mathcal{G}$ there is $H \in \mathcal{H}$ such that $G \in \mathcal{V}_\xi(x)$ for each $x \in H$, equivalently, $H \subset \operatorname{inh}_\xi G$, that is, the second inequality. □

XIX.4. *Partial regularization

If ξ is a free (alias T_1) convergence, then the *partial regularization* $\mathrm{r}\xi$ of ξ is defined by ([8])

$$x \in \lim_{\mathrm{r}\xi} \mathcal{F} \Longleftrightarrow \underset{\mathcal{G} \leq \mathcal{F}}{\exists}\, x \in \lim_\xi \mathcal{G} \wedge \operatorname{adh}_\xi^\natural \mathcal{G} \leq \mathcal{F}. \tag{XIX.4.1}$$

It is obvious that $\mathrm{r}\xi$ is an isotone relation. Moreover, $x \in \lim_\xi x^\uparrow \subset \lim_\xi \operatorname{adh}_\xi^\natural\{x\}^\uparrow \subset \{x\}$, so that $\mathrm{r}\xi$ is a centered relation, hence a convergence.

Notice that (XIX.4.1) amounts to $\lim_{\mathrm{r}\xi}(\operatorname{adh}_\xi^\natural \mathcal{G}) = \lim_\xi \mathcal{G}$. Hence, $\xi \geq \mathrm{r}\xi$, for if $x \in \lim_\xi \mathcal{G}$, then $x \in \lim_{\mathrm{r}\xi}(\operatorname{adh}_\xi^\natural \mathcal{G}) \subset \lim_{\mathrm{r}\xi} \mathcal{G}$.

LEMMA XIX.4.1. *If ξ is a pretopology, then $\mathrm{r}\xi$ is a pretopology.*

PROOF. Indeed, if ξ is a pretopology and $x \in \lim_{\mathrm{r}\xi} \mathcal{F}$, then by (XIX.4.1), there exists $\mathcal{G} = \mathcal{V}_\xi(x)$ such that $\operatorname{adh}_\xi^\natural \mathcal{V}_\xi(x) \leq \mathcal{F}$. Therefore

$$\mathcal{V}_{\mathrm{r}\xi}(x) = \operatorname{adh}_\xi^\natural \mathcal{V}_\xi(x).$$

□

A free convergence ξ is regular if and only if $\lim_\xi \mathcal{F} \subset \lim_\xi(\operatorname{adh}_\xi^\natural \mathcal{F}) = \lim_{\mathrm{r}\xi} \mathcal{F}$, that is, a free convergence is regular if and only if $\mathrm{r}\xi \geq \xi$ equivalently $\mathrm{r}\xi = \xi$.

EXAMPLE XIX.4.2. Consider the standard topology on $\mathbb{R}$. Let $\mathcal{C}_\nu \cup \{\{2^{-n} : n < \omega\}\}$ be a subbase for the closed sets of a topology ζ. Then $\mathcal{N}_\zeta(0) = \mathcal{N}_\nu(0) \setminus \{2^{-n} : n < \omega\}$ but $0 \notin \lim_\zeta \mathcal{N}_\nu(0)$. As $\mathcal{N}_\nu(0) = \operatorname{cl}_\zeta^\natural \mathcal{N}_\zeta(0)$, we infer that $\mathrm{r}\zeta = \nu$.

In Example XIX.4.2, consider the bisequence, where $x_n := 2^{-n}$, and $x_{n,k} := 2^{-n} + 2^{-k}$. This bisequence has the property that $(x_{n,k})_k$ converges to x_n in ζ for each $n < \omega$, and $(x_n)_n$ does not ζ-converge to 0. On the other hand, the contour $\int_{(n)}(x_{n,k})_k$ converges to 0 in ζ.

Since $\xi \geq \mathrm{r}\xi$, for every ordinal α, we define the α-iteration of the partial regularizer r to the effect that $\mathrm{r}^0(\xi) := \xi$, and if $\alpha > 0$,

$$\mathrm{r}^\alpha \xi := \mathrm{r}\big(\bigwedge\nolimits_{\gamma<\alpha} \mathrm{r}^\gamma \xi\big).$$

PROPOSITION XIX.4.3. *For each free convergence ξ, there exists an ordinal number $\rho(\xi)$ such that $\mathrm{R}\xi = \mathrm{r}^{\rho(\xi)}\xi$.*

[8]See S. DOLECKI and D. GAULD [65, 66].

PROOF. By an argument like in Remark XVII.1.7, for every ordinal δ, there exists a transfinite descending sequence $\{\mathrm{r}^{\alpha}\xi : \alpha < \delta\}$. Therefore, there exists a least ordinal $\rho(\xi)$ such that $\mathrm{r}^{\rho(\xi)+1}\xi = \mathrm{r}^{\rho(\xi)}\xi$. Consequently, $\mathrm{r}^{\rho(\xi)}(\mathrm{r}^{\rho(\xi)}\xi) = \mathrm{r}^{\rho(\xi)}\xi$ is a regular convergence, and so the regularization of ξ fulfills

$$\mathrm{R}\xi = \mathrm{r}^{\rho(\xi)}\xi.$$

□

The regularization of a topology need not be diagonal, and the topologization of a regular pretopology need not be regular ([9]).

The ordinal $\rho(\xi)$ is the *irregularity (number)* of ξ. If ξ is regular, then the irregularity of ξ is $\rho(\xi) = 0$. We see that $\rho(\zeta) = 1$, where ζ is from Example XIX.4.2.

PROPOSITION XIX.4.4. *For each ordinal α, there is a Hausdorff pretopology of irregularity α.*

PROOF. Regular pretopologies have the irregularity 0. The pretopology from Example XIX.4.2 has the irregularity 1. Let $\alpha > 1$. By inductive assumption, for each $\gamma < \alpha$, there exists a Hausdorff pretopology π_γ (of irregularity γ) on X_γ, and element x_γ of X_γ and a free filter $\mathcal{F}_\gamma$ on X_γ such that

$$x_\gamma \in \lim_{\mathrm{r}^{\gamma}\pi_\gamma} \mathcal{F}_\gamma \setminus \lim_{\mathrm{r}^{\delta}\pi_\gamma} \mathcal{F}_\gamma \tag{XIX.4.2}$$

for each $\delta < \gamma$. Let $\mathcal{F}_\alpha$ be the image on $\{x_\gamma : \gamma < \alpha\}$ of the coarsest free filter on $\{\gamma : \gamma \leq \alpha\}$ that converges to α in the order topology. On $X := \{x_\alpha\} \sqcup \bigsqcup_{\gamma<\alpha} X_\gamma$, we define a pretopology π_α such that $\pi|_{X_\gamma} = \pi_\gamma$ for each $\gamma < \alpha$, and $x_\alpha \in \lim_{\pi_\alpha} \int_{\mathcal{F}_\alpha}(\mathcal{F}_\gamma)_{\gamma<\alpha}$. Then $\mathrm{r}^{\gamma}\pi_\alpha$ is not regular for each $\gamma < \alpha$ because of (XIX.4.2), but $\mathrm{r}^{\alpha}\pi_\alpha$ is regular, because

$$\mathrm{adh}_{\mathrm{r}^{\alpha}\pi_\alpha}(\int_{\mathcal{F}_\alpha}(\mathcal{F}_\gamma)_{\gamma<\alpha}) \leq \mathcal{F}_\alpha.$$

□

XIX.5. *P-points

DEFINITION XIX.5.1. A free filter $\mathcal{F}$ on an infinite countable set X is said to be a *P-filter* if for each $\{F_n : n < \omega\} \subset \mathcal{F}$ there exists $F_\infty \in \mathcal{F}$ almost included in each F_n, that is, $F_\infty \setminus F_n$ is finite for each $n < \omega$. An ultrafilter which is a P-filter is called a *P-point.*

---

[9] [77].

The existence of *P-points* on countable sets is not assured in ZFC ([10]). On the other hand, on each uncountable set Y, there exist countably complete filters, for example the cocountable filter $(Y)_1$ of Y.

PROPOSITION XIX.5.2. *An ultrafilter $\mathcal{U}$ is a P-point if and only if the β-neighborhood filter of $\mathcal{U}$ on β^*X is countably complete.*

PROOF. Indeed, $U \in \mathcal{U}$ if and only if $\beta U \in \mathcal{N}_\beta(\mathcal{U})$. On the other hand, $\beta^* A_0 \subset \beta^* A_1$ if and only if $A_0 \setminus A_1$ is finite. □

PROPOSITION XIX.5.3. *A free ultrafilter $\mathcal{U}$ on an infinite countable set X is a P-point if and only if for each partition $\{X_n : n < \omega\}$ into infinite sets such that $X_n \notin \mathcal{U}$ for each $n < \omega$, there exists $U \in \mathcal{U}$ such that $U \cap X_n$ is finite for each $n < \omega$.*

Notice that if $\{X_n : n < \omega\}$ is a partition as above, and $F_n := \bigcup_{k>n} X_k$, then the two statements are equivalent:

(1) $A \setminus F_n$ is finite for each $n < \omega$,
(2) $A \cap X_n$ is finite for each $n < \omega$.

PROOF. If $\mathcal{U}$ is a P-point, and $\{X_n : n < \omega\}$ is a partition as above, then $F_n := \bigcup_{k>n} X_k \in \mathcal{U}$ for each $n < \omega$. By assumption, there is $U \in \mathcal{U}$ such that $U \setminus F_n$ is finite for each $n < \omega$, hence $U \cap X_n$ is finite for each $n < \omega$.

Conversely, if a free ultrafilter $\mathcal{U}$ is not a P-point, then there exists $\{U_n : n < \omega\} \subset \mathcal{U}$ such that if A is almost included in each U_n, then $A \notin \mathcal{U}$. It follows that $U_\infty := \bigcap_{n<\omega} U_n \notin \mathcal{U}$, hence $U_n \setminus U_\infty \in \mathcal{U}$ for each $n < \omega$.

It is handy to set $V_n := \bigcap_{k<n} U_k$. Of course, if $A \setminus V_n$ is finite for each $n < \omega$, then $A \setminus U_n$ is finite for each $n < \omega$, hence $A \notin \mathcal{U}$. Notice that there exists a strictly increasing sequence $(n_k)_k$ such that

$$X_k := V_{n_k} \setminus V_{n_{k+1}}$$

is infinite for each $k < \omega$, for otherwise, there would be n such that $V_n \setminus V_{n+k}$ is finite for each $k < \omega$, and thus $V_n \in \mathcal{U}$, and V_n is almost included in each element of the sequence, contrary to the assumption. As a result, $\{X_k : k < \omega\} \cup \{U_\infty\}$ is an infinite partition of ω into infinite sets such that if $A \cap X_n$ is finite for each $n < \omega$, then $A \notin \mathcal{U}$. □

PROPOSITION XIX.5.4. *A free ultrafilter $\mathcal{U}$ is a P-point on an infinite countable set X if and only if for each map $f \in X^X$ there is $U \in \mathcal{U}$ such that the restriction f_U is either constant, or finite-to-one.*

[10] W. RUDIN proved in [227] their existence under the *Continuum Hypothesis.* It was proved by J. KETONES in [159] that P-points exist under $\mathfrak{d} = \mathfrak{c}$ (the dominating number is equal to the continuum), see L. J. HALBEISEN [125, p. 234]. S. SHELAH constructed a model in which P-points do not exist [261].

PROOF. Of course, it is enough to prove this fact for $X = \mathbb{N}$. Let $\mathcal{U}$ be a P-point, and let $f \in \mathbb{N}^{\mathbb{N}}$. Then $\{f^-(n) : n \in \mathbb{N}\}$ is a partition of $\mathbb{N}$. If there is $n < \omega$ such that $f^-(n) \in \mathcal{U}$, then f is constant on $U = f^-(n)$. Otherwise, there is $U \in \mathcal{U}$ such that $U \cap f^-(n)$ is finite for each $n < \omega$, so that the restriction f_U is finite-to-one. Conversely, assume the property, and let $\{X_n : n < \omega\}$ be a partition of $\mathbb{N}$ into infinite sets, such that $X_n \notin \mathcal{U}$ for each $n < \omega$. Let $f \in \mathbb{N}^{\mathbb{N}}$ be defined by $f(x) := n$ whenever $x \in X_n$. Of course, f is constant on $X_n \notin \mathcal{U}$ for each $n < \omega$, hence there is $U \in \mathcal{U}$ such that $U \cap f^-(n) = U \cap X_n$ is finite for each $n < \omega$, thus $\mathcal{U}$ is a P-point by virtue of Proposition XIX.5.3. □

Let $\{X_n : n < \omega\}$ be a partition into infinite sets of a countable set X, and let $\mathcal{F}$ be the *sequential contour associated* with this partition, that is,

$$\mathcal{F} = \bigcup_{n<\omega} \bigcap_{k>n} (X_k)_0,$$

which means that $F \in \mathcal{F}$ whenever there exists $n < \omega$ such that $X_k \setminus F$ is finite for each $k > n$. By definition, $A \in \mathcal{F}$ if and only if there is n_A such that A is cofinite in X_n (equivalently, $A^c \cap X_n$ is finite) for each $n > n_A$. Let

$$F_n := \bigcup_{k>n} X_k.$$

In particular, $F_n \in \mathcal{F}$ and $X_n \notin \mathcal{F}^{\#}$ for each $n < \omega$.

THEOREM XIX.5.5 (Starosolski). *No P-filter on a countable set includes a sequential contour. A free ultrafilter on a countable set is a P-point if and only if it includes no sequential contour* ([11]).

PROOF. A P-filter includes no sequential contour. Indeed, if $\mathcal{G}$ includes a contour $\mathcal{F}$, then $F_n \in \mathcal{G}$ for each $n < \omega$. If $A \setminus F_n$ is finite for each $n < \omega$, then $A^c \in \mathcal{F}$, equivalently $A \notin \mathcal{F}^{\#}$, hence $A \notin \mathcal{G}$.

Conversely, if an ultrafilter $\mathcal{U}$ is not a P-point, then, by Proposition XIX.5.3, there exists a partition $\{X_n : n < \omega\}$ into infinite sets such that $X_n \notin \mathcal{U}$, and if $A \cap X_n$ is finite for each $n < \omega$, then $A \notin \mathcal{U}$.

Let $\mathcal{F}$ be the contour associated with $\{X_n : n < \omega\}$. If $F \in \mathcal{F}$, then there is n_0 such that $X_n \setminus F$ is finite for each $n \geq n_0$. Therefore, $(F_{n_0} \setminus F) \cap X_n$ is finite for each $n < \omega$, hence $F_{n_0} \setminus F \notin \mathcal{U}$, equivalently $\bigcup_{k<n_0} X_k \cup F \in \mathcal{U}$, and thus $F \in \mathcal{U}$, which proves that $\mathcal{F} \subset \mathcal{U}$. □

[11]A. STAROSOLSKI [241].

XIX.6. Supplement

A convergence is called *ultrafilter-diagonal* if (XIX.2.1) holds for each ultrafilter $\mathcal{F}$ and every ξ-selection such that $\mathcal{S}(x)$ is an ultrafilter for each $x \in X$. It is known [83] that diagonal convergences are strictly included in ultrafilter-diagonal convergences and the latter in pseudodiagonal convergences.

DEFINITION XIX.6.1. If ξ and θ are convergences on the same underlying set, then ξ is said to be *θ-regular* whenever

$$\lim_\xi \mathcal{F} \subset \lim_\xi \operatorname{adh}_\theta^\natural \mathcal{F}$$

for each filter $\mathcal{F}$.

EXERCISE XIX.6.2. Show that topological regularity is projective. Show that θ-regularity is projective for each convergence θ. Of course, θ-regularity and $\mathrm{S}_0\theta$-regularity coincide.

REMARK XIX.6.3. A convergence ξ is a regular topology whenever $\xi \in \ker \mathrm{T} \cap \ker \mathrm{Reg}$. However, in general, neither $\mathrm{TReg}\xi$ is regular, nor $\mathrm{RegT}\xi$ is topological. Both TReg and RegT are descending, but are not idempotent.

Actually, there exist Hausdorff pretopologies π such that consecutive regularizations and topologizations can be carried out infinitely many times without yielding a regular topology.

It is proved in [77, Theorem 5.5] that for each $n < \omega$, there exists a pretopology π (on a set of cardinality $\mathfrak{c}$) such that $(\mathrm{RegT})^n\pi < \mathrm{T}(\mathrm{RegT})^{n-1}\pi$ and $(\mathrm{RegT})^n\pi$ is a Hausdorff compact topology ([12]).

EXERCISE XIX.6.4 (P-spaces). Show that in a topological space the intersection of each countable family of open sets is open if and only if each neighborhood filter is countably complete. Such topologies were called *P-spaces* ([13]).

[12]However the proof is very difficult.

[13]L. GILLMAN and M. HENRIKSON [112].

CHAPTER XX

Compactness

We studied compactness and its variants, mostly in the framework of topologies, using either the language of covers (a topological space is compact if its every open cover has a finite subcover) or that of adherence of filters (a topological space is compact if every proper filter is adherent). These definitions, equivalent in the context of topologies, correspond to somewhat different concepts in a broader setting of convergences. Actually, differences appear already for pretopologies. We call the two basic generalizations of compactness, *compactness* and *cover compactness*.

XX.1. Compactness versus cover-compactness

If ξ is a convergence on a set X, then by Proposition VIII.3.1, a subset A of X is ξ-compact if and only if, for every filter $\mathcal{H}$ on X,

$$A \in \mathcal{H}^{\#} \Longrightarrow A \# \operatorname{adh}_{\xi} \mathcal{H}. \tag{XX.1.1}$$

In virtue of Lemma VIII.2.3, a contraposition of (XX.1.1) with $\mathcal{P} := \mathcal{H}_c$ yields: for every ideal $\mathcal{P}$,

$$A \subset \operatorname{inh}_{\xi} \mathcal{P} \Longrightarrow A \in \mathcal{P}, \tag{XX.1.2}$$

which — again by Lemma VIII.2.3 — means that A belongs to each ideal ξ-cover of A.

On the other hand, the cover definition extends for an arbitrary convergence as follows: a set A is called *ξ-cover-compact* whenever

$$A \subset \operatorname{inh}_{\xi} \mathcal{P} \Longrightarrow \underset{P \in \mathcal{P}}{\exists} \quad A \subset \operatorname{inh}_{\xi} P. \tag{XX.1.3}$$

holds for every ideal $\mathcal{P}$.

PROPOSITION XX.1.1. *A set A is ξ-cover-compact if and only if for each ξ-cover $\mathcal{Q}$ of A, there exists a finite subfamily of $\mathcal{Q}$ whose union is a ξ-cover of A.*

PROOF. Let A be ξ-cover-compact and $A \subset \operatorname{inh}_{\xi} \mathcal{Q}$. Then $A \subset \operatorname{inh}_{\xi} \mathcal{Q}^{\cup\downarrow}$ and thus there is a finite subfamily $\mathcal{Q}_0$ of $\mathcal{Q}$ such that $A \subset$

$\operatorname{inh}_\xi(\bigcup \mathcal{Q}_0)$. The converse holds, because each finite union of elements of an ideal belongs to the ideal. $\square$

Recall that, in the topological context, the use of ideal open covers or of open covers is equivalent. This is not the case for general convergences. If we drop the requirement that $\mathcal{P}$ is an ideal, the meaning of the formulae above change.

Each expression involving covers admits an equivalent expression in terms of inherences, and still another equivalent in terms of adherences. For instance, (XX.1.3) holds if and only if, for every filter $\mathcal{H}$,

$$A\# \operatorname{adh}_\xi^\natural \mathcal{H} \Longrightarrow A\# \operatorname{adh}_\xi \mathcal{H}. \tag{XX.1.4}$$

Of course, *cover-compactness* is formally stronger than *compactness*. We have seen in Section VIII.3 that the two concepts are equivalent for topologies. However, they are distinct already for pretopologies.

EXAMPLE XX.1.2 (A bisequence pretopology). Recall the *bisequence pretopology* θ on $X := \{x_\infty\} \sqcup \{x_n : n < \omega\} \sqcup \bigsqcup_{n<\omega}\{x_{n,k} : k < \omega\}$ from Example VI.1.8, for which $\mathcal{V}_\theta(x_\infty) = (X_\infty)_0 \wedge x_\infty^\uparrow$ and $\mathcal{V}_\theta(x_n) = (X_n)_0 \wedge x_n^\uparrow$, and the other elements are isolated. The set $X_\infty := \{x_\infty\} \sqcup$

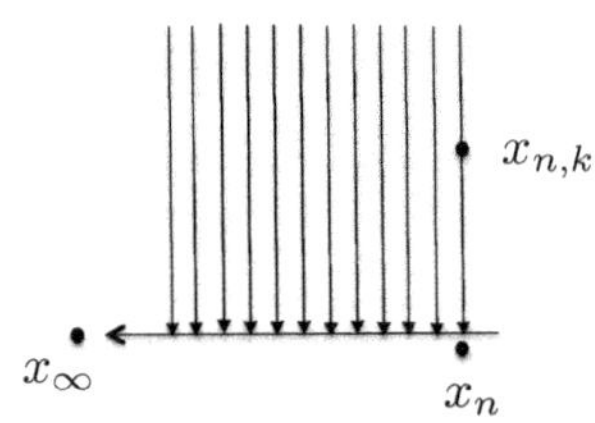

FIGURE XX.1.1. A *bisequence* $x_\infty \xleftarrow{}_n x_n \xleftarrow{}_k x_{n,k}$ defines the *bisequence pretopology*.

$\{x_n : n < \omega\}$ is θ-compact. It is not θ-cover-compact. Indeed, $H_m := \{x_{n,k} : k < \omega, m < n < \omega\}$ with $m < \omega$ defines a filter-base of a filter $\mathcal{H}$ on X. Then $X_\infty \cap \operatorname{adh}_\xi H_m = \{x_n : m < n < \omega\}$ for each $m < \omega$, but $\operatorname{adh}_\xi \mathcal{H} = \bigcap_{m<\infty} \operatorname{adh}_\xi H_m = \varnothing$, so that (XX.1.4) does not hold.

A singleton is compact for every convergence. Indeed, if we set $A = \{x\}$ in (XX.1.1), then the only $\mathcal{H}$ that fulfills the assumption is $\mathcal{H} = x^\uparrow$, and thus $\operatorname{adh}_\xi x^\uparrow \ni x$, so that $\{x\}$ is ξ-compact.

On the other hand, setting $A = \{x\}$ in (XX.1.3), we realize that $\{x\}$ is not ξ-cover-compact when ξ is not a pretopology. Indeed, by Remark XVIII.1.8, if a convergence is not pretopological at a point, then there is an infinite cover of that point without finite subcovers.

One can feel uneasy about a generalization of compactness that does not keep the points compact. And we are still more embarrassed, when we realize that cover-compactness is not preserved by continuous maps. Indeed, take non-cover-compact convergence ξ and the discrete topology ι on the same underlying set. Then, the identity $i \in \mathscr{C}(\iota, \xi)$ is continuous, every singleton is ι-cover-compact, but there is a singleton that is not ξ-cover-compact.

If $A = \{x\}$, then (XX.1.4) becomes

$$\forall_{H \in \mathcal{H}} \, x \in \mathrm{adh}_\xi H \Longrightarrow x \in \mathrm{adh}_\xi \mathcal{H} \tag{XX.1.5}$$

for each filter $\mathcal{H}$.

LEMMA XX.1.3. *If ξ is a convergence such that $\mathrm{S}\xi > \mathrm{S}_0\xi$, then there exists x such that (XX.1.5) does not hold.*

PROOF. By assumption, there exists a filter $\mathcal{F}$ and a point x such that $x \in \lim_{\mathrm{S}_0\xi} \mathcal{F} \setminus \lim_{\mathrm{S}\xi} \mathcal{F}$. By Proposition VI.6.7 and (XVIII.3.5), there is a filter $\mathcal{H}\#\mathcal{F}$ such that $x \notin \mathrm{adh}_\xi \mathcal{H}$, but $x \in \mathrm{adh}_\xi H$ for each $H \in \mathcal{H}$. □

XX.2. From pseudotopologies to compactness

If ξ is a convergence, then a set F is said to be *ξ-compact at* a set B whenever $B\# \mathrm{adh}_\xi \mathcal{H}$ for every filter $\mathcal{H}$ such that $F \in \mathcal{H}^\#$. More generally,

DEFINITION XX.2.1. A filter $\mathcal{F}$ is called *ξ-compact at B* if, for each filter $\mathcal{H}$,

$$\mathcal{H}\#\mathcal{F} \Longrightarrow \mathrm{adh}_\xi \mathcal{H} \cap B \neq \varnothing. \tag{XX.2.1}$$

A filter $\mathcal{F}$ is said to be *ξ-compactoid* if $\mathrm{adh}_\xi \mathcal{H} \neq \varnothing$ for every filter $\mathcal{H}$ such that $\mathcal{H}\#\mathcal{F}$. Finally, a convergence ξ on X is said to be *compact* if $\{X\}$ is ξ-compactoid.

Of course, a set F is ξ-compact (at B) if and only if the filter $\{F\}^\uparrow$ is ξ-compact (at B). It is straightforward that ([1])

PROPOSITION XX.2.2. *A filter $\mathcal{F}$ is ξ-compact at B if and only if $B \cap \lim_\xi \mathcal{U} \neq \varnothing$ for each $\mathcal{U} \in \beta\mathcal{F}$.*

Therefore, compact filters generalize compact sets, but, on the other hand, they generalize convergent filters. Actually,

PROPOSITION XX.2.3. *Every convergent filter is compactoid.*

[1] It is enough to observe that $\mathcal{H}\#\mathcal{F}$ whenever $\beta(\mathcal{H} \vee \mathcal{F}) \neq \varnothing$ and that if $\mathcal{U}$ is an ultrafilter, then $\mathrm{adh}_\xi \mathcal{U} = \lim_\xi \mathcal{U}$.

Indeed, if $\lim_\xi \mathcal{F} \neq \varnothing$ and $\mathcal{H}\#\mathcal{F}$, then $\mathcal{H} \vee \mathcal{F} \geq \mathcal{H}$. Hence, $\varnothing \neq \lim_\xi \mathcal{F} \subset \lim_\xi(\mathcal{H} \vee \mathcal{F}) \subset \operatorname{adh}_\xi \mathcal{H}$.

More precisely,

COROLLARY XX.2.4. *A filter $\mathcal{F}$ is ξ-compact at $\{x\}$ if and only if $x \in \lim_{S\xi} \mathcal{F}$.*

PROOF. By (XVIII.3.5), $x \in \lim_{S\xi} \mathcal{F}$ whenever $x \in \operatorname{adh}_\xi \mathcal{H}$ for each $\mathcal{H}\#\mathcal{F}$, that is, ξ-compact at $\{x\}$ by Definition XX.2.1. □

EXAMPLE XX.2.5 (Non-convergent compactoid filters). Take any Hausdorff pretopology ξ and a finite subset A of $|\xi|$. Then

$$\mathcal{V}_\xi(A) = \bigcap\nolimits_{x\in A} \mathcal{V}_\xi(x)$$

is compactoid. Indeed, by Proposition V.3.8, if a filter $\mathcal{H}$ is such that $\mathcal{H}\#\mathcal{V}_\xi(A)$, then there is $x \in A$ such that $\mathcal{H}\#\mathcal{V}_\xi(x)$, hence $\operatorname{adh}_\xi \mathcal{H} \neq \varnothing$, because $\mathcal{V}_\xi(x)$ is convergent, thus is compactoid. Of course, $\mathcal{V}_\xi(A)$ cannot be convergent if $\operatorname{card} A > 1$, because ξ is Hausdorff.

Here is another instance of a compactoid filter that is neither convergent nor principal.

EXAMPLE XX.2.6. Let ν be the standard topology. Then $\mathcal{N}_\nu[0,1] := \bigcap_{x\in[0,1]} \mathcal{N}_\nu(x)$ is a compactoid non-convergent filter. Indeed, if $\mathcal{U} \in \beta\mathcal{N}_\nu[0,1]$, then $[-1,2] \in \mathcal{U}$ and thus $\lim_\nu \mathcal{U} \neq \varnothing$, more precisely, $\lim_\nu \mathcal{U} \subset [0,1]$. On the other hand, there is no r such that $\mathcal{N}_\nu[0,1] \geq \mathcal{N}_\nu(r)$.

If $\xi \geq \theta$ then ξ-compactness implies θ-compactness. As $\operatorname{adh}_{S\xi} \mathcal{H} = \operatorname{adh}_\xi \mathcal{H}$ for each filter $\mathcal{H}$ and every convergence ξ, it clear that

COROLLARY XX.2.7. *$S\xi$-compactness coincides with ξ-compactness for every convergence ξ.*

We conclude that compact filters constitute an extension of pseudotopologically convergent filters. In other words, the nature of compactness is pseudotopological.

PROPOSITION XX.2.8. *Let $\xi \geq \tau$ be pseudotopologies. If ξ is compact and τ is Hausdorff, then $\xi = \tau$.*

PROOF. As ξ and τ are pseudotopologies such that $\xi \geq \tau$, it is enough to show the equality for ultrafilters. If $x \in \lim_\tau \mathcal{U}$ for an ultrafilter $\mathcal{U}$, then, by compactness of ξ,

$$\varnothing \neq \lim_\xi \mathcal{U} \subset \lim_\tau \mathcal{U} = \{x\},$$

hence $x \in \lim_\xi \mathcal{U}$. □

Of course, each compact topology is an example of compactoid pseudotopology. More significant examples of compact non-topological pseudotopologies will be presented in Section XX.10.

PROPOSITION XX.2.9. *Each convergence on a finite set is compact.*

PROOF. If ξ is a convergence on a finite set X, then each filter is a non-empty principal filter of a finite subset A of X, hence $\varnothing \neq A \subset \operatorname{adh}_\xi A$. □

EXAMPLE XX.2.10. Each prime cofinite convergence is compact. Indeed, let $\pi = \pi[x_\infty, (X)_0]$. Obviously, each finite subset of X is π-compact. If $\mathcal{H}$ is a free filter on X, then $\mathcal{H} \geq (X)_0$, hence $x_\infty \in \lim_\pi \mathcal{H} = \operatorname{adh}_\pi \mathcal{H} \neq \varnothing$.

Compactness is preserved under continuous maps.

PROPOSITION XX.2.11. *If $\mathcal{F}$ is ξ-compact at B and $f \in \mathscr{C}(\xi, \tau)$, then $f[\mathcal{F}]$ is τ-compact at $f(B)$.*

PROOF. Let $\mathcal{H}$ be a filter such that $\mathcal{H}\#f[\mathcal{F}]$, equivalently $f^-[\mathcal{H}]\#\mathcal{F}$. By assumption, $B \cap \operatorname{adh}_\xi f^-[\mathcal{H}] \neq \varnothing$, hence

$$\varnothing \neq f(B) \cap f(\operatorname{adh}_\xi f^-[\mathcal{H}]) = f(B) \cap \operatorname{adh}_{f\xi} \mathcal{H} \subset f(B) \cap \operatorname{adh}_\tau \mathcal{H},$$

because $f \in \mathscr{C}(\xi, \tau)$. □

COROLLARY XX.2.12. *If ξ is a compact convergence, $f \in \mathscr{C}(\xi, \tau)$ is surjective, then τ is compact.*

By Proposition XX.2.8,

COROLLARY XX.2.13. *If ξ is a compact pseudotopology, $f \in \mathscr{C}(\xi, \tau)$ is bijective, and τ is a Hausdorff pseudotopology, then $f \in \mathscr{H}(\xi, \tau)$.*

Let us now generalize the classical Tikhonov theorem on compactness of products.

THEOREM XX.2.14 (generalized Tikhonov theorem). *A filter $\mathcal{F}$ is $\prod \Xi$-compactoid if and only if $p_\xi[\mathcal{F}]$ is ξ-compactoid for each $\xi \in \Xi$.*

PROOF. As $p_\xi \in \mathscr{C}(\prod \Xi, \xi)$ for each $\xi \in \Xi$, if $\mathcal{F}$ is $\prod \Xi$-compactoid, then $p_\xi[\mathcal{F}]$ is ξ-compactoid for each $\xi \in \Xi$ by Proposition XX.2.11.

Conversely, if $\mathcal{F}$ is a filter on $|\prod \Xi|$, and $\mathcal{U} \in \beta\mathcal{F}$, then $p_\xi[\mathcal{U}] \in \beta(p_\xi[\mathcal{F}])$ for each $\xi \in \Xi$. Hence $\lim_\xi p_\xi[\mathcal{U}] \neq \varnothing$, because $p_\xi[\mathcal{F}]$ is ξ-compactoid for each $\xi \in \Xi$. This means that $\lim_{\prod \Xi} \mathcal{U} \neq \varnothing$, that is, $\mathcal{F}$ is $\prod \Xi$-compactoid. □

COROLLARY XX.2.15 (Tikhonov theorem). *If Ξ is a set of convergences, then $\prod \Xi$ is compact if and only if ξ is compact for each $\xi \in \Xi$.*

Let us exhibit the interdependence between compactness and pseudotopologicity even more incisively. To this end, let us recollect characteristic convergences from Example XVII.3.4.

DEFINITION XX.2.16. The *characteristic convergence* χ_ξ of a convergence ξ is defined by

$$\lim_{\chi_\xi} \mathcal{F} := \begin{cases} |\xi| & \lim_\xi \mathcal{F} \neq \varnothing, \\ \varnothing & \lim_\xi \mathcal{F} = \varnothing. \end{cases}$$

Notice that $\operatorname{adh}_{\chi_\xi} \mathcal{H} = |\xi|$ if $\operatorname{adh}_\xi \mathcal{H} \neq \varnothing$, and $\operatorname{adh}_{\chi_\xi} \mathcal{H} = \varnothing$ otherwise. Indeed, $\operatorname{adh}_{\chi_\xi} \mathcal{H} = \bigcup_{\mathcal{F}\#\mathcal{H}} \lim_{\chi_\xi} \mathcal{F}$, hence $\operatorname{adh}_{\chi_\xi} \mathcal{H} \neq \varnothing$ if and only if there exists $\mathcal{F}\#\mathcal{H}$ such that $\lim_{\chi_\xi} \mathcal{F} \neq \varnothing$, equivalently $\lim_\xi \mathcal{F} \neq \varnothing$, that is, $\operatorname{adh}_\xi \mathcal{H} \neq \varnothing$.

LEMMA XX.2.17. *If Ξ is a non-empty set of convergences, then* $\chi_{\prod \Xi} = \prod_{\xi\in\Xi} \chi_\xi$.

PROOF. In fact, $\varnothing \neq \lim_{\chi_{\prod \Xi}} \mathcal{F}$ whenever $\varnothing \neq \lim_{\prod \Xi} \mathcal{F}$, equivalently $\varnothing \neq \lim_\xi p_\xi[\mathcal{F}]$ for each $\xi \in \Xi$. This amounts to $\varnothing \neq \lim_{\chi_\xi} p_\xi[\mathcal{F}]$ for each $\xi \in \Xi$, equivalently $\varnothing \neq \lim_{\prod_{\xi\in\Xi}\chi_\xi} \mathcal{F}$. □

LEMMA XX.2.18. *A filter $\mathcal{F}$ is ξ-compactoid if and only if*

$$\lim_{\mathrm{S}\chi_\xi} \mathcal{F} \neq \varnothing.$$

PROOF. By (XVIII.3.6), $\lim_{\mathrm{S}\chi_\xi} \mathcal{F} \neq \varnothing$ if and only if $\lim_{\chi_\xi} \mathcal{U} \neq \varnothing$ for each $\mathcal{U} \in \beta\mathcal{F}$, equivalently, by Definition XX.2.16, $\lim_\xi \mathcal{U} \neq \varnothing$ for each $\mathcal{U} \in \beta\mathcal{F}$, that is, by Proposition XX.2.2, $\mathcal{F}$ is ξ-compactoid. □

ALTERNATIVE PROOF OF THEOREM XX.2.14. By Lemma XX.2.17 and Theorem XVIII.3.9,

$$\mathrm{S}(\chi_{\prod \Xi}) = \mathrm{S}(\prod_{\xi\in\Xi} \chi_\xi) = \prod_{\xi\in\Xi} \mathrm{S}\chi_\xi. \tag{XX.2.2}$$

In virtue of Lemma XX.2.18, $\mathcal{F}$ is $\prod \Xi$-compactoid if and only if $\lim_{\mathrm{S}(\chi_{\prod \Xi})} \mathcal{F} \neq \varnothing$. By (XX.2.2), this amounts to $\lim_{\mathrm{S}\chi_\xi} p_\xi[\mathcal{F}] \neq \varnothing$ for each $\xi \in \Xi$, that is, $p_\xi[\mathcal{F}]$ is ξ-compactoid for each $\xi \in \Xi$. □

We denote by $\mathcal{K}^\circ(\xi)$ the family of all ξ-compactoid sets, and by $\mathcal{K}(\xi)$ the set of all ξ-compact sets. Evidently, $\mathcal{K}^\circ(\xi)$ is an *ideal*, that is,

$$(K \subset A) \wedge (A \in \mathcal{K}^\circ(\xi)) \Longrightarrow K \in \mathcal{K}^\circ(\xi),$$

$$\mathcal{A} \subset_{\text{fin}} \mathcal{K}^\circ(\xi) \Longrightarrow \bigcup \mathcal{A} \in \mathcal{K}^\circ(\xi),$$

while $\mathcal{K}(\xi)$ is closed under finite unions.

In particular, the empty set is compactoid. The whole underlying set is compactoid if and only if the convergence is compact.

It follows from Proposition XX.2.11 that if A is ξ-compactoid, and $f \in \mathscr{C}(\xi, \tau)$, then $f(A)$ is τ-compactoid. If $\xi \geq \theta$, then $\mathcal{K}^\circ(\xi) \subset \mathcal{K}^\circ(\theta)$. However,

PROPOSITION XX.2.19. *A set is ξ-compactoid if and only if it is $\mathrm{S}\xi$-compactoid, that is,*

$$\mathcal{K}^\circ(\xi) \subset \mathcal{K}^\circ(\mathrm{S}\xi).$$

PROOF. If A is $\mathrm{S}\xi$-compactoid, then $\varnothing \neq \lim_{\mathrm{S}\xi} \mathcal{U} = \lim_\xi \mathcal{U}$, hence A is ξ-compactoid by Proposition XX.2.2. □

In Corollary VIII.3.18, we have observed that compactness of sets in topological spaces is an absolute property. This is still true for arbitrary convergences.

REMARK XX.2.20. Contrary to compactoidness ([2]), compactness is an *absolute* property. In other words, if ζ is a convergence on X_0 and ξ is a convergence on X_1, so that their restrictions to $X_0 \cap X_1$ coincide, then a subset A of $X_0 \cap X_1$ is ζ-compact if and only if it is ξ-compact. Assume that A is ζ-compact, and let $A \in \mathcal{H}^\#$. If $\mathcal{U} \in \beta(\mathcal{H} \vee A)$, then

$$\operatorname{adh}_\xi \mathcal{H} \cap A \supset \lim_\xi \mathcal{U} \cap A = \lim_\zeta \mathcal{U} \cap A \neq \varnothing,$$

because $\zeta|_A = \xi|_A$.

In a Hausdorff convergence space, the intersection of a filter-subbase consisting of compact sets is non-empty. However, without assumption of the Hausdorff property, the intersection of even finitely many compact sets need not be compact.

EXAMPLE XX.2.21. Let X be an infinite set, and let $x_0, x_1 \in X$. Assume that all other the points of X are isolated, and that a free filter $\mathcal{F}$ is θ-convergent provided that $\mathcal{F} \geq (X)_0$, and in this case, $\lim_\theta \mathcal{F} = \{x_0, x_1\}$. Then $X \setminus \{x_0\}$ and $X \setminus \{x_1\}$ are compact, but their intersection $W := X \setminus \{x_0, x_1\}$ is not. In fact, $W \in (X)_0$ and $\operatorname{adh}_\theta (X)_0 = \{x_0, x_1\}$ is disjoint from W.

Finally, we extend the definition of cover compactness to the effect that a filter $\mathcal{F}$ is *ξ-cover-compact at* a set B whenever for each ideal $\mathcal{P}$ such that $B \subset \operatorname{inh}_\xi \mathcal{P}$ there exists $P \in \mathcal{P}$ such that $\operatorname{inh}_\xi P \in \mathcal{F}$. An immediate extension of Proposition XX.13.1 yields

COROLLARY XX.2.22. *A filter $\mathcal{F}$ is ξ-cover-compact at a set B if and only if the contour $\mathcal{V}_\xi\{\mathcal{F}\}$ is ξ-compact at B.*

[2]Exercise XX.13.5.

We conclude that the coincidence of cover compactness with compactness characterizes topologies.

COROLLARY XX.2.23. *A pseudotopology ξ is a topology if and only if for every set B, each filter that is ξ-compact at B is ξ-cover-compact at B.*

PROOF. If ξ is a topology and a filter $\mathcal{F}$ is ξ-compact at a set B, then for every $\mathcal{U} \in \beta\mathcal{F}$, there is $x \in B \cap \lim_\xi \mathcal{U}$. As ξ is diagonal, it follows that $x \in \lim_\xi \mathcal{V}_\xi\{\mathcal{U}\} \subset \operatorname{adh}_\xi \mathcal{V}_\xi\{\mathcal{F}\}$ and thus there is $\mathcal{W} \in \beta\mathcal{V}_\xi\{\mathcal{F}\}$ such that $x \in \lim_\xi \mathcal{W}$.

Conversely, if $x \in \lim_\xi \mathcal{F}$, then by Proposition XX.2.3, $\mathcal{F}$ is ξ-compact at $\{x\}$, and by the assumption, $\mathcal{F}$ is ξ-cover-compact at $\{x\}$, that is $\mathcal{V}_\xi\{\mathcal{F}\}$ is ξ-compact at $\{x\}$. Therefore, $x \in \lim_\xi \mathcal{U}$ for each $\mathcal{U} \in \beta\mathcal{V}_\xi\{\mathcal{F}\}$, and since ξ is a pseudotopology, $x \in \lim_\xi \mathcal{V}_\xi\{\mathcal{F}\}$ by Corollary XX.2.4. In particular, for $\mathcal{F} = x^\uparrow$, we get $x \in \lim_\xi \mathcal{V}_\xi(x)$, which shows that ξ is a diagonal pretopology, hence a topology. □

XX.3. Conditional compactness

Let $\boldsymbol{G}$ be a collection of filters on X.

DEFINITION XX.3.1. A convergence ξ on X is said to be $\boldsymbol{G}$-*compact* provided that $\operatorname{adh}_\xi \mathcal{G} \neq \varnothing$ for each $\mathcal{G} \in \boldsymbol{G}$.

We shall investigate this general concept of conditional compactness for various meaningful collections $\boldsymbol{G}$.

Compactness emerged as an extension of pseudotopologicity. More precisely, the adherence determination of pseudotopologies has been a vehicle of the extension. Following the same scheme, we shall derive other compactness-like properties from other cases of adherence determination.

Recall that, for a composable class $\mathbb{H}$ of filters (XVIII.5.2), that is, such that $f^-[\mathcal{H}] \in \mathbb{H}$ for each $\mathcal{H} \in \mathbb{H}$ and each map f, the modifier $A_\mathbb{H}$ defined by (XVIII.5.3)

$$\lim_{A_\mathbb{H}\xi} \mathcal{F} := \bigcap\nolimits_{\mathbb{H} \ni \mathcal{H} \# \mathcal{F}} \operatorname{adh}_\xi \mathcal{H}$$

is a reflector. In particular, if $\mathbb{H}$ is an initial class, then $\mathbb{F}_0 \subset \mathbb{H} \subset \mathbb{F}$.

We say that a filter $\mathcal{F}$ is $\mathbb{H}(\xi)$-*compact at* B if

$$\text{(XX.3.1)} \qquad \underset{\mathcal{H} \in \mathbb{H}}{\forall} \ \mathcal{H} \# \mathcal{F} \Longrightarrow \operatorname{adh}_\xi \mathcal{H} \cap B \neq \varnothing.$$

On rewriting the properties for complementary ideals ([3]), a filter $\mathcal{F}$ is $\mathbb{H}(\xi)$-compact at B if for each ideal $\mathcal{P}$,

$$\forall_{\mathcal{P}\in\mathbb{H}_\neg} B \subset \mathrm{inh}_\xi \mathcal{P} \Longrightarrow \mathcal{F} \cap \mathcal{P} \neq \varnothing. \tag{XX.3.2}$$

Analogously to Section XX.2, we define $\mathbb{H}$-*compactoid filters* and $\mathbb{H}$-*compact convergences*. Accordingly, a filter $\mathcal{F}$ is $\mathbb{H}(\xi)$-compact at $\{x\}$ if and only if $x \in \lim_{A_{\mathbb{H}}\xi} \mathcal{F}$. In the preceding section, we considered the case of $\mathbb{H}$ being the class $\mathbb{F}$ of all filters. Of course, in that case, $A_{\mathbb{F}} = \mathrm{S}$ is the pseudotopologizer.

Table XVIII.5.1 lists other three fundamental classes of filters and the corresponding classes of adherence-determined convergences: principal (pretopologies), countably based (paratopologies), and countably stable (hypotopologies). Let us complete it here by compactness type.

Class name	reflector	class	$\mathbb{H}$-compactness	filter class name
pseudotopology	S	$\mathbb{F}$	compact	all
hypotopology	$\mathrm{S}_{\wedge 1}$	$\mathbb{F}_{\wedge 1}$	Lindelöf	countably complete
paratopology	S_1	$\mathbb{F}_1$	countably compact	countably based
pretopology	S_0	$\mathbb{F}_0$	finitely compact	principal

TABLE XX.3.1. Basic adherence-determined classes of convergences and the corresponding types of compactness.

The corresponding classes of conditional compactness are listed below:

(1) *Compact* ($\mathbb{F}$-compactness) — pseudotopological property,
(2) *Countably compact* ($\mathbb{F}_1$-compactness) — paratopological property,
(3) *Lindelöf* ($\mathbb{F}_{\wedge 1}$-compactness) — hypotopological property,
(4) *Finitely compact* ($\mathbb{F}_0$-compactness) — pretopological property.

Indeed, the compactness types listed above are characterized in terms of convergence of the characteristic filters with respect to the corresponding reflections.

THEOREM XX.3.2. *A filter $\mathcal{F}$ is $\mathbb{H}(\xi)$-compactoid if and only if*

$$\lim_{A_{\mathbb{H}}(\chi_\xi)} \mathcal{F} \neq \varnothing.$$

[3]By contraposition, (XX.3.1) becomes: for every $\mathcal{H} \in \mathbb{H}$ if $B \subset \mathrm{inh}_\xi \mathcal{H}_c$ then $\mathcal{F} \cap \mathcal{H}_c \neq \varnothing$. Define $\mathbb{H}_\neg$ by setting $\mathcal{P} \in \mathbb{H}_\neg$ whenever $\mathcal{P}_c \in \mathbb{H}$.

PROOF. By definition,

$$\lim_{A_{\mathbb{H}}(\chi_\xi)} \mathcal{F} = \bigcap_{\mathbb{H}\ni\mathcal{H}\#\mathcal{F}} \operatorname{adh}_{\chi_\xi} \mathcal{H}.$$

Therefore $\lim_{A_{\mathbb{H}}(\chi_\xi)} \mathcal{F} \neq \varnothing$ if and only if $\operatorname{adh}_{\chi_\xi} \mathcal{H} \neq \varnothing$ for every $\mathcal{H} \in \mathbb{H}$ such that $\mathcal{F}\#\mathcal{H}$. This means that $\operatorname{adh}_\xi \mathcal{H} \neq \varnothing$ for each $\mathcal{H} \in \mathbb{H}$ such that $\mathcal{H}\#\mathcal{F}$, that is, $\mathcal{F}$ is $\mathbb{H}(\xi)$-compactoid. □

By (XVIII.5.4), $\operatorname{adh}_{A_{\mathbb{H}}\xi} \mathcal{H} = \operatorname{adh}_\xi \mathcal{H}$ for each $\mathcal{H} \in \mathbb{H}$, hence

COROLLARY XX.3.3. *The $\mathbb{H}(\xi)$-compactness and $\mathbb{H}(A_{\mathbb{H}}\xi)$-compactness coincide.*

As in Corollary XX.2.4, it is straightforward that

PROPOSITION XX.3.4. *A filter $\mathcal{F}$ is $\mathbb{H}$-compact at $\{x\}$ if and only if $x \in \lim_{A_{\mathbb{H}}\xi} \mathcal{F}$.*

PROPOSITION XX.3.5. *If $\mathcal{F}$ is $\mathbb{H}(\xi)$-compact at B and $f \in \mathscr{C}(\xi, \tau)$, then $f[\mathcal{F}]$ is τ-$\mathbb{H}$-compact at $f(B)$.*

PROOF. If $\mathcal{H} \in \mathbb{H}$ and $\mathcal{H}\#f[\mathcal{F}]$, then $f^-[\mathcal{H}]\#\mathcal{F}$ and $f^-[\mathcal{H}] \in \mathbb{H}$, because $\mathbb{H}$ is initial. By assumption $B\# \operatorname{adh}_\xi f^-[\mathcal{H}]$. Hence, $f(B)\# f(\operatorname{adh}_\xi f^-[\mathcal{H}])$ and thus $f(B)\# \operatorname{adh}_{f\xi} \mathcal{H}$ by virtue of (VIII.1.11). From $f\xi \geq \tau$ we infer that $f[\mathcal{F}]$ is $\mathbb{H}(\tau)$-compact at $f(B)$. □

The commutation of the pseudotopologizer with arbitrary products entails the (generalized) Tikhnonov Theorem XVII.3.4 on the compactness of products of compact pseudotopologies.

We have observed in Section XVIII.8 that, with the exception of the pseudotopologizer, adherence-determined reflectors do not commute even with finite products. This fact indicates that $\mathbb{H}$-compactness is not preserved under finite products. Nevertheless, we shall later provide some conditions assuring $\mathbb{H}$-compactness of products.

XX.4. Countable and sequential compactness

By (XX.3.1), a filter $\mathcal{F}$ is *countably ξ-compact* at a set B if $\operatorname{adh}_\xi \mathcal{H}\#B$ for every countably based filter $\mathcal{H}$ such that $\mathcal{H}\#\mathcal{F}$.

By (XX.3.2), in terms of covers this means if $\mathcal{P}$ is a countably based ideal such that $B \subset \operatorname{inh}_\xi \mathcal{P}$, then $\mathcal{F} \cap \mathcal{P} \neq \varnothing$. In particular, a convergence ξ on X is countably compact if and only if whenever a countably based ideal $\mathcal{P}$ fulfills $\operatorname{inh}_\xi \mathcal{P} = X$ then $X \in \mathcal{P}$.

Compactness entails countable compactness, because the former is defined in terms of arbitrary filters and ideals. In terms of conditional compactness, countable compactness coincides with $\mathbb{F}_1$-compactness. As $A_{\mathbb{F}_1} = S_1$ is the paratopologizer, we infer that

COROLLARY XX.4.1. *A convergence* ξ *is countably compact if and only if* $\mathrm{S}_1\xi$ *is countably compact.*

On the other hand,

PROPOSITION XX.4.2. *A convergence* ξ *is countably compact if and only if* $\mathrm{adh}_\xi \mathcal{E} \neq \varnothing$ *for each sequential filter* $\mathcal{E}$.

PROOF. If $\mathcal{E} \in \mathbb{S}$, then $\mathrm{adh}_\xi \mathcal{E} \neq \varnothing$, because $\mathbb{S} \subset \mathbb{F}_1$. Conversely, if $\mathcal{H} \in \mathbb{F}_1$, then by Proposition II.3.7 there is $\mathcal{E} \in \mathbb{S}\mathcal{H}$, so that $\varnothing \neq \mathrm{adh}_\xi \mathcal{E} \subset \mathrm{adh}_\xi \mathcal{H}$. □

EXAMPLE XX.4.3 (A product of two countably compact topologies that is not countably compact). We shall specify two countably compact subspaces of $\beta\mathbb{N}$ whose product is not countably compact. They will be the restrictions of the Stone topology β to X and Y that we are going to define by transfinite induction.

For every $A \subset \beta\mathbb{N}$, Let $\mathcal{P}(A)$ be the family of all infinite countable subsets of A. Let $f(P) \in \mathrm{cl}_\beta A \setminus A$ for every $P \in \mathcal{P}(\beta\mathbb{N})$. Let $X_0 = \mathbb{N}$, and, for each $0 < \alpha < \omega_1$,

$$X_\alpha := \bigcup\nolimits_{\gamma<\alpha} X_\gamma \cup f(\mathcal{P}(\bigcup\nolimits_{\gamma<\alpha} X_\gamma)).$$

The set $X := \bigcup_{\alpha<\omega_1} X_\alpha$ is countably compact, because for each $P \in \mathcal{P}(\bigcup_{\gamma<\alpha} X_\gamma)$, there exists $\alpha < \omega_1$ such that $P \subset X_\alpha$, and thus has an accumulation point in $X_{\alpha+1}$, hence in X. By Theorem XXVII.4.5 on transfinite induction, $\mathrm{card}\, X_\alpha \leq \mathfrak{c} \cdot \mathfrak{c} + (\mathfrak{c} \cdot \mathfrak{c})^{\aleph_0} = \mathfrak{c}$, so that $\mathrm{card}\, X \leq \mathfrak{c} \cdot \mathfrak{c} = \mathfrak{c}$.

Let $Y := \mathbb{N} \cup (\beta\mathbb{N} \setminus X)$. By Corollary XII.6.7, $\mathrm{card}\, Y = 2^{\mathfrak{c}}$. As $\mathrm{card}\, \mathrm{cl}_\beta A = 2^{\mathfrak{c}}$ for each infinite subset A of $\beta\mathbb{N}$, we infer that $\mathrm{cl}_\beta A \cap Y \neq \varnothing$ for each infinite countable subset A of Y, and hence Y is countably compact.

As $X \cap Y = \mathbb{N}$, the intersection Δ_0 of $X \times Y$ with the diagonal Δ of $\beta\mathbb{N} \times \beta\mathbb{N}$ is equal to $\Delta_0 = \{(n,n) : n < \omega\}$. As $\{(n,n)\}$ is β^2-open for each $n < \omega$, the set Δ_0 is β^2-open discrete subset of $X \times Y$. As Δ is β^2-closed, Δ_0 is a closed discrete subspace of $X \times Y$. It follows that $X \times Y$ is not countably compact.

A convergence is called *sequentially compact* if every sequence has a convergent subsequence. Therefore, *sequential compactness* is not exactly $\mathbb{S}$-*compactness*, where $\mathbb{S}$ is the class of *sequential filters*.

More explicitly,

DEFINITION XX.4.4. A convergence ξ is *sequentially compact* if for every $\mathcal{E} \in \mathbb{S}$, there exists $\mathcal{H} \in \mathbb{S}$ such that $\mathcal{H} \geq \mathcal{E}$ and $\lim_\xi \mathcal{H} \neq \varnothing$.

On the other hand, it follows from the general definition (XX.3.1) that a convergence ξ is $\mathbb{S}$-*compact* or *sequentially conditionally compact* if $\mathrm{adh}_\xi \mathcal{E} \neq \varnothing$ for every $\mathcal{E} \in \mathbb{S}$.

Of course, each *sequentially compact* convergence is *sequentially conditionally compact*, because for each filter $\mathcal{E}$,

$$\mathrm{adh}_\xi \mathcal{E} = \bigcup\nolimits_{\mathcal{H} \geq \mathcal{E}} \lim\nolimits_\xi \mathcal{H} \supset \bigcup\nolimits_{\mathcal{H} \in \mathbb{S}\mathcal{E}} \lim\nolimits_\xi \mathcal{H},$$

as in (XVI.1.2). And since each countably based has a finer sequential filter,

COROLLARY XX.4.5. *Each sequentially conditionally compact convergence is countably compact.*

On the other hand, if ξ is a *sequentially founded* convergence ([4]), then $\mathrm{adh}_\xi \mathcal{E} = \bigcup_{\mathcal{H} \in \mathbb{S}\mathcal{E}} \lim_\xi \mathcal{H}$ for each filter $\mathcal{E}$, and so

PROPOSITION XX.4.6. *Each sequentially founded, sequentially conditionally compact convergence is sequentially compact.*

PROOF. Although the proposition is obvious, we provide a proof in order to accustom a reader to the terminology. So, let $\mathcal{E} \in \mathbb{S}X$. As ξ is sequentially conditionally compact, $\mathrm{adh}_\xi \mathcal{E} \neq \varnothing$, and since ξ is sequentially founded, there exists $\mathcal{H} \in \mathbb{S}\mathcal{E}$ such that $\lim_\xi \mathcal{H} \neq \varnothing$. $\square$

PROPOSITION XX.4.7. *Sequential compactness with respect to a convergence ξ amounts to countable compactness with respect to* $\mathrm{Seq}\xi$ ([5]).

PROOF. Let A be sequentially ξ-compact at B, and let $\mathcal{H}$ be a countably based filter with $A \in \mathcal{H}^\#$. If $\{H_n : n < \omega\}$ is a filter-base of $\mathcal{H}$, then for each $n < \omega$, there exists $x_n \in A \cap H_n$, hence by sequential compactness there exists a subsequence $(x_{n_k})_k$ such that $B \cap \lim_\xi (x_{n_k})_k \neq \varnothing$. As $(x_{n_k})_k \geq \mathcal{H}$, we infer that $\mathrm{adh}_{\mathrm{Seq}\xi} \mathcal{H} \cap B \neq \varnothing$.

Conversely, if $\mathrm{adh}_{\mathrm{Seq}\xi} \mathcal{H} \cap B \neq \varnothing$ for every countably based $\mathcal{H}$ meshing A, then this holds in particular when $\mathcal{H}$ is sequential. This means that there is a sequential filter $\mathcal{E}$ such that $\mathcal{E} \# \mathcal{H}$ and $\lim_\xi \mathcal{E} \cap B \neq \varnothing$. By Exercise XVI.5.2, $\mathcal{E} \vee \mathcal{H}$ is sequential, then $\lim_\xi (\mathcal{E} \vee \mathcal{H}) \cap B \neq \varnothing$ and $\mathcal{E} \vee \mathcal{H} \geq \mathcal{H}$, which completes the proof. $\square$

COROLLARY XX.4.8. *If $J \geq \mathrm{T}$ is a reflector, and $\xi = J\mathrm{Seq}\xi$, then sequential compactness and countable compactness coincide for ξ.*

[4] See Section XVI.3.

[5] See I. GOTCHEV and H. MINCHEV [115].

PROOF. Sequential ξ-compactness implies countable Seqξ-compactness, hence countable JSeqξ-compactness. Conversely, suppose that $\xi = \mathrm{TSeq}\xi$ is countably compact, and let $(x_n)_n$ be a sequence. By countable compactness, there exists $x \in \mathrm{cl}_\xi(\{x_n : n < \omega\} \setminus \{x\})$, so that the set $C := \{x_n : n < \omega\} \setminus \{x\}$ is not ξ-closed, hence is not Seqξ-closed. This means that there is a free sequential filter $\mathcal{E}$ on C that $\lim_\xi \mathcal{E} \setminus C \neq \varnothing$. Hence $\mathcal{E} \geq (C)_0$, and thus $\mathcal{E} \geq (x_n)_n$. □

XX.5. Lindelöf property

We say that a family $\mathcal{P}$ of sets is a *σ-ideal* if $\bigcup \mathcal{P}_0 \in \mathcal{P}$ for each countable subfamily $\mathcal{P}_0$ of $\mathcal{P}$. Clearly, $\mathcal{P}$ is a σ-ideal if and only if $\mathcal{P}_c$ is a countably complete filter.

By (XX.3.1), a filter $\mathcal{F}$ is *ξ-Lindelöf* at a set B if $\mathrm{adh}_\xi \mathcal{H} \# B$ for every $\mathcal{H} \in \mathbb{F}_{\wedge 1}$ such that $\mathcal{H} \# \mathcal{F}$, where $\mathbb{F}_{\wedge 1}$ stands for the class of *countably complete* filters.

By (XX.3.2), in terms of covers this means that a filter $\mathcal{F}$ is ξ-Lindelöf at B if $\mathcal{P}$ is a σ-ideal such that $B \subset \mathrm{inh}_\xi \mathcal{P}$, then $\mathcal{F} \cap \mathcal{P} \neq \varnothing$. In particular,

PROPOSITION XX.5.1. *A convergence ξ on X is Lindelöf if and only if a σ-ideal $\mathcal{P}$ fulfills $\mathrm{inh}_\xi \mathcal{P} = X$ then $X \in \mathcal{P}$.*

Beacuse $\mathrm{adh}_{\mathrm{S}_{\wedge 1}\xi} \mathcal{H} = \mathrm{adh}_\xi \mathcal{H}$ for each countably complete filter $\mathcal{H}$,

PROPOSITION XX.5.2. *A convergence ξ is Lindelöf if and only if $\mathrm{S}_{\wedge 1}\xi$ is Lindelöf.*

By Remark XI.3.13, the Sorgenfrey line is Lindelöf, but its square is not. Hence, the product of two Lindelöf topologies need not be Lindelöf.

XX.6. Compactoid and compact families

So far, we have been considering filters that are compact at sets. Let us make two steps further, passing from filters to arbitrary families of sets, and from sets also to families of sets.

We do not do it for pure pleasure of generalization! Not only the new setting gains in coherence, but it turns out that the resulting concepts have direct, important applications.

DEFINITION XX.6.1. A family of sets $\mathcal{A}$ is said to be *ξ-compact* at a family of sets $\mathcal{B}$ if, for every filter $\mathcal{H}$,

$$\mathcal{A} \# \mathcal{H} \Longrightarrow \mathrm{adh}_\xi \mathcal{H} \in \mathcal{B}^{\#}. \tag{XX.6.1}$$

$\mathcal{A}$ is called *ξ-compact* if $\mathcal{A}$ is ξ-compact at $\mathcal{A}$; *ξ-compactoid* if it is compact at $\{|\xi|\}$.

Without loss of generality, we assume that $\mathcal{A}$ and $\mathcal{B}$ are isotone families. Indeed, $\mathcal{A}$ is compact at $\mathcal{B}$ if and only if $\mathcal{A}^{\uparrow}$ is compact at $\mathcal{B}^{\uparrow}$ ([6]).

REMARK XX.6.2. It is clear that compactness at $\mathcal{B}$ amounts to compactness at B for each $B \in \mathcal{B}$. Conceptually, this corresponds to the coexistence of multiple conditions. Extension of objects of compactness from filters to arbitrary (isotone) families is somewhat more consequential.

DEFINITION XX.6.3. Let ξ be a convergence. A filter $\mathcal{F}$ is said to be *closed* if $\operatorname{adh}_\xi \mathcal{F} = \ker \mathcal{F}$.

Recall that the kernel $\ker \mathcal{F} = \bigcap_{F \in \mathcal{F}} F$, and thus every filter is closed with respect to the discrete topology ι.

PROPOSITION XX.6.4. *Each compact filter on a Hausdorff convergence space is closed.*

PROOF. If $x \in \operatorname{adh}_\xi \mathcal{F}$, then there is $\mathcal{U} \in \beta\mathcal{F}$ such that $\lim_\xi \mathcal{U} = \operatorname{adh}_\xi \mathcal{U} = \{x\}$, because ξ is Hausdorff. As $\mathcal{F}$ is ξ-compact, $\{x\}\#\mathcal{F}$ equivalently $x \in \ker \mathcal{F}$. □

Applying this proposition to principal filters, we recover an elementary well-known fact that each compact subset of a Hausdorff topology is closed. We shall now extend another basic fact that a closed subset of a compactoid set is compact.

PROPOSITION XX.6.5. *If $\mathcal{F}$ is a ξ-compactoid filter and $\mathcal{G} \geq \mathcal{F}$ and $\mathcal{G}$ is ξ-closed, then $\mathcal{G}$ is ξ-compact.*

PROOF. Let $\mathcal{U} \in \beta\mathcal{G}$, hence $\mathcal{U} \in \beta\mathcal{F}$. Then $\varnothing \neq \operatorname{adh}_\xi \mathcal{U} \subset \ker \mathcal{G}$, hence $\operatorname{adh}_\xi \mathcal{U} \in \mathcal{G}^{\#}$. □

COROLLARY XX.6.6. *A filter on a Hausdorff convergence space is compact if and only if it is compactoid and closed.*

COROLLARY XX.6.7. *If ξ is a Hausdorff convergence and $\mathcal{F}$ is a ξ-compact filter, then* $\operatorname{adh}_\xi \mathcal{F}$ *is a non-empty ξ-compact set.*

PROOF. Let $\mathcal{F}$ be a ξ-compact filter, let and $\mathcal{W} \in \beta(\operatorname{adh}_\xi \mathcal{F})$. Then $\operatorname{adh}_\xi \mathcal{F} = \ker \mathcal{F} \in \mathcal{W}$, because $\mathcal{F}$ is ξ-closed. Thus $\mathcal{W}\#\mathcal{F}$ hence $\varnothing \neq \lim_\xi \mathcal{W} \in \mathcal{F}^{\#}$, and since ξ is Hausdorff, $\lim_\xi \mathcal{W}$ is a singleton, so that $\varnothing \neq \lim_\xi \mathcal{W} \subset \ker \mathcal{F} = \operatorname{adh}_\xi \mathcal{F}$. □

LEMMA XX.6.8. *If $\mathcal{F}$ is a ξ-compactoid filter, then $\mathcal{F} \geq \mathcal{O}_\xi(\operatorname{adh}_\xi \mathcal{F})$.*

[6]Where $\mathcal{A}^{\uparrow}$ and $\mathcal{B}^{\uparrow}$ are the respective isotonizations of $\mathcal{A}$ and $\mathcal{B}$.

PROOF. If $O \in \mathcal{O}_\xi(\operatorname{adh}_\xi \mathcal{F})$, then $O \in \mathcal{U}$ for each $\mathcal{U} \in \beta\mathcal{F}$, because $O \cap \lim_\xi \mathcal{U} \neq \varnothing$ and since O is ξ-open by (VII.2.1). Hence $O \in \mathcal{F} = \bigcap_{\mathcal{U} \in \beta\mathcal{F}} \mathcal{U}$. □

A special case of the preceding lemma appeared already in Lemma XII.10.1. Notice that Lemma XX.6.8 can be viewed as a *finite compactness* property, that is, an $\mathbb{F}_0$-compactness. Indeed,

COROLLARY XX.6.9. *Each ξ-compactoid filter $\mathcal{F}$ is $\mathbb{F}_0(\mathrm{T}\xi)$-compact at* $\operatorname{adh}_\xi \mathcal{F}$.

A converse implication will be discussed in a more general setting in Section XXII.1.

PROPOSITION XX.6.10. *The adherence of a compactoid filter in a regular convergence space is compact.*

PROOF. Let $\mathcal{F}$ be a ξ-compactoid filter. For each $x \in \operatorname{adh}_\xi \mathcal{F}$, let $\boldsymbol{U}(x) := \{\mathcal{U} \in \beta\mathcal{F} : x \in \lim_\xi \mathcal{U}\}$. It follows that $\beta\mathcal{F} = \bigcup_{x \in \operatorname{adh}_\xi \mathcal{F}} \boldsymbol{U}(x)$. If $\mathcal{W} \in \beta(\operatorname{adh}_\xi \mathcal{F})$, then for each choice $\mathcal{U}(x) \in \boldsymbol{U}(x)$, the contour $\mathcal{U}\{\mathcal{W}\} \in \beta\mathcal{F}$. Indeed, $A \in \mathcal{U}\{\mathcal{W}\}$ whenever there is $W \in \mathcal{W}$ such that $A \in \mathcal{U}(x)$ for each $x \in W$. But $F \in \mathcal{U}$ for each $F \in \mathcal{F}$ and every $\mathcal{U} \in \beta\mathcal{F}$. As $\mathcal{F}$ is a ξ-compactoid, $\varnothing \neq \lim_\xi \mathcal{U}\{\mathcal{W}\} \subset \operatorname{adh}_\xi \mathcal{F}$, and thus, by regularity (XIX.3.2), $\lim_\xi \mathcal{U}\{\mathcal{W}\} \subset \lim_\xi \mathcal{W}$. □

COROLLARY XX.6.11. *If ξ is a topology and $K \in \mathcal{K}(\xi)$, then $\mathcal{O}_\xi(K)$ is ξ-compact and* $\operatorname{adh}_\xi \mathcal{O}_\xi(K) = K$.

PROPOSITION XX.6.12. *Let ξ be a convergence. Then the collection $\kappa(\xi)$ of isotone ξ-compact families fulfills:*

$$\varnothing, 2^X \in \kappa(\xi), \tag{XX.6.2}$$

$$\{\mathcal{A}_j : j \in J\} \subset \kappa(\xi) \Longrightarrow \bigcup_{j \in J} \mathcal{A}_j \in \kappa(\xi), \tag{XX.6.3}$$

$$\{\mathcal{A}_j : j \in J\} \subset_{\text{fin}} \kappa(\xi) \Longrightarrow \bigcap_{j \in J} \mathcal{A}_j \in \kappa(\xi). \tag{XX.6.4}$$

In other words, $\kappa(\xi)$ has the properties of a family of open sets of a topology on 2^X (see Exercise XX.13.4). In particular, it is plain that $\{K\}^\uparrow \in \kappa(\xi)$ for every $K \in \mathcal{K}(\xi)$.

A family $\mathcal{A} \subset \mathcal{O}_\xi$ is called of *ξ-openly isotone* if $\mathcal{O} \ni B \supset A \in \mathcal{A}$ implies $B \in \mathcal{A}$.

LEMMA XX.6.13. *Let ξ be a convergence on X. A family $\mathcal{A} \subset 2^X$ is ξ-openly isotone if and only if $\mathcal{O}_\xi\{\mathcal{A}\} \subset \mathcal{A}$.*

PROOF. As $\mathcal{O}_\xi(x) = \{O \in \mathcal{O}_\xi : x \in O\}$, by Definition XIX.1.2 $B \in \mathcal{O}_\xi\{\mathcal{A}\}$ whenever there is $A \in \mathcal{A}$ such that $B \in \mathcal{O}_\xi(A) = \bigcap_{x\in A} \mathcal{O}_\xi(x)$. This completes the proof. □

Recall that $P \in \mathcal{O}_\xi\{\mathcal{A}\}$ whenever there is $A \in \mathcal{A}$ such that $P \in \mathcal{O}_\xi(x)$ for each $x \in A$, that is $A \subset P \in \mathcal{O}_\xi$. Therefore, $\mathcal{A} = \mathcal{O}_\xi\{\mathcal{A}\}$ if and only if $\mathcal{A}$ consists of open sets and if $A \in \mathcal{A}$ and $A \subset P \in \mathcal{O}_\xi$, then $P \in \mathcal{A}$, that is, $\mathcal{A}$ is *openly isotone*.

PROPOSITION XX.6.14. *Let ξ be a topology. An openly isotone family $\mathcal{A}$ is ξ-compact at an openly isotone family $\mathcal{B}$ if and only for every $\mathcal{Q} \subset \mathcal{O}_\xi$ such that $\bigcup \mathcal{Q} \in \mathcal{B}$, there exists a finite subfamily $\mathcal{Q}_0$ of $\mathcal{Q}$ such that $\bigcup \mathcal{Q}_0 \in \mathcal{A}$.*

PROOF. The property (XX.6.1) holds whenever $\operatorname{adh}_\xi \mathcal{H} \# \mathcal{B}$ for every filter $\mathcal{H}$ such that $\mathcal{H} \# \mathcal{A}$. Equivalently, $\operatorname{inh}_\xi \mathcal{H}_c \in \mathcal{B}$ implies that $\mathcal{H}_c \cap \mathcal{A} \neq \varnothing$. In other words, $\bigcup_{H\in\mathcal{H}} \operatorname{int}_\xi H^c \in \mathcal{B}$ entails that there is $H_0 \in \mathcal{H}$ such that $H_0^c \in \mathcal{A}$. By setting, $\mathcal{Q} := \{\operatorname{int}_\xi H^c : H \in \mathcal{H}\}$ and recalling that $\mathcal{H}_c$ is an ideal, we prove the equivalence. □

DEFINITION XX.6.15. An isotone family $\mathcal{A}$ is said to be $\mathbb{H}(\xi)$-*compact*, or simply $\mathbb{H}$-*compact* if (XX.6.1) holds for each filter $\mathcal{H} \in \mathbb{H}$. Denote by $\kappa_{\mathbb{H}}(\xi)$ the set of isotone ξ-$\mathbb{H}$-compact families.

EXAMPLE XX.6.16. $\mathbb{F}_0$-*compactness* or *finite compactness*. On specializing the definition, a family $\mathcal{A}$ is *finitely ξ-compactoid* if, $\operatorname{adh}_\xi H \neq \varnothing$ for every $H \in \mathcal{A}^\#$. In particular, if $\{A\} \subset \mathcal{A} \subset \{A\}^\uparrow$, this means that if $H \# A$ then $\operatorname{adh}_\xi H \neq \varnothing$ for each set H, equivalently, for each non-empty subset H of A. Accordingly, each set is finitely compactoid, so that the condition is trivial for sets, but not for non-principal filters.

PROPOSITION XX.6.17. *Each set of $\mathbb{H}$-compact families constitutes the family of open sets of a topology.*

PROOF. The proof of the case $\kappa(\xi)$ remains unaltered with the additional condition that $\mathcal{H} \in \mathbb{H}$. □

PROPOSITION XX.6.18. *If $\mathcal{A}$ is a ξ-compact family and ξ is a topology, then $\mathcal{O}_\xi\{\mathcal{A}\}$ is ξ-compact.*

PROOF. Let $\mathcal{H} \# \mathcal{O}_\xi\{\mathcal{A}\}$, that is, $\operatorname{cl}_\xi^\natural \mathcal{H} \# \mathcal{A}$ in view of Corollary XIX.1.14. Then $\operatorname{adh}_\xi(\operatorname{cl}_\xi^\natural \mathcal{H}) \in \mathcal{A}^\#$, because $\mathcal{A}$ is ξ-compact. As ξ is a topology, $\operatorname{adh}_\xi(\operatorname{cl}_\xi^\natural \mathcal{H}) = \bigcap_{H\in\mathcal{H}} \operatorname{cl}_\xi H = \operatorname{adh}_\xi \mathcal{H}$, also $\operatorname{adh}_\xi \mathcal{H} \in \mathcal{A}^\# \subset (\mathcal{O}_\xi\{\mathcal{A}\})^\#$. □

PROPOSITION XX.6.19. *If ξ is a topology, $\mathcal{A}$ is an isotone family, and $\mathcal{O}_\xi\{\mathcal{A}\}$ is ξ-compact, then $\mathcal{A}$ is ξ-compact.*

PROOF. Let $\mathcal{H}$ be a filter such that $\mathcal{H}\#\mathcal{A}$, then $\mathcal{O}_\xi\{\mathcal{A}\} = \mathcal{A} \cap \mathcal{O}_\xi$, and $\mathcal{H}\#\mathcal{O}_\xi\{\mathcal{A}\}$. By assumption, $\operatorname{adh}_\xi \mathcal{H} \in (\mathcal{O}_\xi\{\mathcal{A}\})^\#$, that is, $\operatorname{cl}_\xi(\operatorname{adh}_\xi \mathcal{H}) \in \mathcal{A}^\#$. By topologicity, $\operatorname{adh}_\xi \mathcal{H} = \bigcap_{H\in\mathcal{H}} \operatorname{cl}_\xi H$, and thus $\operatorname{adh}_\xi \mathcal{H} \in \mathcal{A}^\#$. □

DEFINITION XX.6.20. A topology ξ is *consonant* if for each ξ-openly isotone ξ-compact family $\mathcal{A}$, there exists $\mathcal{D} \subset \mathcal{K}(\xi)$ such that $\mathcal{A} = \bigcup_{D\in\mathcal{D}} \mathcal{O}_\xi(D)$. We call such a family $\mathcal{D}$ *compactly generated*. A convergent is called *dissonant* if it is not consonant.

DEFINITION XX.6.21. A family $\mathcal{A}$ is said to be ξ-*cover-compact* at $\mathcal{B}$ if for each ideal $\mathcal{P}$ such that $\operatorname{inh}_\xi \mathcal{P} \in \mathcal{B}$, there exists $P \in \mathcal{P}$ such that $\operatorname{inh}_\xi P \in \mathcal{A}$.

It is straightforward that

PROPOSITION XX.6.22. *A family $\mathcal{A}$ is ξ-cover-compact at $\mathcal{B}$ if and only if* $\operatorname{adh}_\xi \mathcal{H} \in \mathcal{B}^\#$ *for each filter $\mathcal{H}$ such that* $(\operatorname{adh}_\xi^\natural \mathcal{H})\#\mathcal{A}$.

DEFINITION XX.6.23. If ξ is a convergence on X, then a family $\mathcal{A}$ is said to be $\mathbb{H}(\xi)$-*compactoid* at $\mathcal{B}$ if for every filter $\mathcal{H}$ on X,

$$(\mathbb{H}\text{-compactoid}) \qquad (\mathcal{H} \in \mathbb{H}) \wedge (\mathcal{A}\#\mathcal{H}) \Longrightarrow \operatorname{adh}_\xi \mathcal{H} \in \mathcal{B}^\#.$$

PROPOSITION XX.6.24. *A family $\mathcal{A}$ is $\mathbb{H}(\xi)$-compactoid at $\mathcal{B}$ if and only if $\mathcal{P} \cap \mathcal{A} \neq \varnothing$ for each ideal $\mathcal{P}$ such that* $\operatorname{inh}_\xi \mathcal{P} \in \mathcal{B}$ *and* $\mathcal{P}_c \in \mathbb{H}$.

DEFINITION XX.6.25. A family $\mathcal{A}$ is $\mathbb{H}(\xi)$-*cover-compactoid* at $\mathcal{B}$ if for each ideal $\mathcal{P}$ such that $\mathcal{P}_c \in \mathbb{H}$ and $\operatorname{inh}_\xi \mathcal{P} \in \mathcal{B}$, there exists $P \in \mathcal{P}$ such that $\operatorname{inh}_\xi P \in \mathcal{A}$.

REMARK XX.6.26. In Remark 19.1.1, we contemplated a concept of compactness with respect to two convergences, $\frac{\tau}{\xi}$-compactness, which generalizes both ξ-cover-compactness (which corresponds to $\frac{\xi}{\xi}$-compactness) and ξ-filter-compactness (which corresponds to $\frac{\xi}{\iota}$-compactness).

We shall extend this concept in a natural way to conditional compactness of families. Let $\mathbb{D}$ and $\mathbb{H}$ be two composable classes of filters, and ξ and τ be two convergences. A family $\mathcal{A}$ is said to be $\frac{\mathbb{H}(\tau)}{\mathbb{D}(\xi)}$*-compact* at a family $\mathcal{B}$ if for each $\mathcal{H} \in \mathbb{H}$ such that $\mathcal{H}_c$ is a ξ-cover of $\mathcal{B}$, there is $\mathcal{D} \in \mathbb{D}$ such that $\mathcal{D} \leq \mathcal{H}$ and $\mathcal{D}_c$ is a τ-cover of $\mathcal{A}$. For instance, $\mathcal{A}$ is $\frac{\mathbb{F}(\xi)}{\mathbb{F}_0(\xi)}$-compact at $\mathcal{B}$ if and only if $\mathcal{A}$ is ξ-cover-compact at $\mathcal{B}$, and $\mathcal{A}$ is $\frac{\mathbb{F}(\xi)}{\mathbb{F}_0(\iota)}$-compact at $\mathcal{B}$ if and only if $\mathcal{A}$ is ξ-compact at $\mathcal{B}$.

It is straightforward that $\mathcal{A}$ is $\frac{\mathbb{H}(\tau)}{\mathbb{D}(\xi)}$-compact at $\mathcal{B}$ if and only if, for each $\mathcal{H} \in \mathbb{H}$ if $\operatorname{adh}_\xi \mathcal{D} \in \mathcal{H}^\#$ for every $\mathcal{D} \in \mathbb{D}$ such that $\mathcal{D} \leq \mathcal{H}$, then $\operatorname{adh}_\tau \mathcal{H} \in \mathcal{B}^\#$. It is a simple observation that if $\mathcal{A}$ is $\frac{\mathbb{H}(\tau)}{\mathbb{D}(\xi)}$-compact at $\mathcal{B}$ and $\mathcal{B}$ is $\frac{\mathbb{J}(\theta)}{\mathbb{H}(\tau)}$-compact at $\mathcal{C}$ ([7]), then $\mathcal{A}$ is $\frac{\mathbb{J}(\theta)}{\mathbb{D}(\xi)}$-compact at $\mathcal{C}$.

XX.7. Conditional compactness of products

We have seen that finite products of countably compact topologies need not be countably compact (Example XX.4.3), and those of Lindelöf topologies need not be Lindelöf (Remark XI.3.13).

Nevertheless, the product of a countably compact convergence with a compact convergence is countably compact. Similarly, the product of a Lindelöf convergence with a compact convergence is Lindelöf. These facts are instances of a principle valid for arbitrary $\mathbb{H}$-compact convergences.

PROPOSITION XX.7.1. *Let $\mathbb{H}$ be a composable class of filters and ξ be a convergence. Then the following statements are equivalent:*

(1) *ξ is $\mathbb{H}$-compact,*
(2) *$\xi \times \tau$ is $\mathbb{H}$-compact for every ($\mathbb{H}$-founded) compact convergence τ,*
(3) *$\xi \times \tau$ is $\mathbb{F}_0$-compact for each $\mathbb{H}$-founded prime topology τ.*

Actually, (1) amounts to (2) also without the assumption that τ be $\mathbb{H}$-founded (Exercise XX.13.2).

Proposition XX.7.1 is a consequence of a more detailed, and consequently, more technical Theorem XX.7.2 below. It is, however, instructive and advisable to prove Proposition XX.7.1 directly as Exercise XX.13.2.

[7]Where $\mathbb{J}$ is a composable class, θ is a convergence, and $\mathcal{C}$ is a family.

In this section, ξ is a convergence on a set X and $\mathcal{A} \subset 2^X$, while τ is a convergence on Y and $\mathcal{B} \subset 2^Y$.

THEOREM XX.7.2. *If $\mathbb{H}$ is a composable class of filters, then the following conditions are equivalent:*

(1) *A filter $\mathcal{F}$ is $\mathbb{H}(\xi)$-compact at $\mathcal{A}$,*
(2) *For each convergence τ and every $\mathbb{H}$-filter $\mathcal{G}$ that is τ-compact at $\mathcal{B}$, the filter $\mathcal{F} \times \mathcal{G}$ is $\mathbb{H}(\xi \times \tau)$-compact at $\mathcal{A} \times \mathcal{B}$,*
(3) *For every $\mathbb{H}$-based prime topology τ and each y and each filter $\mathcal{G}$, if $y \in \lim_\tau \mathcal{G}$, then $\mathcal{F} \times \mathcal{G}$ is $\mathbb{F}_0(\xi \times \tau)$-compact at $\mathcal{A} \times \{y\}$.*

PROOF. (1) implies (2). If $\mathcal{H}$ is an $\mathbb{H}$-filter such that $\mathcal{H}\#(\mathcal{F} \times \mathcal{G})$, then $\mathcal{H}^-[\mathcal{G}]\#\mathcal{F}$ and $\mathcal{H}^-[\mathcal{G}] \in \mathbb{H}$, because of composability. Therefore, by (1) $\operatorname{adh}_\xi \mathcal{H}^-[\mathcal{G}] \in \mathcal{A}^\#$, that is, for every $A \in \mathcal{A}$, there exists $\mathcal{U}_A \in \beta(\mathcal{H}^-[\mathcal{G}])$ such that $A\# \lim_\xi \mathcal{U}_A$. In particular, $\mathcal{G}\#\mathcal{H}[\mathcal{U}_A]$, so that by assumption $\operatorname{adh}_\tau \mathcal{H}[\mathcal{U}_A]\#\mathcal{B}$. This means that for every $B \in \mathcal{B}$, there exists $\mathcal{W}_B \in \beta(\mathcal{H}[\mathcal{U}_A])$ such that $B\# \lim_\tau \mathcal{W}_B$. Consequently, $(\mathcal{U}_A \times \mathcal{W}_B)\#\mathcal{H}$ and thus $(A \times B)\# \operatorname{adh}_{\xi\times\tau} \mathcal{H}$ for each $A \times B \in \mathcal{A} \times \mathcal{B}$, so that $(\mathcal{A} \times \mathcal{B})\# \operatorname{adh}_{\xi\times\tau} \mathcal{H}$.

(2) implies (3), because if $y \in \lim_\tau \mathcal{G}$, then there is an $\mathbb{H}$-filter $\mathcal{Z} \leq \mathcal{G}$ and such that $\{y\} = \lim_\tau \mathcal{Z}$. Accordingly, $\mathcal{Z}\#\mathcal{W}$ for each $\mathbb{H}$-filter such that $\mathcal{W}\#\mathcal{G}$, the filter $\mathcal{Z} \vee \mathcal{W} \in \mathbb{H}$, so that $y \in \lim_\tau(\mathcal{Z} \vee \mathcal{W})$, and $(\mathcal{Z} \vee \mathcal{W})\#\mathcal{G}$.

(3) implies (1). If (1) does not hold for a convergence ξ on X and a filter $\mathcal{F}$ on X, that is, there exists an $\mathbb{H}$-filter $\mathcal{D}$ such that $\mathcal{D}\#\mathcal{F}$ and $\operatorname{adh}_\xi \mathcal{D} \notin \mathcal{A}^\#$. Let y_0 be any element of X, and let τ be a prime topology on X with $\mathcal{N}_\tau(y_0) := \mathcal{D} \wedge \{y_0\}$. Observe that for each free filter $\mathcal{E}$ on $\{(x,x) : x \neq y_0\}$, there exists a free filter $\mathcal{G}$ on X such that $\mathcal{E} = \{(x,x) : x \neq y_0\}[\mathcal{G}]$, so that $(v,w) \in \operatorname{adh}_{\xi\times\tau}\{(x,x) : x \neq y_0\}$ whenever $v = w = y_0$ and there exists $\mathcal{G} \geq \mathcal{D} > \mathcal{N}_\tau(y_0)$. Of course, $\{(x,x) : x \neq y_0\}\#(\mathcal{F} \times \mathcal{N}_\tau(y_0))$, but $\operatorname{adh}_{\xi\times\tau}\{(x,x) : x \neq y_0\} \notin (\mathcal{A} \times \{y_0\})^\#$, for $\operatorname{adh}_\xi \mathcal{D} \notin \mathcal{A}^\#$. Therefore $\mathcal{F} \times \mathcal{D}$ is not $\mathbb{F}_0(\xi \times \tau)$-compact at $\mathcal{A} \times \{y_0\}$. □

REMARK XX.7.3. Note that for (1) $\iff$ (3) only requires $\mathbb{H}$ to be $\mathbb{F}_0$-composable.

It is interesting that conditional compactness of products specializes to some mixed commutation results for the corresponding reflectors. $\mathbb{H}$-compactness was introduced as an extension of $\mathrm{A}_\mathbb{H}$-convergence. Now we will deduce some facts on partial commutation of $\mathrm{A}_\mathbb{H}$ from those on products encompassing $\mathbb{H}$-compact families.

When we apply Theorem XX.7.2 to compactness at singletons, then for $\mathcal{A} := x^\uparrow$, (1) becomes $x \in \lim_{\mathrm{A}_\mathbb{H}\xi} \mathcal{F}$ by Proposition XX.3.4. If moreover $\mathcal{B} := y^\uparrow$, (2) implies that if also $y \in \lim_{\mathrm{B}_{\mathbb{H}}\tau} \mathcal{G}$ then $(x,y) \in \lim_{\mathrm{A}_{\mathbb{H}}(\xi\times\tau)} \mathcal{F} \times \mathcal{G}$.

There result some functorial inequalities involving products.

THEOREM XX.7.4. *Let $\mathbb{H}$ be a composable class of filters and let ξ and θ be convergences on a set X. Then the following statements are equivalent:*

(1) $\theta \geq \mathrm{A}_{\mathbb{H}}\xi$,
(2) $\theta \times \mathrm{B}_{\mathbb{H}}\tau \geq \mathrm{A}_{\mathbb{H}}(\xi \times \tau)$ *for each convergence* τ,
(3) $\theta \times \tau \geq \mathrm{S}_0(\xi \times \tau)$ *for each prime $\mathbb{H}$-founded topology* τ.

PROOF. (1) implies (2). Suppose that $x \in \lim_\theta \mathcal{F} \subset \lim_{\mathrm{A}_{\mathbb{H}}\xi} \mathcal{F}$, that is, $x \in \mathrm{adh}_\xi \mathcal{H}$ for each $\mathcal{H} \in \mathbb{H}$ such that $\mathcal{H}\#\mathcal{F}$. This means that $\mathcal{F}$ is $\mathbb{H}(\xi)$-compact at $\{x\}$ by Proposition XX.3.4, hence by Theorem XX.7.2, for any convergence τ and any filter $\mathcal{G} \in \mathbb{H}$ such that $y \in \lim_\tau \mathcal{G}$, the filter $\mathcal{F} \times \mathcal{G}$ is $\mathbb{H}(\xi \times \tau)$-compact at (x,y), equivalently $(x,y) \in \lim_{\mathrm{A}_{\mathbb{H}}(\xi\times\tau)} \mathcal{F} \times \mathcal{G}$ by Proposition XX.3.4.

(2) implies (3), because if τ is $\mathbb{H}$-founded, then $\mathrm{B}_{\mathbb{H}}\tau = \tau$, and $\mathrm{A}_{\mathbb{H}} \geq \mathrm{A}_{\mathbb{F}_0} = \mathrm{S}_0$, because $\mathbb{F}_0 \subset \mathbb{H}$.

(3) implies (1). Indeed, if $x \in \lim_\theta \mathcal{F}$ and $\mathcal{H} \in \mathbb{H}$ so that $y \in \lim_\tau \mathcal{H}$, then by (3), $(x,y) \in \lim_{\mathrm{S}_0(\xi\times\tau)} \mathcal{F} \times \mathcal{H}$, equivalently $\mathcal{F} \times \mathcal{H}$ is $\mathbb{F}_0(\xi \times \tau)$-compact at (x,y), hence by Theorem XX.7.2, $\mathcal{F}$ is $\mathbb{H}(\xi)$-compact at $\{x\}$, equivalently $x \in \lim_{\mathrm{A}_{\mathbb{H}}\xi} \mathcal{F}$. □

REMARK XX.7.5. In view of Remark XX.7.3, (1) $\iff$ (3) in Theorem XX.7.4 only requires the class $\mathbb{H}$ to be $\mathbb{F}_0$-composable.

If $\mathbb{H}$ is the class of all filters, then $\mathrm{A}_{\mathbb{F}} = \mathrm{S}$ is the pseudotopologizer, and $\mathrm{B}_{\mathbb{F}} = \mathrm{I}$ the identity functor. Hence, it follows from Theorem XX.7.4 that

COROLLARY XX.7.6. *Let ξ and θ be convergences on X. The following statements are equivalent:*

(1) $\theta \geq \mathrm{S}\xi$,
(2) $\theta \times \tau \geq \mathrm{S}(\xi \times \tau)$ *for each convergence* τ,
(3) $\theta \times \tau \geq \mathrm{S}_0(\xi \times \tau)$ *for each prime topology* τ.

Functorial inequalities, obtained in Theorem XX.7.4 as a consequence of considerations on conditional compactness of products, can be applied to product quests for mixed properties in Section XXI.8. In fact,

COROLLARY XX.7.7. *Let $\mathbb{H}$ be a composable class of filters and let $\mathbb{J} \supset \mathbb{H}$ be closed under finite products. Then the following statements are equivalent:*

(1) $\xi \geq \mathrm{A}_{\mathbb{H}}\mathrm{B}_{\mathbb{J}}\xi$,
(2) $\xi \times \tau \geq \mathrm{A}_{\mathbb{H}}\mathrm{B}_{\mathbb{J}}(\xi \times \tau)$ *for each convergence* $\tau \geq \mathrm{SB}_{\mathbb{H}}\tau$,
(3) $\xi \times \tau \geq \mathrm{S}_0\mathrm{B}_{\mathbb{J}}(\xi \times \tau)$ *for each prime* $\mathbb{H}$*-founded topology* τ.

PROOF. In Theorem XX.7.4, set $\xi := \mathrm{B}_{\mathbb{J}}\xi$ and $\theta := \xi$. Then (1) of Theorem XX.7.4 becomes $\xi \geq \mathrm{A}_{\mathbb{H}}\mathrm{B}_{\mathbb{J}}\xi$, and thus $\xi \times \mathrm{B}_{\mathbb{H}}\tau \geq \mathrm{A}_{\mathbb{H}}(\mathrm{B}_{\mathbb{J}}\xi \times \tau)$ for each convergence τ, hence a fortiori for $\tau \geq \mathrm{SB}_{\mathbb{H}}\tau$. As $\mathbb{H} \subset \mathbb{J}$, we infer that $\mathrm{B}_{\mathbb{H}} \geq \mathrm{B}_{\mathbb{J}}$. On the other hand, $\mathrm{A}_{\mathbb{H}}\mathrm{S} = \mathrm{A}_{\mathbb{H}}\mathrm{A}_{\mathbb{F}} = \mathrm{A}_{\mathbb{H}}$. Therefore,

$$\xi \times \tau \geq \xi \times \mathrm{SB}_{\mathbb{H}}\tau \geq \mathrm{A}_{\mathbb{H}}(\mathrm{SB}_{\mathbb{J}}\xi \times \mathrm{SB}_{\mathbb{J}}\tau) \geq \mathrm{A}_{\mathbb{H}}\mathrm{B}_{\mathbb{J}}(\xi \times \tau)$$

for every $\tau \geq \mathrm{SB}_{\mathbb{H}}\tau$, that is, (2) above. A topology τ is $\mathbb{H}$-founded whenever $\tau \geq \mathrm{B}_{\mathbb{H}}\tau$, and on the other hand, $\mathrm{A}_{\mathbb{H}}\mathrm{B}_{\mathbb{J}} \geq \mathrm{A}_{\mathbb{F}_0}\mathrm{B}_{\mathbb{J}} = \mathrm{S}_0\mathrm{B}_{\mathbb{J}}$, (2) implies (3). Assuming (3), we get $\xi \times \tau \geq \mathrm{S}_0(\mathrm{B}_{\mathbb{J}}\xi \times \mathrm{B}_{\mathbb{J}}\tau)$. Since τ is supposed to be a prime $\mathbb{H}$-founded topology, $\tau \geq \mathrm{B}_{\mathbb{J}}\tau \geq \mathrm{B}_{\mathbb{H}}\tau \geq \tau$, hence $\xi \times \tau \geq \mathrm{S}_0(\mathrm{B}_{\mathbb{J}}\xi \times \tau)$ for each prime $\mathbb{H}$-founded topology. Therefore, in virtue of Theorem XX.7.4 (3) to (1) with $\theta := \xi$ and $\xi := \mathrm{B}_{\mathbb{J}}\xi$, we recover (1) above. □

Let us consider a couple of special cases of Corollary XX.7.7. They will provide us ready results in Section XXI.8 when we shall give them a meaning.

COROLLARY XX.7.8. *The following statements are equivalent:*

(1) $\xi \geq \mathrm{SI}_1\xi$,
(2) $\xi \times \tau \geq \mathrm{SI}_1(\xi \times \tau)$ *for each convergence* $\tau \geq \mathrm{S}\tau$,
(3) $\xi \times \tau \geq \mathrm{S}_0\mathrm{I}_1(\xi \times \tau)$ *for each prime topology* τ.

PROOF. Let $\mathbb{H} := \mathbb{F}$ (the class of all filters) and let $\mathbb{J} := \mathbb{F}_1$ (the class of countably based filters). Then $\mathrm{A}_{\mathbb{H}} = \mathrm{S}$ (the pseudotopologizer), $\mathrm{B}_{\mathbb{H}} = \mathrm{I}$ (the identity functor) and $\mathrm{B}_{\mathbb{J}} = \mathrm{I}_1$ (the coreflector on the convergences of countable character). □

COROLLARY XX.7.9. *The following statements are equivalent:*

(1) $\xi \geq \mathrm{S}_1\mathrm{I}_1\xi$,
(2) $\xi \times \tau \geq \mathrm{S}_1\mathrm{I}_1(\xi \times \tau)$ *for each convergence* $\tau \geq \mathrm{SI}_1\tau$,
(3) $\xi \times \tau \geq \mathrm{S}_0\mathrm{I}_1(\xi \times \tau)$ *for each prime topology of countable character* τ.

PROOF. Let $\mathbb{J} = \mathbb{H} := \mathbb{F}_1$, the class of countably based filters. Consequently, $\mathrm{A}_{\mathbb{H}} = \mathrm{S}_1$ is the *paratopologizer*, and $\mathrm{B}_{\mathbb{H}} = \mathrm{B}_{\mathbb{J}} = \mathrm{I}_1$ is the *coreflector* on the class of convergences of *countable character*. □

XX.8. Regularity and topologicity in compact pseudotopologies

Recall Corollary XIX.2.10 that a convergence ξ is subdiagonal whenever

$$\mathrm{T}\xi = \mathrm{S}_0\xi.$$

PROPOSITION XX.8.1. *Each Hausdorff, regular, compact convergence is subdiagonal.*

PROOF. Let ξ be a Hausdorff, regular, compact convergence, and let $x \in \mathrm{adh}_\xi(\mathrm{adh}_\xi A)$. Then there exists $\mathcal{U} \in \beta(\mathrm{adh}_\xi A)$ such that $\{x\} = \lim_\xi \mathcal{U}$, hence by Corollary XIX.1.14, $A \# \mathcal{V}_\xi\{\mathcal{U}\}$. Then there exists $\mathcal{W} \in \beta(A \vee \mathcal{V}_\xi\{\mathcal{U}\})$, hence $\mathcal{W}$ is an ultrafilter on A and by Corollary XIX.1.14 of adherence-vicinity duality, $\mathcal{U} \geq \mathrm{adh}_\xi^\natural \mathcal{W}$. By compactness and regularity,

$$\varnothing \neq \lim_\xi \mathcal{W} \subset \lim_\xi(\mathrm{adh}_\xi^\natural \mathcal{W}) \subset \lim_\xi \mathcal{U} \subset \{x\},$$

that is, $x \in \lim_\xi \mathcal{W}$ and thus $x \in \mathrm{adh}_\xi A$. □

Recall Corollary XIX.3.4 that each regular, subdiagonal convergence is topologically regular. On the other hand,

PROPOSITION XX.8.2. *Every compact topologically regular convergence is normal.*

PROOF. Suppose that on the contrary there exist ξ-closed sets F_0, F_1 such that $F_0 \cap F_1 = \varnothing$ and $\mathcal{N}_\xi(F_0) \# \mathcal{N}_\xi(F_1)$. If $\mathcal{U} \in \beta(\mathcal{N}_\xi(F_0) \vee \mathcal{N}_\xi(F_1))$, then by compactness $\lim_\xi \mathcal{U} \neq \varnothing$. As $\mathcal{U} \# \mathcal{N}_\xi(F_0)$ and $\mathcal{U} \# \mathcal{N}_\xi(F_1)$, we infer that $F_0 \in (\mathrm{cl}_\xi^\natural \mathcal{U})^\#$ and $F_1 \in (\mathrm{cl}_\xi^\natural \mathcal{U})^\#$. By topological regularity and the closedness of F_0 and F_1,

$$\varnothing \neq \lim_\xi \mathcal{U} \subset \lim_\xi \mathrm{cl}_\xi^\natural \mathcal{U} \subset F_0 \cap F_1,$$

which yields a contradiction. □

A Hausdorff normal convergence ξ is topologically Hausdorff, that is, $\mathrm{T}\xi$ is Hausdorff. Therefore ([8]),

THEOREM XX.8.3 (Kac). *Each Hausdorff, regular, compact pseudotopology is a topology.*

PROOF. By Propositions XX.8.1 and XX.8.2, a Hausdorff, regular, compact pseudotopology is in particular topologically Hausdorff. As ξ is a compact pseudotopology, $\xi \geq \mathrm{T}\xi$ and $\mathrm{T}\xi$ is Hausdorff, it follows that $\xi = \mathrm{T}\xi$ by Proposition XX.2.8. □

[8]See M. P. KAC [145].

On the other hand, by Proposition IX.3.3, each Hausdorff compact topology is regular. Therefore,

COROLLARY XX.8.4. *A Hausdorff compact pseudotopology is regular if and only if it is topological.*

In some sense, minimality of Hausdorff compact pseudotopologies entails the coincidence of two antithetic properties, regularity and diagonality.

We considered compactifications in the topological context, more precisely, only in case of functionally regular (alias completely regular) topologies. Compactifications were studied for general convergences ([9]); their interest, however, is limited by the fact that a Hausdorff compact pseudotopology is regular if and only if it is a topology (Corollary XX.8.4).

XX.9. Local compactness

We have considered locally compactoid and locally compact topologies. Let us generalize and refine these notions in case of general convergences. The theme will recur in the context of completeness in Section XXIII.3, where we will see that locally compactoid and finitely complete convergences coincide. Extending Definition XII.3.7, we say that

DEFINITION XX.9.1. A convergence is called *locally compactoid* (*locally compact*) *at* x if each filter convergent to x contains a compactoid (respectively, compact) set. Such x is called a *point of local compactoidness* (*local compactness*). A convergence is said to be *locally compactoid* (*locally compact*) if the property holds at every point.

As each discrete topology is locally compact, a convergence coarser than a locally compact convergence need not be locally compactoid. However,

PROPOSITION XX.9.2. *If ξ is a locally compactoid (locally compact) convergence, then the pseudotopology* $\mathrm{S}\xi$ *is locally compactoid (locally compact).*

PROOF. Indeed, if $x \in \lim_{\mathrm{S}\xi} \mathcal{F}$, hence $x \in \lim_\xi \mathcal{U}$ for each $\mathcal{U} \in \beta\mathcal{F}$. By assumption, there exists $K_\mathcal{U} \in \mathcal{U} \cap \mathcal{K}^\circ(\xi)$. By Proposition V.4.6 on compactness of filters, there exists a finite subcollection $\boldsymbol{F}$ of $\beta\mathcal{F}$ such that $\bigcup_{\mathcal{U}\in\boldsymbol{F}} K_\mathcal{U} \in \mathcal{F}$. As $\bigcup_{\mathcal{U}\in\boldsymbol{F}} K_\mathcal{U} \in \mathcal{K}^\circ(\xi) = \mathcal{K}^\circ(\mathrm{S}\xi)$, the pseudotopology $\mathrm{S}\xi$ is locally compactoid. Of course, the proof is analogous *mutatis mutandis* for local compactness. □

[9]See, for example, D. C. KENT and G. RICHARDSON [**152, 154**].

Recall Corollary XII.3.11 that local compactness is not preserved under continuous maps. On the other hand, it is straightforward that each finite product of locally compactoid convergences is locally compactoid.

Extending Definition XII.3.7 from the framework of topologies, we say that a convergence is *hereditarily locally compact* if it has at each point a pavement of compactly based filters.

Of course, each discrete topology is hereditarily locally compact.

PROPOSITION XX.9.3. *Each Hausdorff sequentially founded convergence is hereditarily locally compact.*

PROOF. Indeed, if ξ is a Hausdorff sequentially founded convergence and $x \in \lim_\xi (x_n)_n$, then $\{x_k : n < k < \omega\} \cup \{x\}$ is ξ-compact for each $n < \omega$, so that $x^\uparrow \wedge (x_n)_n$ has a filter-base of ξ-compact sets. □

EXAMPLE XX.9.4. In Example IX.1.12 we considered a Hausdorff non-regular topology ρ on the real line. This topology is finer than the standard topology $\nu_\mathbb{R}$. We can recall the main difference between these two topologies by noting that $\{2^{-n} : n < \omega\}$ is ρ-closed.

We shall see that ρ is locally compactoid, but not hereditarily locally compact. Of course, $\mathcal{N}_\rho(0) > \mathcal{N}_{\nu_\mathbb{R}}(0)$, but $\mathcal{N}_\rho(x) = \mathcal{N}_{\nu_\mathbb{R}}(x)$ for each $x \neq 0$. A filter-base of $\mathcal{N}_\rho(0)$ consists of $V_r := [-r, r] \setminus \{2^{-n} : n < \omega\}$ for $r > 0$. If $\mathcal{U} \in \beta V_r$ then either $0 \in \lim_\rho \mathcal{U}$, or there is $\varepsilon > 0$ such that $[-\varepsilon, \varepsilon] \notin \mathcal{U}$ and thus $\lim_\rho \mathcal{U} = \lim_{\nu_\mathbb{R}} \mathcal{U} \neq \varnothing$. Consequently, ρ is locally compactoid. However, ρ is not hereditarily locally compact. Indeed if $2^{-n} < r$, then $\lim_\rho \mathcal{U} \cap V_r = \varnothing$ for each $\mathcal{U} \in \beta(\mathcal{N}_\rho(2^{-n}) \vee V_r)$.

There exists a hereditarily locally compact Hausdorff pretopology, whose topologization is not locally compactoid.

EXAMPLE XX.9.5. The bisequence pretopology θ of Example VI.1.8 is locally compact. Indeed, isolated points are θ-compact, and each subset X_n is θ-compact, as is each subset of X_∞. Hence, θ is locally compact.

On the other hand, for each $V \in \mathcal{N}_\theta(x_\infty)$, there exists n_V such that for each $n > n_V$ there is a cofinite subset V_n of X_n such that $\bigcup_{n > n_V} V_n \subset V$. Consequently, if $w_n \in V_n \setminus \{x_n\}$, then the set $\{w_n : n < \omega\}$ is θ-closed and $\mathrm{adh}_\theta \{w_n : n < \omega\} = \varnothing$. Hence $\mathrm{T}\theta$ is not locally compactoid.

Obviously, the identity map is continuous $i \in \mathscr{C}(\theta, \mathrm{T}\theta)$, so that local compactness is not preserved under continuous maps.

Given a convergence ξ on X, we define on X a convergence $\mathrm{K}_\circ\xi$ to the effect that $x \in \lim_{\mathrm{K}_\circ\xi} \mathcal{F}$ whenever $x \in \lim_\xi \mathcal{F}$ and $\mathcal{F} \cap \mathcal{K}_\circ(\xi) \neq \varnothing$.

Notice that $\mathrm{K}_\circ\xi \geq \xi$ for each convergence ξ. Moreover,

$$\zeta \leq \xi \Longrightarrow \mathrm{K}_\circ\zeta \leq \mathrm{K}_\circ\xi.$$

In fact, if $x \in \lim_{\mathrm{K}_\circ\xi} \mathcal{F}$, then $x \in \lim_\xi \mathcal{F} \subset \lim_\zeta \mathcal{F}$ and $\varnothing \neq \mathcal{F} \cap \mathcal{K}_\circ(\xi) \subset \mathcal{F} \cap \mathcal{K}_\circ(\zeta)$. We claim that

$$\mathcal{K}_\circ(\mathrm{K}_\circ\xi) = \mathcal{K}_\circ(\xi).$$

Indeed, if $K \in \mathcal{K}_\circ(\mathrm{K}_\circ\xi)$, then $\lim_{\mathrm{K}_\circ\xi}\mathcal{U} \neq \varnothing$ for every $\mathcal{U} \in \beta K$, because K is $\mathrm{K}_\circ\xi$-compactoid. It follows that $\lim_\xi \mathcal{U} \neq \varnothing$, so that K is ξ-compactoid, that is, $K \in \mathcal{K}_\circ(\xi)$. Finally, for every map f with ξ on its domain,

$$f[\mathcal{K}_\circ(\xi)] \subset \mathcal{K}_\circ(f\xi).$$

In other terms, the class of filters $\mathbb{K}_\circ$ defined by $\mathcal{H} \in \mathbb{K}_\circ\xi$ whenever $\mathcal{H} \in \mathcal{K}_\circ(\xi)$ is a (generalized) foundation of Definition XVIII.2.14. By Corollary XVIII.2.16,

Corollary XX.9.6. *The modifier* $\mathrm{K}_\circ$ *is a coreflector on the class* $\mathbf{K}_\circ$ *of locally compactoid convergences.*

If now we define, for each convergence ξ,

$$\lim_{\mathrm{K}\xi} \mathcal{F} := \begin{cases} \lim_\xi \mathcal{F} & \mathcal{F} \cap \mathcal{K}(\xi) \neq \varnothing, \\ \varnothing & \text{otherwise,} \end{cases}$$

then we easily prove (as for $\mathrm{K}_\circ$) that K is an ascending modifier such that $\mathcal{K}(\mathrm{K}\xi) = \mathcal{K}(\xi)$ for each ξ. As $f[\mathcal{K}(\xi)] \subset \mathcal{K}(f\xi)$ for each map f and every convergence ξ on its domain, we conclude in virtue of Corollary XVIII.2.16 that

Corollary XX.9.7. *The modifier* K *is a coreflector on the class* $\mathbf{K}$ *of locally compact convergences.*

As each locally compact convergence is locally compactoid, $\mathrm{K}^\circ\xi \leq \mathrm{K}\xi$ for every convergence ξ. The class $\mathbf{K}_\circ$ of locally compactoid convergences is not reflective. Indeed,

Example XX.9.8. The topology ν of the real line is locally compactoid and regular, hence hereditarily locally compact. The countable product $\nu^{\mathbb{N}}$ is not locally compactoid. A subset V of $\mathbb{R}^{\mathbb{N}}$ is a neighborhood of the function 0, whenever there is a finite subset F of $\mathbb{N}$ such that $\{\varphi \in \mathbb{R}^{\mathbb{N}} : \forall_{n \in F}\, \varphi(n) = 0\} \subset V$, hence is not compactoid.

In Example XX.9.5, the bisequence pretopology θ is (in particular) locally compactoid, but its topologization $\mathrm{T}\theta$ is not. It is interesting that $\mathrm{K}_\circ(\mathrm{T}\theta) = \theta$. Indeed, we have seen that if K is $\mathrm{T}\theta$-compactoid, then there is a finite subset A of $\mathbb{N} \cup \{\infty\}$ such that $\bigcup_{k \in A} X_k \setminus K$ is finite. Therefore, if $\infty \in \lim_{\mathrm{K}_\circ(\mathrm{T}\theta)} \mathcal{F}$ then $\lim_{\mathrm{K}_\circ(\mathrm{T}\theta)} \mathcal{F} = \lim_\theta \mathcal{F}$.

PROPOSITION XX.9.9. *Each infimum of hereditarily compact convergences is hereditarily compact.*

PROOF. Let Ξ be a set of hereditarily compact convergences on X. If $x \in \lim_{\bigwedge \Xi} \mathcal{F}$, then there exists $\xi \in \Xi$ such that $x \in \lim_{\xi} \mathcal{F}$, so that, by assumption, there exists a filter $\mathcal{H} \leq \mathcal{F}$ which is $\mathcal{K}(\xi)$-based and $x \in \lim_{\xi} \mathcal{H}$. Therefore, $x \in \lim_{\bigwedge \Xi} \mathcal{H}$ and $\mathcal{H}$ is $\mathcal{K}(\bigwedge \Xi)$-based, and thus $\bigwedge \Xi$ is hereditarily compact. □

Since each discrete topology is hereditarily compact, we conclude that

PROPOSITION XX.9.10. *For each convergence ξ, there exists the coarsest convergence $\mathrm{K}_{\mathrm{her}}\xi$ among all hereditarily compact convergences finer than ξ.*

As a result, $\mathrm{K}_{\mathrm{her}}$ is a coprojector.

LEMMA XX.9.11. *For each map f and ξ on its domain,*

$$f(\mathrm{K}_{\mathrm{her}}\xi) \geq \mathrm{K}_{\mathrm{her}}(f\xi).$$

PROOF. If $y \in \lim_{f(\mathrm{K}_{\mathrm{her}}\xi)} \mathcal{G}$, then there exist $x \in f^{-}(y)$ and a filter $\mathcal{F}$ such that $f[\mathcal{F}] \leq \mathcal{G}$ and $x \in \lim_{\mathrm{K}_{\mathrm{her}}\xi} \mathcal{F}$. Therefore, there exists a $\mathcal{K}(\xi)$-based filter $\mathcal{H} \leq \mathcal{F}$ such that $x \in \lim_{\xi} \mathcal{H}$, hence $y \in \lim_{f\xi} \mathcal{F}$. On the other hand, $f[\mathcal{H}] \leq \mathcal{G}$ and $f[\mathcal{H}]$ is $\mathcal{K}(f\xi)$-based, because the continuous image of a compact set is compact. Hence, $y \in \lim_{\mathrm{K}_{\mathrm{her}}(f\xi)} \mathcal{G}$. □

COROLLARY XX.9.12. *The coprojector $\mathrm{K}_{\mathrm{her}}$ is a coreflector.*

We conclude that $\mathrm{K}_{\circ} \leq \mathrm{K} \leq \mathrm{K}_{\mathrm{her}}$.

PROPOSITION XX.9.13. *Each Hausdorff regular locally compact convergence is hereditarily locally compact.*

PROOF. Let ξ be as above. If $x \in \lim_{\xi} \mathcal{F}$, then there exists a ξ-compact set $C \in \mathcal{F}$. Because regularity is reflective, $\xi|_C$ is Hausdorff, regular, and compact, hence subdiagonal in virtue of Proposition XX.8.1, that is, $\mathrm{S}_0(\xi|_C) = \mathrm{T}(\xi|_C)$. As ξ is Hausdorff and C is ξ-compact, C is ξ-closed by Exercise XVIII.10.3, $\mathrm{T}(\xi|_C) = (\mathrm{T}\xi)|_C$. Thus $x \in \lim_{\xi} \mathrm{cl}_{\xi}^{\natural} \mathcal{F}$, showing that ξ is topologically regular. For each $F \in \mathcal{F}$, the set $\mathrm{cl}_{\xi}(F \cap C)$ is a ξ-closed subset of C, hence ξ-compact. Therefore, $\mathrm{cl}_{\xi}^{\natural} \mathcal{F}$ is $\mathcal{K}(\xi)$-based, and $\mathrm{cl}_{\xi}^{\natural} \mathcal{F} \leq \mathcal{F}$, which shows that ξ has a pavement at x consisting of $\mathcal{K}(\xi)$-based filters, and thus is hereditarily locally compact. □

DEFINITION XX.9.14 (Core compactness). Let ξ be a convergence ξ on a set X. A filter $\mathcal{F}$ on X is called *ξ-core-compact* if for each $F \in \mathcal{F}$ there exists $F_0 \in \mathcal{F}$ such that F_0 is ξ-compact at F. A convergence is called *core-compact* if each point has a pavement of core-compact filters.

As each core-compact filter contains a compactoid set, every core-compact convergence is locally compactoid. Examples of core-compact, non-locally-compact topologies are based on sophisticated set-theoretic tools ([10]).

LEMMA XX.9.15. *If $\mathcal{F}$ is ξ-core-compact and $f \in \mathscr{C}(\xi, \tau)$, then $f[\mathcal{F}]$ is τ-core-compact.*

PROOF. If $V \in f[\mathcal{F}]$, then $f^-(V) \in \mathcal{F}$ and thus there is $F \in \mathcal{F}$ that is ξ-compact at $f^-(V)$. Moreover, $f(F) \in f[\mathcal{F}]$ and $f(F) \subset V$. If $\mathcal{W} \in \beta f(F)$, then $f^-[\mathcal{W}]\# F$ so that there exists $\mathcal{U} \in \beta(f^-[\mathcal{W}] \vee F)$ and thus $f[\mathcal{U}] = \mathcal{W}$. Recall that F is ξ-compact at $f^-(V)$, so that $f^-(V)\# \lim_\xi \mathcal{U}$, equivalently $\varnothing \neq V \cap f(\lim_\xi \mathcal{U}) \subset V \cap \lim_\tau \mathcal{W}$, because $f \in \mathscr{C}(\xi, \tau)$. □

LEMMA XX.9.16. *Each infimum of core-compact convergences is core-compact.*

PROOF. Let Ξ be a set of core-compact convergences and let $x \in \lim_{\bigwedge \Xi} \mathcal{F}$ and let $\boldsymbol{D}_\xi(x)$ be a pavement of ξ at x consisting of core-compact filters for each $\xi \in \Xi$. Then there exists $\xi \in \Xi$ such that $x \in \lim_\xi \mathcal{F}$, and so there exists $\mathcal{D} \in \boldsymbol{D}_\xi(x)$ such that $\mathcal{F} \geq \mathcal{D}$. On the other hand, $\boldsymbol{D}(x) := \bigcup_{\xi \in \Xi} \boldsymbol{D}_\xi(x)$ is pavement of $\bigwedge \Xi$ at x. As $\mathcal{K}_\circ(\xi) \subset \mathcal{K}_\circ(\bigwedge \Xi)$, the proof is complete. □

Let Q stand for the corresponding coprojector called the *core-compact modifier*.

LEMMA XX.9.17. *The core-compact modifier is a functor, that is,*

$$\mathrm{Q}(f^- \tau) \geq f^-(\mathrm{Q}\tau)$$

for each f and τ.

PROOF. If $x \in \lim_{\mathrm{Q}(f^-\tau)} \mathcal{F}$, then $f(x) \in \lim_\tau f[\mathcal{F}]$ and there exists a filter $\mathcal{D}_\mathcal{F}$ such that $\mathcal{F} \geq \mathcal{D}_\mathcal{F}, x \in \lim_{f^-\tau} \mathcal{D}_\mathcal{F}$, and $\mathcal{D}_\mathcal{F}$ is $f^-\tau$-core-compact. By Lemma XX.9.15, $f[\mathcal{D}_\mathcal{F}]$ is τ-core-compact, $f(x) \in \lim_\tau f[\mathcal{D}_\mathcal{F}]$ and $f[\mathcal{F}] \geq f[\mathcal{D}_\mathcal{F}]$. On the other hand, $\{f[\mathcal{D}_\mathcal{F}] : \mathcal{F} \in$

[10] K. H. HOFMANN AND J. D. LAWSON [**132**, Section 7]. See also: Electronic supplements of [**116**]. Jean Goubault-Larrecq, 20.02.2019, A core-compact, non-locally compact space.

$(f^-(\tau))^-(x)\}$ is a pavement of τ at $f(x)$, because if $\mathcal{G} \in \tau^-(f(x))$, then $x \in \lim_{f^-\tau} f^-[\mathcal{G}]$ and $f[f^-[\mathcal{G}]] \geq \mathcal{G}$. Therefore, $f(x) \in \lim_{\mathrm{Q}\tau} f[\mathcal{F}]$, equivalently $x \in \lim_{f^-(\mathrm{Q}\tau)} \mathcal{F}$. $\square$

COROLLARY XX.9.18. *The class* $\mathbf{Q}$ *of core-compact convergences is coreflective.*

The corresponding coreflector Q is called the *core-compact modifier*.

As we have said, typically, reflectors do not commute with finite products. This is the case of the topologizer (Example 17.7.3). It turns out that core-compactness is a fundamental additional property enabling a key step towards commutation:

PROPOSITION XX.9.19. *A topology* ξ *is core-compact if and only if*

$$\xi \times \mathrm{S}_0\tau \geq \mathrm{T}(\xi \times \tau)$$

for every convergence τ.

PROOF. Assume ξ is a core-compact topology, and let $(x,y) \in \lim_{\xi \times \mathrm{S}_0\tau} \mathcal{F} \times \mathcal{G}$ where $\mathcal{F}$ is a core-compact filter. As ξ is a topology, we can assume that $\mathcal{F} = \mathcal{N}_\xi(x)$. We want to prove that $(x,y) \in C$ for every $C \in \mathcal{C}_{\xi\times\tau}$ such that $C\#(\mathcal{F} \times \mathcal{G})$, equivalently, $C[\mathcal{F}]\#\mathcal{G}$. For each $F \in \mathcal{F}$, there is $K_F \in \mathcal{F}$ with K_F compact at F. As $C(K_F)\#\mathcal{G}$, $y \in \mathrm{adh}_{\mathrm{S}_0\tau} C(K_F) = \mathrm{adh}_\tau C(K_F)$, so that there is $\mathcal{W}_F\#C(K_F)$ with $y \in \lim_\tau \mathcal{W}_F$. As $C[\mathcal{W}_F]\#K_F$ there are $\mathcal{U}_F\#C[\mathcal{W}_F]$ and $x_F \in \lim_\xi \mathcal{U}_F \cap F$. Thus $(x_F, y) \in C$ because $(\mathcal{U}_F \times \mathcal{W}_F)\#C$. As $\{x_F : F \in \mathcal{F}\}\#\mathcal{F}$, $(x,y) \in \mathrm{cl}_{\xi\times\tau} C \subset C$.

Conversely, if ξ is a topology on X that is not core-compact, then there exists x_0 and $U_0 \in \mathcal{O}_\xi(x_0)$, such that for each $V \in \mathcal{N}_\xi(x_0)$, there is $\mathcal{U}_V \in \beta V$ such that $\lim_\xi \mathcal{U}_V \cap U_0 = \varnothing$. Let τ be a prime convergence on X such that if $x_0 \in \lim_\tau \mathcal{W}$ for $\mathcal{W} \in \beta X$, then either $\mathcal{W} = x_0^\uparrow$ or there is $V \in \mathcal{N}_\xi(x_0)$ such that $\mathcal{W} = \mathcal{U}_V$. Then $x_0 \in \lim_{\mathrm{S}_0\tau} \mathcal{G}$, where $\mathcal{G} := \bigwedge_{V \in \mathcal{N}_\xi(x_0)} \mathcal{U}_V$. On the other hand, if $\mathcal{W} \in \beta^*\{(x,x) : x \neq x_0\}$ is such that $(x,y) \in \lim_{\xi\times\tau} \mathcal{W}$, then there is $\mathcal{W}_0 \in \beta^* X$ such that $p_X[\mathcal{W}] = \mathcal{W}_0$ and thus $y = x_0$, because there is $V \in \mathcal{N}_\xi(x_0)$ such that $\mathcal{W}_0 = \mathcal{U}_V$. Consequently, $\lim_\xi \mathcal{W}_0 \cap U_0 = \varnothing$ and $(x, x_0) \in (X \setminus U_0) \times \{x_0\}$. Hence $\bigcup_{x \neq x_0} \mathrm{cl}_\xi\, x \times \{x\}$ is $(\xi \times \tau)$-closed. Now, it suffices to note that $\{\bigcup_{x \neq x_0} \mathrm{cl}_\xi\, x \times \{x\} \in (\mathcal{N}_\xi(x_0) \times \mathcal{G})^\#$ to conclude that $(x_0, x_0) \in \lim_{\xi \times \mathrm{S}_0\tau}(\mathcal{N}_\xi(x_0) \times \mathcal{G}) \setminus \lim_{\mathrm{T}(\xi\times\tau)}(\mathcal{N}_\xi(x_0) \times \mathcal{G})$. $\square$

We will see in Section XXVI.3 that we could add $\xi \times \mathrm{T}\tau \geq \mathrm{T}(\xi \times \tau)$ for every convergence τ as an equivalent condition in Proposition XX.9.19.

XX.10. Hyperconvergences revisited

In Section XII.9 we delineated hyperconvergences. The term *hyperspace* is used to describe spaces of subsets, typically of closed or open sets. We shall refer to filters on a hyperspace as to *hyperfilters* ([11]).

In this section, we shall illustrate the relation between core compactness of the underlying topology and topologicity of the upper hyperconvergence. Hyperconvergences will be revisited again in full generality in Section XXIV.4 according to Lemma VII.5.1, characterizing hyperspaces as spaces of maps of a special kind.

Let ξ be a topology on a set X. Recall that a filter $\mathcal{F}$ on $\mathcal{C}_\xi$ *upper hyperconverges* to $A_0 \in \mathcal{C}_\xi$, in symbols $A_0 \in \lim_{[\xi,\$_0]} \mathcal{F}$, if

$$\bigcap\nolimits_{F\in\mathcal{F}} \mathrm{cl}_\xi(\bigcup\nolimits_{A\in F} A) \subset A_0. \tag{XX.10.1}$$

It is immediate that the upper hyperconvergence is not only compact, but

PROPOSITION XX.10.1. *The upper hyperconvergence is T_0 and hypercompact (every non-empty proper filter is convergent).*

PROOF. The upper hyperconvergence is T_0 (see Definition III.1.18), because if $A_0, A_1 \in \mathcal{C}_\xi$ and $A_0 \neq A_1$, then there is a point in one of two sets only, say $x \in A_1 \setminus A_0$. Then $A_0 \notin \lim_{[\xi,\$_0]}\{A_1\}^\uparrow$, because $\mathrm{adh}_\xi\,\mathrm{r}(\{A_1\}^\uparrow) = A_1 \setminus A_0 \neq \varnothing$. As for hypercompactness, $X \in \lim_{[\xi,\$_0]} \mathcal{F}$, because $\mathrm{adh}_\xi\,\mathrm{r}(\mathcal{F}) \subset X$ for every hyperfilter $\mathcal{F}$ on $\mathcal{C}_\xi$. $\square$

COROLLARY XX.10.2. *The upper hyperconvergence is not free (alias T_1).*

PROOF. It follows from Exercise XII.10.6 that the undelying set of each free hypercompact convergence is a singleton. On the other hand, if ξ is a topology on a (non-empty set) X, then $\mathrm{card}\,\mathcal{C}_\xi > 1$. In particular,

$$\lim_{[\xi,\$_0]} \mathcal{C}_\xi(A) = \{C \in \mathcal{C}_\xi : A \subset C\}$$

for any $A \in \mathcal{C}_\xi$. $\square$

Another example of hyperconvergence is the *cocompact hypertopology*.

[11] Traditionally, the upper hyperconvergence was called *upper Kuratowski convergence* (for instance, in [**83**]) and sometimes *upper Kuratowski-Painlevé convergence*, until it was rediscovered ([**70, 71**]) that G. PEANO introduced this convergence in Euclidean space ([**215, 218, 219**]) well before K. KURATOWSKI (1896-1980), and applied them routinely in analysis and optimization. L. ZORETTI witnessed in [**263**, p. 8] that P. PAINLEVÉ used the upper limit at the beginning of twentieth century.

DEFINITION XX.10.3. A filter $\mathscr{F}$ on $\mathcal{C}_\xi$ *cocompactly converges* to $A_0 \in \mathcal{C}_\xi$,

$$A_0 \in \lim_{k(\xi,\$_0)} \mathscr{F},$$

if for every ξ-compact set K such that $K \cap A_0 = \varnothing$, there exists $\mathrm{F} \in \mathscr{F}$ such that $K \cap \bigcup_{A \in \mathrm{F}} A = \varnothing$.

PROPOSITION XX.10.4. *The cocompact hyperconvergence is a topology. The collection*

$$\{\mathcal{O}_\xi(K)_c : K \in \mathcal{K}(\xi)\} \tag{XX.10.2}$$

is a base of $k(\xi, \$_0)$*-open sets.*

PROOF. If we take $\mathcal{O}_\xi(K)_c = \{A \in \mathcal{C}_\xi : A \cap K = \varnothing\}$ for each $K \in \mathcal{K}(\xi)$, then we obtain a collection $\boldsymbol{D}$ of families of ξ-closed subsets of X. We observe that, for each $A \in \mathcal{C}_\xi$, there is $\mathcal{D} \in \boldsymbol{D}$ such that $A \in \mathcal{D}$. A trivial choice is that of $\mathcal{C}_\xi = \mathcal{O}_\xi(\varnothing)_c$. We notice also that if $A \in \mathcal{O}_\xi(K_0)_c \cap \mathcal{O}_\xi(K_1)_c$, where $K_0, K_1 \in \mathcal{K}(\xi)$, then $K_0 \cup K_1 \in \mathcal{K}(\xi)$ and $A \in \mathcal{O}_\xi(K_0 \cup K_1)_c$. Therefore, by Proposition IX.2.6, $\boldsymbol{D}$ is a base for the open sets of a hypertopology. On the other hand, $A_0 \in \lim_{k(\xi,\$_0)} \mathscr{F}$ whenever for each $K \in \mathcal{K}(\xi)$ such that $A_0 \in \mathcal{O}_\xi(K)_c$, the family $\mathcal{O}_\xi(K)_c \in \mathscr{F}$. □

PROPOSITION XX.10.5. *If* ξ *is a Hausdorff topology, then* $[\xi, \$_0] \geq k(\xi, \$_0)$.

PROOF. Let $A_0 \in \lim_{[\xi,\$_0]} \mathscr{F}$ and let $K \in \mathcal{K}(\xi)$ such that $K \cap A_0 = \varnothing$. For every $x \in K$, there exists $\mathrm{F}_x \in \mathscr{F}$ and $O_x \in \mathcal{O}_\xi(x)$ such that $O_x \cap \bigcup_{A \in \mathrm{F}_x} A = \varnothing$. As K is ξ-compact, there exists a finite subset D of K such that $K \subset \bigcup_{x \in D} O_x$. On the other hand, $\mathrm{F} := \bigcap_{x \in D} \mathrm{F}_x \in \mathscr{F}$, so that $K \cap \bigcup_{A \in \mathrm{F}} A = \varnothing$, proving that $A_0 \in \lim_{k(\xi,\$_0)} \mathscr{F}$. □

We will see in Section XXV.4 that the upper hyperconvergence is always a pseudotopology. Moreover,

PROPOSITION XX.10.6. *If* ξ *is a topology and* $[\xi, \$_0]$ *is pretopological then* $[\xi, \$_0]$ *is topological.*

PROOF. ([12]). It is enough to show that if $[\xi, \$_0]$ is pretopological, then $A \in \lim \mathcal{V}\{\mathcal{V}(A)\}$ for every $A \in \mathcal{C}_\xi$. To this end, fix a vicinity $\mathcal{A}$ of A and $B \in \mathcal{A}$. As $[\xi, \$_0]$ is pretopological, $A \in \lim \mathcal{V}(A)$. Moreover, in view of (XX.10.1), also $\mathrm{cl}_\xi(\bigcup_{A \in \mathcal{A}} A) \in \lim \mathcal{V}(A)$. As $B \in$

[12]This argument, never published before, was communicated to me by F. Mynard. We will omit the $[\xi, \$_0]$ indices, as this will be the only convergence considered on $\mathcal{C}_\xi$.

$\mathcal{A}$, $B \subset \mathrm{cl}_\xi(\bigcup_{A\in\mathcal{A}} A)$ and $B \in \lim \mathcal{V}(B)$ by pretopologicity, hence $\mathrm{cl}_\xi(\bigcup_{A\in\mathcal{A}} A) \in \lim \mathcal{V}(B)$ for every $B \in \mathcal{A}$. By pretopologicity,

$$\mathrm{cl}_\xi(\bigcup\nolimits_{A\in\mathcal{A}} A) \in \bigcap\nolimits_{B\in\mathcal{A}} \lim \mathcal{V}(B) = \lim \bigcap\nolimits_{B\in\mathcal{A}} \mathcal{V}(B).$$

Moreover, in view of (XX.10.1), $\bigcap_{i\in I} A_i \in \lim \bigvee_{i\in I} \mathcal{F}_i$ whenever $A_i \in \lim \mathcal{F}_i$ for each $i \in I$ and $\bigvee_{i\in I} \mathcal{F}_i$ exists. Thus $\bigcap_{\mathcal{A}\in\mathcal{V}(A)} \mathrm{cl}_\xi(\bigcup_{A\in\mathcal{A}} A) \in \lim \bigvee_{\mathcal{A}\in\mathcal{V}(A)} \bigcap_{B\in\mathcal{A}} \mathcal{V}(B)$. As $\bigvee_{\mathcal{A}\in\mathcal{V}(A)} \bigcap_{B\in\mathcal{A}} \mathcal{V}(B) = \mathcal{V}\{\mathcal{V}(A)\}$ and

$$\bigcap\nolimits_{\mathcal{A}\in\mathcal{V}(A)} \mathrm{cl}_\xi(\bigcup\nolimits_{A\in\mathcal{A}} A) \subset A$$

because $A \in \lim \mathcal{V}(A)$, we conclude that $A \in \lim \mathcal{V}\{\mathcal{V}(A)\}$. □

We will see in Section XXVI.3 that $[\xi, \$_0]$ is pretopological, equivalently topological, if and only if $\xi \times \mathrm{S}_0\tau \geq \mathrm{T}(\xi \times \tau)$ for every τ. In view of Proposition (XX.9.19), a topology ξ is core-compact if and only if $[\xi, \$_0]$ is pretopological if and only if $[\xi, \$_0]$ is topological. For the time being, we prove a particular case:

THEOREM XX.10.7. *If ξ is a Hausdorff regular topology, then the following statements are equivalent:*

(1) *ξ is locally compact,*
(2) *$[\xi, \$_0]$ is a topology,*
(3) *$[\xi, \$_0]$ is a pretopology.*

PROOF. ([13]) (1) $\Rightarrow$ (2). If ξ is locally compact, and let $A_0 \in \lim_{k(\xi,\$_0)} \mathscr{F}$. If $x \notin A_0$, then there exists $O \in \mathcal{O}_\xi(x)$, a ξ-compact set K and $\mathrm{F} \in \mathscr{F}$ such that $O \subset K \cap \bigcup_{A\in\mathrm{F}} A = \varnothing$, proving that $A_0 \in \lim_{[\xi,\$_0]} \mathscr{F}$. Therefore, the topology $k(\xi, \$_0)$ coincides with $[\xi, \$_0]$.

(3) $\Rightarrow$ (1). If ξ is not locally compact, then there exists x_0 such that for each $V \in \mathcal{N}_\xi(x_0)$, there exists $\mathcal{U}_V \in \beta V$ such that $\lim_\xi \mathcal{U}_V = \varnothing$. Let $\mathcal{F} := \bigcap_{V\in\mathcal{N}_\xi(x)} \mathcal{U}_V$. If $F \in \mathcal{F}$, then for each $V \in \mathcal{N}_\xi(x_0)$, there is $U_V \in \mathcal{U}_V$ such that $\bigcup_{V\in\mathcal{N}_\xi(x_0)} U_V \subset F$, so that $\mathcal{N}_\xi(x_0)\#\mathcal{F}$, and thus $x_0 \in \mathrm{adh}_\xi \mathcal{F}$.

Let $\mathscr{U}_V := \{\{x\} : x \in U\}^{\uparrow c_\xi} : U \in \mathcal{U}_V\}$ for each $V \in \mathcal{N}_\xi(x_0)$. Then $\mathrm{r}(\mathscr{U}_V) = \mathcal{U}_V$, hence $\varnothing \in \lim_{[\xi,\$_0]} \mathscr{U}_V$, because $\lim_\xi \mathcal{U}_V = \varnothing$. On the other hand, if

$$\mathscr{F} := \bigcap\nolimits_{V\in\mathcal{N}_\xi(x_0)} \mathscr{U}_V,$$

then $\mathrm{F} \in \mathscr{F}$ whenever there exists $F \in \mathcal{F}$ such that $\{\{x\} : x \in F\} \subset \mathrm{F}$. Therefore, $\mathrm{r}(\mathscr{F}) = \mathcal{F}$, so that $\varnothing \notin \lim_{[\xi,\$_0]} \mathscr{F}$, because $\mathrm{adh}_\xi \mathcal{F} \neq \varnothing$. Consequently, $[\xi, \$_0]$ is not a pretopology. □

[13]Proved by G. CHOQUET, [44].

It is not difficult to describe the pretopologization $\mathrm{S}_0[\xi, \$_0]$ and the topologization $\mathrm{T}[\xi, \$_0]$ in terms of ξ. More precisely, vicinities of $[\xi, \$_0]$ can be related to ξ-compactoid families, and $[\xi, \$_0]$-open sets to ξ-compact families.

We say that a family $\mathcal{A}$ of ξ-closed sets is *stable* if $\mathcal{C}_\xi \ni C \subset A \in \mathcal{A}$, then $C \in \mathcal{A}$.

THEOREM XX.10.8. *If ξ is a topology, then a subfamily $\mathcal{A}$ of $\mathcal{C}_\xi$ is a stable $[\xi, \$_0]$-vicinity of A_0 if and only if $\mathcal{A}_c \subset \mathcal{O}_\xi$ is ξ-openly isotone family ξ-compact at A_0^c.*

PROOF. Let $\mathcal{A}_c = \mathcal{O}_\xi\{\mathcal{A}_c\}$ be ξ-compact at A_0^c. Then $\mathcal{A} \subset \mathcal{C}_\xi$ is stable for ξ-closed subsets. Let $A_0 \in \lim_{[\xi,\$_0]} \mathscr{F}$, that is, $\mathrm{adh}_\xi \mathrm{r}\mathscr{F} \subset A_0$. By the compactness assumption, $\mathrm{r}\mathscr{F} \perp \mathcal{A}_c$, that is, there exist $\mathrm{F} \in \mathcal{F}$ and $A \in \mathcal{A}$ such that $\bigcup_{C \in \mathrm{F}} C \subset A$, that is, $\mathrm{F} \subset \mathcal{A}$ by stability of $\mathcal{A}$, equivalently, $\mathcal{A} \in \mathscr{F}$. In other words, every filter $[\xi, \$]$-convergent to A_0 contains $\mathcal{A}$, hence $\mathcal{A} \in \mathcal{V}_{[\xi,\$_0]}(A_0)$.

Suppose now that ξ is a topology, and $\mathcal{A} \in \mathcal{V}_{[\xi,\$_0]}(A_0)$. In particular, $\mathcal{A}$ is stable for ξ-closed subsets, hence $\mathcal{A}_c = \mathcal{O}_\xi\{\mathcal{A}_c\}$. We need to show that if $\mathcal{H}$ is a filter such that $\mathrm{adh}_\xi \mathcal{H} \subset A_0$, then $\mathcal{H} \perp \mathcal{A}_c$, that is, there exist $A \in \mathcal{A}$ and $H \in \mathcal{H}$ such that $H \subset A$. If $\mathrm{adh}_\xi \mathcal{H} \subset A_0$, and $\widehat{\mathcal{H}} := \{\{C = \mathrm{cl}_\xi C : C \subset H\} : H \in \mathcal{H}\}$, then $A_0 \in \lim_{[\xi,\$_0]} \widehat{\mathcal{H}}$, because $\mathcal{H} \leq \mathrm{r}\widehat{\mathcal{H}}$, and thus $\mathcal{A} \in \widehat{\mathcal{H}}$. Since ξ is a topology, $\mathrm{adh}_\xi \mathcal{H} = \mathrm{adh}_\xi \mathrm{cl}_\xi^\natural \mathcal{H}$, so that there is $H \in \mathcal{H}$ such that $\{C = \mathrm{cl}_\xi C : C \subset \mathrm{cl}_\xi H\} \subset \mathcal{A}$, hence $H \subset \mathrm{cl}_\xi H \in \mathcal{A}$. □

COROLLARY XX.10.9. *If ξ is a topology, then a family $\mathcal{A}$ is a $[\xi, \$_0]$-open set if and only if $\mathcal{A}_c$ is ξ-openly isotone ξ-compact.*

A characterization from Theorem XX.10.8 will be refined and extended.

XX.11. Classification of perfect maps

We say that a map is *compactoid* if the image of each compactoid filter is compactoid. By Proposition XX.2.11, each continuous map is compactoid. On the other hand, each map valued in a compact convergence is compactoid.

PROPOSITION XX.11.1. *Each compactoid graph-closed map is continuous.*

PROOF. Suppose that f is (ξ, τ)-closed and compactoid. If $x \in \lim_\xi \mathcal{F}$, then by graph-closedness, $\mathrm{adh}_\tau f[\mathcal{F}] \subset \{f(x)\}$, and since $f[\mathcal{F}]$ is τ-compactoid, in particular, $\mathrm{adh}_\tau f[\mathcal{F}] \neq \varnothing$. □

EXAMPLE XX.11.2. The map $h \in \mathbb{R}^{\mathbb{R}}$ from Example IV.10.1 defined by $h(x) := \frac{1}{x}$ if $x \neq 0$ and $h(0) := 0$, is graph-closed, but is not compactoid. In fact, as $A(r) := f(]0,r]) \cap [\frac{1}{r}, +\infty[\neq \varnothing$ for each $r > 0$, the filter-base $\{]0,r] : r > 0\}$ converges to 0 in $\nu_{\mathbb{R}}$, but $\mathrm{adh}_{\nu_{\mathbb{R}}}\{A(r) : r > 0\} = \varnothing$.

DEFINITION XX.11.3. Let θ be a convergence on W and σ a convergence on Z. A relation $R \subset W \times Z$ is $\mathbb{H}$-*compactoid* if $\lim_\theta \mathcal{F} \neq \varnothing$ implies that $R[\mathcal{F}]$ is $\mathbb{H}$-compactoid for σ. It is called $\mathbb{H}$-*compact*

$$R \in \mathscr{K}_{\mathbb{H}}(\theta, \sigma)$$

if $w \in \lim_\theta \mathcal{F}$ implies that $R(w) \# \mathrm{adh}_\sigma \mathcal{H}$ for each $\mathcal{H} \# R[\mathcal{F}]$ such that $\mathcal{H} \in \mathbb{H}$.

In other terms, R is $\mathbb{H}$-compact (from θ to σ) if for each $w \in \lim_\theta \mathcal{F}$ the filter $R[\mathcal{F}]$ is $\mathbb{H}(\sigma)$-compact at Rw. In particular, if $\mathbb{H} = \mathbb{F}$, the class of all filters, a relation R is called *compact*: $R \in \mathscr{K}(\theta, \sigma)$.

Let us observe that ([14]) a surjective map $f : |\xi| \longrightarrow |\tau|$ is called $\mathbb{H}$-*perfect* if $f^- \in \mathscr{K}_{\mathbb{H}}(\tau, \xi)$.

PROPOSITION XX.11.4. *A map f from $|\xi|$ to $|\tau|$ is* $\mathbb{H}$-perfect if and only if f is surjective, and for each $\mathcal{H} \in \mathbb{H}$ and every $y \in |\tau|$,

$$y \in \mathrm{adh}_\tau f[\mathcal{H}] \Longrightarrow f^-(y) \cap \mathrm{adh}_\xi \mathcal{H} \neq \varnothing. \tag{XX.11.1}$$

PROOF. Indeed, let $f^- \in \mathscr{K}_{\mathbb{H}}(\tau, \xi)$, and $y \in \mathrm{adh}_\tau f[\mathcal{H}]$ with $\mathcal{H} \in \mathbb{H}$. Consequently, there exists a filter $\mathcal{G}$ such that $y \in \lim_\tau \mathcal{G}$ and $\mathcal{G} \# f[\mathcal{H}]$, equivalently $f^-[\mathcal{G}] \# \mathcal{H}$. Hence by assumption, $\mathrm{adh}_\xi \mathcal{H} \cap f^-(y) \neq \varnothing$. Conversely, if $y \in \lim_\tau \mathcal{G}$ and $\mathcal{H} \in \mathbb{H}$ is such that $\mathcal{H} \# f^-[\mathcal{G}]$, that is, $\mathcal{G} \# f[\mathcal{H}]$ and $y \in \mathrm{adh}_\tau f[\mathcal{H}]$, then by (XX.11.1), $f^-(y) \# \mathrm{adh}_\xi \mathcal{H}$, proving that f^- is compact from τ to ξ. $\square$

It follows immediately from Proposition XX.11.4 that

PROPOSITION XX.11.5. *A surjective map $f : |\xi| \longrightarrow |\tau|$ is $\mathbb{H}$-perfect if and only if*

$$\mathrm{adh}_\tau f[\mathcal{H}] \subset f(\mathrm{adh}_\xi \mathcal{H}) \tag{XX.11.2}$$

for every $\mathcal{H} \in \mathbb{H}$.

Therefore,

[14]Indeed, let $\mathcal{G} \in \mathbb{H}$ be such that $\mathcal{G} \# R[\mathcal{A}]$, equivalently $R^-[\mathcal{G}] \# \mathcal{A}$, hence $\mathrm{adh}\, R^-[\mathcal{G}] \# B$, because $R^-[\mathcal{G}] \in \mathbb{H}$. Therefore, there exists $x \in \mathrm{adh}\, R^-[\mathcal{G}] \cap B$, so that there is a filter $\mathcal{F} \# R^-[\mathcal{G}]$, equivalently, $R[\mathcal{F}] \# \mathcal{G}$, such that $B \# \lim \mathcal{F}$. As R is $\mathbb{H}$-compact, $\mathrm{adh}\, \mathcal{G} \# Rx$, hence $\mathrm{adh}\, \mathcal{G} \# RB$.

PROPOSITION XX.11.6. *A surjective map is $\mathbb{H}$-perfect if and only if*

$$f^{-}\tau \geq \mathrm{A}_{\mathbb{H}}\xi. \tag{XX.11.3}$$

PROOF. Indeed, applying f^{-} to (XX.11.2), we get $\operatorname{adh}_{f^{-}\tau}\mathcal{H} \subset f^{-}(f(\operatorname{adh}_{\xi}\mathcal{H})) \subset \operatorname{adh}_{\xi}\mathcal{H}$, hence

$$\lim_{f^{-}\tau}\mathcal{F} \subset \bigcap_{\mathbb{H}\ni\mathcal{H}\#\mathcal{F}} \operatorname{adh}_{\xi}\mathcal{H} = \lim_{\mathrm{A}_{\mathbb{H}}\xi}\mathcal{F},$$

which is (XX.11.3). □

In particular, $\mathbb{F}$-perfect is called just *perfect*, $\mathbb{F}_1$-perfect is said *countably perfect*, $\mathbb{F}_{\wedge 1}$-perfect, *inversely Lindelöf*, and $\mathbb{F}_0$-perfect is called *adherent.*

The last property specializes to $\operatorname{adh}_{\tau} f(H) \subset f(\operatorname{adh}_{\xi} H)$ for every $H \subset |\xi|$, which, in topological spaces, becomes the closedness of f. More precisely,

PROPOSITION XX.11.7. *If $f^{-} \in \mathscr{K}_{\mathbb{F}_0}(\tau,\xi)$ then $f(H)$ is τ-closed for each ξ-closed set H. If $\mathrm{S}_0\xi = \mathrm{T}\xi$, then the converse is also true.*

PROOF. Suppose that there is $H \subset |\xi|$ such that $f(H)$ is not τ-closed. Then there is $y \notin f(H)$ and a filter $\mathcal{G}$ such that $y \in \lim_{\tau}\mathcal{G}$ and $\mathcal{G}\# f(H)$, equivalently $f^{-}[\mathcal{G}]\# H$. By assumption, $f^{-}[\mathcal{G}]$ is $\mathbb{F}_0(\xi)$-compact at $f^{-}(y)$, so that $\operatorname{adh}_{\xi} H \cap f^{-}(y) \neq \varnothing$. As $f^{-}(y) \cap H = \varnothing$, we infer that H is not ξ-closed.

Conversely, assume that $\mathrm{S}_0\xi = \mathrm{T}\xi$, and $f(H)$ is τ-closed for each ξ-closed set H. As $f(H) \subset f(\operatorname{cl}_{\xi} H)$ and $f(\operatorname{cl}_{\xi} H)$ is τ-closed, $\operatorname{adh}_{\tau} f(H) \subset f(\operatorname{cl}_{\xi} H)$. Then (XX.11.2) holds, because $\operatorname{cl}_{\xi} H = \operatorname{adh}_{\xi} H$. □

EXAMPLE XX.11.8 (A closed, non-adherent map). Let ξ be a non-topological pretopology on X, let $\tau = \mathrm{T}\xi$, and let $i_X : X \longrightarrow X$. Of course, $i_X \in \mathscr{C}(\xi, \mathrm{T}\xi)$ is closed, because $\mathcal{C}_{\xi} = \mathcal{C}_{\mathrm{T}\xi}$. However, i_X is not adherent. Indeed, if $H \subset X$ is such that $\operatorname{adh}_{\xi}^{2} H \setminus \operatorname{adh}_{\xi} H \neq \varnothing$, then $\operatorname{adh}_{\tau} H = \operatorname{cl}_{\tau} H = i_X(\operatorname{cl}_{\xi} H)$ is not included in $i_X(\operatorname{adh}_{\xi} H)$.

Example XX.11.16 is that of an open map that is not closed. As closedness is the weakest variant of perfectness, this is an example for any kind of non-perfectness.

EXAMPLE XX.11.9. Consider the product $\zeta \times \iota_Y$, where ζ is a convergence on X, and Y is an infinite set. The projection map $p_Y : X \times Y \longrightarrow Y$ is continuous $p_Y \in \mathscr{C}(\zeta \times \iota_Y, \iota_Y)$ by the definition of product. Also, p_Y is both open and closed, because any subset of Y is ι_Y-clopen. Using Proposition XX.11.12, we infer that if ζ is compact, then p_Y is a perfect map, that is, $p_Y^- \in \mathscr{K}(\iota_Y, \zeta \times \iota_Y)$; if ζ is countably compact, non-compact, then p_Y is a countably perfect map, non-perfect map; if ζ is not countably compact, then p_Y is not countably perfect; if ζ is Lindelöf, non-compact, then p_Y is inversely Lindelöf, but not perfect.

At this point, a reader might have an impression to have already seen an expression resembling (XX.11.1). Indeed, if $\mathbb{H}$ is a composable class of filters, then a surjective map $f : |\xi| \longrightarrow |\tau|$ is $\mathbb{H}$-*quotient*

$$\tau \geq \mathrm{A}_{\mathbb{H}}(f\xi)$$

whenever (XVIII.9.6)

$$y \in \mathrm{adh}_\tau \mathcal{H} \Longrightarrow f^-(y) \cap \mathrm{adh}_\xi f^-[\mathcal{H}] \neq \varnothing$$

for each $\mathcal{H} \in \mathbb{H}$ and every $y \in |\tau|$.

It turns out that, for every composable class $\mathbb{H}$, each $\mathbb{H}$-perfect map is $\mathbb{H}$-quotient, but not conversely.

THEOREM XX.11.10. *If $\mathbb{H}$ is a composable class of filters, then each $\mathbb{H}$-perfect map is $\mathbb{H}$-quotient.*

PROOF. Let ξ and τ be convergences on X and Y respectively, and let $f \in Y^X$ be surjective. Assume (XX.11.1), and let $\mathcal{G} \in \mathbb{H}Y$ be such that $y \in \mathrm{adh}_\tau \mathcal{G}$. Then $\mathcal{H} := f^-[\mathcal{G}] \in \mathbb{H}X$ by composability. Accordingly, by Lemma IV.1.2, $f[\mathcal{H}] = f[f^-[\mathcal{G}]] = \mathcal{G}$ because f is surjective, so that $y \in \mathrm{adh}_\tau f[\mathcal{H}]$. Hence, by (XX.11.1), $f^-(y) \cap \mathrm{adh}_\xi f^-[\mathcal{G}] \neq \varnothing$. $\square$

Perfect maps. A surjective map $f : |\xi| \longrightarrow |\tau|$ is *perfect* if $y \in \lim_\tau \mathcal{F}$ implies that $f^-[\mathcal{F}]$ is compact at $f^-(y)$. The latter means that $\mathrm{adh}_\xi \mathcal{H} \cap f^-(y) \neq \varnothing$ for each filter $\mathcal{H} \# f^-[\mathcal{F}]$, equivalently $f[\mathcal{H}] \# \mathcal{F}$, that is, $y \in \mathrm{adh}_\tau f[\mathcal{H}]$.

Countably perfect maps. A surjective map $f : |\xi| \longrightarrow |\tau|$ is called *countably perfect*, $f^- \in \mathscr{K}_{\mathbb{F}_1}(\tau, \xi)$, if $y \in \lim_\tau \mathcal{F}$ implies that $f^-[\mathcal{F}]$ is countably compact at $f^-(y)$. In detail, $\mathrm{adh}_\xi \mathcal{H} \cap f^-(y) \neq \varnothing$ for each $\mathcal{H} \in \mathbb{F}_1$ such that $\mathcal{H} \# f^-[\mathcal{F}]$, equivalently $f[\mathcal{H}] \# \mathcal{F}$, that is, $y \in \mathrm{adh}_\tau f[\mathcal{H}]$ for each countably based filter $\mathcal{H}$.

Inversely Lindelöf maps. A surjective map $f : |\xi| \longrightarrow |\tau|$ is called *inversely Lindelöf*, $f^- \in \mathscr{K}_{\mathbb{F}_{\wedge 1}}(\tau, \xi)$, if $y \in \lim_\tau \mathcal{F}$ implies that $f^-[\mathcal{F}]$ is Lindelöf at $f^-(x)$.

Adherent maps. A surjective map $f : |\xi| \longrightarrow |\tau|$ is *adherent*, $f^- \in \mathscr{K}_{\mathbb{F}_0}(\tau, \xi)$, if $y \in \lim_\tau \mathcal{F}$ implies that $f^-[\mathcal{F}]$ is finitely compact at $f^-(y)$. In other terms, $\mathrm{adh}_\xi H \cap f^-(y) \neq \varnothing$ for each $H \in f^-[\mathcal{F}]^\#$, that is, $f(H) \in \mathcal{F}^\#$ implies that $y \in \mathrm{adh}_\tau f(H)$.

Closed maps. A surjective map $f : |\xi| \longrightarrow |\tau|$ is closed, $f^- \in \mathscr{K}_{\mathrm{T}}(\tau,\xi)$, if $\mathrm{adh}_\xi H \subset H$ implies $\mathrm{adh}_\tau f(H) \subset f(H)$.

REMARK XX.11.11. If ξ and τ are topologies, then closed and adherent maps coincide.

To make a link between traditional and current definitions of perfectness, we shall use transitivity of cover compactness. Note that a set is $\frac{\mathbb{H}(\tau)}{\mathbb{F}_0(\tau)}$-compact if and only if it is $\mathbb{H}(\tau)$-cover-compact.

PROPOSITION XX.11.12. *Let $\mathbb{H}$ be a composable class of filters. If R is an $\mathbb{F}_0$-compact relation with $\frac{\mathbb{H}}{\mathbb{F}_0}$-compact values, then R is $\mathbb{H}$-compact.*

PROOF. Let $R \in \mathscr{K}_{\mathbb{F}_0}(\xi,\tau)$ and let $x \in \lim_\xi \mathcal{F}$. Consider an $\mathbb{H}$-filter $\mathcal{H}\#R[\mathcal{F}]$. Then $\mathrm{adh}_\tau H \cap Rx$ for every $H \in \mathcal{H}$. By $\frac{\mathbb{H}(\tau)}{\mathbb{F}_0(\tau)}$-compactness of Rx, we deduce $Rx\#\,\mathrm{adh}_\tau \mathcal{H}$. □

In particular,

PROPOSITION XX.11.13. *A surjective continuous map is perfect if and only if it is adherent and its fibers are cover-compact.*

PROPOSITION XX.11.14. *A surjective continuous map is countably perfect if and only if it is adherent and its fibers are countably cover-compact.*

PROPOSITION XX.11.15. *A surjective continuous map is inversely Lindelöf if and only if it is adherent and its fibers are cover-Lindelöf.*

Reflector	perfect-like	$\Rightarrow$	quotient-like
			open
I			almost open
S	perfect	$\Rightarrow$	biquotient
S_1	countably perfect	$\Rightarrow$	countably biquotient
S_0	adherent	$\Rightarrow$	hereditarily quotient
T	closed	$\Rightarrow$	topologically quotient

TABLE XX.11.1. Comparison of perfect-like and quotient-like maps. No implication can be reversed.

Indeed,

EXAMPLE XX.11.16. Let $f : \mathbb{R} \to S^1$ be given by

$$f(x) := (\cos 2\pi x, \sin 2\pi x),$$

as in Example XX.6.1. Clearly, f is open, hence, has all the quotient-like properties. However, the set $\{n + \frac{1}{n} : n \in \mathbb{N}\}$ is closed, but its image under f is not closed, so that f is not closed, and thus has no perfect-like property.

XX.12. Compactness related properties of maps

Compactoid filters constitute a common generalization of convergent filters and compactoid sets. Similarly, continuity, perfectness, and quotientness and their variants can be seen as special cases of compactness of some relations.

PROPOSITION XX.12.1. *If $\mathbb{H}$ is a composable class of filters, then $f \in \mathscr{C}(\xi, \mathrm{A}_{\mathbb{H}}\tau)$ if and only if $x \in \lim_\xi \mathcal{F}$ implies that $f[\mathcal{F}]$ is $\mathbb{H}(\tau)$-compact at $f(x)$ for each x and $\mathcal{F}$.*

On the other hand, a map is $\mathbb{H}$-perfect if and only if the relation f^- is $\mathbb{H}$-compact.

Recall that a map $f : |\xi| \to |\tau|$ is $\mathbb{H}$-quotient (alias $\mathrm{A}_{\mathbb{H}}$-quotient) if and only if

$$y \in \operatorname{adh}_\tau \mathcal{H} \Longrightarrow f^-(y) \# \operatorname{adh}_\xi f^-[\mathcal{H}] \tag{XX.12.1}$$

for every $\mathcal{H} \in \mathbb{H}$ and each $y \in |\tau|$. In what follows, we show that a map f is $\mathbb{H}$-quotient (alias $\mathrm{A}_{\mathbb{H}}$-quotient) if and only if the relation f is $\mathbb{H}$-compact with respect to f-final and f-initial convergences.

THEOREM XX.12.2. *A map f is $\mathbb{H}$-quotient from ξ to τ if and only if f is $\mathbb{H}$-compact from $f^-\tau$ to $f\xi$, that is, $f \in \mathscr{K}_{\mathbb{H}}(f^-\tau, f\xi)$.*

PROOF. ($\Rightarrow$) If $x \in \lim_{f^-\tau} \mathcal{F}$, then $f(x) \in \lim_\tau f[\mathcal{F}]$. If $\mathcal{H} \# f[\mathcal{F}]$ is such that $\mathcal{H} \in \mathbb{H}$, then $f(x) \in \operatorname{adh}_\tau \mathcal{H}$. By (XX.12.1), $f^-(f(x)) \# \operatorname{adh}_\xi f^-[\mathcal{H}]$, hence

$$f(x) \in f(\operatorname{adh}_\xi f^-[\mathcal{H}]) \subset \operatorname{adh}_{f\xi} \mathcal{H}.$$

($\Leftarrow$) If $y \in \operatorname{adh}_\tau \mathcal{H}$ and $\mathcal{H} \in \mathbb{H}$, then $f^-(y) \subset \operatorname{adh}_{f^-\tau} f^-[\mathcal{H}]$. By surjectivity, $f[f^-[\mathcal{H}]] = \mathcal{H}$, and by $\mathbb{H}$-compactness, $y \in \operatorname{adh}_{f\xi} \mathcal{H} = f(\operatorname{adh}_\xi f^-[\mathcal{H}])$, that is, $f^-(y) \# \operatorname{adh}_\xi f^-[\mathcal{H}]$. □

On the other hand, recall that

PROPOSITION XX.12.3. *A surjective map $f \in \mathscr{C}(\xi, \tau)$ is $\mathbb{H}$-perfect if and only if f^- is $\mathbb{H}$-compact from τ to ξ, that is, $f^- \in \mathscr{K}_{\mathbb{H}}(\tau, \xi)$, if and only if f is adherent (closed) and $f^-(y)$ is $\mathbb{H}(\xi)$-compact for each y in the range.*

Theorem XX.7.2 directly applies to relations, hence to maps and their inverses.

PROPOSITION XX.12.4. *Let* $\mathbb{H}$ *be a composable class of filters, and let* R *be a relation. The following statements are equivalent:*

(1) R *is* $\mathbb{H}$*-compact,*
(2) $R \times S$ *is an* $\mathbb{H}$*-compact relation for each compact relation* S *with an* $\mathbb{H}$*-founded convergence as domain,*
(3) $R \times i_{|\sigma|}$ *is an* $\mathbb{F}_0$*-compact relation, where* $i_{|\sigma|}$ *is the identity on a prime* $\mathbb{H}$*-founded topology* σ*.*

PROOF. Let $R \subset |\xi| \times |\tau|$ be a relation. (1) $\Rightarrow$ (2): let $x \in \lim_\xi \mathcal{F}$ and $\mathcal{H}$ an $\mathbb{H}$-filter such that $y \in \lim_\sigma \mathcal{H}$. By assumptions, $R[\mathcal{F}]$ is $\mathbb{H}(\tau)$-compact at Rx and $S \subset |\sigma \times \theta|$so that the $\mathbb{H}$-filter $S[\mathcal{H}]$ is θ-compact at Sy. By Theorem XX.7.2, $R[\mathcal{F}] \times S[\mathcal{H}]$ is $\mathbb{H}(\tau \times \theta)$-compact at $Rx \times Sy$. Obviously, (2) implies (3). (3) $\Rightarrow$ (1): if $x \in \lim_\xi \mathcal{F}$ and σ is a prime $\mathbb{H}$-founded topology, then, by assumption, $R[\mathcal{F}] \times \mathcal{N}_\sigma(y)$ is $\mathbb{F}_0(\tau \times \sigma)$-compact at $Rx \times \{y\}$. Hence, by Theorem XX.7.2, $R[\mathcal{F}]$ is $\mathbb{H}$-compact at Rx. □

In order to implement Proposition XX.12.4 to $\mathbb{H}$-quotient maps (alias $\mathrm{A}_{\mathbb{H}}$-quotient), we should keep in mind that a map $f \in \mathscr{C}(\xi, \tau)$ is $\mathbb{H}$-quotient if and only if $f \in \mathscr{K}_{\mathbb{H}}(f^- \tau, f\xi)$. Similarly, a map $g \in \mathscr{C}(\sigma, \theta)$ is $\mathbb{H}$-quotient if and only if $f \in \mathscr{K}_{\mathbb{H}}(g^- \theta, g\sigma)$. By composability, $g^- \theta$ is $\mathbb{H}$-founded whenever θ is $\mathbb{H}$-founded. Therefore,

COROLLARY XX.12.5. *Let* $\mathbb{H}$ *be a composable class of filters. For a surjective* $f \in \mathscr{C}(\xi, \tau)$*, the following statements are equivalent:*

(1) f *is* $\mathbb{H}$*-quotient,*
(2) $f \times g$ *is* $\mathbb{H}$*-quotient for every* $\mathbb{H}$*-founded convergence* θ *and each biquotient* $g \in \mathscr{C}(\sigma, \theta)$*,*
(3) $f \times i_{|\sigma|}$ *is hereditarily quotient for each* $\mathbb{H}$*-founded prime topology* σ*.*

PROOF. In order to apply Theorem XX.12.2, let us observe that the final convergence $(f_0 \times f_1)(\xi_0 \times \xi_1)$ is equal to the product of final convergences $f_0\xi_0 \times f_1\xi_1$, and the initial convergence $(f_0 \times f_1)^-(\tau_0 \times \tau_1)$ is equal to the product of initial convergences $f_0^- \tau_0 \times f_1^- \tau_1$. This is because a map f valued in a product is continuous if and only if the compositions $p \circ f$ with all the projections p are continuous. □

Reassuming Corollary XX.12.5, for an $\mathrm{A}_{\mathbb{H}}$-quotient map f, the product $f \times g$ is $\mathrm{A}_{\mathbb{H}}$-quotient if and only if g is biquotient with an additional property of the range depending on $\mathbb{H}$. On the other hand, f is a $\mathrm{A}_{\mathbb{H}}$-quotient map whenever the product of f with the identity

$\mathbb{H}$	$\mathbb{F}$	$\mathbb{F}_1$	$\mathbb{F}_{\wedge 1}$	$\mathbb{F}_0$
f	biquotient	countably biquotient	weakly quotient	$\mathbb{F}_0$-quotient
iff $f \times g$	biquotient	countably biquotient	weakly quotient	$\mathbb{F}_0$-quotient
for each g	biquotient	biquotient	biquotient	biquotient
with the range		countable character	P-space	finitely generated
iff $f \times i_{\lvert\sigma\rvert}$	$\mathbb{F}_0$-quotient	$\mathbb{F}_0$-quotient	$\mathbb{F}_0$-quotient	$\mathbb{F}_0$-quotient
for each prime $\sigma = \mathrm{T}\sigma$		countable character	P-space	finitely generated

TABLE XX.12.1. Products of quotient-like maps. A convergence is said to be a *P-space* if it is founded in countably complete filters.

is hereditarily quotient with the same property of the range depending on $\mathbb{H}$.

For instance, for $\mathbb{H} = \mathbb{F}_1$,

COROLLARY XX.12.6. *A map f is countably biquotient if and only if the product $f \times g$ is countably biquotient for each biquotient map g with the range of countable character, if and only if the product of $f \times i$ is hereditarily quotient for each identity i on a prime countably founded topology.*

On the other hand,

COROLLARY XX.12.7. *Let $\mathbb{H}$ be a composable class of filters. For a surjective $f \in \mathscr{C}(\xi, \tau)$, the following statements are equivalent:*

1. *f is $\mathbb{H}$-perfect,*
2. *$f \times g$ is $\mathbb{H}$-perfect for every $\mathbb{H}$-founded convergence σ and each perfect $g \in \mathscr{C}(\sigma, \theta)$,*
3. *$f \times i_{\lvert\sigma\rvert}$ is adherent for each $\mathbb{H}$-founded prime topology σ.*

$\mathbb{H}$	$\mathbb{F}$	$\mathbb{F}_1$	$\mathbb{F}_{\wedge 1}$	$\mathbb{F}_0$
f	perfect	countably perfect	inversely Lindelöf	adherent
iff $f \times g$	perfect	countably perfect	inversely Lindelöf	adherent
for each g	perfect	perfect	perfect	perfect
with the domain		countable character	P-space	finitely generated
iff $f \times i_{\lvert\sigma\rvert}$	adherent	adherent	adherent	adherent
for each prime $\sigma = \mathrm{T}\sigma$		countable character	P-space	finitely generated

TABLE XX.12.2. Products of perfect-like maps. A convergence is said to be a *P-space* if it is founded in countably complete filters.

Reassuming Corollary XX.12.7, for an $\mathbb{H}$-perfect map f, the product $f \times g$ is $\mathbb{H}$-perfect if and only if g is perfect with an additional property of the domain depending on $\mathbb{H}$. On the other hand, f is a $\mathbb{H}$-perfect map whenever the product of f with the identity is adherent with the same property of the domain depending on $\mathbb{H}$.

For instance, for $\mathbb{H} = \mathbb{F}_1$,

COROLLARY XX.12.8. *A map f is countably perfect if and only if the product $f \times g$ is countably perfect for each perfect map g with the domain of countable character, if and only if the product of $f \times i$ is adherent for each identity i on a prime countably founded topology.*

XX.13. Supplement

In virtue of Corollary XIX.1.14, by (XX.1.4),

PROPOSITION XX.13.1. *A set A is ξ-cover-compact if and only if the filter $\mathcal{V}_\xi(A)$ is ξ-compact.*

EXERCISE XX.13.2. Prove Proposition XX.7.1.

SOLUTION. (1) implies (2). If $\mathcal{H} \in \mathbb{H}(X \times Y)$, then $p_Y[\mathcal{H}] = \mathcal{H}[X] \in \mathbb{H}Y$, because $\mathbb{H}$ is composable. By assumption, there exists $y \in \mathrm{adh}_\tau \mathcal{H}[X]$, and thus there is a filter $\mathcal{G}$ such that $\mathcal{G} \# \mathcal{H}[X]$ and $y \in \lim_\tau \mathcal{G}$. As τ is $\mathbb{H}$-founded, there is $\mathbb{H}Y \ni \mathcal{G}_0 \leq \mathcal{G}$ so that $\mathcal{G}_0 \# \mathcal{H}[X]$ and $y \in \lim_\tau \mathcal{G}_0$.

Accordingly, $\mathcal{H}^-[\mathcal{G}_0] \in \mathbb{H}X$, hence by assumption, there exists a filter $\mathcal{F}$ such that $\mathcal{F} \# \mathcal{H}^-[\mathcal{G}_0]$ and $x \in \lim_\xi \mathcal{F}$. Therefore, $(x, y) \in \lim_{\xi \times \tau}(\mathcal{F} \times \mathcal{G}_0)$ and $(\mathcal{F} \times \mathcal{G}_0) \# \mathcal{H}$, which means that $(x, y) \in \mathrm{adh}_{\xi \times \tau} \mathcal{H}$. Obviously, (2) implies (3).

To see that (3) implies (1), suppose that (1) does not hold: there exists an $\mathbb{H}$-filter $\mathcal{H}$ such that $\operatorname{adh}_\xi \mathcal{H} = \varnothing$. Let y_0 be any element of X, and let τ be a prime topology on X with $\mathcal{N}_\tau(y_0) := \mathcal{H} \wedge \{y_0\}$. Each free filter $\mathcal{E}$ on $\{(x,x) : x \neq y_0\}$ is of the form $\mathcal{E} = \{(x,x) : x \neq y_0\}[\mathcal{G}]$, where $\mathcal{G}$ is a free filter on X. Hence $(v,w) \in \operatorname{adh}_{\xi\times\tau}\{(x,x) : x \neq y_0\}$ whenever $v = w = y_0$, and there exists $\mathcal{G} \geq \mathcal{H} > \mathcal{N}_\tau(y_0)$. But $\operatorname{adh}_{\xi\times\tau}\{(x,x) : x \neq y_0\} = \varnothing$, because $\operatorname{adh}_\xi \mathcal{H} = \varnothing$, so that $\xi \times \tau$ is not $\mathbb{F}_0$-compact.

Exercise XX.13.3. Prove that if ξ is $\mathbb{H}$-compact and τ is compact, then $\xi \times \tau$ is $\mathbb{H}$-compact.

Solution. If $\mathcal{H} \in \mathbb{H}(X \times Y)$, then $p_X[\mathcal{H}] = \mathcal{H}^-[Y] \in \mathbb{H}X$, because $\mathbb{H}$ is composable. By assumption, there exists $x \in \operatorname{adh}_\xi \mathcal{H}^-[Y]$, and thus there is a filter $\mathcal{F}$ such that $x \in \lim_\xi \mathcal{F}$ and $\mathcal{F}\#\mathcal{H}^-[Y]$. It follows that $\mathcal{H}[\mathcal{F}]$ is a proper filter on Y, hence, by compactness of τ, there is $y \in \operatorname{adh}_\tau \mathcal{H}[\mathcal{F}]$, which implies the existence of a filter $\mathcal{G}$ such that $\mathcal{G}\#\mathcal{H}[X]$ and $y \in \lim_\tau \mathcal{G}$. Therefore, $(x,y) \in \lim_{\xi\times\tau}(\mathcal{F}\times\mathcal{G})$ and $(\mathcal{F}\times\mathcal{G})\#\mathcal{H}$, which means that $(x,y) \in \operatorname{adh}_{\xi\times\tau}\mathcal{H}$.

Exercise XX.13.4. Prove that the collection $\kappa(\xi)$ of all isotone ξ-compact families fulfills the axioms of the family of open sets for a topology.

Solution. The empty family does not mesh any proper filter $\mathcal{H}$, that is, $\neg(\varnothing\#\mathcal{H})$, so that the condition is emptily fulfilled. Also $\neg(2^X\#\mathcal{H})$, because $\varnothing \in 2^X$.

If $\mathcal{H} \subset (\bigcup_{j\in J}\mathcal{A}_j)^\# = \bigcap_{j\in J}\mathcal{A}_j^\#$, then by assumption, $\operatorname{adh}_\xi \mathcal{H} \in \mathcal{A}_j^\#$, hence $\operatorname{adh}_\xi \mathcal{H} \in \bigcap_{j\in J}\mathcal{A}_j^\# = (\bigcup_{j\in J}\mathcal{A}_j)^\#$. Finally, let J be finite, and suppose that $\operatorname{adh}_\xi \mathcal{H} \notin (\bigcap_{j\in J}\mathcal{A}_j)^\# = \bigcup_{j\in J}\mathcal{A}_j^\#$, hence $\operatorname{adh}_\xi \mathcal{H} \notin \mathcal{A}_j^\#$ for each $j \in J$, so that there exists $H_j \in \mathcal{H}$ such that $H_j \notin \mathcal{A}_j^\#$, equivalently $H_j^c \in \mathcal{A}_j$, because $\mathcal{A}_j$ is isotone for each $j \in J$. Hence $\bigcup_{j\in J}H_j^c \in \bigcap_{j\in J}\mathcal{A}_j$, and $\bigcap_{j\in J}H_j \in \mathcal{H}$, because $\mathcal{H}$ is a filter.

Exercise XX.13.5. Show that compactoidness is not absolute.

Solution. Consider an infinite set X and let $x_0, x_1 \in X$. Let ζ be a prime topology on $X_0 = X \setminus \{x_1\}$, for which $x_0 \in \lim_\zeta \mathcal{F}$ whenever $\ker \mathcal{F} \subset \{x_0\}$ and $(X)_0 \wedge x_0^\uparrow \leq \mathcal{F}$. Let ξ be the discrete topology on $X_1 = X \setminus \{x_0\}$. Then X is ζ-compact, but not ξ-compact.

Proposition XX.13.6. *Assume that $R^-[\mathbb{H}] \subset \mathbb{H}$. If a family $\mathcal{A}$ is $\mathbb{H}$-compact at B, then $R[\mathcal{A}]$ is $\mathbb{H}$-compact at RB.*

Proof. Let $\mathcal{G} \in \mathbb{H}$ be such that $\mathcal{G}\#R[\mathcal{A}]$, equivalently $R^-[\mathcal{G}]\#\mathcal{A}$, hence $\operatorname{adh} R^-[\mathcal{G}]\#B$, because $R^-[\mathcal{G}] \in \mathbb{H}$. Therefore, there exists $x \in$

$\operatorname{adh} R^{-}[\mathcal{G}] \cap B$, so that there is a filter $\mathcal{F} \# R^{-}[\mathcal{G}]$, equivalently, $R[\mathcal{F}] \# \mathcal{G}$, such that $B \# \lim \mathcal{F}$. As R is $\mathbb{H}$-compact, $\operatorname{adh} \mathcal{G} \# Rx$, hence $\operatorname{adh} \mathcal{G} \# RB$. $\square$

PROPOSITION XX.13.7. *Assume that $R^{-}[\mathbb{H}] \subset \mathbb{H}$. Then the composition of two $\mathbb{H}$-compact relations is $\mathbb{H}$-compact.*

Observe that each ξ-convergent filter $\mathcal{F}$ is ξ-compactoid. More precisely, $\mathcal{F}$ is ξ-compactoid at $\lim_{\xi} \mathcal{F}$. Indeed, if $\mathcal{G} \in \beta\mathcal{F}$, then $\lim_{\xi} \mathcal{F} \subset \lim_{\xi} \mathcal{G} = \operatorname{adh}_{\xi} \mathcal{G}$.

CHAPTER XXI

Mixed properties

We have encountered many instances of reflective classes (e.g., topologies, pretopologies, pseudotopologies) and of coreflective classes (e.g., sequentially founded convergences, convergences of countable character, locally compact convergences). We shall now investigate several mixed properties that result from some interactions of a reflector and a coreflector. More exactly, if J is a reflector and E is a coreflector, then we consider the classes of convergences ξ fulfilling

$$\xi = JE\xi,$$

as well as convergences τ fulfilling

$$\tau = EJ\tau.$$

The former is a subclass of the reflective class $\mathrm{fix}(J)$, and the latter of the coreflective class $\mathrm{fix}(E)$. In general, they are neither reflective nor coreflective.

Nevertheless, by Proposition XVII.1.6, the class of convergences ξ such that

$$\xi \geq JE\xi$$

is coreflective, and the class of convergences τ for which

$$\tau \leq EJ\tau$$

reflective. In general, however, neither the functor JE is a coreflector, nor the functor EJ is a reflector (Remark XVII.1.7).

PROPOSITION XXI.0.8. *If J is a reflector and E is a coreflector, then for any ξ and τ,*

(1) *$\xi = JE\xi$ if and only if $\xi \geq JE\xi$ and $J\xi \geq \xi$,*
(2) *$\tau = EJ\tau$ if and only if $\tau \leq EJ\tau$ and $E\tau \leq \tau$.*

PROOF. (1) Of course, $\xi = JE\xi$ implies $\xi \geq JE\xi$, and since J is idempotent, $J\xi \geq J(JE\xi) = JE\xi = \xi$. Conversely, as $E\xi \geq \xi$, the two inequalities entail $J\xi \geq \xi \geq JE\xi \geq J\xi$. A similar argument shows (2). □

Therefore, in the study of mixed properties it is enough to focus on functorial inequalities.

Before considering a general scheme of interactions between reflective and coreflective classes, let us study a simple special case. Namely, let us consider those pretopologies that are the pretopologizations of sequentially founded convergences, and, conversely, those sequentially founded convergences that are the sequentially founded modifications of pretopologies.

XXI.1. Fréchet pretopologies

A pretopology π is said to be *sequentially induced*, a *Fréchet or* a *Fréchet-Urysohn* pretopology if there exists a sequentially founded convergence θ such that

$$\pi = \mathrm{S}_0\theta.$$

PROPOSITION XXI.1.1. *A pretopology π is sequentially induced (alias Fréchet) if and only if*

$$\pi = \mathrm{S}_0\mathrm{Seq}\,\pi. \tag{XXI.1.1}$$

Moreover, $\mathrm{Seq}\,\pi$ *is the coarsest sequentially founded convergence θ such that $\pi = \mathrm{S}_0\theta$.*

PROOF. It follows immediately from characterizations of the pretopologizer, for instance Corollary VI.6.5, that $\pi = \mathrm{S}_0\theta$ holds whenever

$$\mathcal{V}_\pi(x) = \bigcap\{\mathcal{E} \in \mathbb{S} : x \in \lim_\theta \mathcal{E}\}$$

for each $x \in |\pi|$, where $\mathbb{S}$ stands for the class of sequential filters. Since π is a pretopology, $\mathcal{E} \geq \mathcal{V}_\pi(x)$ if and only if $x \in \lim_\pi \mathcal{E}$, we conclude that $\mathcal{V}_\pi(x) = \mathcal{V}_{\mathrm{Seq}\,\pi}(x)$, that is, $\pi = \mathrm{S}_0\mathrm{Seq}\,\pi$. Moreover, if θ is sequentially founded and $\theta \geq \pi$, then $\theta \geq \mathrm{Seq}\,\pi$. □

Accordingly, in view of Definition II.3.8,

COROLLARY XXI.1.2. *A pretopology is sequentially induced (alias Fréchet) if and only if every vicinity filter is a Fréchet filter.*

It is why we call *sequentially induced pretopologies* alternatively *Fréchet pretopologies.* Consequently, any prime topology $\pi[x_\infty, \mathcal{F}]$, where $\mathcal{F}$ is a Fréchet filter, constitutes an example of a Fréchet topology. In particular, each countably based filter is Fréchet. A non-countably based Fréchet filter was presented in Example XVI.4.5.

Each pretopology is characterized by its vicinity filters, by its adherence, and by its inherence. For instance,

PROPOSITION XXI.1.3. *A pretopology ξ is sequentially induced (alias Fréchet) if and only if for each A and every x,*

$$x \in \operatorname{adh}_\xi A$$

if and only if there exists $\mathcal{E} \in \mathbb{S}$ such that $x \in \lim_\xi \mathcal{E}$ and $A \in \mathcal{E}$.

COROLLARY XXI.1.4. *If $\mathcal{F}$ is a Fréchet filter, then*

$$\mathcal{F}^\# = \bigcup\nolimits_{\mathcal{E} \in \mathbb{S}\mathcal{F}} \mathcal{E}.$$

PROOF. Indeed, as $\mathcal{F} = \bigcap_{\mathcal{E} \in \mathbb{S}\mathcal{F}} \mathcal{E}$, (V.3.9) applies to the effect that $\mathcal{F}^\# = \bigcup_{\mathcal{E} \in \mathbb{S}\mathcal{F}} \mathcal{E}^\#$. Moreover, if $\mathcal{E} \in \mathbb{S}\mathcal{F}$ and $A \in \mathcal{E}^\#$, then $A \in \mathcal{E} \vee A \in \mathbb{S}\mathcal{F}$ and the con □

DEFINITION XXI.1.5. A convergence ξ is called *Fréchet whenever*

$$\xi \geq \mathrm{S}_0\mathrm{Seq}\,\xi. \tag{XXI.1.2}$$

PROPOSITION XXI.1.6. *If a convergence ξ is Fréchet, then $\mathrm{S}_0\xi$ is Fréchet.*

PROOF. As S_0 is descending and idempotent, it follows from (XXI.1.2) that $\xi \geq \mathrm{S}_0\mathrm{Seq}\,\mathrm{S}_0\xi \geq \mathrm{S}_0\xi$, hence $\mathrm{S}_0\xi = \mathrm{S}_0\mathrm{Seq}(\mathrm{S}_0\xi)$. □

The converse is not true.

EXAMPLE XXI.1.7. Let X be an infinite set, $x_\infty \in X$ and η be a prime convergence such that $x_\infty \in \lim_\eta \mathcal{F}$ whenever $\ker \mathcal{F} \subset \{x_\infty\}$ and there is $\boldsymbol{U} \subset_{\text{fin}} \beta X$such that $\bigwedge \boldsymbol{U} = \mathcal{F}$. This is a finitely stable convergence such that the only convergent sequential filters are principal ultrafilters. Therefore, $\mathrm{Seq}\,\eta = \iota_X$ (the discrete topology on X), hence $\mathrm{S}_0\mathrm{Seq}\,\eta = \iota_X \not\leq \eta$. On the other hand, $\mathrm{S}_0\eta$ is the prime cofinite topology $\pi[x_\infty, (X)_0]$, and $\mathcal{N}_\eta(x_\infty) = (X)_0 \wedge x_\infty^\uparrow$ is a Fréchet filter.

Of course, every metrizable topology is Fréchet. The following fact, well-known in metric spaces, is valid in a broader context of pretopologies of countable character.

LEMMA XXI.1.8. *If ξ is a pretopology of countable character, and $x \in \operatorname{adh}_\xi (x_n)_n$, then there exists a subsequence $(x_{n_k})_k$ such that $x \in \lim_\xi (x_{n_k})_k$.*

PROOF. As ξ is a pretopology,

$$x \in \operatorname{adh}_\xi (x_n)_n = \bigcap\nolimits_{k<\omega} \operatorname{adh}_\xi \{x_n : n > k\}.$$

As ξ is of countable character, there exists a countable base $\{V_k : k < \omega\}$ of $\mathcal{V}_\xi(x)$. Therefore, $V_k \cap \{x_n : n > k\} \neq \varnothing$ for each $k < \omega$, that is, there exists $n_k > k$ such that $x_{n_k} \in V_k$. □

By Definition II.3.8, a filter is called *Fréchet* if it is an infimum of sequential filters, and in virtue of Proposition V.5.13, a filter $\mathcal{F}$ is Fréchet whenever it is an infimum of countably based filters, that is, if

$$\mathcal{F} = \bigcap\nolimits_{\mathcal{G}\in\mathbb{F}_1(\mathcal{F})} \mathcal{G}. \tag{XXI.1.3}$$

Thus we can complete Proposition XVI.4.1 by

COROLLARY XXI.1.9. *A filter $\mathcal{F}$ is Fréchet if and only if, for each $H \in \mathcal{F}^{\#}$ there exists a countably based filter $\mathcal{G} \geq \mathcal{F} \vee H$.*

PROOF. If a filter $\mathcal{F}$ is Fréchet, the conclusion follows from Proposition XVI.4.1, because each sequential filter is countably based. Conversely, if a filter $\mathcal{F}$ fulfills the condition, then each $\mathcal{E} \in \mathbb{S}\mathcal{G}$, then $\mathcal{F}$ is Fréchet by Proposition XVI.4.1. $\square$

COROLLARY XXI.1.10. *A pretopology ξ is Fréchet if and only if $\xi = \mathrm{S}_0\mathrm{I}_1\xi$.*

PROOF. It is enough to notice that $\mathrm{adh}_{\mathrm{I}_1\xi} A \subset \mathrm{adh}_{\mathrm{Seq}\xi} A$ for each set A. Indeed, if $\mathcal{G}$ is a countably based filter such that $A \in \mathcal{G}^{\#}$ and $x \in \lim_\xi \mathcal{G}$, then there exists $\mathcal{E} \in \mathbb{S}(\mathcal{G} \vee A)$ such that $x \in \lim_\xi \mathcal{E}$. $\square$

COROLLARY XXI.1.11. *A convergence ξ is Fréchet if and only if*

$$\xi \geq \mathrm{S}_0\mathrm{I}_1\xi.$$

The other way round, a sequentially founded convergence θ is *pretopologically induced* if there exists a pretopology π such that

$$\theta = \mathrm{Seq}\,\pi.$$

PROPOSITION XXI.1.12. *A sequentially founded convergence θ is pretopologically induced if and only if*

$$\theta = \mathrm{Seq}\,\mathrm{S}_0\theta.$$

Moreover, $\mathrm{S}_0\theta$ is the finest pretopology π such that $\theta = \mathrm{Seq}\,\pi$.

PROOF. Indeed, if θ is a pretopologically induced sequentially founded convergence, and π is a pretopology such that $\theta = \mathrm{Seq}\,\pi$, then $\theta \geq \pi$, hence $\mathrm{S}_0\theta \geq \mathrm{S}_0\pi = \pi$, and thus

$$\theta = \mathrm{Seq}\,\theta \geq \mathrm{Seq}\,\mathrm{S}_0\theta \geq \mathrm{Seq}\,\pi = \theta.$$

$\square$

PROPOSITION XXI.1.13. *A sequentially founded convergence θ is pretopologically induced if and only if, for each sequential filter $\mathcal{E}$,*

$$\lim\nolimits_\theta \mathcal{E} \supset \bigcap\nolimits_{H\in\mathcal{E}^{\#}} \mathrm{adh}_\theta H. \tag{XXI.1.4}$$

PROOF. By Proposition XXI.1.12, θ is pretopologically induced whenever $\theta \leq \mathrm{Seq}\,\mathrm{S}_0\theta$, that is whenever $x \in \lim_\theta \mathcal{E}$ for a sequential filter $\mathcal{E}$, provided that $x \in \mathrm{adh}_\theta H$ for each $H \in \mathcal{E}^\#$. □

REMARK XXI.1.14. From $\mathrm{Seq} \geq \mathrm{S}_0\mathrm{Seq}$ it follows that $\mathrm{Seq} \geq \mathrm{Seq}\mathrm{S}_0\mathrm{Seq}$, so that $\mathrm{S}_0\mathrm{Seq} \geq \mathrm{S}_0\mathrm{Seq}\mathrm{S}_0\mathrm{Seq}$. On the other hand, $\mathrm{Seq}\mathrm{S}_0\mathrm{Seq} \geq \mathrm{S}_0\mathrm{Seq}$, hence $\mathrm{S}_0\mathrm{Seq}\mathrm{S}_0\mathrm{Seq} \geq \mathrm{S}_0\mathrm{Seq}$, showing that $\mathrm{S}_0\mathrm{Seq}$ is idempotent.

XXI.2. Galois connections

We shall see that the methods presented in the preceding section constitute a special case of a general pattern of functorial equations and inequalities, which, in turn, follows from an abstract scheme of Galois connections. Galois connections will be also applied in other future chapters, for instance, in Chapter XXIV in duality of convergences.

One of the important features of mathematics is abstraction. It consists in disregarding irrelevant aspects of studied problems, illuminating this way what is essential.

We have already seen that many facts in mathematics hinge on underlying order relations. Often, such facts can be nicely handled with the aid of *Galois connections*. The foundations of this rather abstract theory are in fact very simple.

Consider two ordered sets $(X, \leq)$ and $(Y, \leq)$.

DEFINITION XXI.2.1. A pair of maps $\varphi : X \longrightarrow Y$ and $\psi : Y \longrightarrow X$,

(connection) $$X \underset{\psi}{\overset{\varphi}{\rightleftarrows}} Y$$

is called a *Galois connection* if the maps are decreasing ([1]), and their compositions are *ascending,* that is,

(decreasing) $$x_0 \leq x_1 \Longrightarrow \varphi(x_0) \geq \varphi(x_1),$$

(decreasing) $$y_0 \leq y_1 \Longrightarrow \psi(y_0) \geq \psi(x_1),$$

(ascending) $$x \leq \psi(\varphi(x)), \qquad y \leq \varphi(\psi(y)).$$

EXAMPLE XXI.2.2. In Section I.4, for a relation $R \subset X \times Y$, for $A \subset X$, and for $B \subset Y$, consider the polar $R^\circ A := \bigcap_{x \in A} Rx$, and the inverse polar $R^{-\circ} B := \bigcap_{y \in B} Rx$. By (I.4.2), R° and $R^{-\circ}$ are deceasing, and by (I.4.2) their compositions are ascending.

[1]This is sometimes called an *antitone Galois connection*, as opposed to a monotone Galois connection in which both maps are increasing. We will consider all variants later but we chose this antitone variant as our base case.

PROPOSITION XXI.2.3. *A pair of maps (connection) is a Galois connection if and only if for each $x \in X$ and $y \in Y$,*

$$\varphi(x) \geq y \iff x \leq \psi(y). \tag{XXI.2.1}$$

PROOF. Let (φ, ψ) be a Galois connection. If $\varphi(x) \geq y$, then by (ascending) and (decreasing), $x \leq \psi(\varphi(x)) \leq \psi(y)$. If $x \leq \psi(y)$, then by (ascending) and (decreasing), $\varphi(x) \geq \varphi(\psi(y)) \geq y$.

Conversely, on setting $y = \varphi(x)$ in (XXI.2.1), we get $x \leq \psi(\varphi(x))$; on setting $x = \psi(y)$, we get $\varphi(\psi(y)) \geq y$. Consequently we have established the inequalities in (ascending). If now $x_0 \leq x_1$ and $y \leq \varphi(x_1)$, then $x_0 \leq x_1 \leq \psi(y)$ by virtue of (XXI.2.1), hence, again by (XXI.2.1), $y \leq \varphi(x_0)$. Therefore $\varphi(x_0) \geq \varphi(x_1)$, and similarly the other way round. We conclude that (decreasing) holds. □

DEFINITION XXI.2.4. A map φ between ordered sets is called an *order isomorphism* if φ is increasing and bijective, and φ^{-1} is increasing; an *inverse order isomorphism* if φ is decreasing and bijective, and φ^{-1} is decreasing.

THEOREM XXI.2.5. *If (connection) is a Galois connection, then*

1. *$\psi\varphi$ is a coprojector onto $\psi(Y)$,*
2. *$\varphi\psi$ is a coprojector onto $\varphi(X)$.*

Moreover, the restriction of φ to $\psi(Y)$ is an inverse order isomorphism onto $\varphi(X)$

$$X \supset \psi(Y) \underset{\psi}{\overset{\varphi}{\rightleftarrows}} \varphi(X) \subset Y. \tag{XXI.2.2}$$

PROOF. By definition, $\psi\varphi$ and $\varphi\psi$ are ascending. Each of them is increasing, as a composition of two decreasing maps. By applying φ to the first inequality of (ascending), we get $\varphi(x) \geq \varphi(\psi(\varphi(x))$. On putting $y = \varphi(x)$ into the second inequality of (ascending), we get $\varphi(x) \leq \varphi(\psi(\varphi(x)))$, so that

$$\varphi(x) = \varphi\psi\varphi(x). \tag{XXI.2.3}$$

This implies that $\psi\varphi$ is idempotent. Analogously $\varphi\psi$ is idempotent. Therefore, $\psi\varphi$ and $\varphi\psi$ are coprojectors.

As $\psi\varphi$ is idempotent and ascending, the set of its fixed points coincides with its image $\psi(\varphi(X)) = \{x \in X : \psi\varphi(x) \leq x\}$. We conclude that

$$\text{fix}\,(\psi\varphi) = \psi(Y), \tag{XXI.2.4}$$

because if $x \in \psi(Y)$, then there exists y such that $x = \psi(y)$, hence $\psi\varphi(x) = \psi\varphi\psi(y) = \psi(y) = x$. Likewise for $\varphi\psi$.

By definition φ and ψ are decreasing, and their restrictions in (XXI.2.2) are surjective. To conclude that they are inverse order isomorphisms, we need to show that they are injective. Let $x_0, x_1 \in \psi(Y)$ be such that $\varphi(x_0) = \varphi(x_1)$. By (XXI.2.4), $x_0 = \psi\varphi(x_0) = \psi\varphi(x_1) = x_1$, and similarly for ψ. □

If we simultaneously invert the orders on X and Y, then φ and ψ in (connection) remain *decreasing*, while their compositions $\psi\varphi$ and $\varphi\psi$ become *descending*.

DEFINITION XXI.2.6. We say that (connection) is an *opposite Galois connection* if φ and ψ are decreasing, and $\psi\varphi$ and $\varphi\psi$ are descending.

Of course, all the facts established for *Galois connections* have counterparts for *opposite Galois connections.* In particular, (connection) is an opposite Galois connection if and only if for each x and y,

$$\varphi(x) \leq y \Longleftrightarrow x \geq \psi(y).$$

COROLLARY XXI.2.7. *If (connection) is an opposite Galois connection, then $\psi\varphi$ is a projector onto $\psi(Y)$, $\varphi\psi$ is a projector onto $\varphi(X)$. Moreover, the restriction (XXI.2.2) of φ to $\psi(Y)$ is an inverse order isomorphism onto $\varphi(X)$, and $\psi = \varphi^{-}$.*

Classifying variants of Galois connections in terms of their compositions, we get

- $(+,+)$-(connection) if φ and ψ are decreasing and $\psi\varphi$ and $\varphi\psi$ are ascending, and
- $(-,-)$-(connection) if φ and ψ are decreasing and $\psi\varphi$ and $\varphi\psi$ are descending. If we invert the order on X, then we get a
- $(-,+)$-(connection), in which φ and ψ are both increasing, $\psi\varphi$ is descending, and $\varphi\psi$ is ascending. By inverting the order on Y, we get a
- $(+,-)$-(connection), where φ and ψ are both increasing, $\psi\varphi$ is ascending, and $\varphi\psi$ is descending.

Of course, all the listed properties of a $(+,+)$-(connection) have their obvious reinterpretations in other variants.

XXI.3. Functorial equations and inequalities

Recall from Section XVII.2 that a modifier on the class of convergences is a *functor* if it *preserves continuity*, that is, F is a (concrete) functor whenever

$$
\begin{gathered}
|F\xi| = |\xi|\,, \\
\zeta \leq \xi \Longrightarrow F\zeta \leq F\xi, \\
\mathscr{C}(\xi,\tau) \subset \mathscr{C}(F\xi, F\tau)
\end{gathered}
$$

for any convergences ξ, ζ and τ.

We call ξ an *F-convergence* if $\xi \in \mathrm{fix}(F)$, that is, if $F\xi = \xi$.

It turns out that several interesting classes of convergences materialize as solutions of some *functorial equations* of the type

$$\text{(functorial equation)} \qquad \xi = JE\xi,$$

where J is a reflector, and E is a coreflector, or of *functorial inequalities*, like

$$\text{(functorial inequality)} \qquad \xi \geq JE\xi.$$

By Proposition XVII.1.6, the class of convergences ξ fulfilling (functorial inequality) is coreflective.

In Section XXI.1 we studied the special class of *Fréchet convergences* ξ, characterized by a functorial inequality $\xi \geq \mathrm{S}_0\mathrm{Seq}\,\xi$, and its companion class of *pretopologically induced convergences*, which constitute solutions τ of the functorial inequality $\tau \leq \mathrm{Seq}\,\mathrm{S}_0\tau$.

In general, JE is a functor, but neither ascending, nor descending, nor idempotent. However,

LEMMA XXI.3.1. *If J is a reflector and E is a coreflector, then JE restricted to $\mathrm{fix}J$ is ascending, while EJ restricted to $\mathrm{fix}E$ is descending.*

PROOF. As E is ascending, that is, $E\xi \geq \xi$, hence $JE\xi \geq J\xi$ for each ξ. Therefore, if $\xi \in \mathrm{fix}J$, equivalently $J\xi \geq \xi$, then $JE\xi \geq J\xi \geq \xi$. Similarly, for the restriction of EJ to $\mathrm{fix}E$. □

As functors, JE and EJ are both increasing. Hence, by Lemma XXI.3.1, the restrictions of E to $\mathrm{fix}J$ and of J to $\mathrm{fix}E$,

$$\mathrm{fix}J \underset{J}{\overset{E}{\rightleftarrows}} \mathrm{fix}E$$

form a $(+,-)$-(connection). Hence, by Theorem XXI.2.5,

THEOREM XXI.3.2. *If J is a reflector and E is a coreflector, then*
1. *the restriction of JE to $\mathrm{fix}J$ is a coreflector on $\mathrm{fix}(JE) = J(\mathrm{fix}E)$,*
2. *the restriction of EJ to $\mathrm{fix}E$ is a reflector on $\mathrm{fix}(EJ) = E(\mathrm{fix}J)$,*

(3) *E restricted to* fix(JE) *and J restricted to* fix(EJ) *are reciprocal order isomorphisms:*

$$\mathrm{fix}(JE) \underset{J}{\overset{E}{\rightleftarrows}} \mathrm{fix}(EJ)\ .$$

The connection is represented in Figure XXI.3.1. Typically, fix$J \cap$ fix$E \neq \varnothing$. For example, the discrete convergence belongs to each coprojective class of convergences, and, on the other hand, it is a topology, hence belongs to each projective class that includes topologies.

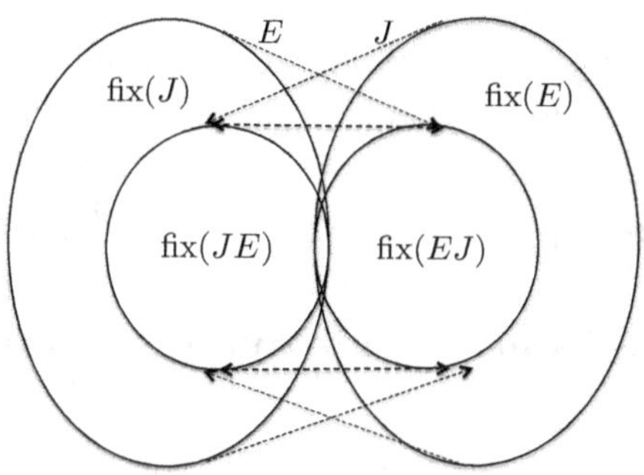

FIGURE XXI.3.1. Galois connection of a reflector J and a coreflector E, restricted to the relative domains.

DEFINITION XXI.3.3. We say that an F-convergence σ is G-*induced* if there is a G-convergence η such that $\sigma = F\eta$.

COROLLARY XXI.3.4. *If J is a reflector, E is a coreflector.*

(1) *A J-convergence ξ is E-induced if and only if $\xi = JE\xi$.*
(2) *An E-convergence τ is J-induced if and only if $\tau = EJ\tau$.*
(3) *If a J-convergence ξ is E-induced, then $\tau := E\xi$ is J-induced, and conversely.*

PROOF. (3) Let ξ be an E-induced J-convergence, that is, $\xi = JE\xi$, then by Theorem XXI.3.2, $\tau := E\xi$ is a J-induced E-convergence, and $J\tau = \xi$. □

Therefore, $\tau \in$ fix(E) and $\xi \in$ fix(J) constitute a pair such that $\tau = E\xi$ and $\xi = J\tau$. In particular,

$$\xi \leq \tau.$$

Indeed, $\xi = JE\xi = J\tau \leq \tau$. As we shall see, it may happen that $\tau = \xi$.

In the particular case discussed in Section XXI.1, we recover from Theorem XXI.3.2 that the subclass of Fréchet pretopologies is coreflective within the class of pretopologies. In particular,

(1) For each pretopology ξ, there exists the coarsest Fréchet pretopology $\mathrm{S_0Seq}\,\xi$ that is finer than ξ.
(2) For each sequentially founded convergence τ, there exists the finest pretopologically induced sequentially founded convergence $\mathrm{SeqS_0}\xi$ that is coarser than τ.
(3) If there exists a sequentially founded convergence τ such that $\xi = \mathrm{S_0}\tau$, then the coarsest sequentially founded convergence with this property is a pretopologically induced sequentially founded convergence $\tau = \mathrm{Seq}\xi$.
(4) If there exists a pretopology ξ such that $\tau = \mathrm{Seq}\xi$, then the finest pretopology with this property is the Fréchet pretopology $\mathrm{S_0}\tau$.

Moreover, by Remark XXI.1.14, $\mathrm{S_0Seq}$ and $\mathrm{SeqS_0}$ are idempotent.

XXI.4. Sequentially induced convergences

In Section XXI.1 we studied the interplay between the pretopologizer $\mathrm{S_0}$ and the coreflectors Seq and $\mathrm{I_1}$. Now, armed with our recent discoveries about functorial equations and inequalities, we shall tackle analogous interactions of the same coreflectors with a larger variety of reflectors.

Let J be a reflector in the class of convergences. In virtue of Definition XXI.3.3 a J-convergence ξ is *sequentially induced* or Seq-*induced* if there exists a sequentially founded convergence θ such that $\xi = J\theta$, that is, whenever $\xi \in J(\mathrm{fix}(\mathrm{Seq}))$. By Proposition XXI.0.8, this is the case if $\xi = J\xi$ and

$$\xi \geq J\mathrm{Seq}\xi.$$

Classically, *sequentially induced topologies* are called *sequential topologies*, while the *topologies* that are *sequentially induced pretopologies* are called *Fréchet topologies.* Of course, such classes were traditionally defined in terms of some topological properties, without aid of convergence structures.

The following formula for adherences of countably based filters will enable us to perceive some of these classes as $\mathrm{I_1}$-induced.

LEMMA XXI.4.1. *For every ξ and each countably based filter $\mathcal{H}$,*

$$\mathrm{adh}_{\mathrm{I_1}\xi}\,\mathcal{H} = \mathrm{adh}_{\mathrm{Seq}\xi}\,\mathcal{H} = \bigcup\nolimits_{\mathcal{E}\in\mathbb{S}\mathcal{H}} \lim\nolimits_{\xi}\mathcal{E}.$$

PROOF. If $x \in \mathrm{adh}_{\mathrm{I}_1\xi}\mathcal{H}$, that is, there exists a countably based filter $\mathcal{G}$ such that $\mathcal{G}\#\mathcal{H}$ and $x \in \lim_\xi \mathcal{G}$, then $\mathcal{G} \vee \mathcal{H}$ is proper and countably based, and thus there is $\mathcal{E} \in \mathbb{S}(\mathcal{G} \vee \mathcal{H})$. Consequently, $x \in \lim_\xi \mathcal{E}$ and, in particular, $\mathcal{E}\#\mathcal{H}$, proving that $x \in \mathrm{adh}_{\mathrm{Seq}\xi}\mathcal{H}$. □

As a result,

COROLLARY XXI.4.2. *For each convergence* ξ,

$$\mathrm{S}_1\mathrm{Seq}\xi = \mathrm{S}_1\mathrm{I}_1\xi, \qquad \mathrm{S}_0\mathrm{Seq}\xi = \mathrm{S}_0\mathrm{I}_1\xi, \qquad \mathrm{T}\mathrm{Seq}\xi = \mathrm{T}\mathrm{I}_1\xi.$$

PROOF. Indeed, by Lemma XXI.4.1,

$$\lim_{\mathrm{S}_1\mathrm{I}_1\xi}\mathcal{F} = \bigcap\nolimits_{\mathbb{F}_1 \ni \mathcal{H}\#\mathcal{F}} \mathrm{adh}_{\mathrm{I}_1\xi}\mathcal{H} = \bigcap\nolimits_{\mathbb{F}_1 \ni \mathcal{H}\#\mathcal{F}} \mathrm{adh}_{\mathrm{Seq}\xi}\mathcal{H} = \lim_{\mathrm{S}_1\mathrm{Seq}\xi}\mathcal{F},$$

and similarly for $\mathrm{S}_0\mathrm{Seq}\xi = \mathrm{S}_0\mathrm{I}_1\xi$ by replacing $\mathbb{F}_1$ by $\mathbb{F}_0$. We show $\mathrm{T}\mathrm{Seq}\xi = \mathrm{T}\mathrm{I}_1\xi$ analogously, on observing that $\mathrm{cl}_{\mathrm{I}_1\xi} H = \mathrm{cl}_{\mathrm{Seq}\xi} H$ for each set H (because a set is closed whenever it includes its adherence). □

On the other hand, by the same Definition XXI.3.3, a sequentially founded convergence τ is J-induced if there is a J-convergence π such that $\tau = \mathrm{Seq}\pi$, that is, whenever $\tau \in \mathrm{Seq}(\mathrm{fix}(J))$. Employing again Proposition XXI.0.8, we deduce that τ fulfills this condition if and only if $\tau = \mathrm{Seq}\tau$ and

$$\tau \leq \mathrm{Seq}J\tau.$$

EXAMPLE XXI.4.3 (Sequential fan topology). In Example III.4.7 we introduced the convergence π on a *sequential fan* $X = \{x_\infty\} \sqcup \bigsqcup_{n<\omega}\{x_{n,k} : k < \omega\}$ for which $x_\infty \in \lim_\pi (x_{n,k})_k$ for each $n < \omega$.

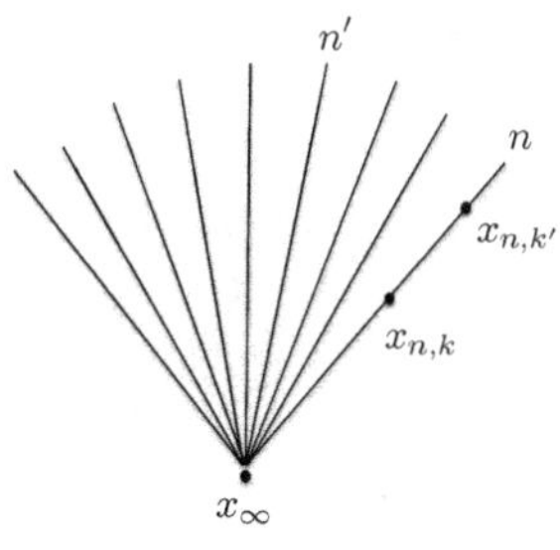

FIGURE XXI.4.1. A sequential fan.

In Example VI.6.11, we observed that $V \in \mathcal{V}_\pi(x_\infty)$ whenever for each $n < \omega$, there is $k_V(n) < \omega$ such that $x_{n,k} \in V$ for every $k < k_V(n)$. The pretopology $\mathrm{S}_0\pi$ is prime, hence is a topology $\mathrm{S}_0\pi = \mathrm{T}\pi$ called the *sequential fan topology*. Let us note that $\mathrm{T}\pi$ is Fréchet. Indeed, if the only non-isolated point $x_\infty \in \mathrm{cl}_\pi A \setminus A$, then there exists $n_A < \omega$ such that $x_\infty \in \mathrm{cl}_\pi(A \cap \{x_{n_A,k} : k < \omega\})$. Otherwise, for each $n < \omega$, there would be $k(n) < \omega$ such that $A \cap \bigcup_{n<\omega}\{x_{n,k} : k > k(n)\} = \varnothing$. It results that a subsequence of $(x_{n_A,k})_k$ converges to x_∞.

We shall review fundamental classes of *sequentially induced* convergences and J-induced sequentially founded convergences, extending their names from the traditional topological context to general convergences. In particular,

(1) a topology θ is called *sequential* (alias *sequentially induced topology*) if $\theta = \mathrm{TSeq}\theta$ (see Examples IV.10.11 and XXI.4.7 of a non-sequential topology); this property extends to arbitrary convergences ξ via

$$\xi \geq \mathrm{TSeq}\xi = \mathrm{TI}_1\xi, \tag{XXI.4.1}$$

(2) a pretopology θ is called *Fréchet* (alias *sequentially induced pretopology*) if $\theta = \mathrm{S}_0\mathrm{Seq}\theta$ (the topologization of the pretopology from Example VI.1.8 for a sequential topology that is not Fréchet); a convergence ξ is called *Fréchet* if

$$\xi \geq \mathrm{S}_0\mathrm{Seq}\xi = \mathrm{S}_0\mathrm{I}_1\xi, \tag{XXI.4.2}$$

(3) a paratopology θ is called *strongly Fréchet* or *countably bisequential* (alias *sequentially induced paratopology*) if $\theta = \mathrm{S}_1\mathrm{Seq}\theta$ (the sequential fan topology from Example XXI.4.3 is Fréchet, but not strongly Fréchet in view of Proposition XXI.8.2 and Lemma XXI.8.3); a convergence ξ is called *countably bisequential* if

$$\xi \geq \mathrm{S}_1\mathrm{Seq}\xi = \mathrm{S}_1\mathrm{I}_1\xi, \tag{XXI.4.3}$$

(4) a pseudotopology θ is called *bisequential* if $\theta = \mathrm{SI}_1\theta$ (see Example XXI.4.10 for a bisequential topology that is not of countable character); a convergence ξ is called *bisequential* if

$$\xi \geq \mathrm{SI}_1\xi. \tag{XXI.4.4}$$

The last inequality has only one term on the right, because in general $\mathrm{SSeq}\xi$ is not equal to $\mathrm{SI}_1\xi$. In fact, (4) does not correspond to sequentially induced pseudotopologies! Actually, we infer from Lemma XXI.5.6 that

THEOREM XXI.4.4. *A pseudotopology is sequentially induced if and only if it is sequentially founded.*

Before justifying Theorem XXI.4.4, let us discuss more in detail the other sequentially induced classes.

DEFINITION XXI.4.5 (Sequential topologies). Let ξ be a convergence on X. A subset A of X is called *sequentially closed* if $\lim_\xi \mathcal{E} \subset A$ for each $\mathcal{E} \in \mathbb{S}$ such that $A \in \mathcal{E}$.

PROPOSITION XXI.4.6. *A topology is sequential if and only each sequentially closed set is closed.*

PROOF. By (VII.1.1) and Corollary VII.3.4, $\mathrm{T}\xi \geq \xi \geq \mathrm{TSeq}\,\xi$ if and only if $\mathcal{C}_{\mathrm{Seq}\xi} \subset \mathcal{C}_\xi$. □

EXAMPLE XXI.4.7 (A non-sequential prime topology). Let X be an uncountable set, $x_\infty \in X$, and let ζ be a prime topology such that $\mathcal{N}_\zeta(x_\infty)$ is the cocountable filter of $X \setminus \{x_\infty\}$. Then $\mathrm{adh}_\zeta A = A$ if A is countable (or empty), and $\mathrm{adh}_\zeta A = A \cup \{x_\infty\}$ if A is uncountable [2]. There is no ζ-convergent free sequential filter (each sequential filter contains a countable set). Therefore, $\mathrm{Seq}\zeta = \iota_X$, hence $\zeta \ngeq \mathrm{TSeq}\zeta = \iota$.

We have discussed in detail the case of Fréchet convergences in Section XXI.1. There exist sequential topologies that are not Fréchet, like that in Example VII.4.9.

Let us provide a classical description of strongly Fréchet topologies.

PROPOSITION XXI.4.8. *A topology ξ is strongly Fréchet (alias countably bisequential) if and only if for each decreasing sequence $(H_n)_{n<\omega}$ of sets such that*

$$x \in \bigcap\nolimits_{n<\omega} \mathrm{adh}_\xi H_n, \tag{XXI.4.5}$$

there exists a sequence $(x_n)_n$ such that $x_n \in H_n$ for each $n < \omega$ and $x \in \lim_\xi (x_n)_n$.

PROOF. By (XVIII.5.4), $\mathrm{adh}_{\mathrm{S}_1\mathrm{Seq}\xi}\,\mathcal{H} = \mathrm{adh}_{\mathrm{Seq}\xi}\,\mathcal{H}$ for each countably based filter $\mathcal{H}$. If a topology $\xi \geq \mathrm{S}_1\mathrm{Seq}\xi$, then $\xi = \mathrm{S}_1\mathrm{Seq}\xi$ [3], and if $(H_n)_n$ is a decreasing filter-base of $\mathcal{H}$, and

$$x \in \bigcap\nolimits_{n<\omega} \mathrm{adh}_\xi H_n = \mathrm{adh}_\xi \mathcal{H} = \mathrm{adh}_{\mathrm{S}_1\mathrm{Seq}\xi}\,\mathcal{H} = \mathrm{adh}_{\mathrm{Seq}\xi}\,\mathcal{H},$$

there exists a sequential filter $\mathcal{E}$ such that $x \in \lim_\xi \mathcal{E}$ and $\mathcal{E} \# \mathcal{H}$. If $(E_n)_n$ is a decreasing filter-base of $\mathcal{E}$, then for every each $n < \omega$, there exists $x_n \in E_n \cap H_n$. Accordingly, $(x_n)_n \geq \mathcal{E}$ and thus $x \in \lim_\xi (x_n)_n$.

Conversely, if the condition holds, then $\mathrm{adh}_\xi \mathcal{H} = \mathrm{adh}_{\mathrm{Seq}\xi}\,\mathcal{H}$ for each countably based filter $\mathcal{H}$, so that $\xi \geq \mathrm{S}_1\mathrm{Seq}\xi$, that is, ξ is strongly Fréchet. □

---

[2] Of course, if A is countable and $x_\infty \in A$, then $A = \mathrm{adh}_\zeta A = A \cup \{x_\infty\}$.

[3] As $\xi = \mathrm{T}\xi \leq \mathrm{TSeq}\xi \leq \mathrm{S}_1\mathrm{Seq}\xi$.

As we have said, a convergence ξ is bisequential if $\xi \geq \mathrm{SI}_1\xi$. As $\mathrm{Seq} \geq \mathrm{I}_1$, each sequentially founded pseudotopology is bisequential.

Explicitly, a convergence ξ is bisequential if $x \in \lim_\xi \mathcal{F}$ whenever for each filter $\mathcal{H}$ such that $\mathcal{H}\#\mathcal{F}$, there exists a countably based filter $\mathcal{G}$ such that $\mathcal{G}\#\mathcal{H}$ and $x \in \lim_\xi \mathcal{G}$. Equivalently,

PROPOSITION XXI.4.9. *A convergence ξ is bisequential if and only if whenever $x \in \lim_\xi \mathcal{F}$ then for every $\mathcal{U} \in \beta\mathcal{F}$, there exists $\mathcal{G} \in \mathbb{F}_1$ such that $\mathcal{U} \geq \mathcal{G}$ and $x \in \lim_\xi \mathcal{G}$.*

A filter $\mathcal{F}$ is *bisequential* if for each filter $\mathcal{U} \in \beta\mathcal{F}$ there is a countably based filter $\mathcal{G}$ such that $\mathcal{F} \leq \mathcal{G} \leq \mathcal{U}$.

EXAMPLE XXI.4.10 (A Hausdorff, compact, bisequential topology of non-countable character). Let ξ be a Hausdorff compact topology on X that is of countable character, but not perfectly normal, for example, that from Proposition XI.5.4. In virtue of Theorem X.7.19 of Vedenisov, there exists a ξ-closed subset C of X that is not $G_{\aleph_0}$. Let $f : X \longrightarrow Y := (X \setminus C) \sqcup \{\infty\}$ be a surjection so that $f(x) := x$ if $x \in X \setminus C$, and $f(x) := \infty$ if $x \in C$. Then f is a perfect map, as a continuous surjection from a compact convergence. Hence it is biquotient (see Table XX.11.1), and the resulting topology $\mathrm{T}(f\xi)$ is bisequential and not of countable character at ∞, because $C = f^-(\infty)$ is not $G_{\aleph_0}$.

On the other hand, there exist non-bisequential, strongly Fréchet topologies.

EXAMPLE XXI.4.11 (A non-bisequential, strongly Fréchet topology). A *Σ-product* of $\{X_j : j \in J\}$ is a subset X_Σ of $X = \prod_{j\in J} X_j$ such that there is $\varphi_0 \in X$ with

$$X_\Sigma = \{\varphi \in X : \mathrm{card}\{j \in J : \varphi(j) \neq \varphi_0(j)\} \leq \aleph_0\}.$$

Let J be an uncountable set and let X_Σ be a subspace of $\{0,1\}^J$ with $\varphi_0 = \overline{0}$ [4]. Take the restriction ψ to X_Σ of the product topology of $\{0,1\}^J$. Then ψ is a strongly Fréchet topology. Indeed, if $\varphi_\infty \in \mathrm{cl}_\psi A$, then there is a countable subset C of J such that $\varphi_\infty \in \mathrm{cl}_\psi(A \cap X_C)$, where $X_C := \{\varphi \in X_\Sigma : j \notin C \Rightarrow \varphi(j) = 0\}$. Indeed, otherwise for each countable $C \subset J$, there would be $V_C \in \mathcal{N}_\psi(\varphi_\infty)$ with $V_C \cap A \cap X_C = \varnothing$. If $C_0 \subset C_1$ then there is $V_{C_1} \supset V_{C_0}$. Therefore, there is a countable family $(V_{C_n})_{n<\omega}$ of neighborhoods of φ_∞ such for each $W \in \mathcal{N}_\psi(\varphi_\infty)$, there is $n < \omega$ such that $W \subset V_{C_n}$. As X_C is homeomorphic to the Cantor cube, hence metrizable, there exists a sequence $(\varphi_k)_k$ in $A \cap X_C$ that ψ-converges to φ_∞.

[4]That is, $\overline{0}(j) := 0$ for each $j \in J$.

If $(A_n)_n$ is a descending sequences of sets such that $\varphi_\infty \in \mathrm{cl}_\psi(A_n)$ for each $n < \omega$, then by the same argument as above, there exists a countable $C \subset J$ such that $\varphi_\infty \in \mathrm{cl}_\psi(A_n \cap X_C)$. We infer that X_Σ is strongly Fréchet.

Let us well order J, so that $j \in J$ whenever $j < \alpha$, where α is an uncountable limit ordinal. For each $j < \alpha$, let $1_j \in X_\Sigma$ be defined by $1_j(k) := 0$ if $k \neq j$ and $1_j(j) := 1$. Let $\mathcal{U}$ be an ultrafilter such that $\{1_j : \gamma < j < \alpha\} \in \mathcal{U}$ for each $\gamma < \alpha$. Then $\overline{0} \in \lim_\psi \mathcal{U}$. Indeed, for every $V \in \mathcal{N}_\psi(\overline{0})$, there is a finite set $F \subset J$ such that $V \supset \{\varphi \in X_\Sigma : j \in F \Rightarrow \varphi(j) = 0\}$, so that $\gamma > \max F$ implies that $\{1_j : \gamma < j < \alpha\} \subset V$. On the other hand, if $(H_n)_n \leq \mathcal{U}$, then there is a sequence $(\gamma_n)_n$ such that $\{1_j : \gamma_n < j < \alpha\} \subset H_n$. Hence, there is γ such that $\sup\{\gamma_n : n < \omega\} < \gamma < \alpha$, and thus $\{1_j : \gamma < j < \alpha\} \subset \bigcap_{n<\omega} H_n$. Let $j_\gamma > \gamma$, then $\{\varphi \in X_\Sigma : \varphi(j_\gamma) = 0\}$ is a ψ-neighborhood of $\overline{0}$ disjoint from $\{1_j : \gamma < j < \alpha\}$. Hence $\overline{0} \notin \lim_\psi (H_n)_n$. Therefore, ψ is not bisequential.

XXI.5. Sequentially founded convergences induced by adherence-determined classes

The class of convergences fulfilling $\tau = \mathrm{SeqT}\tau$ is that of *topologically induced* sequentially founded convergences. It follows from Theorem XXI.3.2 that for each topology ξ, there exists a coarsest sequential topology $\mathrm{TSeq}\,\xi$ finer than ξ. If ξ is a sequential topology, then $\mathrm{Seq}\xi$ is a coarsest sequentially founded convergence θ such that $\xi = \mathrm{T}\theta$.

It follows from Theorem XXI.3.2 (2) that $\mathrm{SeqT} : \mathrm{fix(Seq)} \longrightarrow \mathrm{fix(Seq)}$ is a reflector. From the representation of T (VII.3.2) we can easily derive a formula for SeqT. Indeed, if $\tau = \mathrm{Seq}\tau$, then for each sequential filter $\mathcal{E}$,

$$\lim_{\mathrm{SeqT}\tau} \mathcal{E} = \bigcap_{A \in \mathcal{E}^\#} \mathrm{cl}_\tau A. \tag{XXI.5.1}$$

We have already considered pretopologically induced sequentially founded convergences in Section XXI.1. Let us add here that since $\mathrm{SeqS_0} : \mathrm{fix(Seq)} \longrightarrow \mathrm{fix(Seq)}$ is a reflector, we can use a representation of $\mathrm{S_0}$ given in Proposition VI.6.7 to establish for any sequentially founded convergence τ, for each sequential filter $\mathcal{E}$,

$$\lim_{\mathrm{SeqS_0}\tau} \mathcal{E} = \bigcap_{A \in \mathcal{E}^\#} \mathrm{adh}_\tau A. \tag{XXI.5.2}$$

Therefore, a sequentially founded convergence τ is pretopologically induced if and only if $\bigcap_{A \in \mathcal{E}^\#} \mathrm{adh}_\tau A \subset \lim_\tau \mathcal{E}$.

In Definition XVI.3.15, a sequentially founded convergence θ is called *Urysohn* whenever for each $\mathcal{E} \in \mathbb{S}$, if $x \notin \lim_\theta \mathcal{E}$, then there exists $\mathcal{F} \in \mathbb{S}\mathcal{E}$ such that $x \notin \lim_\theta \mathcal{G}$ for every $\mathcal{G} \in \mathbb{S}\mathcal{F}$, in other terms, whenever for each sequential filter $\mathcal{E}$,

$$\bigcap_{\mathcal{F}\in\mathbb{S}\mathcal{E}} \bigcup_{\mathcal{G}\in\mathbb{S}\mathcal{F}} \lim_\theta \mathcal{G} \subset \lim_\theta \mathcal{E}.$$

PROPOSITION XXI.5.1. *A sequentially founded convergence θ is Urysohn if and only if θ is paratopologically induced:*

$$\theta \leq \mathrm{SeqS}_1\theta. \tag{XXI.5.3}$$

PROOF. If $\theta = \mathrm{Seq}\theta$ and $\mathcal{E} \in \mathbb{S}$, then

$$\lim_{\mathrm{S}_1\theta} \mathcal{E} = \bigcap_{\mathcal{F}\in\mathbb{S}\mathcal{E}} \bigcup_{\mathcal{G}\in\mathbb{S}\mathcal{F}} \lim_\theta \mathcal{G}.$$

Indeed, $\lim_{\mathrm{S}_1\theta} \mathcal{E} = \bigcap_{\mathbb{F}_1 \ni \mathcal{H}\#\mathcal{E}} \mathrm{adh}_{\mathrm{Seq}\theta} \mathcal{H} = \bigcap_{\mathcal{F}\in\mathbb{S}\mathcal{E}} \bigcup_{\mathcal{G}\in\mathbb{S}\mathcal{F}} \lim_\theta \mathcal{G}$. The conclusion follows. □

Example XVI.3.18 exhibits a sequentially founded convergence that is not Urysohn, that is, not paratopologically induced.

EXAMPLE XXI.5.2 (A Urysohn, non-pretopologically-induced convergence). Let X be an infinite countable set and let $x_\infty \in X$. Let π a principal prime convergence such that $x_\infty \in \lim_\pi \{A\}^\uparrow$ if A is finite and non-empty, and $\lim_\pi \{A\}^\uparrow = \varnothing$ if A is infinite. The convergence π is finitely stable, but it is not a pretopology, because $\mathcal{V}_\pi(x_\infty) = \{X\}^\uparrow$ does not converge. It is, however, a paratopology. Indeed, if $x_\infty \notin \lim_\pi \mathcal{F}$, then there is an infinite set $F \in \mathcal{F}$ such that $(F)_0$ is a countably based filter that meshes $\mathcal{F}$, and $x_\infty \notin \mathrm{adh}_\pi (F)_0$.

On the other hand, π is a sequentially based convergence (actually all the convergent filters are sequential), which is not pretopologically induced. However, π is Urysohn. Indeed, if $\mathcal{E}$ is a sequential filter and $x_\infty \notin \lim_\pi \mathcal{E}$, then there is a free filter $\mathcal{F} \in \mathbb{S}\mathcal{E}$, so that $\mathrm{adh}_\pi \mathcal{F} = \mathrm{adh}_{\mathrm{Seq}\pi} \mathcal{F} = \varnothing$.

Nevertheless,

PROPOSITION XXI.5.3. *A free (alias T_1) Urysohn sequentially founded convergence is pretopologically induced.*

PROOF. Let $\mathcal{E}$ be a free sequential filter and let $x \in \bigcap_{A\in\mathcal{E}^\#} \mathrm{adh}_\tau A$. If $A \in \mathcal{E}^\#$, then A is infinite and $\mathcal{E} \vee A$ is the cofinite filter of a countable infinite subset subset D of A. Hence, $x \in \mathrm{adh}_\tau D = \mathrm{adh}_\tau (D)_0 \cup D$, and $x \in \mathrm{adh}_\tau (D)_0 \cup D'$ for each $D' \subset_{\mathrm{cofin}} D$, so that $x \in \mathrm{adh}_\tau (D)_0$ and thus there exists $\mathcal{G} \in \mathbb{S}(D)_0$ such that $x \in \lim_\tau \mathcal{G}$. By setting $\mathcal{D} := (D)_0$, we conclude that $x \in \bigcap_{\mathcal{D}\in\mathbb{S}\mathcal{E}} \bigcup_{\mathcal{G}\in\mathbb{S}\mathcal{D}} \lim_\tau \mathcal{G}$. □

Here is a classical result ([5]).

THEOREM XXI.5.4. *Each Hausdorff Urysohn sequentially founded convergence is topologically induced.*

It will be shown in Example XXII.9.6 that convergence almost everywhere is a non-Urysohn sequentially founded convergence.

It follows from Definition XXI.3.3 that a sequentially founded convergence is *pseudotopologically induced* whenever

$$\lim_{\tau} \mathcal{E} = \bigcap\nolimits_{\mathcal{U} \in \beta\mathcal{E}} \mathrm{adh}_{\mathrm{Seq}\tau} \mathcal{U}$$

for each sequential filter $\mathcal{E}$.

In general, $\mathrm{Seq}\theta \geq \theta$, hence $\mathrm{S}(\mathrm{Seq}\theta) \geq \mathrm{S}\theta$.

LEMMA XXI.5.5. *If a convergence θ is finitely stable, then* $\mathrm{Seq}\theta$ *is finitely stable.*

PROOF. If $x \in \lim_{\mathrm{Seq}\theta} \mathcal{F}_0 \cap \lim_{\mathrm{Seq}\theta} \mathcal{F}_1$, then there are sequential filters $\mathcal{E}_0, \mathcal{E}_1$ such that $x \in \lim_\theta \mathcal{E}_0 \cap \lim_\theta \mathcal{E}_1 = \lim_\theta(\mathcal{E}_0 \wedge \mathcal{E}_1)$ by assumption, and $\mathcal{E}_0 \leq \mathcal{F}_0$ and $\mathcal{E}_1 \leq \mathcal{F}_1$, and thus $\mathcal{E}_0 \wedge \mathcal{E}_1 \leq \mathcal{F}_0 \wedge \mathcal{F}_1$, that is, $x \in \lim_{\mathrm{Seq}\theta} \mathcal{F}_0 \wedge \mathcal{F}_1$. □

LEMMA XXI.5.6. *For every finitely stable convergence θ,*

$$\mathrm{S}(\mathrm{Seq}\theta) \geq \mathrm{Seq}\theta. \tag{XXI.5.4}$$

PROOF. Let $x \in \lim_{\mathrm{S}(\mathrm{Seq}\theta)} \mathcal{F}$. Then, for each $\mathcal{U} \in \beta\mathcal{F}$, there exists $\mathcal{E}_\mathcal{U} \in \mathbb{S}$ such that $\mathcal{E}_\mathcal{U} \leq \mathcal{U}$ and $x \in \lim_\theta \mathcal{E}_\mathcal{U}$. For each $\mathcal{U} \in \beta\mathcal{F}$, let $E_\mathcal{U} \in \mathcal{E}_\mathcal{U}$ be such that $E_\mathcal{U} \setminus A$ is finite for each $A \in \mathcal{E}_\mathcal{U}$. By Proposition V.4.6, there is a finite subset $\boldsymbol{F}$ of $\beta\mathcal{F}$ such that $E := \bigcup_{\mathcal{U} \in \boldsymbol{F}} E_\mathcal{U} \in \mathcal{F}$. Of course, E is countable. If $\mathcal{F}$ is free, then $\bigwedge_{\mathcal{U} \in \boldsymbol{F}} (E_\mathcal{U})_0 = (E)_0 \leq \mathcal{F}$ and $x \in \lim_\theta \bigwedge_{\mathcal{U} \in \boldsymbol{F}} \mathcal{E}_\mathcal{U}$ because θ is finitely stable. As $\mathcal{E}_\mathcal{U} = (E_\mathcal{U})_0$ for each $\mathcal{U}$, we conclude that $x \in \lim_{\mathrm{Seq}\theta} \mathcal{F}$. If $\mathcal{F}$ is principal, then $\mathcal{F} = E^\uparrow$ and E is countable, hence $\mathcal{F} \in \mathbb{S}$, so that $x \in \lim_{\mathrm{Seq}\theta} \mathcal{F}$. If $\mathcal{F}$ is arbitrary, then $\mathcal{F} = \mathcal{F}^* \wedge \mathcal{F}^\bullet$, hence by Lemma XXI.5.5

$$\begin{aligned} \lim_{\mathrm{S}(\mathrm{Seq}\theta)} (\mathcal{F}^* \wedge \mathcal{F}^\bullet) &= \lim_{\mathrm{S}(\mathrm{Seq}\theta)} \mathcal{F}^* \cap \lim_{\mathrm{S}(\mathrm{Seq}\theta)} \mathcal{F}^\bullet \\ &= \lim_{\mathrm{Seq}\theta} \mathcal{F}^* \cap \lim_{\mathrm{Seq}\theta} \mathcal{F}^\bullet = x \in \lim_{\mathrm{Seq}\theta} \mathcal{F}. \end{aligned}$$

□

On setting $\theta = \mathrm{Seq}\theta$, we infer that

COROLLARY XXI.5.7. *A sequentially founded convergence is a pseudotopology if and only if it is finitely stable.*

---

[5]See L. V. KANTOROVICH, B. Z. VULICH and A. G. PINSKER [146], M. DOLCHER [54], and J. KISYŃSKI [160].

Therefore,

COROLLARY XXI.5.8. *If a sequentially founded convergence is pseudotopologically induced, then it is a pseudotopology.*

PROOF. If $\tau = \mathrm{Seq}\tau$ and is pseudotopologically induced if and only if $\tau = \mathrm{SeqS}\tau$, and since $\mathrm{S}\tau$ is, in particular, finitely stable, hence by Proposition XXI.5.5, τ is finitely stable. In virtue of Lemma XXI.5.6, τ is a pseudotopology. □

Therefore a general interaction of Figure XXI.3.1 reduces to Figure XXI.5.1.

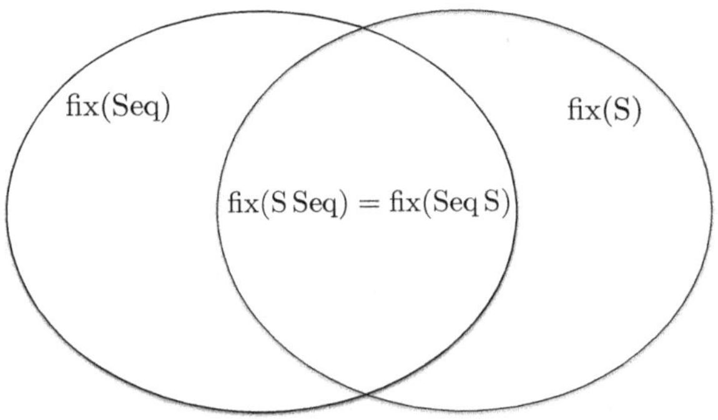

FIGURE XXI.5.1. Coincidence of sequentially induced pseudotopologies and pseudotopologically induced sequentially founded convergences

XXI.6. Subclasses induced by locally compact convergences

As follows from Section XXI.3, functorial equations of the type $\xi = J\mathrm{K}\xi$, where K stands for the *locally compact modifier* and J is a reflector, result in coreflective subclasses of $\mathrm{fix}(J)$ that are determined by ξ-compact sets.

Traditionally ([6]), a convergence ξ is called

(1) *k-topology* whenever $\xi = \mathrm{TK}\xi$,
(2) *k′-pretopology* whenever $\xi = \mathrm{S}_0\mathrm{K}\xi$,
(3) *strongly k′-paratopology* whenever $\xi = \mathrm{S}_1\mathrm{K}\xi$.

On the other hand, $\mathrm{fix(SK)} = \mathrm{fix(K)} \cap \mathrm{fix(S)}$, so that

PROPOSITION XXI.6.1. *A convergence ξ fulfills $\xi = \mathrm{SK}\xi$ if and only if ξ is a locally compact pseudotopology.*

[6] As I have pointed out several times, traditional approach was restricted to the class of topologies.

PROOF. Of course, if ξ is a locally compact pseudotopology, then $\xi \geq \mathrm{K}\xi$, hence $\xi \geq \mathrm{S}\xi \geq \mathrm{SK}\xi$, and $\mathrm{S}(\mathrm{K}\xi) \geq \mathrm{S}\xi \geq \xi$.

If $\xi = \mathrm{SK}\xi$, then obviously ξ is a pseudotopology, that is, $\xi = \mathrm{S}\xi$. If $x \in \lim_\xi \mathcal{F}$, then $x \in \lim_\xi \mathcal{U} = \lim_{\mathrm{K}\xi} \mathcal{U}$ for each ultrafilter $\mathcal{U} \in \beta\mathcal{F}$. Hence, there is $K_\mathcal{U} \in \mathcal{U} \cap \mathcal{K}(\xi)$. By Proposition V.4.6, there is a finite set $\boldsymbol{F} \subset \beta\mathcal{F}$ such that $\bigcup_{\mathcal{U}\in\boldsymbol{F}} K_\mathcal{U} \in \mathcal{F} \cap \mathcal{K}(\xi)$, so that ξ is locally compact. □

As usual, the inequality

$$\xi \geq J\mathrm{K}\xi$$

extends to all convergences from J-convergences any property listed above.

Let ξ be a convergence on a set X. We say that a subfamily $\mathcal{L}$ of subsets of X *determines* the family $\mathcal{C}_\xi$ of ξ-closed sets if

$$F \in \mathcal{C}_\xi \iff \underset{L\in\mathcal{L}}{\forall}\ F \cap L \in \mathcal{C}_{\xi|_L}.$$

PROPOSITION XXI.6.2. *A convergence ξ is a k-convergence, that is,*

$$\xi \geq \mathrm{TK}\xi$$

if and only if $\mathcal{K}(\xi)$ determines ξ.

PROOF. Suppose that $\xi \geq \mathrm{TK}\xi$ and let $H \cap K$ be $\xi|_K$-closed for each $K \in \mathcal{K}(\xi)$, that is, if $\mathcal{U} \in \beta(H \cap K)$, then $\lim_\xi \mathcal{U} \subset H \cap K$. As $\mathcal{U} \cap \mathcal{K}(\xi)$, we infer that $\lim_\xi \mathcal{U} = \lim_{\mathrm{K}\xi} \mathcal{U}$, so that $\operatorname{adh}_{\mathrm{K}\xi} H \subset H$, hence $\operatorname{cl}_{\mathrm{K}\xi} H \subset H$. We need to prove that $\operatorname{cl}_\xi H \subset H$. By (VII.4.1),

$$\lim_{\mathrm{TK}\xi} \mathcal{F} = \bigcap\nolimits_{H\in\mathcal{F}^\#} \operatorname{cl}_{\mathrm{K}\xi} H.$$

Therefore, $\operatorname{cl}_\xi H \subset \operatorname{cl}_{\mathrm{K}\xi} H \subset H$.

Conversely, if ξ is a topology determined by $\mathcal{K}(\xi)$ then every $\mathrm{K}\xi$-closed set is ξ-closed, that is, $\xi \geq \mathrm{TK}\xi$. Indeed, if C is $\mathrm{K}\xi$-closed then $K \cap C$ is ξ-closed for every $K \in \mathcal{K}(\xi)$, because if $K \cap C \in \mathcal{F}$ with $x \in \lim_\xi \mathcal{F}$ then $K \in \mathcal{F}$ and thus $x \in \lim_{\mathrm{K}\xi} \mathcal{F} \subset K \cap C$. □

PROPOSITION XXI.6.3. *A convergence ξ is a k'-convergence, that is,*

$$\xi \geq \mathrm{S}_0\mathrm{K}\xi,$$

if and only if $\operatorname{adh}_\xi H = \operatorname{adh}_{\mathrm{K}\xi} H$ *for each subset H of $|\xi|$.*

PROOF. This follows from (VI.6.5), that is,

$$\lim_\xi \mathcal{F} \subset \bigcap\nolimits_{H\in\mathcal{F}^\#} \operatorname{adh}_{\mathrm{K}\xi} H$$

for each filter $\mathcal{F}$. □

PROPOSITION XXI.6.4. *A convergence* ξ *is a strongly* k'*-convergence, that is*

$$\xi \geq \mathrm{S}_1\mathrm{K}\xi,$$

if and only if $\mathrm{adh}_\xi \mathcal{H} = \mathrm{adh}_{\mathrm{K}\xi} \mathcal{H}$ *for each countably based filter* $\mathcal{H}$ *of* $|\xi|$.

PROOF. Similarly to the preceding proof, by (XVIII.6.1), this holds whenever, for every filter $\mathcal{F}$,

$$\lim_\xi \mathcal{F} \subset \bigcap_{\mathbb{F}_1 \ni \mathcal{H} \# \mathcal{F}} \mathrm{adh}_{\mathrm{K}\xi} \mathcal{H}.$$

□

XXI.7. Preservation of mixed properties under quotient maps

Recall that, given a reflector J, a map f is J*-quotient*, more precisely $f \in Q_J(\xi, \tau)$, whenever

$$\tau \geq J(f\xi).$$

If moreover E is a coreflector, then a J-convergence θ is a JE*-convergence* whenever

$$\theta \geq JE\theta.$$

Similarity of these two formulae is seizing. It is explicated in the following lemma.

LEMMA XXI.7.1. *Let* J *be a reflector, and* E *be a coreflector. Then* $\theta \geq JE\theta$ *if and only if*

$$i \in Q_J(E\theta, \theta). \tag{XXI.7.1}$$

PROOF. Indeed, $\theta \geq JE\theta = J(i(E\theta))$ amounts to (XXI.7.1). □

As a consequence of Lemma XXI.7.1,

THEOREM XXI.7.2. *If* $f \in \mathscr{C}(\xi, \tau)$ *is a* J*-quotient map, and* ξ *is a* JE*-convergence, then* τ *is also a* JE*-convergence. Conversely, each* JE*-convergence is a continuous* J*-quotient image of an* E*-convergence.*

PROOF. By definition, f is a J-quotient map whenever $\tau \geq J(f\xi)$, and ξ is a JE-convergence if and only if $\xi \geq JE\xi$. Putting these two inequalities together, yields

$$\tau \geq J(f\xi) \geq J(f(JE\xi)),$$

and, by (XVII.2.1), $J(f(JE\xi)) \geq JE(f\xi)$, which is greater than $JE\tau$, in view of continuity. By concatenating these formulas, we get $\tau \geq JE\tau$.

Conversely, if $\tau \geq JE\tau = J(i(E\tau))$, then $i \in Q_J(E\tau, \tau)$, as in Lemma XXI.7.1. $\square$

Several classical theorems are corollaries of Theorem XXI.7.2. We shall mention here just those corresponding to the coreflectors Seq, I_1, and K.

We call the second part of Theorem XXI.7.2, the *reconstruction*, because it shows that an E-convergence, which is, so to say, a coreflective ingredient of a JE-convergence, can be recovered from the latter. This reconstruction result can be improved in case of some coreflectors E.

LEMMA XXI.7.3. *Each finitely stable sequentially founded convergence is a continuous convergence quotient image of a topology that is a sequentially founded convergence.*

PROOF. Let $\theta = \mathrm{Seq}\theta$ be a convergence on X. For every $x \in X$, and each sequential filter $\mathcal{E}$ such that $x \in \lim_\theta \mathcal{E}$, consider a convergence $\theta(x, \mathcal{E})$ on X, in which each $w \in X$ different from x is isolated, and $x \in \lim_{\theta(x,\mathcal{E})} \mathcal{F}$ whenever $\mathcal{F} \geq \mathcal{E} \wedge x^\uparrow$. Then, of course, $\theta(x, \mathcal{E}) \geq \theta$ for each $(\mathcal{E}, x) \in \theta$ with $\mathcal{E} \in \mathbb{S}$ (Definition II.2.9). As a prime pretopology, $\theta(x, \mathcal{E})$ is a topology (Proposition VII.3.14), which is a sequentially founded convergence, because the sequential filter $\mathcal{E} \wedge x^\uparrow$ is the vicinity of $\theta(x, \mathcal{E})$ at a unique non-isolated point x.

Considering $\theta(x, \mathcal{E})$ as a convergence on a copy $X_{(\mathcal{E},x)}$ of X, the coproduct $\eta := \bigsqcup_{(\mathcal{E}x) \in \theta} \theta(x, \mathcal{E})$ is a topology and a sequentially founded convergence. Let $i_{(\mathcal{E},x)} : X_{(\mathcal{E},x)} \longrightarrow X$ define the corresponding copy of X. Then, the natural map $\varphi : \bigsqcup_{(\mathcal{E}x) \in \theta} X_{(\mathcal{E},x)} \longrightarrow X$, that is, $\varphi(w) := i_{(\mathcal{E},x)}(w)$ if $w \in X_{(\mathcal{E},x)}$, is a quotient map, and clearly $\varphi\eta = \theta$. $\square$

COROLLARY XXI.7.4. *If J is a reflector, then each JSeq-convergence is a continuous J-quotient of a topology that is a sequentially founded convergence.*

PROOF. Let τ be a JSeq-convergence on X. By the second part of Theorem XXI.7.2, there is a sequentially founded convergence θ on X such that $\tau = J(i\theta)$, where i is the identity. By Lemma XXI.7.3, there exists $\mathrm{T}\eta = \eta = \mathrm{Seq}\eta$, and $\varphi \in CQ(\eta, \theta)$. In particular, $\varphi\eta = \theta$, hence $\tau = J(i\theta) = J(\varphi\eta)$. $\square$

Below we have an instance of a continuous topologically quotient map that is not pretopologically quotient.

EXAMPLE XXI.7.5. Recall the *radial pretopology* ρ of Example VI.1.9 and its topologization, the *radial topology* $\lambda = \mathrm{T}\rho$ from Example

VII.4.9. As we have seen, ρ is a Fréchet pretopology, that is, $\rho = \mathrm{S}_0\mathrm{Seq}\rho$ and thus $\lambda = \mathrm{T}\rho$ is a sequential topology, that is, $\lambda = \mathrm{TSeq}\lambda < \rho$. Moreover, $\eta := \mathrm{Seq}\lambda = \mathrm{Seq}\rho$. Hence, $i \in \mathscr{C}(\eta, \lambda)$ fulfills $\lambda = \mathrm{T}\eta$, but $\lambda < \rho = \mathrm{S}_0\eta$, showing that the identity is a T-quotient, but not an S_0-quotient.

By Theorem XXI.7.2, a continuous map preserves a JE-property whenever it is a J-quotient. Several fundamental cases are gathered below ([7]).

J-quotient	I_1	Seq	K
almost open I	countable character $\xi \geq \mathrm{I}_1\xi$	sequentially founded $\xi \geq \mathrm{Seq}\xi$	locally compact $\xi \geq \mathrm{K}\xi$
biquotient S	bisequential $\xi \geq \mathrm{SI}_1\xi$	sequentially founded $\xi \geq \mathrm{SSeq}\xi$	locally compact $\xi \geq \mathrm{SK}\xi$
countably biquotient S_1	strongly Fréchet $\xi \geq \mathrm{S}_1\mathrm{I}_1\xi$	strongly Fréchet $\xi \geq \mathrm{S}_1\mathrm{Seq}\xi$	strongly k' $\xi \geq \mathrm{S}_1\mathrm{K}\xi$
hereditarily quotient S_0	Fréchet $\xi \geq \mathrm{S}_0\mathrm{I}_1\xi$	Fréchet $\xi \geq \mathrm{S}_0\mathrm{Seq}\xi$	k' $\xi \geq \mathrm{S}_0\mathrm{K}\xi$
quotient T	sequentially induced $\xi \geq \mathrm{TI}_1\xi$	sequentially induced $\xi \geq \mathrm{TSeq}\xi$	k $\xi \geq \mathrm{TK}\xi$

TABLE XXI.7.1. Classes of spaces and functorial inequalities. As $\mathrm{Seq} \geq \mathrm{I}_1$, each property in the third column entails the property in the second column of the same row (actually, three of them even coincide). Since Hausdorff sequentially founded convergences are locally compact, $\mathrm{Seq}\xi \geq \mathrm{K}\xi$ if $\mathrm{Seq}\xi$ is Hausdorff; hence, in this case, the properties in the third column imply those in the fourth column of the same row.

In particular, by Proposition XXI.6.1, a convergence ξ fulfills $\xi = \mathrm{SK}\xi$ if and only if it is a locally compact pseudotopology. Hence,

COROLLARY XXI.7.6. *Locally compact pseudotopologies are preserved under continuous biquotient maps.*

[7]More detailed accounts can be found in [185] by E. MICHAEL, [56] by S. DOLECKI, [83] by S. DOLECKI and F. MYNARD.

XXI.8. Preservation of mixed properties under products

If $\xi \geq JE\xi$ and $\tau \geq JE\tau$, where J is a reflector and E is a coreflector, then a question arises whether

$$\xi \times \tau \geq JE(\xi \times \tau).$$

By assumption, $\xi \times \tau \geq JE\xi \times JE\tau$, so that the inequality would hold, provided that

$$JE\xi \times JE\tau \geq JE(\xi \times \tau).$$

Of course, if $E = \mathrm{I}$ is the identity functor, then the inequality above becomes $J\xi \times J\tau \geq J(\xi \times \tau)$. We know already that this holds if $J = \mathrm{S}$ is the pseudotopologizer, but does not if J is equal to $\mathrm{T}, \mathrm{S}_0, \mathrm{S}_1$ or $\mathrm{S}_{\wedge 1}$.

On the other hand, if $J = \mathrm{I}$ is the identity, then the inequality $E\xi \times E\tau \geq E(\xi \times \tau)$ does not hold if $E = \mathrm{Seq}$ is the sequential modifier for example ([8]), but it does if $E = \mathrm{I}_1$ is the countable character modifier.

This rough analysis implies that the property $\xi \geq \mathrm{SI}_1\xi$ is productive, that is that

PROPOSITION XXI.8.1. *Finite products of bisequential pseudotopologies are bisequential.*

This is an exceptional situation. Among the properties of the type $\xi \geq J\mathrm{I}_1\xi$, this is the only productive one besides $\xi \geq \mathrm{I}_1\xi$, the class of the convergences of *countable character*, which is countably productive in virtue of Corollary XVIII.2.13. As the pseudotopologizer S commutes with arbitrary products by Theorem XVIII.3.9, this means that bisequential pseudotopologies are in fact closed under countable products as well.

It is interesting to first address the productivity quest for Fréchet convergences heuristically ([9]).

Let us start by considering simplest examples, like the product of a sequential fan topology (defined in Example III.4.7 and discussed in III.4.7) with a convergent sequence.

It turns out that

PROPOSITION XXI.8.2. *The product of a sequential fan topology with a free sequence topology is not Fréchet.*

PROOF. Let $A \subset X \times Y$ be given by

$$A := \{(x_{n,k}, y_n) : n, k < \omega\}.$$

[8]Because the product of two sequentially founded convergences need not be sequentially founded.

[9]This way we shall recover the first findings on the topic by A. V. ARHANGEL'SKII, E. MICHAEL, and others.

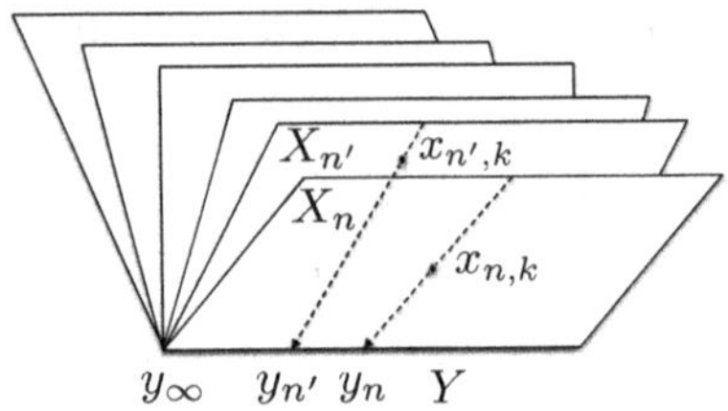

FIGURE XXI.8.1. The product of a sequential fan X with a free sequence Y converging to y_∞. For each $n < \omega$, the segment X_n represents a free sequence converging to x_∞.

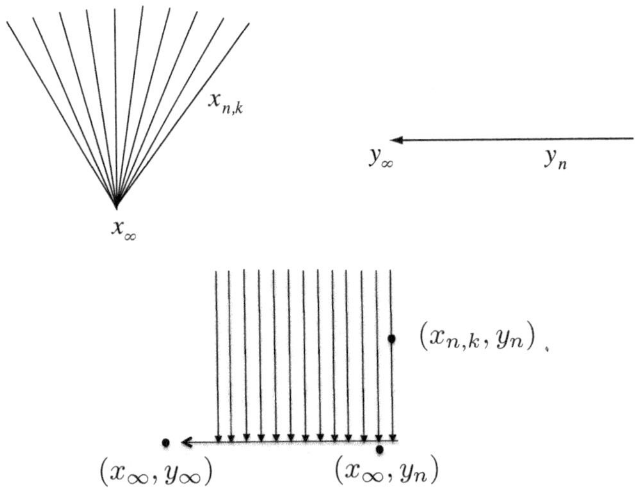

FIGURE XXI.8.2. The product of two Fréchet topologies (a fan with a sequence) is not Fréchet. Indeed, the bisequence above emerges as a subspace of the book from Figure XXI.8.1.

Then $(x_\infty, y_\infty) \in \mathrm{cl}\, A$, because if $O \in \mathcal{O}(x_\infty, y_\infty)$, then there is n_O such that $(x_\infty, y_n) \in O$ for each $n > n_O$, and for each such n, there is $k < \omega$ such that $(x_{n,k}, y_n) \in O$ ([10]).

On the other hand, no sequence on A converges to (x_∞, y_∞). Indeed, if a sequence $(x_{n_p,k_p}, y_{n_p})_p$ converges to (x_∞, y_∞), then $(y_{n_p})_p$ converges to y_∞ since it is a product convergence. Therefore, for each $n < \omega$,

[10]Actually, there is k_n such that $(x_{n,k}, y_n) \in O$ for each $k > k_n$.

there are finitely many p such that $x_{n_p,k_p} \in X_n$, and thus $(x_{n_p,k_p})_p$ does not converge to x_∞. $\square$

Therefore, if the product of a topology ξ with every Fréchet topology τ is Fréchet, then ξ does not include a free sequence (compare with Corollary XXI.8.8). Otherwise, a bisequence appears

$$(x_\infty, y_\infty) \underset{n}{\longleftarrow} (x_\infty, y_n) \underset{k}{\longleftarrow} (x_{n,k}, y_n),$$

contradicting the Fréchetness of the product.

LEMMA XXI.8.3. *If the product of two non-trivial Fréchet topologies is a Fréchet topology, then they are both strongly Fréchet.*

PROOF. If X is not strongly Fréchet and Y contains a free sequence $(y_n)_n$ converging to y_∞, then there exists a decreasing sequence $(H_n)_n$ of subsets of X such that $x_\infty \in \operatorname{cl} H_n$ for each $n < \omega$, but no sequence $x_n \in H_n$ converges to x_∞. Let

$$A := \bigcup\nolimits_{n<\omega} (H_n \times \{y_n\}).$$

Then $(x_\infty, y_\infty) \in \operatorname{cl} A$, because if $O \in \mathcal{O}(x_\infty, y_\infty)$, then there is n_O such that $(x_\infty, y_n) \in O$ for each $n > n_O$, and for each such n, there is $x_n \in H_n$ such that $(x_n, y_n) \in O$.

If there is a sequence (v_k, w_k) in A converging to (x_∞, y_∞), then there is a subsequence $(n_k)_k$ such that $y_{n_k} = w_k$ and $v_k \in H_{n_k}$ and $(n_k)_k$ tends to ∞. Then $x_n := v_k$ if $n_k \leq n < n_{k-1}$ is like in the assumption, a contradiction. $\square$

Therefore, strong Fréchetness of factors is necessary for Fréchetness of the product. It is, however, not sufficient ([11]).

A decisive feature of Fréchet factors for the Fréchetness of their product is *transversality*. It was possible in Proposition XXI.8.2 to construct a bisequence without any convergent diagonal sequence, because there was no convergent transversal sequence in the sequential fan. It is clear that an additional transversality condition is needed for the Fréchetness of products. Namely, ξ is called α_4 *at* x if for every countable collection of (disjoint) sequences $(x_{n,k})_k$ converging to x there are sequences $(n_p)_p$ tending to x and $(k_p)_p$ such that $(x_{n_p,k_p})_p$ ξ-converges to x. A convergence is α_4 if it is α_4 at every point.

PROPOSITION XXI.8.4. *A Fréchet pretopology is strongly Fréchet if and only if it is* α_4.

[11]See F. JORDAN and F. MYNARD [141]. P. SIMON gave in [236] an example of a Hausdorff compact Fréchet topology, the square of which is not Fréchet. This example is presented in [83, p. 468] by S. DOLECKI and F. MYNARD.

PROOF. Let ξ be an α_4 Fréchet convergence and let $(H_n)_n$ be a descending sequence of sets such that $x \in \bigcap_{n<\omega} \operatorname{adh}_\xi H_n$. By Fréchetness, for every $n < \omega$, there exists a sequence $(x_{n,k})_k$ of elements of H_n such that $x_\infty \in \lim_\xi (x_{n,k})_k$. By α_4, there exists $(n_p)_p$ tending to ∞ and $(k_p)_p$ such that $x_\infty \in \lim_\xi (x_{n_p,k_p})_p$, and $(x_{n_p,k_p})_p \geq \{H_n : n < \omega\}$, showing that ξ is strongly Fréchet.

Conversely, let ξ be strongly Fréchet, and let $\{x_\infty\} \sqcup \{x_{n,k} : n, k < \omega\}$ is a ξ-convergent fan. Let $H_m := \{x_{n,k} : m \leq n < \omega, k < \omega\}$. Hence $x \in \bigcap_{m<\omega} \operatorname{adh}_\xi H_m$. By strong Fréchetness, there exists a selection $x_{n_m,k_m} \in H_m$ such that $x_\infty \in \lim_\xi (x_{n_m,k_m})_m$ showing α_4. □

An interested reader will find more details on this direct approach in Section XXI.11, while we now pass to a more systematic approach. In Section XX.7, we established conditions assuring countable compactness of finite products, and we pointed out that they readily apply to products of various mixed properties.

Let us start by interpreting Corollary XX.7.8.

COROLLARY XXI.8.5. *The following statements are equivalent:*

(1) $\xi \geq \mathrm{SI}_1\xi$ *(bisequential),*
(2) $\xi \times \tau \geq \mathrm{SI}_1(\xi \times \tau) = \mathrm{SI}_1\xi \times \mathrm{SI}_1\tau$ *for each pseudotopology* $\tau = \mathrm{SI}_1\tau$,
(3) $\xi \times \tau \geq \mathrm{S}_0\mathrm{I}_1(\xi \times \tau) = \mathrm{S}_0(\mathrm{I}_1\xi \times \mathrm{I}_1\tau)$ *for each prime topology* $\tau = \mathrm{I}_1\tau$.

In other words, a convergence is bisequential if and only if its product with each bisequential pseudotopology is bisequential if and only if its product with a prime topology of countable character is Fréchet.

Therefore,

COROLLARY XXI.8.6. *Finite products of bisequential convergences are bisequential.*

In fact, as mentioned before, products of countably many bisequential convergences are bisequential.

We are now in a position to interpret Corollary XX.7.9 in terms of products of strongly Fréchet convergences.

COROLLARY XXI.8.7. *The following statements are equivalent:*

(1) $\xi \geq \mathrm{S}_1\mathrm{I}_1\xi$ *(strongly Fréchet),*
(2) $\xi \times \tau \geq \mathrm{S}_1\mathrm{I}_1(\xi \times \tau)$ *for each convergence* $\tau \geq \mathrm{SI}_1\tau$,
(3) $\xi \times \tau \geq \mathrm{S}_0\mathrm{I}_1(\xi \times \tau)$ *for each prime topology* τ *of countable character.*

In other words, a convergence is strongly Fréchet if and only if its product with each bisequential convergence (equivalently, with each prime topology of countable character) is strongly Fréchet (equivalently, Fréchet).

Similarly, we characterize Fréchet convergences. Recall Table XVIII.2.1 that a convergence is called *finitely generated* if it $\mathbb{F}_0$-founded.

COROLLARY XXI.8.8. *The following statements are equivalent:*

(1) $\xi \geq \mathrm{S}_0\mathrm{I}_1\xi$ *(Fréchet),*
(2) $\xi \times \tau \geq \mathrm{S}_0\mathrm{I}_1(\xi \times \tau)$ *for each convergence* $\tau \geq \mathrm{SI}_0\tau$,
(3) $\xi \times \tau \geq \mathrm{S}_0\mathrm{I}_1(\xi \times \tau)$ *for each prime* $\mathbb{F}_0$*-founded topology.*

In other words, a convergence is Fréchet if and only if its product with each finitely generated convergence (equivalently, with each prime finitely generated topology) is Fréchet.

PROOF. Set $\mathbb{H} = \mathbb{F}_0$ and $\mathbb{J} = \mathbb{F}_1$ in Corollary XX.7.7. □

W shall characterize in Section XXI.10 convergences whose product with each strongly Fréchet convergence is strongly Fréchet ([12]).

DEFINITION XXI.8.9. A convergence ξ is called *productively Fréchet* if $\xi \times \tau$ is strongly Fréchet for each strongly Fréchet convergence τ.

Let us observe right now that

PROPOSITION XXI.8.10. *Finite products of productively Fréchet convergences are productively Fréchet.*

PROOF. Let ζ and ξ be productively Fréchet. If θ is strongly Fréchet, then $\theta \times (\zeta \times \xi) = (\theta \times \zeta) \times \xi$. By definition, $\theta \times \zeta$ is strongly Fréchet, hence $(\theta \times \zeta) \times \xi$ is strongly Fréchet, which proves that $\zeta \times \xi$ is productively Fréchet. □

We shall characterize productively Fréchet convergences later. We shall also see in Theorem XXII.8.4 that there exists a Hausdorff compact Fréchet topology whose square is not Fréchet.

XXI.9. *Relations and polarities in classes of filters

The relations that we considered so far were *binary*, that is, identified with subsets of products of two sets. On one hand, it is often useful to extend this concept to n-ary relations, that is, subsets of products of

[12]Such convergences were characterized by F. JORDAN and F. MYNARD in [**141**]. It is not known if each countable product of productively Fréchet convergences is productively Fréchet. An analogous characterization of productively sequential convergences was presented by S. DOLECKI and F. MYNARD in [**84**].

n sets ([13]). On the other, the notion of relation is trivially generalized to classes. The idea of polar comes along with that of a relation. Rather than to adapt the idea of inverse relation in case of several variables, it is more practical to convene that if $R \subset \prod_{j\in J} X_j$ and $A \subset \prod_{j\in J_0} X_j$ for $\varnothing \neq J_0 \subset J$, then $RA \subset \prod_{j\in J\setminus J_0} X_j$, and $\psi \in RA$ if there is $\varphi \in R$ such that $\varphi|_{J_0} \in A$ and $\psi = \varphi|_{J\setminus J_0}$.

EXAMPLE XXI.9.1. If $R \subset X \times Y \times Z$, then a triple (x, y, z) is R-related whenever $(x, y, z) \in R$. In this situation several variants of images and polars, as well as of their combinations, are conceivable.

If $A \subset X$, then $RA := \bigcup_{x\in A}\{(y, z) \in Y \times Z : (x, y, z) \in R\}$, so that $RA \subset Y \times Y$ constitutes now a binary relation. Consequently, if $B \subset Y$, then

$$(RA)B = \bigcup\nolimits_{y\in B}\{z \in Z : (y, z) \in RA\}$$
$$= \bigcup\nolimits_{y\in B}\bigcup\nolimits_{x\in A}\{z \in Z : (x, y, z) \in R\} = R(A \times B) = (RB)A.$$

We can also alternate images and polars, for instance if $A \subset X$ and $B \subset Y$, then

$$(RA)^{\circ}B = \bigcap\nolimits_{y\in B}\bigcup\nolimits_{x\in A} R(x, y), \quad (R^{\circ}A)B = \bigcup\nolimits_{y\in B}\bigcap\nolimits_{x\in A} R(x, y).$$

Given two classes $\mathbb{L}$ and $\mathbb{H}$ of filters, the class $\mathbb{H}^{\mathbb{L}}$ is defined by

$$\mathcal{F} \in \mathbb{H}^{\mathbb{L}} \iff \underset{\mathcal{H}\in\mathbb{H}}{\forall}\left(\mathcal{F}\#\mathcal{H} \Longrightarrow \underset{\mathcal{L}\in\mathbb{L}}{\exists}(\mathcal{L} \geq \mathcal{F} \vee \mathcal{H})\right). \tag{XXI.9.1}$$

See the first figure of Figure XXI.9.1.

EXAMPLE XXI.9.2. If $\mathbb{H} = \mathbb{F}_0$ is the class of principal filters, and $\mathbb{L} = \mathbb{F}_1$ is the class of countably based filters, then

$$\mathcal{F} \in \mathbb{F}_0^{\mathbb{F}_1}$$

if and only if, for each principal filter $\mathcal{H}$, that is, of the form $\mathcal{H} = \{H\}^{\uparrow}$ for some non-empty set H, such that $\mathcal{F}\#\mathcal{H}$, equivalently, $H \in \mathcal{F}^{\#}$, there is a countably based filter $\mathcal{L}$ such that (XXI.9.1). We recognize a characterization of *Fréchet* filters.

EXAMPLE XXI.9.3. If $\mathbb{L} = \mathbb{F}_1 = \mathbb{H}$ is the class of countably based filters, then

$$\mathcal{F} \in \mathbb{F}_1^{\mathbb{F}_1}$$

if and only if, for each countably based filter $\mathcal{H}$ such that $\mathcal{F}\#\mathcal{H}$, there is a countably based filter $\mathcal{L}$ such that (XXI.9.1). We recognize a characterization of *strongly Fréchet* filters.

[13] Of course, one can contemplate relations with infinitely many arguments.

Let us observe that (XXI.9.1) represents a relation between three filters, say $R \subset \mathbb{F} \times \mathbb{F} \times \mathbb{F}$ is defined by $(\mathcal{F}, \mathcal{H}, \mathcal{L}) \in R$ whenever (XXI.9.1) holds. Therefore, for a given class $\mathbb{L}$ of filters, the following relation $(\mathcal{F}, \mathcal{H}) \in R\mathbb{L}$

$$\exists_{\mathcal{L} \in \mathbb{L}} \mathcal{L} \geq \mathcal{F} \vee \mathcal{H} \tag{XXI.9.2}$$

results from (XXI.9.1) as the image of the class $\mathbb{L}$ under R. Accordingly, given classes $\mathbb{H}$ and $\mathbb{L}$ of filters, $\mathcal{F} \in (R\mathbb{L})^{\circ}\mathbb{H}$ if and only if $\mathcal{F} \in \mathbb{H}^{\mathbb{L}}$, so that,

$$\mathbb{H}^{\mathbb{L}} := (R\mathbb{L})^{\circ}\mathbb{H}$$

is the $\mathbb{L}$-*polar* of $\mathbb{H}$.

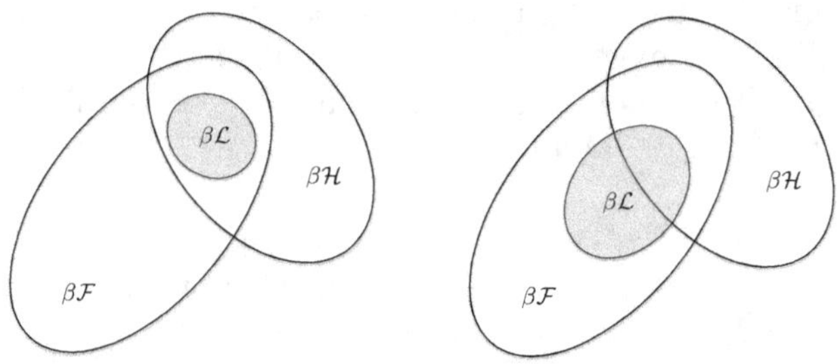

FIGURE XXI.9.1. The Stone representation of $\mathcal{F} \in \mathbb{H}^{\mathbb{L}}$ (for each $\mathcal{H} \in \mathbb{H}$ such that $\mathcal{H}\#\mathcal{F}$, there is $\mathcal{L} \in \mathbb{L}$ such that $\mathcal{L} \geq \mathcal{F}$ and $\mathcal{L}\#\mathcal{H}$) and of $\mathcal{F} \in \mathbb{H}^{\sharp\mathbb{L}}$ (for each $\mathcal{H} \in \mathbb{H}$ such that $\mathcal{H}\#\mathcal{F}$, there is $\mathcal{L} \in \mathbb{L}$ such that $\mathcal{L} \geq \mathcal{F} \vee \mathcal{H}$).

EXAMPLE XXI.9.4. Another similar characterization as in Example XXI.9.2 is that $\mathcal{F}$ is Fréchet if and only if, for each $\mathcal{H} \in \mathbb{F}_0$ such that $\mathcal{F}\#\mathcal{H}$, there exists $\mathcal{L} \in \mathbb{F}_1$ such that

$$(\mathcal{L} \geq \mathcal{F}) \wedge (\mathcal{H}\#\mathcal{L}). \tag{XXI.9.3}$$

This is a consequence of the foregoing characterization, because if $\mathcal{L} \in \mathbb{F}_1$ fulfills (XXI.9.1), then *a fortiori* $\mathcal{L} \geq \mathcal{F}$ and $\mathcal{L}\#\mathcal{H}$. The converse is also true, because if $\mathcal{L}\#\mathcal{H}$, then $\mathcal{L} \vee \mathcal{H} \in \mathbb{F}_1$ and $\mathcal{L} \vee \mathcal{H} \geq \mathcal{F} \vee \mathcal{H}$.

More generally, let

$$\mathcal{F} \in \mathbb{H}^{\sharp\mathbb{L}} \iff \forall_{\mathcal{H} \in \mathbb{H}} \left(\mathcal{F}\#\mathcal{H} \Longrightarrow \exists_{\mathcal{L} \in \mathbb{L}} (\mathcal{L} \geq \mathcal{F}) \wedge (\mathcal{L}\#\mathcal{H}) \right). \tag{XXI.9.4}$$

For a given class $\mathbb{L}$ of filters, $(\mathcal{F}, \mathcal{H}) \in S\mathbb{L}$ if and only if

$$\exists_{\mathcal{L} \in \mathbb{L}} (\mathcal{L} \geq \mathcal{F}) \wedge (\mathcal{L}\#\mathcal{H}). \tag{XXI.9.5}$$

Analogously, $\mathcal{F} \in \mathbb{F}_1^{\mathbb{F}_1}$ is strongly Fréchet if and only if, for each $\mathcal{H} \in \mathbb{F}_1$ such that $\mathcal{F}\#\mathcal{H}$, there exists $\mathcal{L} \in \mathbb{F}_1$ such that (XXI.9.2).

Therefore, if $S \subset \mathbb{F} \times \mathbb{F} \times \mathbb{F}$ is defined $(\mathcal{F}, \mathcal{H}, \mathcal{L}) \in S$ by (XXI.9.3), then $\mathcal{F} \in (S\mathbb{L})^\circ\mathbb{H}$ if and only if, for each filter $\mathcal{H} \in \mathbb{H}$ such that $\mathcal{H}\#\mathcal{F}$, there exists $\mathcal{L} \in \mathbb{L}$ so that (XXI.9.3).

As we have seen, $(S\mathbb{F}_1)^\circ\mathbb{F}_0 = (R\mathbb{F}_1)^\circ\mathbb{F}_0$ and $(S\mathbb{F}_1)^\circ\mathbb{F}_1 = (R\mathbb{F}_1)^\circ\mathbb{F}_1$. Actually, $(R\mathbb{L})^\circ\mathbb{H} \subset (S\mathbb{L})^\circ\mathbb{H}$, and

$$(R\mathbb{L})^\circ\mathbb{H} \supset (S\mathbb{L})^\circ\mathbb{H} \iff \mathbb{L} \vee \mathbb{H} \subset \mathbb{L}. \tag{XXI.9.6}$$

EXAMPLE XXI.9.5. A filter $\mathcal{F}$ is called *bisequential* if for every filter $\mathcal{H}$ with $\mathcal{F}\#\mathcal{H}$, there exists $\mathcal{L} \in \mathbb{F}_1$ such that (XXI.9.3) holds. In other terms, $(S\mathbb{F}_1)^\circ\mathbb{F}$ represents the class of bisequential filters.

To see this, let us recall that $\mathcal{F}$ is bisequential if and only if for each $\mathcal{U} \in \beta\mathcal{F}$, there exists $\mathcal{L} \in \mathbb{F}_1$ such that $\mathcal{U} \geq \mathcal{L}$. One implication is obvious. If the last condition holds, and $\mathcal{H} \in \mathbb{F}$ fulfills $\mathcal{F}\#\mathcal{H}$, then there exists $\mathcal{U} \in \beta(\mathcal{F} \vee \mathcal{H})$ and thus there is $\mathcal{L} \in \mathbb{F}_1$ with $\mathcal{U} \geq \mathcal{L}$, hence $\mathcal{L}\#\mathcal{H}$.

However, $(R\mathbb{F}_1)^\circ\mathbb{F} \neq (S\mathbb{F}_1)^\circ\mathbb{F}$ because $(R\mathbb{F}_1)^\circ\mathbb{F}$ does not contain any free filter: if $\mathcal{F}$ is free and $\mathcal{H} \in \beta\mathcal{F}$, then there is no countably-based filter finer than $\mathcal{H}$.

We use the abbreviations

$$\mathbb{H}^{\sharp\mathbb{L}} := (S\mathbb{L})^\circ\mathbb{H} \text{ and } \mathbb{H}^{\mathbb{L}} := (R\mathbb{L})^\circ\mathbb{H},$$

and call the class $\mathbb{H}^{\sharp\mathbb{L}}$ the *strict* $\mathbb{L}$*-polar of* $\mathbb{H}$*, and* the class $\mathbb{H}^{\mathbb{L}}$ is called the $\mathbb{L}$*-polar.*

Notice that

$$\mathbb{L} \subset \mathbb{H}^{\sharp\mathbb{L}} \tag{XXI.9.7}$$

for every classes $\mathbb{H}$ and $\mathbb{L}$. Indeed, if $\mathcal{F} \in \mathbb{L}$ and $\mathcal{H} \in \mathbb{H}$ is such that $\mathcal{F}\#\mathcal{H}$, then $\mathcal{L} := \mathcal{F} \geq \mathcal{F}$ and $\mathcal{L}\#\mathcal{H}$.

Notice also that $\mathbb{H}^{\mathbb{L}} \subset \mathbb{H}^{\sharp\mathbb{L}}$ and (XXI.9.6) becomes

$$\mathbb{H}^{\sharp\mathbb{L}} = \mathbb{H}^{\mathbb{L}} \iff \mathbb{L} \vee \mathbb{H} \subset \mathbb{L}. \tag{XXI.9.8}$$

In particular, $\mathbb{F}_1 \vee \mathbb{F}_1 \subset \mathbb{F}_1$, and, as $\mathbb{F}_0 \subset \mathbb{F}_1$, also $\mathbb{F}_0 \vee \mathbb{F}_1 \subset \mathbb{F}_1$. Therefore, in view of (XXI.9.8):

$$\mathbb{F}_1^{\sharp\mathbb{F}_1} = \mathbb{F}_1^{\mathbb{F}_1} \text{ and } \mathbb{F}_0^{\sharp\mathbb{F}_1} = \mathbb{F}_0^{\mathbb{F}_1}. \tag{XXI.9.9}$$

REMARK XXI.9.6. The *countably based polar* of $\mathbb{H}$ has been denoted by $\mathbb{H}^\triangle = \mathbb{H}^{\mathbb{F}_1}$ ([14]).

[14] See F. JORDAN and F. MYNARD [141], S. DOLECKI and F. MYNARD [83].

The relations $R\mathbb{L}$ and $S\mathbb{L}$ are binary. Hence, their inverse relations, $(R\mathbb{L})^-$ and $(S\mathbb{L})^-$ are well defined, as well as the corresponding inverse polars. Accordingly, $\mathbb{H} \subset \mathbb{H}^{\mathbb{L}\mathbb{L}}$ and $\mathbb{H}^{\mathbb{L}\mathbb{L}\mathbb{L}} = \mathbb{H}^{\mathbb{L}}$.

PROPOSITION XXI.9.7. $\mathbb{H}^{\sharp\mathbb{L}}$-compactness coincides with $\mathbb{L}$-compactness on (filters from) $\mathbb{H}$.

PROOF. Let ξ be a convergence. As by (XXI.9.7), $\mathbb{L} \subset \mathbb{H}^{\sharp\mathbb{L}}$, every $\mathbb{H}^{\sharp\mathbb{L}}$-compact filter is $\mathbb{L}$-compact. Conversely, assume that a filter $\mathcal{H} \in \mathbb{H}$ is $\mathbb{L}(\xi)$-compact at $\mathcal{B}$ and let $\mathcal{F} \in \mathbb{H}^{\sharp\mathbb{L}}$ be such that $\mathcal{F}\#\mathcal{H}$. By definition, there exists $\mathcal{G} \in \mathbb{L}$ such that $\mathcal{G} \geq \mathcal{F}$ and $\mathcal{G}\#\mathcal{H}$. It follows that $\operatorname{adh}_\xi \mathcal{G} \in \mathcal{B}^\#$, because $\mathcal{H}$ is $\mathbb{L}(\xi)$-compact at $\mathcal{B}$. As $\mathcal{G} \geq \mathcal{F}$, also $\operatorname{adh}_\xi \mathcal{F} \in \mathcal{B}^\#$, so that $\mathcal{H}$ is $\mathbb{H}^{\sharp\mathbb{L}}(\xi)$-compact at $\mathcal{B}$. □

XXI.10. Coreflectors for mixed properties

Proposition XVII.1.6 states that, for every functor F, the class of convergences ξ defined by the functorial inequality $\xi \geq F\xi$ is coreflective. Consequently, F is not in general the corresponding coreflector. Nevertheless, it is possible to construct such a coreflector within $\mathrm{I}X$, the set of all convergences on X, by iterating the (ascending) functor

$$F \vee \mathrm{I}$$

(see Remark XVII.1.7).

On the other hand, by Proposition XVIII.2.5, if $\mathbb{F}_0 \subset \mathbb{D} \subset \mathbb{F}$ and $\mathbb{D}$ is a final class of filters, then the class of $\mathbb{D}$-founded convergences is coreflective, and the corresponding coreflector is denoted by $\mathrm{B}_\mathbb{D}$.

In this section we shall describe coreflectors $\mathrm{B}_\mathbb{D}$ for a mixed property of the form $\{\xi : \xi \geq \mathrm{A}_\mathbb{H}\mathrm{B}_\mathbb{M}\xi\}$ by constructing a class $\mathbb{D}$ of filters in terms of classes $\mathbb{H}$ and $\mathbb{M}$, so that

$$\xi \geq \mathrm{B}_\mathbb{D}\xi \iff \xi \geq \mathrm{A}_\mathbb{H}\mathrm{B}_\mathbb{M}\xi \tag{XXI.10.1}$$

for every pretopology ξ. Namely,

THEOREM XXI.10.1. *Let $\mathbb{H}$ and $\mathbb{M}$ be classes of filters and let $\mathbb{D} := \mathbb{H}^{\sharp\mathbb{M}}$. If $\xi \geq \mathrm{B}_\mathbb{D}\xi$ then $\xi \geq \mathrm{A}_\mathbb{H}\mathrm{B}_\mathbb{M}\xi$. Conversely, if ξ is a pretopology and $\xi \geq \mathrm{A}_\mathbb{H}\mathrm{B}_\mathbb{M}\xi$ then $\xi \geq \mathrm{B}_\mathbb{D}\xi$.*

PROOF. $\xi \geq \mathrm{B}_\mathbb{D}\xi$ and $x \in \lim_\xi \mathcal{F}$ there is $\mathcal{G} \leq \mathcal{F}$ with $\mathcal{G} \in \mathbb{D}$ and $x \in \lim_\xi \mathcal{G}$. If $\mathcal{H} \in \mathbb{H}$ with $\mathcal{H}\#\mathcal{G}$ there is $\mathcal{M} \in \mathbb{M}$ with $\mathcal{M} \geq \mathcal{G}$ and $\mathcal{M}\#\mathcal{H}$ because $\mathcal{G} \in \mathbb{H}^{\sharp\mathbb{M}}$. As $x \in \lim_\xi \mathcal{M}$, we conclude that $x \in \operatorname{adh}_{\mathrm{B}_\mathbb{M}\xi} \mathcal{H}$, that is, $x \in \lim_{\mathrm{A}_\mathbb{H}\mathrm{B}_\mathbb{M}\xi} \mathcal{G} \subset \lim_{\mathrm{A}_\mathbb{H}\mathrm{B}_\mathbb{M}\xi} \mathcal{F}$.

If $\xi \geq \mathrm{A}_\mathbb{H}\mathrm{B}_\mathbb{M}\xi$ is a pretopology then $x \in \lim_{\mathrm{A}_\mathbb{H}\mathrm{B}_\mathbb{M}\xi} \mathcal{V}_\xi(x)$ and thus $\mathcal{V}_\xi(x) \in \mathbb{H}^{\sharp\mathbb{M}}$. □

Moreover, the assumption that ξ be a pretopology is essential for the converse part, as shows the following:

EXAMPLE XXI.10.2 ($\xi \geq \mathrm{S}_0\mathrm{I}_1\xi$ but $\xi \ngeq \mathrm{B}_{\mathbb{F}_0^{\mathbb{F}_1}}\xi$). Consider the sequential fan convergence π of Example III.4.7. The vicinity filter $\mathcal{V}_\pi(x_\infty)$ is a Fréchet filter. The countably-based filter $\mathcal{H}$ generated by $\left\{\bigcup_{n\geq p}\{x_{n,k} : k < \omega\} : p < \omega\right\}$ meshes with $\mathcal{V}_\pi(x_\infty)$ but $x_\infty \notin \mathrm{adh}_\pi \mathcal{H}$. Let $\mathcal{U} \in \beta(\mathcal{V}_\pi(x_\infty) \vee \mathcal{H})$. Then $x_\infty \in \lim_{\mathrm{S}_0\pi} \mathcal{U}$ and $\lim_\pi \mathcal{U} = \emptyset$. Let ξ be the convergence on X in which $\lim_\xi \mathcal{F} = \lim_\pi \mathcal{F}$ if $\mathcal{U} \perp \mathcal{F}$, $\lim_\xi \mathcal{F} = \emptyset$ if $\mathcal{F} < \mathcal{U}$ and $\lim_\xi \mathcal{U} = \{\infty\}$. Then $\xi \geq \mathrm{S}_0\mathrm{I}_1\xi$ because $\pi \geq \mathrm{S}_0\pi = \mathrm{S}_0\mathrm{I}_1\pi$ and if $A\#\mathcal{U}$ then $A\#\mathcal{V}_\pi(x_\infty)$ so that there is $\mathcal{F} \perp \mathcal{U}$ with $A \in \mathcal{F}$ and $\infty \in \lim_\pi \mathcal{F} = \lim_{\mathrm{I}_1\xi} \mathcal{F}$. However, $\mathcal{U}$ is not Fréchet and its coarser filters are non-convergent, so $\xi \ngeq \mathrm{B}_{\mathbb{F}_0^{\mathbb{F}_1}}\xi$.

Here is a list of some cases of $\mathbb{F}_1$-polars.

(1) The class $\mathbb{F}_0^{\mathbb{F}_1}$ $(= \mathbb{F}_0^{\sharp\mathbb{F}_1})$ is the class of *Fréchet filters.*
(2) The class $\mathbb{F}_1^{\mathbb{F}_1}(= \mathbb{F}_1^{\sharp\mathbb{F}_1})$ is the class of *strongly Fréchet filters* (alias *countably bisequential filters*).
(3) The class $\mathbb{F}^{\sharp\mathbb{F}_1}$ is the class of *bisequential filters.*
(4) The elements of the class

$$(\mathbb{F}_1^{\mathbb{F}_1})^{\mathbb{F}_1}$$

are called *productively Fréchet filters.* In other words, a filter $\mathcal{F}$ is productively Fréchet whenever, for each strongly Fréchet filter $\mathcal{H}$ such that $\mathcal{F}\#\mathcal{H}$, there exists $\mathcal{G} \in \mathbb{F}_1$ such that $\mathcal{G} \geq \mathcal{F} \vee \mathcal{H}$.

As a result of the difference exhibited in Example XXI.10.2, unless we are talking about pretopologies, we will distinguish between a *Fréchet-founded convergence* (i.e., $\xi \geq \mathrm{B}_{\mathbb{F}_0^{\mathbb{F}_1}}\xi$) and a *Fréchet convergence* (i.e., $\xi \geq \mathrm{S}_0\mathrm{I}_1\xi$), and we make similar distinctions between the two variants of strongly Fréchet, productively Fréchet and bisequential.

Accordingly, some mixed properties corresponding to functorial inequalities have a stronger variant as classes of convergences founded in appropriate classes of filters, which coincides for pretopologies. The class of $\mathbb{H}$-founded convergences is coreflective.

PROPOSITION XXI.10.3. *The product of a strongly Fréchet filter with a countably based filter is strongly Fréchet.*

PROOF. Let $\mathcal{F}$ be a strongly Fréchet filter on X, that is, $\mathcal{F} \in \mathbb{F}_1^{\mathbb{F}_1}X$, let $\mathcal{M}$ be a countably based filter on Y, that is, $\mathcal{M} \in \mathbb{F}_1 Y$. To prove that $\mathcal{F}\times\mathcal{M}$ is strongly Fréchet, let $\mathcal{H} \in \mathbb{F}_1(X\times Y)$ be such that $\mathcal{H}\#(\mathcal{F}\times\mathcal{M})$,

filter class $\mathbb{D}$	$\mathbb{D}$-founded	inequality version	
$\mathbb{F}_0$	principal-founded	$\xi \geq \mathrm{I}_0\xi$	finitely generated
$\mathbb{F}_1$	of countable character	$\xi \geq \mathrm{I}_1\xi$	of countable character
$\mathbb{F}^{\#\mathbb{F}_1}$	bisequential-founded	$\xi \geq \mathrm{SI}_1\xi$	bisequential
$(\mathbb{F}_1^{\mathbb{F}_1})^{\mathbb{F}_1}$	productively Fréchet-founded	$\xi \geq \mathrm{A}_{\mathbb{F}_1^{\mathbb{F}_1}}\mathrm{I}_1\xi$	productively Fréchet
$\mathbb{F}_1^{\mathbb{F}_1}$	strongly Fréchet-founded	$\xi \geq \mathrm{S}_1\mathrm{I}_1\xi$	strongly Fréchet
$\mathbb{F}_0^{\mathbb{F}_1}$	Fréchet-founded	$\xi \geq \mathrm{S}_0\mathrm{I}_1\xi$	Fréchet
		$\xi \geq \mathrm{TI}_1\xi$	sequential

TABLE XXI.10.1. Examples of mixed properties expressed in terms of their foundations. With the exception of first two, the functors in the last column are not coreflectors, although they determine coreflective classes.

equivalently $\mathcal{H}^-[\mathcal{M}]\#\mathcal{F}$. It is plain that $\mathcal{H}^-[\mathcal{M}]$ is countably based, that is, $\mathcal{H}^-[\mathcal{M}] \in \mathbb{F}_1 X$. As $\mathcal{F} \in \mathbb{F}_1^{\mathbb{F}_1} X$, by definition, there exist $\mathcal{L} \in \mathbb{F}_1$ such that $\mathcal{L} \geq \mathcal{F}$ and $\mathcal{L}\#\mathcal{H}^-[\mathcal{M}]$. Therefore, $\mathcal{H}\#(\mathcal{L}\times\mathcal{M})$ and $\mathcal{L}\times\mathcal{M} \geq \mathcal{F} \times \mathcal{M}$. Now it is enough to note that $\mathcal{L} \times \mathcal{M} \in \mathbb{F}_1$. □

LEMMA XXI.10.4. *The class $\mathbb{F}_1^{\mathbb{F}_1}$ (of strongly Fréchet filters) is $\mathbb{F}_1$-composable.*

PROOF. Let $\mathcal{F} \in \mathbb{F}_1^{\mathbb{F}_1}$ and $\mathcal{H} \in \mathbb{F}_1$. We need to show that $\mathcal{H}[\mathcal{F}] \in \mathbb{F}_1^{\mathbb{F}_1}$. Let $\mathcal{G} \in \mathbb{F}_1$ and $\mathcal{G}\#\mathcal{H}[\mathcal{F}]$, equivalently $\mathcal{F}\#\mathcal{H}^-[\mathcal{G}]$. As $\mathcal{F} \in \mathbb{F}_1^{\mathbb{F}_1}$ and $\mathcal{H}^-[\mathcal{G}] \in \mathbb{F}_1$, there exists $\mathcal{M} \in \mathbb{F}_1$ such that $\mathcal{M} \geq \mathcal{F}$ and $\mathcal{M}\#\mathcal{H}^-[\mathcal{G}]$. Therefore, $\mathcal{H}[\mathcal{M}] \geq \mathcal{H}[\mathcal{F}]$ and $\mathcal{H}[\mathcal{M}]\#\mathcal{G}$ and $\mathcal{H}[\mathcal{M}] \in \mathbb{F}_1$, so that $\mathcal{H}[\mathcal{F}] \in \mathbb{F}_1^{\mathbb{F}_1}$. □

Hence, we have an analog of Corollary XXI.8.7 for the stronger variant of the strongly Fréchet property:

COROLLARY XXI.10.5. *For each ξ and τ,*

$$\mathrm{B}_{\mathbb{F}_1^{\mathbb{F}_1}}\xi \times \mathrm{I}_1\tau \geq \mathrm{B}_{\mathbb{F}_1^{\mathbb{F}_1}}(\xi \times \tau).$$

In other words, the product of a strongly Fréchet-founded convergence with a countably based convergence is strongly Fréchet-founded.

PROOF. If $x \in \lim_{\mathrm{B}_{\mathbb{F}_1^{\mathbb{F}_1}}\xi} \mathcal{F}$ and $y \in \lim_{\mathrm{I}_1\tau} \mathcal{G}$, then there exist $\mathcal{F}_0 \in \mathbb{F}_1^{\mathbb{F}_1}$ and $\mathcal{G}_0 \in \mathbb{F}_1$ such that $x \in \lim_\xi \mathcal{F}_0$ and $\mathcal{F}_0 \leq \mathcal{F}$, and $y \in \lim_\tau \mathcal{G}_0$ and $\mathcal{G}_0 \leq \mathcal{G}$. Now, $\mathcal{F}_0 \times \mathcal{G}_0 \leq \mathcal{F} \times \mathcal{G}$, and by Proposition XXI.10.3, $\mathcal{F}_0\times\mathcal{G}_0 \in \mathbb{F}_1^{\mathbb{F}_1}$ and $(x, y) \in \lim_{\xi\times\tau}(\mathcal{F}_0\times\mathcal{G}_0)$, and thus $(x, y) \in \lim_{\mathrm{B}_{\mathbb{F}_1^{\mathbb{F}_1}}(\xi\times\tau)} (\mathcal{F} \times \mathcal{G})$. □

Note that the arguments for the latest group of results can be repeated almost verbatim for a polar $\mathbb{H}^{\sharp\mathbb{M}}$ as in the following. Note that under the assumption that $\mathbb{F}_0 \subset \mathbb{H} \subset \mathbb{M}$ and $\mathbb{M}$ is composable, Proposition XVIII.2.8 applies to the effect that $\mathbb{H} \vee \mathbb{M} \subset \mathbb{M}$, so that, in view of (XXI.9.8),

$$\mathbb{H}^{\sharp\mathbb{M}} = \mathbb{H}^{\mathbb{M}}.$$

THEOREM XXI.10.6. *Let $\mathbb{F}_0 \subset \mathbb{H} \subset \mathbb{M}$, where $\mathbb{H}, \mathbb{M}$ are two composable classes of filters.*

(1) *The class $\mathbb{H}^{\sharp\mathbb{M}}$ is $\mathbb{H}$-composable;*
(2) *The product of a filter of $\mathbb{H}^{\sharp\mathbb{M}}$ with a filter of $\mathbb{H}$ is a filter of $\mathbb{H}^{\sharp\mathbb{M}}$;*
(3) $\mathrm{B}_{\mathbb{H}^{\sharp\mathbb{M}}}\xi \times \mathrm{B}_{\mathbb{H}}\tau \geq \mathrm{B}_{\mathbb{H}^{\sharp\mathbb{M}}}(\xi \times \tau)$ *for every ξ and τ.*

PROOF. (1): Let $\mathcal{F} \in \mathbb{H}^{\sharp\mathbb{M}}(X)$ and let $\mathcal{H} \in \mathbb{H}(X \times Y)$ and let $\mathcal{G} \in \mathbb{H}(Y)$ with $\mathcal{G}\#\mathcal{H}[\mathcal{F}]$. Then $\mathcal{H}^-[\mathcal{G}]\#\mathcal{F}$ and $\mathcal{H}^-[\mathcal{G}] \in \mathbb{H}(X)$ because $\mathbb{H}$ is composable. Hence there is $\mathcal{M} \in \mathbb{M}(X)$ with $\mathcal{M} \geq \mathcal{F}$ and $\mathcal{M}\#\mathcal{H}^-[\mathcal{G}]$, so that $\mathcal{H}[\mathcal{M}]\#\mathcal{G}$ and $\mathcal{H}[\mathcal{M}] \geq \mathcal{H}[\mathcal{F}]$. Moreover, $\mathcal{H}[\mathcal{M}] \in \mathbb{M}(Y)$ because $\mathbb{M}$ is composable.

(2): Let $\mathcal{F} \in \mathbb{H}^{\sharp\mathbb{M}}(X)$, $\mathcal{G} \in \mathbb{H}(Y)$, and $\mathcal{H} \in \mathbb{H}(X \times Y)$ with $\mathcal{H}\#(\mathcal{F} \times \mathcal{G})$, equivalently, $\mathcal{H}^-[\mathcal{G}]\#\mathcal{F}$. Moreover, $\mathcal{H}^-[\mathcal{G}] \in \mathbb{H}(X)$ by composability of $\mathbb{H}$. Hence, there is $\mathcal{M} \in \mathbb{M}(X)$ with $\mathcal{M} \geq \mathcal{F}$ and $\mathcal{M}\#\mathcal{H}^-[\mathcal{G}]$. Hence $\mathcal{M}\times\mathcal{G} \geq \mathcal{F}\times\mathcal{G}$ and $(\mathcal{M}\times\mathcal{G})\#\mathcal{H}$, while $\mathcal{M}\times\mathcal{G} \in \mathbb{M}(X\times Y)$ because $\mathcal{G} \in \mathbb{H}(Y) \subset \mathbb{M}(Y)$, and $\mathbb{M}$ is composable, hence closed under finite product by Proposition XVIII.2.8. (3) follows immediately from (2). □

THEOREM XXI.10.7. *The following statements are equivalent:*

(1) $\mathcal{F} \in (\mathbb{F}_1^{\mathbb{F}_1})^{\mathbb{F}_1}$ *(productively Fréchet)*
(2) $\mathcal{F} \times \mathcal{G} \in \mathbb{F}_1^{\mathbb{F}_1}$ *for every* $\mathcal{G} \in \mathbb{F}_1^{\mathbb{F}_1}$,
(3) $\mathcal{F} \times \mathcal{G} \in \mathbb{F}_0^{\mathbb{F}_1}$ *for every* $\mathcal{G} \in \mathbb{F}_1^{\mathbb{F}_1}$.

In other words, $\mathcal{F}$ is productively Fréchet if and only if $\mathcal{F}\times\mathcal{G}$ is strongly Fréchet for every strongly Fréchet filter $\mathcal{G}$, if and only if $\mathcal{F}\times\mathcal{G}$ is Fréchet for every strongly Fréchet filter $\mathcal{G}$.

PROOF. Assume (1) and let $\mathcal{G} \in \mathbb{F}_1^{\mathbb{F}_1}$ and $\mathcal{H} \in \mathbb{F}_1$ be such that $\mathcal{H}\#(\mathcal{F} \times \mathcal{G})$, equivalently $\mathcal{H}^-[\mathcal{G}]\#\mathcal{F}$. It follows that $\mathcal{H}^-[\mathcal{G}] \in \mathbb{F}_1^{\mathbb{F}_1}$, because $\mathbb{F}_1^{\mathbb{F}_1}$ is $\mathbb{F}_1$-composable by Lemma XXI.10.4. Since $\mathcal{F} \in (\mathbb{F}_1^{\mathbb{F}_1})^{\mathbb{F}_1}$, there exists $\mathcal{L} \in \mathbb{F}_1$ such that $\mathcal{L} \geq \mathcal{H}^-[\mathcal{G}] \vee \mathcal{F}$. Consequently, $\mathcal{H}[\mathcal{L}] \in \mathbb{F}_1$ (because $\mathbb{F}_1$ is self-composable) and $\mathcal{H}[\mathcal{L}]\#\mathcal{G}$, and since $\mathcal{G} \in \mathbb{F}_1^{\mathbb{F}_1}$, there exists $\mathcal{M} \in \mathbb{F}_1$ such that $\mathcal{M} \geq \mathcal{H}[\mathcal{L}] \vee \mathcal{G}$.

Accordingly, $\mathcal{L}\times\mathcal{M}\in\mathbb{F}_1$ and $\mathcal{L}\times\mathcal{M}\geq\mathcal{F}\times\mathcal{G}$. On the other hand, $\mathcal{M}\geq\mathcal{H}[\mathcal{L}]$ and *a fortiori* $\mathcal{M}\#\mathcal{H}[\mathcal{L}]$, hence $(\mathcal{L}\times\mathcal{M})\#\mathcal{H}$, proving that $\mathcal{F}\times\mathcal{G}\in\mathbb{F}_1^{\mathbb{F}_1}$, that is (2).

That (2) implies (3) is clear.

Assume (3), let $\mathcal{F}\in\mathbb{F}(X)$ let $\mathcal{G}\in\mathbb{F}_1^{\mathbb{F}_1}(X)$ be such that $\mathcal{F}\#\mathcal{G}$. Then the diagonal subset $\triangle(X)$ on $X\times X$ meshes $\mathcal{F}\times\mathcal{G}$. In other words, $\triangle(X)\in\mathbb{F}_0(X\times X)$ and $\triangle(X)\in(\mathcal{F}\times\mathcal{G})^{\#}$. By (3), there exists $\mathcal{H}\in\mathbb{F}_1(X\times X)$ such that $\mathcal{H}\geq(\mathcal{F}\times\mathcal{G})\vee\triangle(X)$. Accordingly, the countably based filter $p_X[\mathcal{H}]$ is finer than $\mathcal{F}\vee\mathcal{G}$, proving that $\mathcal{F}\in(\mathbb{F}_1^{\mathbb{F}_1})^{\mathbb{F}_1}$. □

Analogously to Corollary XXI.10.5, we obtain the following generalization to convergences of the Jordan-Mynard Theorem [141, Theorem 9] initially proved for topologies only:

COROLLARY XXI.10.8 (Jordan-Mynard, foundation variant). *A convergence ξ is productively Fréchet-founded if and only if $\xi\times\tau$ is strongly Fréchet-founded for every strongly Fréchet-founded convergence τ.*

On the other hand, the results of this section also yield a variant of Theorem XX.7.4, which in turn will give us another variant of the Jordan-Mynard Theorem.

THEOREM XXI.10.9. *Let $\mathbb{H}$ be a composable class of filters containing $\mathbb{F}_0$. The following are equivalent:*

(1) $\theta\geq \mathrm{A}_{\mathbb{H}^{\mathbb{H}}}\,\xi$;
(2) $\theta\times \mathrm{B}_{\mathbb{H}^{\mathbb{H}}}\tau\geq \mathrm{A}_{\mathbb{H}}(\xi\times\tau)$ *for every* τ;
(3) $\theta\times\tau\geq \mathrm{S}_0(\xi\times\tau)$ *for every prime $\mathbb{H}^{\mathbb{H}}$-founded topology* τ.

PROOF. In view of Remark XX.7.5 and given that $\mathbb{H}^{\mathbb{H}}$ is $\mathbb{H}$-composable, hence $\mathbb{F}_0$-composable by Theorem XXI.10.6, we conclude that (1) and (3) are equivalent. Moreover, (2) implies (3) is clear.

To see that (1) implies (2), let $x\in\lim_\theta\mathcal{F}$, let $y\lim_\tau\mathcal{G}$ for $\mathcal{G}\in\mathbb{H}^{\mathbb{H}}$ and let $\mathcal{H}\in\mathbb{H}$ with $\mathcal{H}\#(\mathcal{F}\times\mathcal{G})$. In view of Theorem XXI.10.6, $\mathcal{H}^-[\mathcal{G}]\in\mathbb{H}^{\mathbb{H}}$. Moreover $\mathcal{H}^-[\mathcal{G}]\#\mathcal{F}$ and $x\in\lim_{\mathrm{A}_{\mathbb{H}^{\mathbb{H}}}\xi}\mathcal{F}$. Hence

$$x\in\mathrm{adh}_{\mathrm{A}_{\mathbb{H}^{\mathbb{H}}}\xi}\mathcal{H}^-[\mathcal{G}]=\mathrm{adh}_\xi\mathcal{H}^-[\mathcal{G}]$$

by (XVIII.5.4). Hence, there is $\mathcal{U}\#\mathcal{H}^-[\mathcal{G}]$ with $x\in\lim_\xi\mathcal{U}$. As $(\mathcal{U}\times\mathcal{G})\#\mathcal{H}$, we conclude that $(x,y)\in\mathrm{adh}_{\xi\times\tau}\mathcal{H}$ so that $(x,y)\in\lim_{\mathrm{A}_{\mathbb{H}}(\xi\times\tau)}(\mathcal{F}\times\mathcal{G})$. □

Taking $\mathbb{H}=\mathbb{F}_1$, $\theta=\xi$ and $\xi=\mathrm{I}_1\,\xi$ in Theorem XXI.10.9 yields:

COROLLARY XXI.10.10 (Jordan-Mynard, inequality variant). *The following are equivalent:*

(1) *ξ is productively Fréchet* ($\xi \geq \mathrm{A}_{\mathbb{F}_1^{\mathbb{F}_1}} \mathrm{I}_1 \xi$);
(2) *$\xi \times \tau$ is strongly Fréchet* ($\xi \times \tau \geq \mathrm{S}_1 \mathrm{I}_1(\xi \times \tau)$) *for every strongly Fréchet-founded convergence* τ ($\tau = \mathrm{B}_{\mathbb{F}_1^{\mathbb{F}_1}} \tau$);
(3) *$\xi \times \tau$ is Fréchet for every strongly Fréchet prime topology.*

PROOF. Theorem XXI.10.9 with $\mathbb{H} = \mathbb{F}_1$, $\theta = \xi$ and $\xi = \mathrm{I}_1 \xi$ yields

$$\xi \geq \mathrm{A}_{\mathbb{F}_1^{\mathbb{F}_1}} \mathrm{I}_1 \xi \iff \underset{\tau}{\forall}\ \xi \times \mathrm{B}_{\mathbb{F}_1^{\mathbb{F}_1}} \tau \geq \mathrm{S}_1(\mathrm{I}_1 \xi \times \tau)$$
$$\iff \underset{\tau = \mathrm{B}_{\mathbb{F}_1^{\mathbb{F}_1}} \tau}{\forall}\ \xi \times \tau \geq \mathrm{S}_0(\mathrm{I}_1 \xi \times \tau).$$

As a result, (1) implies (2) because if $\tau = \mathrm{B}_{\mathbb{F}_1^{\mathbb{F}_1}} \tau$ then $\tau \geq \mathrm{S}_1 \mathrm{I}_1 \tau$ by Theorem XXI.10.1 and therefore

$$\mathrm{I}_1 \xi \times \tau \geq \mathrm{I}_1 \xi \times \mathrm{S}_1 \mathrm{I}_1 \tau \geq \mathrm{S}_1 (\mathrm{I}_1 \xi \times \mathrm{I}_1 \tau),$$

by applying Theorem XX.7.4 with $\theta = \tau \geq \mathrm{S}_1(\mathrm{I}_1 \tau)$, $\xi = \mathrm{I}_1 \tau$ and $\tau = \mathrm{I}_1 \xi$. Thus,

$$\xi \times \tau \geq \mathrm{S}_1(\mathrm{I}_1 \xi \times \tau) \geq \mathrm{S}_1(\mathrm{S}_1 (\mathrm{I}_1 \xi \times \mathrm{I}_1 \tau)) = \mathrm{S}_1 \mathrm{I}_1(\xi \times \tau).$$

(2) implies (3) is clear and (3) implies (1) because if $\xi \times \tau \geq \mathrm{S}_0 \mathrm{I}_1(\xi \times \tau)$ for every $\tau = \mathrm{B}_{\mathbb{F}_1^{\mathbb{F}_1}} \tau$ then also $\xi \times \tau \geq \mathrm{S}_0(\mathrm{I}_1 \xi \times \tau)$ and we conclude from (3) implies (1) in Theorem XXI.10.9. □

XXI.11. Supplement

The α_4 property is the weakest diagonal property of convergent sequential fans that is contemplated. Several stronger diagonal properties have been studied and applied to productivity quests, namely $\alpha_3, \alpha_2, \alpha_{1\frac{1}{2}}$, and α_1 ([15]).

If for every countable collection of (disjoint) sequences $(x_{n,k})_k$ ξ-converging to x, there is

$$D \subset \{x_{n,k} : n, k < \omega\}$$

with $(D)_0$ ξ-converging to x then ξ is called:

- α_4 at x if $\{n : D \cap \{x_{n,k} : k < \omega\} \neq \varnothing\}$ is infinite;
- α_3 at x if $\{n : D \# (x_{n,k})_k\}$ is infinite;
- α_2 at x if $\{n : D \# (x_{n,k})_k\}$ is cofinite;
- $\alpha_{1\frac{1}{2}}$ at x if $\{n : (D)_0 \leq (x_{n,k})_k\}$ is infinite;
- α_1 at x if $\{n : (D)_0 \leq (x_{n,k})_k\}$ is cofinite.

[15]They were introduced by A. V. ARHANGEL'SKII in [13], and $\alpha_{1\frac{1}{2}}$ by NYIKOS in [212]; similar properties α, β, γ, and δ were studied by Novák [211].

PROPOSITION XXI.11.1. *If ξ is a Fréchet hereditarily locally countably compact topology, and τ is an α_3 Fréchet topology, then $\xi \times \tau$ is Fréchet.*

PROOF. Let ξ be a topology on X, and let τ be a topology on Y fulfilling the assumptions. Let $A \subset X \times Y$ and $(x, y) \in \mathrm{adh}_{\xi\times\tau} A$. Then $y \in \mathrm{adh}_\tau AV$ for each countably compact $V \in \mathcal{N}_\xi(x)$. As τ is Fréchet, for every $V \in \mathcal{V}_\xi(x)$, there exists a sequence ψ_V on AV such that $y \in \lim_\tau \psi_V$. Let φ_V be a sequence on V such that $(\varphi_V(n), \psi_V(n)) \in A$ for each $n < \omega$. Then there exist a map $\gamma_V \in \mathbb{N}^{\mathbb{N}}$ tending to infinity and $\varphi_V^0 = \varphi_V \circ \gamma_V$ and $x_V \in V$ such that $x_V \in \lim_\xi \varphi_V^0$. Consequently, $x \in \mathrm{cl}_\xi\{x_V : V \in \mathcal{N}_\xi(x)\}$ and since ξ is Fréchet, there exists a sequence $(V_n)_n$ such that $x \in \lim_\xi (x_{V_n})_n$.

On the other hand, let $\psi_{V_n}^0 := \psi_{V_n} \circ \gamma_{V_n}$ for each $n < \omega$. Of course, $y \in \lim_\tau \psi_{V_n}^0$ for each $n < \omega$, and since τ is an α_3 Fréchet topology, there exists a sequence $(\gamma, \delta) : \mathbb{N} \longrightarrow \mathbb{N} \times \mathbb{N}$ such that if $\psi(k) := \psi_{V_{\gamma(k)}}^0(\delta(k))$ for each $k < \omega$, then ψ is frequent-frequent and $y \in \lim_\tau \psi$.

Let $\varphi(k) := \varphi_{V_{\gamma(k)}}^0(\delta(k))$ for each $k < \omega$. Then $(\varphi(k), \psi(k)) \in A$ for each $k < \omega$. By topologicity, $x \in \mathrm{cl}_\xi\{\varphi(k) : k < \omega\}$, and since ξ is Fréchet, there exists $\alpha \in \mathbb{N}^{\mathbb{N}}$ tending to infinity such that $x \in \lim_\xi \varphi \circ \alpha$. On the other hand, $y \in \lim_\tau \psi \circ \alpha$, and the proof is complete. □

By Lemma XXI.8.3,

COROLLARY XXI.11.2. *Each Fréchet, hereditarily locally countably compact topology is strongly Fréchet.*

REMARK XXI.11.3. We say that a filter $\mathcal{F}$ is *AB-diagonal*, where $A, B \in \{\#, \geq\}$, if for every sequence $\boldsymbol{S} := \{\mathcal{S}_n : n < \omega\}$ of cofinite filters such that the contour $\int_{\mathcal{N}} \boldsymbol{S} \geq \mathcal{F}$, where $\mathcal{N}$ stands for the cofinite filter on $\{n \in N : n < \omega\}$, there exists a cofinite filter $\mathcal{D} \geq \mathcal{F}$ that is (A, B) with respect to $\boldsymbol{S}$.

Diagonal properties of these types were called $\beta_3, \beta_2, \beta_{1\frac{1}{2}}$, and β_1 by analogy the properties from the series of transversal properties $\alpha_3, \alpha_2, \alpha_{1\frac{1}{2}}$, and α_1, used in the case of fans ([16]).

There are two symmetric sets of sufficient conditions for the Fréchetness of the product of two Fréchet convergences, either one is transversal and the other frequent-frequent diagonal, or one is diagonal and the other frequent-frequent transversal ([17]).

COROLLARY XXI.11.4. *If ξ is a countably core-compact, Fréchet and β_3, and τ is strongly Fréchet, then $\xi \times \tau$ is Fréchet.*

[16] S. DOLECKI and T. NOGURA [85].

[17] There is also a supplementary compactness assumption.

PROOF. At the large stage of the proof, use ## for the bisequence, and then strong Fréchetness for the corresponding subfan. □

It follows from Lemma XXI.8.3 that each Fréchet, β_3 and countably core-compact convergence is strongly Fréchet.

CHAPTER XXII

Implementations and refinements

XXII.1. Active boundary

Lemma XX.6.8 asserts that, for any convergence ξ, a ξ-compactoid filter $\mathcal{F}$ fulfills $\mathcal{F} \geq \mathcal{N}_\xi(\mathrm{adh}_\xi \mathcal{F})$. Corollary XX.6.9 points out, that this property is in fact finite compactness of $\mathcal{F}$ at $\mathrm{adh}_\xi \mathcal{F}$ with respect to the topologization of ξ. More precisely, recall the following:

LEMMA XXII.1.1. *Let ξ be a convergence, $A \subset |\xi|$ and $\mathcal{F} \in \mathbb{F}|\xi|$. Then $\mathcal{F}$ is $\mathbb{F}_0$-compact at A if and only if $\mathcal{F} \geq \mathcal{V}_\xi(A)$.*

PROOF. If $\mathcal{F}$ is $\mathbb{F}_0$-compact at A and $V \in \mathcal{V}_\xi(A)$ with $V \notin \mathcal{F}$ then $V^c \# \mathcal{F}$ by (V.3.4) so that $\mathrm{adh}_\xi V^c \cap A \neq \emptyset$ by $\mathbb{F}_0$-compactness of $\mathcal{F}$ at A, in contradiction with $A \subset \mathrm{inh}_\xi V = (\mathrm{adh}_\xi V^c)^c$.

Conversely, if $\mathcal{F} \geq \mathcal{V}_\xi(A)$ and $H \# \mathcal{F}$ then $H \# \mathcal{V}_\xi(A)$ and $\mathcal{V}_\xi(A) = \bigcap_{x \in A} \bigcap_{\mathcal{G} \in \xi^-(x)} \mathcal{G}$ so that, in view of (V.3.9), there is $x \in A$ and $\mathcal{G} \in \xi^-(x)$ with $H \in \mathcal{G}^\#$, that is, $\mathrm{adh}_\xi H \cap A \neq \varnothing$. □

If ξ is a topology $\mathcal{V}_\xi(A) = \mathcal{N}_\xi(A)$ and the condition $\mathcal{F} \geq \mathcal{N}_\xi(A)$ describes the *upper Vietoris topology* on the hyperspace of subsets of a topological space ([1]) in terms of the reduced filter.

Let us suppose that $\mathcal{F}$ is out of A, that is,

(XXII.1.1) $$A^c \in \mathcal{F} \text{ and } \mathcal{F} \geq \mathcal{V}_\xi(A).$$

The *active boundary* of A with respect to $\mathcal{F}$ is defined by

$$\partial_\xi^{\mathcal{F}} A := \mathrm{adh}_\xi(\mathcal{F} \vee A^c).$$

LEMMA XXII.1.2. *If ξ is a free (alias T_1) convergence and (XXII.1.1) holds, then $\mathrm{adh}_\xi(x_n)_n \cap A \neq \varnothing$ for each sequence $(x_n)_n \geq \mathcal{F}$.*

PROOF. Suppose that, on the contrary, there exists a sequence $(x_n)_n \geq \mathcal{F}$ such that $\mathrm{adh}(x_n)_n \cap A = \varnothing$. As $A^c \in \mathcal{F}$, we may assume $\{x_n : n < \omega\} \cap A = \varnothing$. By Lemma VIII.4.1,

$$\mathrm{adh}\{x_n : n < \omega\} = \{x_n : n < \omega\} \cup \mathrm{adh}(x_n)_n$$

is disjoint from A, hence $\{x_n : n < \omega\} \notin \mathcal{F}^\#$ by $\mathbb{F}_0$-compactness of $\mathcal{F}$ at A, which yields a contradiction. □

[1]See [183], [89], [78], [170], [171], and [57].

PROPOSITION XXII.1.3. *If ξ is a free convergence and $\mathcal{F}$ is a strongly Fréchet filter such that (XXII.1.1) holds, then $\mathcal{F}$ is countably compact at A.*

PROOF. Let $\mathcal{H}$ be a countably based filter such that $\mathcal{H}\#\mathcal{F}$. As $\mathcal{F}$ is strongly Fréchet, there exists a sequential filter $\mathcal{E} \geq \mathcal{H} \vee \mathcal{F}$, and thus, by Lemma XXII.1.2, $\varnothing \neq A \cap \operatorname{adh}_\xi \mathcal{E} \subset A \cap \operatorname{adh}_\xi \mathcal{H}$, completing the proof. □

Note however that, in contrast to Corollary XX.6.7 and Proposition XX.6.10, the adherence of a countably compact filter does not need to be countably compact, but it is functionally compactoid.

To see this, let us first define a filter $\mathcal{F}$ on (X, ξ) *to be functional* if there is a continuous real-valued function $f \in C(\xi, \nu)$ such that $\{\{|f| \geq n\} : n \in \mathbb{N}\}^\uparrow = \mathcal{F}$. Note that a functional filter has empty ξ-adherence ([2]) and that on a functionally compact (alias pseudocompact) convergence, there is no non-degenerate functional filter.

Hence, a filter is $\mathbb{D}$-compactoid for the class $\mathbb{D}$ of functional filters if and only if it does not mesh with any functional filter, that is, for every $f \in C(\xi, \nu)$ there is an element of the filter on which f is bounded. Such filters are called *functionally compactoid.* As the class of functional filters is a subclass of $\mathbb{F}_1$, every countably compactoid filter is functionally compactoid. □

LEMMA XXII.1.4. *If $\mathcal{F}$ is a filter that is functionally compactoid, then* $\operatorname{adh}_\xi \mathcal{F}$ *is a (possibly empty) functionally compactoid set.*

PROOF. Suppose $\mathcal{F}$ is functionally compactoid and let $f \in C(\xi, \nu)$. There is $F \in \mathcal{F}$ on which f is bounded. Then f is also bounded on $\operatorname{adh}_\xi F$ ([3]) hence on its subset $\operatorname{adh}_\xi \mathcal{F}$, which is therefore functionally compactoid. □

PROPOSITION XXII.1.5. *Let* $\mathrm{Seq}\xi$ *be a Hausdorff convergence. If a strongly Fréchet filter $\mathcal{F}$ fulfills (XXII.1.1), then* $\operatorname{adh}_{\mathrm{Seq}\xi} \mathcal{F} \subset A$.

PROOF. If $x \in \operatorname{adh}_{\mathrm{Seq}\xi} \mathcal{F}$, then there exists a sequential filter $\mathcal{E}\#\mathcal{F}$ such that $x \in \lim_\xi \mathcal{E}$. As $\mathcal{F}$ is strongly Fréchet, there exists a sequential filter $\mathcal{G} \in \mathbb{S}(\mathcal{F} \vee \mathcal{E})$, hence $x \in \lim_\xi \mathcal{G}$. By Lemma XXII.1.2, $\operatorname{adh}_\xi \mathcal{G} \cap A \neq \varnothing$. Therefore, by Proposition VIII.1.3 on coincidence of non-empty limit with adherence in Hausdorff spaces, $\varnothing \neq A \cap \operatorname{adh}_\xi \mathcal{G} = A \cap \lim_\xi \mathcal{G} \subset \{x\}$. □

[2]Indeed, if $\mathcal{U} \in \beta(\{\{|f| \geq n\} : n \in \mathbb{N}\})$ with $\lim_\xi \mathcal{U} \neq \varnothing$ then by continuity $\lim_\nu f[\mathcal{U}] \neq \varnothing$ in contradiction to $f[\mathcal{U}] \in \beta(\{\{x : |x| \geq n\} : n \in \mathbb{N}\})$.

[3]Indeed, if $f(F) \subset]-n, n[$ then by continuity $f(\operatorname{adh}_\xi F) \subset \operatorname{adh}_\nu f(F) \subset [-n, n]$

We are looking for additional conditions under which (XXII.1.1) implies $\mathrm{adh}_\xi \mathcal{F} \subset A$. Recall that if $\mathcal{F}$ is a free filter on a set Y and $\infty \in Y$, then $\pi[\infty, \mathcal{F}]$ denotes the prime topology determined by $\mathcal{F}$ and ∞.

PROPOSITION XXII.1.6. *Let ξ be a convergence on X and let $\mathcal{F}$ be a free strongly Fréchet filter on X and let $\pi := \pi[\infty, \mathcal{F}]$. Then the following two statements are equivalent:*

(1) $\mathrm{adh}_\xi R^-[\mathcal{F}] \subset \mathrm{adh}_{\mathrm{Seq}\xi} R^-[\mathcal{F}]$ *for every* $R \subset (X, \xi) \times (X, \pi)$*;*
(2) $\xi \times \pi$ *is Fréchet.*

PROOF. Let $\pi = \pi[\infty, \mathcal{F}]$ be the prime topology on X. Assume (1) and let $R \subset X \times X$ such that $(x, \infty) \in \mathrm{adh}_{\xi\times\pi} R$. Then there exists a filter $\mathcal{G}$ such that $x \in \lim_\xi \mathcal{G}$ and $R \in (\mathcal{G} \times \mathcal{F})^\#$, equivalently $\mathcal{G} \# R^-[\mathcal{F}]$, that is $x \in \mathrm{adh}_\xi R^-[\mathcal{F}]$. By assumption, $x \in \mathrm{adh}_{\mathrm{Seq}\xi} R^-[\mathcal{F}]$, and thus there exists $\mathcal{E} \in \mathbb{S}(R^-[\mathcal{F}])$ such that $x \in \lim_\xi \mathcal{E}$. Hence $R[\mathcal{E}] \# \mathcal{F}$, and since $\mathcal{F}$ is strongly Fréchet, there exists $\mathcal{D} \in \mathbb{S}(R[\mathcal{E}] \vee \mathcal{F})$. Consequently, $\infty \in \lim_\pi \mathcal{D}$, $R \in (\mathcal{E} \times \mathcal{D})^\#$, and $(x, \infty) \in \lim_{\xi\times\pi}(\mathcal{E} \times \mathcal{D})$. On the other hand, $\mathcal{E} \times \mathcal{D} \in \mathbb{F}_1$, showing that $\xi \times \pi$ is Fréchet.

Assume (2) and let $R \subset X \times X$ If $x \in \mathrm{adh}_\xi R^-[\mathcal{F}]$, then there exists a filter $\mathcal{G}$ such that $\mathcal{G} \# R^-[\mathcal{F}]$, equivalently, $R[\mathcal{G}] \# \mathcal{F}$, and $x \in \lim_\xi \mathcal{G}$. As $(x, \infty) \in \lim_{\xi\times\pi} \mathcal{G} \times \mathcal{F}$, $(x, \infty) \in \mathrm{adh}_{\xi\times\pi} R$. By assumption, there exists $\mathcal{E} \in \mathbb{S}(R)$ such that $(x, \infty) \in \lim_{\xi\times\pi} \mathcal{E}$, thus $\infty \in \lim_\pi p_2[\mathcal{E}]$ so that $p_2[\mathcal{E}] \geq \mathcal{F}$, and $x \in \lim_\xi p_1[\mathcal{E}]$ where p_1 and p_2 are the projections on the first and second factors respectively. If $\mathcal{E} = (e_n)_n$, the condition $p_2[\mathcal{E}] \geq \mathcal{F}$ with $\mathcal{E} \in \mathbb{S}(R)$ implies that for each $F \in \mathcal{F}$ there is n with $\{p_2(e_k) : k \geq n\} \subset F$, and $(p_1(e_k), p_2(e_k)) \in R$ for each k. Hence $p_1[\mathcal{E}] \geq R^-[\mathcal{F}]$. Of course, $p_1[\mathcal{E}] \in \mathbb{S}$ and thus $x \in \mathrm{adh}_{\mathrm{Seq}\xi} R^-[\mathcal{F}]$. □

Consequently, by Theorem XXI.10.7,

COROLLARY XXII.1.7. *If ξ is a Hausdorff productively Fréchet convergence, $A \subset |\xi|$ and $\mathcal{F}$ is a strongly Fréchet filter such that (XXII.1.1) holds, then $\mathrm{adh}_\xi \mathcal{F}$ is a non-empty ξ-functionally compactoid subset of A.*

PROOF. By Proposition XXII.1.3, $\mathcal{F}$ is countably compact, hence functionally compact and thus, in view of Lemma XXII.1.4, $\mathrm{adh}_\xi \mathcal{F}$ is ξ-functionally compactoid. It is non-empty because $\varnothing \neq \mathrm{adh}_\xi \mathcal{E} \subset \mathrm{adh}_\xi \mathcal{F}$ for every $\mathcal{E} \in \mathbb{S}(\mathcal{F}) \neq \varnothing$. By Proposition XXII.1.5, $\mathrm{adh}_{\mathrm{Seq}\xi} \mathcal{F} \subset A$. Since ξ is productively Fréchet and $\pi = \pi[\infty, \mathcal{F}]$ is strongly Fréchet, $\xi \times \pi$ is Fréchet. In view of Proposition XXII.1.6 using $B = \{(x, x) : x \in X\}$, we conclude that $\mathrm{adh}_\xi \mathcal{F} \subset A$. □

Or, the other way around

COROLLARY XXII.1.8. *If ξ is a strongly Fréchet Hausdorff convergence and $\mathcal{F}$ is a productively Fréchet filter such that (XXII.1.1) holds. Then* $\operatorname{adh}_\xi \mathcal{F}$ *is a non-empty ξ-functionally compactoid subset of A.*

Because functionally compactoid (alias pseudocompact) subsets and countably compact subsets coincide in normal spaces, and countably compact subsets of a metrizable space are compact, functionally compactoid subsets of a metrizable space are compact.

Notice that A has no prescribed convergence-theoretic properties; in particular, it need not be closed. As a consequence we recover a result on upper semicontinuity of G. CHOQUET [44], which translated into our terminology becomes:

PROPOSITION XXII.1.9. *If $\mathcal{F}$ is a countably based filter on a metrizable space ξ and A is an arbitrary set, then the active boundary $\partial_\xi^{\mathcal{F}} A$ is compact and $\partial_\xi^{\mathcal{F}} A \subset A$.*

REMARK XXII.1.10. In view of our results, $\mathcal{F}$ only need be strongly Fréchet in Proposition XXII.1.9.

Let $f : X \longrightarrow Y$ and consider a topology ξ on X and a topology τ on Y. Recall that f is called *closed* if $f(C)$ is τ-closed for every ξ-closed subset C of X. As a special case of Propositions IX.4.8 and XX.11.7,

PROPOSITION XXII.1.11. *A map f between topological spaces is closed if and only if the relation f^- is upper semicontinuous, if and only if f^- is an $\mathbb{F}_0$-compact relation.*

It follows that [4]

LEMMA XXII.1.12 (Vainštein). *If ξ be a metrizable topology and τ be a topology of countable character, and $f \in \mathscr{C}(\xi, \tau)$ is a surjective closed map, then the boundary $\partial f^-(y)$ is ξ-compact for each $y \in |\tau|$.*

In fact, we can show the following generalization:

PROPOSITION XXII.1.13. *If ξ be a metrizable topology and τ be a strongly Fréchet topology, and $f \in \mathscr{C}(\xi, \tau)$ is a surjective closed map, then the boundary $\partial f^-(y)$ is ξ-compact for each $y \in |\tau|$.*

PROOF. For a given y in the domain, the filter generated by $\mathcal{F} := f^-[\mathcal{N}_\tau(y)] \setminus f^-(y)$ is strongly Fréchet. As ξ is in particular strongly Fréchet Hausdorff, we can use either Corollary XXII.1.7 or Corollary XXII.1.8 to the effect that $\operatorname{adh}_\xi \mathcal{F}$ is functionally compactoid and $\operatorname{adh}_\xi \mathcal{F} \subset \partial f^-(y)$. As ξ is metrizable, $\operatorname{adh}_\xi \mathcal{F}$ is ξ-compact. In

[4] I. A. VAINŠTEIN [253], see also [97, 4.4.16].

fact, the converse inclusion above also holds. Indeed, as $f \in \mathscr{C}(\xi, \tau)$, if $x \in \partial f^-(y)$ and $V \in \mathcal{N}_\tau(y)$, then there exists $W \in \mathcal{N}_\xi(x)$ such that $W \subset f^-(V)$. Therefore, $(W \setminus f^-(y)) \# f^-(V)$ for each $W \in \mathcal{N}_\xi(x)$ and $V \in \mathcal{N}_\tau(y)$, that is, $x \in \operatorname{adh}_\xi \mathcal{F}$. □

It follows that the relation ∂f^- is compact-valued and upper semicontinuous. Therefore, the map f restricted to the image of ∂f^- is a perfect map ([5]). As a result, we recover a result of E. MICHAEL [185] to the effect that a closed continuous image of a metric space (often called a *Lašnev space*) is metrizable if and only if it is strongly Fréchet ([6]):

THEOREM XXII.1.14 (Michael). *If ξ is metrizable, $f \in \mathscr{C}(\xi, \tau)$ is surjective and closed, and τ is strongly Fréchet, then τ is metrizable.*

PROOF. By Vainštein Lemma XXII.1.12, the map f restricted to $\bigcup_{y \in Y} \partial f^-(y)$ is perfect, hence by Theorem XI.5.6, τ is metrizable. □

XXII.2. *Topological defect and sequential order

We have already considered convergences ξ, for which certain reflections coincide, but the convergence is equal to none of them. For instance, $\xi > \mathrm{S}\xi = \mathrm{T}\xi$, or $\xi > \mathrm{S}_0\xi = \mathrm{T}\xi$. In the second example, the condition amounts to $\operatorname{adh}_\xi = \operatorname{cl}_\xi$.

Here we shall investigate in more detail relations between adh_ξ and cl_ξ. As we well know, $\operatorname{adh}_\xi H \subset \operatorname{cl}_\xi H$ for every subset H of $|\xi|$. They coincide if the adherence is idempotent: $\operatorname{adh}_\xi(\operatorname{adh}_\xi H) = \operatorname{adh}_\xi H$ for each set H, but in general, $\operatorname{adh}_\xi(\operatorname{adh}_\xi H) \subset \operatorname{cl}_\xi H$ does not entail the equality.

Given two pretopologies σ, τ on X, let us define a *composition of pretopologies* $\sigma\tau$ on X by

$$\mathcal{V}_{\sigma\tau}(x) := \textstyle\int_{\mathcal{V}_\tau(x)} \mathcal{V}_\sigma(\cdot).$$

PROPOSITION XXII.2.1. *For each A,*

$$\operatorname{adh}_{\sigma\tau} A = \operatorname{adh}_\tau(\operatorname{adh}_\sigma A), \tag{XXII.2.1}$$

equivalently, $\operatorname{inh}_{\sigma\tau} A = \operatorname{inh}_\tau(\operatorname{inh}_\sigma A)$.

---

[5]Such maps are called *inductively perfect*.

[6]This is a particular case of a more general result [185, Corollary 9.10]. This particular case in turn is a generalization of a classical metrization theorem of Hanai-Morita-Stone for the case of a codomain of countable character: See K. MORITA and S. HANAI [190], A. H. STONE [245].

PROOF. Indeed, by definition, $x \in \operatorname{inh}_{\sigma\tau} A$ whenever $A \in \mathcal{V}_{\sigma\tau}(x)$, that is, there exists $V \in \mathcal{V}_\tau(x)$ such that $A \in \mathcal{V}_\sigma(v)$ for each $v \in V$. Equivalently, there is V such that $x \in \operatorname{inh}_\tau V$ and $v \in \operatorname{inh}_\sigma A$ for each $v \in V$. This means that $x \in \operatorname{inh}_\tau(\operatorname{inh}_\sigma A)$. □

We notice that $\pi(\sigma\tau) = (\pi\sigma)\tau$, but, in general, $\sigma\tau \neq \tau\sigma$.

EXAMPLE XXII.2.2. Let ν be the standard topology on $I = [0, 1]$ and by ρ the following pretopology: $\mathcal{V}_\rho(x) := \mathcal{V}_\iota(x)$ if $x \in \mathbb{Q}$ and $\mathcal{V}_\rho(x) := \mathcal{V}_o(x)$ otherwise. Then $\rho\nu = o$, because $W \in \mathcal{V}_{\rho\nu}(x)$ if there is $V \in \mathcal{V}_\nu(x)$ such that $W \in \mathcal{V}_\rho(v)$ for each $v \in V$. There exists $z \in V \setminus \mathbb{Q}$, hence if $W \in \mathcal{V}_\rho(z)$ then $W = I$. On the other hand, $U \in \mathcal{V}_{\nu\rho}(x)$ and $x \in \mathbb{Q}$ imply that $U \in \mathcal{V}_\nu(x)$; if $x \in I \setminus \mathbb{Q}$ then $U = I$.

EXERCISE XXII.2.3. Check that $o\sigma = o = \sigma o$, and $\sigma\iota = \sigma = \iota\sigma$. Check that the product preserves order in each variable. Prove that $\sigma\tau \leq \sigma \wedge \tau$.

EXAMPLE XXII.2.4 (Some pretopologies of the plane). We shall consider several simple pretopologies on $\mathbb{R}^2$. Let $\nu = \nu_{\mathbb{R}}$ be the standard topology of $\mathbb{R}$, and $\iota = \iota_{\mathbb{R}}$ be the discrete topology on $\mathbb{R}$. Then $\nu \times \nu$ is the standard topology of $\mathbb{R}^2$, $\nu \times \iota$ is the *horizontal topology* of $\mathbb{R}^2$, and $\iota \times \nu$ is the *vertical topology* of $\mathbb{R}^2$. The infimum $\xi := (\nu \times \iota) \wedge (\iota \times \nu)$ is a pretopology, called the *Féron cross* ([7]). A set O is ξ-open if and only if for each $(x, y) \in \mathbb{R}^2$ there exists $\varepsilon > 0$ such that

$$C^\varepsilon_{(x,y)} := (\{x\} \times [y - \varepsilon, y + \varepsilon]) \cup ([x - \varepsilon, x + \varepsilon] \times \{y\}) \subset O.$$

Of course, $C^\varepsilon_{(x,y)} \in \mathcal{V}_\xi(x, y)$ and $\operatorname{inh}_\xi C^\varepsilon_{(x,y)} = \varnothing$, so that ξ is not a topology.

The compositions $\xi_h := (\iota \times \nu)(\nu \times \iota)$ and $\xi_v := (\nu \times \iota)(\iota \times \nu)$ are two distinct non-topological pretopologies. Indeed, $V \in \mathcal{V}_{\xi_h}(0, 0)$ whenever there is $\varepsilon > 0$ such that for each $x \in [-\varepsilon, \varepsilon]$, there exists $\delta_x > 0$ such that $\{x\} \times [-\delta_x, \delta_x] \subset V$, while $W \in \mathcal{V}_{\xi_v}(0, 0)$ whenever there is $\delta > 0$ such that for each $y \in [-\delta, \delta]$, there exists $\varepsilon_y > 0$ such that $[-\varepsilon_y, \varepsilon_y] \subset W$. Of course, $\xi^2 \leq \xi_h \wedge \xi_v \leq \xi$, hence $\mathrm{T}\xi = \mathrm{T}\xi_h = \mathrm{T}\xi_v$.

Let ξ be a pretopology. We denote $\xi^0 := \iota$, and $\xi^{n+1} := \xi\xi^n$ for each $n < \omega$. Accordingly, $\xi^n \geq \xi^{n+1}$ for each $n < \omega$. In general, an infimum of pretopologies is not a pretopology, so that

$$\bigwedge\nolimits_{\mathrm{S}_0} \{\xi^n : n < \omega\} = \mathrm{S}_0 \bigwedge \{\xi^n : n < \omega\},$$

that is,

$$\operatorname{adh}_{\bigwedge_{n<\omega} \xi^n} A = \operatorname{adh}_\xi\Big(\bigcup\nolimits_{n<\omega} \operatorname{adh}_{\xi^n} A\Big).$$

[7] See R. FÉRON [100]. The terminology is that of [67].

We will show that $\mathrm{adh}_{\bigwedge_{n<\omega}\xi^n}$ need not to be idempotent. Therefore the iteration can be continued.

More generally, we set $\xi^0 := \iota$, and for any ordinal number $\alpha > 0$, we define the α-th iteration of a pretopology ξ by

$$\text{(XXII.2.2)} \qquad \mathrm{adh}_\xi^\alpha A = \mathrm{adh}_{\xi^\alpha} A := \mathrm{adh}_\xi(\bigcup\nolimits_{\gamma<\alpha} \mathrm{adh}_{\xi^\gamma} A).$$

In particular, $\xi^1 = \xi$. It is plain that

PROPOSITION XXII.2.5. *A pretopology ξ is a topology if and only if $\xi^2 = \xi$.*

PROOF. By (XXII.2.1), $\xi^2 = \xi$ if and only if adh_ξ is idempotent. □

LEMMA XXII.2.6. *If ξ and τ are free pretopologies, then $\xi\tau$ is free.*

By Theorem XXVII.4.6 of Zermelo, every set can be well-ordered. If ξ is a pretopology on a set X, then the cardinality of the family 2^X of subsets of X is $2^{\mathrm{card}\,X}$.

Therefore, there exists a least ordinal ζ such that $2^{\mathrm{card}\,X} < \mathrm{card}\,\zeta$, and thus the transfinite non-decreasing sequence $\{\mathrm{adh}_\xi^\alpha A : \alpha < \zeta\}$ must have a greatest element for each $A \subset X$. Consequently,

PROPOSITION XXII.2.7. *For every pretopology ξ there exists a least ordinal $\mathrm{t}(\xi)$ such that $\xi^{\mathrm{t}(\xi)+1} = \xi^{\mathrm{t}(\xi)}$.*

We call $\mathrm{t}(\xi)$ the *topological defect* of ξ. It can be characterized by

$$\mathrm{adh}_\xi(\mathrm{adh}_{\xi^{\mathrm{t}(\xi)}} A) \subset \mathrm{adh}_{\xi^{\mathrm{t}(\xi)}} A$$

for every A. If there exists x and A such that $x \in \mathrm{adh}_\xi^{\mathrm{t}(\xi)} A \setminus \bigcup_{\gamma<\mathrm{t}(\xi)} \mathrm{adh}_\xi^\gamma A$, then we say that the topological defect is *attained*.

PROPOSITION XXII.2.8. *If ξ be a pretopology, then $\mathrm{T}\xi = \xi^{\mathrm{t}(\xi)}$.*

PROOF. As $\xi \geq \mathrm{T}\xi$, and $\xi(\mathrm{T}\xi) \geq (\mathrm{T}\xi)^2 = \mathrm{T}\xi$, it follows that $\xi^\gamma \geq \mathrm{T}\xi$ for every ordinal γ. Therefore,

$$\xi^{\mathrm{t}(\xi)}\xi^{\mathrm{t}(\xi)} = \xi^{\mathrm{t}(\xi)} \geq \mathrm{T}\xi.$$

Hence $\xi^{\mathrm{t}(\xi)}$ is a pretopology with idempotent adherence, that is, a topology coarser than ξ. Since $\mathrm{T}\xi$ is the finest topology coarser than ξ, it follows that $\mathrm{T}\xi = \xi^{\mathrm{t}(\xi)}$. □

PROPOSITION XXII.2.9. *For each ordinal α, there exists a pretopology ξ such that $\mathrm{t}(\xi) = \alpha$ and the topological defect is attained.*

PROOF. We proceed by transfinite induction. The claim is true for $\alpha = 0$ (a discrete topology) and $\alpha \leq 1$ (a topology). Let $\alpha > 1$, and suppose that for each $\gamma < \alpha$, there exists a pretopology ξ_γ on a set X_γ, a subset $A_\gamma \subset X_\gamma$, and $x_\gamma \in X_\gamma$ such that $\mathrm{t}(\xi_\gamma) = \gamma$, and

$$x_\gamma \in \mathrm{adh}^{\gamma}_{\xi_\gamma} A_\gamma \setminus \bigcup\nolimits_{\delta<\gamma} \mathrm{adh}^{\delta}_{\xi_\gamma} A_\gamma. \tag{XXII.2.3}$$

Let $X := \bigsqcup_{\gamma<\alpha} X_\gamma \sqcup \{x\}$, and let ξ be defined on X by $\xi|_{X_\gamma} := \xi_\gamma$, and by $V \in \mathcal{V}_\xi(x)$ whenever there exists $\delta < \alpha$ such that $\{x_\gamma : \delta < \gamma < \alpha\} \cup \{x\} \subset V$. Therefore, $x \in \mathrm{adh}_\xi\{x_\gamma : \gamma < \alpha\}$. If we set $A := \bigcup_{\gamma<\alpha} A_\gamma$, then $\{x_\gamma : \gamma < \eta\} \subset \mathrm{adh}^{\eta}_{\xi} A$ and $\{x_\gamma : \eta \leq \gamma < \alpha\} \cap \mathrm{adh}^{\eta}_{\xi} A = \varnothing$, because of (XXII.2.3). Therefore, by (XXII.2.2),

$$\mathrm{adh}^{\alpha}_{\xi} A = \mathrm{adh}_\xi(\bigcup\nolimits_{\gamma<\alpha} \mathrm{adh}^{\gamma}_{\xi} A) \supset \mathrm{adh}_\xi\{x_\gamma : \gamma < \alpha\} \ni x.$$

It remains to show that $\mathrm{adh}_\xi(\mathrm{adh}^{\alpha}_{\xi} B) \subset \mathrm{adh}^{\alpha}_{\xi} B$ for any $B \subset X$. If $\gamma_B := \sup\{\gamma : B \cap X_\gamma \neq \varnothing\} < \alpha$, then $\mathrm{adh}^{\eta}_{\xi} B \cap \{x_\gamma : \gamma_B < \gamma < \alpha\} = \varnothing$ for every η, and thus $\mathrm{adh}_\xi(\mathrm{adh}^{\gamma_B}_{\xi} B) = \mathrm{adh}^{\gamma_B}_{\xi} B = \mathrm{adh}^{\alpha}_{\xi} B$. Otherwise there exists a cofinal subset of $\{\gamma : B \cap X_\gamma \neq \varnothing\}$, and if $x_\gamma \in \mathrm{adh}^{\eta}_{\xi}(B \cap X_\gamma)$, then $\eta \leq \mathrm{t}(\xi_\gamma) = \gamma$, and thus if $\mathrm{adh}^{\eta+1}_{\xi} B = \mathrm{adh}^{\eta}_{\xi} B$, then $\eta \leq \alpha$. □

If θ is a sequentially founded convergence, that is, if $\mathrm{S}_0\theta$ is a sequentially induced pretopology (alias sequentially pretopological), in other terms, a Fréchet pretopology, then $\mathrm{t}(\mathrm{S}_0\theta)$ is called the *sequential order* of θ or of $\mathrm{S}_0\theta$.

PROPOSITION XXII.2.10. *The topological defect of a Fréchet pretopology does not exceed* ω_1.

PROOF. Let ξ be a Fréchet pretopology on X, and let $A \subset X$. We shall show that the set $\bigcup_{\gamma<\omega_1} \mathrm{adh}^{\gamma}_{\xi} A$ is ξ-closed. Let $x \in \mathrm{adh}^{\omega_1}_{\xi} A$, the ξ-adherence of the considered set. By the definition (XXII.2.2), there exists a sequence $(x_n)_n$ of elements of $\bigcup_{\gamma<\omega_1} \mathrm{adh}^{\gamma}_{\xi} A$ such that $x \in \lim_\xi (x_n)_n$. Accordingly, for every $n < \omega$, there is $\gamma_n < \omega_1$ such that $x_n \in \mathrm{adh}^{\gamma_n}_{\xi} A$. As each supremum of countable sequence of countable ordinals is countable, $\gamma := \sup\{\gamma_n : n < \omega\} < \omega_1$, and thus $x \in \mathrm{adh}^{\gamma+1}_{\xi} A \subset \bigcup_{\gamma<\omega_1} \mathrm{adh}^{\gamma}_{\xi} A$. It follows that $\mathrm{adh}_{\mathrm{T}\xi} A = \mathrm{cl}_\xi A \subset \mathrm{adh}^{\omega_1}_{\xi} A$, hence $\mathrm{adh}_{\mathrm{T}\xi} A = \mathrm{adh}^{\omega_1}_{\xi} A$. □

EXAMPLE XXII.2.11. The *radial pretopology* was introduced in Example VI.1.9 and investigated in Example VII.4.9. On the real plane, $x \in \mathrm{adh}_\rho A$ if there exists a vector h and a strictly decreasing real sequence $(t_n)_n$ converging to 0 such that $x + t_n h \in A$.

It follows that ρ is a Fréchet pretopology. We have already checked that ρ is not topological. Its topologization $\lambda = \mathrm{T}\rho$ has been characterized to the effect that a set A is λ-closed if and only if $A \cap L$ is L-closed for each straight line L.

It turns out that the topological defect of ρ is equal to ω_1. It was already shown in Example VI.1.9 that the set $A := \{(r,s) \in \mathbb{R}^2 : 0 < s \leq r^2, 0 < r\}$ fulfills $(0,0) \in \operatorname{adh}_\rho(\operatorname{adh}_\rho A) \backslash \operatorname{adh}_\rho A$. Therefore, $\operatorname{adh}_\rho^2 A \backslash \operatorname{adh}_\rho A \neq \varnothing$. We shall treat A as a module, on transforming it with the aid of rotations, translations and homotheties onto homeomorphic sets to be employed in our construction.

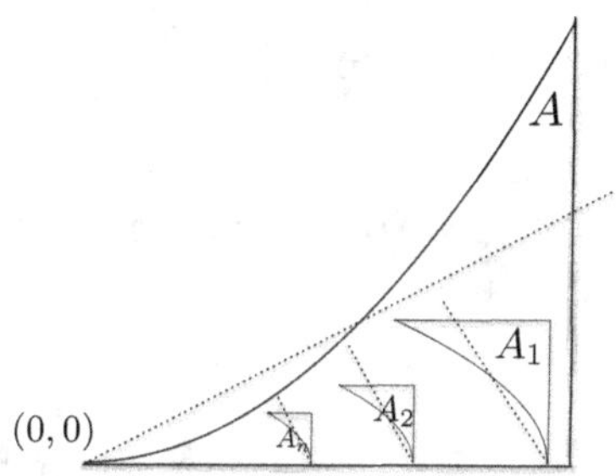

FIGURE XXII.2.1. Topological defect of the radial pretopology is ω_1.

We shall show that for each $0 < \gamma < \omega_1$, there exists a (bounded) subset A_γ and x_γ such that $\{x_\gamma\} = \operatorname{adh}_\rho^\gamma A_\gamma \setminus \bigcup_{\delta<\gamma} \operatorname{adh}_\rho^\delta A_\gamma$. This is obvious for $\gamma = 1$, and has been proved to be true for $\gamma = 2$. Suppose that the property holds for each $\delta < \gamma < \omega_1$. Then there exists a sequence $(\delta_n)_n$ of increasing ordinals such that $\gamma = \sup\{\delta_n + 1 : n < \omega\}$. By inductive assumption, and since ρ is homogeneous, for each $n < \omega$, there exists a bounded $A_n \subset A$ such that $\{(0, \frac{1}{n})\} = \operatorname{adh}_\rho^{\delta_n} A_n \setminus \bigcup_{\delta<\delta_n} \operatorname{adh}_\rho^\delta A_n$ and $\operatorname{diam} A_n < 2^{-n}$.

We have seen in Proposition XXI.8.2 that the product of the sequential fan and a free sequence is not a Fréchet filter. It turns out that certain disposition of sequential fans and free sequences in factor spaces influence the sequential order of the product.

In Figure XXI.8.2, $x_\infty \in \lim_{k\to\infty} x_{n,k}$ for each $n < \omega$, and, on the other hand, $y \in \lim_{n\to\infty} y_n$. Therefore, bisequences appear

$$(x_\infty, y_\infty) \underset{n}{\longleftarrow} (x_\infty, y_n) \underset{k}{\longleftarrow} (x_{n,k}, y_n).$$

For example, the sequential order of a product of a sequential fan with a sequence of sequential fans is strictly greater than 2.

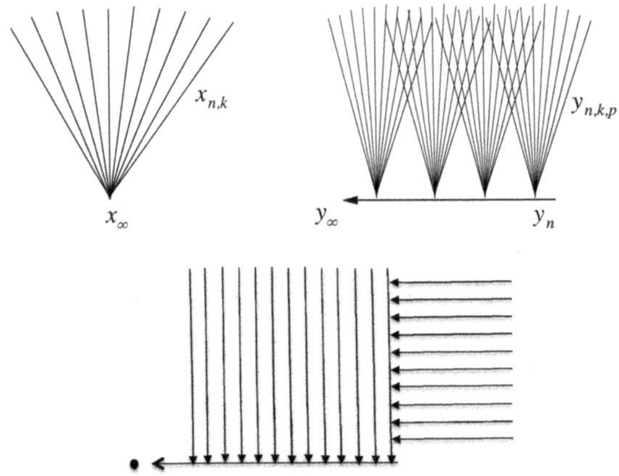

FIGURE XXII.2.2. The sequential order of products of two Fréchet topologies (one including a sequential fan, and the other a sequence of sequential fans) equals 3 (a trisequence).

In Figure XXII.2.2, a fan is defined as before, that is, $x_\infty \in \lim_{k\to\infty} x_{n,k}$ for each $n < \omega$. On the other hand, $y \in \lim_{n\to\infty} y_n$, and for every $n < \omega$, there is a sequential fan converging to y_n, that is, $y_n \in \lim_{p\to\infty} y_{n,k,p}$ for each $k < \omega$. There results a *trisequence*

$$(x_\infty, y_\infty) \underset{n}{\longleftarrow} (x_\infty, y_n) \underset{k}{\longleftarrow} (x_{n,k}, y_n) \underset{p}{\longleftarrow} (x_{n,k}, y_{n,k,p}),$$

so that the sequential order is at least 3.

One can iterate such constructions of Fréchet topologies, so that their product has the sequential order equal to any countable ordinal ([8]). For example, to get the sequential order equal to 4, consider the product of a sequence of sequential fans with a sequential fan of sequences of sequential fans.

XXII.3. *Sequential cascades

An ordered set is called *well-ordered* if its every non-empty subset has a least element. It is plain that every well-ordered set is linearly ordered (see Section XXVII.3).

An ordered set $(T, \leq)$ is called a *tree* if, for every $t \in T$, the set

$$T^{\downarrow}(t) := \{s \in T : s \leq t\}$$

is well-ordered.

[8]See S. DOLECKI and S. SITOU [90, Theorem 6.2].

A tree $(\Sigma, \leq)$ is called *sequential* if there exists a least element $\varnothing \in \Sigma$, and for every $t \in \Sigma$, the set

$$\Sigma^+(t) := \mathrm{Min}\{s \in \Sigma : t < s\}$$

of immediate successors of t is infinite countable. It follows that if $t_0 \neq t_1$, then $\Sigma^+(t_0) \cap \Sigma^+(t_1) = \varnothing$ and that the elements of $\Sigma^+(t)$ are incomparable for each $t \in \Sigma$.

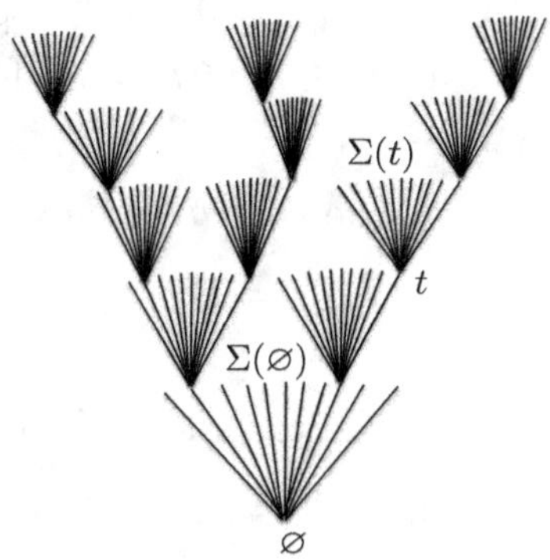

FIGURE XXII.3.1. A visualization of the sequential tree: $\Sigma(t)$ is the set of immediate successors of t for each $t \in \Sigma$.

Denote by $\Sigma^{\uparrow}(t) := \{s \in \Sigma : s \geq t\}$, the set of all *successors* of t in Σ, and notice that

PROPOSITION XXII.3.1. *$(\Sigma^{\uparrow}(t), \leq)$ is order-isomorphic with $(\Sigma, \leq)$ for every $t \in \Sigma$.*

Two sequential trees are order-isomorphic. Let us construct a sequential tree. Let $\Sigma := \{(n_1, \ldots, n_k) \in \mathbb{N}^k : k < \omega\}$ be ordered by inclusion.

As $\Sigma(\varnothing)$ is assumed to be infinite and countable, let $\Sigma(\varnothing) := \{(n) : n < \omega\}$, where (n) stands for a sequence of length 1, whose single term is n. We proceed by induction on the length of finite sequences. The least element $\varnothing$ is a sequence of length 0, that is, an empty sequence. Consequently, for each sequence $(n_1, n_2, \ldots, n_k) \in \Sigma$, of length $k < \omega$, we define

$$\Sigma(n_1, n_2, \ldots, n_k) := \{(n_1, n_2, \ldots, n_k, p) : p < \omega\}.$$

Accordingly, each $t \in \Sigma$ there exist $k < \omega$ and a sequence $(n_1, n_2, \ldots, n_k)$ such that $t = (n_1, n_2, \ldots, n_k)$, and the construction is complete.

EXAMPLE XXII.3.2 (a Hausdorff, regular, non-topologically regular pretopology). Let Σ stand for the *sequential tree*, to which we add an extra element ∞. Let γ be the following convergence: Σ is the set of finite sequences of natural numbers ordered by inclusion, and for each $s \in \Sigma$,

$$s \in \lim_\gamma \mathcal{F} \iff (\ker\mathcal{F} \subset \{s\}) \wedge (\Sigma^+(s) \in \mathcal{F}).$$

On the other hand, $\infty \in \lim_\gamma \mathcal{F}$ if $\ker\mathcal{F} \subset \{\infty\}$ and $\{\infty\} \cup \{t \in \Sigma : l(t) > n\} \in \mathcal{F}$ for each $n < \omega$, where $l(t)$ denotes the length of t. We see that, in particular, $0 \in \mathrm{cl}_\gamma\{t \in \Sigma : l(t) > n\}$ for each $n < \omega$, hence γ is not topologically regular.

If $t, s \in \Sigma$, then we define the *concatenation* $t \frown s$ of t with s by

$$t=(n_1,\ldots,n_k) \wedge s = (m_1,\ldots,m_p) \Longrightarrow t \frown s=(n_1,\ldots,n_k,m_1,\ldots,m_p),$$

abridging $t \frown n := t \frown (n)$.

The *natural convergence* γ on Σ is defined by

$$t \in \lim_\gamma \mathcal{F} \Longrightarrow \mathcal{F} \geq t^\uparrow \wedge (\Sigma^+(t))_0 \tag{XXII.3.1}$$

for every $t \in \Sigma$. Therefore, for each $t \in \Sigma$,

$$t \in \lim_\gamma (t \frown n)_n.$$

Notice that, if $t \in \Sigma$, then $V \in \mathcal{V}_\gamma(t)$ whenever $t \in V$ and $\Sigma^+(t) \setminus V$ is finite. Accordingly, $\Sigma^\uparrow(t)$ is γ-open for each $t \in T$. If $t_0 \neq t_1$, then either t_0, t_1 are incomparable and there exist elements $s_0 \neq s_1$ of $\Sigma^+(t_0 \wedge t_1)$ such that $s_0 \leq t_0$ and $s_1 \leq t_1$. Therefore $t_0 \in \Sigma^\uparrow(s_0), t_1 \in \Sigma^\uparrow(s_1)$, and $\Sigma^\uparrow(s_0)$ and $\Sigma^\uparrow(s_1)$ are γ-open and disjoint, or, say $t_0 < t_1$, then there is $s_1 \in \Sigma^+(t_0)$ such that $s_1 \leq t_1$. The set $\Sigma^\uparrow(t_0) \setminus \Sigma^\uparrow(s_1)$ is γ-open and disjoint from $\Sigma^\uparrow(t_1)$, so that γ is topologically Hausdorff. Moreover, $\Sigma^\uparrow(t)$ is γ-clopen for each $t \in \Sigma$. As a result, each cofinite subset of $\Sigma^+(t)$ is γ-closed, hence γ is topologically regular.

It follows that

PROPOSITION XXII.3.3. *The natural convergence γ of a sequential tree is a topologically Hausdorff, topologically regular, sequentially founded, homogeneous pretopology.*

COROLLARY XXII.3.4. *The natural topology $\mathrm{T}\gamma$ of a sequential tree is Hausdorff, regular, sequential, and homogeneous.*

A tree $(T, \leq)$ is called a *sequential cascade* if T is an infinite countable set, there is a least element $\varnothing_T$ of $(T, \leq)$, each non-empty subset of T has a maximal element, and, for each non-maximal $t \in T$, the set $T^+(t)$ of immediate successors of t is infinite.

On anticipating the formal Definition XXVII.8.1, we say that an ordered set $(X, \leq)$ is *well-founded* if for each $\varnothing \neq A \subset X$ there exists a minimal $a_0 \in A$. Of course, each tree is well-founded. On the other hand, each sequential cascade is well-founded with respect to the inverse order. Therefore,

PROPOSITION XXII.3.5. *Each sequential cascade $(T, \leq)$ is well-founded with respect to $\leq$ and to the inverse order $\leq^-$.*

It follows from the definition that every *branch* of $(T, \leq)$ ([9]) is finite. Hence, for each $t \in T$, let *level* $l_T(t)$, defined by

$$l_T(t) := \text{card}\{s \in T : \varnothing_T < s \leq t\},$$

is finite. In particular, $l_T(\varnothing_T) = 0$.

The *rank* $r_T(t)$ of an element of a sequential cascade $(T, \leq)$ can be defined by Theorem XXVII.8.4 to the effect that $r_T(t) = 0$ if t is maximal, and otherwise let

$$r_T(t) := \sup\{r_T(s) + 1 : s > t\}. \tag{XXII.3.2}$$

It follows that $r_T(t) := \sup\{r_T(s) + 1 : s \in T^+(t)\}$.

Consequently, the rank of an element of a sequential cascade is an ordinal number. We say that a sequential cascade T is of rank α, in symbols $r(T) = \alpha$ whenever $r_T(\varnothing_T) = \alpha$.

A sequential cascade T is called *essentially monotone* if

$$r_T(t) = \sup_{F \subset_{\text{fin}} T^+(t)} \inf\{r_T(s) + 1 : s \in T^+(t) \setminus F\}, \tag{XXII.3.3}$$

that is, if $r_T(t)$ is equal to the lower limit of $r_T(\cdot) + 1$ along the cofinite filter of $T^+(t)$.

PROPOSITION XXII.3.6. *Every sequential cascade is order-isomorphic with a subset of the sequential tree.*

PROOF. Let $(T, \leq)$ be a sequential cascade. We shall define an injection $f : T \longrightarrow \Sigma$ by induction on the *level*. If $\varnothing_T$ is the least element of T, then let $f(\varnothing_T) := \varnothing$. If $\varnothing_T$ is not maximal, then $T^+(\varnothing_T)$ is infinite countable, define $f|_{T^+(\varnothing_T)}$ to be any bijection onto $\Sigma^+(\varnothing_T)$. We have defined f for $t \in T$ with $l_T(t) \leq 1$. Suppose that we have already defined $f(t)$ for each $t \in T$ with $l_T(t) \leq r$. If $l_T(t) = r$ and t is not maximal, then let $f|_{T^+(t)}$ be any bijection between $T^+(t)$ and $\Sigma^+(f(t))$. As $l_T(t)$ is finite for each $t \in T$, the construction is complete. It is obvious that f is an order-isomorphism. □

[9]That is, a linearly ordered subset of T.

Let $(V, \leq)$ be an ordered set. A subset S of V is said to be *closed downwards* whenever if $V \ni v < s \in S$, then $v \in S$. A subset S of Σ is said to be a *full* whenever if $S \cap \Sigma^+(s) \neq \varnothing$ then $\Sigma^+(s) \subset S$ for each $s \in \Sigma$.

Notice that if $f : T \longrightarrow \Sigma$ is the order-isomorphism just constructed, $s \in f(T)$ and $r \in \Sigma$ and $r \leq s$, then $r \in f(T)$.

Accordingly, it is practical to consider merely sequential cascades that are subsets of the sequential tree.

PROPOSITION XXII.3.7. *Every sequential cascade is of countable rank (or of rank 0).*

PROOF. Indeed, the rank of maximal elements is equal to 0. By well-founded induction (Proposition XXVII.8.3), if we assume that for $t \in T$, the rank of every $s \in T$ with $s > t$ is countable, then by (XXII.3.2), $r_T(t)$ is countable, as the supremum countably many countable ordinals $\{r_T(s) + 1 : s \in T, s > t\}$. □

EXAMPLE XXII.3.8. If $\operatorname{card} T = 1$, then $r(T) = 0$. If $T = \{\varnothing\} \sqcup \{(n) : n < \omega\}$ then $r(T) = 1$. If $T = \{\varnothing\} \sqcup \{(n) : n < \omega\} \sqcup \{(n, k) : n, k < \omega\}\}$ ordered by inclusion, then $r(T) = 2$. If T_n is a sequential cascade of rank n for each $n < \omega$, then the sequential cascade

$$T = \{\varnothing\} \sqcup \bigsqcup\nolimits_{n<\omega} (n) \frown T_n, \tag{XXII.3.4}$$

is of rank $r(T) = \omega_0$. If T_n is a sequential cascade of rank ω_0 for every $n < \omega$, then (XXII.3.2) is of rank $\omega_0 + 1$.

PROPOSITION XXII.3.9. *There exist sequential cascades of arbitrary countable rank.*

PROOF. $\{\varnothing\}$ is a sequential cascade of rank 0. Proceeding by transfinite induction, let $0 < \alpha < \omega_0$ and assume that for every $\gamma < \alpha$, there exists a sequential cascade T_γ such that $r(T_\gamma) = \gamma$. As there exists a non-decreasing sequence $(\alpha_n)_n$ of ordinals such that $\alpha = \sup\{\alpha_n + 1 : n < \omega\}$, let T be of the form (XXII.3.4) with $r(T_n) = \alpha_n$ for every $n < \omega$. As $r_T(n) = \alpha_n$ for each $n < \omega$, by (XXII.3.2), $r_T(\varnothing_T) = \alpha$. □

DEFINITION XXII.3.10. The *natural pretopology* γ_T on a sequential cascade $(T, \leq)$ is defined like in (XXII.3.1), substituting $T^+(t)$ for $\Sigma(t)$.

It is plain that $\gamma_T = \gamma|_T$, where γ is the natural pretopology on the sequential tree Σ, when we view T as a full closed downwards subset of the sequential tree Σ.

We know that the topologizer is not hereditary. However, in this case $\mathrm{T}(\gamma_T) = (\mathrm{T}\gamma)|_T$. Let $O \subset T$ be $\gamma|_T$-open. Thus $T \cap \Sigma^\uparrow(x) =$

$T^{\uparrow}(x) \subset O$ for each $x \in O$. As $\Sigma^{\uparrow}(x)$ is γ-open for each $x \in \Sigma$, the set $O^{\uparrow} := \Sigma^{\uparrow}(x)$ is γ-open and $O^{\uparrow} \cap T = O$, hence O is $(\mathrm{T}\gamma)|_T$-open.

COROLLARY XXII.3.11. *The natural pretopology of a sequential cascade is topologically Hausdorff, topologically regular, and sequentially founded. Hence, its topologization is a Hausdorff regular sequential topology.*

If $(T, \leq)$ is a sequential cascade, then

$$\mathrm{ext}T := \min T \cup \max T$$

denotes the set of *extremal* elements, that is, either maximal or minimal elements. We know that $\min T = \{\varnothing_T\}$ and that $\max T \neq \varnothing$, because $(T, \leq)$ is well founded for the inverse order of $\leq$. A sequential cascade is degenerate if $\min T = \max T$, which happens when the rank of T is null.

The restriction $\mathrm{T}\gamma|_{\mathrm{ext}T}$ of the natural topology of a sequential tree Σ, equivalently of T is a prime topology. The restriction

$$\mathcal{N}_{\gamma}(\varnothing_T) \vee \max T$$

is called the *sequential contour* of T, and is denoted by $\int T$. Here is a proposition that justifies the terminology.

PROPOSITION XXII.3.12. *The contour of sequential contours along a sequential contour is a sequential contour.*

PROOF. Let T be a sequential cascade, and let T_s be a sequential cascade for each $s \in \max T$. Let $W := T \cup \bigcup_{s \in \max T} T_s$ so that $T \cap T_s = \{s\}$ and $T_s \cap T_r = \varnothing$ if $s \neq r$. Then each $w \in W \setminus T$ is of the form $s \frown r$, where $s \in \max T$ and $r \in T_s \setminus \{s\}$. Then W with the natural order is a sequential cascade, and by topologicity,

$$\textstyle\int W = \int_{s \in \int T} \int T_s.$$

$\square$

DEFINITION XXII.3.13. We name a *subsequential contour* the image of a sequential contour under a map.

A sequential cascade T is said to ξ-converge to x, in symbols,

$$x \in \lim_{\xi} T$$

if $i|_T \in \mathscr{C}(\gamma_T, \xi)$.

PROPOSITION XXII.3.14. *If ξ is a sequential topology, that is, $\xi \geq \mathrm{TSeq}\xi$, then $x \in \mathrm{cl}_{\xi} A$ if and only if there exists a sequential cascade T so that $\max T \subset A$ and $x \in \lim_{\xi} T$.*

PROOF. Let $x \in \mathrm{cl}_\xi A$. By Proposition XXII.2.10, there exists a countable ordinal $\alpha \leq \mathrm{t}(\mathrm{Seq}\xi)$ such that $\mathrm{cl}_\xi A = \mathrm{adh}^\alpha_{\mathrm{Seq}\xi} A$. If $n = 0$, then $\mathrm{adh}^0_{\mathrm{Seq}\xi} A = A$, so that $x \in A$. Therefore $\{\varnothing\}$ is a sequential cascade of rank 0, and $T := \{x\}$ is the corresponding sequential cascade. If $\alpha = 1$, that is, $\mathrm{cl}_\xi A = \mathrm{adh}_{\mathrm{Seq}\xi} A$, then there exists a sequence $(x_n)_n$ of elements of A such that $x \in \lim_\xi (x_n)_n$, then we define $T := \{x\} \sqcup \{(n) : n < \omega\}$. Let $\alpha > 0$ and suppose that the claim holds for $\eta < \alpha$. If

$$x \in \mathrm{adh}^\alpha_{\mathrm{Seq}\xi} A = \mathrm{adh}_{\mathrm{Seq}\xi}(\bigcup\nolimits_{\eta<\alpha} \mathrm{adh}^\eta_{\mathrm{Seq}\xi} A),$$

then there exist sequences $(\eta_n)_{n<\omega}$ and $(x_n)_{n<\omega}$ such that $x_n \in \mathrm{adh}^{\eta_n}_{\mathrm{Seq}\xi} A$ and $x \in \lim_\xi (x_n)_n$. By inductive assumption, for each $n < \omega$, there exists on A a sequential cascade T_n of rank η_n such that $x_n \in \lim_\xi T_n$. Let T be given by (XXII.3.4), and $\varnothing := x$. Then $x \in \lim_\xi T$. The converse holds, because if $\max T \subset A$ and $x \in \lim_\xi T$, then $i|_T \in \mathscr{C}(\gamma_T, \xi)$, hence $x \in \mathrm{cl}_\xi A$. □

An easy modification of the proof above yields

COROLLARY XXII.3.15. *If ξ is a sequential topology, $\alpha < \omega_0$ and $x \in \mathrm{adh}^\alpha_\xi A$, then there exists a sequential cascade T of rank α such that $x \in \lim_\xi T$ and $\max T \subset A$.*

REMARK XXII.3.16. In preceding occurrences Proposition XXII.3.14 was formulated in terms of multisequences to the effect that $x \in \mathrm{cl}_\xi A$ whenever there exists a multisequence on A that ξ-converges to x. A map $f : \mathrm{ext}T \longrightarrow X$ is said to be a *multisequence* on X. If ξ is a convergence on X, then a multisequence f is said to ξ-converge to $x \in X$, whenever there exists $F \in \mathscr{C}(\gamma_T, \xi)$ such that $F(\varnothing_T) = x$ and $F|_{\max T} = f$.

In preceding sections we studied sequential fans, sequences of sequential fans, and so on. All these objects can be easily described with the aid of multisequences.

EXAMPLE XXII.3.17. A *sequential fan* from Example III.4.7 can be seen as a convergent multisequence $f : T \longrightarrow X$ from a sequential cascade of rank 2, so that $f(\varnothing) = f((n)) := x_\infty$ for each $n < \omega$, and $f((n,k)) := x_{n,k}$ for every $n, k < \omega$.

The sequence of sequential fans from Figure XXII.2.2, is a convergent multisequence $f : T \longrightarrow X$ from a sequential cascade of rank 3 defined by $f(\varnothing) := x_\infty$, and $f((n)) = f((n,k)) := x_n$ for every $n, k < \omega$, and $f((n,k,p)) := x_{n,k,p}$ for every $n, k, p < \omega$.

EXAMPLE XXII.3.18. Slightly more generally, if T is a sequential cascade of finite even rank m (equal to the length), then a multisequence $f : T \longrightarrow X$ is called a *multifan* if for each $t \in T$ of even length $k < m$,

$$f(t \frown n) = f(t),$$

and if T is a sequential of finite odd rank $m + 1$ (equal to the length), then a multisequence $f : T \longrightarrow X$ is called a *multiarrow* if the equality above holds for each $t \in T$ of odd length $k < m$. It is not difficult to check that the product of a multifan of rank m with a multiarrow of rank $m + 1$ yields a convergent multisequence of sequential order $m + 1$ ([10]).

XXII.4. Subsequential topologies

Because the topologizer is not hereditary, a subconvergence of a sequential topology is not, in general, sequential, like the Arens topology in Example VII.4.6 (see below).

A topology ξ is called *subsequential* if it is a subconvergence of a sequential topology, that is, if there exists $\tau = \mathrm{TSeq}\tau$ so that $\xi \subset \tau$.

EXAMPLE XXII.4.1. The *Arens topology* of Example VII.4.6 is subsequential. It is a subspace of the *bisequence topology* from Example VI.1.8. Actually, the bisequence topology is homeomorphic to the natural topology of the (essentially monotone) sequential cascade of rank 2. The Arens topology is not sequential, because the only convergent sequences are almost constant.

PROPOSITION XXII.4.2. *A topology ξ is subsequential if and only if $x \in \mathrm{cl}_\xi A$ implies the existence of a subsequential contour $\mathcal{F}$ on A such that $x \in \lim_\xi \mathcal{F}$.*

PROOF. Let τ be a sequential topology, $\xi \subset \tau$ and let $x \in \mathrm{cl}_\xi A$. Hence $x \in \mathrm{cl}_\tau A$ and thus by Proposition XXII.3.14, there exists a sequential cascade T and a multisequence $f \in \mathscr{C}(\gamma_T, \tau)$ such that $f(\max T) \subset A$ and $f(\varnothing_T) = x$. In particular, the sequential contour $\int T$ on $\max T$ is mapped into A, and this $A \in f(\int T)$. As τ is a topology, $f \in \mathscr{C}(\mathrm{T}\gamma_T, \tau)$, hence $x \in \lim_\tau f(\int T)$.

Conversely, if the condition holds for a topology ξ on X. Let M be the set of $(x, A) \in X \times 2^X$ such that $x \in \mathrm{cl}_\xi A$, let $\mathcal{F}_{(x,A)}$ be a

[10]Multifans and multiarrows of arbitrary countable rank give rise to ordinal numbers (called *fascicularity* and *sagittality*) associated with Fréchet topologies. There exist formulae that enable to estimate, or in certain cases, calculate the sequential order of their product.

subsequential filter such that $A \in \mathcal{F}_{(x,A)}$ and $x \in \lim_\xi \mathcal{F}_{(x,A)}$. By assumption, there exists a sequential cascade $T_{(x,A)}$ such that $x \approx \varnothing_{T_{(x,A)}}$, and $\max T_{(x,A)} \subset A$ and $\int T_{(x,A)} = \mathcal{F}_{(x,A)}$, and moreover let ξ coincide with the natural topology $\gamma_{(x,A)}$ of $T_{(x,A)}$ on $X \cap T_{(x,A)}$. Let $Y := X \sqcup \bigsqcup_{(x,A)\in M}(T_{(x,A)} \setminus X)$ endowed with $\tau := \xi \oplus \bigoplus_{(x,A)\in M} \gamma_{(x,A)}|_{T_{(x,A)}\setminus X}$. To show that τ is sequential, let $y \in \mathrm{cl}_\tau B$. Then either there is $(x, A) \in M$ such that $y \in T_{(x,A)}$ and $B \subset T_{(x,A)}$ and $\gamma_{(x,A)}$ is sequential, or $y \in X$ and $B \subset X$. In the latter case, $(y, B) \in M$ and thus $\max T_{(y,B)} \subset B$ and $y \approx \varnothing_{T_{(y,B)}}$, so that the proof is complete. □

PROPOSITION XXII.4.3. *A topology is subsequential if and only if a set is closed provided that it is subsequentially closed.*

DEFINITION XXII.4.4. A filter is designated as *subsequential* whenever it is an infimum of subsequential contours.

COROLLARY XXII.4.5. *A topology is subsequential if and only if every neighborhood filter is a subsequential filter.*

EXAMPLE XXII.4.6. Let $\mathcal{A} = \{A_y : y \in Y\}$ be a *MAD* family of subsets of a countable set X, and let $\zeta_\mathcal{A}^\infty$ be a compact topology on Y_∞ from Section XII.8. Then $\zeta_\mathcal{A}^\infty$ is a sequential topology of sequential rank 2. Its restriction to $X \sqcup \{\infty\}$ is therefore a subsequential prime topology, which is not sequential, because each convergent sequential filter is almost constant. Indeed, if B is an infinite countable subset of X, then by the maximality of $\mathcal{A}$, there is $A \in \mathcal{A}$ such that $B \cap A$ is infinite. Therefore the cofinite filter $(B)_0$ does not converge to ∞, because $(A)_0$ and $\mathcal{N}_{\zeta_\mathcal{A}^\infty}(\infty)$ are dissociated for each $A \in \mathcal{A}$.

PROPOSITION XXII.4.7. *There exists a prime subsequential topology that cannot be extended to a Hausdorff sequential topology on a countable set* ([11]).

PROOF. Let $\zeta_\mathcal{A}^\infty$ be the sequential topology from Example XXII.4.6. Then $\xi = \zeta_\mathcal{A}^\infty|_{X\sqcup\{\infty\}}$ is subsequential and prime on a countable set X. I claim that ξ cannot be extended to a Hausdorff sequential topology. Suppose that, on the contrary, such an extension exists. Let W be a countable set, and τ a sequential topology on $X \sqcup W \sqcup \{\infty\}$ such that $\xi \subset \tau$. By assumption, $\mathcal{N}_\tau(\infty)|_X = \mathcal{N}_{\zeta_\infty}(\infty)|_X = \mathrm{c}\mathcal{A}$, which the residual filter of $\mathcal{A}$.

As τ is Hausdorff, for each $w \in W$, the restriction $\mathcal{F}_w$ to X of $\mathcal{N}_\tau(w)$ is dissociated from $\mathrm{c}\mathcal{A}$, hence by the Stone transform, $\beta_X^*\mathcal{F}_w$ is

[11]This is a negative answer given in [61] to a question of S. GARCIA-FERREIRA asked at the International Conference on Topology and Applications in December 2007 in Kyoto (see [108]).

a $\beta_{\iota X}$-closed subset of $\bigsqcup_{A\in\mathcal{A}}\beta^*_X A = \beta^* X \setminus \beta^*_X(\mathcal{F}_\mathcal{A})$. As each $\beta^*_X A$ is $\beta_{\iota X}$-clopen, there exists a finite subfamily $\mathcal{A}_w$ of $\mathcal{A}$, such that $\beta^*_X\mathcal{F}_w \subset \bigcup_{A\in\mathcal{A}_w}\beta^*_X A$. Therefore, there exists a countable subfamily $\mathcal{A}_W$ of $\mathcal{A}$ such that

$$\bigcup\nolimits_{w\in W}\beta^*_X\mathcal{F}_w \subset \bigcup\nolimits_{A\in\mathcal{A}_W}\beta^*_X A.$$

Hence, $\int_{(W)_0}\mathcal{F}_w < \mathrm{c}\mathcal{A}$, which is a contradiction by Corollary XII.8.16. $\square$

XXII.5. Topologically maximal pretopologies

Given a topology τ on a set X, we consider $\mathbf{S}_0(\tau) := \{\xi \in \mathbf{S}_0 : \mathrm{T}\xi = \tau\}$. Accordingly, for each $\xi \in \mathbf{S}_0(\tau)$, there exists the topological defect $\mathrm{t}(\xi)$, and of course, $\xi^{\mathrm{t}(\xi)} = \tau$.

Which are the maximal elements of $\mathbf{S}_0(\tau)$? A pretopology ξ is called *topologically maximal* if whenever a pretopology $\theta \geq \xi$ is such that $\mathrm{T}\theta = \mathrm{T}\xi$, then $\theta = \xi$. We shall see that this property is actually pointwise. We say that a pretopology is *topologically maximal at* x if whenever a pretopology $\theta \geq \xi$ is such that $\mathrm{T}\theta = \mathrm{T}\xi$, then $\mathcal{V}_\theta(x) = \mathcal{V}_\xi(x)$.

THEOREM XXII.5.1. *A pretopology ξ on X is topologically maximal at x if and only if for every $A \subset X$ such that $x \in \operatorname{adh}_\xi A \setminus A$, there exists $\Omega \subset X$ such that $\{x\} = \operatorname{adh}_\xi \Omega \setminus \Omega$ and $x \notin \operatorname{adh}_\xi(\Omega \setminus A)$.*

PROOF. Suppose that the condition does not hold: there is $A \subset X$ such that $x \in \operatorname{adh}_\xi A \setminus A$ and for each $\Omega \subset X$,

$$\{x\} = \operatorname{adh}_\xi \Omega \setminus \Omega \Longrightarrow x \in \operatorname{adh}_\xi(\Omega \setminus A). \tag{XXII.5.1}$$

Define on X a pretopology σ by setting $\mathcal{V}_\sigma(x) := \mathcal{V}_\xi(x)\setminus A$ and $\mathcal{V}_\sigma(w) := \mathcal{V}_\xi(w)$ for each $w \neq x$. Obviously, $\sigma \geq \xi$ and $\{x\} = \operatorname{adh}_\xi A \setminus \operatorname{adh}_\sigma A$. We need to show that if $\operatorname{adh}_\sigma \Omega \subset \Omega$, then $\operatorname{adh}_\xi \Omega \subset \Omega$. Let

$$\begin{aligned}\operatorname{adh}_\sigma \Omega &= \operatorname{adh}_\sigma(\Omega \cap A) \cup \operatorname{adh}_\sigma(\Omega \setminus A)\\ &= \operatorname{adh}_\xi(\Omega \cap A) \setminus \{x\} \cup \operatorname{adh}_\xi(\Omega \setminus A).\end{aligned}$$

Accordingly, if $\operatorname{adh}_\sigma \Omega \subset \Omega$, then $\operatorname{adh}_\xi \Omega \subset \Omega \cup \{x\}$. Then either $\operatorname{adh}_\xi \Omega \subset \Omega$, or $\{x\} = \operatorname{adh}_\xi \Omega \setminus \Omega$, and by (XXII.5.1), $x \in \operatorname{adh}_\xi(\Omega \setminus A) = \operatorname{adh}_\sigma(\Omega \setminus A) \subset \Omega$, a contradiction.

Suppose that a pretopology ξ on X is not topologically maximal at x. This means that there exists a pretopology $\theta \geq \xi$ such that $\mathrm{T}\theta = \mathrm{T}\xi$ and there is $W \in \mathcal{V}_\theta(x) \setminus \mathcal{V}_\xi(x)$. Consequently, $A := X \setminus W \in \mathcal{V}_\xi(x)^\#$. Let σ be as in the first part of the proof, that is, $\mathcal{V}_\sigma(x) := \mathcal{V}_\xi(x) \setminus A = \mathcal{V}_\xi(x) \vee W$. As $W \in \mathcal{V}_\theta(x)$, we infer that $\theta \geq \sigma \geq \xi$. Now, $x \in \operatorname{adh}_\xi A \setminus A$ and if $\{x\} = \operatorname{adh}_\xi \Omega \setminus \Omega$, then Ω is not ξ-closed, hence it is not

σ-closed, because $\mathrm{T}\theta \geq \mathrm{T}\sigma \geq \mathrm{T}\xi$. Finally, since $\mathrm{adh}_\xi \Omega \setminus \mathrm{adh}_\sigma \Omega \subset \{x\}$, we conclude that $x \in \mathrm{adh}_\xi \Omega$, hence by (XXII.5.1) $x \in \mathrm{adh}_\xi(\Omega \setminus A)$, completing the proof. □

COROLLARY XXII.5.2. *Every Hausdorff Fréchet pretopology is topologically maximal.*

PROOF. Indeed, if ξ is a Fréchet pretopology and $x \in \mathrm{adh}_\xi A \setminus A$, then there exists a sequence $(x_n)_n$ of elements of A such that $x \in \lim_\xi (x_n)_n$. If we set $\Omega := \{x_n : n < \omega\}$, then $\{x\} = \mathrm{adh}_\xi \Omega \setminus \Omega$, because ξ is Hausdorff, and on the other hand, $\Omega \setminus A = \varnothing$. □

EXAMPLE XXII.5.3. Examples of topologically non-maximal pretopologies are ξ_h and ξ_v from Example XXII.2.4. Indeed, from $\xi > \xi_h > \xi^2$, we infer that $\mathrm{T}\xi = \mathrm{T}\xi_h$, where ξ is the Féron cross pretopology, which is a Hausdorff Fréchet pretopology, hence topologically maximal in virtue of Theorem XXII.5.1.

XXII.6. Applications of the Stone topology

It turns out that numerous non-topological convergence properties can be characterized with the aid of the Stone topology ([12]).

We know that if ξ is a pseudotopology on X, then $x \in \lim_\xi \mathcal{F}$ if and only if $x \in \lim_\xi \mathcal{U}$ for every $\mathcal{U} \in \beta\mathcal{F}$, equivalently if $\beta\mathcal{F} \subset \xi^-(x)$.

Let us define

$$\mathbb{U}_\xi(x) := \xi^-(x) \cap \beta X$$

the set of ξ-convergent ultrafilters. In other words, for a pseudotopology ξ,

$$x \in \lim_\xi \mathcal{F} \iff \beta\mathcal{F} \subset \mathbb{U}_\xi(x).$$

Notice that for any map $S : X \longrightarrow 2^{\beta X}$ such that $x^\uparrow \in S(x)$, there is a pseudotopology σ_S on X such that $\mathbb{U}_{\sigma_S}(x) = S(x)$. A first simple observation is that

PROPOSITION XXII.6.1. *A point x is pretopological for a pseudotopology ξ if and only if $\mathbb{U}_\xi(x)$ is β-*closed.

PROOF. Indeed, a pseudotopology ξ is pretopological at x whenever $\mathcal{V}_\xi(x) = \bigcap_{\mathcal{U} \in \mathbb{U}_\xi(x)} \mathcal{U}$, that is, $\beta\mathcal{V}_\xi(x) = \mathbb{U}_\xi(x)$, which is β-closed as the Stone image of a filter $\mathcal{V}_\xi(x)$. □

---

[12] A Stone-topological characterization of Fréchetness is due to V. I. MALYHIN [**182**]. First results in this sense appeared in [**56**]. A systematic study was carried out in [**250**] by A. TRAN-MYNARD and F. MYNARD, and developed in [**200**] by F. MYNARD.

Recall Example VI.1.6, which illustrates Proposition XXII.6.1. Let us observe that if τ is a topology on a set Y, then the collection

$$\{\bigcap \mathcal{Q} : \mathcal{Q} \subset \mathcal{O}_\tau, \operatorname{card} \mathcal{Q} \leq \aleph_0\},$$

of $G_{\aleph_0}\tau$-sets, that is, G_δ-sets in traditional terminology, is a base for the open sets of a topology on X. In the case of the Stone topology $\beta = \beta\iota_X$ we shall denote this topology by $\delta = \delta_X$.

LEMMA XXII.6.2. *A filter $\mathcal{F}$ on X is countably based if and only if $\beta\mathcal{F}$ is a β-closed $G_{\aleph_0}$ subset of βX.*

PROOF. If $\{F_n : n < \omega\}$ is a filter-base of $\mathcal{F}$, then $\mathcal{F} = \bigvee_{n<\omega}\{F_n\}^\uparrow$, so that $\beta\mathcal{F} = \bigcap_{n<\omega} \beta F_n$. □

PROPOSITION XXII.6.3. *Let ξ be a pseudotopology. Then x is paratopological if and only if $\mathbb{U}_\xi(x)$ is δ_X-*closed.

PROOF. Indeed, if a point x is paratopological in a pseudotopology ξ, if $\mathcal{U} \in \beta X$ and $x \notin \lim_\xi \mathcal{U}$, then there exists a countably based filter $\mathcal{D} \leq \mathcal{U}$ such that $x \notin \operatorname{adh}_\xi \mathcal{D}$, that is, $\beta\mathcal{D} \cap \mathbb{U}_\xi(x) = \varnothing$. If $\{D_n : n < \omega\}$ is a filter-base of $\mathcal{D}$, then $\mathcal{D} = \bigvee_{n<\omega} D_n^\uparrow$, hence $\beta\mathcal{D} = \bigcap_{n<\omega} \beta D_n$ is a $G_{\aleph_0}\beta$-set disjoint from $\mathbb{U}_\xi(x)$, which completes the proof. □

More generally, a class $\mathbb{D}$ of filters stable under finite infima, that is such that $\mathcal{D}_0, \mathcal{D}_1 \in \mathbb{D}$ implies that $\mathcal{D}_0 \wedge \mathcal{D}_1 \in \mathbb{D}$, defines a topology on βX for each set X to the effect that

$$\{\beta\mathcal{D} : \mathcal{D} \in \mathbb{D}X\}$$

is a base for the open sets of a topology on βX denoted by $\beta\mathbb{D}$. Of course, $\beta \leq \beta\mathbb{D}$ if $\mathbb{F}_0 \subset \mathbb{D}$. Notice that $\beta\mathbb{F}$ coincides with the discrete.

The case of countably complete filters is a little more complicated.

LEMMA XXII.6.4. *A filter $\mathcal{F}$ on X is countably complete if and only if*

$$\mathcal{N}_\beta(\beta\mathcal{F}) = \mathcal{N}_\delta(\beta\mathcal{F}).$$

PROOF. As $\delta \geq \beta$, the inclusion $\subset$ always holds. If $\mathcal{F} \in \mathbb{F}_{\wedge 1}$ and $\{F_n : n < \omega\} \subset \mathcal{F}$, then $F := \bigcap_{n<\omega} F_n \in \mathcal{F}$, that is, $\{F\}^\uparrow = \bigvee_{n<\omega}\{F_n\}^\uparrow \leq \mathcal{F}$ and $\beta F = \bigcap_{n<\omega} \beta F_n$. Therefore $\beta\mathcal{F} = \bigcap_{D\in\mathcal{D}(\mathcal{F})} \beta D$, where $\mathcal{D}(\mathcal{F})$ is the set of countably based filters coarser than $\mathcal{F}$. □

PROPOSITION XXII.6.5. $\mathrm{S}_0 \geq \mathrm{S}_{\wedge 1}\mathrm{S}_1$.

PROOF. Suppose that $\mathcal{U} \notin \operatorname{cl}_{\mathbb{F}_{\wedge 1}} \operatorname{cl}_{\mathbb{F}_1} \mathbb{U}_\xi(x)$, that is, there exists $\mathcal{H} \in \mathbb{F}_{\wedge 1}$ such that $\mathcal{U} \in \beta\mathcal{H}$ and $\beta\mathcal{H} \cap \operatorname{cl}_\delta \mathbb{U}_\xi(x) = \varnothing$, equivalently $\mathbb{U}_\xi(x) \notin (\mathcal{N}_\delta(\beta\mathcal{H}))^\#$. By Lemma XXII.6.4, $\mathcal{N}_\delta(\beta\mathcal{H}) = \mathcal{N}_\beta(\beta\mathcal{H})$, so that $\beta\mathcal{H} \cap \operatorname{cl}_\beta \mathbb{U}_\xi(x) = \varnothing$, and in particular, $\mathcal{U} \notin \operatorname{cl}_\beta \mathbb{U}_\xi(x)$. □

One proves the following proposition by mimicking the proof of Proposition XXII.6.3.

PROPOSITION XXII.6.6. *If ξ is a pseudotopology on X, then for each $x \in X$,*

$$\mathbb{U}_{\mathrm{A}_{\mathbb{D}}\xi}(x) = \mathrm{cl}_{\beta\mathbb{D}} \mathbb{U}_{\xi}(x).$$

Recall a missing-a-filter convergence from Example XVIII.5.5, which illustrates Proposition XXII.6.6. Let $x^{\uparrow} \in \mathbb{U}_{\xi}(x)$ for each $x \in X$. Then

(1) a convergence ξ on X is a pseudotopology whenever $\mathbb{U}_{\xi}(x)$ is $\beta\mathbb{F}$-closed for each $x \in X$, hence arbitrary, because any subset of βX is $\beta\mathbb{F}$-closed (for the discrete topology).
(2) a pseudotopology ξ is a paratopology at x whenever $\mathbb{U}_{\xi}(x)$ is $\beta\mathbb{F}_1$-closed, that is, δ-closed.
(3) a pseudotopology ξ is a pretopology at x whenever $\mathbb{U}_{\xi}(x)$ is $\beta\mathbb{F}_0$-closed, that is, β-closed.

COROLLARY XXII.6.7. *The pseudotopological infimum of countably many paratopologies is a paratopology.*

PROOF. Let $\Pi = \{\pi_n : n < \omega\}$ be a sequence of paratopologies on X and $x \in X$. Then $\mathbb{U}_{\mathrm{S}(\bigwedge \Pi)}(x) = \bigcup_{n<\omega} \mathbb{U}_{\pi_n}(x)$ in virtue of (XVIII.3.10). If $\mathcal{U} \notin \bigcup_{n<\omega} \mathbb{U}_{\pi_n}(x)$, then for each $n < \omega$, there exists a sequence $\{D_{n,k} : k < \omega\}$ of subsets of X such that $\mathcal{U} \in \bigcap_{k<\omega} \beta_X D_{n,k}$ and $\bigcap_{k<\omega} \beta_X D_{n,k} \cap \mathbb{U}_{\pi_n}(x) = \varnothing$. Therefore $\mathcal{U} \in \bigcap_{n<\omega} \bigcap_{k<\omega} \beta_X D_{n,k}$ and

$$\bigcap_{n,k<\omega} \beta_X D_{n,k} \cap \bigcup_{n<\omega} \mathbb{U}_{\pi_n}(x) = \varnothing.$$

On the other hand, $\bigcap_{n,k<\omega} \beta_X D_{n,k}$ is a δ-neighborhood of $\mathcal{U}$. $\square$

One can also characterize mixed properties of convergences of the type

$$\xi \geq \mathrm{A}_{\mathbb{D}}\mathrm{B}_{\mathbb{H}}\xi.$$

THEOREM XXII.6.8. *If ξ is a pseudotopology, then for each x,*

$$\mathbb{U}_{\mathrm{A}_{\mathbb{D}}\mathrm{B}_{\mathbb{H}}\xi}(x) = \mathrm{cl}_{\beta\mathbb{D}} \mathrm{int}_{\beta\mathbb{H}} \mathbb{U}_{\xi}(x).$$

PROOF. Indeed, if $\mathcal{U} \notin \mathbb{U}_{\mathrm{A}_{\mathbb{D}}\mathrm{B}_{\mathbb{H}}\xi}(x)$, then by Proposition XXII.6.6, $\mathbb{U}_{\mathrm{A}_{\mathbb{D}}\mathrm{B}_{\mathbb{H}}\xi}(x) = \mathrm{cl}_{\beta\mathbb{D}} \mathbb{U}_{\mathrm{B}_{\mathbb{H}}\xi}(x)$. On the other hand, $\mathcal{U} \in \mathbb{U}_{\mathrm{B}_{\mathbb{H}}\xi}(x)$, then $x \in \lim_{\mathrm{B}_{\mathbb{H}}\xi} \mathcal{U}$, that is, there exists $\mathcal{H} \in \mathbb{H}$ such that $x \in \lim_{\xi} \mathcal{H}$ and $\mathcal{U} \geq \mathcal{H}$. In other words, $\mathcal{U} \in \beta\mathcal{H} \subset \mathbb{U}_{\xi}(x)$. Therefore, $\mathbb{U}_{\mathrm{B}_{\mathbb{H}}\xi}(x) = \mathrm{int}_{\beta\mathbb{H}} \mathbb{U}_{\xi}(x)$, and the proof is complete. $\square$

Consequently,

COROLLARY XXII.6.9. *A pseudotopology ξ is*

(1) *bisequential at x if and only if* $\mathbb{U}_\xi(x) \subset \mathrm{int}_{\beta\mathbb{F}_1} \mathbb{U}_\xi(x)$, *that is,* $\mathbb{U}_\xi(x)$ *is* δ*-open,*
(2) *strongly Fréchet at x if and only if* $\mathbb{U}_\xi(x) \subset \mathrm{cl}_{\beta\mathbb{F}_1} \mathrm{int}_{\beta\mathbb{F}_1} \mathbb{U}_\xi(x)$, *that is,* $\mathbb{U}_\xi(x)$ *is the* δ*-closure of a* δ*-open set,*
(3) *Fréchet at x if and only if* $\mathbb{U}_\xi(x) \subset \mathrm{cl}_{\beta\mathbb{F}_0} \mathrm{int}_{\beta\mathbb{F}_1} \mathbb{U}_\xi(x)$, *that is,* $\mathbb{U}_\xi(x)$ *is the* β*-closure of a* δ*-open set.*

Of course, if ξ is a pretopology, then $\mathbb{U}_\xi(x) = \beta\mathcal{V}_\xi(x)$ for every $x \in |\xi|$. As we have seen in Corollary XXI.1.2, a pretopology is Fréchet if and only if every vicinity filter is Fréchet. Accordingly, we can recover the Stone topology characterization of Fréchet pretopologies by applying the Stone transform to (XXI.1.3)

$$\mathcal{F} = \bigcap\nolimits_{\mathcal{G}\in\mathbb{F}_1(\mathcal{F})} \mathcal{G}$$

to the effect $\beta\mathcal{F} = \mathrm{cl}_\beta(\bigcup_{\mathcal{G}\in\mathbb{F}_1(\mathcal{F})} \beta\mathcal{G})$. Now, if a countably based filter $\mathcal{G}$ is finer than $\mathcal{F}$, by taking any filter-base $(G_n)_n$ of $\mathcal{G}$, we obtain $\mathcal{G} = \bigvee_{n<\omega}\{G_n\}^\uparrow$ (see Proposition II.3.5). Hence, $\beta\mathcal{G} = \bigcap_{n<\omega} \beta G_n \subset \beta\mathcal{F}$, that is, $\beta\mathcal{G}$ is a $G_{\aleph_0}(\beta)$-subset of $\beta\mathcal{F}$, equivalently a δ-open subset of $\beta\mathcal{F}$. Consequently, a filter $\mathcal{F}$ is Fréchet if and only if

$$\beta\mathcal{F} = \mathrm{cl}_\beta(\mathrm{int}_\delta\, \beta\mathcal{F}),$$

as in Corollary XXII.6.9.

If X is countable, then $\beta^* D = \beta^*(D)_0$ for each infinite subset D of X, and thus ([13])

COROLLARY XXII.6.10. *A free filter $\mathcal{F}$ on a countable set is Fréchet if and only if*

$$\beta\mathcal{F} = \mathrm{cl}_\beta(\mathrm{int}_\beta\, \beta\mathcal{F}).$$

These are merely a few basic examples of applications of the Stone topology ([14]).

XXII.7. *Homogeneity quest

A convergence ξ on a set X is called *homogeneous* if for any pair (x_0, x_1) of elements of X, there exists a homeomorphism $h \in \mathscr{H}(\xi, \xi)$ so that $h(x_0) = x_1$ (Definition IV.10.14).

The discrete and the chaotic topologies ι_X and o_X are homogeneous for every set X. On the other hand, a prime convergence is homogeneous if and only if it is discrete.

Loosely speaking, a convergence is homogeneous whenever it looks alike at every point.

[13] V. I. MALYHIN [182]

[14] Many other utilizations are presented in [200].

As ι_X is homogeneous, β^*X is made of free ultrafilters on X, and since the Stone topology β is defined without making distinction between the points in β^*X, one might suspect that the restriction β^* of β to β^*X is homogeneous.

In fact, this is not the case! We will see in this section that β^* is very heterogeneous ([15]).

Let N be an infinite countable set and β^*N the set of free ultrafilters on N. By Theorem XII.7.17, every $f \in N^N$ has a unique extension $f^* \in \mathscr{C}(\beta^*N, \beta^*N)$.

Two free ultrafilters p_0 and p_1 are of the same *type*,

$$\mathrm{t}(p_0) = \mathrm{t}(p_1),$$

if there exists a permutation $f \in N^N$ such that $f^*(p_0) = p_1$. If f is a permutation, then $f^* \in \mathscr{H}(\beta^*, \beta^*)$. Indeed, by Example XII.7.18, $\mathcal{U}(f^*(p)) = f[\mathcal{U}(p)]$ and, since f^{-1} is a permutation, $\mathcal{U}((f^{-1})^*(f^*(p))) = f^{-1}[f[\mathcal{U}(p)]] = \mathcal{U}(p)$. The type of ultrafilters is an equivalence relation.

As there are $2^{\aleph_0} = \mathfrak{c}$ of permutations of N, and $2^{2^{\aleph_0}} = 2^{\mathfrak{c}}$ of free ultrafilters on N,

PROPOSITION XXII.7.1. *There exist* $2^{\mathfrak{c}}$ *of types, and* $\mathfrak{c}$ *of ultrafilters of every type.*

DEFINITION XXII.7.2. A set $\boldsymbol{F}$ of filters on N is said to be *discrete* if there is a disjoint family $\{N_{\mathcal{F}} : \mathcal{F} \in \boldsymbol{F}\}$ of subsets of N such that $N_{\mathcal{F}} \in \mathcal{F}$ for every $\mathcal{F} \in \boldsymbol{F}$.

If $\{p_n : n < \omega\}$ and $\{q_n : n < \omega\}$ are discrete sequences of ultrafilters such that $\mathrm{t}(p_n) = \mathrm{t}(q_n)$ for each $n < \omega$, and $\mathcal{W}$ is an ultrafilter on $\mathbb{N}$, then the types of the corresponding contours are equal, that is,

$$\mathrm{t}(\textstyle\int_{\mathcal{W}} (p_n)_n) = \mathrm{t}(\textstyle\int_{\mathcal{W}} (q_n)_n).$$

Indeed, let $\{M_n : n < \omega\}, \{N_n : n < \omega\}$ be partitions of N such that $M_n \in \mathcal{U}(p_n)$ and $N_n \in \mathcal{U}(q_n)$ for each $n < \omega$. For each $n < \omega$, let $f_n : M_n \longrightarrow N_n$ be a bijection such that $f^*(p_n) = q_n$. Then $f : N \longrightarrow N$ given by $f(m) := f_n(m)$ if $m \in M_n$, is a permutation. By definition, $A \in \mathcal{U}(\int_{\mathcal{W}} (p_n)_n)$ whenever there exists $W \in \mathcal{W}$ such that $A \in \mathcal{U}(p_n)$

[15]The first indication in this sense was provided by WALTER RUDIN in his celebrated paper [**227**] *Homogeneity problems in the theory of Čech compactifications.* He proved that $\beta^*\mathbb{N}$ is not homogenous under the *Continuum Hypothesis* (that the first non-countable cardinal number is equal to the continuum $\mathfrak{c} = \operatorname{card}\mathbb{R}$), which is a supplementary assumption with respect to ZFC. ZDENĚK FROLÍK, in an equally famous paper [**106**] *Sums of ultrafilters* proved without any set-theoretic assumption that $\beta^*\mathbb{N}$ is very heterogeneous.

for each $n \in W$. Thus $f(A) \in f_n[\mathcal{U}(p_n)] = \mathcal{U}(q_n)$ for each $n \in W$, that is, $f(A) \in \mathcal{U}(\int_{\mathcal{W}}(q_n)_n)$.

Therefore, for any sequence of types $(\tau_n)_n$ and an ultrafilter $\mathcal{W}$, the type $\int_{\mathcal{W}}(\tau_n)_n$ is the type of the contour $\int_{\mathcal{W}}(p_n)_n$, where $\{p_n : n < \omega\}$ is discrete, $\mathrm{t}(p_n) = \tau_n$.

DEFINITION XXII.7.3. We say that a type τ *produces* a type σ, in symbols $\sigma \in \Phi(\tau)$, if there exist an infinite countable discrete subset $X = \{x_n : n < \omega\}$ of β^*N and $\mathcal{W} \in \beta^*X$ such that $\mathrm{t}(\mathcal{W}) = \tau$ and $\mathrm{t}(\int_{\mathcal{W}}(x_n)_n) = \sigma$.

Each infinite countable set X of β^*N is a discrete subset of βN if and only if $\{\mathcal{U}(x) : x \in X\}$ is discrete in the sense of Definition XXII.7.2, in which case $\mathrm{cl}_\beta X$ is homeomorphic to βN in virtue of Theorem XII.7.21.

Accordingly, if $z \in \mathrm{cl}_{\beta^*} X \setminus X$, then there is on X a free ultrafilter $\mathcal{U}_X(z)$ such that $z \in \lim_{\beta^*} \mathcal{U}_X(z)$, that is, by Proposition XII.7.15,

$$\mathcal{U}(z) = \int_{\mathcal{U}_X(z)} \mathcal{U}(\cdot). \tag{XXII.7.1}$$

If $l \in \mathscr{H}(\beta^*, \beta^*)$, then

(XXII.7.2)

$$\mathcal{U}(l(z)) = l[\mathcal{U}(z)] = l[\int_{\mathcal{U}_X(z)} \mathcal{U}(\cdot)] = \int_{\mathcal{U}_X(z)} l[\mathcal{U}(\cdot)] = \int_{\mathcal{U}_{l(X)}(l(z))} \mathcal{U}(\cdot).$$

We can consider the types of ultrafilters with respect to X. Two free ultrafilters z_0 and z_1 on X are of the same X-*type*, $\mathrm{t}_X(z_0) = \mathrm{t}_X(z_1)$, if there is a permutation $f \in X^X$ such that the Stone extension $f^* \in \mathscr{C}(\beta^*X, \beta^*X)$ fulfills $f^*(z_0) = z_1$.

Therefore, by (XXII.7.1),

$$\mathrm{t}(z) \in \Phi(\mathrm{t}_X(z)).$$

It follows from (XXII.7.2) that, for any homeomorphism l,

$$t_{l(X)}(l(z)) = t_X(z). \tag{XXII.7.3}$$

To describe $\Phi(\tau)$ we consider all possible infinite countable discrete subsets X of β^*N, and for each such X, the set $\Phi_X(\tau) := \{\mathrm{t}(z) : z \in \mathrm{cl}_\beta X \setminus X, \mathrm{t}_X(z) = \tau\}$. This yields

$$\Phi(\tau) = \bigcup\nolimits_{X \looparrowright \beta^*N} \Phi_X(\tau),$$

where $X \looparrowright \beta^*N$ means that X is an infinite countable discrete subset of β^*N.

THEOREM XXII.7.4. *Every type of ultrafilters on an infinite countable set produces* $2^{\mathfrak{c}}$ *types.*

PROOF. Let τ be a type. For each infinite countable discrete subset X of β^*N, there is at least one $z \in \mathrm{cl}_{\beta^*} X \setminus X$ such that $\mathrm{t}_X(z) = \tau$. By Proposition XII.8.5, there are $2^{\mathfrak{c}}$ infinite disjoint closed subsets of β^*N (each of which includes an infinite countable discrete subset), hence $2^{\mathfrak{c}}$ different z are produced by τ. As there are $\mathfrak{c}$ of ultrafilters of the same type, there are $2^{\mathfrak{c}}$ types produced by τ. □

LEMMA XXII.7.5. *For each $y \in \beta^*N$, there exists a family $\mathcal{X}$ of cardinality $\mathfrak{c}$ of infinite countable discrete subsets X of β^*N such that $y \in \mathrm{cl}_{\beta^*} X \setminus X$, and if Y is an infinite countable discrete subset of β^*N with $y \in \beta^*Y$, then there is $X \in \mathcal{X}$ such that $X \subset Y$.*

PROOF. For a given $y \in \beta^*N$, consider all infinite countable partitions $\mathcal{M} = \{M_n : n < \omega\}$ of N into infinite countable sets, for which there is a sequence $x_n \in \mathrm{cl}_{\beta^*} M_n \setminus M_n$ such that $y \in \mathrm{cl}_{\beta^*}\{x_n : n < \omega\} \setminus \{x_n : n < \omega\}$. The number of such partitions is (not greater than) the cardinality of N^N, that is, $\aleph_0^{\aleph_0} = 2^{\aleph_0} = \mathfrak{c}$.

Let $\mathcal{X}$ consist of all infinite subsets X of such $\{x_n : n < \omega\}$ for all partitions $\mathcal{M}$, for which $y \in \mathrm{cl}_{\beta^*} X \setminus X$. Then any $Y = \{y_n : n < \omega\}$ as in the claim defines a partition $\mathcal{M}$ of N such that $y_n \in \mathrm{cl}_{\beta^*} M_n \setminus M_n$ and $y \in \mathrm{cl}_{\beta^*} Y \setminus Y$. Let K be the set of $n \in N$ such that $y_n \neq x_n$. If $n \in K$ then M_n can be split into M_n^0 and M_n^1 so that $M_n^0 \in \mathcal{U}(y_n)$ and $M_n^1 \in \mathcal{U}(x_n)$. We claim that $y \notin \mathrm{cl}_{\beta^*}\{y_n : n \in K\}$, hence $y \in \mathrm{cl}_{\beta^*}\{y_n : n \in N\setminus K\}$, and thus Y includes an $X \in \mathcal{X}$. Indeed, if $y \in \mathrm{cl}_{\beta^*}\{y_n : n \in K\}$, then $\bigcup_{n\in K} M_n^0 \in \mathcal{U}(y)$, hence $\bigcup_{n\in K} M_n^1 \cup \bigcup_{n\in N\setminus K} M_n \notin \mathcal{U}(y)$, which is in contradiction with $y \in \mathrm{cl}_{\beta^*}\{x_n : n < \omega\}$. For each $\{x_n : n < \omega\}$, the number of subsets is $2^{\aleph_0} = \mathfrak{c}$, so that $\mathrm{card}\, \mathcal{X} \leq \mathfrak{c} \cdot \mathfrak{c} = \mathfrak{c}$. □

Notice that if X and Y are related as in Lemma XXII.7.5, then $\mathrm{t}_X(y) = \mathrm{t}_Y(y)$. Hence, recalling that there are $\mathfrak{c}$ ultrafilters of each type, it follows that

PROPOSITION XXII.7.6. *Every type of ultrafilters on an infinite countable set is produced by at most $\mathfrak{c}$ types.*

THEOREM XXII.7.7. *The Stone topology on the set of free ultrafilters is not homogeneous.*

PROOF. Let $l \in \mathscr{H}(\beta^*, \beta^*)$, let $y \in \beta^*N$. If $\mathcal{X}_y$ and $\mathcal{X}_{l(y)}$ correspond to y and to $l(y)$ as in Lemma XXII.7.5, then $l(\mathcal{X}_y) = \mathcal{X}_{l(y)}$. As $\Phi^-(\mathrm{t}(y)) = \{\mathrm{t}_X(y) : X \in \mathcal{X}_y\}$ is of cardinality at most $\mathfrak{c}$, and there exist $\mathfrak{c}$ of homeomorphisms l. On a given infinite countable V subset of β^*N, there exists an ultrafilter whose relative type with respect to V is different from all the types producing $\mathrm{t}(l(y))$ for all homeomorphisms

l, because the cardinality of all types is $2^{\mathfrak{c}}$. If $z \in \beta^* N$ is the resulting from this type, then $z \neq l(y)$ for any $l \in \mathscr{H}(\beta^*, \beta^*)$. □

XXII.8. *A compact Fréchet topology with non-Fréchet square

We know already that the product of two Fréchet topologies need not be Fréchet, and if it is Fréchet, then, in virtue of Lemma XXI.8.3, both are necessarily strongly Fréchet. However, this is not a sufficient condition.

On the other hand, it follows from Corollary XXVI.6.11 that a Hausdorff compact Fréchet topology is strongly Fréchet. In this section, we shall see that the product of two Hausdorff compact Fréchet topologies need not be Fréchet.

This fact will be derived from some facts on almost disjoint families, which were defined in Section XII.8.

Recall that if $\mathcal{A}$ is an almost disjoint family on an infinite set X, then the residual filter $\mathrm{c}\mathcal{A}$ of $\mathcal{A}$ is equal to the restriction to X of the neighborhood filter at ∞ of the compact AD topology $\zeta_{\mathcal{A}}^{\infty}$. Accordingly, the filter $\mathrm{c}\mathcal{A}$ is Fréchet if and only if for an infinite set B, the restriction

$$\mathcal{A} \vee_{\infty} B := \{A \cap B : A \in \mathcal{A}, \mathrm{card}(A \cap B) = \infty\}$$

is maximal, then it is finite. Formally,

PROPOSITION XXII.8.1. *The residual filter* $\mathrm{c}\mathcal{A}$ *of an almost disjoint family* $\mathcal{A}$ *is Fréchet if and only if the maximality of* $\mathcal{A} \vee_{\infty} H$ *implies the finiteness of* $\mathcal{A} \vee_{\infty} H$ *for each infinite* $H \subset X$.

PROOF. The residual filter $\mathrm{c}\mathcal{A}$ is Fréchet if and only if for every infinite subset H of X such that $H \in (\mathrm{c}\mathcal{A})^{\#}$, there exists $E \subset H$ such that $(E)_0 \geq \mathrm{c}\mathcal{A}$, which means that the family $\mathcal{A} \vee_{\infty} H \cup \{E\}$ of subsets of H is almost disjoint, that is, $\mathcal{A} \vee_{\infty} H$ not maximal. Hence, $\mathrm{c}\mathcal{A}$ is Fréchet if and only if $\mathcal{A} \vee_{\infty} H$ is MAD amounts to $H \notin (\mathrm{c}\mathcal{A})^{\#}$, that is, by (XII.8.1), $\beta^* H \subset \bigcup_{A \in \mathcal{A}} \beta^* A$. As $\beta^* H$ is β-compact, $\mathcal{A} \vee_{\infty} H$ is finite. □

Let us call *Fréchet* an almost disjoint family the residual filter of which is Fréchet.

From Lemma XVI.4.6 it follows that, for each decreasing sequence of sequential filters $(\mathcal{E}_n)_n$, there exists a sequential filter $\mathcal{H}$ such that $\mathcal{H} \leq \bigwedge_{n<\omega} \mathcal{E}_n$ (although $\bigwedge_{n<\omega} \mathcal{E}_n$ is sequential only if $(\mathcal{E}_n)_n$ is stationary). The following lemma improves this result.

LEMMA XXII.8.2 (Dočkalková). *Let* $\mathcal{A}$ *be a MAD family on infinite countable set, and let* $(E_n)_n$ *be a sequence of infinite sets such that*

$\mathcal{A} \vee_\infty E_n$ is infinite for each $n < \omega$. Then there exists an infinite set H such that $H \setminus E_n$ is finite for each $n < \omega$ and $\mathcal{A} \vee_\infty H$ is infinite.

PROOF. Let $H_0 := \{x_n^0 : n < \omega\}$ be a set of distinct terms such that $x_n^0 \in E_n$ for every $n < \omega$. As $\mathcal{A}$ is MAD, there exists $A_0 \in \mathcal{A}$ such that $A_0 \cap H_0$ is infinite. We will construct by induction, a family $\{A_k : k < \omega\}$ of distinct elements of $\mathcal{A}$, and $H_k := \{x_n^k : k \leq n < \omega\}$ such that for each $k < \omega$ and $n < \omega$,

$$x_n^k \in E_n \setminus \bigcup\nolimits_{0 \leq j \leq k} A_j, \qquad \operatorname{card}(H_k \cap A_k) = \aleph_0. \tag{XXII.8.1}$$

We have proved (XXII.8.1) for $k = 0$. Let $l > 0$ and the claim holds for $k \leq l$. Then since $\mathcal{A} \vee_\infty E_l$ is infinite, there exists a set $H_l = \{x_n^l : l \leq n < \omega\}$ fulfilling (XXII.8.1) for $k = l$, there is $A_l \in \mathcal{A}$ such that $H_l \cap A_l$ is infinite, and since $H_l \cap \bigcup_{0 \leq j < l} A_j = \varnothing$, the sets $A_0, \ldots, A_l$ are distinct. Hence, $H := \bigcup_{k<\omega} H_k$ and $\mathcal{A} \vee_\infty H$ is infinite, while $H \setminus E_n$ is finite for each $n < \omega$. □

LEMMA XXII.8.3 (Simon). *On any infinite countable set there exists a MAD family admitting a partition into two Fréchet families.*

PROOF. Let X be an infinite countable set. Suppose that the claim is false, that is, for each infinite MAD family $\mathcal{A}$ on X and for each partition $\mathcal{A} = \mathcal{A}_0 \sqcup \mathcal{A}_1$, either $\mathcal{A}_0$ or $\mathcal{A}_1$ is not a Fréchet family. By Proposition XII.8.4, there exists on X a MAD family $\mathcal{A}$ of cardinality continuum. Let us index

$$\mathcal{A} = \{A_f : f \in \{0,1\}^{\mathbb{N}}\}.$$

If $\mathcal{A}_{n,0} := \{A_f : f(n) = 0\}$ and $\mathcal{A}_{n,1} := \{A_f : f(n) = 1\}$., then $\mathcal{A}_{n,0} \sqcup \mathcal{A}_{n,1} = \mathcal{A}$ for each $n \in \mathbb{N}$. By assumption, there exists $k_0 \in \{0,1\}$ and an infinite set E_0 such that $\mathcal{A}_{0,k_0} \vee_\infty E_0$ is maximal and infinite.

If $k_0, \ldots, k_p \in \{0,1\}$ and infinite sets $E_0 \supset \ldots \supset E_p$ are such that

$$\mathcal{A}_{n,k_n} \vee_\infty E_n \tag{XXII.8.2}$$

is an infinite MAD family for every $0 \leq n \leq p$, then by inductive assumption there exists $k_{p+1} \in \{0,1\}$ and an infinite subset E_{p+1} of E_p such that (XXII.8.2) with $n = p+1$ is an infinite MAD family. Hence there exist a sequence $(k_n)_n \subset \{0,1\}$ and a decreasing sequence $(E_n)_n$ such that (XXII.8.2) is infinite and MAD for each $n \in \mathbb{N}$.

In particular, $\{A \in \mathcal{A} : \operatorname{card}(A \cap E_n) = \aleph_0\}$ is infinite for every $n \in \mathbb{N}$. Therefore, by Lemma XXII.8.2, there exists an infinite set H so that $H \setminus E_n$ is finite for every $n \in \mathbb{N}$ and such that $\mathcal{A} \vee_\infty H$ is infinite.

Let $f \in \{0,1\}^{\mathbb{N}}$ be such that $f(n) = k_n$ for each $n \in \mathbb{N}$. Then there is $h \neq f$ such that $A_h \in \mathcal{A} \vee_\infty H$. Let n_h be such that $h(n_h) \neq f(n_h)$. Accordingly, $A_h \cap E_{n_h}$ is infinite. Now, $A_h \notin \mathcal{A}_{n_h, f(n_h)}$ so that $A_h \cap A$

is finite for every $A \in \mathcal{A}_{n_h,f(n_h)}$, yet $\mathcal{A}_{n_h,f(n_h)} \vee_\infty E_{n_h}$ is MAD on E_{n_h}, a contradiction. □

THEOREM XXII.8.4 (Simon). *There exists a Hausdorff compact Fréchet topology ξ such that $\xi \times \xi$ is not Fréchet.*

PROOF. The AD topologies $\zeta_0 := \zeta_{\mathcal{A}_0}^\infty$ and $\zeta_1 := \zeta_{\mathcal{A}_1}^\infty$ with $\mathcal{A}_0$ and $\mathcal{A}_1$ on an infinite countable set X from Lemma XXII.8.3 are Hausdorff compact Fréchet. Let $\zeta_0 \times \zeta_1$ be their product on $X^2 = \{(x_0, x_1) : x \in X\}$. Then the free part of the neighborhood filter of $\zeta_0 \times \zeta_1$ at (∞_0, ∞_1) is equal to $\mathcal{F} := \mathcal{F}_{\mathcal{A}_0} \times \mathcal{F}_{\mathcal{A}_1}$. Hence,

$$\mathcal{F}|_{\triangle X} \approx \mathcal{F}_{\mathcal{A}_0} \vee \mathcal{F}_{\mathcal{A}_1} = \mathcal{F}_{\mathcal{A}_0 \cup \mathcal{A}_1},$$

that is, $\mathcal{F}|_{\triangle X}$ is the residual filter of the infinite MAD family $\mathcal{A}$ from Lemma XXII.8.3, hence is not Fréchet, and thus the topology $\zeta_0 \times \zeta_1$ is not Fréchet. To conclude that proof, let $\xi := \zeta_0 \oplus \zeta_1$ is Fréchet, as the coproduct of two Fréchet topologies. Then $\zeta_0 \times \zeta_1$ is a subspace of ξ^2, hence ξ^2 is not Fréchet. □

XXII.9. Measure-theoretic convergences

Measure theory studies various specific convergences of sequences. As measures are countably additive, it is not surprising that convergences related to measure theory are defined in terms of sequences. Of course, convergences of sequences can be be easily extended to sequentially founded convergences.

I shall delineate classical theory of convergences in measure theory ([16]) before embedding it into convergence theory of filters.

A family $\mathscr{S}$ of subsets of a given non-empty set Ω is called a *σ-algebra* if $\Omega \in \mathscr{S}$ and $\Omega \setminus S \in \mathscr{S}$ if $S \in \mathscr{S}$, and any countable union of elements of $\mathscr{S}$ belongs to $\mathscr{S}$. It follows that $\varnothing \in \mathscr{S}$ and any countable intersection of elements of $\mathscr{S}$ belongs to $\mathscr{S}$. A function $\mu : \mathscr{S} \longrightarrow \mathbb{R}_+$ is called a *measure* provided that $\mu(\varnothing) = 0$ and

$$\mu(\bigsqcup\nolimits_{n<\omega} S_n) = \sum\nolimits_{n<\omega} \mu(S_n)$$

for each sequence $\{S_n : n < \omega\}$ of (mutually disjoints) elements of $\mathscr{S}$. A real-valued function f on Ω is called *measurable* if $f^-(O) \in \mathscr{S}$ for each open set O, and the set of all measurable functions is denoted by $\mathfrak{M}$. In examples, we shall use the *Lebesgue measure* on $[0, 1]$ ([17]).

[16]Following their presentations in [**126**] by P. R. HALMOS and in [**228**].

[17]The Lebesgue measure μ on $\Omega = [0, 1]$ is defined the least σ-algebra including intervals and subsets of sets of measure 0, and such that if $b \geq a$ then $\mu([a, b]) = b - a$.

In measure theory, the following basic convergences are defined in terms of sequences of measurable functions.

DEFINITION XXII.9.1. A sequence $(f_n)_n$ of measurable functions converges to a function f

(1) *in measure* whenever, for each $\eta > 0$,

$$\lim_{n\to\infty} \mu\{\,|f_n - f| \geq \eta\} = 0.$$

(2) *almost uniformly* whenever, for each $\lambda > 0$, there exists A such that $\mu(\Omega \setminus A) < \lambda$ and

$$\lim_{n\to\infty} \sup_{x\in A} |f_n(x) - f(x)| = 0.$$

(3) *almost everywhere* whenever there exists a set A such that $\mu(\Omega \setminus A) = 0$ and for each $x \in A$,

$$\lim_{n\to\infty} |f_n(x) - f(x)| = 0.$$

Therefore, any of these convergences can be represented by a relation between $\mathfrak{M}^N$ and $\mathfrak{M}$ (see Section XVI.3).

PROPOSITION XXII.9.2. *Almost uniform convergence entails convergence almost everywhere.*

PROOF. Let $(h_n)_n$ almost uniformly converge to 0, in particular, for each $k < \omega$, there exists E_k such that $\mu(E_k) < 2^{-k}$ and $(h_n(x))_n$ converges to 0 for each $x \in \Omega \setminus E_k$. Hence, $\mu(\bigcap_{k<\omega} E_k) = 0$ and $(h_n)_n$ pointwise converges to 0 on $\Omega \setminus \bigcap_{k<\omega} E_k$. □

REMARK XXII.9.3. Notice that in the proof above, the convergence on $\Omega \setminus \bigcap_{k<\omega} E_k$ is, in general, not uniform.

PROPOSITION XXII.9.4. *If the measure is finite, then convergence almost everywhere implies almost uniform convergence.*

PROOF. Let $\mu(\Omega) < \infty$, and let $(h_n)_n$ converge to 0 almost everywhere. For simplicity sake, we assume that it *converges everywhere* ([18]). Let

$$Q_{n,\varepsilon} := \bigcup_{m\geq n} \{|h_m| \geq \varepsilon\}. \tag{XXII.9.1}$$

Observe that $\bigcap_{n<\omega} Q_{n,\varepsilon} = \varnothing$, because by assumption $(h_n)_n$ converges to 0 everywhere. As $\mu(\Omega) < \infty$, we infer that $\lim_{n\to\infty} \mu(Q_{n,\varepsilon}) = 0$ for each $\varepsilon > 0$.

[18]This is done without loss of generality, as we can redefine Ω so that the sequence converge everywhere.

Fix $\delta > 0$ and let $\varepsilon = 2^{-k}$. Then there exists n_k such that $\mu(Q_{n_k,2^{-k}}) < \delta 2^{-k-1}$. If $Q := \bigcup_{k<\omega} Q_{n_k,2^{-k}}$, then

$$\mu(Q) < \sum_{k<\omega} \delta 2^{-k-1} = \delta.$$

If $x \notin Q$, then $x \notin Q_{n_k,2^{-k}}$ for each $k < \omega$, hence by definition (XXII.9.1), $|h_m(x)| < 2^{-k}$ for each $m \geq n_k$. □

LEMMA XXII.9.5. *Almost uniform convergence implies convergence in measure.*

PROOF. Suppose that $(h_n)_n$ of measurable positive functions *almost uniformly converges* to 0. Let $\varepsilon > 0$ and $\eta > 0$. Then there exist A_ε and $n_\varepsilon < \omega$ such that $\mu(\Omega \setminus A_\varepsilon) < \varepsilon$, and $\sup_{A_\varepsilon} |h_n| < \eta$ for $n > n_\varepsilon$. On the other hand, $A_\varepsilon \cap \{|h_n| \geq \eta\} = \varnothing$ for $n > n_\varepsilon$, because $\sup_{A_\varepsilon} |h_n| < \eta$. Accordingly, $\mu\{|h_n| \geq \eta\} < \varepsilon$ for each $n > n_\varepsilon$. □

EXAMPLE XXII.9.6 (Convergent in measure, but not almost everywhere). Let $f_{k,n} \in [0,1]^{[0,1]}$ be the characteristic function $[\frac{n-1}{k}, \frac{n}{k}]$ for $0 < n \leq k, k \in \mathbb{N}_1$.

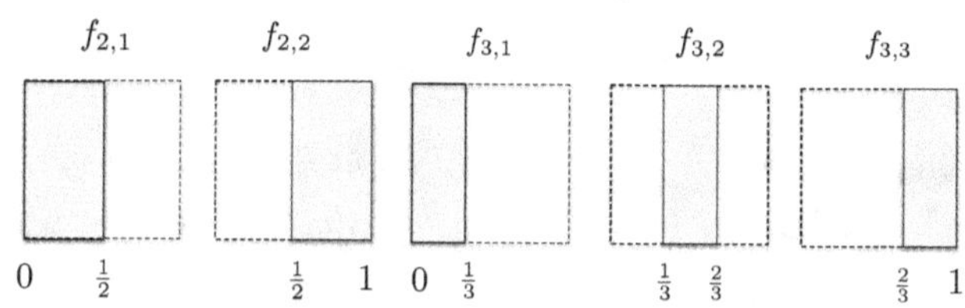

FIGURE XXII.9.1. Several terms of a sequence of characteristic functions converging to 0 in measure, but not almost everywhere.

Then if $0 < \eta \leq 1$, then $\{x : \chi_E(x) \geq \eta\} = E$, hence $\mu\{f_{k,n} \geq \eta\} = \frac{1}{k}$ for each $0 < n \leq k$, which means that $\lim(f_{k,n})_{k,n} = 0$, where the index k, n denotes the cofinite filter on $\{(n,k) : 0 < n \leq k, k \in \mathbb{N}_1\}$. On the other hand, $(f_{k,n})_{k,n}$ does not converge almost everywhere, because the set $\{(n,k) : f_{k,n}(x) = 1\}$ is infinite for each $x \in [0,1]$.

LEMMA XXII.9.7. *A sequence that is fundamental in measure has a subsequence convergent almost uniformly.*

PROOF. Let $(f_n)_n$ be fundamental in measure. If $\varepsilon > 0$, then for each $k < \omega$, there exists n_k such that $\mu\{|f_n - f_m| \geq \frac{1}{2^k}\} < \frac{\varepsilon}{2^k}$ for each $n, m > n_k$.

Let $E_k := \{|f_{n_{k+1}} - f_{n_k}| \geq \frac{1}{2^k}\}$, and $E := \bigcup_{k<\omega} E_k$. Then $\mu(E) \leq \sum_{k<\omega} \mu(E_k) < \varepsilon$. We claim that $(f_{n_k})_k$ converges uniformly on $\Omega \setminus E$. Consider

(XXII.9.2) $$f(x) := f_{n_0}(x) + \sum_{k<\omega} f_{n_{k+1}}(x) - f_{n_k}(x).$$

If $x \notin E$, then $x \notin E_k$ for each $k < \omega$, hence $|f_{n_{k+1}}(x) - f_{n_k}(x)| < \frac{1}{2^k}$ for each $k < \omega$. Therefore, by Exercise X.4.14, the series is convergent uniformly on $\Omega \setminus E$. □

PROPOSITION XXII.9.8. *Convergence in measure is complete.*

PROOF. By Lemma XXII.9.7, each measure fundamental sequence has an almost uniformly convergent subsequence, which is convergent in measure by Lemma XXII.9.5. By Corollary X.4.2, a fundamental sequence with a convergent subsequence is convergent. □

In Section XVI.3, we have seen that any convergence of sequences can be extended to a relation $\theta \subset \mathbb{F}\mathfrak{M} \times \mathfrak{M}$ such that $\theta^{-}\mathfrak{M} \subset \mathbb{S}\mathfrak{M}$, that is, a sequential convergence. Therefore, by Lemma III.7.10, θ has an extension to the finest convergence $\mathrm{I}\theta$ coarser than θ. As for all these convergences, each subsequence of a convergence sequence is convergent (to the same point), θ is (isotone) restricted to sequential filters, and (centered). In fact,

PROPOSITION XXII.9.9. *If θ is a sequential convergence, then its convergence modification $\mathrm{I}\theta$ is a sequentially founded convergence.*

A convergence (of filters) is called *measure-theoretic* if whenever a filter $\mathcal{F}$ defined on a set of measurable functions $\mathfrak{M}$ converges to $f \in \mathfrak{M}$, and $\varphi : \mathfrak{M} \longrightarrow \mathfrak{M}$ is such that $\varphi(h) = h$ a.e., then $\varphi(f)$ is a limit of $\varphi[\mathcal{F}]$. This gives rise to a concept of convergence on equivalence classes $\mathfrak{M}_\mu$ of functions equal μ-a.e.

We proceed to embed into convergence-theoretic setting convergences occurring in measure theory. Let $(\Omega, \mathscr{S}, \mu)$ be a measure space, where $\mathscr{S}$ a σ-algebra of subsets of Ω, and μ a finite measure defined on elements of $\mathscr{S}$. The set of measurable functions is denoted by $\mathfrak{M}$, while $\mathfrak{M}_\mu$ stands for the set of equivalence classes μ-a.e. Let

$$[f]_\mu := \{g \in \mathfrak{M} : \mu\{g \neq f\} = 0\},$$

be the class of functions μ-equivalent to f.

Recall Proposition IV.3.3 that $\Phi^{-}\tau$ is a convergence of fibers of Φ. In general, $\xi \geq \Phi^{-}(\Phi\xi)$. Let $\Phi : \mathfrak{M} \longrightarrow \mathfrak{M}_\mu$ be the quotient map with respect to the equivalence relation μ-a.e., that is,

(XXII.9.3) $$\Phi(f) := [f]_\mu.$$

If ξ is a convergence on $\mathfrak{M}$, then let

$$[\xi]_\mu := \Phi\xi,$$

the final convergence of ξ with respect to Φ ([19]).

PROPOSITION XXII.9.10. *A convergence ξ on $\mathfrak{M}$ is measure-theoretic if and only if $\xi = \Phi^-(\Phi\xi)$.*

PROOF. By definition, $g \in [f]_\mu$ whenever $\mu\{f \neq g\} = 0$. Therefore, μ is measure-theoretic if and only if $\varphi : \mathfrak{M} \longrightarrow \mathfrak{M}$ such that $\mu\{\varphi(f) \neq f\} = 0$ for each $f \in \mathfrak{M}$, then

$$f_\infty \in \lim_\xi \mathcal{F} \Longrightarrow \varphi(f_\infty) \in \lim_\xi \varphi[\mathcal{F}],$$

that is, by (XXII.9.3), $\Phi(f) = \Phi(\varphi(f))$ as well as $\Phi[\mathcal{F}] = \Phi[\varphi[\mathcal{F}]]$, that is, by Proposition IV.3.3, $\xi = \Phi^-(\Phi\xi)$. $\square$

Accordingly, a convergence ξ on $\mathfrak{M}$ is measure-theoretic if and only if $\xi = \Phi^-[\xi]_\mu$. We observe that if ξ and θ are measure-theoretic convergences on $\mathfrak{M}$, then

$$\xi \leq \theta \Longleftrightarrow [\xi]_\mu \leq [\theta]_\mu.$$

Therefore, all the relations on measure-theoretic convergences on $\mathfrak{M}$ defined in terms of order are tantamount to the corresponding relations on $\mathfrak{M}_\mu$. For instance, if ξ is measure-theoretic, then $[\mathrm{Seq}\xi]_\mu = \mathrm{Seq}([\xi]_\mu)$.

Here is a natural extension to filters of convergence *almost everywhere*. Given a filter $\mathcal{F}$ on $\mathfrak{M}$ and $x \in \Omega$, let

$$\mathcal{F}(x) := \{\{f(x) : f \in F\} : F \in \mathcal{F}\}.$$

It is clear that $\mathcal{F}(x)$ is a filter on the real line.

DEFINITION XXII.9.11. A filter $\mathcal{F}$ on $\mathfrak{M}$ *converges almost everywhere* to $f_\infty \in \mathfrak{M}$, in symbols

$$f_\infty \in \lim_{e_\mu} \mathcal{F},$$

whenever there exists $S \in \mathscr{S}$ such that $\mu(\Omega \setminus S) = 0$ and $f_\infty(x) \in \lim \mathcal{F}(x)$ for each $x \in S$.

In particular, if $\mathcal{F}$ is a sequential filter on $\mathfrak{M}$, say $\{\{f_n : n > k\} : k < \omega\}$ is a filter-base of $\mathcal{F}$, then $\mathcal{F}$ converges to f_∞ in the sense of Definition XXII.9.11 if and only if $(f_n)_n$ converges to f_∞ in almost everywhere in the usual sense.

The convergence from Definition XXII.9.11, however, is not measure-theoretic.

[19] As Φ is a quotient map, $\Phi\xi$ is called the *quotient convergence.* I do not use the term *quotient* in order to avoid confusion with the notion of *quotient topology.*

EXAMPLE XXII.9.12. Consider on $[0,1]$ the relation $\approx_{\mathbb{Q}}$ defined by $x_0 \approx_{\mathbb{Q}} x_1$ whenever $x_1 - x_0 \in \mathbb{Q}$. This is clearly an equivalence relation. We denote by $\mathcal{R}$ the family of equivalence classes. Of course, each $R \in \mathcal{R}$ is countable, and $\operatorname{card} \mathcal{R} = \mathfrak{c}$. Let $F_r := \{s\chi_R : 0 < s < r, R \in \mathcal{R}\}$ for $r > 0$. Then $\{F_{2^{-n}} : n < \omega\}$ is a filter-base of a filter $\mathcal{F}$ on $\mathfrak{M}$. It is clear that $\mathcal{F}$ converges to the function 0 everywhere. On the other hand, let $\varphi(s\chi_R) := \chi_R$ for each $R \in \mathcal{R}$ and every $s > 0$. Then $\varphi[\mathcal{F}]$ is the principal filter, and $\{\chi_R : R \in \mathcal{R}\}$ is its filter-base. Note that $\varphi(s\chi_R) = \chi_R$ almost everywhere, and that $\varphi[\mathcal{F}]$ does not converge almost everywhere. Indeed, for every $x \in [0,1]$, there is $R_x \in \mathcal{R}$ such that $x \in R_x$ and $x \notin R$. Therefore, for any $S \in \mathscr{S}$ with $\mu(S) = 1$, there is $x \in S$ such that χ_{R_x} belongs to every element of $\varphi[\mathcal{F}]$, and thus 1 belongs to every element of the filter $\varphi[\mathcal{F}](x)$. Accordingly, $\varphi[\mathcal{F}]$ does not converge to 0 almost everywhere.

Nevertheless,

PROPOSITION XXII.9.13. *The countably carried modification of convergence almost everywhere is measure-theoretic.*

PROOF. Indeed, if $f_\infty \in \lim_{\mathrm{Ee}_\mu} \mathcal{F}$ and $\varphi : \mathfrak{M} \longrightarrow \mathfrak{M}$ is such that $\mu\{h \neq \varphi(h)\} = 0$ for each $h \in \mathfrak{M}$, then there is $S \in \mathscr{S}$ such that $\mu(\Omega \setminus S) = 0$ and $f_\infty(x) \in \lim \mathcal{F}(x)$ for each $x \in S$. As there is a countable set $F_0 \in \mathcal{F}$, also $\varphi(f_\infty)(x) \in \lim \varphi[\mathcal{F}](x)$ for each $x \in S_0 := S \setminus \bigcup_{f \in F_0 \cup \{f_\infty\}} \{f \neq \varphi(f)\}$, and $\mu(\Omega \setminus S_0) = 0$. $\square$

At this point we recollect that convergences *almost everywhere* and *almost uniform* coincide on sequential filters. As we shall see, *almost uniform convergence* naturally extends to a measure-theoretic convergence on all filters.

Here $f|_A$ stands for the restriction of f to a set A, and

$$\|h\|_\infty := \inf\{c : \mu\{|h| > c\} = 0\}$$

is the *essential supremum* of $|h|$.

DEFINITION XXII.9.14. A filter $\mathcal{F}$ on $\mathfrak{M}$ converges *almost uniformly* to a function 0,

$$0 \in \lim_{a_\mu} \mathcal{F},$$

if for each $\varepsilon > 0$, there exist $S \in \mathscr{S}$ such that $\mu(\Omega \setminus S) < \varepsilon$, and for each $\lambda > 0$, there exists $F \in \mathcal{F}$ so that $\| f|_S \|_\infty < \lambda$ for each $f \in F$.

Almost uniform convergence is obviously measure-theoretic, because it is defined in terms of $\|\cdot\|_\infty$ which is constant on each $[f]_\mu$.

PROPOSITION XXII.9.15. *Almost uniform convergence is finer than convergence almost everywhere, that is,*

$$a_\mu \geq e_\mu.$$

PROOF. If $0 \in \lim_{a_\mu} \mathcal{F}$, then, in particular, for each $k < \omega$, there is $S_k \in \mathscr{S}$ such that $\mu(\Omega \setminus S_k) < 2^{-k}$ and $0 \in \lim \mathcal{F}(x)$ for each $x \in S_k$. Hence, $0 \in \lim \mathcal{F}(x)$ for each $x \in S := \bigcup_{k<\omega} S_k$ and $\mu(\Omega \setminus S) = 0$. □

DEFINITION XXII.9.16. A filter $\mathcal{F}$ on $\mathfrak{M}$ *converges in measure* to $f_\infty \in \mathfrak{M}$, in symbols

$$f_\infty \in \lim_{m_\mu} \mathcal{F}, \tag{XXII.9.4}$$

if for all real numbers $\lambda > 0$ and $\varepsilon > 0$, there exists $F \in \mathcal{F}$ such that $\mu\{|f - f_\infty| \geq \lambda\} < \varepsilon$ for each $f \in F$.

We promptly observe that the convergence m_μ is measure-theoretic. Indeed, if $\varphi : \mathfrak{M} \longrightarrow \mathfrak{M}$ is such that $\varphi(h) = h$ a.e. for every $h \in \mathfrak{M}$, then $\mu\{|\varphi(f) - \varphi(f_\infty)| \geq \lambda\} = \mu\{|f - f_\infty| \geq \lambda\}$ for any f and f_∞.

Also, m_μ is μ-a.e. Hausdorff, that is, $[m_\mu]_\mu$ is Hausdorff. Indeed, if $f \neq g$ μ-a.e, that is, there is $S \in \mathscr{S}$ with $\mu(S) > 0$, for which $S \cap \{f = g\} = \varnothing$, then $S \subset \bigcap_{n<\omega}\{|f - g| > 2^{-n}\}$ and thus there is $n < \omega$ such that $\mu\{|f - g| > 2^{-n}\} > 0$. Hence a filter cannot m_μ-converge both to f and g.

PROPOSITION XXII.9.17. *Almost uniform convergence is finer than convergence in measure (for arbitrary filters).*

PROOF. Suppose that $0 \in \lim_{a_\mu} \mathcal{F}$ and let $\varepsilon > 0$. By assumption, there is a measurable set $S \subset \Omega$ such that $\mu(\Omega \setminus S) < \varepsilon$ and for each $\lambda > 0$, there exists $F \in \mathcal{F}$ such that $\| f|_S\|_\infty < \lambda$ for each $f \in F$.

This implies that for each $\lambda > 0$ and $\varepsilon > 0$, there exist $S \subset \Omega$ and $F \in \mathcal{F}$ such that $\mu(\Omega \setminus S) < \varepsilon$ and $S \subset \{|f| \geq \lambda\}$ for each $f \in F$, that is, $\mu(\{|f| \geq \lambda\}) < \varepsilon$ for each $f \in F$. □

We notice that *convergence in measure* m_μ of Definition XXII.9.16 is equivalent to continuity of a family of real-valued functions. Indeed, given measurable functions f, g and a real number $\lambda > 0$, we define

$$e_\lambda(f, g) := \mu\{|f - g| \geq \lambda\}. \tag{XXII.9.5}$$

Notice that $e_\lambda(f, g) = 0$ for each $\lambda > 0$ if and only if $f = g$ μ-a.e. Then $f_\infty \in \lim_{m_\mu} \mathcal{F}$ if and only if, for each $\lambda > 0$, the filter $e_\lambda[\mathcal{F}, f_\infty] = \{\{e_\lambda(f, f_\infty) : f \in F\} : F \in \mathcal{F}\}$ on $\mathbb{R}_+$ tends to 0. Of course, $e_\lambda([f]_\mu, [g]_\mu) := e_\lambda(f, g)$ is well-defined.

Consequently,

PROPOSITION XXII.9.18. *Convergence in measure is the finest convergence, for which the functionals* $\{e_\lambda : \lambda > 0\}$ *are continuous.*

In other words, m_μ is the initial convergence with respect to a family of real-valued functions $\{e_\lambda(\cdot, f) : \lambda > 0, f \in \mathfrak{M}\}$, and thus $[m_\mu]_\mu$ is a Hausdorff, functionally regular (alias completely regular) topology of countable character (see, e.g., [83]).

If μ is the Lebesgue measure on a bounded closed interval I, then the topology m_μ is separable. Indeed, if $\mu\{|h| > \lambda\} < \varepsilon$, then there is a simple function s valued in $\mathbb{Q}$ and with the intervals with rational ends such that $\mu\{|s| > \lambda\} < \varepsilon$. There are countably many of them. As we can also take $\lambda, \varepsilon \in \mathbb{Q}$, we conclude that the topology m_μ is separable. It follows from the Urysohn metrization theorem (e. g., [83, p. 204]) that m_μ is metrizable.

COROLLARY XXII.9.19. *If* μ *is the Lebesgue measure on a bounded closed interval, then convergence in measure* $[m_\mu]_\mu$ *(on equivalence classes) is a metrizable topology.*

Recall the *Urysohn modification* of ξ, denoted by $\mathrm{U}\xi$ (XVI.3.6): for each sequential filter $\mathcal{E}$,

$$\lim\nolimits_{\mathrm{U}\xi} \mathcal{E} := \bigcap\nolimits_{\mathcal{G} \in \mathbb{S}\mathcal{E}} \bigcup\nolimits_{\mathcal{H} \in \mathbb{S}\mathcal{G}} \lim\nolimits_\xi \mathcal{H}. \tag{XXII.9.6}$$

If $\mathcal{E}$ is a sequential filter and $x \in \lim_\xi \mathcal{E}$, then $x \in \lim_\xi \mathcal{H}$ for every filter $\mathcal{H} \supset \mathcal{E}$, hence $\lim_\xi \mathcal{E} \subset \lim_{\mathrm{U}\xi} \mathcal{E}$ for each $\mathcal{E}$ and ξ, so that $\xi \geq \mathrm{U}\xi$ for each sequentially founded convergence ξ. Accordingly, ξ is a Urysohn convergence whenever $\mathrm{U}\xi \geq \xi$.

By Lemma XXII.9.7,

COROLLARY XXII.9.20. *Sequentially founded convergence in measure is the Urysohn modification of the sequentially founded convergence almost everywhere.*

PROOF. Indeed, if a sequential filter $\mathcal{E}$ converges to f in measure, then there exists $\mathcal{F} \in \mathbb{S}\mathcal{E}$ that converges to f a.e. Hence each $\mathcal{G} \in \mathbb{S}\mathcal{F}$ also converges to f a.e., that is, if $\mathcal{E}$ converges to f in measure, then it also does in the Urysohn modification of the convergence a.e. On the other hand, convergence a.e. entails convergence in measure. □

It follows from Corollary XXI.5.7 that each sequentially founded convergence almost everywhere is a pseudotopology. Hence,

$$\mathrm{Seq}e_\mu = \mathrm{S}(\mathrm{Seq}e_\mu) > \mathrm{U}(\mathrm{Seq}e_\mu) = \mathrm{Seq}m_\mu. \tag{XXII.9.7}$$

However,

PROPOSITION XXII.9.21. *Let μ be the Lebesgue measure on $[0,1]$. The countably carried modification of almost uniform convergence a_μ is not a pseudotopology, that is,* $\mathrm{E}e_\mu > \mathrm{S}(\mathrm{E}e_\mu)$.

PROOF. Consider the closed unit interval with Lebesgue measure. Following [**126**, p. 94], let $f_{k,n} \in [0,1]^{[0,1]}$ be the characteristic function of $[\frac{n-1}{k}, \frac{n}{k}]$ for $0 < n \leq k, 0 \neq k \in \mathbb{N}$. Let $\mathcal{F}$ be a family of subsets of $\{0,1\}^{[0,1]}$ defined by $F \in \mathcal{F}$ whenever $\{(k,n) : f_{k,n} \notin F\}$ is finite ([20]). Then $\mathcal{F}$ is a sequential filter that does not converge to 0 almost everywhere, equivalently almost uniformly, because $\{(k,n) : f_{k,n}(x) = 1\}$ is infinite for each $x \in [0,1]$ ([21]).

I claim that each ultrafilter $\mathcal{U}$ finer than $\mathcal{F}$ converges to 0 almost uniformly. Let $\pi : \{0,1\}^{[0,1]} \longrightarrow [0,1]$ be any selection $\pi(\chi_A) \in A$. Then $\pi(U) := \{\pi(f) : f \in U\}$, and $\pi[\mathcal{U}] := \{\pi(U) : U \in \mathcal{U}\}\}$ is an ultrafilter-base on $[0,1]$. By compactness of $[0,1]$, there exists $x_\mathcal{U}$ such that $\pi[\mathcal{U}]$ converges to $x_\mathcal{U}$. Hence, for each $\varepsilon > 0$, there is $U \in \mathcal{U}$ such that $\pi(U) \subset (x_\mathcal{U} - \frac{\varepsilon}{2}, x_\mathcal{U} + \frac{\varepsilon}{2})$.

On the other hand, there is $F \in \mathcal{F}$ such that if $f_{n,k} \in F$, then $\frac{1}{k} < \frac{\varepsilon}{2}$ and $\pi(f_{n,k}) \in [\frac{n-1}{k}, \frac{n}{k}]$. Consequently, $[\frac{n-1}{k}, \frac{n}{k}] \subset]x_\mathcal{U} - \varepsilon, x_\mathcal{U} + \varepsilon[$ for each (k,n) such that $f_{k,n} \in U \cap F$. Therefore $\mathcal{U}$ converges uniformly to 0 out of $]x_\mathcal{U} - \varepsilon, x_\mathcal{U} + \varepsilon[$ for each $\varepsilon > 0$, hence converges to 0 almost uniformly. Therefore, $\mathcal{F}$ converges to 0 in the pseudotopological modification of the countably carried modification of almost uniform convergence. □

REMARK XXII.9.22. In the proof, we construct a filter $\mathcal{F}$ such that $0 \in \lim_{\mathrm{S}e_\mu} \mathcal{F} \setminus \lim_{e_\mu} \mathcal{F}$. As $\mathcal{F}$ is sequential, $\lim_{e_\mu} \mathcal{F} = \lim_{\mathrm{Seq}e_\mu} \mathcal{F}$. However, it follows from the construction that $\lim_{\mathrm{S}e_\mu} \mathcal{F} \neq \lim_{\mathrm{SSeq}e_\mu} \mathcal{F}$. Of course, the considered ultrafilters are countably carried, because they include the sequential filter $\mathcal{F}$, but none includes a sequential filter convergent to 0.

REMARK XXII.9.23. Notice that a filter is countably carried if and only if every finer ultrafilter is countably carried. Therefore, SE = ES. Consequently, $\mathrm{E}a_\mu > \mathrm{S}(\mathrm{E}a_\mu)$ implies $a_\mu > \mathrm{S}a_\mu$. Hence, by Proposition XXII.9.17 and Corollary XXII.9.19, if μ is the Lebesgue measure on $[0,1]$, then

$$a_\mu > \mathrm{S}a_\mu \geq m_\mu = \mathrm{T}m_\mu.$$

By Propositions XXII.9.2 and XXII.9.4, then $\mathrm{Seq}\, a_\mu$ is a common restriction to sequentially founded convergences of convergences *almost uniform* and *almost everywhere*, and thus is measure-theoretic.

---

[20]In other words, $\mathcal{F}$ is the cofinite filter on the set $X := \{f_{n,k} : 0 < n \leq k, 0 \neq k \in \mathbb{N}\}$, or else corresponds to any order of type ω on X.

[21]In fact, it converges nowhere.

This can be also see directly, because each $\mathrm{Seq}\, a_\mu$-convergent filter $\mathcal{F}$ contains a countable set, say F_0. Therefore, if there is $S \in \mathscr{S}$ such that $\mu(\Omega \setminus S) = 0$ so that $0 \in \lim \mathcal{F}(x)$ for each $x \in S$, and $\varphi(h) = h$ μ-a.e. for each $h \in \mathfrak{M}$, then $0 \in \lim \varphi[\mathcal{F}](x)$ for each $x \in S_0 := S \cup \bigcup_{f \in F_0} \{|\varphi(f) - f|\}$, and $\mu(\Omega \setminus 0) = 0$.

The following example highlights the difference between $\mathrm{Seq} e_\mu$ and $\mathrm{E} e_\mu$. It shows that while the former is a pseudotopology finer than $\mathrm{Seq}\, m_\mu$ (XXII.9.7), the latter is not, and $\mathrm{S}(\mathrm{E} e_\mu)$ is not finer than $\mathrm{E}\, m_\mu$. In particular, the countably carried modification of convergence almost everywhere is not sequentially founded.

EXAMPLE XXII.9.24 ([22]). Consider the Lebesgue measure on $[0, 1]$. For each $n < \omega$, let $\mathcal{K}_n$ consist of the sets of the form $\bigcup_{i \in I} [i2^{-n}, (i + 1)2^{-n}]$, where $I \subset \{0, \dots, 2^n - 1\}$ such that $\mathrm{card}\, I = n$. Accordingly, $\mu(K) = \frac{n}{2^n}$ for each $K \in \mathcal{K}_n$. Let $X := \bigcup_{n < \omega} \{\chi_K : K \in \mathcal{K}_n\}$. It is plain that the cofinite filter of X converges *in measure* to the function 0, hence every free ultrafilter also does.

On the other hand, for each $A \subset [0, 1]$ ([23]), there exists an ultrafilter $\mathcal{U}_A$ on X that converges almost everywhere to χ_A. Indeed, each neighborhood of χ_A in the pointwise convergence intersects the cofinite filter of X, equivalently, for each finite subset F of $[0, 1]$, there are infinitely many $f \in X$ that coincide with χ_A on F, that is, $f(x) = 1$ if $x \in F \cap A$ and $f(x) = 0$ if $x \in F \setminus A$.

Since all the free filters on X from Example XXII.9.24 are finer than the cofinite filter of X, they converge to 0 in measure. As convergence in measure is μ-a.e.-Hausdorff, we infer that

COROLLARY XXII.9.25. *The pseudotopological modification of the countably carried modification of convergence almost everywhere is not finer than convergence in measure.*

By Propositions XXII.9.2 and XXII.9.4, $\mathrm{Seq} e_\mu = \mathrm{Seq} a_\mu$, but $\mathrm{E} e_\mu$ is not necessarily finer than $\mathrm{E} a_\mu$. Indeed, it follows from Proposition XXII.9.15 that $\mathrm{E} a_\mu \geq \mathrm{E} e_\mu$, and from Proposition XXII.9.17 that $\mathrm{E} a_\mu \geq \mathrm{E} m_\mu$. As, by Corollary XXII.9.25, $\mathrm{E} e_\mu \not\geq \mathrm{E} m_\mu$, we infer that $\mathrm{E} e_\mu \not\geq \mathrm{E} a_\mu$.

REMARK XXII.9.26. Let X be the set from Example XXII.9.24. Endow $\beta^* X$, the set of all free ultrafilters on X, with the Stone topology: $\{\beta^* W : W \subset X, \mathrm{card}\, W = \infty\}$ is a base for the topology β and

[22] Communicated by Professor D. Fremlin.

[23] In order to stay in our setting, we suppose that A is measurable, but the property holds for any A.

consists of β-clopen sets. It turns out that the subset $\boldsymbol{A}$ of β^*X of those ultrafilters that converge to 0 in Seqe_μ is open and dense. This means that $\{\mathcal{U}_A : A \subset [0,1]\}$ is nowhere dense. Indeed, as the cofinite filter of X converges to 0 is measure, for every infinite subset W of X there is an infinite subset V such that the cofinite filter of V converges to 0 almost everywhere by Proposition XXII.9.24. Therefore, each $\mathcal{U} \in \beta^*V$ converges to 0 almost everywhere. This shows that $\boldsymbol{A}$ is β^*-dense. But if $\mathcal{U} \in \boldsymbol{A}$, then by definition there exists a free sequential filter $\mathcal{E}$ included in $\mathcal{U}$ that converges to 0 almost everywhere. Hence there is an infinite subset E of X such that $\mathcal{E}$ is the cofinite filter of E, and thus each $\mathcal{U} \in \beta^*E$ converges to 0 almost everywhere, proving that $\boldsymbol{A}$ is β^*-open.

CHAPTER XXIII

Completeness

Completeness is a common generalization of compactness, local compactness, metric completeness, and of topological completeness. Completeness number of a convergence indicates how far it is from being compact (a convergence is compact whenever its completeness number is zero).

In Section X.4 we encountered completely metrizable topologies. A metric is complete if each Cauchy sequence (equivalently, fundamental filter) is convergent. A metrizable topology can admit both complete and non-complete metrics; it is completely metrizable if at least one of compatible metrics is complete. If a metrizable space is compact, then all its compatible metrics are complete.

Topological completeness, also known as *Čech completeness*, is defined in terms of collections of covers, which become collections of balls in the metric case. Traditionally topological completeness was defined for Hausdorff, functionally regular (alias completely regular) topologies ([1]). As we shall see, a natural framework for investigation of completeness is that of pseudotopologies.

XXIII.1. Fundamental filters and completeness

Let ξ be a convergence on X, and let $\boldsymbol{P}$ be a collection of families of subsets of X. We shall say that ξ is $\boldsymbol{P}$*-complete* if $\operatorname{adh}_\xi \mathcal{F} \neq \varnothing$ for each filter $\mathcal{F}$ such that $\mathcal{F} \cap \mathcal{P} \neq \varnothing$ for every $\mathcal{P} \in \boldsymbol{P}$.

It is useful to rephrase this definition splitting it into two.

DEFINITION XXIII.1.1 (Fundamental filter). A filter $\mathcal{F}$ is called $\boldsymbol{P}$*-fundamental* or $\boldsymbol{P}$*-Cauchy* if

$$\mathcal{F} \cap \mathcal{P} \neq \varnothing$$

for every $\mathcal{P} \in \boldsymbol{P}$.

EXAMPLE XXIII.1.2. Let (X, d) be a metric space, and let $\mathrm{t}d$ be the corresponding convergence. Let $\mathcal{P}_\varepsilon := \{B_d(x, \varepsilon) : x \in X\}$ be the family of all d-balls of radius $\varepsilon > 0$, and $\boldsymbol{P} := \{\mathcal{P}_\varepsilon : \varepsilon > 0\}$. A filter

[1]For example, R. ENGELKING, [97, Theorem 3.9.2].

$\mathcal{F}$ is $\boldsymbol{P}$-fundamental if and only if for each $\varepsilon > 0$, there is $x \in X$ such that $B_d(x, \varepsilon) \in \mathcal{F}$, that is, if and only if $\mathcal{F}$ is a d-Cauchy filter:

$$\inf_{F \in \mathcal{F}} \operatorname{diam} F = 0. \tag{XXIII.1.1}$$

DEFINITION XXIII.1.3 (Completeness). A convergence ξ is said to be $\boldsymbol{P}$-*complete* whenever each $\boldsymbol{P}$-fundamental filter is ξ-adherent.

Of particular interest are convergences that are complete with respect to a collection of convergence covers. The following example goes back to the origins of completeness.

EXAMPLE XXIII.1.4. A metric d is *complete* if each d-Cauchy filter is $\mathrm{t}d$-adherent. By Proposition X.4.3, $\mathrm{t}d$ constitutes $\boldsymbol{P}$-complete if and only if d constitutes a complete metric. By Lemma X.4.1, a $\mathrm{t}d$-adherent d-Cauchy filter is $\mathrm{t}d$-convergent.

EXAMPLE XXIII.1.5. A convergence is compact if and only if it is $\varnothing$-complete. Here $\varnothing$ stands for the empty collection of families of subsets of X. Therefore, every filter $\mathcal{F}$ on X is $\varnothing$-fundamental, because there is no $\mathcal{P}$ in $\varnothing$. In other terms, ξ is $\varnothing$-complete whenever $\operatorname{adh}_\xi \mathcal{F} \neq \varnothing$ for each filter $\mathcal{F}$.

Let $\boldsymbol{P}$ and $\boldsymbol{Q}$ be collections of families of subsets of X.

LEMMA XXIII.1.6. *If $\mathcal{F}$ is $\boldsymbol{Q}$-fundamental, and for every $\mathcal{P} \in \boldsymbol{P}$, there exists a refinement $\mathcal{Q} \in \boldsymbol{Q}$ of $\mathcal{P}$, then $\mathcal{F}$ is $\boldsymbol{P}$-fundamental.*

PROOF. Let $\mathcal{P} \in \boldsymbol{P}$, and let $\mathcal{Q}_\mathcal{P} \in \boldsymbol{Q}$ be a refinement of $\mathcal{P}$. As $\mathcal{F}$ is $\boldsymbol{Q}$-fundamental, there exists $Q_\mathcal{P} \in \mathcal{F} \cap \mathcal{Q}_\mathcal{P}$, hence there exists $P \in \mathcal{P}$ such that $Q_\mathcal{P} \subset P$, and thus $P \in \mathcal{F}$. □

COROLLARY XXIII.1.7. *If a convergence ξ is $\boldsymbol{P}$-complete, and for every $\mathcal{P} \in \boldsymbol{P}$, there exists a refinement $\mathcal{Q} \in \boldsymbol{Q}$ of $\mathcal{P}$, then ξ is $\boldsymbol{Q}$-complete.*

PROOF. If $\mathcal{F}$ is $\boldsymbol{Q}$-fundamental, then by Lemma XXIII.1.6, $\mathcal{F}$ is $\boldsymbol{P}$-fundamental, hence $\operatorname{adh}_\xi \mathcal{F} \neq \varnothing$ by $\boldsymbol{P}$-completeness of ξ. Thus $\boldsymbol{Q}$-completeness follows. □

If $\boldsymbol{P}$ is a collection of families of sets, then let

$$\boldsymbol{P}_{\cup\downarrow} := \{\mathcal{P}_{\cup\downarrow} : \mathcal{P} \in \boldsymbol{P}\}.$$

COROLLARY XXIII.1.8. *An ultrafilter $\mathcal{U}$ is $\boldsymbol{P}$-fundamental if and only if it is $\boldsymbol{P}_{\cup\downarrow}$-fundamental.*

PROOF. Necessity follows from Lemma XXIII.1.6. The converse holds, because if an ultrafilter contains a finite union, then it contains at least one of its components. □

Actually, it is enough to consider completeness with respect to collections of ideals.

PROPOSITION XXIII.1.9. *A convergence is $\boldsymbol{P}$-complete if and only if it is $\boldsymbol{P}_{\cup\downarrow}$-complete.*

PROOF. If ξ is $\boldsymbol{P}$-complete, and a filter $\mathcal{F}$ is $\boldsymbol{P}_{\cup\downarrow}$-fundamental, then each $\mathcal{U} \in \beta\mathcal{F}$ is $\boldsymbol{P}$-fundamental by Corollary XXIII.1.8. Therefore, $\mathcal{F}$ is ξ-compactoid, so that ξ is $\boldsymbol{P}_{\cup\downarrow}$-complete. Conversely, if ξ is $\boldsymbol{P}_{\cup\downarrow}$-complete, and $\mathcal{F}$ is $\boldsymbol{P}$-fundamental, then by Lemma XXIII.1.7, $\mathcal{F}$ is $\boldsymbol{P}_{\cup\downarrow}$-fundamental, because $\boldsymbol{P}$ is a refinement of $\boldsymbol{P}_{\cup\downarrow}$. □

Recollect Definition XIX.3.14 that $\mathcal{S}$ is a ξ-regular refinement of $\mathcal{P}$ if $\operatorname{adh}^{\natural}_{\xi} \mathcal{S}$ is a refinement of $\mathcal{P}$. We have already observed in XIX.3.15 that

LEMMA XXIII.1.10. *Each cover of a regular convergence has a regular covering refinement.*

XXIII.2. Cocompleteness

Let $\boldsymbol{H}$ be a collection of families of subsets of X.

DEFINITION XXIII.2.1. A convergence ξ on X is called *$\boldsymbol{H}$-cocomplete* if $\operatorname{adh}_{\xi} \mathcal{F} \neq \varnothing$ for each filter $\mathcal{F}$ such that $\mathcal{F} \perp \mathcal{H}$ for each $\mathcal{H} \in \boldsymbol{H}$.

By our convention,

$$\mathcal{H}_c := \{H^c : H \in \mathcal{H}\}, \quad \boldsymbol{H}_{\neg} := \{\mathcal{H}_c : \mathcal{H} \in \boldsymbol{H}\}. \tag{XXIII.2.1}$$

PROPOSITION XXIII.2.2. *A convergence is $\boldsymbol{H}$-cocomplete if and only if it is $\boldsymbol{H}_{\neg}$-complete.*

PROOF. By definition, a convergence ξ is $\boldsymbol{H}$-cocomplete if and only if $\operatorname{adh}_{\xi} \mathcal{F} \neq \varnothing$, provided that $\mathcal{F}$ does not mesh any $\mathcal{H} \in \boldsymbol{H}$. Now, $\mathcal{F} \perp \mathcal{H}$ means that there exists $H \in \mathcal{H}$ such that $H^c \in \mathcal{F}$, equivalently $\mathcal{F} \cap \mathcal{H}_c \neq \varnothing$. □

Consequently, completeness with respect to a collection $\boldsymbol{P}$ is tantamount to cocompleteness with respect to the collection of the complements of families from $\boldsymbol{P}$. Both formulations can be useful, as one or another becomes more incisive in different situations.

Because by (XVIII.3.8), $\operatorname{adh}_{\mathrm{S}\xi} \mathcal{G} = \operatorname{adh}_{\xi} \mathcal{G}$, a convergence ξ is $\boldsymbol{H}$-cocomplete if and only if $\mathrm{S}\xi$ is $\boldsymbol{H}$-cocomplete.

DEFINITION XXIII.2.3. A filter $\mathcal{F}$ is called *$\boldsymbol{H}$-cofundamental if* $\mathcal{F} \perp \mathcal{H}$ for each $\mathcal{H} \in \boldsymbol{H}$.

PROPOSITION XXIII.2.4. *A filter $\mathcal{F}$ is $\boldsymbol{H}$-cofundamental if and only if it is $\boldsymbol{H}_{\neg}$-fundamental.*

PROOF. Indeed, by definition $\boldsymbol{H}$-cofundamental whenever $\mathcal{F}\perp\mathcal{H}$ for each $\mathcal{H} \in \boldsymbol{H}$, equivalently if $\mathcal{F} \cap \mathcal{H}_c \neq \varnothing$ for each $\mathcal{H} \in \boldsymbol{H}$. □

It follows immediately that

PROPOSITION XXIII.2.5. *A convergence ξ is $\boldsymbol{H}$-cocomplete if and only if one of the following equivalent conditions holds:*

(1) *each $\boldsymbol{H}$-cofundamental filter is ξ-adherent,*
(2) *each $\boldsymbol{H}$-cofundamental ultrafilter is ξ-convergent,*
(3) *each $\boldsymbol{H}$-cofundamental filter is ξ-compactoid.*

PROOF. Each ultrafilter is a filter, and if an ultrafilter is adherent, then it is convergent. On the other hand, a filter is compactoid whenever each finer ultrafilter is convergent. □

By Propositions XXIII.1.9 and XXIII.2.4, cocompleteness with respect to a collection of families with the finite intersection property is equivalent to cocompleteness with respect to the corresponding filters.

If $\boldsymbol{H}$ is a collection of families of sets, then let

$$\boldsymbol{H}_{\cap\uparrow} := \{\mathcal{H}_{\cap\uparrow} : \mathcal{H} \in \boldsymbol{H}\}.$$

COROLLARY XXIII.2.6. *A convergence is $\boldsymbol{H}$-cocomplete if and only if it is $\boldsymbol{H}_{\cap\uparrow}$-cocomplete.*

Notice that both the definitions of $\boldsymbol{H}$-cocompleteness and of $\mathbb{H}$-compactness follow a similar general pattern. Let $\mathfrak{C}_{\boldsymbol{H}}$ stand for the family of *$\boldsymbol{H}$-cofundamental filters* on a set X. Then, in view of Definition XX.3.1,

PROPOSITION XXIII.2.7. *A convergence is $\boldsymbol{H}$-cocomplete if and only if it is $\mathfrak{C}_{\boldsymbol{H}}$-compact*

Accordingly, $\boldsymbol{H}$-cocomplete and $\mathbb{H}$-compact convergences coincide if $\mathfrak{C}_{\boldsymbol{H}} = \mathbb{H}X$. For instance, $\mathfrak{C}_{\varnothing} = \mathbb{F}X$, so that $\varnothing$-complete and compact convergences coincide.

By Lemma VIII.2.3, a family $\mathcal{P}$ is a ξ-cover if and only if $\mathrm{adh}_\xi \mathcal{P}_c = \varnothing$. Therefore,

COROLLARY XXIII.2.8. *A convergence ξ is $\boldsymbol{P}$-complete and $\boldsymbol{P}$ is a ξ-cover if and only if ξ is $\boldsymbol{P}_{\neg}$-cocomplete and* $\mathrm{adh}_\xi \mathcal{H} = \varnothing$ *for each $\mathcal{H} \in \boldsymbol{P}_{\neg}$.*

Recapitulating, the same situations can be described either in terms of fundamental filters and completeness, or of cofundamental filters and cocompleteness. The former language is that traditionally used in topology. The latter has been introduced recently ([2]), and has an advantage of enabling representations in terms of spaces of ultrafilters. Namely, Definition XXIII.2.3 can be reformulated in terms ultrafilters, as follows

LEMMA XXIII.2.9. *A filter $\mathcal{F}$ is $\boldsymbol{H}$-cofundamental whenever*

$$\beta\mathcal{F}\cap\bigcup\nolimits_{\mathcal{H}\in\boldsymbol{H}}\beta\mathcal{H}=\varnothing.$$

Therefore,

COROLLARY XXIII.2.10. *A convergence ξ on X is $\boldsymbol{H}$-cocomplete if and only if*

$$\beta X\setminus\xi^{-}(X)\subset\bigcup\nolimits_{\mathcal{H}\in\boldsymbol{H}}\beta\mathcal{H}.$$

If, moreover, $\boldsymbol{H}$ is composed of non-ξ-adherent filters, then

$$\beta X\setminus\xi^{-}(X)=\bigcup\nolimits_{\mathcal{H}\in\boldsymbol{H}}\beta\mathcal{H}.$$

A convergence ξ is $\boldsymbol{H}$-cocomplete if and only if all the ultrafilters of $\beta\mathcal{F}$ are ξ-convergent, provided that $\mathcal{F}$ is $\boldsymbol{H}$-cofundamental. Equivalently, if $\lim_\xi\mathcal{U}=\varnothing$ for an ultrafilter $\mathcal{U}$, then there exists $\mathcal{H}\in\boldsymbol{H}$ such that $\mathcal{U}\in\beta\mathcal{H}$.

COROLLARY XXIII.2.11. *If ξ is a convergence on X, and $\boldsymbol{H}$ is a collection of non-ξ-adherent filters, then ξ is $\boldsymbol{H}$-cocomplete if and only if $\{\beta\mathcal{H}:\mathcal{H}\in\boldsymbol{H}\}$ is a β-compact covering of $\beta X\setminus\xi^{-}(X)$.*

By Corollary VIII.1.10,

LEMMA XXIII.2.12. *If $f\in\mathscr{C}(\xi,\tau)$, then*

$$\operatorname{adh}_\xi\mathcal{H}\neq\varnothing\Longrightarrow\operatorname{adh}_\tau f[\mathcal{H}]\neq\varnothing.$$

COROLLARY XXIII.2.13. *If ξ is $\boldsymbol{H}$-cocomplete and $f\in\mathscr{C}(\xi,\tau)$, then τ is $\{f[\mathcal{H}]:\mathcal{H}\in\boldsymbol{H}\}$-cocomplete.*

PROOF. Let $\mathcal{G}\perp f[\mathcal{H}]$, equivalently $f^{-}[\mathcal{G}]\perp\mathcal{H}$, for each $\mathcal{H}\in\boldsymbol{H}$. As ξ is $\boldsymbol{H}$-cocomplete, $\operatorname{adh}_\xi f^{-}[\mathcal{G}]\neq\varnothing$, thus by Lemma XXIII.2.12, $\varnothing\neq\operatorname{adh}_\tau f[f^{-}[\mathcal{G}]]\subset\operatorname{adh}_\tau\mathcal{G}$. □

XXIII.3. Finitely complete convergences

In this section we study in detail the case of finite completeness. Be aware that the results presented here constitute a special case of general facts discussed in Sections XXIII.5, XXIII.6 and XXIII.7.

Because of Proposition XXIII.1.9 and Corollary XXIII.2.6, completeness can be defined either in terms of collections of ideals or of filters.

[2]See [60].

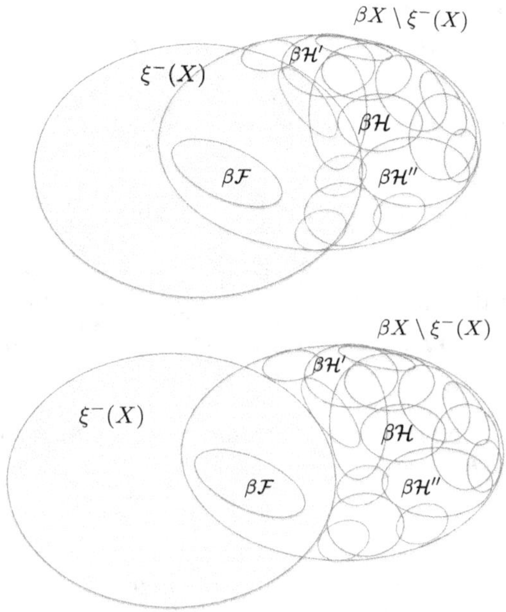

FIGURE XXIII.2.1. Here $\xi^{-}(X)$ stands for the set of ξ-convergent ultrafilters. A convergence ξ is $\boldsymbol{H}$-cocomplete provided that a filter $\mathcal{F}$ is ξ-compactoid if $\mathcal{F} \perp \mathcal{H}$, that is $\beta\mathcal{F} \cap \beta\mathcal{H} = \varnothing$, for each $\mathcal{H} \in \boldsymbol{H}$. Beneath, $\boldsymbol{H}$ consists of ξ-non-adherent filters, so that $\beta\mathcal{H} \cap \xi^{-}(X) = \varnothing$ for every $\mathcal{H} \in \boldsymbol{H}$.

DEFINITION XXIII.3.1. A convergence ξ is called *finitely complete* if there exists a finite collection $\boldsymbol{P}$ of ideal ξ-covers such that ξ is $\boldsymbol{P}$-complete.

In particular, in view of Example XXIII.1.5, a convergence is compact if and only if it is $\varnothing$-complete.

PROPOSITION XXIII.3.2. *A convergence ξ is finitely complete if and only if there exists a finite collection $\boldsymbol{H}$ of ξ-non-adherent filters such that ξ is $\boldsymbol{H}$-cocomplete.*

PROOF. By Proposition XXIII.2.4, a filter is $\boldsymbol{H}$-cofundamental whenever it is $\boldsymbol{H}_{\neg}$-fundamental. Of course, $\boldsymbol{H}$ and $\boldsymbol{H}_{\neg}$ are equipotent, hence one is finite whenever the other is finite. By Lemma VIII.2.3, a family $\mathcal{H} \in \boldsymbol{H}$ is ξ-non-adherent if and only if $\mathcal{H}_c \in \boldsymbol{H}_{\neg}$ is a ξ-cover. $\square$

PROPOSITION XXIII.3.3. *A non-compact convergence ξ is finitely complete if and only if there exists a ξ-non-adherent filter $\mathcal{H}$ such that each $\mathcal{H}$-cofundamental filter is ξ-compactoid.*

PROOF. If $\boldsymbol{H}$ is a finite collection of filters such that $\operatorname{adh}_\xi \mathcal{H} = \varnothing$ for each $\mathcal{H} \in \boldsymbol{H}$, then

$$\operatorname{adh}_\xi(\bigcap\nolimits_{\mathcal{H}\in\boldsymbol{H}} \mathcal{H}) = \bigcup\nolimits_{\mathcal{H}\in\boldsymbol{H}} \operatorname{adh}_\xi \mathcal{H} = \varnothing,$$

by virtue of (VIII.1.4). By Definition XXIII.2.3, a filter is $\boldsymbol{H}$-cofundamental if and only if it is $\bigcap_{\mathcal{H}\in\boldsymbol{H}} \mathcal{H}$-cofundamental. □

In particular, if $\boldsymbol{H} = \varnothing$, then each filter is ξ-compactoid, that is, ξ is compact.

COROLLARY XXIII.3.4. *A convergence ξ is finitely complete if and only if there exists an ideal ξ-cover $\mathcal{P}$ such that each $\mathcal{P}$-fundamental filter is ξ-compactoid.*

Recall that $\mathcal{K}_\circ(\xi)$ stands for the ideal of all ξ-compactoid sets, and, consequently, $\mathcal{K}_{\circ c}(\xi)$ is the filter of their complements, that is, of ξ-cocompactoid sets.

Given a convergence ξ on X, the family

$$\mathcal{K}_{\circ c}(\xi) = \{F \subset X : X \setminus F \in \mathcal{K}_\circ(\xi)\}$$

is a filter, called the *ξ-cocompactoid filter*.

Of course, a convergence ξ is locally compactoid if and only if it has a cover composed of ξ-compactoid sets, equivalently whenever $\mathcal{K}_\circ(\xi)$ is a ξ-cover.

THEOREM XXIII.3.5. *A convergence is finitely complete if and only if it is locally compactoid.*

PROOF. If ξ is finitely complete, then by Proposition XXIII.3.3, there exists an ideal ξ-cover $\mathcal{P}$ such that $\operatorname{adh}_\xi \mathcal{F} \neq \varnothing$ for each filter such that $\mathcal{F} \cap \mathcal{P} \neq \varnothing$. In particular, each $P \in \mathcal{P}$ is ξ-compactoid ([3]). Therefore $\mathcal{K}_\circ(\xi)$ is a ξ-cover, so that ξ is locally compactoid. Conversely, if ξ is locally compactoid, hence $\mathcal{K}_\circ(\xi)$ is a ξ-cover, then if $\mathcal{F} \cap \mathcal{K}_\circ(\xi) \neq \varnothing$, that is, if $\mathcal{F}$ contains a ξ-compactoid set, then $\operatorname{adh}_\xi \mathcal{F} \neq \varnothing$. □

Accordingly, each finite product of finitely complete convergences is finitely complete.

COROLLARY XXIII.3.6. *A convergence ξ is locally compactoid if and only if* $\operatorname{adh}_\xi \mathcal{K}_{\circ c}(\xi) = \varnothing$.

[3]Because if $P \in \mathcal{P}$, then each filter $\mathcal{F}$ such that $P \in \mathcal{F}$ is ξ-fundamental, hence $\operatorname{adh}_\xi \mathcal{F} \neq \varnothing$.

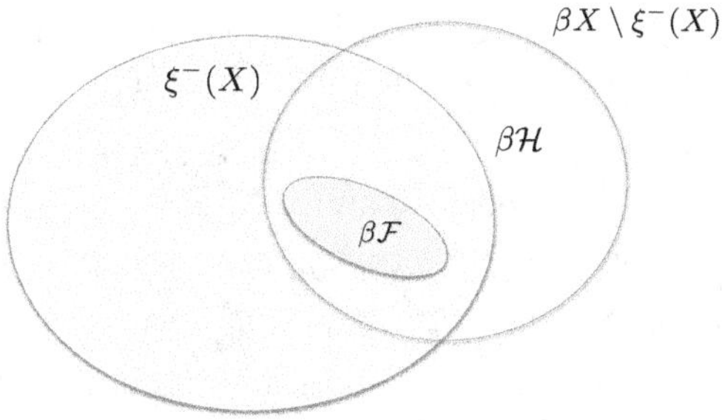

FIGURE XXIII.3.1. A convergence ξ is finitely complete whenever there exists a filter $\mathcal{H}$ such that all finer ultrafilters are precisely those that are not ξ-convergent. Such a filter consists of the complements of all ξ-compactoid sets, that is, is equal to $\mathcal{K}_{\circ c}(\xi)$. Consequently, a filter $\mathcal{F}$ is ξ-compactoid if and only if it does not mesh $\mathcal{K}_{\circ c}(\xi)$.

In particular, ξ is compact if and only if the ideal $\mathcal{K}_{\circ}(\xi)$ is improper, that is, $|\xi| \in \mathcal{K}_{\circ}(\xi)$. Therefore,

COROLLARY XXIII.3.7. *A convergence ξ is compact if and only if the filter $\mathcal{K}_{\circ c}(\xi)$ is improper.*

EXAMPLE XXIII.3.8. Consider the standard topology $\nu_{\mathbb{R}}$ on $\mathbb{R}$. We shall see that $\nu_{\mathbb{R}}$ is locally compactoid, equivalently finitely complete. By Corollary XII.1.10, a subset of $\mathbb{R}$ is $\nu_{\mathbb{R}}$-compactoid if and only if it is bounded. On the other hand, each $\nu_{\mathbb{R}}$-convergent filter contains a bounded set ([4]), hence $\nu_{\mathbb{R}}$ is locally compactoid. Accordingly, the family $\mathcal{K}_{\circ}(\nu_{\mathbb{R}})$ of $\nu_{\mathbb{R}}$-compactoid sets is a cover of $\nu_{\mathbb{R}}$, equivalently the $\nu_{\mathbb{R}}$-cocompactoid filter $\mathcal{K}_{\circ c}(\nu_{\mathbb{R}})$ is not $\nu_{\mathbb{R}}$-adherent.

It is straightforward that each closed subset of a finitely complete convergence is finitely complete. On the other hand, each open subset of a locally compact convergence is locally compact, equivalently

PROPOSITION XXIII.3.9. *Each open subspace of a regular, finitely complete convergence is finitely complete.*

PROOF. Let $\mathcal{P}$ be a ξ-cover such that ξ is $\mathcal{P}$-complete, and let $O \in \mathcal{O}_{\xi}$. By Lemma VIII.2.5, $\mathcal{P}$ is a ξ-cover of O, hence $\mathcal{P}_O := \{P \cap O : P \in \mathcal{P}\}$ is a $\xi|_O$-cover. As ξ is regular, by Lemma XXIII.1.10, there exists a $\xi|_O$-cover $\mathcal{S}$ such that $\operatorname{adh}^{\natural}_{\xi} \mathcal{S}$ is a refinement of $\mathcal{P}_O$, in other words, for each $S \in \mathcal{S}$ there is $P \in \mathcal{P}$ such that $\operatorname{adh}_{\xi} S \subset P \cap O$. By

[4]Indeed, $x \in \lim_{\nu_{\mathbb{R}}} \mathcal{F}$ whenever $[x-\varepsilon, x+\varepsilon] \in \mathcal{F}$ for each $\varepsilon > 0$.

Lemma XXIII.1.7, ξ is $\operatorname{adh}^{\natural}_{\xi}\mathcal{S}$-complete, that is, if $\mathcal{U} \in \beta O$ and there is $S \in \mathcal{S}$ such that $\operatorname{adh}_{\xi} S \in \mathcal{U}$, then $\varnothing \neq \lim_{\xi} \mathcal{U} \subset \operatorname{adh}_{\xi} S \subset O$. □

Recall Definition XIX.2.6 that a convergence ξ is called *pseudodiagonal* if $\operatorname{adh}_{\xi} \mathcal{H}$ is ξ-closed for each filter $\mathcal{H}$.

PROPOSITION XXIII.3.10. *Each dense, finitely complete subset of a Hausdorff pseudodiagonal convergence is open.*

PROOF. Let ξ be a Hausdorff pseudodiagonal convergence on X, and let A be a subset of X such that ξ_A is finitely complete. Let $\mathcal{H}$ be a filter on A, for which ξ_A is cocomplete. Accordingly, $A \cap \operatorname{adh}_{\xi} \mathcal{H} = \operatorname{adh}_{\xi_A} \mathcal{H} = \varnothing$. We shall show that $X \setminus A = \operatorname{adh}_{\xi} \mathcal{H}$, hence A is ξ-open, because ξ is pseudodiagonal. Let $y \in X \setminus A$. As A is ξ-dense, there exists $\mathcal{U} \in \beta A$ such that $\lim_{\xi} \mathcal{U} = \{y\}$, because of Hausdorffness, and thus $\lim_{\xi_A} \mathcal{U} = \varnothing$. As ξ_A is $\mathcal{H}$-complete, $\mathcal{U} \in \beta \mathcal{H}$, and thus $y \in \lim_{\xi} \mathcal{U} \subset \operatorname{adh}_{\xi} \mathcal{H}$, which is what we needed to prove. □

Each topology is diagonal, *a fortiori* pseudodiagonal. On the other hand, a regular locally compactoid topology is hereditarily locally compact. Therefore

COROLLARY XXIII.3.11. *A dense subset of a Hausdorff compact topology is locally compact if and only if it is open.*

We know from Example XII.3.10 that finite completeness is not preserved under continuous maps. However, by Corollary XXI.7.6,

PROPOSITION XXIII.3.12. *Finite completeness is preserved under continuous biquotient maps.*

XXIII.4. Countably complete convergences

Let me remind that the facts presented in this section will be investigated independently in full generality in Sections XXIII.5, XXIII.6 and XXIII.7.

DEFINITION XXIII.4.1. A convergence ξ is called *countably complete* if there exists a countable collection $\boldsymbol{P}$ of ideal ξ-covers such that ξ is $\boldsymbol{P}$-complete.

In particular, each finitely complete convergence is countably complete.

REMARK XXIII.4.2. Let us remind that traditionally a topology is called *topologically complete* or *Čech complete* ([5]) if it is countably complete (in the sense of Definition XXIII.4.1), Hausdorff and functionally regular (alias completely regular).

[5]R. ENGELKING [97]

By Lemma VIII.2.3,

PROPOSITION XXIII.4.3. *A convergence ξ is countably complete if and only if there exists a countable collection $\boldsymbol{H}$ of ξ-non-adherent filters such that ξ is $\boldsymbol{H}$-cocomplete.*

Accordingly, a countable collection $\boldsymbol{H}$ from the proposition above consists of ξ-non-adherent filters, that is, the set $\xi^{-}(X)$ of each ξ-convergent ultrafilters is disjoint from $\beta\mathcal{H}$ for each $\mathcal{H} \in \boldsymbol{H}$:

$$\xi^{-}(X) \cap \bigcup_{\mathcal{H}\in\boldsymbol{H}} \beta\mathcal{H} = \varnothing.$$

A filter $\mathcal{F}$ is $\boldsymbol{H}$-cofundamental whenever $\mathcal{F}$ does not mesh any $\mathcal{H}$ from $\boldsymbol{H}$, in particular, an ultrafilter $\mathcal{U}$ is $\boldsymbol{H}$-cofundamental whenever $\mathcal{U} \notin \bigcup_{\mathcal{H}\in\boldsymbol{H}} \beta\mathcal{H}$. By Proposition XXIII.4.3, each $\boldsymbol{H}$-cofundamental is ξ-convergent, that is, $\beta X \setminus \xi^{-}(X) = \bigcup_{\mathcal{H}\in\boldsymbol{H}} \beta\mathcal{H}$.

The standard topology of the real line is countably complete, because it is locally compact.

EXAMPLE XXIII.4.4 (A countably complete functionally regular topology, that is not locally compact). The set $\mathbb{R} \setminus \mathbb{Q}$ of irrational numbers with the standard convergence restricted from $\mathbb{R}$ is countably complete, but not locally compactoid.

Indeed, we have seen that the standard topology ν of $\mathbb{R}$ is homeomorphic to its restriction $\nu|_{]0,1[}$. Hence ν restricted to $\mathbb{R} \setminus \mathbb{Q}$ is homeomorphic to ν restricted to $X :=]0,1[\setminus\mathbb{Q}$. On the other hand, $Y := [0,1]$ with ν is compact, and $\mathrm{cl}_\nu X = Y$. To see that $\nu|_X$ is countably complete, let $\boldsymbol{H} := \{\mathcal{N}_{\nu|_Y}(y) \vee X : y \in \mathbb{Q} \cap [0,1]\}$. This collection of filters on X is countable, and $\mathrm{adh}_{\nu|_Y}(\mathcal{N}_{\nu|_Y}(y) \vee X) = \{y\}$, hence $\mathrm{adh}_{\nu|_X}(\mathcal{N}_{\nu|_Y}(y) \vee X) = \varnothing$. Let $\mathcal{U}$ be a $\boldsymbol{H}$-cofundamental ultrafilter on $X \subset Y$. By compactness, there exists $w \in Y$ such that $w \in \lim_{\nu|_Y} \mathcal{U}$. If $w \in \mathbb{Q}$, then $\mathcal{U} \geq \mathcal{N}_{\nu|_Y}(w) \vee X$, which is impossible, because $\mathcal{U}$ is $\boldsymbol{H}$-cofundamental. Therefore, $w \in X$, proving countable completeness.

That ν_X is not locally compactoid, notice that for each $0 < r < s < 1$, there is $y \in \mathbb{Q}\cap]r,s[$, hence $\mathrm{adh}_{\nu|_X}(\mathcal{N}_{\nu|_Y}(y) \vee X) = \varnothing$ and $]r,s[\setminus\mathbb{Q} \in \mathcal{N}_{\nu|_Y}(y) \vee X$, which show that $]r,s[\setminus\mathbb{Q}$ is not $\nu|_X$-compactoid.

In Section X.4 we studied *completely metrizable* topologies, that is, those that admit a compatible complete metric. It turns out that they coincide with countably complete metrizable topologies ([6]).

PROPOSITION XXIII.4.5. *A metrizable topology is completely metrizable if and only if it is countably complete.*

[6] E. ČECH [39].

PROOF. Let d be a complete metric on X. Then $\mathcal{B}_n := \{B(x, 2^{-n}) : x \in X\}$ is an td-open cover. The countable collection $\{\mathcal{B}_n : n < \omega\}$ is complete, because if a filter $\mathcal{F}$ is d-fundamental, then $\varnothing \neq \lim_{\mathrm{t}d} \mathcal{F} \subset \mathrm{adh}_{\mathrm{t}d} \mathcal{F}$ by the assumption of completeness of d. Conversely, let ξ be a countably complete metrizable topology on X, and let d be a metric compatible with ξ. By Theorem X.5.2, there exists a completion $(\widehat{X}, \widehat{d})$ of (X, d). By the first part of the proof, $\mathrm{t}\widehat{d}$ is countably complete, and by assumption τ is a countably complete topology densely embedded in $\mathrm{t}\widehat{d}$, hence a $G_{\aleph_0}$ in virtue of Theorem XXIII.6.3.

By Lemma X.10.5, τ is homeomorphic to a closed subspace of $\mathrm{t}\widehat{d} \times \prod_{n<\omega} \nu_{\mathbb{R}}$. This product is completely metrizable in view of Proposition X.4.10, as well as its closed subspaces. Therefore τ is completely metrizable. □

Proposition XXIII.4.5 helps us to supply an example of a non-countably complete convergence.

EXAMPLE XXIII.4.6. The (metrizable) standard topology $\nu|_{\mathbb{Q}}$ of the set of *rational numbers* is not completely metrizable in virtue of Example X.4.17. By Proposition XXIII.4.5, $\nu|_{\mathbb{Q}}$ is not countably complete.

It is easy to see that a closed subset of a countably complete convergence is countably complete (Proposition XXIII.6.1).

An open subset of a Hausdorff regular countably complete convergence is countably complete. We will deduce this fact from the next theorem. Recall that a subset of a convergence space is called $G_{\aleph_0}$ (alias G_δ) if it is a countable intersection of open subsets.

THEOREM XXIII.4.7. *Each $G_{\aleph_0}$-subset of a regular countably complete convergence is countably complete.*

PROOF. Let ξ be a regular convergence on X and let $\boldsymbol{P} = \{\mathcal{P}_k : k < \omega\}$ be a collection of ξ-covers of X such that ξ is $\boldsymbol{P}$-complete. Let $W := \bigcap_{n<\omega} O_n$ where $O_n \in \mathcal{O}_\xi$. For each $n < \omega$, the collection

$$\boldsymbol{P}_n := \{\mathcal{P}_k \vee O_n : k < \omega\}$$

is a collection of ξ-covers of O_n. By Lemma XXIII.1.10, there exists a ξ-regular refinement $\mathcal{S}_{n,k}$ of $\mathcal{P}_k \vee O_n$ so that $\mathcal{S}_{n,k}$ is a ξ-cover of O_n. Then

$$\boldsymbol{S}_n := \{\mathcal{S}_{n,k} : k < \omega\}$$

is a collection of $\xi|_{O_n}$-covers such that $\xi|_{O_n}$ is $\boldsymbol{S}_n$-complete. Indeed, let $\mathcal{U} \in \beta O_n$ be $\boldsymbol{S}_n$-fundamental, that is, for each $k < \omega$, there is $U_k \in \mathcal{U} \cap \mathcal{S}_{n,k}$. As $\mathcal{S}_{n,k}$ is a ξ-regular refinement of $\mathcal{P}_k \vee O_n$, there is $P \in \mathcal{P}_k$

such that $\operatorname{adh}_\xi U_{\mathcal{P}} \subset P$, and since ξ is $\boldsymbol{P}$-complete, $\operatorname{adh}_\xi(\operatorname{adh}^\natural_\xi \mathcal{U}) \neq \varnothing$. Let

$$\boldsymbol{S} := \{\mathcal{S}_{n,k} \vee W : n, k < \omega\}.$$

We shall show that $\xi|_W$ is $\boldsymbol{S}$-complete. Let $\mathcal{U} \in \beta W$ be $\boldsymbol{S}$-fundamental, hence it is a $\boldsymbol{S}_n$-fundamental ultrafilter-base, and thus $\varnothing \neq \lim_\xi \mathcal{U} \subset \lim_\xi(\operatorname{adh}^\natural_\xi \mathcal{U}) \subset \operatorname{adh}_\xi S_n \subset O_n$ for each $n < \infty$. Hence $\varnothing \neq \lim_\xi \mathcal{U} \subset \bigcap_{n<\omega} O_n = W$. □

THEOREM XXIII.4.8. *Every dense, countably complete subconvergence of a Hausdorff pseudodiagonal convergence is* $G_{\aleph_0}$.

PROOF. Let ξ be a Hausdorff, pseudodiagonal convergence on X, and let A be a dense subset of X, for which $\xi|_A$ is countably complete. Let $\{\mathcal{H}_n : n < \omega\}$ be a collection of filters on A, for which $\xi|_A$ is complete, and such that $\operatorname{adh}_{\xi|_A} \mathcal{H}_n = \varnothing$ for each $n < \omega$. We shall show that under the conditions of the theorem,

$$X \setminus A = \bigcup_{n<\omega} \operatorname{adh}_\xi \mathcal{H}_n.$$

Indeed, as $\varnothing = \operatorname{adh}_{\xi|_A} \mathcal{H}_n = A \cap \operatorname{adh}_\xi \mathcal{H}_n$, we need to show that $X \setminus A \subset \bigcup_{n<\omega} \operatorname{adh}_\xi \mathcal{H}_n$. So let $y \in X \setminus A$. As A is ξ-dense, there exists an ultrafilter $\mathcal{U}$ on A such that $\{y\} = \lim_\xi \mathcal{U}$, because ξ is Hausdorff. Therefore, $\lim_{\xi|_A} \mathcal{U} = \varnothing$, and since $\xi|_A$ is complete with respect to $\{\mathcal{H}_n : n < \omega\}$, there exists $n < \omega$ such that $\mathcal{U} \in \beta\mathcal{H}_n$, hence $y \in \lim_\xi \mathcal{U} \subset \operatorname{adh}_\xi \mathcal{H}_n$, completing the proof. □

Each topology is diagonal, *a fortiori* pseudodiagonal. Therefore,

COROLLARY XXIII.4.9. *A dense subconvergence of a Hausdorff regular countably complete topology is countably complete if and only if it is* $G_{\aleph_0}$.

Recall that a map $f : |\xi| \longrightarrow |\tau|$ is perfect (XX.11.1)

$$f^- \in \mathscr{K}(\tau, \xi)$$

whenever $f^-(y) \cap \operatorname{adh}_\xi \mathcal{H} \neq \varnothing$ for each filter $\mathcal{H}$ such that $y \in \operatorname{adh}_\tau f[\mathcal{H}]$. Therefore,

LEMMA XXIII.4.10. *If* $f^- \in \mathscr{K}(\tau, \xi)$, *then for each filter* $\mathcal{H}$,

$$\operatorname{adh}_\xi \mathcal{H} = \varnothing \Longrightarrow \operatorname{adh}_\tau f[\mathcal{H}] = \varnothing.$$

PROPOSITION XXIII.4.11. *Countable completeness is preserved under continuous perfect maps.*

Proof. If $\boldsymbol{H}$ is a collection of filters such that ξ is $\boldsymbol{H}$-cocomplete, then by Lemma XXIII.2.13 τ is $\{f[\mathcal{H}] : \mathcal{H} \in \boldsymbol{H}\}$-cocomplete, because $f \in \mathscr{C}(\xi, \tau)$. By Corollary XXIII.4.10, if $\operatorname{adh}_\xi \mathcal{H} = \varnothing$ then $\operatorname{adh}_\tau f[\mathcal{H}] = \varnothing$ for each $\mathcal{H} \in \boldsymbol{H}$, because $f^- \in \mathscr{K}(\tau, \xi)$. By taking $\boldsymbol{H}$ countable, we complete the proof. □

Remark XXIII.4.12. One can read in some books ([7]) that *Čech completeness is not preserved under (continuous) perfect maps,* which is not contradictory with Proposition XXIII.4.11, because a topology is said to be *Čech complete* if it is *countably complete* and *completely regular* (that is, *functionally regular* in our terminology). It is *functional regularity* that is not preserved under continuous perfect maps.

To see that countable completeness is not preserved under continuous maps, it is enough to take a non-countably complete convergence ξ on X, and use the fact that the identity map $i \in \mathscr{C}(\iota_X, \xi)$ is continuous, and the discrete topology ι_X is locally compact, hence finitely complete, and *a fortiori* countably complete.

Proposition XXIII.4.13. *A product $\prod_{j \in J} \xi_j$ is countably complete if and only if there exists a countable or empty subset J_0 of J such that ξ_j is countably complete for $j \in J_0$, and ξ_j is compact for $j \notin J_0$.*

Proof. Let ξ_n be a countably complete convergence fo each $n < \omega$. Let $\boldsymbol{G}_n$ be a collection of non-ξ_n-adherent filters such that ξ_n is $\boldsymbol{G}_n$-cocomplete. Let $\xi := \prod_{n<\omega} \xi_n$. Then

$$\boldsymbol{G} := \{p_n^-[\mathcal{G}] : \mathcal{G} \in \boldsymbol{G}_n, n < \omega\}$$

is a countable collection of non-ξ-adherent filters, becuase $p_n \in \mathscr{C}(\xi, \xi_n)$ for every $n < \omega$. On the other hand, ξ is $\boldsymbol{G}$-cocomplete. Indeed, let $\mathcal{U}$ be an ultrafilter dissociated from $\boldsymbol{G}$, that is, $\mathcal{U} \perp p_n^-[\mathcal{G}]$ for each $\mathcal{G} \in \boldsymbol{G}_n$ and every $n < \omega$. Equivalently, $p_n[\mathcal{U}] \perp \boldsymbol{G}_j$ for each $n < \omega$, and thus $\lim_{\xi_n} p_n[\mathcal{U}] \neq \varnothing$ for every $n < \omega$. As a result, $\lim_\xi \mathcal{U} \neq \varnothing$. It follows that ξ is countably complete. If we consider a general case, then $\zeta := \prod_{j \in J \setminus J_0} \xi_j$ is compact by Corollary XX.2.15, hence countably complete, thus $\xi \times \zeta$ is countably complete by the first part of the proof. Let J be uncountable, and let ξ_j be a countably complete non-compact convergence for each $j \in J$. □

XXIII.5. Completeness number

Let κ be a cardinal number. A convergence ξ is called *κ-complete* if there exists a collection $\boldsymbol{P}$ of ideal *ξ-covers* such that $\operatorname{card} \boldsymbol{P} \leq \kappa$ and ξ is *$\boldsymbol{P}$-complete.*

[7]R. Engelking [97].

DEFINITION XXIII.5.1. The *completeness number* or *completeness* of a convergence ξ is the least cardinal κ, for which ξ is κ-complete.

We denote it by $\mathrm{compl}(\xi)$. Because of Proposition XXIII.2.2,

PROPOSITION XXIII.5.2. *The completeness number of ξ is equal to the least cardinality of a collection of ξ-non-adherent filters, with respect to which ξ is cocomplete.*

As completeness number is defined in terms of adherences,

$$\mathrm{compl}(\xi) = \mathrm{compl}(\mathrm{S}\xi).$$

Figure XXIII.2.1 represents a collection $\boldsymbol{H}$ consisting of ξ-non-adherent filters $\mathcal{H}$ such that each filter $\mathcal{F}$, dissociated from each filter $\mathcal{H} \in \boldsymbol{H}$, is ξ-adherent. On recalling that $\xi^{-}(X)$ stands for the set of all ξ-convergent ultrafilters, this amounts to

$$\beta X \setminus \xi^{-}(X) = \bigcup\nolimits_{\mathcal{H} \in \boldsymbol{H}} \beta\mathcal{H}. \tag{XXIII.5.1}$$

The *compact covering number* of a convergence ξ is the least cardinality of compact covers of ξ. As each β-closed set is β-compact (Section XII.7),

PROPOSITION XXIII.5.3. *A convergence ξ is κ-complete if and only if the set of non-ξ-convergent ultrafilters has a β-compact covering of cardinality κ.*

Therefore,

COROLLARY XXIII.5.4. *The completeness number of ξ is equal to the β-compact covering number of the set of non-ξ-convergent ultrafilters*

$$\mathrm{compl}(\xi) = \mathrm{kc}(\beta X \setminus \xi^{-}(X)).$$

A convergence ξ on X is compact whenever $\beta X \subset \xi^{-}(X)$, that is, ξ is $\boldsymbol{H}$-cocomplete whenever $\bigcup_{\mathcal{H} \in \boldsymbol{H}} \beta\mathcal{H} = \varnothing$. This holds for $\boldsymbol{H} = \varnothing$, equivalently, $\mathrm{compl}(\xi) = 0$.

The completeness number of ξ is finite whenever $\mathrm{compl}(\xi) \leq 1$, equivalently if there exists a filter $\mathcal{H}$ such that

$$\beta\mathcal{H} = \beta X \setminus \xi^{-}(X).$$

In other terms, $H \in \mathcal{H}$ if and only if $\lim_{\xi} \mathcal{U} \neq \varnothing$ for each $\mathcal{U} \in \beta(X \setminus H)$. This means that $\mathcal{H}$ is a ξ-cocompactoid filter, or else ξ is locally compactoid. Finally, $\mathrm{compl}(\xi) = \aleph_0$ whenever ξ is countably complete.

Completeness number exists for every convergence ξ. It is not greater than $2^{2^{\mathrm{card}|\xi|}}$, the number of all filters on X. Indeed, the class of $\boldsymbol{H}$

fulfilling (XXIII.5.1) is a set, so that there exists $\boldsymbol{H}$ with the least cardinality $\mathrm{compl}(\xi)$.

The following fact is a consequence of theorems from Section XXIII.6, but it is interesting, and perhaps motivating, to prove it directly right now.

PROPOSITION XXIII.5.5. *Let τ be a Hausdorff compact topology on X, and let A be a τ-dense subset of X. Then*

$$\mathrm{compl}(\tau|_A) = \mathrm{kc}(\tau|_{X\setminus A}).$$

PROOF. Let $\mathcal{D} \subset \mathcal{K}(\tau)$ such that $\bigcup_{D\in\mathcal{D}} D = X \setminus A$ and $\mathrm{card}\,\mathcal{D} = \kappa$. By Remark XX.2.20, $\mathcal{D} \subset \mathcal{K}(\tau|_{X\setminus A})$. By Corollary XX.6.11, the filter $\mathcal{N}_\tau(D)$ is τ-compact and $\mathrm{adh}_\tau \mathcal{N}_\tau(D) = D$ for each $D \in \mathcal{D}$. Therefore, $\mathrm{adh}_{\tau|_A} \mathcal{N}_\tau(D) = \varnothing$, so that by density, $\{\mathcal{N}_\tau(D) \vee A : D \in \mathcal{D}\}$ is a collection of cardinality κ of $\tau|_A$-non-adherent filters. To see that this collection is $\tau|_A$-cocomplete, let $\mathcal{U} \in \beta A$ be such that $\mathcal{U} \perp \mathcal{N}_\tau(D)$ for every $D \in \mathcal{D}$. As τ is compact and Hausdorff, there exists $x \in X$ such that $\{x\} = \mathrm{adh}_\tau \mathcal{U}$, hence and $\mathcal{U} \geq \mathcal{N}_\tau(x)$. If $x \notin A$, then there exists $D \in \mathcal{D}$ with $x \in D$, so that $\mathcal{U} \geq \mathcal{N}_\tau(x) \geq \mathcal{N}_\tau(D)$, a contradiction.

Conversely, let $\boldsymbol{H}$ be a $\tau|_A$-cocomplete collection of $\tau|_A$-non-adherent filters. By compactness of τ, each $\mathcal{H} \in \boldsymbol{H}$ is τ-compact, $\mathrm{adh}_\tau \mathcal{H} \cap A = \varnothing$, and by Lemma XX.6.8, $\mathcal{H} \geq \mathcal{N}_\tau(\mathrm{adh}_\tau \mathcal{H})$. If $x \notin \bigcup_{\mathcal{H}\in\boldsymbol{H}} \mathrm{adh}_\tau \mathcal{H}$, then by density, there exists $\mathcal{U} \in \beta(\mathcal{N}_\tau(x) \vee A)$, thus $\{x\} = \lim_\tau \mathcal{U}$. Now, by normality, $\mathcal{U} \perp \mathcal{H}$ for each $\mathcal{H} \in \boldsymbol{H}$. Consequently, $\lim_\tau \mathcal{U} \cap A \neq \varnothing$ by the $\tau|_A$-cocompleteness of $\boldsymbol{H}$, so that $x \in A$, showing that $\{\mathrm{adh}_\tau \mathcal{H} : \mathcal{H} \in \boldsymbol{H}\}$ is a compact covering of $\tau|_{X\setminus A}$. □

This proposition recovers from Example XXIII.4.4 the fact that the space of *irrational numbers* is *countably complete.* In fact, by taking for τ a one-point compactification of $\nu_{\mathbb{R}}$ and setting $A := \mathbb{R} \setminus \mathbb{Q}$, the complementary set of A is countable, hence a countable union of singletons. On the other hand, as $\mathbb{R} \setminus \mathbb{Q}$ is not locally compact, hence is not finitely complete, the *compact covering* of $\mathbb{Q}$ is precisely $\aleph_0$.

XXIII.6. Completeness of subspaces

PROPOSITION XXIII.6.1. *Completeness is invariant with respect to closed subspaces.*

PROOF. Let ξ be a convergence on X, and let W be a ξ-closed subset of X. We shall show that

$$\mathrm{compl}(\xi|_W) \leq \mathrm{card}(\xi).$$

Let $\boldsymbol{H}$ be a collection of ξ-non-adherent filters on X, so that ξ is $\boldsymbol{H}$-cocomplete. As $\operatorname{adh}_{\xi|_W} \mathcal{G} = W \cap \operatorname{adh}_\xi \mathcal{G}$ for a filter $\mathcal{G}$ on W, the collection $\boldsymbol{H}_W := \{\mathcal{H} \vee W : \mathcal{H} \in \boldsymbol{H}\}$ consists of $\xi|_W$-non-adherent filters. If $\mathcal{U} \in \beta W \setminus \beta(\mathcal{H} \vee W)$ for each $\mathcal{H} \in \boldsymbol{H}$, then $\mathcal{U}^\uparrow$ is $\boldsymbol{H}$-cofundamental, and thus $\lim_\xi \mathcal{U} \neq \varnothing$, because ξ is $\boldsymbol{H}$-cocomplete. Consequently $\lim_{\xi|_W} \mathcal{U} = W \cap \lim_\xi \mathcal{U} \neq \varnothing$, because $\operatorname{adh}_\xi W \subset W$. We wind up the proof by observing that $\operatorname{card} \boldsymbol{H}_W \leq \operatorname{card} \boldsymbol{H}$. □

A convergence is compact whenever its completeness is 0. Hence, we recover a very elementary fact that each closed subset of a compact convergence is compact.

Recall Proposition XXIII.3.9 that each open subset of a regular finitely complete convergence is finitely complete. In other words, every open subset of a locally compact convergence is locally compact.

Similarly, by Theorem XXIII.4.7, each countable intersection of open sets of a regular countably complete convergence is countably complete.

The following theorem is a common generalization of these facts.

THEOREM XXIII.6.2. *Let κ be $0, 1$ or an infinite cardinal. Every G_κ-subset of a regular κ-complete convergence is κ-complete.*

PROOF. Let ξ be a regular κ-complete convergence on X, and let W be a G_κ-subset of X. Consequently, there exists a collection $\boldsymbol{P}$ of ξ-covers of X such that $\operatorname{card} \boldsymbol{P} = \kappa$ and ξ is $\boldsymbol{P}$-complete. Let $W := \bigcap_{\alpha<\kappa} O_\alpha$ where $O_\alpha \in \mathcal{O}_\xi$. For each $\alpha < \kappa$, the collection

$$\boldsymbol{P}_\alpha := \{\mathcal{P} \vee O_\alpha : \mathcal{P} \in \boldsymbol{P}\}$$

is a collection of $\xi|_{O_\alpha}$-covers. By Lemma XXIII.1.10, there exists a ξ-regular refinement $\mathcal{S}_{\alpha,\mathcal{P}}$ of $\mathcal{P} \vee O_\alpha$ so that $\mathcal{S}_{\alpha,\mathcal{P}}$ is a $\xi|_{O_\alpha}$-cover of O_α. Then

$$\boldsymbol{S}_\alpha := \{\mathcal{S}_{\alpha,\mathcal{P}} : \mathcal{P} \in \boldsymbol{P}\}$$

is a collection of $\xi|_{O_\alpha}$-covers such that $\xi|_{O_\alpha}$ is $\boldsymbol{S}_\alpha$-complete. Indeed, let $\mathcal{U} \in \beta O_\alpha$ be $\boldsymbol{S}_\alpha$-fundamental, that is, for each $\mathcal{P} \in \boldsymbol{P}$ there is $U_\mathcal{P} \in \mathcal{U} \cap \mathcal{S}_{\alpha,\mathcal{P}}$. As $\mathcal{S}_{\alpha,\mathcal{P}}$ is a ξ-regular refinement of $\mathcal{P} \vee O_\alpha$, there is $P \in \mathcal{P}$ such that $\operatorname{adh}_\xi U_\mathcal{P} \subset P$, and since ξ is $\boldsymbol{P}$-complete, $\operatorname{adh}_\xi(\operatorname{adh}_\xi^\natural \mathcal{U}) \neq \varnothing$.

Let

$$\boldsymbol{S} := \{\mathcal{S}_{\alpha,\mathcal{P}} \vee W : \mathcal{P} \in \boldsymbol{P}, \alpha < \kappa\}.$$

We shall show that $\xi|_W$ is $\boldsymbol{S}$-complete. Let $\mathcal{U} \in \beta W$ be $\boldsymbol{S}$-fundamental, thus it is a $\boldsymbol{S}_\alpha$-fundamental ultrafilter-base, and thus $\varnothing \neq \lim_\xi \mathcal{U} \subset \lim_\xi(\operatorname{adh}_\xi^\natural \mathcal{U}) \subset \operatorname{adh}_\xi S_\alpha \subset O_\alpha$ for each $\alpha < \kappa$. Hence $\varnothing \neq \lim_\xi \mathcal{U} \subset \bigcap_{\alpha<\kappa} O_\alpha = W$. □

THEOREM XXIII.6.3. *Let κ be $0, 1$ or an infinite cardinal. Each dense κ-complete subset of a Hausdorff pseudodiagonal convergence is G_κ.*

PROOF. Let ξ be a Hausdorff convergence on X such that $\operatorname{adh}_\xi \mathcal{G}$ is ξ-closed for each filter $\mathcal{G}$ on X. Let W be a subset of X such that $\operatorname{adh}_\xi W = X$. Suppose that there is a collection $\boldsymbol{H}$ of filters on W such that $\operatorname{card} \boldsymbol{H} \leq \kappa$ and $W \cap \operatorname{adh}_\xi \mathcal{H} = \operatorname{adh}_{\xi|_W} \mathcal{H} = \varnothing$ for each $\mathcal{H} \in \boldsymbol{H}$, so that $\lim_{\xi|_W} \mathcal{U} \neq \varnothing$ for each $\mathcal{U} \in \beta W$ that is dissociated from $\boldsymbol{H}$. It follows that $W \cap \bigcup_{\mathcal{H} \in \boldsymbol{H}} \operatorname{adh}_\xi \mathcal{H} = \varnothing$. On the other hand, for each $x \in X$ there exists $\mathcal{U} \in \beta W$ such that $\{x\} = \lim_\xi \mathcal{U}$, hence $\lim_{\xi|_W} \mathcal{U} = \varnothing$. As $\xi|_W$ is $\boldsymbol{H}$-cocomplete, there exists $\mathcal{H} \in \boldsymbol{H}$ such that $\mathcal{U} \geq \mathcal{H}$, which proves that $X \setminus W = \bigcup_{\mathcal{H} \in \boldsymbol{H}} \operatorname{adh}_\xi \mathcal{H}$. Since $\operatorname{card} \boldsymbol{H} \leq \kappa$, W is an intersection κ many ξ-open sets, and thus is $G_\kappa(\xi)$. □

COROLLARY XXIII.6.4 (Arhangel'skii-Frolík). *A Hausdorff, functionally regular (alias completely regular) topological space is Čech-complete if and only if it is a $G_{\aleph_0}$-subspace of any of its compactifications.*

PROOF. A Hausdorff compact topology is regular and κ-complete for each $\kappa \geq 0$, because its completeness number is 0, in particular is $\aleph_0$-complete. Hence, by Theorem XXIII.6.2, its $G_{\aleph_0}$-subset is $\aleph_0$-complete. Conversely, each space is dense in its compactification, and every topology is diagonal. Therefore, by Theorem XXIII.6.3, a $G_{\aleph_0}$-subspace is countably complete. □

The dominating number $\mathfrak{d}$ specified in Definition XXVII.9.2 is a small cardinal, that is, such that $\aleph_1 \leq \mathfrak{d} \leq \mathfrak{c}$. It is shown in Proposition XXVII.9.6 that the compact covering number of the space $\mathbb{R} \setminus \mathbb{Q}$ is equal to $\mathfrak{d}$, that is, $\operatorname{kc}(\nu_{\mathbb{R}\setminus\mathbb{Q}}) = \mathfrak{d}$.

COROLLARY XXIII.6.5. *The completeness of the space of rational numbers is equal to the dominating number $\mathfrak{d}$.*

PROOF. Indeed, $\mathbb{R}$ is countably complete, thus *a fortiori* $\mathfrak{d}$-complete, and the space $\mathbb{R} \setminus \mathbb{Q}$ of irrational numbers is $G_\mathfrak{d}$ and (functionally) regular. Therefore, $\mathbb{Q}$ is $\mathfrak{d}$-complete in virtue of Theorem XXIII.6.2. Suppose $\kappa \leq \mathfrak{d}$ and $\mathbb{Q}$ is κ-complete. As $\mathbb{Q}$ is dense, and $\mathbb{R}$ is a Hausdorff topological space, by Theorem XXIII.6.3, $\mathbb{Q}$ is G_κ, that is, $]0,1[\setminus\mathbb{Q}$ (homeomorphic to $\mathbb{R} \setminus \mathbb{Q}$) is a union of κ many compact sets. As its compact covering number is $\mathfrak{d}$, it follows that $\mathfrak{d} \leq \kappa$. □

XXIII.7. Operations preserving completeness number

Continuity preserves compactness, but not local compactness (see XX.9.5). This means that there exist convergences ξ and τ and $f \in \mathscr{C}(\xi, \tau)$ so that $\mathrm{compl}(\xi) = 1$ and $\mathrm{compl}(\tau)$ is infinite.

THEOREM XXIII.7.1. *If* $f \in \mathscr{C}(\xi, \tau)$ *is perfect, i.e.* $f^- \in \mathscr{K}(\tau, \xi)$, *then* $\mathrm{compl}(\xi) = \mathrm{compl}(\tau)$.

PROOF. ($\geq$) Let ξ be $\boldsymbol{G}$-cocomplete, where $\mathrm{card}\, \boldsymbol{G} \leq \mathrm{compl}(\xi)$ and let $\mathrm{adh}_\xi\, \mathcal{G} = \varnothing$ for each $\mathcal{G} \in \boldsymbol{G}$. By Lemma XXIII.4.10, $\mathrm{adh}_\tau\, f[\mathcal{G}] = \varnothing$ for each $\mathcal{G} \in \boldsymbol{G}$. If a filter $\mathcal{H}$ is $\{f[\mathcal{G}] : \mathcal{G} \in \boldsymbol{G}\}$-cofundamental, that is, $\mathcal{H} \perp f[\mathcal{G}]$, equivalently $f^-[\mathcal{H}] \perp \mathcal{G}$, for each $\mathcal{G} \in \boldsymbol{G}$, then $\mathrm{adh}_\xi\, f^-[\mathcal{H}] \neq \varnothing$, because ξ is assumed to be $\boldsymbol{G}$-cocomplete. By Lemma XXIII.2.13, $\mathrm{adh}_\tau\, \mathcal{H} \neq \varnothing$, that is, τ is $\{f[\mathcal{G}] : \mathcal{G} \in \boldsymbol{G}\}$-cocomplete. It follows that

$$\mathrm{comp}(\tau) \leq \mathrm{card}\{f[\mathcal{G}] : \mathcal{G} \in \boldsymbol{G}\} \leq \mathrm{card}\, \boldsymbol{G} \leq \mathrm{comp}(\xi).$$

Conversely, let τ be $\boldsymbol{H}$-cocomplete, $\mathrm{card}\, \boldsymbol{H} \leq \mathrm{comp}(\tau)$, and $\mathrm{adh}_\tau\, \mathcal{H} = \varnothing$ for every $\mathcal{H} \in \boldsymbol{H}$. Then by Lemma XXIII.2.13, $\mathrm{adh}_\xi\, f^-[\mathcal{H}] = \varnothing$ for each $\mathcal{H} \in \boldsymbol{H}$. To see that ξ is $\{f^-[\mathcal{H}] : \mathcal{H} \in \boldsymbol{H}\}$-cocomplete, let $\mathcal{F} \perp f^-[\mathcal{H}]$ equivalently $f[\mathcal{F}] \perp \mathcal{H}$ for every $\mathcal{H} \in \boldsymbol{H}$, then $\mathrm{adh}_\tau\, f[\mathcal{H}] \neq \varnothing$. By Lemma XXIII.4.10, $\mathrm{adh}_\xi\, \mathcal{F} \neq \varnothing$, because $f^- \in \mathscr{K}(\tau, \xi)$. Finally,

$$\mathrm{comp}(\xi) \leq \mathrm{card}\{f^-[\mathcal{H}] : \mathcal{H} \in \boldsymbol{H}\} \leq \mathrm{card}\, \boldsymbol{H} \leq \mathrm{comp}(\tau).$$

□

In particular, a continuous perfect image of a finitely complete convergence is finitely complete. As, by Theorem XX.11.10, continuous perfect maps are biquotient, Proposition XXIII.3.12 is an improvement, in this special case, of the general result above.

LEMMA XXIII.7.2. *Let* λ *be an infinite cardinal. If* $\mathrm{compl}(\prod_{j \in J} \xi_j) < \lambda$, *then there exists* $J_0 \subset J$ *such that* $\mathrm{card}\, J_0 < \lambda$ *and* ξ_j *is compact for each* $j \in J \setminus J_0$.

PROOF. Let ξ_j be a convergence on X_j and let $\xi := \prod_{j \in J} \xi_j$ and let $\boldsymbol{P}$ be a collection of ξ-covers with $\mathrm{card}\, \boldsymbol{P} < \lambda$, for which ξ is $\boldsymbol{P}$-complete. Since each product convergence has a pavement made of polyhedral filters, let $\mathcal{H}$ be any polyhedral ξ-convergent filter. Then, for every $\mathcal{P} \in \boldsymbol{P}$, there exists a polyhedron $H_\mathcal{P} \in \mathcal{H} \cap \mathcal{P}$, that is, there is a finite subset $J_\mathcal{P}$ of J such that $H_\mathcal{P} = \prod_{j \in J} D_j$ and $D_j = X_j$ for each $j \in J \setminus J_\mathcal{P}$. Set $J_0 := \bigcup_{\mathcal{P} \in \boldsymbol{P}} J_\mathcal{P}$ and observe that $\mathrm{card}\, J_0 < \lambda$, because each $J_\mathcal{P}$ is finite. The family

$$\mathcal{K} := \{H_\mathcal{P} : \mathcal{P} \in \boldsymbol{P}\}^{\cap\uparrow}$$

is a ξ-compactoid filter, because ξ is $\boldsymbol{P}$-complete. Hence $p_j[\mathcal{K}]$ is ξ_j-compactoid for each $j \in J$. In particular, if $j \notin J_0$, then $p_j[\mathcal{K}] = \{X_j\}$, hence ξ_j is compact. $\square$

COROLLARY XXIII.7.3. *If* $\mathrm{compl}(\prod_{j\in J}\xi_j)$ *is finite, then*

$$\mathrm{compl}(\prod\nolimits_{j\in J}\xi_j) \leq 1,$$

and there exists a finite subset J_0 of J such that $\mathrm{compl}(\xi_j) = 0$ for each $j \in J \setminus J_0$.

PROOF. If $\mathrm{compl}(\prod_{j\in J}\xi_j) < \infty$, then by Proposition XXIII.3.3,

$$\mathrm{compl}(\prod\nolimits_{j\in J}\xi_j) \leq 1.$$

Since ξ_j is homeomorphic to a ξ-closed subset of $\prod_{j\in J} X_j$, by XXIII.6.1, $\mathrm{compl}(\xi_j) \leq 1$ for each $j \in J$, and by Lemma XXIII.7.2, there is a finite subset J_0 of J such that ξ_j is compact for each $j \notin J_0$, that is, $\mathrm{compl}(\xi_j) = 0$ for each $j \in J \setminus J_0$. $\square$

LEMMA XXIII.7.4. *Let* κ *be an infinite ordinal, and let* $\lambda_\alpha > 0$ *for each* $\alpha < \kappa$. *Then*

$$\sum\nolimits_{\alpha<\kappa}\lambda_\alpha = \kappa \cdot \sup\nolimits_{\alpha<\kappa}\lambda_\alpha.$$

PROOF. If $\lambda := \sup_{\alpha<\kappa}\lambda_\alpha \leq \sum_{\alpha<\kappa}\lambda_\alpha$, then

$$\sum\nolimits_{\alpha<\kappa}\lambda_\alpha \leq \sum\nolimits_{\alpha<\kappa}\lambda = \kappa\cdot\lambda.$$

On the other hand, $\kappa = \sum_{\alpha<\kappa} 1 \leq \sum_{\alpha<\kappa}\lambda_\alpha$. Therefore,

$$\lambda\cdot\kappa \leq \lambda\cdot\sum\nolimits_{\alpha<\kappa}\lambda_\alpha \leq \sum\nolimits_{\alpha<\kappa}\lambda_\alpha.$$

$\square$

THEOREM XXIII.7.5. *Let* ξ_j *be a convergence on* X_j *for each* $j \in J$. *If* $\mathrm{compl}(\prod_{j\in J}\xi_j)$ *is infinite, then*

$$\mathrm{compl}(\prod\nolimits_{j\in J}\xi_j) = \sum\nolimits_{j\in J}\mathrm{compl}(\xi_j). \tag{XXIII.7.1}$$

PROOF. Set $\xi := \prod_{j\in J}\xi_j$. For each $j \in J$, let $\boldsymbol{G}_j$ be a collection of ξ_j-non-adherent filters (on X_j) such that ξ_j is $\boldsymbol{G}_j$-cocomplete and $\mathrm{card}\,\boldsymbol{G}_j = \mathrm{compl}(\xi_j)$. Let

$$\boldsymbol{G} := \{p_j^-[\mathcal{G}] : \mathcal{G} \in \boldsymbol{G}_j, j \in J\}.$$

By Lemma XXIII.2.12, every element of $\boldsymbol{G}$ is ξ-non-adherent, because $p_j \in \mathscr{C}(\xi, \xi_j)$. On the other hand, ξ is $\boldsymbol{G}$-cocomplete. Indeed, let $\mathcal{U}$ be an ultrafilter dissociated from $\boldsymbol{G}$, that is, $\mathcal{U} \perp \boldsymbol{G}$. Hence, $\mathcal{U} \perp p_j^-[\mathcal{G}]$ for each

$\mathcal{G} \in \boldsymbol{G}_j$ and every $j \in J$. Equivalently, $p_j[\mathcal{U}] \perp \boldsymbol{G}_j$ for each $j \in J$, and thus $\lim_{\xi_j} p_j[\mathcal{U}] \neq \varnothing$ for each $j \in J$. As a result, $\lim_\xi \mathcal{U} \neq \varnothing$, proving the claim. As $\operatorname{card} \boldsymbol{G} \leq \sum_{j \in J} \operatorname{card} \boldsymbol{G}_j$, the inequality $\leq$ is proved in (XXIII.7.1).

On the other hand, since ξ_j is homeomorphic to a ξ-closed subset of $\prod_{j \in J} X_j$, by XXIII.6.1, $\sup_{j \in J} \operatorname{compl}(\xi_j) \leq \operatorname{compl}(\xi)$. By Lemma XXIII.7.4,

$$\sum_{j \in J} \operatorname{compl}(\xi_j) = \operatorname{card}\{j \in J : \operatorname{compl}(\xi_j) > 0\} \cdot \sup_{j \in J} \operatorname{compl}(\xi_j).$$

As $\operatorname{card}\{j \in J : \operatorname{compl}(\xi_j) > 0\} \subset J_0$, by Lemma XXIII.7.2, $\operatorname{card} J_0 \leq \operatorname{compl}(\xi)$, hence $\sum_{j \in J} \operatorname{compl}(\xi_j) \leq \operatorname{compl}(\xi)^2 = \operatorname{compl}(\xi)$. □

Completeness equal to 0 is equivalent to compactness. Consequently, we recover Tikhonov Theorem XX.2.15.

COROLLARY XXIII.7.6. *A product of convergences is compact if and only if each factor is compact.*

Finite completeness implies that completeness is less than or equal to 1, which is equivalent to local compactoidness. Therefore,

COROLLARY XXIII.7.7. *A product of convergences is locally compactoid if and only if finitely many factors are locally compactoid and all others are compact.*

Finally,

COROLLARY XXIII.7.8. *A product of convergences is countably complete if and only if countably many factors are countably complete and all others are compact.*

XXIII.8. Baire category theorem

It follows from Theorems XXIII.6.2 and XXIII.6.3 that a $G_{\aleph_0}$-subspace of a regular, countably complete topology is countably complete, and a dense countably complete subset of a topology is a $G_{\aleph_0}$-subspace. By Corollary X.9.12, each functionally regular topology can be (densely) embedded in a compact topology. It follows that

COROLLARY XXIII.8.1. *A Hausdorff, functionally regular (alias completely regular) topology is countably complete if and only if it is a $G_{\aleph_0}$-subspace of its each compactification, equivalently of one of its compactifications.*

Baire Category Theorem X.4.15, which usefulness was already proved, extends to countably complete regular topologies. It is useful to generalize furthermore to pseudo-complete quasi-regular topologies.

Recall that a topology has the *Baire property* if each intersection of countably many open dense subsets is dense. Let ξ be a topology on X.

DEFINITION XXIII.8.2. A family $\mathcal{B}$ of ξ-open subsets of X is called a *pseudobase* of ξ if every ξ-open set includes an element of $\mathcal{B}$.

Observe that a ξ-pseudobase need not to be a ξ-open cover.

EXAMPLE XXIII.8.3. Consider the standard topology of $\mathbb{R}$. Arrange $\mathbb{Q} = \{q_k : k \in \mathbb{N}\}$ and let $\mathcal{B} := \{B(q_k, 2^{-m-k}) : k, m \in \mathbb{N}\}$. This is a pseudobase, because for each open set O, there exist $k \in \mathbb{N}$ such that $q_k \in O$, and thus $m \in \mathbb{N}$ such that $B(q_k, 2^{-m-k}) \subset O$. On the other hand, $\mathcal{B}$ is not a cover, because $B := \bigcup_{k,n\in\mathbb{N}} B(q_k, 2^{-m-k}) = \bigcup_{k\in\mathbb{N}} B(q_k, 2^{-k})$, and $\sum_{k\in\mathbb{N}} 2^{-k} = 1$ by (X.1.3); therefore, $B \neq \mathbb{R}$.

On the other hand, for each ideal ξ-cover $\mathcal{P}$, the family $\mathcal{P} \cap \mathcal{O}_\xi$ is a ξ-base, hence a ξ-pseudobase.

DEFINITION XXIII.8.4. A topology ξ is said to be *pseudo-complete* if it has a sequence $(\mathcal{B}_n)_{n<\omega}$ of ξ-pseudobases so that $\bigcap_{n<\omega} \mathrm{cl}_\xi F_n \neq \varnothing$ for each $F_n \in \mathcal{B}_n$ with $F_n \supset \mathrm{cl}_\xi F_{n+1}$ for each $n < \omega$.

REMARK XXIII.8.5. Each countably complete topology is pseudo-complete. Indeed, as we have noticed, open sets of any ideal form a pseudobase. Therefore, if $\mathcal{B}_n := \mathcal{P}_n \cap \mathcal{O}_\xi$ and ξ is $(\mathcal{P}_n)_n$-complete, then each filter base $(F_n)_n$ from Definition XXIII.8.4 is $(\mathcal{P}_n)_n$-fundamental, hence $\mathrm{adh}_\xi (F_n)_n = \bigcap_{n<\omega} \mathrm{cl}_\xi F_n \neq \varnothing$.

A topology is called *quasi-regular* if each non-empty open set includes the closure of a non-empty open set.

THEOREM XXIII.8.6 (Baire category theorem). *Each pseudo-complete, quasi-regular topology has the Baire property.*

PROOF. Consider a pseudo-complete, quasi-regular topology, and let $(\mathcal{B}_n)_n$ be a sequence of pseudobases that witnesses pseudo-completeness. Let $(O_n)_n$ be a sequence of open, dense sets. To prove that $\bigcap_{n<\omega} O_n$ is dense, let $Q \in \mathcal{O}_\xi$. As O_0 is dense, $Q \cap O_0 \neq \varnothing$, and since O_0 is open, hence $Q \cap O_0$ is open, non empty. By quasi-regularity, there exists a non-empty open set $F_0 \in \mathcal{B}_0$ such that $\mathrm{cl}\, F_0 \subset Q \cap O_0$. If $F_0, \ldots, F_n$ of elements of $\mathcal{B}_0, \ldots, \mathcal{B}_n$ has been defined so that $\mathrm{cl}\, F_{k+1} \subset F_k$ for $0 \leq k < n$, then there is $F_{n+1} \in \mathcal{B}_{n+1}$ such that $\mathrm{cl}\, F_{n+1} \subset F_n \cap O_n$. Hence $\varnothing \neq \bigcap_{n<\omega} \mathrm{cl}_\xi F_n \subset Q \cap \bigcap_{n<\omega} O_n$. □

COROLLARY XXIII.8.7. *Each countably complete, regular topology has the Baire property.*

THEOREM XXIII.8.8 (Oxtoby). *Each product of pseudo-complete, quasi-regular topologies is pseudo-complete and quasi-regular.*

PROOF. For each $j \in J$, let ξ_j be a pseudo-complete quasi-regular topology on a set X_j. Denote $\xi := \prod_{j\in J} \xi_j$. By assumption, for each $j \in J$, there exists a sequence $\{\mathcal{B}_n^j : n < \omega\}$ of ξ_j-pseudobases that witnesses pseudo-completeness. By Proposition IV.6.5, ξ has a base of polyhedral sets. For each $n < \omega$, let $\mathcal{B}_n$ consist of all the polyhedral products of the elements of $\{\mathcal{B}_n^j : j \in J\}$, and thus is a ξ-pseudobase. If $(F_n)_n$ is a sequence such that $F_n \in \mathcal{B}_n$ and $F_n \supset \mathrm{cl}_\xi F_{n+1}$, then of course each F_n is a polyhedron, say $F_n = \prod_{j\in J} B_n^j$ so that $B_n^j \in \mathcal{B}_n^j$ and there is a finite $J_n \subset J$ such that $B_n^j = X_j$ for each $j \in J \setminus J_n$. Then necessarily $J_n \subset J_{n+1}$. It follows that for each $j \in J$, there is $x_j \in \bigcap_{n<\omega} \mathrm{cl}_{\xi_j} B_n^j$, so that if $\varphi \in \prod_{j\in J} X_j$ fulfills $\varphi(j) := x_j$ for each $j \in J$, then $\varphi \in \bigcap_{n<\omega} \mathrm{cl}_\xi F_n$. Quasi-regularity follows from the fact that $\mathrm{cl}_\xi \prod_{j\in J} D^j = \prod_{j\in J} \mathrm{cl}_{\xi_j} D^j$ for each polyhedron $\prod_{j\in J} D^j$. □

COROLLARY XXIII.8.9. *Each product of regular, countably complete topologies has the Baire property.*

Such a topology is not necessarily countably complete. Indeed, as we remember, if a product of topologies is countably complete, then only countably many factors are countably complete non-compact.

On the other hand, it is known ([8]) that the Baire property is not preserved by finite products.

EXAMPLE XXIII.8.10. Consider $\mathbb{R}^{\mathbb{R}}$ with the topology of pointwise convergence, equivalently with the product topology $\nu^{\mathbb{R}}$. By Theorem XXIII.8.8, this a pseudo-complete, regular topology, hence has the Baire property. This is not a countably complete space, because it has uncountably many non-compact factors.

XXIII.9. *Ultracompleteness

We will consider now ultracompleteness, which is a stronger variant of completeness ([9]). We are motivated by forthcoming duality investigations, in which some global properties of convergences correspond to certain local properties of their dual convergences.

Consider a collection $\boldsymbol{P}$ of families of subsets of X. A filter $\mathcal{F}$ is called $\boldsymbol{P}$-*subfundamental* if $\mathcal{F}^\# \cap \mathcal{P} \neq \varnothing$ for every $\mathcal{P} \in \boldsymbol{P}$. By Definition

[8] See J. C. OXTOBY [214]

[9] Ultracompletness was studied by F. MYNARD [199], S. ROMAGUERA [225], D. BUHAGIAR and I. YOSHIOKA [35], V. I. PONOMAREV and V. V. TKACHUK [221]. See a survey of D. JARDÓN [135].

XXIII.1.1, a filter is $\boldsymbol{P}$-fundamental if $\mathcal{F} \cap \mathcal{P} \neq \varnothing$ for every $\mathcal{P} \in \boldsymbol{P}$. Therefore each $\boldsymbol{P}$-fundamental filter is $\boldsymbol{P}$-subfundamental.

As a filter $\mathcal{F}$ is an ultrafilter if and only if $\mathcal{F}^{\#} = \mathcal{F}$, an ultrafilter is $\boldsymbol{P}$-subfundamental if and only if it is $\boldsymbol{P}$-fundamental.

A convergence ξ on X is said to be $\boldsymbol{P}$-*ultracomplete* if $\operatorname{adh}_\xi \mathcal{F} \neq \varnothing$ for every $\boldsymbol{P}$-subfundamental filter. Consequently, each $\boldsymbol{P}$-ultracomplete convergence is $\boldsymbol{P}$-complete.

Likewise in Proposition XXIII.1.9, recalling that

$$\boldsymbol{P}_{\cup\downarrow} = \{\mathcal{P}^{\cup\downarrow} : \mathcal{P} \in \boldsymbol{P}\},$$

COROLLARY XXIII.9.1. *A convergence is* $\boldsymbol{P}$-*ultracomplete if and only if it is* $\boldsymbol{P}_{\cup\downarrow}$-*ultracomplete.*

PROOF. Because an ultrafilter is $\boldsymbol{P}_{\cup\downarrow}$-fundamental if and only if it is $\boldsymbol{P}$-fundamental. □

A convergence is said to be κ-*ultracomplete* if there exists a collection $\boldsymbol{P}$ of ξ-covers of cardinality κ, for which a convergence ξ is $\boldsymbol{P}$-ultracomplete. The least cardinal κ such that ξ is $\boldsymbol{P}$-ultracomplete is called the *ultracompletness number*, and is denoted by $\operatorname{compl}_{\#}(\xi)$. Since $\boldsymbol{P}$-ultracompleteness entails $\boldsymbol{P}$-completeness,

$$\operatorname{compl}(\xi) \leq \operatorname{compl}_{\#}(\xi)$$

for every convergence ξ.

PROPOSITION XXIII.9.2. *If* $\operatorname{compl}_{\#}(\xi) < \infty$, *then*

$$\operatorname{compl}(\xi) = \operatorname{compl}_{\#}(\xi) \leq 1.$$

PROOF. Indeed, if $n < \omega$ and ξ is $\{\mathcal{P}_0, \ldots, \mathcal{P}_n\}$-ultracomplete, hence it is $\{\mathcal{P}_0, \ldots, \mathcal{P}_n\}$-complete, and thus $\mathcal{P}$-complete, where $\mathcal{P} := (\mathcal{P}_0 \cup \ldots \cup \mathcal{P}_n)^{\cup\downarrow}$. Conversely, if ξ is $\mathcal{P}$-complete, and $\mathcal{F}$ is $\mathcal{P}$-subfundamental, that is, there is $P \in \mathcal{P}$ such that $P \in \mathcal{F}^{\#}$, then there is $\mathcal{U} \in \beta(\mathcal{F} \vee P)$. By $\mathcal{P}$-completeness, $\varnothing \neq \lim_\xi \mathcal{U} \subset \operatorname{adh}_\xi \mathcal{F}$, hence ξ is $\mathcal{P}$-ultracomplete. □

As we shall see, there exist (Hausdorff, functionally regular) topologies ξ such that $\aleph_0 = \operatorname{compl}(\xi) < \operatorname{compl}_{\#}(\xi)$, more precisely, the space $\mathbb{R} \setminus \mathbb{Q}$ of irrational number is countably complete, but not countably ultracomplete.

Let $\boldsymbol{H}$ be a collection of filters. A filter is called $\boldsymbol{H}$-*subcofundamental* if it is $\boldsymbol{H}_{\neg}$-subfundamental, that is,

$$\operatorname{adh}_\xi \mathcal{G} = \varnothing \Longrightarrow \underset{\mathcal{H} \in \boldsymbol{H}}{\exists}\, \mathcal{G} \geq \mathcal{H}. \tag{XXIII.9.1}$$

PROPOSITION XXIII.9.3. *A convergence ξ is κ-ultracomplete if and only if there exists a collection* $\boldsymbol{H}$ *of non-adherent filters such that* $\operatorname{card}\boldsymbol{H} \leq \kappa$ *and* $\operatorname{adh}_\xi \mathcal{G} \neq \varnothing$ *for each* $\boldsymbol{H}$-subcofundamental filter $\mathcal{G}$.

On using again the Stone space representation, we get that a convergence ξ is $\boldsymbol{H}$-ultracomplete whenever for each filter $\mathcal{H}$ such that $\operatorname{adh}_\xi \mathcal{F}$, there exists $\mathcal{H} \in \boldsymbol{H}$ such that

$$\beta\mathcal{G} \subset \beta\mathcal{H}. \tag{XXIII.9.2}$$

PROPOSITION XXIII.9.4. *The ultracompleteness of ξ on X is equal to the β-compact cofinality number of the set of ξ-non-convergent ultrafilters.*

$$\operatorname{compl}_{\#}(\xi) = \operatorname{kcf}(\beta X \setminus \xi^{-}X).$$

PROOF. A collection $\boldsymbol{H}$ consists of non-ξ-adherent filters whenever $\bigcup_{\mathcal{H}\in\boldsymbol{H}} \beta\mathcal{H} \subset \beta X \setminus \xi^{-}X$. A convergence ξ is $\boldsymbol{H}$-ultracomplete whenever $\operatorname{adh}_\xi \mathcal{G} = \varnothing$ implies the existence of $\mathcal{H} \in \boldsymbol{H}$ such (XXIII.9.2) holds, equivalently $\{\beta\mathcal{H} : \mathcal{H} \in \boldsymbol{H}\}$ is cofinal in $\mathcal{K}(\beta X \setminus \xi^{-}X)$. □

The *character* $\chi_\xi(A)$ of $A \subset X$ of a convergence ξ on X is defined as the character of the filter $\mathcal{N}_\xi(A)$ (see Definition II.2.17), that is, the least cardinality of the filter-base of neighborhoods of A.

COROLLARY XXIII.9.5. *The ultracompleteness number of a convergence ξ on X is equal to the character of the set of ξ-convergent ultrafilters in βX*

$$\operatorname{compl}_{\#}(\xi) = \chi_\beta(\xi^{-}X \cap \beta X).$$

We shall provide a condition for a metrizable topology to be countably ultracomplete. As we know, a convergence is *locally compact* whenever each convergent filter contains a compact set; *hereditarily locally compact* whenever each convergent filter has a base of compact sets.

In particular, a pretopology is (hereditarily) locally compact if each vicinity filter contains a compact set (has a base of compact sets).

DEFINITION XXIII.9.6. We say that x is a point of *local non-compactness* of a topology ξ, or a ξ-*LNC* point, if $\mathcal{V}_\xi(x) \cap \mathcal{K}(\xi) = \varnothing$. A point x is said to be of *local compactness* of ξ if $\mathcal{V}_\xi(x) \cap \mathcal{K}(\xi) \neq \varnothing$.

It is straightforward that the set of points of local compactness of a topological space is open (Exercise XII.10.7).

THEOREM XXIII.9.7 (Romaguera). *A metrizable topology is countably ultracomplete if and only if the set of its points of local non-compactness is compact.*

PROOF. Suppose that X is a metrizable space and K is a compact subset such that for each $x \in X \setminus K$ there exists a compact neighborhood V_x of x. Fix a metric and let

$$\mathcal{P}_n := \{\operatorname{int} V_x : x \in X \setminus K\} \cup \{B(x, \tfrac{1}{n}) : x \in K\}.$$

Let $\mathcal{H}$ be a filter such that for each $n < \omega$ there exists $P_n \in \mathcal{P}_n \cap \mathcal{H}^{\#}$. If $P_n = \operatorname{int} V_x$ for some $n < \omega$, then $\operatorname{adh} \mathcal{H} \cap V_x \neq \varnothing$, because V_x is compact. Otherwise, for each $n < \omega$ there exists $x_n \in K$ such that $B(x_n, \frac{1}{n}) \in \mathcal{H}^{\#}$. By the compactness of K, there exists an adherence point $x \in K$ of $\{x_n : n < \omega\}$. Consequently, for each $V \in \mathcal{N}(x)$ there exists $n < \omega$ such that $B(x_n, \frac{1}{n}) \subset V$, and thus $V \in \mathcal{H}^{\#}$, so that $x \in \operatorname{adh} \mathcal{H}$.

Conversely, let $\{x_n : n < \omega\}$ be a set of distinct LNC elements without accumulation point. For a given metric, there exists $\{V_n : n < \omega\}$ of disjoint closed non-compact sets such that $V_n \in \mathcal{N}(x_n)$ such that $V_n \subset B(x_n, \frac{1}{n})$ and

$$\operatorname{adh}\{V_n : n < \omega\} := \bigcap_{n<\omega} \operatorname{cl}(\bigcup_{k>n} V_n) = \varnothing.$$

For each $n < \omega$, let $\mathcal{U}_n \in \beta V_n$ such that $\lim \mathcal{U}_n = \varnothing$. Let $\mathcal{H} := \bigcap_{n<\omega} \mathcal{U}_n$. Suppose that $\{\mathcal{P}_k : k < \omega\}$ be an ultracomplete system of covers, and for each $k < \omega$ there is $P_k \in \mathcal{P}_k$ such that

$$P_k \in \mathcal{H}^{\#} = (\bigcap_{n<\omega} \mathcal{U}_n)^{\#} = \bigcup_{n<\omega} \mathcal{U}_n.$$

Therefore there exists a sequence $(n_k)_k$ such that $P_k \in \mathcal{U}_{n_k}$. For each $n < \omega$,

$$\begin{aligned} \operatorname{adh} \mathcal{H} &= \operatorname{adh}(\bigcap_{k \leq n} \mathcal{U}_k) \cup \operatorname{adh}(\bigcap_{k>n} \mathcal{U}_k) \\ &= \operatorname{adh}(\bigcap_{k>n} \mathcal{U}_k) \subset \operatorname{adh}\{V_n : n < \omega\} = \varnothing, \end{aligned}$$

which is a contradiction. □

COROLLARY XXIII.9.8. *The space of irrational numbers is not countably ultracomplete.*

PROOF. No point has a compact neighborhood, that is, the whole space is of local non-compactness and it is not compact. □

REMARK XXIII.9.9. Actually, $\operatorname{compl}_{\#}(\mathbb{R} \setminus \mathbb{Q}) = \mathfrak{d}$ (the dominating number). This follows from the fact that the cofinality of the set of compact subsets of $\mathbb{Q}$ of rational numbers is $\mathfrak{d}$, that is, $\operatorname{kcf}(\mathbb{Q}) = \mathfrak{d}$ ([10]).

[10] E. K. VAN DOUWEN [**255**, p. 139].

EXAMPLE XXIII.9.10. The restriction of $\nu_{\mathbb{R}}$ to $\mathbb{R} \setminus \{2^{-n} : n < \omega\}$ is countably ultracomplete. Indeed, the only point of local non-compactness is 0. We conclude by Theorem XXIII.9.7.

EXAMPLE XXIII.9.11. Let $\gamma = \pi[x_\infty, \mathcal{G}]$ be a prime topology on an infinite set X from Examples III.4.6 and XII.3.12 such that $\mathcal{G}$ is countably based, and there is a filter-base $\{G_n : n < \omega\}$ of $\mathcal{G}$ such that $G_n \setminus G_{n+1}$ is infinite for each $n < \omega$. As in every prime convergence, all but (possibly) one points are isolated, hence they are points of local compactness. Therefore, γ is not finitely complete. For every $n < \omega$, let $P \in \mathcal{P}_n$ if either there is $x \in$ such that $P = \{x\}$, or $P = G_n$. Let $\mathcal{F}$ be a filter such that $\mathcal{F}^\# \cap \mathcal{P}_n \neq \varnothing$ for each $n < \omega$. Consequently, either $\{x\} \in \mathcal{F}^\#$ and thus $x \in \ker \mathcal{F} \subset \operatorname{adh}_\gamma \mathcal{F}$, or $\mathcal{F}\#\mathcal{G}$ and thus $x_\infty \in \operatorname{adh}_\gamma \mathcal{F}$. We conclude that γ is-countably ultracomplete.

XXIII.10. Supplement

LEMMA XXIII.10.1. *Let ξ be a Hausdorff compact topology on X, and let $W \subset X$. Then* $\operatorname{compl}(\xi|_W) \leq \operatorname{card}(X \setminus W)$.

PROOF. The collection $\boldsymbol{H} := \{\mathcal{V}_\xi(x) \vee W : x \in X \setminus W\}$ consists of filters that are $\xi|_W$-non-adherent. Indeed, if $\mathcal{U} \in \beta(\mathcal{V}_\xi(x) \vee W)$ then $\{x\} = \lim_\xi \mathcal{U}$, and $\lim_{\xi|_W} \mathcal{U} = W \cap \lim_\xi \mathcal{U} = \varnothing$.

If $\mathcal{W} \in \beta W$, then $\lim_\xi \mathcal{W} \neq \varnothing$ by the compactness of ξ, that is there exists $x \in X$ such that $\mathcal{W} \geq \mathcal{V}_\xi(x) \vee W$. As $\boldsymbol{H}$ consists of $\xi|_W$-non-adherent filters, $x \in W$. □

CHAPTER XXIV

Spaces of maps

Convergences on spaces of maps are interesting from a theoretic viewpoint and because of their numerous applications in various branches of mathematics, like analysis or measure theory. They are defined in terms of convergences already present on the domain and the codomain of these maps ([1]). Moreover, they are often restricted to the subspaces of continuous maps, or even linear continuous maps, as occurs in functional analysis. Hyperconvergences constitute a special case of convergences of maps, because subsets can be identified with maps valued in a two-point set.

The most familiar is the pointwise convergence, in which Z is a convergence space, and a filter $\mathcal{F}$ on a set Z^X of maps converges to f_∞ whenever

$$\mathcal{F}(x) := \{\{f(x) : f \in F\} : F \in \mathcal{F}\}$$

converges to $f_\infty(x)$ for each $x \in X$. More precisely, if σ is a convergence on Z, then the *σ-pointwise convergence* $p(X, \sigma)$ on Z^X is defined by $f_\infty \in \lim_{p(X,\sigma)} \mathcal{F}$ whenever $f_\infty(x) \in \lim_\sigma \mathcal{F}(x)$ for each $x \in X$.

In this chapter we shall focus on convergences on sets of continuous maps, say $\mathscr{C}(\xi, \sigma)$. The σ-pointwise convergence belongs to this category, when we consider the space $\mathscr{C}(\iota, \sigma)$, where ι is the discrete topology.

Traditionally, σ-pointwise convergence has been studied on $\mathscr{C}(\xi, \sigma)$, where ξ is a topology on X, which is not necessarily discrete, and σ is a topology on Z. The term C_p-theory is deployed to design studies of pointwise convergence on spaces of continuous maps ([2]).

A convergent filter on a set of continuous maps does not need to converge pointwise to a continuous map, as witness many elementary examples.

EXAMPLE XXIV.0.2. Let $\mathscr{C}(I, I) \subset I^I$ be the set of continuous functions from a closed unit interval to itself. The sequence $(f_n)_n$

[1] They can also be defined in terms of other structures, like metrics or uniformities (see for instance *uniform convergence* defined in (X.3.7) with the aid of a metric).
[2] See A. V. ARHANGEL'SKII [14], D. FREMLIN [104].

defined by $f_n(x) := x^n$, converges pointwise to $\chi_{\{1\}}$ (the characteristic function of $\{1\}$), which is manifestly discontinuous.

In other words, $\mathscr{C}(\xi, \sigma)$ is not $p(\iota, \nu)$-closed.

A convergence on $\mathscr{C}(\xi, \sigma)$ that is most intimately related to both σ and ξ, is the *σ-dual convergence* $[\xi, \sigma]$ of ξ. It is the coarsest convergence on $\mathscr{C}(\xi, \sigma)$, for which the natural coupling between points and maps is jointly continuous (a formal definition will be given in the next section).

In the special case, where $\xi = \iota$, the discrete topology, the σ-pointwise and the σ-dual convergences coincide

$$[\iota, \sigma] = p(\iota, \sigma).$$

By (IV.3.4), $\mathscr{C}(\xi, \sigma) \subset \mathscr{C}(\iota, \sigma)$. The restriction of the convergence $p(\iota, \sigma) = [\iota, \sigma]$ to $\mathscr{C}(\xi, \sigma)$ will be denoted by

(XXIV.0.1) $$p(\xi, \sigma) := [\iota, \sigma]|_{\mathscr{C}(\xi,\sigma)} .$$

XXIV.1. Dual (natural) convergence

Let ξ be a convergence on X and σ a convergence on Z. The *evaluation* $\mathrm{ev} : X \times \mathscr{C}(\xi, \sigma) \longrightarrow Z$ ([3]) is an obvious restriction of $\mathrm{ev} : X \times Z^X \longrightarrow Z$ given by

(XXIV.1.1) $$\langle x, f\rangle = \mathrm{ev}(x, f) := f(x)$$

for each $x \in X$ and $f \in Z^X$. If $G \subset X$ and $F \subset Z^X$, then $\langle G \times F\rangle := F(G) = \{f(x) : x \in G, f \in F\}$ and if $\mathcal{G} \subset 2^X$ and $\mathcal{F} \subset 2^{(Z^X)}$, then

(XXIV.1.2) $$\langle \mathcal{G}, \mathcal{F}\rangle := \mathcal{F}[\mathcal{G}] = \{F(G) : G \in \mathcal{G}, F \in \mathcal{F}\}.$$

Let σ^ξ be a subset of $\mathbb{F}(Z^X) \times Z^X$ defined by

(dual) $$f \in \lim\nolimits_{\sigma^\xi} \mathcal{F}$$

if $f(x) = \langle x, f\rangle \in \lim_\sigma \langle \mathcal{G}, \mathcal{F}\rangle$ for each $x \in X$ and every filter $\mathcal{G}$ such that $x \in \lim_\xi \mathcal{G}$.

The relation σ^ξ is a preconvergence. Indeed, if $\mathcal{F}_0 \le \mathcal{F}_1$ then $\langle \mathcal{G}, \mathcal{F}_0\rangle \le \langle \mathcal{G}, \mathcal{F}_1\rangle$ for each $\mathcal{G}$, hence σ^ξ is isotone.

If $\mathcal{F} = f^\uparrow$, then, by the definition of (dual), $f \in \lim_{\sigma^\xi} f^\uparrow$ if and only if $f(x) \in \lim_\sigma \langle \mathcal{G}, f^\uparrow\rangle = \lim_\sigma f[\mathcal{G}]$ for each $x \in X$ and every filter $\mathcal{G}$ such that $x \in \lim_\xi \mathcal{G}$, which means that $f \in \mathscr{C}(\xi, \sigma)$.

Therefore, the preconvergence σ^ξ restricted to $\mathscr{C}(\xi, \sigma)$ is a convergence. We denote

$$[\xi, \sigma] := \sigma^\xi|_{\mathscr{C}(\xi,\sigma)} .$$

[3]The set $\mathscr{C}(\xi, \sigma)$ is non-empty, because it includes all the constant maps.

PROPOSITION XXIV.1.1. *The convergence* $[\xi,\sigma]$ *is the coarsest among the convergences* θ *on* $\mathscr{C}(\xi,\sigma)$ *for which* $\mathrm{ev}\in\mathscr{C}(\xi\times\theta,\sigma)$.

PROOF. Indeed, it follows from the definition that $\mathrm{ev}\in\mathscr{C}(\xi\times[\xi,\sigma],\sigma)$, and if $\mathrm{ev}\in\mathscr{C}(\xi\times\theta,\sigma)$, then $\theta\geq[\xi,\sigma]$. □

By Proposition IV.3.15, $\mathrm{ev}\in\mathscr{C}(\xi\times[\xi,\sigma],\sigma)$ amounts to

$$\xi\times[\xi,\sigma]\geq\mathrm{ev}^{-}\sigma. \tag{XXIV.1.3}$$

DEFINITION XXIV.1.2. The convergence $[\xi,\sigma]$ is called the σ*-dual convergence* of ξ ([4]), while ξ is said to be σ*-primal convergence* of $[\xi,\sigma]$.

It turns out that

THEOREM XXIV.1.3. *If a convergence* σ *is regular, then* $\mathscr{C}(\xi,\sigma)$ *is* σ^{ξ}*-closed.*

PROOF. Let ξ be a convergence on X. Assume that σ is a regular convergence, let $f\in\lim_{\sigma^{\xi}}\mathcal{F}$ and $\mathscr{C}(\xi,\sigma)\in\mathcal{F}$. By definition of dual convergence and regularity, for each x_0 and every filter $\mathcal{G}$ such that $x_0\in\lim_{\xi}\mathcal{G}$,

$$f(x_0)\in\lim_{\sigma}\langle\mathcal{G},\mathcal{F}\rangle\subset\lim_{\sigma}\mathrm{adh}_{\sigma}^{\natural}\langle\mathcal{G},\mathcal{F}\rangle\,.$$

Now, it is enough to prove that $f[\mathcal{G}]\geq\mathrm{adh}_{\sigma}^{\natural}\langle\mathcal{G},\mathcal{F}\rangle$. Indeed, $f(x)\in\lim_{\sigma}\langle x,\mathcal{F}\rangle$ for each $x\in X$, because $x\in\lim_{\xi}x^{\uparrow}$. Take $G\in\mathcal{G}$ and $F\in\mathcal{F}$. Then $\langle G,F\rangle\in\langle x,\mathcal{F}\rangle$ for every $x\in G$ and $F\in\mathcal{F}$, hence $f(x)\in\mathrm{adh}_{\sigma}\langle G,F\rangle$ for every $x\in G$, and thus $f(G)\subset\mathrm{adh}_{\sigma}\langle G,F\rangle$, that is, $f[\mathcal{G}]\geq\mathrm{adh}_{\sigma}^{\natural}\langle\mathcal{G},\mathcal{F}\rangle$. □

Actually, the converse also holds, that is, a convergence σ is regular if and only if for each convergence (equivalent, each topology) ξ, the set $\mathscr{C}(\xi,\sigma)$ is σ^{ξ}-closed. A proof ([5]) is rather tedious and lengthy.

It is clear from the definition that the finer ξ is, the coarser σ^{ξ} is. However, by (IV.3.4),

$$\zeta\geq\xi\Longrightarrow\mathscr{C}(\xi,\sigma)\subset\mathscr{C}(\zeta,\sigma).$$

Therefore, $\left|[\xi,\sigma]\right|=\mathscr{C}(\xi,\sigma)$ and $\left|[\zeta,\sigma]\right|=\mathscr{C}(\zeta,\sigma)$ are *a priori* different, so that we cannot compare $[\xi,\sigma]$ with $[\zeta,\sigma]$ directly via $\geq$.

As a continuation of Remark III.3.10,

REMARK XXIV.1.4. If θ and τ are (pre)convergences, recall from Remark III.3.10 that we say that θ is a *finer subspace* of τ, in symbols,

$$\theta\trianglerighteq\tau,$$

[4]Or the *natural convergence,* also traditionally called *continuous convergence.*

[5]It can be consulted in [**83**, Theorem XVI.5.2].

if $|\theta| \subset |\tau|$ and $\theta \geq \tau|_{|\theta|}$ ([6]).

Of course, if $\theta \geq \tau$, then $\theta \trianglerighteq \tau$. It follows from the definition that $\theta_0 = \bigvee^{\trianglerighteq} \Theta$ if and only if $|\theta_0| \subset \bigcap_{\theta\in\Theta} |\theta|$ and $\theta_0 \geq \theta|_{|\theta_0|}$ for each $\theta \in \Theta$, while $\theta_1 = \bigwedge^{\trianglerighteq} \Theta$ if and only if $\bigcup_{\theta\in\Theta} |\theta| \subset |\theta_1|$ and $\theta \geq \theta_1|_{|\theta|}$.

We are now in a position to compare dual convergences. Let ξ and ζ be convergences on X, and let σ and τ be convergences on Z. Then

(XXIV.1.4) $$\zeta \geq \xi \Longrightarrow [\zeta,\sigma] \trianglelefteq [\xi,\sigma], \qquad \sigma \geq \tau \Longrightarrow [\theta,\sigma] \trianglerighteq [\theta,\tau].$$

PROOF. Indeed, if $\zeta \geq \xi$, then $\mathscr{C}(\xi,\sigma) \subset \mathscr{C}(\zeta,\sigma)$, so that (XXIV.1.3) yields

$$\mathrm{ev}^{-}\sigma \leq \xi \times [\xi,\sigma] \leq \zeta \times [\xi,\sigma].$$

As $[\zeta,\sigma]$ is the coarsest convergence on $\mathscr{C}(\zeta,\sigma)$, for which $\mathrm{ev}^{-}\sigma \leq \zeta \times [\zeta,\sigma]$, this is also true, when restricted to $X \times \mathscr{C}(\xi,\sigma)$, hence

$$[\zeta,\sigma]|_{\mathscr{C}(\xi,\sigma)} \leq [\xi,\sigma],$$

that is, $[\zeta,\sigma] \trianglelefteq [\xi,\sigma]$. If now $\sigma \geq \tau$, then $\theta \times [\theta,\sigma] \geq \mathrm{ev}^{-}\sigma \trianglerighteq \mathrm{ev}^{-}\tau$. Hence $[\theta,\sigma] \trianglerighteq [\theta,\tau]$ by definition of $[\theta,\tau]$. □

In particular, as the discrete convergence ι is finer than any other convergence ξ on the same underlying set,

$$[\xi,\sigma] \trianglerighteq [\iota,\sigma] = p(|\xi|,\sigma),$$

that is, the σ-dual convergence is finer than the σ-pointwise convergence.

PROPOSITION XXIV.1.5. *If Ξ and Σ are non-empty sets of convergences, then*

$$[\bigwedge \Xi,\sigma] = \bigvee^{\trianglerighteq}_{\xi\in\Xi}[\xi,\sigma], \quad [\xi,\bigvee \Sigma] = \bigvee^{\trianglerighteq}_{\sigma\in\Sigma}[\xi,\sigma].$$

PROOF. By (XXIV.1.4), $[\bigwedge \Xi,\sigma] \trianglerighteq \bigvee^{\trianglerighteq}_{\xi\in\Xi}[\xi,\sigma]$. Conversely, by Remark IV.10.10, $\bigvee^{\trianglerighteq}_{\xi\in\Xi}[\xi,\sigma]$ is a preconvergence on $Y := \bigcap_{\xi\in\Xi}\mathscr{C}(\xi,\sigma) = \mathscr{C}(\bigwedge \Xi,\sigma)$. For each ξ, on restricting ev to $X \times Y$,

$$\xi \times \bigvee^{\trianglerighteq}_{\xi\in\Xi}[\xi,\sigma] \geq \xi \times [\xi,\sigma] \geq \mathrm{ev}^{-}\sigma,$$

hence $\bigwedge \Xi \times \bigvee^{\trianglerighteq}_{\xi\in\Xi}[\xi,\sigma] \geq \mathrm{ev}^{-}\sigma$, thus $[\bigwedge \Xi,\sigma] \leq \bigvee^{\trianglerighteq}_{\xi\in\Xi}[\xi,\sigma]$ by definition of $[\bigwedge \Xi,\sigma]$.

[6] Mind that in [83] we considered the same order, but in terms of convergences, which obliged us to introduce injections as an ingredient of the definition. A recourse to preconvergences simplifies the approach, situating the concept in set-theoretic terms.

By (XXIV.1.4), $[\xi, \bigvee \Sigma] \trianglerighteq \bigvee^{\trianglerighteq}_{\sigma\in\Sigma}[\xi,\sigma]$. Conversely, as $\xi \times [\xi,\sigma] \geq \mathrm{ev}^{-}\sigma$ for each $\sigma \in \Sigma$, we get $\xi \times \bigvee^{\trianglerighteq}_{\sigma\in\Sigma}[\xi,\sigma] \geq \mathrm{ev}^{-}\bigvee\Sigma$, hence $[\xi, \bigvee\Sigma] \leq \bigvee^{\trianglerighteq}_{\sigma\in\Sigma}[\xi,\sigma]$. □

XXIV.2. Exponentiation, transposition, and adjoint maps

Be patient! The beginning of this section requires some extra concentration, because certain objects are highlighted in several different ways. It is therefore important not to miss here the point.

We shall now describe two canonical reciprocal bijections between $Z^{X\times Y}$ and $(Z^X)^Y$, the *exponentiation* exp and the *transposition* tp:

$$Z^{X\times Y} \underset{\mathrm{tp}}{\overset{\exp}{\rightleftarrows}} (Z^X)^Y$$

With each $g \in Z^{X\times Y}$, we associate an element $\exp(g)$ of $(Z^X)^Y$ (if $g : X \times Y \longrightarrow Z$, then $\exp(g) : Y \longrightarrow Z^X$) by ([7])

$$\exp(g)(y)(x) := g(x,y).$$

Conversely, with $h \in (Z^X)^Y$, we associate $\mathrm{tp}(h) = \widehat{h} \in Z^{X\times Y}$ by

$$\mathrm{tp}(h)(x,y) = \widehat{h}(x,y) := h(y)(x).$$

The map $h \mapsto \widehat{h} = \mathrm{tp}(h)$ is called the *transposition* ([8]).

Clearly, $\mathrm{tp}\circ\exp$ and $\exp\circ\mathrm{tp}$ are the identities on the corresponding sets, that is,

$$\widehat{\exp(g)} = \mathrm{tp}(\exp(g)) = g \text{ and } \exp(\mathrm{tp}(h)) = h$$

for each $g \in Z^{X\times Y}$ and every $h \in (Z^X)^Y$.

If $F \subset X, G \subset Y$ and $H \subset Z^{X\times Y}$, then we set

$$\begin{aligned} H(F\times G) &= \exp(H)(G)(F) \\ &:= \{h(x,y) : h \in H, y \in G, x \in F\}. \end{aligned}$$

Thus, if $\mathcal{F} \subset 2^X, \mathcal{G} \subset 2^Y$, and $\mathcal{H} \subset 2^{Z^{X\times Y}}$, then

$$\begin{aligned} \mathcal{H}[\mathcal{F}\times\mathcal{G}] &= \exp[\mathcal{H}][\mathcal{G}][\mathcal{F}] \\ &= \{H(F\times G) : H \in \mathcal{H}, G \in \mathcal{G}, F \in \mathcal{F}\}. \end{aligned}$$

---

[7]As $\exp(g)(y) \in Z^X$, we conclude that $\langle x, \exp(g)(y)\rangle = \exp(g)(y)(x) = g(x,y)$, which means that $\mathrm{ev}\circ(i_X \times \exp(g)) = g$, where i_X is the identity on X.

[8]It follows that $\langle x, h(y)\rangle = h(y)(x) = \widehat{h}(x,y)$, hence $\widehat{h} = \mathrm{ev}\circ(i_X \times h)$.

For a map $f : X \longrightarrow Y$, we define two types of *adjoint map*, *upper* and *lower*. Let Z be another set. The *upper adjoint* map $f^* : Z^Y \longrightarrow Z^X$ is defined by

$$f^*(h) := h \circ f, \tag{XXIV.2.1}$$

where $h \in Z^Y$. More explicitly, $f^*(h) \in Z^X$ and, by the definition above, $f^*(h)(x) = h(f(x))$ for each $x \in X$.

Accordingly, the diagram

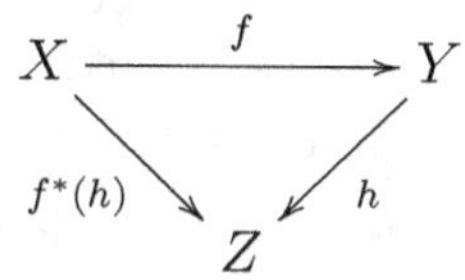

commutes. In the sequel, it will be instrumental to unfold (XXIV.2.1) in terms of the evaluation map ev, namely

$$\langle x, f^*(h)\rangle = \langle f(x), h\rangle .$$

Mind, however, that the two occurrences of evaluation in the formula above represent two, generally different, objects: the first $\mathrm{ev} : X \times Z^X \longrightarrow Z$, and the second $\mathrm{ev} : Y \times Z^Y \longrightarrow Z$. Therefore, by the usual convention (XXIV.1.2), for $\mathcal{F} \subset 2^X$ and $\mathcal{H} \subset 2^{Z^Y}$,

$$\langle \mathcal{F}, f^*[\mathcal{H}]\rangle = \langle f[\mathcal{F}], \mathcal{H}\rangle ,$$

the resulting family being a subset of 2^Z.

We will need to iterate the operation $f \mapsto f^*$, to the effect that $f^{**} := (f^*)^*$. More precisely, starting from $f : X \longrightarrow Y$, we have defined by (XXIV.2.1) the upper adjoint map (with respect to a set Z) $f^* : Z^Y \longrightarrow Z^X$. Now, by (XXIV.2.1), the upper adjoint map $f^{**} : Z^{Z^X} \longrightarrow Z^{Z^Y}$ of f^*, with respect to the same set Z, associates with each $s : Z^X \longrightarrow Z$,

$$f^{**}(s) = s \circ f^*.$$

In other terms, for $h : Z^Y \longrightarrow Z$,

$$\langle h, f^{**}(s)\rangle = \langle f^*(h), s\rangle ,$$

that is,

$$f^{**}(s)(h) = s \circ f^*(h) = s \circ h \circ f.$$

Hence,

$$f^{**}(s)(h)(x) = (s \circ h \circ f)(x) = s(h(f(x))).$$

DEFINITION XXIV.2.1. The map $j = j_X : X \longrightarrow Z^{Z^X}$, defined by

$$j_X(x)(s) := s(x)$$

for each $s \in Z^X$, is called the *point-evaluation* (of X with respect to Z).

PROPOSITION XXIV.2.2. *If* $\operatorname{card} X > 1$, *then the point-evaluation with respect to* Z *is injective if and only if* $\operatorname{card} Z > 1$.

PROOF. Let $\operatorname{card} Z > 1$, that is, there exist two distinct elements z_0 and z_1 of Z. Let $x_0 \neq x_1$ be elements of X. Then there exists $s \in Z^X$ such that $j(x_0)(s) = s(x_0) = z_0$ and $j(x_1)(s) = s(x_1) = z_1$, that is, $j(x_0) \neq j(x_1)$. If $Z = \{z\}$, then $s(x_0) = s(x_1)$ for each $s \in \{z\}^X$, so that j is constant. □

We shall see that the double adjoint map f^{**} coincides on X with f. More precisely,

PROPOSITION XXIV.2.3. *Let* $j_X : X \longrightarrow Z^{Z^X}$ *and* $j_Y : Y \longrightarrow Z^{Z^Y}$ *be the point-evaluation maps, and let* $f : X \longrightarrow Y$. *Then*

$$f^{**} \circ j_X = j_Y \circ f.$$

PROOF. Indeed, $f^{**}(j_X(x)) = j_X(x) \circ f^* : Z^Y \to Z$. So, if $h \in Z^Y$ then

$$f^{**}(j_X(x))(h) = j_X(x)(f^*(h)) = j_X(x)(h \circ f)).$$

By definition, $j_X(x)(h \circ f) = h(f(x)) = (j_Y \circ f)(x)(h)$. □

We shall now consider restrictions of adjoint maps to spaces of continuous maps. If $f \in \mathscr{C}(\xi, \tau)$ and $h \in \mathscr{C}(\tau, \sigma)$, then $f^*(h) = h \circ f \in \mathscr{C}(\xi, \sigma)$ by Proposition IV.2.2. In the diagram below, an arrow between two convergences means continuity, for instance, $\xi \xrightarrow{f} \tau$ means $f \in \mathscr{C}(\xi, \tau)$.

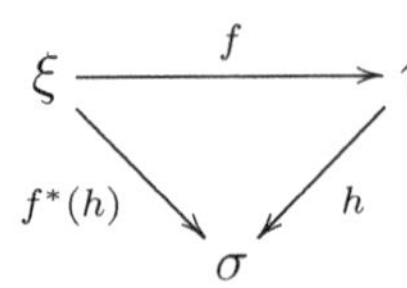

PROPOSITION XXIV.2.4. *If* $f \in \mathscr{C}(\xi, \tau)$, *then* $f^* \in \mathscr{C}([\tau, \sigma], [\xi, \sigma])$.

PROOF. Let $h \in \lim_{[\tau,\sigma]} \mathcal{H}$. To show that $f^*(h) \in \lim_{[\xi,\sigma]} f^*[\mathcal{H}]$, let $x \in \lim_\xi \mathcal{F}$. By continuity, $f(x) \in \lim_\tau f[\mathcal{F}]$, thus $\langle f(x), h \rangle \in \lim_\sigma \langle f[\mathcal{F}], \mathcal{H} \rangle$, that is, $\langle x, f^*(h) \rangle \in \lim_\sigma \langle \mathcal{F}, f^*[\mathcal{H}] \rangle$, which completes the proof. □

Proposition XXIV.2.5. *For every ξ, σ, and a surjective map $f : |\xi| \longrightarrow Y$,*

$$[f\xi, \sigma] = (f^*)^-[\xi, \sigma].$$

Proof. From Proposition XXIV.2.4 applied to $f \in \mathscr{C}(\xi, f\xi)$, we infer that $[f\xi, \sigma] \geq (f^*)^-[\xi, \sigma]$. Conversely, let $h \in \lim_{(f^*)^-[\xi,\sigma]} \mathcal{H}$, that is, $f^*(h) \in \lim_{[\xi,\sigma]} f^*[\mathcal{H}]$, which means that

$$\langle x, f^*(h) \rangle \in \lim_\sigma \langle \mathcal{F}, f^*[\mathcal{H}] \rangle,$$

whenever $x \in \lim_\xi \mathcal{F}$. This amounts to $\langle f(x), h \rangle \in \lim_\sigma \langle f[\mathcal{F}], \mathcal{H} \rangle$. For each $y \in \lim_{f\xi} \mathcal{G}$, there exist x and $\mathcal{F}$ such that $y = f(x)$, $x \in \lim_\xi \mathcal{F}$ and $\mathcal{G} = f[\mathcal{F}]$. Hence, $\langle f(x), h \rangle \in \lim_\sigma \langle f[\mathcal{F}], \mathcal{H} \rangle = \langle y, h \rangle \in \lim_\sigma \langle \mathcal{G}, \mathcal{H} \rangle$ and thus, $h \in \lim_{[f\xi,\sigma]} \mathcal{H}$. □

The *lower adjoint* map $f_* : X^W \longrightarrow Y^W$ of a map $f : X \longrightarrow Y$, with respect to a set W, is defined by

$$f_*(g) := f \circ g, \tag{XXIV.2.2}$$

where $g \in X^W$. Hence, $f_*(g)(w) = f(g(w))$ for each $w \in W$. Therefore, in terms of the evaluation map ev,

$$\langle w, f_*(g) \rangle = \langle g(w), f \rangle,$$

where the first instance of evaluation is $\mathrm{ev} : W \times Y^W \longrightarrow Y$, while the second appearance is $\mathrm{ev} : X \times Y^X \longrightarrow Y$. Of course, the diagram below

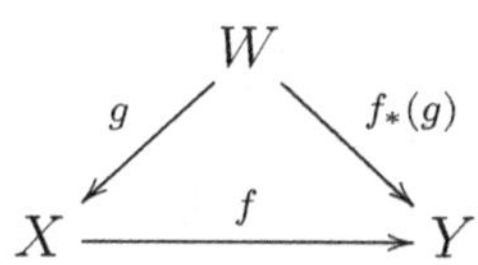

commutes. Finally, by the usual convention, for $\mathcal{W} \subset 2^W$ and $\mathcal{G} \subset 2^{X^W}$,

$$\langle \mathcal{W}, f_*[\mathcal{G}] \rangle = f[\langle \mathcal{W}, \mathcal{G} \rangle],$$

the resulting family is included in 2^Y.

An analogous iteration of $f \mapsto f_*$ can be carried out, but we will not need it for our considerations.

As for the lower adjoints restricted to continuous maps,

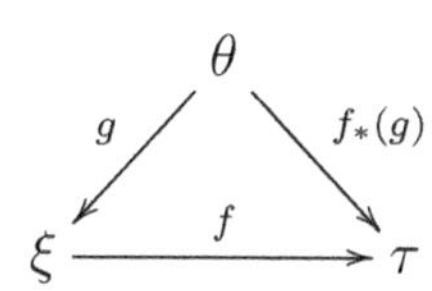

Proposition XXIV.2.6. *If $f \in \mathscr{C}(\xi, \tau)$, then $f_* \in \mathscr{C}([\theta, \xi], [\theta, \tau])$.*

Proof. Let $f \in \mathscr{C}(\xi,\tau)$, and let $g \in \lim_{[\theta,\xi]} \mathcal{G}$, that is,

$$g(w) = \langle w, g\rangle \in \lim_{\xi} \langle \mathcal{W}, \mathcal{G}\rangle = \lim_{\xi} \mathcal{G}[\mathcal{W}]$$

for each $w \in \lim_{\theta} \mathcal{W}$. Thus $f(g(w)) \in \lim_{\tau} f[\mathcal{G}[\mathcal{W}]]$ by continuity of f. In other terms,

$$\langle w, f_*(g)\rangle = \langle g(w), f\rangle \in \lim_{\tau} f[\langle \mathcal{W}, \mathcal{G}\rangle] = \lim_{\tau} \langle \mathcal{W}, f_*[\mathcal{G}]\rangle\,,$$

that is, $f_*(g) \in \lim_{[\theta,\tau]} f_*[\mathcal{G}]$. □

Proposition XXIV.2.7. *For every θ, σ, and $f : |\theta| \longrightarrow |\sigma|$,*

$$[\theta, f^{-}\sigma] = (f_*)^{-}[\theta,\sigma].$$

Proof. From Proposition XXIV.2.6 applied to $f \in \mathscr{C}(f^{-}\sigma,\sigma)$,we infer that $[\theta, f^{-}\sigma] \geq (f_*)^{-}[\theta,\sigma]$. Conversely, let $g \in \lim_{(f_*)^{-}[\theta,\sigma]} \mathcal{G}$, equivalently $f_*(g) \in \lim_{[\theta,\sigma]} f_*[\mathcal{G}]$, which is equivalent to $\langle w, f_*(g)\rangle \in \lim_{\sigma} \langle \mathcal{W}, f_*[\mathcal{G}]\rangle$ whenever $w \in \lim_{\theta} \mathcal{W}$. This amounts to $\langle g(w), f\rangle \in \lim_{\sigma} f[\langle \mathcal{W}, \mathcal{G}\rangle]$ (equivalently to $\langle w, g\rangle \in \lim_{f^{-}\sigma} \langle \mathcal{W}, \mathcal{G}\rangle$) for each $w \in \lim_{\theta} \mathcal{W}$, that is, $g \in \lim_{[\theta, f^{-}\sigma]} \mathcal{G}$. □

With a view to applications in Chapter XXVI, let us now consider a situation that will be typical in the quest for stability under products of mixed properties considered in Chapter XXI: for two convergences θ and ξ on the same set, a reflector J and a coreflector E, we want to understand how θ and ξ are related for

$$\theta \times \tau \geq J(\xi \times \tau)$$

to hold for every $\tau = E\tau$. The following lemma will be instrumental in reducing such problems to an initially dense subclass of $\mathrm{fix}(J)$ and a finally dense subclass of $\mathrm{fix}(E)$.

Lemma XXIV.2.8. *Let J be a reflector, E be a coreflector, let $\mathbf{W}$ be finally dense in $\mathrm{fix}(E)$ and let $\mathbf{D}$ be initially dense in $\mathrm{fix}(J)$. The following are equivalent:*

(1)
$$\underset{\tau=E\tau}{\forall}\ \theta \times \tau \geq J(\xi \times \tau);$$

(2)
$$\underset{\tau=E\tau}{\forall}\ \underset{\sigma=J\sigma}{\forall}\ \mathscr{C}(\xi \times \tau, \sigma) \subset \mathscr{C}(\theta \times \tau, \sigma);$$

(3)
$$\underset{\tau\in\mathbf{W}}{\forall}\ \underset{\sigma\in\mathbf{D}}{\forall}\ \mathscr{C}(\xi \times \tau, \sigma) \subset \mathscr{C}(\theta \times \tau, \sigma);$$

(4)
$$\underset{\tau\in\mathbf{W}}{\forall}\ \theta \times \tau \geq J(\xi \times \tau).$$

PROOF. (1) $\iff$ (2) and (3) $\iff$ (4) are instances of Proposition XVII.4.8 and (2)$\Longrightarrow$(3) is clear. To see that (3)$\Longrightarrow$(2), assume (3) and let $\tau = E\tau$ and $\sigma = J\sigma$. Then

$$\tau = E\tau = \mathrm{C}^{\mathbf{W}}\tau = \bigwedge\nolimits_{\alpha\in\mathbf{W}} \bigwedge\nolimits_{f\in\mathscr{C}(\alpha,\tau)} f\alpha$$

and

$$\sigma = J\sigma = \mathrm{R}^{\mathbf{D}}\sigma = \bigvee\nolimits_{\delta\in\mathbf{D}} \bigvee\nolimits_{g\in\mathscr{C}(\sigma,\delta)} g^{-}\delta,$$

so that, given that

$$\begin{aligned} \xi\times\tau &= \xi\times\bigwedge\nolimits_{\alpha\in\mathbf{W}}\bigwedge\nolimits_{f\in\mathscr{C}(\alpha,\tau)} f\alpha \\ &= \bigwedge\nolimits_{\alpha\in\mathbf{W}}\bigwedge\nolimits_{f\in\mathscr{C}(\alpha,\tau)} \xi\times f\alpha \\ &= \bigwedge\nolimits_{\alpha\in\mathbf{W}}\bigwedge\nolimits_{f\in\mathscr{C}(\alpha,\tau)} (\mathrm{id}_{|\xi|}\times f)(\xi\times\alpha), \end{aligned}$$

we conclude from Propositions XXIV.1.5, XXIV.2.5 and XXIV.2.7 that (XXIV.2.3)

$$[\xi\times\tau,\sigma]=\bigvee\nolimits^{\trianglerighteq}_{\alpha\in\mathbf{W}}\bigvee\nolimits^{\trianglerighteq}_{f\in\mathscr{C}(\alpha,\tau)}\bigvee\nolimits^{\trianglerighteq}_{\delta\in\mathbf{D}}\bigvee\nolimits^{\trianglerighteq}_{g\in\mathscr{C}(\sigma,\delta)}\left((\mathrm{id}_{|\xi|}\times f)^{*}\right)^{-}(g_{*})^{-}[\xi\times\alpha,\delta].$$

By assumption, $\mathscr{C}(\xi\times\alpha,\delta)\subset\mathscr{C}(\theta\times\alpha,\delta)$ for each $\alpha\in\mathbf{W}$ and $\delta\in\mathbf{D}$. Hence, considering only underlying sets in (XXIV.2.3), we obtain

$$\begin{aligned} \mathscr{C}(\xi\times\tau,\sigma) &\subset \bigcap\nolimits_{\alpha\in\mathbf{W},f\in\mathscr{C}(\alpha,\tau)}\bigcap\nolimits_{\delta\in\mathbf{D},g\in\mathscr{C}(\sigma,\delta)}\left(g_{*}\circ(\mathrm{id}_{|\xi|}\times f)^{*}\right)^{-}(\mathscr{C}(\xi\times\alpha,\delta)) \\ &\subset \bigcap\nolimits_{\alpha\in\mathbf{W},f\in\mathscr{C}(\alpha,\tau)}\bigcap\nolimits_{\delta\in\mathbf{D},g\in\mathscr{C}(\sigma,\delta)}\left(g_{*}\circ(\mathrm{id}_{|\xi|}\times f)^{*}\right)^{-}(\mathscr{C}(\theta\times\alpha,\delta)) \\ &\subset \mathscr{C}(\theta\times\tau,\sigma). \end{aligned}$$

□

XXIV.3. Exponential law

We shall now restrict the *exponential map* and the *transposition* to sets of continuous maps.

We have already defined in Definition XXIV.1.2 the *natural convergence* $[\xi,\sigma]$ on $\mathscr{C}(\xi,\sigma)$ to be the coarsest convergence, for which

$$\mathrm{ev}\in\mathscr{C}(\xi\times[\xi,\sigma],\sigma),$$

where ev is the *evaluation* map, that is, $\mathrm{ev}(x,f)=\langle x,f\rangle=f(x)$ for every $x\in X$ and $f\in Y^X$.

Recall from Section XXIV.2 that the *exponential map* $\exp: Z^{X\times Y}\longrightarrow(Z^X)^Y$ associates with each $f\in Z^{X\times Y}$ a map $\exp(f)\in(Z^X)^Y$, defined by

$$\exp(f)(y)(x)=f(x,y).$$

Moreover, exp is a natural bijection between $Z^{X\times Y}$ and $(Z^X)^Y$.

Let ξ, τ, and σ be convergences on X, Y, and Z, respectively. If $f \in \mathscr{C}(\xi \times \tau, \sigma) \subset Z^{X \times Y}$, then $\exp(f) \in (Z^X)^Y$. It turns out that

LEMMA XXIV.3.1. *The restriction of the exponential map* exp *to* $\mathscr{C}(\xi \times \tau, \sigma)$ *is a bijection onto* $\mathscr{C}(\tau, [\xi, \sigma])$.

PROOF. We show that exp maps $\mathscr{C}(\xi \times \tau, \sigma)$ to $\mathscr{C}(\tau, [\xi, \sigma])$. As $\mathrm{ev} \circ (i_X \times \exp(f)) = f$ by ([9]), the condition $f \in \mathscr{C}(\xi \times \tau, \sigma)$ is equivalent to

$$\xi \times \exp(f)\tau \geq e^{-}\sigma.$$

Because $[\xi, \sigma]$ is the coarsest convergence θ on $\mathscr{C}(\xi, \sigma)$, for which the evaluation $\mathrm{ev} \in \mathscr{C}(\xi \times \theta, \sigma)$, we conclude that $\exp(f)\tau \geq [\xi, \sigma]$, that is, $\exp(f) \in \mathscr{C}(\tau, [\xi, \sigma])$.

It remains to show that exp is surjective, equivalently that the transposition maps $\mathscr{C}(\tau, [\xi, \sigma])$ to $\mathscr{C}(\xi \times \tau, \sigma)$. So let $h \in \mathscr{C}(\tau, [\xi, \sigma])$, equivalently, $h\tau \geq [\xi, \sigma]$. As $\widehat{h}(x, y) = h(y)(x)$,

$$\widehat{h}(\xi \times \tau) = \mathrm{ev}(\xi \times h\tau) \geq \mathrm{ev}(\xi \times [\xi, \sigma]) \geq \sigma,$$

that is, $\widehat{h} \in \mathscr{C}(\xi \times \tau, \sigma)$. □

THEOREM XXIV.3.2. *The exponential map is a homeomorphism, that is, for each* ξ, τ, *and* σ,

(exponential law) $$[\xi \times \tau, \sigma] \approx [\tau, [\xi, \sigma]].$$

PROOF. If $h \in \mathscr{C}(\xi \times \tau, \sigma)$, then by Lemma XXIV.3.1, $\exp(h) \in \mathscr{C}(\tau, [\xi, \sigma])$. To see that exp is continuous, let $h \in \lim_{[\xi \times \tau, \sigma]} \mathcal{H}$, that is,

(XXIV.3.1) $$h(x, y) \in \lim_\sigma \langle \mathcal{F} \times \mathcal{G}, \mathcal{H} \rangle = \lim_\sigma \mathcal{H}[\mathcal{F} \times \mathcal{G}]$$

for each $x \in \lim_\xi \mathcal{F}$ and $y \in \lim_\tau \mathcal{G}$.

On the other hand, by definition, $\exp(h) \in \lim_{[\tau, [\xi, \sigma]]} \exp[\mathcal{H}]$ whenever $\exp(h)(y) \in \lim_{[\xi, \sigma]} \exp[\mathcal{H}][\mathcal{G}]$ for each $y \in \lim_\tau \mathcal{G}$. This amounts to

$$h(x, y) = \exp(h)(y)(x) \in \lim_\sigma \exp[\mathcal{H}][\mathcal{G}][\mathcal{F}]$$

for each $y \in \lim_\tau \mathcal{G}$ and every $x \in \lim_\xi \mathcal{F}$. Because $\mathcal{H}[\mathcal{F} \times \mathcal{G}] = \exp[\mathcal{H}][\mathcal{G}][\mathcal{F}]$, the two conditions above are the same.

Conversely, let $d \in \mathscr{C}(\tau, [\xi, \sigma])$ and let $\mathcal{D}$ be a filter on $\mathscr{C}(\tau, [\xi, \sigma])$ such that $d \in \lim_{[\tau, [\xi, \sigma]]} \mathcal{D}$. By definition, if $y \in \lim_\tau \mathcal{G}$, then $d(y) \in \lim_{[\xi, \sigma]} \langle \mathcal{G}, \mathcal{D} \rangle$, that is, if $x \in \lim_\xi \mathcal{F}$ then $\widehat{d}(x, y) = d(y)(x) \in \lim_\sigma \langle \mathcal{F}, \langle \mathcal{G}, \mathcal{D} \rangle \rangle = \lim_\sigma \widehat{\mathcal{G}}[\mathcal{F} \times \mathcal{D}]$. Therefore, (XXIV.3.1), exp is a homeomorphism. □

XXIV.4. Hyperconvergences as dual convergences

One of the most important special cases is the $-dual convergence of ξ, which appears in two homeomorphic versions, $[\xi, \$_0]$ and $[\xi, \$_1]$. The version dual with respect to the upper Sierpiński topology $\$_0$ can be represented as a hyperconvergence on $\mathcal{C}_\xi$ ([9]), the set of ξ-closed sets; the $\$_1$-dual can be seen as a hyperconvergence on $\mathcal{O}_\xi$, the set of ξ-open sets ([10]).

It turns out that $[\xi, \$_0]$ coincides with the *upper hyperconvergence* $[\xi, \$_0]$ of ξ, as defined in Section XX.10 for a topology ξ. We anticipated this fact using the same notation.

Indeed, if ξ is an arbitrary convergence, then by the definition (dual) of the dual preconvergence $(\$_0)^\xi$, which is equal to $[\xi, \$_0]$ if $A_0 \in \mathcal{C}_\xi$ and $\mathscr{F} \in \mathbb{F}\mathcal{C}_\xi$,

$$A_0 \in \lim_{[\xi,\, \$_0]} \mathscr{F}$$

if and only if, for each $x_0 \notin A_0$ and each $\mathcal{G}$ such that $x_0 \in \lim_\xi \mathcal{G}$, there exists $\mathrm{F} \in \mathscr{F}$ and $G \in \mathcal{G}$ so that $\bigcup_{A \in \mathrm{F}} A \cap G = \varnothing$. Equivalently,

$$\operatorname{adh}_\xi \mathrm{r}\mathscr{F} \subset A_0.$$

If ξ is a pretopology, then the formula above amounts to

$$\bigcap\nolimits_{\mathrm{F} \in \mathscr{F}} \operatorname{adh}_\xi(\bigcup\nolimits_{A \in \mathrm{F}} A) \subset A_0. \tag{XXIV.4.1}$$

Let us write down the variant $[\xi, \$_1]$. By definition, if $O_0 \in \mathcal{O}_\xi$ and $\mathscr{D} \in \mathbb{F}\mathcal{O}_\xi$, then

$$O_0 \in \lim_{[\xi, \$_1]} \mathscr{D} \tag{XXIV.4.2}$$

if and only if, for every x_0 and each filter $\mathcal{G}$,

$$x_0 \in \lim_\xi \mathcal{G} \Longrightarrow \langle x_0, O_0 \rangle \in \lim_{\$_1} \langle \mathcal{G}, \mathscr{D} \rangle .$$

In other words, if $\langle x_0, O_0 \rangle = \chi_{O_0}(x_0) = 1$, then there exists $G \in \mathcal{G}$ and $\mathrm{D} \in \mathscr{D}$ such that $1 = \langle x, O \rangle$ for each $x \in G$ and $O \in \mathrm{D}$. Equivalently, if $x_0 \in O_0$ and $x_0 \in \lim_\xi \mathcal{G}$, then there exists $G \in \mathcal{G}$ and $\mathrm{D} \in \mathscr{D}$ such that $G \subset \bigcap_{O \in \mathrm{D}} O$. If ξ is a pretopology, then (XXIV.4.2) amounts to

$$O_0 \subset \bigcup\nolimits_{\mathrm{D} \in \mathscr{D}} \operatorname{inh}_\xi(\bigcap\nolimits_{O \in \mathrm{D}} O). \tag{XXIV.4.3}$$

[9] A subset A of X is ξ-closed whenever $\chi_A \in \mathscr{C}(\xi, \$_0)$. The evaluation ev is defined by $\operatorname{ev}(x, A) := \chi_A(x)$ and restricted to $X \times \mathscr{C}(\xi, \$_0) \to \{0, 1\}$, and $\{1\}$ is the only proper $\$_0$-closed set. Accordingly, $A = \chi_A^-(1)$ is ξ-closed.

[10] Traditionally, $[\xi, \$_0]$ was called the *upper Kuratowski convergence*, and $[\xi, \$_1]$ the *Scott convergence*.

Given a filter $\mathcal{H}$ on X, the *ξ-erected filter* on $\mathcal{C}_\xi$ is defined by

$$\mathrm{e}_\xi^\natural \mathcal{H} := \{\mathrm{e}_\xi H : H \in \mathcal{H}\}^\uparrow,$$

where $\mathrm{e}_\xi H := \{C \in \mathcal{C}_\xi : C \subset H\}$. Clearly, if $\mathcal{G} \subset \mathcal{H}$, then $\mathrm{e}_\xi^\natural \mathcal{G} \subset \mathrm{e}_\xi^\natural \mathcal{H}$. On observing that $\bigcup \mathrm{e}_\xi H \subset H$, we deduce that

$$\mathrm{r}(\mathrm{e}_\xi^\natural \mathcal{H}) \geq \mathcal{H}.$$

Conversely, if $\mathscr{F}$ is a filter on $\mathcal{C}_\xi$, then

$$\mathscr{F} \geq \mathrm{e}_\xi^\natural(\mathrm{r}\mathscr{F}).$$

Indeed, if $\mathrm{F} \in \mathscr{F}$, then $\mathrm{r(F)} = \bigcup\{F : F \in \mathrm{F}\}$, and $\mathrm{e}_\xi(\mathrm{r(F)}) = \{C : C = \mathrm{cl}_\xi C \subset \mathrm{r(F)}\} \supset \mathrm{F}$. Therefore

$$\mathbb{F}X \underset{\mathrm{r}}{\overset{\mathrm{e}_\xi^\natural}{\rightleftarrows}} \mathbb{F}\mathcal{C}_\xi$$

constitutes a Galois connection of type $(+,-)$ (see Section XXI.2), that is, both $\mathrm{e}_\xi^\natural$ and r are order-preserving, $\mathrm{r}\mathrm{e}_\xi^\natural$ is ascending, and $\mathrm{e}_\xi^\natural \mathrm{r}$ is descending ([11]). It follows that that $\mathrm{e}_\xi^\natural(\mathrm{r}(\mathrm{e}_\xi^\natural \mathcal{H})) = \mathrm{e}_\xi^\natural \mathcal{H}$, and $\mathrm{r}\mathscr{F} = \mathrm{r}(\mathrm{e}_\xi^\natural(\mathrm{r}\mathscr{F}))$.

A hyperfilter $\mathscr{F}$ on $\mathcal{C}_\xi$ is called *ξ-saturated* if $\mathscr{F} = \mathrm{e}_\xi^\natural(\mathrm{r}\mathscr{F})$. It is immediate that $\mathscr{F}$ is ξ-saturated whenever it has a filter-base of ξ-saturated hypersets F, that is, such that if $C = \mathrm{cl}_\xi C \subset \bigcup_{F \in \mathrm{F}} F$, then $C \in \mathrm{F}$. Notice that $[\xi, \$_0]$ has a pavement of ξ-saturated hyperfilters. A filter on X is called *ξ-reduced* if $\mathrm{r}(\mathrm{e}_\xi^\natural \mathcal{H}) = \mathcal{H}$.

With a view to characterizing reduced filters, we define

$$\mathrm{cl}_{\xi^\bullet} A := \bigcup\nolimits_{x \in A} \mathrm{cl}_\xi\{x\}.$$

The map $\mathrm{cl}_{\xi^\bullet}$ on 2^X is expansive, idempotent, such that $\mathrm{cl}_{\xi^\bullet} \varnothing = \varnothing$, and additive, that is

$$\mathrm{cl}_{\xi^\bullet}(\bigcup\nolimits_{j \in J} A_j) = \bigcup\nolimits_{j \in J} \mathrm{cl}_{\xi^\bullet} A_j. \tag{XXIV.4.4}$$

Consequently, the closure $\mathrm{cl}_{\xi^\bullet}$ defines a topology $\xi^\bullet$, called the *point topology* of ξ, for which each union of closed sets is closed (hence each intersection of open sets is open). Evidently, $\xi^\bullet \geq \xi$. Observe that if ξ is free (alias T_1), then $\xi^\bullet = \iota$.

By (XXIV.4.4), $\mathrm{cl}_{\xi^\bullet}$ is a relation on X, hence

$$x \in \mathrm{cl}_{\xi^\circ}\{y\} \iff y \in \mathrm{cl}_{\xi^\bullet}\{x\}$$

[11]Observed by A. Lechicki in [172]

defines the inverse relation, which is expansive, idempotent, such that $\mathrm{cl}_{\xi^\circ} \varnothing = \varnothing$, and additive. Therefore, the closure cl_{ξ° defines a topology ξ°, called the *inverse point topology*, for which each union of closed sets is closed.

PROPOSITION XXIV.4.1. *For any convergence ξ, the topologies $\xi^\bullet$ and ξ° are complementary, that is,*

$$\mathcal{C}_{\xi^\bullet} = \mathcal{O}_{\xi^\circ}.$$

PROOF. Let A be ξ°-open and let $x \in A$, and let $y \in \lim_{\xi^\bullet} x^\uparrow$, that is, $x \in \lim_{\xi^\circ} y^\uparrow$, hence $A \in y^\uparrow$, equivalently $y \in A$, showing that A is $\xi^\bullet$-closed. □

If ξ is free, then $\xi^\circ = \iota$. However, ξ° need not be finer than ξ.

EXAMPLE XXIV.4.2. Let ι be the discrete topology on $\{0,1\}$, and let $\xi := [\iota, \$_0]$. Then $\mathrm{cl}_\xi\{0\} = \mathrm{cl}_{\xi^\bullet}\{0\} = \{\{0\},\{0,1\}\}$, and $\mathrm{cl}_{\xi^\circ}\{0\} = \{\varnothing, \{0\}\}$.

PROPOSITION XXIV.4.3. *A filter is reduced if and only if it is regular for the point topology if and only if it is openly based with respect to the inverse point topology.*

PROOF. Let $\mathscr{F}$ be a hyperfilter on $\mathscr{C}(\xi, \$_0)$, and let $x \in \bigcup_{A\in \boldsymbol{F}} A$ for a $\boldsymbol{F} \in \mathscr{F}$. As $A \in \mathscr{C}(\xi, \$_0)$ for each $A \in \boldsymbol{F}$, that is, A is ξ-closed, there is $A \in \boldsymbol{F}$ such that $x \in A$, hence $x \in \mathrm{cl}_\xi\{x\} \subset A \subset \bigcup_{A\in\boldsymbol{F}} A$. It follows that $\mathrm{r}\mathscr{F}$ is $\xi^\bullet$-regular. Conversely, if $\mathrm{cl}_{\xi^\bullet} H \subset H$, then $\bigcup\{A \subset H : \mathrm{cl}_\xi A \subset A\} \subset \bigcup_{x\in H} \mathrm{cl}_\xi\{x\} = \mathrm{cl}_{\xi^\bullet} H = H$. In other terms, if $\mathcal{H}$ is a $\xi^\bullet$-regular filter, then $\mathrm{r}(\mathrm{e}_\xi^\natural \mathcal{H}) = \mathcal{H}$. As ξ°-open sets coincide with $\xi^\bullet$-closed sets, the proof is complete. □

XXIV.5. Other convergences on function spaces

In the introduction to this chapter, we defined the *pointwise convergence* $p(\xi, \sigma)$ as the restriction to $\mathscr{C}(\xi,\sigma)$ of $[\iota, \sigma]$.

The pointwise convergence $p(\iota, \sigma)$ is homeomorphic to the product convergence $\prod_{x\in X} \sigma$, as the coarsest convergence on Z^X, for which the projection (or point-evaluation) $p_x : Z^X \longrightarrow Z$ is continuous for every $x \in X$. If σ is a topology, then $p(\xi, \sigma)$ is a topology. In general, $\mathscr{C}(\xi,\sigma)$ is not $p(\xi,\sigma)$-closed, as witnesses Example XXIV.0.2.

EXAMPLE XXIV.5.1. When $\sigma = \$$, then the hypertopology $p(\xi, \$)$ has two homeomorphic variants $p(\xi, \$_0)$ and $p(\xi, \$_1)$, on hyperspaces of ξ-closed and of ξ-open subsets of X respectively. Explicitly, $A_0 \in$

$\lim_{p(\xi,\$_0)} \mathscr{F}$ if and only if, for each $C \subset_{\text{fin}} X$ such that $C \cap A_0 = \varnothing$, there exists $\mathrm{F} \in \mathscr{F}$ such that $C \cap \bigcup_{A \in \mathrm{F}} A = \varnothing$. Therefore, if we set

$$C^{\neg\#} := \{A \in \mathscr{C}(\xi, \$_0) : C \cap A = \varnothing\},$$

then $\{C^{\neg\#} : C \subset_{\text{fin}} X\}$ is a subbase for the open sets of $p(\xi, \$_0)$. In fact, it is a base, because $(C_0 \cup C_1)^{\neg\#} = C_0^{\neg\#} \cap C_1^{\neg\#}$. We notice that $\varnothing^{\neg\#}$ is the whole space, and $X^{\neg\#} = \varnothing$. The collection

$$p(\xi) := \{\mathcal{O}_\xi(F) : F \subset_{\text{fin}} X\} \tag{XXIV.5.1}$$

is a base for the open sets of $p(\xi, \$_1)$.

As we will see, if σ is a topology, then the topology $p(\xi, \sigma)$ can be reconstructed preimage-wise from the corresponding hypertopology $p(\xi, \$)$.

For any $A \subset X$ and $O \in \mathcal{O}_\sigma$, let

$$[A, O] := \{f \in \mathscr{C}(\xi, \sigma) : A \subset f^-(O)\}, \tag{XXIV.5.2}$$

and if $\mathcal{A} \subset 2^X$, then let

$$[\mathcal{A}, O] := \bigcup_{A \in \mathcal{A}} [A, O]. \tag{XXIV.5.3}$$

It follows from $f^-(O) \in \mathcal{O}_\xi$ that $[A, O] = [\mathcal{O}_\xi(A), O]$ and thus also $[\mathcal{A}, O] = [\mathcal{O}_\xi\{\mathcal{A}\}, O]$.

It is straightforward that

$$\{[\mathcal{A}, O] : \mathcal{A} \in p(\xi), O \in \mathcal{O}_\sigma\},$$

where $p(\xi)$ is defined in (XXIV.5.1), is a subbase for the open sets of $p(\xi, \sigma)$.

If $\mathcal{F}$ is a filter on $\mathscr{C}(\xi, \sigma)$ and $O \in \mathcal{O}_\sigma$, then

$$\mathcal{F}^-(O) := \{\{f^-(O) : f \in F\} : F \in \mathcal{F}\}$$

is a filter-base on $\mathcal{O}_\xi$.

LEMMA XXIV.5.2. *If $\mathcal{F}$ is a filter on $\mathscr{C}(\xi, \sigma)$, $O \in \mathcal{O}_\sigma$, and $\mathcal{A} = \mathcal{O}_\xi(\mathcal{A})$ is an openly-isotone family of subsets of $|\xi|$, then*

$$[\mathcal{A}, O] \in \mathcal{F} \iff \mathcal{A} \in \mathcal{F}^-(O)^\uparrow.$$

PROOF. $[\mathcal{A}, O] \in \mathcal{F}$ if and only if there is $F \in \mathcal{F}$ with $F \subset [\mathcal{A}, O]$, that is, there is $A \in \mathcal{A}$ with $A \subset f^-(O)$ for every $f \in F$. As $\mathcal{A}$ is openly isotone, this is equivalent to $f^-(O) \in \mathcal{A}$ for every $f \in F$, that is, to $\{f^-(O) : f \in F\} \subset \mathcal{A}$. In other words, this is equivalent to $\mathcal{A} \in \mathcal{F}^-(O)^\uparrow$. □

As $f_0 \in \lim_{p(\xi,\sigma)} \mathcal{F}$ if and only if $[\mathcal{A}, O] \in \mathcal{F}$ for every $\mathcal{A} \in p(\xi)$ with $f_0 \in [\mathcal{A}, O]$, this condition rephrases in view of Lemma XXIV.5.2 into $\mathcal{A} \in \mathcal{F}^-(O)^\uparrow$ for every $\mathcal{A} \in p(\xi)$ with $f_0^-(O) \in \mathcal{A}$, that is, $f_0^-(O) \in \lim_{p(\xi,\$_1)} \mathcal{F}^-(O)$:

PROPOSITION XXIV.5.3. *If σ is a topology, then $f_0 \in \lim_{p(\xi,\sigma)} \mathcal{F}$ if and only if*

$$f_0^-(O) \in \lim_{p(\xi,\$_1)} \mathcal{F}^-(O)$$

for each $O \in \mathcal{O}_\sigma$.

This proposition follows from a general fact formulated in Proposition XXIV.5.6 to come. Another well-known hyperconvergence is the so-called *compact-open* topology $k(\xi, \sigma)$.

DEFINITION XXIV.5.4. If σ is a topology, then $f_0 \in \lim_{k(\xi,\sigma)} \mathcal{F}$ if for every $K \in \mathcal{K}(\xi)$ and each $O \in \mathcal{O}_\sigma$ such that $f_0(K) \subset O$, there is $F \in \mathcal{F}$ such that $f(K) \subset O$ for every $f \in F$.

EXAMPLE XXIV.5.5. When $\sigma = \$$, then the hyperconvergence $k(\xi, \$)$ is the *cocompact topology* (or the *upper Wijsman topology*) on the hyperspace of ξ-closed subsets of X. Analogously to Example XXIV.5.1, $A_0 \in \lim_{k(\xi,\$_0)} \mathscr{F}$ if and only if, for each $K \in \mathcal{K}(\xi)$ such that $K \cap A_0 = \varnothing$, there exists $\mathrm{F} \in \mathscr{F}$ such that $K \cap \bigcup_{A\in \mathrm{F}} A = \varnothing$. Hence, $\{K^{\neg\#} : K \in \mathcal{K}(\xi)\}$ is a base for the open sets of $k(\xi, \$_0)$. The homeomorphic variant $k(\xi, \$_1)$ on the hyperspace of ξ-open sets has a base for the open sets of $k(\xi, \$_1)$ has $\{\mathcal{O}_\xi(K) : K \in \mathcal{K}(\xi)\}$ for a base for the open sets.

Let us reword Definition XXIV.5.4: a filter $\mathcal{F}$ converges to a map f_0 in $k(\xi, \sigma)$ if and only if $[K, O] \in \mathcal{F}$ for each $K \in \mathcal{K}(\xi)$ and each $O \in \mathcal{O}_\sigma$ with $f_0 \in [K, O]$. Let us note that by definition, sets $[K, O]$ are $k(\xi, \sigma)$-open, so that in view of Lemma IX.2.5, $k(\xi, \sigma)$ is a topology on $\mathscr{C}(\xi, \sigma)$ and

$$\{[K, O] : K \in \mathcal{K}(\xi), O \in \mathcal{O}_\xi\}$$

is a subbase for this topology.

As $f^-(O) \in \mathcal{O}_\xi$ if $f \in \mathscr{C}(\xi, \sigma)$ and $O \in \mathcal{O}_\sigma$, we see that $[K, O] = [\mathcal{O}_\xi(K), O]$. It follows that

$$k(\xi) := \{\mathcal{O}_\xi(K) : K \in \mathcal{K}(\xi)\},$$

is a base for the open sets of $k(\xi, \$_1)$, and that $\{[\mathcal{A}, O] : \mathcal{A} \in k(\xi), O \in \mathcal{O}_\sigma\}$ is the same subbase displayed above. As before for the pointwise convergence, Lemma XXIV.5.2 leads to:

PROPOSITION XXIV.5.6. *If σ is a topology, then $f_0 \in \lim_{k(\xi,\sigma)} \mathcal{F}$ if and only if*

$$f_0^-(O) \in \lim_{k(\xi,\$_1)} \mathcal{F}^-(O)$$

for each $O \in \mathcal{O}_\sigma$.

More generally, let α be a collection of ξ-openly isotone families of ξ-open sets that is a base for the open sets of a topology on $\mathcal{O}_\xi$. This means that, for each $A \in \mathcal{O}_\xi$, there is $\mathcal{A} \in \alpha$ such that $A \in \mathcal{A}$, and if $\mathcal{A}_0, \mathcal{A}_1 \in \alpha$ and $A \in \mathcal{A}_0 \cap \mathcal{A}_1$, then there is $\mathcal{A} \in \alpha$ such that $A \in \mathcal{A} \subset \mathcal{A}_0 \cap \mathcal{A}_1$. Then we define a convergence $\alpha(\xi,\sigma)$ on $\mathscr{C}(\xi,\sigma)$ by

$$f_0 \in \lim_{\alpha(\xi,\sigma)} \mathcal{F} \iff \forall_{O\in\mathcal{O}_\sigma} f_0^-(O) \in \lim_\alpha \mathcal{F}^-(O), \tag{XXIV.5.4}$$

where $\lim_\alpha$ denotes convergence in the topology with base α. In the special case where $\sigma = \$_1$, this means that $A_0 \in \lim_{\alpha(\xi,\$_1)} \mathcal{F}$ if and only if

$$\forall_{\mathcal{A}\in\alpha} A_0 \in \mathcal{A} \Longrightarrow \mathcal{A} \in \mathcal{F}.$$

Therefore, $\alpha(\xi,\$_1)$ is the topology with base α and $\lim_\alpha = \lim_{\alpha(\xi,\$_1)}$.

Moreover, as observed for the particular cases of $p(\xi,\sigma)$ and $k(\xi,\sigma)$,

PROPOSITION XXIV.5.7. *let α be a collection of ξ-openly isotone families of ξ-open sets that is a base for the open sets of a topology on $\mathcal{O}_\xi$. Then $\alpha(\xi,\sigma)$ is a topology on $\mathscr{C}(\xi,\sigma)$ and*

$$\{[\mathcal{A}, O] : \mathcal{A} \in \alpha, O \in \mathcal{O}_\sigma\} \tag{XXIV.5.5}$$

is a subbase for its open sets.

PROOF. By definition, $f_0 \in \lim_{\alpha(\xi,\sigma)} \mathcal{F}$ if and only if $f_0^-(O) \in \lim_\alpha \mathcal{F}^-(O)$ for each $O \in \mathcal{O}_\sigma$, that is, if for every $O \in \mathcal{O}_\sigma$, $\mathcal{A} \in \mathcal{F}^-(O)^\uparrow$ for every $\mathcal{A} \in \alpha$ with $f_0^-(O) \in \mathcal{A}$. By Lemma XXIV.5.2, this is equivalent to $[\mathcal{A}, O] \in \mathcal{F}$ for every $\mathcal{A} \in \alpha$ with $f_0 \in [\mathcal{A}, O]$. Hence sets $[\mathcal{A}, O]$ are $\alpha(\xi,\sigma)$-open and the conclusion follows from Lemma IX.2.5. □

By Proposition XX.6.12, the collection $\kappa(\xi)$ of isotone ξ-compact families is a base for the open sets of a topology. This is also the case for the collection of ξ-openly isotone ξ-compact families.

In view of Proposition XXIV.5.7, the collection

$$\{[\mathcal{A}, O] : \mathcal{A} \in \kappa(\xi), O \in \mathcal{O}_\sigma\} \tag{XXIV.5.6}$$

is a subbase for the open sets of $\kappa(\xi,\sigma)$ on $\mathscr{C}(\xi,\sigma)$.

REMARK XXIV.5.8. If ξ is a topology and $\mathcal{A}$ is a ξ-compact family, then $\mathcal{O}_\xi\{\mathcal{A}\}$ is also a ξ-compact family in virtue of Proposition XX.6.18.

The resulting topology $\kappa(\xi,\$_1)$ on $\mathcal{O}_\xi \approx \mathscr{C}(\xi,\$_1)$ is called the *Scott topology*, and $\kappa(\xi,\sigma)$ is called the *Isbell topology*. If ξ is a topology and $K \in \mathcal{K}(\xi)$, then $\mathcal{O}_\xi(K)$ is a ξ-openly isotone ξ-compact family, we infer that $k(\xi,\sigma) \leq \kappa(\xi,\sigma)$.

It turns out that the preimage-wise way of constructing a topology $\alpha(\xi,\sigma)$ on $\mathscr{C}(\xi,\sigma)$ from the topology $\alpha(\xi,\$_1)$ on $\mathcal{O}_\xi \approx \mathscr{C}(\xi,\$_1)$ can be extended to more general convergences.

PROPOSITION XXIV.5.9. *Let ξ be a convergence and σ a topology. Then $f_0 \in \lim_{[\xi,\sigma]} \mathcal{F}$ if and only if, for every $O \in \mathcal{O}_\sigma$,*

$$f_0^-(O) \in \lim_{[\xi,\$_1]} \mathcal{F}^-(O).$$

PROOF. By definition, $f_0 \in \lim_{[\xi,\sigma]} \mathcal{F}$ if and only if $f_0(x_0) \in \lim_\sigma \langle \mathcal{G}, \mathcal{F}\rangle$ for each x_0 and every filter $\mathcal{G}$ such that $x_0 \in \lim_\xi \mathcal{G}$, that is, whenever for each $O \in \mathcal{O}_\sigma$ such that $f_0(x_0) \in O$, there exist $G \in \mathcal{G}$ and $F \in \mathcal{F}$ such that $f(x) \in O$ for each $x \in G$ and $f \in F$. In other words, for every $O \in \mathcal{O}_\sigma$ with $x_0 \in f_0^-(O)$, there exist $G \in \mathcal{G}$ and $F \in \mathcal{F}$ so that $G \subset \bigcap_{f \in F} f^-(O)$. □

By Corollary XX.10.9, a family $\mathcal{A} \subset \mathcal{O}_\xi$ is $[\xi,\$_1]$-open if and only if $\mathcal{A} \in \kappa(\xi)$. Therefore, $\kappa(\xi,\$) = \mathrm{T}[\xi,\$]$, and in particular $\kappa(\xi,\$) \leq [\xi,\$]$.

COROLLARY XXIV.5.10. *The Isbell topology $\kappa(\xi,\sigma)$ is coarser than $[\xi,\sigma]$, the σ-dual convergence of ξ.*

Therefore, if σ and ξ are topologies (the topologicity of ξ is used for the second inequality), then

$$p(\xi,\sigma) \leq k(\xi,\sigma) \leq \kappa(\xi,\sigma) \leq \mathrm{T}[\xi,\sigma] \leq [\xi,\sigma], \tag{XXIV.5.7}$$

and in the special case where $\sigma = \$$,

$$p(\xi,\$) \leq k(\xi,\$) \leq \kappa(\xi,\$) = \mathrm{T}[\xi,\$] \leq [\xi,\$], \tag{XXIV.5.8}$$

corresponding to any of two variants $\$_0$ and $\$_1$ of $\$$. In view of Definition XX.6.20,

COROLLARY XXIV.5.11. *A topology ξ is consonant if and only if $k(\xi,\$) = \kappa(\xi,\$)$.*

REMARK XXIV.5.12. It is enough to take a subbase of $\mathcal{O}_\xi$ and ξ-openly isotone ξ-compact families in order that (XXIV.5.6) constitute a subbase for the open sets of the same topology.

Let us yet provide another description of pointwise convergence. The *discretization modifier* Dis, introduced in Example XVII.3.13, associates the discrete topology ι_X with each convergence ξ on X, that is, $[\mathrm{Dis}\xi,\sigma]=[\iota,\sigma]$ so that, as a consequence of (XXIV.0.1),

$$p(\xi,\sigma)=[\mathrm{Dis}\xi,\sigma]|_{\mathscr{C}(\xi,\sigma)}.$$

The discretization modifier Dis is a coreflector. Similarly, K is the locally compact modifier, and $\mathrm{K}_{\mathrm{her}}$ is the locally hereditary compact modifier. They are both coreflectors, and $\mathrm{K}_{\mathrm{her}}\geq \mathrm{K}$, so that $[\mathrm{K}_{\mathrm{her}}\xi,\sigma]\trianglelefteq[\mathrm{K}\xi,\sigma]$ for every ξ and σ.

We shall see that ([12])

PROPOSITION XXIV.5.13. *If σ is a topology, then*

$$[\mathrm{K}_{\mathrm{her}}\xi,\sigma]|_{\mathscr{C}(\xi,\sigma)}\leq k(\xi,\sigma)\leq[\mathrm{K}\xi,\sigma]|_{\mathscr{C}(\xi,\sigma)}.$$

PROOF. If $f_0\notin\lim_{[\mathrm{K}_{\mathrm{her}}\xi,\sigma]}\mathcal{F}$, then there is $x_0\in\lim_{\mathrm{K}_{\mathrm{her}}\xi}\mathcal{G}$ such that $f_0(x_0)\notin\lim_\sigma\langle\mathcal{G},\mathcal{F}\rangle$. Hence there is $O_0\in\mathcal{O}_\sigma(f(x_0))$ such that $\langle G,F\rangle\setminus O_0\neq\varnothing$ for every $G\in\mathcal{G}$ and $F\in\mathcal{F}$. As $f_0\in\mathscr{C}(\xi,\sigma)$, there is $K_0\in\mathcal{G}\cap\mathcal{K}(\xi)$ such that $f_0(K_0)\subset O_0$ and thus $[K_0,O_0]\in\mathcal{N}_{k(\xi,\sigma)}(f_0)\setminus\mathcal{F}$. Therefore, $f_0\notin\lim_{k(\xi,\sigma)}\mathcal{F}$.

If $f_0\notin\lim_{k(\xi,\sigma)}\mathcal{F}$, then there exists $\varnothing\neq K_0\in\mathcal{K}(\xi)$ and $O_0\in\mathcal{O}_\sigma$ such that $f_0(K_0)\subset O_0$ and $\langle K_0,F\rangle\setminus O_0\neq\varnothing$ for every $F\in\mathcal{F}$. We show by contradiction that there is an ultrafilter $\mathcal{U}$ on K_0, which is of course $\mathrm{K}\xi$-convergent to a point of $K_0\subset f_0^-(O_0)$, with $O_0\notin\langle\mathcal{U},\mathcal{F}\rangle$, so that $f_0\notin\lim_{[\mathrm{K}\xi,\sigma]}\mathcal{F}$.

Otherwise, for every $\mathcal{U}\in\beta(K_0)$ there is $U_\mathcal{U}\in\mathcal{U}$ and $F_\mathcal{U}\in\mathcal{F}$ with $\langle U_\mathcal{U},F_\mathcal{U}\rangle\subset O_0$. In view of Proposition V.4.6, there is a finite subset $\mathbb{D}$ of $\beta(K_0)$ with $K_0\subset\bigcup_{\mathcal{U}\in\mathbb{D}}U_\mathcal{U}$ and thus

$$\left\langle K_0,\bigcap\nolimits_{\mathcal{U}\in\mathbb{D}}F_\mathcal{U}\right\rangle\subset\left\langle\bigcup\nolimits_{\mathcal{U}\in\mathbb{D}}U_\mathcal{U},\bigcap\nolimits_{\mathcal{U}\in\mathbb{D}}F_\mathcal{U}\right\rangle\subset O_0,$$

which is a contradiction because $\bigcap_{\mathcal{U}\in\mathbb{D}}F_\mathcal{U}\in\mathcal{F}$, as $\mathbb{D}$ is finite. □

Recall that $\mathrm{K}\xi=\mathrm{K}_{\mathrm{her}}\xi$ provided that ξ is Hausdorff and regular by Proposition XX.9.13. Hence

COROLLARY XXIV.5.14. *If a convergence ξ is Hausdorff and regular, then*

$$k(\xi,\sigma)=[\mathrm{K}\xi,\sigma]|_{\mathscr{C}(\xi,\sigma)}.$$

We will see in Section XXIV.6 that each convergence of the type (XXIV.5.7) is initial with respect to the corresponding hyperconvergence of the type (XXIV.5.8).

[12] See F. MYNARD [194, Proposition 4.3].

XXIV.6. A unification of functional convergences and hyperconvergences

Let ξ be a convergence on a set X, and let τ be a convergence on the hyperspace of ξ-open sets $\mathcal{O}_\xi \approx \mathscr{C}(\xi,\$_1)$. Let now σ be a topology on a set Z. We shall construct on $\mathscr{C}(\xi,\sigma)$ a convergence $\tau^{\Uparrow\sigma}$ in terms of the hyperconvergence τ of the preimages under continuous maps of σ-open sets.

We have already implemented this method of construction in Propositions XXIV.5.3, XXIV.5.6, and XXIV.5.9.

The *preimagewise convergence* $\tau^{\Uparrow\sigma}$ on $\mathscr{C}(\xi,\sigma)$ is defined by

$$f_0 \in \lim_{\tau^{\Uparrow\sigma}} \mathcal{F} \Longleftrightarrow \underset{O\in\mathcal{O}_\sigma}{\forall}\ f_0^-(O) \in \lim_\tau \mathcal{F}^-(O). \tag{XXIV.6.1}$$

In particular, $\tau^{\Uparrow\$_1}$ is a convergence on $\mathcal{O}_\xi \approx \mathscr{C}(\xi,\$_1)$. It follows that $\tau = \tau^{\Uparrow\$_1}$. Indeed, the only non-trivial $\$_1$-open set is $\{1\}$, while each $A \in \mathcal{O}_\xi$ is identified with $\chi_A \in \mathscr{C}(\xi,\$_1)$. On the other hand, $x \in \chi_A^-(1)$ if and only if $x \in A$.

In virtue of Proposition XXIV.5.7, for each topology σ and for every base α for the open sets on $\mathcal{O}_\xi$,

$$\alpha(\xi,\sigma) = \alpha(\xi,\$_1)^{\Uparrow\sigma},$$

and by Proposition XXIV.5.9,

$$[\xi,\sigma] = [\xi,\$_1]^{\Uparrow\sigma}.$$

Formula (XXIV.6.1) defines a convergence on a space of maps in terms of hyperconvergences on the set of open sets. As open sets can be interpreted as functions valued in a Sierpiński space, one can reformulate the mentioned hyperconvergences functionally. If $f \in Z^X$ and $O \subset Z$, then $f^-(O) \subset X$, and, in terms of characteristic functions ([13]),

$$\chi_{f^-(O)} = \chi_O \circ f. \tag{XXIV.6.2}$$

In other words, $f_0 \in \lim_{\tau^{\Uparrow\sigma}} \mathcal{F}$ if and only if, for every $O \in \mathcal{O}_\sigma$,

$$\chi_O \circ f_0 \in \lim_\tau \chi_O \circ \mathcal{F}, \tag{XXIV.6.3}$$

where $\chi_O \circ \mathcal{F} = \{\{\chi_O \circ f : f \in F\} : F \in \mathcal{F}\}$.

PROPOSITION XXIV.6.1. *Each convergence τ on $\mathscr{C}(\xi,\$)$ is initially dense in the corresponding class of preimagewise convergences $\tau^{\Uparrow\sigma}$*, that is,

$$\tau^{\Uparrow\sigma} := \bigvee_{h\in\mathscr{C}(\sigma,\$)} h_*^-\tau. \tag{XXIV.6.4}$$

[13] Indeed, $\chi_O(f(x)) = 1$ if and only if $f(x) \in O$ if and only if $x \in f^-(O)$ if and only if $\chi_{f^-(O)}(x) = 1$.

PROOF. By (XXIV.6.2) and (XXIV.2.2), we can apply the lower adjoint map $(\chi_O)_*(f) = \chi_O \circ f$ of χ_O to the effect that (XXIV.6.3) becomes $(\chi_O)_*(f_0) \in \lim_\tau (\chi_O)_*[\mathcal{F}]$, that is, $f_0 \in \lim_{\tau^{\Uparrow\sigma}} \mathcal{F}$ if and only if $h_*(f_0) \in \lim_\tau h_*[\mathcal{F}]$ for every $h \in \mathscr{C}(\sigma, \$)$. In other terms, (XXIV.6.4) holds. □

It follows that all the reflective properties of τ are transferred to $\tau^{\Uparrow\sigma}$.

XXIV.7. Consonance

By Definition XX.6.20, a convergence ξ is *consonant* if for each ξ-openly isotone ξ-compact family $\mathcal{A}$, there exists a family $\mathcal{D}$ of ξ-compact sets such that

$$\mathcal{A} = \bigcup_{D \in \mathcal{D}} \mathcal{O}_\xi(D).$$

Otherwise, it is *dissonant*.

PROPOSITION XXIV.7.1. *A topology ξ is consonant if and only if*

$$k(\xi, \$_0) = \mathrm{T}[\xi, \$_0].$$

PROOF. By Proposition XX.10.5, $[\xi, \$_0] \geq k(\xi, \$_0)$. As the latter convergence is a topology, $\mathrm{T}[\xi, \$_0] \geq k(\xi, \$_0)$. By definition, ξ is consonant whenever each ξ-compact ξ-openly family is ξ-compactly generated. We conclude by Corollary XX.10.9 and Proposition XX.10.4. □

THEOREM XXIV.7.2. *Every regular countably complete convergence is consonant.*

PROOF. Let ξ be a regular topology and let $(\mathcal{P}_n)_n$ be a sequence of open covers such that ξ is $(\mathcal{P}_n)_n$-complete. Let $\mathcal{A}$ be a ξ-openly isotone compact family. We need to show that for each $A \in \mathcal{A}$, there exists $K \in \mathcal{K}(\xi)$ such that $A \in \mathcal{O}_\xi(K) \subset \mathcal{A}$.

Let $A_0 := A$ and $\mathcal{Q}_0 := \{A_0 \cap P : P \in \mathcal{P}_0\}$. Then $A_0 = \bigcup \mathcal{Q}_0 \in \mathcal{A}$, hence by compactness, there exists a finite subfamily $\mathcal{S}_0$ of $\mathcal{Q}_0$ such that $A_1 := \bigcup \mathcal{S}_0 \in \mathcal{A}$. Let $\mathcal{Q}_1 := \{A_1 \cap P : P \in \mathcal{P}_1\}$, and thus $A_1 = \bigcup \mathcal{Q}_1 \in \mathcal{A}$, hence by compactness, there is a finite subfamily $\mathcal{S}_1$ of $\mathcal{Q}_1$ such that $A_2 := \bigcup \mathcal{S}_1 \in \mathcal{A}$. By induction, we define a filter $\mathcal{F} := \{A_n : n < \omega\}^\uparrow$ such that each $\mathcal{U} \in \beta\mathcal{F}$ contains an element of $\mathcal{P}_n$ for each $n < \omega$, and thus $\lim_\xi \mathcal{U} \neq \varnothing$, because ξ is $(\mathcal{P}_n)_n$-complete. In other words, $\mathcal{F}$ is ξ-compactoid. As ξ is regular, $\varnothing \neq K := \mathrm{adh}_\xi \mathcal{F}$ is ξ-compact in virtue of Proposition XX.6.10. By Lemma XX.6.8, $\mathcal{O}_\xi(K) \subset \mathcal{F}$, hence $A \in \mathcal{O}_\xi(K) \subset \mathcal{A}$. □

COROLLARY XXIV.7.3. *Completely metrizable topologies are consonant.*

The space $\mathbb{Q}$ of rational numbers (with the standard topology) is dissonant ([14]).

With a view to give examples of dissonant topologies, let us observe that

LEMMA XXIV.7.4. *If* $(\mathcal{A}_n)_n$ *is a sequence of* ξ*-compact families and* $\mathcal{A}$ *is a* ξ*-compact proper family such that*

$$\mathcal{A} \subset \bigcup\nolimits_{n<\omega} \bigcap\nolimits_{k>n} \mathcal{A}_k, \tag{XXIV.7.1}$$

and for each $K \in \mathcal{K}(\xi)$ *such that* $\mathcal{O}_\xi(K) \setminus \mathcal{A} \neq \varnothing$, *there is* $n < \omega$, *for which* $\mathcal{O}_\xi(K) \subset \mathcal{A}_n$, *then* ξ *is dissonant.*

PROOF. We shall see that $\mathcal{A}_\infty := \mathcal{A} \cap \bigcap_{n<\omega} \mathcal{A}_n$ is ξ-compact. Let $\mathcal{P}$ be a family of ξ-open sets such that $\bigcup_{P\in\mathcal{P}} P \in \mathcal{A}_\infty$. In particular, $\bigcup_{P\in\mathcal{P}} P \in \mathcal{A}$, then, by the ξ-compactness of $\mathcal{A}$, there is $\mathcal{P}_\infty \subset_{\text{fin}} \mathcal{P}$ such that $\bigcup_{P\in\mathcal{P}_\infty} P \in \mathcal{A}$. Hence, by (XXIV.7.1), there is $n < \omega$ such that $\bigcup_{P\in\mathcal{P}_\infty} P \in \mathcal{A}_k$ for each $k > n$. On the other hand, $\bigcup_{P\in\mathcal{P}} P \in \mathcal{A}_k$ for each $k \leq n$, so that there is $\mathcal{P}_k \subset_{\text{fin}} \mathcal{P}$ such that $\bigcup_{P\in\mathcal{P}_k} P \in \mathcal{A}_k$. If $\mathcal{R} := \mathcal{P}_\infty \cup \bigcup_{k\leq n} \mathcal{P}_k$ then $\bigcup_{P\in\mathcal{R}} P \in \mathcal{A}_\infty$, showing that $\mathcal{A}_\infty$ is ξ-compact. It is not compactly generated, because if $K \in \mathcal{K}(\xi)$ the family $\mathcal{O}_\xi(K)$ cannot be simultaneously included in $\mathcal{A}$ and in $\bigcap_{n<\omega} \mathcal{A}_n$. □

EXAMPLE XXIV.7.5 (Bisequence topology). Consider the bisequence topology from Example VI.1.8. Compact subsets are convergent sequences with limits, and their finite unions. On the other hand, the neighborhood filters are compact families. As $\mathcal{N}_\theta(x_\infty) = \bigcup_{n<\omega} \bigcap_{k>n} \mathcal{N}_\theta(x_k)$, the condition (XXIV.7.1) is fulfilled. Moreover, the set $\{\mathcal{N}_\theta(x_n) : n < \omega\}$ is discrete, and each finite infimum is dissociated from $\mathcal{N}_\theta(x_\infty)$, hence Lemma XXIV.7.4 can be applied. Therefore, $\mathrm{T}\theta$ is a dissonant topology.

EXAMPLE XXIV.7.6 (Arens topology). In Example VII.4.6, we considered the Arens topology, that is, the restriction of the bisequence topology from Example XXIV.7.5 to $\{x_\infty\} \cup \{x_{n,k} : n, k < \omega\}$. Every compact set is finite, because there is a neighborhood base of the single non-isolated point x_∞ consisting of non-cofinite sets. Consequently, for every compact set K, there is $n < \omega$ such that $K \cap \{x_{n,k} : k < \omega\} = \varnothing$. The family $\mathcal{A}_n := \bigcup_{k<\omega} \mathcal{O}(x_{n,k})$ is compact for each $n < \omega$. Let $\mathcal{A}$ stand for the neighborhood filter of x_∞, which is compact, as a compactoid

[14] Consonance was defined by G. H. GRECO, A. LECHICKI and the present author in [74], where they asked if $\mathbb{Q}$ is consonant. A. BOUZIAD and D. FREMLIN proved independently in Summer 1995 that $\mathbb{Q}$ is dissonant [32]. See also C. COSTANTINI and S. WATSON [49], and J. SAINT-RAYMOND [229].

and closed filter in a Hausdorff space. Moreover, (XXIV.7.1) holds, so that by Lemma XXIV.7.4 the Arens topology is dissonant.

The radial topology $\lambda = \mathrm{T}\rho$ on $\mathbb{R}^2$ defined in Example VII.4.9 is dissonant ([15]).

XXIV.8. Compactness in function spaces

We have considered equicontinuous families of functions (Definition X.2.5). The classical *Ascoli-Arzelà theorem* signifies that each equicontinuous sequence of real-valued bounded functions on a closed interval has a uniformly convergent subsequence. The Ascoli-Arzelà theorem is a criterion for sequential compactness in function spaces endowed with the topology of uniform convergence. Equicontinuity requires a metric on the set of values. Even continuity replaces equicontinuity in the context of convergence spaces ([16]).

Let ξ be a convergence on X and σ be a convergence on Y.

DEFINITION XXIV.8.1. A subset H of Y^X is said to be *evenly continuous* with respect to (ξ, σ) or $\|\xi, \sigma\|$-*continuous* at $x \in X$ if for each $y \in Y$, filters $\mathcal{G}$ and $\mathcal{F}$ such that $H \in \mathcal{F}$,

$$(y \in \lim_\sigma \langle x, \mathcal{F}\rangle) \wedge (x \in \lim_\xi \mathcal{G}) \Longrightarrow y \in \lim_\sigma \langle \mathcal{G}, \mathcal{F}\rangle \,. \tag{XXIV.8.1}$$

For those who remember the traditional definition of even continuity in the topological context, let us check that in that special case the notions coincide.

LEMMA XXIV.8.2. *If ξ and σ are topologies, then H is evenly continuous with respect to (ξ, σ) if and only if for every $y \in Y$ and each $U \in \mathcal{O}_\sigma(y)$, there exist $W \in \mathcal{O}_\sigma(y)$ and $V \in \mathcal{O}_\xi(x)$ such that*

$$(f \in H) \wedge (f(x) \in W) \Longrightarrow f(V) \subset U.$$

PROOF. Suppose that the condition holds, and let (XXIV.8.1) holds. Then there is $F \in \mathcal{F}$ such that $F \subset H$ such that $\langle x, F\rangle \subset W$. Then by assumption, there is $V \in \mathcal{O}_\xi(x) \leq \mathcal{G}$ such that $\langle V, F\rangle \subset U$. Hence, $y \in \lim_\sigma \langle \mathcal{G}, \mathcal{F}\rangle$.

[15] A justification of this fact is rather involved [74].

[16] See G. ASCOLI in [20] and C. ARZELÀ in [19]. J. KELLEY introduced the notion of *even continuity* in [147], which is a consequence of equicontinuity. The convergence-theoretic extension of the Ascoli-Arzelà theory to convergence spaces is due H. POPPE and his collaborators [23, 222]. It was revisited by F. MYNARD [197].

Conversely, by way of contradiction, let $U \in \mathcal{O}_\sigma(y)$ such that for each $W \in \mathcal{O}_\sigma(y)$ and $V \in \mathcal{O}_\xi(x)$, there exists $f_{V,W} \in H$ such that $f_{V,W}(x) \in W$ but $f_{V,W}(V) \setminus U \neq \varnothing$. For every $W \in \mathcal{O}_\sigma(y)$, let

$$N_W := \{f_{V,W} : V \in \mathcal{O}_\xi(x), P \in \mathcal{O}_\sigma(y), P \subset W\}.$$

Then $\{N_W : W \in \mathcal{O}_\sigma(y)\}$ is a filter-base. The resulting filter $\mathcal{F}$ is such that $y \in \lim_\sigma \langle x, \mathcal{F}\rangle$, but $y \notin \lim_\sigma \langle \mathcal{O}_\xi(x), \mathcal{F}\rangle$. □

It follows directly from the definition that

COROLLARY XXIV.8.3. *If H is $\|\xi,\sigma\|$-continuous, then for each filter $\mathcal{F}$ such that $H \in \mathcal{F}$,*

$$\lim_{p(\xi,\sigma)} \mathcal{F} \subset \lim_{[\xi,\sigma]} \mathcal{F}.$$

In other terms, the pointwise convergence coincides with the natural convergence on evenly continuous sets.

PROOF. If $f \in \lim_{p(\xi,\sigma)} \mathcal{F}$, that is, $f(x) \in \lim_\sigma \langle x, \mathcal{F}\rangle$ for every $x \in X$, then by (XXIV.8.1), $f(x) \in \lim_\sigma \langle \mathcal{G}, \mathcal{F}\rangle$ for each $x \in \lim_\xi \mathcal{G}$, hence $f \in \lim_{[\xi,\sigma]} \mathcal{F}$. □

In Definition X.2.5, equicontinuity was defined for families of real-valued functions on a topological space. The concepts readily extends to families of maps from a finitely stable convergence ξ to a metric d.

A subset H of $\mathscr{C}(\xi, \mathrm{t}d)$ is said to be *(ξ, d)-equicontinuous* at x if for every $\varepsilon > 0$ and each filter $\mathcal{G}$ such that $x \in \lim_\xi \mathcal{G}$, there exists $G \in \mathcal{G}$ such that

$$d(f(v), f(x)) < \varepsilon$$

for each $v \in G$ and $f \in H$. Plainly, if ξ is a pretopology, then it is enough to take $\mathcal{G} := \mathcal{V}_\xi(x)$, the vicinity filter of ξ at x.

PROPOSITION XXIV.8.4. *If σ is metrizable and ξ is finitely stable, then for every metric d such that $\sigma = \mathrm{t}d$, each (ξ, d)-equicontinuous set H (at x) is evenly continuous with respect to (ξ, σ) (at x). If moreover $\langle x, H\rangle$ is σ-compactoid, then the converse also holds.*

PROOF. Assume that H is (ξ, d)-equicontinuous. We need to show (XXIV.8.1) in the case of $\sigma = \mathrm{t}d$. Let $H \in \mathcal{F}, x \in \lim_\xi \mathcal{G}$, and $\varepsilon > 0$. Accordingly, there exists $G \in \mathcal{G}$ such that $d(f(x), f(v)) < \frac{\varepsilon}{2}$ for every $v \in G$ and $f \in H$. On the other hand, by assumption, there exists $F \in \mathcal{F}$ such that $F \subset H$ and $d(y, f(x)) < \frac{\varepsilon}{2}$ for each $f \in F$. Therefore, for each $f \in F$ and $v \in G$,

$$d(y, f(v)) \leq d(y, f(x)) + d(f(x), f(v)) < \varepsilon,$$

that is, $y \in \lim_{\mathrm{t}d} \langle \mathcal{G}, \mathcal{F}\rangle$.

If $\langle x, H\rangle$ is td-compactoid, and by way of contradiction, H is not (ξ, d)-equicontinuous at x, then there exists $\varepsilon > 0$ and $\mathcal{G}$ such that $x \in \lim_\xi \mathcal{G}$ and for each $G \in \mathcal{G}$, there is $f_G \in H$ and $x_G \in G$ such that $d(f(x), f_G(x_G)) \geq \varepsilon$. Then the family

$$\{\{f_G : G \subset V\} : V \in \mathcal{G}\} \tag{XXIV.8.2}$$

is a filter-base on H. If $\mathcal{U}$ is an ultrafilter including (XXIV.8.2), then there exists $y \in \lim_{td} \langle x, \mathcal{U}\rangle$. However, $\lim_{td} \langle \mathcal{G}, \mathcal{U}\rangle = \varnothing$, because for each $V \in \mathcal{G}$ and $U \in \mathcal{U}$,

$$\operatorname{diam} \langle V, U \cap \{f_G : G \subset V\}\rangle \geq \varepsilon,$$

so that H is not evenly continuous. □

PROPOSITION XXIV.8.5. *If σ is a regular convergence and H is $\|\xi, \sigma\|$-continuous, then $H \subset \mathscr{C}(\xi, \sigma)$.*

PROOF. By Corollary XXIV.8.3, if H is $\|\xi, \sigma\|$-continuous, then $p(\xi, \sigma)|_H \geq [\xi, \sigma]$, and since $[\xi, \sigma] \geq p(\xi, \sigma)$, we deduce that $p(\xi, \sigma)|_H = [\xi, \sigma]|_H$. By Theorem XXIV.1.3, then $\mathscr{C}(\xi, \sigma)$ is $[\xi, \sigma]$-closed. □

THEOREM XXIV.8.6. *Let ξ and σ be convergences. If σ is regular, H is $\|\xi, \sigma\|$-continuous, and $\langle x, H\rangle$ is σ-compactoid for every $x \in |\xi|$, then H is $[\xi, \sigma]$-compactoid.*

PROOF. Let H be $\|\xi, \sigma\|$-continuous, and let $\mathcal{U}$ be an ultrafilter on $\mathscr{C}(\xi, \sigma)$ such that $H \in \mathcal{U}$. As $\langle x, H\rangle \in \langle x, \mathcal{U}\rangle$ and $\langle x, H\rangle$ is σ-compactoid, there exists $f(x) \in \lim_\sigma \langle x, \mathcal{U}\rangle$ for every $x \in |\xi|$, that is, $f \in \lim_{p(\xi,\sigma)} \mathcal{U}$. By Corollary XXIV.8.3 $f \in \lim_{[\xi,\sigma]} \mathcal{U}$, and by Proposition XXIV.8.5 $f \in \mathscr{C}(\xi, \sigma)$. □

Conversely,

THEOREM XXIV.8.7. *Let ξ and σ be convergences. If σ is a Hausdorff pseudotopology and H is $[\xi, \sigma]$-compactoid, then H is $\|\xi, \sigma\|$-continuous and $\langle x, H\rangle$ is σ-compactoid for every $x \in |\xi|$.*

PROOF. Let H be $[\xi, \sigma]$-compactoid. By Definition XXIV.1.2 of the natural convergence, $\langle x, \cdot\rangle \in \mathscr{C}([\xi, \sigma], \sigma)$ and thus $\langle x, H\rangle$ is σ-compactoid, as the continuous image of a $[\xi, \sigma]$-compactoid set.

To see that H is evenly continuous with respect to (ξ, σ), let $H \in \mathcal{F}$ and $\mathcal{G}$ be filters such that $y \in \lim_\sigma \langle x, \mathcal{F}\rangle$ and $x \in \lim_\xi \mathcal{G}$. To prove that $y \in \lim_\sigma \langle \mathcal{G}, \mathcal{F}\rangle$, it is enough to show that $y \in \lim_\sigma \mathcal{W}$ for every $\mathcal{W} \in \beta \langle \mathcal{G}, \mathcal{F}\rangle$, because σ is a pseudotopology. By definition (XXIV.1.1) of evaluation map ev, this amounts to $\mathcal{W}\#\mathrm{ev}(\mathcal{G} \times \mathcal{F})$, hence by (V.3.6), to $\mathrm{ev}^-[\mathcal{W}]\#(\mathcal{G} \times \mathcal{F})$, equivalently, to $\mathrm{ev}^-[\mathcal{W}][\mathcal{G}]\#\mathcal{F}$. Consequently, $H \in \mathrm{ev}^-[\mathcal{W}][\mathcal{G}] \vee \mathcal{F}$. As H is $[\xi, \sigma]$-compactoid, for every ultrafilter $\mathcal{U}$ finer than $\operatorname{ev}^{-}[\mathcal{W}][\mathcal{G}] \vee \mathcal{F}$, there exists $f \in \mathscr{C}(\xi,\sigma)$ such that $f \in \lim_{[\xi,\sigma]} \mathcal{U}$. Consequently, $f(x) \in \lim_{\sigma} \langle x,\mathcal{U}\rangle$ and $\langle x,\mathcal{U}\rangle \geq \langle x,\mathcal{F}\rangle$. By assumption, $y \in \lim_{\sigma} \langle x,\mathcal{F}\rangle \subset \lim_{\sigma} \langle x,\mathcal{U}\rangle$. Since σ is Hausdorff, $y = f(x)$ for each $x \in |\xi|$, hence $y = f(x) \in \lim_{\sigma} \langle \mathcal{G},\mathcal{U}\rangle$, because $f \in \lim_{[\xi,\sigma]} \mathcal{U}$.

On the other hand, $\mathcal{U} \# \operatorname{ev}^{-}[\mathcal{W}][\mathcal{G}]$, so that $\mathcal{W} \geq \langle \mathcal{G},\mathcal{U}\rangle \geq \langle \mathcal{G},\mathcal{F}\rangle$, because $\mathcal{W}$ is an ultrafilter, hence $y = f(x) \in \lim_{\sigma} \mathcal{W}$, which completes the proof. $\square$

CHAPTER XXV

Duality

Duality investigates interrelations between a given convergence and its dual with respect to another convergence. Consequently, the dual of the dual convergence is defined on a space of maps over a space of maps. Its restriction to the underlying set of the primal convergence gives rise to a bidual convergence, which is always coarser than, but generally not equal to, the primal one. However, due to a Galois connection that underlies these interrelations, the dual of the bidual is already equal to the dual.

XXV.1. Dual and bidual convergences

If ξ and σ are convergences on X and Z respectively, then the *σ-dual convergence* $[\xi, \sigma]$ of ξ is defined on $\mathscr{C}(\xi, \sigma) \subset Z^X$. Hence the σ-dual of $[\xi, \sigma]$,

$$[[\xi, \sigma], \sigma],$$

is a convergence on $\mathscr{C}([\xi, \sigma], \sigma) \subset Z^{Z^X}$.

The *point-evaluation* j from Definition XXIV.2.1 (defined by $j(x)(f) = f(x)$) is an injection

$$j = j_X : X \longrightarrow Z^{Z^X}. \tag{XXV.1.1}$$

It turns out that

PROPOSITION XXV.1.1. *For all convergences ξ and σ, the point-evaluation (XXV.1.1) is continuous, that is,*

$$j_{|\xi|} \in \mathscr{C}(\xi, [[\xi, \sigma], \sigma]).$$

PROOF. Let $x \in \lim_\xi \mathcal{F}$. Then $f(x) = \langle x, f \rangle \in \lim_\sigma \langle \mathcal{F}, \mathcal{H} \rangle$ if $f \in \lim_{[\xi,\sigma]} \mathcal{H}$, by definition of $[\xi, \sigma]$. As $j(x)(f) = f(x)$ and $\langle \mathcal{H}, j[\mathcal{F}] \rangle = \langle \mathcal{F}, \mathcal{H} \rangle$, we infer that $j(x) \in \lim_{[[\xi,\sigma],\sigma]} j[\mathcal{F}]$. □

Fixing a convergence σ, for any convergence ξ, we define a σ-bidual convergence of ξ with the aid of a Galois $(-,-)$-connection.

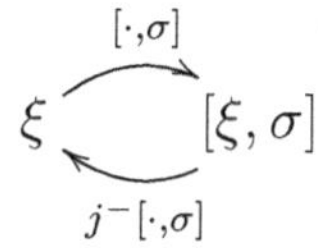

Therefore, the resulting modifier is a reflector.

DEFINITION XXV.1.2. If ξ is a convergence on X, then the convergence

$$\mathrm{Epi}^{\sigma}\xi := j_X^{-}[[\xi,\sigma],\sigma] \tag{XXV.1.2}$$

is called the *σ-bidual convergence.*

By Proposition XXV.1.1, $\xi \geq \mathrm{Epi}^{\sigma}\xi$.

PROPOSITION XXV.1.3. *For every convergence σ, the modifier* Epi^{σ} *is a projector. Moreover,* $\mathrm{Epi}^{\sigma}\xi$ *is the coarsest among convergences ζ such that $[\zeta,\sigma]=[\xi,\sigma]$. In particular*

$$[\mathrm{Epi}^{\sigma}\xi,\sigma]=[\xi,\sigma]. \tag{XXV.1.3}$$

PROOF. By (XXIV.1.4), the maps $\varphi(\xi)=[\xi,\sigma]$ and $\psi(\theta)=j^{-}[\theta,\sigma]$ are decreasing between the ordered set $(\mathbf{I}(X),\leq)$ of convergences on X, and the ordered set $\mathbf{I}(|\sigma|^{X},\trianglelefteq)$ of subspaces. Moreover, $\psi\varphi=\mathrm{Epi}^{\sigma}$ and $\varphi\psi$ are descending, so that (φ,ψ) is a $(-,-)$-connection. Thus, by Corollary XXI.2.7 and (XXI.2.3), $\varphi\psi\varphi=\varphi$ amounts to (XXV.1.3). Hence, by Corollary XXI.2.7, $\mathrm{Epi}^{\sigma}\xi=\psi\varphi(\xi)$ defines a projector. Therefore, if $[\zeta,\sigma]=[\xi,\sigma]$, then $\zeta\geq\mathrm{Epi}^{\sigma}\zeta=\mathrm{Epi}^{\sigma}\xi$. $\square$

In particular, $\mathrm{Epi}^{\sigma}(\mathrm{Epi}^{\sigma}\xi)=\mathrm{Epi}^{\sigma}\xi$, and

$$\xi\geq\theta\geq\mathrm{Epi}^{\sigma}\xi \Longrightarrow \mathrm{Epi}^{\sigma}\theta=\mathrm{Epi}^{\sigma}\xi. \tag{XXV.1.4}$$

THEOREM XXV.1.4. *For each σ, the modifier* Epi^{σ} *is a reflector.*

PROOF. By Proposition XXV.1.3, Epi^{σ} is a projector for any convergence σ. To see that Epi^{σ} is a functor, we need to show that $\mathscr{C}(\xi,\tau)\subset\mathscr{C}(\mathrm{Epi}^{\sigma}\xi,\mathrm{Epi}^{\sigma}\tau)$. Let $f\in\mathscr{C}(\xi,\tau)$. By applying Proposition XXIV.2.4, we get $f^{*}\in\mathscr{C}([\tau,\sigma],[\xi,\sigma])$, and then

$$f^{**}\in\mathscr{C}([[\xi,\sigma],\sigma],[[\tau,\sigma],\sigma]).$$

Since $j_X\in\mathscr{C}(\mathrm{Epi}^{\sigma}\xi,[[\xi,\sigma],\sigma])$, we infer that

$$(f^{**}\circ j_X)(\mathrm{Epi}^{\sigma}\xi)\geq[[\tau,\sigma],\sigma],$$

and as $f^{**}\circ j_X=j_Y\circ f$, we conclude that $f(\mathrm{Epi}^{\sigma}\xi)\geq j_Y^{-}[[\tau,\sigma],\sigma]=\mathrm{Epi}^{\sigma}\tau$ and thus $f\in\mathscr{C}(\mathrm{Epi}^{\sigma}\xi,\mathrm{Epi}^{\sigma}\tau)$. $\square$

Therefore, every σ-bidual convergence belongs to the reflective class $\mathrm{fix}(\mathrm{Epi}^\sigma)$. As we shall see below, also each σ-dual convergence belongs to $\mathrm{fix}(\mathrm{Epi}^\sigma)$.

PROPOSITION XXV.1.5. *For every convergences σ and ξ,*

$$[\xi, \sigma] \leq \mathrm{Epi}^\sigma [\xi, \sigma].$$

PROOF. It is enough to show that the point-evaluation $j : |[\xi, \sigma]| \to |[[[\xi, \sigma], \sigma], \sigma]|$ is initial, which follows from the fact that if $j(f) \in \lim_{[[[\xi,\sigma],\sigma],\sigma]} j[\mathcal{F}]$ then for every filter $\mathcal{G}$ with $g \in \lim_{[[\xi,\sigma],\sigma]} \mathcal{G}$ then $\langle \mathcal{G}, j[\mathcal{F}] \rangle = \langle \mathcal{F}, \mathcal{G} \rangle$ converges to $g(f) = j(f)(g)$ in σ and thus $f \in \lim_{[\xi,\sigma]} \mathcal{F}$. □

Thus (XXV.1.3) signifies that any investigation of σ-dual convergences can be restricted to primal convergences from $\mathrm{fix}(\mathrm{Epi}^\sigma)$.

Recall (XVII.4.1) that for a convergence σ,

$$\mathrm{R}^\sigma \xi = \bigvee\nolimits_{f \in \mathscr{C}(\xi,\sigma)} f^- \sigma$$

is the *σ-initial reflection* of ξ.

PROPOSITION XXV.1.6. *For every pair ξ and σ of convergences,*

$$\xi \geq \mathrm{Epi}^\sigma \xi \geq \mathrm{R}^\sigma \xi. \tag{XXV.1.5}$$

PROOF. Indeed, as $[\xi, \sigma] = [\mathrm{Epi}^\sigma \xi, \sigma]$ is a convergence on $\mathscr{C}(\xi, \sigma)$, for each $f \in \mathscr{C}(\xi, \sigma)$,

$$f \in \lim\nolimits_{[\mathrm{Epi}^\sigma \xi, \sigma]} f^\uparrow,$$

that is, $x \in \lim_{\mathrm{Epi}^\sigma \xi} \mathcal{F}$ implies $f(x) \in \lim_\sigma \langle \mathcal{F}, f^\uparrow \rangle = \lim_\sigma f[\mathcal{F}]$, which means that $f \in \mathscr{C}(\mathrm{Epi}^\sigma \xi, \sigma)$. Therefore, $\mathrm{Epi}^\sigma \xi \geq \bigvee_{f \in \mathscr{C}(\xi,\sigma)} f^- \sigma = \mathrm{R}^\sigma \xi$. □

Recapitulating, the reflective hull of a convergence σ is a subclass of the reflective class of σ-bidual convergences:

$$\mathrm{fix}(\mathrm{R}^\sigma) \subset \mathrm{fix}(\mathrm{Epi}^\sigma).$$

An important case is that of the standard topology $\nu_\mathbb{R}$ on $\mathbb{R}$. Recall that a convergence ξ is called *functionally initial* whenever $\xi = \mathrm{R}^\nu \xi$, and that functionally initial convergences coincide with *functionally regular* (alias completely regular) topologies.

DEFINITION XXV.1.7. A convergence is called *functionally embedded* if it is a ν-bidual convergence:

$$\xi \leq \mathrm{Epi}^\nu \xi.$$

XXV.2. Bidual reflector

We have so far investigated duality with respect to a single convergence. We shall now extend our exploration to duality with respect to a reflective class of convergences. Our findings will be then applied to basic reflective classes of topologies, pretopologies, and pseudotopologies.

If Σ is a set of convergences, then the Σ-*bidual* reflection is defined by

$$\mathrm{Epi}^{\Sigma}\xi := \bigvee\nolimits_{\sigma\in\Sigma} \mathrm{Epi}^{\sigma}\xi.$$

In particular, if J is a modifier, then

$$\mathrm{Epi}^{J}\xi = \bigvee\nolimits_{\sigma=J\sigma} \mathrm{Epi}^{\sigma}\xi \tag{XXV.2.1}$$

is called the J-*bidual reflection* of ξ.

REMARK XXV.2.1. Sure enough, if J is a modifier, then $\mathrm{fix}(J)$ is not a set, but a class of convergences. However, the supremum in (XXV.2.1) is in fact taken over a set of convergences. Indeed, if $f \in \mathscr{C}(\xi,\sigma)$, then $f \in \mathscr{C}(\xi,\sigma|_{f(X)})$. Hence, up to homeomorphisms, we can consider in (XXV.2.1) only convergences on subsets of X, which constitutes a set.

It follows from Proposition XVII.3.10 that Epi^{J} is a reflector for every functor J, and, of course, $\mathrm{Epi}^{J}\xi \geq \mathrm{Epi}^{\sigma}\xi$ for each $\sigma = J\sigma$. If J_0 and J_1 are reflectors, then it is plain that

$$J_0 \leq J_1 \Longrightarrow \mathrm{Epi}^{J_0}\xi \leq \mathrm{Epi}^{J_1}\xi. \tag{XXV.2.2}$$

Accordingly, $\mathrm{fix}(\mathrm{Epi}^{J})$ is a reflective class of convergences that includes $\mathrm{fix}(\mathrm{Epi}^{\sigma})$ for each $\sigma \in \mathrm{fix}(J)$. Proposition XXV.1.3 can be extended as follows.

PROPOSITION XXV.2.2. *If J is a reflector, then $\mathrm{Epi}^{J}\xi$ is the coarsest among convergences θ such that $[\theta,\sigma] = [\xi,\sigma]$ for each $\sigma = J\sigma$. In particular, for each $\sigma = J\sigma$,*

$$[\mathrm{Epi}^{J}\xi,\sigma] = [\xi,\sigma]. \tag{XXV.2.3}$$

PROOF. As $\xi \geq \mathrm{Epi}^{J}\xi \geq \mathrm{Epi}^{\sigma}\xi$ for each $\sigma = J\sigma$, by (XXIV.1.4), $[\xi,\sigma] \trianglelefteq [\mathrm{Epi}^{J}\xi,\sigma] \trianglelefteq [\mathrm{Epi}^{\sigma}\xi,\sigma]$ for each $\sigma = J\sigma$. By Proposition XXV.1.3, $[\xi,\sigma] = [\mathrm{Epi}^{\sigma}\xi,\sigma]$, so that all three terms above are equal. If θ is a convergence on $|\xi|$ such that $[\theta,\sigma] = [\xi,\sigma]$ for every $\sigma = J\sigma$, then by Proposition XXV.1.3, $\theta \geq \mathrm{Epi}^{\sigma}\xi$ for every $\sigma = J\sigma$, hence $\theta \geq \mathrm{Epi}^{J}\xi$ by definition. □

Hence,

$$\xi \geq \theta \geq \mathrm{Epi}^J\xi \Longrightarrow \mathrm{Epi}^J\theta = \mathrm{Epi}^J\xi. \tag{XXV.2.4}$$

From (XXV.1.5), it follows that, $\xi \geq \mathrm{Epi}^J\xi \geq \mathrm{R}^J\xi$ for every convergence ξ, that is, a *bidual reflection* is always finer than the corresponding *initial reflection.*

By Proposition XVII.4.7,

COROLLARY XXV.2.3. *If J is a reflector, then*

$$\mathrm{Epi}^J \geq \mathrm{R}^J = J. \tag{XXV.2.5}$$

By (XXV.2.3), it is enough to restrict our study to primal convergences from the reflective class $\mathrm{fix}(\mathrm{Epi}^J)$.

It follows from the definitions that $\mathrm{Epi}^{\mathrm{R}^\sigma}\xi \geq \mathrm{Epi}^\sigma\xi$ for every σ and ξ. Actually,

PROPOSITION XXV.2.4. *If J is a reflector and $\mathbf{D}$ is an initially dense subclass of $\mathrm{fix}(J)$, then the J-bidual reflector fulfills*

$$\mathrm{Epi}^J\xi = \bigvee_{\sigma\in\mathbf{D}} \mathrm{Epi}^\sigma\xi.$$

PROOF. Of course,

$$\mathrm{Epi}^J\xi = \bigvee_{\sigma=J\sigma} \mathrm{Epi}^\sigma\xi \geq \bigvee_{\sigma\in\mathbf{D}} \mathrm{Epi}^\sigma\xi.$$

On the other hand, let us show that $\theta = \bigvee_{\sigma\in\mathbf{D}} \mathrm{Epi}^\sigma\xi$ satisfies $[\theta,\tau] = [\xi,\tau]$ for each $\tau = J\tau$, and the conclusion follows from Proposition XXV.2.2. To this end, note that if $\tau = J\tau$, then $\tau = \bigvee_{\sigma\in\mathbf{D}} \bigvee_{f\in\mathscr{C}(\tau,\sigma)} f^-\sigma$. By Propositions XXIV.1.5 and XXIV.2.7, and by Proposition XXV.1.3,

$$[\theta,\tau] = [\theta, \bigvee_{\sigma\in\mathbf{D}, f\in\mathscr{C}(\tau,\sigma)} f^-\sigma] = \bigvee^{\trianglerighteq}_{\sigma\in\mathbf{D}, f\in\mathscr{C}(\tau,\sigma)} f^-_*[\theta,\sigma].$$

Moreover, as $\xi \geq \theta \geq \mathrm{Epi}^\sigma\xi$ for each $\sigma \in \mathbf{D}$, by (XXIV.1.4), $[\xi,\sigma] \trianglelefteq [\theta,\sigma] \trianglelefteq [\mathrm{Epi}^\sigma\xi,\sigma]$ for each $\sigma \in \mathbf{D}$. By Proposition XXV.1.3, $[\theta,\sigma] = [\mathrm{Epi}^\sigma\xi,\sigma] = [\xi,\sigma]$, so that

$$[\theta,\tau] = \bigvee^{\trianglerighteq}_{\sigma\in\mathbf{D}, f\in\mathscr{C}(\tau,\sigma)} f^-_*[\xi,\sigma] = [\xi, \bigvee_{\sigma\in\mathbf{D}, f\in\mathscr{C}(\tau,\sigma)} f^-\sigma] = [\xi,\tau].$$

□

In particular, applying this result to $\mathbf{D} = \{\sigma\}$ and $J = \mathrm{R}^\sigma$ yields

$$\mathrm{Epi}^{\mathrm{R}^\sigma}\xi = \mathrm{Epi}^\sigma\xi, \tag{XXV.2.6}$$

for all convergences σ and ξ.

Recall that the *Sierpiński topology* \$ is initially dense in the class of *topologies* ($\mathrm{R}^{\$} = \mathrm{T}$), the *Bourdaud pretopology* ¥ is initially dense in the class of *pretopologies* ($\mathrm{R}^{¥} = \mathrm{S}_0$). Therefore by Proposition

XXV.2.4, $\mathrm{Epi}^{\mathrm{T}} = \mathrm{Epi}^{\$}$ and $\mathrm{Epi}^{\mathrm{S}_0} = \mathrm{Epi}^{¥}$. As the *standard topology* ν of the real line is initially dense in the class of *functionally regular* (alias completely regular) *topologies* $\mathrm{fix}(\mathrm{R}^{\nu})$, by Proposition XXV.2.4, $\mathrm{Epi}^{\mathrm{R}^{\nu}}\xi = \mathrm{Epi}^{\nu}\xi$.

XXV.3. Reflective inequalities in duality theory

Some of the functorial inequalities studied in Section XX.7 have applications to duality theory. For example, consider Corollary XX.7.6, and for simplicity of comparison, replace in Corollary XX.7.6 (3) *"for each prime topology τ"* by *"for each convergence τ"*. Then we get

COROLLARY XXV.3.1. *The following statements are equivalent for any convergences θ and ξ:*

(1) $\theta \geq \mathrm{S}\xi$,
(2) $\theta \times \tau \geq \mathrm{S}(\xi \times \tau)$ *for each convergence* τ,
(3) $\theta \times \tau \geq \mathrm{S}_0(\xi \times \tau)$ *for each convergence* τ.

Analogous inequalities can be substantiated in terms of bidual reflectors. A comparison of respective inequalities will enable us to settle various quests on exponential hulls and objects.

THEOREM XXV.3.2. *Let J be a reflector, and let $\xi \geq \theta$. Then the following conditions are equivalent:*

(1) $\theta \geq \mathrm{Epi}^J\xi$,
(2) $\theta \times \tau \geq \mathrm{Epi}^J(\xi \times \tau)$ *for each convergence* τ,
(3) $\theta \times \tau \geq J(\xi \times \tau)$ *for each convergence* τ.

PROOF. By Proposition XXV.2.2, if $\xi \geq \theta$, then (1) amounts to $[\theta, \sigma] = [\xi, \sigma]$ for each $\sigma = J\sigma$. If τ is any convergence, then by (exponential law) and Proposition XXV.2.2,

$$[\mathrm{Epi}^J\xi \times \tau, \sigma] = [\tau, [\mathrm{Epi}^J\xi, \sigma]] = [\tau, [\xi, \sigma]] = [\xi \times \tau, \sigma]. \tag{XXV.3.1}$$

Proposition XXV.2.2 applies to the effect that $\mathrm{Epi}^J(\xi \times \tau)$ is the coarsest convergence such that $[\xi \times \tau, \sigma] = [\mathrm{Epi}^J(\xi \times \tau), \sigma]$ for each $\sigma = J\sigma$. Hence, by (XXV.3.1),

$$\theta \times \tau \geq \mathrm{Epi}^J\xi \times \tau \geq \mathrm{Epi}^J(\xi \times \tau),$$

and (2) is proved ([1]). (2) implies (3) by (XXV.2.5).

By setting in (3) $\tau = [\xi, \sigma]$, then for each $\sigma = J\sigma$,

$$\theta \times [\xi, \sigma] \geq J(\xi \times [\xi, \sigma]) \geq J(\mathrm{ev}^{-}\sigma) = \mathrm{ev}^{-}\sigma,$$

[1]By the way, if we set in (2), τ to be a (unique) convergence on a singleton, then (1) follows immediately.

thus $[\theta,\sigma] \leq [\xi,\sigma]$. On the other hand, by assumption $\xi \geq \theta$, hence $[\xi,\sigma] \leq [\theta,\sigma]$. Hence, by Proposition XXV.2.2, $\theta \geq \mathrm{Epi}^J\xi$. □

Proposition XXV.3.5 will complete Theorem XXV.3.2 further.

COROLLARY XXV.3.3. *The pretopologically bidual reflector is equal to the pseudotopologizer, which is the pseudotopologically bidual reflector:*

$$\mathrm{Epi}^{\mathrm{S_0}} = \mathrm{S} = \mathrm{Epi}^{\mathrm{S}}. \tag{XXV.3.2}$$

PROOF. On setting $J = \mathrm{S_0}$ in Theorem XXV.3.2, it follows from Corollary XXV.3.1 that $\theta \geq \mathrm{S}\xi$ is equivalent to $\theta \geq \mathrm{Epi}^{\mathrm{S_0}}\xi$ for any $\theta \leq \xi$. Therefore, $\mathrm{S}\xi \geq \mathrm{Epi}^{\mathrm{S_0}}\xi$ and $\mathrm{Epi}^{\mathrm{S_0}}\xi \geq \mathrm{S}\xi$. On setting $J = \mathrm{S}$ in Theorem XXV.3.2, we exhibit the equivalence of $\theta \geq \mathrm{Epi}^{\mathrm{S}}\xi$ for any $\theta \leq \xi$ and $\theta \times \tau \geq \mathrm{S}(\xi \times \tau)$ for each convergence τ. By Corollary XXV.3.1, the latter is equivalent to $\theta \geq \mathrm{S}\xi$. Therefore, $\mathrm{Epi}^{\mathrm{S}}\xi = \mathrm{S}\xi$. □

Hence,

$$\mathrm{Epi}^{\nu_{\mathbb{R}}} \leq \mathrm{Epi}^{\mathrm{T}} \leq \mathrm{Epi}^{\mathrm{S_0}} = \mathrm{S} = \mathrm{Epi}^{\mathrm{S}}. \tag{XXV.3.3}$$

Another important conclusion of Theorem XXV.3.2 is that if J is a reflector, then

$$\mathrm{Epi}^J\xi \times \tau \geq \mathrm{Epi}^J(\xi \times \tau) \tag{XXV.3.4}$$

for every convergences ξ and τ.

COROLLARY XXV.3.4. *If J is a reflector, then the reflector* Epi^J *commutes with finite products.*

PROOF. By applying Epi^J to (XXV.3.4), we get $\mathrm{Epi}^J(\mathrm{Epi}^J\xi \times \tau) \geq \mathrm{Epi}^J(\xi \times \tau)$. Applying again (XXV.3.4), we get $\mathrm{Epi}^J\xi \times \mathrm{Epi}^J\tau \geq \mathrm{Epi}^J(\mathrm{Epi}^J\xi \times \tau)$, and thus

$$\mathrm{Epi}^J\xi \times \mathrm{Epi}^J\tau \geq \mathrm{Epi}^J(\xi \times \tau).$$

□

Recall from (XXV.2.6) that, for each convergence σ,

$$\mathrm{Epi}^{\mathrm{R}^\sigma} = \mathrm{Epi}^\sigma,$$

where $\mathrm{fix}(\mathrm{R}^\sigma)$ is the reflective hull of σ, that is, the least reflective class of convergences containing σ. Consequently, by Corollary XXV.3.4, for each convergence σ,

$$\mathrm{Epi}^\sigma\xi \times \mathrm{Epi}^\sigma\tau \geq \mathrm{Epi}^\sigma(\xi \times \tau) \tag{XXV.3.5}$$

for each pair ξ and τ of convergences. This fact provides another easy mode to prove that every σ-dual convergence is an Epi^{σ}-convergence (Proposition XXV.1.5), that is

$$[\xi, \sigma] \leq \mathrm{Epi}^{\sigma}[\xi, \sigma]. \tag{XXV.3.6}$$

PROOF. Indeed, by definition, $[\xi, \sigma]$ is the coarsest convergence such that $\xi \times [\xi, \sigma] \geq \mathrm{ev}^{-}\sigma$. On applying Epi^{σ} to this inequality, by (XXV.3.5) we get

$$\xi \times \mathrm{Epi}^{\sigma}[\xi, \sigma] \geq \mathrm{Epi}^{\sigma}\xi \times \mathrm{Epi}^{\sigma}[\xi, \sigma] \geq \mathrm{Epi}^{\sigma}(\mathrm{ev}^{-}\sigma) = \mathrm{ev}^{-}\sigma,$$

hence $\mathrm{Epi}^{\sigma}[\xi, \sigma] \geq [\xi, \sigma]$. □

Recall that by Definition VI.7.4, a subclass $\mathbf{D}$ is *initially dense* in a class $\mathbf{J}$ if $\xi = \mathrm{R}^{\mathbf{D}}\xi = \bigvee_{\sigma \in \mathbf{D}} \bigvee_{f \in \mathscr{C}(\xi, \sigma)} f^{-}\sigma$ for each $\xi \in \mathbf{J}$.

The following proposition completes Theorem XXV.3.2.

PROPOSITION XXV.3.5. *If J is a reflector, $\mathbf{D}$ is initially dense in* $\mathrm{fix}(J)$, *and $\xi \geq \theta$, then the following conditions are equivalent:*

(1) $\theta \geq \mathrm{Epi}^{J}\xi$,
(2) $[\xi, \sigma] \geq [\theta, \sigma]$ *for each* $\sigma = \mathrm{Epi}^{J}\sigma$,
(3) $[\xi, \sigma] \geq [\theta, \sigma]$ *for each* $\sigma = J\sigma$,
(4) $[\xi, \sigma] \geq [\theta, \sigma]$ *for each* $\sigma \in \mathbf{D}$.

PROOF. Obviously, (2) ⇒ (3) ⇒ (4). Assume (4). As $\xi \geq \theta$, by (XXV.2.4), $\mathrm{Epi}^{\sigma}\xi = \mathrm{Epi}^{\sigma}\theta$ for each $\sigma \in \mathbf{D}$, and thus $\theta \geq \bigvee_{\sigma \in \mathbf{D}} \mathrm{Epi}^{\sigma}\theta = \bigvee_{\sigma \in \mathbf{D}} \mathrm{Epi}^{\sigma}\xi = \mathrm{Epi}^{J}\xi$, proving (1). If (1) holds, then by (2) of Theorem XXV.3.2 and by Proposition XVII.4.8, $\mathscr{C}(\xi \times \tau, \sigma) \subset \mathscr{C}(\theta \times \tau, \sigma)$ for each $\sigma = \mathrm{Epi}^{J}\sigma$ and each τ. In particular for $\tau = [\xi, \sigma]$, the inequality $\xi \times [\xi, \sigma] \geq \mathrm{ev}^{-}\sigma$ implies $\theta \times [\xi, \sigma] \geq \mathrm{ev}^{-}\sigma$, and thus by Proposition XXIV.1.1, $[\xi, \sigma] \geq [\theta, \sigma]$ for each $\sigma = \mathrm{Epi}^{J}\sigma$, proving (2). □

Therefore, $[\xi, \sigma] = [\mathrm{Epi}^{\sigma}\xi, \sigma]$ for each $\sigma = \mathrm{Epi}^{J}\sigma$.

XXV.4. Exponential classes and hulls

DEFINITION XXV.4.1. We say that a reflective subclass $\mathbf{J}$ of convergences is *exponential* (called also *Cartesian closed*) if $[\xi, \sigma] \in \mathbf{J}$ for every $\sigma \in \mathbf{J}$ and each convergence ξ. It is handy to call a reflector J *exponential* whenever $\mathrm{fix}J$ is exponential.

If J is a reflector (onto $\mathbf{J}$), then by Proposition XVII.2.5,

$$J\xi \times J\tau \leq J(\xi \times \tau)$$

for each pair ξ and τ of convergences. As we have repeated several times, the converse inequality does not hold in general.

THEOREM XXV.4.2. *A reflector J is exponential if and only if, for each ξ and each τ,*

$$\xi \times J\tau \geq J(\xi \times \tau). \tag{XXV.4.1}$$

PROOF. Suppose that (XXV.4.1) holds, and let $\sigma = J\sigma$. By definition, for any convergence ξ,

$$\xi \times [\xi, \sigma] \geq \mathrm{ev}^{-}\sigma.$$

Hence, on setting $\tau = [\xi, \sigma]$ in (XXV.4.1), in view of the definition above, we get

$$\xi \times J[\xi, \sigma] \geq J(\xi \times [\xi, \sigma]) \geq J(\mathrm{ev}^{-}\sigma) = \mathrm{ev}^{-}\sigma.$$

By definition, $[\xi, \sigma]$ is the coarsest convergence η, for which $\mathrm{ev} \in \mathscr{C}(\xi \times \eta, \sigma)$, hence $J[\xi, \sigma] = [\xi, \sigma]$.

Conversely if J is exponential and $f \in \mathscr{C}(\xi \times \tau, \sigma)$ for $\sigma = J\sigma$ then

$$\exp(f) \in \mathscr{C}(\tau, [\xi, \sigma]) \subset \mathscr{C}(J\tau, J[\xi, \sigma]) = \mathscr{C}(J\tau, [\xi, \sigma])$$

because J is exponential. Hence, $f \in \mathscr{C}(\xi \times J\tau, \sigma)$. In view of Proposition XVII.4.8, $\xi \times J\tau \geq J(\xi \times \tau)$. □

Let us observe that if (XXV.4.1) holds for each pair ξ and τ of convergences, then also

$$J\xi \times J\tau \geq J(\xi \times \tau) \tag{XXV.4.2}$$

holds.

Of course, (XXV.4.1) amounts to

$$J\xi \times \tau \geq J(\xi \times \tau) \tag{XXV.4.3}$$

for each ξ and every τ. Then, by (XXV.4.1), $J\xi \times J\tau \geq J(J\xi \times \tau)$, hence, by the inequality above, $J(J\xi \times \tau) \geq JJ(\xi \times \tau) = J(\xi \times \tau)$, because J is idempotent.

Obviously (XXV.4.2) implies (XXV.4.1) and (XXV.4.3).

In particular, the class of pseudotopologies is exponential. Finally,

COROLLARY XXV.4.3. *A reflector J is exponential if and only if $J = \mathrm{Epi}^{J}$.*

PROOF. By Corollary XXV.3.4 and Theorem XXV.4.2, Epi^{J} is exponential. Thus if $J = \mathrm{Epi}^{J}$, then J is exponential. If J is exponential, and $\sigma = J\sigma$, then, in particular, $[[\xi, \sigma], \sigma] = J[[\xi, \sigma], \sigma]$ for every convergence ξ. Hence, by definitions and (XVII.2.1),

$$\begin{aligned}\mathrm{Epi}^J\xi = \bigvee_{\sigma=J\sigma} j^-[[\xi,\sigma],\sigma] = \bigvee_{\sigma=J\sigma} j^-(J[[\xi,\sigma],\sigma]) &\leq \bigvee_{\sigma=J\sigma} J(j^-[[\xi,\sigma],\sigma]) \\ &\leq J(\bigvee_{\sigma=J\sigma}(j^-[[\xi,\sigma],\sigma])) \\ &\leq J\xi.\end{aligned}$$

As, by (XXV.2.5), $J \leq \mathrm{Epi}^J$, the inequality above implies that $J = \mathrm{Epi}^J$. $\square$

DEFINITION XXV.4.4. The least exponential reflective class including a reflective class $\mathbf{J}$ is called the *exponential hull* of $\mathbf{J}$.

THEOREM XXV.4.5. *If a reflective class* $\mathbf{J}$ *is finally dense in* $\mathrm{Epi}^{\mathbf{J}}$, *then* $\mathrm{Epi}^{\mathbf{J}}$ *is the exponential hull of* $\mathbf{J}$.

PROOF. By (XXV.2.5), $\mathbf{J} \subset \mathrm{Epi}^{\mathbf{J}}$. Let $\mathbf{R}$ be an exponential reflective class including $\mathbf{J}$. Then $[\theta,\sigma] \in \mathbf{R}$ for each $\theta,\sigma \in \mathbf{J}$.

Now, $\mathrm{Epi}^J[\xi,\sigma] \geq \mathrm{Epi}^\sigma[\xi,\sigma] = [\xi,\sigma]$ for each convergence ξ, and $\sigma \in \mathbf{J} \subset \mathrm{Epi}^{\mathbf{J}}$. By final density, there is a set $\{\eta_i : i \in I\} \subset \mathbf{J}$, and a set of surjective maps $\{h_i : i \in I\}$ such that $[\xi,\sigma] = \bigwedge_{i\in I} h_i\eta_i$. By Proposition XXIV.1.5,

$$[[\xi,\sigma],\sigma] = [\bigwedge\nolimits_{i\in I} h_i\eta_i,\sigma] = \bigvee\nolimits_{i\in I}^{\trianglerighteq}(h_i^*)^-[\eta_i,\sigma].$$

By assumption, $[\eta_i,\sigma] \in \mathbf{R}$, because $\mathbf{R}$ is exponential, thus $\mathrm{Epi}^J\xi = \bigvee_{\sigma=J\sigma} j^-[[\xi,\sigma],\sigma] \in \mathbf{R}$, because $\mathbf{R}$ is reflective. Thus, $\mathrm{Epi}^{\mathbf{J}} \subset \mathbf{R}$. $\square$

THEOREM XXV.4.6. *The class of pseudotopologies is the exponential hull of the class of pretopologies, that is,*

$$\mathrm{Epi}^{\mathbf{S}_0} = \mathbf{S}.$$

PROOF. By Theorem XXV.3.2, $\theta \geq \mathrm{Epi}^J\xi$ whenever $\theta\times\tau \geq J(\xi\times\tau)$ for each convergence τ. By Corollary XXV.3.1, $\theta\times\tau \geq \mathrm{S}_0(\xi\times\tau)$ for each convergence τ whenever $\theta \geq \mathrm{S}\xi$. By setting $J = \mathrm{S}_0$, we recover $\mathrm{Epi}^{\mathrm{S}_0} = \mathrm{S}$. As $\mathbf{S}_0$ is finally dense in $\mathbf{S}$, by Theorem XXV.4.5, $\mathbf{S} = \mathrm{Epi}^{\mathbf{S}_0}$ is the exponential hull of $\mathbf{S}_0$. $\square$

We shall postpone a characterization of the exponential hull of the class of topologies to Section XXVI.1.

XXV.5. Functionally embedded convergences

Theorem X.9.5 affirms that functionally regular (alias completely regular) topologies constitute the reflective hull of the standard topology of the real line $\nu_{\mathbb{R}}$, equivalently of the standard topology of the closed unit interval.

In other terms, $\xi = \mathrm{R}^{\nu}\xi$ whenever ξ is a functionally regular topology. Therefore, by (XXV.2.6),

$$\mathrm{Epi}^{\mathrm{R}^{\nu}}\xi = \mathrm{Epi}^{\nu}\xi,$$

which means that all the class of ν-bidual convergences coincides with that of functionally embedded *convergences* (Definition XXV.1.7).

THEOREM XXV.5.1. *The class of functionally embedded convergences constitutes the exponential hull of functionally regular (alias completely regular) topologies.*

PROOF. In virtue of Theorem XXV.4.5, it is enough to show that the class of functionally regular topologies is finally dense in the class of functionally embedded convergences. Let us note that, by (XXV.3.3), each functionally embedded convergence is a pseudotopology, hence a finitely stable convergence. On the other hand, by Proposition VII.3.15, each finitely stable convergence is an infimum of prime topologies. As each prime topology is functionally regular by Proposition XIII.4.2, the proof is concluded. □

We are going to characterize internally functionally embedded convergences. Here is a useful result not only for the $\nu_{\mathbb{R}}$-initial topologies, but for σ-initial convergences for any regular convergence σ.

Recall that Definition XIX.6.1 designates *θ-regular* a convergence ξ such that

$$\lim_{\xi}\mathcal{F} \subset \lim_{\xi}(\mathrm{adh}_{\theta}^{\natural}\mathcal{F})$$

for each filter $\mathcal{F}$.

LEMMA XXV.5.2. *Let σ be a regular convergence. If ξ is a convergence, then* $\mathrm{Epi}^{\sigma}\xi$ *is* $\mathrm{R}^{\sigma}\xi$*-regular.*

PROOF. Let $x \in \lim_{\mathrm{Epi}^{\sigma}\xi}\mathcal{F}$, equivalently, $f(x) \in \lim_{\sigma}\langle\mathcal{F},\mathcal{G}\rangle$ whenever $f \in \lim_{[\xi,\sigma]}\mathcal{G}$. Hence, by regularity, $f(x) \in \lim_{\sigma}\mathrm{adh}_{\sigma}^{\natural}\langle\mathcal{F},\mathcal{G}\rangle$. The proof will be concluded if we show that

$$\mathrm{adh}_{\sigma}^{\natural}\langle\mathcal{F},\mathcal{G}\rangle \leq \left\langle \mathrm{adh}_{\mathrm{R}^{\sigma}\xi}^{\natural}\mathcal{F},\mathcal{G}\right\rangle$$

for any filters $\mathcal{F}$ and $\mathcal{G}$.

Indeed, $w \in \mathrm{adh}_{\mathrm{R}^{\sigma}\xi}F$ whenever there exists a filter $\mathcal{H}$ on F such that $\varphi(w) \in \lim_{\sigma}\varphi[\mathcal{H}]$ for each $\varphi \in \mathscr{C}(\xi,\sigma)$. In particular, $\varphi(w) \in$

$\mathrm{adh}_\sigma \varphi(F)$ for each $\varphi \in \mathscr{C}(\xi,\sigma)$. Hence, if $G \in \mathcal{G} \in \mathbb{F}|[\xi,\sigma]|$, $\langle \mathrm{adh}_{\mathrm{R}^\sigma\xi} F, G\rangle \subset \mathrm{adh}_\sigma \langle F, G\rangle$, which ends the proof. □

As a result, $\mathrm{Epi}^\nu\xi$ is $\mathrm{R}^\nu\xi$-regular. On the other hand, it follows from (XXV.3.3) that $\mathrm{Epi}^\nu\xi$ is a pseudotopology. Therefore,

COROLLARY XXV.5.3. *Each functionally embedded convergence ξ is a $\mathrm{R}^\nu\xi$-regular pseudotopology.*

In order to demonstrate that this is in fact a characterization, we shall need a lemma.

LEMMA XXV.5.4. *If $\mathcal{S}$ is an (ideal) ξ-cover, then the filter-base*

$$\mathcal{S}^{=0} := \{\{f \in \mathscr{C} : S \subset f^-(0)\} : S \in \mathcal{S}\}$$

$[\xi,\sigma]$-converges to the constant function $\widehat{0}$.

PROOF. To see that $\mathcal{S}^{=0}$ is a filter-base, let $F_0, F_1 \in \mathcal{S}^-0$. Then there exist $S_0, S_1 \in \mathcal{S}$ such that if $S_0 \subset f^-(0)$, then $f \in F_0$, and if $S_1 \subset f^-(0)$, then $f \in F_1$. Hence if $S_0 \cup S_1 \subset f^-(0)$, then $f \in F_0 \cap F_1$. If now $x \in \lim_\xi \mathcal{G}$, then there exists $S \in \mathcal{S} \cap \mathcal{G}$. By the definition of $\mathcal{S}^{=0}$, there exists $F \in \mathcal{S}^{=0}$ such that $\langle S, F\rangle = 0$, which proves that $\widehat{0} \in \lim_{[\xi,\sigma]} \mathcal{S}^{=0}$. □

THEOREM XXV.5.5. *A Hausdorff convergence ξ is functionally embedded if and only if it is a $\mathrm{R}^\nu\xi$-regular pseudotopology.*

PROOF. One implication follows from Corollary XXV.5.3. Assume that ξ is a $\mathrm{R}^\nu\xi$-regular pseudotopology on X. It is enough to show that $\lim_{\mathrm{Epi}^\nu\xi} \mathcal{U} \subset \lim_\xi \mathcal{U}$ for each ultrafilter $\mathcal{U}$. Assume that $x \notin \lim_\xi \mathcal{U}$, so that $\lim_\xi \mathcal{U} = \varnothing$. By $\mathrm{R}^\nu\xi$-regularity, $\mathcal{U} \notin \beta(\mathrm{adh}^\natural_{\mathrm{R}^\nu\xi} \mathcal{F})$ for each ξ-convergent filter $\mathcal{F}$, that is, there exists $F_\mathcal{F} \in \mathcal{F}$ such that $F_\mathcal{F} = \mathrm{cl}_{\mathrm{R}^\nu\xi} F_\mathcal{F} \notin \mathcal{U}$. Of course, the family $\mathcal{S} := \{F_\mathcal{F} : \mathcal{F} \in \xi^-(X)\}$ is a ξ-cover of X. By Lemma XXV.5.4, $\widehat{0} \in \lim_{[\xi,\sigma]}(\mathcal{S}^{=0})^{\cap\uparrow}$. Suppose that $x \in \lim_{\mathrm{Epi}^\nu\xi} \mathcal{U}$. Then $0 \in \lim_\nu \langle \mathcal{U}, (\mathcal{S}^{=0})^{\cap\uparrow}\rangle$, that is, for each $\varepsilon > 0$, there exist $U \in \mathcal{U}$ and $H \in (\mathcal{S}^{=0})^{\cap\uparrow}$ such that $\langle U, H\rangle \subset]-\varepsilon,\varepsilon[$. On the other hand, there exists $S \in \mathcal{S}^\cup$ such that

$$\{f \in \mathscr{C}(\xi,\nu) : S \subset f^-(0)\} \subset H.$$

Since $S \notin \mathcal{U}$, there is $t \in U \setminus S$. Since S is $\mathrm{R}^\nu\xi$-closed, there exists $f \in \mathscr{C}(\xi,\nu)$ such that $S \subset f^-(0)$ and $f(t) = 1$. Therefore, $f \in H$ in contradiction with $\langle U, H\rangle \subset]-\varepsilon,\varepsilon[$. □

XXV.6. Epitopologies

Before we characterize $\mathrm{Epi}^{\mathbf{T}}$, recall that

$$\mathbf{T} \subset \mathrm{Epi}^{\mathbf{T}} \subset \mathrm{Epi}^{\mathbf{S}_0} = \mathbf{S}.$$

Although the topologizer is not adherence-determined, it has a representation analogous to those of adherence-determined reflectors, that is,

$$\lim_{\mathrm{T}\xi} \mathcal{F} = \bigcap\nolimits_{H \in \mathcal{F}^{\#}} \mathrm{cl}_\xi H.$$

For this reason, it used to be presented as generalized adherence-determined ([2]). Given a convergence θ, let

$$\mathbb{F}_\theta := \{\mathcal{F} \in \mathbb{F} : \mathrm{adh}_\theta^\natural \mathcal{F} = \mathcal{F}\}$$

be the set of θ-regular filters ([3]).

LEMMA XXV.6.1.

$$\lim_{\mathrm{Epi}^{\mathrm{T}}\xi} \mathcal{F} = \bigcap\nolimits_{\mathbb{F}_{\xi^\bullet} \ni \mathcal{H} \# \mathcal{F}} \mathrm{cl}_\xi(\mathrm{adh}_\xi \mathcal{H}). \tag{XXV.6.1}$$

PROOF. Since $\$_0$ is initially dense in the class of topologies, $\mathrm{Epi}^{\mathrm{T}} = \mathrm{Epi}^{\$_0}$, so that $x \in \lim_{\mathrm{Epi}^{\mathrm{T}}\xi} \mathcal{F}$ whenever for each $\mathscr{H} \in \mathbb{F}([\xi, \$_0])$ and $A \in \lim_{[\xi,\$_0]} \mathscr{H}$,

$$\chi_A(x) \in \lim_{\$_0} \langle \mathcal{F}, \mathscr{H} \rangle ,$$

equivalently, if $0 \in \lim_{\$_0} \langle \mathcal{F}, \mathscr{H} \rangle$ then $\{0\} \in \langle \mathcal{F}, \mathscr{H} \rangle$ by the definition of $\$_0$. This means that if $x \notin A$ then $\mathcal{F}$ and $\mathrm{r}\mathscr{H}$ do not mesh. In other words, $x \in A$ whenever $\mathcal{F} \# \mathrm{r}\mathscr{H}$ and $\mathrm{adh}_\xi \mathrm{r}\mathscr{H} \subset A$, that is, $\mathcal{F} \# \mathrm{r}\mathscr{H}$ implies $x \in \mathrm{cl}_\xi(\mathrm{adh}_\xi \mathrm{r}\mathscr{H})$. But $\mathcal{H} \in \mathbb{F}_{\xi^\bullet}$ whenever there exists $\mathscr{H} \in \mathbb{F}\,|[\xi, \$_0]|$ such that $\mathcal{H} = \mathrm{r}\mathscr{H}$, by Proposition XXIV.4.3. $\square$

A convergence ξ is called an *epitopology* if $\xi = \mathrm{Epi}^{\mathrm{T}}\xi$. We observe that (XXV.6.1) amounts to

$$\lim_{\mathrm{Epi}^{\mathrm{T}}\xi} \mathcal{F} = \bigcap\nolimits_{\mathcal{U} \in \beta\mathcal{F}} \mathrm{cl}_\xi(\mathrm{adh}_\xi(\mathrm{cl}_{\xi^\bullet}^\natural \mathcal{U})). \tag{XXV.6.2}$$

Indeed, instead of considering $\mathcal{H} \in \mathbb{F}_{\xi^\bullet}$, one can take arbitrary $\mathcal{H} \in \mathbb{F}$ but replace $\mathcal{H}$ by $\mathrm{cl}_{\xi^\bullet}^\natural \mathcal{H}$ in (XXV.6.1). Then for each $\mathcal{H} \# \mathcal{F}$, for any $\mathcal{U} \in \beta(\mathcal{H} \vee \mathcal{F})$, the inclusion $\mathrm{adh}_\xi(\mathrm{cl}_{\xi^\bullet}^\natural \mathcal{U}) \subset \mathrm{adh}_\xi(\mathrm{cl}_{\xi^\bullet}^\natural \mathcal{H})$ allows to conclude.

Notice that

$$\mathrm{cl}_{\xi^\circ}^\natural \mathcal{F} \# \mathcal{H} \iff \mathcal{F} \# \mathrm{cl}_{\xi^\bullet}^\natural \mathcal{H}. \tag{XXV.6.3}$$

[2]For instance, S. DOLECKI and F. MYNARD, [83].

[3]It is implicit that if θ is a convergence on X, then $\mathcal{F} \in \mathbb{F}X$.

Indeed, $F\#\,\mathrm{cl}_{\xi^\bullet} H$ whenever there exists $x \in H$ such that $\mathrm{cl}_\xi\{x\}\#F$, equivalently $x \in \mathrm{cl}_{\xi^\circ} F$.

LEMMA XXV.6.2. *If a convergence ξ is $\circ$-regular, then*

$$\mathrm{adh}_\xi(\mathrm{cl}^\natural_{\xi^\bullet}\mathcal{H}) = \mathrm{adh}_\xi \mathcal{H}$$

for each filter $\mathcal{H}$.

PROOF. If $w \in \mathrm{adh}_\xi(\mathrm{cl}^\natural_{\xi^\bullet}\mathcal{H})$, then there exists $\mathcal{W} \in \beta(\mathrm{cl}^\natural_{\xi^\bullet}\mathcal{H})$ such that $w \in \lim_\xi \mathcal{W} = \lim_\xi \mathrm{cl}^\natural_{\xi^\circ}\mathcal{W}$, because ξ is $\circ$-regular. By (XXV.6.3), $\mathcal{H}\#\,\mathrm{cl}^\natural_{\xi^\circ}\mathcal{W}$, and thus $x \in \mathrm{adh}_\xi \mathcal{H}$. □

PROPOSITION XXV.6.3. *A convergence is an epitopology if and only if it is a $\circ$-regular pseudotopology with closed limits.*

PROOF. If ξ is an epitopology, then $\xi \leq \mathrm{Epi}^{\mathrm{T}}\xi \leq \mathrm{Epi}^{\mathrm{So}}\xi = \mathrm{S}\xi$, hence ξ is a pseudotopology. By (XXV.6.1), each ξ-limit is ξ-closed. Moreover, ξ is ξ°-regular, for if $x \in \lim_\xi \mathcal{F}$ and $\mathcal{H} \in \mathbb{F}_{\xi^\bullet}$ and $\mathcal{H}\#\,\mathrm{cl}^\natural_{\xi^\circ}\mathcal{F}$, then by (XXV.6.3) $\mathcal{H}\#\mathcal{F}$, so that $x \in \mathrm{cl}_\xi(\mathrm{adh}_\xi \mathcal{H})$ and thus $x \in \lim_\xi \mathrm{cl}^\natural_{\xi^\circ}\mathcal{F}$ by (XXV.6.3).

Let ξ be a $\circ$-regular pseudotopology with closed limits, and let $x \in \lim_{\mathrm{Epi}^{\mathrm{T}}\xi}\mathcal{F}$. We need to show that if $\mathcal{U} \in \beta(\mathrm{cl}^\natural_{\xi^\circ}\mathcal{F})$, then $x \in \lim_\xi \mathcal{U}$, because ξ is a $\circ$-regular pseudotopology. By (XXV.6.3), $\mathcal{F}\#\,\mathrm{cl}^\natural_{\xi^\bullet}\mathcal{U}$ so that $x \in \mathrm{cl}_\xi(\mathrm{adh}_\xi(\mathrm{cl}^\natural_{\xi^\bullet}\mathcal{U}))$ by (XXV.6.2). By Lemma XXV.6.2,

$$\mathrm{adh}_\xi(\mathrm{cl}^\natural_{\xi^\bullet}\mathcal{U}) = \mathrm{adh}_\xi \mathcal{U} = \lim_\xi \mathcal{U},$$

hence $\mathrm{cl}_\xi \mathrm{adh}_\xi(\mathrm{cl}^\natural_{\xi^\bullet}\mathcal{U}) = \mathrm{cl}_\xi \lim_\xi \mathcal{U}$, and since $\lim_\xi \mathcal{U}$ is ξ-closed, $x \in \lim_\xi \mathcal{U}$. □

XXV.7. Duality in completeness

Lemma XXIII.10.1 asserts that the completeness number of a subspace of a Hausdorff compact topology does not exceed the cardinality of its complement. More precisely, Theorem XXIII.6.2 says that the covering number of the complementary subspace is not greater than the completeness of the original subspace.

THEOREM XXV.7.1. *Let ξ be a Hausdorff topology. Then*

$$\mathrm{compl}_\#(\xi) = \mathfrak{p}([\xi, \$_0], \varnothing),$$

under the convention that $\mathfrak{p}(\theta, y) = 0$ whenever y is θ-isolated.

PROOF. Let $\boldsymbol{H}$ be a collection of non-ξ-adherent filters such that ξ is $\boldsymbol{H}$-ultracocomplete. Then $\{\mathrm{e}_\xi^\natural \mathcal{H} : \mathcal{H} \in \boldsymbol{H}\}$ is a pavement of $[\xi, \$_0]$ at $\varnothing$. Indeed, as $\mathrm{r}(\mathrm{e}_\xi^\natural \mathcal{H}) \geq \mathcal{H}$, we infer that $\operatorname{adh}_\xi \mathrm{r}(\mathrm{e}_\xi^\natural \mathcal{H}) \subset \operatorname{adh}_\xi \mathcal{H} = \varnothing$, hence $\varnothing \in \lim_{[\xi,\$_0]} \mathrm{e}_\xi^\natural \mathcal{H}$. If now $\varnothing \in \lim_{[\xi,\$_0]} \mathscr{F}$, then by Lemma XII.9.1, $\operatorname{adh}_\xi \mathrm{r}\mathscr{F} = \varnothing$, hence by $\boldsymbol{H}$-ultracocompleteness of ξ, there exists $\mathcal{H} \in \boldsymbol{H}$ such that $\mathrm{r}\mathscr{F} \geq \mathcal{H}$, hence $\mathscr{F} \geq \mathrm{e}_\xi^\natural(\mathrm{r}\mathscr{F}) \geq \mathrm{e}_\xi^\natural \mathcal{H}$, which proves the claim. Hence $\mathrm{compl}_\#(\xi) \geq \mathfrak{p}([\xi, \$_0], \varnothing)$.

Conversely, if $\boldsymbol{D}$ is a pavement of $[\xi, \$_0]$ at $\varnothing$, then $\{\mathrm{r}\mathscr{D} : \mathscr{D} \in \boldsymbol{D}\}$ is a collection of non-ξ-convergent filters, for which ξ is ultracocomplete. Indeed, if $\operatorname{adh}_\xi \mathcal{F} = \varnothing$, then $\varnothing \in \lim_{[\xi,\$_0]} \mathrm{e}_\xi^\natural \mathcal{F}$, because $\mathrm{r}(\mathrm{e}_\xi^\natural \mathcal{F}) \geq \mathcal{F}$, and thus $\operatorname{adh}_\xi \mathrm{r}(\mathrm{e}_\xi^\natural \mathcal{F}) = \varnothing$. By assumption, there is $\mathscr{D} \in \boldsymbol{D}$ such that $\mathscr{D} \leq \mathrm{e}_\xi^\natural \mathcal{F}$, hence $\mathrm{r}\mathscr{D} \leq \mathrm{r}(\mathrm{e}_\xi{}^\natural \mathcal{F}) \leq \mathcal{F}$. Therefore, $\mathrm{compl}_\#(\xi) \leq \mathfrak{p}([\xi, \$_0], \varnothing)$. □

THEOREM XXV.7.2. *Let ξ be a Hausdorff topology. Then*

$$\mathrm{compl}(\xi) \geq \mathfrak{p}_\#([\xi, \$_0], \varnothing),$$

under the convention that $\mathfrak{p}_\#(\theta, y) = 0$ whenever y is θ-isolated.

PROOF. Indeed, let $\boldsymbol{H}$ be a collection of non-ξ-adherent filters such that ξ is $\boldsymbol{H}$-cocomplete. Then $\{\mathrm{e}_\xi^\natural \mathcal{H} : \mathcal{H} \in \boldsymbol{H}\}$ is a pseudopavement of $[\xi, \$_0]$ at $\varnothing$. We have already proved that if $\operatorname{adh}_\xi \mathcal{H} = \varnothing$, then $\varnothing \in \lim_{[\xi,\$_0]} \mathrm{e}_\xi^\natural \mathcal{H}$. If now $\mathscr{U}$ is an ultrafilter such that $\varnothing \in \lim_{[\xi,\$_0]} \mathscr{U}$, then by Lemma XII.9.1, $\operatorname{adh}_\xi \mathrm{r}\mathscr{U} = \varnothing$, hence by $\boldsymbol{H}$-cocompleteness of ξ, there exists $\mathcal{H} \in \boldsymbol{H}$ such that $\mathrm{r}\mathscr{U} \# \mathcal{H}$, hence there exists $\mathcal{W} \in \beta(\mathrm{r}\mathscr{U} \vee \mathcal{H})$, and thus $\mathrm{e}_\xi^\natural(\mathrm{r}\mathscr{U}) \leq \mathrm{e}_\xi^\natural \mathcal{W}$, and $\mathrm{e}_\xi^\natural \mathcal{H} \leq \mathrm{e}_\xi^\natural \mathcal{W}$, so that $\mathrm{e}_\xi^\natural(\mathrm{r}\mathscr{U}) \# \mathrm{e}_\xi^\natural \mathcal{H}$, hence $\mathscr{U} \geq \mathrm{e}_\xi^\natural(\mathrm{r}\mathscr{U}) \vee \mathrm{e}_\xi^\natural \mathcal{H}$. Hence $\mathrm{compl}(\xi) \geq \mathfrak{p}_\#([\xi, \$_0], \varnothing)$. □

REMARK XXV.7.3. The inverse equality does not hold in general. Actually, the completeness number of ξ is equal to a modified pseudopaving number ([4]).

As a result, under the convention concerning paving and pseudopaving numbers 0,

$$\mathfrak{p}_\#([\xi, \$_0], \varnothing) \leq \mathrm{compl}(\xi) \leq \mathrm{compl}_\#(\xi) = \mathfrak{p}([\xi, \$_0], \varnothing).$$

As we know, if the paving and pseudopaving numbers, respectively, completeness and ultracompleteness numbers, are finite, then they respectively coincide, and then are less than or equal to 1.

[4]See F. Mynard [**199**].

Example XXV.7.4. The space of irrational numbers is countably complete, but is not countably ultracomplete by Corollary XXIII.9.8. Therefore the upper convergence on the hyperspace of closed subsets of the irrational numbers is countably pseudopaved at $\varnothing$, but is not countably paved.

Corollary XXV.7.5. *If* ξ *is a Hausdorff topology, then* $[\xi, \$_0]$ *is pretopological if and only if* ξ *is locally compactoid.*

Compare this corollary with Theorem XX.10.7.

CHAPTER XXVI

Modified duality

We know that the class of topologies is not exponential. In other terms, if σ is a topology, then the σ-dual convergence $[\xi, \sigma]$ of a convergence (a topology) ξ is in general not topological.

A question arises under which conditions on ξ, and maybe on a topology σ, the inequality $\mathrm{T}[\xi, \sigma] \geq [\xi, \sigma]$ holds. We know already from Sections XX.10 and XXV.7 that $\mathrm{S}_0[\xi, \$] \geq [\xi, \$]$ is intimately related to the local compactness of ξ. More generally, under what conditions on a primal convergence, a given functorial inequality holds for the dual convergence, and conversely? It turns out that modified duality is a way of addressing this quest.

XXVI.1. Functorial modifications of duality

For any functor F, define the *F-modified σ-bidual* convergence of a convergence ξ on X by

$$\text{(XXVI.1.1)} \qquad \mathrm{Epi}_F^{\sigma}\xi := j_X^{-}[F[\xi, \sigma], \sigma].$$

In comparison to Definition XXV.1.2 of the σ-bidual convergence, we interpose F in the process of building the bidual. On recalling that I stands for the identity functor, $\mathrm{Epi}_{\mathrm{I}}^{\sigma}\xi = \mathrm{Epi}^{\sigma}\xi$, that is, the I-modified σ-bidual convergence of ξ coincides with the σ-bidual reflection of ξ.

REMARK XXVI.1.1. It is interesting that the σ-initial reflection of ξ can be represented as a specially modified σ-bidual convergence of ξ. Indeed,

$$\mathrm{R}^{\sigma}\xi = \mathrm{Epi}_{\mathrm{Dis}}^{\sigma}\xi,$$

where Dis is the discretization coreflector introduced in Example XVII.3.13. As $\mathrm{Dis}\,\theta = \iota_{|\theta|}$ for every convergence θ, we see that

$$\mathrm{Epi}_{\mathrm{Dis}}^{\sigma}\xi := j_{|\xi|}^{-}[\mathrm{Dis}[\xi, \sigma], \sigma] = j_{|\xi|}^{-}[\iota_{\mathscr{C}(\xi,\sigma)}, \sigma].$$

Accordingly, $\xi = \mathrm{Epi}_{\mathrm{Dis}}^{\sigma}\xi$ whenever $x \in \lim_{\xi} \mathcal{F}$ amounts to $f(x) \in \lim_{\sigma} f[\mathcal{F}]$ for every $f \in \mathscr{C}(\xi, \sigma)$. Hence, $\mathrm{R}^{\sigma} \leq \mathrm{Epi}_F^{\sigma}$ for every functor F.

As in Chapter XXV, if Σ is a set of convergences, then the *F-modified Σ-bidual* convergence is defined by

$$\mathrm{Epi}_F^\Sigma \xi := \bigvee\nolimits_{\sigma \in \Sigma} \mathrm{Epi}_F^\sigma \xi.$$

A special case of this concept is obtained for a modifier J, on recalling Remark XXV.2.1, by

$$\mathrm{Epi}_F^J \xi = \bigvee\nolimits_{\sigma = J\sigma} \mathrm{Epi}_F^\sigma \xi. \tag{XXVI.1.2}$$

Because of (XXIV.1.4), it is plain that

$$F_0 \leq F_1 \Longrightarrow \mathrm{Epi}_{F_0}^J \geq \mathrm{Epi}_{F_1}^J. \tag{XXVI.1.3}$$

Consequently, if D is a reflector, and E is a coreflector, then

$$\mathrm{R}^J \leq \mathrm{Epi}_E^J \leq \mathrm{Epi}^J \leq \mathrm{Epi}_D^J. \tag{XXVI.1.4}$$

PROPOSITION XXVI.1.2. *If F and J are functors, then Epi_F^J is a functor.*

PROOF. Let ξ and τ be convergences on X and Y, respectively, and $f \in \mathscr{C}(\xi, \tau)$. We need to show that $f \in \mathscr{C}(\mathrm{Epi}_F^J \xi, \mathrm{Epi}_F^J \tau)$ to get the functorial condition (functorial), hence that Epi_F^J is (increasing).

Let $\sigma = J\sigma$. By Proposition XXIV.2.4, $f^* \in \mathscr{C}([\tau, \sigma], [\xi, \sigma])$, hence $f^* \in \mathscr{C}(F[\tau, \sigma], F[\xi, \sigma])$, because F is a functor. Applying once more Proposition XXIV.2.4, we get $f^{**} \in \mathscr{C}([F[\xi, \sigma], \sigma], [F[\tau, \sigma], \sigma])$, that is,

$$[F[\xi, \sigma], \sigma] \geq (f^{**})^- [F[\tau, \sigma], \sigma].$$

If $j_X : X \longrightarrow Z^{Z^X}$ and $j_Y : Y \longrightarrow Z^{Z^Y}$ are the point-evaluation maps, then $f^{**} \circ j_X = j_Y \circ f$ by Proposition XXIV.2.3, and hence

$$\begin{aligned}\mathrm{Epi}_F^\sigma \xi &= j_X^- [F[\xi, \sigma], \sigma] \geq j_X^- (f^{**})^- [F[\tau, \sigma], \sigma] \\ &= (j_Y f)^- [F[\tau, \sigma], \sigma] = f^- \mathrm{Epi}_F^\sigma \tau.\end{aligned}$$

Finally, by Proposition XVII.2.4, Epi_F^J is a functor as a supremum of functors. □

COROLLARY XXVI.1.3. *If F is a coreflector and J is a reflector, then Epi_F^J is a reflector.*

PROOF. Indeed, if in the proof of Proposition XXV.1.3, we set $\varphi(\xi) = [\xi, \sigma]$ and $\psi(\theta) = j^- F[\theta, \sigma]$, then (φ, ψ) is a $(-,-)$-connection. Thus, by Corollary XXI.2.7, Epi_F^J is a reflector. □

If J is a reflector, then by Proposition XVII.4.7 and by (XXVI.1.3),

$$J = \mathrm{R}^J \leq \mathrm{Epi}_F^J \tag{XXVI.1.5}$$

for any functor F.

REMARK XXVI.1.4. Let us stress that, in general, the point-evaluations considered above are not continuous. More precisely, by Proposition XXV.1.1 $j \in \mathscr{C}(\xi, [[\xi, \sigma], \sigma])$, but this does not imply that $j \in \mathscr{C}(\xi, [F[\xi, \sigma], \sigma])$ for a given functor.

Here is a generalization of the equivalence of (1) $\theta \geq \mathrm{Epi}^J \xi$ and (3) $\theta \times \tau \geq J(\xi \times \tau)$ for every convergence τ of Theorem XXV.3.2 (We defer a discussion of (2) of Theorem XXV.3.2 to Corollary XXVI.1.12).

THEOREM XXVI.1.5. *Let F be a functor, J a reflector, and let $\theta \geq J\xi$. Then the following statements are equivalent:*

(1) $\theta \geq \mathrm{Epi}_F^J \xi$,
(2) $F[\xi, \sigma] \trianglerighteq [\theta, \sigma]$ *for each* $\sigma = J\sigma$,
(3) $\theta \times F\tau \geq J(\xi \times \tau)$ *for every convergence* τ.

PROOF. Assume (1), that is, in view of (IV.3.6), $j\theta \geq [F[\xi, \sigma], \sigma]$ for every $\sigma = J\sigma$. In order to prove (3), by Proposition XVII.4.8, it is enough to show that if $\sigma = J\sigma$, then $\mathscr{C}(\xi \times \tau, \sigma) \subset \mathscr{C}(\theta \times F\tau, \sigma)$. By Lemma XXIV.3.1, $f \in \mathscr{C}(\xi \times \tau, \sigma)$ implies that $\exp f \in \mathscr{C}(\tau, [\xi, \sigma]) \subset \mathscr{C}(F\tau, F[\xi, \sigma])$, because F is a functor. By Proposition XXIV.2.4, $(\exp f)^* \in \mathscr{C}([F[\xi, \sigma], \sigma], [F\tau, \sigma])$, hence by assumption, $(\exp f)^* \circ j \in \mathscr{C}(\theta, [F\tau, \sigma])$, that is, $f \in \mathscr{C}(\theta \times F\tau, \sigma)$.

(3) implies (2). Let $\sigma = J\sigma$. By definition, $\xi \times [\xi, \sigma] \geq \mathrm{ev}^- \sigma$. As J is a functor,

$$J(\xi \times [\xi, \sigma]) \geq J(\mathrm{ev}^- \sigma) \geq \mathrm{ev}^-(J\sigma) = \mathrm{ev}^- \sigma.$$

Therefore, (3) with $\tau = [\xi, \sigma]$ implies $\theta \times F[\xi, \sigma] \geq \mathrm{ev}^- \sigma$. From $\theta \geq J\xi$ it follows that $\mathscr{C}(\xi, \sigma) \subset \mathscr{C}(\theta, \sigma)$, and thus $F[\xi, \sigma]$ is a convergence on $\mathscr{C}(\xi, \sigma)$. On the other hand, by definition, $[\theta, \sigma]$ is the coarsest convergence on $\mathscr{C}(\theta, \sigma)$ for which $\theta \times [\theta, \sigma] \geq \mathrm{ev}^- \sigma$, so that $F[\xi, \sigma] \trianglerighteq [\theta, \sigma]$.

(2) implies (1). Indeed, by (2), $[[\theta, \sigma], \sigma] \trianglerighteq [F[\xi, \sigma], \sigma]$ for each $\sigma = J\sigma$, hence

$$\theta \geq \mathrm{Epi}^\sigma \theta = j_X^-[[\theta, \sigma], \sigma] \geq j_X^-[F[\xi, \sigma], \sigma] = \mathrm{Epi}_F^\sigma \xi, \tag{XXVI.1.6}$$

so that, $\theta \geq \mathrm{Epi}_F^J \xi$ by (XXVI.1.2). □

Proposition XXV.2.4 implies that $\mathrm{Epi}^{\mathrm{R}\sigma} = \mathrm{Epi}^\sigma$ for every convergence σ. This equality continues to hold in modified duality. Namely,

LEMMA XXVI.1.6. *If F is a functor and σ is a convergence, then*

$$\mathrm{Epi}_F^{\mathrm{R}\sigma} = \mathrm{Epi}_F^\sigma.$$

PROOF. It follows from the definition that $\mathrm{Epi}_F^{\mathrm{R}^\sigma}\xi \geq \mathrm{Epi}_F^\sigma\xi$. By Theorem XXVI.1.5, $\mathrm{Epi}_F^{\mathrm{R}^\sigma}\xi$ is the coarsest θ such that $F[\xi,\tau] \trianglerighteq [\theta,\tau]$ for each $\tau = \mathrm{R}^\sigma\tau$. By Propositions XXIV.1.5 and XXIV.2.7, for every $\tau = \mathrm{R}^\sigma\tau$,

$$[\mathrm{Epi}_F^\sigma\xi,\tau] = [\mathrm{Epi}_F^\sigma\xi, \bigvee_{f\in\mathscr{C}(\tau,\sigma)} f^-\sigma] = \bigvee^{\trianglerighteq}_{f\in\mathscr{C}(\tau,\sigma)} f_*^-[\mathrm{Epi}_F^\sigma\xi,\sigma].$$

In virtue of Theorem XXVI.1.5 and by (XVII.2.2),

$$[\mathrm{Epi}_F^\sigma\xi,\tau] \trianglelefteq \bigvee^{\trianglerighteq}_{f\in\mathscr{C}(\tau,\sigma)} f_*^- F[\xi,\sigma] \leq \bigvee^{\trianglerighteq}_{f\in\mathscr{C}(\tau,\sigma)} Ff_*^-[\xi,\sigma]$$
$$\leq F(\bigvee^{\trianglerighteq}_{f\in\mathscr{C}(\tau,\sigma)} f_*^-[\xi,\sigma]) = F[\xi, \bigvee_{f\in\mathscr{C}(\tau,\sigma)} f^-\sigma] = F[\xi,\tau].$$

Therefore, $[\mathrm{Epi}_F^\sigma\xi,\tau] \trianglelefteq F[\xi,\tau]$ for each $\tau = \mathrm{R}^\sigma\tau$, and thus $\mathrm{Epi}_F^\sigma\xi \geq \mathrm{Epi}_F^{\mathrm{R}^\sigma}\xi$ in virtue of Theorem XXVI.1.5. □

COROLLARY XXVI.1.7. *If F is a functor, J a reflector, and* $\mathbf{D}$ *is an initially dense subclass of* $\mathrm{fix}(J)$, *then*

$$\mathrm{Epi}_F^J\xi = \bigvee_{\sigma\in\mathbf{D}} \mathrm{Epi}_F^\sigma\xi. \tag{XXVI.1.7}$$

We are now ready to complete Theorem XXVI.1.5.

COROLLARY XXVI.1.8. *Let F be a functor, J a reflector, and let* $\mathbf{D}$ *be initially dense in* $\mathrm{fix}(J)$. *If $\theta \geq J\xi$, then $F[\xi,\sigma] \trianglerighteq [\theta,\sigma]$ for each $\sigma \in \mathbf{D}$ if and only if $\theta \geq \mathrm{Epi}_F^J\xi$.*

PROOF. Theorem XXVI.1.5 (3) implies that $F[\xi,\sigma] \trianglerighteq [\theta,\sigma]$ for each $\sigma \in \mathbf{D}$. Conversely, the latter condition entails (XXVI.1.6), which together with (XXVI.1.7) implies that $\theta \geq \mathrm{Epi}_F^J\xi$. □

LEMMA XXVI.1.9. *If F and B are functors such that, for each σ* and τ,

$$F\sigma \times B\tau \geq F(\sigma \times \tau), \tag{XXVI.1.8}$$

and if J is a reflector, then the following are equivalent:

(1) $\theta \geq \mathrm{Epi}_F^J\xi$;
(2) $\theta \times F\tau \geq \mathrm{Epi}_B^J(\xi \times \tau)$ *for every* τ;
(3) $\theta \times F\tau \geq J(\xi \times \tau)$ *for every* τ.

PROOF. By (XXVI.1.8) and Theorem XXVI.1.5,

$$\mathrm{Epi}_F^J\xi \times (F\sigma \times B\tau) \geq \mathrm{Epi}_F^J\xi \times F(\sigma \times \tau) \geq J(\xi \times (\sigma \times \tau)).$$

Let $\theta := \mathrm{Epi}_F^J\xi \times F\sigma$ and $F := B$ in Theorem XXVI.1.5 (3). Then (1) becomes $\mathrm{Epi}_F^J\xi \times F\sigma \geq \mathrm{Epi}_B^J(\xi \times \sigma)$ for every σ, which proves (1) implies (2). That (2) implies (3) follows from the fact that $\mathrm{Epi}_B^J \geq J$ by (XXVI.1.3).Finally (3) implies (1) follows from Theorem XXVI.1.5. □

We are now in a position to generalize (XXV.3.4) from $F = \mathrm{I}$ (identity) to an arbitrary functor F that commutes with products.

COROLLARY XXVI.1.10. *If J is a reflector and F is a functor that commutes with finite products, then*

$$\mathrm{Epi}^J_F \xi \times F\tau \geq \mathrm{Epi}^J_F(\xi \times \tau) \tag{XXVI.1.9}$$

for every ξ and τ.

It follows from Theorem XXVI.1.5 and Corollary XXVI.1.10 that

COROLLARY XXVI.1.11. *If J is a reflector and F is a functor that commutes with finite products, then*

$$F[\xi, \sigma] \trianglerighteq [\mathrm{Epi}^J_F \xi, \sigma] \tag{XXVI.1.10}$$

for every $\sigma \leq \mathrm{Epi}^J_F \sigma$.

PROOF. If we set $\tau = [\xi, \sigma]$ in (XXVI.1.9), then by Proposition XVII.4.8, $\mathscr{C}(\xi \times [\xi, \sigma], \sigma) \subset \mathscr{C}(\mathrm{Epi}^J_F \xi \times F[\xi, \sigma], \sigma)$ for each $\sigma = \mathrm{Epi}^J_F \sigma$. In particular, the evaluation $\mathrm{ev} \in \mathscr{C}(\xi \times [\xi, \sigma], \sigma)$ by the definition of $[\xi, \sigma]$. Therefore, $\mathrm{Epi}^J_F \xi \times F[\xi, \sigma] \geq \mathrm{ev}^- \sigma$, hence $[\mathrm{Epi}^J_F \xi, \sigma] \leq F[\xi, \sigma]$ for each $\sigma \leq \mathrm{Epi}^J_F \sigma$. □

If E is a coreflector and J is a reflector, then by Corollary XXVI.1.3 Epi^J_E is a reflector. If, moreover, E commutes with finite products, then from Corollary XXVI.1.11 and Theorem XXVI.1.5 it follows that

COROLLARY XXVI.1.12. *If J is a reflector and E is a coreflector that commutes with finite products, then the following statements are equivalent for $\theta \geq J\xi$:*
1. *$\theta \geq \mathrm{Epi}^J_E \xi$,*
2. *$\theta \times E\tau \geq \mathrm{Epi}^J_E(\xi \times \tau)$ for each convergence τ,*
3. *$\theta \times E\tau \geq J(\xi \times \tau)$ for each convergence τ.*

Hence, if J is a reflector, E is a coreflector, it follows from (3) that ξ and $\mathrm{Epi}^J_E \xi$ are equivalent with respect to E-modified J-duality. In other terms, if J is a reflector and E is a coreflector, then

$$(\tau = E\tau) \Longrightarrow J(\mathrm{Epi}^J_E \xi \times \tau) = J(\xi \times \tau) \tag{XXVI.1.11}$$

for all convergences ξ and τ.

COROLLARY XXVI.1.13. *If J is a reflector and E is a coreflector that commutes with finite products, then*

$$\mathrm{Epi}_E^{\mathrm{Epi}^J_E} = \mathrm{Epi}^J_E.$$

In particular, $\mathrm{Epi}^{\mathrm{Epi}^J} = \mathrm{Epi}^J$ for each reflector J.

With a view to forthcoming applications to Theorem XXVI.6.14, let us note that

LEMMA XXVI.1.14. *If J is a reflector, E is a coreflector that commutes with finite products, and F is a reflector such that*

$$F \geq \mathrm{Epi}_E^J, \tag{XXVI.1.12}$$

then for each $\xi = E\xi$,

$$\theta \times F\sigma \geq J(\xi \times \sigma) \tag{XXVI.1.13}$$

holds for each $\sigma = E\sigma$ if and only if it does for every $\sigma = EF\sigma$.

PROOF. As E is idempotent, it is clear that if $\sigma = EF\sigma$, then $\sigma = E\sigma$. Assume that (XXVI.1.13) holds for every $\sigma = EF\sigma$, and let $\tau = E\tau$. As $\xi = E\xi$, we use (XXVI.1.11) interchanging the roles of ξ and τ to the effect that

$$J(\xi \times \tau) = J(\xi \times \mathrm{Epi}_E^J \tau) \leq J(\xi \times F\tau) \leq J(\xi \times EF\tau).$$

By setting $\sigma := EF\tau$ and applying (XXVI.1.13), we get

$$J(\xi \times \tau) \leq J(\xi \times \sigma) \leq \theta \times F\sigma = \theta \times FEF\tau \leq \theta \times FE\tau = \theta \times F\tau,$$

because we assumed that $\tau = E\tau$. □

XXVI.2. Applications to mixed properties

It turns out that functorial inequalities of modified duality can be implemented to the quest of preservation of mixed properties of the form $\xi \geq JE\xi$, where J is reflector and E a coreflector (see Chapter XXI).

COROLLARY XXVI.2.1. *If J is a reflector and E and F are coreflectors that commute with finite products, $F \geq E$ and $\mathbf{W}$ is finally dense in $\mathbf{F} = \mathrm{fix}\, F$, then the following statements are equivalent:*

(1) $\xi \geq \mathrm{Epi}_F^J E\xi$,
(2) $\xi \times \tau \geq \mathrm{Epi}_F^J E(\xi \times \tau)$ *for each convergence* $\tau = F\tau$,
(3) $\xi \times \tau \geq JE(\xi \times \tau)$ *for each convergence* τ *of* $\mathbf{W}$.

PROOF. By setting $\xi := E\xi$ and $\theta := \xi$ in Corollary XXVI.1.12, we get the equivalence of

$$\xi \geq \mathrm{Epi}_F^J E\xi$$

with $\xi \times F\tau \geq \mathrm{Epi}_F^J(E\xi \times \tau)$ for each convergence τ, and with $\xi \times F\tau \geq J(E\xi \times \tau)$ for each convergence τ. If L is any modifier, then $\xi \times F\tau \geq L(E\xi \times \tau)$ for each convergence τ implies, for every $\tau = F\tau$,

$$\xi \times \tau \geq L(E\xi \times F\tau) \geq L(E\xi \times E\tau) \geq LE(\xi \times \tau).$$

This gives the implications (1) $\Rightarrow$ (2) $\Rightarrow$ (3), because $\mathrm{Epi}_F^J \geq J$.

Now, assume (3). In view of Lemma XXIV.2.8, $\xi \times \tau \geq JE(\xi \times \tau)$ for each convergence $\tau = F\tau$. Let τ be an arbitrary convergence. Because $F \geq E$, $E(F\tau) = F\tau$, so that

$$JE(\xi \times F\tau) = J(E\xi \times F\tau) \geq J(E\xi \times \tau),$$

and $\xi \times F\tau \geq JE(\xi \times F\tau)$ by (3). Hence $\xi \times F\tau \geq J(E\xi \times \tau)$ for all τ. Therefore, by (3) $\Rightarrow$ (1) of Corollary XXVI.1.12 applied to $\xi := E\xi$, we get (1). □

For $E = \mathrm{B}_{\mathbb{H}}$, $F = \mathrm{B}_{\mathbb{D}}$ and $\mathbf{W}$ the class of $\mathbb{D}$-based prime topologies (which is finally dense in that of $\mathbb{D}$-based convergences), we get the following corollary.

Corollary XXVI.2.2. *Let J be a reflector, $\mathbb{D} \subset \mathbb{H}$ be composable classes, and let $\mathbb{H}$ be closed under finite products. Then the following statements are equivalent:*

(1) $\xi \geq \mathrm{Epi}_{\mathrm{B}_{\mathbb{D}}}^J \mathrm{B}_{\mathbb{H}}\xi$,
(2) $\xi \times \tau \geq \mathrm{Epi}_{\mathrm{B}_{\mathbb{D}}}^J \mathrm{B}_{\mathbb{H}}(\xi \times \tau)$ *for each $\mathbb{D}$-based convergence τ,*
(3) $\xi \times \tau \geq J\mathrm{B}_{\mathbb{H}}(\xi \times \tau)$ *for each $\mathbb{D}$-based prime topology τ.*

Proof. On setting $E := \mathrm{B}_{\mathbb{H}}$ and $F := \mathrm{B}_{\mathbb{D}}$, and observing that $F \geq E$ and $\mathrm{B}_{\mathbb{H}}\xi \times \tau \geq \mathrm{B}_{\mathbb{H}}(\xi \times \tau)$ because $\mathbb{H}$ closed under finite products, we apply Corollary XXVI.2.1 for each convergence τ. □

Theorem XXVI.2.3. *If $\mathbb{D}$ is a composable class of filters, then*

$$\mathrm{Epi}_{\mathrm{B}_{\mathbb{D}}}^{\mathrm{S}_0} = \mathrm{A}_{\mathbb{D}}.$$

Proof. By Corollary XVIII.2.9, $\mathrm{B}_{\mathbb{D}}$ is a coreflector that commutes with finite products. Then, by Corollary XXVI.1.12, $\theta \geq \mathrm{Epi}_{\mathrm{B}_{\mathbb{D}}}^{\mathrm{S}_0}\xi$ whenever $\theta \times \mathrm{B}_{\mathbb{D}}\tau \geq \mathrm{S}_0(\xi \times \tau)$ for each convergence τ. On the other hand, by Theorem XX.7.4, $\theta \geq \mathrm{A}_{\mathbb{D}}\xi$ is equivalent to $\theta \times \mathrm{B}_{\mathbb{D}}\tau \geq \mathrm{S}_0(\xi \times \tau)$ for each convergence τ. The equality follows. □

If $J = \mathrm{S}_0$ is the pretopologizer, then Corollary XXVI.2.2 in view of Theorem XXVI.2.3, specializes to

Proposition XXVI.2.4. *Let $\mathbb{D} \subset \mathbb{H}$ be composable classes so that $\mathbb{H}$ is closed under finite products. Let $\mathbf{W}$ be a finally dense class in $\mathrm{B}_{\mathbb{D}}$. Then the following statements are equivalent:*

(1) $\xi \geq \mathrm{A}_{\mathbb{D}}\mathrm{B}_{\mathbb{H}}\xi$,
(2) $\xi \times \tau \geq \mathrm{A}_{\mathbb{D}}\mathrm{B}_{\mathbb{H}}(\xi \times \tau)$ *for each $\mathbb{D}$-based convergence τ,*
(3) $\xi \times \tau \geq \mathrm{S}_0\mathrm{B}_{\mathbb{H}}(\xi \times \tau)$ *for every $\tau \in \mathbf{W}$.*

In particular, we recover Proposition XXI.8.7, because the pseudotopologizer S commutes with product:

COROLLARY XXVI.2.5. *The following statements are equivalent:*

(1) $\xi \geq \mathrm{S}_1\mathrm{I}_1\xi$,
(2) $\xi \times \tau \geq \mathrm{S}_1\mathrm{I}_1(\xi \times \tau)$ *for each convergence* $\tau \geq \mathrm{SI}_1\tau$,
(3) $\xi \times \tau \geq \mathrm{S}_0\mathrm{I}_1(\xi \times \tau)$ *for each prime topology* τ *of countable character.*

In other terms, a convergence ξ *is strongly Fréchet if and only if* $\xi \times \tau$ *is strongly Fréchet for each bisequential convergence* τ *if and only if* $\xi \times \tau$ *is Fréchet for each prime topology* τ *of countable character.*

PROOF. In Proposition XXVI.2.4, set $\mathbb{H} = \mathbb{F}_1$ and $\mathbb{D} = \mathbb{F}_1$. Then $\mathrm{A}_{\mathbb{D}} = \mathrm{S}_1$ (the paratopologizer) and $\mathrm{B}_{\mathbb{H}} = \mathrm{I}_1$ (the coreflector on the class of convergence of countable character). □

REMARK XXVI.2.6. Note that, as a product of two strongly Fréchet filters may fail to be strongly Fréchet, the class $\mathbb{D} = \mathbb{F}_1^{\mathbb{F}_1}$ of strongly Fréchet filters is not composable and therefore Proposition XXVI.2.4 and Theorem XXVI.2.3 do not apply directly to this choice of class $\mathbb{D}$. However, in view of Corollary XXI.10.5, Lemma XXVI.1.9 applies to $F = \mathrm{B}_{\mathbb{F}_1^{\mathbb{F}_1}}$, $J = \mathrm{S}_0$ and $B = \mathrm{I}_1$ to the effect that

$$\text{(XXVI.2.1)}\ \theta \geq \mathrm{Epi}^{\mathrm{S}_0}_{\mathrm{B}_{\mathbb{F}_1^{\mathbb{F}_1}}}\xi \iff \underset{\tau}{\forall}\ \theta \times \mathrm{B}_{\mathbb{F}_1^{\mathbb{F}_1}}\tau \geq \mathrm{S}_1(\xi \times \tau) \iff \underset{\tau}{\forall}\ \theta \times \mathrm{B}_{\mathbb{F}_1^{\mathbb{F}_1}}\tau \geq \mathrm{S}_0(\xi \times \tau).$$

Comparing with Theorem XXI.10.9, we conclude that

$$\mathrm{Epi}^{\mathrm{S}_0}_{\mathrm{B}_{\mathbb{F}_1^{\mathbb{F}_1}}} = \mathrm{A}_{\mathbb{F}_1^{\mathbb{F}_1}}\,. \tag{XXVI.2.2}$$

As a result of (XXVI.2.1) and (XXVI.2.2), Corollary XXI.10.10 can also be recovered as a corollary of the general results of modified duality.

XXVI.3. Exponential objects

By Definition XXV.4.1, a reflective class $\mathbf{J}$ of convergences is called *exponential* if $[\xi, \sigma] \in \mathbf{J}$ for every convergence ξ and every $\sigma \in \mathbf{J}$.

DEFINITION XXVI.3.1. Let J be a reflector. A convergence ξ is said to be *J-exponential* if $[\xi, \sigma] = J[\xi, \sigma]$ for each $\sigma = J\sigma$.

T-*exponential* convergences are also called *topologically exponential*, S_0-*exponential* convergences are also called *pretopologically exponential*, and so on.

Accordingly, a reflective class $\mathrm{fix}(J)$ is exponential if and only if every convergence ξ is J-exponential. Notice that we do not assume that $\xi \in \mathrm{fix}(J)$!

For each reflector J, there always exist some ξ, for which $[\xi, \sigma] = J[\xi, \sigma]$ for every $\sigma = J\sigma$, for example, the discrete convergence ι, for which $[\iota, \sigma] = p(\iota, \sigma)$ is the product convergence with factors in σ. As products preserve reflective properties, $[\iota, \sigma] \in \mathbf{J}$ provided that $\sigma \in \mathbf{J}$.

Since $[\xi, \sigma] = [\mathrm{Epi}^J \xi, \sigma]$ for each $\sigma = J\sigma$, we can study without loss of generality J-exponential convergences ξ from $\mathrm{fix}(\mathrm{Epi}^J)$.

It follows from Theorem XXVI.1.5 and Corollary XXVI.1.8 that

THEOREM XXVI.3.2. *If J is a reflector, $\mathbf{D}$ is an initially dense class in* $\mathrm{fix}(J)$*, and $\xi \in \mathrm{fix}(\mathrm{Epi}^J)$, then the following statements are equivalent:*

(1) *ξ is J-exponential,*
(2) *$[\xi, \sigma] = J[\xi, \sigma]$ for each $\sigma \in \mathbf{D}$,*
(3) *$\xi \geq \mathrm{Epi}_J^J \xi$,*
(4) *$\xi \times J\tau \geq J(\xi \times \tau)$ for every convergence τ.*

PROOF. (1) $\Leftrightarrow$ (2) by Definition XXVI.3.1. The equivalences of (2), (3), and (4) follows from Theorem XXVI.1.5 and Corollary XXVI.1.8. □

By Propositions XX.9.19 and XX.10.6, for a topology ξ, the upper hyperconvergence $[\xi, \$_0]$ is topological if and only if ξ is core-compact. We conclude that

COROLLARY XXVI.3.3. *A topology is topologically exponential if and only if it is core-compact.*

PROOF. As the Sierpiński topology $\$_0$ is initially dense in the class of topologies (see Corollary VII.5.3), on setting $J := \mathrm{T}$ and $\mathbf{D} := \{\$_0\}$ in Theorem XXVI.3.2, (2) specializes to $[\xi, \$_0] = \mathrm{T}[\xi, \$_0]$. □

REMARK XXVI.3.4. The assumption of topologicity enables a simple characterization of topologically exponential convergences. A full characterization of topologically exponential *convergences*, presented in Theorem XXVI.4.6, is rather involuted ([1]).

In order to be able to recognize J-exponential objects, we need to give a practical meaning to the condition $\xi \geq \mathrm{Epi}_J^J \xi$. If J is an exponential reflector, then each ξ is J-exponential. Otherwise,

PROPOSITION XXVI.3.5. *If J is a reflector, then the restriction*

$$\mathrm{Epi}_J^J : \mathrm{fix}(\mathrm{Epi}^J) \longrightarrow \mathrm{fix}(\mathrm{Epi}^J)$$

is an ascending functor.

[1]A characterization was given by S. DOLECKI and F. MYNARD, [80, Theorem 16.5]. See also F. Jordan and F. Mynard, [144].

PROOF. We have already observed that Epi_J^J is a functor. By (XXVI.1.3), if J is a reflector, thus $J \leq \mathrm{I}$, then $\mathrm{Epi}_J^J \geq \mathrm{Epi}^J = \mathrm{Epi}_{\mathrm{I}}^J$. Therefore, if $\xi \in \mathrm{fix}(\mathrm{Epi}^J)$, that is, $\mathrm{Epi}^J\xi \geq \xi$, then $\mathrm{Epi}_J^J\xi \geq \xi$, hence Epi_J^J is ascending on $\mathrm{fix}(\mathrm{Epi}^J)$.

Let us show that $\mathrm{Epi}_J^J\xi \in \mathrm{fix}(\mathrm{Epi}^J)$, that is, $\mathrm{Epi}^J(\mathrm{Epi}_J^J\xi) \geq \mathrm{Epi}_J^J\xi$. To this end, recall that

$$\theta \geq \mathrm{Epi}_J^J\xi \iff \forall_{\tau \in \mathbf{I}}\ \theta \times J\tau \geq J(\xi \times \tau). \tag{XXVI.3.1}$$

As Epi^J commutes with finite products, it follows that

$$\mathrm{Epi}^J\theta \times J\tau = \mathrm{Epi}^J\theta \times \mathrm{Epi}^J J\tau \geq \mathrm{Epi}^J J(\xi \times \tau) = J(\xi \times \tau).$$

By setting $\theta := \mathrm{Epi}_J^J\xi$, we get $\mathrm{Epi}^J(\mathrm{Epi}_J^J\xi) \times J\tau \geq J(\xi \times \tau)$ for each τ. Hence by (XXVI.3.1), $\mathrm{Epi}^J(\mathrm{Epi}_J^J\xi) \geq \mathrm{Epi}_J^J\xi$ for each $\xi \in \mathrm{fix}(\mathrm{Epi}^J)$. □

Therefore,

COROLLARY XXVI.3.6. *If J is a reflector, then the class*

$$\mathrm{fix}(\mathrm{Epi}_J^J)$$

(of J-exponential convergences) is a coreflective subclass of

$$\mathrm{fix}(\mathrm{Epi}^J)$$

(of the exponential hull of J). Moreover, the corresponding coreflector D^J commutes with finite products.

PROOF. By Propositions XXVI.3.5 and XVII.1.6, the class

$$\{\xi \in \mathrm{Epi}^J\xi : \xi \geq \mathrm{Epi}_J^J\xi\}$$

is coreflective in $\mathrm{fix}(\mathrm{Epi}^J)$. On the other hand, by Remark XVII.3.12, the idempotent iteration D^J of Epi_J^J is a coreflector on $\mathrm{fix}(\mathrm{Epi}_J^J)$ within $\mathrm{fix}(\mathrm{Epi}^J)$.

To see that D^J commutes with finite products, it is enough to prove that the product of two J-exponential convergences is J-exponential. Indeed, if ξ and ζ are arbitrary convergences, then $\mathrm{D}^J\xi$ and $\mathrm{D}^J\zeta$ are J-exponential, and since $\mathrm{D}^J\xi \times \mathrm{D}^J\zeta \geq \xi \times \zeta$, we infer that $\mathrm{D}^J\xi \times \mathrm{D}^J\zeta \geq \mathrm{D}^J(\xi \times \zeta) \geq \mathrm{D}^J\xi \times \mathrm{D}^J\zeta$, the second inequality holding for every functor.

Assume that ξ and ζ are J-exponential and that $\sigma = J\sigma$. Then, by Theorem XXIV.3.2, $[\xi \times \zeta, \sigma] \approx [\zeta, [\xi, \sigma]]$,and $[\xi, \sigma] = J[\xi, \sigma]$, because ξ is J-exponential. Hence, $[\zeta, [\xi, \sigma]] = J[\zeta, [\xi, \sigma]]$, because ζ is J-exponential (and $[\xi, \sigma] = J[\xi, \sigma]$). It follows that $[\xi \times \zeta, \sigma] = J[\xi \times \zeta, \sigma]$, that is, $\xi \times \zeta$ is J-exponential. □

PROPOSITION XXVI.3.7. *If ξ is a Hausdorff, functionally regular (alias completely regular) topology and $[\xi, \nu]$ is a pretopology, then ξ is locally compact.*

PROOF. Let $x \in \lim_\xi \mathcal{F}$, hence $x \in \lim_\xi \mathrm{cl}_\xi^\natural \mathcal{F}$. By definition, there exists a functionally closed F_0 and $V_0 \in \mathcal{V}_{[\xi,\nu]}(\widehat{0})$ such that

$$\text{(XXVI.3.2)} \qquad \langle F_0, V_0\rangle \subset]-\tfrac{1}{2},\tfrac{1}{2}[.$$

If F_0 were not ξ-compact, there would be $\mathcal{U} \in \beta F_0$ with $\varnothing = \lim_\xi \mathcal{U} = \mathrm{adh}_\xi \mathcal{U} \supset \mathrm{adh}_\xi \int_\mathcal{U} \mathcal{N}_\xi$, because of Proposition XIX.3.13. By Lemma VIII.2.3, the complementary family $\mathrm{cl}_\xi^\natural \mathcal{U}_c$ is a ξ-cover refining $\mathcal{U}_c$. By Lemma XXV.5.4, $\widehat{0} \in \lim_{[\xi,\nu]} \mathcal{U}_c^{=0}$, and as $[\xi,\nu]$ is a pretopology, $V_0 \in \mathcal{U}_c^{=0}$, that is, there exists a functionally ξ-open set $U \in \mathcal{U}$ such that

$$\{f > 0\} \subset U \Longrightarrow f \in V_0.$$

If $x \in U \subset F_0$, then there exists $h \in \mathscr{C}(\xi,\nu)$ such that $h(x) = 1$ and $\{h > 0\} = U$, in contradiction with (XXVI.3.2). □

THEOREM XXVI.3.8. *If ξ is a Hausdorff, functionally regular (alias completely regular) topology, then the following statements are equivalent:*

(1) $[\xi,\nu] \le \mathrm{R}^\nu[\xi,\nu]$,
(2) $[\xi,\nu] \le \mathrm{T}[\xi,\nu]$,
(3) $[\xi,\nu] \le \mathrm{S}_0[\xi,\nu]$,
(4) *ξ is locally compact.*

PROOF. Obviously, (1) $\Rightarrow$ (2) $\Rightarrow$ (3), and (3) $\Rightarrow$ (4) in virtue of Proposition XXVI.3.7. If ξ is locally compact, then by Corollary XXVI.3.3, $[\xi,\nu]$ is a topological, Hausdorff, and functionally embedded by Proposition XXV.1.5, hence functionally regular. □

COROLLARY XXVI.3.9. *A Hausdorff, functionally regular (alias completely regular) topology is R^ν-exponential if and only if it is locally compact.*

THEOREM XXVI.3.10. *Let $\mathbb{D}$ be a composable class of filters. Then a pseudotopology τ is $\mathrm{A}_\mathbb{D}$-exponential if and only if*

$$\tau \ge \mathrm{SB}_\mathbb{D}\tau.$$

PROOF. By Theorem XX.7.4,

$$\mathrm{A}_\mathbb{D}\xi \times \mathrm{B}_\mathbb{D}\tau \ge \mathrm{A}_\mathbb{D}(\xi \times \tau),$$

for each ξ and τ. As $\mathrm{S} = \mathrm{A}_\mathbb{F} \ge \mathrm{A}_\mathbb{D}$, and S commutes with products, we deduce that $\mathrm{A}_\mathbb{D}\xi \times \mathrm{SB}_\mathbb{D}\tau \ge \mathrm{A}_\mathbb{D}(\xi \times \tau)$, so that if $\tau \ge \mathrm{SB}_\mathbb{D}\tau$, then $\mathrm{A}_\mathbb{D}\xi \times \tau \ge \mathrm{A}_\mathbb{D}(\xi \times \tau)$ for every ξ, hence τ is $\mathrm{A}_\mathbb{D}$-exponential in virtue of Theorem XXVI.3.2.

Conversely, if the condition does not hold for a pseudotopology τ on X, then there exist $x_0 \in X$ and $\mathcal{U} \in \beta X$ such that $x_0 \in \lim_\tau \mathcal{U} \setminus$

$\lim_{\mathrm{B}_{\mathbb{D}}\tau}\mathcal{U}$, so that if $\mathcal{D}\in\mathbb{D}$ and $\mathcal{D}\leq\mathcal{U}$, then $x_0\notin\lim_\tau\mathcal{D}$. By the pseudotopologicity of τ, there exists $\mathcal{W}_\mathcal{D}\in\beta\mathcal{D}$ such that $x_0\notin\lim_\tau\mathcal{W}_\mathcal{D}$.

Let ξ be the prime convergence on X defined by

$$x_0\in\lim_\xi\mathcal{F}\iff\underset{\mathcal{D}\in\mathbb{D}}{\exists}(\mathcal{D}\leq\mathcal{U})\wedge(\mathcal{F}\geq\mathcal{W}_\mathcal{D}\wedge x_0^\uparrow).$$

Then $(x_0,x_0)\in\lim_{\mathrm{A}_{\mathbb{D}}\xi\times\tau}(\mathcal{U}\times\mathcal{U})$. However, $(x_0,x_0)\notin\lim_{\mathrm{A}_{\mathbb{D}}(\xi\times\tau)}(\mathcal{U}\times\mathcal{U})$, because the diagonal $\triangle=\{(x,x):x\in X,x\neq x_0\}\in(\mathcal{U}\times\mathcal{U})^\#$, but $(x_0,x_0)\notin\mathrm{adh}_{\xi\times\tau}\triangle$. Indeed, if $\triangle\in\mathcal{H}$ and $x_0\in\lim_\xi p_X[\mathcal{H}]$, then by definition of ξ, there is $\mathcal{D}\in\mathbb{D}$ with $\mathcal{D}\leq\mathcal{U}$ and such that $p_X[\mathcal{H}]\geq\mathcal{W}_\mathcal{D}$, hence $p_X[\mathcal{H}]=\mathcal{W}_\mathcal{D}$, in contradiction with $x_0\notin\lim_\tau\mathcal{W}_\mathcal{D}$. □

COROLLARY XXVI.3.11. *A pseudotopology is*

(1) *pretopologically exponential if and only if it is finitely generated,*
(2) *paratopologically exponential if and only if it is bisequential.*

PROOF. (1) As the pretopologizer is $\mathbb{F}_0$-adherence determined: $\mathrm{S}_0=\mathrm{A}_{\mathbb{F}_0}$, by Theorem XXVI.3.10, a pseudotopology τ is pretopologically exponential whenever $\tau\geq\mathrm{SB}_{\mathbb{F}_0}\tau=\mathrm{SI}_0\tau$. Actually, $\mathrm{SI}_0\tau=\mathrm{I}_0\tau$, because if $x\in\lim_{\mathrm{SI}_0\tau}\mathcal{F}$, then for each $\mathcal{U}\in\beta\mathcal{F}$ there is $U_\mathcal{U}\in\mathcal{U}$ such that $x\in\lim_\tau\{U_\mathcal{U}\}^\uparrow$. By Proposition V.4.6, there is a finite set $\boldsymbol{F}\subset\beta\mathcal{F}$ such that $U_{\boldsymbol{F}}:=\bigcup_{\mathcal{U}\in\boldsymbol{F}}U_\mathcal{U}\in\mathcal{F}$. Each pseudotopology is finitely stable, so that, $x\in\lim_\tau\{U_{\boldsymbol{F}}\}^\uparrow$ and $\{U_{\boldsymbol{F}}\}^\uparrow\leq\mathcal{F}$, completing the argument.

(2) As $\mathrm{S}_1=\mathrm{A}_{\mathbb{F}_1}$, a pseudotopology τ is pseudotopologically exponential whenever $\tau\geq\mathrm{SB}_{\mathbb{F}_1}\tau=\mathrm{SI}_1\tau$, that is, τ is bisequential (see for instance Table XXI.7.1). □

XXVI.4. Topologicity of dual convergences

We say that a family $\mathcal{H}$ *ξ-rigidly meshes* a family $\mathcal{D}$ if for each $H\in\mathcal{H}$ there exists $A_H\in\mathcal{D}^\#$ such that $\mathrm{cl}_\xi A_H=A_H\subset H$. Of course, if $\mathcal{H}$ ξ-rigidly meshes $\mathcal{D}$, then $\mathcal{H}\#\mathcal{D}$. It is plain that the converse holds provided that $\mathcal{H}=\mathrm{cl}_\xi^\natural\mathcal{H}$.

We say that a family $\mathcal{D}$ is *ξ-rigidly compactoid* at a family $\mathcal{B}$ if $\mathrm{adh}_\xi\mathcal{H}\in\mathcal{B}^\#$ for each filter $\mathcal{H}$ that ξ-rigidly meshes $\mathcal{D}$; *ξ-rigidly compact* if $\mathrm{adh}_\xi\mathcal{H}\in\mathcal{D}^\#$ for each filter $\mathcal{H}$ that ξ-rigidly meshes $\mathcal{D}$.

Notice that if $\mathcal{D}$ is ξ-compactoid at $\mathcal{B}$, then $\mathcal{D}$ is ξ-rigidly compactoid at $\mathcal{B}$ for every convergence ξ. The converse also holds provided that ξ is a topology.

LEMMA XXVI.4.1. *A set F is $\mathbb{F}_{\xi^\bullet}(\xi)$-compact at B if and only if $\mathcal{O}_\xi(F)$ is ξ-rigidly compact at B.*

PROOF. A filter $\mathcal{H}$ ξ-rigidly meshes with $\mathcal{O}_\xi(F)$ if and only if for each $H \in \mathcal{H}$ there exists $\mathrm{cl}_\xi D = D \subset H$ such that $D \in \mathcal{O}_\xi(F)^\#$, equivalently, $\mathrm{cl}_\xi D \# F$, that is, $F \# \mathrm{int}^\natural_{\xi^\circ} \mathcal{H}$. Therefore, if $\mathcal{O}_\xi(F)$ is ξ-rigidly compact at B and $\mathrm{int}^\natural_{\xi^\circ} \mathcal{H} = \mathcal{H} = \mathrm{cl}^\natural_{\xi^\bullet} \mathcal{H}$ meshes F, then $\mathcal{H}$ ξ-rigidly meshes $\mathcal{O}_\xi(F)$, and thus $\mathrm{adh}_\xi \mathcal{H} \# B$. Conversely, if F is $\mathbb{F}_{\xi^\bullet}(\xi)$-compact at B, and $\mathcal{H}$ ξ-rigidly meshes $\mathcal{O}_\xi(F)$, then $\mathrm{int}^\natural_{\xi^\circ} \mathcal{H} = \mathcal{H} = \mathrm{cl}^\natural_{\xi^\bullet} \mathcal{H}$ meshes F, hence $\mathrm{adh}_\xi \mathcal{H} \# B$. □

The proof of Theorem XX.10.8 can now be easily adapted to yield the following generalization for arbitrary convergences.

THEOREM XXVI.4.2. *A subfamily $\mathcal{A}$ of $\mathcal{C}_\xi$ is a stable $[\xi, \$_0]$-vicinity of A_0 if and only if $\mathcal{A}_c = \mathcal{O}_\xi\{\mathcal{A}_c\}$ is ξ-rigidly compact at A_0^c.*

COROLLARY XXVI.4.3. *A subfamily $\mathcal{A}$ of $\mathcal{C}_\xi$ is $[\xi, \$_0]$-open if and only if $\mathcal{A}_c = \mathcal{O}_\xi\{\mathcal{A}_c\}$ is ξ-rigidly compact.*

LEMMA XXVI.4.4. *A family $\mathcal{A}$ is* $\mathrm{Epi}^{\mathrm{T}}\xi$*-compact at $\mathcal{B} = \mathcal{O}_\xi\{\mathcal{B}\}$ if and only if $\mathcal{A}$ is $\mathbb{F}_{\xi^\bullet}(\xi)$-compact at $\mathcal{B}$.*

PROOF. Let $\mathcal{A}$ be $\mathrm{Epi}^{\mathrm{T}}\xi$-compact at $\mathcal{B}$ and let $\mathcal{H} \in \mathbb{F}_{\xi^\bullet}$ be such that $\mathcal{H} \# \mathcal{A}$. Then $\mathcal{B} \# \mathrm{adh}_{\mathrm{Epi}^{\mathrm{T}}\xi} \mathcal{H}$, equivalently, $\mathcal{B} \# \mathrm{cl}_\xi(\mathrm{adh}_{\mathrm{Epi}^{\mathrm{T}}\xi} \mathcal{H})$, because $\mathcal{B} = \mathcal{O}_\xi\{\mathcal{B}\}$. As $\mathrm{Epi}^{\mathrm{T}}$ is idempotent, in view of Lemma XXV.6.1, $\mathrm{cl}_\xi(\mathrm{adh}_{\mathrm{Epi}^{\mathrm{T}}\xi} \mathcal{H}) = \mathrm{cl}_\xi(\mathrm{adh}_\xi \mathcal{H})$, because $\mathcal{H} \in \mathbb{F}_{\xi^\bullet}$. It follows that $\mathcal{B} \# \mathrm{cl}_\xi(\mathrm{adh}_\xi \mathcal{H})$, hence $\mathcal{B} \# \mathrm{adh}_\xi \mathcal{H}$, because $\mathcal{B} = \mathcal{O}_\xi\{\mathcal{B}\}$. Conversely, if $\mathcal{A}$ is $\mathbb{F}_{\xi^\bullet}(\xi)$-compact at $\mathcal{B}$ and $\mathcal{H} \# \mathcal{A}$, then *a fortiori* $\mathrm{cl}^\natural_{\xi^\bullet} \mathcal{H} \# \mathcal{A}$, hence by assumption $\mathcal{B} \# \mathrm{adh}_\xi \mathrm{cl}^\natural_{\xi^\bullet} \mathcal{H}$, equivalently $\mathcal{B} \# \mathrm{adh}_{\mathrm{Epi}^{\mathrm{T}}\xi} \mathrm{cl}^\natural_{\xi^\bullet} \mathcal{H}$, because $\mathcal{B} = \mathcal{O}_\xi\{\mathcal{B}\}$ and $\mathrm{cl}^\natural_{\xi^\bullet} \mathcal{H} \in \mathbb{F}_{\xi^\bullet}$. Finally, in view of Proposition XXV.6.3, $\mathcal{B} \# \mathrm{adh}_{\mathrm{Epi}^{\mathrm{T}}\xi} \mathcal{H}$ by $\circ$-regularity of $\mathrm{Epi}^{\mathrm{T}}$. □

We call a convergence ξ *topologically core compact* if whenever $x \in \lim_\xi \mathcal{F}$ and $O \in \mathcal{O}_\xi(x)$ there is $F \in \mathcal{F}$ that is ξ-compact at O. As $\mathcal{O}_\xi(x) \subset \mathcal{F}$ whenever $x \in \lim_\xi \mathcal{F}$, a core compact convergence is also topologically core compact, but not conversely (see [**144**, Example 5.8]).

THEOREM XXVI.4.5. *A hyperconvergence $[\xi, \$_0]$ is a pretopology if and only if* $\mathrm{Epi}^{\mathrm{T}}\xi$ *is topologically core compact.*

PROOF. A hyperconvergence $[\xi, \$_0]$ is a pretopology if and only if $A_0 \in \lim_{[\xi,\$_0]} \mathcal{V}_{[\xi,\$_0]}(A_0)$ for every $A_0 \in \mathcal{C}_\xi$, that is, $\mathrm{adh}_\xi \mathrm{r}\mathcal{V}_{[\xi,\$_0]}(A_0) \subset A_0$. Equivalently, if $\lim_\xi \mathcal{F} \cap A_0^c \neq \varnothing$ then $\mathcal{F} \perp \mathrm{r}\mathcal{V}_{[\xi,\$_0]}(A_0)$, that is, there exist $F \in \mathcal{F}$ and a stable $\mathcal{A} \in \mathcal{V}_{[\xi,\$_0]}(A_0)$ such that $F \cap \bigcup_{A \in \mathcal{A}} A = \varnothing$, equivalently, $F \subset \bigcap_{A \in \mathcal{A}} A^c$. By Theorem XXVI.4.2, this means that for each $O \in \mathcal{O}_\xi$ and each filter $\mathcal{F}$ such that $\lim_\xi \mathcal{F} \cap O \neq \varnothing$, there exists a family $\mathcal{D} = \mathcal{O}_\xi\{\mathcal{D}\}$ that is ξ-rigidly compact at O and such that $\bigcap_{D \in \mathcal{D}} D \in \mathcal{F}$.

Since $\mathcal{D} = \mathcal{O}_\xi\{\mathcal{D}\}$ is ξ-rigidly compact at O, and $\mathcal{D} \subset \mathcal{O}_\xi(F)$, the latter family is also ξ-rigidly compact at O, thus by Lemma XXVI.4.1, F is $\mathbb{F}_{\xi^\bullet}(\xi)$-compact at O, equivalently, F is $\mathrm{Epi}^{\mathrm{T}}\xi$-compact at O, by Lemma XXVI.4.4. In other words, $\mathrm{Epi}^{\mathrm{T}}\xi$ is topologically core compact.

Conversely, we need to show $\mathrm{adh}_\xi \, \mathrm{r}\mathcal{V}_{[\xi,\$_0]}(A_0) \subset A_0$ if ξ $\mathrm{Epi}^{\mathrm{T}}\xi$ is topologically core compact. In other words, letting $O = A_0^c$, we need to show that if $\lim_\xi \mathcal{F} \cap O \neq \varnothing$, $\mathcal{F} \perp \mathrm{r}\mathcal{V}_{[\xi,\$_0]}(A_0)$, that is, there is $\mathcal{A} \in \mathrm{r}\mathcal{V}(A_0)$ with $\bigcap \mathcal{A}_c \in \mathcal{F}$. Under our assumption, there exists $F \in \mathcal{F}$ that is $\mathrm{Epi}^{\mathrm{T}}\xi$-compact at O, equivalently $\mathcal{O}_\xi(F)$ is ξ-rigidly compact at O, by Lemmas XXVI.4.1 and XXVI.4.4. In view of Theorem XXVI.4.2, $\mathcal{A} = \mathcal{O}_\xi(F)$ is as desired because $\bigcap_{Q \in \mathcal{O}_\xi(F)} Q \in \mathcal{F}$. □

Of course, $[\xi, \$]$ is a topology if and only if a hyperfilter $\mathscr{F}$ contains every element of $\mathcal{O}_{[\xi,\$]}(A_0)$, then $A_0 \in \lim_{[\xi,\$]} \mathscr{F}$. This can be rephrased as follows:

THEOREM XXVI.4.6. *The hyperconvergence* $[\xi, \$]$ *is a topology if and only if for every* $O \in \mathcal{O}_\xi$ *and each filter* $\mathcal{F}$ *such that* $\lim_\xi \mathcal{F} \cap O \neq \varnothing$, *there exists a family* $\mathcal{B} = \mathcal{O}_\xi\{\mathcal{B}\}$ *that is* $\mathrm{Epi}\xi$*-rigidly compact,* $O \in \mathcal{B}$ *and* $\bigcap_{B \in \mathcal{B}} B \in \mathcal{F}$.

XXVI.5. Conditional versions of core compactness

Analogously to local and core compactness, we consider *conditional* variants of these notions. If $\mathbb{J}$ is a class of filters, then a convergence ξ is called *locally* $\mathbb{J}$*-compact* if each ξ-convergent filter contains a $\mathbb{J}$-compact set. Similarly, *locally* $\mathbb{J}$*-compactoid* and *locally hereditarily* $\mathbb{J}$*-compact* convergences are defined. Structural properties of these classes of convergences are similar to those of local compactness.

Core compactness came to light in the quest of (partial) commutation of the topologizer with finite products in Proposition XX.9.19. In order to tackle commutation of products with other reflectors, we need a somewhat more refined concepts of conditional core compactness.

Let $\mathbb{D}$ and $\mathbb{J}$ be two composable classes of filters. Define a modifier $\mathrm{Q}_{\mathbb{D}}^{\mathbb{J}}$ to the effect that

$$x \in \lim_{\mathrm{Q}_{\mathbb{D}}^{\mathbb{J}}\xi} \mathcal{F} \tag{XXVI.5.1}$$

if $x \in \lim_\xi \mathcal{F}$ and for every $V \in \mathcal{O}_\xi(x)$, there exists $\mathcal{D}_V \in \mathbb{D}$ such that $\mathcal{D}_V \leq \mathcal{F}$ and $\mathcal{D}_V$ is $\mathbb{J}(\xi)$-compact at V.

For instance, if $\mathbb{D} = \mathbb{F}_0$ and $\mathbb{J} = \mathbb{F}$, then (XXVI.5.1) if and only if ξ is topologically core-compact; if $\mathbb{D} = \mathbb{F}_0$ and $\mathbb{J} = \mathbb{F}_1$, then (XXVI.5.1) if and only if ξ is *topologically countably core-compact.*

PROPOSITION XXVI.5.1. *The modifier* $\mathrm{Q}_{\mathbb{D}}^{\mathbb{J}}$ *is a functor.*

PROOF. By Proposition XVII.2.3, we need to prove that $f(\mathrm{Q}_{\mathbb{D}}^{\mathbb{J}}\xi) \geq \mathrm{Q}_{\mathbb{D}}^{\mathbb{J}}(f\xi)$ for any surjective map $f : X \longrightarrow Y$ and any convergence ξ on X. Let $y \in \lim_{f(\mathrm{Q}_{\mathbb{D}}^{\mathbb{J}}\xi)} \mathcal{F}$ and $V \in \mathcal{O}_{f\xi}(y)$. By Proposition VII.3.10, $f^-(V) \in \mathcal{N}_\xi(x)$ for each $x \in f^-(y)$. On the other hand, there exists a filter $\mathcal{G}$ such that $f[\mathcal{G}] = \mathcal{F}$ and $x \in \lim_{\mathrm{Q}_{\mathbb{D}}^{\mathbb{J}}\xi} \mathcal{G}$. Thus there exists a $\mathbb{D}$-filter $\mathcal{D} \leq \mathcal{G}$ that is $\mathbb{J}(\xi)$-compact at $f^-(V)$. By composability, $f[\mathcal{D}]$ is a $\mathbb{D}$-filter $\mathbb{J}(f\xi)$-compact at V and $f[\mathcal{D}] \leq \mathcal{F}$. We conclude that $y \in \lim_{\mathrm{Q}_{\mathbb{D}}^{\mathbb{J}}(f\xi)} \mathcal{F}$. □

It is clear that $\mathrm{Q}_{\mathbb{D}}^{\mathbb{J}}$ is isotone and ascending. Therefore, the class $\mathrm{fix}(\mathrm{Q}_{\mathbb{D}}^{\mathbb{J}})$ is coreflective. It can be shown that $\mathrm{Q}_{\mathbb{D}}^{\mathbb{J}}$ is not idempotent.

Note that Lemma XXV.6.1 is easily modified to the following more general statement, with the obvious variant of the proof:

LEMMA XXVI.5.2. *Let* $\mathbb{J}$ *be a composable class of filters. Then*

$$\lim_{Epi^T_{\mathrm{B}_\mathbb{J}}\xi} \mathcal{F} = \bigcap\nolimits_{\mathbb{J}_{\xi^\bullet} \ni \mathcal{H} \# \mathcal{F}} \mathrm{cl}_\xi(\mathrm{adh}_\xi \mathcal{H}), \tag{XXVI.5.2}$$

where $\mathbb{J}_{\xi^\bullet} = \{\mathcal{H} \in \mathbb{J} : \mathcal{H} = \mathrm{cl}^\natural_{\xi^\bullet} \mathcal{H}\}$.

THEOREM XXVI.5.3. *If* $\mathbb{D}$ *and* $\mathbb{J}$ *are composable classes of filters, then*

$$\theta \times \mathrm{A}_\mathbb{D}\tau \geq \mathrm{T}(\xi \times \tau)$$

for every $\mathbb{J}$*-founded convergence* τ *if and only if*

$$\theta \geq \mathrm{S}\mathrm{Q}_{\mathbb{D}}^{\mathbb{J}}\mathrm{Epi}^{\mathrm{T}}_{\mathrm{B}_\mathbb{J}}\xi.$$

PROOF. Assume that τ is $\mathbb{J}$-founded, $\theta = \mathrm{Q}_{\mathbb{D}}^{\mathbb{J}}\mathrm{Epi}^{\mathrm{T}}_{\mathrm{B}_\mathbb{J}}\xi$, and let $x \in \lim_\theta \mathcal{F}$ and $y \in \lim_{\mathrm{A}_\mathbb{D}\tau} \mathcal{G}$. Let H be a $(\xi \times \tau)$-closed set such that $H \#(\mathcal{F} \times \mathcal{G})$. If we show that $(x, y) \in H$, then this holds *a fortiori* for $\theta \geq \mathrm{S}\mathrm{Q}_{\mathbb{D}}^{\mathbb{J}}\mathrm{Epi}^{\mathrm{T}}_{\mathrm{B}_\mathbb{J}}\xi$. For every $V \in \mathcal{O}_\xi(x)$, there is a $\mathbb{D}$-filter $\mathcal{D}_V \leq \mathcal{F}$ so that $\mathcal{D}_V$ is $\mathbb{J}(\mathrm{Epi}^{\mathrm{T}}_{\mathrm{B}_\mathbb{J}}\xi)$-compact at V. As $\mathbb{D}$ is composable, $H\mathcal{D}_V$ is a $\mathbb{D}$-filter and $H\mathcal{D}_V \# \mathcal{G}$, so that $y \in \mathrm{adh}_\tau H\mathcal{D}_V$. As τ is $\mathbb{J}$-founded, there exists a $\mathbb{J}$-filter $\mathcal{L}_V \# H\mathcal{D}_V$ such that $y \in \lim_\tau \mathcal{L}_V$. By composability, $H^-\mathcal{L}_V$ is a $\mathbb{J}$-filter and $\mathcal{D}_V \# H^-\mathcal{L}_V$, and by assumption on θ, there exists $x_V \in V \cap \mathrm{adh}_{\mathrm{Epi}^{\mathrm{T}}_{\mathrm{B}_\mathbb{J}}\xi} H^-\mathcal{L}_V$. Accordingly, $(x_V, y) \in \mathrm{adh}_{\mathrm{Epi}^{\mathrm{T}}_{\mathrm{B}_\mathbb{J}}\xi \times \tau} H \subset H$, because $\mathrm{T}(\mathrm{Epi}^{\mathrm{T}}_{\mathrm{B}_{\mathbb{J}}}\xi \times \tau) = \mathrm{T}(\xi \times \tau)$ for each $\tau = \mathrm{B}_{\mathbb{J}}\tau$. Since $x \in \mathrm{cl}_\xi\{x_V : V \in \mathcal{O}_\xi(x)\}$, we infer that

$$(x, y) \in \mathrm{cl}_{\xi\times\tau}\{(x_V, y) : V \in \mathcal{O}_\xi(x)\} \subset H.$$

Conversely, let $X := |\theta|$ and assume that $\theta \ngeq \mathrm{SQ}^{\mathbb{J}}_{\mathbb{D}}\mathrm{Epi}^{\mathrm{T}}_{\mathrm{B}_{\mathbb{J}}}\xi$, that is, there exists an ultrafilter $\mathcal{U}$ and $x_0 \in X$ such that

$$x_0 \in \lim_\theta \mathcal{U} \setminus \lim_{\mathrm{Q}^{\mathbb{J}}_{\mathbb{D}}\mathrm{Epi}^{\mathrm{T}}_{\mathrm{B}_{\mathbb{J}}}\xi} \mathcal{U}.$$

Hence, there exists $V \in \mathcal{O}_\xi(x)$ such that for every $\mathbb{D}$-filter $\mathcal{H}\#\mathcal{U}$, there is a $\mathbb{J}$-filter $\mathcal{L}_\mathcal{H}\#\mathcal{H}$ such that $\mathrm{adh}_{\mathrm{Epi}^{\mathrm{T}}_{\mathrm{B}_{\mathbb{J}}}\xi} \mathcal{L}_\mathcal{H} \cap V = \varnothing$. Let τ be a prime convergence on X such that $x_0 \in \lim_\tau \mathcal{F}$ if and only if there is a $\mathbb{D}$-filter $\mathcal{H}\#\mathcal{U}$ such that $\mathcal{F} \geq \mathcal{L}_\mathcal{H} \wedge x_0^\uparrow$, so that τ is a $\mathbb{J}$-founded convergence. We observe that $(x_0, x_0) \in \lim_{\theta\times\mathrm{A}_{\mathbb{D}}\tau}(\mathcal{U}\times\mathcal{U})$, but $(x_0, x_0) \notin \lim_{\mathrm{T}(\xi\times\tau)}(\mathcal{U}\times\mathcal{U})$, because, if

$$A := \{(x, y) : y \neq x_0, x \in \lim_{\mathrm{Epi}^{\mathrm{T}}_{\mathrm{B}_{\mathbb{J}}}\xi}\{y\}^\uparrow\},$$

then $A\#(\mathcal{U}\times\mathcal{U})$, but $(x_0, x_0) \notin \mathrm{cl}_{\xi\times\tau} A$. Indeed, if $(x, y) \in \lim_{\xi\times\tau}(\mathcal{M} \times \mathcal{G})$ and $y \neq x_0$, then $\mathcal{G} = \{y\}^\uparrow$, and since $A \in \mathcal{M} \times \{y\}^\uparrow$, we infer that $\mathcal{M}$ is a filter on $\lim_{\mathrm{Epi}^{\mathrm{T}}_{\mathrm{B}_{\mathbb{J}}}\xi}\{y\}^\uparrow$. As $\mathrm{Epi}^{\mathrm{T}}_{\mathbb{J}}\xi$ has ξ-closed limits by Lemma XXVI.5.2, $(x, y) \in A$. If now $y = x_0$, then there is a $\mathbb{D}$-filter $\mathcal{H}$ such that $\mathcal{G} \geq \mathcal{L}_\mathcal{H}$. As $A\#(\mathcal{M} \times \mathcal{L}_\mathcal{H})$, we conclude that $\mathcal{L}_\mathcal{H}\# \mathrm{cl}^\natural_{\xi^\circ} \mathcal{M}$, so that $\lim_{\mathrm{Epi}^{\mathrm{T}}_{\mathrm{B}_{\mathbb{J}}}\xi} \mathrm{cl}^\natural_{\xi^\circ} \mathcal{M} = \lim_{\mathrm{Epi}^{\mathrm{T}}_{\mathrm{B}_{\mathbb{J}}}\xi} \mathcal{M}$ and thus $(x, x_0) \in V^c \times \{x_0\}$. Since V^c is ξ-closed, $(x_0, x_0) \notin \mathrm{cl}_{\xi\times\tau} A$. □

In particular, if $\mathbb{J} = \mathbb{F}$ and $\mathbb{D} = \mathbb{F}_0$, we obtain a generalization of Proposition XX.9.19, without the assumption that ξ be topological:

Corollary XXVI.5.4. *An epitopology ξ is topologically core-compact (equivalently, $\xi \geq \mathrm{SQ}^{\mathbb{F}}_{\mathbb{F}_0}\xi$) if and only if*

$$\xi \times \mathrm{S}_0\tau \geq \mathrm{T}(\xi \times \tau)$$

for every convergence τ.

If now $\mathbb{J} = \mathbb{F}_1$ and $\mathbb{D} = \mathbb{F}_0$, then

Corollary XXVI.5.5. *The following statements are equivalent:*

(1) *$\xi \times \mathrm{S}_0\tau \geq \mathrm{T}(\xi \times \tau)$ for each convergence τ of countable character,*
(2) *$\xi \geq \mathrm{SQ}^{\mathbb{F}_1}_{\mathbb{F}_0}\mathrm{Epi}^{\mathrm{T}}_{\mathrm{I}_1}\xi$.*

Finally, if $\mathbb{J} = \mathbb{F}_1$ and $\mathbb{D} = \mathbb{F}_1$, then

Corollary XXVI.5.6. *The following statements are equivalent:*

(1) $\xi \times \mathrm{S}_1\tau \geq \mathrm{T}(\xi \times \tau)$ *for each convergence* τ *of countable character,*

(2) $\xi \geq \mathrm{SQ}_{\mathbb{F}_1}^{\mathbb{F}_1}\mathrm{Epi}_{\mathrm{I}_1}^{\mathrm{T}}\xi$.

COROLLARY XXVI.5.7. *Let* $\mathbb{D}$ *and* $\mathbb{J}$ *be composable classes of filters. The following conditions are equivalent:*

(1) $\xi \geq \mathrm{SQ}_{\mathbb{D}}^{\mathbb{J}}\mathrm{Epi}_{\mathrm{B}_{\mathbb{J}}}^{\mathrm{T}}\xi$,

(2) $\xi \times \mathrm{A}_{\mathbb{D}}\tau \geq \mathrm{T}(\xi \times \tau)$ *for every* $\mathbb{J}$*-founded convergence* τ,

(3) $\xi \times \mathrm{A}_{\mathbb{D}}\tau \geq \mathrm{T}(\xi \times \tau)$ *for every* $\tau \geq \mathrm{A}_{\mathbb{D}}\mathrm{B}_{\mathbb{J}}\tau$,

(4) $\mathrm{A}_{\mathbb{D}}\mathrm{B}_{\mathbb{J}}[\xi, \sigma] \geq [\xi, \sigma]$ *for every* $\sigma = \mathrm{T}\sigma$,

(5) $\mathrm{A}_{\mathbb{D}}\mathrm{B}_{\mathbb{J}}[\xi, \$] \geq [\xi, \$]$.

There exist long-established classes of topologies that can be characterized with the aid of the modifier $\mathrm{Q}_{\mathbb{D}}^{\mathbb{J}}$ for particular classes $\mathbb{D}$ and $\mathbb{J}$.

For example, let ξ be a topology on X. A *q-sequence* is a descending sequence $(A_n)_n$ of subsets of X such that there is a countably ξ-compact set K with $(A_n)_n \geq \mathcal{N}_\xi(K)$ and $K \subset \bigcap_{n<\omega} A_n$. A topology is called *bi-quasi-q* if every convergent filter includes a q-sequence.

PROPOSITION XXVI.5.8. *A regular topology* ξ *is bi-quasi-q if and only if*

$$\xi \geq \mathrm{SQ}_{\mathbb{F}_1}^{\mathbb{F}_1}\xi.$$

PROOF. Suppose that ξ is a regular bi-quasi-q topology, and let $\mathcal{U}$ be an ultrafilter such that $x \in \lim_\xi \mathcal{U}$. Then $\mathcal{U} \geq \mathcal{N}_\xi(x)$ and there is a q-sequence $(A_n)_n$ such that $\mathcal{U} \geq (A_n)_n \geq \mathcal{N}_\xi(K)$, where K is a countably ξ-compact set included in $\bigcap_{n<\omega} A_n$. If V is a ξ-closed element of $\mathcal{U}$, then $V \# \mathcal{N}_\xi(K)$, equivalently $\mathrm{cl}_\xi V \cap K = V \cap K \neq \varnothing$. Then $(A_n \cap V)_n$ is countably ξ-compact at V. Indeed, if $\mathcal{D} \in \mathbb{F}_1$ meshes $(A_n \cap V)_n$, then $(\mathcal{D} \vee V) \# \mathcal{N}_\xi(K)$, so that $\mathrm{cl}_\xi^\natural(\mathcal{D} \vee V) \# K$, and thus $\mathrm{adh}_\xi(\mathcal{D} \vee V) \cap V \neq \varnothing$, because $\mathrm{cl}_\xi V \subset V$.

Conversely, assume $\xi \geq \mathrm{SQ}_{\mathbb{F}_1}^{\mathbb{F}_1}\xi$ and let $\mathcal{U}$ be an ultrafilter such that $x \in \lim_\xi \mathcal{U}$. For each ξ-closed $V \in \mathcal{N}_\xi(x)$, there a countably based filter $\mathcal{C}_V \leq \mathcal{U}$ that is countably ξ-compact at V. Then $\mathrm{cl}_\xi^\natural \mathcal{C}_V$ is a q-sequence included in $\mathcal{U}$ with $K := V \cap \mathrm{adh}_\xi \mathcal{C}_V = V \cap \mathrm{adh}_\xi(\mathrm{cl}_\xi^\natural \mathcal{C}_V)$. By assumption, K is non-empty.

To see that K is countably ξ-compact, let $\mathcal{D}$ be countably based with $K \in \mathcal{D}^\#$. As $(\mathcal{D} \vee K) \# \mathrm{cl}_\xi^\natural \mathcal{C}_V$, equivalently $\mathcal{N}_\xi\{\mathcal{D} \vee K\} \# \mathcal{C}_V$, we conclude that $\mathrm{adh}_\xi \mathcal{N}_\xi\{\mathcal{D} \vee K\} \cap V \neq \varnothing$. By regularity and in view of Proposition XIX.3.13, $\varnothing \neq \mathrm{adh}_\xi \mathcal{N}_\xi\{\mathcal{D} \vee K\} \subset \mathrm{adh}_\xi \mathcal{D} \vee K \subset K$, because K is ξ-closed, and $\mathrm{adh}_\xi(\mathcal{D} \vee K) \subset \mathrm{adh}_\xi \mathcal{D}$. Hence, $\mathrm{adh}_\xi \mathcal{D} \cap K \neq \varnothing$.

If O is ξ-open such that $K \subset O$, then $O \in \mathrm{cl}_\xi^\natural \mathcal{C}_V$, for otherwise $O^c \in (\mathrm{cl}_\xi^\natural \mathcal{C}_V)^\#$, and thus $\mathrm{cl}_\xi^\natural \mathcal{C}_V \vee O^c$ is countably based and countably

ξ-compact at V. Hence $\operatorname{adh}_\xi(\operatorname{cl}_\xi^\natural \mathcal{C}_V \vee O^c) \cap V \neq \varnothing$. On the other hand, $\operatorname{adh}_\xi \operatorname{cl}_\xi^\natural \mathcal{C}_V \cap O^c \cap V = \varnothing$, a contradiction. □

XXVI.6. Strongly and productively sequential convergences

Recall that a convergence ξ is called *sequential* if each sequentially ξ-closed set is ξ-closed, which amounts to (XXI.4.1)

$$\text{(sequential)} \qquad \xi \geq \mathrm{TI}_1\xi.$$

If ξ is assumed to be a topology, that is, $\mathrm{T}\xi \geq \xi$, this inequality is equivalent to the equality $\xi = \mathrm{TI}_1\xi = \mathrm{TSeq}\xi$.

LEMMA XXVI.6.1. *If* $\mathbb{H} \subset \mathbb{F}_1^{\mathbb{F}_1}$, *then* $\mathrm{A}_{\mathbb{H}}\mathrm{Seq} = \mathrm{A}_{\mathbb{H}}\mathrm{I}_1$.

PROOF. As $\mathrm{I}_1 \leq \mathrm{Seq}$, we need only to prove that $\mathrm{A}_{\mathbb{H}}\mathrm{Seq} \leq \mathrm{A}_{\mathbb{H}}\mathrm{I}_1$. So let $x \in \lim_{\mathrm{A}_{\mathbb{H}}\mathrm{I}_1\xi} \mathcal{F}$ and $\mathcal{H} \in \mathbb{H}$ be such that $\mathcal{H}\#\mathcal{F}$. Then $x \in \operatorname{adh}_{\mathrm{I}_1\xi} \mathcal{H}$, that is, there exists $\mathcal{L} \in \mathbb{F}_1$ such that $\mathcal{L}\#\mathcal{H}$ and $x \in \lim_\xi \mathcal{L}$. As $\mathcal{H} \in \mathbb{F}_1^{\mathbb{F}_1}$, there exists $\mathcal{E} \in \mathbb{S}(\mathcal{L} \vee \mathcal{H})$. Consequently, $x \in \lim_\xi \mathcal{E}$ and thus $x \in \operatorname{adh}_{\mathrm{Seq}\xi} \mathcal{H}$, and thus $x \in \lim_{\mathrm{A}_{\mathbb{H}}\mathrm{Seq}\xi} \mathcal{F}$. □

We are now in a position to tackle the productivity problem of sequential topologies. A convergence ξ is called *strongly sequential* if

$$\text{(strongly sequential)} \qquad \xi \geq \mathrm{Epi}_{\mathrm{I}_1}^{\mathrm{T}}\mathrm{I}_1\xi,$$

where $\mathrm{Epi}_{\mathrm{I}_1}^{\mathrm{T}}$ is the I_1-modified T-bidual functor, defined by (XXVI.1.1) and (XXVI.1.2).

Given a class $\mathbb{H}$ of filters, w denote by $\mathbb{H}_\theta := \{\mathcal{H} \in \mathbb{H} : \operatorname{adh}_\theta^\natural \mathcal{H} = \mathcal{H}\}$ the set of θ-regular filters from the class $\mathbb{H}$.

PROPOSITION XXVI.6.2. *A topology* ξ *is strongly sequential if and only if*

$$\text{(XXVI.6.1)} \qquad \operatorname{adh}_\xi \mathcal{H} \subset \operatorname{cl}_{\mathrm{Seq}\xi}(\operatorname{adh}_{\mathrm{Seq}\xi} \mathcal{H})$$

for every $\mathcal{H} \in (\mathbb{F}_1)_{\xi^\bullet}$.

PROOF. A topology ξ is strongly sequential if and only if $\xi = \mathrm{Epi}_{\mathrm{I}_1}^{\mathrm{T}}\mathrm{I}_1\xi$. By Lemma XXVI.5.2 applied with $\mathbb{J} = \mathbb{F}_1$,

$$\lim_{\mathrm{Epi}_{\mathrm{I}_1}^{\mathrm{T}}\mathrm{I}_1\xi} \mathcal{F} = \bigcap_{(\mathbb{F}_1)_{\xi^\bullet} \ni \mathcal{H}\#\mathcal{F}} \operatorname{cl}_{\mathrm{I}_1\xi}(\operatorname{adh}_{\mathrm{I}_1\xi} \mathcal{H}).$$

Therefore, $\xi = \mathrm{Epi}_{\mathrm{I}_1}^{\mathrm{T}}\mathrm{I}_1\xi$ if and only if $\operatorname{adh}_\xi \mathcal{H} \subset \operatorname{cl}_{\mathrm{I}_1\xi}(\operatorname{adh}_{\mathrm{I}_1\xi} \mathcal{H}) = \operatorname{cl}_{\mathrm{Seq}\xi}(\operatorname{adh}_{\mathrm{Seq}\xi} \mathcal{H})$ for every countably based filter $\mathcal{H}$ such that $\operatorname{cl}_{\xi^\bullet}^\natural \mathcal{H} \geq \mathcal{H}$, where $\operatorname{cl}_{\xi^\bullet} H = \bigcup_{x \in H} \operatorname{cl}_\xi\{x\}$ (see Section XXV.6). From Lemma XXVI.6.1 applied to $\mathbb{H} = (\mathbb{F}_1)_{\xi^\bullet}$, we deduce that $\operatorname{cl}_{\mathrm{I}_1\xi}(\operatorname{adh}_{\mathrm{I}_1\xi} \mathcal{H}) = \operatorname{cl}_{\mathrm{Seq}\xi}(\operatorname{adh}_{\mathrm{Seq}\xi} \mathcal{H})$. □

COROLLARY XXVI.6.3. *A strongly Fréchet convergence is strongly sequential.*

PROOF. Indeed, it follows from $\mathrm{S}_0 \geq \mathrm{T}$ that $\mathrm{Epi}_F^{\mathrm{S}_0} \geq \mathrm{Epi}_F^{\mathrm{T}}$ for every functor F. In particular, $\mathrm{S}_1 = \mathrm{Epi}_{\mathrm{I}_1}^{\mathrm{S}_0} \geq \mathrm{Epi}_{\mathrm{I}_1}^{\mathrm{T}}$, so that $\xi \geq \mathrm{S}_1\mathrm{I}_1\xi \geq \mathrm{Epi}_{\mathrm{I}_1}^{\mathrm{T}}\mathrm{I}_1\xi$. Hence, if ξ is strongly Fréchet, it is also strongly sequential. $\square$

On the other hand, among free Fréchet convergences, the classes of strongly sequential and of strongly Fréchet convergences coincide:

PROPOSITION XXVI.6.4. *A free strongly sequential and Fréchet convergence is strongly Fréchet.*

PROOF. If ξ is strongly sequential and free then (XXVI.6.1) holds for every $\mathcal{H} \in \mathbb{F}_1$. To show that $\mathrm{adh}_\xi \mathcal{H} \subset \mathrm{adh}_{\mathrm{Seq}\xi}\mathcal{H}$, let $(H_n)_n$ be a decreasing filter base of $\mathcal{H}$ and let $x \in \mathrm{adh}_\xi \mathcal{H}$. In view of (XXVI.6.1) and since ξ is Fréchet, there is a sequence $(x_n)_n$ of distinct terms of $\mathrm{adh}_{\mathrm{Seq}\xi}\mathcal{H}$ converging to x. For each n, there is a sequence $(x_{n,k})_k$ finer than $\mathcal{H}$ and converging to x_n. As ξ is Fréchet and $x \in \mathrm{cl}_{\mathrm{Seq}\xi}\{x_{n,k} : n,k < \omega\}$ there is a sequence $(x_{n_j,k_j})_j$ converging to x, which is finer than $\mathcal{H}$, hence $x \in adh_{\mathrm{Seq}\xi}\mathcal{H}$. $\square$

THEOREM XXVI.6.5 (Mynard). *A convergence ξ is strongly sequential if and only if $\xi \times \tau$ is strongly sequential for every $\tau = \mathrm{I}_1\tau$ if and only if $\xi \times \tau$ is sequential for every prime topology of countable character τ.*

PROOF. Corollary XXVI.2.1 with $J = \mathrm{T}$ and $F = E = \mathrm{I}_1$ amounts to the equivalence of

(1) $\xi \geq \mathrm{Epi}_{\mathrm{I}_1}^{\mathrm{T}}\mathrm{I}_1\xi$,
(2) $\xi \times \tau \geq \mathrm{Epi}_{\mathrm{I}_1}^{\mathrm{T}}\mathrm{I}_1(\xi \times \tau)$ for each convergence $\tau = \mathrm{I}_1\tau$,
(3) $\xi \times \tau \geq \mathrm{T}\mathrm{I}_1(\xi \times \tau)$ for each prime topology $\tau = \mathrm{I}_1\tau$.

$\square$

We shall see that countably compact, sequential topologies are strongly sequential. Actually, we shall provide a more precise statement.

We say that a topology ξ has the *Tanaka property* if $\mathrm{adh}_\xi \mathcal{H} \neq \varnothing$ implies that $\mathrm{adh}_{\mathrm{Seq}\xi} \mathcal{H} \neq \varnothing$ for every countably based filter $\mathcal{H}$.

Notice that $\mathrm{adh}_\xi \mathcal{H} \subset \mathrm{adh}_\xi(\mathrm{cl}_{\xi\bullet}^{\natural} \mathcal{H})$, and if ξ is free (alias T_1), then $\mathrm{cl}_{\xi\bullet}^{\natural} \mathcal{H} = \mathcal{H}$, for every filter $\mathcal{H}$. Therefore, each free strongly sequential topology has the Tanaka property by (XXVI.6.1).

EXAMPLE XXVI.6.6. A bisequence topology is sequential but does not have the Tanaka property. As described in Examples VI.1.8, VII.4.6, the *bisequence pretopology* θ is defined on $X = \{x_\infty\} \sqcup \{x_n : n < \omega\} \sqcup \{x_{n,k} : n, k < \omega\}$ (of distinct points) to the effect that $x_\infty \in \lim_\theta (x_n)_n$, and $x_n \in \lim_\theta (x_{n,k})_k$ for each $n < \omega$, while each $x_{n,k}$ is isolated. The *bisequence topology* $\mathrm{T}\theta$ is sequential (but not Fréchet). It is not a Tanaka topology. Indeed, let $\mathcal{H}$ be a filter such that $\{\{x_{n,k} : n > m\} : m < \omega\}$ is the filter-base, then $\mathcal{H}$ is countably based, $x_\infty \in \operatorname{adh}_{\mathrm{T}\theta} \mathcal{H}$, but $\operatorname{adh}_{\mathrm{SeqT}\theta} \mathcal{H} = \varnothing$.

PROPOSITION XXVI.6.7. *Each regular sequential topology with the Tanaka property is strongly sequential.*

PROOF. Let $\mathcal{H}$ be a countably based filter such that $x \in \operatorname{adh}_\xi \mathcal{H}$. Therefore, $x \in \operatorname{adh}_\xi(\mathcal{H} \vee W)$ for each $W \in \mathcal{N}_\xi(x)$. By Tanaka, there is a sequential filter $\mathcal{E}_W \# (\mathcal{H} \vee W)$ such that $\lim_\xi \mathcal{E}_W \cap \operatorname{cl}_\xi W \neq \varnothing$. Therefore, $\bigcup_{W \in \mathcal{N}_\xi(x)} \lim_\xi \mathcal{E}_W \in (\operatorname{cl}_\xi^\natural \mathcal{N}_\xi(x))^\#$, hence $\operatorname{adh}_{\mathrm{Seq}\xi} \mathcal{H} \in (\operatorname{cl}_\xi^\natural \mathcal{N}_\xi(x))^\#$, and since ξ is regular, $x \in \operatorname{cl}_\xi(\operatorname{adh}_{\mathrm{Seq}\xi} \mathcal{H})$. Finally, as ξ is a sequential topology, $x \in \operatorname{cl}_{\mathrm{Seq}\xi}(\operatorname{adh}_{\mathrm{Seq}\xi} \mathcal{H})$. □

EXAMPLE XXVI.6.8 (Tanaka, non-strongly-sequential topology). Consider X as in Example XXVI.6.6, and let $X_n := \{x_{n,k} : k < \omega\}$ for each $n < \omega$. Let $\mathcal{A} = \{A_y : y \in Y\}$ be a family of infinite subsets of $\{x_{n,k} : n, k < \omega\}$ such that $\mathcal{A} \sqcup \{X_n : n < \omega\}$ is *MAD*. On $X \sqcup Y$ let τ be the topology such that $\tau|_X = \mathrm{T}\theta$, and $y \in \lim_\tau (A_y)_0$ for each $y \in Y$. The topology τ is Hausdorff and sequential. Moreover, τ is a Tanaka topology. As all the points but x_∞ are of countable character, it is enough to consider a filter $\mathcal{H} \approx \{H_p : p < \omega\}$ on $\{x_{n,k} : n, k < \omega\}$ such that $x_\infty \in \operatorname{adh}_\tau \mathcal{H}$. Then for each $p < \omega$, there is $w_p \in H_p \setminus \bigcup_{n \leq p} X_n$ so that $\{w_p : p < \omega\}$ is infinite. By maximality, there is either $y \in Y$ such that $A_y \cap \{w_p : p < \omega\}$ is infinite, and thus $y \in \operatorname{adh}_{\mathrm{Seq}\tau} \mathcal{H}$. However, τ is not strongly sequential. Indeed, if $\mathcal{X} \approx \{\bigcup_{p<n<\omega} X_n : p < \omega\}$, then $x_\infty \in \operatorname{adh}_\tau \mathcal{X}$, but $\operatorname{adh}_{\mathrm{Seq}\tau} \mathcal{X} \subset Y$, which is a τ-discrete set, so that $x_\infty \notin \operatorname{cl}_\tau \operatorname{adh}_{\mathrm{Seq}\tau} \mathcal{X} \supset \operatorname{cl}_{\mathrm{Seq}\tau} \operatorname{adh}_{\mathrm{Seq}\tau} \mathcal{X}$.

PROPOSITION XXVI.6.9. *Each regular sequential bi-quasi-q topology is strongly sequential.*

PROOF. Assume that ξ fulfills the assumptions, and consider a filter $\mathcal{H} \approx \{H_n : n < \omega\}$, where $H_n \supset H_{n+1}$ for each $n < \omega$. Let $x \in \operatorname{adh}_\xi \mathcal{H}$. Let $(A_n)_n$ be descending sequence and K be a countably ξ-compact set such that $(A_n)_n \geq \mathcal{N}_\xi(K)$ and $K \subset \bigcap_{n<\omega} A_n$. If $x_n \in A_n \cap H_n$, then $\operatorname{adh}_\xi (x_n)_n \neq \varnothing$, and thus $\{x_n : n < \omega\}$ is not ξ-closed, hence not sequentially ξ-closed, so that $(x_n)_n$ has a ξ-convergent subsequence, and thus $\mathrm{adh}_{\mathrm{Seq}\xi}\,\mathcal{H} \neq \varnothing$. Therefore, ξ is a Tanaka topology, so that by Proposition XXVI.6.7, ξ is strongly sequential. □

In particular,

COROLLARY XXVI.6.10. *Every regular, locally countably compact, free (alias T_1) sequential topology is strongly sequential.*

Hence, in virtue of Proposition XXVI.6.4,

COROLLARY XXVI.6.11. *Every regular, locally countably compact, free Fréchet topology is strongly Fréchet.*

A convergence ξ is called *productively sequential* if $\xi \times \tau$ is strongly sequential (equivalently, sequential) for every strongly sequential convergence τ.

Each productively sequential convergence is strongly sequential, because, if ξ is productively sequential, then, in particular, $\xi \times \tau$ is sequential for every convergence τ of countable character.

PROPOSITION XXVI.6.12. *Finite products of productively sequential convergences are productively sequential.*

PROOF. If ξ_0 and ξ_1 are productively sequential and τ is strongly sequential, then

$$(\xi_0 \times \xi_1) \times \tau = \xi_0 \times (\xi_1 \times \tau), \tag{XXVI.6.2}$$

where $\xi_1 \times \tau$ is strongly sequential, because ξ_1 is productively sequential and τ is strongly sequential. Hence, (XXVI.6.2) is strongly sequential, which proves that $\xi_0 \times \xi_1$ is productively sequential. Conclude the proof by obvious induction. □

LEMMA XXVI.6.13. *If ξ is productively sequential, then*

$$\xi \times \mathrm{Epi}^{\mathrm{T}}_{\mathrm{I}_1}\sigma \geq \mathrm{T}(\mathrm{I}_1\xi \times \sigma) \tag{XXVI.6.3}$$

for each $\sigma = \mathrm{I}_1\sigma$.

PROOF. Let us show this first for $\sigma = \mathrm{I}_1\mathrm{Epi}^{\mathrm{T}}_{\mathrm{I}_1}\sigma$. Let $\tau = \mathrm{Epi}^{\mathrm{T}}_{\mathrm{I}_1}\sigma$, and thus $\sigma = \mathrm{I}_1\tau$. Hence $\tau = \mathrm{Epi}^{\mathrm{T}}_{\mathrm{I}_1}\mathrm{I}_1\tau$, that is, τ is strongly sequential. By assumption, ξ is productively sequential, that is,

$$\xi \times \tau \geq \mathrm{TI}_1(\xi \times \tau) = \mathrm{T}(\mathrm{I}_1\xi \times \mathrm{I}_1\tau).$$

As $\tau = \mathrm{Epi}^{\mathrm{T}}_{\mathrm{I}_1}\sigma$, and $\sigma = \mathrm{I}_1\tau$, we get (XXVI.6.3). We apply now Lemma XXVI.1.14 with $E = \mathrm{I}_1, J = \mathrm{T}$ and $F = \mathrm{Epi}^{\mathrm{T}}_{\mathrm{I}_1}$, on replacing θ by ξ, and ξ by $\mathrm{I}_1\xi$. □

THEOREM XXVI.6.14. *If ξ is a regular topology, then the following statements are equivalent:*

(1) ξ *is productively sequential,*
(2) $\xi \times \tau$ *is strongly sequential for each strongly sequential convergence* τ,
(3) $\xi \times \tau$ *is sequential for each strongly Fréchet convergence* τ,
(4) $\xi \geq \mathrm{SQ}_{\mathbb{F}_1}^{\mathbb{F}_1} \mathrm{Epi}_{\mathrm{I}_1}^{\mathrm{T}} \mathrm{I}_1 \xi$,
(5) ξ *is bi-quasi-q and strongly sequential,*
(6) ξ *is bi-quasi-q and sequential.*

PROOF. (1) $\Leftrightarrow$ (2), by definition, and clearly imply (3).

(3) $\Rightarrow$ (4). We notice that (3) implies

$$\tau = \mathrm{S}_1 \mathrm{I}_1 \tau \Longrightarrow \xi \times \tau \geq \mathrm{T}\mathrm{I}_1(\xi \times \tau) = \mathrm{T}(\mathrm{I}_1 \xi \times \mathrm{I}_1 \tau). \tag{XXVI.6.4}$$

Assume that $\sigma = \mathrm{I}_1 \mathrm{S}_1 \sigma$ and let $\tau := \mathrm{S}_1 \sigma$. Then $\tau = \mathrm{S}_1 \mathrm{I}_1 \mathrm{S}_1 \sigma = \mathrm{S}_1 \mathrm{I}_1 \tau$, hence by (XXVI.6.4), $\xi \times \tau \geq \mathrm{T}(\mathrm{I}_1 \xi \times \mathrm{I}_1 \tau)$, that is,

$$\sigma = \mathrm{I}_1 \mathrm{S}_1 \sigma \Longrightarrow \xi \times \mathrm{S}_1 \sigma \geq \mathrm{T}(\mathrm{I}_1 \xi \times \sigma),$$

because $\sigma = \mathrm{I}_1 \mathrm{S}_1 \sigma$. By Lemma XXVI.1.14, this is equivalent to

$$\sigma = \mathrm{I}_1 \sigma \Longrightarrow \xi \times \mathrm{S}_1 \sigma \geq \mathrm{T}(\mathrm{I}_1 \xi \times \sigma). \tag{XXVI.6.5}$$

By Theorem XXVI.5.3, applied with $\mathbb{D} = \mathbb{J} = \mathbb{F}_1$, ξ replaced by $\mathrm{I}_1 \xi$, $\theta := \xi$ and $\tau := \sigma$, (XXVI.6.5) amounts to

$$\xi \geq \mathrm{SQ}_{\mathbb{F}_1}^{\mathbb{F}_1} \mathrm{Epi}_{\mathrm{I}_1}^{\mathrm{T}} \mathrm{I}_1 \xi.$$

(4) $\Rightarrow$ (5). As $\mathrm{SQ}_{\mathbb{F}_1}^{\mathbb{F}_1}$ and $\mathrm{Epi}_{\mathrm{I}_1}^{\mathrm{T}} \mathrm{I}_1$ are ascending on topologies, $\xi \geq \mathrm{SQ}_{\mathbb{F}_1}^{\mathbb{F}_1} \mathrm{Epi}_{\mathrm{I}_1}^{\mathrm{T}} \mathrm{I}_1 \xi$ implies $\xi \geq \mathrm{SQ}_{\mathbb{F}_1}^{\mathbb{F}_1} \xi$ and $\xi \geq \mathrm{Epi}_{\mathrm{I}_1}^{\mathrm{T}} \mathrm{I}_1 \xi$ and the result follows from Proposition XXVI.5.8.

Of course, (5) $\Rightarrow$ (6). Finally (6) $\Rightarrow$ (2) is rather technical and we refer the reader to [**249**, Theorem 12-(2)i]. □

CHAPTER XXVII

Outline of principles

Here I gather most fundamental facts from set theory ([1]).

Set theory is a foundation of modern mathematics. Axiomatic set theory was developed to avoid ambiguities and contradictions that occurred as a consequence of the lack of precision of informal language used in mathematics in the past.

An intuitive notion of *set* that is determined by its *elements* is vague and gives rise to paradoxes. For instance, consider $Y := \{x : x \notin x\}$. If Y were a set, then we should be able to decide whether a given set belongs to Y. In particular, if $Y \in Y$, then it follows from the description of Y that $Y \notin Y$; on the other hand, if $Y \notin Y$, then it follows from the description of Y that $Y \in Y$. This is the famous *Russell's paradox*.

Axiomatic set theory delimits objects that can be qualified as sets in order to avoid contradictions. This theory is deductive, that is, based on a collection of *primitive notions* and *axioms*, that is, *primitive propositions*.

The primitive notions are undefined, and the axioms are assumed to be valid. A meaning of a primitive notions is derived from the set of all axioms. All the other notions are defined from the primitive notions, and all the other propositions are inferred from the axioms with the aid of rules of logic.

The primitive notions of set theory are the *set* and the *membership* relation $\in$. Actually, each object of the theory is made up of sets. Propositions are built with the aid of formulae, which are constructed from two *atomic formulae*

$$x \in y, \qquad x = y$$

by means of logical connectives,

$$\neg, \vee, \wedge, \Rightarrow, \Leftrightarrow$$

[1]A principal, classical reference book is that of THOMAS JECH ([**136**]), which is written in a concise way, is suitable for a rather advanced reader; this chapter represents a fraction of its first chapter.

(negation, alternative, conjunction, implication, and equivalence) and quantifiers $\exists, \forall$ (existential and universal). The set of connective displayed above is redundant; for example, the connective $\Rightarrow$ can be derived from $\neg$ and $\vee$. A really smart irreducible set of connectives, however, is

$$\neg, \Rightarrow .$$

Effectively, $\alpha \vee \beta$ is equivalent to $\neg\alpha \Rightarrow \beta$, then $\alpha \wedge \beta$ amounts to $\neg(\neg\alpha \vee \neg\beta)$, and $\alpha \Leftrightarrow \beta$ to $(\alpha \Rightarrow \beta) \wedge (\beta \Rightarrow \alpha)$, for example.

The system of axioms *ZFC* (*Zermelo* and *Fraenkel* with the *Axiom of Choice*) presented here is generally adopted.

The notion of *universe* V that consists of all sets is not a set, but a *class*, that is, is *external* to the set theory. Therefore, the relation $X \in V$, meaning that X is a set, is also external.

XXVII.1. Axioms of set theory

(1) AXIOMS OF EXTENSIONALITY states that X, Y are sets, then $X = Y$ if and only if, for each z,

$$z \in X \Longleftrightarrow z \in Y.$$

(2) AXIOM SCHEMA OF REPLACEMENT. Let $\varphi(x, p)$ be a formula (with parameter p), then for each p and every set X, there exists a set Y such that

$$Y = \{x : x \in X \wedge \varphi(x, p)\}.$$

If $\varphi(x, p_0) = \varphi(x, p_1)$ implies that $p_0 = p_1$, then $\varphi(x, y)$ defines a map Φ by $y = \Phi(x)$ if and only if $\varphi(x, y)$. In other terms, for every set X,

$$\Phi(X) := \{\Phi(x) : x \in X\}$$

is a set.

(3) AXIOM OF SEPARATION ([2]) stipulates that for every formula $\psi(x)$ and for every set X, there exists a set Y such that

$$Y = \{x \in X : \psi(x)\}.$$

(4) AXIOM OF UNION states that for every set X, there exists a set $\bigcup X$ such that

$$y \in \bigcup X \Longleftrightarrow \underset{A \in X}{\exists}\, y \in A.$$

[2] If $\psi(x) = \varphi(x, p)$ is a formula independent of p, then Axiom of replacement and the existence of a set (see later) imply Axiom of separation.

(5) AXIOM OF POWER asserts that for every set X, there exists a set Y such that

$$Y \in 2^X \Longleftrightarrow Y \subset X.$$

(6) AXIOM OF REGULARITY affirms that each non-empty set has an $\in$-minimal element.
It implies that no set is an element of itself. In other words, $x \neq \{x\}$ for every set x. It follows from the Axioms of regularity and separation that the set of all sets does not exist.

(7) AXIOM OF INFINITY postulates the existence of an infinite set.
The axiom of infinity implies the existence of a set, which implies the existence of an empty set, denoted by $\varnothing$. Indeed, if X is a set, then $\{x \in X : x \neq x\}$ is a set by Axiom of Separation, and is empty by Axiom of extensionality.

(8) AXIOM OF CHOICE declares that for every set X of non-empty elements, there exists a map $f : X \longrightarrow \bigcup X$ such that $f(A) \in A$ for each $A \in X$.

REMARK XXVII.1.1. Axiom 3 (Separation) prevents from paradoxes as above. By taking $\psi(x) = x \notin x$, we conclude that if X is a set, then $Y = \{x \in X : x \notin x\}$ is a set. By Axiom 6 (Regularity) $Y \notin Y$ yields no contradiction, because $Y \notin X$.

In some situations, we consider classes *definable* by a formula

$$\mathbf{A} := \{x : \varphi(x)\}.$$

For instance, while the class $\mathbb{F}$ of all filters is not a set, while $\mathbb{F}X$ is a set of filters on X for every set X. The *universal class* or the *universe* is defined by

$$V := \{x : x = x\}.$$

Operations on classes are defined naturally

$$\mathbf{A} \subset \mathbf{B} \Longleftrightarrow \forall x\, (x \in \mathbf{A} \Rightarrow x \in \mathbf{B}),$$
$$\mathbf{A} \cap \mathbf{B} := \{x : x \in \mathbf{A} \wedge x \in \mathbf{B}\},$$
$$\mathbf{A} \cup \mathbf{B} := \{x : x \in \mathbf{A} \vee x \in \mathbf{B}\},$$
$$\mathbf{A}^c := \{x : x \notin \mathbf{A}\}.$$

XXVII.2. Cardinality

Two sets X, Y are said to be *equipotent* $X \sim Y$ if there exists a bijection $f : X \longrightarrow Y$. For any sets X, Y, Z,

$$X \sim X,$$
$$X \sim Y \Longrightarrow Y \sim X,$$
$$X \sim Y \wedge Y \sim Z \Longrightarrow X \sim Z.$$

Indeed, the identity map $i_X : X \longrightarrow X$ is a bijection; if $f : X \longrightarrow Y$ is a bijection, then $f^{-1} : Y \longrightarrow X$ is also a bijection; finally, a composition of two bijections is a bijection.

Therefore, the relation $\sim$ of *equipotence* is an equivalence relation on the universe of sets.

I shall use this equivalence to introduce *cardinal numbers*. Consequently, because of the use of classes of sets, this introduction will not be purely set-theoretic. However, very likely, this way of presenting cardinal numbers is more accessible than rigorous set-theoretic procedure passing through ordinals (see Section XXVII.6) that is commonly adopted nowadays.

DEFINITION XXVII.2.1. An equivalence class with respect to equipotence $\sim$ is called a *cardinal number* or a *cardinal*.

We denote by $\operatorname{card} X$ the equivalence class containing X, called the *cardinality* of X. Traditionally, small Greek letters $\kappa, \lambda, \mu, \ldots$ are used to denote cardinalities.

DEFINITION XXVII.2.2. Let X, Y be sets. Then $\operatorname{card} X \leq \operatorname{card} Y$ whenever there is an injection $f : X \longrightarrow Y$.

In other words, if κ, λ are cardinals, then $\kappa \leq \lambda$ means that if X, Y are sets such that $\kappa = \operatorname{card} X$ and $\lambda = \operatorname{card} Y$, then there exists an injection $f : X \longrightarrow Y$.

It is clear that the relation $\leq$ is reflexive and transitive. Next theorem shows that it is also antisymmetric, hence an order relation.

THEOREM XXVII.2.3 (Cantor-Bernstein). *Let κ, λ be cardinal numbers. If $\kappa \leq \lambda$ and $\lambda \leq \kappa$, then $\kappa = \lambda$.*

PROOF.[3] Let X, Y be sets such that $\kappa = \operatorname{card} X$ and $\lambda = \operatorname{card} Y$. As $\kappa \leq \lambda$, there exists an injective map $f_0 : X \longrightarrow Y$, and since $\lambda \leq \kappa$, there exists an injective map $f_1 : Y \longrightarrow X$. Then $f = f_1 \circ f_0 : X \longrightarrow X$ is an injection, and $X \supset f_1(Y) \supset f(X)$.

[3] This proof follows that of [**136**, Theorem 7] by T. JECH, detailing the arguments that were sketched therein.

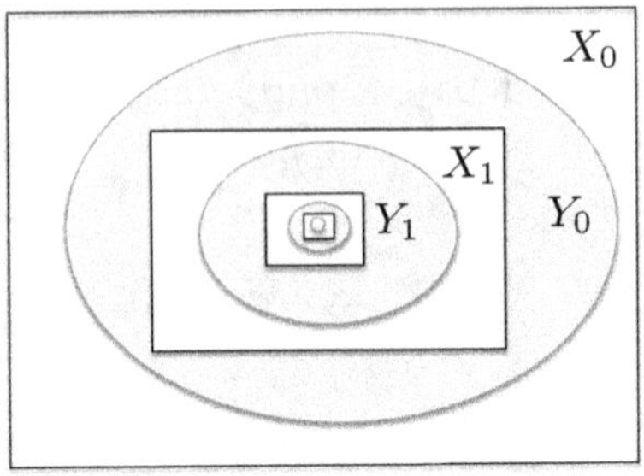

FIGURE XXVII.2.1. Here Z is represented by the white area.

Let $X_0 := X, Y_0 := f_1(Y)$, and by induction,

$$X_{n+1} := f(X_n), \text{ and } Y_{n+1} := f(Y_n).$$

Accordingly, $f : X_0 \longrightarrow X_1$ is a bijection. Let $Z := \bigcup_{n<\omega}(X_n \setminus Y_n)$ and notice that $f(Z) \subset Z$. Now let $h : X \longrightarrow Y$ be given by

$$h(x) := \begin{cases} f(x) & x \in Z, \\ x & x \in X \setminus Z. \end{cases}$$

The map h is injective, because $f(Z) \subset Z$, and the restriction $f|$of an injective map f is injective; on the other hand, the restriction $h|_{X\setminus Z}$ of h is the identity $i_{X\setminus Z}$. Finally, h is surjective, because $h(Z) = X_1 \cap Z$ and $h(X \setminus Z) = Y \setminus Z$. Hence $h : X \longrightarrow Y$ is a bijection, so that $\kappa = \lambda$. □

We write $\kappa < \lambda$ if $\kappa \leq \lambda$ and $\lambda \neq \kappa$. By Cantor-Bernstein Theorem XXVII.2.3, this means that if $\kappa = \operatorname{card} X$ and $\lambda = \operatorname{card} Y$, then there is an injection $f : X \longrightarrow Y$, and there is no injection from Y to X.

We denote by $0 = \operatorname{card}(\varnothing)$, the cardinality of the empty set, and by 1, the cardinality of a singleton, that is $1 = \operatorname{card}\{\varnothing\}$.

REMARK XXVII.2.4. The cardinal 0 is the least cardinal number, that is, $0 \leq \kappa$ for each cardinal κ. To justify this claim, we need to show that there exists an injective map from $\varnothing$ to any other set X. By definition, a map $f : \varnothing \longrightarrow X$ is a relation $f \subset \varnothing \times X$ such that f^- is injective and surjective. As $\varnothing \times X = \varnothing$, we infer that $f = \varnothing$. As $f^-(X) = \varnothing^- X = \varnothing$, we conclude that $\varnothing^-$ is surjective. As $\varnothing^- x_0 \cap \varnothing^- x_1 = \varnothing$, it is also injective, hence an application, necessarily injective, for there is no $w \in \varnothing$.

PROPOSITION XXVII.2.5. *If* $\varnothing \neq X$, *then* $\operatorname{card} X \leq \operatorname{card} Y$ *if and only if there exists a surjection* $g : Y \longrightarrow X$.

PROOF. If $g : Y \longrightarrow X$ is a surjection, then by the Axiom of Choice, for each $x \in X$, there exists $f(y) \in g^{-}(y)$. Obviously, f is injective.

Conversely, if $f : X \to Y$ is an injection and $x_0 \in X$, then let $g : Y \longrightarrow X$ be defined by

$$g(y) := \begin{cases} f^{-1}(y) & y \in f(X), \\ x_0 & y \notin f(X). \end{cases}$$

Then g is surjective. □

Three arithmetic operations (addition, multiplication and power) are defined. If $\kappa = \operatorname{card} X$ and $\lambda = \operatorname{card} Y$, then

$$\begin{aligned} \kappa + \lambda &:= \operatorname{card}(X \cup Y) \text{ if } X \cap Y = \varnothing, \\ \kappa \cdot \lambda &:= \operatorname{card}(X \times Y), \\ \lambda^{\kappa} &:= \operatorname{card}(Y^X). \end{aligned}$$

DEFINITION XXVII.2.6. A set X is said to be *finite* if each injection $f : X \to X$ is a surjection.

REMARK XXVII.2.7. The empty set is finite, because $\varnothing \subset \varnothing \times \varnothing$, the only relation, is a bijective application. By induction, if κ is finite, then $\kappa + 1$ too. The least infinite cardinal is $\aleph_0$, defined in Chapter I, that is, every infinite subset of $\aleph_0$ is equipotent to it.

In fact, the shift $s : \mathbb{N} \longrightarrow \mathbb{N}$, defined by $s(n) := n + 1$ is injective, but not surjective, because $0 \notin s(\mathbb{N})$. It follows from the *Peano's axioms* ([4]) of natural numbers that it is the least one.

On the other hand,

PROPOSITION XXVII.2.8. *If κ is a cardinal, then $\kappa^0 = 1, 1^\kappa = 1$, and*

$$\kappa > 0 \Longrightarrow 0^\kappa = 0.$$

PROOF. By definition, $\kappa^0 = \operatorname{card}(Y^\varnothing)$, where $\operatorname{card} Y = \kappa$. The number of all subsets in $\varnothing \times Y = \varnothing$ is 1, and the only subset is $\varnothing$, and it is a map. Indeed, $\varnothing^-$ is injective, because $\varnothing^- y_0 \cap \varnothing^- y_0 = \varnothing$, and surjective, because $\varnothing^- Y = \varnothing$ is equal to the first factor, so that by Corollary I.1.6, $\varnothing \in Y^\varnothing$.

[4] G. PEANO in [**216**] adopts $\mathbb{N}, 0$ and an operation σ on $\mathbb{N}$ *primitive notions* and proposes six axioms: (P0) $\mathbb{N}$ is a set, (P1) $0 \in \mathbb{N}$, (P2) if $m \in \mathbb{N}$ then $\sigma(m) \in \mathbb{N}$, (P3) if S is a set such that $0 \in S$ and $\sigma(S) \subset S$, then $\mathbb{N} \subset S$, (P4) σ is injective, (P5) $0 \notin \sigma(\mathbb{N})$ (see also S. Dolecki and G. H. Greco [**71**, pp. 25–26]).

By definition, $1^\kappa = \operatorname{card}(\{\varnothing\}^X)$, where $\operatorname{card} X = \kappa$. There exists the only subset R of $X\times\{\varnothing\}$, for which R^- is surjective (it is $R = X\times\{\varnothing\}$). It is also injective, $(X \times \{\varnothing\})^- x = \{x\}$ for each $x \in X$. As a result, there exists a single map from any set to a singleton.

Finally, $0^\kappa = \operatorname{card}(\varnothing^X)$, where $\operatorname{card} X = \kappa$. As $X \times \varnothing = \varnothing$, the only subset of this product is $\varnothing$, but if $X \neq \varnothing$, then it is not a map. Indeed, $\varnothing^-\varnothing = \varnothing \neq X$. Hence, there is no map from a non-empty set to $\varnothing$. □

We deduce that, in particular,

$$0^0 = 1.$$

Here are some basic rules of arithmetics of cardinal numbers, that can be routinely verified. The operations of *addition* $+$ and of *multiplication* $\cdot$ are associative, commutative and multiplication is *distributive* with respect to addition: $\kappa(\lambda + \mu) = \kappa\lambda + \kappa\mu$. Moreover,

$$(\lambda \cdot \kappa)^\mu = \lambda^\mu \cdot \kappa^\mu, \quad \lambda^{\kappa+\mu} = \lambda^\kappa\lambda^\mu, \quad (\lambda^\kappa)^\mu = \lambda^{\kappa\mu}.$$

If $\kappa \leq \lambda$, then $\kappa + \mu \leq \lambda + \mu$ and $\kappa \cdot \mu \leq \lambda \cdot \mu$, as well as $\kappa^\mu \leq \lambda^\mu$. On the other hand,

$$0 < \kappa \leq \lambda \Longrightarrow \mu^\kappa \leq \mu^\lambda.$$

In general, $0 < \kappa < \lambda$ does not entail $\mu^\kappa < \mu^\lambda$, as we can see below.

We know already from Corollary I.5.6 that

$$\aleph_0 = \aleph_0 + \aleph_0 = \aleph_0\aleph_0.$$

On the other hand, we know that $\mathfrak{c} = 2^{\aleph_0}$, and $\mathfrak{c} < 2^\mathfrak{c}$ by the Cantor Theorem I.5.8. It follows that

$$\mathfrak{c} = 2^{\aleph_0} = \aleph_0^{\aleph_0} = \mathfrak{c}^{\aleph_0},$$

because $\mathfrak{c}^{\aleph_0} = (2^{\aleph_0})^{\aleph_0} = 2^{\aleph_0\aleph_0} = 2^{\aleph_0}$. On the other hand,

$$\mathfrak{c} = \mathfrak{c} + \mathfrak{c} = \mathfrak{c}\mathfrak{c} = \mathfrak{c}^{\aleph_0},$$

because $\mathfrak{c}\mathfrak{c} = 2^{\aleph_0}2^{\aleph_0} = 2^{\aleph_0+\aleph_0} = 2^{\aleph_0} = \mathfrak{c}$.

We shall see (Corollary XXVII.6.5) that for each $\kappa \geq \aleph_0$,

$$\kappa = \kappa + \kappa = \kappa \cdot \kappa.$$

XXVII.3. Well-ordered sets

An ordered set is said to be *well-ordered* if each non-empty subset has a least element. Of course, each well-ordering is linear, because a subset of two elements has a least element.

The set of natural numbers with its natural order $\mathbb{N}$ is well-ordered; neither the set of integers $\mathbb{Z}$, nor that of positive reals $\mathbb{R}_+$ are not.

LEMMA XXVII.3.1. *Each strictly increasing bijection between linearly ordered sets is an isomorphism* ([5]).

PROOF. Let X, Y be linearly ordered sets and let $f : X \longrightarrow Y$ be a strictly increasing bijection. We shall show that f^{-1} is strictly increasing: let $y_0 < y_1$ and suppose that, on the contrary, $f^{-1}(y_0) \geq f^{-1}(y_1)$, then $y_0 = f(f^{-1}(y_0)) \geq f(f^{-1}(y_1)) = y_1$, which is a contradiction. □

LEMMA XXVII.3.2. *If W is well-ordered, and $f : W \longrightarrow W$ is strictly increasing, then $w \leq f(w)$ for each $w \in W$.*

PROOF. Otherwise, the set $A := \{w \in W : f(w) < w\} \neq \varnothing$, hence, by the definition of well-ordered set, there exists a least element w_0 of A. Consequently, if $w_1 = f(w_0) < w_0$, and since f is strictly increasing, $f(w_1) < f(w_0) = w_1$, showing that $w_1 \in A$ and $w_1 < w_0$, a contradiction. □

Recall that a map between two ordered sets is an (*order*) *isomorphism* if it is strictly increasing and if its inverse map is also strictly increasing.

COROLLARY XXVII.3.3. *The only isomorphism of a well-ordered set with itself is the identity.*

PROOF. If W is well-ordered, and $f : W \longrightarrow W$ is an isomorphism, then $f^{-1} : W \longrightarrow W$ is too. Therefore, $w \leq f(w)$ and $f^{-1}(w) \geq w$ for $w \in W$. Therefore $w = f(f^{-1}(w)) \geq f(w)$, and thus $w = f(w)$ for $w \in W$. □

Accordingly,

COROLLARY XXVII.3.4. *If two well-ordered sets are isomorphic, then there is a unique isomorphism.*

If W is well-ordered and $x \in W$, then the set

$$W(x) := \{w \in W : w < x\}$$

is called an *initial segment* of W.

LEMMA XXVII.3.5. *No well-ordered set is isomorphic with its initial segment.*

PROOF. If $f : W \longrightarrow W(x) = \{w \in W : w < x\}$ for some $x \in W$ is strictly increasing, then $f(x) < x$, a contradiction. □

[5]An example of a strictly increasing bijection that is not an isomorphism: let $f : \{0, \ldots, 3\} \to 2^{\{0,1\}}$ be given by $f(0) = \varnothing, f(1) = \{0\}, f(2) = 1$, and $f(3) = \{0, 1\}$.

THEOREM XXVII.3.6 (Trichotomy). *If X, Y are well-ordered sets, then either they are isomorphic, or X is isomorphic with an initial segment of Y, or else Y is isomorphic with an initial segment of X.*

PROOF. Let us write $X \simeq Y$ to indicate that the ordered sets X, Y are isomorphic. Consider

$$F := \{(x, y) \in X \times Y : X(x) \simeq Y(y)\}.$$

We see that the relations F and F^- are injective. Indeed, if $y_0, y_1 \in Fx$, then $X(y_0) \simeq X(x) \simeq Y(y_1)$, hence by Corollaries XXVII.3.3 and XXVII.3.5, $y_0 = y_1$, and likewise for F^-.

If $h : X(x) \longrightarrow Y(y)$ is an isomorphism and $v < x$, then $X(v) \simeq Y(h(v))$. Therefore F and F^- are order-preserving, and F^-Y is equal either to X or to its initial segment, and likewise for FX.

If $FX = Y$, then $X \simeq Y$ by Lemma XXVII.3.1, which is the first case considered.

If $FX \neq Y$, then there exists a least element y_0 of $Y \setminus f(X)$, and thus $FX = Y(y_0)$, and F defines a strictly increasing bijection between X and $Y(y_0)$, hence $X \simeq Y(y_0)$ by Lemma XXVII.3.1, which is the second case. Similarly, if $F^-Y \neq X$, getting the third case. □

Two ordered sets W_0 and W_1 have the same *order-type*

$$t(W_0) = t(W_1)$$

if they are order-isomorphic. A *type of well-ordering* ([6]) is the equivalence class of all isomorphic well-orderings.

XXVII.4. Ordinal numbers

Cardinal numbers evaluate the size of sets. Ordinal numbers assess well-orderings of sets. They are both respective generalizations of natural numbers to infinite sets.

With each type of well-ordering we will associate an ordinal number, that is to say, a representative of that class in such a way that, for two distinct ordinal numbers, one is an initial segment of another.

A set T is called *transitive* if $x \in T$ then $x \subset T$. The empty set $\varnothing$ is transitive, because the transitivity condition is emptily fulfilled. As $\varnothing \subset x$ for every set x, a transitive set T of cardinality 1, say $T = \{x\}$, fulfills $x \subset \{x\}$, hence $T = \{\varnothing\}$. Consequently,

$$\varnothing, \{\varnothing\}, \{\varnothing, \{\varnothing\}\}, \{\varnothing, \{\varnothing, \{\varnothing\}\}\}, \ldots \tag{XXVII.4.1}$$

are transitive sets.

[6] A formal definition in terms of transitive sets well-ordered by inclusion is given in [**136**, p. 14].

An *ordinal number* (or an *ordinal*) is a transitive set well-ordered by the (strict) order $\in$. Therefore, if α and β are ordinals, then

$$\alpha < \beta \Longleftrightarrow \alpha \in \beta.$$

If $\alpha < \beta$, then we say that α is a *predecessor* of β Consequently, (XXVII.4.1) are ordinals. More generally, for each ordinal α,

$$\alpha = \{\gamma : \gamma < \alpha\}.$$

It follows from Theorem XXVII.3.6 that

THEOREM XXVII.4.1. *Every well-ordered set is order-isomorphic with a unique ordinal number.*

Ordinals form a class, which is not a set, denoted by Ord. Traditionally, to denote ordinals, one uses small Greek letters from the very beginning of the alphabet, $\alpha, \beta, \gamma, \delta$, and son on.

All linear orders on a finite set are well-orderings ([7]), and they are isomorphic. The situation changes radically in case of infinite sets. Already on an infinite countable set there exist infinitely many orders that are not isomorphic one with another.

The ordinal *zero* 0 is the type of well-ordering of the empty set $\varnothing$. A unique relation on $\varnothing$ is empty ($\varnothing \subset \varnothing \times \varnothing$). This relation is of well-ordering, because there are no elements, for which the conditions, using general quantifiers $\forall$, need to hold.

On denoting by t the order type,

$$t(\varnothing) = 0, t(\{\varnothing\}) = 1, t(\{\varnothing, \{\varnothing\}\} = 2,$$

and so on. This way, each finite ordinal is an order type on the set of its predecessors. The standard order of the set $\mathbb{N}$ natural numbers is a well-ordering. It is denoted by ω_0 or ω.

More generally, if α is an ordinal, then the *cardinality* of α is

$$\operatorname{card} \alpha := \operatorname{card}\{\gamma : \gamma < \alpha\}.$$

PROPOSITION XXVII.4.2. *For each ordinal α, there exists the least ordinal that is greater than α.*

PROOF. Let X be well-ordered so that $\alpha = t(X)$. By Cantor Theorem I.5.8, $X < 2^X$ and by Zermelo Theorem XXVII.4.6, there is a well-ordering on 2^X. If β is a type of ordering of 2^X, then $\alpha < \beta$, and there exists

$$\alpha + 1 := \min\{\beta \in \operatorname{Ord} : \alpha < \beta\}.$$

□

[7]We know that there are $n!$ such orders on a set of cardinality n.

The least ordinal greater than α is called the *successor* of α and is denoted by $\alpha + 1$.

Therefore, there exists a map $s : \mathrm{Ord} \longrightarrow \mathrm{Ord}$ such that $s(\alpha) := \alpha + 1$. An ordinal β is called a *successor ordinal* if there is α such that $\alpha + 1 = \beta$. Otherwise, it said to be a *limit ordinal*. For example, 0 is a limit ordinal.

PROPOSITION XXVII.4.3. *An ordinal β is a limit ordinal if and only if $\beta = \sup\{\alpha : \alpha < \beta\}$.*

PROOF. By definition, $\sup\{\alpha : \alpha < \beta\} = \min\{\gamma \in \mathrm{Ord} : \underset{\alpha<\beta}{\forall}\ \alpha < \gamma\} \leq \beta$. If β is a successor ordinal, then $\sup\{\alpha : \alpha < \beta\} + 1 = \beta$; otherwise, $\sup\{\alpha : \alpha < \beta\} = \beta$. □

An ordinal is called finite if its well-ordering is on a finite set. It follows from Zermelo Theorem XXVII.4.6 that there exists a least infinite ordinal ω_0. Consequently, ω_0 is a limit ordinal. By Proposition XXVII.4.2, there exists $\omega_0 + 1$.

A map $f : \alpha \longrightarrow X$ is called a *transfinite sequence* of *length* α (or shortly, a *sequence of length* α). A sequence in the usual sense is a transfinite sequence of length ω_0 ([8]).

If $\{\gamma_\xi : \xi < \alpha\}$ is a non-decreasing sequence, then

$$\lim_{\xi\to\alpha} \gamma_\xi := \sup\{\gamma_\xi : \xi < \alpha\}.$$

A sequence $\{\gamma_\xi : \xi \in \mathrm{Ord}\}$ is called *normal* if it is strictly increasing and $\lim_{\xi\to\alpha} \gamma_\xi = \gamma_\alpha$ for every α.

PROPOSITION XXVII.4.4. *Each decreasing sequence of ordinals is almost constant.*

PROOF. Let $\{\delta_n : n < \omega\}$ be such that $\delta_n \geq \delta_{n+1}$. If $\delta_m = \min\{\delta_n : n < \omega\}$, then $\delta_{m+k} = \delta_m$ for each $k < \omega$. □

An ordinal α is said to be *countable* if the set $\{\gamma \in \mathrm{Ord} : \gamma < \alpha\}$ is countable. Notice that if $\{\gamma_\xi : \xi < \alpha\}$ is a countable sequence of countable ordinals, then $\sup\{\gamma_\xi : \xi < \alpha\}$ is countable, because a countable union of countable sets is countable.

Mathematical induction stipulates that if $P \subset \mathbb{N}$ is such that $0 \in P$, and $n \in P$ implies $n + 1 \in P$, then $P = \mathbb{N}$.

[8]We showed that form the convergence viewpoint a particular well-ordering on $\mathbb{N}$ is insignificant. This is not the case for transfinite sequences of length greater than ω_0.

In fact, mathematical induction constitutes a part of *Peano's axioms* of natural numbers. As we know, mathematical induction constitutes a tool used in numerous mathematical proofs and definitions. We can interpret it by saying if a property $P(n)$ holds for $n = 0$, and $P(n)$ implies $P(n+1)$ for each $n \in \mathbb{N}$, then $P(n)$ holds for each $n \in \mathbb{N}$. The following theorem extends *mathematical induction* to *transfinite induction.*

THEOREM XXVII.4.5 (transfinite induction). *Let* $P \subset \{\gamma \in \text{Ord} : \gamma < \alpha\}$ *such that* $0 \in P$, *and*

(transfinite induction) $$\{\gamma : \gamma < \delta\} \subset P \Longrightarrow \delta \in P,$$

then $P = \{\gamma \in \text{Ord} : \gamma < \alpha\}$.

PROOF. If $\{\gamma \in \text{Ord} : \gamma < \alpha\} \setminus P \neq \varnothing$, then there exists a least $\gamma_0 < \alpha$ such that $\gamma_0 \notin P$. By assumption, $\gamma_0 \neq 0$. Therefore, $\varnothing \neq \{\gamma : \gamma < \gamma_0\} \subset P$, hence by inductive assumption (transfinite induction), $\gamma_0 \in P$, which yields a contradiction. □

THEOREM XXVII.4.6 (Zermelo). *Every set can be well-ordered.*

PROOF. We have already showed that $\varnothing$ is well-ordered by (the order type) 0. If $X \neq \varnothing$, then by the *Axiom of Choice*, there exists a map $f : 2^X \longrightarrow X$ such that $f(A) \in A$ for each $\varnothing \neq A \subset X$. Let $x_0 := f(X)$ and suppose that $\{x_\gamma : \gamma < \alpha\} \subset X$ has been determined. If it is not empty, then let $x_\alpha := f(X \setminus \{x_\gamma : \gamma < \alpha\})$. As f is defined on $2^X \setminus \{\varnothing\}$, there is α such that $X = \{x_\gamma : \gamma < \alpha\}$, and thus, by the definition of well-ordering, there exists $\delta = \min\{\alpha : X = \{x_\gamma : \gamma < \alpha\}\}$. Consequently, δ is a type of well-ordering on X. □

We shall see that if G is a map valued in a set X associating to every transfinite sequence $\{x_\gamma : \gamma < \alpha\}$ and element $G(\{x_\gamma : \gamma < \alpha\})$ of X, then for each ordinal θ, there exists a unique transfinite sequence $\{a_\alpha : \alpha < \theta\}$ such that

$$a_\alpha = G(\{a_\gamma : \gamma < \alpha\}).$$

Recall that an element u of an ordered set X is *maximal* if $x \geq u$, then $x = u$ for each $x \in X$.

This fact follows from *transfinite recursion*, that is, *definition by transfinite induction.*

THEOREM XXVII.4.7 (Transfinite recursion). *If G maps transfinite sequences valued in a set X to elements of X, then there exists a unique map $F : \mathrm{Ord} \longrightarrow X$ such that*

$$F(\alpha) = G(\{F(\gamma) : \gamma < \alpha\}). \tag{XXVII.4.2}$$

PROOF. Let $x := F(\alpha)$ whenever there exists a transfinite sequence $\{a_\gamma : \gamma < \alpha\}$ such that

$$\underset{\gamma<\alpha}{\forall}\ a_\gamma = G(\{a_\xi : \xi < \gamma\}), \tag{1}$$

$$x = G(\{a_\gamma : \gamma < \alpha\}). \tag{2}$$

It is clear that if for every α, there exists such sequence, then it is unique. Hence F is determined by (1). We use Theorem XXVII.4.5 to justify (2). Therefore, $F(\alpha)$ is defined for every ordinal α, and obviously fulfills (XXVII.4.2). □

THEOREM XXVII.4.8 (Kuratowski-Zorn Lemma). *Let $(X, \leq)$ be an ordered set such that, for each linearly ordered subset L of X, there exists $w \in X$ such that $l \leq w$ for each $l \in L$. Then for each $x \in X$, there exists a maximal element u of X such that $x \leq u$.*

PROOF. Given $x_0 \in X$, we construct by transfinite induction a maximal element $u \in X$ such that $x_0 \leq u$. If $\{x_\gamma : \gamma < \alpha\}$ is a strictly increasing sequence in X. Let

$$A_\alpha := \{x \in X : \underset{\gamma<\alpha}{\forall}\ x_\gamma < x\}. \tag{XXVII.4.3}$$

If $A_\alpha \neq \varnothing$, then by Axiom of Choice, there exists $x_\alpha \in A_\alpha$. As $\operatorname{card}\alpha \leq \operatorname{card} X$, there exists δ such that $A_\delta = \varnothing$. As $\{x_\gamma : \gamma < \delta\}$ is linearly ordered, there exists $u \in X$ such that $x_\gamma \leq u$ for each $\gamma < \delta$. But since $A_\delta = \varnothing$, we infer that $u \in \{x_\gamma : \gamma < \delta\}$, hence by (XXVII.4.3), there is no $x > u$. Therefore, u is maximal and $x_0 \leq u$. □

COROLLARY XXVII.4.9. *For each filter $\mathcal{F}$ on a set X, there exists an ultrafilter $\mathcal{U} \geq \mathcal{F}$.*

PROOF. If $\boldsymbol{L}$ is a set of filters on X linearly ordered by inclusion, then $\bigcup_{\mathcal{F}\in\boldsymbol{L}} \mathcal{F}$ is a filter. Indeed, if $F \in \bigcup_{\mathcal{F}\in\boldsymbol{L}} \mathcal{F}$, and $G \supset F$, then there exists $\mathcal{G} \in \boldsymbol{L}$ such that $F \in \mathcal{G}$, hence $G \in \mathcal{G} \subset \bigcup_{\mathcal{F}\in\boldsymbol{L}} \mathcal{F}$. If $A_0, A_1 \in \bigcup_{\mathcal{F}\in\boldsymbol{L}} \mathcal{F}$, then there exist $\mathcal{F}_0, \mathcal{F}_1 \in \boldsymbol{L}$ such that $A_0 \in \mathcal{F}_0$ and $A_1 \in \mathcal{F}_1$. As, by assumption, $\mathcal{A} := \mathcal{F}_0 \cup \mathcal{F}_1 \in \boldsymbol{L}$, we infer that $A_0 \cap A_1 \in \mathcal{A} \subset \bigcup_{\mathcal{F}\in\boldsymbol{L}} \mathcal{F}$. Therefore, by the Kuratowski-Zorn Theorem XXVII.4.8, for each filter $\mathcal{F}$ on X, there exists a maximal filter $\mathcal{U} \geq \mathcal{F}$. □

XXVII.5. Ordinal arithmetic

The *ordinal sum* $\alpha + \beta$ is defined as a unique ordinal $\delta \geq \alpha$ such that β is isomorphic with $\{\gamma : \alpha \leq \gamma < \delta\}$.

In other terms, if α is the order type of X, and β is the order type of Y, then $\alpha + \beta$ is the type of well-ordering of $X \sqcup Y$ such that $x < y$ for each $x \in X$ and $y \in Y$.

The ordinal sum is *associative* $\alpha + (\beta + \gamma) = (\alpha + \beta) + \gamma$, but is not *commutative*, for example,

$$1 + \omega_0 = \omega_0 < \omega_0 + 1.$$

The *ordinal product* $\alpha\beta$ is the unique ordinal, isomorphic with $\alpha \times \beta = \{\gamma : \gamma < \alpha\} \times \{\delta : \delta < \beta\}$ under the horizontal lexicographic order (see Example I.3.5), that is, that $(\gamma_0, \delta_0) < (\gamma_1, \delta_1)$ whenever either $\delta_0 < \delta_1$ or $\delta_0 = \delta_1$ and $\gamma_0 < \gamma_1$.

For instance,

$$2 \cdot \omega_0 = \omega_0 < \omega_0 + \omega_0 = \omega_0 \cdot 2.$$

In fact, $2 \cdot \omega_0$ is the order type of $\{0, 1\} \times \omega_0 = \{0, 1\} \times \{\delta : \delta < \omega_0\}$, arranged as follows

$$2 \cdot \omega_0 \approx (0, \delta_0), (1, \delta_0), (0, \delta_1), (1, \delta_1), (0, \delta_2), \ldots,$$

while $\omega_0 \cdot 2 \approx (0, \delta_0), (0, \delta_1), 0, \delta_2), \ldots (1, \delta_0), (1, \delta_1), \ldots$ In other terms, in $\omega_0 \cdot 2$ all the elements of one copy ω_0 precede all the elements of the second copy. Therefore, there is a limit ordinal in the middle.

The ordinal product is *distributive on the left* with respect to the ordinal sum, that is,

$$\gamma \cdot (\alpha + \beta) = \gamma\alpha + \gamma\beta,$$

but not *on the right*, because

$$(\omega_0 + 1) \cdot 2 = (\omega_0 + 1) + (\omega_0 + 1) = \omega_0 + (1 + \omega_0) + 1 = \omega_0 \cdot 2 + 1.$$

By definition,

$$\alpha^0 := 1, \text{ and } \alpha^{\gamma+1} := \alpha^\gamma \cdot \alpha,$$

$$\beta \text{ is a limit ordinal } \Rightarrow \alpha^\beta := \lim_{\xi \to \beta} \alpha^\xi.$$

We notice that if $\alpha > 0$, then $\xi \mapsto \alpha^\xi$ is normal.

XXVII.6. Cardinal ordinals (Alephs)

An ordinal α is called *cardinal* or an *aleph* if $\alpha = \min\{\gamma : \operatorname{card}\gamma = \operatorname{card}\alpha\}$. In other words, α is a cardinal ordinal if and only if $\operatorname{card}\gamma < \operatorname{card}\alpha$ for each $\gamma < \alpha$.

Every finite ordinal is cardinal, and ω_0 — the first infinite ordinal $\operatorname{card}\omega_0 = \aleph_0$ — too. The first uncountable ordinal ω_1 is also cardinal, and $\operatorname{card}\omega_1 = \aleph_1$.

LEMMA XXVII.6.1. *For each ordinal α, there exists a least cardinal α^+ such that* $\operatorname{card}\alpha < \alpha^+$.

PROOF. If $\operatorname{card}\alpha = \operatorname{card}X$, then the ordinals γ such that $\operatorname{card}\gamma \leq \operatorname{card}2^X$ constitutes a set, and by Cantor Theorem I.5.8, $\operatorname{card}\alpha < \operatorname{card}2^X$. Hence $\min\{\gamma \in \mathrm{Ord} : \operatorname{card}\alpha < \gamma\}$ exists, and is a cardinal ordinal. □

COROLLARY XXVII.6.2. *The ordinal supremum of a set of cardinal ordinals is a cardinal ordinal.*

By definition, $\aleph_0 := \operatorname{card}\omega_0$. We are now in a position to define consecutive alephs:

$$\aleph_{\alpha+1} := (\omega_\alpha)^+ = \min\{\beta : \operatorname{card}\beta > \operatorname{card}(\omega_\alpha)\},$$

and for each limit ordinal α,

$$\aleph_\alpha := \sup\{\aleph_\gamma : \gamma < \alpha\}.$$

Therefore, for each infinite cardinal κ, there exists an ordinal α such that $\kappa = \aleph_\alpha$, that is,

COROLLARY XXVII.6.3. *Each infinite cardinal is an aleph.*

We shall define a *canonical well-ordering* on the class $\mathrm{Ord} \times \mathrm{Ord}$. Let

$$(\zeta_0, \eta_0) \prec (\zeta_1, \eta_1)$$

whenever $\max(\zeta_0, \eta_0) < \max(\zeta_1, \eta_1)$, or if $\max(\zeta_0, \eta_0) = \max(\zeta_1, \eta_1)$ and $\zeta_0 < \zeta_1$, or if $\max(\zeta_0, \eta_0) = \max(\zeta_1, \eta_1)$ and $\zeta_0 = \zeta_1$ and $\eta_0 < \eta_1$.

Therefore, the set $\alpha \times \alpha = \{(\zeta, \eta) : \zeta < \alpha, \eta < \alpha\}$ is an initial segment of $(0, \alpha)$.

The relation $\prec$ is a linear strict order on $\mathrm{Ord} \times \mathrm{Ord}$. Moreover, each non-empty set of $\mathrm{Ord} \times \mathrm{Ord}$ has a least element.

The order type $\Gamma(\zeta, \eta)$ of the set $\{(\xi, \theta) : (\xi, \theta) \prec (\zeta, \eta)\}$ is well-ordered, hence by Theorem XXVII.3.6, $\Gamma(\zeta, \eta)$ is an initial segment of Ord, and conversely each $\alpha = \{\zeta \in \mathrm{Ord} : \zeta < \alpha\}$ is an initial segment of $\mathrm{Ord} \times \mathrm{Ord}$. Therefore $\Gamma : \mathrm{Ord} \times \mathrm{Ord} \longrightarrow \mathrm{Ord}$ is an order isomorphism.

The map $\gamma(\zeta) := \Gamma(\zeta, \zeta)$ is an increasing function such that $\zeta \leq \gamma(\zeta)$. It is easily shown to be normal, and thus $\Gamma(\zeta, \zeta) = \zeta$ for arbitrarily large ζ.

THEOREM XXVII.6.4. *For each ordinal α,*

$$\aleph_\alpha \cdot \aleph_\alpha = \aleph_\alpha.$$

PROOF. We shall show that $\Gamma(\omega_\alpha, \omega_\alpha) = \omega_\alpha$ for each ordinal α. This is true for $\alpha = 0$. Suppose that α is the least ordinal such that ω_α is an initial segment of $\Gamma(\omega_\alpha, \omega_\alpha)$. Let $\eta < \omega_\alpha$ and $\theta < \omega_\alpha$ such that $\Gamma(\eta, \theta) = \omega_\alpha$, and thus there exists δ such that $\max(\eta, \theta) < \delta < \omega_\alpha$, so that $\omega_\alpha < \Gamma(\delta, \delta)$. On the other hand, $\delta \times \delta$ is an initial segment of $\mathrm{Ord} \times \mathrm{Ord}$ in the canonical well-ordering, and $(\eta, \theta) \in \delta \times \delta$. Therefore, $\omega_\alpha \subset \Gamma(\delta, \delta)$, hence, by inductive assumption, $\aleph_\alpha \leq \operatorname{card}(\delta \times \delta) = \operatorname{card} \delta$. But, since ω_α is a cardinal ordinal greater than δ, we conclude that $\operatorname{card} \delta < \aleph_\alpha$, a contradiction. □

As a result,

COROLLARY XXVII.6.5. *For any infinite cardinal numbers κ and λ,*

$$\kappa + \lambda = \kappa \cdot \lambda = \max(\kappa, \lambda). \tag{XXVII.6.1}$$

XXVII.7. Closed unbounded sets

Let $\alpha > 0$ be a limit ordinal. A subset A of α is called *cofinal* in α if $\sup A = \alpha$. An increasing β-sequence $\{\alpha_\xi : \xi < \beta\}$ is called *cofinal* in α if β is a limit ordinal and

$$\sup\{\alpha_\xi : \xi < \beta\} = \alpha. \tag{XXVII.7.1}$$

The least ordinal β such that (XXVII.7.1) holds is called the *cofinality* of α, and is denoted by $\mathrm{cf}\alpha$.

It follows from the definition that $\mathrm{cf}\alpha$ is limit for every (limit) ordinal α, that $\mathrm{cf}\alpha \leq \alpha$ and

$$\mathrm{cf}(\mathrm{cf}\alpha) = \mathrm{cf}\alpha. \tag{XXVII.7.2}$$

Indeed, if β is cofinal in α and γ is cofinal in β, then there are sequences such that $\sup\{\alpha_\xi : \xi < \beta\} = \alpha$ and $\sup\{\xi_\eta : \eta < \gamma\} = \beta$. Therefore, $\sup\{\alpha_{\xi_\eta} : \eta < \gamma\} = \alpha$.

An infinite cardinal $\aleph_\theta$ is called *regular* if $\mathrm{cf}\omega_\theta = \omega_\theta$; *singular* if $\mathrm{cf}\omega_\theta < \omega_\theta$.

A cofinality is always a regular cardinal. Indeed, if α is not a cardinal, then there exists $\gamma < \alpha$ and such that $\operatorname{card} \gamma = \operatorname{card} \alpha$, so that γ is cofinal in α. By (XXVII.7.2), $\mathrm{cf}\alpha$ is a cardinal, and is regular.

A subset C of a regular uncountable cardinal λ is called *closed* if $\lim_{\xi \to \gamma} \alpha_\xi \in C$ for each increasing sequence $\{\alpha_\xi : \xi < \gamma\}$ included in C with $\gamma < \lambda$; it is called *unbounded* if for each $\alpha < \lambda$, there exists $\beta > \alpha$ such that $\beta \in C$.

LEMMA XXVII.7.1. *The family of all closed unbounded subsets of a regular uncountable cardinal is a filter-base.*

PROOF. Let $\mathcal{C}_\lambda$ be the family of all closed and unbounded subsets of a regular uncountable cardinal λ. Of course, $\varnothing \notin \mathcal{C}_\lambda$, because the empty set is not unbounded. If $C_0 \in \mathcal{C}_\lambda$ and $C_1 \in \mathcal{C}_\lambda$, then obviously $C_0 \cap C_1$ is closed. To see that it is unbounded, let $\alpha < \lambda$. As C_0 is unbounded, there exists $\alpha_0 \in C_0$ such that $\alpha < \alpha_0$, and since C_1 is unbounded, there exists $\alpha_1 \in C_1$ such that $\alpha_0 < \alpha_1$. By induction, we construct a strictly increasing sequence $\{\alpha_n : n < \omega\}$ such that $\alpha_{2n} \in C_0$ and $\alpha_{2n+1} \in C_1$ for every $n < \omega$. If $\beta := \lim_{n\to\omega_0} \alpha_n$, then $\beta \in C_0 \cap C_1$, because C_0 and C_1 are closed. □

THEOREM XXVII.7.2. *If λ is a regular uncountable cardinal, then each intersection of less than λ closed unbounded subsets of λ is closed and unbounded.*

PROOF. Let $\mathcal{C}_\lambda$ be the family of all closed and unbounded subsets of a regular uncountable cardinal λ. We shall show by induction on $\gamma < \lambda$ that if $\{C_\xi : \xi < \gamma\} \subset \mathcal{C}_\lambda$, then $\bigcap_{\xi<\gamma} C_\xi \in \mathcal{C}_\lambda$. In virtue of Lemma XXVII.7.1, it is enough to consider the case of limit ordinals γ. The closedness of $\bigcap_{\xi<\gamma} C_\xi$ is obvious. By inductive assumption, $D_\xi := \bigcap_{\zeta<\xi} C_\zeta \in \mathcal{C}_\lambda$, we assume, without loss of generality, that $\{C_\xi : \xi < \gamma\}$ is decreasing. If $\alpha < \lambda$, then let $\beta_0 > \alpha$ be such that $\beta_0 \in C_0$, and for any $\xi < \gamma$, let $\beta_\xi > \sup\{\beta_\zeta : \zeta < \xi\}$ be such that $\beta_\xi \in C_\xi$ by the assumption that C_ξ is unbounded. For each $\eta < \gamma$, the limit $\beta := \lim_{\xi\to\gamma} \beta_\xi \in C_\eta$, because $\{\beta_\xi : \eta < \xi < \gamma\} \subset C_\eta$, and thus $\beta := \lim_{\xi\to\gamma} \beta_\xi \in \bigcap_{\xi<\gamma} C_\xi$. □

XXVII.8. Well-founded relations

Let $R \subset X \times X$ be a relation. Recall that $(x, y) \in R$ if and only if $y \in Rx$ if and only if xRy.

An element a_0 of a subset A of X is called *R-minimal element* if for each $a \in A$,

$$aRa_0 \Longrightarrow a = a_0.$$

DEFINITION XXVII.8.1. A relation R on X is said to be *well-founded* if for any non-empty subset A of X, there exists an R-minimal element $a_0 \in A$.

REMARK XXVII.8.2. Let us specialize these concepts to the case where R is an order relation $\leq$, just to see how they look like in that case. A ordered set $(X, \leq)$ is well-founded whenever for each non-empty subset A of X, there exists $x_A \in A$ such that if $x \leq x_A$, then $x = x_A$. In particular, if $(X, \leq)$ is well-ordered, then $\leq$ is well-founded.

Denote by $\min_R T$ the set of R-minimal elements of T.

PROPOSITION XXVII.8.3 (Well-founded induction). *If R is a well-founded relation on T and $P \subset T$ is such that*

(1) $\min_R T \subset P$ *and*
(2) $Rt \setminus \{t\} \subset P$ *then* $t \in P$.

Then $T = P$.

PROOF. Suppose that $S = T \setminus P \neq \varnothing$. Then there exists an R-minimal $s_0 \in S$. By assumption, $s_0 \notin \min_R T$ and $Rs_0 \setminus \{s_0\} \subset P$. Hence, by assumption, $s_0 \in P$, a contradiction. □

Notice that Proposition XXVII.8.3 generalizes Theorem XXVII.4.5.

THEOREM XXVII.8.4. *If R is a well-founded relation on a set T, then there exists a unique function function $l_R : T \longrightarrow \mathrm{Ord}$ such that $t_0 R t_1$ implies $l_R(t_0) \le l_R(t_1)$, and*

$$l_R(t) = \sup\{l_R(s) + 1 : sRt\}. \tag{XXVII.8.1}$$

PROOF. By induction, $T_0 := \varnothing$, and

$$T_{\alpha+1} := \{t \in T : R^{-}t \subset T_\alpha\},$$

$$\alpha = \sup\{\gamma : \gamma < \alpha\} \Longrightarrow T_\alpha := \bigcup_{\gamma<\alpha} T_\gamma.$$

Notice that $T_\alpha \subset T_{\alpha+1}$. By Replacement (Axiom 2), there exists the least ordinal θ such that $T_{\theta+1} = T_\theta$. Then $T = T_\theta$, for otherwise there would be an R-minimal element m of $T \setminus T_\theta$. Consequently, $t \in T_\theta$ for each tRm, and thus $t \in T_{\theta+1}$ by inductive hypothesis. This yields a contradiction. Let

$$l_R(t) := \min\{\alpha : t \in T_{\alpha+1}\}.$$

It is plain that l_R just defined fulfills (XXVII.8.1). To see the uniqueness, suppose $a : T \longrightarrow \mathrm{Ord}$ fulfills $a(t) = \sup\{a(s) + 1 : sRt\}$. Then consider an R-minimal element t_0 of the set $\{t \in T : l_R(t) \neq a(t)\}$. Let $\alpha < \theta$ be the least ordinal such that $t_0 \in T_\alpha$. Hence if $\gamma < \alpha$ and $s \in T_\gamma$, then $l_R(s) = a(s)$, so that by inductive assumption $l_R(t_0) = a(t_0)$, a contradiction. □

The ordinal θ from the proof above is called the *length* of R and is denoted by $l(R)$

XXVII.9. Small cardinals

We have proved that the cardinality $\mathfrak{c}$ of the real line is not countable. On the other hand, by Lemma XXVII.6.1, there exists a least uncountable cardinal. It is denoted by $\aleph_1$ (*aleph* 1). Accordingly,

$$\aleph_0 < \aleph_1 \le \mathfrak{c} = 2^{\aleph_0}.$$

The *Continuum Hypothesis* states that $\aleph_1 = 2^{\aleph_0}$. It turned out that *CH* cannot be settled within the system *ZFC*.

Several uncountable cardinals less or equal to $\mathfrak{c}$, called often *small cardinals*, are related to some elementary properties of subsets of ω. A detailed account of them can be found in **[255]** and in **[29]**. Some of them are related to our considerations.

If $\mathbb{A} \subset \mathbb{B}$ are classes, then

$$\#\mathbb{A} := \min\{\operatorname{card} \mathcal{A} : \mathcal{A} \in \mathbb{A}\} \geq \min\{\operatorname{card} \mathcal{B} : \mathcal{B} \in \mathbb{B}\} =: \#\mathbb{B},$$

because $\mathbb{B}$ has more elements than $\mathbb{A}$.

If for each $\mathcal{A} \in \mathbb{A}$, there is $\mathcal{B} \in \mathbb{B}$ such that $\operatorname{card} \mathcal{A} \geq \operatorname{card} \mathcal{B}$, then also

$$\#\mathbb{A} \geq \#\mathbb{B}.$$

Maximal almost disjoint number $\mathfrak{a}$. Namely, in Definition XII.8.1, a family $\mathcal{A}$ of infinite subsets of X is called *maximal almost disjoint* or *MAD* if $A_0 \cap A_1$ is finite for each $A_0, A_1 \in \mathcal{A}$ and if an infinite $B \notin \mathcal{A}$ then $\mathcal{A} \cup \{B\}$ is not almost disjoint.

$$\mathfrak{a} := \min\{\operatorname{card} \mathcal{A} : \mathcal{A} \text{ is an infinite } MAD \text{ on } \mathbb{N}\}.$$

By Proposition XII.8.2, an almost disjoint family $\mathcal{A}$ on X is *MAD* if and only if for each infinite subset B of X, there exists $A \in \mathcal{A}$ such that $A \cap B$ is infinite.

We know from Propositions XII.8.3 and XII.8.4 that $\aleph_1 \leq \mathfrak{a} \leq \mathfrak{c}$.

PROPOSITION XXVII.9.1. *If $\mathcal{D}$ is an infinite partition of $\mathbb{N}$ into infinite subsets, then the least cardinality of a family $\mathcal{A}$ such that $\mathcal{A} \cup \mathcal{D}$ is* MAD, *is $\mathfrak{a}$.*

PROOF. Let us call $\mathfrak{a}_0$ the least cardinal of a family $\mathcal{A}$ fulfilling the conditions above. For each such a family $\mathcal{A}$, the family $\mathcal{A} \cup \mathcal{D}$ is *MAD* family and $\operatorname{card} \mathcal{D} = \aleph_0$, we infer that $\mathfrak{a} \leq \mathfrak{a}_0 + \aleph_0 = \mathfrak{a}_0$.

Conversely, let $\mathcal{B}$ be any infinite *MAD* family. Let $\{B_n : n < \omega\} \subset \mathcal{B}$ be a subfamily of distinct elements. Define by induction, $B_0^\circ := B_0$ and $B_n^\circ := (B_n \cup \{n\}) \setminus \bigcup_{k<n} B_k^\circ$ for each $0 < n < \omega$. Then $B_n \subset_{\text{fin}} B_n^\circ \subset_{\text{fin}} B_n$, hence $B_n \cap X$ is infinite if and only if $B_n^\circ \cap X$ is infinite for any infinite set X and each $n < \omega$. By setting $\mathcal{D} := \{B_n^\circ : n < \omega\}$ and $\mathcal{A} := \mathcal{B} \setminus \mathcal{D}$, we notice that $\operatorname{card} \mathcal{B} \geq \operatorname{card} \mathcal{A}$, hence $\mathfrak{a} \geq \mathfrak{a}_0$. □

Dominating number $\mathfrak{d}$. For any pair f, g of elements of $\mathbb{N}^{\mathbb{N}}$, let

$$f \leq_0 g \tag{XXVII.9.1}$$

mean that $f(n) \leq g(n)$ for almost all n, that is, that there exists $n(f,g) < \omega$ such that $f(n) \leq g(n)$ for every $n \geq n(f,g)$. This relation is obviously reflexive and transitive. If $f \leq_0 g$ and $g \leq_0 f$, then, of course, there is $n(f,g) < \omega$ such that $f(n) = g(n)$ for every $n \geq n(f,g)$, that is, f and g are equal almost everywhere $f =_0 g$. Accordingly, (XII.8.4) constitutes an order on the class of almost equal elements of $\mathbb{N}^{\mathbb{N}}$.

DEFINITION XXVII.9.2. A family $\mathcal{D} \subset \mathbb{N}^{\mathbb{N}}$ is called *dominating* if for every $f \in \mathbb{N}^{\mathbb{N}}$ there exists $d \in \mathcal{D}$ such that $f \leq_0 d$. The *dominating number* is defined by

$$\mathfrak{d} := \min\{\operatorname{card} \mathcal{D} : \mathcal{D} \text{ is a } \leq_0\text{-dominating subset of } \mathbb{N}^{\mathbb{N}}\}.$$

Bounding number $\mathfrak{b}$. A family $\mathcal{B}$ is called *unbounded* if there is no $f \in \mathbb{N}^{\mathbb{N}}$ such that $b \leq_0 f$ for each $b \in \mathcal{B}$. In other words, $\mathcal{B}$ is called unbounded whenever, for every $f \in \mathbb{N}^{\mathbb{N}}$, there exists $b \in \mathcal{B}$ such that $b \not\leq_0 f$, that is, $f(n) < b(n)$ for infinitely many $n < \omega$.

DEFINITION XXVII.9.3. The *bounding number* is defined by

$$\mathfrak{b} := \min\{\operatorname{card} \mathcal{B} : \mathcal{B} \text{ is an unbounded subset of } \mathbb{N}^{\mathbb{N}}\}.$$

Clearly, each dominating family is unbounded. Therefore,

$$\mathfrak{b} \leq \mathfrak{d}.$$

REMARK XXVII.9.4. Let us observe that the dominating number $\mathfrak{d}$ can be obtained if we replaced $\leq_0$ by the usual pointwise order $\leq$ on $\mathbb{N}^{\mathbb{N}}$ ([9]). Indeed, the set of finite modifications of all elements of a dominating family $\mathcal{D}$ has the same cardinality of $\mathcal{D}$. However, $\leq$-unbounded families lead to a different number.

Families $\mathcal{A}$ and $\mathcal{B}$ of infinite subsets of X are called *almost dissociated*

$$\mathcal{A} \perp_0 \mathcal{B}$$

whenever if $A \cap B$ is finite for each $A \in \mathcal{A}$ and $B \in \mathcal{B}$. If $\mathcal{A} = \{A\}$ then we abridge $A \perp_0 \mathcal{B}$.

Families $\mathcal{A}$ and $\mathcal{B}$ of infinite subsets of X are separated if there exists $S \subset X$ such that $A \subset_0 S$ for each $A \in \mathcal{A}$, and $B \subset_0 X \setminus S$ for each $B \in \mathcal{B}$ (equivalently, $S \perp_0 \mathcal{B}$).

For each $D \subset \mathbb{N} \times \mathbb{N}$, let $(n,k) \in D^{\downarrow}$ whenever there exists $l \geq k$ such that $(n,l) \in D$, and $(n,k) \in D^{\uparrow}$ whenever there exists $l \leq k$ such that $(n,l) \in D$. In particular, if $f \in \mathbb{N}^{\mathbb{N}}$, then $f^{\downarrow}$ is the hypograph, and $f^{\uparrow}$ is the epigraph of f.

THEOREM XXVII.9.5. $\mathfrak{a} \geq \mathfrak{b}$.

[9]That is, $f \leq g$ if and only if $f(n) \leq g(n)$ for each $n < \omega$.

PROOF. Let $\mathcal{V}$ be a family of subsets of $\mathbb{N}\times\mathbb{N}$ given by $\mathcal{V} := \{V_n : n<\omega\}$, where $V_n := \{n\}\times\mathbb{N}$. In particular, it follows from Proposition XXVII.9.1 that there exists a family $\mathcal{B}$ on $\mathbb{N}\times\mathbb{N}$ such that $\operatorname{card}\mathcal{B} = \mathfrak{a}$ such that $\mathcal{B}\perp_0\mathcal{V}$ and, for each infinite $W\perp_0\mathcal{V}$, there exists $B\in\mathcal{B}$ such that $W\cap B$ is infinite ([10]).

A fortiori $\mathcal{B}_\downarrow := \{B^\downarrow : B\in\mathcal{B}\}$ fulfills this condition. It follows that $\mathcal{B}_\downarrow$ and $\mathcal{V}$ are not separated. Indeed, if $V_n \subset_0 S$ for each $n<\omega$, then there is $f_S(n)<\omega$ such that $(n,k)\in S$ for each $k\geq f_S(n)$ for each $n<\omega$. In particular $f_S\perp_0\mathcal{V}$, hence there is $B\in\mathcal{B}$ such that $f_S\cap B^\downarrow$ is infinite, so that $S\cap B^\downarrow$ is infinite.

As $\mathcal{B}_\downarrow$ and $\mathcal{V}$ are not separated, if $f\in\mathbb{N}^\mathbb{N}$, then $V_n\subset_0 f^\uparrow$ for each $n<\omega$, then there is $B\in\mathcal{B}$ such that $B^\downarrow$ is not almost included in $\mathbb{N}\times\mathbb{N}\setminus f^\uparrow$, equivalently $f^\uparrow\cap B^\downarrow$ is infinite, thus $f(n)<\max_B(n)+1$ for infinitely many n, where

$$\max_B(n) := \max\{k : (n,k)\in B\}.$$

The subfamily $\{\max_B : B\in\mathcal{B}\}$ of $\mathbb{N}^\mathbb{N}$ is unbounded and its cardinality does not exceed $\mathfrak{a}$. Therefore, $\mathfrak{a}\geq\mathfrak{b}$. □

Applications. Let ξ be a convergence on X. Recall that $\mathcal{K}(\xi)$ stands for the family of all ξ-compact sets. A family $\mathcal{L}\subset\mathcal{K}(\xi)$ is called a *compact cover* if $X\subset\mathcal{L}$. The *compact covering number* of ξ is the least cardinality of compact covers of ξ, that is,

$$\mathrm{kc}(\xi) := \min\{\operatorname{card}\mathcal{L} : \mathcal{L}\subset\mathcal{K}(\xi)\wedge X\subset\bigcup\mathcal{L}\}.$$

The *compact cofinality* $\mathrm{kcf}(\xi)$ of ξ is the least cardinality of a family $\mathcal{L}\subset\mathcal{K}(\xi)$ such that for each $K\in\mathcal{K}(\xi)$ there is $L\in\mathcal{L}$ such that $K\subset L$, that is,

$$\mathrm{kcf}(\xi) := \min\{\operatorname{card}\mathcal{L} : \mathcal{L}\subset\mathcal{K}(\xi)\wedge X\subset\bigcup\mathcal{L}\wedge \underset{K\in\mathcal{K}(\xi)}{\forall}\ \underset{L\in\mathcal{L}}{\exists} K\subset L\}.$$

The *k-ness* or *compact determination* of ξ denoted by $\mathrm{k}(\xi)$ is the least cardinality of a family $\mathcal{L}$ such that F is ξ-closed if $F\cap L$ is $\xi|_L$-closed for each $L\in\mathcal{L}$, that is,

$$\mathrm{k}(\xi) := \min\{\operatorname{card}\mathcal{L} : \mathcal{L}\subset\mathcal{K}(\xi)\wedge X\subset\bigcup\mathcal{L}\wedge \underset{F\subset X}{\forall}(F\cap L\in\mathcal{C}_{\xi|_L}\Rightarrow F\in\mathcal{C}_\xi)\}.$$

Let us observe that if ξ is a k-topology, then

$$\mathrm{kc}(\xi)\leq\mathrm{k}(\xi)\leq\mathrm{kcf}(\xi).$$

The three numbers coincide in the case of the space of irrational numbers. Indeed,

[10]We drop an extra assumption that $\mathcal{B}$ is *MAD*.

PROPOSITION XXVII.9.6. *The compact covering number and the compact cofinality of the space of irrational numbers are equal to* $\mathfrak{d}$.

PROOF. By Theorem XIII.5.16, the Baire space $\mathbb{B} = \mathbb{N}^{\mathbb{N}}$ is homeomorphic to the space of irrational numbers $\mathbb{P}$. Let $\varphi \in \mathbb{N}^{\mathbb{N}}$ be a generic element of the Baire space, and let $\Phi \subset \mathbb{N}^{\mathbb{N}}$ stand for a subset of the Baire space. Then, denote as usually $\Phi(n) := \{\varphi(n) : \varphi \in \Phi\}$, and $\sup_{\Phi} : \mathbb{N} \longrightarrow \mathbb{N}$ be defined by

$$\sup_{\Phi}(n) := \sup \Phi(n) = \sup\{\varphi(n) : \varphi \in \Phi\}.$$

By Tikhonov Theorem VIII.3.4, a subset Φ of $\mathbb{B}$ is compact if and only if $\Phi(n) \subset \mathbb{N}$ is finite for each $n < \omega$, that is, whenever $\sup_{\Phi} \in \mathbb{N}^{\mathbb{N}}$. Therefore, a family $\mathcal{A}$ of compact sets is a cover if for each $f \in \mathbb{N}^{\mathbb{N}}$ there exists $\Phi \in \mathcal{A}$ such that $f \leq \sup_{\Phi}$. Equivalently, whenever $\{\sup_{\Phi} : \Phi \in \mathcal{A}\}$ is dominating. It follows that $\mathrm{kc}(\mathbb{B}) = \mathfrak{d}$. Likewise, $\mathcal{A} \subset \mathcal{K}(\mathbb{B})$ is cofinal, if for every $K \in \mathcal{K}(\mathbb{B})$, there is $\Phi \in \mathcal{A}$ such that $\sup_K \leq \sup_{\Phi}$, that is, if $\mathcal{A}$ is dominating. Thus $\mathrm{kcf}(\mathbb{B}) = \mathfrak{d}$. Since the space of irrational numbers is homeomorphic to the Baire space, these numbers are the same. □

In the case of the space of rational numbers $\mathbb{Q}$,

$$\mathrm{kc}(\mathbb{Q}) = \aleph_0 < \mathrm{k}(\mathbb{Q}) = \mathfrak{b} \leq \mathrm{kcf}(\mathbb{Q}) = \mathfrak{d}. \tag{XXVII.9.2}$$

While the first equality is obvious, because $\mathbb{Q}$ is countable, the other equalities are more difficult to prove; proofs can be found in ([11]).

XXVII.10. Supplement

EXERCISE XXVII.10.1. What is the cardinality of

(1) the set of maps $f : \mathbb{R} \longrightarrow \mathbb{R}$?
(2) the set sequences on $\{0, 1\}$?
(3) the set sequences on $\mathbb{N}$?
(4) the set sequences on $\mathbb{R}$?
(5) the set of sequences converging to 0 on $\mathbb{R}$?
(6) What is the paving number of $\sigma_{\mathbb{R}}$ at any point?

SOLUTION. (1) $\mathrm{card}(\mathbb{R}^{\mathbb{R}}) = \mathfrak{c}^{\mathfrak{c}} = (2^{\aleph_0})^{\mathfrak{c}} = 2^{\aleph_0 \cdot \mathfrak{c}} = 2^{\mathfrak{c}}$ by Corollary XXVII.6.5. (2) $2^{\aleph_0} = \mathfrak{c}$. (4) $\mathrm{card}(\mathbb{R}^{\mathbb{N}}) = \mathfrak{c}^{\aleph_0} = (2^{\aleph_0})^{\aleph_0} = 2^{\aleph_0} = \mathfrak{c}$. (3) follows from (2) and (4).

(6) As $\sigma_{\mathbb{R}}$ is homogeneous (see Definition IV.10.14) it is enough to show this at 0. By (4), $\mathfrak{p}(\sigma_{\mathbb{R}}, 0) \leq \mathfrak{c}$. On the other hand, by Proposition XII.8.3, each infinite almost disjoint family of countable sets is

[11] [255] by E. K. van Douwen, who qualified the proof of $\mathrm{kcf}(\mathbb{Q}) = \mathfrak{d}$ as *rather unpleasant*.

uncountable. In particular, every almost disjoint family of countable subsets (of rational numbers) adherent to 0 is uncountable. This means that each sequence $\sigma_{\mathbb{R}}$-convergent to 0 (on rational numbers) has a common subsequence with such a family, so that $\mathfrak{p}_{\#}(\sigma_{\mathbb{R}}, 0)$ is uncountable, hence $\mathfrak{p}(\sigma_{\mathbb{R}}, 0)$ is uncountable. Hence $\mathfrak{p}(\sigma_{\mathbb{R}}, 0) = \mathfrak{c}$ under the Continuum Hypothesis.

CHAPTER XXVIII

Category theoretic perspective

The founding idea of *category theory* ([1]) is to investigate those objects and laws that are common to a large variety of mathematical theories, abstracting from specific aspects of these theories. Primitive notions of category theory are *objects* and *morphisms. Sets* and *maps, topologies* and *continuous maps, convergences* and *continuous maps, groups* and *homomorphisms, vector spaces* and *linear maps* are all instances of objects and morphisms. Therefore, it is rational to explore these common traits simultaneously. At this level of abstraction one can contemplate, for example, *products* and *quotients* of categories, and *functors* between categories. All the mentioned instances are the so called *concrete categories*, that is, carried by the single underlying category, that of sets and maps. At this point, one can check if they admit *initial* and *final* objects, and so on.

The aim of this chapter is to sketch some most basic concepts of category theory, and to describe in this context the category of convergences and continuous maps, and of its subcategories.

It turns out that most of them are *topological categories*, that is, are closed under initial and final objects. This feature enables us to simplify the presentation of categorical aspects of convergences, and thus, to reveal an intelligible underlying structure and a resulting calculus ([2]).

[1] *Category theory* was introduced by S. EILENBERG and S. MACLANE [**96**]. An exhaustive reference book on categories is [**1**] of J. ADÁMEK, H. HERRLICH, and E. STRECKER. Its complexity, however, makes a discourse very arduous for a non-initiated public. See also [**178**] of R. LOWEN, M. SIOEN, and S. VERWULGEN for topological categories.

[2] This calculus was initiated in [**68, 69**], developed in [**56**], and subsequently has been refined and implemented by the author and his collaborators, e.g. in [**83**]. An interesting example of employment of category theory in a somewhat broader context than convergence theory is *convergence without points* developed by J. GOUBAULT-LARRECQ and F. MYNARD in [**117**].

XXVIII.1. Objects, morphisms, and functors

DEFINITION XXVIII.1.1. A *category* $\boldsymbol{C}$ consists of a class of objects Ob_C and, for each pair of $\boldsymbol{C}$-objects ξ, τ, of a set $\mathrm{Hom}_{\boldsymbol{C}}(\xi, \tau)$ of morphisms, so that for all ξ, τ, and θ,

(1) if $\mathfrak{f} \in \mathrm{Hom}_{\boldsymbol{C}}(\xi, \tau)$ and $\mathfrak{h} \in \mathrm{Hom}_{\boldsymbol{C}}(\tau, \theta)$, then $\mathfrak{h} \circ \mathfrak{f} \in \mathrm{Hom}_{\boldsymbol{C}}(\xi, \theta)$,
(2) the *composition* $\circ$ is associative,
(3) for each object ξ, there exists $1_\xi \in \mathrm{Hom}_{\boldsymbol{C}}(\xi, \xi)$ so that, for each $\mathfrak{f} \in \mathrm{Hom}_{\boldsymbol{C}}(\xi, \tau)$,

$$\mathfrak{f} \circ 1_\xi = \mathfrak{f}, \quad 1_\tau \circ \mathfrak{f} = \mathfrak{f}.$$

The morphism 1_ξ is called a $\boldsymbol{C}$-*identity* of ξ. If a category $\boldsymbol{C}$ has been fixed, then one often abridges

$$\mathfrak{f} : \xi \longrightarrow \tau \Longleftrightarrow \mathfrak{f} \in \mathrm{Hom}_{\boldsymbol{C}}(\xi, \tau), \tag{XXVIII.1.1}$$

and say that $\mathfrak{f}$ is a morphism *from* (an object) ξ *to* (an object) τ.

It is tacitly assumed that the sets of morphisms corresponding to distinct pairs of objects are disjoint. It may happen that $\mathrm{Hom}_{\boldsymbol{C}}(\xi, \tau)$ is empty. However, $\mathrm{Hom}_{\boldsymbol{C}}(\xi, \xi) \neq \varnothing$ for each $\xi \in \mathrm{Ob}_{\boldsymbol{C}}$ ([3]).

EXAMPLE XXVIII.1.2. The category **Set** (of sets) consists of sets as objects with maps between sets as morphisms. Accordingly, for any $X, Y \in \mathrm{Ob}_{\mathbf{Set}}$,

$$\mathrm{Hom}_{\mathbf{Set}}(X, Y) = Y^X.$$

Therefore, if $f \in Y^X$ and $x \in X$, then $f(x) \in Y$. The identity $1_X \in \mathrm{Hom}_{\mathbf{Set}}(X, Y)$ is equal to the identity map $i_X : X \longrightarrow X$.

REMARK XXVIII.1.3. Note that Definition XXVIII.1.1 does not postulate that objects of a given category contain elements!

A morphism $\mathfrak{f} \in \mathrm{Hom}_{\boldsymbol{C}}(\xi, \tau)$ is called a *monomorphism* if, for each object ζ and every $\mathfrak{l}, \mathfrak{m} \in \mathrm{Hom}_{\boldsymbol{C}}(\zeta, \xi)$,

$$\mathfrak{f} \circ \mathfrak{l} = \mathfrak{f} \circ \mathfrak{m} \Longrightarrow \mathfrak{l} = \mathfrak{m}.$$

Monomorphisms generalize injective maps, because $f \in Y^X$ is injective if and only if $f \in \mathrm{Hom}_{\mathbf{Set}}(X, Y)$ is a monomorphism.

A morphism $\mathfrak{f} \in \mathrm{Hom}_{\boldsymbol{C}}(\xi, \tau)$ is called an *epimorphism* if, for each object θ and every $\mathfrak{l}, \mathfrak{m} \in \mathrm{Hom}_{\boldsymbol{C}}(\tau, \theta)$,

$$\mathfrak{l} \circ \mathfrak{f} = \mathfrak{m} \circ \mathfrak{f} \Longrightarrow \mathfrak{l} = \mathfrak{m}.$$

---

[3] For example, the empty category $\boldsymbol{C}_\varnothing$, that is, such that $\mathrm{Ob}_{\boldsymbol{C}_\varnothing} = \varnothing$ fulfills the axioms of a category. Another example is given for any class $\mathbf{X}$, by defining $\boldsymbol{C}_{\mathbf{X}}$ so that $\mathbf{X}$ is the class of objects, and $\mathrm{Hom}_{\boldsymbol{C}_{\mathbf{X}}}(\xi, \tau) = \varnothing$ if $\xi \neq \tau$, and $\mathrm{Hom}_{\boldsymbol{C}_{\mathbf{X}}}(\xi, \xi) = \{1_\xi\}$ for each $\xi \in \mathbf{X}$. Then $\boldsymbol{C}_{\mathbf{X}}$ fulfills the axioms of a category.

Epimorphisms generalize surjective maps, because $f \in Y^X$ is surjective if and only if $f \in \mathrm{Hom}_{\mathbf{Set}}(X,Y)$ is an epimorphism.

EXAMPLE XXVIII.1.4. The *category of convergences* **Conv** consists of all convergences as objects, and of all continuous maps between convergences as morphisms. Recall that each convergence ξ determines the underlying set $|\xi|$. For convergences ξ and τ, each morphism is identified with a continuous map:

$$\mathrm{Hom}_{\mathbf{Conv}}(\xi,\tau) \approx \mathscr{C}(\xi,\tau)$$

means that $\mathfrak{f} \in \mathrm{Hom}_{\mathbf{Conv}}(\xi,\tau)$ if and only if there is $f \in |\tau|^{|\xi|}$ such that $f \in \mathscr{C}(\xi,\tau)$. Then we write $|\mathfrak{f}| = f$. In particular, for every convergence ξ, the identity morphism 1_ξ of ξ is identified with the identity map $|1_\xi| = i_{|\xi|} : |\xi| \longrightarrow |\xi|$; of course, $i_{|\xi|} \in \mathscr{C}(\xi,\xi)$.

Topologies, pretopologies, and pseudotopologies, considered with continuous maps, are other examples of categories, denoted by **Top**, **PrTop**, and **PsTop**, respectively. Actually, all the subclasses of convergences that we have individuated in this book, considered with continuous maps, become categories. One should distinguish the category of topologies from that of Hausdorff topologies, both considered with continuous maps. Metric spaces considered give rise to several categories when considered with continuous maps, isometries, or contractions.

A map F from a category $\boldsymbol{C}$ to a category $\boldsymbol{D}$ is called a *functor* (or, a *covariant functor*) if $F : \mathrm{Ob}_{\boldsymbol{C}} \longrightarrow \mathrm{Ob}_{\boldsymbol{D}}$, and for every $\xi,\tau \in \mathrm{Ob}_{\boldsymbol{C}}$,

$$\text{(functor)} \qquad F : \mathrm{Hom}_{\boldsymbol{C}}(\xi,\tau) \to \mathrm{Hom}_{\boldsymbol{D}}(F\xi, F\tau)$$

so that

$$F(\mathfrak{h} \circ \mathfrak{f}) = F\mathfrak{h} \circ F\mathfrak{f}, \quad F(1_\xi) = 1_{F\xi}$$

whenever $\mathfrak{h} \circ \mathfrak{f}$ is defined and for every $\xi \in \mathrm{Ob}_{\boldsymbol{C}}$.

EXAMPLE XXVIII.1.5. In Section VII.4, the topologizer T was defined only on objects, associating to each convergence ξ, a topology $\mathrm{T}\xi$. We extend $\mathrm{T} : \mathrm{Hom}_{\mathbf{Conv}}(\xi,\tau) \longrightarrow \mathrm{Hom}_{\mathbf{Top}}(\mathrm{T}\xi, \mathrm{T}\tau)$ by

$$|\mathrm{T}(\mathfrak{f})| = |\mathfrak{f}| \in \mathscr{C}(\xi,\tau) \subset \mathscr{C}(\mathrm{T}\xi, \mathrm{T}\tau).$$

A functor $F : \boldsymbol{C} \to \boldsymbol{D}$ is called *full* (respectively, *faithful*), if for each pair ξ,τ of $\boldsymbol{C}$-objects, (functor) is *surjective* (respectively, *injective*).

Mind that a faithful functor F need not be injective on objects,

$$F : \mathrm{Ob}_{\boldsymbol{C}} \longrightarrow \mathrm{Ob}_{\boldsymbol{D}},$$

that is, $F\xi = F\tau$ need not imply that $\xi = \tau$.

Indeed, for example, the topologizer $\mathrm{T} : \mathrm{PrTop} \longrightarrow \mathrm{Top}$ is a faithful functor. In fact, if $\mathfrak{f} \in \mathrm{Hom}_{\mathbf{PrTop}}(\xi, \tau)$, then $\mathfrak{f}$ and $\mathrm{T}\mathfrak{f}$ correspond to the same map, because $\mathscr{C}(\xi, \tau) \subset \mathscr{C}(\mathrm{T}\xi, \mathrm{T}\tau)$. On the other hand, in Example XXII.2.4, the Féron cross topology $\mathrm{T}\xi = \mathrm{T}\xi_h$, but $\xi > \xi_h$. Observe that $\mathscr{C}(\xi_h, \xi_h) \ni 1_{\xi_h} \neq 1_\xi \in \mathscr{C}(\xi, \xi)$.

A map F from the class of morphisms of a category $\boldsymbol{C}$ to the class of morphisms of a category $\boldsymbol{D}$ is called a *cofunctor* (or, *contravariant functor*) if

$$F(\mathfrak{h} \circ \mathfrak{f}) = F\mathfrak{f} \circ F\mathfrak{h}, \quad F(1_\xi) = 1_{F\xi}$$

for every $\xi \in \boldsymbol{C}$, and whenever $\mathfrak{h} \circ \mathfrak{f}$ is defined ([4]). Let us only note that the composition of two cofunctors is a functor.

A category $\boldsymbol{D}$ is called a *subcategory* of a category $\boldsymbol{C}$ if $\mathrm{Ob}_{\boldsymbol{D}} \subset \mathrm{Ob}_{\boldsymbol{C}}$ and if ξ and τ are $\boldsymbol{D}$-objects, then $\mathrm{Hom}_{\boldsymbol{D}}(\xi, \tau) \subset \mathrm{Hom}_{\boldsymbol{C}}(\xi, \tau)$, and the composition of $\boldsymbol{D}$-morphims is the restriction of the composition of $\boldsymbol{C}$-morphims. In particular, the $\boldsymbol{D}$-identity 1_ξ of ξ is also the $\boldsymbol{C}$-identity of ξ. A subcategory $\boldsymbol{D}$ of $\boldsymbol{C}$ is *full* provided that

$$\mathrm{Hom}_{\boldsymbol{D}}(\xi, \tau) = \mathrm{Hom}_{\boldsymbol{C}}(\xi, \tau)$$

for every $\boldsymbol{D}$-objects ξ and τ.

Each subcategory $\boldsymbol{D}$ of $\boldsymbol{C}$ specifies a functor $I : \boldsymbol{D} \longrightarrow \boldsymbol{C}$, called the *inclusion* such that if $I\xi := \xi$ for each $\boldsymbol{D}$-object ξ and $I\mathfrak{f} := \mathfrak{f}$ for each morphism.

For instance, the categories of topologies, pretopologies, paratopologies, hypotopologies, pseudotopologies, convergences of countable character, sequentially founded convergences, countably carried convergences (with continuous maps as morphisms) are full subcategories of the category of convergences. On the other hand, the category of pretopologies is a full subcategory of that of paratopologies, that of sequentially founded convergences of that of countably carried convergences, and so on.

XXVIII.2. Constructs

Recall that the class of all sets (with maps between them as morphisms) constitutes a category, denoted by **Set**. More precisely,

$$\mathrm{Hom}_{\mathrm{Set}}(X, Y) = Y^X$$

for any objects X and Y of **Set**.

[4]Every cofunctor can be seen as a functor when of the categories is replaced by its opposite category.

DEFINITION XXVIII.2.1. A category $\boldsymbol{C}$ is called *concrete over* **Set** if there exists a faithful functor $A : \boldsymbol{C} \to \mathbf{Set}$. Then the pair $(\boldsymbol{C}, A)$ is called a *construct*. A faithful functor $A : \boldsymbol{C} \to \mathbf{Set}$ is often called a *forgetful* functor ([5]).

EXAMPLE XXVIII.2.2. The category **Conv** of convergences is concrete. In Section III.1, $|\xi|$ was defined as the underlying set of a convergence ξ, and thus in this case the forgetful functor $A : \mathbf{I} \to \mathbf{Set}$ is defined by

$$A\xi := |\xi|\,,\ \mathfrak{f} \in \mathrm{Hom}_{\mathbf{I}}(\xi, \tau) \Longrightarrow A\mathfrak{f} \in (A\tau)^{A\xi} = |\tau|^{|\xi|}$$

Still in Section III.1, we used $\mathbf{I}$ to design the class of all convergences, and $\mathbf{I}X$ the set of convergences on a set X. In the context of concrete categories,

$$\mathbf{I}X = \{\xi \in \mathbf{I} : |\xi| = X\},$$

is the fiber of X. Accordingly, $\mathfrak{f} \in \mathrm{Hom}_{\mathbf{I}}(\xi, \tau)$ and $|\mathfrak{f}| = f \in |\tau|^{|\xi|}$ amount to $f \in \mathscr{C}(\xi, \tau)$. Of course, all the subclasses of convergences defined by fix$F$, where F is a functor described in Definition XVII.2.1, are concrete categories with continuous maps as morphisms.

If $(\boldsymbol{C}, A)$ is a construct, then $A\xi$ is a set for every $\boldsymbol{C}$-object ξ (the *underlying set* of ξ), and $A\mathfrak{f} \in \mathrm{Hom}_{\mathbf{Set}}(A\xi, A\tau) \subset (A\tau)^{A\xi}$ for every morphism $\mathfrak{f} \in \mathrm{Hom}_{\boldsymbol{C}}(\xi, \tau)$. In other words,

$$A : \mathrm{Hom}_{\boldsymbol{C}}(\xi, \tau) \longrightarrow Y^X \tag{XXVIII.2.1}$$

is an injection with $Y = A\tau$ and $X = A\xi$. Hence if $A\mathfrak{f} = f$, then $\mathfrak{f}$ is unique. Therefore, if $\xi, \tau \in \mathrm{Ob}_{\boldsymbol{C}}$, let

$$f \in \mathscr{C}_{\boldsymbol{C}}(\xi, \tau) \tag{XXVIII.2.2}$$

if $f \in (A\tau)^{A\xi}$ and $\mathfrak{f} \in \mathrm{Hom}_{\boldsymbol{C}}(\xi, \tau)$ is the unique morphism such that $A\mathfrak{f} = f$. In particular, if $\xi \in \boldsymbol{C}X$, then

$$A(1_\xi) = i_X : X \longrightarrow X$$

is the *identity map* on X. Accordingly, in the case of constructs, we can forget about morphisms, substituting them with their underlying maps!

Given a set X, the class

$$\boldsymbol{C}X := A^-(X) = \{\xi \in \boldsymbol{C} : A\xi = X\} \tag{XXVIII.2.3}$$

[5]More generally, a category $\boldsymbol{C}$ is said to be *concrete* over a category $\boldsymbol{A}$ (not necessarily over **Set**) if there is a faithful functor $A : \boldsymbol{C} \to \boldsymbol{A}$.

is called the *fiber* of X for the category $\boldsymbol{C}$. A $\boldsymbol{C}$-fiber is said to be *small* if it is a set. If $\zeta, \xi \in \boldsymbol{C}X$, then we define the following relation

$$\zeta \geq_{\boldsymbol{C}} \xi \Longleftrightarrow i_X \in \mathscr{C}_{\boldsymbol{C}}(\zeta, \xi). \tag{XXVIII.2.4}$$

If a category $\boldsymbol{C}$ has been fixed, then we abridge $\zeta \geq \xi$ whenever $\zeta \geq_{\boldsymbol{C}} \xi$.

PROPOSITION XXVIII.2.3. *The relation (XXVIII.2.4) is a preorder on each fiber of the concrete category.*

PROOF. Indeed, $\geq$ is reflexive, because $i_X \in \mathscr{C}_{\boldsymbol{C}}(\xi, \xi)$ for every $A\xi = X$. It is transitive, because $i_X \in \mathscr{C}_{\boldsymbol{C}}(\zeta, \xi)$ and $i_X \in \mathscr{C}_{\boldsymbol{C}}(\xi, \tau)$ implies that $i_X = i_X \circ i_X \in \mathscr{C}_{\boldsymbol{C}}(\zeta, \tau)$. □

A concrete category is said to be *amnestic* if the preorder on each fiber is an order, that is, whenever $\leq$ is antisymmetric.

Let $(\boldsymbol{C}, A_{\boldsymbol{C}})$ and $(\boldsymbol{D}, A_D)$ be constructs. A functor $F : \boldsymbol{C} \to \boldsymbol{D}$ is said to be *concrete* whenever $A_{\boldsymbol{C}} = A_{\boldsymbol{D}} \circ F$, that is,

$$A_{\boldsymbol{C}}(\mathfrak{f}) = A_{\boldsymbol{D}}(F\mathfrak{f})$$

for every $\boldsymbol{C}$-morphism $\mathfrak{f}$. In other terms, a morphism and its image by a concrete functor correspond to the same map: if $\mathfrak{f} = A_{\boldsymbol{C}}^{-1}(f)$ then $F\mathfrak{f} = A_{\boldsymbol{D}}^{-1}(f)$.

Therefore, if $F : \boldsymbol{C} \to \boldsymbol{D}$ is a concrete functor, then

$$\mathscr{C}_{\boldsymbol{C}}(\xi, \tau) \subset \mathscr{C}_{\boldsymbol{D}}(F\xi, F\tau) \subset \mathscr{C}_{\mathbf{Set}}(A\xi, A\tau) = (A\tau)^{A\xi}. \tag{XXVIII.2.5}$$

Every concrete functor is determined by its values on identities, that is, on objects. Indeed, suppose that $F\xi = G\xi$ for every $\boldsymbol{C}$-object ξ. If $\mathfrak{f} \in \mathrm{Hom}_{\boldsymbol{C}}(\xi, \tau)$, then

$$F\mathfrak{f} \in \mathrm{Hom}_{\boldsymbol{D}}(F\xi, F\tau) = \mathrm{Hom}_{\boldsymbol{D}}(G\xi, G\tau) \ni G\mathfrak{f},$$

and since F and G are concrete, $A(F\mathfrak{f}) = A\mathfrak{f} = A(G\mathfrak{f})$, and as A is faithful, $F\mathfrak{f} = \mathfrak{f} = G\mathfrak{f}$.

In particular, (XXVIII.2.5) implies that

$$\xi \geq_{\boldsymbol{C}} \tau \Longrightarrow F\xi \geq_{\boldsymbol{D}} F\tau, \tag{XXVIII.2.6}$$

which means that every (covariant) functor is increasing.

A subcategory $\boldsymbol{D}$ of a concrete category $\boldsymbol{C}$ is called *concrete* if $(\boldsymbol{D}, A_{\boldsymbol{C}} \circ I)$ is a construct, where $A_{\boldsymbol{C}}$ is the forgetful functor of $\boldsymbol{C}$, and I is the inclusion functor of $\boldsymbol{D}$ into $\boldsymbol{C}$. Then $A_{\boldsymbol{D}} := A_{\boldsymbol{C}} \circ I$ is a forgetful functor. Accordingly, for every set X,

$$\boldsymbol{D}X \subset \boldsymbol{C}X.$$

Indeed, if $\xi \in \boldsymbol{D}X$, then $A_{\boldsymbol{C}}(I\xi) = (A_{\boldsymbol{C}} \circ I)\xi = A_{\boldsymbol{D}}\xi = X$.

XXVIII.3. Topological categories

A *source* is an object ξ with a collection of morphisms from ξ, while a *sink* is an object ξ with a collection of morphisms to ξ.

Formally, a *source* in a category $\boldsymbol{C}$ consists of a $\boldsymbol{C}$-object ξ and of a collection of morphisms $\{\mathfrak{g}_j : \xi \longrightarrow \tau_j\}_{j\in J}$. A *sink* in a category $\boldsymbol{C}$ consists of a $\boldsymbol{C}$-object ξ and of a collection of morphisms $\{\mathfrak{f}_m : \zeta_m \longrightarrow \xi\}_{m\in M}$ ([6]).

Let us assume now that $\boldsymbol{C}$ is a concrete category. Then each morphism is determined by a map, so that a source of ξ is determined by a collection of maps $\{g_j : j \in J\}$ such that $g_j \in \mathscr{C}_{\boldsymbol{C}}(\xi, \tau_j)$ for each $j \in J$, and a sink of ξ by a collection of maps $\{f_m : m \in M\}$ such that $f_m \in \mathscr{C}_{\boldsymbol{C}}(\zeta_m, \xi)$ for each $m \in M$.

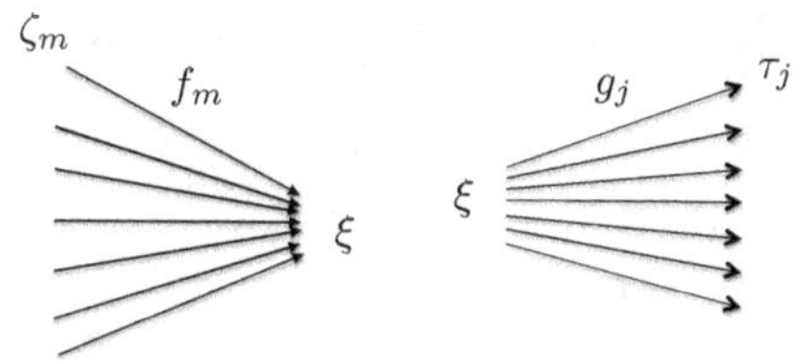

FIGURE XXVIII.3.1. A sink and a source

We have seen instances of sources and sinks in Chapter IV, where a convergence ξ is called *initial* with respect to a set of pairs $\{(g_j, \tau_j) : j \in J\}$ when it is the coarsest convergence such $g_j \in \mathscr{C}(\xi, \tau_j)$ for every $j \in J$, and similarly a convergence ξ is called *final* with respect to a set of pairs $\{(\xi_m, f_m) : m \in M\}$ when it is the finest convergence such $f_m \in \mathscr{C}(\zeta_m, \xi)$ for every $m \in M$.

We shall now examine the same notions in a greater generality of concrete categories. Let $(\boldsymbol{C}, A)$ be a construct ([7]). In (XXVIII.2.2), for two $\boldsymbol{C}$-objects ξ, τ such that $A\xi = X$ and $A\tau = Y$, we have defined the symbol

$$f \in \mathscr{C}_{\boldsymbol{C}}(\xi, \tau)$$

by the existence of $\mathfrak{f} \in \mathrm{Hom}_{\boldsymbol{C}}(\xi, \tau)$ such that $A\mathfrak{f} = f \in Y^X$. By definition, the functor A is faithful, so that the morphism $\mathfrak{f}$ is determined by the map f. The other way round, if $f \in Y^X$, that is, $f \in \mathrm{Hom}_{\mathbf{Set}}(X, Y) = \mathscr{C}_{\mathbf{Set}}(X, Y)$, then for $\xi \in \boldsymbol{C}X$ and $\tau \in \boldsymbol{C}Y$,

[6]We use different symbols for morphisms and indices in the case of sources and sinks with a view to distinguish them in arguments using both of them.

[7]Then $f_m := A\mathfrak{f}_m \in \mathscr{C}_{\boldsymbol{C}}(\xi, \tau_m)$ for each $m \in M$, each f_m determines $\mathfrak{f}_m$.

either $f \in \mathscr{C}_{\boldsymbol{C}}(\xi, \tau)$ or not, and if it does, then there is a unique $\mathfrak{f} \in \mathrm{Hom}_{\boldsymbol{C}}(\xi, \tau)$ such that $A\mathfrak{f} = f$.

Topological categories are usually defined in terms of the so-called *structured sources*, or, equivalently, of the so-called *structured sinks*. These are **Set**-morphisms comprising dormant $\boldsymbol{C}$-objects. I find it handy to denote them as follows

$$f \in \mathscr{C}_{\boldsymbol{C}}(X, \tau) \Longleftrightarrow f : X \longrightarrow A\tau, \tag{XXVIII.3.1}$$

$$f \in \mathscr{C}_{\boldsymbol{C}}(\xi, Y) \Longleftrightarrow f : A\xi \longrightarrow Y. \tag{XXVIII.3.2}$$

We can conceive these maps as *hybrid morphisms*. This notation will be formally justified after the introduction of topological categories.

An *A-structured source* of a set X is a collection of maps

$$\{g_j \in \mathscr{C}_{\boldsymbol{C}}(X, \tau_j) : j \in J\}, \tag{structured source}$$

where τ_j is a $\boldsymbol{C}$-object for every $j \in J$. A source of $\xi \in \boldsymbol{C}X$

$$\{g_j \in \mathscr{C}_{\boldsymbol{C}}(\xi, \tau_j) : j \in J\} \tag{source}$$

is called *initial* provided that if $f \in \mathscr{C}_{\boldsymbol{C}}(\zeta, X)$ is such that $g_j \circ f \in \mathscr{C}_{\boldsymbol{C}}(\zeta, \tau_j)$ for each $j \in J$, then

$$f \in \mathscr{C}_{\boldsymbol{C}}(\zeta, \xi).$$

DEFINITION XXVIII.3.1. A concrete category $\boldsymbol{C}$ is called *topological* if each A-structured source (structured source) has a unique initial lift, that is, $\xi \in \boldsymbol{C}X$ such that (source) is initial.

As mentioned, the category of convergences with continuous maps is topological by virtue of Proposition XXVIII.2.2. On the other hand, there exist non-topological convergences. Therefore, one should be cautious not to mix up the meanings of the term *topological.* To make things even more complicated, there exist subcategories of topologies with continuous maps that are not topological categories. For instance, the category of *Hausdorff topologies* with continuous maps is not topological.

EXAMPLE XXVIII.3.2. Let $f \in \{0\}^{\{0,1\}}$ be a (unique) map. Then $g : \{0,1\} \longrightarrow \iota_{\{0\}}$ is a structured source in the category of Hausdorff topologies with continuous maps. The only Hausdorff topology on $\{0,1\}$ is the discrete topology $\iota_{\{0,1\}}$, but $g : \iota_{\{0,1\}} \longrightarrow \iota_{\{0\}}$ is not an initial source. Indeed, let ζ be a prime, hence Hausdorff, topology on an infinite set X such that $x_\infty \in \lim_\zeta \mathcal{F}$ whenever $\ker \mathcal{F} \subset \{x_\infty\}$ and $\mathcal{F} \setminus \{x_\infty\}$ is a free filter on X. Then if $f \in \{0,1\}^X$ is given by $f(x_\infty) := 0$ and $f(x) := 1$ if $x \neq x_\infty$, then $g \circ f \in \mathscr{C}(\zeta, \iota_{\{0\}})$, but $f \notin \mathscr{C}(\zeta, \iota_{\{0,1\}})$, because $0 \notin \lim_\iota f[\mathcal{F}]$ for any free filter $\mathcal{F}$.

PROPOSITION XXVIII.3.3. *Each topological category is amnestic.*

PROOF. Consider a topological category $\boldsymbol{C}$, and two objects such that $\tau \geq_{\boldsymbol{C}} \xi \geq_{\boldsymbol{C}} \tau$. It follows that there is a set X such that $\xi, \tau \in \boldsymbol{C}X$. As $\boldsymbol{C}$ is a topological category, a structured source $i_X \in \mathscr{C}_{\boldsymbol{C}}(X, \xi)$ has a unique initial source. The source $i_X \in \mathscr{C}_{\boldsymbol{C}}(\xi, \xi)$ is initial. Indeed, if $f \in \mathscr{C}_{\boldsymbol{C}}(\eta, X)$ is such that $i_X \circ f \in \mathscr{C}_{\boldsymbol{C}}(\eta, \xi)$, then obviously $f \in \mathscr{C}_{\boldsymbol{C}}(\eta, \xi)$. On the hand, the source $i_X \in \mathscr{C}_{\boldsymbol{C}}(\tau, \xi)$ is also initial. Indeed, if $f \in \mathscr{C}_{\boldsymbol{C}}(\eta, X)$ is such that $i_X \circ f \in \mathscr{C}_{\boldsymbol{C}}(\eta, \xi)$, then $f \in \mathscr{C}_{\boldsymbol{C}}(\eta, \xi)$, hence $f \in \mathscr{C}_{\boldsymbol{C}}(\eta, \tau)$, because $\xi \geq_{\boldsymbol{C}} \tau$, equivalently $i_X \in \mathscr{C}_{\boldsymbol{C}}(\xi, \tau)$. Hence $\xi = \tau$. □

By Proposition XXVIII.2.4, if $\boldsymbol{C}$ is a concrete category, then for every set X, the fiber $\boldsymbol{C}X$ is endowed with the preorder $\geq$ defined by $\zeta \geq \xi$ if and only if $i_X \in \mathscr{C}_{\boldsymbol{C}}(\zeta, \xi)$. If the category is topological, then it is plain that $\geq$ is an order, and that $(\boldsymbol{C}X, \geq)$ is a complete lattice. Nevertheless, I will detail the proof.

COROLLARY XXVIII.3.4. *Each fiber in a topological category is a complete lattice.*

PROOF. If $\varnothing \neq \Xi \subset \boldsymbol{C}X$, then the source $\{i_X \in \mathscr{C}_{\boldsymbol{C}}(X, \xi) : \xi \in \Xi\}$ has a unique initial element ζ. In particular, $i_X \in \mathscr{C}_{\boldsymbol{C}}(\zeta, \xi)$ for each $\xi \in \Xi$, and if $i_X \in \mathscr{C}_{\boldsymbol{C}}(\eta, \xi)$ for each $\xi \in \Xi$, then $i_X \in \mathscr{C}_{\boldsymbol{C}}(\eta, \zeta)$. In other words, $\zeta \geq \xi$ for each $\xi \in \Xi$, and if $\eta \geq \xi$ for each $\xi \in \Xi$, then $\eta \geq \zeta$, that is, $\zeta = \bigvee \Xi$. The existence of infima of Ξ can be proved analogously. □

In particular, if $f \in \mathscr{C}_{\boldsymbol{C}}(\iota_X, \tau)$, then the least ξ such that $f \in \mathscr{C}_{\boldsymbol{C}}(\xi, \tau)$ is denoted by $f^- \tau$, and called *initial object* of τ under f. Similarly, if $f \in \mathscr{C}_{\boldsymbol{C}}(\xi, o_X)$, then the greatest τ such that $f \in \mathscr{C}_{\boldsymbol{C}}(\xi, \tau)$ is denoted by $f\xi$, and called *final object* of ξ under f. In other words, if a construct $(\boldsymbol{C}, A)$ is topological, then (XXVIII.3.1) and (XXVIII.3.2) fulfill

$$\mathscr{C}_{\boldsymbol{C}}(X, \tau) = \mathscr{C}_{\boldsymbol{C}}(\iota_X, \tau), \text{ and } \mathscr{C}_{\boldsymbol{C}}(\xi, Y) = \mathscr{C}_{\boldsymbol{C}}(\xi, o_Y).$$

Therefore,

$$f\zeta \geq_{\boldsymbol{C}} \tau \iff f \in \mathscr{C}_{\boldsymbol{C}}(\zeta, \tau) \iff \zeta \geq_{\boldsymbol{C}} f^- \tau. \tag{XXVIII.3.3}$$

Accordingly, if a source (source) is initial then

$$\xi = \bigvee\nolimits_{j \in J} f_j^- \tau_j,$$

and if (sink) is final, then

$$\tau = \bigwedge\nolimits_{j \in J} g_j \xi_j.$$

Therefore, in topological categories, morphisms can be characterized by inequalities, as it was the case for convergences. Many of the rules established for convergences still hold for topological categories. For instance, for all objects ξ and τ, and any suitable map f,

$$\xi \geq f^{-}(f\xi), \quad f(f^{-}\tau) \geq \tau. \tag{XXVIII.3.4}$$

Indeed, as $f\xi$ is the greatest object τ for which $f \in \mathscr{C}_{\boldsymbol{C}}(\xi, \tau)$, we infer that $f \in \mathscr{C}_{\boldsymbol{C}}(\xi, f\xi)$, hence $\xi \geq f^{-}(f\xi)$. And since $f^{-}\tau$ is the least object θ for which $f \in \mathscr{C}_{\boldsymbol{C}}(\theta, \tau)$, we deduce that $f \in \mathscr{C}_{\boldsymbol{C}}(f^{-}\tau, \tau)$, hence $f(f^{-}\tau) \geq \tau$.

Given a set Y and $\tau \in \boldsymbol{C}Y$, a sink

$$\{f_m \in \mathscr{C}_{\boldsymbol{C}}(\zeta_m, \xi) : m \in M\} \tag{sink}$$

is called *final* provided that if a map $h \in \mathscr{C}_{\boldsymbol{C}}(Y, \theta)$ is such that $h \circ f_m \in \mathscr{C}_{\boldsymbol{C}}(\xi_m, \theta)$ for every $m \in M$, then

$$h \in \mathscr{C}_{\boldsymbol{C}}(\tau, \theta).$$

Like for initial objects, the finest convergence τ such that $f_j \in \mathscr{C}(\xi_j, \tau)$ exists for every $\{(f_j, \tau) : j \in J\}$. It is a unique object τ of **Conv** that fulfills the conditions above.

A class of maps

$$\{f_m \in \mathscr{C}_{\boldsymbol{C}}(\zeta_m, Y)\}_{m \in M},$$

where each ζ_m is a $\boldsymbol{C}$-object for each $m \in M$, is called an *A-structured sink* in $\boldsymbol{C}$.

PROPOSITION XXVIII.3.5. *A concrete category is topological if and only if each structured sink has a unique final lift.*

PROOF. Let $\{f_m \in \mathscr{C}_{\boldsymbol{C}}(\zeta_m, X) : m \in M\}$ be a structured sink in $\boldsymbol{C}$. Consider the structured source $\{g_j \in \mathscr{C}_{\boldsymbol{C}}(X, \tau_j) : j \in J\}$ consisting of all maps such that

$$g_j \circ f_m \in \mathscr{C}_{\boldsymbol{C}}(\zeta_m, \tau_j) \tag{XXVIII.3.5}$$

for every $m \in m$ and $j \in J$. As $\boldsymbol{C}$ is topological, there exists ξ such that $\{g_j \in \mathscr{C}_{\boldsymbol{C}}(\xi, \tau_j) : j \in J\}$ is an initial source. Hence $f_m \in \mathscr{C}_{\boldsymbol{C}}(\zeta_m, \xi)$ for every $m \in M$, because (XXVIII.3.5) holds for each $j \in J$. If h is a map such that $f_m \circ h$ is a morphism, then there is $j \in J$ such that $h = g_j \in \mathscr{C}_{\boldsymbol{C}}(\zeta, \tau_j)$, then $g_j \circ f_m \in \mathscr{C}_{\boldsymbol{C}}(\zeta_m, \tau_j)$. Therefore the sink $\{f_m \in \mathscr{C}_{\boldsymbol{C}}(\zeta_m, \xi) : m \in M\}$ is final. The converse is proved analogously. $\square$

We have seen that for concrete categories, covariant functors are increasing (XXVIII.2.6). In topological categories, we use (XXVIII.3.3) to obtain equivalences for (XXVIII.2.5)

$$\mathscr{C}_{\boldsymbol{C}}(\xi, \tau) \subset \mathscr{C}_{\boldsymbol{D}}(F\xi, F\tau)$$

in terms of inequalities.

PROPOSITION XXVIII.3.6. *In topological categories, an increasing concrete map F defined on objects is a functor if and only if the following equivalent statements hold for all ξ, τ and f,*

$$\begin{gathered}\operatorname{Hom}_{\boldsymbol{C}}(\xi, \tau) \subset \operatorname{Hom}_{\boldsymbol{D}}(F\xi, F\tau), \\ f(F\xi) \geq_{\boldsymbol{D}} F(f\xi), \\ F(f^{-}\tau) \geq_{\boldsymbol{C}} f^{-}(F\tau).\end{gathered}$$

PROOF. The same as that of Proposition XVII.2.3. □

XXVIII.4. Reflections and coreflections in concrete categories

DEFINITION XXVIII.4.1. A subcategory $\boldsymbol{H}$ of a category $\boldsymbol{C}$ is said to be *reflective* if for every $\boldsymbol{C}$-object ξ, there is a unique $\boldsymbol{H}$-object $H\xi$ and a unique $\mathfrak{h} \in \operatorname{Hom}_{\boldsymbol{C}}(\xi, H\xi)$ such that, for every $\boldsymbol{H}$-object τ and each $\mathfrak{f} \in \operatorname{Hom}_{\boldsymbol{C}}(\xi, \tau)$, there is $\mathfrak{g} \in \operatorname{Hom}_{\boldsymbol{H}}(H\xi, \tau)$ such that $\mathfrak{f} = \mathfrak{g} \circ \mathfrak{h}$, that is, the following diagram commutes:

$$\begin{array}{ccc} \xi & \xrightarrow{\mathfrak{h}} & H\xi \\ & \searrow^{\mathfrak{f}} & \downarrow{\mathfrak{g}} \\ & & \tau \end{array}$$

If moreover each $\mathfrak{h}$ in the definition above is an epimorphism, then the subcategory is called *epireflective.*

If $\boldsymbol{H}$ is a concrete subcategory of a topological category $\boldsymbol{C}$, then $\boldsymbol{H}X \subset \boldsymbol{C}X$ for every set X and $(\boldsymbol{C}X, \leq)$ is a complete lattice. Therefore, for each set X,

$$H : \boldsymbol{C}X \longrightarrow \boldsymbol{H}X.$$

Let us interpret the general definition in this special case.

LEMMA XXVIII.4.2. *A concrete subcategory $\boldsymbol{H}$ of a topological category $\boldsymbol{C}$ is reflective if and only if for each set X and every $\xi \in \boldsymbol{C}X$, there is a unique $H\xi \in \boldsymbol{H}X$ such that $\mathscr{C}_{\boldsymbol{C}}(\xi, \tau) \subset \mathscr{C}_{\boldsymbol{H}}(H\xi, \tau)$ for every Y and $\tau \in \boldsymbol{H}Y$.*

Proof. By concreteness, for each set X, for each $\xi \in \boldsymbol{C}X$ there is a unique $H\xi \in \boldsymbol{H}X$ such that if $\tau \in \boldsymbol{H}Y$ and $f \in \mathscr{C}_{\boldsymbol{C}}(\xi, \tau)$, then $f \in \mathscr{C}_{\boldsymbol{H}}(H\xi, \tau)$. In other words, if $\xi \geq H\xi$ and $\xi \geq f^{-}\tau$, then $H\xi \geq f^{-}\tau$. □

In Definition XVII.3.3 a concrete functor in the category of convergences is called a *reflector* if it is idempotent and descending. We shall see that H from Lemma XXVIII.4.2 is a concrete reflector in the same sense. This fact can be easily proved with the aid of inequalities, but it will be more instructive to carry out a detailed proof below.

Proposition XXVIII.4.3. *Let $\boldsymbol{H}$ be a reflective concrete subcategory of a topological category. Then H is a reflector such that* $\mathrm{fix}H = \boldsymbol{H}$.

Proof. By definition $H\xi \in \boldsymbol{H}X$, so that if $\xi = H\xi$, then $\xi \in \boldsymbol{H}X$. Conversely, if $\xi \in \boldsymbol{H}X \subset \boldsymbol{C}X$, then there is a unique $H\xi \in \boldsymbol{H}X$ such that $i_X \in \mathscr{C}_{\boldsymbol{C}}(\xi, H\xi)$. Apply the definition for $\tau = \xi$ and $f = i_X$ to the effect that if $i_X \in \mathscr{C}_{\boldsymbol{C}}(\xi, \xi)$, then $i_X \in \mathscr{C}_{\boldsymbol{H}}(H\xi, \xi)$, hence $\xi = H\xi$, because $\boldsymbol{C}$ is topological and thus amnestic.

By definition, $i_X \in \mathscr{C}_{\boldsymbol{C}}(\xi, H\xi)$, that is, $\xi \geq H\xi$, so that H is descending. To see that H is increasing, let $\xi \geq \tau$, hence $\xi \geq H\tau$. Then let $f = i_X \in \mathscr{C}_{\boldsymbol{C}}(\xi, H\tau)$. Therefore, by definition, $i_X \in \mathscr{C}_{\boldsymbol{C}}(H\xi, H\tau)$, that is, $H\xi \geq H\tau$.

If $\xi = H\xi$, then there is a unique $\boldsymbol{H}$-object $H(H\xi)$ such that $i_X \in \mathscr{C}_{\boldsymbol{C}}(\xi, H(H\xi))$, and setting $\tau = H\xi$ and $f = i_X$, we infer that $i_X \in \mathscr{C}_{\boldsymbol{H}}(H(H\xi); H\xi)$, that is $H(H\xi) = H\xi$, showing that H is idempotent.

We need to show that $H : \boldsymbol{C} \longrightarrow \boldsymbol{H}$ is a functor. If $\tau \in \boldsymbol{H}Y$ and $f \in Y^X$, then $f^{-}\tau \in \boldsymbol{H}X$. Let $\xi = f^{-}\tau$, that is, $f \in \mathscr{C}_{\boldsymbol{C}}(f^{-}\tau, \tau)$ and thus $f \in \mathscr{C}_{\boldsymbol{H}}(H(f^{-}\tau), \tau)$, equivalently $H(f^{-}\tau) \geq f^{-}\tau = f^{-}(H\tau)$. By Proposition XXVIII.3.6, this amounts to $\mathscr{C}_{\boldsymbol{C}}(\xi, \tau) \subset \mathscr{C}_{\boldsymbol{C}}(H\xi, H\tau)$. □

Conversely, if H is a (concrete) reflector in the sense of Definition XVII.3.3, that is, a functor in the sense of Proposition XXVIII.3.6 such that, for every non-empty set X, a map $H : \mathbf{I}X \longrightarrow \mathbf{I}X$ that is increasing, idempotent, and descending ($H\xi \leq \xi$ for every $\xi \in \mathbf{I}X$), then $\mathfrak{h} \in \mathrm{Hom}_{\boldsymbol{C}}(\xi, H\xi)$ such that $A(\mathfrak{h})(\xi) = H\xi$ is the morphism associated with the reflective subcategory $\mathbf{H}$ in Definition XXVIII.4.1.

Dually,

Definition XXVIII.4.4. A subcategory $\boldsymbol{E}$ of a category $\boldsymbol{C}$ is said to be *coreflective* if for every $\boldsymbol{C}$-object ξ, there is a unique $\boldsymbol{E}$-object $E\xi$ and a unique $\mathfrak{e} \in \mathrm{Hom}_{\boldsymbol{C}}(E\xi, \xi)$ such that, for every $\boldsymbol{E}$-object ζ and each $\mathfrak{f} \in \mathrm{Hom}_{\boldsymbol{C}}(\zeta, \xi)$, there is $\mathfrak{h} \in \mathrm{Hom}_{\boldsymbol{E}}(\zeta, E\xi)$ such that $\mathfrak{f} = \mathfrak{e} \circ \mathfrak{h}$, that is, the following diagram commutes:

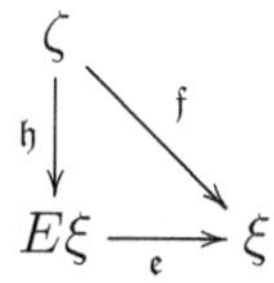

Let us interpret the general definition in this special case.

LEMMA XXVIII.4.5. *A concrete subcategory* $\boldsymbol{E}$ *of a topological category* $\boldsymbol{C}$ *is coreflective if and only if for each set* X *and every* $\xi \in \boldsymbol{C}X$, *there is a unique* $E\xi \in \boldsymbol{E}X$ *such that* $\mathscr{C}_{\boldsymbol{C}}(\zeta,\xi) \subset \mathscr{C}_{\boldsymbol{E}}(\zeta, E\xi)$ *for every set* W *and* $\zeta \in \boldsymbol{E}W$.

PROOF. By concreteness, for each set X, for each $\xi \in \boldsymbol{C}X$ there is a unique $E\xi \in \boldsymbol{E}X$ such that if $\zeta \in \boldsymbol{E}W$ and $f \in \mathscr{C}_{\boldsymbol{C}}(\zeta,\xi)$, then $f \in \mathscr{C}_{\boldsymbol{E}}(\zeta, E\xi)$. In other words, if $E\xi \geq \xi$ and $f\zeta \geq \xi$, then $f\zeta \geq E\xi$. $\square$

PROPOSITION XXVIII.4.6. *Let* $\boldsymbol{E}$ *be a coreflective concrete subcategory of a topological category. Then* E *is a coreflector such that* $\mathrm{fix}E = \boldsymbol{E}$.

Similarly as for reflectors, in Definition XVII.3.6 a concrete functor in the category of convergences is called a *coreflector* if it is idempotent and ascending. The functor E from Lemma XXVIII.4.5 is a concrete coreflector in the same sense.

XXVIII.5. Conclusion

Topological categories, concrete subcategories, and concrete functors can be apprehended object-wise, in terms of order on fibers. Our approach to the class of convergences was originally conceived independently of category theory. Later we realized that, to a great extend, it amounts to the category theoretic approach. Our considerations on duality, reflective hulls, and exponential hulls are also instances of category theory. A presentation of these aspects in full generality would be rather onerous.

Category theory in its full generality is very vast and rich in subtleties, which make it intricate. Topological categories with concrete functors, however, constitute a relatively simple but powerful instrument of investigation.

Postface

First non-topological convergences considered were sequential convergences, which became disused with the advent of abstract topological spaces. In 1914 FELIX HAUSDORFF defined axiomatically, with the aid of neighborhoods, *topologies* fulfilling the separation axiom now called the *Hausdorff* property. In 1922 KAZIMIERZ KURATOWSKI introduced in [**165**] a set of axioms (Section VII.1) of the closed sets for a topology. Another set of axioms was introduced by R. L. MOORE in [**188**].

The emergence of *pretopologies* in [**223, 224**] by FRIGYES RIESZ in 1907 (in terms of accumulations points) passed almost unnoticed as "too general [... and ...] too complicated" according to R. ENGELKING [**97**, p. 36].

In 1935 FELIX HAUSDORFF [**128**] introduced *pretopologies* in terms of set-adherence. They reappeared as well in [**235**] of 1961 by WACŁAW SIERPIŃSKI and in [**40**] of 1966 by EDUARD ČECH.

The decisive innovation in the founding process of convergence theory was, in my opinion, the appearance of pseudotopologies in [**44**] in 1947-48 by GUSTAVE CHOQUET.

Gustave Choquet (1915–2006)

In a study of various concepts of tangency, CHOQUET deployed hyperconvergences ([8],[9]). Then he realized that the upper hyperconvergences are non-topological unless the underlying topology is locally compact. They are, however, always *pseudotopological*, the name introduced by CHOQUET. Actually, he introduced pseudotopologies as an extension to all filters of ultrafilter convergences. In doing so, he evoked the *Urysohn condition* regarding sequential convergences, that is, if $x \notin \lim(x_n)_n$ then there exists a subsequence $(x_{n_k})_k$ such that $x \notin \mathrm{adh}_{\mathrm{Seq}}(x_{n_k})_k$. We do not know whether this was an inspiration or an *a posteriori* association of ideas for the pseudotopologicity condition for a convergence ξ that if $x \notin \lim_\xi \mathcal{F}$, then there is a filter $\mathcal{G} \geq \mathcal{F}$ such that $x \notin \mathrm{adh}_\xi \mathcal{G}$. Anyway, this is the condition enabling one to extend a convergence from ultrafilters to all filters to the effect that

$$\lim_\xi \mathcal{F} = \bigcap\nolimits_{\mathcal{U} \in \beta\mathcal{F}} \lim_\xi \mathcal{U}.$$

Then CHOQUET moved to other areas, like his theory of capacities, that enhanced his celebrity. But he remained interested in developments of convergence theory.

[8]CHOQUET apparently ignored the theory of tangency in Euclidean spaces developed in [215] by GIUSEPPE PEANO, carried out with the aid of hyperlimits, introduced therein well before KAZIMIERZ KURATOWSKI (see [55] and [71] for details).
[9]It turns out that, three decades later, the author studied various tangent cones in topological vector spaces [55] on using hyperlimits of the homothety difference quotient. Needless to say that he realized that the upper hyperlimit was not of topological nature, what lead him to the fundamental paper [44] of CHOQUET.

Bibliography

[1] J. Adámek, H. Herrlich, and E. Strecker. *Abstract and Concrete Categories.* John Wiley and Sons, Inc., 1990.

[2] P. Alexandroff. On the metrization of topological spaces (in Russian). *Bull. Acad. pol. Sc., Série Math.*, 8:135–140, 1960.

[3] P. Alexandroff and P. Urysohn. Une condition nécessaire et suffisante pour qu'une classe (L) soit une classe (D). *C.R. Acad. Sc. Paris*, 177:1274–1276, 1923.

[4] B. Alleche and J. Calbrix. Semi-continuous multifunctions and bases of countable order. *Topology Appl.*, 104(1-3):3–12, 2000.

[5] E. M. Aniskovič. On subspaces of sequential spaces. *Soviet Math. Dokl.*, 28:202–205, 1981.

[6] P. Antosik. On a topology of convergence. *Coll. Math.*, 21:205–209, 1970.

[7] A. V. Arhangel'skii. Some types of factor mappings and the relations between classes of topological spaces. *Dokl. Akad. Nauk SSSR*, 153:743–763, 1963.

[8] A. V. Arhangel'skii. On quotient mappings defined on metric spaces. *Soviet Math. Dokl.*, 5:368–371, 1964.

[9] A. V. Arhangel'skii. Bicompact sets and the topology of spaces. *Trudy Moskov. Mat. Obsc.*, 13:3–55, 1965.

[10] A. V. Arhangel'skii. Mappings and spaces. *Russian Math. Surveys*, 21:115–162, 1966.

[11] A. V. Arhangel'skii. Open and closed-to-open mappings. Relations among spaces. *Trudy Moskov. Mat. Obsc.*, 15:181–223, 1966.

[12] A. V. Arhangel'skii. The frequency spectrum of a topological space and the classification of spaces. *Math. Dokl.*, 13:1185–1189, 1972.

[13] A. V. Arhangel'skii. The frequency spectrum of a topological space and the product operation. *Trans. Moscow Soc.*, 40:164–200, 1981.

[14] A. V. Arhangel'skii. *Topological Function Spaces.* Kluwer Academic, Dordrecht, 1992.

[15] A. V. Arhangel'skii. *Topological Function Spaces.* Mathematics and Its Applications. Kluwer, 1992.

[16] A. V. Arhangel'skii. Bisequential spaces, tightness of products, and metrizability conditions in topological groups. *Trans. Moscow Math. Soc.*, 55:207–219, 1994.

[17] A. V. Arhangel'skii and S. P. Franklin. Ordinal invariants for topological spaces. *Michigan Math. J.*, 19:295–298, 1988.

[18] A. V. Arkhangel'skiĭ. New criteria for the paracompactness and metrizability of an arbitrary T_1-space (in Russian). *Soviet Math. Dokl.*, 2:1367–1369, 1961.

[19] C. Arzelà. Funzioni di linee. *Atti della Reale Accademia dei Lincei. Serie quarta*, 5:342–348, 1889.

[20] G. Ascoli. Le curve limiti di una varietà data di curve. *Atti della R. Accad. Dei Lincei. Memorie della Cl. Sci. Fis. Mat. Nat.*, 18(3):521–586, 1884.

[21] K. Austin and J. Dydak. Partitions of unity and coverings. *Topology Appl.*, 173:74–82, 2014.

[22] A. Barría Comicheo and K. Shamseddine. Summary on non-Archimedean valued fields. In *Advances in ultrametric analysis*, Contemporary Mathematics 704, pages 1–36. A.M.S., 2018.

[23] R. Bartsh, P. Dencker, and H. Poppe. Ascoli-Arzelà-theory based on continuous convergence in an (almost) non-Haudorff setting. In E. Giuli, editor, *Categorical Topology*, pages 221–240. Kluwer Academic, 1996.

[24] R. Beattie and H. P. Butzmann. *Convergence Structures and Applications to Functional Analysis.* Kluwer Academic, 2002.

[25] G. Beer. *Topologies on Closed and Closed Convex Sets.* Kluwer Academic, 1993.

[26] G. Beer and J. Rodríguez-López. Topologies associated with Kuratowski-Painlevé convergence of closed sets. *J. Convex Anal.*, 17(3-4):805–826, 2010.

[27] H. L. Bentley, E. Collebunders, and E. Vandersmissen. A convenient setting for completions and function spaces. In F. Mynard and E. Pearl, editors, *Beyond Topology*, volume 486 of *Contemporary Mathematics*, pages 37–88. A.M.S., 2009.

[28] R. H. Bing. Metrization of topological spaces. *Canad. J. Math.*, 3:175–186, 1951.

[29] A. Blass. Combinatorial cardinal characteristics of the continuum. In *Handbook of Set Theory*, pages 395–489, Dortrecht, 2010. Springer-Verlag.

[30] B. Bolzano. *Rein analytischer Beweis des Lehrsatzes, ...* Prag, 1817.

[31] N. Bourbaki. *Éléments de Mathémathiques, Topologie Générale.* Hermann, 1951.

[32] A. Bouziad. Borel measures in consonant spaces. *Topology Appl.*, 70:125–132, 1996.

[33] D. Brian and I. Labuda. Inherent compactness of upper continuous set valued maps. *Rocky Mountain J. of Math.*, 39(2):463–484, 2009.

[34] H. D. Brunk and R. P. Boas. Solution 4813. *Amer. Math. Monthly*, 66(7):599, 1959.

[35] D. Buhagiar and I. Yoshioka. Ultracomplete topological spaces. *Acta Math. Hungar.*, 92(1-2):19–26, 2001.

[36] G. Cantor. Über unendliche, lineare Punktmannigfaltigkeiten. *Mathematische Annalen*, 23:453–488, 1884.

[37] H. Cartan. Filtres et ultrafiltres. *C. R. Acad. Sc. Paris*, 205:777–779, 1937.

[38] H. Cartan. Théorie des filtres. *C. R. Acad. Sc. Paris*, 205:595–598, 1937.

[39] E. Čech. On bicompact spaces. *Ann. Math.*, 38:823–844, 1937.

[40] E. Čech. *Topological Spaces.* J. Wiley & Sons, 1966.

[41] E. Čech. *Topological papers of Eduard Čech.* Academia, Prague, 1968.

[42] M. Chiccourat. Convergences, pretopologies and proximities induced by Cauchy and Riesz screens. *Acta Math. Hungar.*, 128(3):199–220, 2010.

[43] G. Choquet. Sur les notions de filtre et de grille. *C. R. Acad. Sci. Paris*, 224:171–173, 1947.

[44] G. Choquet. Convergences. *Ann. Univ. Grenoble*, 23:55–112, 1947-48.

[45] K. C. Ciesielski and J. Wojciechowski. Cardinality of regular spaces admitting only constant continuous functions. *Topology Proceedings*, 47:313–329, 2016.

[46] A. C. Cochran and R. B. Trail. *Regularity and Complete Regularity for Convergence Spaces*, volume 375 of *Lecture Notes in Math.*, pages 64–70. Springer-Verlag, 1975.

[47] E. Collebunders and R. Lowen. Metrically generated theories. *Proc. Amer. Math. Soc.*, 133(5):1547–1556, 2005.

[48] C. Cook and H. Fischer. Uniform convergence structures. *Math. Ann.*, 173:290–306, 1967.

[49] C. Costantini and S. Watson. On the dissonance of some metrizable spaces. *Topology Appl.*, 84:259–268, 1998.

[50] Á. Császár. *Fondements de la topologie générale*. Gautier-Villars, 1960.

[51] G. Darboux. Sur un théorème relatif à la continuité des fonctions. *Bull. Soc. Math. Astronom.*, 3:307–313, 1872.

[52] J. de Groot. Non-Archimedean metrics in topology. *Proc. Amer. Math. Soc.*, 7(5):948–953, 1956.

[53] D. Dikranjan and I. Gotchev. Sequentially closed and absolutely closed spaces. *Bolletino U. M. I.*, 7(1B):849–860, 1987.

[54] M. Dolcher. Topologie e strutture di convergenza. *Ann. Sc. Norm. Pisa*, 14:63–92, 1960.

[55] S. Dolecki. Tangency and differentiation: some applications of convergence theory. *Ann. Mat. pura e appl.*, 130(4):223–255, 1982.

[56] S. Dolecki. Convergence-theoretic methods in quotient quest. *Topology Appl.*, 73:1–21, 1996.

[57] S. Dolecki. Active boundaries of upper semicontinuous and compactoid relations; closed and inductively perfect maps. *Rostock. Math. Coll.*, 54:51–68, 2000.

[58] S. Dolecki. Convergence-theoretic characterizations of compactness. *Topology Appl.*, 125:393–417, 2002.

[59] S. Dolecki. Method of multisequences. *Not. S. Afr. Math. Soc.*, 34(1):1–12, 2003.

[60] S. Dolecki. Elimination of covers in completeness. *Topology Proceedings*, 28:445–465, 2004.

[61] S. Dolecki. A subsequential filter that cannot be extended to a countable Hausdorff sequetial space. unpublished, December 2007.

[62] S. Dolecki. Erratum to "Convergence-theoretic characterizations of compactness". *Topology and Appl.*, 154:1216–1217, 2007.

[63] S. Dolecki. An initiation into convergence theory. In F. Mynard and E. Pearl, editors, *Beyond Topology*, volume 486 of *Contemporary Mathematics*, pages 115–161. A.M.S., 2009.

[64] S. Dolecki. *Analyse Fondamentale. Espaces métriques, topologiques et normés.* Hermann, 2nd edition, 2013.

[65] S. Dolecki and D. Gauld. Irregularity. *Topology Appl.*, 154(8):1565–1580, 2007.

[66] S. Dolecki and D. Gauld. Erratum to "Irregularity". *Topology Appl.*, 159(17):3658–3659, 2012.

[67] S. Dolecki and G. H. Greco. Topologically maximal pretopologies. *Studia Math.*, 77:265–281, 1984.

[68] S. Dolecki and G. H. Greco. Cyrtologies of convergences, I. *Math. Nachr.*, 126:327–348, 1986.

[69] S. Dolecki and G. H. Greco. Cyrtologies of convergences, II: Sequential convergences. *Math. Nachr.*, 127:317–334, 1986.

[70] S. Dolecki and G. H. Greco. Towards historical roots of necessary conditions of optimality: Regula of Peano. *Control and Cybernetics*, 36:491–518, 2007.

[71] S. Dolecki and G. H. Greco. The forgotten mathematical legacy of Peano. *Dissert. Math.*, 537:1–77, 2019.

[72] S. Dolecki, G. H. Greco, and A. Lechicki. Compactoid and compact filters. *Pacific J. Math.*, 117:69–98, 1985.

[73] S. Dolecki, G. H. Greco, and A. Lechicki. Sur la topologie de la convergence supérieure de Kuratowski. *C.R. Acad. Sci. Paris*, 312:923–926, 1991.

[74] S. Dolecki, G. H. Greco, and A. Lechicki. When do the upper Kuratowski topology (homeomorphically, Scott topology) and the cocompact topology coincide? *Trans. Amer. Math. Soc.*, 347:2869–2884, 1995.

[75] S. Dolecki, F. Jordan, and F. Mynard. Reflective classes of sequentially based convergence spaces, sequential continuity and sequence-rich filters. *Topology Proc.*, 31(2):457–479, 2007.

[76] S. Dolecki, F. Jordan, and F. Mynard. Group topologies coarser than the Isbell topology. *Topology Appl.*, 158(15):1962–1968, 2011.

[77] S. Dolecki, H.-P. Künzi, and T. Nogura. Weak regularity and consecutive topologizations and regularizations of pretopologies. *Topology and Appl.*, 156(7):1306–1314, 2009.

[78] S. Dolecki and A. Lechicki. On structure of upper semicontinuity. *J. Math. Anal. Appl.*, 88:547–554, 1982.

[79] S. Dolecki and F. Mynard. Convergence-theoretic mechanisms behind product theorems. *Topology Appl.*, 104:67–99, 2000.

[80] S. Dolecki and F. Mynard. Hyperconvergences. *Appl. Gen. Top.*, 4(2):391–419, 2003.

[81] S. Dolecki and F. Mynard. When is the Isbell topology a group topology? *Topology and Appl.*, 157(10):1370–1378, 2010.

[82] S. Dolecki and F. Mynard. A unified approach to function spaces and hyperspaces: local properties. *Houston J. of Math.*, 40(1):285–318, 2014.

[83] S. Dolecki and F. Mynard. *Convergence Foundations of Topology.* World Scientific, 2016.

[84] S. Dolecki and F. Mynard. Productively sequential spaces. *Math. Slovaca*, 68(3):667–676, 2018.

[85] S. Dolecki and T. Nogura. Two-fold theorem on Fréchetness of products. *Czech. Math. J.*, 49 (124):421–429, 1999.

[86] S. Dolecki, T. Nogura, R. Peirone, and G. M. Reed. Non-normality numbers. *Topology Proceedings*, 27(1):111–123, 2003.

[87] S. Dolecki and M. Pillot. Relatively maximal convergences. *Bolletino U. M. I.*, (8) 1-B:699–704, 1998.

[88] S. Dolecki and M. Pillot. Topologically maximal convergences and types of accessibility. *Math. Bohemica*, 123:371–384, 1998.

[89] S. Dolecki and S. Rolewicz. Metric characterizations of upper semicontinuity. *J. Math. Anal. Appl.*, 69:146–152, 1979.

[90] S. Dolecki and S. Sitou. Precise bounds for sequential order of products of some Fréchet topologies. *Topology Appl.*, 84:61–75, 1998.

[91] S. Dolecki and A. Starosolski. Continuous extension of maps between sequential cascades. *Ann. Pure Appl. Logic*, 172(4), 2021.

[92] S. Dolecki, A. Starosolski, and S. Watson. Extensions of multisequences and countably uniradial classes of topologies. *Comment. Math. Univ. Carolin.*, 44(1):165–181, 2003.

[93] S. Dolecki and S. Watson. Internal characterizations of subsequential topologies. preprint, 2002.

[94] J. Dugundji. *Topology.* Allyn and Bacon, 1978.

[95] J. Dydak. Partitions of unity. *Topology Proceedings*, 27(1):125–171, 2003.

[96] S. Eilenberg and S. MacLane. General theory of natural equivalences. *Trans. Amer. Math. Soc.*, 58(2):231–294, 1945.

[97] R. Engelking. *General Topology.* Heldermann Verlag, 1989.

[98] M. Erné. The ABC of order and topology. In H. Herrlich and H.-E. Porst, editors, *Category Theory at Work*, pages 57–83. Heldermann, 1991.

[99] M. Escardó, J. Lawson, and A. Simpson. Comparing Cartesian closed categories of (core) compactly generated spaces. *Topology Appl.*, 143(1-3):105–145, 2004.

[100] R. Féron. Espaces à écart de Fréchet. *C. R. Acad. Sc. Paris. Série A*, (262):278–280, 1960.

[101] P. Fletcher and W. F. Lindgren. *Quasi-Uniform Spaces.* Marcel Dekker, 1982.

[102] S. Franklin and M. Rajagopalan. On subsequential spaces. *Topology Appl.*, 35:1–19, 1990.

[103] M. Fréchet. Sur quelques points du calcul fonctionnel. *Rend. Circ. Mat. Palermo*, 22:1–74, 1906.

[104] D. Fremlin. Sequential convergence in $C_p(X)$. *Comment. Math. Univ. Carolin.*, 35:371–382, 1994.

[105] Z. Frolík. Generalizations of the G_δ-property of complete metric spaces. *Czechoslovak Math. J.*, 10:359–379, 1960.

[106] Z. Frolík. Sums of ultrafilters. *Bull. Amer. Math. Soc.*, 73:87–91, 1967.

[107] W. Gähler. *Grundstrukturen der Analysis*, volume I and II. Akademie-Verlag, Berlin, 1977.

[108] S. García-Ferreira and C. Uzcátegui. Subsequential filters. *Topology Appl.*, 156(18):2949–2959, 2009.

[109] R. J. Gazik. A hyperspace for convergence spaces. *Proc. Amer. Math. Soc.*, 37:234–240, 1973.

[110] G. Gierz, K. H. Hofmann, K. Keimel, J. Lawson, M. Mislove, and D. Scott. *A Compedium of Continuous Lattices.* Springer-Verlag, Berlin, 1980.

[111] G. Gierz, K.H. Hofmann, K. Keimel, J. Lawson, M. Mislove, and D. Scott. *Continuous Lattices and Domains*, volume 93 of *Encyclopedia of Mathematics.* Cambridge University Press, 2003.

[112] L. Gillman and M. Henriksen. Concerning rings of continuous functions. *Trans. Amer. Math. Soc.*, 77(2):330–362, 1954.

[113] L. Gillman and M. Jerison. *Rings of Continuous Functions.* Van Nostrand, 1960.

[114] I. Gotchev. Sequentially P-closed spaces. *Rend. Istit. Mat. Univ. Trieste*, 20(1):1–17, 1988.

[115] I. Gotchev and H. Minchev. On sequential properties of Noetherian topological spaces. *Topology Proc.*, 28(2):487–501, 2004.

[116] J. Goubault-Larrecq. *Non-Hausdorff Topology and Domain Theory.* Number 22 in New Mathematical Monographs. Cambridge University Press, 2013.

[117] J. Goubault-Larrecq and F. Mynard. Convergence without points. *Houston J. of Math.*, 46(1):227–282, 2020.

[118] G. H. Greco. Limitoidi e reticoli completi. *Ann. Univ. Ferrara*, 29:153–164, 1983.

[119] G. H. Greco. Decomposizioni di semifiltri e Γ-limiti sequenziali in reticoli completamente distributivi. *Ann. Mat. pura e appl.*, 137:61–82, 1984.

[120] G. H. Greco. Limites et fonctions d'ensemble. *Rend. Sem. Mat. Padova*, 72:89–97, 1984.

[121] G. H. Greco. The sequential defect of the cross topology is ω_1. *Topology and Appl*, 19:91–94, 1985.

[122] G. H. Greco. *Calcolo differenziale e integrale. In occasione del 150° anniversario della nascita di Giuseppe Peano (1858-1934)*, volume 1. Aracne, 2008.

[123] G. Gruenhage. Products of Fréchet spaces. *Top. Proc.*, 30(2):475–499, 2006.

[124] V. Gutev. Consonance and Cantor set-selectors. *Cent. Eur. J. Math.*, 11(2):41–348, 20133.

[125] L. J. Halbeisen. *Combinatorial Set Theory; with a gentle introduction to forcing.* Springer, 2017.

[126] P. R. Halmos. *Measure Theory.* Springer-Verlag, 1950.

[127] F. Hausdorff. *Grundzüge der Mengenlehre.* Veit & Comp., Leipzig, 1914.

[128] F. Hausdorff. Gestufte Räume. *Fund. Math.*, 25:486–506, 1935.

[129] E. Heine. Die Elemente der Functionenlehre. *J. Reine Angew. Math.*, 74, 1872.

[130] H. Herrlich. Wann sind alle stetigen Abbildungen in Y konstant. *Math. Zeitschrift*, 90:152–154, 1965.

[131] D. Hilbert. Über die stetige Abbildung einer Linie auf ein Flächenstück. *Mathematische Annalen*, 38:459–460, 1891.

[132] K. H. Hofmann and J. D. Lawson. The spectral theory of distributive continuous lattices. *Trans. Amer. Math. Soc.*, 246:285–309, 1978.

[133] J. Isbell. Function spaces and adjoints. *Math. Scand.*, 36:317–339, 1975.

[134] J. R. Isbell. General function spaces, products and continuous lattices. *Math. Proc. Cambridge Phil. Soc.*, 100:193–205, 1986.

[135] D. Jardón. Ultracomplete spaces. In *Workshop on Applied Topological Structures*, pages 67–76, 2016.

[136] T. Jech. *Set Theory.* Academic Press, 1978.

[137] F. Jordan. Productive local properties of function spaces. *Topology Appl.*, 154:870–883, 2007.

[138] F. Jordan. Coincidence of function space topologies. *Top. Appl.*, 157(2):336–351, 2010.

[139] F. Jordan, I. Labuda, and F. Mynard. Finite products of filters that are compact relative to a class of filters. *Applied Gen. Top.*, 8(2):161–170, 2007.

[140] F. Jordan and F. Mynard. Espaces productivement de Fréchet. *C. R. Acad. Sci. Paris, Ser I*, 335:259–262, 2002.

[141] F. Jordan and F. Mynard. Productively Fréchet spaces. *Czech. Math. J.*, 54(129):981–990, 2004.

[142] F. Jordan and F. Mynard. Compatible relations of filters and stability of local topological properties under supremum and product. *Topology Appl.*, 153:2386–2412, 2006.

[143] F. Jordan and F. Mynard. Compatible relations on filters and stability of local topological properties under supremum and product. *Top. Appl.*, 153:2386–2412, 2006.

[144] F. Jordan and F. Mynard. Core compactness and diagonality in spaces of open sets. *Applied General Topology*, 12(2):143–162, 2011.

[145] M. P. Kac. Characterization of some classes of pseudotopological linear spaces. In *Convergence Structures and Applications to Analysis*. Akademie-Verlag, 1980.

[146] L. V. Kantorovich, B. Z. Vulich, and A. G. Pinsker. *Functional Analysis in Preordered Spaces*. GITTL (in Russian), 1950.

[147] J. Kelley. *General Topology*. Van Nostrand, 1955.

[148] D. C. Kent. Convergence quotient maps. *Fund. Math.*, 65:197–205, 1969.

[149] D. C. Kent and R. Frič. The finite product theorem for certain epireflections. *Math. Nachr.*, 150:7–14, 1991.

[150] D. C. Kent, K. McKennon, G. Richardson, and M. Schroder. Continuous convergence in $C(X)$. *Pacific J. Math.*, 52:457–465, 1974.

[151] D. C. Kent and Won Keun Min. Neighborhood spaces. *Inter. J. Math. Math. Sci.*, 32(7):387–399, 2002.

[152] D. C. Kent and G. Richardson. Regular compactification of convergence spaces. *Proc. Amer. Math. Soc.*, 31:571–573, 1972.

[153] D. C. Kent and G. Richardson. Locally compact convergence spaces. *Michigan Math. J.*, 22:353–360, 1975.

[154] D. C. Kent and G. Richardson. Compactifications of convergence spaces. *Internat. J. Math. and Math. Sci.*, 2(3):345–368, 1979.

[155] D. C. Kent and G. Richardson. Some product theorems for convergence spaces. *Math. Nachr.*, 87:43–51, 1979.

[156] D. C. Kent and G. Richardson. Two generalizations of Novak's sequential envelope. *Math. Nachr.*, 91:77–85, 1979.

[157] D. C. Kent and G. Richardson. Convergence spaces and diagonal conditions. *Topology Appl.*, 70:167–174, 1996.

[158] D. C. Kent and G.D. Richardson. Minimal convergence spaces. *Trans. Amer. Math. Soc.*, 160:487–499, 1971.

[159] J. Ketonen. On the existence of P-points in the Stone-Čech compactification of integers. *Fund. Math.*, 92(2):91–94, 1976.

[160] J. Kisyński. Convergence du type L. *Colloq. Math.*, 7:205–211, 1960.

[161] H. J. Kowalsky. Limesräume und Kompletierung. *Math. Nachr.*, 12:302–340, 1954.

[162] P. Kratochvil. Multisequences and their structure in sequential spaces. In *Convergence Structures*. Akademie-Verlag, 1985.

[163] K. Kunen and J.E. Vaughan, editors. *Handbook of Set-Theoretic Topology*. Elsevier, 1984.

[164] H.-P. Künzi. An introduction to quasi-uniform spaces. In E. Pearl and F. Mynard, editors, *Beyond Topology*, volume 486 of *Contemporary Mathematics*, pages 239–304. A.M.S., 2009.

[165] K. Kuratowski. Une méthode d'élimination des nombres transfinis des raisonnements mathématiques. *Fund. Math.*, 3, 1922.

[166] K. Kuratowski. *Topologie*, volume I and II of *Monografie Matematyczne*. PWN Polish Scientific Publishing, 1933.

[167] K. Kuratowski. Quelques problèmes concernant les espaces métriques non-séparables. *Fund. Math.*, 25:534–545, 1935.

[168] K. Kuratowski. *Introduction to set theory and topology (in Polish, Wstęp do teorii mnogości i topologii)*. PWN Polish Scientific Publishers, 1965.

[169] K. Kuratowski. *Topology*. Academic Press — PWN, 1966.

[170] I. Labuda. On a theorem of Choquet and Dolecki. *J. Math. Anal. Appl.*, 126:1–10, 1987.

[171] I. Labuda. Boundaries of upper semicontinuous set valued maps. *Commentationes Mathematicae (Prace Mat.)*, 55(2):225–235, 2005.

[172] A. Lechicki. The Vietoris hyperspace for convergence spaces. *Bolletino U. M. I.*, 7(2-B):41–52, 1988.

[173] A. Lechicki, S. Levi, and A. Spakowski. Bornological convergences. *J. Math. Anal. Appl.*, 297(2):751–770, 2004.

[174] A. Lechicki and J. Ziemińska. Closed filters and graph-closed multifunctions in convergence spaces. *Rocky Mountain J. of Math.*, 17(4):815–827, 1987.

[175] A. Lechicki and J. Ziemińska. On uniform convergence structures. *Math. Nachr.*, 134:169–187, 1987.

[176] Chuan Liu. On weakly bisequential spaces. *Comment. Math. Univ. Carolinae*, 41(3):611–617, 2000.

[177] E. Lowen and R. Lowen. On the nonsimplicity of some convergence categories. *Proc. Amer. Math. Soc.*, 105(2):305–308, 1989.

[178] R. Lowen, M. Sioen, and S. Verwulgen. Categorical topology. In F. Mynard and E. Pearl, editors, *Beyond Topology*, volume 486 of *Contemporary Mathematics*, pages 1–35. A.M.S., 2009.

[179] E. Lowen-Colebunders. An internal and an external characterization of convergence spaces in which adherences of filters are closed. *Proc. Amer. Math. Soc.*, 72:205–210, 1978.

[180] E. Lowen-Colebunders. *Function Classes of Cauchy Continuous Maps*. Number 123 in Monographs and Textbooks in Pure and Applied Math. Marcel Dekker, 1989.

[181] E. Lowen-Colebunders. *Function Classes of Cauchy Continuous Maps*. Dekker, 1989.

[182] V. I. Malyhin. Sequential and Fréchet-Uryson bicompacta (in Russian). *Vestnik Moskov. Univ. Ser. I Mat. Mekh.*, 31:42–48, 1976.

[183] E. Michael. Topologies on spaces of subsets. *Trans. Amer. Math. Soc.*, 71:152–182, 1951.

[184] E. Michael. A survey of continuous selections. In *Set-Valued Mappings, Selections and Topological Properties of* 2^X, volume 171 of *Lecture Notes in Mathematics*, pages 54–58. Springer, 1970.

[185] E. Michael. A quintuple quotient quest. *Gen. Topology Appl.*, 2:91–138, 1972.

[186] E. Michael, R. C. Olson, and F. Siwiec. A-spaces and countably biquotient maps. *Dissert. Math.*, 133, 1976.

[187] E. H. Moore and H. L. Smith. A general theory of limits. *Amer. J. Math.*, 44:102–121, 1922.

[188] R. L. Moore. A set of axioms for plane analysis situs. *Fund. Math.*, 25:13–28, 1935.

[189] K. Morita. On the simple extension of a space with respect to the uniformity. *Proc. Japan Acad.*, 27:632–636, 1951.

[190] K. Morita and S. Hanai. Closed mappings and metric spaces. *Proc. Japan Acad.*, 32:10–14, 1956.

[191] J. R. Munkres. *Topology.* Prentice-Hall, 1975.

[192] M. G. Murdeshwar and S. A. Naimpally. *Quasi-uniform topological spaces.* P. Noordhoff Ltd., Groningen, 1966.

[193] F. Mynard. When are compact, countably compact and Lindelöf convergences topological? *Top. Proc.*, 26(1):283–296, 2002.

[194] F. Mynard. First-countability, sequentiality and tightness of the upper Kuratowski convergence. *Rocky Mountain J. of Math.*, 33(3):1011–1038, 2003.

[195] F. Mynard. Products of compact filters and applications to classical product theorems. *Top. Appl.*, 154:953–968, 2007.

[196] F. Mynard. Relations that preserve compact filters. *Applied Gen. Top.*, 8(2):171–185, 2007.

[197] F. Mynard. A convergence-theoretic viewpoint on the Arzelà-Ascoli theorem. *Real Analysis Exchange*, 38(2):431–444, 2013.

[198] F. Mynard. *An Introduction to the Language of Mathematics.* Springer, 2018.

[199] F. Mynard. (Ultra-) completeness numbers and (pseudo-) paving numbers. *Topology Appl.*, 256:86–103, 2019.

[200] F. Mynard. Comparison of countability conditions within three fundamental classifications of convergences. *Topology and Appl.*, 278(107233), 2020.

[201] F. Mynard and J. Wojciechowski. When is a category of adherence-determined convergences simple? *Topology and Appl.*, 322:108319, 2022.

[202] A. Mysior. A regular space which is not completely regular. *Proc. Amer. Math. Soc.*, 81(4):652–653, 1981.

[203] L. Nachbin. *Topology and Order.* Van Nostrand, 1965.

[204] L. D. Nel. *Continuity Theory.* Springer, 2016.

[205] T. Nogura. The product of α_i-spaces. *Topology Appl.*, 21:251–259, 1985.

[206] T. Nogura. A counterexample for a problem of Arhangel'skiĭ concerning the product of Fréchet spaces. *Topology Appl.*, 25:75–80, 1987.

[207] T. Nogura. Products of sequential convergence properties. *Czechoslovak Math. J*, 39:262–279, 1989.

[208] T. Nogura and D. Shakhmatov. When does the Fell topology on a hyperspace of closed sets coincide with the meet of the upper Kuratowski and the lower Vietoris topologies? *Topology Appl.*, 70:213–243, 1996.

[209] T. Nogura and A. Shibakov. Sequential order of product spaces. *Topology and Appl.*, 65:271–285, 1995.

[210] T. Nogura and A. Shibakov. Sequential order of products of Fréchet spaces. *Topology and Appl*, 70:245–253, 1996.

[211] J. Novák. On convergence groups. *Czechoslovak Math. J.*, 20:357–374, 1970.

[212] P. Nyikos. Subsets of ω^ω and the Fréchet-Urysohn and α_i-properties. *Topology Appl.*, 48:91–116, 1992.

[213] R. C. Olson. Biquotient maps, countably bisequential spaces and related topics. *Topology Appl.*, 4:1–28, 1974.

[214] J. C. Oxtoby. Cartesian products of Baire spaces. *Fund. Math.*, 49:157–166, 1960/1961.

[215] G. Peano. *Applicazioni Geometriche del calcolo infinitesimale.* Fratelli Bocca Editori, 1887.

[216] G. Peano. *Arithmetices principia, nova methodo exposita.* Bocca, Augustae Taurinorum, 1889.

[217] G. Peano. Sur une courbe, qui remplit toute une aire plane. *Mathematische Annalen*, 36:157–160, 1890.

[218] G. Peano. *Formulaire Mathématique.* Fratelli Bocca Editori, 4th edition, 1902–1903.

[219] G. Peano. *Formulario Mathematico.* Fratelli Bocca Editori, 1908.

[220] W. J. Pervin. Quasi-uniformities of topological spaces. *Math. Ann.*, 147:316–317, 1962.

[221] V. I. Ponomarev and V. V. Tkachuk. The countable character of X in βX compared with the countable character of the diagonal in $X \times X$. *Vestnik Moskov. Univ. Ser. I Mat. Mekh.*, 42(5):16–19, 1987.

[222] H. Poppe. *Compactness in General Function Spaces.* VEB Deutscher Verlag der Wissenschaften, Berlin, 1974.

[223] F. Riesz. Die Genesis des Raumbegriffes. *Math. und Naturwiss. Berichte aus Ungarn*, 24:309–353, 1907.

[224] F. Riesz. Stetigkeitsbegriff und abstrakte Mengenlehre. In *Atti del IV Congresso Internazionale dei Matematici*, volume I, pages 18–24, Roma, 1908.

[225] S. Romaguera. On cofinally complete metric spaces. *Questions and Answers in General Topology*, 16(165-170), 1998.

[226] M. E. Rudin. A new proof that metric spaces are paracompact. *Proc. Amer. Math. Soc.*, 20:603, 1969.

[227] W. Rudin. Homogeneity problems in the theory of Čech compactifications. *Duke Math. J.*, 23:409–419, 1956.

[228] C. Ryll-Nardzewski. Real functions (In Polish, Funkcje rzeczywiste). Handwritten notes (by S. Dolecki) of a course on measure theory held at Wrocław University, 1970-71.

[229] J. Saint-Raymond. Espaces métrisables dissonants. Technical report, Université Paris 6, 1995.

[230] M. Schroder. Solid convergences spaces. *Bull. Australian Math. Soc.*, 8:443–459, 1973.

[231] M. Schroder. Compactness theorems. In *Categorical Topology*, pages 566–577. Springer-Verlag, 1974.

[232] F. Schwarz. Powers and exponential objects in initially structured categories and application to categories of limits spaces. *Quaest. Math.*, 6:227–254, 1983.

[233] V. E. Shneider. Continuous images of Suslin and Borel sets. Metrization theorems (in Russian). *Dokl. Akad. Nauk SSSR*, 50:77–79, 1945.

[234] W. Sierpiński. Sur quelques propriétés topologiques du plan. *Fund. Math.*, 4:1–6, 1923.

[235] W. Sierpiński. *General Topology.* Univ. Toronto, 1961.

[236] P. Simon. A compact Fréchet space whose square is not Fréchet. *Comment. Math. Univ. Carolin.*, 21:749–753, 1980.

[237] S. Sitou. On multisequences and their application to product of sequential spaces. *Math. Slovaca*, 49:235–241, 1999.

[238] F. Siwiec. Sequence-covering and countably bi-quotient mappings. *Gen. Topology Appl.*, 1:143–154, 1971.

[239] F. Siwiec. On defining a space by a weak base. *Pacific J. Math.*, 52(1):233–245, 1974.

[240] F. Siwiec and V. J. Mancuso. Relations among certain mappings and conditions for their equivalence. *Gen. Topology Appl.*, 1:34–41, 1971.

[241] A. Starosolski. P-hierarchy on $\beta\omega$. *J. Symbolic Logic*, 73(4):1202–1214, 2008.

[242] A. Starosolski. Cascades, order, and ultrafilters. *Annals of Pure and Applied Logic*, 165:1626–1638, 2014.

[243] L. A. Steen and J. A. Seebach. *Counterexamples in Topology.* Dover, 1995.

[244] A. H. Stone. Paracompactness and product spaces. *Bulletin Amer. Math. Soc.*, 54:977–982, 1948.

[245] A. H. Stone. Metrizability of decomposition spaces. *Proc. Amer. Math. Soc.*, 7:690–700, 1956.

[246] A. H. Stone. Sequences of coverings. *Pacific J. Math.*, 10:689–691, 1960.

[247] M. H. Stone. Application of the theory of Boolean rings to general topology. *Trans. Amer. Math. Soc.*, 41(375-481), 1937.

[248] Y. Tanaka. Products of sequential spaces. *Proc. Amer. Math. Soc.*, 54:371–375, 1976.

[249] Y. Tanaka. Tanaka spaces and products of sequential spaces. *Comm. Math. Univ. Carol.*, 48(3):529–540, 2007.

[250] A. Tran-Mynard and F. Mynard. Closure spaces and characterizations of filters in terms of their Stone images. *Czech. Math. J.*, 57(3):1025–1034, 2007.

[251] A. Tychonoff. Über die topologische Erweiterung von Räumen. *Math. Annalen*, 102(1):544–561, 1930.

[252] P. Urysohn. Über die Mächtigkeit der zusammenhängenden Mengen. *Mathematische Annalen*, 94:262–295, 1925.

[253] I. A. Vainštein. On closed mappings of metric spaces (in Russian). *Dokl. Akad. Nauk SSSR*, 57:319–321, 1947, in Russian.

[254] J. H. van der Walt. The order convergence structure. *Indag. Math.*, 21(3-5):138–15, 2011.

[255] E. K. van Douwen. The integers and topology. In K. Kunen and J. E. Vaughan, editors, *Handbook of Set-Theoretic Topology*, pages 111–167. North-Holland, 1984.

[256] A. C. M. van Rooij. *Non-Archimedean functional analysis*, volume 51 of *Monographs and Textbooks in Pure and Applied Math.* Marcel Dekker, 1978.

[257] J. E. Vaughan. Countably compact and sequentially compact spaces. In K. Kunen and J. E. Vaughan, editors, *Handbook of Set-Theoretic Topology*, pages 569–602. Elsevier Science, 1984.

[258] J. E. Vaughan. An introduction to images of the irrational numbers. IV Taller de Investigacion de Topologia, UNAM, Oaxaca (preprint, Topology Atlas), 1996.

[259] L. Vietoris. Stetige Mengen. *Monatsh. Math.*, 31:173–204, 1921.

[260] A. Weil. *Sur les espaces à structure uniforme et sur la topologie générale.* Hermann, 1938.

[261] E. Wimmers. The Shelah P-point independence theorem. *Israel J. Math.*, 43(1):28–48, 1982.

[262] J. Wojciechowski. Three problems in convergence theory. *Topology Proceedings*, 60:137–145, 2022.

[263] L. Zoretti. Sur les fonctions analytiques uniformes qui possèdent un ensemble parfait discontinu de points singuliers. *J. Math. Pures Appl.*, 1:1–51, 1905.

List of Symbols

$\zeta_{\mathcal{A}}$	almost disjoint topology of $\mathcal{A}$
$\operatorname{adh}_\xi \mathcal{A}$	adherence of a family $\mathcal{A}$ with respect to a convergence ξ
$\operatorname{adh}^\natural_\xi \mathcal{A}$	$\{\operatorname{adh}_\xi A : A \in \mathcal{A}\}$
$\operatorname{adh}_\xi A$	adherence of a set A with respect to a convergence ξ
$\bigoplus$	coproduct
$\boldsymbol{A}_\neg$	the set of complementary families of those from $\boldsymbol{A}$
$\perp$	dissociated
$\perp_0$	almost dissociated
$\mathbb{A}$	class of antitone families
$\mathbb{A}X$	the set of antitone families of subsets of X
$\mathcal{A}\#\mathcal{B}$	$\mathcal{A}$ meshes $\mathcal{B}$
$\mathcal{A}\perp\mathcal{B}$	$\mathcal{A}$ is dissociated from $\mathcal{B}$ ($\mathcal{A}\neg\#\mathcal{B}$)
$\mathcal{A}^\#$	the grill of $\mathcal{A}$
$\mathcal{A}^\star$	the conjugate family of $\mathcal{A}$
$\mathcal{A}^\uparrow$	$G \in \mathcal{A}^\uparrow$ if there is $A \in \mathcal{A}$ such that $A \subset G$
$\mathcal{A}_c$	family of complements of the elements of $\mathcal{A}$
$\sqcup, \bigsqcup$	disjoint union
$\times, \prod$	product
$A\#B$	$A \cap B \neq \varnothing$
$B_r(x)$	ball of radius r centered at x (when a metric is implicit)
βX	set of all ultrafilters on X (also $\mathbb{U}X$)
$\beta, \beta\iota_X$	the Stone topology on βX
$\beta\mathcal{F}, \beta_X(\mathcal{F})$	set of all ultrafilters finer than a filter $\mathcal{F}$ (on X)
$\beta^* X$	set of all free ultrafilters on X
$\beta_X A$	set of all ultrafilters $\mathcal{U}$ on X such that $A \in \mathcal{U}$
$B_d(x,r)$	ball of radius r (with respect to a metric d) centered at x
$\mathfrak{C}(\tau)$	the class of compactifications of τ
$\aleph_0$	aleph zero (the cardinality of $\mathbb{N}$)
$\operatorname{cl}_{\xi^\bullet}$	pointwise closure
$\operatorname{cl}_{\xi^\circ}$	inverse pointwise closure
$\mathcal{C}_\xi$	family of all ξ-closed sets
$\mathcal{C}^I_\xi$	family of functionally ξ-closed sets
$\mathfrak{c}$	continuum (the cardinality of $\mathbb{R}$)

$\mathrm{compl}(\xi)$	completeness number of ξ
$\mathrm{compl}_{\#}(\xi)$	ultracompleteness number of ξ
$\mathscr{C}(\mathcal{F},\mathcal{G})$	the set of relations continuous from $\mathcal{F}$ to $\mathcal{G}$
$\mathscr{C}(\xi,\tau)$	set of continuous functions from ξ to τ
$\mathrm{dist}(\cdot,\cdot)$	distance between two sets
$\partial\mathscr{F}$	the support of a family $\mathscr{F}$ of functions
$\triangle(X)$	diagonal subset of a power set of X
$\triangle$	diagonal map
$\triangle_X$	identity relation on X
$\exp(g)$	exponentiation of the map g
$\mathbb{E}$	class of countably carried filters
ev	evaluation map
$f^{\sim 1}$	the semi-inverse map of f
$\mathbb{F}$	class of all filters
$\mathbb{F}(\mathcal{F})$	the set of all filters that are finer than a filter $\mathcal{F}$
$\mathbb{F}_0$	class of principal filters
$\mathbb{F}_*$	class of all free filters
$\mathbb{F}_1$	class of countably based filters
$\mathbb{F}_{\#1}$	class of countanly tight filters
$\mathbb{F}_\theta$	set of θ-regular filters
$\mathbb{F}_{\wedge 1}$	class of countably complete filters
$\mathbb{F}X$	set of all filters on X
$\mathbb{H}_\theta$	set of θ-regular $\mathbb{H}$-filters
$f\|_V$	restriction of a map f to a subset V of the domain
f^*	upper adjoint map of f
f^{-1}	the inverse map of a map f
f^-	the inverse relation of a map f
f_*	lower adjoint map of f
$F_{\aleph_0} = \mathrm{P}_0$	finitely stable modifier
$F_{\aleph_1} = \mathrm{P}_1$	countably stable modifier
R^-	the inverse relation of a relation R
$[X]^{<\omega}$	the set of finite subsets of X
$\subset_{\mathrm{cofin}}$	is a cofinite subset of
$\subset_{\mathrm{fin}}$	is a finite subset of
$\mathcal{G} \in \mathbb{H}(\mathcal{F})$	$\mathcal{G}$ is an $\mathbb{H}$-filter finer than $\mathcal{F}$
$\mathrm{gr}(f)$	the graph of f
$\mathbb{H}^{\mathbb{M}}$	$\mathbb{M}$-polar of $\mathbb{H}$
$\mathbb{H}^{\sharp\mathbb{M}}$	strict $\mathbb{M}$-polar of $\mathbb{H}$
$\mathbb{H}_*$	the subclass of $\mathbb{H}$ consisting of free filters
$\mathbb{H}_\neg$	the class of complementary families of those from $\mathbb{H}$
$\mathbf{A}_{\mathbb{H}}$	class of $\mathbb{H}$-adherence-determined convergences

$\mathbf{B}_{\mathbb{H}}$	class of $\mathbb{H}$-founded convergences
$\mathrm{A}_{\mathbb{H}}$	reflector on $\mathbb{H}$-adherence-determined convergences
$\mathrm{B}_{\mathbb{H}}$	coreflector on $\mathbb{H}$-founded convergences
$\mathscr{H}(\xi,\tau)$	set of homeomorphisms from ξ to τ
inh_{ξ}	inherence with respect to a convergence ξ
$\mathbb{I}$	class of isotone families of sets
$\mathbb{I}X$	the set of isotone families of subsets of X
$\mathbf{I}$	class of all convergences
$\mathbf{I}_{\circ}X$	set of all preconvergences on X
$\mathbf{I}_{\circ}$	class of all preconvergences
$\mathbf{I}X$	set of all convergences on X
I	convergence modifier
$\mathrm{I}_{\circ}$	preconvergence modifier
$\jmath, \jmath_X$	point-evaluation map (of X)
$\kappa(\xi)$	collection of isotone ξ-compact families
$\mathcal{K}(\xi)$	set of ξ-compact sets
$\mathcal{K}_{oc}(\xi)$	the ξ-cocompactoid filter
$\mathcal{K}_{\circ}(\xi)$	set of ξ-compactoid sets
$\mathrm{kcf}(\xi)$	compact cofinality number of ξ
$\mathrm{kc}(\xi)$	compact covering number of ξ
$\mathrm{k}(\xi)$	compact determination number of ξ
$\mathscr{K}_{\mathbb{H}}(\theta,\sigma)$	the set of $\mathbb{H}$-compact relations from θ to σ
$\lim_{\xi}\mathcal{F}$	limit of a filter $\mathcal{F}$ for a convergence ξ
$\mathrm{Max}A$	the set of maximal elements of A
$\mathrm{Min}A$	the set of minimal elements of A
$\min_R T$	the set of R-mininimal elements pf T
$\mathbb{N}$	set of natural numbers
$\mathbb{N}_m$	the set of integers n such that $n \geq m$
$\mathbb{P}$	set of irrational numbers
$\mathbb{Q}$	set of rational numbers
$\mathbb{R}$	set of real numbers
$\mathbb{Z}$	set of integers
$\mathcal{N}_{\xi}(A)$	nighborhood filter of ξ at a set A
$\mathcal{N}_{\xi}(x)$	neighborhood filter of ξ at x
$\nu, \nu_{\mathbb{Q}}$	standard topology of $\mathbb{Q}$
$\nu, \nu_{\mathbb{R}}$	standard convergence (topology) of $\mathbb{R}$
$\nu_{\mathbb{I}}, I$	standard topology of $I = [0,1]$
ν_C	standard topology of the Cantor set
$\sigma, \sigma_{\mathbb{R}}$	standard sequential convergence of $\mathbb{R}$
$\mathcal{O}_{\xi}$	family of all ξ-open sets
$\mathcal{O}_{\xi}(A)$	family of ξ-open sets including A
$\mathcal{O}_{\xi}(x)$	family of ξ-open sets containing x

$\mathcal{O}_\xi\{\mathcal{A}\}$	the contour of $\mathcal{O}_\xi$ along $\mathcal{A}$
$\mathcal{O}_\xi^I$	family of functionally ξ-open sets
$t(W)$	the order-type of an ordered set W
$\boldsymbol{P}_{\cup\downarrow}$	collections of ideals of families from $\boldsymbol{P}$
$\mathbf{P}_0$	class of finitely stable convergences
$\mathbf{P}_1$	class of countably stable convergences
$\mathfrak{p}_\#(\xi, x)$	pseudopaving number of ξ at x
$\mathfrak{p}(\xi, x)$	paving number of ξ at x
$\pi[x, \boldsymbol{D}]$	prime convergence, where $\boldsymbol{D}$ is a pavement at x
$\pi[x_\infty, \mathcal{F}]$	prime convergence, where $\{\mathcal{F}\}$ is a pavement at x_∞
$\mathrm{Q}_{\mathbb{D}}^{\mathbb{J}}$	Modifier of local $\mathbb{J}$-compactness with respect to $\mathbb{D}$-based filters
$Q_J(\xi, \tau)$	the set of J-quotient maps from ξ to τ
$\mathrm{fix}(\mathrm{R}^\sigma)$	reflective hull of a convergence σ
$\mathrm{r}\mathscr{F}$	the reduced filter of a hyperfilter $\mathscr{F}$
$\mathrm{R}^\sigma\xi$	σ-inital reflection of ξ
$\widehat{r}$	the constant function equal to r
$\mathbb{S}$	class of sequential filters
$\mathbb{S}\mathcal{A}$	set of sequential filters finer than $\mathcal{A}$
$\mathbb{S}X$	set of sequential filters on X
$\mathbf{Seq}X$	set of sequentially founded convergences on X
$\mathbf{Seq}$	class of sequentially founded convergences
$\mathbf{S}$	class of pseudotopologies
$\mathbf{S}_0$	class of pretopologies
$\mathbf{S}_0X$	set of pretopologies with the underlying set X
$\mathbf{S}_1$	class of paratopologies
$\mathbf{S}_1X$	set of paratopologies with the underlying set X
$\mathbf{S}_{\wedge 1}$	class of hypotopologies
$\mathbf{S}_{\wedge 1}X$	set of hypotopologies with the underlying set X
$\mathbf{S}X$	set of pseudotopologies with the underlying set X
S	pseudotopologizer
S_0	pretopologizer
S_1	paratopologizer
$\mathrm{S}_{\wedge 1}$	hypotopologizer
Σ	sequential tree
$\$$	Sierpiński convergence
$\mathbf{T}$	class of topologies
$\mathbf{T}X$	set of topologies with the underlying set X
$\mathrm{ext}T$	the set of extremal elements of T
$\mathrm{tp}(h) = \widehat{h}$	transposition of the map h
T	topologizer

$\mathrm{t}d$	the topology determined by a metric d
$\max T$	the set of maximal elements of T
$\min T$	the set of minimal elements of T
$\int T$	the contour of a cascade T
$\tau \succeq \theta$	τ is a larger compactification than θ
$\tau \looparrowright \theta$	τ is densely embedded in θ
$\theta \trianglerighteq \tau$	θ is a finer subspace of τ
$\xi \hookrightarrow \tau$	ξ is embedded in τ
$l_T(t)$	the level of t in a cascade T
$r_T(t)$	the rank of t in a cascade T
$\mathbb{U}_*$	class of free ultrafilters
$\mathbb{U}$	class of ultrafilters
$\mathscr{U}(\mathcal{P}, \mathcal{R})$	the set of uniformly continuous maps between $\mathcal{R}$ and $\mathcal{P}$
$\mathscr{U}(d_X, d_Y)$	the set of uniformly continuous maps from d_X to d_Y
$\mathcal{V}_\xi(x)$	vicinity filter of ξ at x
$\mathcal{V}_\xi(A)$	vicinity filter of ξ at a set A
$V(v)$	the set of immediate successors of v
$V^\uparrow(v)$	the set of successors of v
$x^\uparrow$	the principal ultrafilter $\{F \subset X : x \in F\}$ of x
¥	Bourdaud pretopology

Index

www.ingramcontent.com/pod-product-compliance
Lightning Source LLC
LaVergne TN
LVHW052012200926
848481LV00005B/1471